The Natural History of Primates

The Natural History of Primates

A Systematic Survey of Ecology and Behavior

EDITED BY
ROBERT W. SUSSMAN, DONNA HART, AND IAN C. COLQUHOUN

ROWMAN & LITTLEFIELD
Lanham • Boulder • New York • London

Acquisitions Editor: Alyssa Palazzo
Assistant Acquisitions Editor: Samantha Delwarte
Sales and Marketing Inquiries: textbooks@rowman.com

Credits and acknowledgments for material borrowed from other sources, and
reproduced with permission, appear on the appropriate pages within the text.

Published by Rowman & Littlefield
An imprint of The Rowman & Littlefield Publishing Group, Inc.
4501 Forbes Boulevard, Suite 200, Lanham, Maryland 20706
www.rowman.com

86-90 Paul Street, London EC2A 4NE

Copyright © 2023 by The Rowman & Littlefield Publishing Group, Inc.

British Library Cataloguing in Publication Information Available

Library of Congress Cataloging-in-Publication Data

Names: Sussman, Robert W., 1941- editor. | Hart, Donna, editor. | Colquhoun, Ian C., editor.
Title: The natural history of primates : a systematic survey of ecology and behavior /
    edited by Robert W. Sussman, Donna Hart, and Ian C. Colquhoun.
Description: Lanham : Rowman & Littlefield Publishers, [2022] | Includes bibliographical references and index.
Identifiers: LCCN 2022010163 (print) | LCCN 2022010164 (ebook) | ISBN 9781442248991 (cloth) |
    ISBN 9781442249004 (ebook)
Subjects: LCSH: Primates—Behavior. | Primates—Ecology.
Classification: LCC QL737.P9 N356 2022 (print) | LCC QL737.P9 (ebook) | DDC 599.815—dc23/eng/20220325
LC record available at https://lccn.loc.gov/2022010163
LC ebook record available at https://lccn.loc.gov/2022010164

♾™ The paper used in this publication meets the minimum requirements of
American National Standard for Information Sciences—Permanence of Paper
for Printed Library Materials, ANSI/NISO Z39.48-1992.

Dedication

This book is dedicated to Robert W. Sussman (1941–2016) in appreciation of his contributions to primatology.

Contents

Foreword

IAN TATTERSALL

ALTHOUGH BANG UP TO DATE ON ITS PUBLICATION, this volume has been a very long time in gestation. That's because, for almost as long as I can remember, Bob Sussman was dreaming of a single book that would synthesize what's most important in our knowledge of all the 200–300 or more species of primates (let's not get into that one). In the beginning that seemed a plausible ambition, as relatively little was known. But as the field of primatology burgeoned in the last quarter of the twentieth century, the task of realizing this vision became exponentially more daunting for a single individual, even one as energetic as Bob. And as a result, it's not very surprising that, although he repeatedly returned to the idea, Bob never got around to writing his synthesis himself, or even with a single co-author (he brought the subject up almost every time I saw him). After all, he faced basically the same problem that faces the painters of the George Washington Bridge: by the time they reach the far end, the paint is beginning to peel back where they started. In Bob's case, the target of a comprehensive account of primates was just receding too rapidly, as new information flooded in from all sides; and besides, there was always a welter of other projects to attend to.

Happily, all along Bob was energetically training a large cohort of talented students, each one of whom he encouraged to follow his or her intellectual nose wherever it might lead, and many of whom consequently went on to become leaders in diverse subfields of primatology. He thus had a huge pool of expertise to draw from when, toward the unexpected and tragically early end of his busy life, he concluded that his long-envisaged single-volume survey of the primates would have to be a corporate effort. Bob it was, then, who drew up the initial proposal for the book you are holding; and he would have been enormously gratified to see how successfully his vision has been implemented by his students and editorial successors, Donna Hart and Ian Colquhoun, and by the colleagues—many also trained by Bob—who contributed the individual chapters.

As we are reminded by the authors of the book's section commentaries, primatology has come a very long way since Bob entered it in the late 1960s. In those days the young science was just beginning to feel its way, propelled largely by a handful of groundbreaking ape and baboon studies in eastern Africa—studies that placed their greatest emphasis on social behaviors, and on how those behaviors related to their human equivalents. Bob, in contrast, took a more bottom-up approach—one that was grounded in his long-standing concern with the evolution of behavior among the primates as a whole. One consequence was that he was just as much interested in his primate subjects' relationships with their environments as in those they had with each other; and he was also a pioneer in the use of timed sampling methods to quantify those interactions.

I was fortunate to have been, in a way, Bob's first student. In 1969, while preparing a thesis on the extinct "subfossil" lemurs, I had the good fortune of being introduced to Madagascar by Bob's supervisor John Buettner-Janusch and his wife Vina. One of their many aims at the time was to pave the way for Bob's planned thesis work on sympatric and allopatric ring-tailed and brown lemurs in the island's southwest. I left Madagascar shortly before Bob and his medical anthropologist wife, Linda, were scheduled to arrive to begin their research; but I was back the next year, and my first glimpse of them was entirely unexpected. It came in the middle of nowhere, as I had just cleaned out the carburetor of my stalled Renault 4L at the side of the dreadful dirt road to Madagascar's far south. With Bob at the wheel, the Buettner-Januschs' Land Rover suddenly appeared around a curve at high speed and vanished in a cloud of dust before I could register who its occupants were. That dust slowly settled into the 4L's open carburetor—which I had to remove, cursing, in order to blow it out once more. We finally met under happier circumstances in New York City when, at the end of his fieldwork, Bob came to take up an appointment at Hunter College. By that time my allegiance had long since shifted from the extinct lemurs to the living ones, and we duly began planning joint fieldwork in Madagascar, where we passed the summer of 1973 studying mongoose lemurs at Ampijoroa in the northwest. To our great surprise, the supposedly diurnal lemurs turned out to be on a nocturnal schedule, so that during those memorable weeks I learned from Bob not only how to go about doing a lemur field study, but how to deal with the unexpected when doing so.

I think Bob would have been as happy, as I certainly would have, to spend his entire professional life in Madagascar. But fate intervened in the form of the 1973–1975 revolution, which slowed field research there to a creep for almost a decade. In 1974, I was forced to leave the island and transfer my attention to the Comoro archipelago; and soon Bob, too, had to look elsewhere—though not before he and Alison Richard had set up the unique Beza-Mahafaly Reserve. We both wanted to continue working on lemurs, and accordingly we planned further joint field research in the Comoros, the only place besides Madagascar with wild-living lemur populations. But in 1977, political events caught up with us in the Comoros too—just as we happened to be on our way down there to begin a field season—and we were obliged to divert to Mauritius, where I had fled after my involuntary exit from Madagascar three years earlier. On that occasion I had been able to observe some of the introduced Mauritian long-tailed macaques, albeit from a great distance because these monkeys were persecuted as pests in the sugar plantations. And for other reasons too: "curry number two" was a commonly heard euphemism for a popular local dish that did not entirely lack macaque associations.

Trying to observe shy, skittish, and very fast monkeys under those conditions was an intimidating prospect. It proved equally daunting in practice; so, it was in Mauritius that I learned from Bob the value of perseverance in the pursuit of primatology. By the time I left a couple of months later to go to the newly reopened Comoros (a big mistake, but that's another story), the basis had already been laid for a long-term study of the Mauritian macaques, one that Bob and his students intermittently carried out over several subsequent years. Study conditions in Mauritius were very different from those in Madagascar; but in both cases Bob was scrupulous about being as rigorous as conditions permitted and about extracting the broadest spectrum of information possible. He and Linda also made the most of the personal opportunities that presented themselves and, over the years, they became fixtures on the Mauritian scene, just as they had integrated themselves into the isolated local village near their study site of Antserananomby in Madagascar's

southwest. The macaques also eventually cooperated, proving to exploit remarkably similar diets in both undisturbed forest and degraded settings—diets that, moreover, echoed those preferred by the parental populations in Indonesia from which they were almost certainly derived as a result of sixteenth-century Portuguese explorations. These findings amply confirmed Bob's abiding belief, initially acquired in the forests of southwest Madagascar, in species-specific dietary adaptation and the importance of evolutionary legacy in the determination of foraging behaviors.

The contents of this book are in a very real sense an homage to Bob's amazingly productive career following those early formative days in Madagascar and Mauritius. Its descriptive chapters compactly cover all of the major primate taxonomic groups, each one hewing to a similar format and asking broadly the same questions; and all of them are written by Bob's students or close associates. Indeed, in an eloquent accolade to the breadth of his influence, well over a third of the descriptive contributions acknowledge Bob as co-author. All of the writers are distinguished independent investigators; and perhaps the book's most remarkable feature is that, even with its restricted reservoir of authorship, the end result is as comprehensive, authoritative, and unified an account of the primates as you could ever wish to find in a single manageable volume. The reason is not far to seek: Bob never took his eye off the big picture, and he wanted to know where each and every primate species fit into it. One of the great mentors in primatology, he never hesitated to send his students into parts (and species) unknown, to ask the questions that he felt needed answering.

This book is, then, a tribute to Bob Sussman, both personally and professionally, as well as an authoritative, current, and disciplined synthesis of findings in the field of primatology that he loved and did so much to help advance. It is additionally leavened, and given larger context, by the reminiscences, historical perspectives, and research insights of the seasoned observers who wrote the section commentaries. The end result is a unified whole that has a coherence one rarely encounters in an edited volume, especially one that is also an unofficial Festschrift. The editors and authors are to be congratulated; and Bob would have been proud to see his vision so beautifully realized.

Introduction

Donna Hart

IT WOULD BE IMPOSSIBLE TO INTRODUCE THE CONTENTS of this volume without declaring that it was the brainchild of Robert W. Sussman. Those of us who had the good fortune to be schooled in the science of primatology by Bob Sussman were, indeed, privileged. Both Ian Colquhoun, my co-editor, and I earned our PhDs at Washington University in St. Louis under Bob's guidance and scrutiny. I remember being enlisted (probably along with many other grad students) to participate in literature searches for "the book." That was back in the 1990s; sometime in the 2000-teens, Bob decided that an edited volume was likely a better idea than a sole-authored one (see Ian Tattersall's explanation in the Foreword regarding the wisdom behind that decision). Thus, the publication of this volume was a decades-long goal for Bob Sussman, and it is an honor to have inherited it and brought it to completion. Bob's footprint has been large on present and future generations of primatologists. Many of the contributors to this volume were Bob's graduate students (as shown in Figure 1). His teaching mission still continues through the hundreds of students that his graduates have taught.

The Natural History of Primates: A Systematic Survey of Ecology and Behavior incorporates a distinctly anthropological, holistic perspective to the summary and synthesis of field investigations carried out on wild primates. This volume contains overviews of primate clades from experts with field experience spanning all primate taxa. Chapters were written by an ensemble of contributors with extraordinary depth and breadth of research experience, both through time (often decades) and space (everywhere that primate species inhabit). Each chapter focuses on particular primate taxa—infraorders, families, subfamilies, tribes, or genera. The organization incorporates a systematic approach in the sense of taxonomic groupings, from strepsirrhines and tarsiers, to New World monkeys, to Old World monkeys, to apes.

Chapters bring ecological and behavioral material together to provide a comprehensive overview of primates from different parts of the world—Malagasy lemuriforms, small Asian and African nocturnal primates, Neotropical arboreal monkeys, African and Asian cercopithecines and colobines, Asian hylobatids and pongids, as well as African apes. Consequently, readers wanting to learn about the ecology and behavior of a primate species and its relatives (perhaps, slow lorises, or titi monkeys, or mandrills, or bonobos) are able to obtain this information in one inclusive chapter. The addition of expert commentaries at the end of each section provides topical perspectives that speak to related issues and are written by authorities on the primates within that segment of the book. Robert Martin writes of his own prosimian research and reflects on the foundations of lemuriform studies in Madagascar. Anthony Rylands and Russell Mittermeier, both with Re:wild (formerly Global Wildlife Conservation), provide a historical timeline of research, exploitation, and conservation of platyrrhines, the Neotropical monkeys. Larissa Swedell and W. Scott McGraw deliver a phylogenetic overview of African and Asian monkeys and their socioecological adaptations. Michele Goldsmith gives engaging insight into her exhilarating personal experiences in the field as a gorilla researcher.

Each chapter of *The Natural History of Primates* has a similar configuration with important ecological and behavioral topics covered for each taxon. Chapters sort primates into systematic hierarchies through up-to-date taxonomy, phylogeny, and geographic distribution. Then, aspects of ecology and behavior are examined and reviewed: locomotion and habitat preferences, home range and ranging behavior, diet and feeding, activity cycles and sleeping sites, predation and competition, social structure and organization, patterns of communication, reproduction and infant development, and anthropogenic threats causing population declines and habitat destruction. Chapter contributors present thorough and rigorous overviews of ecology and behavior. However, there are various depths of data available for each taxon—in other words, there is diversity in longitudinal field data and unequal datasets due to research emphasis on different socioecological aspects. Unique habitats and research sites, diverse lengths of time between earlier studies and current ones, and changing environmental conditions are all at play to skew what can be discerned. Classic ecological and behavioral studies are cited; nonetheless, no theory or finding applies to all primates or even

Figure 1   The "Academic Tree" of PhD students supervised by Professor Robert W. Sussman (and including his own dissertation supervisor, John Buettner-Janusch).
(Credit: www.academictree.org)

the primates in one genus or one species—there is always inter- and intraspecific variation implicit in complex animals such as primates (see Strier, 2017). Still, to the extent that each chapter was able to present material in a similar configuration, inter-taxa comparisons across chapters can be readily made.

For purely organizational purposes we have divided the Order Primates into four sections in this volume: strepsirrhines (lemurs, galagos, lorises) and tarsiers, New World monkeys (platyrrhines with broad, flat noses and sideways-facing nostrils that are indigenous to Mexico, Central, and South America), Old World monkeys (indigenous to Africa and Asia), and the apes (tail-less hominoids native to Africa and Southeast Asia in the same superfamily as humans). So-called traditional taxonomy divides the Order Primates into two suborders, Prosimii and Anthropoidea, based on anatomical traits. The prosimians (the lemurs, galagos, lorises, and tarsiers under this categorization) were considered to exhibit more ancestral traits of the early, now-extinct, primates appearing 65 million years ago with, for example, their relatively high dependence on sense of smell. The more recent taxonomic division into Suborders Strepsirrhini and Haplorrhini was proposed to better accommodate the tarsiers. While the strepsirrhines have wet noses (i.e.,

rhinaria, an ancestral mammalian trait exhibited by all lorisiform and lemuriform primates), the haplorrhines have dry noses, and tarsiers have dry noses. The gross morphology and ecology of the tarsier is similar in many ways to the lorises, galagos, and lemurs. Still, it has features that link it to the monkeys and apes—a dry nose, an enhanced sense of vision with forward-facing eyes, and a relatively larger brain-body ratio.

Because there are no other primatology books organized solely in this systematic fashion, e.g., focusing not on cross-species comparisons of socioecological topics but on the topics as they apply across taxa in the Order Primates, we hope this volume serves as a reference book for those wanting an accessible and comprehensive treatment of any particular taxon of primate. It is designed to focus on primate taxa so that readers can acquire background on their biogeography, the ecological communities in which they are found, and their basic ecology, social organization, and behavior. Fascinating and unusual features of particular taxa are also covered. For example, torpor in some nocturnal lemurs, hunting behavior in capuchins, new tool variants in chimpanzees, the speed (and noise) with which slow lorises can use their teeth to gouge exudate holes in trees, wide-spread gummivory across primate taxa, and primate seed dispersal to rejuvenate forests are all put into context and perspective given the total adaptive pattern of the taxon and of primates in general.

Each chapter in *The Natural History of Primates* introduces recent changes in primate taxonomy since DNA and mitochondrial DNA analyses have expanded the ability to separate species that might have been previously lumped together based on morphology or distribution. There is no doubt taxonomy is an ever-evolving discipline. Fueled by new DNA and mtDNA analyses, geneticists have confirmed a wealth of new genera and species. For instance, there is recognition now that 24 species of *Microcebus* exist (Rasoloarison et al., 2000; Yoder et al., 2000; Andriantomphohavana et al., 2006; Louis et al., 2006, 2008; Olivieri et al., 2007; Schüßler et al., 2020) when once the mouse lemur was thought to be monospecific with two subspecies (Petter et al., 1977). In a similar proliferation, the nocturnal genus *Lepilemur* (sportive or weasel lemurs) now includes 26 species placed in the family Lepilemuridae (IUCN, 2020; but see Tattersall, 2007). The numbers of New World platyrrhines also have expanded, with additional genera and species identified. A case in point involves the recent recognition of 35 titi monkey species, along with a reclassification by Byrne et al. (2016) dividing the previously considered monogeneric titis into three genera (*Callicebus*, *Cheracebus*, and *Plecturocebus*) based on molecular evidence that has been corroborated by biogeographical and morphological data. Even the primate clades in Asia, particularly the langurs and tarsiers, are in the midst of taxonomic turmoil—and, thus, the list goes on and on.

As recently as the first decades of the twenty-first century, new species of primates have been discovered and are discussed in their respective chapters: examples include the novel genus and species of an African monkey, *Rungwecebus kipunji* (Jones et al., 2005; Davenport et al., 2006); a new species of titi monkey (*Callicebus miltoni*) in Brazil named after Milton Thiago de Mello, professor at the University of Brasília (Dalponte et al., 2014); the Arunachal macaque (*Macaca munzala*) found at high altitudes in northeastern India as described by Sinha et al. (2005); the white-cheeked macaque (*M. leucogenys*) discovered in southeastern Tibet (Li et al., 2015) and identified from 738 photos and DNA extracted from skin samples; the black, or Burmese, snub-nosed monkey (*Rhinopithecus strykii*) found in Myanmar (Geissmann et al., 2011; Meyer et al., 2017); and a new guenon, *Cercopithecus lomamiensis*, from the Democratic Republic of Congo (Hart et al., 2012). Although one would think that every mammalian species is already known to science, discoveries continue to mount. Even as we go to press with this volume, one heretofore unknown species, the Popa langur (*Trachypithecus popa*), has been found

in Myanmar, although published data are not yet available (Roos et al., 2020). The recognition of a new primate species can be exciting, but a word of caution should be linked to each discovery. New sightings are often directly correlated with areas of the world that had been fairly devoid of human influence until recent incursions into pristine and impenetrable environments. In other words, new species are literally being uncovered and revealed because their surrounding habitats have been penetrated by human activity.

Three significant components make this volume unique: (1) the information presented in each chapter, for all taxa, is grounded in and focused on field research of wild primates in their natural environments, (2) each chapter includes a discussion of predation on primates in the context of predators as a primary and natural ecological factor, and (3) each chapter gives a comprehensive update on the population status of the respective primate species in the wild and conservation efforts on their behalf.

Observing animals in the wild has always been the backbone of primatology. Collecting data in a methodical and organized manner, with confounding variables acknowledged or accounted for, while recording the unfolding of natural events, both ecological and social, is the gold standard for fieldwork on primate species. Yet, reality intrudes on the best-laid fieldwork plans, whereby some primates can be habituated to human observers, while others cannot. Some habitats are hospitable, while others make fieldwork extremely difficult. For example, terrain such as deep, submontane rainforest where drills (*Mandrillus leucophaeus*) reside (Fa et al., 2005; Oates & Butynski, 2008) may defy typical observational methodology. Drills are shy and wary; thus, diet information must be obtained mainly from the examination of fecal samples (Astaras et al., 2008). Another illustration is the rugged mountain home of the black snub-nosed monkey, *Rhinopithecus strykeri*. No complete data on its feeding habits have been gathered due to the steep mountainous habitat (Yang et al., 2019). While much important and interesting information has been gleaned from captive primate studies, laboratory testing involves manipulation of the animals involved and has the inherent bias of information extracted from stressed wild animals. Studies on groups of zoo or free-ranging animals, a description that is slightly euphemistic for primates that are not in cages but still confined outside their natural habitat, have also added to scientific information about primate behavior. However, such data may be fraught with stresses or even be unnaturally devoid of them. Predation or seasonal changes in fruit availability, for example, may occur but not necessarily be the same as what transpires in a wild state. For these reasons, this volume has little reference to captive or free-ranging groups of primates.

All chapters in *The Natural History of Primates* focus on fieldwork. Precise scientific methodologies may be taken to the field, but unpredictable occurrences are almost guaranteed. If the habitat in which primates are studied is a functioning ecosystem and not overly altered by human activity, there will be predators of all sorts. Raptors (eagles, hawks, buzzards, owls), other predatory birds (crows, toucans, shrike-like vangas), reptiles (pythons, boa constrictors, anacondas, tegus, monitor lizards, crocodiles), hyaenids, felids, canids, ursids, and small carnivores (fossa, civets, genets, tayra, mongoose) may all be a threat since all have been documented to be predators of primate taxa (Hart, 2000; Miller & Treves, 2011). Predation is termed "natural mortality," just like sickness or old age, rather than additive mortality which occurs through hunting or other human-inflicted death. Predation is a part of the ecological reality of prey species. It befalls the smallest of the mouse lemur species in Madagascar (Schäffler & Kappeler, 2014) that weigh as little as 33 g, all the way to gorilla subspecies in Central Africa, weighing in at 169.5 kg (Fay et al., 1995). Predation is a force interwoven into primate evolution and adaptations (e.g., Headland & Greene, 2011; McGraw & Berger, 2013), and it is reflected in their morphology, ecology, behavior, overall activity patterns, and social structure. Selective pressure exerted by predation is advanced as an explanation for group living in primates (van Schaik & van Hooff, 1983), as well as the size of social groups (Janson & Goldsmith, 1995; Hill & Lee, 1998), the type of social system found in arboreal primates (Treves, 1999), and multifaceted and complex vocal repertoires in primates (Zuberbühler et al., 1999; Schei et al., 2010; Adams & Kitchen, 2018).

While predation as an impact (or force) on primates was often minimized in the past (Hall, 1966; Hausfater & Hrdy, 1984; Cheney & Wrangham, 1987), there is now a recognition that the ecological parameters for primates (i.e., all of the interacting communities of plants and animals as part of a functioning ecosystem) include those species for which they are prey (see, for example, Isbell, 1994; Hart, 2000; Zuberbühler & Jenny, 2002; Miller & Treves, 2011; Burnham et al., 2013; McGraw & Berger, 2013; Bidner, 2014; Farris et al., 2014). Some primate researchers have documented, as much as possible, the circumstances surrounding an attempted or successful killing of a primate by a predator. However, many field researchers are in the process of studying a particular aspect of primate ecology or behavior when the unexpected drama of predation unfolds or, alternatively, the puzzling disappearance of a study animal happens. Given the instantaneous reaction time necessary to process the former phenomenon and the mysterious situation of the latter, records of predation are frequently brief and anecdotal asides to the main body of the published research. However, the wider scientific literature encompassing field research on mammalian carnivores, birds of prey, and large reptiles contains both quantitative and qualitative discussions of primates as prey. Quantitative data are presented through sampling methodologies such as fecal or regurgitation examination, or analyses of nest and den remains and prey carcasses (e.g., McGraw, 2006). Descriptions of hunting behavior on the part of the predators reflect adaptations that enable them to successfully hunt primates (e.g., Schultz & Thomsett, 2007; Headland & Greene, 2011). Combining the two mirror images of predation—the predator and the prey—offers a holistic view of ecological relationships (Hart & Sussman, 2009).

Of course, predation by other wild animals is *not* the force driving primates to extinction. Estrada et al. (2017, 2018, 2019, 2020) reported that approximately 60% of primate species are threatened with extinction and about 75% of primate species are experiencing decline in population numbers. This dire situation affects 504 extant species in 79 genera in all regions inhabited by nonhuman primates. Just a few years since those analyses, the International Union for the Conservation of Nature has been forced to increase the percentage of primate species threatened with extinction to 65% (IUCN, 2020). Circumstances leading to extinction are solely the result of escalating anthropogenic pressures on primates and their habitats, and they continue to accelerate (Estrada et al., 2017, 2018). Besides climate change and its mounting effects on habitat, human encroachment into once-pristine rainforests, and hunting for bushmeat, primates are involved in the pet and "trinket" trade, both legal and illegal. (The illegal trade in all wildlife species is estimated to be worth $8–$10 billion per year, right alongside drugs, guns, and people trafficking [Redmond, 2019].) Americans may not perceive themselves as primary agents of primate destruction, but statistics tell a different story. The most recent data that can be accessed, obtained from the U.S. Fish and Wildlife Service for 2010–2014, list 2,068,328 primates and primate parts imported into the U.S., including "dead and alive individual primates, primate skulls, products (such as boots), milliliters of blood, skins and [hunting] trophies.... Commercially, the U.S. also imported primate skulls and primate 'bodies' as décor, including baboons, macaques and slow lorises" (Sanerib & Uhlemann, 2020: p.ii). Humans in both rich and poor nations are, indeed, at the center of primate devastation worldwide.

Raising awareness in scientific circles and within the general public about the global annihilation of our closest relatives and what their loss will mean to ecosystem health is a staggering, but imperative, task (Estrada et al., 2017, 2018, 2019, 2020). Conservation is a field spanning many specialties, and to be successful, it must realistically include many components. It follows that everything from the collection of scientific data, to amazing photography, cultural relativity, and wide ecosystemic views, along with written words that unite emotion with facts, plus development of pedagogical applications, activism, and the stirring of public opinion by scientists and dedicated environmentalists, to legal acumen, international treaties, and contacts with government leaders, lawmakers, and local peoples across the globe, must be a part of and must integrate into active political strategies to preserve wildlife and wild environments.

Michael Soulé, the founder of "conservation biology" (a scientific field distinct from ecology), declared, "Decision makers responsible for conservation policy and practice must balance a desire for knowledge... with the need to act" (Cook et al., 2013: p.669, from Soulé, 1985). Soulé proposed that the traditional sciences were neutral pursuits, but conservation biology, on the other hand, was actively devoted to saving species from extinction and ecosystems from destruction. Soulé was known to have once told an audience of scientists that if not a single ecology paper was ever published again, the knowledge already existed to stop the extinction crisis. Action is what is missing, and in order to change the direction of the extraordinary meltdown we are experiencing, those who care need to become activists. "We are dealing with a diminishing asset—wildness—whether it's the disappearance of a single species or an entire fauna, such as Africa's wildlife. Lurking in the minds of most conservation biologists is a singular realization: our subject matter is disappearing," Soulé warned (Tonino, 2018: p.12).

The spectrum of activism is wide and primatologists are involved in many areas of conservation. Dr. Jane Goodall is a prime example of the scientist-activist, giving speeches on both primatology and activism to audiences throughout the world. Perhaps less well-known to the general public, is the innovative and imaginative activism by nongovernmental organizations (NGOs) which provides unique niche solutions to conservation dilemmas. Indeed, many of the contributors to this volume are deeply committed to and involved in primate and biodiversity protection through conservation NGOs. While there are numerous examples, two described here serve to highlight the very wide diversity and unusual tactics found in activism on behalf of primates: Twenty-thousand young plants of *Yushania alpine* collected from Kenya's Aberdare Range are currently being raised in a special bamboo nursery by African Wildlife Foundation.

These seedlings, the foliage of choice for mountain gorillas in the montane rainforest ecosystem, are intended to restore mountain gorilla habitat in Rwanda (Sehmi, 2020). Another, but very different, example of activism involves animal welfare and wild primates. People for the Ethical Treatment of Animals (PETA) recently launched a campaign targeting the capture of wild juvenile *Macaca nemestrina* (categorized as a Vulnerable species by IUCN). Thirty years ago, Oi (1990) identified the exploitation of southern pig-tailed macaques in Southeast Asia on coconut farms and raised concerns about how the capture of wild *M. nemestrina* might be depressing the population size. Yet, it wasn't until the burgeoning Western market for coconut milk emerged that an uber-activist organization became involved. According to PETA (2020), their campaign has resulted in large grocery chains, with a total of 26,000 stores, banning purchases of brand-name coconut milk products from the corporation that captures, chains, and coerces young macaques to climb trees and pick coconuts.

In some ways, it is hard to see how this volume could not be a new way to present primates, given that Bob Sussman always brought new perspectives to primatology. One of his breakthrough theories relating to primate taxa concerned species-specific dietary adaptations. As Peter Ungar (2019) explained in a tribute to Bob in the *American Journal of Primatology*, the theory, first and foremost, is that diet is related to dental and digestive tract adaptations that are part of a primate species' evolutionary legacy. Another illustration is Bob's angiosperm theory which posited a co-evolutionary link between primates and flowering plants (Sussman, 1991). But, certainly, the most groundbreaking is his primate cooperation theory, one which presents a major challenge to cultural stereotyping. Intended to dispute the Western paradigm that competition and aggression are at the roots of primate and, therefore, human evolution, Sussman et al. (2005) presented a data-driven survey of social interactions from 81 studies published in the primate literature, concluding that cooperative and affiliative behaviors are overwhelmingly more common in all primate species than agonistic conflicts. (Taking this one step further, if our primate relatives and our pre-hominin ancestors were more cooperative than they were aggressive, then why are humans so often labeled as inherently violent?)

Much has happened in primatology over the course of its history as a scientific discipline. Field primatology began in earnest more than 60 years ago, and since then the research on our closest living relatives has progressed from anecdotal records to long-term studies that demand statistically sophisticated analyses. But the lure of primatology has always been the complexity of the species encountered and the familiarity we humans recognize in the natural behavior of our primate relatives. Bob Sussman's work embodied these concepts. He loved to be in the field with primates and urged others to pursue knowledge about them that would add to our understanding of their life histories. Perhaps the best way to end this introduction to *The Natural History of Primates* is with a reminder of what Michael Soulé said (Tonino, 2018: p.12), and what Bob Sussman believed: "We only protect what we love."

REFERENCES CITED—INTRODUCTION

Adams, D., & Kitchen, D. (2018). Experimental evidence that titi and saki monkey alarm calls deter an ambush predator. *Animal Behaviour*, 145, 141-147.

Andriantomphohavana, R., Zaonarivelo, J. R., Engberg, S. E., Randriamampionona, R., McGuire S. M., Shore, G. D., Rakotonomenjanahary, R., Brenneman, R. A., & Louis, E. E. Jr. (2006). Mouse lemurs of northwestern Madagascar with a description of a new species at Lokobe Special Reserve. *Occasional Papers: Museum of Texas Tech University*, 259, 1-24.

Astaras, C., Mühlenberg, M., & Waltert, M. (2008). Note on drill (*Mandrillus leucophaeus*) ecology and conservation status in Korup National Park, southwestern Cameroon. *American Journal of Primatology*, 70, 306-310.

Bidner, L. (2014). Primates on the menu: Direct and indirect effects of predation on primate communities. *International Journal of Primatology*, 35, 1164-1177.

Burnham, D., Bearder, S., Cheyne, S., Dunbar, R., & Macdonald, D. (2013). Predation by mammalian carnivores on nocturnal primates: Is the lack of evidence support for the effectiveness of nocturnality as an antipredator strategy? *Folia Primatologica*, 83(3-6), 236-251.

Byrne, H., Rylands, A. B., Carneiro, J. C., Alfaro, J. W., Bertuol, F., da Silva, M. N. F., Messias, M., Groves, C., Mittermeier, R., Farias, I., Hrbek, T., Schneider, H., Sampaio, I., & Boubli, J. (2016). Phylogenetic relationships of the New World titi monkeys (*Callicebus*): First appraisal of taxonomy based on molecular evidence. *Frontiers in Zoology*, 13(10). https://doi.org/10.1186/s12983-016-0142-4.

Cheney, D, & Wrangham, R. (1987). Predation. In B. B. Smuts, D. L. Cheney, R. M. Seyfarth, R. W. Wrangham, & T. T. Struhsaker (Eds.), *Primate societies* (pp. 227-239). University of Chicago Press.

Cook, C. N., Mascia, M. B., Schwartz, M. W., Possingham, H. P., & Fuller, R. A. (2013). Achieving conservation science that bridges the knowledge-action boundary. *Conservation Biology*, 27(4), 669-678.

Dalponte J., Silve, F., & Jùnior, J. (2014). New species of titi monkey, genus *Callicebus* Thomas, 1902 (Primates, Pitheciidae) from southern Amazonia, Brazil. *Papéis Avulsos de Zoologia*, 54, 457-472.

Davenport, T. R. B., Stanley, W. T., Sargis, E. J., de Luca, D. W., Mpunga, N. E., Machaga, S. J., & Olson, L. E. (2006). A new genus of African monkey, *Rungwecebus*: Morphology, ecology, and molecular phylogenetics. *Science*, 312, 1378-1381.

Estrada, A., Garber, P. A., Rylands, A. B., Roos, C., Fernandez-Duque, E., Di Fiore, A., Nekaris, K. A. I., Nijman, V., Heymann, E. W., Lambert, J. E., Rovero, F., Barelli, C., Setchell, J. M., Gillespie, T. R., Mittermeier, R. A., Arregoitia, L. V., de Guinea, M., Gouveia, S., Dobrovolski, R.,... & Li, B. (2017). Impending extinction crisis of the world's primates: Why primates matter. *Science Advances*, 3, e1600946.

Estrada, A., Garber, P. A., Mittermeier, R. A., Wich, S., Gouveia, S., Dobrovolski, R., Nekaris, K. A. I., Nijman, V., Rylands, A. B., Maisels, F., Williamson, E., Bicca-Marques, J., Fuentes, A., Jerusalinsky, L., Johnson, S., Rodrigues de Melo, F., Oliveira, L., Schwitzer, C.,... & Setiawan, A. (2018). Primates in peril: The significance of Brazil, Madagascar, Indonesia and the Democratic Republic of the Congo for global primate conservation. *PeerJ*, 6, e4869.

Estrada, A., Garber, P.A., Chaudhary, A. (2019). Expanding global commodities trade and consumption place the world's primates at risk of extinction. *PeerJ*, 7, e7068

Estrada, A., Garber, P.A., Chaudhary, A. (2020). Current and future trends in socio-economic, demographic and governance factors affecting global primate conservation. *PeerJ*, 8, e9816

Fa, J. E., Ryan, S., & Bell, D. J. (2005). Hunting vulnerability, ecological characteristics and harvest rates of bushmeat species in Afrotropical forests. *Biological Conservation*, 121, 167-176.

Farris, Z. J., Karpanty, S. M., Ratelolahy, F., & Kelly, M. J. (2014). Predator-primate distribution, activity, and co-occurrence in relation to habitat and human activity across fragmented and contiguous forest in northeastern Madagascar. *International Journal of Primatology*, 35(5), 859-880.

Fay, J., Carroll, R., Kerbis Peterhans, J., & Harris, D. (1995). Leopard attack on and consumption of gorillas in the Central African Republic. *Journal of Human Evolution*, 29(1), 93-99.

Geissmann, T., Lwin, N., Aung, S., Aung, T., Aung, Z., Hla, T., Grindley, M., & Momberg, F. (2011). A new species of snub-nosed monkey, genus *Rhinopithecus* Milne-Edwards, 1872 (Primates, Colobinae), from northern Kachin state, northeastern Myanmar. *American Journal of Primatology*, 73, 96-107.

Hall, K. (1966). Distribution and adaptation of baboons. *Symposium of the Zoological Society of London*, 17, 49-73.

Hart, D. (2000). *Primates as prey: Ecological, morphological and behavioral relationships between primate species and their predators.* [PhD dissertation, Washington University].

Hart, D., & Sussman, R. W. (2009). *Man the hunted: Primates, predators, and human evolution.* Westview Press.

Hart, J. A., Detwiler, K. M., Gilbert, C. C., Burrell, A. S., Fuller, J. L., Emetshu, M., Hart, T. B., Vosper, A., Sargis, E. J., & Tosi, A. J. (2012). Lesula: A new species of *Cercopithecus* monkey endemic to the Democratic Republic of Congo and implications for conservation of Congo's central basin. *PLoS ONE*, 7(9), e44271.

Hausfater, G., & Hrdy, S. (1984). *Infanticide: Comparative and evolutionary perspectives*. Aldine Publishing.

Headland, T. N., & Greene, H. W. (2011). Hunter-gatherers and other primates as prey, predators, and competitors of snakes. *Proceedings of the National Academy of Sciences*, 108(52), E1470-E1474.

Hill, R., & Lee, P. (1998). Predation risk as an influence on group size in cercopithecoid primates: Implications for social structure. *Journal of Zoology*, 245(4), 447-456.

IUCN. (2020). *International Union for the Conservation of Nature Red List of Threatened Species*. 2020-Version 3. www.iucnredlist.org.

Isbell, L. (1994). Predation on primates: Ecological patterns and evolutionary consequences. *Evolutionary Anthropology*, 3(2), 61-71.

Janson, C. H., & Goldsmith, M. L. (1995). Predicting group size in primates: Foraging costs and predation risks. *Behavioral Ecology*, 6(3), 326-336.

Jones, T., Ehardt, C., Butynski, T., Davenport, T., Mpunga, N., Machaga, S., & De Luca, D. (2005). The highland mangabey *Lophocebus kipunji*: A new species of African monkey. *Science*, 308(5725), 1161-1164.

Li, C., Zhao, C., & Fan, P. (2015). White-cheeked macaque (*Macaca leucogenys*): A new macaque species from Medog, southeastern Tibet. *American Journal of Primatology*, 77, 753-766.

Louis, E. E. Jr., Coles, M. S., Andriantompohavana, R., Sommer, J. A., Engberg, S. E., Zaonarivelo, J. R., & Mayor, M. I. (2006). Revision of the mouse lemurs (*Microcebus*) of eastern Madagascar. *International Journal of Primatology*, 27, 347-389.

Louis, E. E. Jr., Engberg, S. E., McGuire, S. M., McCormick, M. J., Randriamampionona, R., Ranaivoarisoa, J. F., Bailey, C. A., Mittermeier, R. A., & Lei, R. (2008). Revision of the mouse lemurs, *Microcebus* (Primates, Lemuriformes) of northern and northwestern Madagascar with descriptions of two new species at Montagne d'Ambre National Park and Antafondro Classified Forest. *Primate Conservation*, 23, 19-38.

McGraw, W. S. (2006). Primate remains from African crowned eagle (*Stephanoaetus coronatus*) nests in Ivory Coast's Taï Forest: Implications for primate predation and early hominid taphonomy in South Africa. *American Journal of Physical Anthropology*, 131(2), 151-165.

McGraw, W. S., & Berger, L. R. (2013). Raptors and primate evolution. *Evolutionary Anthropology*, 22(6), 280-293.

Meyer, D., Momberg, F., Matauschek, C., Oswald, P., Lwin, N., Aung, S., Yang, Y., Xiao, W., Long, Y-C., Grueter, C., & Roos, C. (2017). *Conservation status of the Myanmar or black snub-nosed monkey Rhinopithecus strykeri*. Fauna & Flora International, Yangon, Myanmar; Institute of Eastern-Himalaya Biodiversity Research, Dali, China; German Primate Center, Göttingen, Germany.

Miller, L., & Treves, A. (2011). Predation on primates: Past studies, current challenges, and directions for the future.

In C. Campbell, A. Fuentes, K. MacKinnon, M. Panger, & S. Bearder (Eds.), *Primates in perspective* (2nd Ed., pp. 525-543). Oxford University Press.

Oates, J., & Butynski, T. (2008). *Mandrillus leucophaeus. The International Union for the Conservation of Nature Red List of Threatened Species*. www.iucnredlist.org.

Oi, T. (1990). Mating systems of macaques. *13th Congress of the International Primatological Society, Nagoya and Kyoto, Japan*. [Abstract].

Olivieri, G., Zimmerman, E., Randrianambinina, B., Rasoloharijaona, S., Rakotondravony, D., Guschanski, K., & Radespiel, U. (2007). The ever-increasing diversity in mouse lemurs: Three new species in north and northwestern Madagascar. *Molecular Phylogenetics and Evolution*, 43, 309-327.

PETA. (2020). People for the Ethical Treatment of Animals cracks down on forced monkey labor. *PETA Global*, Autumn 4, 5.

Petter, J. J., Albignac, R., & Rumpler, Y. (1977). *Faune de Madagascar, 44: Mammifères Lémuriens (Primates Prosimiens)*. Ostrom, CNRS.

Rasoloarison, R., Goodman, S., & Ganzhorn, J. (2000). Taxonomic revision of mouse lemurs (*Microcebus*) in the western portions of Madagascar. *International Journal of Primatology*, 21, 963-1019.

Redmond, I. (2019). Highlights of the CITES CoP 18, Geneva, Switzerland. *International Primate Protection League News*, 46(3), 5-7.

Roos, C., Helgen, K., Portela Miguez, R., Thant, N., Lwin, N., Lin, A., Lin, A., Yi, K., Soe, P., Hein, Z., Myint, M., Ahmed, T., Chetry, D., Urh, M., Veatch, M., Duncan, N., Kamminga, P., Chua, M., Yao, L., Matauschek, C., Meyer, D., Liu, Z., Li, M., Nadler, T., Fan, P., Quyet, L., Hofreiter, M., Zinner, D., & Momberg, F. (2020). Mitogenomic phylogeny of the Asian colobine genus *Trachypithecus* with special focus on *Trachypithecus phayrei* (Blyth, 1847) and description of a new species. *Zoological Research*, 41(6), 656-669.

Sanerib, T., & Uhlemann, S. (2020). *Dealing in disease: How U.S. wildlife imports fuel global pandemic risks*. Center for Biological Diversity Report. www.biologicaldiversity.org.

Schäffler, L., & Kappeler, P. (2014). Distribution and abundance of the world's smallest primate, *Microcebus berthae*, in central western Madagascar. *International Journal of Primatology*, 35(2), 557-572.

Schei, A., Candiotti, A., & Zuberbühler, K. (2010). Predator-deterring alarm call sequences in guereza colobus monkeys are meaningful to conspecifics. *Animal Behaviour*, 80(5), 799-808.

Schüßler, D., Blanco, M. B., Salmona, J., Poelstra, J., Andriambeloson, J. B., Miller, A., Randrianambinina, B., Rasolofoson, D. W., Mantilla-Contreras, J., Chikhi, L., Louis Jr., E. E., Yoder, A. D., & Radespiel, U. (2020). Ecology and morphology of mouse lemurs (*Microcebus* spp.) in a hotspot of microendemism in northeastern Madagascar, with the description of a new species. *American Journal of Primatology*, 82(9), e23180.

Schultz, S., & Thomsett, S. (2007). Interactions between African crowned eagles and their prey community. In W. S. McGraw, K. Zuberbühler, & R. Noë (Eds.), *Monkeys of the Taï Forest: An African primate community* (pp. 171-193). Cambridge University Press.

Sehmi, H. 2020. *Baby bamboo plants restore mountain gorilla habitat in Rwanda*. African Wildlife Foundation. www.awf.org/blog/baby-bamboo-plants-restore.

Sinha, A., Datta, A., Madhusudan, M., & Mishra, C. (2005). *Macaca munzala*: A new species from Western Arunachal Pradesh, Northeastern India. *International Journal of Primatology,* 26(4), 977-989.

Soulé, M. E. (1985). What is conservation biology? *BioScience,* 35, 727-734.

Strier, K. (2017). What does variation in primate behavior mean? *American Journal of Physical Anthropology*, 162(S63), 4-14.

Sussman, R. W. (1991). Primate origins and the evolution of angiosperms. *American Journal of Primatology*, 23(4), 209-223.

Sussman, R. W., Garber, P. A., & Cheverud, J. M. (2005). Importance of cooperation and affiliation in the evolution of primate sociality. *American Journal of Physical Anthropology*, 128, 84-97.

Tattersall, I. (2007). Madagascar's lemurs: Cryptic diversity or taxonomic inflation? *Evolutionary Anthropology*, 16(1), 12-23.

Tonino, L. (2018). We only protect what we love: Michael Soulé on the vanishing wilderness. *The Sun Magazine.* www.thesunmagazine.org/issues/508/we-only-protect-what-we-love.

Treves, A. (1999). Has predation shaped the social systems of arboreal primates? *International Journal of Primatology,* 20(1), 35-67.

Ungar, P. (2019). Bob Sussman and the concept of species-specific dietary adaptations. *Primate Conservation*, 33, 1-4.

van Schaik, C. P., & van Hooff, J. (1983). On the ultimate causes of primate social systems. *Behaviour*, 1, 91-117.

Yang, Y., Groves, C., Garber, P., Wang, X., Li, H., Long, Y., Li, G., Tian, Y., Dong, S.,Yang, S., Behie, A., & Xiao, W. (2019). First insights into the feeding habits of the critically endangered black snub-nosed monkey, *Rhinopithecus strykeri* (Colobinae, Primates). *Primates*, 60(2), 143-153.

Yoder, A. D., Rasoloarison, R. M., Goodman, S. M., Irwin, J. A., Atsalis, S., Ravosa, M. J., & Ganzhorn, J. U. (2000). Remarkable species diversity in Malagasy mouse lemurs (Primates, *Microcebus*). *Proceedings of the National Academy of Sciences*, 97, 11325-11330.

Zuberbühler, K., & Jenny, D. (2002). Leopard predation and primate evolution. *Journal of Human Evolution*, 43(6), 873-886.

Zuberbühler, K., Jenny, D., & Bshary, R. (1999). The predator deterrence function of primate alarm calls. *Ethology*, 105(6), 477-490.

About the Contributors

Angela Achorn graduated from Rhode Island College in 2016 with a BA in Anthropology and a minor in Environmental Studies. She earned her MA in Anthropology at Texas A&M in 2018 and is now a PhD candidate there. Angela is currently in Indonesia on a Fulbright fellowship collecting data for her dissertation, which explores relationships between coloration, mating behaviors, parasite infections, and hormones in Sulawesi crested macaques.

Pamela Ashmore earned a PhD from Washington University in St. Louis with Bob Sussman as her advisor. Her dissertation focused on the behavioral ecology of macaques. She is co-author of *The Life of Primates* and is currently researching the private ownership of primates in the United States. She served as chair of the Department of Anthropology at the University of Missouri–St. Louis; presently, she recently retired as professor of biological anthropology at the University of Tennessee–Chattanooga.

Sylvia Atsalis is the founder of "Professional Development for Good," supporting young adults in STEM, service, and policy careers. Known for her research on the smallest living primates, mouse lemurs, and her study of reproductive aging and sexual behavior in female gorillas in North American zoos, Atsalis's publications include *Primate Reproductive Aging: Cross-Taxon Perspectives*. More recently Atsalis has focused on building conservation leadership in countries with fragile ecosystems and received a 2019 Fulbright Scholar award to train students in Suriname in primate ecology, conservation biology, professional development, and leadership.

Thad Q. Bartlett is a biological anthropologist at the University of Texas at San Antonio where he teaches undergraduate and graduate courses in human origins, human nature, and primate behavior and ecology. His primary research interest is in the behavioral ecology of gibbons. More recently he has begun to focus on the applied dimensions of primatology with a focus on documenting the distribution and population density of gibbons in Thailand and Peninsular Malaysia.

Michelle Bezanson is a primatologist and the chair of the Department of Anthropology at Santa Clara University. She conducts field studies on living nonhuman primates and has focused on conservation, natural history, positional behavior, ontogeny, and field ethics. She is also a distinguished artist whose unique illustrations have been featured in a number of prominent publications.

Ian C. Colquhoun first conducted behavioral observations on lemurs while an undergraduate at the University of Western Ontario. Fieldwork on lemurs continued during MA (McMaster University) and PhD studies (Washington University in St. Louis). He is an associate professor in the Department of Anthropology at the University of Western Ontario, past president of the Canadian Association for Biological Anthropology/L'Association canadienne d'anthropologie biologique (CABA-ACAB), and member of the Madagascar Section of the IUCN/SSC Primate Specialist Group.

Kristena Cooksey is a graduate student in Biological Anthropology at Washington University in St. Louis, Missouri. Prior to pursuing her doctoral studies, Cooksey worked at several accredited zoological parks in the United States. Her doctoral research focuses on gorilla sociality and health implications of group living in northern Republic of Congo.

Camille Coudrat is the founder-director of Association Anoulak, a nongovernmental organization based in Lao People's Democratic Republic (Lao PDR) dedicated to biodiversity conservation and human community resilience in the Annamite Mountains, Southeast Asia. Dr. Coudrat focuses on scientific research of primates, otters, small carnivores, ungulates, and elephants. She is a member of several IUCN SSC Specialist Groups and has co-authored publications on biodiversity in Southeast Asia.

Jennifer Cramer is a professor and program director of the Sociology and Interdisciplinary Studies Programs at American Public University System. Working in the Caribbean and in several African countries, her research started with a focus on monkey coloration and behavior. More recently, Jennifer's work in Africa has shifted to studying connections between primate and human health, and partnering with local communities to focus on primate conservation.

Roberto Delgado is a program director at the National Science Foundation. Known for his work on orangutan behavioral ecology, geographic variation, primate communication, and sexual selection, Delgado's publications include articles in the *American Journal of Physical Anthropology*, the *American Journal of Primatology*, *Current Biology*, *Ethology*, *Evolutionary Anthropology*, *Folia Primatologica*, and the *International Journal of Primatology*; and chapters in *Orangutans: Geographic Variation in Behavioral Ecology and Conservation* (Oxford University Press), and *Indonesian Primates* (Springer Academic Press).

Anneke DeLuycker is associate professor of Conservation Biology at the Smithsonian-Mason School of Conservation and Affiliate Faculty in the Department of Biology at George Mason University. Known for her work on the first long-term field study on the critically endangered Andean titi monkey in Peru (*Plecturocebus oenanthe*), she continues to contribute to primatological research concerning feeding plasticity, social structure dynamics, and behavior in fragmented landscapes. She witnessed and filmed the only scientifically documented birthing event of a titi monkey in the wild. Her recent publications have been featured in the *International Journal of Primatology*, the *American Journal of Primatology*, *Primates*, and *Oryx*. She is passionate about developing local and community-based initiatives to enhance conservation methodology.

Paul A. Garber is a professor emeritus in the Department of Anthropology and the Program in Ecology, Evolution, and Conservation Biology at the University of Illinois, Urbana. He has studied wild nonhuman primates in Mexico, Nicaragua, Costa Rica, Panama, Peru, Brazil, Bolivia, and China. Dr. Garber was the American Society of Primatologists' 2017 Distinguished Primatologist and was the executive editor of the *American Journal of Primatology* from 2008–2017.

Michele L. Goldsmith is professor and program director of Environmental Science at Southern New Hampshire University. Known for her work on western lowland and mountain gorilla behavioral ecology, she most recently studied impacts of ecotourism. She publishes her fieldwork in peer-reviewed journals and edited volumes, such as *Primate Ecotourism* and *Ethics in Biological Anthropology*. Goldsmith has co-edited two books: *Gorilla Biology: A Multidisciplinary Perspective* (with T. A. Taylor) and *Sacred Commerce: Environment, Ethics and Innovation* (with J. Chryssavgis).

Sharon Gursky has studied wild tarsiers in Indonesia for most of her career. She has published 6 books and over 50 peer-reviewed articles. Dr. Gursky received her BA from Hartwick College in Oneonta, New York, her MS from the University of New Mexico in Albuquerque, and her PhD from SUNY Stony Brook, New York. She currently is professor of Anthropology at Texas A&M University.

Donna Hart was recently retired from the Department of Anthropology, University of Missouri–St. Louis. She earned a PhD at Washington University in St. Louis. Her research areas include predation on primates and the ramifications of predation on early hominins. The latter topic is explored in *Man the Hunted*, the book for which Hart, and her co-author Bob Sussman, were awarded the 2006 W. W. Howells Book Prize. Preceding academics, Hart worked in the field of wildlife conservation, specializing in international wildlife treaties.

Kathryn Judson is a PhD student in the Department of Anthropology at Washington University in St. Louis. Prior to her doctoral studies, she was a research assistant for the Max Planck Institute for Evolutionary Anthropology for three years, working at the Loango Gorilla Project in Gabon and the Luikotale Bonobo Project in Democratic Republic of Congo.

Elizabeth (Lisa) Kelley is the executive director of the Saint Louis Zoo WildCare Institute. Through her position, Lisa provides institutional oversight for the zoo's 17 wildlife conservation WildCare Centers and 13 WildCare programs, which are located throughout the globe. She received her PhD in Anthropology from Washington University in St. Louis in 2011, under the mentorship of the late Dr. Robert W. Sussman. Her research background is in primatology, with a focus on lemurs in Madagascar.

Joanna E. Lambert is a professor in the Program of Environmental Studies and faculty of Ecology and Evolutionary Biology at the University of Colorado–Boulder. She has engaged in field research on wild mammals for over 30 years in various remote localities around the world, including primates in Equatorial Africa.

Martha Lyke is a lecturer in the Department of Anthropology at the University of Texas at San Antonio. Dr. Lyke is a broadly trained biological anthropologist and molecular ecologist and has published in *Molecular Ecology* and *PLoS ONE*.

Katherine C. MacKinnon is associate professor of Anthropology at Saint Louis University. She holds a BA and PhD in Anthropology from the University of California at Berkeley, and an MA in Anthropology from the University of Alberta. Dr. MacKinnon has done fieldwork in Central and South America and Africa. Research interests include primate social behavior, evolution of social complexity, conservation/management issues, and ethics in field primatology. She has published on biological anthropology and primate behavioral ecology and co-edited the volume *Primates in Perspective*.

Robert (Bob) Martin currently holds an honorary post-retirement appointment as Academic Guest at the Institute for Evolutionary Medicine at the University of Zürich (Switzerland) and is emeritus curator at The Field Museum, Chicago. His prior academic appointments were professor in the Anthropology Department at University College London (1969–1986), professor and director of the Anthropological Institute & Museum at the University of Zürich (1986–2001), and curator for Biological Anthropology at The Field Museum, Chicago (2001–2013). Over the past 50 years he has devoted his career to a wide-ranging exploration of the origins and evolution of primates, extending from morphology to genes. From the outset, special research interest was devoted to reproductive biology, brain evolution, and the theory and practice of allometric scaling. A mid-career exemplar of this approach is provided by his 1990 textbook *Primate Origins and Evolution: A Phylogenetic Reconstruction*.

W. Scott McGraw received his BA from Northwestern University in 1987 and his PhD from Stony Brook University in 1996. Dr. McGraw has been studying the behavior, ecology, and functional morphology of African cercopithecoid monkeys for over three decades. He co-directs the Taï Monkey Project at Taï, Côte d'Ivoire, one of the most species-rich primate field sites in the world, and is currently professor of Anthropology at Ohio State University.

Russell A. Mittermeier, PhD, is chief conservation officer of Re:wild (formerly Global Wildlife Conservation, GWC) and chair of the IUCN SSC Primate Specialist Group. Prior to joining GWC/Re:wild, he served as president of Conservation International for 25 years. He has been working on biodiversity conservation in tropical forests for the past 50 years, with a special emphasis on primate conservation. He also created the concept of primate-watching and primate life-listing, based on a bird-watching model. His publication list includes 41 books and more than 700 scientific and popular publications.

David Morgan is a research fellow of the Lester E. Fisher Center for the Study and Conservation of Apes at Lincoln Park Zoo and co-director of the Goualougo Triangle Ape Research Project. Dr. Morgan has spent over 25 years conducting scientific and conservation initiatives in Central Africa. He collaborates with other field biologists and scientists on research projects aimed at improving the conservation status of chimpanzees, gorillas, and their habitats throughout Africa.

Stephanie Musgrave is an assistant professor of Anthropology at the University of Miami. Her research focuses on the social, cognitive, and ecological underpinnings of primate tool use. She is particularly interested in understanding what factors shape the diverse tool traditions of wild chimpanzees in order to help illuminate the evolutionary origins of culture and technology.

Anna Nekaris is professor of Anthropology and Primate Conservation at Oxford Brookes University, UK. She studies the evolution, conservation, and ecology of lorises and pottos, and has led long-term field projects in India, Sri Lanka, Cambodia, and Indonesia. Her work involves understanding loris and potto taxonomy and behavior and using this information to improve captive management, develop conservation strategies, and help to mitigate both their illegal wildlife trade and persecution.

Pia Nystrom became interested in baboons as an undergraduate when she was invited to visit Awash National Park, Ethiopia. This led to a PhD study at Washington University in St. Louis with a focus on the reproductive success of male anubis, hamadryas, and hybrid males living in the same social group. After a post-doc at Harvard Medical School, she has taught at the University of Sheffield, UK, and has co-authored *The Life of Primates*, a comprehensive textbook. Her interest in the survival and well-being of primates has been the central theme throughout her career.

Patricia T. Ormond earned a BS in Anthropology from the University of Tennessee at Chattanooga. She is president of Lambda Alpha Gamma, a member of the Alpha Scholastic Honor Society, and won the Outstanding Senior Award in Anthropology in 2020. She has worked on a range of archaeological research projects, including broader efforts to archive archaeological records and collections housed at the Jeffrey L. Brown Institute of Archaeology (JBIA). She is working on a GIS-based project to record new and old archaeological sites based on historic JBIA data, applying new technology to old data.

Juan Ortega Peralejo is a primatologist with extensive experience studying great apes in the wild. He has conducted long-term monitoring of western lowland gorillas in Central African Republic, assisted with studies of bonobo behavioral ecology near Salonga National Park in the Democratic Republic of Congo, habituated groups of western lowland gorillas in the Moukalaba-Doudou National Park of Gabon, and managed research on chimpanzees and gorillas in the Goualougo Triangle of northern Republic of Congo.

Stephanie A. Poindexter is an assistant professor of Evolutionary Anthropology at the State University of New York at Buffalo. She studies behavioral evolution, movement, and sensory ecology of nocturnal primates. Her research is based on wild and captive populations to address theoretical and practical primatology and conservation questions.

Joyce Powzyk pursued doctoral studies in the Department of Biological Anthropology and Anatomy at Duke University. Her dissertation research was conducted in Mantadia National Park, Madagascar, where she studied feeding ecologies of sympatric indri (*Indri indri*) and diademed sifaka (*Propithecus diadema*) populations. She is currently an Associate Professor of the Practice in Biology in the Department of Biology at Wesleyan University, Middletown, Connecticut. She has also authored and illustrated several natural history books for young readers.

Jill Pruetz is professor of Anthropology and honorary professor of International Studies at Texas State University. She received a PhD in Anthropology at the University of Illinois at Urbana-Champaign and studied the socioecology of patas monkeys and vervets in Kenya for two years as part of her doctoral research. Since 2000, she has focused on the behavioral ecology of chimpanzees living in a savanna landscape at the Fongoli site in Senegal, West Africa.

Anthony B. Rylands is primate conservation director at Re:wild (formerly Global Wildlife Conservation, GWC). He began his career in 1976, at the National Institute for Amazon Research (INPA) in Manaus. From 1986 to 2003, he was professor of Vertebrate Zoology at the Federal University of Minas Gerais, Brazil, and from 2000 to 2017, a researcher at Conservation International, Washington, DC. He is a member of the Brazilian Academy of Sciences and Deputy Chair of the IUCN/SSC Primate Specialist Group.

Crickette Sanz is a professor of Anthropology at Washington University in St. Louis and co-director of the Goualougo Triangle and Mondika Ape Projects, both affiliated with the Nouabalé-Ndoki National Park in Republic of Congo. Her research career began with studies of ape language and animal welfare issues in captive populations but expanded to include studies of behavioral ecology, sociality, and cognition among wild chimpanzees and western lowland gorillas. Sanz is also dedicated to conducting research that can aid in mitigating threats to the preservation of the remaining great apes and their habitats.

Christopher Shaffer is an assistant professor of anthropology at Grand Valley State University in Michigan with research interests in the ecological and social interactions between humans and nonhuman primates. Dr. Shaffer is the principal investigator of the Konashen Ecosystem Health Project, a partnership between a multidisciplinary team of researchers and indigenous Waiwai in Guyana, South America, focused on human-wildlife interactions, biodiversity conservation, and the prevention of zoonotic emergence. Dr. Shaffer's work also focuses on the behavioral ecology of pitheciine primates,

particularly movement ecology, fission-fusion dynamics, and quantifying resource dispersion.

Myron Shekelle received a PhD at Washington University in St. Louis, conducting his dissertation research on the taxonomy and biogeography of the Eastern (Sulawesi) tarsiers. He is a member of the IUCN/SSC Primate Specialist Group and pursues initiatives to mitigate biodiversity loss, while simultaneously fostering sustainable development to support the livelihoods of people. Research activities and projects have been carried out in Brunei, Cambodia, Indonesia, Korea, Laos, Malaysia, Philippines, Singapore, Thailand, United States, and Vietnam.

Karen Strier is the Vilas research professor and Irven DeVore professor of Anthropology at the University of Wisconsin–Madison. She is an international authority on the endangered northern muriqui monkey, which she has been studying in the Brazilian Atlantic forest since 1982. Her pioneering, long-term field research has been critical to conservation efforts on behalf of this species, and has been influential in broadening comparative perspectives on primate behavioral and ecological diversity.

Robert W. Sussman was a professor of Anthropology at Washington University, St. Louis, Missouri, and until his death in 2016 was considered one of the most outstanding field primatologists of his generation. He was the author of almost 200 journal articles, book chapters, books, and edited volumes on a wide range of issues in primate behavior, ecology, and conservation. In 2015, Bob published a landmark volume titled *The Myth of Race: The Troubling Persistence of an Unscientific Idea* (Harvard University Press). His death represents a major loss to the world's primate community.

Larissa Swedell received her BA from Cornell University in 1991 and her PhD from Columbia University in 2000. Dr. Swedell has been studying the behavior and ecology of hamadryas baboons in Ethiopia since 1996 and chacma baboons in South Africa since 2006. She co-directs the Filoha Hamadryas Project in Ethiopia, the only currently operational field study of hamadryas baboons, and is professor of Anthropology at Queens College and the Graduate Center of the City University of New York.

Nelson Ting is an associate professor in the Department of Anthropology and Institute of Ecology and Evolution at the University of Oregon. He obtained his PhD in Anthropology (2008) from the City University of New York Graduate Center through the New York Consortium in Evolutionary Primatology. His expertise is in genetics, conservation, ecology, and evolution of colobus monkeys, and he helped lead the first conservation action plan for a colobus monkey (Red Colobus Conservation Action Plan 2021–2026).

Rajnish Vandercone is a senior lecturer attached to the Department of Biological Sciences of the Faculty of Applied Sciences, Rajarata University of Sri Lanka, and an honorary research associate at the Department of Anthropology, Washington University in St Louis. His research focuses on the ecology and behavior of sympatric primates, with special emphasis on the mechanisms of co-existence and traits that make some species more resilient to anthropogenic habitat change than others.

Eva Wikberg is an assistant professor in the Department of Anthropology at the University of Texas at San Antonio. She received her PhD in Biological Anthropology from the University of Calgary in 2012. Eva's research focuses on behavioral and molecular ecology of colobus monkeys, and she is co-principal investigator of the Boabeng-Fiema colobus project and the molecular anthropology lab at the University of Texas at San Antonio.

Acknowledgments

THIS BOOK IS THE RESULT OF THE COMBINED efforts of many people. As such, there are numerous individuals we want to recognize to express our heartfelt thanks. Dr. Linda Sussman was the initial source of encouragement to restart this project following its period in limbo after Bob Sussman's untimely passing. We are happy to have been able to bring about the completion of a volume that Bob had begun and had foreseen as an important addition to the primate literature. We would also like to recognize the support of Nancy Roberts during the early phases of our editorial work. We salute all the lead authors and co-authors who, in making their chapter contributions, drew on their years of field research experience and their primatological expertise. Our sincere gratitude goes to Dr. Ian Tattersall who was happily on board with the project as soon as he was approached, signaling his deep friendship with Bob Sussman by writing the Foreword to this volume. His perceptive contribution provides an insider's view of the backstory, when the book existed only on the drawing board. Similarly, the distinguished primatologists we asked to contribute commentaries also graciously accepted our invitations. We thank them for providing detailed, engaging observations following each taxonomic section of the volume: Dr. Robert Martin ("Strepsirrhini and Tarsiers [Prosimians]"), Dr. Anthony Rylands and Dr. Russell Mittermeier ("Neotropical Monkeys"), Dr. Larissa Swedell and Dr. W. Scott McGraw ("African and Asian Monkeys"), and Dr. Michele Goldsmith ("Apes").

Patricia Ormond, at the University of Tennessee–Chattanooga, not only made contributions to two of the volume's chapters, but also took on the meticulous task of standardizing the reference sections of all the chapters. We thank Debbie Bruns, at St. Charles Community College, for the countless hours she spent on the bibliographic and in-text citation proofing of each chapter and multiple other tasks, as well as for the technical assistance she provided. We also extend our thanks to Dr. Mary Willis, University of Nebraska–Lincoln, for her valuable feedback and suggestions. Thanks to Sonia Wolf who was always available for consultations concerning the book's graphics. Still with graphics, we are most appreciative to everyone who shared their images of primates in the field, enriching the volume as a result. Finally, we would like to acknowledge and thank Kate Powers, Associate Acquisitions Editor for Anthropology, at Rowman & Littlefield Publishers for working with us as we all coped with the COVID-19 pandemic through most of 2020 and into 2021. Her interest in the project and editorial assistance is very much appreciated. Also at Rowman & Littlefield, Alyssa Palazzo (Senior Acquisitions Editor, Sociology & Anthropology) and Hannah Fisher (Production Editor) helped us wrap up the volume.

To the primate species that have inspired the studies presented in this volume, we acknowledge our concern over their future and our fervent hope that primatologists will play an important role in guaranteeing their continued existence.

A TRIBUTE TO OUR COLLEAGUE, DR. DONNA LEE HART (1944–2022)

Just as this volume was about to go to press (we had just finalized the cover design), Dr. Donna Hart suffered a sudden illness and passed away. She had taken on this project at the behest of Dr. Linda Sussman and as a tribute to her PhD advisor, Dr. Robert Sussman. Bob had started working on an edited volume about primate ecology and behavior before his death in 2016. So, Donna picked up where Bob had left off and, along with her co-editor, Dr. Ian Colquhoun, they achieved the herculean task of reviving Bob Sussman's vision for an edited volume. Ian and the many contributing authors were devastated to learn of Donna's death; all understood that without her collegiality, professionalism, and dogged determination, this project would never have been completed.

Dr. Hart earned an MA and PhD in Biological Anthropology from Washington University in St. Louis. Prior to beginning graduate school, Donna had been a lifelong advocate for animals, supporting species both domestic and wild. She spent nearly 35 years working in conservation, holding positions such as Senior Program Director and Vice President for the International Wildlife Coalition, Legislative Liaison and Project Coordinator for the International Fund for Animal Welfare, and

Conservation Coordinator for the Wild Canid Survival and Research Center in St. Louis, Missouri, an NGO started by the late Marlin Perkins. Due to her skill and insight, Donna was invited by governments and NGOs in places such as Zimbabwe, Dominica, Hawaii, Tanzania, Sri Lanka, Belgium, Sweden, France, Philippines, Australia, and Norway to assess a plethora of wildlife issues and policies. Similarly, she conducted field site inspections involving human-wildlife conflicts in numerous international locales, including: Canada, Zimbabwe, Japan, Tanzania, Kenya, Sri Lanka, and Norway, to name a few. She was a non-governmental observer and advisor on a variety of government delegations that would later develop multi-lateral treaties to the Convention on International Trade in Endangered Species (CITES). Donna was an unyielding champion for primates, birds, jaguars, leopards, pangolins, walrus, seals, and numerous other marine mammals. But she was equally devoted to the lone turtle on the road or the abandoned cat or dog. On both international and national scales, Donna advocated for the humane and respectful treatment of all living species. In retirement, she provided expert testimony against having primates as pets and worked with diligence to break up a notorious breeding operation for dog-fighting in Missouri. She and her late husband, Al Bruns, adopted retired race horses, worked to protect endangered bat species and preserve land, along with providing a home for countless numbers of canine and feline rescues.

Donna's dissertation research on primates as prey revealed the role that predation has played in the evolution of primates—including humans. She uncovered ecological, morphological, and behavioral relationships between primate species and their predators, ultimately upending long-held views of "man the hunter" for a more accurate vision of "man the hunted"; humans, like other primates, were "just another item on the menu."[1] Co-authored by Robert Wald Sussman, her 2005 book, *Man the Hunted: Primates, Predators, and Human Evolution*, made this data-based view of early human evolution accessible to a general reading audience.[2] Their work was later acknowledged with the W. W. Howells Award from the Biological Anthropology Section (BAS) of the American Anthropological Association (AAA). Donna also published on methods for teaching about human variation and sharing strategies that challenged and changed students' previously learned perceptions and understanding of race.[3] She had a nuanced, yet practical, approach that resonated with college students, exposing them to science-based facts about the way in which humans vary worldwide. Donna retired in 2015 as an Associate Teaching Professor and Director of Undergraduate Research at the University of Missouri–St. Louis (UMSL). Along the way, she mentored dozens of undergraduates, many of whom received Fulbright Awards, graduated with honors, and/or went on to graduate and professional school.

Donna Lee Hart was a generous, kind, and nurturing person. Her great concern for the environment, and the fauna and flora which depend upon their ecosystems for survival, began in her childhood in St. Louis, and was strengthened as she worked for fellow Missourian and conservationist, Marlin Perkins. Donna would want to see her work carried on by a large network, including colleagues such as those who contributed to this volume, as well as the many friends, relatives, children, and grandchildren whom she educated each day on topics of import to conservation. She had urged all of us to contribute to this essential task, writing: "*Our earliest evolutionary history is not pushing us to be awful bullies. Instead, our millions of years as prey suggest that we should be able to take our ancestral tool kit of sociality, cooperation, interdependency, and mutual protection and use it to make a brighter future for ourselves and our planet*" [*Man the Hunted* 2009 p. 285].

NOTES

1. Hart, D. & Sussman, R. (2009). Man the hunted: just another item on the menu. In E. Angeloni (Ed.), *Annual Editions: Physical Anthropology 09/10*. Dushkin McGraw-Hill.
2. Hart, D. & Sussman, R. (2009). *Man the Hunted: Primates, Predators, and Human Evolution*. Expanded 2nd edition. Westview Press.
3. Ashmore, P. & Hart, D. (2010). Challenging students' concepts about race. In G. Strkali (Ed.), *Teaching Human Variation: Issues, Trends, and Challenges* (pp. 81-100). Nova Science Publishers.

PART I

STREPSIRRHINI AND TARSIERS (PROSIMIANS)

CHAPTER 1

Lorises and Galagos
The Lorisiform Primates

K. A. I. Nekaris, Stephanie A. Poindexter, Robert W. Sussman, and Ian C. Colquhoun

THE LORISIFORM PRIMATES ARE ONE OF THE MOST intriguing, yet understudied, groups in the Order Primates. In the past, the lorisiform species were summed up as being either vertical clingers and leapers or slow climbers (Masters, 2020). We now know that the 39 currently recognized species listed on Table 1.1 exhibit a wide range of feeding ecologies (from faunivory/insectivory, to frugivory, to specialized exudativory), diverse locomotor patterns (from slow climbing, to arboreal quadrupedalism, to vertical clinging and leaping), varying forms of social organization (from promiscuity to social monogamy), and interspecific differences in patterns of infant care (including care of offspring by adult males and siblings). In many cases, our knowledge of lorisiform behavioral ecology is quite incomplete, with long-term data on some genera being very scarce (e.g., *Arctocebus*, *Euoticus*: Pimley, 2002; *Sciurocheirus*: Ambrose, 2003, 2013). Within other genera (i.e., *Galagoides*, *Paragalago*, *Galago*, *Nycticebus*, and *Loris*), numerous species and even yet unrecognized taxa, remain to be studied in detail (e.g., Butynski & de Jong, 2004; Butynski et al., 2006; Kumara et al., 2009; Munds et al., 2013; Pozzi et al., 2015; Gamage et al., 2017; Svensson et al., 2017, 2020a; Luhrs et al., 2018; Rosti et al., 2020).

TAXONOMY, EVOLUTION, AND DISTRIBUTION

The living Lorisiformes are widely distributed throughout sub-Saharan Africa, and South and Southeast Asia (Figure 1.1). Lorisiform taxonomic classification has been in flux for a number of years. A conservative classificatory approach sees the Infraorder Lorisiformes comprising a single superfamily—the Lorisoidea, with one family, the Lorisidae (Nekaris & Bearder, 2011). There is general consensus that three distinct lorisiform subfamilies are represented: Galaginae (the galagos), Perodicticinae (the pottos and angwantibos), and Lorisinae (the Asian lorises) (e.g., Rasmussen & Nekaris, 1998; Grubb et al., 2003; Hartwig, 2011). Pottos and angwantibos are restricted to the remaining forests of Equatorial Central and West Africa, while the galagos occur more widely across sub-Saharan Africa. African lorisiforms have adult body masses ranging from 45–1,510 g (Table 1.1). The Asian lorises are found in South and Southeast Asia, with adult body weights ranging from 105–2,100 g.

Recent genetic evidence has helped to resolve that galagines diverged from the lorisines and perodicticines during the late Eocene, which was followed by a split between the latter two subfamilies by approximately 37–38 mya (Pozzi et al., 2014a,b, 2015; dos Reis et al., 2018). Clearly, at that time, the environment selected for two different morphs, with the Asian lorises and African pottos specializing in non-leaping, "slow" climbing (Walker, 1969; Oxnard et al., 1990), and the galagos specializing in rapid leaping (Crompton, 1980, 1984). (Therefore, in this chapter, any generalization referring to Asian lorises and African pottos *together* will refer to them collectively as lorises due to their slow locomotion.) Rasmussen and Nekaris (1998) suggest that a primary cause for the lineage divergences of these groups may have been a separation in foraging strategies, with the galagos specializing on fast-moving, elusive insect prey, resulting in an emphasis on hearing and leaping, and the lorises concentrating on less mobile, noxious insect prey species, with a subsequent reliance on olfaction and a reduced basal metabolic rate coinciding with slow locomotion and slow life history (Nekaris, 2014). These specialized locomotor patterns pervade all aspects of the morphology and behavior of both galagos and lorises (see Oxnard et al., 1990), and are the major determinants of their habitat preferences. Eleven genera (6 galagos and 5 lorises) and 39 species of lorisiform primates are recognized (Table 1.1). It is important to note that the behavior and anatomy of many of these species has been described previously using biological names different from current taxonomic nomenclature due to subsequent revisions.

Galagos

Prior to 1979, it was accepted that the subfamily Galaginae had a monogeneric taxonomic structure, with one genus, *Galago*, containing six species (Petter & Petter-Rousseau, 1979). For the galagos, we adopt the current taxonomy published by Rowe and Myers (2016), which recognizes five galagine genera, together with a

Table 1.1 Scientific Names, Common Names, Distributions, and Body Weights of the Lorisiformes

Scientific Name	Common Name	Distribution	Body Weight (g)
Galaginae			
Galagoides demidoff	Demidoff's dwarf galago	Bioko, Cameroon, Gabon, Ivory Coast, Nigeria, Uganda	45-75
Gd. thomasi	Thomas's dwarf galago	Bioko, Cameroon, Gabon, Ivory Coast, Nigeria, Uganda	55-90
Gd. kumbirensis	Angolan dwarf galago	Angola	?
Paragalago orinus	Mountain dwarf galago	Tanzania	70-100
Pg. zanzibaricus	Zanzibar lesser galago	Tanzania	149
Pg. rondoensis	Rondo dwarf galago	Tanzania	60-75
Pg. cocos	Kenya coastal galago	Kenya, Tanzania	115-185
Pg. granti	Mozambique lesser galago	Malawi, Mozambique, Tanzania, Zimbabwe (?)	134
Pg. nyasae (=Pg. granti?)	Malawi lesser galago	Malawi	?
Galago senegalensis	Senegal lesser galago	Cameroon, Kenya, Tanzania, Uganda	265-360
G. gallarum	Somali lesser galago	Kenya	200
G. moholi	Southern lesser galago	Botswana, Malawi, Namibia, South Africa	175-250
G. matschiei	Spectacled galago	Uganda	195-225
Euoticus elegantulus	Southern needle-clawed galago	Cameroon, Gabon	270-360
E. pallidus	Northern needle-clawed galago	Bioko, Cameroon	180-210
Sciurocheirus alleni	Allen's squirrel galago	Bioko, Cameroon	240-355
S. gabonensis	Gabon squirrel galago	Cameroon, Gabon	185-340
S. cameronensis	Cross River squirrel galago	Cameroon, Nigeria	?
S. makandensis	Makandé squirrel galago	Gabon	?
Otolemur garnettii	Garnett's, or sm.-eared, greater galago	Kenya, Tanzania	600-1060
O. crassicaudatus	Thick-tailed greater galago	Malawi, South Africa, Tanzania, Zimbabwe	1250-1510
O. monteiri	Silvery greater galago	Kenya	?
Perodicticinae			
Perodicticus potto	Western potto	Guinea, Guinea Bissau, Nigeria	600
P. edwardsi	Milne-Edwards', or Central, potto	Central African Republic, Nigeria, Zaire	850-1600
P. juju	S. Nigerian, or Benin, potto	Guinea Coast of Nigeria	?
P. ibeanus	Bosman's, or Eastern, potto	Burundi, Kenya, Rwanda, Zaire	850-920
Arctocebus aureus	Golden angwantibo	Gabon	150-270
A. calabarensis	Calabar angwantibo	Cameroon, Congo, Gabon	270-325
Lorisinae			
Loris lydekkerianus	Gray slender loris	South India, Sri Lanka	225-320
Loris tardigradus	Red slender loris	Sri Lanka	105-170
Nycticebus bengalensis	Bengal slow loris	Bhutan, Burma, Cambodia, China, India, Laos, Thailand, Vietnam	1140-2100
N. coucang	Greater slow loris	Peninsular Malaysia, Singapore, Sumatra, Thailand	635-850
N. javanicus	Javan slow loris	Indonesia (Java)	750-1150
N. menagensis	Philippine slow loris	Brunei, Indonesia and Malaysia (Borneo), Philippines	265-800
N. borneanus	Bornean slow loris		360-580
N. kayan	Kayan slow loris	Borneo	500-700
N. bancanus	Sody's slow loris	Borneo, Malaysia	?
		Banka and Belitung Islands, Borneo, Indonesia	
N. hilleri	Sumatran slow loris	Sumatra	688
Xanthonycticebus (N.) pygmaeus	Pygmy loris	Cambodia, China, Laos, Vietnam	360-580

subsequent revision introduced by Masters et al. (2017), to deal with paraphyly in the classification of the western and eastern dwarf galagos (see Masters et al., 2007; Pozzi et al., 2015). The Masters et al. (2017) classification moved five *Galagoides* species to a new, sixth galagine genus, *Paragalago*, which is comprised of the eastern dwarf galagos (see Butynski et al., 2006; Table 1.1). The dwarf galago genera (Figure 1.2) represent adaptive radiations as distinct as any of the clades of Madagascar's lemuriforms. Phylogenetic analysis indicates the *Galagoides* and *Paragalago* lineages diverged from one another as far back as 33 mya (Pozzi et al., 2014b, 2015). Despite these

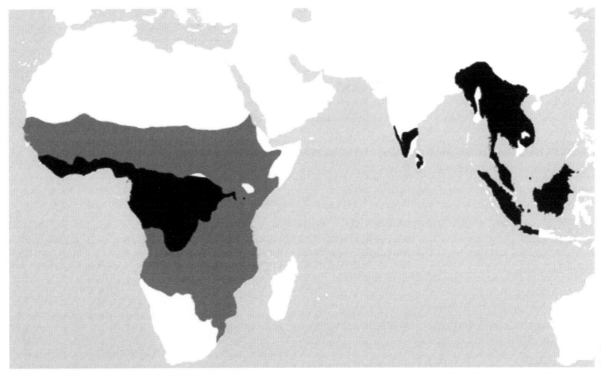

Figure 1.1 Geographic distribution map for all genera of African pottos and Asian lorises (black), and galagos (dark gray + sympatry with pottos).
(Based on spatial data from IUCN Red List; IUCN, 2020)

deep phylogenetic splits at the generic level (Pozzi et al., 2014a; Masters et al., 2017), many galagine taxa are difficult to morphologically differentiate from each other in the field due to their similar external appearances.

In a taxonomic sense, because many galagine taxa (e.g., both dwarf galago genera and taxa in the genus *Galago*) can be regarded as "cryptic species," bioacoustics studies (using Paterson's mate recognition concept of species;

Figure 1.2 Demidoff's dwarf galago, *Galagoides demidoff.*
(Photo credit: Bruce Davidson/www.arkive.org; licensed under Creative Commons Attribution/Share-Alike License)

Paterson, 1985; Bearder et al., 1995) have been of critical importance in validating galagine taxonomic revisions (e.g., Masters, 1988, 1991, 1998; Honess, 1996; Bearder, 1999; Perkin et al., 2002; Ambrose, 2003, 2013; Bearder et al., 2013; Masters et al., 2017; Pozzi et al., 2019). Taxonomic revisions have also been supported through several other independent streams of research: behavioral, nonvocal, studies (Bearder & Martin, 1980b; Harcourt & Bearder, 1989); analyses of hand, foot, and penile morphology; comparative assessment of hair-scale patterns (Dixson, 1989, 1995, 1998; Anderson, 1998, 2000, 2001; Anderson et al., 2000; Butynski et al., 2006; Perkin, 2007), as well as comparative genetic research (Masters et al., 1994, 2007; Bayes, 1998; Roos, 2003; Pozzi et al., 2019).

Galagos are distributed throughout forested regions of Africa, south of the Sahara, with the exception of the southernmost regions of South Africa. Galagos occupy a wide range of habitat types (Chapman et al., 1999), from arid *Acacia-Commiphora* thornbush formations at the northern fringes of their distribution in Somalia and northern Kenya (some of the driest habitat utilized by any African primate; Butynski & de Jong, 2004; de Jong & Butynski, 2004), through subtropical miombo woodland-savannas (Svensson & Bearder, 2013; Schneiderová et al., 2020), to riverine forests, to dense tropical rainforests and coastal forest formations (Charles-Dominique, 1974b, 1977a; Svensson et al., 2017). In elevational occurrence, they range from coastal forests at/near sea level to montane forest zones (Butynski et al., 2006). Some species live allopatrically or parapatrically (e.g., Butynski et al., 2006), but in many areas up to four galago species can occur in sympatry, in addition to as many as two potto species (Charles-Dominique, 1974b, 1977a).

Pottos, Angwantibos, and Lorises

The Perodicticinae of Africa and the Lorisinae of Asia were once thought to each comprise two monospecific genera, each containing a gracile and robust form: *Perodicticus potto* (the robust form) and *Arctocebus calabarensis* (the gracile form) in the perodicticines, and *Nycticebus coucang* (the robust form) and *Loris tardigradus* (the gracile form) in the lorisines (Yoder et al., 2001). Differences in behavior, morphology, facial markings, and comparative genetics have all contributed to revisions of the taxonomic diversity found in this nocturnal primate infraorder (see Figure 1.3; Nekaris & Jayewardene, 2003; Roos, 2003; Nekaris & Jaffe, 2007; Nekaris & Munds, 2010; Nekaris, 2014).

In Africa, two gracile perodicticine species are allopatric and confined to the rainforests of western Equatorial Africa (Svensson & Luhrs, 2020)—the golden angwantibo

Figure 1.3   Adult female Bengal slow loris (*Nycticebus bengalensis*) with six-week-old infant clinging to mother.
(Photo credit: Helena Snyder; GNU Free Documentation License, Version 1.2)

(*Arctocebus aureus*) and the Calabar angwantibo (*A. calabarensis*). The robustly built potto (*Perodicticus*) is represented by four species across Equatorial Africa (cf. Butynski & de Jong, 2017; Luhrs et al. 2018)—*P. edwardsi*, *P. ibeanus*, *P. potto*, and *P. juju*. However, the potto populations presently classified as *P. juju*, from Nigeria to Benin, may well represent more than one distinct species, with *P. juju* populations being allopatrically distributed to the north and to the south of the Cross River, which forms a biogeographic barrier between them (Oates et al., 2020).

The Asian gracile lorisines (genus *Loris*) also appear to have allopatric distributions. What was initially thought to be a narrow zone of sympatry between two subspecies, the Mysore (*Loris lydekkerianus lydekkerianus*) and Malabar (*L. l. malabaricus*) gray slender lorises on the eastern slopes of the southern Western Ghats Mountains in India (Kumara et al., 2006, 2009; Sasi & Kumara, 2014), has more recently been assessed as the extent of occurrence of a previously unrecognized (and still not formally described) subspecies of gray slender loris (Kumara et al., 2009, 2013; Sasi & Kumara, 2014). The distribution of this gray slender loris form is limited to a small region of southern Kerala state and far southwestern Tamil Nadu state (Kumara et al., 2009, 2013; Sasi & Kumara, 2014). The gray slender loris also occurs in northern Sri Lanka, where a further two subspecies exist (*L. l. nordicus* and *L. l. grandis*). The red slender loris (*Loris tardigradus*), the smallest of the lorisines, is endemic to Sri Lanka's lowland and montane rainforests in the island's southwestern wet zone (Gamage et al., 2016, 2017).

Nine species of the robust slow loris (*Nycticebus* spp., plus the pygmy loris, *Xanthonycticebus* (*Nycticebus*) *pygmaeus*—see below) are recognized (Rowe & Myers, 2016; see Table 1.1). By and large these species are allopatric, apart from regions of Indochina where pygmy and Bengal slow lorises occur in sympatry. Variation in Bornean slow loris facemask markings and the geographic distribution of that variation was recently analyzed (Nekaris & Munds, 2010), which has led to Borneo being recognized as home to four slow loris species (*N. borneanus, N. bancanus, N. menagensis,* and *N. kayan*), with the Kayan slow loris being recognized as a species new to science (Munds et al., 2013). Recently, Nekaris and Nijman (2022) proposed removing the pygmy slow loris from the genus *Nycticebus* and placing it in the newly erected genus *Xanthonycticebus*. This taxonomic revision draws on multiple biological variables that distinguish pygmy from slow lorises (i.e., genetic distance between pygmy and slow loris species based on comparative analysis of complete mtDNA sequences, sympatric occurrence of pygmy lorises with *Nycticebus*, twins births being usual in pygmy lorises, unique seasonal changes in pygmy loris body mass and coat color, and a multi-male, multi-female form of social organization).

ECOLOGY OF THE LORISIFORMES

Despite their wide geographic range (from tropical West African forests to South and Southeast Asia), species of the Lorisiformes remain among the least studied of primates. In the early 1970s, Charles-Dominique (1974b, 1977a) carried out his now-classic study of sympatric lorisiform primates at Makokou, in Gabon (Chapman et al., 1999). The study highlights a long-standing problem occurring in field research on this poorly studied group of nocturnal primates. Charles-Dominique's field study spanned 42 months, during which time he thought he was studying five sympatric lorisiform species, but he was, in fact, studying six sympatric species. The western dwarf galagos, *Galagoides thomasi* and *Gd. demidoff*, are sympatric across many parts of their respective ranges (Masters et al., 2017) and are nearly identical in external appearance. Only comparison of their calls, inspection of subtle differences in pelage, and species-distinguishing penile morphology (e.g., Butynski et al., 2006) can help

field researchers tell these species apart. Still, Charles-Dominique's field study (1972, 1974a,b, 1977a,b) remains a landmark piece of research, not only in the study of lorisiforms but also in field primatology more broadly. All of the subsequent studies of lorisiforms have drawn on his valuable field data as a basis for comparison.

Locomotion and Habitat Preferences

Generally, locomotion is often used to characterize the Lorisiformes and is possibly the best-studied aspect of their behavior, forming the basis for a number of field studies (e.g., Crompton, 1980, 1983, 1984) and captive studies (e.g., Dykyj, 1980; Glassman & Wells, 1984; Oxnard et al., 1990; Ishida et al., 1992; Demes et al., 1998). A complex suite of morphological traits, linked to locomotion, differentiates the galagos from the lorises (Charles-Dominique & Bearder, 1979; Oxnard et al., 1990). All galagos have long tails and elongated feet, particularly elongation of the navicular and calcaneus (Stevens et al., 1971), adaptations consistent with leaping. Galagos are characterized by intermembral indices considerably less than 100, reflecting that legs are appreciably longer than arms, which enables hindlimb-dominated locomotion such as leaping (Ankel-Simons, 2000). Alternatively, lorises have tails that are significantly reduced in length, and their intermembral indices are close to 100, indicating that the arms and legs are of approximately equal lengths (the general pattern of most quadrupeds) (Martin, 1990; Ankel-Simons, 2000). As a result of these differing limb length ratios, galagos can cross arboreal pathway gaps by hopping or leaping, while lorises bridge across arboreal gaps by grabbing handfuls of small branches, stretching their limbs and trunk, and "cantilevering" between supports to cross the gap (Figure 1.4) (Oxnard et al., 1990; Sellers, 1996; Nekaris, 2014).

Although vertical clinging and leaping (Napier & Walker, 1967) has been considered the locomotor mode typical of galago species, it is actually only used by most galago species to negotiate sizable gaps in arboreal pathways. A few species, such as *Sciurocheirus alleni* and *Paraalago rondoensis*, use it as their stereotypic mode of locomotion (Honess, 1996; Perkin, 2002; Pimley et al., 2005a). However, the greater galagos (*Otolemur* spp.), largest of the galagines, rarely use this mode of locomotion even though they are capable of bipedal hopping and leaping (Crompton, 1983; Harcourt & Nash, 1986a). These larger galagos are more monkey-like in their arboreal progressions, regularly traveling quadrupedally through the trees on larger horizontal and oblique branches. Oxnard et al. (1990), following Oates (1984), describe *Otolemur* as the most "potto-like" of the galagines, both in terms

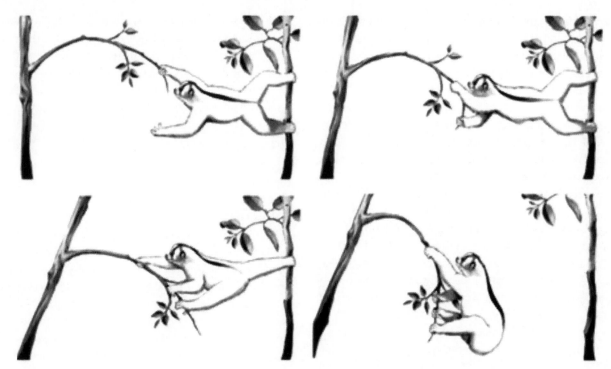

Figure 1.4 Javan slow loris using typical "cantilevering" locomotion to bridge across a gap in the canopy.
(Illustration credit: A. Setiad)

of their general slow arboreal quadrupedalism, and their gummivory. Many of the smaller galagos (i.e., *Galagoides*, *Paragalago*, and *Galago*) nimbly use quadrupedal running, climbing, and agile, often ricochetal, leaping to maneuver through the three-dimensional network of supports that tiny branches form—the so-called fine-branch niche (Charles-Dominique, 1972; Martin, 1972, 1973; Ambrose, 1999; Nekaris, 2014). Both pottos (*Arctocebus* spp., *Perodicticus* spp.) and lorises (*Loris* spp., *Nycticebus* spp., and *Xanthonycticebus*) use their long bodies, flexible limbs, and mobile wrist and ankle joints (Ankel-Simons, 2000) to stretch and cantilever across gaps in the canopy (Oxnard et al., 1990; Gamage et al., 2016). Branches must be of a threshold diameter to support their body weights during gap crossings (Charles-Dominique, 1974b; Nekaris, 2001, 2014). Such crossings might take multiple attempts to accomplish, and lorises do this with utmost patience. Unlike galagos, which typically move quickly through the forest, the "slowness" of loris movement comes, in part, from testing the strength of branches and perhaps having to change position in the canopy to find an accessible arboreal pathway (Charles-Dominique, 1977a; Nekaris, 2001). Nonetheless, this progression need not be slow-paced; Asian lorises are capable of a swift locomotor mode (Ankel-Simons, 2000), which has been termed the "race walk" (Demes et al., 1998; Nekaris & Stevens, 2007). Wild *L. tardigradus* and *X. (N.) pygmaeus* both exhibit race-walk locomotion at times (e.g., when seizing insect prey, or avoiding potential predators;

Masters, 2020), and Nekaris and Stevens (2007) suggest that *L. tardigradus* uses rapid quadrupedalism more often than does *L. lydekkerianus*, but further field observations are needed to confirm these species differences. As Nekaris (2014) points out, rapid quadrupedal race-walking locomotion bears little similarity to the cautious arboreal locomotion lorises utilize most of the time. *L. tardigradus* and *X. (N.) pygmaeus* have also both been observed to negotiate arboreal pathway gaps with "mini leaps," rearing up on their hind legs and launching their bodies over gaps of several inches (Duckworth, 1994; Nekaris & Stevens, 2007; Starr et al., 2011). *Perodicticus edwardsi* has also been described to have a mini leap, which is used when an individual cannot negotiate a gap regardless of any amount of stretching in efforts to cantilever (Charles-Dominique, 1977a).

Before long-term field studies on lorisiform ranging behavior were conducted (Bearder, 1999), it was thought that the slow locomotion of lorises would mean that quick galagos would be able to cross larger home-ranges and return to dispersed sleeping sites with greater ability than the lorises (Charles-Dominique, 1977a,b; Oates, 1984). In fact, home ranges of similar-sized galago and loris species are comparable in area. Studies conducted with all-night follows (Nekaris, 2003a; Bearder et al., 2006) and with radio-tracking (Wiens & Zitzmann, 2003; Pimley et al., 2005b; Nekaris & Bearder, 2011; Nekaris, 2014) have shown there is no support for early suggestions that lorises may move as little as 10 m per night (Nekaris, 2014).

These supposedly slow animals may actually have larger night ranges than the quick galagos (Table 1.2). Milne-Edwards' pottos, Javan slow lorises, and pygmy slow lorises may move up to 6 km per night; gray slender lorises move several hundred meters nightly; red slender lorises travel up to 1 km per night, and greater slow lorises can travel up to 400 m in an hour. Based on available comparative data, the only ecological relationship that appears linked to home range size is that the more faunivorous lorisiforms (Figure 1.5) are also those with the smaller home ranges (Nekaris, 2014).

Habitat Use

The lorisiforms show preference for using a wide range of both substrates and strata in the forest (e.g., Crompton, 1983; Honess, 1996; Ambrose, 1999; Nekaris, 2001; Nekaris et al., 2005; Pimley et al., 2005a). Substrate size selection is almost always related to the body weight of the animal, with smaller animals being able to move on smaller-gauged twigs, branches, and lianas, and larger animals utilizing sturdier supports with greater girth. An exception is seen with *Euoticus*, the needle-clawed galagos, which make more use of large vertical tree trunk supports due to their largely exudativorous feeding ecology (Charles-Dominique, 1977a). A number of species (e.g., *Sciurocheirus gabonensis*, *Galagoides demidoff*, *Arctocebus aureus*, *Loris lydekkerianus*) thrive in the forest shrub layer undergrowth and also utilize the zones of vigorous plant regrowth that occur in light gaps created by tree falls, whereas a preference for the canopy is exhibited by other species (e.g., *Paragalago orinus* and *Nycticebus coucang*; Butynski et al., 1998). This ecological stratification (Napier, 1966) is what allows the African lorisiforms, in particular, to occur in sympatry in many places throughout their respective ranges (cf. Charles-Dominique, 1977a), and may also influence the distribution of sympatric Asian lorises (Duckworth, 1994).

Activity Cycles and Sleeping Sites

All lorisiforms are nocturnal in their activity patterns. There is no diurnal or cathemeral lorisiform species, although animals may occasionally become active during daylight hours to change sleeping position for thermoregulatory purposes, to eat during periods of pronounced food scarcity, or to avoid predators (Bearder et al., 2006). Across species, ambient light levels do not necessarily seem to be linked to the amount of nocturnal activity that species exhibit. *Galago moholi*, for example, was reported to increase its behavior and range of travel patterns during the light moon phases of the lunar cycle and during twilight periods, while *Loris lydekkerianus* maintained activity regardless of moon phase (Bearder et al., 2002). Galago species living in closed forest, on the other hand, do not appear to be influenced by changes in moonlight levels (Nash, 1986), whereas *Xanthonycticebus (N.) pygmaeus*, the pygmy loris, are consistently active on dark nights regardless of temperature but decrease their activity on bright moonlit nights with temperatures below 26°C (Starr et al., 2012). Pygmy lorises in northern Vietnam

Table 1.2 Home Range, Population Density, Sleeping Association, and Social Organization of Lorisiform Taxa.

Taxa	Home Range (ha)	Population Density (animals/km^2)	Sleeping Associations (male; female; offspring)	Social Organization
Galagoides demidoff	0.5-2.7	50-80	(1 m; 2-10 f; o)	Dispersed multi-male
Gd. thomasi	0.5-2.7	50-80	?	Dispersed multi-male
Paragalago cocos	1.8-2.2	170-180	(1 m; 1-2 f; o)	Spatial monogamy
Galago moholi	4.4-22.9	?	(1 m; ≥1 f; o)	Dispersed multi-male
Euoticus elegantulus	?	10-15	(1 m; 2-7 f; o)	?
Sciurocheirus alleni	2.0-2.8	15	(≥2-3)	Dispersed multi-male
S. gabonensis	8.0-50.0	15-20	?	Dispersed multi-male
Perodicticus potto	7.5-31.5	2-18	(1 m) (1 f; o)	Semi-dispersed uni-male/uni-female
Arctocebus aureus	?	?	(1 m) (1 f; o)	?
A. calabarensis	?	?	(1 m) (1 f; o)	?
Loris l. lydekkerianus	1.6-3.6	28	(≥1 m; 1 f; o)	Semi-dispersed multi-male
L. l. malabaricus	0.9-35.2	?	(1 m; 1 f; o)	Semi-dispersed multi-male
L. tardigradus	1.2-6.9	43	(≥1 m; 1 f; o)	Semi-dispersed multi-male
Nycticebus coucang	0.8-10.4	4-26	(1 m) (1 f; o) / (1 m; 1 f; o)	Semi-dispersed uni-male/uni-female
N. javanicus	25.0	16	(1 m; 1 f; o)	Semi-dispersed uni-male/uni-female
N. bengalensis	30.0	23	(1 m; 1 f; o)	Semi-dispersed uni-male/uni-female
Xanthonycticebus (N.) pygmaeus	12.0-22.0	9	(1 m; 1 f; o)	Semi-dispersed uni-male/uni-female

Sources: Nekaris and Bearder (2011); Nekaris (unpub.); for home range size and density, ranges are presented; this may include degraded or fragmented areas with unusually high or low values; (?) indicates no available data.

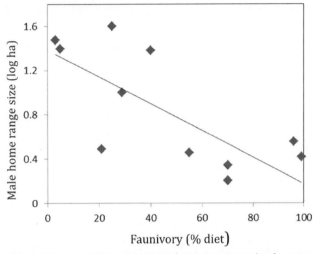

Figure 1.5 Lorisiform primates that eat more animal matter have smaller home ranges (based on 5 galagos and 7 lorises). Relationship between the degree of faunivory and the size of a male home range (log transformed, ranging from 1.6–24.2 ha). The relationship is significant, with 55% of the variation in home range size being explained by diet.

experience winter temperatures as low as 5–6°C and have recently been shown to enter hibernation (i.e., periods of torpor > 24 hours) under such ambient conditions; periods of hibernation lasted up to 63 hours. During periods of hibernation, pygmy loris minimum body temperatures dropped to as low as 8–9°C (Ruf et al., 2015). Recent analyses by Campera and colleagues (2022) of a large comparative 5-year data set found that Javan slow lorises are also lunarphobic, exhibit reduced levels of activity on chilly nights (i.e., < 12°C), while higher activity levels were exhibited when overnight temperatures ranged between 20–30°C; activity levels also increased with increasing relative humidity (especially > 90% humidity).

Galagos and lorises differ dramatically in their use of sleeping sites. Following a comparative study of diurnal sleep site choice/use of the Lorisiformes that included 36 species, Svensson et al. (2018) found that across the infraorder there was a variety of nests (i.e., nests fashioned from leaves and thin branches and located in tree crowns), tree holes, dense vegetation tangles (such as liana tangles), and branch forks. *Galagoides, Paragalago, Galago,* and *Otolemur* are canopy nest users; *Galago, Paragalago, Sciurocheirus,* and *Perodicticus* use tree holes; *Nycticebus, Loris, Galagoides, Galago, Euoticus, Otolemur, Perodicticus,* and *Arctocebus* all utilize dense vegetation tangles as sleeping sites, while both *Sciurocheirus* and *Otolemur* sleep in branch forks (Svensson et al., 2018). *Galago* will also use tree branch forks and has even been reported to sleep sheltered among palm leaves, a day sleeping site choice perhaps driven by scarcity of preferred, better hidden sleeping sites (Ellison et al., 2019). Some lorisiform species share sleeping sites, and the age-sex composition

of these sleeping groups may vary considerably (Radhakrishna & Singh, 2002). For those species that sleep socially, the sleeping group can represent an important aspect of lorisiform social behavior since being in a sleeping group may take up approximately 12 hours of each day. Allogrooming bouts at dawn and dusk may last up to 3 hours, and intermittent grooming may also occur among sleeping group members during daylight hours.

Diet and Feeding Behavior

Given the new understanding of the diversity in lorisiform primates, it is not surprising that a variety of dietary specialists can be found among this group, including gummivores/exudativores, highly frugivorous taxa, and some of the most faunivorous of all primates (e.g., Charles-Dominique, 1977a; Oates, 1984; Happold & Happold, 1992; Nekaris & Rasmussen, 2003; Wiens et al., 2006; Starr & Nekaris, 2013; Das et al., 2014; Poindexter & Nekaris, 2017) (see Table 1.3). Some taxa also consume floral nectar, fungus, and bamboo, exemplifying again past misconceptions that these primates are uniform or limited in their feeding ecologies.

The ability to consume and digest gum may be the most fundamental feeding adaptation of galagos, as all include some gum in their diets (Harcourt, 1980, 1984; Bearder & Martin, 1980a; Nash, 1989; Nash & Whitten, 1989). Besides the adaptations linked to gummivory/exudativory, body size is the key predictor of diet among the galagos. The small-bodied *Galagoides* and *Paragalago* rely more on insects, while medium-sized species (e.g., *Galago moholi, Euoticus elegantulus*) add more exudates to their diets, and the largest of the galagos (e.g., *Sciurocheirus* and *Otolemur*) will increase their fruit intake with seasonal availability. In the case of *O. crassicaudatus*, however, gum is a year-round staple of the diet and an essential food source during the austral winter in the southern parts of its geographic distribution (Masters et al., 1988; Oxnard et al., 1990). Nontoxic insects, such as grasshoppers, cicadas, and beetles, comprise the invertebrate part of galago diets (Bearder & Doyle, 1974b; Harcourt & Nash, 1986a), and the fruits eaten by galagines are sweet and soft (Charles-Dominique, 1977a). Some *Galagoides* species also consume floral nectar, suggesting a potentially important role in forest tree pollination (Rylands et al., 2004).

Galagos have adapted to their varied diets via a range of morphological and behavioral means. All galagos are capable of localizing animal prey with their large and independently mobile ears; hearing is frequently used to detect cryptic and concealed insect prey. They also search for insects visually and find sources of gum using their

Table 1.3 Diet of Wild Lorisiform Primates Based on Long-Term Studies.

Taxa	Diet					References
	Animal Prey	Fruit	Gum	Nectar	Other	
Galaginae						
Galagoides demidoff/thomasi [a]	70%	19%	10%			Charles-Dominique (1977a)
Paragalago cocos	70%	30%				Harcourt and Nash (1986a)
Galago moholi	52%		48%			Harcourt (1986b)
Euoticus elegantulus	20%	5%	75%			Charles-Dominique (1977a)
Sciurocheirus gabonensis	25%	73%		2%		Charles-Dominique (1977a)
S. alleni cameronensis	55%	55%				Pimley (2002)
Otolemur garnettii	50%	50%				Harcourt and Nash (1986b)
O. crassicaudatus	5%	33%	62%			Harcourt (1986b)
Perodicticinae						
Perodicticus (potto) edwardsi	40%	50%			10%	Pimley (2002)
P. (potto) edwardsi	11%	67%	22%			Charles-Dominique (1977a)
Arctocebus aureus	87%	13%				Charles-Dominique (1977a)
Lorisinae						
Loris lydekkerianus lydekkerianus	96%	1%	3%			Nekaris and Rasmussen (2003)
L. l. nordicus	95%	5%				Nekaris and Jayewardene (2003)
L. tardigradus tardigradus	98%	2%				Nekaris (unpub.)
Nycticebus coucang coucang	3%	22%	43%	32%		Wiens (2002)
N.coucang coucang	29%	71%				Barrett (1984)
N. javanicus	7%	3%	56%	32%		Rode-Margono et al. (2014)
N. bengalensis	3%	<1%	87%	6%	4%	Swapna et al. (2010)
Xanthonycticebus (N.) pygmaeus	18%	19%	46%	13%	6%	Starr and Nekaris (2013)
X. (N.) pygmaeus	33%		63%		4%	Streicher et al. (2012)

[a] Charles-Dominique (1977a) did not recognize *Gd. thomasi* as a distinct species, thus all data combined.

keen sense of smell (Bearder, 1969; Charles-Dominique, 1977a; Hladik, 1979; Pariente, 1979). As is the case for all insectivorous strepsirrhines, insects are secured by galagos in a stereotypic fashion involving both hands; insect prey is pressed firmly between the palmar pads of each hand with the fingers wrapping around the insect, since the individual fingers cannot be moved independently (Hladik, 1979; Ankel-Simons, 2000; Martin, 1990). The mandibular toothcomb plays an important role in scraping gum from trees, and gum is cleaned from between the teeth using the serrated cartilaginous sublingual—a sort of secondary tongue with denticulate notching at its tip located underneath the main tongue (Nekaris, 2014). Gum is digested in an elongated cecum containing symbiotic gastrointestinal bacteria capable of digesting the complex polymerized sugars found in tree gums. *Euoticus* spp., that primarily eat gum, have additional specializations in the form of enlarged canines and caniniform premolars for exposing sources of gum, and keeled (pointed) nails that the animals use to cling to large, vertical tree trunks in order to reach exudates that would otherwise be inaccessible (Osman Hill, 1953; Charles-Dominique, 1977a; Ambrose, 1999). Galagos living in seasonal environments in South Africa may rely almost completely on carbohydrate-rich gum during the cold months of the austral winter (Masters et al., 1988; Oxnard et al., 1990); they reduce their activity levels accordingly, adopting an "energy minimizer" activity pattern in which energy expenditure is limited (Bearder & Martin, 1980a). Squirrel galagos (*Sciurocheirus* spp.) will descend to the forest floor to feed on fallen fruits, even swallowing them whole, which allows them to eat their fill quickly and then withdraw to other areas away from potential predators (Charles-Dominique, 1977a).

Pottos (*Perodicticus* spp.) are mainly frugivorous, but supplement their diet with (1) relatively large amounts of gums, utilizing gum secretions from natural tree wounds, rather than by tree-gouging as slow lorises do (Nekaris et al., 2010b), and (2) animal prey, including ants, slow-moving arthropods, as well as birds, and bats (Jewell & Oates, 1969; Charles-Dominique, 1977a; Oates, 1984; Pimley, 2002). Pottos have more powerful jaws than galagos. Their insect prey tends to be less active species, and they are able to eat caterpillars and noxious beetles that most other insectivores avoid. The angwantibos (*Arctocebus* spp.), sympatric with pottos in Equatorial West-Central Africa (Pozzi et al., 2015; Svensson et al., 2016), forage mainly in the thick vegetation of the forest shrub layer, with the bulk of their diet being invertebrates, especially caterpillars (Jewell & Oates, 1969; Charles-Dominique, 1971, 1977a; Oates, 1984). "Dietary conditioning" is exhibited by all the African strepsirrhines, whereby a young animal learns what to eat by snatching food from its parent and examining novel food items with curiosity as indicated by head-cocking (Bearder, 1969; Charles-Dominique, 1977a). This developmental behavior seems to be particularly important for the angwantibos; they process certain lepidopteran species of caterpillars by removing stinging "hairs" before consuming them.

In his groundbreaking study of sympatric galagos (*Galagoides demidoff*, *Gd. thomasi*, *Euoticus elegantulus*, *Sciurocheirus gabonensis*) and pottos (*Perodicticus edwardsi*, *Arctocebus aureus*) in Gabon, Charles-Dominique (1974a; 1977a) revealed classic dietary niche partitioning between nocturnal lorisiform primate species, meaning that even though some overlap in resource use may occur, these sympatric species are not in direct feeding competition. Species that spent the most time in the forest canopy concentrated mainly on insects (*Galagoides* and *Paragalago*), gums (*Euoticus*), and/or fruits (*Perodicticus*). Species preferring the undergrowth of the forest shrub layer subsisted on caterpillars (*Arctocebus*) or fallen fruits (*Sciurocheirus*). As noted above, subsequent to Charles-Dominique's (1974a, 1977a) study, it was discovered that the dwarf galagos in Gabon were not a single species but, in fact, two different species (i.e., *Galagoides* [*demidovii*] *demidoff* and *Gd. thomasi*) that occur sympatrically throughout much of their respective ranges in the tropical forests of West-Central Africa (Wickings et al., 1998; Masters et al., 2017). Both of these species prefer insects but, as would be predicted by econiche partitioning theory and ecological stratification in three-dimensional forest habitats (Napier, 1966), one species (*Gd. thomasi*) forages mainly in the canopy whereas the other species (*Gd. demidoff*) forages in the forest undergrowth where it consumes fast-moving insects, in contrast to the slower, noxious insect prey taken by angwantibos (*Arctocebus* spp.) in the forest shrub layer. Similar ecological separations occur between sympatric lorisiform species in other parts of Africa. For example, in the Rondo Forest of southeastern Tanzania, Garnett's galagos (*Otolemur garnettii*) forage in the canopy, Grant's galagos (*Paragalago granti*) use the forest middle-story strata, and Rondo dwarf galagos (*Pg. rondoensis*) remain approximately one meter above the ground and feed almost exclusively on insects and grubs from the leaf litter (Honess, 1996).

The Asian lorises, although perhaps less complex in their econiche partitioning than the African lorisiforms, exhibit some of the most fascinating extreme dietary adaptations seen among primates (Nekaris, 2014). The slender lorises (*Loris* spp.) are considered, along with tarsiers, to be the most faunivorous primates (Petter & Hladik, 1970; Nekaris, 2002; Nekaris & Jayewardene, 2003; Nekaris & Rasmussen, 2003). Lorisine species specialize on prey of small size-classes and are highly tolerant of toxic prey, such as ants and noxious beetles (Rode-Margono et al., 2015). They are unusual among primates in their ability to consume large quantities

of ants (Nekaris, 2014) and have been reported to engage in something akin to a primate analog of the "anting behavior" reported in many bird species (Morozov, 2015). *Loris* seem to facilitate ant consumption by licking their hands as a means of ensnaring ants (Kumara et al., 2005), or by "urine-washing," i.e., urinating on their hands and feet prior to approaching an ant colony. The urine-washing could be to anoint themselves with the chemical profile of their intended prey (Nekaris et al., 2007; Nekaris, 2014) through secondary metabolites in their urine from previous ant predations in order to prevent or minimize attack (Nekaris et al., 2007; Nekaris, 2014). Alternatively, the urine-washing could possibly provide some counteractive defense (Nekaris et al., 2007) to dangerous secretions many ants can produce, such as ant species in the widespread subfamily Formicinae that secrete aqueous formic acid at concentrations of up to 60% (Attygalle & Morgan, 1984; Torres et al., 2013; Nekaris, 2014). In a brief field study of a radio-collared female *Galago* [*senegalensis*] *moholi*, Harcourt (1981) did not discern any function of urine-washing for olfactory sign-marking of home range boundaries; rather, she suggested that urine-washing in this species served to facilitate the gripping of arboreal substrates.

When foraging at night, slender lorises stalk silently through the branches, not unlike chameleons, scanning

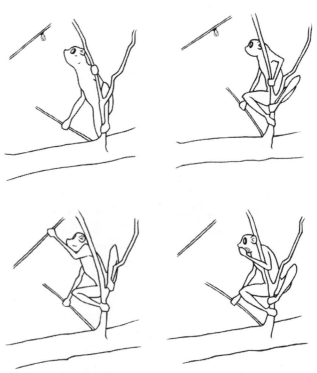

Figure 1.6 Slender loris focusing in on a prey item visually before it coils, snatches the insect with lightning speed, and then devours it.
(Illustration credit: Andrew Brady)

for potential prey as they move. Slender lorises stealthily pounce upon their insect prey with precision, rarely missing the mark, and they consume captured prey head-first (Nekaris, 2005) (Figure 1.6).

Quick and efficient prey capture seems to characterize the Asian lorises, but for the larger-bodied slow lorises, animal prey is a supplement (see Table 1.3). Possible physiological connections between noxious insects in the overall diet of *Nycticebus* and sequestering insect toxins to incorporate in their own venom are unclear (Rode-Margono et al., 2015). Exudate feeding has now been identified as a critical component of the specialized feeding ecology of *Nycticebus coucang, N. bengalensis, N. javanicus,* and *Xanthonycticebus (N.) pygmaeus,* as well as *Perodicticus potto* (Tan & Drake, 2001; Wiens et al., 2006; Swapna et al., 2010; Nekaris et al., 2010b; Starr & Nekaris, 2013; Das et al., 2014; Nekaris, 2014; Nekaris & Starr, 2015; Cabana et al., 2017; Poindexter & Nekaris, 2017). These recent insights on loris feeding ecology clearly show that this is a highly specialized group of gum/exudate feeders and that exudates are not fallback foods for these species. Slow lorises actively search for gum sources, employing a head-down searching mode, with the sniffing that accompanies searching being both audible and visible to observers (Nekaris et al., 2010b). Using their mandibular toothcomb, slow lorises are capable, within a matter of seconds, of creating large gouge holes in tree trunks to produce gum flow; the sound of crunching bark and wood from a foraging slow loris can be audible from as far away as 10 m (Nekaris et al., 2010b; Nekaris, 2014). Das et al. (2014) showed that the preferred gums consumed by slow lorises are also those that contain significant medicinal properties and are commonly used locally for traditional medicinal purposes. Floral nectar is another important food resource for slow lorises (Cabana et al., 2017). Wiens (2002) found nectar from the flowers of the Bertram palm (*Eugeissona tristis*) to be the food most often consumed by *N. coucang*; he attributed secondary metabolites in this food as the evolutionary driver of slow life history typical of lorisines (see also Wiens et al., 2006). Javan slow lorises subsist primarily on gum during both the wet and dry seasons, but insects and nectar are both consumed more in the wet season than the dry (Cabana et al., 2017). When foraging for nectar, slow lorises use both vision and precise manual inspection of individual flowers (Rode-Margono et al., 2014). *X. (N.) pygmaeus* and *N. bengalensis,* sympatric in many parts of their respective ranges, are known from preliminary observations to share feeding sites. At present, however, nothing is known about how they partition their econiches (Duckworth, 1994).

PREDATION

With their saltatory locomotor abilities, galagos can escape from predators by leaping away, whereas lorises and pottos have developed a suite of morphological characteristics that allow them to remain motionless for prolonged periods, providing camouflage and crypsis, which serve as protection from potential predators (Charles-Dominique, 1977a; Bearder, 1987; Nekaris, 2001, 2014; Poindexter & Nekaris, 2017). Because of the relatively small size of many galagine species, galagos are likely subject to significant predation pressure; however, field data on predation of galagos are scarce. Predators thought likely to be threats to galagos include small felids, mongoose, genets, jackals, large snakes, and owls (Ballenger, 2001; de Jong & Butynski, 2004). Galagos are also preyed upon by several larger primate species, including yellow baboons (*Papio cynocephalus*) that prey on *Galago senegalensis* (Altmann & Altmann, 1970), blue monkeys (*Cercopithecus mitis*) that prey on either *Galagoides demidoff* or *Galago senegalensis* (Butynski, 1982), bonobos (*Pan paniscus*) that have been seen preying on *Galagoides demidoff* (Hohmann & Fruth, 2008), and eastern chimpanzees (*Pan troglodytes schweinfurthii*) in Gombe Stream National Park, Tanzania, that have been observed to kill and eat *Galago senegalensis*, although only rarely (O'Malley, 2010). Notably, western chimpanzees (*P. t. verus*), inhabiting the mosaic of savanna, open woodland, bamboo, and gallery forest habitats at Fongoli, Senegal, use stick tools to hunt *G. senegalensis* in their day-nest tree holes, repeatedly jabbing the stick into the tree hole to disable or kill the galago before extracting it (Pruetz & Bertolani, 2007; Newton-Fisher, 2014; Preutz et al., 2015).

Galagos give intense alarm calls and often mob potential predators at distances of several meters, following them while constantly calling. They also direct alarm calls at arboreal snakes (Bearder, 1987). These alarm calls act as a warning system to all other galagos in the vicinity, identifying the location of the potential predator. Some galagos move more during moonlit nights and, since they use visual detection, mobbing, and escape, rather than concealment, Nash (1986) suggests that galagos may be safer from nocturnal predators in full moonlight.

Risk of predation is greater for young galagids and lorisids. During the daylight period, lorisiform species will utilize a variety of sleeping sites, including nests of branches and leaves located in dense tree crown foliage, tree holes, natural vegetation/liana tangles, or tree branch forks (Svensson et al., 2018). When galagids and lorisids are active at night, a mother will carry her young infant in her mouth and "park" the youngster on a thin branch close to where she will be foraging. The infant remains completely immobile in the absence of its mother. If the branch to which it clings shakes, the infant may drop to the forest floor and utter a distress call. The call attracts not only the mother, but also other adults in the vicinity. If a predator is also attracted, the adults will band together and "mob" the predator, issuing alarm calls. This defense mechanism is quite successful in the tropical rainforest. During Charles-Dominique's (1972, 1974a,b, 1977a) study in Gabon, only two young bushbabies disappeared from his study population.

Lorises, with their slow locomotion, exhibit several morphological adaptations against predators (Ankel-Simons, 2000; Nekaris et al., 2007). An important defense is their cryptic locomotion—usually, these silent primates can remain still and undetected, or silently slip away from potential danger. They can also cling to arboreal supports and remain motionless for long periods; pottos and lorises both exhibit features, such as shortened second digits on the hands and feet, highly mobile ankles and wrists, and *retia miribalia* of the distal limb vessels, that allow for prolonged gripping by their upper and lower appendages with no fatigue (Oates, 1984; Rasmussen & Nekaris, 1998; Ankel-Simons, 2000). *Retia miribalia* are fine networks of capillaries in the arms and legs that optimize blood flow to the musculature, permitting pottos and lorises to maintain very strong grips on arboreal substrates for long periods without losing sensation in their extremities (Ankel-Simons, 2000). Thus, the exchange of oxygen and physiological by-product materials in the muscles continues even under prolonged muscular contraction (preventing, for example, a buildup of lactic acid in the limb musculature).

In situations where *Perodicticus* is confronted by a potential predator, and there may appear to be no option for escape, pottos will turn and face the danger. They adopt a hunched defensive posture and make sudden lunges at the predator, while also growling and hissing (Daschbach et al., 1981; Zimmermann, 1985; see also Nekaris et al., 2007, 2013). Three physical adaptations make this potto/lorisine defensive hunch behavioral display a formidable defense: (1) pottos possess a "scapular shield," formed by the shoulder blades (scapulae) and, in the upper vertebral column, a series of raised apophyseal cervical vertebral spines, some of which protrude above the skin in the form of tubercles covered by thick, cornified epithelial tissue surrounded by sensory bristles that are distributed across the upper surface (Walker, 1970; Oates, 1984; Ankle-Simons, 2000). This scapular shield provides protection and defense, as well as acute sensitivity (Walker, 1970); (2) muscular hands and feet allow the potto to grip arboreal supports firmly (Ankel-Simons, 2000); (3) the *retia mirabalia* allow lorises to remain on their arboreal

supports in stationary postures for extended periods. When threatened, the potto, and other lorises, will sometimes drop to the ground and rapidly move through the undergrowth to safety (Jewell & Oates, 1969; Oates, 1984; Nekaris et al., 2007). Still, such defensive actions and adaptations may not suffice. Hockings et al. (2012) describe a case from Bossou, Guinea, of a potto being killed and devoured by two male chimpanzees (*P. t. verus*).

Although cryptic and stealthy behaviors are their main defenses, pottos, angwantibos, and Asian lorises all assume a characteristic defensive posture in which the arms are held up and wrapped around the head (see Nekaris et al., 2007, 2013; Nekaris, 2014; Rode-Margono & Nekaris, 2015). If the potential predator gets too close, the threatened individual can lift its arm and bite the attacker on the muzzle or forelimb. The "arms over head" defensive posture and the potential for bites are significant in the case of *Nycticebus* (Fuller et al., 2016). Slow lorises are venomous, a rare trait among mammals; indeed, *Nycticebus* species are the only venomous primates (Nekaris et al., 2007, 2013; Nekaris, 2014). Their bite can be serious, even fatal, both intra-specifically (e.g., due to severe tissue damage in wounds that result from aggressive encounters between males), and also for humans who might get bitten as it can cause anaphylactic shock (Wilde, 1972; Alterman, 1995; Hagey et al., 2007; Nekaris et al., 2013; Madani & Nekaris, 2014; Nekaris & Starr, 2015; Rode-Margono & Nekaris, 2015). The roots of this unique defense mechanism are not fully known, but thus far researchers have shown that loris venom will repel ectoparisites and helps to immobilize arthropod prey (Grow & Nekaris, 2015). Before biting prey items, predators, or conspecifics, slow lorises combine a secretion from their brachial sebaceous glands with their saliva in order to produce a numbing poison (Nekaris et al., 2013; Nekaris, 2014). After biting, they then keep their arms tightly coiled above their head, protecting this vulnerable part of their body.

Known predators of slender and slow lorises include pythons (genus *Python*: Wiens & Zitzmann, 1999), monitor lizards (genus *Varanus*), orangutans (genus *Pongo* in the case of *Nycticebus* spp.: Utami & van Hooff, 1997; Hardus et al., 2012), changeable hawk eagles (*Nisaetus cirrhatus*), and domestic dogs (*Canis familiaris*: Nekaris et al., 2007). Other possible predators of slender lorises include: Indian palm civets (*Paradoxurus hermaphroditus*), jungle cats (*Felis chaus*), rusty-spotted cats (*Prionailurus rubiginosus*), feral domestic cats (*F.* cf. *domesticus*), serpent eagles (*Spilornis cheela*), and barn owls (*Tyto alba*) (Nekaris, 2001; Nekaris et al., 2007; Gnanaolivu & Singh, 2019). Gnanaolivu and Singh (2019) also provide a unique report on predation of an adult female Malabar slender loris (*L. lydekkerianus malabaricus*) by one of a pair of brown palm civets (*Paradoxurus jerdoni*), a species that, based on scat analysis, is primarily frugivorous but may include animal prey in its diet when fruit is scarce.

When in face-to-face confrontations with a potential predator, Malabar slender lorises have been observed/heard to give either loud whistles or shrieks directed at the predator (Nekaris et al., 2007; Gnanaolivu & Singh, 2019). Slender lorises will also sometimes face-off against potential predators by performing a "cobra display" in which they raise their arms to ear-level or over their heads while making serpent-like sways of their body and giving hissing vocalizations (Nekaris et al., 2007). In fact, it has been suggested that slender and slow lorises co-evolved with cobras (e.g., spectacled cobra, *Naja naja*) during the dispersal of both groups into South and Southeast Asia, and that venom in slow lorises and the "cobra display" of slender lorises may be rare examples of mimicry between mammals and reptiles (Nekaris et al., 2007, 2013; Nekaris, 2014).

SOCIAL BEHAVIOR OF THE LORISIFORMES

To help dispel the idea that the lorisiform primates lead solitary existences (Charles-Dominique, 1978; Bearder, 1987), Sterling (1993) recommends three components be used to emphasize the diversity of nocturnal primate social organization: *social system*, *spacing system*, and *mating system*. Social system relates to social behavior and relationships within a group, and most lorisiforms engage in considerable amounts of social behavior. Many adult nocturnal primates, including lesser galagos, thick-tailed galagos, and slender lorises, spend time together outside the breeding season, grooming, foraging, and feeding together (Bearder & Martin, 1980b; Clark, 1985; Radhakrishna & Singh, 2002). Some taxa spend up to 50% of their time in social proximity with adult conspecifics. Variability also exists in choice of companions. Among *Galago moholi*, for example, females were the most common social partners (Bearder & Doyle, 1974b), whereas in *Loris lydekkerianus*, females only formed positive affiliations with multiple adult males (Nekaris, 2002, 2003a; Radhakrishna & Singh, 2004). There is some relationship with social behavior and body size, with the smaller taxa being somewhat more gregarious, but this may be an artifact of larger-bodied species occurring at much lower population densities.

Physical contact alone does not define sociality—home range overlap is also very important, and animals that are physically dispersed still maintain regular vocal and olfactory contact. Determination of home range overlap via radio-tracking, or simply by all-night follows of known individuals, further elucidates the varied social relationships of the lorisiforms and defines the spacing system. Building on pioneering work by Bearder (1987), Müller and Thalmann (2000) constructed a framework by which home range overlap, or *spacing system*, can be used to illustrate the diversity among nocturnal mammal social organizations. In this framework, grouping systems can be cohesive and gregarious, dispersed yet social, or solitary—meaning no social contacts are made outside the mating system (Müller & Thalmann, 2000; Sterling et al., 2000). Adult sex ratio compositions mirror those seen among the social organizations of diurnal primates, with single male and female units, single male and multiple female groups, single female with multiple male groups, and groups with multi-male/multi-female structure (Table 1.4). In the case of rainforest primates, for which observational conditions may be very difficult (if not impossible), radio-tracking has become invaluable to infer social organization based on spacing patterns.

Table 1.2 contains a summary of the inferred social organizations of those lorisiforms that are best-studied to date (Bearder, 1987; Müller & Thalmann, 2000). Most galagos appear to exhibit a dispersed multi-male system, whereby males have larger home ranges than females, and females form matrilocal clusters of related females who may sleep together. These related females tend to be aggressive toward females from other groups, whereas males may be aggressive toward one another (Bearder & Doyle, 1974a; Charles-Dominique, 1974a; Bearder & Martin, 1980b). Male social behavior may correlate with age and status; for example, smaller resident males may be tolerated by the larger territorial males, whereas other males may be constantly on the move (so-called floaters) and, finally, some males remain solitary during the process of dispersing from their natal groups (Charles-Dominique, 1972; Bearder, 1987). A one-male/multiple-female system may be present in *Sciurocheirus gabonensis*, the Gabon squirrel galago, in which males are exclusively associated with small groups of females and only have aggressive contact with other males (Charles-Dominique, 1974a, 1977a,b). Another exception is found in *Paragalago cocos*, the Kenya coastal galago, in which one male/one female or one male/two–three female associations form, although variability between study sites shows some convergence with the general dispersed multi-male social organization of other galagos (Nash & Harcourt, 1986; Harcourt & Nash, 1986b).

Two different systems of social organization have been observed in *Perodicticus edwardsi*, Milne-Edwards' potto. Charles-Dominique (1977a) studied this species in a restricted forest environment where no matriarchies were present, and female home ranges were isolated from one another. Males may overlap their ranges with more than

Table 1.4 Reproductive and Life History Parameters of Lorisiform Primates

Taxa	Infants/ Year	Litter Size	Parking	Infant Carriage	Gestation (days)	Weaning (days)	References
Galaginae							
Galagoides demidoff	1-2	1	yes	mouth	111-114	40-50	Charles-Dominique (1977a)
Paragalago cocos	2-4	1-2	yes	mouth	120	49	Harcourt (1986a)
Sciurocheirus gabonensis	1-2	1	yes	mouth	133	?	Charles-Dominique (1977a)
Galago moholi	1-2	2	yes	mouth	120-126	?	Bearder and Doyle (1974a,b)
G. senegalensis	1-2	1	yes	mouth	141±2	70-98	Nash (1993)
Otolemur garnettii	1	1	yes	mouth/fur	126-138	140	Bearder (1987)
O. crassicaudatus	2-3	1	rare	mouth/fur	136	70-134	Bearder et al. (2003)
Perodicticinae							
P. (potto) edwardsi	1/(2)	1	rare	fur	197 (193-205)	120-180	Charles-Dominique (1977a)
Arctocebus aureus	1	1	yes	fur	131-136	100-130	Charles-Dominique (1977a)
A. calabarensis	1-2	1	yes	fur	130	115	Charles-Dominique (1977a)
Lorisinae							
Loris l. lydekkerianus	1-4	1-2	yes	fur	164 (160-166)	120-150	Rhadakrishna and Singh (2004)
L. l. malabaricus	1-4	1-2	yes	fur	166-169	120-150	Izard and Rasmussen (1985)
Loris l. nordicus	1-4	1-2	yes	fur	157-172	148-175	Fitch-Snyder and Schulze (2001)
Loris t. tardigradus	1-4	1-2	yes	fur	167-175	153-219	Fitch-Snyder and Schulze (2001)
Nycticebus coucang	1	1	yes	fur	165-175	85-180	Rasmussen (1986)
Xanthonycticebus (N.) pygmaeus	1-4	1/2	yes	fur	176-199	123-146	Fitch-Snyder and Ehrlich (2003)

one female, but tend to avoid one another, probably using scent. A second study found that males and females shared their home ranges to the exclusion of other male-female pairs. These same pairs also slept together or very near to one another on most occasions, suggesting a single male/single female spacing system (Pimley et al., 2005b).

Mysore slender lorises (*L. l. lydekkerianus*) exhibit limited range overlap between females, who were aggressive toward each other at home range boundaries. Male ranges are much larger than those of females. One or more adult males share sleeping sites with females; males are aggressive only to those males from other sleeping groups. The spacing indicates a single male/single female, and single male/multiple female system, but is also combined with promiscuous mating, suggesting a multimale/multi-female social organization (Nekaris, 2003a). Most slow lorises (*Nycticebus* spp.) and red slender lorises (*L. tardigradus*) appear to exhibit a single male/single female social organization, with the most common groupings being an adult male and female pair and their dependent offspring (Nekaris, 2014). This assessment corresponds with low testes volume for the genera (Wiens & Zitzmann, 2003). Nevertheless, a polygynous mating system is likely (Elliot & Elliot, 1967).

Mating system is defined as determining which animals actually mate and produce offspring. Pullen (2000) and Pullen et al. (2000) show that despite their spatial advantages, and despite fathering a majority of offspring in the study population, alpha male lesser galagos (*Galago moholi*) were not always the fathers of infants. Furthermore, not all twins were fathered by the same individuals (Pullen et al., 2000). These results are in line with both the testicular and copulatory evidence for this species, which suggests polygynandry. Pimley's (2002) molecular data for *P. edwardsi* show that offspring of mothers were not fathered by the male with whom those females were spatially paired, suggesting the social system differs from the mating system (Pimley et al., 2005b).

Acoustic and Olfactory Communication

Perhaps more than any other aspects of their behavior, calls have been invaluable for understanding the social behavior and taxonomy of galagos and, to a lesser degree, also of lorises (Zimmermann et al., 1988; Zimmerman, 1990; Bearder et al., 1995, 2013; Honess, 1996, 1995a; Anderson et al., 2000; Bearder et al., 2002; Coultas, 2002). Since animals can always remain silent, their calls invariably reflect circumstances where they may benefit in some way and, therefore, provide strong clues to important aspects of their socioecology. Calls are given in situations where it is advantageous (1) to attract and maintain contact with conspecifics, (2) to increase distance between competing individuals, (3) to warn kin of the presence of dangers, or (4) to advertise to potential predators that they have been detected. In the case of galagos, the safety provided by their arboreal, nocturnal activity patterns, and the ability to escape rapidly if detected, means they can communicate effectively by sound even when they appear to be alone. They have rich vocal repertoires of between 8–25 structurally distinct calls, including sounds that are relatively invariable and others that continuously change from one species to another (e.g., see Butynski et al., 2006, 2013). Added to this, galagos are able to mix different calls into bouts of rapidly changing vocalization sequences that can sometimes last for over 30 minutes at a time. Calls are used during short-range social interactions, with some variation between animals of different ages and sexes, but each species also has some calls that are loud and are used when mobbing predators, attracting partners, or repelling rivals. Every species has one particular loud call that is common to both sexes and used to advertise their presence to companions and rivals (Bearder et al., 2013). Since the advertisement call helps to bring mates together, it is invariably species-specific, remaining more or less constant across the entire geographical range of each species, thereby providing a convenient diagnostic tool for identifying new species (Masters, 1991; Zimmermann, 1995a,b; Anderson et al., 2000; Bearder et al., 2013; Svensson et al., 2017; Schneiderová et al., 2020).

The less agile pottos and lorises, as a group, are less audibly vocal, but unlike galagos, some of their calls are at frequencies in the ultrasonic range and are inaudible to humans (Zimmerman, 1985; Schulze & Meier, 1995; Nekaris & Jayewardene, 2004, Geerah et al., 2019). Both of the slender loris species call throughout the night (their calls are sometimes uttered over 50 times in one hour), and although not all parts of their loud calls are audible to humans, most of the call can be heard. The calls have several functions, including social spacing, aggression, affiliation, and dawn assembly (Bearder et al., 2002). At least six loud whistles with different functions have been identified for *L. tardigradus* (Coultas, 2002). Slow lorises, too, produce at least eight call types, and some calls, including mother-infant contact calls, are species-specific. Occurring at much higher frequency wavelengths, and often monotonal in sonic structure, these calls are much more difficult to hear in a noisy forest environment, especially for the human observer. In captivity, Javan slow lorises regularly counter-call to each other up to four times an hour. Javan slow lorises being reintroduced into the wild were also observed calling to wild conspecifics (Geerah et al., 2019).

One of the most understudied areas of lorisiform social behavior is that of olfactory communication. Aided by an acute sense of smell and the Jacobsen's organ (or vomeronasal organ) in the roof of the mouth, which senses volatile chemicals transferred from the rhinarium, or moist nose (Schilling, 1979; Martin, 1990), the lorisiforms are in constant delayed communication with each other via olfactory channels in the environment. Nocturnal lorisiforms communicate both with a number of specialized scent glands, as well as with urine (Harcourt, 1981). The visual systems of nocturnal lorisiform are highly sensitive, and are supplemented by olfactory communication (Bearder et al., 2006). The main advantage of olfactory communication via scent gland and urine marking in general is that it conveys information that is indirect and deferred in time, with a result that individuals do not have to come together in order to communicate. Captive studies of pygmy slow lorises (Fisher et al., 2003a,b), Senegal galagos (Nash, 1993), and thick-tailed galagos (Clark, 1978a,b, 1982a,b) have shown the ability of nocturnal primates not only to differentiate the state of sexual receptivity of conspecifics using scent, but also to recognize specific individuals of different age-sex classes. In fact, Clark (1985) suggests that the ability for fine olfactory differentiation contributed to increased gregariousness among *Otolemur crassicaudatus*. In the only systematic study of olfactory behavior in free-ranging nocturnal lorisiforms, Charles-Dominique (1974b, 1977b) showed that rather than using scent as trails, the taxa he studied would deposit scent marks in specific areas, conveying clear signals for both sexual attraction and avoidance.

Reproduction and Infant Development
The life history strategies of the lorisiforms have previously been the focus of a number of captive studies (e.g., Manley, 1966, 1967; Ehrlich & Musicant, 1977; Doyle, 1979; Izard & Rasmussen, 1985; Rasmussen, 1986; Rasmussen & Izard, 1988, Ehrlich & Macbride, 1989; Nash, 1993; Weisenseel et al., 1998; Fitch-Snyder & Schulze, 2001; Fitch-Snyder & Ehrlich, 2003), though field data are becoming available (Gursky & Nekaris, 2003). All taxa, with few exceptions, give birth to either singletons or twins, with twin births being known from more than half the taxa studied at present. A number of taxa have two litters per year. Little is known about survivorship ratios of lorisiforms in the wild, but when it is mentioned, it is not uncommon for only one infant to reach sexual maturity. Infant "parking" is common among the lorisiforms (Charles-Dominique, 1977b; Nekaris, 2003b, 2014). In general, the practice is for a mother to leave her infant on a branch or in a tree hole while she forages. The only variation seems to be whether infants are parked throughout the night (as is the case for most pottos and lorises), or whether the mother carries the infant with her for short distances, caching it in multiple sites throughout the night (which is the pattern for most galagos). Variation

across taxa also exists in whether or not infants cling to the fur while being carried or are transported by being held in the mother's mouth. Contrasting rates of life history among the galagos, pottos, and lorises are considered by some authors to be related to other locomotor and ecological differences among the three subfamilies (see Rasmussen & Nekaris, 1998). The pottos and lorises are noted for having some of the longest (i.e., slowest) life histories of any primate taxa of their body size, including long gestation lengths followed by low birth weights, and long periods of lactation, in contrast to galagos, which fall more in line with other primates of their body size (Martin, 1990).

Dixson (1995, 1998) has pointed out that a number of features of the genital morphology and the copulatory behavior of nocturnal lorisiforms may provide evidence that the spacing system does not necessarily coincide with the mating system. For example, larger testes size, or increase of testes size during a breeding season, should be linked with multiple male/multiple female mating systems (polygynandry). The penile morphology of most lorisiforms might also serve to identify species (e.g., Butynski et al., 1998), to enhance female receptivity and genital lock, or to break up copulatory plugs left by other males, and might also provide a clue to the mating systems of these primates.

LORISIFORM CONSERVATION

The general threats to lorisiform species are those that threaten primate populations wherever they occur (Estrada et al., 2017), anthropogenic activities producing habitat degradation and loss of suitable environmental conditions (Junker et al., 2012), hunting and trade in live animals as pets or their body parts for ornaments or for medicinal use (Ripple et al., 2016), and climate change (Estrada et al., 2017). These anthropogenic activities and their impacts can be expected to create latent extinction risks and future conservation concerns (Cardillo et al., 2006).

Lorisiform species in both Africa and Asia are subject to all the above pressures, but there are important regional differences in the types of anthropogenic pressures to which lorisiform taxa are subject (Nekaris et al., 2010a). Habitat degradation and loss are clearly proximate conservation threats (Gamage et al., 2017), but illegal trade of both live animals and bushmeat are persistent and considerable threats to lorisiforms (Svensson & Friant, 2014; Nekaris & Starr, 2015; Svensson et al., 2016; Nijman et al., 2017). Accurate monitoring of populations/ taxa of concern is necessary so that unexpected declines in conservation status do not occur unnoticed (Bland et al., 2015).

Table 1.5 summarizes the current International Union for the Conservation of Nature (IUCN) status for species and subspecies of the lorisiform primates. In terms of conservation category designation, 44.7% of all lorisiform taxa (17 species) are currently considered of Least Concern; 13.2% (5 species) are classified as Near Threatened; 10.5% of lorisiform taxa (4 species) are considered Vulnerable; 21.1% (8 species) are listed as Endangered; and 5.2% (2 species) are classified as Critically Endangered (IUCN, 2020). Inspection of Table 1.5 shows that conservation concern surrounding lorisiform species is not equally distributed across the subfamilies of the infraorder. Among the galagines, 13 species of currently recognized taxa are listed as Least Concern; 4 species are regarded as Near Threatened, with 1 species being considered either Vulnerable or Data Deficient (IUCN, 2020). Among the perodicticines, 3 species are currently listed as Least Concern; 2 species are classified as Near Threatened; and 1 species is listed as Endangered (IUCN, 2020). The picture for the lorisines is, however, dramatically different (Nekaris & Starr, 2015). Currently, a single lorisine species is classified in each of the Data Deficient and Least Concern categories; 3 species are considered Vulnerable; 5 species are listed as Endangered; and another 2 species are now considered Critically Endangered (IUCN, 2020). Thus, 90% of all lorisine taxa are currently classified as either Vulnerable, Endangered, or Critically Endangered; only the gray slender loris (*L. lydekerianus*) is currently classified as Least Concern. There are sufficient field data on the gray and red slender lorises to permit assessments of the conservation status of subspecies, but a clear exception is the undescribed gray slender loris form reported from the southern Western Ghats Mountains in India (Kumara et al., 2009, 2013; Sasi & Kumara, 2014), which for now is classified as Data Deficient. However, this is not the case with the slow lorises (Nekaris & Starr, 2015). Using the available IUCN (2020) conservation status assessments for both lorisine species and subspecies, 100% of known lorisine taxa face an extremely precarious future. This state of affairs makes the subfamily Lorisinae as endangered a taxonomic group of primates as any of Madagascar's highly endangered lemurform subfamily lineages.

In some cases (e.g., Javan slow lorises, in particular), where traditional local beliefs about leaving slow lorises alone and not disturbing or capturing them continue to be followed, slow loris populations can continue to live in close proximity to human rural communities (Nijman & Nekaris, 2014; Nekaris & Starr, 2015). However, the trade in slow lorises has become more popular over time (Nijman et al., 2017). Animals confiscated from the pet trade are often in poor health—it is a common practice for

Table 1.5 Conservation Status of the Lorisiform Taxa.

Scientific Name	Common Name	IUCN Red List Status[a]	References
Galaginae			
Galagoides demidoff	Demidoff's dwarf galago	Least Concern	Svensson et al. (2019)
Gd. thomasi	Thomas's dwarf galago	Least Concern	Svensson and Bearder (2019)
Gd. kumbirensis	Angolan dwarf galago	Near Threatened	Svensson et al. (2020a)
Paragalago orinus	Mountain dwarf galago	Vulnerable	Perkin (2020a)
Pg. zanzibaricus	Zanzibar lesser galago	Endangered	Perkin et al. (2020a)
Pg. rondoensis	Rondo dwarf galago	Endangered	Perkin (2020b)
Pg. cocos	Kenya coastal galago	Least Concern	Butynski and de Jong (2019a)
Pg. granti	Mozambique lesser galago	Least Concern	de Jong et al. (2019a)
Pg. nyasae (?)	Malawi lesser galago	[Conspecific with *Pg. granti*?]	Rovero et al. (2009)
Galago senegalensis	Senegal lesser galago	Least Concern	de Jong et al. (2019b)
G. gallarum	Somali lesser galago	Least Concern	de Jong and Butynski (2019)
G. moholi	Southern lesser galago	Least Concern	Bearder et al. (2019)
G. matschiei	Spectacled galago	Least Concern	Butynski and de Jong (2019b)
Euoticus elegantulus	Southern needle-clawed galago	Least Concern	Oates and Butynski (2019)
E. pallidus	Northern needle-clawed galago	Near Threatened	Cronin et al. (2020)
Sciurocheirus alleni	Allen's squirrel galago	Near Threatened	Perkin et al. (2020b)
S. gabonensis	Gabon squirrel galago	Least Concern	Oates (2019)
S. cameronensis	Cross River squirrel galago	Near Threatened	Pimley and Oates (2020)
S. makandensis	Makandé squirrel galago	Data Deficient	Svensson et al. (2020b)
Otolemur garnettii	Garnett's (sm.-eared) greater galago	Least Concern	de Jong et al. (2019c)
O. crassicaudatus	Thick-tailed greater galago	Least Concern	Masters and Bearder (2019a)
O. monteiri	Silvery greater galago	Least Concern	Masters and Bearder (2019b)
Perodicticinae			
Perodicticus potto	Western potto	Near Threatened	Svensson et al. (2020c)
P. edwardsi	Milne-Edwards', or Central, potto	Least Concern	Svensson and Pimley (2019)
P. ibeanus	Bosman's, or Eastern, potto	Least Concern	de Jong et al. (2019d)
P. (potto) juju (?)	S. Nigerian, or Benin, potto	Endangered	Oates et al. (2020)
Arctocebus aureus	Golden angwantibo	Least Concern	Svennson and Nekaris (2019)
A. calabarensis	Calabar angwantibo	Near Threatened	Oates and Svensson (2019)
Lorisinae			
Loris lydekkerianus	Gray slender loris	Least Concern	Nekaris et al. (2008)
L. l. lydekkerianus	Mysore gray slender loris	Near Threatened	Nekaris et al. (2008)
L. l. malabaricus	Malabar gray slender loris	Near Threatened	Bayani (2020)
L. lydekkerianus ssp.	Undescribed *L. lydekkerianus* ssp. (?)	Near Threatened	Nekaris et al. (2008)
L. l. nordicus	N. Sri Lanka gray slender loris	Near Threatened	Bayani (2020)
L. l. grandis	Highland gray slender loris	Data Deficient	Kumara et al. (2009)
Loris tardigradus	Red slender loris	Endangered	Nekaris et al. (2008)
		Vulnerable	Gamage et al. (2017)
		Endangered	Nekaris et al. (2008)
		Endangered	Gamage et al. (2017)
		Endangered	Nekaris (2008a)
L. t. tardigradus	Southwestern red slender loris	Endangered	Nekaris (2008b)
L. t. nycticeboides	Highland/Horton Plains slender loris	Endangered	Gamage et al. (2017)
L. t. parvus	Northwestern red slender loris	Endangered	Nekaris (2008c)
Nycticebus bengalensis	Bengal slow loris	Critically Endangered	Gamage et al. (2017)
		Critically Endangered	Gamage et al. (2017)
		Endangered	Nekaris et al. (2020a)
N. coucang	Greater slow loris	Endangered	Nekaris et al. (2020b)
N. javanicus	Javan slow loris	Critically Endangered	Nekaris et al. (2020c)
N. menagensis	Philippine slow loris	Vulnerable	Nekaris et al. (2020d)
N. borneanus	Bornean slow loris	Vulnerable	Nekaris and Miard (2020a)
N. kayan	Kayan slow loris	Vulnerable	Nekaris and Miard (2020b)
N. bancanus	Sody's slow loris	Critically Endangered	Nekaris and Marsh (2020)
N. hilleri	Sumatran slow loris	Endangered	Nekaris and Poindexter (2020)
Xenonycticebus (N.) pygmaeus	Pygmy loris	Endangered	Blair et al. (2020)

[a] IUCN (2020).

animal traders to remove the lower incisors and canines of lorises (which forms their mandibular toothcomb, essential for tree-gouging and exudate feeding); if the lorises survive, such individuals cannot be reintroduced to the wild, as they would be unable to exudate-feed naturally (Nijman & Nekaris, 2014; Nekaris & Starr, 2015). While some small-scale, controlled reintroductions of lorises have been conducted with wild-born individuals confiscated from the pet trade (e.g., Kenyon et al., 2014), the effectiveness of, and loris survivorship from, these reintroductions require further study (Nekaris & Starr, 2015). More problematic is the widespread and unmonitored practice of so-called hard releases of confiscated lorises (Nekaris & Starr, 2015), with no pre-release assessments of the suitability of habitat surrounding release sites nor post-release follow-up on the health and status of released animals (Moore et al., 2014; Nekaris & Starr, 2015). Several studies advocate for future conservation research into the anthropogenic pressures on lorisiform species to adopt an ethnoprimatological approach to understanding the cultural drivers behind the exploitation of lorisiform species and, thus, inform and improve lorisiform conservation efforts (Nekaris et al., 2010a; Nijman & Nekaris, 2014; Svensson & Friant, 2014; Nekaris & Starr, 2015; Svensson et al., 2016; Nijman et al., 2017).

LORISIFORM BEHAVIORAL ECOLOGY—A SUMMARY

Field research over the past 20 years (with significant reports published within just the last decade) has revealed previously unrecognized complexity in lorisiform feeding ecology (e.g., Nekaris et al., 2010b; Starr & Nekaris, 2013; Das et al., 2014; Cabana et al., 2017) and social organization (e.g., Radhakrishna & Singh, 2002; Geerah et al., 2019), clearly reflecting that the lorisiform primates are more than just an assemblage of leapers and slow climbers (Bearder, 1999; Masters, 2020). At the same time, however, an informed conceptualization of the full range of conservation perils facing lorisiform species now and in the immediate future is only beginning to take shape (e.g., Nekaris & Starr, 2015). An examination of Table 1.5 is sufficient to appreciate that many lorisiform species are of conservation concern with populations in decline, particularly among the lorisine taxa (with habitat loss and fragmentation, plus bushmeat hunting and the burgeoning pet trade among the major drivers of those declines; Nekaris et al., 2010a; Estrada et al., 2017). In both Africa and Asia, lorisiform species that are new to science continue to be identified (e.g., Ambrose, 2013; Munds et al., 2013; Svensson et al., 2017). It stands to reason, then, that there are still interspecific patterns of lorisiform similarity and difference which we currently do not appreciate, let alone understand.

Such lacunae do not diminish, however, the degree of ecological diversity that has been documented from recent studies across the taxonomic subfamilies Galaginae, Perodicticinae, and Lorisinae. But, with many lorisiform species still being only poorly studied or relatively unstudied, significant gaps in our accumulated understanding of the Infraorder Lorisiformes remain. Such blind spots in our knowledge base are problematic as there are also concurrent conservation concerns surrounding many lorisiform taxa arising from severe human-induced threats (Estrada et al., 2017). The bushmeat trade in Africa (Svensson & Friant, 2014; Svensson et al, 2016), and the pet and medicinal trades in Southeast Asia (Nekaris & Jaffe, 2007; Nekaris et al., 2010a; Nijman et al., 2017) are having detrimental effects on lorisiform populations. Habitat loss in both Africa and Asia as a result of human population pressures also poses a severe threat to these species (Erdelen, 1988; Butynski, 1996/97; Ratajszczak, 1998). In Africa and Asia, the multiple factors of human population pressures (Butynski, 1996/97), conversion of land to agriculture, including industrial-scale agriculture (Estrada et al., 2017), and deforestation from timber extractions in logging concessions are the chief causes of forest loss (Mill, 1995). The tendency for lorises to cling to, rather than flee out of, trees that are being felled, makes them easy targets for capture for the pet trade. Thus, whereas other animals may more readily escape fronts of human activity in forested habitats, capture of lorises for the pet trade can completely remove all individuals from forest areas adjacent to localities of human activity (Streicher, 2004; Nekaris & Jaffe, 2007). Furthermore, logging and human disturbance have been shown to adversely affect lorisiform population densities (Weisenseel et al., 1993; Nekaris & Jayewardene, 2004).

New and ongoing field research projects will continue to generate contributions to our understanding of lorisiform behavioral ecology. With the possibility that there are cryptic lorisiform species that still remain unrecognized, even while forest fragmentation and habitat loss continue apace, there is a genuine chance that the taxonomic description of any "new" lorisiform species discovered in the future will include the recommendation that the species also be immediately classified as Critically Endangered (Bearder, 1999). Future studies of individual lorisiform species, particularly those that are today either poorly known or for which no long-term data are available, will provide additional valuable comparative data on these biologically diverse and intriguing primates.

REFERENCES CITED—CHAPTER 1

Alterman, L. (1995). Toxins and toothcombs: Potential allospecific chemical defenses in *Nycticebus* and *Perodicticus*. In L. Alterman, G. A. Doyle, & M. K. Izard (Eds.), *Creatures of the dark: The nocturnal prosimians* (pp. 413-424). Plenum Press.

Altmann, S. A., & Altmann, J. (1970). *Baboon ecology: African field research*. University of Chicago Press.

Ambrose, L. (1999). *Species diversity in West and Central African galagos (Primates, Galagonidae): The use of acoustic analysis*. [Doctoral dissertation, Oxford Brookes University].

Ambrose, L. (2003). Three acoustic forms of Allen's galagos (Primates; Galagonidae) in the Central African region. *Primates, 44*, 25-39.

Ambrose, L. (2013). *Sciurocheirus makandensis sp. nov.* Makandé Squirrel Galago. In T. M. Butynski, J. Kingdon, & J. Kalina (Eds.), *Mammals of Africa, Vol. II—Primates* (pp. 421-422). Bloomsbury Publishing.

Anderson, M. J. (1998). Comparative morphology and speciation in galagos. *Folia Primatologica, 69*(Suppl. 1), 325-331.

Anderson, M. J. (2000). Penile morphology and classification of bush babies (subfamily Galagoninae). *International Journal of Primatology, 21*(5), 815-836.

Anderson, M. J. (2001). The use of hair morphology in the classification of galagos (Primates, subfamily Galagoninae). *Primates, 42*(2), 113-121.

Anderson, M. J., Ambrose, L., Bearder, S. K., Dixson, A. F., & Pullen, S. (2000). Intraspecific variation in the vocalizations and hand pad morphology of southern lesser bush babies (*Galago moholi*): A comparison with *G. senegalensis*. *International Journal of Primatology, 21*(3), 537-555.

Ankel-Simons, F. (2000). *Primate anatomy: An introduction* (2nd ed.). Academic Press.

Attygalle, A. B., & Morgan, E. D. (1984). Chemicals from the glands of ants. *Chemical Society Reviews, 13*, 245-278.

Ballenger, L. (2001). "*Galago senegalensis.*" Animal Diversity Web. https://animaldiversity.org/accounts/Galago_senegalensis/

Barrett, E. (1984). *The ecology of some nocturnal, arboreal mammals in the rainforests of peninsular Malaysia*. [Doctoral dissertation, Cambridge University].

Bayani, A. 2020. *Loris lydekkerianus* Cabrera, 1908—Slender loris. In V. Ramachandran, A. Bayani, R. Chakravarty, P. Roy, & K. Kunte (Eds.), *Mammals of India, v. 1.13*.http://www.mammalsofindia.org/sp/287/Loris-lydekkerianus

Bayes, M. K. (1998). *A molecular phylogenetic study of the galagos, strepsirhine primates and archontan mammals*. [Doctoral dissertation, Oxford Brookes University].

Bearder, S. K. (1969). *Territorial and intergroup behaviour of the lesser bushbaby, Galago senegalensis moholi, in semi-natural conditions and in the field*. [Master's thesis, University of the Witwatersrand].

Bearder, S. K. (1974). *Aspects of the ecology and behaviour of the thick-tailed bushbaby galago crassicaudatus*. [Doctoral dissertation, University of Witwatersrand].

Bearder, S. K. (1987). Lorises, bushbabies and tarsiers: Diverse societies in solitary foragers. In B. B. Smuts, D. L. Cheney, R. M. Seyfarth, R. W. Wrangham, & T. T. Struhsaker (Eds.), *Primate societies* (pp.11-24). University of Chicago Press.

Bearder, S. K. (1999). Physical and social diversity among nocturnal primates: A new view based on long term research. *Primates, 40*(1), 267-282.

Bearder, S. K., Ambrose, L., Harcourt, C., Honess, P., Perkin, A., Pimley, E., Pullen, S., & Svoboda, N. (2003). Species-typical patterns of infant contact, sleeping site use and social cohesion among nocturnal primates in Africa. *Folia Primatologica, 74*(5-6), 337-354.

Bearder, S. K., Butynski, T. M., & de Jong, Y. A. (2013). Vocal profiles for the galagos: A tool for identification. *Primate Conservation, 27*, 75.

Bearder, S. K., & Doyle, G. A. (1974a). Field and laboratory studies of social organisation in bushbabies (*Galago senegalensis*). *Journal of Human Evolution, 3*, 37-50.

Bearder, S. K., & Doyle, G. A. (1974b). Ecology of bushbabies *Galago senegalensis* and *Galago crassicaudatus*, with some notes on their behaviour in the field. In R. D. Martin, G. A. Doyle, & A. C. Walker (Eds.), *Prosimian Biology* (pp.109-130). University of Pittsburgh Press.

Bearder, S. K., Honess, P. E., & Ambrose, L. (1995). Species diversity among galagos with special reference to mate recognition. In L. Alterman, G. A. Doyle, & M. K. Izard (Eds.), *Creatures of the dark: The nocturnal prosimians* (pp. 331-352). Plenum Press.

Bearder, S. K., & Martin, R. D. (1980a). Acacia gum and its use by bushbabies, *Galago senegalensis* (Primates: Lorisidae). *International Journal of Primatology, 1*, 103-128.

Bearder, S. K., & Martin, R. D. (1980b). The social organization of a nocturnal primate revealed by radio tracking. In C. J. Amlaner, & D. W. MacDonald (Eds.), *A handbook on biotelemetry and radio tracking* (pp. 633-648). Pergamon Press.

Bearder, S. K., Nekaris, K. A. I., & Buzzell, C. A. (2002). Dangers of the night: Are some primates afraid of the dark? In L. E. Miller (Ed.), *Eat or be eaten: Predator sensitive foraging in primates* (pp. 21-43). Cambridge University Press.

Bearder, S. K., Nekaris, K. A. I., & Curtis, D. J. (2006). A re-evaluation of the role of vision in the activity and communication of nocturnal primates. *Folia Primatologica, 77*(1-2), 50-71.

Bearder, S. K., Svensson, M., & Butynski, T. M. (2019). *Galago moholi. The IUCN Red List of Threatened Species* 2019: e.T8788A17963285. https://dx.doi.org/10.2305/IUCN.UK.2019-3.RLTS.T8788A17963285.en.

Blair, M., Nadler, T., Ni, O., Samun, E., Streicher, U., & Nekaris, K. A. I. (2020). *Nycticebus pygmaeus. The IUCN Red List of Threatened Species* 2020: e.T14941A17971417. https://dx.doi.org/10.2305/IUCN.UK.2020-2.RLTS.T14941A17971417.en.

Bland, L. M., Collen, B., Orme, C. D., & Bielby, J. (2015). Predicting the conservation status of data-deficient species. *Conservervation Biology, 29*, 250-259.

Butynski, T. M. (1982). Blue monkey (*Cercopithecus mitis stuhlmanni*) predation on galagos. *Primates, 23,* 563-566.

Butynski, T. M. (1996/97). African primate conservation—the species and the IUCN/SSC primate specialist group network. *Primate Conservation, 17,* 87-100.

Butynski, T. M., & de Jong, Y. A. (2004). Natural history of the Somali lesser galago (*Galago gallarum*). *Journal of East African Natural History, 93*(1), 23-38.

Butynski, T. M., & de Jong, Y. A. (2017). The Mount Kenya potto is a subspecies of the eastern potto *Perodicticus ibeanus. Primate Conservation, 31,* 49-52.

Butynski, T. M., & de Jong, Y. A. (2019a). *Paragalago cocos. The IUCN Red List of Threatened Species* 2019: e.T136212A17963050. https://dx.doi.org/10.2305/IUCN .UK.2019-3.RLTS.T136212A17963050.en.

Butynski, T. M., & de Jong, Y. A. (2019b). *Galago matschiei. The IUCN Red List of Threatened Species* 2019: e.T8787A17963414. https://dx.doi.org/10.2305/IUCN.UK .2019-3.RLTS.T8787A17963414.en.

Butynski, T. M., de Jong, Y. A., Perkin, A. W., Bearder, S. K., & Honess, P. E. (2006). Taxonomy, distribution, and conservation status of three species of dwarf galagos (Galagoides) in eastern Africa. *Primate Conservation, 21,* 63-79.

Butynski, T. M., Ehardt, C. L., & Struhsaker, T. T. (1998). Notes on two dwarf galagos (*Galagoides udzungwensis* and *Galagoides orinus*) in the Udzungwa Mountains, Tanzania. *Primate Conservation, 18,* 69-75.

Cabana, F., Dierenfeld, E., Wirdateti, W., Donati, G., & Nekaris, K. A. I. (2017). The seasonal feeding ecology of the Javan slow loris (*Nycticebus javanicus*). *American Journal of Physical Anthropology, 162*(4), 768-781.

Campera, M., Balestri, M., Stewart, A. N., & Nekaris, K. A. I. (2022). Influence of moon luminosity, seasonality, sex and weather conditions on the activity levels of the nocturnal Javan slow loris. *Ecologies, 3,* 257-266.

Cardillo, M., Mace, G. M., Gittleman, J. L., & Purvis, A. (2006). Latent extinction risk and the future battlegrounds of mammal conservation. *Proceedings of the National Academy of Sciences, 103*(11), 4157-4161.

Chapman, C. A., Gautier-Hion, A., Oates, J. F., & Onderdonk, D. A. (1999). African primate communities: Determinants of structure and threats to survival. In J. G. Fleagle, C. Janson, & K. E. Reed (Eds.), *Primate Communities* (pp. 1-37). Cambridge University Press.

Charles-Dominique, P. (1971). Eco-ethologie des prosimiens du Gabon. *Biol. Gabon, 7,* 121-228.

Charles-Dominique, P. (1972). Ecologie et vie sociale de *Galago demidovii* (Fisher 1808, Prosimii). *Zeitschrift für Tierpsychologie Suppl., 9,* 7-41.

Charles-Dominique, P. (1974a). Vie sociale de *Perodicticus potto* (Primates, Lorisides). Étude de terrain en forêt equatorial de l'ouest africain au Gabon. *Mammalia, 38,* 355-379.

Charles-Dominique, P. (1974b). Ecology and feeding behaviour of five sympatric lorisids in Gabon. In R. D. Martin., G. A. Doyle, & A. C. Walker (Eds.), *Prosimian biology* (pp. 131-150). University of Pittsburgh Press.

Charles-Dominique, P. (1977a). *Ecology and behaviour of nocturnal primates: Prosimians of equatorial West Africa.* Columbia University Press.

Charles-Dominique, P. (1977b). Urine marking and territoriality in *Galago alleni* (Waterhouse, 1837—Lorisoidea, Primates)—A field study by radio-telemetry. *Zeitschrift für Tierpsychologie, 43,* 113-138.

Charles-Dominique, P. (1978). Solitary and gregarious prosimians: Evolution of social structures in primates. In D. J. Chivers, & K. A. Joysey (Eds.), *Recent advances in primatology, Volume Three—Evolution* (pp. 139-149). Academic Press.

Charles-Dominique, P., & Bearder, S. K. (1979). Field studies of lorisid behavior: Methodological aspects. In G. A. Doyle, & R. D. Martin (Eds.), *The Study of Prosimian Behavior* (pp. 567-629). Academic Press.

Clark, A. B. (1978a). Olfactory communication, *Galago crassicaudatus,* and the social life of prosimians. In D. J. Chivers, & K. A. Joysey (Eds.), *Recent Advances in Primatology, Volume Three—Evolution* (pp. 109-117). Academic Press.

Clark, A. B. (1978b). Sex ratio and local resource competition in a prosimian primate. *Science, 201,* 163-165.

Clark, A. B. (1982a). Scent marks as social signals in *Galago crassicaudatus* I. Sex and reproductive status as factors in signals and responses. *Journal of Chemical Ecology, 8*(8), 1133-1151.

Clark, A. B. (1982b). Scent marks as social signals in *Galago crassicaudatus* II. Discrimination between individuals by scent. *Journal of Chemical Ecology, 8*(8), 1153-1165.

Clark, A. B. (1985). Sociality in a nocturnal "solitary" prosimian: *Galago crassicaudatus. International Journal of Primatology, 6,* 581-600.

Coultas, D. S. (2002). *Bioacoustic analysis of the loud call of two species of slender loris (Loris tardigradus and L. lydekkerianus nordicus) from Sri Lanka.* [Master's thesis, Oxford Brookes University].

Crompton, R. H. (1980). *A leap in the dark: Locomotor behaviour and ecology in* Galago senegalensis *and* Galago crassicaudatus. [Doctoral dissertation, Harvard University].

Crompton, R. H. (1983). Age differences in locomotion in two subtropical galaginae. *Primates, 24,* 241-259.

Crompton, R. H. (1984). Foraging, habitat structure, and locomotion in two species of galago. In P. S. Rodman, & J. G. H. Cant (Eds.), *Adaptations for foraging in non-human primates* (pp. 73-111). Columbia University Press.

Cronin, E. S., Oates, J. F., & Butynski, T. M. (2020). *Euoticus pallidus. The IUCN Red List of Threatened Species* 2020: e.T8266A17961858. https://dx.doi.org/10.2305/IUCN.UK .2020-2.RLTS.T8266A17961858.en.

Das, N., Nekaris, K. A. I., & Bhattacharjee, P. C. (2014). Medicinal plant exudativory by the Bengal slow loris *Nycticebus bengalensis. Endangered Species Research, 23,* 149-157.

Daschbach, N., Schein, M., & Haines, D. (1981). Vocalizations of the slow loris, *Nycticebus coucang* (Primates, Lorisidae). *International Journal of Primatology, 2*, 71-80.

de Jong, Y. A, & Butynski, T. M. (2004). Life in the thornbush— the Somali bushbaby. *Swara, 27*, 22.

de Jong, Y. A., & Butynski, T. M. (2019). *Galago gallarum. The IUCN Red List of Threatened Species* 2019: e.T8786A17963185. https://dx.doi.org/10.2305/IUCN.UK .2019-3.RLTS.T8786A17963185.en.

de Jong, Y. A., Butynski, T. M., & Perkin, A. (2019a). *Paragalago granti. The IUCN Red List of Threatened Species* 2019: e.T91970347A17962454. https://dx.doi.org/10.2305/IUCN .UK.2019-3.RLTS.T91970347A17962454.en.

de Jong, Y. A., Butynski, T. M., Perkin, A., & Svensson, M. (2019c). *Otolemur garnettii. The IUCN Red List of Threatened Species* 2019: e.T15644A17963837. https://dx.doi .org/10.2305/IUCN.UK.2019-3.RLTS.T15644A17963837 .en.

de Jong, Y. A., Butynski, T. M., Perkin, A., Svensson, M., & Pimley, E. (2019d). *Perodicticus ibeanus. The IUCN Red List of Threatened Species* 2019: e.T136875A91996195. https://dx.doi.org/10.2305/IUCN.UK.2019-3.RLTS .T136875A91996195.en.

de Jong, Y. A., Butynski, T. M., Svensson, M., & Perkin, A. (2019b). *Galago senegalensis. The IUCN Red List of Threatened Species* 2019: e.T8789A17963505. https://dx.doi .org/10.2305/IUCN.UK.2019-3.RLTS.T8789A17963505.en.

Demes, B., Fleagle, J. G., & Lemelin, P. (1998). Myological correlates of prosimian leaping. *Journal of Human Evolution, 34*(4), 385-399.

Dixson, A. F. (1989). Effects of sexual selection upon the genitalia and copulatory behaviour in male primates. *International Journal of Primatology, 10*, 47-55.

Dixson, A. F. (1995). Sexual selection and the evolution of copulatory behavior in nocturnal prosimians. In L. Alterman, G. A. Doyle, & M. K. Izard (Eds.), *Creatures of the dark: The nocturnal prosimians* (pp. 93-118). Plenum Press.

Dixson, A. F. (1998). *Primate sexuality: Comparative studies of the prosimians, monkeys, apes and human beings*. Oxford University Press.

dos Reis, M., Gunnell, G. F., Baeba-Montoya, Wilkins, A., Yang, Z., & Yoder, A. D. (2018). Using phylogenomic data to explore the effects of relaxed clocks and calibration strategies on divergence time estimation: Primates as a test case. *Systematic Biology, 67*(4), 594-615.

Doyle, G. A. (1979). Development of behaviour in prosimians with special reference to the lesser bushbaby, *Galago senegalensis moholi*. In G. A. Doyle, & R. D. Martin (Eds.), *The study of prosimian behaviour* (pp. 157-189). Academic Press.

Duckworth, J. W. (1994). Field sighting of the pygmy loris (*Nycticebus pygmaeus*) in Laos. *Folia Primatologica, 63*, 99-101.

Dykyj, D. (1980). Locomotion of the slow loris in a designed substrate context. *American Journal of Physical Anthropology, 52*, 577-586.

Ehrlich, A., & Macbride, L. (1989). Mother-infant interactions in captive slow lorises (*Nycticebus coucang*). *American Journal of Primatology, 19*, 217-228.

Ehrlich, A., & Musicant, A. (1977). Social and individual behaviors in captive slow lorises (*Nycticebus coucang*). *Behaviour, 60*, 195-220.

Elliot, O., & Elliot, M. (1967). Field notes on the slow loris in Malaya. *Journal of Mammalogy, 48*, 497-498.

Ellison, G., Wolfenden, A., Kahana, L., Kisingo, A., Jamieson, J., Jones, M., & Bettridge, C. M. (2019). Sleeping site selection in the nocturnal northern lesser galago (*Galago senegalensis*) supports antipredator and thermoregulatory hypotheses. *International Journal of Primatology, 40*, 276-296.

Erdelen, W. (1988). Forest ecosystems and nature conservation in Sri Lanka. *Biological Conservation, 43*, 115-135.

Estrada, A., Garber, P. A., Rylands, A. B., Roos, C., Fernandez-Duque, E., Di Fiore, A., Nekaris, K. A., Nijman, V., Heymann, E. W., Lambert, J. E., Rovero, F., Barelli, C., Setchell, J. M., Gillespie, T. R., Mittermeier, R. A., Arregoitia, L. V., de Guinea, M., Gouveia, S., Dobrovolski, R.,... & Li, B. (2017). Impending extinction crisis of the world's primates: Why primates matter. *Science Advances, 3*(1), e1600946. https:// doi.org/10.1126/sciadv.1600946.

Fisher, H. S., Swaisgood, R. R., & Fitch-Snyder, H. (2003a). Odor familiarity and female preferences for males in a threatened primate, the pygmy loris *Nycticebus pygmaeus*: Applications for genetic management of small populations. *Naturwissenschaften, 90*(11), 509-512.

Fisher, H. S., Swaisgood, R. R., & Fitch-Snyder, H. (2003b). Countermarking by male pygmy lorises (*Nycticebus pygmaeus*): Do females use odor cues to select mates with high competitive ability? *Behavioral Ecology and Sociobiology, 53*(2), 123-130.

Fitch-Snyder, H., & Ehrlich, A. (2003). Mother-infant interactions in slow lorises (*Nycticebus bengalensis*) and pygmy lorises (*Nycticebus pygmaeus*). *Folia Primatologica, 74*(5-6), 259-271.

Fitch-Snyder, H., & Schulze, H. (Eds.). (2001). *Management of lorises in captivity. A husbandry manual for Asian Lorisines (Nycticebus & Loris spp.)*. Center for Reproduction of Endangered Species.

Fuller, G., Nijman, V., Wirdateti, W., & Nekaris, K. A. I. (2016). Do chemical cues in the venom of slow lorises repel avian predators? *Emu, 116*(4), 435-439.

Gamage, S. N., Groves, C. P., Marikar, F. M. M. T., Turner, C. S., Padmalal, U. K. G. K., & Kotagama, S. W. (2017). Taxonomy, distribution, and conservation status of the slender loris (Primates, Lorisidae: Loris) in Sri Lanka. *Primate Conservation, 31*, 83-106.

Gamage, S. N., Hettiarachchi, C. J., Mahanayakage, C. A., Padmalal, U. K. G. K., & Kotagama, S. W. (2016). The red slender loris (*Loris tardigradus*). *Wildlanka, 4*(2), 66-74.

Geerah, D. R., O'Hagan, R. P., Wirdateti, W., & Nekaris, K. A. I. (2019). The use of ultrasonic communication to maintain social cohesion in the Javan slow loris (*Nycticebus javanicus*). *Folia Primatologica, 90*(5), 392-403.

Glassman, D. M., & Wells, J. P. (1984). Positional and activity behavior in a captive slow loris: a quantitative assessment. *American Journal of Primatology*, 7(2), 121-132.

Gnanaolivu, S. D., & Singh, M. (2019). First sighting of predatory attack on a Malabar grey slender loris (*Loris lydekkerianus malabaricus*) by brown palm civet (*Paradoxurus jerdoni*). *Asian Primates Journal*, 8(1), 37-40.

Groves, C. P. (2001). *Primate taxonomy*. Smithsonian Institution Press.

Groves, C. P. (2005). Order primates. In D. E. Wilson, & D. M. Reeder (Eds.), *Mammal species of the world: A taxonomic and geographic reference* (3rd ed., pp. 111-184). Johns Hopkins University Press.

Grow, N. B., & Nekaris, K. A. I. (2015). Does toxic defence in *Nycticebus* spp. relate to ectoparasites? The lethal effects of slow loris venom on arthropods. *Toxicon*, 95, 1-5.

Grubb, P., Butynski, T. M., Oates, J. F., Bearder, S. K., Disotell, T. R., Groves, C., & Struhsaker, T. (2003). An assessment of the diversity of African primates. *International Journal of Primatology*, 24(6), 1301-1357.

Gursky, S., & Nekaris, K. A. I. (2003). An introduction to mating, birthing and rearing systems of nocturnal prosimians. *Folia Primatologica*, 74(5-6), 272-284.

Hagey, L., Fry, B., & Fitch-Snyder, H. (2007). Talking defensively, a dual use for the brachial gland exudate of slow and pygmy lorises. In S. Gursky, & K. A. I. Nekaris (Eds.), *Primate anti-predator strategies* (pp. 253-272). Springer.

Happold, D., & Happold, M. (1992). Termites as food for the thick-tailed bushbaby (*Otolemur crassicaudatus*) in Malawi. *Folia Primatologica*, 58, 118-120.

Harcourt, C. S. (1980). *Behavioural adaptations of South African galagos.* [Master's thesis, University of the Witwatersrand].

Harcourt, C. S. (1981). An examination of the function of urine washing in *Galago senegalensis*. *Zeitschrift für Tierpsychologie*, 55, 119-128.

Harcourt, C. S. (1984). *The behaviour and ecology of galagos in Kenyan Coastal Forest.* [Unpublished doctoral dissertation, University of Cambridge].

Harcourt, C. S. (1986a). *Galago zanzibaricus*: Birth seasonality, litter size and perinatal behaviour of females. *Journal of Zoology*, 210(3), 451-457.

Harcourt, C. S. (1986b). Seasonal variation in the diet of South African galagos. *International Journal of Primatology*, 7(5), 491-506.

Harcourt, C. S. & Bearder, S. K. (1989). A comparison of *Galago moholi* in South Africa with *Galago zanzibaricus* in Kenya. *International Journal of Primatology*, 10(1), 35-45.

Harcourt, C. S., & Nash, L. T. (1986a). Species differences in substrate use and diet between sympatric galagos in two Kenyan coastal forests. *Primates*, 27, 41-52.

Harcourt, C. S. & Nash, L. T. (1986b). Social organization of galagos in Kenyan coastal forests, I. *Galago zanzibaricus*. *American Journal of Primatology*, 10, 339-356.

Hardus, M. E., Lameira, A. R., Zulfa, A., Atmoko, S. S. U., de Vries, H., & Wich, S. A. (2012). Behavioral, ecological, and evolutionary aspects of meat-eating by Sumatran orangutans (*Pongo abelii*). *International Journal of Primatology*, 33, 287-304.

Hartwig, W. (2011). Primate evolution. In C. J. Campbell, A. Fuentes, K. C. MacKinnon, S. K. Bearder, & R. M. Stumpf (Eds.), *Primates in perspective* (2nd ed., pp. 19-31). Oxford University Press.

Hladik, C. M. (1979). Diet and ecology of prosimians. In G. A. Doyle, & R. D. Martin (Eds.), *The study of prosimian behavior* (pp. 307-357). Academic Press.

Hockings, K. J., Humle, T., Carvalho, S., & Matsuzawa, T. (2012). Chimpanzee interactions with nonhuman species in an anthropogenic habitat. *Behaviour*, 149(3-4), 299-324.

Hohmann, G., & Fruth, B. (2008). New records on prey capture and meat eating by bonobos at Lui Kotale, Salonga National Park, Democratic Republic of Congo. *Folia Primatologica*, 79, 103-110.

Honess, P. E. (1996). *Speciation among galagos (Primates, Galagidae) in Tanzanian forests.* [Doctoral dissertation, Oxford Brookes University].

Ishida, H., Hirasaki, E., & Matano, S. (1992). Locomotion of the slow loris between discontinuous substrates. In S. Matano, R. H. Tuttle, & H. Ishida (Eds.), *Topics in primatology. Vol. 3—Evolutionary biology, reproductive endocrinology and virology* (pp. 139-152). University of Tokyo Press.

IUCN. (2020). *The International Union for the Conservation of Nature Red List of Threatened Species.* Version 2020-2. (www.iucnredlist.org). IUCN: Gland, Switzerland, and Cambridge, U.K.

Izard, M. K., & Rasmussen, D. T. (1985). Reproduction in the slender loris (*Loris tardigradus malabaricus*). *American Journal of Primatology*, 8, 153-165.

Jewell, P. A., & Oates, J. F. (1969). Ecological observations of the lorisoid primates of African lowland forest. *Zoologica Africana*, 4, 231-248.

Junker, J., Blake, S., Boesch, C., Campbell, G., du Toit, L., Duvall, C., Ekobo, A., Etoga, G., Galat-Luong, A., Gamys, J., Ganas-Swaray, J., Gatti, S., Ghiurghi, A., Granier, N., Hart, J., Head, J., Herbinger, I., Cleveland Hicks, T., Huijbregts, B.,...& Kuehl, H. S. (2012). Recent decline in suitable environmental conditions for African great apes. *Diversity and Distributions*, 18(11), 1-15. https://doi.org/10.1111/ddi.12005.

Kenyon, M., Streicher, U., Loung, H., Tran, T., Tran, M., Vo, B., & Cronin, A. (2014). Survival of reintroduced pygmy slow loris *Nycticebus pygmaeus* in south Vietnam. *Endangered Species Research*, 25, 185-195.

Kumara, H. N., Irfan-Ullah, M., & Kumar, S. (2009). Mapping potential distribution of slender loris subspecies in peninsular India. *Endangered Species Research*, 7, 29-38.

Kumara, H. N., Kumar, S., & Singh, M. (2005). A novel foraging technique observed in slender loris (*Loris lydekkerianus malabaricus*) feeding on red ants in the western Ghats, India. *Folia Primatologica*, 76, 116-118.

Kumara, H. N., Singh, M., Irfan-Ullah, M., & Kumar, S. (2013). Status, distribution, and conservation of slender lorises in India. In J. Masters, M. Gamba, & F. Génin (Eds.),

Leaping ahead: Advances in prosimian biology (pp. 343-352). Springer.

Kumara, H. N., Singh, M., & Kumar, S. (2006). Distribution, habitat correlates and conservation of *Loris lydekkerianus* in Karnataka, India. *International Journal of Primatology*, 27, 941-969.

Luhrs, A. M., Svensson, M. S., & Kekaris, K. A. I. (2018). Comparative ecology and behaviour of eastern potto (*Nycticebus kayan*): Implications for slow loris conservation. *P. Edwardsi* in Angola, Cameroon, Kenya, Nigeria, Rwanda and Uganda. *Journal of East African Natural History*, 107, 17-30.

Madani, G., & Nekaris, K. A. I. (2014). Anaphylactic shock following the bite of a wild Kayan slow loris (*Perodicticus ibeanus*) and central potto *Journal of Venomous Animals and Toxins Including Tropical Diseases*, 20, 43.

Manley, G. H. (1966). Reproduction in lorisoid primates. *Symposia of the Zoological Society of London*, 15, 493-509.

Manley, G. H. (1967). Gestation periods in the lorisidae. *International Zoo Yearbook*, 7, 80-81.

Martin, R. D. (1972). A preliminary field study of the lesser mouse lemur (*Microcebus murinus* J. F. Miller 1777). *Zeitschrift für Tierpsychologie*. 9, 43-89.

Martin, R. D. (1973). A review of the behaviour and ecology of the lesser mouse lemur (*Microcebus murinus* J. F. Miller 1777). In R. P. Michael, & J. H. Crook (Eds.), *Comparative ecology and behaviour of primates* (pp. 1-68). Academic Press.

Martin, R. D. (1990). *Primate origins and evolution: A Phylogenetic Reconstruction*. Chapman and Hall.

Masters, J. C. (1988). Speciation in the greater galagos (Prosimii: Galaginae): A review and synthesis. *Biological Journal of the Linnean Society*, 34, 149-174.

Masters, J. C. (1991). Loud calls of *Galago crassicaudatus* and *G. garnettii* and their relation to habitat structure. *Primates*, 32, 153-167.

Masters, J. C. (1998). Speciation in the lesser galagos. *Folia Primatologica*, 69(Suppl. 1), 357-370.

Masters, J. C. (2020). Sluggards and drunkards?: A history of the discovery and description of the Afro-Asian lorisidae. In K. A. I. Nekaris, & A. Burrows (Eds.), *Evolution, ecology and conservation of lorises and pottos* (pp. 19-32). Cambridge University Press.

Masters, J. C., & Bearder, S. (2019a). *Otolemur crassicaudatus* ssp. *crassicaudatus*. *The IUCN Red List of Threatened Species* 2019: e.T136881A17989293. https://dx.doi.org/10.2305/IUCN.UK.2019-3.RLTS.T136881A17989293.en.

Masters, J. C., & Bearder, S. (2019b). *Otolemur crassicaudatus* ssp. *monteiri*. *The IUCN Red List of Threatened Species* 2019: e.T91991178A17989322. https://dx.doi.org/10.2305/IUCN.UK.2019-3.RLTS.T91991178A17989322.en.

Masters, J. C., Boniotto, M., Crovella, S., Roos, C., Pozzi, L., & Delpero, M. (2007). Phylogenetic relationships among the lorisoidea as indicated by craniodental morphology and mitochondrial sequence data. *American Journal of Primatology*, 69, 6-15.

Masters, J. C., Génin, F., Couette, S., Groves, C., Nash, S. D., Delpero, M., & Pozzi, L. (2017). A new genus for the eastern dwarf galagos (Primates: Galagidae). *Zoological Journal of the Linnean Society*, 181(1), 229-241.

Masters, J. C., Lumsden, W. H. R., & Young, D. A. (1988). Reproductive and dietary parameters in wild greater galago populations. *International Journal of Primatology*, 9(6), 573-592.

Masters, J. C., Rayner, H., Ludewig, H., Zimmermann, E., Molez-Verriere, F., Vincent, F., & Nash, L. T. (1994). Phylogenetic relationships among the galaginae as indicated by erythrocytic allozymes. *Primates*, 35, 177-190.

Mill, R. R. (1995). Regional overview: Indian subcontinent. In V. H. Heywood, & S. D. Davis (Eds.), *Centres of plant diversity: A guide to strategy for their conservation. Asia, Australia and the Pacific* (Vol. 2, pp. 62-135). Cambridge: World Wildlife Fund for Nature and IUCN Press.

Moore, R. S., Wihermento, S., & Nekaris, K. A. I. (2014). Compassionate conservation, rehabilitation and translocation of Indonesian slow lorises. *Endangered Species Research*, 26, 93-102.

Morozov, N. (2015). Why do birds practice anting? *Biology Bulletin Reviews*, 5(4), 353-365.

Müller, A. E., & Thalmann, U. (2000). Origin and evolution of primate social organisation: A reconstruction. *Biological Review of the Cambridge Philosophical. Society*, 75(3), 405-435.

Munds, R. A., Nekaris, K. A. I., & Ford, S. M. (2013). Taxonomy of the Bornean slow loris, with new species *Nycticebus kayan* (Primates, Lorisidae). *American Journal of Primatology*, 75, 46-56.

Napier, J. R. (1966). Stratification and primate ecology. *Journal of Animal Ecology*, 35, 411-412.

Napier, J. R., & Walker, A. C. (1967). Vertical clinging and leaping—a newly recognized category of locomotor behaviour of primates. *Folia Primatologica*, 6(3), 204-219.

Nash, L. T. (1986). Influence of moonlight level on traveling and calling patterns in two sympatric species of galago in Kenya. In D. M. Taub & F. A. King (Eds.), *Current perspectives in primate social dynamics* (pp. 357-367). Van Nostrand Reinhold Co.

Nash, L. T. (1989). Galagos and gummivory. *Human Evolution*, 4(2-3), 199-206.

Nash, L. T. (1993). Juveniles in nongregarious primates. In M. E. Pereira & L. A. Fairbanks (Eds.), *Juvenile primates: Life history, development, and behavior* (pp. 119-137). Oxford University Press.

Nash, L. T. & Harcourt, C. H. (1986). Social organization of galagos in Kenyan coastal forest II: *Galago garnettii*. *American Journal of Primatology*, 10, 357-369.

Nash, L. T., & Whitten, P. L. (1989). Preliminary observations on the role of *Acacia* gum chemistry in *Acacia* utilization by *Galago senegalensis* in Kenya. *American Journal of Primatology*, 17(1), 27-39.

Nekaris, K. A. I. (2001). Activity budget and positional behavior of the Mysore slender loris (*Loris tardigradus lydekkarianus*): Implications for "slow climbing" locomotion. *Folia Primatologica, 72*, 228-241.

Nekaris, K. A. I. (2002). Slender in the night. *Natural History, 111*(2), 54-59.

Nekaris, K. A. I. (2003a). Spacing system of the Mysore slender loris (*Loris lydekkerianus lydekkerianus*). *American Journal of Physical Anthropology, 121*, 86-96.

Nekaris, K. A. I. (2003b). Observations on mating, birthing and parental care in three taxa of slender loris in India and Sri Lanka (*Loris tardigradus* and *Loris lydekkerianus*). *Folia Primatologica, 74*(Suppl.), 312-336.

Nekaris, K. A. I. (2005). Visual predation in the slender loris. *Journal of Human Evolution, 49*, 289-300.

Nekaris, K. A. I. (2008a). *Loris tardigradus. The IUCN Red List of Threatened Species* 2008: e.T12375A3338689. https://dx.doi.org/10.2305/IUCN.UK.2008.RLTS.T12375A3338689.en.

Nekaris, K. A. I. (2008b). *Loris tardigradus* ssp. *tardigradus. The IUCN Red List of Threatened Species* 2008: e.T39757A10262945. https://dx.doi.org/10.2305/IUCN.UK.2008.RLTS.T39757A10262945.en.

Nekaris, K. A. I. (2008c). *Loris tardigradus* ssp. *nycticeboides. The IUCN Red List of Threatened Species* 2008: e.T39756A10262765. https://dx.doi.org/10.2305/IUCN.UK.2008.RLTS.T39756A10262765.en.

Nekaris K. A. I. (2014). Extreme primates: Ecology and evolution of Asian lorises. *Evolutionary Anthropology, 23*(5), 177-187.

Nekaris, K. A. I., Al-Razi, H., Blair, M., Das, J., Ni, Q., Samun, E., Streicher, U., Xue-long, J., & Yongcheng, L. (2020a). *Nycticebus bengalensis. The IUCN Red List of Threatened Species* 2020: e.T39758A17970536. https://dx.doi.org/10.2305/IUCN.UK.2020-2.RLTS.T39758A17970536.en.

Nekaris, K. A. I., & Bearder, S. K. (2011). The strepsirrhine primates of Asia and mainland Africa: Diversity shrouded in darkness. In C. Campbell, A. Fuentes, K. MacKinnon, S. K. Bearder, & R. Stumpf (Eds.), *Primates in perspective* (2nd ed., pp. 34-54). Oxford University Press.

Nekaris, K. A. I., & Jaffe, S. (2007). Unexpected diversity of slow lorises (*Nycticebus* spp.) within the Javan pet trade: Implications for slow loris taxonomy. *Contributions to Zoology, 76*(3), 187-196.

Nekaris, K. A. I., & Jayewardene, J. (2003). Pilot study and conservation status of the slender loris (*Loris tardigradus* and *Loris lydekkerianus*) in Sri Lanka. *Primate Conservation, 19*, 83-90.

Nekaris, K. A. I., & Jayewardene, J. (2004). Distribution of slender lorises in four ecological zones in Sri Lanka. *Journal of Zoology, 262*, 1-12.

Nekaris, K. A. I., Liyanage, W. K. D. D., & Gamage, S. (2005). Relationship between forest structure and floristic composition and population density of the southwestern Ceylon slender loris (*Loris tardigradus tardigradus*) in Masmullah Forest, Sri Lanka. *Mammalia, 69*(2), 1-10.

Nekaris, K.A.I., & Marsh, C. (2020). *Nycticebus bancanus. The IUCN Red List of Threatened Species* 2020: e.T163015864A163015867. https://dx.doi.org/10.2305/IUCN.UK.2020-2.RLTS.T163015864A163015867.en.

Nekaris, K. A. I., & Miard, P. (2020a). *Nycticebus borneanus. The IUCN Red List of Threatened Species* 2020: e.T163015906A163015915. https://dx.doi.org/10.2305/IUCN.UK.2020-2.RLTS.T163015906A163015915.en.

Nekaris, K. A. I., & Miard, P. (2020b). *Nycticebus kayan. The IUCN Red List of Threatened Species* 2020: e.T163015583A163015849. https://dx.doi.org/10.2305/IUCN.UK.2020-2.RLTS.T163015583A163015849.en.

Nekaris, K. A. I., Miard, P., & Streicher, U. (2020d). *Nycticebus menagensis. The IUCN Red List of Threatened Species* 2020: e.T163013860A17970781. https://dx.doi.org/10.2305/IUCN.UK.2020-2.RLTS.T163013860A17970781.en.

Nekaris K. A. I., Moore, R. S., Rode, E. J., & Fry, B. G. (2013). Mad, bad and dangerous to know: The biochemistry, ecology and evolution of slow loris venom. *Journal of Venomous Animals and Toxins including Tropical Diseases, 19*, 21.

Nekaris, K. A. I., & Munds, R. (2010). Using facial markings to unmask diversity: The slow lorises (Primates: Lorisidae: *Nycticebus*) of Indonesia. In S. Gursky, & J. Supriatna (Eds.), *The primates of Indonesia* (pp. 383-396). Springer.

Nekaris, K. A. I., & Nijman, V. (2022). A new genus name for pygmy lorises, *Xanthonycticebus* gen. nov. (Mammalia, Primates). *Zoosystematics and Evolution, 98*(1), 87-92.

Nekaris, K. A. I., Pimley, E. R., & Ablard, K. (2007). Anti-predator behaviour of lorises and pottos. In S. G. Gursky, & K. A. I. Nekaris (Eds.), *Primate anti-predator strategies* (pp. 220-238). Springer.

Nekaris, K. A. I., & Poindexter, S. (2020). *Nycticebus hilleri. The IUCN Red List of Threatened Species* 2020: e.T163019804A163020000. https://dx.doi.org/10.2305/IUCN.UK.2020-2.RLTS.T163019804A163020000.en.

Nekaris, K. A. I., Poindexter, S., & Streicher, U. (2020b). *Nycticebus coucang. The IUCN Red List of Threatened Species* 2020: e.T163017685A17970966. https://dx.doi.org/10.2305/IUCN.UK.2020-2.RLTS.T163017685A17970966.en.

Nekaris, K. A. I., & Rasmussen, D. T. (2003). Diet and feeding behavior of Mysore slender lorises. *International Journal of Primatology, 24*(1), 33-46.

Nekaris, K. A. I., Shekelle, M., Wirdateti, Rode-Margono, E. J., & Nijman, V. (2020c). *Nycticebus javanicus. The IUCN Red List of Threatened Species* 2020: e.T39761A86050473. https://dx.doi.org/10.2305/IUCN.UK.2020-2.RLTS.T39761A86050473.en.

Nekaris, K. A. I., Shepherd, C. R., Starr, C. R., & Nijman, V. (2010a). Exploring cultural drivers for wildlife trade via an ethnoprimatological approach: A case study of slender and slow lorises (*Loris* and *Nycticebus*) in South and Southeast Asia. *American Journal of Primatology, 72*, 877-886.

Nekaris, K. A. I., Singh, M., & Kumar Chhangani, A. (2008). *Loris lydekkerianus. The IUCN Red List of Threatened Species* 2008: e.T44722A10942453. https://dx

.doi.org/10.2305/IUCN.UK.2008.RLTS.T44722A10942453
.en.

Nekaris, K. A. I., & Starr, C. R. (2015). Conservation and ecology of the neglected slow loris: Priorities and prospects. *Endangered Species Research, 28*, 87-95.

Nekaris K. A. I., Starr, C. R., Collins, R. L., & Wilson, A. (2010b). Comparative ecology of exudate feeding by lorises (*Nycticebus, Loris*) and pottos (*Perodicticus, Arctocebus*). In A. M. Burrows, & L. T. Nash (Eds.), *The Evolution of exudativory in primates* (pp. 155-168). Springer.

Nekaris, K. A. I., & Stevens, N. J. (2007). Not all lorises are slow: Rapid arboreal locomotion in *Loris tardigradus* of southwestern Sri Lanka. *American Journal of Primatology, 69*, 112-120.

Newton-Fisher, N. (2014). The hunting behavior and carnivory of wild chimpanzees. In W. Henke, & I. Tattersall (Eds.), *Handbook of paleoanthropology* (Vol. 2, pp. 1661-1691). Springer.

Nijman, V., & Nekaris, K. A. I. (2014). Traditions, taboos and trade in slow lorises in Sundanese communities in southern Java, Indonesia. *Endangered Species Research, 25*, 79-88.

Nijman, V., Spaan, D., Rode-Margono, E. J., Wirdateti, K. A., & Nekaris, K. A. I. (2017). Changes in the primate trade in Indonesian wildlife markets over a 25-year period: Fewer apes and langurs, more macaques, and slow lorises. *American Journal of Primatology, 79*(11), e22517.

Oates, J. F. (1984). The niche of the potto, *Perodicticus potto. International Journal of Primatology, 5*(1), 51-61.

Oates, J. F. (2019). *Sciurocheirus gabonensis. The IUCN Red List of Threatened Species* 2019: e.T136214A17961659. https://dx.doi.org/10.2305/IUCN.UK.2019-3.RLTS .T136214A17961659.en.

Oates, J. F., & Butynski, T. M. (2019). *Euoticus elegantulus. The IUCN Red List of Threatened Species* 2019: e.T8265A17961768. https://dx.doi.org/10.2305/IUCN.UK .2019-3.RLTS.T8265A17961768.en.

Oates, J. F., Ikemeh, R., Nobimè, G., & Svensson, M. (2020). *Perodicticus potto* ssp. *juju. The IUCN Red List of Threatened Species* 2020: e.T91995465A91995471. https://dx.doi.org/10 .2305/IUCN.UK.2020-2.RLTS.T91995465A91995471.en.

Oates, J. F., & Svensson, M. (2019). *Arctocebus calabarensis. The IUCN Red List of Threatened Species* 2019: e.T2054A17969996. https://dx.doi.org/10.2305/IUCN.UK .2019-1.RLTS.T2054A17969996.en.

O'Malley, R. C. O. (2010). Two observations of galago predation by the Kasekela chimpanzees of Gombe Stream National Park, Tanzania. *Pan Africa News, 17*(2), 17-19.

Osman Hill, W. C. (1953). *Primates—Comparative anatomy and taxonomy. I. Strepsirhini.* Edinburgh University Press.

Oxnard, C. E., Crompton, R. H., & Lieberman, S. S. (1990). *Animal lifestyles and anatomies.* University of Washington Press.

Pariente, G. (1979). The role of vision in prosimian behaviour. In G. A. Doyle, & R. D. Martin (Eds.), *The study of prosimian behaviour* (pp. 411-459). Academic Press.

Paterson, H. E. H. (1985). The recognition concept of species. In E. S. Vrba (Ed.), *Species and speciation* (pp. 21-29). Transvaal Museum.

Perkin, A. W. (2002). The rondo galago *Galagoides rondoensis* (Honess & Bearder, 1996): A primate conservation priority. *Primate Eye, 77*, 14-15.

Perkin, A. W. (2007). Comparative penile morphology of East African galagos of the genus *Galagoides* (Family Galagidae): Implications for taxonomy. *American Journal of Primatology, 69*(1), 16-26.

Perkin, A. W. (2020a). *Paragalago orinus. The IUCN Red List of Threatened Species* 2020: e.T40651A17961996. https://dx.doi .org/10.2305/IUCN.UK.2020-2.RLTS.T40651A17961996 .en.

Perkin, A. W. (2020b). *Paragalago rondoensis. The IUCN Red List of Threatened Species* 2020: e.T40652A17962115. https://dx.doi.org/10.2305/IUCN.UK.2020-2.RLTS .T40652A17962115.en.

Perkin, A. W., Bearder, S. K., Butynski, T. M., Agwanda, B., & Bytebier, B. (2002). The Taita Mountain dwarf galago *Galagoides* Sp: A new primate for Kenya. *Journal of East African Natural History, 91*(1), 1-13.

Perkin, A., Butynski, T. M., Cronin, D. T., Masters, J., Oates, J. F., & Pimley, E. (2020b). *Sciurocheirus alleni. The IUCN Red List of Threatened Species* 2020: e.T8785A95509640. https://dx.doi.org/10.2305/IUCN.UK.2020-2.RLTS .T8785A95509640.en.

Perkin, A. W., Butynski, T. M., & de Jong, Y. A. (2020a). *Paragalago zanzibaricus* ssp. *zanzibaricus. The IUCN Red List of Threatened Species* 2020: e.T136924A17989117. https://dx.doi .org/10.2305/IUCN.UK.2020-2.RLTS.T136924A17989117.en.

Petter, J. J., & Hladik C. M. (1970). Observations sur le domaine vital et la densité de population de *Loris tardigradus* dans les forêts de Ceylan. *Mammalia, 34*, 394-409.

Petter, J. J., & Petter-Rousseaux, A. (1979). Classification of the prosimians. In G. A. Doyle, & R. D. Martin (Eds.), *The study of prosimian behaviour* (pp. 1-44). Academic Press.

Pimley, E. R. (2002). *The behavioural ecology and genetics of the potto* (Perodicticus potto edwardsi) *and Allen's bushbaby* (Galago alleni cameronensis). [Doctoral dissertation, University of Cambridge].

Pimley, E. R., Bearder, S. K., & Dixson, A. F. (2005a). Home range analysis of *Perodicticus potto edwardsi* and *Sciurocheirus cameronensis. International Journal of Primatology, 26*(1), 191-206.

Pimley, E. R., Bearder, S. K., & Dixson, A. F. (2005b). Examining the social organization of the Milne-Edwards's potto *Perodicticus potto edwardsi. American Journal of Primatology, 66*(4), 317-330.

Pimley, E., & Oates, J. F. (2020). *Sciurocheirus alleni* ssp. *cameronensis. The IUCN Red List of Threatened Species* 2020: e.T136854A91978997. https://dx.doi.org/10 .2305/IUCN.UK.2020-2.RLTS.T136854A91978997.en.

Poindexter, S. A., & Nekaris, K. A. I. (2017). Vertical clingers and gougers: Rapid acquisition of adult limb proportions facilitates feeding behaviours in young Javan slow lorises (*Nycticebus javanicus*). *Mammalian Biology, 87*, 40-49.

Pozzi, L., Disotell, T. R., Bearder, S. K., Karlsson, I., Perkin, A., & Gamba, M. (2019). Species boundaries within morphologically cryptic galagos: Evidence from acoustic and genetic data. *Folia Primatologica, 90*, 279-299.

Pozzi, L., Disotell, T. R., & Masters, J. C. (2014a). A multilocus phylogeny reveals deep lineages within African galagids (Primates: Galagidae). *BMC Evolutionary Biology, 14*, 72.

Pozzi, L., Hodgson, J. A., Burrell, A. S., Sterner, K. N., Raaum, R. L., & Disotell, T. R. (2014b). Primate phylogenetic relationships and divergence dates inferred from complete mitochondrial genomes. *Molecular Phylogenetics and Evolution, 75*, 165-183.

Pozzi, L., Nekaris, K. A., Perkin, A., Bearder, S. K., Pimley, E. R., Schulze, H., Streicher, U., Nadler, T., Kitchener, A., Zischler, H., Zinner, D., & Roos, C. (2015). Remarkable ancient divergences amongst neglected lorisiform primates. *Zoological Journal of the Linnean Society, 175*(3), 661-674.

Pruetz, J. D., & Bertolani, P. (2007). Savanna chimpanzees, *Pan troglodytes verus*, hunt with tools. *Current Biology, 17*, 412-417.

Pruetz, J. D., Bertolani, P., Ontl, K. B., Lindshield, S., Shelley, M., & Wessling, E. G. (2015). New evidence on the tool-assisted hunting exhibited by chimpanzees (*Pan troglodytes verus*) in a savannah habitat at Fongoli, Sénégal. *Royal Society Open Science, 2*(4), 140507.

Pullen, S. L. (2000). *Behavioural and genetic studies of the mating system in a nocturnal primate: The lesser galago (*Galago moholi*)*. [Doctoral dissertation, University of Cambridge].

Pullen, S.L., Bearder, S. K., & Dixson, A. F. (2000). Preliminary observations on sexual behavior and the mating system in free-ranging lesser galagos (*Galago moholi*). *American Journal of Primatology, 51*(1), 79-88.

Radhakrishna, S., & Singh, M. (2002). Social behaviour of the slender loris (*Loris tardigradus lydekkerianus*). *Folia Primatologica, 73*(4), 181-196.

Radhakrishna, S., & Singh, M. (2004). Reproductive biology of the slender loris (*Loris lydekkerianus lydekkerianus*). *Folia Primatologica, 75*(1), 1-13.

Rasmussen, D. T. (1986). *Life history and behavior of slow lorises and slender lorises.* [Doctoral dissertation, Duke University].

Rasmussen, D. T., & Izard, M. K. (1988). Scaling of growth and life-history traits relative to body size, brain size and metabolic rate in lorises and galagos (Lorisidae, primates). *American Journal of Physical Anthropology, 75*, 357-367.

Rasmussen, D. T., & Nekaris, K. A. I. (1998). Evolutionary history of the lorisiform primates. *Folia Primatologica, 69*(Suppl. 1), 250-285.

Ratajszczak, R. (1998). Taxonomy, distribution and status of the lesser slow loris *Nycticebus pygmaeus* and their implications for captive management. *Folia Primatologica, 69*(Suppl. 1), 171-174.

Ripple W. J., Abernethy, K., Betts, M. G., Chapron, G., Dirzo, R., Galetti, M., Levi1, T., Lindsey, P. A., Macdonald, D. W., Machovina, B., Newsome, T. M., Peres, C. A., Wallach, A. D., Wolf, C., & Young, H. (2016). Bushmeat hunting and extinction risk to the world's mammals. *Royal Society Open Science, 3*(10), 160498.

Rode-Margono E. J., & Nekaris K. A. I. (2015). Cabinet of curiosities: Venom systems and their ecological function in mammals, with a focus on primates. *Toxins, 7*, 2639-2658.

Rode-Margono, E. J., Nijman, V., Wirdateti, & Nekaris, K. A. I. (2014). Ethology of the critically endangered Javan slow loris *Nycticebus javanicus* É. Geoffroy Saint-Hilaire in West Java. *Asian Primates, 4*(2), 27-41.

Rode-Margono, E. J., Rademaker, M., Wirdateti, Strijkstra, A., & Nekaris, K. A. I. (2015). Noxious arthropods as potential prey of the venomous Javan slow loris (*Nycticebus javanicus*) in a West Javan volcanic agricultural system. *Journal of Natural History, 49*(31-32), 1949-1959.

Roos, C. (2003). Molekulare phylogenie der Halbaffen, Schlankaffen, und Gibbons. [Doctoral dissertation, Technischen Universitaet München].

Rosti, H., Rikkinen, J., Pellikka, P., Bearder, S. K., & Mwamodenyi, J. M. (2020). Taita Mountain dwarf galago is extant in the Taita Hills of Kenya. *Oryx, 54*(2), 152-153.

Rovero, F., Marshall, A. R., Jones, T., & Perkin, A. (2009). The primates of the Udzungwa Mountains: Diversity, ecology and conservation. *Journal of Anthropological Sciences, 87*, 93-126.

Rowe, N. (2016). *All the world's primates.* (M. Myers, Ed.) Pogonius Press.

Ruf, T., Streicher, U., Stalder, G. L., Nadler, T., & Walzer, C. (2015). Hibernation in the pygmy slow loris (*Nycticebus pygmaeus*): multiday torpor in primates is not restricted to Madagascar. *Scientific Reports, 5*, 17392.

Rylands, A. B., Mittermeier, R. A., & Konstant, B. R. (2004). *IUCN/ SSC Primate Specialist Group Report 2001-2004.* Conservation International Unpublished Report.

Sasi, R., & Kumara, H. N. (2014). Distribution and relative abundance of the slender loris *Loris lydekkerianus* in Southern Kerala, India. *Primate Conservation, 28*, 165-170.

Schilling, A. (1979). Olfactory communication in prosimians. In G. A. Doyle, & R. D. Martin (Eds.), *The study of prosimian behaviour* (pp. 461-542). Academic Press.

Schneiderová, I., Singh, N. J., Baklová, A., Smetanová, M., Nicolas Benty Gomis, N. B., & Lhota, S. (2020). Northern lesser galagos (*Galago senegalensis*) increase the production of loud calls before and at dawn. *Primates, 61*, 331-338.

Schulze, H., & Meier, B. (1995). Behaviour of captive *Loris tardigradus nordicus*: A qualitative description including some information about morphological bases of behavior. In L. Alterman, G. A. Doyle, & M. K. Izard (Eds.), *Creatures of the dark: The nocturnal prosimians* (pp. 221-250). Springer.

Sellers, W. (1996). A biomechanical investigation into the absence of leaping in the locomotor repertoire of the slender loris (*Loris tardigradus*). *Folia Primatologica, 67*, 1-14.

Starr, C. R., & Nekaris, K. A. I. (2013). Obligate exudativory characterizes the diet of the pygmy slow loris *Nycticebus pygmaeus. American Journal of Primatology, 75, 1054-1061.

Starr, C. R., Nekaris, K. A. I., & Leung, L. (2012). Hiding from the moonlight: Luminosity and temperature affect activity of Asian nocturnal primates in a highly seasonal forest. *PLoS ONE* 7(4): e36396. https://doi.org/10.1371/journal.pone.0036396.

Starr, C. R., Nekaris, K. A. I., Streicher, U., & Leung, L. (2011). Field surveys of the threatened pygmy slow loris (*Nycticebus pygmaeus*) using local knowledge in Mondulkiri Province, Cambodia. *Oryx,* 45, 135-142.

Sterling, E. J. (1993). Patterns of range use and social organization in aye-ayes (*Daubentonia madagascariensis*) on Nosy Mangabe. In P. M. Kappeler, & J. U. Ganzhorn (Eds.), *Lemur social systems and their ecological basis* (pp. 1-10). Plenum Press.

Sterling, E. J., Nguyen, N., & Fashing, P. (2000). Spatial patterning in nocturnal prosimians: A review of methods and relevance to studies of sociality. *American Journal of Primatology,* 51, 3-19.

Stevens, J. L., Egerton, V. R., & Mitton, S. (1971). Gross anatomy of the hindlimb skeletal system of the *Galago senegalensis. Primates,* 12, 313-321.

Streicher, U. (2004). *Aspects of the ecology and conservation of the pygmy loris* Nycticebus pygmaeus *in Vietnam.* [Doctoral dissertation, Ludwig-Maximilians Universität, München].

Streicher, U., Collins, R., Wilson, A., & Nekaris, K. A. I. (2012). Observations on the feeding preferences of slow lorises (*N. pygmaeus, N. javanicus, N. coucang*) confiscated from the trade. In J. Masters, M. Gamba, & F. Génin (Eds.), *Leaping ahead: Advances in prosimian biology* (pp. 165-172). Springer.

Svensson, M. S., Ambrose, L., & Bearder, S. (2020b). *Sciurocheirus makandensis*, Makandé Squirrel Galago. The IUCN Red List of Threatened Species 2020: e.T91979463A91979703. https://dx.doi.org/10.2305/IUCN.UK.2020-2.RLTS.T91979463A91979703.en.

Svensson, M., & Bearder, S. (2013). Sightings and habitat use of the northern lesser galago (*Galago senegalensis senegalensis*) in Niumi National Park, The Gambia. *African Primates,* 8, 51-58.

Svensson, M., & Bearder, S. (2019). *Galagoides thomasi.* The IUCN Red List of Threatened Species 2019: e.T40653A17962691. https://dx.doi.org/10.2305/IUCN.UK.2019-3.RLTS.T40653A17962691.en.

Svensson, M., Bersacola, E., & Bearder, S. (2019). *Galagoides demidoff.* The IUCN Red List of Threatened Species 2019: e.T40649A17962255. https://dx.doi.org/10.2305/IUCN.UK.2019-3.RLTS.T40649A17962255.en.

Svensson, M. S., Bersacola, E., Mills, M. S., Munds, R. A., Nijman, V., Perkin, A., Masters, J. C., Couette, S., Nekaris, K. A. I., & Bearder, S. K. (2017). A giant among dwarfs: A new species of galago (Primates: Galagidae) from Angola. *American Journal of Physical Anthropology,* 163(1), 30-43.

Svensson, M., Bersacola, E., Nijman, V., Mills, S. L., Munds, R., Perkin, A., & Bearder, S. (2020a). *Galagoides kumbirensis.* The IUCN Red List of Threatened Species 2020: e.T164378198A164378551. https://dx.doi.org/10.2305/IUCN.UK.2020-2.RLTS.T164378198A164378551.en.

Svensson, M. S., & Friant, S. C. (2014). Threats from trading and hunting of pottos and angwantibos in Africa resemble those faced by slow lorises in Asia. *Endangered Species Research,* 23, 107-114.

Svensson, M. S., Ingram, D. J., Nekaris, K. A. I., & Nijman, V. (2016). Trade and ethnozoological use of African lorisiforms in the last 20 years. *Hystrix—Italian Journal of Mammalogy,* 26(2), 153-161.

Svensson, M. S., & Luhrs, A. (2020). Behaviour of pottos and angwantibos. In K. Nekaris & A. Burrows (Eds.), *Evolution, ecology and conservation of lorises and pottos* (pp. 204-209). Cambridge University Press.

Svensson, M. S. & Nekaris, K. A. I. (2019). *Arctocebus aureus.* The IUCN Red List of Threatened Species 2019: e.T2053A17969875. https://dx.doi.org/10.2305/IUCN.UK.2019-3.RLTS.T2053A17969875.en.

Svensson, M. S., Nekaris, K. A. I., Bearder, S. K., Bettridge, C. M., Butynski, T. M., Cheyne, S. M., Das, N., de Jong, Y. A., Luhrs, A. M., Luncz, L. V., Maddock, S. T., Perkin, A., Pimley, E., Poindexter, S. A., Reinhardt, K. D., Spaan, D., Stark, D. J., Starr, C., & Nijman, V. (2018). Sleep patterns, daytime predation, and the evolution of diurnal sleep site selection in lorisiforms. *American Journal of Physical Anthropology,* 166(3), 563-577.

Svensson, M. S., Oates, J. F., Pimley, E., & Gonedelé Bi, S. (2020c). *Perodicticus potto.* The IUCN Red List of Threatened Species 2020: e.T91995408A92248699. https://dx.doi.org/10.2305/IUCN.UK.2020-2.RLTS.T91995408A92248699.en.

Svensson, M. S., & Pimley, E. (2019). *Perodicticus edwardsi.* The IUCN Red List of Threatened Species 2019: e.T136852A91996061. https://dx.doi.org/10.2305/IUCN.UK.2019-3.RLTS.T136852A91996061.en.

Swapna, N., Radhakrishna, S., Gupta, A. K., & Kumar, A. (2010). Exudativory in the Bengal slow loris (*Nycticebus bengalensis*) in Trishna Wildlife Sanctuary, Tripura, Northeast India. *American Journal of Primatology,* 72, 113-121.

Tan, C. L., & Drake, J. H. (2001). Evidence of tree gouging and exudate eating in pygmy slow lorises (*Nycticebus pygmaeus*). *Folia Primatologica,* 72, 37-39.

Torres, A. F. C., Quinet, Y. P., Havt, A., Rádis-Baptista, G., & Martins, A. M. C. (2013). Molecular pharmacology and toxinology of venom from ants. In G. R. Baptista (Ed.), *An integrated view of the molecular recognition and toxinology—From analytical procedures to biomedical applications* (pp. 207-222). Intech Open.

Utami, S. C., & van Hooff, J. A. R. A. M. (1997). Meat-eating by adult female Sumatran orangutans (*Pongo pygmæus abelii*). *American Journal of Primatology,* 43, 159-165.

Walker, A. C. (1969). The locomotion of the lorises, with special reference to the potto. *East African Wildlife Journal,* 7, 1-5.

Walker, A. C. (1970). Nuchal adaptations in *Perodicticus potto*. *Primates, 11*, 135-144.

Weisenseel, K., Chapman, C. A., & Chapman, L. J. (1993). Nocturnal primates of Kibale Forest: Effects of selective logging on prosimian densities. *Primates, 34*, 445-450.

Weisenseel, K. A., Izard, M. K., Nash, L. T., Ange, R. L., & Poorman-Allen, P. (1998). A comparison of reproduction in two species of *Nycticebus. Folia Primatologica, 69*(Suppl. 1), 321-324.

Wickings, E. J., Ambrose, L., & Bearder, S. K. (1998). Sympatric populations of *Galagoides demidoff* and *G. thomasi* in the Haut Ogooué region of Gabon. *Folia Primatologica, 69*(Suppl.1), 389-393.

Wiens, F. (2002). *Behavior and ecology of wild slow lorises (*Nycticebus coucang*): Social organisation, infant care system and diet.* [Doctoral dissertation, Bayreuth].

Wiens, F., & Zitzmann, A. (1999). Predation on a wild slow loris (*Nycticebus coucang*) by a reticulated python (*Python reticulatus*). *Folia Primatologica, 70*, 362-364.

Wiens, F., & Zitzmann, A. (2003). Social structure of the solitary slow loris *Nycticebus coucang* (Lorisidae). *Journal of Zoology, 261*(1), 35-46.

Wiens, F., Zitzmann, A., & Hussein, N. A. (2006). Fast food for slow lorises: Is low metabolism related to secondary compounds in high-energy plant diet? *Journal of Mammalogy, 87*(4), 790-798.

Wilde, H. (1972). Anaphylactic shock following bite by a "slow loris," *Nycticebus coucang. American Journal of Tropical Medicine and Hygiene, 21*, 592.

Yoder, A. D., Irwin, J. D., & Payseur, B. A. (2001). Failure of the ILD to determine data combinability for slow loris phylogeny. *Systematic Biology, 50*(3), 408-424.

Zimmermann, E. (1985). Vocalisations and associated behaviours in adult slow loris (*Nycticebus coucang*). *Folia Primatologica, 44*, 52-64.

Zimmermann, E. (1990). Differentiation of vocalisations in bushbabies [Galaginae, Prosimii, Primates] and the significance for assessing phylogenetic relationships. *Zeitschrift für zoologische Systematik und Evolutionsforschung, 28*, 217-239.

Zimmermann, E. (1995a). Loud calls in nocturnal prosimians: Structure, evolution and ontogeny. In E. Zimmermann, J. D. Newman, & U. Jürgens (Eds.), *Current topics in primate vocal communication* (pp. 47-72). Plenum Press.

Zimmermann, E. (1995b). Acoustic communication in nocturnal prosimians. In L. Alterman, G. A. Doyle, & M. K. Izard (Eds.), *Creatures of the dark: The nocturnal prosimians* (pp. 311-330). Plenum Press.

Zimmermann, E., Bearder, S. K., Doyle, G. A., & Anderson, A. B. (1988). Variations in vocal patterns of Senegal and South African lesser bushbabies and their implications for taxonomic relationships. *Folia Primatologica, 51*(2-3), 87-105.

CHAPTER 2

Ecology and Life History of the Nocturnal Lemurs

SYLVIA ATSALIS AND ROBERT W. SUSSMAN

NOCTURNALITY, THE COMMON FEATURE OF THE LEMURS IN this chapter, impacts their ecology, social behavior and organization, and life history. With research activity intensifying in Madagascar—one of the world's biodiversity hotspots—an understanding of the nocturnal lemuriforms has been considerably enhanced through the addition of intensive observational, as well as geographical, morphological, and molecular data. In the last two decades, in particular, knowledge of lemurs has increased appreciably, with *Microcebus* and *Lepilemur* both being genera with increasing numbers of recognized species. The dynamic constellation of research on these and the other nocturnal lemuriform taxa, while providing improved understanding of Madagascar's biodiversity, has also created a complex body of literature. We now recognize that each species of nocturnal lemuriform has evolved its own repertoire of socioecological adaptations.

NOCTURNAL LEMUR DIVERSITY AND DISTRIBUTION

A principle driving factor behind the high species diversity and local endemism among all lemurs, including the nocturnal lemuriforms, is Madagascar's topography. The island is composed of a central plateau surrounded by regional zones of coastal forest. The plateau is highest in the east, which, with the predominant east to west movement of weather systems off the Indian Ocean, gives rise to a north–south chain of humid forested highlands. In the west the plateau slopes more gradually toward the Mozambique Channel. The western forests are dry and deciduous. The topographical pattern of the two major phytogeographic zones, the humid eastern and the dry western, together with Madagascar's large river systems, are associated with the microendemic speciation and geographic distribution of many of Madagascar's endemic taxa, including the present distributions of the island's lemurs (Wilmé et al., 2006).

The genus *Cheirogaleus* (dwarf lemurs) of the family Cheirogaleidae, previously included only two species: *C. major* (the greater dwarf lemur) in eastern rainforests and *C. medius* (the fat-tailed dwarf lemur) of western dry forests. Today, however, at least an additional 8 species are recognized, for a total of 10 (Table 2.1), and additional

species may still be described in the future (e.g., Frasier et al., 2016). The genus is distributed widely across the island with some *Cheirogaleus* species living sympatrically (Blanco et al., 2009).

Microcebus (mouse lemurs) of the family Cheirogaleidae, with average adult body weights between 30–60 g, are the smallest living primates (Figure 2.1). Mouse lemurs exhibit an exceptional level of species diversity (Rasoloarison et al., 2000; Yoder et al., 2000) and are widely distributed across Madagascar (Radespiel, 2006), inhabiting almost all forest types (Figure 2.2). Like dwarf lemurs, the mouse lemurs also have been discovered to be species-rich taxa. Molecular techniques, morphological indicators, and shifts in conceptual approaches (e.g., an increased application of the Phylogenetic Species Concept; see Groves, 2001) have been used to clarify new species status, resulting in dramatic increase in the number of species described. Genus *Microcebus* was at one time considered to be monospecific, with two subspecies (Petter et al., 1977). Today, by contrast, at least 24 species of mouse lemur are recognized, but with the pace of new discoveries of mouse lemur species, this number cannot be considered definitive (Rasoloarison et al, 2000; Yoder et al., 2000; Andriantomphohavana et al., 2006; Louis et al., 2006, 2008; Olivieri et al., 2007; Schüßler et al., 2020) (Table 2.2). Ecologically, as many as three *Microcebus* species can live in sympatry (Schmid & Kappeler, 1994; Atsalis et al., 1996; Zimmermann et al., 1998; Heckman et al., 2006; Louis et al., 2006; Radespiel et al., 2006; Olivieri et al., 2007; Rakotondranary & Ganzhorn, 2012).

Species of the genus *Phaner* (the forked-marked lemurs), in the family Cheirogaleidae, are discontinuously distributed on Madagascar (Groves & Tattersall, 1991; Colquhoun, 1998; Hending et al., 2018, 2020; Salmona et al., 2018; Webber et al., 2020). The four species (cf. Groves & Tattersall, 1991) occur in the south and west (*P. pallescens*), the northwest (*P. parienti*), the north (*P. electromontis*), and in the northeast (*P. furcifer*).

The genus *Mirza* (the giant mouse lemurs), in the family Cheirogaleidae, is represented by *Mirza coquereli* in western lowland forests and by *M. zaza* in northwestern Madagascar (Kappeler et al., 2005). Prior to

Table 2.1 *Cheirogaleus, Phaner, Mirza, Allocebus* (Family Cheirogaleidae): Habitat, Distribution, and IUCN Conservation Status

Species	Common Name	Habitat and Areas of Occupation	IUCN Red List[a]
Cheirogaleus medius	Western fat-tailed dwarf lemur	Dry deciduous forests, gallery, evergreen humid, transitional sub-humid forest; Petriky and Mandena, remnants in Lavasoa-Ambatotsirongorongo Mountains	Vulnerable (Decreasing)
C. major	Geoffroy's dwarf lemur	Restricted range. Humid and littoral forests in southeastern and western-central Madagascar; near Tolagnaro	Vulnerable (Decreasing)
C. crossleyi	Furry-eared dwarf lemur	Humid forests in northern and eastern Madagascar; Vohima, Lac Alaotra, Andasibe, Vohémar, Talatakely and Vatateza, Andasivodihazo, Ranomafana National Park	Vulnerable (Decreasing)
C. sibreei	Sibree's dwarf lemur	Restricted to Eastern high-altitude rainforests; Ranomafana National Park (Mount Maharira at 1,400 m), and Tsinjoarivo (around Onive and Mangoro Rivers)	Critically Endangered (Decreasing)
C. minusculus	Lesser iron gray dwarf lemur	Limited to one locale in east-central Madagascar; Ambostra—known from one museum specimen	Data Deficient
C. andysabini	Montagne d'Ambre dwarf lemur	Restricted range in N Madagascar; Amber Mountain National Park	Endangered (Decreasing)
C. grovesi	Groves' dwarf lemur	Restricted to SE rainforests; Ranomafana National Park, Andringitra National Park	Data Deficient (Decreasing)
C. lavasoensis	Lavasoa dwarf lemur	Limited to S Madagascar; occurs in transitional areas between dry, spiny, humid littoral forest, and humid forest; found in remnant forest on slopes of Lavasoa Mountains (approx. 50 individuals); Kalambatritra forest population unassessed	Endangered (Decreasing)
C. shethi	Ankarana dwarf lemur	Limited to N Madagascar; Ankarana Special Reserve; Analamerana Special Reserve	Endangered (Decreasing)
C. thomasi	Thomas' dwarf lemur	Restricted to SE Madagascar in littoral forest fragments (from Petriky to Ste. Luce); seems restricted to coastal plain below 40 m elevation. Need for tree-holes in large-sized trees might explain dependency on intact forests—these would be rare in degraded forests. The minimum estimated Area of Occupancy is 12 km²; the maximum estimated AOO is 260 km².	Endangered (Decreasing)
Phaner pallescens	Western fork-marked lemur	W Madagascar; south of the Fiherenana River to the region of Soalala; Morondava region	Endangered (Decreasing)
P. furcifer	Eastern/Masoala fork-marked lemur	Lowland humid forests in NE Madagasar; Parc National de Masoala	Endangered (Decreasing)
P. parienti	Sambirano fork-marked lemur	Limited. Lowland and mid-altitude humid forests in NW Madagascar; Sambirano region	Endangered (Decreasing)
P. electromontis	Amber Mt. fork-marked lemur	Limited; far northern Madagascar; Mt. d'Ambre N.P. and Ankarana Reserve	Endangered (Decreasing)
Mirza coquereli	Coquerel's giant mouse lemur	Western lowland dry deciduous forests, gallery forests, plantations; between the Onilahy River to the south and the Tsiribinha River to the north, including Vohibasia, Zombitse, and Isalo National Parks	Endangered (Decreasing)
M. zaza	Northern giant mouse lemur	NW lowland dry and seasonally humid forests, plantations, gallery forests; Ampasindava Peninsula (from Ambato and Pasandava); Ankiabe and Andranobe near Befotaka; Sahamalaza region (Ambendrana and Ankarafa forest); Nosy Be	Vulnerable (Decreasing)
Allocebus trichotis	Hairy-eared dwarf lemur	Intact moist forest, NE Madagascar; Marojejy National Park, Anjanaharibe-Sud Special Reserve, Masoala National Park, Mananara-Nord National Park	Endangered (Decreasing)

[a] IUCN (2020).

the description of *M. zaza* in 2005, *Mirza* was considered to be a monospecific genus. Historically, Coquerel's mouse lemur has, at times, been variously classified as a member of genus *Cheirogaleus* or as a member of genus *Microcebus* (Tattersall, 1982). Gray (1870) proposed that *Mirza* be split away from *Microcebus* and placed in its own genus—that classification was largely not used during much of the twentieth century; however, Tattersall

Table 2.2 *Microcebus* (Family Cheirogaleidae): Habitat, Distribution, and IUCN Conservation Status

Species	Common Name	Habitat and Areas of Occupation	IUCN Red List[a]
Microcebus murinus	gray mouse lemur	Widespread: dry deciduous, semi-humid, moist lowland, transitional, littoral, and spiny forest; Ampijoroa, Kirindy, Mandena, Beza Mahafaly, Vohimena, Manamby, Berenty, Ankarafantsika	Least Concern
M. rufus	brown mouse lemur	Widespread: eastern lowland, montane rainforests; Ranomafana, Nosy Mangabe, Tampolo	Vulnerable
M. myoxinus	pygmy mouse lemur	Limited to dry deciduous, degraded forests; Bemaraha, Kirindy, Aboalimena, Bekopaka	Vulnerable
M. griseorufus	reddish-gray mouse lemur	Limited to spiny, gallery, dry deciduous western forests; Beza Mahafaly, Berenty, Tsimananampetsotsa	Least Concern
M. ravelobensis	golden-brown mouse lemur	Dry deciduous forests; Ankarafantsika, Ampijoroa	Vulnerable
M. berthae	Madame Berthe's mouse lemur	Limited to dry deciduous western forests; Kirindy forest	Critically Endangered
M. sambiranensis	Sambirano mouse lemur	Limited to humid lowland forest; Sambirano region—Manongarivo Special Reserve	Endangered
M. tavaratra	northern brown mouse lemur	Dry deciduous, gallery forests; Ankarana Massif	Vulnerable
M. lehilahytsara	Goodman's mouse lemur	Mid-altitude humid forest; Andasibe	Vulnerable
M. jollyae	Jolly's mouse lemur	Coastal, lowland mid-altitude forest; Manombo, Mananjary, Kianjavato	Endangered
M. mittermeieri	Mittermeier's mouse lemur	Mid-altitude rainforests; Anjanaharibe-Sud	Endangered
M. simmonsi	Simmon's mouse lemur	Lowland, mid-altitude rainforests; Betampona, Zahamena, Ranomafana	Endangered
M. danfossi	Danfoss' mouse lemur	Deciduous dry forests; Mahajanga, Bora, Anjiamangirana	Vulnerable
M. bongolavensis	Bongolava mouse lemur	Western deciduous forests; Mahajanga	Endangered
M. macarthurii	MacArthur's mouse lemur	Humid eastern rainforests; Makira	Endangered
M. mamiratra (= *M. lokobensis*)	Claire's mouse lemur	Humid primary and secondary forest fragments; Nosy Be (Lokobe Special Reserve)	Endangered
M. margotmarshae	Margot Marsh's mouse lemur	Limited. Humid forests; Andranomalaza Classified Forest (south of Andranomalaza R. and north of Maevarano R.); recently recorded in riparian forest on lower section of Mahavavy Nord R.	Endangered
M. marohita	Marohita mouse lemur	Limited. Eastern lowland rainforests; Marohita forest	Critically Endangered
M. gerpi	Gerp's mouse lemur	Limited. Eastern lowland primary and secondary rainforest, 29-230 m ASL; Sahafina Forest in eastern Madagascar	Critically Endangered
M. arnholdi	Arnhold's mouse lemur	Limited. Northern montane rainforest; Montagne d'Ambre National Park and Montagne d'Ambre Special Reserve	Vulnerable
M. tanosi	Anosy mouse lemur	Limited. Moist forests and forest remnants of SE Madagascar. Northern portion of Tsitongambarika Protected Area; littoral forest fragments of Sainte-Luce Conservation Area	Endangered
M. boraha	Nosy Boraha mouse lemur	Limited. Isolated on Nosy Boraha (Ile St. Marie)	Data Deficient
M. bongolavensis	Bongolava mouse lemur	Limited. Northwestern Madagascar—only known to occur in three small forest fragments in the region of Port-Bergé, between the Mahajamba R. and the Sofia R.	Endangered
M. ganzhorni	Ganzhorn's mouse lemur	Limited. SE Madagascar—occurs only in littoral forests in immediate vicinity of Tolagnaro (Fort Dauphin), from dry forest (Petriky) W of Tolagnaro, to wet forest (Mandena) E of Tolagnaro	Endangered

[a] IUCN (2020).

Figure 2.1 *Microcebus rufus*, the brown mouse lemur.
(Photo credit: Alain Mafart-Renodier; arkive.org/Biosphoto)

(1982) resurrected Gray's genus-level reclassification, placing *Mirza* in its own genus because of its larger size and other differences.

Genus *Allocebus* (the hairy-eared dwarf lemur), of the family Cheirogaleidae, contains a single species, *A. trichotis*. The species is similar in color to *Microcebus* but, with an adult weight range of 65–98 g, *Allocebus* can be approximately 50% larger than the largest *Microcebus* taxa and two to three times as large as the smallest *Microcebus* species. As its common name indicates, and unlike *Microcebus*, *Allocebus* also sports long, wavy hairs on its ears. *A. trichotis* was thought to be extinct until it was reported in 1965 (Petter-Rousseaux & Petter, 1967), and rediscovered again in 1989 (Meier & Albignac, 1991). Since that time, researchers have now recorded the presence of *Allocebus* in several national parks and reserves in northeastern Madagascar (Rakotoarison, 1998; Schütz & Goodman, 1998; Goodman & Raselimanana, 2002; Biebouw, 2009; Miller et al., 2015).

Mirroring the recent proliferation of species discovered for mouse lemurs, the genus *Lepilemur*, sportive or weasel lemurs (see Figure 2.3), now includes 26 species placed within their own family, the Lepilemuridae (IUCN, 2020; but see Tattersall, 2007). *Lepilemur* species occur in all forested areas of Madagascar (Table 2.3 and Figure 2.4).

Daubentonia madagascariensis (the aye-aye) is the only living member of the family Daubentoniidae (Figure 2.5). Aye-ayes have bat-like ears and ever-growing, rodent-like incisors; in the late eighteenth century, initial accounts of

the species led to questions about its correct taxonomic classification—the species was, for example, at one time classified as an endemic Madagascar squirrel (see Tattersall, 1982; see also Jenkins, 1987; Harcourt & Thornback, 1990; Sterling & McCreless, 2006). The history of the taxonomic debates surrounding this unique lemur species is detailed by Sterling (1993a, 1994a). Aye-ayes occur at low densities but the species has a wider geographic distribution than any other lemur species (Mittermeier et al., 2010). Aye-ayes are, thus, able to adapt to a variety of Malagasy forest habitats, ranging from the western deciduous forests to the rainforests of the east (Sterling & McCreless, 2006) (Figure 2.6). Prior to field research beginning in the 1960's, little was known of aye-aye behavior and ecology, and the species was even feared to be near extinction; initial field observations began with the release of individuals on the Special Reserve of Nosy Mangabe, an island off the east coast of Madagascar (Petter & Petter-Rousseaux, 1967; Petter & Peyriéras, 1970; Petter, 1977; Sterling, 1993a). More recently, the species has been found to have a more extensive geographic distribution than had been thought initially (Simons, 1993; Simons & Meyers, 2001).

ECOLOGY AND LIFE HISTORY

Among primates, including other lemurs, the nocturnal lemuriforms are characterized by a unique set of traits, including nocturnality and, in some families, hibernation and the production of litters. A comparative approach has been helpful in understanding this group. Specific

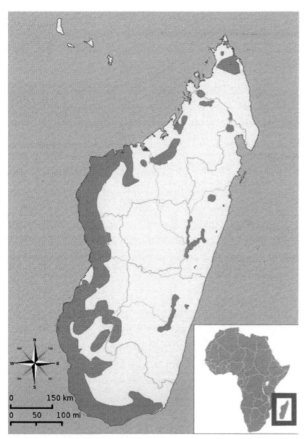

Figure 2.2 Geographic distribution (dark gray) of mouse lemur species (genus *Microcebus*), based on spatial data from IUCN Red List (IUCN, 2020).
(Map by Alex Dunkel/Maky; image licensed under Creative Commons Attribution-Share Alike 3.0 Unported license—cc by-sa 3.0)

studies have focused on *Cheirogaleus* and *Microcebus*. For a primate, entering a period of significant fattening followed by hibernation to reduce energy requirements is unique (McKechnie & Mzilikazi, 2011), but it is not unusual among other small mammals especially those in temperate climates (Atsalis, 2008). Thus, although tropical and subtropical in distribution, cheirogaleid species constitute distinctive examples of how energy management, through daily and seasonal periods of hypothermia, can occur in primates just as it does in mammals that live in temperate latitudes. These nocturnal lemurs adapted to save energy by becoming inactive and lowering their body temperatures (T_b) when food resources are low and the climate is unpredictable. The first comparative study of five nocturnal Malagasy lemurs was carried out over 30 years ago in the seasonal dry forest of Marosalaza in western Madagascar and included observations on *Cheirogaleus medius*, *Microcebus murinus*, *Mirza coquereli*, *Phaner furcifer*, and *Lepilemur mustelinus* (Petter, 1978; Hladik et al., 1980). The study, conducted in short periods of time between October 1973–June 1974, revealed that insect availability and production of

flowers, fruit, and leaves are low during the dry season, which corresponds to the austral winter in Madagascar and spans about eight months beginning in April–May (Hladik et al., 1980). Specific dietary and physiological adaptations of these five small nocturnal species, which will be described in this chapter, constitute responses to the extreme seasonality characteristic of west and southwest Madagascar.

As with diurnal lemurs, most nocturnal lemurs are seasonal breeders, likely to avoid reproducing when resources are scarce. Photoperiod, through day length, serves to synchronize reproduction. Mating and gestation typically occur in the dry season, and lactation and weaning in the wet season when resource abundance is greatest (Martin, 1972). Some genera, like *Microcebus*, produce litters with young developing quickly, ready to reproduce in the first year of their lives. Alternatively, *Phaner*, *Lepilemur*, and *Daubentonia*, commonly produce singletons, and at least in the latter, the maturation rate is relatively slow.

Family Cheirogaleidae
Cheirogaleus

The most extreme adaptation to seasonal reduction in resources is seen among species in the genus *Cheirogaleus* which, unlike any other primates (but see Ruf et al., 2015), exhibit the ability to hibernate for six to eight months every year during the austral winter (Hladik et al., 1980; Foerg & Hoffman, 1982; McKechnie & Mzilikazi, 2011). While hibernating, three to five individuals may sleep together in deep hollows of tree trunks (Hladik et al., 1980). In the dry forests of Marosalaza and Kirindy, *C. medius*, fat-tailed dwarf lemurs, sustain themselves while hibernating by utilizing large fat reserves, especially in the tail, that were accumulated during the pre-hibernation wet season (Fietz & Ganzhorn, 1999); body masses of individuals captured at Marosalaza varied from 142 g in November to 217 g in March (Hladik et al., 1980). Fat stored in the tail caused swelling, from a minimum of 9 cm³ in November (mean = 15 cm³) to an astonishing maximum of 56 cm³ (mean = 42 cm³) in May. Traveling quadrupedally on tree branches, the thickened tail slows the agility and movement of dwarf lemurs compared to other nocturnal lemurs (Petter et al., 1977). Emergence from hibernation in November is timed to coincide with maximum food production in the rainy season.

The diet of the dwarf lemur was studied in Marosalaza by observing animals feeding in the forest and by sifting through fecal material. Primarily an "opportunistic" frugivore, the dwarf lemurs of Marosalaza feed on fruits and other foods as they become seasonally available (Hladik et al., 1980). Specifically, from October–January, after

Figure 2.3 *Lepilemur ankaranensis*, photographed in Ankarana National Park, northern Madagascar.
(Photo credit: Leonora Enking; image licensed under Creative Commons Attribution-Share Alike 2.0 Generic license)

February–April, animals accumulated fat by consuming fruits (high in sugar) but also by decreasing activity by 50%. Animals trapped in April had gained 87.8% of their former body mass, fattening up from an average of 124 g to 234 g in just a few weeks. One female gained 118% of her body mass. During this time, dwarf lemurs were observed feeding on 34 different kinds of plants. Animal prey was also eaten seasonally, with highest consumption occurring in the rainy season.

The littoral forest of Mandena near the island's southeast coast is a very different habitat from Kirindy, with precipitation reaching 1,000 mm or more in the rainy season (Lahann, 2007). Several species of the family Cheirogaleidae live there sympatrically: *C. medius*, *C. major*, and *Microcebus murinus*. All are mostly frugivorous, with 91% of plant foods in common, and they are rarely seen consuming arthropods or licking gums (Lahann, 2007). They avoid competition through vertical stratification (Napier, 1966) by feeding at different heights of the forest, separated in the vertical dimension; *C. medius* feeds at 5 m above ground level, the larger *C. major* at 7 m, and the mouse lemur is typically located at 4 m. All three species are seed dispersers. *C. medius* was even observed smearing its feces laden with seeds of mistletoe species on tree branches (Ganzhorn & Kappeler, 1996). Eastern dwarf lemurs (*C. crossleyi* and *C. sibreei*) also hibernate in Ranomafana National Park (RNP), another southeastern forest, and Tsinjoarivo, a fragmented highland rainforest site in southeast Madagascar—while some animals sleep in tree hollows, others burrow into the leaf litter, such as at the bases of hollow trees and hibernate in underground hibernacula (Blanco & Rahalinarivo, 2010; Blanco et al., 2013, 2018; Dausmann & Warnecke, 2016).

At Kirindy, fat-tailed dwarf lemurs were fitted with temperature-sensitive collar transmitters to monitor changes in body temperature (T_b), as they hibernate in tree holes. In the non-hibernating period when animals sleep during the day, T_b decreases only slightly, from about 37° C to about 36° C (Dausmann et al., 2005). But by the end of May and over the following dry season months, when all animals are hibernating, body temperatures may fluctuate greatly, from 10° C to 30° C, passively tracking the temperature in the tree hole. Such extreme fluctuations in daily body temperature are rare in other hibernating mammals (Dausmann et al., 2000; McKechnie & Mzilikazi, 2011).

Microcebus

Mouse lemurs exhibit great flexibility in adapting to Madagascar's diverse habitats (Kappeler & Rasoloarison, 2003). *Microcebus* species occur in nearly all Madagascar's remaining forests, from very humid evergreen

animals emerged from hibernation, the majority of the diet was nectar from the abundance of tree species flowering at that time of year. In December, one individual, fitted with a radio-tracking device, fed on nectar for six consecutive days, suggesting that the dwarf lemur plays a role in pollination (Hladik et al., 1980). From December–January, the pre-hibernation period and the time of rainy season resource abundance, a higher proportion of fruits and an increasing proportion of insects, especially Coleoptera (beetles), are consumed (Hladik et al., 1980; Fietz & Ganzhorn, 1999). Gums and other plant and insect secretions make up a small part of the diet. Interestingly, even when given a choice of food, the same food patterns are mirrored by captive dwarf lemurs (Petter-Rousseaux, 1980).

The fat-tailed dwarf lemur has also been studied in the Kirindy forest, a northern dry, west coast forest with annual precipitation averaging 800 mm, mostly falling between December–February (Fietz & Ganzhorn, 1999). In one research study, 36 individuals were radio-tracked between 1–8 nights (Fietz & Ganzhorn, 1999). From

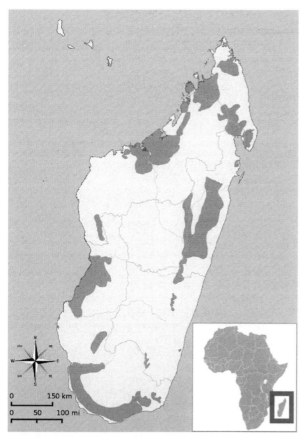

Figure 2.4   Geographic distribution (dark gray) of the sportive, or weasel, lemurs (genus *Lepilemur*), based on spatial data from IUCN Red List (IUCN, 2020). While the genus *Lepilemur* is widespread, many of the individual species in the genus are of elevated conservation concern.

(Map by Alex Dunkel/Maky; image licensed under Creative Commons Attribution-Share Alike 3.0 Unported license)

2000, 2001); *M. berthae*, Madame Berthe's mouse lemur (Schmid et al., 2000); *M. rufus*, the brown mouse lemur (Randrianambinina et al., 2003b; Atsalis, 2008); *M. ravelobensis*, the golden-brown mouse lemur (Radespiel et al., 2003); and *M. griseorufus*, the reddish-gray mouse lemur (sometimes called the gray-brown mouse lemur) (Génin, 2008). Whereas seasonal hibernation, triggered by short photoperiod, is part of a time-fixed regime for surviving in unfavorable environmental conditions, daily torpor is a flexible way to adjust to energy requirements as needed.

To understand torpor patterns, researchers at Kirindy kept *M. murinus* and sympatric *M. berthae* in outdoor enclosures. Even with food and water readily available, both species entered states of torpor on cold and warmer days, averaging 9.3 hours of torpor (Schmid, 2000). When in torpor, metabolic rate was reduced by up to 90% (Ortmann et al., 1997; Schmid, 2000, 2001), with body temperatures dipping to as low as 6.8°C, down from 38°C (Schmid, 1996). Madame Berthe's mouse lemurs engaged in daily torpor but not in seasonal torpor (Ortmann et al., 1996, 1997; Schmid, 1996). In both species, arousal from torpor began with body temperature passively rising to 24°–25° C, then through active production of body heat to reach normal body temperature. Passive arousal may help to conserve energy (Schmid, 1996).

During prolonged seasonal torpor, animals rest in their nests for weeks or even months. Photoperiod serves as the predictable environmental cue that triggers seasonal torpor. Even in captivity, changes in photoperiod can be used to trigger torpor (Perret, 1992; Génin, 2000; Génin & Perret, 2003). However, only populations of dry forest-dwelling *M. murinus* and the rainforest-dwelling *M. rufus* have been found to enter seasonal torpor. *M. ravelobensis*, *M. berthae*, and *M. griseorufus* are inhabitants of dry forests and were not observed to hibernate seasonally (Ortmann et al., 1996, 1997; Schmid, 1996; Schmid et al., 2000; Génin, 2008). Despite seasonal torpor, mouse lemurs are observed to be active throughout the year in the forest since torpor patterns vary widely. Even within a species, some populations may hibernate while others do not; for instance, at Ampijoroa, *M. murinus* sympatric with *M. ravelobensis* did not exhibit seasonal fattening and hibernation (Schmelting et al., 2000; Lutermann & Zimmermann, 2001). Finally, within the same population, some individuals may hibernate some years while others do not (Schmid & Kappeler, 1998a; Atsalis, 1999a, 2008; Rasoazanabary, 1999; Schmid, 1999). This variety in seasonal torpor patterns is unlike what occurs with *Cheirogaleus* where individuals are not sighted in the forest in the dry season. Genetic variability and differences in local habitats, including micro-climatic

habitats in the east and northeast regions of the island to the very extreme conditions that characterize the dry spiny forests in southern Madagascar (Kappeler & Rasoloarison, 2003; Radespiel, 2006). They are also found in secondary forests affected by human activity, in plantations, and along forest edges (Lehman et al., 2006; Radespiel, 2006). Differences in the responses of *Microcebus* species to habitat fragmentation are known—e.g., *M. murinus* appears quite resilient to habitat disturbance and fragmentation, whereas abundance of the sympatric *M. ravelobensis* is negatively affected by habitat fragmentation (Andriatsitohaina et al., 2019).

To manage the seasonal changes in resources characteristic of Madagascar, and reduce energy expenditure, mouse lemurs, like dwarf lemurs, exhibit heterothermic adaptations by entering states of hypothermia and inactivity, either through daily torpor, or through deep and extended torpor during the dry season. During bouts of daily torpor, body temperatures decrease and mouse lemurs become exceptionally lethargic. Evidence of daily torpor has been reported for *M. murinus* (Schmid,

Table 2.3 *Lepilemur* (Family Lepilemuridae): Habitat, Distribution, and IUCN Conservation Status

Species	Common Name	Habitat and Areas of Occupation	IUCN Red List[a]
L. mustelinus	Weasel sportive lemur	Primary/secondary lowland and montane rainforests; eastern Madagascar	Vulnerable
L. microdon	small-toothed sportive lemur	Primary/secondary rainforests; Ranomafana National Park, Andringitra Ranomafana National Park, Andringitra National Park	Endangered
L. jamesorum	James' sportive lemur	Moist lowland rainforest in southeastern Madagascar; Manombo Special Reserve region	Critically Endangered
L. wrightae	Wright's sportive lemur	Restricted range. Lowland rainforest in southeastern Madagascar; Kalambatritra Special Reserve in southeastern Madagascar, west of the Mananara River and north of the Mandrare River (Louis et al., 2006).	Endangered
L. fleuretae	Madame Fleurette's sportive lemur	Restricted range. Southeastern rainforest; Andohahela National Park between the Mandrare and Mananara Rivers	Endangered
L. seali	Seal's sportive lemur	Dry deciduous/gallery/semi-evergreen forests in northern Madagascar; Daraina region south of the Loky River; Analamerana Forest	Vulnerable
L. milanoii	Daraina sportive lemur	Dry deciduous/gallery/semi-evergreen forests in northern Madagascar; Daraina region south of the Loky River; Analamerana Forest	Endangered
L. septentrionalis	Sahafary or northern sportive lemur	Restricted range. Dry deciduous and gallery forest fragments in far northern Madagascar. Patches of forest near the villages of Madirobe and Ankarongana in the Sahafary region.	Critically Endangered
L. ankaranensis	Ankara sportive lemur	Dry deciduous/some humid evergreen forests in northern Madagascar. Forests of Ankarana, Andrafiamena, and Analamerana, from low elevations to 1,500 m.	Endangered
L. dorsalis	gray-backed sportive lemur	Restricted range. Humid and sub-humid rainforest/degraded forests in northwestern Madagascar; Sambirano region	Endangered
L. tymerlachsoni	Hawk's sportive lemur	Humid Sambirano-type forests, some modified secondary habitats; Nosy Be	Critically Endangered
L. mittermeieri	Mittermeier's sportive lemur	Restricted range. Dry deciduous primary and secondary forests in northwestern Madagascar; Ampasindava Peninsula	Critically Endangered
L. sahamalaza	Sahamalaza sportive lemur	Limited. Primary and mature secondary dry deciduous forests; Sahamalaza Peninsula	Critically Endangered
L. grewcockorum	Grewcock's sportive lemur	Limited. Dense primary forest in mountainous and coastal western forests; Southern Sambirano	Critically Endangered
L. edwardsi	Milne-Edwards' sportive lemur	Limited. Dry deciduous lowland western forests; Ankarafantsika National Park	Endangered
L. aeeclis	red-shouldered, or Antafia, sportive lemur	Restricted range. Western dry deciduous forests; the Betsiboka and Mahavavy du Sud Rivers	Endangered
L. ahmansoni	Ahmanson's sportive lemur	Restricted range. Western dry deciduous forests; Tsiombikibo region, west of the Mahavavy River	Critically Endangered
L. randrianasoloi	Randrianasolo's sportive lemur	Restricted range. Western dry deciduous forests; Andramasay, Bemaraha	Endangered
L. ruficaudatus	red-tailed sportive lemur	Dry deciduous western forests; Morondava	Critically Endangered
L. betsileo	Betsileo sportive lemur	Restricted range. Central-eastern rainforests; Fandriana region	Endangered
L. otto	Otto's sportive lemur	Restricted range. Northwestern dry deciduous forests; Ambodimahabibo	Endangered
L. scottorum	Scott's sportive lemur	Restricted range. Primary lowland rainforest in northeastern Madagascar; Masoala National Park	Endangered
L. hubbardorum	Hubbard's sportive lemur	Restricted range. Dry transitional forests in southwestern Madagascar; Zombitse-Vohibasia National Park region, north of the Onilahy River and south of the Fiherena River	Endangered
L. petteri	Petter's sportive lemur	Southwestern spiny and gallery forests; Beza-Mahafaly	Endangered
L. leucopus	white-footed sportive lemur	Southeastern spiny Didiereaceae forest and bushy areas, gallery, riverine, and subtropical dry lowland forest; Berenty Private Reserve	Endangered
L. hollandorum	Holland's sportive lemur	Restricted range. Lowland rainforest in northeastern Madagascar. Known only from the Ivontaka-Sud and Verezanantsoro (Ambinanibeorana) parcels of the Mananara-Nord Biosphere Reserve; S. limit of distribution not yet defined	Critically Endangered

[a] IUCN (2020).

Figure 2.5 Adult aye-aye, *Daubentonia madagascariensis.* Clearly visible are the specialized, long middle fingers on both hands used during foraging and feeding, particularly to reach wood-boring insect grubs.

(Photo credit: Edward E. Louis Jr.; image published as cover of *Genome Biology and Evolution* 4(2), 2012; distributed by Creative Commons Attribution Non-Commercial License 3.0, permitting unrestricted use in any medium)

conditions, may drive variation in mouse lemur seasonal torpor patterns, among other life history traits linked to thermoregulation and resource seasonality/availability (Lahann et al., 2006). Schmid (1999) speculated that females in the wild entered the seasonal period of inactivity once they had reached a critical mass of at least 50 g.

Live-trapping, a safe method of capture traditionally used by small mammal biologists, is suited to collecting information on mouse lemurs, including data on torpor, reproduction, and food eaten (Atsalis, 1998, 2008). Using this method, researchers have found that, like *Cheirogaleus*, mouse lemurs were found to undergo a period of pre-hibernation increase in body mass. Specifically, in May–June, individual *M. murinus* and *M. rufus* begin storing fat, including substantial fat gain in the tail (Hladik, 1979; Fietz, 1998a; Schmid & Kappeler, 1998a; Atsalis, 1999a; Schmid, 1999). Female gray mouse lemurs stored more fat than males; in one case a male increased body mass by 18%, whereas females increased their mass by 26% (Schmid, 1999). Rufus mouse lemurs at RNP weighed between 30–50 g pre-hibernation. Males gained more, increasing body mass by 15–91% with one male reaching ~90 g, while females increased body mass by 15–51%, with one reaching 65 g (Atsalis, 1999a, 2008). A similar pattern was noticed at Mantadia National Park, another

eastern rainforest site, where a male *M. rufus* reached 70 g and a female 55 g (Randrianambinina et al., 2003a).

The reddish-gray mouse lemur, *M. griseorufus*, survives in the dry spiny forest of Berenty Reserve in southern Madagascar, an unpredictable environment with up to seven months of dry season annually and highly irregular rainfall patterns (Génin, 2008). Yet, in this arid environment, there was no evidence of seasonal prolonged torpor, although there were hints that daily torpor may occur occasionally when ambient temperatures were low (Génin, 2008). Despite the lack of seasonal hibernation, body mass of females trapped in March–April increased, although male body mass did not, suggesting the presence of intersexual competition for food (Génin, 2008).

When inactive, either as part of the daily cycle or when in seasonal hibernation, mouse lemurs typically rest in tree holes and leaf nests (Radespiel et al., 1998, 2003; Atsalis, 2008; Thorén et al., 2010). At one site, gray mouse lemur females selected tree holes that averaged 4 m above the ground, significantly higher than the location of nests chosen by males, which averaged 0.77 m (Radespiel et al., 1998). Females choose well-insulated tree holes, which may protect from predators and the elements. Males often choose dead tree hollows and leaf nests, changing these poorer-quality sleeping sites frequently (Radespiel et al., 1998). Compared to the gray mouse lemur, sympatric golden-brown mouse lemurs use a broader range of sites, sometimes nesting in collections of branches (Radespiel et al., 2003). As in gray mouse lemurs, female golden-brown mouse lemurs are concerned with safety; they use tree holes in 46% of cases, some located at the base of trees (Radespiel et al., 2003) but choose to construct leaf nests higher in trees during the rearing season when offspring require added protection from predators (Thorén et al., 2010).

The diet of mouse lemurs has been studied through direct observation of animals feeding, examination of fecal matter deposited in live traps, and, in the past, through analysis of stomach contents. Although mouse lemurs are the smallest primates and expected to be highly insectivorous like other small species, in fact less than half of their diet is made up of insects. They consume a broad range of dietary items that change seasonally.

Mouse lemur behavior is highly seasonal, revealed in detail through the first long-term continuous study on wild mouse lemurs by Atsalis (1998, 1999a, 1999b, 2000, 2008) (Figure 2.7). During the 16-month study, feeding ecology of the brown mouse lemur at RNP was examined intensively, revealing how dietary choices changed in association with seasonal resource availability (Atsalis, 1999a, 2008). Tagged plants were checked monthly for flower and fruit production (phenological patterns), and

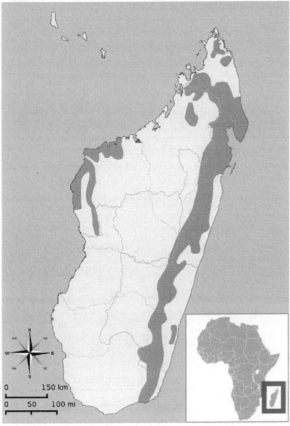

Figure 2.6 Geographic distribution (dark gray) of the aye-aye (*Daubentonia madagascariensis*), based on spatial data from IUCN Red List (IUCN, 2020). Aye-ayes occur at low densities, but the species has a wide geographic distribution.
(Map by Hogweard; image is licensed under Creative Commons Attribution-Share Alike 3.0 Unported license)

typically birds (McKey, 1975; Howe & Estabrook, 1977; Godschalk, 1983). *Bakerella* fruits were found to be high in fat, which has twice the energy content of the carbohydrates typical of fruits (Atsalis, 1999a, 2008). Atsalis (1999a) proposed that for this population of mouse lemurs, *Bakerella* was both a keystone resource eaten during the dry season when resources were low, and a staple dietary item eaten as a year-round resource.

In western Madagascar, gray mouse lemurs eat fruits, insects, gums, and insect secretions, in addition to flowers, buds, tree sap, and even chameleons and tree frogs (Martin, 1972, 1973; Barre et al., 1988; Corbin & Schmid, 1995; Reimann et al., 2003; Dammhahn & Kappeler, 2006; Radespiel et al., 2006). Fruits are considered to be a staple, complemented by insects, mostly beetles (Hladik et al., 1980). A high level of frugivory was further confirmed for this species in Mandena where 63% of feeding was on fruit (Lahann, 2007). Mouse lemurs were also seen to feed on flowers (22% of observations), and on arthropods and gums (7% and 1% of observations, respectively). They consume fruit (62 species) and flowers (8 species); small fruits are preferred, especially berries and drupes. Like the brown mouse lemur, gray mouse lemurs at Mandena feed on the semiparasitic, epiphytic mistletoe *Bakerella*. Gray mouse lemurs also eat maggots, spiders, stick insects, crickets, cockroaches, moths, and beetles, using their relatively large ear pinnae to locate insects before dashing to seize them, sometimes in the ground leaf litter (Martin, 1972). They have also been observed to feed on floral nectar (Martin, 1973; Hladik, 1980), very likely serving as plant pollinators (Nilsson et al., 1993).

Whereas the early study at Marosalaza focused on understanding how nocturnal lemuriform species coexist within the same forest (e.g., Hladik et al., 1980), recent studies have concentrated specifically on sympatric mouse lemur species. Given their similar body size and ecology, how do mouse lemur congeners partition their niches? *M. murinus* and *M. ravelobensis* both feed mainly on insect secretions and gum in the dry season, sharing 70% or more of plant species. Co-existence is facilitated through specialized selection of plant species by *M. murinus* compared to the more generalized diet of *M. ravelobensis* (Radespiel et al., 2006). Similarly, in a multiyear study, Dammhahn and Kappeler (2006, 2008b, 2009) discovered that *M. murinus* and *M. berthae* differ in the proportions of their dietary items: *M. murinus* eats more fruit, whereas for *M. berthae*, up to 81% of the diet was made up of sugary secretions produced by Homoptera larvae. Insect secretions are an important food resource of *M. ravelobensis* as well, particularly in the dry season (Corbin & Schmid, 1995; Reimann et al., 2003; Radespiel

insect abundance was monitored through various collection methods. Hundreds of fecal samples were examined, and radio-collared individuals were tracked in the forest. Even in a rainforest with annual precipitation of over 4,000 mm, the dry season is characterized by relative scarcity in fruit and insect abundance (Overdorff, 1991; Hemingway, 1995; Atsalis, 2008). Mouse lemurs rely heavily on 75+ different fruit species, increasing the quantity and diversity eaten during rainy season when fruit availability is high (Atsalis, 1999a). In contrast, reliance on insects does not increase during the rainy season when insect abundance is at its highest. As in gray mouse lemurs, beetles are a regular part of the diet. In some locations mouse lemurs concentrate on one or two locally available fruit-bearing species at a time, particularly berries (Martin, 1972; Petter, 1978). At RNP, the brown mouse lemur relies heavily on the fruits of several varieties of the mistletoe, *Bakerella*, which occurs in small patches but is widely distributed in the forest (Atsalis, 1999b, 2008). Mistletoe produces highly nutritious berries that are dispersed by specialized frugivores,

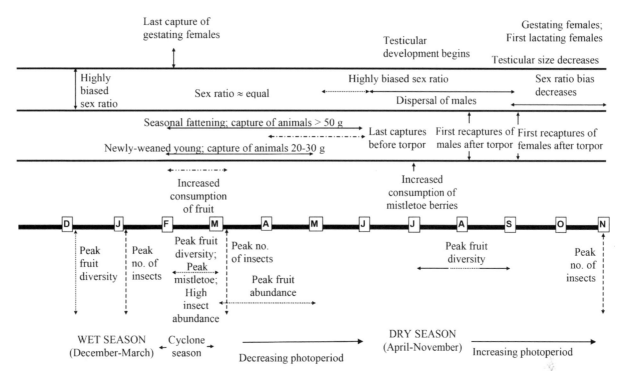

Figure 2.7 Schematic representation of variation in resource availability for *Microcebus rufus* individuals across the annual cycle, along with co-occurring changes that take place in *M. rufus* physiology and life history events.

et al., 2006). Produced by colonies of larvae, the secretions drip onto branches drying into a crystalline form that mouse lemurs lick (Corbin & Schmid, 1995). To supplement with protein, *M. murinus* consumed a wider variety of arthropods than *M. berthae*, though beetles figured importantly in the diets of both (Dammhahn & Kappeler, 2009). The gray mouse lemur was more opportunistic in its food habits with a feeding niche that completely encompassed the more specialized niche of Madame Berthe's mouse lemur. As a dietary specialist, *M. berthae* may need a larger home range to overcome the challenges of competing with sympatric lemurs, hence its exceptionally large range averaging 2 ha in contrast to that of *M. murinus* at 0.26 ha.

Génin's study (2008) on the reddish-gray mouse lemur in the spiny forest of Berenty Reserve in southern Madagascar—one of the driest habitats in the Afrotropical biogeographic realm—took place during a very arid year in which only 46 mm of rain fell. Animals relied heavily on gums, foraging on them for up to 58% of total activity time, as opposed to only 25% during a year with 168 mm of rainfall. When available, individuals preferred fruit over gums, expanding their home ranges to twice the size of the dry year, and increasing nightly activity in order to gain access to newly produced fruits. Gums, just like homopteran secretions, are an important food type in the seasonal diet of mouse lemurs studied in dry forests, leading Génin (2001) to conclude that gum trees are keystone resources for mouse lemurs.

The importance of nectar for some lemurs, including mouse lemurs, as well as the role of lemurs as pollinators has been widely recognized (Sussman & Raven, 1978; Kress, 1993; Nilsson et al., 1993; Kress et al., 1994; Birkinshaw & Colquhoun, 1998). We have also noted examples of the importance of mouse lemurs as seed dispersers (Atsalis, 2008). Seed dispersal may be among the most important components of plant–animal interactions. In the New and Old World, birds are important seed dispersers, and mistletoe fruits may be exclusively consumed by birds in some ecological communities (Godschalk, 1985; Reid et al., 1995). In Madagascar, the diversity of birds and other mammals besides lemurs that eat fruits is low (Wright, 1997; Hawkins & Goodman, 2003), leaving lemurs as the primary seed dispersers (Smith & Ganzhorn, 1996; Dew & Wright, 1998). Brown mouse lemurs likely disperse mistletoe seeds; the characteristically sticky feces of mouse lemurs, containing intact mistletoe seeds, often being detected adhering to trunks of trees (Atsalis, 2008) where epiphytes could establish themselves. Because mammals are rarely dispersers of mistletoes (Amico & Aizen, 2000), the discovery of a close relationship between mistletoes and brown mouse lemurs is significant (Atsalis, 2008), since epiphytes play an important role in forest nutrient cycling (Nadkarni, 1983). As a result, mouse lemurs, too, likely play an important role in the ecological dynamics of forests in Madagascar.

Mouse lemurs travel quadrupedally on branches, running and leaping short distances, descending to the forest floor to hunt for insects or to cross short open areas (Martin, 1973; Pagès-Feuillade, 1988). Although they roam at all heights of the forest (Radespiel et al., 2006; Atsalis, 2008), seasonal preferences have been observed; e.g., at the end of the dry season, with little available plant food, some mouse lemurs spend 70% of their time searching for insects below 3 m above the ground (Pagès-Feuillade, 1988). Locomotor differences described between *M. ravelobensis*, the golden-brown mouse lemur, which moves through its environment by leaping, and *M. murinus*, the gray mouse lemur, which is predominantly quadrupedal, are likely associated with differences in body morphology and preferred forest strata (Zimmermann et al., 1998).

Phaner

Phaner species, the fork-marked lemurs, with average adult size ranges of 350–500 g (Tattersall, 1982), are quick-moving, agile, quadrupedal animals that spend most of their time on the horizontal branches of the canopy as well as utilizing the vertical strata of the forest, from the ground to above 10 m (Petter et al., 1975). Fork-marked lemurs have keeled, pointed nails allowing them to cling vertically on tree trunks to feed on gums of *Terminalia* tree species. Trees produce gums to seal wounds when beetle larvae burrow tunnels in the bark of the tree. In the daytime, *Terminalia* trees are visited for gums by the crested coua (*Coua cristata*), an endemic bird species, and, at night, the fork-marked lemur feeds on the gums (Charles-Dominique, 1978). *Phaner* survive mainly on this resource, which is available throughout the year. The genus has evolved additional dental traits for easy access to the resource; the upper first premolars resemble canines, while the lower canines and incisors are nearly horizontal (i.e., procumbent), excellent for gouging to release exudates (Petter et al., 1975). *Phaner* also have long, pointed tongues and guts specialized for digestion of gums (Petter et al., 1975). Other gum-feeding primates, such as *Allocebus*, the African lorisoid needle-clawed bushbabies (*Euoticus elegantulus* and *E. pallidus*), and the marmosets and tamarins of Central and South America have also evolved similar adaptations.

In the wet season, *Phaner* consumes floral nectar and fruit, clinging to fine terminal branches and licking clusters of flowers (Petter et al., 1975). *Phaner* also feeds on insect secretions, primarily those from homopteran larvae. Although some gums are proteinaceous, the fork-marked lemur complements its diet with animal food throughout the year. Approximately 10% of the diet is composed of insects (Hladik et al., 1980).

Studies of *P. pallescens* in Marosalaza Forest and Kirindy Forest provide some of the most detailed field data available on fork-marked lemurs (Hladik et al., 1980;

Schülke, 2003a). At these sites, interspecific competition with other sympatric nocturnal lemurs for food and sleeping sites is reduced through spatial separation of animals in the forest. Members of fork-marked lemur populations usually sleep in tree hollows that are located at ~8.5 m (Schülke, 2003a). During the dry season, *P. pallescens* successfully compete with *M. murinus* and *M. coquereli*, using their anatomical specializations for gum feeding. The gums/exudates of more than 20 tree species make up to 90% of monthly feeding time by *Phaner*. Despite the year-round presence of gums, fork-marked lemurs may suffer from resource shortages in the dry season as body mass drops by 20% during this time (Schülke, 2003a).

Mirza

Mirza coquereli, Coquerel's giant mouse lemur, is typically found in the densest part of the forest (Pagès, 1980), moving quadrupedally primarily on small branches and at all forest heights (Hladik et al., 1980; Kappeler, 2003). By chewing off small twigs, giant mouse lemurs build their own nests, which are typically found high in the canopy often surrounded by other nests (Andrianarivo, 1981; Sarikaya & Kappeler, 1997); similar nest site preference has been reported for *M. zaza* (Rode et al., 2013). Multiple nest sites may be used, with *M. coquereli* rotating between as many as 10–12 at any one time (Kappeler, 2003). *M. zaza* has been described as using fewer (1–3) day nest sleeping sites, with males and females sharing day nests (Rode et al., 2013). By contrast, individual adult *M. coquereli* generally have solitary day nests (Kappeler et al., 2005; Rode et al., 2013).

As an omnivore, Coquerel's giant mouse lemur feeds primarily on fruit and insects, supplementing the diet with chameleons, frogs, floral nectar, and gums (Hladik et al., 1980; Pagès, 1980). Additionally, 50% of feeding observations in the dry season were on secretions of insect larvae, mainly aggregates of the endemic insect *Flatida coccinea*; lemurs visited these insect colonies nightly to lick the liquid secreted (Pagès, 1980).

Unlike dwarf and mouse lemurs, giant mouse lemurs do not undergo seasonal weight gain or enter hibernation, but they may reduce activity during the austral winter by retiring to their nests in the second half of the night (Kappeler, 2003). The ability to exploit the secretions of insects enables both *M. coquereli* and *M. zaza* to be active throughout the year (Hladik et al., 1980; Rode, 2010). Although this food source is constant, it occurs in low density, thus influencing *Mirza* population density. Giant mouse lemurs are able to survive in secondary forests. In one degraded forest located in northern Madagascar, the population density of northern giant mouse lemurs

(*M. zaza*) was much higher than expected because of the presence of cashew trees (*Anacardium occidentale*), which supply ripe fruit and gums in the dry season (Andrianarivo, 1981; Kappeler et al., 2005).

Allocebus trichotis
Biebouw (2006, 2009) studied *A. trichotis* for one year in central eastern Madagascar, shedding light on the behavior and ecology of this little-known species. Biebouw radio-collared members of a group that slept together and followed them from April–November, doing the same for another sleeping group whose members she followed from September to the end of the year. *Allocebus* sleep in tree holes, approximately 1–9 m above ground and lined with fresh leaves (Biebouw et al., 2009). Each animal uses four to five different tree holes but favored one or two. They were sometimes observed in sleeping associations with white-tailed tree rats (*Brachytarsomys albicauda*), which may confer thermoregulatory or antipredatory advantages. As in *Microcebus*, large trees are preferred nesting sites, but hairy-eared dwarf lemurs do not hibernate, although the animals occasionally enter short periods of torpor (Biebouw et al., 2009).

Like *Phaner*, hairy-eared dwarf lemurs feed on gums of *Terminalia*, using their keeled nails, relatively long tongues, and dental specializations (Rakotoarison et al., 1997; Fleagle, 1999). They also consume small moths, fruit, flowers, and leaves (Biebouw, 2009).

Family Lepilemuridae
Lepilemur
Lepilemur, the sportive lemurs, are unusual in being relatively small-bodied primates that are predominantly folivorous. (Folivorous animals tend to be large-bodied to accommodate the increased gut capacity needed to digest leafy matter.) To process and digest their leafy diet, *Lepilemur* has molars with well-developed shearing crests (Seligson & Szalay, 1978), a specialized alimentary tract with a large cecum, and a wide colon (Charles-Dominique & Hladik, 1971; Chivers & Hladik, 1980). They have been observed to feed on cecotropes (Charles-Dominique & Hladik, 1971; Hladik, 1979), which are produced in the cecum and excreted. Cecotrophic behavior benefits individuals by providing them access to the nutrients available in semi-digested leafy matter that form the cecotropes, thereby maintaining energy balance (Charles-Dominique & Hladik, 1971; Ratsirarson & Rumpler, 1988).

The sportive lemurs are vertical clingers and leapers (Napier & Walker, 1967), jumping from one vertical trunk to another (see Charles-Dominique & Hladik, 1971), usually maintaining the same horizontal height (Nash, 1998;

Warren & Crompton, 1998a; but see Blanchard et al., 2015). Sportive lemurs ascend and descend tree trunks by limb-over-limb slow climbing, or by upward hops (Walker, 1967; Russell, 1977; Warren & Crompton, 1998a). Typically observed in the larger branches of the canopy, sportive lemurs are most often found from 4–15 m high in the forest, rarely if ever descending to the ground (Hladik et al, 1980; Nash, 1998). They are among the few lemurs that inhabit the arid *Didierea* spiny forests of southern Madagascar where vertical clinging and leaping is a necessary mode of locomotion for an arboreal species.

Russell (1977) noted that *L. mustelinus*, the weasel sportive lemurs, spend 91% of their feeding time ingesting leaves, 5.9% ingesting fruit, and 3.2% ingesting latex. They feed on 24 species of plants in a *Didierea* forest, concentrating on 2–14 species per month. Despite some diversity in the plant species consumed, only two plant species accounted for 64% of the diet: *Alluaudia procera* (Madagascan ocotillo = 52.8%) and *Grewia* (11.1%). Nash's (1998) findings on *L. leucopus*, the white-footed sportive lemur, were similar; they are almost exclusively folivorous, and two tree species and one vine species account for 75–85% of the diet of five animals that she radio-tracked.

Nash (1998) reported that, depending on the season, 79–85% of active time was devoted to feeding (31–35%) or resting (44–54%). After spending hours patiently watching them rest, Nash refers to sportive lemurs as the sloths of the primate world and, as seen in the title of Nash (1998), "vertical clingers and sleepers" (cf. the "lazy leapers" description of *Lepilemur* by Warren & Crompton, 1998b). The sportive lemurs rest more and travel less in the cold dry season of the austral winter, even though time spent feeding does not change seasonally. Sportive lemurs move little while eating as much as possible at all times (Nash, 1998). The increased time spent resting is typical of folivores since they spend long periods digesting their food. Also related to the low nutritional quality of leaves may be the exceptionally low resting metabolic rate (RMR) of the sportive lemur—among the lowest of all folivorous mammals thus far measured (Schmid & Ganzhorn, 1996). *L. leucopus* lives in Madagascar's spiny forest, where they rely mainly on *Alluaudia procera*, a succulent that produces potentially toxic secondary compounds (Dröscher et al., 2016). The low RMR of *Lepilemur* may be effective in minimizing the possible toxicity of their diet.

Ganzhorn (1988, 1993) compared the chemical composition of the diet of *L. mustelinus* with that of the ecologically similar nocturnal species *Avahi laniger* (family Indriidae). The diet of sportive lemurs was more restricted when the two taxa were sympatric; when allopatric, however, sportive lemurs choose higher-quality leaves with higher protein and lower fiber content (see also Dröscher et al., 2016). They feed on leaves containing alkaloids, chemicals that are generally toxic, and not eaten by *Avahi*. Additional research revealed that sympatrically living *Avahi occidentalis* and *L. edwardsi* feed on leaves in approximately 75% of feeding bouts, each consuming leaves from 25 tree species and 5 liana species (Thalmann, 2001, 2002, 2006). Interspecies competition was avoided because the overlap in the diets was, in fact, minor; of the two, *Lepilemur* was a generalist, feeding on the more common plant species, whereas *Avahi* specialized on a narrower selection of foods. Competition between *Avahi* and *Lepilemur* can also be avoided by feeding in different forest strata (Napier, 1966) or on trees with different crown diameters (Ganzhorn, 1988, 1989), engaging in different locomotor behaviors (Warren & Crompton, 1998a,b; Blanchard et al., 2015), or through different activity schedules, as in the case of *Avahi*, which exhibit high crepuscular activity and short day bouts of activity when sympatric with *L. fleuretae* (Campera et al., 2019).

Although *Lepilemur* do not enter torpor, they do adjust their behavior to seasonal conditions. To illustrate, data collected over dry and wet seasons on 14 *L. sahamalaza* reveal that individuals significantly decrease time spent traveling and increase their time spent resting in the dry season compared to the wet season, and lemurs seek more confined sleeping locations in colder periods (Mandl et al., 2018). Isolated on the offshore islet of Nosy Be, the newly recognized and little-known *L. tymerlachsoni* was studied by Sawyer and colleagues (2015) who discovered that the animals occasionally eat fruit but, like other lepilemurs, are predominantly folivorous.

Family Daubentoniidae
Daubentonia madagascariensis

The aye-aye has unusual anatomical traits, most notably: rodent-like dentition with continuously growing, enlarged incisors; greatly reduced posterior dentition; an extremely long, thin, and mobile third manual digit that can rotate 360°; claws rather than nails except on the first toe; and an oddly shaped head with large bat-like, mobile ears. Aye-ayes weigh ~2,700 g and are relatively slow and deliberate arboreal quadrupeds. Their claws enable them to move on large vertical and angled branches, but they are also adept at both leaping and moving in the fine terminal branches of the forest canopy and bushes. When the aye-aye is on the ground, it moves slowly on its palms with the fingers of the hand raised, not touching the ground (Petter, 1977).

In an intensive study on Nosy Mangabe, an island reserve off Madagascar's northeast coast, two males and two females were radio-tracked for 46 nights (Sterling, 1993a,b, 1994b,c; Sterling et al., 1994). Aye-ayes choose a mix of habitats that includes large trees, small trees, and shrubs. Animals use all levels of the forest, including tree crowns reaching up to 35 m in height; traveling typically occurs at about 7 m above the ground and feeding, on average, at approximately 11 m. On Nosy Mangabe, aye-ayes use the ground 25% of the time (Sterling, 1993a) but, on the mainland, they spend >80% in the two highest forest levels, rarely venturing to the ground (Ancrenaz et al., 1994). They sleep in well-constructed day nests, high in the trees in locations with a high concentration of lianas (Sterling, 1993a; Ancrenaz et al., 1994).

On Nosy Mangabe, aye-ayes eat seeds, larvae, fungus, nectar, and adult insects, each from only a few species. *Ramy* seeds (the hard-shelled nut of the endemic tree, *Canarium* sp., cf. *madagascariensis*, Burseraceae), larvae, a fungus growing on leguminous *Intsia* trees, and nectar from the Madagascar traveler's palm (*Ravenala madagascariensis*) together make up over 90% of the diet (Sterling, 1993a,b, 1994c; Sterling et al., 1994). Food choice is seasonal on Nosy Mangabe. Aye-ayes eat *Canarium* seeds most of the year, except during the cool-wet season (mid-May through mid-September), when they eat more insects and fungus; *Ravenala* nectar is consumed predominantly in the hot-wet season (January to mid-May). On the mainland, however, Ancrenaz et al. (1994) found *Ravenala* nectar to be the main food source in the cool dry season. At the Mananara-Nord Biosphere Reserve on the mainland south of Nosy Mangabe, Andriamasimanana (1994) found that other foods are part of the diet, including wood-boring insects, the fruit of various fig species, nectar from banana flowers, seeds of *Terminalia* trees, coconut flesh and milk, the flesh and seeds of breadfruit, and ripe bananas and sugarcane grown on plantations. At the Mananara-Nord site, aye-ayes rub the stalks of bamboo in the gap between their incisors and molars, possibly to sharpen their teeth.

Aye-ayes use their long, middle finger or their incisors to exploit several food sources, but they only use both to gain access to foods that require "extractive foraging" (e.g., hard-shelled seeds or wood-boring insect larvae). These foods are structurally defended either by hard outer coats or by wood or bark and are fairly abundant but rarely exploited by other species. The aye-ayes gnaw the endocarp of hard seeds with their incisors and extract the cotyledon with their long middle finger, employing its 360° rotational adaptation (Petter, 1977; Iwano & Iwakawa, 1988; Sterling, 1993a). Sterling (1993a) discovered that endocarp of seeds eaten by aye-ayes is considerably harder than those eaten by South American primates specializing in seed predation. The ability to consume hard seeds and nuts prompted Iwano and Iwakawa (1988) to speculate that aye-ayes fill the ecological niche of squirrels, which are absent in Madagascar.

Wood-boring adaptations provide nurture and protection for insect larvae and pupae, such as those of the Lepidoptera and Coleoptera. The aye-aye's unusual morphological traits are used to exploit these wood-boring larvae, which serve as an important dietary component year-round (Sterling, 1993b). Woodpeckers are common vertebrate predators on wood-boring insect larvae. Where woodpeckers are absent, other birds have developed adaptations analogous to those of the woodpeckers. In Madagascar, the aye-aye fills the unique ecological niche-space that combines some aspects of both the squirrel and the wood-boring woodpecker niches. Cartmill (1974) points out that New Guinea and Madagascar are the only extensive regions in the world which have not been colonized by woodpeckers or other bird species that prey on wood-boring insects. He hypothesizes that the aye-aye fills the woodpecker niche and can be considered a member of a specialized foraging guild shared by *Dactylopsila*—a marsupial genus found in New Guinea and northern Queensland, Australia, that also hunts for insects by tapping branches (i.e., using "percussive foraging"), and has procumbent incisors—and by wood-boring birds found elsewhere in the world. There are, in fact, several avian species in the family Vangidae (e.g., the hook-billed vanga, *Vanga curvirostris*), endemic to Madagascar and the Comoros Islands, that probe for insect larvae found in the bark of trees (Erickson, 1991). It is not known to what degree these small to medium-sized passerine birds compete with aye-ayes for food.

Studies in captivity shed light on how aye-ayes use their combination of traits to locate prey in woody cavities in the absence of prey movement or visual/olfactory cues (Erickson, 1994, 1998; see also Goix, 1993). Tapping the surface of a branch with its long, thin middle finger, a *specialized* form of percussive foraging, the aye-aye receives either auditory or surface vibration cues about the hollow spaces within and stimulates larvae to move, giving away their location. In the wild, aye-ayes use both live and dead wood for percussive foraging of invertebrates (Sefczek et al., 2017) and select trees with dense interiors, likely because they are better sounding boards for locating grubs (Thompson et al., 2016).

SOCIAL ORGANIZATION

There has been continued interest in the social organization of the nocturnal Lemuriformes as reflected in numerous studies: *Microcebus* species (e.g., Martin, 1972; Atsalis, 2008), *Mirza coquereli* (e.g., Kappeler, 2003),

Cheirogaleus medius (e.g., Müller, 1999c), *Lepilemur* species (e.g., Thalmann, 2006), genus *Phaner* (e.g., Schülke, 2003b), and *Daubentonia madagascariensis* (e.g., Sterling & Richard, 1995). Members of these species most often forage alone but may sleep in pairs or groups during the day. They communicate through chemical signals transmitted via scent-producing glands, saliva, urine, and even feces (e.g., Braune et al., 2001), and are characterized by substantial variation in social behavior and organization (Table 2.4). Births and maturation rates differ across species. Some genera, like *Microcebus*, produce litters with young that develop quickly and are ready to reproduce in the first year of their lives. *Phaner*, *Lepilemur*, and *Daubentonia* produce single offspring with slower maturation rates.

Family Cheirogaleidae
Cheirogaleus

The social organization of *Cheirogaleus medius*, the fat-tailed dwarf lemur, whose annual activity is interrupted by a prolonged period of hibernation, has been the subject of several studies resulting in significant insights on social behavior. Müller (1999a) and Fietz (1999a, 2003a) studied this species in northwestern and western Madagascar, respectively, each over several study periods. Their research revealed that although *C. medius* are solitary foragers, male-female pairs live together in monogamous pairs that are likely permanently bonded. Pairs reside with their offspring of one or more breeding seasons in a system characterized as dispersed small family groups (Fietz, 1998b, 1999a, 2003a,b; Müller, 1999b; Müller & Thalmann, 2000) (Table 2.5). Müller (1999a,b,c) calculated a home range size that varied from 2.0–2.7 ha, reporting that overlap between ranges of adjacent groups vary from 5.2–21.9%.

Fietz (1999b, 2003a,b) discovered the presence of surplus males, which she named "floaters." Floaters do not have mates but instead roam in relatively large ranges. Fietz (1999a) reported that floater males have ranges of up to 11 ha that overlap the home ranges of other floaters as well as the territories of established groups.

Each fat-tailed dwarf lemur uses between 4–22 different sleeping sites (Müller, 1999b). Most sites are used by more than one member of the same group and some are used by all members of the group. Floater males sleep at sites used by other groups. Bonded pairs sleep together significantly more than expected by chance, although they always sleep apart during the birthing season. Overall, the composition of sleeping associations is made up of mothers with their offspring, but also fathers and offspring, as well as brothers and sisters (Müller, 1999c).

Table 2.4 Body Mass in Families Cheirogaleidae, Lepilemuridae, and Daubentoniidae

Species	Mass (g)	Notes	References
Cheirogaleus medius	165-195 g (n=4 males); 270 g highest 142-157 g (n=3 females); 214 g highest	Range of body mass following hibernation and highest body mass measured	Müller, 1999b
Microcebus murinus	39-84 g; avg. 57.9 g (n=52); (33 males, 19 females)	-	Schmid and Kappeler, 1994
Microcebus rufus	37.3-88.0 g; avg. 47.9 g (n=6 males; 32 recaptures in 15 mo.) 30–62 g; avg. 44.2 g (n=10 females; 45 recaptures in 15 mo.)	-	Atsalis, 1999
Microcebus rufus	30-90 g (n=220 males)[a] 27-72 g (n=82 females)[a]	June-May data (incl. pre-hibernation fattening period)	Atsalis, 2008
Microcebus rufus	male avg. 50.38 g +/- 4.75 g (range: 39.0-73.5 g) female avg. 44.6 g +/- 5.3 g (range: 31.5-55.0 g)	-	Blanco, 2010
Microcebus griseorufus	51.4 ± 4 g (n=4 males) 59 ± 3 g (n=6 females)	Average body mass in dry season (March-May)	Génin, 2008
Mirza coquereli	280-335 g (n=14 males & females)	-	Kappeler, 1997
Phaner furcifer	327 ± 20 g (n=18 males & females)	-	Schülke, 2003a
Allocebus trichotis	75.5-84.5 g (n=2 females) 81.5-78.0 g (n=2 males)	Average body mass from dry to rainy season	Biebouw, 2009
Lepilemur edwardsi	804-1020 g (n=10 males) 816-1131 g (n=10 females)	Male & female body mass ranges from dry season (May)	Rasoloharijaona et al., 2003
Daubentonia madagascariensis	2350-3000 g (n=6 males, 2 females)	-	Sterling, 1993b

[a] Sample size reflects all data points, not averages, from repeated captures of each individual.

Pair-bonded males and females scent-mark and actively defend the boundaries of their territories, taking turns protecting infants in the nests (Müller, 1998; Fietz, 1999a, 2003a). Closely related females sometimes raise their infants together (Fietz, 1999a). Offspring leave the family in the second year of their life (Fietz, 2003a), but there is evidence that males may disperse at a younger age than females, seeking a mate and home range of their own (Müller, 1999c).

Male extra-pair copulations are not uncommon. Therefore, pairs are said to be socially, but not reproductively, monogamous (Müller, 1998, 1999a; Fietz et al., 2000). Indeed, in 44% of cases analyzed, offspring are sired through extra-pair copulations, although not by floater males who have smaller testes and do not sire offspring (Fietz et al., 2000). In *C. medius*, timing of reproduction is usually dependent on the onset of the rainy season, which occurs in December (Fietz, 1999a). Gestation is 61–64 days (Foerg, 1982), although *C. major* gestation may be closer to three months (Fietz & Dausmann, 2003). *Cheirogaleus* produce two young (Fietz, 1999a). Fathers actively participate in infant care; they take turns with the mother "babysitting" offspring, and they accompany the young in their initial forays outside the nest, playing with them, defending them, and responding to

their calls (Fietz & Dausmann, 2003). The paternal role is critical to offspring survival because even though females may be able to defend their sleeping nest sites and fend off predators, infant care by male partners lessens the energetic costs incurred by mothers (Ross, 2003); females who lose their male partners are not successful in raising young, since infants are at greater risk of injury and death (Fietz, 1999a,b; Fietz & Dausmann, 2003). This extensive degree of parental care benefits young through improved thermoregulation of offspring during sleep and through safety from predators (Fietz & Dausmann, 2003). Finally, a comparison of fathers and non-fathers indicates a tendency toward decreased fat storage in the tail in the former (Fietz & Dausmann, 2003). Mothers, too, differ significantly from non-mothers in body mass, tail fat storage, and body condition (Fietz & Dausmann, 2003).

Less is known about eastern dwarf lemurs living in Madagascar's rainforests, but social organization generally mirrors that of the fat-tailed dwarf lemur. Lahann (2008) studied dwarf lemurs in the littoral rainforest of southeast Madagascar. Lahann identified two family groups, each composed of an adult mated pair and their offspring of previous years. Females were pregnant in December and infants were seen traveling away from a nest site for the first time in February (Lahann, 2007).

Table 2.5   Social Organization Characteristics of Select Nocturnal Lemuriformes

Species	Home Range[a]	Social Structure	Notes	References
Cheirogaleus medius	Relatively small: 2.0–2.7 ha	Monogamous pair bonds living with offspring in dispersed small family groups	Solitary males (floaters) range further up to 11 ha; extra-pair copulations	Müller, 1999b,c; Fietz, 1999a
Cheirogaleus major	4.0–4.4 ha	Adult mating pair with offspring	-	Lahann, 2007; Lahann, 2008
Microcebus murinus	-	Dispersed multimale-multifemale networks with female philopatry and resident versus dispersing males	-	Kappeler, 2000
Microcebus griseorufus	~0.93/0.54 ha (n = 10 males and females; rainy/dry season)	Dispersed multimale-multifemale networks with female philopatry and resident versus dispersing males	-	Génin, 2008
Microcebus ravelobensis	0.52/0.56 (n = 14 females; before mating season/ during mating season) 0.55/0.65 (n = 15 males; before mating season/ during mating season)	Dispersed multimale-multifemale networks with female philopatry and resident versus dispersing males		Weidt et al., 2004
Mirza coquereli	4.0–17 ha (males) 2.5–4.5 ha (females)	Dispersed multimale-multifemale networks	Male home range increased during mating season	Kappeler, 1997
Phaner furcifer	~4.7 ha (n = 8 females), ~5 ha (n = 7 males)	Pair living with low cohesion	-	Shülke and Kappeler, 2003
Allocebus trichotis	~5 ha (n = 4 females, 2 males)	Males ranges overlap with those of females	-	Biebouw, 2009
Lepilemur edwardsi	1 ha (combining both studies: n = 7 females, n = 6 males)	Pair-living sometimes with adult offspring; dispersed (social?) monogamy	-	Thalmann, 2001; Rasoloharijaona et al., 2003, 2006
Daubentonia madagascariensis	~36 ha (n = 2 females), n = ~170 (n = 2 males)	Males highly overlapping Females non-overlapping	- -	Sterling, 1993a,b
Daubentonia madagascariensis	7.5 ha (n = 1 female), n = 11 ha (n = 1 male)	-		Andriamasimanana, 1994

[a] Intraspecific differences may reflect differences in methodologies, environmental conditions, and duration of study periods. For example, see Biebouw (2009) for how different methodological protocols result in widely differing results for the same individuals.

The home ranges of members within a group overlap extensively (from 78–94%), and group members sleep together. As expected, overlap between group home ranges was minor (0–5.9%). Adults were noted to begin hibernation as early as April, even though young remain active until May (Fietz & Dausmann, 2003).

Microcebus

Despite the diminutive size of mouse lemurs and their solitary habits, the social organization of the genus has been the subject of several excellent field studies. By combining results from multiple field research methods, including trail censusing, live-trapping, radio-tracking, genetic analyses, and investigation of nesting groups with infrared cameras, important strides have been made in revealing the complexity of mouse lemur social behavior and relationships. From these methods, it is possible to summarize several important points. Not unlike

other small mammals, mouse lemurs live in dispersed multi-male/multi-female social networks characterized by considerable home-range overlap between the sexes (e.g., Barre et al., 1988; Andrès et al., 2001; Radespiel et al., 2001a; Atsalis, 2008). Males compete for access to females, their home ranges overlapping those of females to varying degrees (e.g. Radespiel, 2000; Schmelting et al., 2000; Weidt et al., 2004; Dammhahn & Kappeler, 2009); it is unlikely that males are able to monopolize females (Radespiel et al., 2001a,b; but see Génin, 2008), and cases of mixed paternity in infants born in the same litter have been reported—a rather rare occurrence in mammals (Radespiel et al., 2002; Andrès et al., 2003). Mouse lemurs are generally solitary foragers, and the degree of social behavior exhibited when they are active at night differs among species (e.g., Pagès-Feuillade, 1988; Schwab, 2000; Atsalis, 2008). Females remain in their natal group establishing relatively matrilinear lines, whereas males tend to disperse from their natal group (e.g., Martin, 1973; Atsalis, 2000; Eberle & Kappeler, 2003; Schmelting et al., 2007). Finally, sleeping group associations are important to the social life of mouse lemurs (e.g., Fietz, 1999c; Radespiel et al., 2003; Eberle & Kappeler, 2006).

Atsalis (1999b, 2000, 2008) sheds light on social spacing, highlighting seasonal changes in population dynamics of *M. rufus*. During the 16-month study, more males were live-trapped than females (102 versus 72), with a strongly seasonal sex bias that was highest in the dry season and in the months associated with the early mating period. Some males and females remained active throughout the dry season without changes in body mass, whereas other males and females fattened and disappeared from the live-traps for weeks during the dry season (Atsalis, 1998, 1999a, 2008). The sex ratio of trapped animals began to favor males in May and became highly biased in favor of males from May–September. During this period, many new males entered the population and were trapped, leading Atsalis (2000, 2008) to hypothesize that young males were dispersing from natal areas. Other males, termed residents, were stable members of the population, maintaining relatively larger ranges, each with access to several females (Atsalis, 2008). In a different study at RNP (which tracked the transfer of lice between individual mouse lemurs), Zohdy and colleagues (2012) suggested that male-only transfers implied the occurrence of fights over females for mating rights. Atsalis (2000) inferred that brown mouse lemur females are more philopatric than males since three times more females trapped at the start of the study were still being trapped at the end. Finally, over the course of the latter study, as many as 19 mouse lemurs, both males and females, were captured at a single trap location, suggesting considerable range overlap.

Of all mouse lemur species, the gray mouse lemur (*M. murinus*) has been the most extensively studied. At Kirindy, males occupy large overlapping home ranges, also overlapping those of females, and in the mating season, heavier males range closer to females, suggesting physical competition for access to females (Fietz, 1998a). Based on data from trapping, radio-tracking, and DNA analysis collected over successive years during the mating period, male gray mouse lemurs at Ampijoroa are also found to have larger home ranges than females, which increase in size during the mating period (Radespiel, 2000). Radio-tracked males have spatial access to three to four females and those females have access to an average of four males, suggesting that males are unlikely to monopolize females as mates (Radespiel, 2000). Male home ranges increase in size during the mating season potentially allowing access to more females (Radespiel, 2000; Schmelting et al.,

2000). Males wait by the entrance of nest sites before dusk in order to copulate with emerging females (Radespiel, 2000). The occasional presence of "sperm plugs" (formed by a post-ejaculatory gelatinous secretion deposited by the male which coagulates in the female vagina, forming a temporary plug the female can later expel, but which prevents another male's sperm from vaginal entry) provides further support that mouse lemurs are a promiscuous species (Schmid & Kappeler, 1998b; Schwab, 2000). Sex ratios remain consistently male-biased with genetic analyses confirming the findings at RNP, i.e., that males dispersed away from their natal ranges (Radespiel et al., 1998, 2001a,b; Radespiel, 2000). Closely related females (likely of the same matriline) commonly sleep in groups that remain stable even when the nest location changes (Radespiel, 2000; Radespiel et al., 2001b).

Gray mouse lemurs at Ampijoroa also experience very high turnover rate from year to year with 25–45% of the yearly population recaptured in the following year (Schmelting et al., 2007). New males account for 60–81% of yearly captures, and resident males have significantly larger body mass and larger home ranges. Although both resident and new males contribute equally to the next generation in terms of absolute numbers, resident males have more than 100% higher probability to reproduce than new males; only 22% of recaptured new males contributed to the next generation, but 50–100% of recaptured resident males reproduced successfully within their second to fourth season (Schmelting et al., 2007). On the other hand, at Kirindy, both resident and new males sired offspring according to their proportion in the population, likely because the density of females was much higher than at Ampijoroa (Eberle & Kappeler, 2004).

Sexually active in the first year of life, mouse lemur females produce 1 to 3 offspring per parturition (Radespiel et al., 2001a; Atsalis, 2008). Reproductive activity is triggered by photoperiod. As day length begins to increase in August–September at the end of the dry season, the vulval area of females, which is otherwise sealed by a membrane, begins to redden and swell leading to the vaginal opening; the testes of males also undergo considerable enlargement at this time (e.g., Schmid & Kappeler, 1998b; Radespiel, 2000; Schwab, 2000; Randrianambinina et al., 2003b; Dammhahm & Kappeler, 2005; Atsalis, 2008). At RNP, 71% of *M. rufus* gestating females sighted were seen in mid- to late November; lactating females were sighted in January–April (Atsalis, 2008; Blanco, 2008, 2011). In other mouse lemur species, lactating females were seen as late as April, possibly because some females bear second litters (Schmelting et al., 2000). The timing of reproductive events—mating, parturition, and weaning—in species that hibernate is critical. Timing should allow mothers the ability to prepare for seasonal torpor by increasing body fat during the period of resource abundance and for young mouse lemurs to attain independence. Where mouse lemur species are sympatric, however, they maintain reproductive isolation by staggering their estrus periods, e.g., female *M. murinus* enter estrus in August, while *M. ravelobensis* females undergo estrus in November (Randrianambinina et al., 2003a,b). Seasonal breeding sets the stage for closely related gray mouse lemur females to raise their young cooperatively and even to adopt the young of relatives who die before weaning is complete (Eberle & Kappeler, 2003). Communal rearing can improve infant survival through social thermoregulation (Radespiel & Zimmermann, 2001b).

Long-term observation of Madame Berthe's mouse lemur (*M. berthae*) has been conducted at Kirindy (Dammhahn & Kappeler, 2005, 2009). Compared to other mouse lemur species, home ranges are considerably larger (Dammhahn & Kappeler, 2009). Males share home ranges with up to nine other males and with several females (Dammhahn & Kappeler, 2003, 2005). The ranges of females overlap with an average of two other females compared to *M. murinus* females who overlap spatially with, on average, five (but up to 16) other females (Dammhahn & Kappeler, 2009). They rarely share sleeping sites and have unstable sleeping group compositions (also true for recently studied *M. sambiranensis*, Hending et al., 2017). Like other species, the average relatedness of females with overlapping ranges is higher than that of males, hinting that females are philopatric whereas males disperse (Dammhahn & Kappeler, 2005). As in other mouse lemur species, sex ratio of live-trap captures is also male-biased, coinciding with increased male activity during the mating season (Schwab, 2000; Dammhahn & Kappeler, 2005).

Génin (2008), who studied *M. griseorufus* in the extremely dry spiny forest of southern Madagascar, noted that resident males form long-term relationships with the philopatric females. Males engage in mate guarding by staying close to the female, sniffing her frequently, and scent marking. Other males are chased away.

A consistent pattern across species reveals that mouse lemurs have few social interactions while foraging. In gray mouse lemurs, Radespiel (2000) observed what amounted to 0.53 social encounters/hour, some of which were female-female encounters among the same sleeping group. In a different study, although social behavior occurred only in 7–8% of observation intervals, long grooming and huddling bouts (some lasting up to 23 minutes) were observed at the onset and end of nightly activity periods (Dammhahn & Kappeler, 2005). At RNP, brown mouse lemurs were seen in close proximity of one another in only 10% of sightings (Atsalis, 2008). Peaceful social interactions were observed in April–May, when young mouse lemurs are expected to make their appearance from nesting sites (Atsalis, 2008). Sleeping groups are usually small, only one to two occupants were counted in 69% of occupied nests, but in 31% of nests there were three to five mouse lemurs, often including young ones (Atsalis, 2008). In a follow-up study, Karanewsky (2013) reported that males and females share sleeping sites. In communal nests, gray mouse lemur females can raise young together maintaining matrilineal relationships (Radespiel et al., 1998, 2001a; Fietz, 1999c; Radespiel, 2000; Eberle & Kappeler, 2003, 2006). Up to 16 gray mouse lemurs were found sleeping together in the dry season (Radespiel et al., 2001b; Rasoazanabary, 2006), but males mostly sleep alone (Radespiel, 2000). In contrast, members of both sexes typically sleep together in the sympatric *M. ravelobensis* (Radespiel et al., 2003). Madame Berthe's mouse lemur females tend to sleep alone, but notably, males sometimes sleep in single-sex groups, which contrasts with the solo-sleeping gray mouse lemur males (Dammhahn & Kappeler, 2005).

M. ravelobensis, the golden-brown mouse lemur, was studied in the dry deciduous forest of Ankarafantsika in northwestern Madagascar (Schmelting et al., 2000; Weidt et al., 2004). Typical of mouse lemurs, individuals forage alone, with just 1.7 mostly neutral social encounters/hour occurring (Weidt et al., 2004). Mixed sleeping groups are typical of the golden-brown mouse lemur, but unlike sympatric gray mouse lemurs, which tend to use better-insulated tree holes, they tend to sleep in branches and tree nests (Weidt et al., 2004). The presence of males in sleeping groups may help to improve thermoregulation (Weidt et al., 2004). Sleeping group partners huddle and groom during social encounters at night, and at dawn members use a gathering vocalization before entering the sleeping site together (Braune et al., 2001).

Mouse lemurs typically communicate through deferred and remote social communication (i.e., scent marking and vocalizations). They lack true scent glands, instead using saliva, urine, feces, and genital secretions to recognize individuals, attract mates, and mark territories (Glatston, 1979; Perret, 1995; Braune et al., 2005). Mouse lemurs are also quite vocal; 11 different vocalizations were discovered for *M. murinus* and *M. rufus* (Zimmermann, 1995a,b). Mother-infant vocal exchanges are ritualized, are essential to bonding, and may have deep evolutionary roots (Scheumann et al., 2017). Ultrasonic calls, inaudible to humans, are also emitted (Cherry et al., 1987; Schilling, 2000; Atsalis, pers.obs.). Males are attracted to the breeding calls of females (Buesching et al., 1998). Sympatric species of mouse lemur differ in their reproductive advertisement calls (Braune et al., 2001). Adjacent, genetically related populations of mouse lemurs exhibit specific dialects of advertisement trill calls, most often emitted by males; young males migrating into the population may gain acceptance more easily by adopting these dialect calls (Zimmermann, 1995a; Hafen et al., 1998).

Phaner

Phaner furcifer typically lives in pairs (Charles-Dominique & Petter, 1980; Schülke, 2003a), although variations, such as one-male/multi-female groups, have been reported (Charles-Dominique & Petter, 1980; Schülke & Kappeler, 2003). These variations may be the result of fully grown offspring delaying dispersal as they wait for a vacancy to open up in their natal territory (Schülke & Ostner, 2005). A long-term study at Kirindy revealed that pair members live in small, stable territories that are actively defended by both sexes (Schülke & Kappeler, 2003). The respective ranging areas of pair members overlap extensively; pair members sit together, often grooming each other (Schülke & Kappeler, 2003). Yet, they spend more than 75% of their active time apart, sleeping together on average only every third day (Schülke & Kappeler, 2003). Paternity analyses revealed that a high proportion of offspring are fathered by non-pair males (Schülke et al., 2004).

To explain the evolution of pair-living in the fork-marked lemur, Schülke (2005) draws attention to the relative scarcity and patchy distribution of the exudate trees upon which the lemurs rely for food. Under these resource-sparse conditions, females can only share their range with one male, and males cannot defend access to more than one female against rivals (Schülke, 2003b, 2005). Members of a pair keep in contact primarily through loud calls but not frequently enough to know the whereabouts of the other on a constant basis (Schülke, 2003a). Schülke (2005) believes that males are unable to increase their foraging efficiency by coordinating their movements with females, and since females are physically dominant to the submissive males, they can chase and displace the males at feeding sites (Schülke, 2003b). In fact, the majority of interactions between a pair are conflicts over access to food resources (Schülke, 2003a).

The pair-bonded fork-marked lemur exhibits highly seasonal reproduction (Schülke, 2003a). Although females have two pairs of mammae, there is no evidence that they deliver litters like other members of the Cheirogaleidae (Kappeler, 1998; Schülke, 2003b). The pair has one infant per season, born in November–December (Harcourt & Thornback, 1990). Infants are initially cached in the tree hole in which their mother rests during the day. Later, mothers park their infants in vegetation at variable distances from the tree hole; infants are carried in the mother's mouth from the day shelter to a parking place (Schülke, 2005). At seven weeks the young begin moving independently (Schülke, 2005). Both males and females engage in extra-pair copulations (Schülke, 2005).

Mirza

Mirza coquereli, Coquerel's mouse lemur, has been the subject of an ongoing study in the Kirindy Forest since 1993, with more than 100 individuals marked, regularly captured, and observed within the 60-ha study area (Kappeler, 2003). Females occupy stable home ranges of between 2.5–4.5 ha (Kappeler, 1997). Each female's home range overlaps with as many as 8 other females. In contrast, male home ranges change in size but only during the mating season when home range size increases to more than 4 times their non-mating season size (from ~4+ to 17+ ha). It is only during this time that male home ranges overlap. During the mating season, male ranges also overlapped with those of at least 4–15 of the 19 known females in the study area (Kappeler, 1997); after the mating season, male ranges overlapped with those of only two to six females. These giant mouse lemurs do not defend their home ranges.

At Kirindy, results from genetic analyses suggest that Coquerel's mouse lemur lives in matrilinear clusters maintained through female philopatry (Kappeler et al., 2002). Some males tend to move over large distances, whereas others stay in the natal area for more than one year. Adults are very solitary, typically sleeping and foraging alone (Kappeler, 1997). Social interactions are extremely rare (Pagès, 1978, 1980; Kappeler, 1997), though male injuries during the mating season suggest that agonistic encounters occur with some frequency, and this competition explains the occurrence of mixed paternity litters (Kappeler, 1997; Kappeler et al., 2002). Young are foraging independently by three months of age (Kappeler, 1998).

In contrast to *M. coquereli*, *M. zaza* sleeps in groups of two to eight animals that can include several males (Rode, 2010). Individual home ranges are 0.5–2.2 ha (Rode, 2010). *M. zaza* may have the largest relative testis size of all primates and seems to be highly promiscuous (Kappeler et al., 2005; Rode, 2010).

Allocebus trichotis

Little was known about the behavior of *A. trichotis*, the hairy-eared dwarf lemur, until its rediscovery in 1989 (Meier & Albignac, 1991). Since then, individuals have been observed traveling alone or in pairs, and up to six animals have been found sleeping together in tree holes (Albignac et al., 1991; Meier & Albignac, 1991; Rakotoarison et al., 1997; Goodman & Raselimanana, 2002). Biebouw's (2006, 2009) yearlong study shed light on male and female ranging habits. Individuals whose home ranges overlap often are the ones that sleep with each other, typically in groups of three of mixed-sex composition, suggesting that the sleeping group represents the basic social unit (Biebouw et al., 2009). By radio-tracking members of two different sleeping groups, Biebouw (2009) discovered that home range size for each group differed widely, one ranged over approximately 7 ha, while the other ranged over 15 ha. Within each sleeping group, the home ranges of individuals also differ widely from each other.

Male and female home ranges overlap; males ranges overlap with those of two to four females, and female ranges overlap with those of up to two males. Social encounters occur between adults, as well as between mother and offspring; individuals also call to each other, usually near home range edges. Seasonal fluctuations occur in range use by both males and females during the reproductive period and at times when shifts in diet are expected (Biebouw, 2009). Not unlike mouse lemurs, estrus in *Allocebus* females seems to coincide with the wet season in November–December, with births occurring in January–February (Meier & Albignac, 1991). Pairs have been seen with a single young, suggesting that females give birth to singletons following a two-month gestation period (Meier & Albignac, 1991). Young are found nesting in a tree hole by March (Biebouw et al., 2009).

Family Lepilemuridae
Lepilemur
Across species, sportive lemurs live in dispersed pair-living situations (Russell, 1977; Ganzhorn & Kappeler, 1996; Thalmann, 2001, 2006; Rasoloharijaona et al., 2003, 2006; Zinner et al., 2003). Parents may share their range with adult offspring (Thalmann, 2001).

In an early study, Russell (1977) discovered that the home ranges of male *L. (mustelinus) leucopus*, the white-footed sportive lemur, in southern Madagascar did not overlap, but that those of females, as well as of males and females sometimes with offspring, did. Range-mates come together in the same tree for five minutes to an hour to feed, play, groom, or simply to rest together. During the day, range-mates often sleep together. Subsequent research showed that *L. ruficaudatus*, the red-tailed sportive lemur of Kirindy forest in western Madagascar, was also observed to live in pairs with overlapping ranges (Ganzhorn & Kappeler, 1996; Zinner et al., 2003). Pairs share sleeping sites, but the frequency of simultaneous use of sleeping trees between pair members varies considerably, from 1.7–83% (Zinner et al., 2003). A minority of males share their range with more than one female, and female–female associations observed within a home range may be mothers and daughters (Zinner et al., 2003). Based on follows of 46 individuals, Zinner and colleagues (2003) established that home ranges are quite small, averaging 0.8 ha.

Pair-living in sportive lemurs was confirmed from a multiple-year study of *L. edwardsi* (Thalmann, 2001, 2006). The home ranges of the radio-tracked male and female sportive lemurs were small (approximately 1 ha each), remained constant over time, and were occupied to the exclusion of neighboring groups (Thalmann, 2001, 2006). Members of a pair kept in contact by vocalizing, especially after the birth of their single young (Thalmann, 2001, 2006).

L. edwardsi, the Milne Edwards's sportive lemur, was studied in the dry forest of Ampijoroa (Rasoloharijaona et al., 2006). As in other locations, pairs use stable, small home ranges (about 1 ha) almost exclusively, suggesting that pairs are territorial and that they defend their sleeping and foraging sites cooperatively. Latrine behavior, the preferential use of specific sites for defecation, may also be used by *Lepilemur* for intergroup resource defense (Irwin et al., 2004). Loud calls, emitted by pairs in the morning and night, regulate the spacing between the pairs and identify ownership of territories (Rasoloharijaona et al., 2006). One notable finding of the study was that members of four of nine pairs followed slept together on average 92% of the days, and the other pairs usually slept in close proximity (Rasoloharijaona et at., 2006). These lemurs exhibited seasonal reproduction restricted to two months (Randrianambinina et al., 2007). Testicular volume increases in May and adult females are in estrus in June. After a four to five-month gestation, females produce single infants, which they initially transport by mouth. Young thought to be yearlings from the previous birth season show no signs of reproductive activity (Randrianambinina et al., 2007).

L. mustelinus, the rainforest-dwelling greater weasel sportive lemur, was much more solitary than *L. edwardsi*, both foraging and sleeping primarily alone (Rasoloharijaona et al., 2008). Ecological differences between the two broad habitat types supporting *Lepilemur* populations (i.e., dry western versus wet eastern forests), such as the availability of food resources and sleeping sites, may drive reported intra-generic differences in *Lepilemur* social organization (Rasoloharijaona et al., 2008).

The little-known Sahamalaza sportive lemur, *L. saha-malaza*, recently studied over a 10-month period, has also been determined to be solitary with no evidence of pair-specific home ranges, no pattern of social interactions between the individuals that were followed (only 0.32 interactions/hour), and no sleeping associations between adults (Mandl et al., 2019). The researchers concluded that this species was characterized by solitary social organization and structure with potential for a polygamous mating system (Mandl et al., 2019).

Family Daubentoniidae
Daubentonia madagascariensis

The social organization of the aye-aye is far more complex than originally thought. Although aye-ayes are typically solitary outside the mating season (Ancrenaz et al., 1994; Sterling, 2003), during Sterling's two-year study of aye-aye behavioral ecology on Nosy Mangabe, up to four individuals were noticed traveling and foraging together outside of the mating period (Sterling & Richard, 1995). Sterling (1993b) reported associations between males and females, and between males whose home ranges overlap considerably. Two aye-aye males she followed had large home ranges (126 and 215 ha), approximately four to five times larger than those of females. Furthermore, females were more solitary than males, rarely interacting with each other except through aggressive interactions, although they were never seen within each other's ranges (Sterling & Richard, 1995; Sterling, 2003; see also Ancrenaz et al., 1994). In contrast, males engage in many more affiliative, as well as aggressive, interactions. Sterling (1993b) noted that two to three aye-ayes occasionally forage as a unit, exchanging calls as they move. These foraging units included groupings of two adult males, young and adult males, and adult males and females. Animals remained within 20 m of each other for less than 14% of the time, but patterns of inter-individual interactions were still discernable. For example, some males avoided each other, and an adult male and female pair was observed foraging and traveling together several times when the female was not in estrus. Typically, aye-ayes sleep alone, but males (Sterling, 1993a,b) and mothers with young (Petter et al., 1977) have been observed to sleep together. During her study on Nosy Mangabe, Sterling (1993a,b) did not observe aye-ayes to engage in mutual grooming (see also Ancrenaz et al., 1994).

Andriamasimanana (1994) studied aye-ayes in the Mananara-Nord Biosphere Reserve in northeastern Madagascar. She found that for approximately 26% of observation time, the male and female under observation stayed in close proximity for periods ranging from five minutes up to one hour. They engaged in mutual nose-touching, sniffing, mouth-licking, and "chewing" on one another. Aye-ayes participate in what appeared to be play behavior—a male-female pair was seen running and jumping together along branches, possibly as a preamble to copulation (Andriamasimanana, 1994). This pair remained in close proximity after the birth of an infant, and later the mother and infant were seen grooming and playing together. Wild female aye-ayes can park their infants in the nest while they forage (Sterling & Feistner, 2000).

Female aye-ayes appear to be sexually receptive year-round, exhibiting genital swelling and color change (Sterling, 1993b; Winn, 1994a). Females begin reproducing at approximately 3.5 years of age and males at 2.5 years (Winn, 1994a,b). In the mating season, both males and females engage in scent-marking activity, and females vocalize to attract mates (Sterling & Richard, 1995). Males compete for access to females, with up to six males observed openly competing for a female (Barrows, 2001). Copulations can last for an hour between a pair; prolonged copulation prevents other males from gaining access to a sexually receptive female (Sterling & Richard, 1995). In captivity, sperm plugs have been found in the vaginas of females, which function as a form of passive mate defense in male-male reproductive competition (Haring et al., 1994).

PREDATION ON NOCTURNAL LEMURS

Predation exerts strong selective pressure on the behavior of primates, influencing and potentially shaping behavior, ecology, and overall activity patterns (van Schaik & van Hooff, 1983). With their small body sizes, nocturnal lemurs may be preferred prey for numerous Malagasy predators. Although there are relatively few large extant predators in Madagascar, carnivore scats and raptor prey remains indicate that predators of nocturnal lemurs include a number of endemic mongoose-like carnivoran species of the family Eupleridae, birds of prey, feral cats, domestic dogs, large snakes, and, in rare instances, even other lemurs.

Endemic to Madagascar, the cat-like fossa (*Cryptoprocta ferox*, family Eupleridae), weighing 6–9 kg (Dollar et al., 2007), is the largest of Madagascar's extant carnivores. Active both at night and during the day, the fossa utilizes auditory, visual, and olfactory cues to prey upon species across the island. Documented prey includes both eastern and western dwarf lemurs (Rasoloarison et al., 1995; Goodman et al., 1997; Rasamison, 1997; Hawkins, 2003; Hawkins & Racey, 2008). Other mammalian predators of dwarf lemurs are the ring-tailed mongoose, *Galidia elegans* (Wright & Martin, 1995; Deppe et al., 2008) and the narrow-striped mongoose, *Mungotictis* decemlineata (Rabeantoandro, 1997), like the fossa, both members of the family Eupleridae. Raptor predators include the Madagascar buzzard, *Buteo brachypterus* (Goodman et al., 1993a), the Madagascar cuckoo-falcon, *Aviceda madagascariensis* (Rasoloarison et al., 1995), the Madagascar harrier-hawk, *Polyboroides radiatus* (Gilbert & Tingay, 2001), and Henst's goshawk, *Accipiter henstii* (Karpanty & Wright, 2007). Reptilian predators include two large boids, the Madagascar ground boa, *Acrantophis madagascariensis*, and the Madagascar tree boa, *Sanzinia madagascariensis* (Wright & Martin, 1995).

Mammalian predators of mouse lemurs include the narrow-striped mongoose, common mongoose, and the domestic dog (Albignac, 1976; Petter et al., 1977; Russell, 1977; Goodman et al., 1993a; Rasoloarison et al., 1995; Rabeantoandro, 1997; Dollar et al., 2007). Species of snakes, such as the colubrid snake, *Ithycyphys miniatus,* and the Madagascar tree boa, *Sanzinia madagascariensis*, have been observed preying on western gray mouse lemurs (Richard, 1978; Randrianarivo, 1979; Goodman et al., 1993a; Dammhahn & Kappeler, 2008a).

Most birds of prey in Madagascar are diurnal and hunt by visually spotting prey. These birds include the previously mentioned Henst's goshawk, *Accipiter henstii*, and the Madagascar harrier hawk, *Polyboroides radiatus*, the Madagascar buzzard, *Buteo brachypterus*, and the Madagascar cuckoo-falcon, *Aviceda madagascariensis*; the hook-billed vanga, *Vanga curvirostris*, a shrike-like passerine, also preys on small lemurs (Goodman et al., 1993a). Russell (1977) observed the Madagascar harrier hawk, clinging to trees and tearing out sections of bark with their talons in pursuit of reptiles and perhaps nesting mouse lemur prey, and *Lepilemur* has been seen to be hunted in the same manner (Goodman et al., 1993a). At least six species of raptors prey on various mouse lemur species, including: *Asio madagascariensis*, the Madagascar long-eared owl; *Tyto alba*, the barn owl; *Tyto soumagnei*, the Madagascar red owl; *Accipiter henstii*, Henst's goshawk; and *Polyboroides radiatus*, the Madagascar harrier hawk (Langrand, 1990; Goodman et al., 1991, 1993a,b,c; Goodman, 1994; Rasoloarison et al., 1995, 2000; Goodman & Thorstrom, 1998; Karpanty & Goodman, 1999; Sterling & McFadden, 2000; Thorstrom & La Marca, 2000; Génin, 2010). Direct observations of raptor nest sites in Ranomafana National Park indicated *Microcebus rufus* are among the primary prey of both Henst's goshawk and the Madagascar harrier hawk (Karpanty & Wright, 2007).

In addition to diurnal birds of prey, the nocturnal barn owl and long-eared owl are among the most skilled predators of small nocturnal lemurs. In a yearlong study of predation on *M. murinus* by these two species, the remains of at least 74 mouse lemurs were identified in the regurgitated pellets of three individual owls. The researchers estimated a predation rate (percent of population taken by predators per year) of approximately 25%, the highest known for any primate species (Goodman et al., 1993a,b,c). Moreover, the long-eared owl has been observed feeding on at least four *Microcebus* species (Goodman et al., 1991, 1993b; Rasoloarison et al., 1995; Sterling & McFadden, 2000), while the barn owl preys on three (Langrand, 1990; Goodman et al., 1993c; Rasoloarison et al., 1995, 2000). The small, cat-like *Mungotictis decemlineata* in the west and *Galidia elegans* in the east (both in the family Eupleridae) search in the dense vegetation for prey, often discovering lemur tree hole nests and eating the occupants after enlarging the opening (Petter et al., 1977; Rasoloarison et al., 1995; Deppe et al., 2008). Overall, the intensity of predation on mouse lemurs suggests strong evolutionary impact; captive mouse lemur males live on average twice as long, and females 50% longer, than wild animals (Hämäläinen et al., 2014).

Predation on *Mirza* by two birds of prey, *Buteo brachypterus* and *Asio madagascariensis*, and two viverrid carnivores, *Cryptoprocta ferox* and *Mungotictis decemlineata*, has been reported (Kappeler, 1997; Rabeantoandro, 1997; Rasamison, 1997). Several predators have been identified for *Phaner* species, including the Madagascar buzzard (Goodman et al., 1993a,b), the Madagascar cuckoo-falcon, *A. madagascariensis* (Charles-Dominique & Petter, 1980), and the fossa (Rasolonandrasana, 1994; Rasoloarison et al., 1995). *Lepilemur*, while larger-bodied, are preyed upon by several predators—fossa, long-eared owls, Madagascar harrier hawks, and constricting snakes (*Acrantophis*) at night, as well as during the day when they rest in their tree holes (Ratsirarson, 1986; Goodman et al., 1993a; Rasoloarison et al., 1995; Schülke & Ostner, 2001; Hilgartner et al., 2008). Remains of *Lepilemur* have been found in pellets of the long-eared owl (Rasoloarison et al., 1995), though not in Ranomafana National Park, possibly because they are relatively rare or because they sleep deep in tree holes (Porter, 1998).

Joint mobbing, an anti-predator strategy by *Mirza zaza* and *L. sahamalazensis* against the colubrid snake, *Ithycyphus perineti*, has been observed with several individuals of each species partaking (Mandl et al., 2017). However, because they are small and relatively solitary, nocturnal lemurs are more likely to use evasion or stealth to avoid predation. They may choose well-protected nest sites and spend more time in dense vegetation to minimize predation (Goodman et al., 1993a). Thus, the depth and diameter of the opening of nest holes may play a role in protecting mouse lemurs. To illustrate this, a fossa was observed unsuccessfully attempting to claw into a mouse lemur tree hole (Martin, 1972; Russell, 1977); as well, *Mirza coquereli* nests are located within the crowns of trees in areas associated with thick liana growth where, presumably, they are well hidden (Sarikaya & Kappeler, 1997). Other anti-predatory behaviors have been noted in mouse lemurs: *Microcebus griseorufus* individuals encountering snakes stood up and exhibited lateral head movements, and a radio-tracked female bit a colubrid, *Madagascarophis colubrinus*, which fell down from a tree. A female mouse lemur who encountered an arboreal boa emitted several sequences of short whistles, a possible alarm call presumably aimed at a second female observed 10 m away, who responded with a short whistle. Another female exhibited evasive behavior when attacked by a Madagascar long-eared owl (Génin, 2010). Sportive lemurs produce loud distress calls, or "barks," which are also given during disturbances at the tree hole (Schülke & Ostner, 2001).

Commonly preyed-upon species, such as mouse and dwarf lemurs, may gain reprieve from predation when in torpor or hibernation (Karpanty & Wright, 2007). Torpor, however, does not assure protection; diurnal raptors prey upon sleeping nocturnal lemurs, while fossa and boas hunt both by day and night (Karpanty & Wright, 2007). Moreover, there are risks when emerging from torpor. Atsalis (2008) observed one hibernating female mouse lemur who, after emerging from her nest, remained in a lethargic state without seeking cover for nearly three hours. Unable to stay upright, she slowly rotated while holding on to a vine with her hind feet until she was suspended entirely upside down! Clearly, the behavior of lethargic animals may place them at increased risk of predation.

In summary, the presence of multiple predators that seek out nocturnal lemurs throughout the 24-hour cycle has significant impact on their behavioral ecology. These key ecological relationships are ripe as an important area of future research.

CONSERVATION EFFORTS

When population numbers in the wild exhibit indications of decline, the conservation status of threatened species is assessed by the International Union for the Conservation of Nature (IUCN) on a Red List of Threatened Species. Categories in ascending order of peril are Near-Threatened (a taxon is likely at risk in the near future), Vulnerable (assessment of a species indicates that it is facing high risk), Endangered (a species is considered to be at risk of extinction), and Critically Endangered (the best available evidence indicates that a species is facing extremely high risk of extinction in the near future unless drastic measures are implemented to ensure survival) (IUCN,

2020; see Tables 2.1–2.3). The vast majority of lemur species are now considered to be on the brink of extinction (Vaughan, 2020); 98% of all lemur species are classified by IUCN as either Vulnerable, Endangered, or Critically Endangered. Nearly one-third (31%) of all lemur species are now considered Critically Endangered, according to detailed reports by primate researchers (IUCN, 2020).

Nocturnal lemurs face pressures from deforestation associated with expansion of human settlements (Gillespie et al., 2008), parasitic diseases (Rasambinarivo et al., 2013; Ehlers et al., 2019), hunting for bushmeat associated with need, erosion of taboos against hunting due to increase of illegal gold mining (Jenkins et al., 2011), and a growing illegal pet trade in lemurs (Reuter et al., 2017). Research by Borgerson and colleagues (2016) emphasizes that illegal hunting will not decrease without attempts to improve the welfare of local people. Although remote areas of Madagascar are increasingly being surveyed, resulting in the discovery of new populations of nocturnal lemurs, these remote areas may be slated for exploitation, requiring special management to protect small and isolated lemur populations (e.g., Miller et al., 2015; Salmona et al., 2018).

How conservation threats are manifested may not be immediately obvious. The percussive foraging of the aye-aye requires wood and potentially causes resource competition between aye-ayes and humans, who also rely on dead trees as sources of fuel (Miller et al., 2017; Vieilledent et al., 2018). Because of the necessity to use vertical supports while traveling, the ever-increasing distance between trees that results from logging activities (both selective and intensive) has resulted in declining *Lepilemur* populations throughout Madagascar (Ganzhorn, 1993). *Lepilemur leucopus* occurs in the extremely dry and hot spiny forest, making it particularly vulnerable to climate change (Bethge et al., 2017). As forests change under human exploitation and in response to changing climatic conditions, it is important to identify criteria associated with future suitable habitat (Brown & Yoder, 2015; Morelli et al., 2019; Tinsman, 2020). For example, logistic regression models implemented for *Microcebus griseorufus* revealed that optimal habitat required a specific threshold for tree density and mean tree diameter (Steffens et al., 2017). Related to this requirement is the need for tree hollows that provide shelter for nocturnal lemurs, and reduced availability of tree hollows may lead to declines in lemurs, including sportive lemurs. In one extensive survey, *Strychnos madagascariensis* (Loganiaceae), the spineless monkey orange tree, was found to be a principal sleeping site of *Lepilemur ankaranensis* and *L. milanoii*, accounting for 32.5% of 458 recorded sleeping sites (Salmona et al., 2015). Finally, for all nocturnal lemurs, improved understanding of seasonal reproductive patterns (Schmelting et al., 2000) and the special requirements of sleeping ecology (Hending et al., 2017) are needed to deliver a better plan for long-term habitat management including reforestation plans.

Many of the newly discovered nocturnal species are especially vulnerable because they occupy severely restricted ranges of forest where destruction spells extinction for rare species (Blanco et al., 2020; Tables 2.1–2.3). Density estimates may be deceptive—e.g., higher habitat fragmentation in western dry forests may partially explain the higher densities of mouse lemurs in western versus eastern humid forests (Setash et al., 2017). Overall, nocturnal lemurs highlight the need for immediate protective conservation measures. To illustrate, *Microcebus berthae*, the smallest species in the Order Primates, requires nondegraded habitat in the rainy season, such as thick vegetation for nesting. During the dry season, however, their need to feed on homopteran secretions pushes them into degraded areas near human habitation which they normally avoid (Schäffler & Kappeler, 2014). *M. berthae* is an example of how a geographically restricted species responds to seasonal and habitat variation and must also somehow adjust to anthropogenic pressures (Schäffler & Kappeler, 2014). Thus, preserving exceptional biodiversity should include not only protection of habitat fragments but also of forest corridors to prevent population isolation while also permitting animals to move away from human disturbance. Researchers have called attention to the enormous conservation challenge inherent in preserving species with small-scale distributions and have signaled the need for legal protection of additional forest cover (Louis et al., 2008). Yet, to understand the minimal habitat needs of so many species requires an intensity of research that has not yet been reached.

In the past, regional taboos to harming lemurs were a critical element to their survival. One prominent example is the aye-aye, a species considered taboo to harm by the cultural traditions of some villagers. However, other villagers respond to the plantation raiding on the part of aye-ayes by hunting and killing them with fishing spears and dogs (Andriamasimanana, 1994). Regrettably, although the aye-aye is now known to be more widespread than was previously believed, actual densities may be low and the habitat that aye-ayes need for continued survival is rapidly disappearing (Green & Sussman, 1990; Farris et al., 2011). The IUCN now classifies the aye-aye in the Endangered designation (Louis et al., 2020).

To date, given the less topographically challenging terrain and drier climatic conditions of western Madagascar compared to eastern Madagascar, most nocturnal species have been more thoroughly studied at research sites in the west. Nevertheless, excellent research has taken place in eastern rainforests, making for interesting ecological comparisons between western and eastern habitats. Research that concentrates on comparing the lifestyles of sympatric nocturnal species is particularly needed in order to understand how species partition resources for optimal survival. Among all biologists, the common plea is for quick conservation measures to be implemented on the island of Madagascar in order to secure the island's diverse habitats, the genetic diversity of lemurs and other endemic fauna and flora, and what is a significant part of the world's biological heritage. Toward this common purpose, concern and resources should be rallied from all partners, including dedicated conservationists, researchers, and the public, both in Madagascar and internationally (Schwitzer et al., 2014).

SUMMARY

The nocturnal lemuriforms are as highly diverse, taxonomically and ecologically, as any clade of diurnal primates. Within the Order Primates, they exhibit unusual traits, encompassing the ability of some species to produce litters and to hibernate for lengthy periods of time. Although much remains to be learned about their ecology and social organization—particularly with the ongoing discovery of new species—results from research conducted to date provide solid comparative background on behavioral diversity. Combining traditional methods with creative use of modern technologies (e.g., Blanco et al., 2020) and intensified efforts of researchers willing to work at night, sometimes under difficult conditions of climate and terrain, has resulted in many exciting discoveries.

That most nocturnal lemurs are omnivorous to one degree or another may come as no surprise, given that they are primates. They consume fruits for carbohydrates that provide the sugars necessary for energy, and for protein they hunt for insects, other arthropods, or even small vertebrates. There is increasing evidence that some species may play particularly important roles in pollination and seed dispersal (e.g., Sussman & Raven, 1978; Atsalis, 2008). Dietary and habitat choices are sometimes directly related to morphological and physiological traits. For example, *Phaner* is specifically adapted for feeding on gums. On the other hand, *Lepilemur* is an atypical example of a small mammal that consumes appreciable quantities of leaves. As in diurnal primates, it is now clear that several nocturnal primates can live in sympatry and that there is a complex realm of nocturnal ecological interspecies networks within which adjustments are made to minimize feeding competition. Niche separation may occur in subtle ways, understood only through detailed long-term studies focusing on microhabitat traits. Niche separation may be supported through differences in dietary choices (*Avahi occidentalis* and *Lepilemur edwardsi*: Thalmann, 2001, 2002, 2006; *Microcebus murinus* and *M. berthae*: Dammhahn & Kappeler, 2006, 2008a,b, 2009; *M. murinus* and *M. ravelobensis*: Radespiel et al., 2006), or through the quality of foods eaten (*Lepilemur mustelinus* and *Avahi laniger*: Ganzhorn, 1988, 1993), differences in spacing including vertical strata (*M. berthe* and *M. murinus*: Dammhahn & Kappeler, 2009; *Avahi occidentalis* and *L. edwardsi*: Thalmann, 2001, 2006; *Cheirogaleus medius, C. major*, and *M. murinus*: Lahann, 2008), and even choice of sleeping sites (*M. murinus* and *M. ravelobensis*: Radespiel et al., 2003; Weidt et al., 2004).

As a group, the nocturnal lemuriforms occupy a broad range of habitats. The ecological niches of these small, nocturnal primates seem to be the result of adaptations to the seasonal resource shortages of the unstable island climate of Madagascar, which is characterized by sharply defined seasonality in many regions of the island, rainfall patterns that vary considerably from one year to another, and dramatic—often catastrophic—annual cyclone activity (Wright, 1999). Hibernation, limited to only a few of the nocturnal Lemuriformes, is undoubtedly an adaptation to the volatile climatic conditions. As such, it seems likely that mouse lemurs may well have evolved under conditions of unpredictability and high seasonality, conditions that required strategies for energy conservation (Génin, 2008).

Although nocturnal lemurs usually travel and forage alone at night, they are not solitary but, rather, display diversity in behavior and social organization. Thought to live in dispersed social networks, their social organizations are far from uniform. For instance, whereas matrilines form the core of many of the species studied, when compared to other primate groups, pair-living is relatively frequent in this group (occurring in *Phaner, Cheirogaleus*, and *Lepilemur*). Despite the relative larger body sizes, these taxa are also associated with extremely small home ranges compared to the exceptionally large home ranges of the much smaller *Allocebus trichotis* and *Microcebus berthae*. Finally, sleeping groups constitute an important component of social life in nocturnal lemuriforms. Research into species-specific patterns of sleeping associations can lead to a deeper comprehension of how individuals space themselves and communicate with each other.

Current socioecological theory assumes that the distribution of females is related to the distribution of major resources used by a species, whereas males distribute themselves according to the locations of females (Trivers, 1972; Dammhahn & Kappeler, 2009). Nocturnal prosimians are suitable for testing hypotheses concerning relationships between resource distribution and individual spacing patterns because, even in pair-living species, males and females do not travel socially while foraging. The distribution of males and females must also be related to seasonally occurring male-biased sex ratios and changes in home range size and overlap, the results of males migrating and mating activity, as well as other seasonal behaviors such as fattening and hibernation. The strong seasonal component in the behavior of some species underscores the importance of not drawing conclusions based on data from short-term studies that do not cover more than one season.

Comparing the African galagos and the cheirogaleids of Madagascar, we find that both groups are basically nocturnal, arboreal, fast-moving, quadrupedal, and omnivorous, characteristics proposed to be primitive primate evolutionary characters (Charles-Dominique, 1978; Martin, 1990). They share traits, such as the ability to enter torpor/hibernate, with other small mammals (Atsalis, 2008). Thus, evolutionary biologists can look to the small nocturnal lemurs of Madagascar as possible models for the earliest primates (Sussman & Raven, 1978; Martin, 1990; Sussman, 1999; Rasmussen & Sussman, 2007). We anticipate that understanding their feeding ecology, how they distribute themselves in space, how they communicate in their dispersed social networks, and why, when, and how they use torpor/hibernation to decrease energetic needs will open windows to comprehending the evolutionary biology of the earliest primates.

REFERENCES CITED—CHAPTER 2

Albignac, R. (1976). L'ecologie de *Mungotictis decemlineata* dans le forets decidues de l'ouest de Madagascar. *Terre et Vie, 30,* 347-376.

Albignac, R., Justin, R., & Meier, B. (1991). Study of the first behavior of *Allocebus trichotis* Günther 1875 (hairy-eared dwarf lemurs): Prosimian lemur rediscovered in 1989 in the northeast of Madagascar (Biosphere Reserve of Mananara-Nord). In A. Ehara, T. Kimura, O. Takenaka, & M. Iwamonto (Eds.), *Primatology today: Proceedings of the 13th Congress International Primatological Society* (pp. 85-88). Elsevier Science Publishing.

Amico, G., & Aizen, M. A. (2000). Mistletoe seed dispersal by a marsupial. *Nature, 408,* 929-930.

Ancrenaz, M., Lackman-Ancrenaz, I., & Mundy, N. (1994). Field observations of Aye-ayes (*Daubentonia madagascariensis*) in Madagascar. *Folia Primatologica, 62,* 22-36.

Andrès, M., Gachot-Neveu, H., & Perret, M. (2001). Genetic determination of paternity in captive grey mouse lemurs: Pre-copulatory sexual competition rather than sperm competition in a nocturnal prosimian? *Behaviour, 138(8),* 1047-1063.

Andrès, M., Solignac, M., & Perret, M. (2003). Mating system in mouse lemurs: Theories and facts, using analysis of paternity. *Folia Primatologica, 74,* 355-366.

Andriamasimanana, M. (1994). Ecoethological study of free-ranging aye-ayes (*Daubentonia madagascariensis*) in Madagascar. *Folia Primatologica, 62,* 37-45.

Andrianarivo, A. J. (1981). *Etude comparee de l'organisation sociale chez* Microcebus coquereli. [Unpublished doctoral dissertation, Université de Madagascar].

Andriantompohavana, R., Zaonarivelo, J. R., Engberg, S. E., Randriamampionona, R., McGuire S. M., Shore, G. D., Rakotonomenjanahary, R., Brenneman, R. A., & Louis, E. E., Jr. (2006). Mouse lemurs of northwestern Madagascar with a description of a new species at Lokobe Special Reserve. *Occasional Papers: Museum of Texas Tech University, 259,* 1-24.

Andriatsitohaina, B., Ramsay, M. S., Kiene, F., Lehman, S. M., Rasoloharijaona, S., Rakotondravony, R., & Radespiel, U. (2019). Ecological fragmentation effects in mouse lemurs and small mammals in northwestern Madagascar. *American Journal of Primatology, 82,* e23059.

Atsalis, S. (1998). *Feeding ecology and aspects of life history in* Microcebus rufus. [Unpublished doctoral dissertation, City University of New York].

Atsalis, S. (1999a). Diet of the brown mouse lemur, *Microcebus rufus* (Family Cheirogaleidae), in Ranomafana National Park, Madagascar. *International Journal of Primatology,* 20(2), 193-229.

Atsalis, S. (1999b). Seasonal fluctuations in body fat and activity levels in a rainforest species of mouse lemur, *Microcebus rufus. International Journal of Primatology,* 20(6), 883-909.

Atsalis, S. (2000). Spatial distribution and population composition of the brown mouse lemur (*Microcebus rufus*) in Ranomafana National Park, Madagascar, and its implications for social organization. *International Journal of Primatology,* 51(1), 61-78.

Atsalis, S. (2008). *A natural history of the brown mouse lemur: Primate field studies.* Prentice Hall Publishing.

Atsalis, S., Schmid, J., & Kappeler, P. M. (1996). Metrical comparisons of three species of mouse lemur. *Journal of Human Evolution, 31,* 61-68.

Barre, V., Levec, A., Petter, J. J., & Albignac, R. (1988). Etude du microcèbe par radiotracking dans la Forêt de l'Akarafantsika. In L. Rakotovao, V. Barre, & J. Sayer (Eds.), *L'equilibre des ecosystemes forestièrs à Madagascar: Actes d'un semiaire international* (pp. 61-71). IUCN.

Barrows, E. M. (2001). *Animal behavior desk reference: A dictionary of animal behavior, ecology, and evolution* (2nd ed.). CRC Press.

Bethge, J., Wist, B., Stalenberg, E., & Dausmann, K. (2017). Seasonal adaptations in energy budgeting in the primate *Lepilemur leucopus. Journal of Comparative Physiology B, 187,* 827-834.

Biebouw, K. (2006). Pilot study on the conservation status of the hairy-eared dwarf lemur (*Allocebus trichotis*) in eastern Madagascar. *Primate Eye, 90,* 22.

Biebouw, K. (2009). Home range size and use in *Allocebus trichotis* in Analamazaotra Special Reserve, Central Eastern Madagascar. *International Journal of Primatology, 30,* 367-386.

Biebouw, K., Bearder, S., & Nekaris, A. (2009). Tree hole utilization by the hairy-eared dwarf lemur (*Allocebus trichotis*) in Analamazaotra Special Reserve, Central Eastern Madagascar. *Folia Primatologica, 80,* 89-103.

Birkinshaw, C. R., & Colquhoun, I. C. (1998). Pollination of *Ravenala madagascariensis* and *Parkia madagascariensis* by *Eulemur macaco* in Madagascar. *Folia Primatologica, 69,* 252-259.

Blanchard, M. L., Furnell, S., Sellers, W. I., & Crompton, R. H. (2015). Locomotor flexibility in *Lepilemur:* Explained by habitat and biomechanics. *American Journal of Physical Anthropology, 156,* 58-66.

Blanco, M. B. (2008.) Reproductive schedules of female *Microcebus rufus* at Ranomafana National Park, Madagascar. *International Journal of Primatology, 29,* 323-338.

Blanco, M. B. (2010). *Reproductive biology of mouse and dwarf lemurs of eastern Madagascar, with an emphasis on brown mouse lemurs* (Microcebus rufus) *at Ranomafana National Park, a southeastern rainforest.* [Doctoral dissertation, University of Massachusetts].

Blanco, M. B. (2011). Timely estrus in wild brown mouse lemur females at Ranomafana National Park, southeastern Madagascar. *American Journal of Physical Anthropology, 145,* 311-317.

Blanco, M. B., Dausmann, K. H., Faherty, S. L., & Yoder, A. D. (2018). Tropical heterothermy is "cool": The expression of daily torpor and hibernation in primates. *Evolutionary Anthropology, 27,* 147-161.

Blanco, M. B., Dausmann, K. H., Ranaivoarisoa, J. F., & Yoder, A. D. (2013). Underground hibernation in a primate. *Scientific Reports, 3,* 1768.

Blanco, M. B, Godfrey L. R., Rakotondratsima, M., Rahalinarivo, V., Samonds, K. E., Raharison, J.-L., & Irwin, M. T. (2009). Discovery of sympatric dwarf lemur species in the high-altitude rain forest of Tsinjoarivo, eastern Madagascar: Implications for biogeography and conservation. *Folia Primatologica, 80,* 1-17.

Blanco, M. B., Greene, L. K., Rasambainarivo, F., Toomey, E., Williams, R. C., Andrianandrasana, L., Larsen, P. A., & Yoder, A. D. (2020). Next-generation technologies applied to age-old challenges in Madagascar. *Conservation Genetics,* 21(5), 785-793.

Blanco, M. B., & Rahalinarivo, V. (2010). First direct evidence of hibernation in an eastern dwarf lemur species (*Cheirogaleus crossleyi*) from the high-altitude forest of Tsinjoarivo, central-eastern Madagascar. *Naturwissenschaften,* 97(10), 945-950.

Borgerson, C., McKean, M. A., Sutherland, M. R., & Godfrey, L. R. (2016). Who hunts lemurs and why they hunt them. *Biological Conservation, 197,* 124-130.

Braune, P., Polenz, S., Zietemann, V., & Zimmermann, E. (2001). Species-specific signaling in two sympatrically living nocturnal primates, the grey and the golden-brown mouse lemurs (*Microcebus murinus* and *Microcebus ravelobensis*), in northwestern Madagascar. *Advances in Ethology, 36,* 126.

Braune, P., Schmid, S., & Zimmermann, E. (2005). Spacing and group coordination in a nocturnal primate, the golden-brown mouse lemur (*Microcebus ravelobensis*): The role of olfactory and acoustic signals. *Behavioral Ecology and Sociobiology,* 58 (6), 587-596.

Brown, J. L., & Yoder, A. D. (2015). Shifting ranges and conservation challenges for lemurs in the face of climate change. *Ecology and Evolution,* 5(6), 1131-1142.

Buesching, C. D., Heistermann, M., Hodges, J. K., & Zimmermann, E. (1998). Multimodal oestrus advertisement in a small nocturnal prosimian, *Microcebus murinus. Folia Primatologica,* 69(1), 295-308.

Campera, M., Balestri, M., Chimienti, M., Nijman, V., Nekaris, K. A. I., & Donati, G. (2019). Temporal niche separation between the two ecologically similar nocturnal primates *Avahi meridionalis* and *Lepilemur fleuretae. Behavioral Ecology and Sociobiology,* 73, 55.

Cartmill, M. (1974). *Daubentonia, Dactylopsila,* woodpeckers and klinorhynchy. In R. D. Martin, G. A. Doyle, & A. C. Walker (Eds.), *Prosimian biology* (pp. 655-670). University of Pittsburgh Press.

Charles-Dominique, P. (1978). Solitary and gregarious prosimians: Evolution of social behavior in primates. In D. Chivers, & K. Joysey (Eds.), *Recent advances in primatology* (pp. 139-149). Academic Press.

Charles-Dominique, P., & Hladik, M. (1971). Le lépilemur du sud de Madagascar: Écologie, alimentation, et vie sociale. *Terre et Vie,* 25, 3-66.

Charles-Dominique, P., & Petter, J. J. (1980). Ecology and social life of *Phaner furcifer.* In H. Cooper, A. Hladik, C. Hladik, E. Pages, G. Pariente, A. Petter-Rousseaux, J. Petter, A. Schilling, & P. Charles-Dominique (Eds.), *Nocturnal Malagasy primates: Ecology, physiology and behavior* (pp. 75-95). Academic Press.

Cherry, J. A., Izard, M. K., & Simons, E. L. (1987). Description of ultrasonic vocalizations of the mouse lemur (*Microcebus murinus*) and the fat-tailed dwarf lemur (*Cheirogaleus medius*). *American Journal of Primatology,* 13(2), 181-185.

Chivers, D., & Hladik, M. (1980). Morphology of the gastrointestinal tract in primates: Comparisons with other mammals in relation to diet. *Journal of Morphology,* 166, 337-386.

Colquhoun, I. C. (1998). The lemur community of Ambato Massif: An example of the species richness of Madagascar's classified forests. *Lemur News,* 3, 11-14.

Corbin, G. D., & Schmid, J. (1995). Insect secretions determine habitat use patterns by a female lesser mouse lemur (*Microcebus murinus*). *American Journal of Primatology,* 37, 317-324.

Dammhahn, M., & Kappeler, P. M. (2003). The social system of the world's smallest primate, the pygmy mouse lemur (*Microcebus berthae*). *Folia Primatologica,* 74(4), 188.

Dammhahn, M., & Kappeler, P. M. (2005). Social system of *Microcebus berthae,* the world's smallest primate. *International Journal of Primatology,* 26(2), 407-435.

Dammhahn M., & Kappeler P. M. (2006). Feeding ecology and activity patterns of female *Microcebus berthae* and sympatric *M. murinus* (Cheirogaleidae) [abstract]. *International Journal of Primatology,* 27(Suppl. 1), 44.

Dammhahn, M., & Kappeler, P. M. (2008a). Small-scale coexistence of two mouse lemur species (*Microcebus berthae* and *M. murinus*) within a homogeneous competitive environment. *Oecologia,* 157, 473-483.

Dammhahn, M., & Kappeler, P. M. (2008b). Comparative feeding ecology of sympatric *Microcebus berthae* and *M. murinus. International Journal of Primatology,* 29(6), 1567-1589.

Dammhahn, M., & Kappeler, P. M. (2009). Females go where the food is: Does the socio-ecological model explain variation in social organization of solitary foragers? *Behavioral Ecology and Sociobiology,* 63(6), 939-952.

Dausmann, K. H., Ganzhorn, J. U., & Heldmaier, G. (2000). Body temperature and metabolic rate of a hibernating primate in Madagascar: Preliminary results from a field study. In G. Heldmaier, & M. Klingenspor (Eds.), *Life in the cold: Eleventh International Hibernation Symposium* (pp. 41-47). Springer.

Dausmann, K. H., Glos, J., Ganzhorn, J. U., & Heldmaier, G. (2005). Hibernation in the tropics: lessons from a primate. *Journal of Comparative Physiology B,* 175(3), 147-155.

Dausmann, K. H., & Warnecke, L. (2016). Primate torpor expression: Ghost of the climatic past. *Physiology,* 31, 398-408.

Deppe, A. M., Randriamiarisoa, M., Kasprak, A. H., & Wright, P. C. (2008). Predation on the brown mouse lemur (*Microcebus rufus*) by a diurnal carnivore, the ring-tailed mongoose (*Galidia elegans*). *Lemur News,* 13, 19.

Dew, J. L., & Wright, P. (1998). Frugivory and seed dispersal by four species of primates in Madagascar's eastern rain forest. *Biotropica,* 30(3), 425-437.

Dollar, L., Ganzhorn, J. U., & Goodman, S. M. (2007). Primates and other prey in the seasonally variable diet of *Cryptoprocta ferox* in the dry deciduous forests of western Madagascar. In S. L. Gursky, & K. A. I. Nekaris (Eds.), *Primate anti-predator strategies; Developments in primatology—Progress and prospects* (pp. 63-76). Springer.

Dröscher, I., Rothman, J. M., Ganzhorn, J. U., & Kappeler, P. M. (2016). Nutritional consequences of folivory in a small-bodied lemur (*Lepilemur leucopus*): Effects of season and reproduction on nutrient balancing. *American Journal of Physical Anthropology,* 160(2), 197-207.

Eberle, M., & Kappeler, P. M. (2003). Cooperative breeding in grey mouse lemurs (*Microcebus murinus*). *Folia Primatologica,* 74(5-6), 367.

Eberle, M., & Kappeler, P. M. (2004). Sex in the dark: Determinants and consequences of mixed male mating tactics in *Microcebus murinus*, a small solitary nocturnal primate. *Behavioral Ecology and Sociobiology, 57*(1), 77-90.

Eberle, M., & Kappeler, P. M. (2006). Family insurance: Kin selection and cooperative breeding in a solitary primate (*Microcebus murinus*). *Behavioral Ecology and Sociobiology, 60*(4), 582-588.

Ehlers, J., Poppert, S., Ratovonamanaa, R. Y., Ganzhorn, J. U., Tappe, D., & Krüger, A. (2019). Ectoparasites of endemic and domestic animals in southwest Madagascar. *Acta Tropica, 196*, 83-92.

Erickson, C. J. (1991). Percussive foraging in the aye-aye (*Daubentonia madagascariensis*). *Animal Behaviour, 41*, 793-801.

Erickson, C. J., (1994). Tap-scanning and extractive foraging in aye-ayes, *Daubentonia madagascariensis*. *Folia Primatologica, 62*, 125-135.

Erickson, C. J., (1998). Cues for prey location by aye-ayes (*Daubentonia madagascariensis*). *Folia Primatologica, 69*(1), 35-40.

Farris, Z. J., Morelli, T. L., Sefczek, T., & Wright, P. C. (2011). Comparing aye-aye (*Daubentonia madagascariensis*) presence and distribution between degraded and non-degraded forest within Ranomafana National Park, Madagascar. *Folia Primatologica, 82*(2), 94-106.

Fietz, J. (1998a). Body mass in wild *Microcebus murinus* over the dry season. *Folia Primatologica, 69*, 183-190.

Fietz, J. (1998b). Parental care and monogamy in a nocturnal lemur (*Cheirogaleus medius*). *Folia Primatologica, 69*, 210.

Fietz, J. (1999a). Monogamy as a rule rather than exception in nocturnal lemurs: The case of the fat-tailed dwarf lemur, *Cheirogaleus medius*. *Ethology, 105*(3), 255-272.

Fietz, J. (1999b). Demography and floating males in a population of *Cheirogaleus medius*. In B. Rakotosamimanana, H. Rasamimanana, J. Ganzhorn, & S. Goodman (Eds.), *New directions in lemur studies* (pp. 159-172). Plenum Press.

Fietz, J. (1999c), Mating system of *Microcebus murinus*. *American Journal of Primatology, 48*, 127-133.

Fietz, J. (2003a). Pair living and mating strategies in the fat-tailed dwarf lemur (*Cheirogaleus medius*). In U. H. Reichard, & C. Boesch (Eds.), *Monogamy: Mating strategies and partnerships in birds, humans and other mammals* (pp. 214-231). Cambridge University Press.

Fietz, J. (2003b). Primates: *Cheirogaleus*, Dwarf lemurs or fat-tailed lemurs. In S. M. Goodman, & J. P. Benstead (Eds.), *The natural history of Madagascar* (pp. 1307-1309). University of Chicago Press.

Fietz, J., & Dausmann, K. H. (2003). Costs and potential benefits of parental care in the nocturnal fat-tailed dwarf lemur (*Cheirogaleus medius*). *Folia Primatologica, 74*(5-6), 246-258.

Fietz, J., & Ganzhorn, J. U. (1999). Feeding ecology of the hibernating primate *Cheirogaleus medius*: How does it get so fat? *Oecologia, 121*(2), 157-164.

Fietz, J., Zischler, H., Schwiegk, C., Tomiuk, J., Dausmann, K. H., & Ganzhorn, J. U. (2000). High rates of extra-pair young in the pair-living fat-tailed dwarf lemur, *Cheirogaleus medius*. *Behavioral Ecology and Sociobiology, 49*(1), 8-17.

Fleagle, J. G. (1999). *Primate adaptation and evolution* (2nd ed.). Academic Press.

Foerg, R. (1982). Reproduction in *Cheirogaleus medius*. *Folia Primatologica, 39*, 49-62.

Foerg, R., & Hoffmann, R. (1982). Seasonal and daily activity changes in captive *Cheirogaleus medius*. *Folia Primatologica, 38*, 259-268.

Frasier, C. L., Lei, R., McLain, A. T., Taylor, J. M., Bailey, C. A., Ginter, A. L., Nash, S. D., Randriamampionona, R., Groves, C. P., Mittermeier, R. A., & Louis, E. E., Jr. (2016). A new species of dwarf lemur (Cheirogaleidae: *Cheirogaleus medius* group) from the Ankarana and Andrafiamena-Andavakoera Massifs, Madagascar. *Primate Conservation, 30*, 59-72.

Ganzhorn, J. U. (1988). Food partitioning among Malagasy primates. *Oecologia, 75*, 436-450.

Ganzhorn, J. U. (1989). Niche separation of seven lemur species in the eastern rainforest of Madagascar. *Oecologia, 79*, 279-286.

Ganzhorn, J. U. (1993). Flexibility and constraints of *Lepilemur* ecology. In P. Kappeler, & J. Ganzhorn (Eds.), *Lemur social systems and their ecological basis* (pp. 153-165). Plenum Press.

Ganzhorn, J. U., & Kappeler, P. M. (1996). Lemurs of the Kirindy Forest. *Primate Report, 46* (1), 257-274.

Génin, F. (2000). Food restriction enhances deep torpor bouts in the grey mouse lemur. *Folia Primatologica, 71*(4), 258-259.

Génin, F. (2001). Gumnivory in mouse lemurs during the dry season in Madagascar. *Folia Primatologica, 72*(3), 119-120.

Génin, F. (2008). Life in unpredictable environments: First investigation of the natural history of *Microcebus griseorufus*. *International Journal of Primatology, 29*(2), 289-302.

Génin, F. (2010). Who sleeps with whom? Sleeping association and socio-territoriality in *Microcebus griseorufus*. *Journal of Mammalogy, 91*(4), 942-951.

Génin, F., & Perret, M. (2003). Daily hypothermia in captive grey mouse lemurs (*Microcebus murinus*): Effects of photoperiod and food restriction. *Comparative Biochemistry and Physiology B, 136*(1), 71-81.

Gilbert, M., & Tingay, R. E. (2001). Predation of a fat-tailed dwarf lemur *Cheirogaleus medius* by a Madagascar Harrier-Hawk (*Polyboroides radiatus*): An incidental observation. *IUCN Primate Specialist Newsletter, 6*, 6.

Gillespie, T. R., Nunn, C. L., & Leendertz, F. H. (2008). Integrative approaches to the study of primate infectious disease: Implications for biodiversity conservation and global health. *American Journal of Physical Anthropology, 47*, 53-69.

Glatston, A. R. H. (1979). *Reproduction and behavior of the lesser mouse lemur* (Microcebus murinus, *Miller 1777*). [Doctoral dissertation, University of London].

Godschalk, S. K. B. (1983). Mistletoe dispersal by birds in South Africa. In M. D. Calder & P. Bernhardt (Eds.), *The biology of mistletoes* (pp. 117-128). Academic Press.

Godschalk, S. K. B. (1985). Feeding behavior of avian dispersers of mistletoe fruit in the Loskop Dam Nature Reserve, South Africa. *South African Journal of Zoology, 20*(3), 136-146.

Goix, E. (1993). L'utilisation de la main chez le aye-aye en captivité (*Daubentonia madagascariensis*) (Prosimiens, Daubentoniidés). *Mammalia*, 57, 171-188.

Goodman, S. M. (1994). The enigma of antipredator behavior in lemurs: Evidence of a large extinct eagle on Madagascar. *International Journal of Primatology*, 15, 129-134.

Goodman, S. M., Creighton, G. K., & Raxworthy, C. (1991). The food habits of the Madagascar long-eared owl *Asio madagascariensis* in southeastern Madagascar. *Bonn. zool. Beitr.* 42(1), 21-26.

Goodman, S. M., Langrand, O., & Rasolonandrasana, B. P. N. (1997). The food habits of *Cryptoprocta ferox* in the high mountain zone of the Andringitra Massif, Madagascar (Carnivora, Viverridae). *Mammalia*, 61, 185-192.

Goodman, S. M., Langrand, O., & Raxworthy, C. J. (1993b). Food habits of the Madagascar long-eared owl *Asio madagascariensis* in two habitats in southern Madagascar. *Ostrich*, 64, 79-85.

Goodman, S. M., Langrand, O., & Raxworthy, C. J. (1993c). Food habits of the barn owl *Tyto alba* at three sites on Madagascar. *Ostrich*, 64, 160-171.

Goodman, S. M., O'Conner, S., & Langrand, O. (1993a). A review of predation on lemurs: Implications for the evolution of social behavior in small, nocturnal primates. In P. M. Kappeler, & J. U. Ganzhorn (Eds.), *Lemur social systems and their ecological basis* (pp. 51-66). Plenum Press.

Goodman, S. M., & Raselimanana, A. P. (2002). The occurrence of *Allocebus trichotis* in the Parc National de Marojejy. *Lemur News*, 7, 21-22.

Goodman, S. M., & Thorstrom, R. (1998). The diet of the Madagascar red owl (*Tyto soumagnei*) on the Masoala Peninsula, Madagascar. *Wilson Bulletin*, 110(3), 417-421.

Gray, J. E. (1870). *Catalogue of monkeys, lemurs, and fruit-eating bats in British Museum*. Natural History Museum Publications.

Green, G. M., & Sussman, R. W. (1990). Deforestation history of the eastern rain forests of Madagascar from satellite images. *Science*, 248, 212-215.

Groves, C. P. (2001). *Primate taxonomy*. Smithsonian Institution Press.

Groves, C. P., & Tattersall, I. (1991). Geographical variation in the fork-marked lemur, *Phaner furcifer* (Primates, Cheirogaleidae). *Folia Primatologica*, 56(1), 39-49.

Hafen, T., Heveu, H., Rumpler, Y., Wilden, I, & Zimmermann, E. (1998). Acoustically dimorphic advertisement calls separate morphologically and genetically homogenous populations of the grey mouse lemur (*Microcebus murinus*). *Folia Primatologica*, 69(1), 342-356.

Hämäläinen, A., Dammhahn, M., Aujard, F., Eberle, M., Hardy, I., Kappeler, P. M., Perret, M., Schliehe-Diecks, S., & Kraus, C. (2014). Senescence or selective disappearance? Age trajectories of body mass in wild and captive populations of a small-bodied primate. *Proceedings of the Royal Society, B*, 281(1791), 20140830.

Harcourt, C., & Thornback, J. (1990). *Lemurs of Madagascar and the Comoros: The IUCN Red Data Book*. IUCN.

Haring, D. M., Hess, W. R., Coffman, B. S., Simons, E. L., & Owens, T. M. (1994). Natural history and captive management of the aye-aye (*Daubentonia madagascariensis*) at the Duke University Primate Center, Durham. *International Zoo Yearbook*, 33, 201-219.

Hawkins, A. F. A., & Goodman, S. M. (2003). Introduction to the birds. In S. M. Goodman, & J. P. Benstead (Eds.), *The natural history of Madagascar* (pp. 1019-1044). University of Chicago Press.

Hawkins, C. E. (2003). *Cryptoprocta ferox*, fossa. In S. M. Goodman, & J. P. Benstead (Eds.), *The natural history of Madagascar* (pp. 1361-1363). University of Chicago Press.

Hawkins, C. E., & Racey, P. A. (2008). Food habits of an endangered carnivore, *Cryptoprocta ferox*, in the dry deciduous forest of western Madagascar. *Journal of Mammalogy*, 89(1), 64-74.

Heckman, K. L., Rasoazanabary, E., Machlin, E., Godfrey, L. R., & Yoder, A. D. (2006). Incongruence between genetic and morphological diversity in *Microcebus griseorufus* of Beza Mahafaly. *BMC Evolutionary Biology*, 6 (1), 98.

Hemingway, C. A. (1995*). Feeding and reproductive strategies of the Milne-Edwards' sifaka*, Propithecus diadema edwardsi. [Unpublished doctoral dissertation, Duke University].

Hending, D., Andrianiaina, A., Rakotomalala, Z., & Cotton, S. (2018). The use of vanilla plantations by lemurs: Encouraging findings for both lemur conservation and sustainable agroforestry in the Sava Region, northeast Madagascar. *International Journal of Primatology*, 39, 141-153.

Hending, D., McCabe, G., & Holderied, M. (2017). Sleeping and ranging behavior of the Sambirano mouse lemur, *Microcebus sambiranensis*. *International Journal of Primatology*, 38, 1072-1089.

Hending, D., Sgarlata, G. M., Le Pors, B., Rasolondraibe, E., Jan, F., Rakotonanahary, A. N., Ralantoharijaona, T. N., Debulois, S., Andrianiaina, A., Cotton, S., Rasoloharijaona, S., Zaonarivelo, J. R., Andriaholinirina, N. V., Chikhi, L., & Salmona, J. (2020). Distribution and conservation status of the endangered Montagne d'Ambre fork-marked lemur (*Phaner electromontis*). *Journal of Mammalogy*, 101(4), 1049-1060.

Hilgartner, R., Zinner, D., & Kappeler, P. M. (2008). Life history traits and parental care in *Lepilemur ruficaudatus*. *American Journal of Primatology*, 69, 1-15.

Hladik, A. (1980). The dry forest of the west coast of Madagascar: Climate, phenology, and food availability. In P. Charles-Dominique, H. Cooper, A. Hladik, C. Hladik, E. Pages, G. Pariente, A. Petter-Rousseaux, J. Petter, & A. Schilling (Eds.), *Nocturnal Malagasy primates: Ecology, physiology and behavior* (pp. 3-40). Academic Press.

Hladik, P. (1979). Diet and ecology of prosimians. In G. Doyle, & R. Martin (Eds.), *The study of prosimian behavior* (pp. 307-339). Academic Press.

Hladik, P., Charles-Dominique, P., & Petter, J. J. (1980). Feeding strategies of five nocturnal prosimians in the dry forest of the west coast of Madagascar. In P. Charles-Dominique, H. Cooper, A. Hladik, C. Hladik, E. Pages,

G. Pariente, A. Petter-Rousseaux, J. Petter, & A. Schilling (Eds.), *Nocturnal Malagasy primates: Ecology, physiology and behavior* (pp. 41-73). Academic Press.

Howe, H. F., & Estabrook, G. F. (1977). On intra-specific competition for avian dispersers in tropical trees. *American Naturalist, 111,* 817-832.

Irwin, M. T., Samonds, K. E., Raharison, J. L., & Wright, P. C. (2004). Lemur latrines: Observations of latrine behavior in wild primates and possible ecological significance. *Journal of Mammalogy, 85*(3), 420-427.

IUCN. (2020). *The International Union for the Conservation of Nature Red List of Threatened Species.* Version 2020-2. www.iucnredlist.org.

Iwano, T., & Iwakawa, C. (1988). Feeding behaviour of the aye-aye (*Daubentonia madagascariensis*) on nuts of ramy (*Canarium madagascariensis*). *Folia Primatologica, 50,* 136-142.

Jenkins, P. D. (1987). *Catalogue of primates in the British Museum (Natural History) and elsewhere in the British Isles. Part 14: Suborder Strepsirrhini, including the subfossil Madagascan lemurs and family Tarsiidae.* Natural History Museum Publications.

Jenkins, R. K. B., Keane, A., Rakotoarivelo, A. R., Rakotomboavonjy, V., Randrianandrianina, F. H., Razafimanahaka, H. J., Ralaiarimalala, S. R., & Jones, J. P. G. (2011). Analysis of patterns of bushmeat consumption reveals extensive exploitation of protected species in eastern Madagascar. *PLoS ONE, 6,* e27570.

Kappeler, P. M. (1997). Intrasexual selection *Mirza coquereli*: Evidence for scramble competition polygyny in a solitary primate. *Behavioral Ecology and Sociobiology, 45,* 115-127.

Kappeler, P. M. (1998). Nests, tree holes and the evolution of primate life histories. *American Journal of Primatology, 46,* 7-33.

Kappeler, P. M. (2000). Ecologie des microcèbes. *Primatologie, 3,* 145-171.

Kappeler, P. M. (2003). *Mirza coquereli,* Coquerel's dwarf lemur. In S. Goodman, & J. Benstead (Eds.), *The natural history of Madagascar* (pp. 1316-1318). University of Chicago Press.

Kappeler, P. M., & Rasoloarison, R. M. (2003). *Microcebus,* mouse lemurs, *Tsidy.* In S. Goodman, & J. Benstead (Eds.), *The natural history of Madagascar* (pp. 1310-1315). University of Chicago Press.

Kappeler, P. M., Rasoloarison, R. M., Razafimanantsoa, L., Walter, L., & Roos, C. (2005). Morphology, behaviour and molecular evolution of giant mouse lemurs (*Mirza* spp.) Gray, 1870, with description of a new species. *Primate Report, 71,* 3-26.

Kappeler, P. M., Wimmer, B., Zinner, D., & Tautz, D. (2002). The hidden matrilineal structure of a solitary lemur: Implications for primate social evolution. *Proceedings of the Royal Society, B,* 269(1502), 1755-1763.

Karanewsky C. (2013). *The ecology of hibernation in a rain forest primate,* Microcebus rufus. [Unpublished doctoral dissertation, Stony Brook University].

Karpanty S. M., & Goodman S. M. (1999). Diet of the Madagascar harrier-hawk, *Polyboroides radiatus,* in southeastern Madagascar. *Journal of Raptor Research, 33*(4), 313-316.

Karpanty, S. M., & Wright P. C. (2007). Predation on lemurs in the rainforest of Madagascar by multiple predator species: Observations and experiments. In S. L. Gursky, & A. Nekaris (Eds.), *Primate anti-predator strategies developments in primatology: Progress and prospects* (pp. 77-99). Springer.

Kress, W. J. (1993). Coevolution of plants and animals: Pollination of flowers by primates in Madagascar. *Current Science, 65,* 253-257.

Kress, W. J., Schatz, G. E., Andrianifahanana, M., & Morland, H. S. (1994). Pollination of *Ravenala madagascariensis* (Strelitziaceae) by lemurs in Madagascar: Evidence for an archaic coevolutionary system? *American Journal of Botany, 81,* 542-551.

Lahann, P. (2007). Biology of *Cheirogaleus major* in a littoral rainforest in southeast Madagascar. *International Journal of Primatology, 28,* 895-905.

Lahann, P. (2008). Habitat utilization of three sympatric cheirogaleid lemur species in a littoral rainforest of southeastern Madagascar. *International Journal of Primatology, 29,* 117-134.

Lahann, P., Schmid, J., & Ganzhorn, J. U. (2006). Geographic variation in populations of *Microcebus murinus* in Madagascar: Resource seasonality or Bergmann's Rule? *International Journal of Primatology, 27*(4), 983-999.

Langrand, O. (1990). *Guide to the birds of Madagascar.* Yale University Press.

Lehman, S. M., Rajaonson, A., & Day, S. (2006). Edge effects and their influence on lemur density and distribution in southeast Madagascar. *American Journal of Physical Anthropology, 129*(20), 232-241.

Louis, E. E., Jr., Coles, M. S., Andriantompohavana, R., Sommer, J. A., Engberg, S. E., Zaonarivelo, J. R., & Mayor, M. I. (2006). Revision of the mouse lemurs (*Microcebus*) of eastern Madagascar. *International Journal of Primatology, 27,* 347-389.

Louis, E. E., Jr., Engberg, S. E., McGuire, S. M., McCormick, M. J., Randriamampionona, R., Ranaivoarisoa, J. F., Bailey, C. A., Mittermeier, R. A., & Lei, R. (2008). Revision of the mouse lemurs, *Microcebus* (Primates, Lemuriformes), of northern and northwestern Madagascar with descriptions of two new species at Montagne d'Ambre National Park and Antafondro Classified Forest. *Primate Conservation, 23,* 19-38.

Louis, E. E., Jr., Sefczek, T. M., Randimbiharinirina, D. R., Raharivololona, B., Rakotondrazandry, J. N., Manjary, D., Aylward, M., & Ravelomandrato, F. (2020). *Daubentonia madagascariensis. IUCN Red List of Threatened Species* 2020: e.T6302A115560793. https://dx.doi.org/10.2305/IUCN.UK.2020-2.RLTS.T6302A115560793.en.

Lutermann, H., & Zimmermann, E. (2001). Nesting and resting in a small nocturnal primate: Behavioural ecology of inactivity from a female perspective. *Folia Primatologica, 72* (3), 172.

Mandl, I., Holderied, M., & Schwitzer, C. (2018). The effects of climate seasonality on behavior and sleeping site choice in Sahamalaza sportive lemurs, *Lepilemur sahamalaza*. *International Journal of Primatology, 39*, 1039-1067.

Mandl, I., Holderied, M., & Schwitzer, C. (2019). Spatiotemporal distribution of individuals as an indicator for the social system of *Lepilemur sahamalaza*. *American Journal of Primatology, 81*, e22984.

Mandl, I., Rabemanan-jara, N. R., Rakotomalala, A. N. A., Sorlin, M., Holderied, M., & Schwitzer, C. (2017). A case of mobbing observed in two species of nocturnal lemur, *Mirza zaza* and *Lepilemur sahamalazensis* in north-west Madagascar. *Lemur News, 20*, 6-7.

Martin, R. D. (1972). A preliminary field-study of the lesser mouse lemur (*Microcebus murinus* J. F. Miller 1777). *Journal of Comparative Ethology, 9*, 43-89.

Martin, R. D. (1973). A review of the behavior and ecology of the lesser mouse lemur (*Microcebus murinus* J. F. Miller 1777). In R. Michael, & J. Crook (Eds.), *Comparative ecology and behavior of primates* (pp. 1-68). Academic Press.

Martin, R. D. (1990). *Primate origins and evolution: A phylogenetic reconstruction*. Princeton University Press.

McKechnie, A. E., & Mzilikazi, N. (2011). Heterothermy in afrotropical mammals and birds: A review. *Integrative and Comparative Biology, 51*(3), 349-363.

McKey, D. (1975). The ecology of coevolved seed dispersal systems. In L. Gilbert, & P. Raven (Eds.), *Coevolution of animals and plants* (pp. 151-191). University of Texas Press.

Meier, B., & Albignac, R. (1991). Rediscovery of *Allocebus trichotis* Gunther 1875 (Primates) in northeast Madagascar. *Folia Primatologica, 56*, 57-63.

Miller, M., Ralantoharijaona, T., Misandeau, C., Volasoa, Z. A., Mills, H., Bencini, R., Chikhi, L., & Salmona, J. (2015). A biological survey of Antsahanadraitry forest (Alan'Antanetivy corridor, Manompana) reveals the presence of the hairy-eared dwarf lemur (*Allocebus trichotis*). *Lemur News, 19*, 4-6.

Miller, R. T., Raharison, J., & Irwin, M. T. (2017). Competition for dead trees between humans and aye-ayes (*Daubentonia madagascariensis*) in central eastern Madagascar. *Primates, 58*, 367-375.

Mittermeier, R. A., Louis, E. E., Jr., Richardson, M. J., Schwitzer, C., Langrand, O., Rylands, A. B. & Hawkins, F., et al. (2010). *Lemurs of Madagascar* (3rd Ed.). Conservation International.

Morelli, T. L., Smith, A. B., Mancini, A. N., Balko, E. A., Borgerson, C., Dolch, R., Farris, Z., Federman, S., Golden, C. D., Holmes, S. M., Irwin, M., Jacobs, R. L., Johnson, S., King, T., Lehman, S. M., Louis, E. E., Jr., Murphy, A., Randriahaingo, H. N. T., Randrianarimanana, H. L. L.,... & Baden, A. L.(2019). The fate of Madagascar's rainforest habitat. *Nature Climate Change, 10*, 89-96.

Müller, A. E. (1998). A preliminary report on the social organization of *Cheirogaleus medius* (Cheirogaleidae; Primates) in northwest Madagascar. *Folia Primatolologica, 69* (Suppl. 1), 160-166.

Müller, A. E. (1999a). *The social organisation of the fat-tailed dwarf lemur, Cheirogaleus medius (Lemuriformes; Primates)*. [Unpublished doctoral dissertation, University of Zurich].

Müller, A. E. (1999b). Aspects of social life in the fat-tailed dwarf lemur (*Cheirogaleus medius*): Inferences from body weights and trapping data. *American Journal of Primatology, 49*(3), 265-280.

Müller, A. E. (1999c). Social organization of the fat-tailed dwarf lemur (*Cheirogaleus medius*) in northwestern Madagascar. In H. Rasamimanana, B. Rakotosamimanana, J. U. Ganzhorn, & S. M. Goodman (Eds.), *New directions in lemur studies* (pp. 139-158). Plenum Press.

Müller, A. E., & Thalmann, U. (2000). Origin and evolution of primate social organization: A reconstruction. *Biological Review of the Cambridge Philosophical Society, 75* (3), 405-435.

Nadkarni, N. M. (1983). *The effects of epiphytes on nutrient cycles within temperate and tropical rainforest tree canopies*. [Unpublished doctoral dissertation, University of Washington].

Napier, J. R. (1966). Stratification and primate ecology. *Journal of Animal Ecology, 25*, 411-412.

Napier, J. R., & Walker, A. C. (1967). Vertical clinging and leaping, a newly recognised category of locomotor behaviour among primates. *Folia Primatologica, 6*, 204-219.

Nash, L. T. (1998). Vertical clingers and sleepers: Seasonal influences on the activities and substrate use of *Lepilemur leucopus* at Beza Mahafaly Special Reserve, Madagascar. *Folia Primatologica, 69*(1), 204-217.

Nilsson, L. A., Rabakonandrianina, E., & Pettersson, B. (1993). Lemur pollination in the Malagasy rainforest liana *Strongylodon craveniae* (Leguminosae). *Evolutionary Trends in Plants, 7*, 49-56.

Olivieri, G., Zimmerman, E., Randrianambinina B., Rasoloharijaona, S., Rakotondravony, D., Guschanski, K., & Radespiel, U. (2007). The ever-increasing diversity in mouse lemurs: Three new species in north and northwestern Madagascar. *Molecular Phylogenetics and Evolution, 43*, 309-327.

Ortmann, S., Heldmaier, G., Schmid, J., & Ganzhorn, J. U. (1997). Spontaneous daily torpor in Malagasy mouse lemurs. *Naturwissenschaften, 84*, 28-32.

Ortmann, S., Schmid, J., Ganzhorn, J. U., & Heldmaier, G. (1996). Body temperature and torpor in a Malagasy small primate, the mouse lemur. In F. Geiser, A. Hulbert, & S. Nicol (Eds.), *Adaptations to the cold: Tenth International Hibernation Symposium* (pp. 55-61). University of New England Press.

Overdorff, D. J. (1991). *Ecological correlates of social structure in two prosimian primates: Eulemur fulvus and Eulemur rubriventer in Madagascar*. [Unpublished doctoral dissertation, Duke University].

Pagès, E. (1978). Home range, behaviour and tactile communication in a nocturnal Malagasy lemur *Microcebus coquereli*. In D. A. Chivers, & K. A. Joysey (Eds.), *Recent advances in primatology* (pp. 171-177). Academic Press.

Pagès, E. (1980). Ethoecology of *Microcebus coquereli* during the dry season. In P. Charles-Dominique, H. Cooper, A. Hladik, C. Hladik, E. Pages, G. Pariente, A. Petter-Rousseaux, J. Petter, & A. Schilling (Eds.), *Nocturnal Malagasy primates: Ecology, physiology and behavior* (pp. 97-116). Academic Press.

Pagès-Feuillade, E. (1988). Modalités de l'occupation de l'espace et relations interindividuelles chez un prosimien nocturne malgache (*Microcebus murinus*). *Folia Primatologica, 50,* 204-220.

Perret, M. (1992). Environmental and social determinants of sexual function in the male lesser mouse lemur (*Microcebus murinus*). *Folia Primatologica, 59,* 1-25.

Perret, M. (1995). Chemocommunication in the reproductive function of mouse lemurs. In L. Alterman, G. A. Doyle, & M. K. Izard (Eds.), *Creatures of the dark: The nocturnal prosimians* (pp. 377-392). Plenum Press.

Petter, J. J. (1977). The aye-aye. In H. Prince Rainier III, & G. Bourne (Eds.), *Primate conservation* (pp. 38-59). Academic Press.

Petter, J. J. (1978). Ecological and physiological adaptations of five sympatric nocturnal lemurs to seasonal variations in food production. In D. Chivers, & K. Joysey (Eds.), *Recent advances in primatology* (pp. 211-223). Academic Press.

Petter, J. J, Albignac, R., & Rumpler, Y. (1977). *Faune de Madagascar, 44: Mammifères Lémuriens (Primates Prosimiens).* Ostrom, CNRS.

Petter J. J., & Petter-Rousseaux, A. (1967). The aye-aye of Madagascar. In S. Altmann (Ed.), *Social communication among primates* (pp. 195-205). University of Chicago Press.

Petter, J. J., & Peyrieras, A. (1970). Nouvelle contribution à l'étude d'un lémurien malgache, le aye-aye (*Daubentonia madagascariensis* E. Geoffroy). *Mammalia, 34,* 167-193.

Petter, J. J., Schilling, A., & Pariente, G. (1975). Observations on the behavior and ecology of *Phaner furcifer*. In I. Tattersall, & R. W. Sussman (Eds.), *Lemur biology* (pp. 209-218). Plenum Press.

Petter-Rousseaux, A. (1980). Seasonal activity rhythms, reproduction, and body weight variations in five sympatric nocturnal prosimians, in simulated light and climatic conditions. In P. Charles-Dominique, H. M. Cooper, A. Hladik, C. M. Hladik, E. Pages, G. F. Pariente, A. Petter-Rousseaux, J. J. Petter, & A. Schilling (Eds.), *Nocturnal Malagasy primates: Ecology, physiology and behavior* (pp. 137-152). Academic Press.

Petter-Rousseaux, A., & Petter, J. J. (1967). Contribution a la systematique des Cheirogaleinae (*lemuriens malgaches*), *Allocebus*, gen. nov., pour *Cheirogaleus trichotis* Günther 1875. *Mammalia, 31,* 574-582.

Porter, L. (1998). Influences on the distribution of *Lepilemur microdon* in Ranomafana National Park, Madagascar. *Folia Primatologica, 69,* 172-176.

Rabeantoandro, Z. S. (1997). Contribution a l'étude du *Mungotictis decemlineata* (Grandidier 1867) de la forest de Kirindy, Morondava. [D. E. A. Memoire, Université d'Antananarivo].

Radespiel, U., (2000). Sociality in the gray mouse lemur (*Microcebus murinus*) in northwestern Madagascar. *American Journal of Primatology, 51*(1), 21-40.

Radespiel, U. (2006). Ecological diversity and seasonal adaptations of mouse lemurs (*Microcebus* spp.). In L. Gould, & M. Sauther (Eds.), *Lemurs: Ecology and adaption* (pp. 211-234). Springer.

Radespiel, U., Cepok, S., Zimmermann E., & Zietemann, V. (1998). Sex-specific usage patterns of sleeping sites in grey mouse lemurs (*Microcebus murinus*) in northwestern Madagascar. *American Journal Primatology, 46,* 77-84.

Radespiel, U., Dal Secco, V., Drögemüller, C., Braune, P., Labes, E., & Zimmermann, E. (2002). Sexual selection, multiple mating and paternity in grey mouse lemurs, *Microcebus murinus. Animal Behavior, 63*(2), 259-268.

Radespiel, U., Ehresmann, P., & Zimmermann, E. (2001b). Contest versus scramble competition for mates: The composition and spatial structure of a population of gray mouse lemurs (*Microcebus murinus*) in north-west Madagascar. *Primates, 42*(3), 207-220.

Radespiel, U., Ehresmann, P., & Zimmermann, E. (2003). Species-specific usage of sleeping sites in two sympatric mouse lemur species (*Microcebus murinus* and *M. ravelobensis*) in northwestern Madagascar. *American Journal of Primatology, 59,* 139-151.

Radespiel, U., Reimann, W., Rahelinirina, M., & Zimmermann, E. (2006). Feeding ecology of sympatric mouse lemur species in northwestern Madagascar. *International Journal of Primatology, 27,* 311-321.

Radespiel, U., Sarikaya, Z., Zimmermann, E., & Bruford, M. W. (2001a). Sociogenetic structure in a free-living nocturnal primate population: Sex-specific differences in the grey mouse lemur (*Microcebus murinus*). *Behavioral Ecology and Sociobiology, 50*(6), 493-502.

Radespiel, U., & Zimmermann, E. (2001a). Female dominance in captive gray mouse lemurs (*Microcebus murinus*). *American Journal of Primatology, 54,* 181-192.

Radespiel, U., & Zimmermann, E. (2001b). Dynamics of estrous synchrony in captive gray mouse lemurs (*Microcebus murinus*). *International Journal of Primatology, 22,* 71-90.

Rakotoarison, N. (1998). Recent discoveries of the hairy-eared dwarf lemur (*Allocebus trichotis*). *Lemur News, 3,* 21.

Rakotoarison, N., Zimmermann, H., & Zimmermann, E. (1997). First discovery of the hairy-eared dwarf lemur (*Allocebus trichotis*) in a highland rain forest of eastern Madagascar. *Folia Primatologica, 68,* 86-94.

Rakotondranary, S. J., & Ganzhorn, J. U. (2012). Habitat separation of sympatric *Microcebus* spp. in the dry spiny forest of south-eastern Madagascar. *Folia Primatologica, 82,* 212-223.

Randrianambinina, B., Mbotizafy, S., Rasoloharijaona, S., Ravoahangimalala, R O., & Zimmermann, E. (2007). Seasonality in reproduction of *Lepilemur edwardsi*. *International Journal of Primatology, 28,* 783-790.

Randrianambinina, B., Raktondravony, D., Radespiel, U., & Zimmermann, E. (2003b). Seasonal changes in general activity, body mass and reproduction of two small nocturnal primates: A comparison of the golden-brown mouse lemur (*Microcebus ravelobensis*) in northwestern Madagascar and the brown mouse lemur (*Microcebus rufus*) in eastern Madagascar. *Primates*, 44, 321-331.

Randrianambinina, B., Rasoloharijaona, S., Rakotosamimanana, B., & Zimmermann, E. (2003a) Inventaires des communautés lémuriennes dans la Réserve Spéciale de Bora au nordouest et la forêt domaniale de Mahilaka-Maromandia au nord de Madagascar. *Lemur News*, 8, 15–18.

Randrianarivo, R. (1979). *Essai d'inventaire des Lemuriens de la future reserve de Beza Mahafaly.* [Memoire de fin d'etude, Université d'Antananarivo].

Rasambinarivo, F. T., Gillespie, T. R., Wright, P. C., Arsenault, J., Villeneuve, A., & Lair, S. (2013). Survey of *Giardia* and *Cryptosporidium* in lemurs from the Ranomafana National Park, Madagascar. *Journal of Wildlife Diseases*, 49(3), 741-743.

Rasamison, A. A. (1997). *Contribution à l'étude biologique, écologique et éthologique de* Cryptoprocta ferox *(Bennett, 1833) dans la forêt de Kirindy à Madagascar.* [D. E. A. memoire, Université d'Antananarivo].

Rasmussen, D. T., & Sussman, R.W. (2007). Parallelisms among primates and possums. In M. J. Ravosa, & M. Dagosto (Eds.), *Primate origins and adaptations* (pp. 775-803). Plenum Press.

Rasoazanabary, E. (1999). Do male mouse lemurs (*Microcebus murinus*) face a trade-off between survival and reproduction during the dry season? *Primate Report*, 54 (1), 27.

Rasoazanabary, E. (2006). Male and female activity patterns in *Microcebus murinus* during the dry season at Kirindy forest, western Madagascar. *International Journal of Primatology*, 27(2), 437-464.

Rasoloarison, R., Rasolonandrasana, B., Ganzhorn, J., & Goodman, S. (1995). Predation on vertebrates in the Kirindy forest, western Madagascar. *Ecotropica*, 1, 59-65.

Rasoloarison, R., Goodman, S., & Ganzhorn, J. (2000). Taxonomic revision of mouse lemurs (*Microcebus*) in the western portions of Madagascar. *International Journal of Primatology*, 21, 963-1019.

Rasoloharijaona, S., Rakotosamimanana, B., Randrianambinina, B., & Zimmermann, E. (2003). Pair-specific usage of sleeping sites and their implications for social organization in a nocturnal Malagasy primate, the Milne-Edwards' sportive lemur (*Lepilemur edwardsi*). *American Journal of Physical Anthropology*, 122, 251-258.

Rasoloharijaona, S., Randrianambinina, B., & Zimmermann, E. (2008). Sleeping site ecology in a rain-forest dwelling nocturnal lemur (*Lepilemur mustelinus*): Implications for sociality and conservation. *American Journal of Physical Anthropology*, 70(3), 247-253.

Rasolonandrasana, B. P. N. (1994). *Contribution à l'étude de l'alimentation de* Cryptoprocta ferox Bennett *(1833) dans son milieu naturel.* [Mémoire de D.E.A., Université d'Antananarivo].

Ratsirarson, J. (1986). *Contribution à l'etude compare du l'eco-ethologie de deux expéces de lemuriens:* Lepilemur mustelinus *(I. GEOFFROY 1850) et* Lepilemur septentrionalis *(Rumpler et Albignac 1975).* [Doctoral dissertation, Université Louis Pasteur].

Ratsirarson, J., & Rumpler, Y. (1988). Contribution à l'étude comparée de l'eco-ethologie de deux espéces de lemuriens, *Lepilemur mustelinus* (I. Geoffroy 1850), *Lepilemur septentrionalis* (Rumpler and Albignac 1975). In L. Rakotovao, V. Barre, & J. Sayer (Eds.), *L'equilibre des ecosystèmes forestiers à Madagascar, actes d'un séminaire international* (pp. 100-102). IUCN.

Reid, N., Stafford Smith, M., & Yan, Z. (1995). Ecology and population biology of mistletoes. In M. Lowman, & N. Nadkarni (Eds.), *Forest canopies* (pp. 285-310). Academic Press.

Reimann, W., Radespiel, U., & Zimmermann, E. (2003). Feeding regimes of two sympatric mouse lemurs in north-western Madagascar (*Microcebus murinus* and *M. ravelobensis*): No clear evidence for niche separation. *Folia Primatologica*, 74, 215.

Reuter, K. E., LaFleur, M., & Clarke, T. A. (2017). Illegal lemur trade grows in Madagascar. *Nature*, 541, 157.

Richard, A. F. (1978). *Behavioral variation: Case study of a Malagasy prosimian.* Bucknell University Press.

Rode, E. J. (2010). *Conservation ecology, morphology and reproduction of the nocturnal northern giant mouse lemur* Mirza zaza *in Sahamalaza National Park, northwestern Madagascar* [Master's thesis, Oxford Brookes University].

Rode, E. J., Nekaris, K. A. I., Markolf, M., Schliehe-Diecks, S., Seiler, M., Radespiel, U., & Schwitzer, C. (2013). Social organisation of the northern giant mouse lemur *Mirza zaza* in Sahamalaza, north western Madagascar, inferred from nest group composition and genetic relatedness. *Contributions to Zoology*, 82(2), 71-83.

Ross, C. (2003). Life history, infant care strategies, and brain size in primates. In P. M. Kappeler, & M. E. Pereira (Eds.), *Primate life histories and socioecology* (pp. 266-284). University of Chicago Press.

Ruf, T., Streicher, U., Stalder, G. L., Nadler, T., & Walzer, C. (2015). Hibernation in the pygmy slow loris (*Nycticebus pygmaeus*): multiday torpor in primates is not restricted to Madagascar. *Scientific Reports*, 5, 17392.

Russell, R. J. (1977). *The behavior, ecology, and environmental physiology of a nocturnal primate,* Lepilemur mustelinus. [Unpublished doctoral dissertation, Duke University].

Salmona, J., Banks, M., Ralantoharijaona, T. N., Rasolondraibe, E., Zaranaina, R., Rakotonanahary, A., Wohlhauser, S., Sewall, B. J., & Chikhi, L. (2015). The value of the spineless monkey orange tree (*Strychnos madagascariensis*) for conservation of northern sportive lemurs (*Lepilemur milanoii* and *L. ankaranensis*). *Madagascar Conservation and Development*, 10(2), 53-59.

Salmona, J., Rasolondraibe, E., Jan, F., Rakotonanahary, A. N., Ralantoharijaona, T., Pors, B. L., Dhurham, S. A. O., Ousseni, A., Aleixo-Pais, I., Marques, A. D. J., Sgarlata, G.

M., Teixeira, H., Gabillaud, V., Miller, A., Ibouroi, M. T., Zaonarivelo, J. R., Andriaholinirina, N. V., & Chikhi, L. (2018). Re-discovering the forgotten *Phaner* population of the small and isolated Analafiana forest (Vohémar, SAVA). *Lemur News,* 21, 31-36.

Sarikaya, Z., & Kappeler, P. M. (1997). Nest building behavior of Coquerel's dwarf lemur (*Mirza coquereli*). *Primate Report,* 47, 3-9.

Sawyer, R. M., Mena, H. E., & Donati, G. (2015). Habitat use, diet and sleeping site selection of *Lepilemur tymerlachsoni* in a disturbed forest of Nosy Be: Preliminary observations. *Lemur News,* 19, 25-30.

Schäffler, L., & Kappeler, P. M. (2014). Distribution and abundance of the world's smallest primate, *Microcebus berthae*, in central western Madagascar. *International Journal of Primatology,* 35, 557-572.

Scheumann, M., Linn, S., & Zimmermann, E. (2017). Vocal greeting during mother-infant reunions in a nocturnal primate, the gray mouse lemur (*Microcebus murinus*). *Scientific Reports,* 7(1), 10321.

Schilling, A. (2000). Sensory organs and communication in *Microcebus murinus*. *Primatologie,* 3, 85-143.

Schmelting, B., Ehresmann, P., Lutermann, H., Randrianambinina, B., & Zimmermann, E. (2000). Reproduction of two sympatric mouse lemur species (*Microcebus murinus* and *M. ravelobensis*) in north-west Madagascar: First results of a long-term study. In W. R. Lourenco, & S. M. Goodman (Eds.), *Diversité et endemisme à Madagascar* (pp. 165-175). Societe de Biogeographie.

Schmelting, B., Zimmermann, E., Berke, O., Bruford, M. W., & Radespiel, U. (2007). Experience-dependent recapture rates and reproductive success in male grey mouse lemurs (*Microcebus murinus*). *American Journal of Physical Anthropology,* 133(1), 743-752.

Schmid, J. (1996). Oxygen consumption and torpor in mouse lemurs (*Microcebus murinus* and *M. myoxinus*): Preliminary results of a study in western Madagascar. In F. Geiser, A. Hulbert, & S. Nicol (Eds.), *Adaptations to the cold: Tenth International Hibernation Symposium* (pp. 47-54). University of New England Press.

Schmid, J. (1999). Sex-specific differences in activity patterns and fattening in the gray mouse lemur (*Microcebus murinus*) in Madagascar. *Journal of Mammalogy,* 80(3), 749-757.

Schmid, J. (2000). Daily torpor in the gray mouse lemur (*Microcebus murinus*) in Madagascar: Energetic consequences and biological significance. *Oecologia,* 123, 175-183.

Schmid, J. (2001). Daily torpor in free-ranging gray mouse lemurs (*Microcebus murinus*) in Madagascar. *International Journal of Primatology,* 22(6), 1021-1031.

Schmid, J., & Ganzhorn, J. U. (1996). Resting metabolic rates of *Lepilemur ruficaudatus. American Journal of Primatology,* 38, 169-174.

Schmid, J., & Kappeler, P. M. (1994). Sympatric mouse lemurs (*Microcebus* spp.) in western Madagascar. *Folia Primatologica,* 63, 162-170.

Schmid, J., & Kappeler, P. M. (1998a). Fluctuating sexual dimorphism and differential hibernation by sex in a primate, the gray mouse lemur (*Microcebus murinus*). *Behavioral Ecology and Sociobiology,* 43, 125-132.

Schmid, J., & Kappeler, P. M. (1998b). Intrasexual selection in *Microcebus murinus. Folia Primatologica,* 69, 211.

Schmid, J., Ruf, T., & Heldmaier, G. (2000). Metabolism and temperature regulation during daily torpor in the smallest primate, the pygmy mouse lemur (*Microcebus myoxinus*) in Madagascar. *Journal of Comparative Physiology B,* 170(1), 59-68.

Schülke, O. (2003a). To breed or not to breed—food competition and other factors involved in female breeding decisions in the pair-living nocturnal fork-marked lemur (*Phaner furcifer*). *Behavioral Ecology and Sociobiology,* 55, 11-21.

Schülke, O. (2003b). *Phaner furcifer*, fork-marked lemur, *Vakihandry, Tanta*. In S. Goodman, & J. Benstead (Eds.), *The natural history of Madagascar* (pp. 1318-1320). University of Chicago Press.

Schülke, O. (2005). Evolution of pair-living in *Phaner fucifer. International Journal of Primatology,* 26, 903-919.

Schülke, O., & Kappeler, P. M. (2003). So near and yet so far: Territorial pairs but low cohesion between pair partners in a nocturnal lemur, *Phaner furcifer. Animal Behaviour,* 65(2), 331-343.

Schülke, O. Kappeler, P. M., & Zischler, H. (2004). Small testes size despite high extra-pair paternity in the pair-living nocturnal primate *Phaner furcifer. Behavioral Ecology and Sociobiology,* 55, 293-301.

Schülke, O., & Ostner, J. (2001). Predation on a *Lepilemur* by a harrier hawk and implications of sleeping site quality. *Lemur News,* 6, 5.

Schülke, O., & Ostner, J. (2005). Big times for dwarfs: Social organization, sexual selection, and cooperation in the Cheirogaleidae. *Evolutionary Anthropology,* 14(5), 170-185.

Schüßler, D., Blanco, M. B., Salmona, J., Poelstra, J., Andriambeloson, J. B., Miller, A., Randrianambinina, B., Rasolofoson, D. W., Mantilla-Contreras, J., Chikhi, L., Louis, E. E., Jr., Yoder, A. D., & Radespiel, U. (2020). Ecology and morphology of mouse lemurs (*Microcebus* spp.) in a hotspot of microendemism in northeastern Madagascar, with the description of a new species. *American Journal of Primatology,* 82(9), e23180.

Schütz, H., & Goodman, S. M. (1998). Photographic evidence of *Allocebus trichotis* in the Reserve Spéciale d'Anjanaharibe-Sud. *Lemur News,* 3, 21-22.

Schwab, D. (2000). A preliminary study of spatial distribution and mating system of pygmy mouse lemur (*Microcebus cf. myoxinus*). *American Journal of Primatology,* 51, 41-60.

Schwitzer, C., Mittermeier, R., Johnson, S., Donati, G., Irwin, M., Peacock, H., Ratsimbazafy, J., Razafindramanana, J., Louis, E. E., Chikhi, L., Colquhoun, I. C., Tinsman, J., Dolch, R., LaFleur, M., Nash, S., Patel, E., Randrianambinina, B., Rasolofoharivelo, T., & Wright, P. C. (2014). Averting lemur extinctions amid Madagascar's political crisis. *Science,* 343(6173), 842-843.

Sefczek, T. M., Randimbiharinirina, D., Raharivololona, B. M., Rabekianja, J. D., & Louis, E. E. (2017). Comparing the use of live trees and deadwood for larval foraging by aye-ayes (*Daubentonia madagascariensis*) at Kianjavato and Torotorofotsy, Madagascar. *Primates,* 58, 535-546.

Seligson, D., & Szalay, F. S. (1978). Relationship between natural selection and dental morphology: Tooth function and diet in *Lepilemur* and *Hapalemur*. In P. M. Butler, & K. A. Joysey (Eds.), *Studies in the development, function and evolution of teeth* (pp. 289-307). Academic Press.

Setash, C. M., Zohdy, S., Gerber, B. D., & Karanewsky, C. J. (2017). A biogeographical perspective on the variation in mouse lemur density throughout Madagascar. *Mammal Review,* 47, 212-229.

Simons, E. L. (1993). Discovery of the western aye-aye. *Lemur News,* 1, 6.

Simons, E. L., & Meyers, D. M. (2001). Folklore and beliefs about the aye-aye (*Daubentonia madagascariensis*). *Lemur News,* 6, 11-16.

Smith, A. P., & Ganzhorn, J. U. (1996). Convergence in community structure and dietary adaptation in Australian possums and gliders and Malagasy lemurs. *Australian Journal of Ecology,* 21(1), 31-46.

Steffens, K. J. E., Rakotondranary, S. J., Ratovonamana, Y. R., & Ganzhorn, J. U. (2017). Vegetation thresholds for the occurrence and dispersal of *Microcebus griseorufus* in southwestern Madagascar. *International Journal of Primatology,* 38, 1138-1153.

Sterling, E. J. (1993a). *Behavioral ecology of the aye-aye (*Daubentonia madagascariensis*) on Nosy Mangabe, Madagascar.* [Unpublished doctoral dissertation, Yale University].

Sterling, E. J. (1993b). Patterns of range use and social organization in aye-ayes (*Daubentonia madagascariensis*) on Nosy Mangabe. In P. M. Kappeler, & J. U. Ganzhorn (Eds.), *Lemur social systems and their ecological basis* (pp. 1-10). Plenum Press.

Sterling, E. J. (1994a). Taxonomy and distribution of *Daubentonia*: A historical perspective. *Folia Primatologica,* 62, 8-13.

Sterling, E. J. (1994b). Evidence for nonseasonal reproduction in wild aye-ayes (*Daubentonia madagascariensis*). *Folia Primatologica,* 62, 45-53.

Sterling, E. J. (1994c). Aye-ayes: Specialists on structurally defended resources. *Folia Primatologica,* 62, 142-154.

Sterling, E. J. (2003). *Daubentonia madagascariensis,* aye-aye. In S. M. Goodman, & J. P. Benstead (Eds.), *The natural history of Madagascar* (pp. 1348-1351). University of Chicago Press.

Sterling, E. J., Dierenfeld, E. S., Ashbourne, C. J., & Feistner, A. T. C. (1994). Dietary intake, food composition and nutrient intake in wild and captive populations of *Daubentonia madagascariensis*. *Folia Primatologica,* 62, 115-124.

Sterling, E. J., & Feistner, A. T. C. (2000). *Aye-aye. Endangered animals: A reference guide to conflicting issues.* Greenwood Press.

Sterling, E. J., & McCreless, E. M. (2006). Adaptations in the aye-aye: A review. In L. Gould, & M. L. Sauther (Eds.), *Lemurs: Ecology and adaptation (Developments in primatology: Progress and prospect)* (pp. 159-184). Springer.

Sterling, E. J., & McFadden, K. (2000). Rapid census of lemur populations in the Parc National de Marojejy, Madagascar. *Field Zoology,* 97, 265-274.

Sterling, E. J., & Richard, A. F. (1995). Social organization in the aye-aye (*Daubentonia madagascariensis*) and the perceived distinctiveness of nocturnal primates. In K. Izard, L. Anderson, & G. A. Doyle (Eds.), *Creatures of the dark: The nocturnal prosimians* (pp. 439-451). Plenum Press.

Sussman, R. W. (1999). *Primate ecology and social structure. Vol. 1: Lorises, lemurs and tarsiers.* Pearson Custom Publishing.

Sussman, R. W., & Raven, P. H. (1978). Pollination by lemurs and marsupials: An archaic coevolutionary system. *Science,* 200, 731-736.

Tattersall, I. (1982). *The primates of Madagascar.* Columbia University Press.

Tattersall, I. (2007). Madagascar's lemurs: Cryptic diversity or taxonomic inflation? *Evolutionary Anthropology,* 16(1), 12-23.

Thalmann, U. (2001). Food resource characteristics in two nocturnal lemurs with different social behavior: *Avahi occidentalis* and *Lepilemur edwardsi*. *International Journal of Primatology,* 22, 287-324.

Thalmann, U. (2002). Contrasts between two nocturnal leaf-eating lemurs. *Evolutionary Anthropology,* 11, 105-107.

Thalmann, U. (2006). Behavioral and ecological adaptations in two small folivorous lemurs with different social organization: *Avahi* and *Lepilemur*. In L. Gould, & M. Sauther (Eds.), *Lemurs: Ecology and adaption* (pp. 327-352). Springer.

Thompson, K. E. T., Bankoff, R. J., Louis, E. E., Jr., & Perry, G. H. (2016). Deadwood structural properties may influence aye-aye (*Daubentonia madagascariensis*) extractive foraging behavior. *International Journal of Primatology,* 37, 281-295.

Thoren, S., Quietzsch, F., & Radespiel, U. (2010). Leaf nest use and construction in the golden-brown mouse lemur (*Microcebus ravelobensis*) in the Ankarafantsika National Park. *American Journal of Primatology,* 72(1), 48-55.

Thorstrom, R., & La Marca, G. (2000). Nesting biology and behavior of the Madagascar harrier-hawk (*Polyboroides radiatus*) in northeastern Madagascar. *Journal of Raptor Research,* 34, 120-125.

Tinsman, J. (2020). *Geospatial and genomic tools for conserving the critically endangered blue-eyed black lemur (*Eulemur flavifrons*) and the sportive lemurs (genus* Lepilemur). [Doctoral dissertation, Columbia University].

Trivers, R. L. (1972). Parental investment and sexual selection. In B. Campbell (Ed.), *Sexual selection and the descent of man* (pp. 1871-1971). Aldine Publishing.

van Schaik, C. P., & van Hooff, J. A. (1983). On the ultimate causes of primate social systems. *Behaviour, 85*(1-2), 91-117.

Vaughan, A. (2020). Almost all lemur species are now officially endangered. *NewScientist, 3291.*

Vieilledent, G., Grinandc, C., Rakotomalalac, F. A., Ranaivosoad, R., Rakotoarijaonad, J.-R., Allnutt, T. F., & Achard, F. (2018). Combining global tree cover loss data with historical national forest cover maps to look at six decades of deforestation and forest fragmentation in Madagascar. *Biological Conservation, 222,* 189-197.

Walker, A. C. (1967*). Locomotor adaptation in recent and fossil Madagascan lemurs.* [Doctoral dissertation, University of London].

Warren, R. D., & Crompton. R. H. (1998a). Diet, body size and the energy costs of locomotion in saltatory primates. *Folia Primatologica, 69*(Suppl. 1), 86-100.

Warren, R. D., & Crompton, R. H. (1998b). Lazy leapers: Locomotor behaviour and ecology of *Lepilemur edwardsi* and *Avahi occidentalis. American Journal of Physical Anthropology, 104,* 471-486.

Webber, A. D., Solofondranohatra, J. S., Razafindramoana, S., Fernández, D., Parker, C. A., Steer, M., Abrahams, M., & Allainguillaume, J. (2020). Lemurs in cacao: Presence and abundance within the shade plantations of northern Madagascar. *Folia Primatologica, 91,* 96-107.

Weidt, A., Hagenah, N., Randrianambinina, B., Radespiel, U., & Zimmermann, E. (2004). Social organization of the golden-brown mouse lemur (*Microcebus ravelobensis*). *American Journal of Physical Anthropology, 123*(1), 40-51.

Wilmé, L., Goodman, S. M., & Ganzhorn, J. U. (2006). Biogeographic evolution of Madagascar's microendemic biota. *Science, 312*(5776), 1063-1065.

Winn, R. M. (1994a). Preliminary study of the sexual behaviour of three aye-ayes (*Daubentonia madagascariensis*) in captivity. *Folia Primatologica, 62*(1-3), 63-73.

Winn, R. M. (1994b). Development of behaviour in a young aye-aye (*Daubentonia madagascariensis*) in captivity. *Folia Primatologica, 62*(1-3), 93-107.

Wright, P. C. (1997). The future of biodiversity in Madagascar: A view from Ranomafana National Park. In S. Goodman, & B. Patterson (Eds.), *Natural change and human impact in Madagascar* (pp. 381-405). Smithsonian Institution Press.

Wright, P. C. (1999). Lemur traits and Madagascar ecology: Coping with an island environment. *Yearbook of Physical Anthropology, 42,* 31-72.

Wright, P. C., & Martin, L. B. (1995). Predation, pollination, and torpor in two nocturnal prosimians: *Cheirogaleus major* and *Microcebus rufus* in the rain forest of Madagascar. In K. Izard, L. Anderson, & G. A. Doyle (Eds.), *Creatures of the dark: The nocturnal prosimians* (pp. 325-334). Plenum Press.

Yoder, A. D., Rasoloarison, R. M., Goodman, S. M., Irwin, J. A., Atsalis, S., Ravosa, M. J., & Ganzhorn, J. U. (2000). Remarkable species diversity in Malagasy mouse lemurs (Primates, *Microcebus*). *Proceedings of the National Academy of Sciences, 97,* 11325-11330.

Zimmermann, E. (1995a). Acoustic communication in nocturnal prosimians. In K. Izard, L. Anderson, & G. A. Doyle (Eds.), *Creatures of the dark: The nocturnal prosimians* (pp. 311-330). Plenum Press.

Zimmermann, E. (1995b). Loud calls in nocturnal prosimians: Structure, evolution and ontogeny. In E. Zimmermann, J. D. Newman, & U. Jürgens (Eds.), *Current topics in primate vocal communication* (pp. 47-72). Plenum Press.

Zimmermann, E., Cepok, S., Rakotoarison, N., Zietemann, V., & Radespiel, U. (1998). Sympatric mouse lemurs in north-west Madagascar: A new rufous mouse lemur species (*Microcebus ravelobensis*). *Folia Primatologica, 69,* 106-114.

Zinner, D., Hilgartner, R. D., Kappeler, P. M., Pietsch, T., & Ganzhorn, J. U. (2003). Social organization of *Lepilemur ruficaudatus. International Journal of Primatology, 24*(4), 869-888.

Zohdy, S., Kemp, A., Durden, L., Wright, P. C., & Jernvall, J. (2012). Mapping the social network: Tracking lice in a wild primate population (*Microcebus rufus*) to infer social contacts and vector potential. *BMC Ecology, 12*(1), 4.

CHAPTER 3

Diurnal and Cathemeral Lemurs

Ian C. Colquhoun and Joyce Powzyk

THE DIURNAL AND CATHEMERAL LEMURS, I.E., THE LEM-URIDS and indriids (families Lemuridae and Indriidae), are iconic species—both in Madagascar and internationally. A fascinating branch of primate biodiversity, lemurid and indriid species exhibit a spectrum of activity patterns, from diurnal (day active), to cathemeral (activity at irregular but significant intervals throughout a 24-hour cycle; Tattersall, 1987), as well as nocturnal (night active) patterns in the woolly lemurs (genus *Avahi*, Indriidae). The lemurids and indriids exhibit diverse social organizations, and have varied and flexible feeding ecologies, making them an interesting group of primates to contrast with monkey and ape species. But lemurs also attract attention beyond the primate research community. This is embodied in the ring-tailed lemur (*Lemur catta*), or *maki*, and to some extent genus *Propithecus*, the *sifakas*. Today, these taxa have high public recognition in Madagascar and the Western world. The unmistakable *L. catta* and the "dancing" sifaka are species that, for many, have become synonymous with the word "lemur." Lemurs seem to have great appeal to humans. There are now conservation-themed annual lemur festivals held in communities across Madagascar and, since 2014, an International Lemur Day has been recognized annually. Communities around the world hold their own lemur festivals in the weeks around each year's International Lemur Day. But it was not always so. The situation was strikingly different in the late 1950s when the first fieldwork on lemur behavior and ecology was carried out—at that time, wild lemur populations were almost completely unstudied.

Groundbreaking survey work in 1956–1957 by Jean-Jacques Petter on multiple lemur species across Madagascar provided the first detailed field data on wild lemur populations (Petter, 1962). In short order thereafter, the first in-depth field study of lemuriform behavioral ecology was conducted by Alison Jolly during 1963–1964 on the ring-tailed lemur (*L. catta*, family Lemuridae) and Verreaux's sifaka (*P. verreauxi,* family Indriidae) at Berenty Reserve, in the extreme southeast of the island (Jolly, 1966). Since then, research on the genera *Lemur, Eulemur, Varecia, Hapalemur, Prolemur, Propithecus,* and *Indri* has dominated field studies on the lemuriforms. For

taxonomic consistency, the woolly lemurs of the nocturnal indriid genus *Avahi* are also discussed in this chapter.

There are several valuable historical overviews of field studies on lemuriform species. These cover the expansion of the number of species that have been the focus of field research, along with several long-term research projects on day-active lemurs that have generated decades of data on selected taxa, as well as histories of some of the study sites themselves. Notable historical overviews of field research on diurnal and cathemeral lemur species have been presented by Buettner-Janusch et al. (1975), Jolly et al. (2006), Jolly and Sussman (2006), Sussman and Rat-sirarson (2006), Jolly (2012), Kappeler and Fichtel (2012), Kappeler et al. (2012, 2017), Sussman et al. (2012), and Wright et al. (2012). See Sussman (2011) for a concise overview on the history of conceptual developments leading to today's problem-oriented primatological studies.

TAXONOMY, DISTRIBUTION, AND HABITAT

The lemuriform species are a major primate lineage, representing over 21% of all species in the Order Primates (Rowe & Myers, 2016). Madagascar, home to the lemuriforms, is a large island—an obvious geographic fact, but its size is an important piece in building an understanding of Madagascar's endemic primate radiation, the lemurs. Madagascar is often referred to as a "megadiversity" country (Estrada et al., 2017), not just for its lemur biodiversity but for other fauna and flora groups as well. The world's fourth largest island (after Greenland, New Guinea, and Borneo; Leigh et al., 2007), Madagascar is also a very old landmass, having been an oceanic island for approximately 90 million years (de Wit, 2003; Mittermeier et al., 2010). Over that time, Madagascar has experienced considerable climatic changes (Ohba et al., 2016; Godfrey et al., 2020), which produced widespread changes in the island's plant communities, including the forest habitats on which lemurs are dependent (Ganzhorn, 1995; Ohba et al., 2016). Another key factor affecting lemur distributions, in addition to variation in climate and plant communities, is the island's topography (MacArthur, 1972; Tattersall, 1982; Brown, 1995; Ganzhorn et al., 2006, 2014; Wilmé et al., 2006), which has remained relatively stable over the past 65 million years (Ohba et al., 2016).

Madagascar's physical geography is marked by asymmetric eastern and western slopes (Leigh et al., 2007; Ganzhorn et al., 2014), with a narrow east coastal plain giving way within 100 km of the coast to an escarpment that runs parallel to the coast and rises steeply to 1,500–2,500 masl. The high-relief topography in the east contrasts with a much broader and less steep slope that descends gradually to the western coast.

Madagascar's large size and topographic features combine to give its climate and regional environments continental qualities. Those qualities include several distinct climatic zones (and associated plant communities) occurring in different regions of the island, from the arid southwest to the rainforests of the east and northeast (de Wit, 2003; Leigh et al., 2007). These climatic, or phytogeographic, zones can be generally described as incorporating moist evergreen lowland and montane forests in the east, deciduous and semi-deciduous forests across much of the western and northwestern regions of the island, and endemic dry-adapted plant communities in the arid and highly seasonal southwestern zone of the island (Jolly, 1966; Tattersall & Sussman, 1975; Tattersall, 1982; Mittermeier et al., 1992, 2010).

Lemurs represent an ancient lineage (Yoder et al., 1996; Herrera, 2016; Godfrey et al., 2020). However, Madagascar is older than the date reconstructed for the divergences of lemuriforms and lorisiforms lineages from a common ancestral line—between 50–60 mya, although dating this lineage divergence has produced a range of estimates (Mittermeier et al., 2014). In other words, there were no ancestral lemurs on Madagascar when it became an island; lemurs had to cross a water barrier to colonize the island. Thus, colonization of Madagascar by many of the terrestrial vertebrate groups found there, including the lemuriform species, is considered to have involved dispersal from Africa across the Mozambique Channel on natural vegetation "rafts" washed out to sea following intense tropical storms (Yoder & Nowak, 2006; Samonds et al., 2012; Godfrey et al., 2020).

Madagascar's climatic, phytogeographic, and topographic variation have contributed to high levels of regional lemur biodiversity (see Ganzhorn et al., 2006, 2014; Wilmé et al., 2006; Mittermeier et al., 2008, 2010). The extant lemurid and indriid species (see Table 3.1) occur in each of Madagascar's phytogeographic zones (Mittermeier et al., 1992). However, most forest cover on the central plateau has been lost, effectively meaning that the plateau's forest biome, and the biota that once occurred there, have disappeared (Ganzhorn et al., 2001). While lemurids and indriids do occur in multiple mid-altitude rainforest sites in the eastern escarpment rainforests (e.g.,

Goodman & Ganzhorn, 2004; Irwin, 2007; Everson et al., 2020), *L. catta* is the only species from either family known to occur in one of the island's highest elevation habitats (ranging above the tree line to 2,520 m) on the Andringitra Massif (Goodman & Langrand, 1996; Goodman & Ganzhorn, 2004; Goodman et al., 2006).

In the time since lemuriform field studies were initiated over 60 years ago, lemurid and indriid taxonomic classifications have both undergone significant revisions as field data on these taxa have accumulated. Petter (1962) and Jolly (1966) employed taxonomies in which just 19–20 species comprised the entire Infraorder Lemuriformes. Increased fieldwork has uncovered lemur taxa that were new to science, and during the last 20 years, the increased application of the phylogenetic species concept in taxonomic matters has also produced growing numbers of recognized species, mostly through subspecies being elevated to full species status (Groves, 2001).

Lemuridae
Lemur

The genus *Lemur* was erected by Linnaeus in 1758 in reference to the ring-tailed lemur, which he christened *L. catta* (Tattersall, 1982; Groves, 2001). Through much of the twentieth century, the genus *Lemur* encompassed multiple species beyond the ring-tailed lemur and included, among others, the "brown lemur" complex of taxa (Johnson, 2006; Tattersall & Sussman, 2016), that due to shifting views on taxonomic synonymies, were variously referred to as *L. macaco* subspecies and then as subspecies of *L. fulvus*. *Lemur* became a monospecific genus with resurrection of the genus *Eulemur* (Simons & Rumpler, 1988) for species other than the ring-tailed that had been classified previously in the genus *Lemur* (Yoder, 1997; Wilson & Hanlon, 2010). Comparative genetic data were concordant with this taxonomic revision since they indicated that, despite morphological resemblances between *L. catta* and the *Eulemur* taxa, the ring-tailed lemur lineage actually forms a clade with *Hapalemur* and *Prolemur* (the gentle and bamboo lemurs) to the exclusion of *Eulemur* (Yoder, 1997; Groves, 2001). The ring-tailed lemur has a patchy distribution across the dry forest and spiny forest formations of the southwestern portion of the island as far east as where climatic patterns shift to more humid conditions. And, as mentioned previously, a population of *L. catta* also even occurs at high elevations on the Andringitra Massif at the eastern edge of ring-tailed lemur distribution (Goodman & Langrand, 1996; Yoder et al., 1999).

Lemur catta (Figure 3.1)—the most terrestrial species among the extant lemurs (Mittermeier et al., 2008)—form the largest social groups of any lemuriform species (Jolly,

Table 3.1   Scientific Names, Common Names, Distributions, and Adult Body Weights of the Lemuridae and Indriidae

Scientific Name	Common Name	Distribution	Adult Body Weight (kg)[a]
Lemuridae	Lemurids		
Lemur catta	Ring-tailed lemur	Southwestern and southern Madagascar	2.2 (mean)
Eulemur fulvus	Brown lemur	Northwestern and central eastern Madagascar	1.7-2.1
E. rufus	Rufous brown lemur	Central western Madagascar	~2.0 (mean)
E. rufifrons	Red-fronted brown lemur	Southwestern and southeastern Madagascar	2.2-2.3
E. albifrons	White-fronted brown lemur	Northeastern Madagascar	~2.0
E. sanfordi	Sanford's brown lemur	Northern Madagascar	1.8-1.9
E. cinereiceps	Gray-headed, or white-collared brown lemur	Southeastern Madagascar	2.0-2.5
E. collaris	Red-collared brown lemur	Southeastern Madagascar	2.25-2.5
E. coronatus	Crowned lemur	Northern Madagascar	1.1-1.3
E. macaco	Black lemur	Northwestern Madagascar	1.8-1.9
E. flavifrons	Blue-eyed, or Sclater's, black lemur	Northwestern Madagascar	1.8-1.9
E. rubriventer	Red-bellied lemur	Eastern Madagascar (from NE to SE)	1.6-2.4
E. mongoz	Mongoose lemur	Northwestern Madagascar	1.1-1.6
Varecia variegata	Black-and-white ruffed lemur	Eastern Madagascar	3.1-3.6
V. rubra	Red-ruffed lemur	Northeastern Madagascar (Masoala Peninsula only)	3.3-3.6
Hapalemur griseus	Gray gentle lemur	Central eastern and central western Madagascar	0.7-0.85
H. occidentalis	Northern, or western, gentle lemur	NW, N, and NE Madagascar	0.84-0.87
H. meridionalis	Southern gentle lemur	Southeastern Madagascar	0.85 (m), 1.19 (f)
H. alaotrensis	Lac Alaotra gentle lemur	Western margins of Lac Alaotra	1.1-1.55
H. aureus	Golden bamboo lemur	Southeastern Madagascar	1.3-1.7
Prolemur simus	Greater bamboo lemur	Central and southeastern Madagascar	2.2-2.5
Indriidae	Indriids		
Indri indri	Indri	Northeastern and central eastern Madagascar	6.0-9.5
Propithecus.diadema	Diademed sifaka	Northeastern and central eastern Madagascar	6.0-8.5
P. candidus	Silky sifaka	Northeastern Madagascar	5.0-6.0
P. edwardsi	Milne-Edwards' sifaka	Southeastern Madagascar	5.0-6.5
P. perrieri	Perrier's sifaka	Northern Madagascar	4.3-5.0
P. tattersalli	Tattersall's, or golden-crowned, sifaka	Northern Madagascar	3.4-3.6
P. verreauxi	Verreaux's sifaka	Southwestern and southern Madagascar	3.0-3.5
P. deckenii	van der Decken's sifaka	Central western Madagascar	3.0-4.5
P. coronatus	Crowned sifaka	Central western Madagascar	3.5-4.3
P. coquereli	Coquerel's sifaka	Central western Madagascar	3.7-4.3
Avahi laniger	Eastern, or Gmelin's, woolly lemur	Northeastern and central eastern Madagascar	1.0-1.4
A. mooreorum	Moore's woolly lemur	Northeastern Madagascar (Masoala Peninsula only)	~0.920
A. peyrierasi	Peyrieras' woolly lemur	Southeastern Madagascar	0.9-1.2
A. betsileo	Betsileo woolly lemur	Central eastern Madagascar	0.9-1.2
A. ramanantsoavanai	Manombo woolly lemur	Southeastern Madagascar	0.9-1.2
A. meridionalis	Southern woolly lemur	Southeastern Madagascar Borneo	0.95-1.4
A. occidentalis	Western, or von Liburnau's, woolly lemur	Western Madagascar	0.8-1.1
A. unicolor	Sambirano woolly lemur	Northwestern Madagascar	0.7-1.0
A. cleesei	Cleese's, or Bemaraha, woolly lemur	Western Madagascar (Tsingy d'Bemaraha; Beanka forest)	0.75-1.3

[a]Adult lemurid and indriid body weights drawn from Mittermeier et al. (2010) and references therein.

Figure 3.1 Adult male ring-tailed lemur (*Lemur catta*). Dark spot on the lemur's right forearm is the antebrachial gland used in scent-marking by males. Both male and female ring-tailed lemurs scent-mark extensively with anogenital glands, but only males possess antebrachial glands.
(Photo credit: Mathius Appel; image made available under Creative Commons CC0 1.0 Universal Public Domain Dedication)

2003), are the most-studied lemuriform (Mittermeier et al., 2008), and are also the nonhuman primate most often seen in zoological gardens (LaFleur et al., 2016a). In the southwestern portion of their geographic range, in the region of the Beza Mahafaly Special Reserve, local *fadys* (taboos) accord culturally protected status to both *L. catta* and *P. verreauxi*. Traditional beliefs of the local Mahafaly people hold that both species originated from humans and that harming, killing, or eating them is forbidden since such actions would bring bad luck (Loudon et al., 2006).

Eulemur

Species of the genus *Eulemur* (Simons & Rumpler, 1988) occur across the remaining forested areas in Madagascar's eastern, western, and northern regions. Over the last decades, *Eulemur* taxa have been the focus of a large

number of field studies (Jolly & Sussman, 2006). With separation of the genus *Lemur* from the *Eulemur* taxa (Simons & Rumpler, 1988), the latter was generally understood at that time to include five species—the brown lemur complex (*E. fulvus*, including 6 subspecies), the black lemurs (*E. macaco,* including 2 subspecies), and the crowned, mongoose, and red-bellied lemurs (*E. coronatus, E. mongoz,* and *E. rubriventer,* respectively) (Overdorff & Johnson, 2003; Johnson, 2006). Groves (2001) introduced further taxonomic revisions by raising all subspecies of the brown lemur complex to full species status, while retaining the black lemur subspecies. Subsequently, the blue-eyed black lemur (*E. flavifrons*) was also raised to full species status, a taxonomic revision consistent with the genetic distance between *E. macaco* and *E. flavifrons* (Pastorini et al., 2002; Mittermeier et al., 2008; Schwitzer et al., 2014a). With those taxonomic revisions, the genus *Eulemur* is presently composed of 12 species (Mittermeier et al., 2008, 2010). The geographic distributions of these *Eulemur* taxa correspond closely to the island's eco-geographic regions (Markolf & Kappeler, 2013). It is notable that the branching patterns of the genus *Eulemur* sub-clades presented in Figure 3.2 reflect major observable aspects of *Eulemur* phylogeography and behavioral ecology. *E. albifrons, E. sanfordi, E. fulvus, E. rufus,* and *E. rufifrons* represent closely related species in the brown lemur complex. *E. cinereiceps* and *E. collaris* are also part of the complex but exhibit unique genetic characteristics; their geographic ranges in southeast Madagascar are at one end of brown lemur distribution. *E. rubriventer* and *E. mongoz* are the two *Eulemur* species exhibiting pair-bonded social organization and are sister lineages to each other. *E. coronatus, E. macaco,* and *E. flavifrons* occur in a north to south array of parapatric distributions, with a narrow hybrid zone at the margins of the northeastern distribution of *E. flavifrons* and the southern distribution limit of *E. macaco*. Across the distributions of *Eulemur* species, several cases occur where a member of the brown lemur complex occurs in sympatry with a *Eulemur* taxon that is *not* a brown lemur (i.e., the crowned, mongoose, red-bellied, and black lemurs), and some populations of *E. rufifrons* in southwestern Madagascar are sympatric with *L. catta* (Mittermeier et al., 2008, 2010; Markolf & Kappeler, 2013).

Varecia

The ruffed lemurs (genus *Varecia*) are the largest members of the Lemuridae (see Figure 3.3). While the *Eulemur* species have adult body weights that range between 2–3 kg (Tattersall, 1982), the ruffed lemurs have adult body weights that are approximately 50% greater (mean of ~3.8–3.9 kg; Tattersall, 1982). As early as the

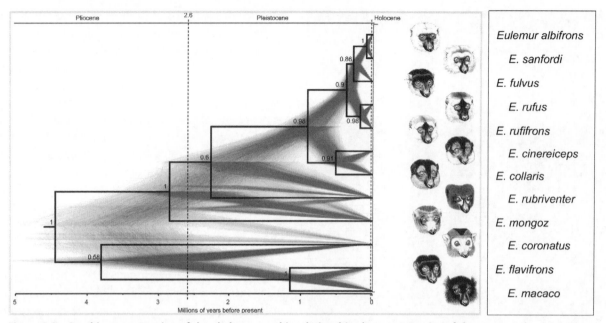

Figure 3.2 Graphic reconstruction of the phylogeographic relationships between species of the genus *Eulemur*. Darkest strands of the reconstructed lineages represent regional species clusters for which interspecies relationships are best resolved genetically. See text for description of these closely related regional clusters of *Eulemur* taxa.
(Image derived from Markolf and Kappeler, 2013-Fig. 3; Creative Commons Attribution License 2.0, permitting unrestricted reproduction)

late eighteenth century, ruffed lemurs were being classified in the (formerly expansive) genus *Lemur* (Tattersall, 1982). Over the last several decades, taxonomic consensus shifted from recognizing one species of ruffed lemur (*V. variegata*) with two subspecies, to the two forms being elevated to full species status: *V. variegata*, the black-and-white ruffed lemur, and *V. rubra*, the red-ruffed lemur (Tattersall, 1982; Groves, 2001; Vasey & Tattersall, 2002; Vasey, 2003; Mittermeier et al., 2008, 2010). Both ruffed lemur species are found only in the eastern evergreen rainforests. The red-ruffed lemur is restricted to the Masoala Peninsula in northeastern Madagascar (Hekkala et al., 2007), while the three subspecific forms of *V. variegata* are distributed in a north–south geographic array over the eastern rainforests. Mittermeier et al. (2010) discuss the problem that observed variability in pelage patterning among black-and-white ruffed lemurs is not easily accommodated in this three-subspecies taxonomy.

Geographic occurrence of black-and-white ruffed lemurs stretches from the northern limit of their distribution at the Antainambalana River (west of the Masoala Peninsula), to the Mananara River in the southeastern part of the island (Hekkala et al., 2007; Mittermeier et al., 2008, 2010). More recently, Rakotonirina and colleagues (2013) have reported *V. variegata* from south of the Mananara River, which would define a southern expansion of geographic range for this species. Historically, there appears to have been a narrow zone of sympatry/parapatry (with occasional natural hybridizations) between *V. rubra* and *V. variegata*

at the northwestern edge of the Masoala Peninsula, but field surveys to identify this potential hybrid zone were unsuccessful (Vasey & Tattersall, 2002; Vasey, 2003; Hekkala et al., 2007). Results from fieldwork in the region by Hekkala and colleagues (2007) show that only about 22% of the habitat they surveyed can be considered either primary or secondary forest. Remaining forest habitat in the survey area is fragmented and under considerable anthropogenic pressure; forest clearance for subsistence agricultural plots is widespread as is hunting pressure on *V. rubra*. Results of the survey led Hekkala et al. (2007) to conclude that *V. rubra* habitat immediately west of Masoala National Park is rapidly disappearing, and red-ruffed lemur populations in that region are at high risk of extirpation.

Ruffed lemurs, with their frugivorous feeding ecology, are dietary specialists with a preference for and reliance on intact primary rainforest (e.g., Ratsimbazafy, 2006; Vasey, 2006; Hekkala et al., 2007). However, Ratsimbazafy (2006) also found that, following extensive cyclone damage to forest structure (toppled emergent trees, significant loss of branches, and defoliation in the forest canopy), ruffed lemurs are able to maintain a high degree of frugivory by shifting their food species choices and even incorporating exotic invasive species into their diet.

Hapalemur *and* Prolemur
The genera *Hapalemur* and *Prolemur* are behaviorally (Petter & Peyriéras, 1970, 1975), morphologically (Milton, 1978; Eronen et al., 2017), and physiologically

Figure 3.3 Young adult male black-and-white ruffed lemur (*Varecia variegata*) engaging in the sloth-like below-branch posture that ruffed lemurs often assume during arboreal foraging. The coloration of *V. variegata* helps to obscure and disguise the lemur's physical outline, providing some camouflaging against the background of ambient light and tree canopy shadows.
(Photo credit: Sonia Wolf)

(Glander et al., 1989; Eppley et al., 2017b) specialized bamboo or grass eaters. There are two species of bamboo lemurs—*Prolemur simus* (the greater, or broad-nosed, bamboo lemur) and *Hapelemur aureus* (the golden bamboo lemur); four species of gentle lemurs are recognized—*H. griseus* (the eastern, or gray, gentle lemur), *H. occidentalis* (the western, or northern, gentle lemur), *H. meridionalis* (the southern gentle lemur), and *H. alaotrensis* (the Lac Alaotra gentle lemur recently elevated from a subspecies of *H. griseus* to species level) (Pastorini et al., 2002; Andrianandrasana et al., 2005; Groves, 2005; Ralainasolo et al., 2006; Fleagle, 2013; Ratsimbazafy et al., 2013; Mittermeier et al., 2014; but see debate about *H. alaotrensis* in Fausser et al., 2002; Rabarivola et al., 2007).

The gentle lemurs are smaller than the bamboo lemurs. Tan (1999, 2000) gives an average weight of 900 g for *H. griseus*. Fleagle (2013) gives weights for male *H. griseus* of 748 g and 670 g for females, mean = 709 g. Male body weights for *H. alaotrensis* = 1,228–1,350 g

and females = 1,251–1,550 g (Rabarivola et al., 2007; Fleagle, 2013). *H. meridionalis* and *H. occidentalis* are estimated to fall within the weight range of *H. griseus* (Mutschler & Tan, 2003). *H. aureus* is of intermediate size among the gentle and bamboo lemurs with male body weight = 1,514 g and female weight = 1,355 g, mean = 1,434.5 g (Tan, 2000; Wright & Tan, 2016a). *P. simus* is the largest species among the gentle and bamboo lemurs with males weighing 2,238 g and females 2,250 g, mean = 2,244 g (Tan, 2000; Mittermeier et al., 2008; Wright & Tan, 2016b). These body mass data reflect a major determining dimension of bamboo and gentle lemur synecology, i.e., body size and biological scaling (Schmidt-Nielsen, 1975, 1984; Tan, 1999, 2000). *H. griseus*, on average, is about 75% the body size of *H. aureus*, and *H. aureus* is, on average, about 64% the body size of *P. simus*. These morpho-ecological parameters and differentiations are played out in the econiche partitioning that underpins how the bamboo-eating species can exist sympatrically (Tan, 1999, 2000).

The gentle lemurs exhibit wide geographic distributions, from west and northwest Madagascar, throughout the eastern rainforests, to the southeast of the island. The geographic ranges of the gentle lemurs follow a north–south biogeographic gradient, with the range of western and northern populations of *H. occidentalis* being replaced by *H. griseus* in the eastern rainforests to the region of Ranomafana National Park in the southeast, where *H. griseus* is then replaced by the range of *H. meridionalis* farther to the southeast. The southern limit of the species range for *H. griseus* and the northern extent of the distribution of *H. meridionalis* are both unclear. However, *H. meridionalis* is known to inhabit the littoral rainforest fragment of Mandena in southeastern Madagascar (Rabarivola et al., 2007; Eppley et al., 2011; Eppley, 2015). *H. alaotra* is endemic to the reed beds in and around Lac Alaotra in northeast Madagascar. The western, also called the northern, gentle lemur, *H. occidentalis*, occupies dry and sub-humid semi-deciduous forests of west and northwest Madagascar (Colquhoun, 1998a; Mutschler & Tan, 2003; Tan, 2006) and, to date, is among the least-studied of the *Hapalemur* species. While *H. occidentalis* originally gained its taxonomic name due to its purported geographic distribution in western Madagascar, the common name of the taxon no longer accurately reflects its more recently determined extent of occurrence in northern Madagascar. In the northwestern Forêt Classée of Ambato Massif, Colquhoun (1998a) reported that *H. occidentalis* occurs at a relatively low population density of ~ 0.35–0.40 animals/ha in his study area.

The golden bamboo lemurs (*H. aureus*) were first discovered in 1986 (Meier et al., 1987). They can be identified by their distinctive golden ventrum and circumfacial ring, with a dorsal pelage that is a rich reddish-brown color (Mittermeier et al., 2008; Wright & Tan, 2016a; Wright et al., 2020b). In the wake of their discovery, *H. aureus* were initially thought to have a highly restricted geographic range consisting of the area in and around Ranomafana National Park, where they are sympatric with *H. griseus* and *P. simus* (Tan, 1999, 2000). Subsequent field surveys, however, have revealed a second population to the south in Andringitra National Park (see Rakotondravony & Razafindramahatra, 2004); it is also possible that the species still occurs in the Ranomafana-Andringitra forest corridor (Mittermeier et al., 2008; Rakotonirina et al., 2013; Wright et al., 2020b; but see Lehman et al., 2005b, 2006, for surveys that produced no sightings).

P. simus has a scattered, fragmented distribution; it is extremely rare and is considered among the most critically endangered of primate species (Mittermeier et al., 2008; Wright et al., 2008; Ravaloharimanitra et al., 2011, 2020). The species was thought to be extinct until being rediscovered in 1964 and again in 1972 (Petter & Peyriéras, 1970; Petter et al., 1977). Following rediscovery, *P. simus* was thought to be restricted to a 200 km stretch of the Ambositra-Vondrozo Corridor, but recent surveys have extended the range of the species 345 km northward (Rakotonirina et al., 2011; Ravaloharimanitra et al., 2011; Ramiadantsoa et al., 2015; Eronen et al., 2017). The current known geographic distribution of *P. simus* covers approximately 1,700 km^2 (Hawkins et al., 2018); however, it is now well established that *P. simus* is only patchily distributed within its geographic range (Mittermeier et al., 2008; Olson et al., 2013).

Indriidae

Three genera constitute the taxonomic family Indriidae: *Indri* (the indri), *Propithecus* (the sifakas), and *Avahi* (the woolly lemurs). Indriids are medium- to large-bodied lemurs; body sizes vary greatly between the genera, with adult body weights of *Indri* being in the range of 6.5–8.8 kg (Powzyk & Mowry, 2006), while adult *Avahi* average just 0.8 kg (Thalmann, 2003). The indriids include the two largest lemur species, the indri (*I. indri*) and the diademed sifaka (*P. diadema*). Field data indicate that the heaviest lemurs are adult female *Indri* (Powzyk, 1997; Glander & Powzyk, 1998; Powzyk & Thalmann, 2003; Oliver, 2017). Sexual dimorphism is low among indriids, but females are typically slightly heavier than males. Powzyk (1997), working in Mantadia National Park, found that adult weight of *Indri* females averaged 7.14 kg (n = 2), while the average weight of adult males was 5.83 kg (n = 2). Average adult body weights for

P. diadema were virtually identical for females, 6.51 kg (n = 6), and males, 6.50 kg (n = 5).

All indriid species are highly arboreal. All taxa in the family are characterized by their vertical clinging and leaping locomotor pattern to propel themselves between vertical substrates (Napier & Walker, 1967). As an adaptation to this style of locomotion, indriids have the longest legs in proportion to their arms of all lemurs, with an average intermembral index of 62 (Ankel-Simons, 2007). Prior to leaping, indriids cling to tree trunks or large branches, the body held vertically with the legs in a tightly flexed position; rapid, full extension of those powerful flexed legs push the animal off into its leap. Indriid locomotion can thus be characterized as strongly hindlimb dominated. When leaping, indriids also typically hold their trunk upright while in mid-leap between supports with their arms raised in preparation for contacting the landing site (Oxnard et al., 1990).

Indri are capable of 10 m leaps (Pollock, 1975), while *Avahi*, the smallest in the family, leap an average distance of about 1.5 m (Warren, 1997). Sifakas in the dry forests of southwestern Madagascar (e.g., *P. verreauxi*) will occasionally travel short distances on the ground. In these instances, they spring bipedally, often with the shoulders and trunk rotated in the direction of progression (e.g., Loudon et al., 2006). Perhaps uniquely among its congeners, *P. perrieri* in northern Madagascar will occasionally descend to the ground during the dry season and move as far as a kilometer across open habitat in order to access water in riverbeds (Mayor & Lehman, 1999). Indriids possess long fingers and toes, strong hands, and particularly large, strong feet that allow them to powerfully grasp arboreal substrates when foraging (Quinn & Wilson, 2002). They readily employ suspensory (i.e., below-branch) feeding postures.

Indri

Indri are strictly diurnal and rarely move during the night or during times of low light (dawn and dusk) unless disturbed (Powzyk, 1997). *Indri* occur only in the evergreen rainforests of northeastern and central eastern Madagascar (White, 1983; Quinn & Wilson, 2002). Habitats that are preferred include montane evergreen forests and lowland rainforests. *Indri* have been observed on mountain slopes up to 1,800 masl and also in forests near sea level (Goodman & Ganzhorn, 2004), and they can survive in selectively logged moist forests (Pollock, 1975). Subfossil remains indicate that *Indri* once had a much broader geographic extent of occurrence (Jungers et al., 1995).

Indri males and females are relatively monomorphic. Head and body length measures about 61 cm; only

P. diadema approach them in size with a head and body length of 53 cm (Glander & Powzyk, 1998). *Indri* possess a short, rudimentary tail measuring 5–7 cm (Glander & Powzyk, 1998), unlike other lemuriforms which are all long-tailed (Quinn & Wilson, 2002).

The pelage of *Indri* is black with white patches on the forearms and white to gray areas running down the lateral aspect of the legs. The majority of individuals also have a triangular white pygal region extending up the lower back. The head and face are predominantly black with large tufted ears; some individuals have a distinct crown of white fur. Overall, *Indri* show highly variable markings even within a regional population. Individuals that are almost completely black and individuals with an abundance of white were once considered to be two subspecies of *Indri*, but genetic sampling has not substantiated this view (Thalmann et al., 1993; Mittermeier et al., 2008, 2010). Infant *Indri* are born black and develop a grayish pygal region as they mature. *Indri* eyes have arresting yellow-green irises. When illuminated in their sleeping trees at night, the reflective "eye-shine" that is clearly visible indicates that the eyes of *Indri* possess a functional *tapedum lucidum* (an adaptation to activity under nocturnal/low-light conditions), which is indicative of species that have arisen from nocturnal ancestors.

Propithecus

The sifakas (genus *Propithecus*) form a diverse genus with nine recognized species. Congeners vary in their habitat preferences, behavior, and phenotypic traits. *Propithecus* species have the widest geographic distributions among the indriid genera, occurring across much of Madagascar's remaining forest habitats—ranging from the southern spiny forest, to eastern evergreen rainforests, to seasonally deciduous forests in the west and north of the island (Richard, 2003). A simple dichotomy of dry-forest versus wet-forest sifaka species is sometimes used, although these characterizations are not entirely precise. Taxonomic clusters of species in the genus can be recognized on the basis of geography, i.e., the eastern sifakas (*P. diadema*, *P. edwardsi*, and *P. candidus* all occur in rainforest habitats, plus *P. perrieri* and *P. tattersalli* that occur in more seasonal forests in far northeastern Madagascar; Irwin, 2007) and the western sifakas (*P. verreauxi*, *P. deckenii*, *P. coronatus*, and *P. coquereli* found in the drier forests of the west and south). The eastern sifakas tend to exhibit larger body sizes than the western sifakas (Albrecht et al., 1990; Godfrey et al., 2004; Lehman et al., 2005a), although *P. tattersalli*, with its distribution in far northeastern Madagascar, is closer in body size to the western species. Regionally, western *Propithecus* are referred to

by the Malagasy as "sifaka" (pronounced *shee-fak*), while the eastern *Propithecus* are known as "simpona" (pronounced *shim-poon*). Both of these common names are onomatopoeic words based on mimicking the alarm calls of western and eastern *Propithecus*.

Among lemuriform species, *Propithecus* are large-bodied with weights ranging from 3–6.5 kg (Powzyk, 1997; Lehman et al., 2005a). *Propithecus* have long, slender tails that equal or exceed their combined head and body length. The head is round and plushly furred, while the face is short and mostly hairless. Species in the western and southern dry forest group of *Propithecus* tend to be lighter in color and have lower body weights compared to their eastern congeners in wet-forest habitats (Mittermeier et al., 2010). *Propithecus* utilize all levels of the forest habitats they occupy; they will also descend to the ground to forage, including seeking terrestrial sources of water (Mayor & Lehman, 1999), to engage in geophagy (Norscia et al., 2005), and to play. *Propithecus* descend from trees to the ground by backing down a tree trunk hand over hand. When moving on the ground, they hop and bound bipedally (Figure 3.4), holding their arms raised to the level of their shoulders (or higher) as they move forward.

As in the *Indri*, the eyes of *Propithecus* possess the *tapetum lucidum* (Ankel-Simons, 2007). The geographic ranges of *Propithecus* taxa are allopatric and the current distributions are arrayed in the form of a ring around the island's dry central plateau. Verreaux's sifaka (*P. verreauxi*), found in southwestern and southern Madagascar, are arguably the second most-studied lemuriform species, after the ring-tailed lemur, with which it is sympatric.

Avahi

The woolly lemurs (*Avahi* spp.) inhabit both Madagascar's eastern rainforests as well as dry, deciduous forests of the west. Similar to taxa in the genus *Propithecus*, the *Avahi* species are distributed in both evergreen rainforest and seasonally deciduous forest habitats. The woolly lemurs are the smallest members of family Indriidae, with adult body weights ranging between 700–1,600 g. In the latter part of the twentieth century, the genus *Avahi* was considered to either be monospecific with two subspecies (one eastern and one western; Tattersall, 1982) or to be represented by two species, one eastern and one western (Groves, 2001; Thalmann & Geissmann, 2000). Today, at least nine species of *Avahi* are recognized (see Table 3.1). An interesting observed pattern of sympatry and allopatry occurs with *Avahi* populations and the nocturnal *Lepilemur*. Where *Avahi* are present, a *Lepilemur* species

Figure 3.4 Adult Verreaux's sifaka (*Propithecus verreauxi*), Beza Mahafaly Special Reserve, engaging in terrestrial bipedal locomotion.
(Photo credit: Michelle Sauther, published in Loudon et al., 2006, under Environmental Sciences Commons by University of Nebraska–Lincoln)

will also always be a part of that lemuriform ecological community; however, the reverse is not true—where a lemur ecological community includes a *Lepilemur* species, *Avahi* are not necessarily present (Petter et al., 1977; Thalmann & Geissmann, 2000).

ACTIVITY CYCLES, DIET, AND FEEDING ECOLOGY
Lemuridae
Lemur, Eulemur, *and* Varecia

Since *L. catta* is the most-studied lemur species, we now have decades of fieldwork on this species and a data-rich basis for discerning emergent patterns in ring-tailed lemur socioecology (Sauther et al., 1999) and identifying key questions to pursue in future research (Gould, 2006a). While Jolly (1966) reported hearing *L. catta* anti-predator "yap" vocalizations at night, and considered it possible that *L. catta* could be engaging in nocturnal activity,

ring-tailed lemurs have traditionally been considered strictly diurnal. It has only been relatively recently that investigation of *L. catta* activity patterns has led to the view that the species should be regarded as cathemeral (Donati et al., 2013; LaFleur et al., 2014). Recognition of this shifting diurnal/nocturnal activity pattern adds another layer of complexity to our knowledge of the flexible behavioral ecology of ring-tailed lemurs (Sauther et al., 1999). Indeed, behavioral flexibility is a hallmark of the diet and feeding ecology of *L. catta*—ring-tailed lemurs being very accurately described as opportunistic omnivores. The broad diet of the species includes fruits (pods produced by the tamarind tree, *Tamarindus indica*, are a particularly key resource for *L. catta*; Mertl-Millhollen et al., 2011), leaves, leaf buds, leaf stems, flowers, flower stems, insects and other arthropods, passerine birds (occasionally), and soil, i.e., cases of geophagy (Sauther, 1992).

Cathemeral activity cycles have now been reported in all species of the genus *Eulemur* (Tattersall, 1979, 1987; Curtis, 1997, 2006; Colquhoun, 1998b, 2007; Donati et al., 1999; Rasmussen, 1999; Curtis & Rasmussen, 2002, 2006). Rasmussen (1999) and Curtis and Rasmussen (2002), recognized a trimodal structure in *Eulemur* cathemerality patterns. Their Pattern A refers to the seasonal switch in temporal activity exhibited by *E. mongoz* from being significantly more diurnal during the wet season to being primarily nocturnal during the dry season (Curtis, 1997). Pattern B is exhibited by *E. fulvus* and *E. rufifrons* in western dry-forest habitats and involves a shift from wet season diurnality to 24-hour activity in the dry season. Finally, Pattern C is a year-round 24-hour activity pattern which has been observed in *Eulemur* taxa, as well as *H. alaotrensis* and *H. meridionalis*, found in humid forest habitats (Curtis & Rasmussen, 2002; Eppley, 2015; Eppley et al., 2015a; see also Curtis, 2006, listing reports of lemurid cathemerality from the mid-1970s to 2006). Further, Eppley et al. (2015a) note that cathemerality reported in *Eulemur, Hapalemur, Prolemur,* and *Lemur* is also associated with lunarphilia—that is, nocturnal activity in these taxa increases with increased lunar luminosity around the time of the full moon. This taxonomic pattern of occurrence would be consistent with cathemerality being an ancestral trait of the Lemuridae radiation (Tattersall, 1987; Eppley et al., 2015a) and subsequently being lost in the largely diurnal *Varecia* (Curtis & Rasmussen, 2006; Santini et al., 2015).

Fruits tend to constitute the largest proportion of the diets of *Eulemur* species on an annual basis, but their broad feeding ecologies also see seasonal variability in diets, which include leaves, seed pods, flowers, nectar (both through nondestructive nectar feeding and flower predation, depending on the species being utilized), mushrooms and other fungi, arthropods (occasionally, e.g., millipedes), and intermittent bouts of geophagy. The nondestructive nectar feeding of some *Eulemur* taxa also means that they are potential pollinators of some tree species (e.g., Birkinshaw & Colquhoun, 1998). Because *Eulemur* diets have been characterized by their breadth and seasonal flexibility, there has been a general interpretation that *Eulemur* are ecologically generalized frugivore-folivores with good resilience to ecological disturbances, such as that resulting from anthropogenic factors (Colquhoun, 1997; Ralainasolo et al., 2008). Nevertheless, a meta-analysis comparing dietary flexibility across 10 *Eulemur* and 7 *Propithecus* populations point to *Eulemur* being somewhat specialized toward frugivory, with fruit constituting 65–99% of the total diet, leaves making up < 25% of the annual diet, and seasonal food sources like flowers and nectar also forming part of the *Eulemur* diet (Sato et al., 2016). Both *V. variegata* (Kress et al., 1994) and *E. macaco* (Birkinshaw & Colquhoun, 1998) have been recorded feeding nondestructively on nectar from the endemic tree *Ravenala madagascariensis*, acting as a pollinator in the process. In contrast, *Propithecus* diets are composed of more equal amounts of leaves (particularly young leaves and leaf buds) and fruit, with a good amount of dietary flexibility exhibited across the annual cycle (Sato et al., 2016). Populations of genera with heavier body weights have higher proportions of fruit in their diets, while the *Eulemur* populations in drier habitats exhibit more frequent dietary-switching from fruit to alternative food resources. *Eulemur* feeding strategies include "power-feeding" in exploiting fruit resources—increasing energy expenditure in order to utilize scattered fruit resources (Sato et al., 2016). Alternatively, *Eulemur* populations that experience dry season conditions can exhibit "energy minimizer" patterns of inactivity, reducing levels of activity and energy expenditure (i.e., resting for long periods through the daylight period, reducing daily path length distances, etc.). Dry season conditions can also induce *Eulemur* to feed on food items that they do not utilize at other times of the year (Colquhoun, 1997, 2005).

Freed (1996, 2006) reported that sympatric *Eulemur* species (*E. coronatus* and *E. sanfordi*) in Mt. d'Ambre National Park regularly form polyspecific associations throughout the year, but more frequently during the wet season. When the species are not in polyspecific associations, they each largely utilize separate strata in the forest; *E. sanfordi* prefer foraging in the forest canopy, while *E. coronatus* make greater use of the forest midstory (Freed, 1996, 2006). There is little agonism related to polyspecific associations, but when agonistic interactions arise, it is often related to groups of each species foraging in the same large fruiting trees.

There are some limited data to indicate that, like the other members of the Lemuridae, *Varecia* also exhibits cathemerality. However, definitive data on cathemeral activity in *Varecia* have not been collected thus far. Vasey (2003) reported that while *Varecia* populations exhibit inter-site variation in home range area, community size, and territorial behavior, there are also some consistent patterns that clearly differentiate *Varecia* from the other lemurid taxa. Whereas *Eulemur* clearly prefer fruit, both *Varecia* species are obligate frugivore specialists that occur at highest densities in primary rainforest (Vasey, 1997, 2003). Moses and Semple (2011) examined primary seed dispersal by *V. variegata* at Manombo Forest Reserve (see also Ratsimbazafy, 2006). The larger body size of *V. variegata*, and its rapid gut passage time (mean = 4.4 hours), makes them efficient dispersers of large-seed tree species (Moses & Semple, 2011). Seeds dispersed by *Varecia* show superior germination rates compared to seeds removed from fruits or seeds in whole fruit, reflecting that ruffed lemurs are likely critically important (and possibly the sole) dispersers of tree species with large seeds >10 mm in length that smaller-bodied frugivores cannot swallow (Moses & Semple, 2011; but see Sato, 2012, who suggests that *E. fulvus* also could be a seed disperser of tree species with seeds that are >10 mm in length). Martinez and Razafindratsima (2014) present commensurate data on seed dispersal by *V. rubra* (mean gut passage time = 3.75 hours) in Masoala National Park and reach similar conclusions. As the largest lemur species on the Masoala Peninsula, *V. rubra* are dispersing seeds that are likely too large for most smaller lemur species to ingest, as well as performing other ecological services, such as pollination, that affect forest structure and floristic composition.

In studying the synecology of *V. rubra* and *E. albifrons* at Andranobe, Vasey (2000, 2002) reported that *V. rubra* are canopy foragers making particular utilization of large-crowned fruiting trees since the species' diet is composed primarily of ripe fruit. Vasey (2000, 2002) found that red-ruffed lemurs forage mainly in tree crowns above a height of 15 m; *E. albifrons*, on the other hand, utilize forest strata below 15 m. While both species are primarily frugivorous, their foraging and harvesting of fruit occurs in different-sized trees, in differing quantities, and in different forest strata. Vasey (2000, 2002, 2003, 2006) also reported that both *V. rubra* and *E. albifrons* females have more diversified diets than the males of either species. This trend for diet diversification was most pronounced in female *V. rubra*; Vasey (2000, 2002) showed that the reproductive needs of the female red-ruffed lemurs are the prime driver in niche partitioning between *V. rubra* and *E. albifrons* because of the high energetic investments in reproduction and infant rearing by female red-ruffed lemurs.

Hapalemur *and* Prolemur

No long-term studies have been conducted on *H. occidentalis* to date, but brief surveys suggest that they are cathemeral and are most active at night during the dry season, July–September (Mutschler & Tan, 2003; Martinez, 2008). Group size is often three or four individuals but can be as large as six (Raxworthy & Rakotondraparany, 1988; Martinez, 2008). *H. occidentalis* have been reported to feed primarily on giant bamboo (*Cathariostachys madagascarensis*) in small (about 1 hectare) forest patches but also include ripe fruits in their diet (Birkinshaw & Colquhoun, 2003; Martinez, 2008).

The Alaotran gentle lemur (*H. alaotrensis*) only inhabits the marshland and reed beds along the shores of Madagascar's largest lake, Lac Alaotra, in the northeast of the island. Although bamboo is absent, the "bandro" are able to subsist on abundant papyrus (Cryperaceae) and reed (Poaceae) species (Mutschler, 1999). Dietary diversity was found to be very low. In a 15-month study, Mutschler (1998) found that the Alaotran gentle lemur fed on a total of 11 different plant species and also engaged in geophagy. Moreover, >95% of feeding time was invested in just 4 species of grass from the families Cyperaceae and Poaceae (Mutschler, 1998). Gentle lemurs usually move slowly in the dense vegetation or from one reed stem to another, often over the water. While apparently not a common behavior, Alaotran gentle lemurs are reported to be excellent swimmers (Petter & Peyriéras, 1970, 1975).

The most extensive studies of *H. griseus* have been conducted at Ranomafana National Park, where it is sympatric with *Prolemur simus* and *H. aureus*. At Talatakely, a site within Ranomafana, bamboo comprises >80% of *H. griseus'* overall diet throughout the year. Although liana bamboo species comprise more of the dry season diet, consumption of giant bamboo (*C. madagascarensis*) increases in the wet season (Overdorff et al., 1997; Tan, 1999). The remainder of their diet includes non-bamboo leaves, fruits, and flowers (Overdorff et al., 1997; Tan, 1999; Grassi, 2006). *H. griseus* demonstrate some ecological flexibility as evidenced by the feeding and ranging patterns of a group at Ranomafana for which the density of bamboo groves in their home range was low. When a species of introduced guava (*Psidium cattleyanum*) was fruiting, this *H. griseus* group spent much of their time feeding in the guava stands (Tan, 2000; Grassi, 2006).

Eastern gentle lemurs appear to exhibit intra-specifc variability in their activity patterns. At Ranomafana

National Park, Tan (1999, 2000) found that *H. griseus* are diurnal and spend almost half of their time feeding. However, Vasey (1997, 2000) found that at Andranobe, on the southwest coast of the Masoala Peninsula, *H. griseus* are largely nocturnal. Grassi (2002) found that females exhibit a higher degree of dietary diversity than males, spending more time lower in the forest understory strata (<5 m), whereas males often occupy higher (>11 m) canopy levels (Grassi, 2002). Group movements are cohesive and primarily led by adult females.

The southern gentle lemurs (*H. meridionalis*) are characterized by a darker pelage than their congeners. They occupy marsh habitats and littoral forest fragments in southeast Madagascar (Warter et al., 1987; Eppley, 2015). In a three-month pilot study of this species, Eppley et al. (2011) discovered that, similar to the Alaotran gentle lemur, the diet of *H. meridionalis* consists primarily of species from the grass family Poaceae. *H. meridionalis* exhibit a considerable amount of terrestrial feeding behavior while foraging for grasses (Eppley & Donati, 2009; Eppley et al., 2011, 2016a; Eppley, 2015). Groups have been observed to include as many as seven individuals, but usually consist of one adult male, one or two adult females, and one to two infants or juveniles (Eppley & Donati, 2009). Although huddling has been found to be more important for southern gentle lemur thermoregulation than microhabitat selection, because of the typically small family group size of *H. meridionalis* groups, huddling behavior (especially during the austral winter) likely only confers marginal benefits of social thermoregulation (Eppley et al., 2017a).

Golden bamboo lemurs (*H. aureus*) feed almost exclusively on giant bamboo; 91% of the diet consists of shoots, leaf bases, pith, and the viny tendrils of this bamboo. They forage in a "lawn mower" pattern by depleting the food resources at a specific bamboo culm before moving on to another (Tan, 2000). The bamboo parts that *H. aureus* consume are high in protein but also contain naturally high quantities of cyanide—levels of which would kill most mammals (Glander et al., 1989). In fact, golden bamboo lemurs daily ingest about half their body weight in bamboo parts. This daily amount of cyanogenic bamboo contains ~10–50 times the normal lethal dose of cyanide (Eppley et al., 2017b). While the physiological mechanisms that permit *H. aureus* to avoid cyanide poisoning are not entirely clear, a recent field study found cyanide in the urine, but only rarely in feces, of *Hapalemur* and *Prolemur* at Ranomafana (Yamashita et al., 2010). This supports the hypothesis that the kidneys play a role in filtering out and excreting dietary cyanide (Yamashita et al., 2010), possibly coupled with synchronistic effects from a metabolic pathway involving sulfur-containing amino acids to detoxify cyanide by converting it to thiocyanate (Eppley et al., 2017b).

At Ranomafana, *P. simus* feed primarily (95% of overall feeding time) on giant bamboo with a focus on ground shoots in the rainy months of November–December, and leaves, shoots, and inner pith of the bamboo culm throughout the remainder of the year (Wright et al., 1987; Glander et al., 1989; Tan, 1999, 2000). Olson and colleagues (2013) report that *P. simus* appear to prefer patches of primary forest exhibiting low-to-moderate levels of natural or anthropogenic disturbance; the highest densities of giant bamboo recorded by Olson et al. (2013) were in areas that exhibit moderate anthropogenic disturbance. *P. simus* tend to feed on parts of *C. madagascarensis* that are more mechanically challenging to masticate in comparison to the bamboo parts utilized by *Hapalemur* species—the larger jaws and jaw musculature of *P. simus* likely facilitate this econiche partitioning (Yamashita et al., 2009; Gron, 2010; Fleagle, 2013; Eronen et al., 2017). The greater bamboo lemur is cathemeral, which may be an adaptation to maintain high feeding frequencies in response to their energy-poor diets (Mutschler, 1999; cf. Eppley et al., 2011).

The bamboo and gentle lemurs have exploited what is essentially an open econiche in Madagascar. It is quite remarkable that these six species of bamboo-eating lemurs coexist in some Madagascar forests, having subdivided the bamboo econiche. Niche partitioning is accomplished partly by the differing body sizes of *H. griseus*, *H. aureus*, and *P. simus*, as well as by different bamboo and gentle lemur species specializing in feeding on different species of bamboo, or feeding on different parts of the same bamboo species (Tan 1999, 2000; Wright & Tan, 2016a,b), with food selection being mediated by the chemical and nutrient content of particular food items (Ganzhorn, 1988; Glander et al., 1989; Gron, 2010; Eronen et al., 2017).

Several studies, both long-term fieldwork and new field surveys over the past 20 years, have yielded previously unknown information regarding the dietary, ecological, and social variation within and between the gentle and bamboo lemur species (Dolch et al., 2008; Mittermeier et al., 2008; Yamashita et al., 2010; Ravaloharimanitra et al., 2011; Olson et al., 2013; Eppley et al., 2017b). Further research is still needed, however, to gain a more comprehensive understanding of the behavioral ecology and social structure of these unique primates, particularly the relatively under-studied taxa *H. occidentalis*, *H. meridionalis*, *H. aureus*, and *Prolemur simus*.

Indriidae
Indri, Propithecus, *and* Avahi

The indriids are highly folivorous, with expansive salivary glands and capacious stomachs. With a comparatively long gastrointestinal tract, they also possess an elongated cecum in their mid-gut, which allows microbial processing of plant fiber by symbiotic gut bacteria (hind-gut fermentation), releasing plant nutrients that are then absorbed through the gut wall (Langer, 2003; Richard, 2003).

In habit, *Indri* are highly diurnal lemuriforms; indri rarely move during the night or during times of low light (dawn and dusk) unless disturbed (Powzyk, 1997). They can be described as having a limited daily activity period compared to other lemur species. They are known to be late to rise in the morning and early to retire for the night. During the cooler months of the annual cycle, *Indri* show a reduction in periods of activity compared to the warmer months. By taking long rest periods, *Indri* may be facilitating a more efficient fermentation of leaf fiber in the gut. Although passage rates are unknown, *Indri* defecate approximately twice a day, indicating a lengthy gut retention time (compared to frugivorous *Eulemur* species, for example). Sleeping sites are often in food trees; individuals will either rest on a horizontal substrate or tuck themselves into a forked tree branch. Male and female pairs typically sleep apart, with the female maintaining a higher position in the sleeping tree along with any dependent offspring (Powzyk, 1997).

The most preferred food choices for *Indri* are immature leaves, followed by fruit and/or fruit seeds, flowers, mature leaves, plant galls, and bark. While *Indri* are considered folivores (and certainly exhibit the dental and gastrointestinal adaptations associated with folivory), the breadth of their diet shows their feeding ecology is not strictly folivorous (Pollock, 1975; Powzyk, 1997; Britt et al., 2002; Powzyk & Mowry, 2006). The most preferred plant family for *Indri* is the Lauraceae (laurels) in both montane forests of Mantadia National Park and low altitude forests of Betampona Reserve (Britt et al., 2002; Powzyk & Mowry, 2006). *Indri* were found to consume soil on 23% of the research days in Mantadia National Park and 75% of the research days in Analamazaotra Special Reserve. Soil was typically eaten from areas around tree falls (Pollock, 1977; Powzyk, 1997). When eating leaves, an indri pulls a branch with its outstretched hand toward its mouth (Rigamonti et al., 2005), sometimes selectively nipping off just the leaf petiole and letting the leaf blade drop. Fruit is also picked with the mouth and lips and then transferred to a clutched hand while the indri bites or scoops out seeds from leathery fruits. Feeding bouts for *Indri* can be long—for example, one female was observed feeding continuously on a Lauraceae fruit (*Cryptocarya* spp.) for 2.5 hours (Powzyk, 1997; Powzyk & Mowry, 2006). Trees are the preferred source of plant material (98%), followed by lianas and parasitic plants

(Powzyk, 1997). Data show that *Indri* females spend significantly more time feeding than their male partners (Powzyk, 1997). *Indri* show considerable variability in their use of vertical space within forests. In addition to descending to within one meter of the forest floor (e.g., to engage in geophagy from the bases of uprooted trees or at the forest floor/tree bole interface; Pollock, 1975, 1977), they will also forage throughout the forest canopy, including up into the crowns of emergent trees >30 m in height (Powzyk, 1997).

All *Propithecus* species are diurnal (but not all are as exclusively so as *Indri*). They have been reported to be active into the early evening and may depart their sleeping site before dawn (e.g., *P. diadema*; Powzyk, 1997). Activity data recorded for *P. verreauxi* in Kirindy Forest (collected over an annual cycle through the use of automatic data-loggers) show that the sifakas exhibit a bimodal AM/PM activity pattern that is strictly diurnal (Erkert & Kappeler, 2004). The diets of *Propithecus* species are high in immature leaves, fruits, and flowers. They show a preference for fruits when available and are considered seed predators since they also masticate whole seeds. Other food items in the diets of *Propithecus* taxa include leaf petioles, plant galls, bark, new stems, and soil (Pebsworth et al., 2019; Semel et al., 2019). *Propithecus* feed on a variety of plant types, including trees in all levels of their forest habitats, climbing vines, parasitic plants, ferns, herbs, and fungi. Their gut specializations reveal an extensive cecum for plant fiber fermentation together with an elongated small intestine that provides extensive surface area for sugar and fatty acid absorption, components typical of fruit seeds (Hill, 1953). *Propithecus* show early and relatively rapid dental eruption to facilitate processing of their foliose diet, i.e., high or rich in leafy matter, yet not exclusively (Schwartz, 1974; Godfrey et al., 2004), with crested molars for leaf shearing, dentition which may also be quite effective in seed predation (Yamashita, 2002). Most *Propithecus* species can get enough water from the plant resources they consume, but *P. perrieri* in the seasonally dry forest of the Analamerana Reserve in northeastern Madagascar have been reported to come to the ground and travel up to a kilometer to access local riverbeds in order to obtain water (Mayor & Lehman, 1999), which makes them vulnerable to terrestrial predators (Banks, 2013). Group sizes among *Propithecus* taxa typically range from 2–6 individuals, but larger groups up to 10 or more can occur (Patel, 2006; Lewis, 2008; Irwin et al., 2019). Home range sizes vary in area from 4 ha for dry-forest living *P. verreauxi* (Petter, 1962; Richard, 1978), to 42 ha for *P. diadema* (Powzyk & Mowry, 2006). Wet forest *Propithecus* tend to have larger home ranges than dry forest species.

Wet forest species include *P. diadema, P. edwardsi,* and *P. candidus,* species living along the eastern escarpment in humid evergreen forests that receive up to 3.5 m of rainfall annually. Dry forest *Propithecus* species reside in lower altitude dry scrub and/or dry forest along the western margin of the island and experience 1.5 m or less of rainfall. The most northern species are *P. perrieri* and *P. tattersalli. P. perrieri* reside in semi-humid forests and dry deciduous forests on karst limestone between sea level and 500 m. This species has a highly seasonal diet with fruit constituting 70–90% of food consumption diet during the wet season, with a shift to leaves and flowers in the dry season (Irwin, 2006). Researchers note a lack of intergroup aggression when animals meet during feeding bouts (Lehman & Mayor, 2004). *P. perrieri* have a highly restricted range of fragmented forest constituting ~40 sq km (Salmona et al., 2013). Found in both the Andrafiamena Protected Area and the Analamerana Special Reserve, it is believed extirpated from the Ankarana National Park (Salmona et al., 2013).

The nine currently recognized woolly lemur species are all strictly nocturnal, making the genus *Avahi* an outlier in the activity patterns seen among members of the Indriidae (although, the possibility of cathemerality in *Avahi* species has not been fully explored). The genus is also less studied than either *Propithecus* or *Indri*. Several of the more recently described *Avahi* species have not yet been the subjects of detailed, long-term study (e.g., Thalmann & Geissmann, 2000, 2005; Mittermeier et al., 2010). From the available data compiled during field surveys and short-term studies on *Avahi* taxa, however, some emergent patterns in comparison with the feeding ecologies and diets of other indriids can be identified. Despite their small size (adult body weights across the genus range from approximately 900 g to 1.5 kg (Tattersall, 1982; Mittermeier et al., 2010), the diets of woolly lemurs are composed primarily of leaves, but flowers and fruits are also incorporated seasonally. The feeding ecology of woolly lemurs could be broadly characterized as one of folivory (as are *Indri* and *Propithecus* diets). However, the selective and mixed nature of *Avahi* diets (from what we know) might best be labeled foliose rather than strict folivory.

SOCIAL STRUCTURE AND SOCIAL ORGANIZATION
Lemuridae
Lemur, Eulemur, *and* Varecia

Across these three genera, the composition and organization of social groups show a good deal of variability, both interspecifically and intraspecifically. The multi-male/multi-female social groups of *L. catta* are the largest social groups of any lemuriform species (group size mean = ~13 individuals, group size range = 5–27; Sauther & Sussman, 1993). Female philopatry is the general rule in these social groups; typically, males disperse from their natal groups as they reach young adulthood (Sussman, 1991, 1992), with individual males sometimes transferring between groups in the company of male relatives or age-mates (Gould, 2006b; Parga et al., 2015). Males may change groups multiple times over their adult life; however, female intergroup transfers do occur rarely (Parga et al., 2015). Ring-tailed lemurs are also a lemuriform species that exhibits female social dominance (e.g., group progressions are led by a dominant adult female, and females have priority of access to food sources and can displace males from preferred resting or sleeping sites).

There are two forms of social organization exhibited by the 12 species in the genus *Eulemur*—multi-male/multi-female groups and pair-bonded family groups. The majority of lemur taxa form mid-sized multi-male/multi-female social groups that average around 8–10 individuals but can range in size from as small as 3–5 individuals to as large as 29 (Overdorff & Johnson, 2003). In other words, there can be a considerable intra-population range of social group sizes and compositions. Additionally, flexible fission-fusion social organization has been recorded in some *Eulemur* taxa, with the temporary fissioning of groups into subgroups occurring on an unpredictable basis and with subgroups exhibiting variable compositions (e.g., *E. coronatus*; Freed, 1996). In a field study of *E. macaco* at Ambato Massif, Colquhoun (1997) found that larger black lemur groups fissioned more frequently than smaller groups and that groups in anthropogenically disturbed forest would subgroup more often than groups with home ranges that were in more intact forest. At least some *Eulemur* species exhibit female social dominance. Overdorff and Johnson (2003) list *E. coronatus*, *E. macaco*, and *E. rubriventer* as species for which there is sufficient evidence to support female social dominance (i.e., adult females leading group progressions, often besting males in agonistic encounters, and having priority of access to food sources) as a feature of *Eulemur* social grouping. Female priority of access becomes most apparent when competition arises for quite localized resources or those with limited accessibility (such as water reservoirs in tree holes which only can be accessed by one animal at a time, or nectar-feeding from the large nectar-producing flowers of the traveler's palm (*Ravenala madagascariensis*) (Colquhoun pers.obs. of *E. macaco*).

There are just two *Eulemur* species that form pair-bonded family groups—the mongoose lemur (*E. mongoz*) and the red-bellied lemur (*E. rubriventer*); the family groups in both these species are territorial, with the male-female pair cooperatively defending border areas of their territory. From a review of the literature concerning *Eulemur* scent-marking behavior and olfactory communication, Colquhoun (2011) discerned differing scent-marking patterns between males and females of the two pair-bonded *Eulemur* species (*E. mongoz* and *E. rubriventer*) compared to *Eulemur* taxa that live in multi-male/multi-female social groups. All of the multi-male/multi-female *Eulemur* species exhibit male scent-marking directly on the female members of their groups, but this is a one-way behavior with no *Eulemur* females scent-marking males. However, the two pair-bonded *Eulemur* species both exhibit a scent-marking pattern in which not only the male scent-marks the female, but the female scent-marks the male (especially during intergroup encounters).

The social organizations of *Hapalemur* and *Prolemur* are less studied, but from the available information, there seems to be some degree of intra- and inter-generic variation. The two best-studied gentle lemur species (*H. griseus* and *H. meridionals*) reflect this variability. Social group sizes and compositions reported for *H. griseus* range from 2–7 individuals, with some groups having as many as 11 members (Muschler & Tan, 2003). Eppley et al. (2015a) describe the southern bamboo lemur (*H. meridionalis*) as characterized by a female-dominated social structure similar to congeners.

Many lemurs exhibit suites of traits belonging to the so-called lemur syndrome—a constellation of characteristics which largely differentiate them from all other primates. Wright (1999) recognized the following defining features of the lemur syndrome: (1) female social dominance, (2) targeted female-female aggression, (3) lack of sexual dimorphism across mating systems, (4) occurrence of monogamy in multiple taxa, (5) both sperm competition and male-male contest competition, (6) high infant mortality, (7) presence of cathemerality (especially *L. catta*, all taxa in genus *Eulemur*, and most taxa in genus *Hapalemur*), (8) low metabolic rates, and (9) temporally constrained breeding seasons triggered by photoperiod cues. A number of these features relate directly to, and affect, lemur breeding systems. Wright's (1999) *energy frugality hypothesis* asserts that these traits demonstrate adaptations either to conserve energy or to maximize use of scarce resources in response to Madagascar's sometimes harsh and often unpredictable climatic conditions. Despite the majority of lemurs showing no sexual dimorphism (Glander et al., 1992), female lemurs in several lemurid species typically occupy a dominant position in their social groups (Richard & Nicoll, 1987). Since reproduction—ovulatory cycling, gestation, birth, infant care, and lactation—is a major energetic cost for females, female lemurs take feeding priority in an environment where access to resources is not consistent (Young et al., 1990; Wright, 1999). Indeed, during their research period, Baden and colleagues (2013) found that female black-and-white ruffed lemurs (*V. variegata*) reproduced only once out of six annual cycles, a result attributed to lower available resources in their environment. This emphasizes how resource availability can drastically affect fecundity in Madagascar.

Sharply defined birth seasons are relatively rare across primate species, but across lemuriform species, they are the general pattern (Janson & Verdolin, 2005). The photoperiodicity, or daily sunlight, of Madagascar's distinct annual dry and wet seasons drives reproductive seasonality; this zeitgeber is so strong that even captive lemurs in the northern hemisphere latitudes adjust their mating season to the ambient light levels of their nonnative environment (Rasmussen, 1985). Breeding and gestation, across lemur species, are restricted to the dry season (roughly March–April through July–September, depending on the region of the island where a species is located). Birth and lactation occur during the wet months (with births typically occurring from late August through October, while lactation and, eventually, weaning occur through the latter part of the wet season from February through April) when more food is available and energy needs of a mother are high. This timing of early infant development also means that infants will reach weaning during the wet season when resources are available.

The breeding season is highly synchronized and limited to roughly a two-week window (Wright, 1999). Individual female receptivity is even more concentrated. Females are in estrus annually for as short a period as 4–32 hours, and in the ruffed lemur species (as well as all cheirogaleids and the aye-aye), the vaginal opening is sealed outside of the female's estrus period (Wright, 1999). Male testicular development also tracks female reproductive receptivity, exhibiting reduction in size outside of the breeding season (Glander et al., 1992). Lemurs, being strepsirrhines, have very sensitive olfactory systems (compared to haplorrhine primate taxa), allowing males a way to monitor the estrus status of females. Females may also advertise sexual receptivity via labial swelling and reddening, as in *Lemur, Eulemur, Varecia,* and *Propithecus*.

Tecot (2010) reported 100% mortality for infant red-bellied lemurs (*E. rubriventer*) that were born outside the peak period of births from late August–October. This clearly indicates the presence of environmentally determined limits to the best birth period in order to guarantee high infant survival. Seasonal conditions in Madagascar signify that longer birth seasons are not adaptive. Births

that occur outside the peak periods for specific lemur taxa could lead to scenarios where female lactation and infant weaning would occur during periods of the year when food availability is limited (e.g., dry season), having the effect of reducing infant survival and maturation.

In general, infant mortality among Lemuriformes species is also higher (40–80%; see summary in Wright, 1999) than that of other primate infraorders. In lemur species that are relatively large-bodied, such as *Indri, Propithecus,* and *Eulemur*, females typically give birth to singletons annually or at intervals of two to three years. This contrasts with the reproductive biology of small-bodied lemurs, such as the mouse and dwarf lemurs (genera *Microcebus* and *Cheirogaleus*, respectively), that are more *r*-selected and usually produce annual small litters of two to three offspring, a pattern similar to small non-primate mammals (Wright, 1999). Ruffed lemurs (*Varecia*) represent an exception—although ruffed lemurs are the largest lemurid species, female *Varecia* also give birth to litters of two to three infants and possess three pairs of nipples (two located pectorally, and one pair inguinally). The added maternal reproductive costs of additional offspring are offset by the unusual infant care system that *Varecia* employ involving "crèching," the placement of young in communal nests with communal care. Individuals of all age-sex classes in a group participate in guarding, carrying, and even allonursing group infants (Baden et al., 2013). Thus, lemur mothers offset the risk of infant mortality with a variety of strategies: longer birth intervals to expend more energy on single offspring, large litters, or communal care. Variation in ruffed lemur group structure and social organization from nine field studies is summarized by Vasey (2006). Most *Varecia* study populations have been interpreted as multi-male/multi-female social structures, although a handful of studies have interpreted small *Varecia* groups as exhibiting pair-bonded group structures. Multi-male/multi-female *Varecia* groups exhibit a pattern of fission-fusion/dispersed social organization. In short, *Varecia* have a complex social organization which involves large dispersed communities (or "nuclear neighborhoods"; Baden et al., 2021), composed of multiple smaller core groups—a very fluid and dynamic system (Vasey, 2006; Baden et al., 2021). Additionally, Baden and colleagues report that female *V. variegata* occupy home ranges that are significantly larger than those of males; moderate home range overlap occurs at equal levels for both sexes.

Hapalemur *and* Prolemur

H. griseus live in small groups of three to seven individuals (mean = 4.5), and groups typically contain one adult male and one to two adult females. The average home range size is 15 ha, and average day ranges are 425 m (range = 375–495 m) (Wright, 1986; Tan, 2000). *H. griseus* groups use "latrine" behavior (i.e., sites of habitual defecation) as a mode of intergroup resource defense (Irwin et al., 2004) as do groups of *H. meridionalis* (Eppley et al., 2016b). (There is one report of *L. catta* at Beza Mahafaly Reserve in southwestern Madagascar also using latrine behavior; Loudon et al., 2006). At Ranomafana, population density of *H. griseus* is estimated at 9.8–26.7 individuals/km². Higher population density was correlated with a greater degree of habitat disturbance (Grassi, 2006). *H. griseus* usually give birth to one infant following a gestation period of 137–140 days. Infants are born October–January (Petter & Peyriéras, 1970; Tan, 2000). The infant clings to the mother's back from birth and, based on observations in captivity, both the male and female will carry the infant (Petter & Peyriéras, 1970, 1975). Both sexes disperse from their natal group, and female social dominance has been reported (Grassi, 2001; Tan, 2006).

H. alaotrensis live in small multi-male/multi-female groups of two to nine individuals (mean = 4.3; Mutschler, 1998; Nievergelt et al., 2002). Adult sex ratios approximate 1:1 on average, but group compositions are variable. Groups of *H. alaotrensis* may contain one adult male and one adult female, or one adult male and two adult females, or one adult female and two adult males, or several adults of both sexes (Nievergelt et al., 2002). Despite this variability in group composition, Nievergelt and colleagues (2002) found only one breeding male and one or two breeding females per group. Although Petter and Peyriéras (1975) reported that, at certain times of the year, 30–40 animals often gather together, Mutschler (1998) did not observe any large seasonal aggregations. In fact, more recent data indicate that *H. alaotrensis* are territorial and defend large areas of their home ranges (Nievergelt et al., 1998). Mutschler (1998) observed both male and female dispersal, with females leaving their natal groups as subadults and males transferring after reaching adulthood. Waeber and Hemelrjk (2003) reported female dominance in Alaotran gentle lemurs evidenced by males exhibiting submissive behavior toward females in 83% of intersexual social encounters. Additionally, females lead group movement more often than males. Unlike all other lemuriforms except the aye-aye (*Daubentonia madagascariensis*), the Alaotran gentle lemur does not have a highly restricted birth season; births are spread over a six-month period, and 40% of births are twins. Reproductive maturity is reached at two years of age in females and three years in males. This gentle lemur species was found to have a crepuscular activity cycle which does not differ significantly between seasons (Mutschler, 1998; Mutschler et al., 2000).

The social system of *H. aureus* appears to be similar to that of *H. griseus*, with groups that contain two to four individuals. Golden bamboo lemur groups spend 53.6% of their daily activity budget resting, 36.9% feeding, and 8.4% traveling (Tan, 2000). Average home range size is 26 ha but can be as large as 80 ha (Tan, 1999, 2000). Like *H. griseus* at Ranomafana, *H. aureus* is diurnal. Females give birth to a single infant during November–December, and the birth interval is one year (Mutschler & Tan, 2003). *H. aureus* has been described as extremely territorial. Their vocal repertoire includes a complex duetting with sex-specific call segments used as a territorial defense mechanism against conspecifics. Golden bamboo lemur groups often begin their days with male-female territorial vocal duets (Tan, 2006).

Greater bamboo lemur (*P. simus*) groups, ranging in size from 7–11 individuals, occupy home ranges of >100 ha, with an average size of 60 ha (Meier et al.,1987; Wright et al., 1987; Tan,1999, 2000). However, at some sites, larger groups numbering up to 28 individuals have been observed (Tan, 1999, 2000, 2006; Mutschler & Tan, 2003; Wright et al., 2008). Within social groups, in contrast to many other lemur species, males have been reported to exhibit social dominance (Andriaholinirina et al., 2003; Tan, 2006). The breeding season occurs in May–June, and the reported gestation length for *P. simus* is 149 days (Tan, 2000). Sexual maturity is reached by both sexes at two years, and males disperse from their natal group between three to four years of age (Tan, 2000).

Indri, Propithecus, and Avahi

Indri family groups generally consist of a pair-bonded adult male and female with their immature offspring. This species is noted for its remarkable (and loud!) vocal communication. *Indri* are the only lemur species with a duetting long call between a pair-bonded male and female. When calling, indri utilize their large laryngeal throat sac. The song commences with a common "roar" from the pair and then the duet ensues, which can last over two minutes and be heard over a radius of several kilometers depending upon the terrain (Pollock, 1975, 1986). Male indri exhibit higher frequencies in their songs than females, and the sexes have periods of co-singing (Gamba et al., 2016). Within pair-bonded social groups, the pair may have maturing offspring that also sing during the mating pair's duet. The singing behavior is used as a territorial defense/advertisement and can assist groups in monitoring their spacing with neighboring groups. The duets are also contagious, with a neighboring group commencing its own duet as the initiating group completes its call. During the mating season, *Indri* perform their singing duet more frequently in the day but do occasionally duet after sunset (Powzyk, 1997).

Pair-bonded male and female *Indri* were, in the past, interpreted as being monogamous and were thought to mate for life, but extra-pair copulations have been observed when a female trespasses into an adjoining territory and mates. In one study the female bore an offspring after extra-pair copulations, but her resident male was revealed to be the father through genetic testing (Bonadonna et al., 2014). When *Indri* mate, the male sniffs the female's urogenital region and mounts her from behind, often disregarding her aggression (Bonadonna, pers.comm.). Mating is highly seasonal and occurs during the hot rainy season (starting in November) with infants born June–August during the cooler dry season. Gestation is approximately 150 days (Bonadonna et al., 2014). Newborns cling to the lower ventrum of the mother and move to riding on the mother's back at four to five months of age; young begin to move independently at around 8 months, always maintaining close proximity to the mother. The reproductive biology of *Indri* is relatively *K*-selected (for a lemuriform species). Whereas *Eulemur* and *Lemur* females can have an offspring annually after a gestation of about 124–125 days, adult female *Indri* have birth intervals of two to three years under good resource conditions. A young *Indri* will nurse for about a year, and during the subsequent weaning stage, the offspring is vulnerable (in a life history survivorship sense); juvenile individuals may be at greater risk of predation. *Indri* groups are generally small, composed of two to six individuals—a breeding-age adult male and female pair and any maturing offspring (Garbutt, 1999; Bonadonna et al., 2014). Larger groups, with multiple offspring of varying ages, have been observed in fragmented forest habitat of the Analamazaotra Special Reserve. Territory size also varies; *Indri* living in primary evergreen forest in Mantadia National Park utilize territories of 34–40 ha (Powzyk, 1997), while territories in Analamazaotra Special Reserve are only 17.7–18 ha (Pollock, 1979). In the Maromizaha Forest, *Indri* maintain exclusive territories of 12 ha in an area that reaches 800–1,200 m in elevation (Bonadonna et al., 2014). Territories appear to be exclusive, maintained by the duetting of the resident pairs, as well as their scent-marking.

All *Propithecus* are known for their conspicuous scent-marking habits using scent glands and urine. Females anogenital scent-mark throughout their territories as do males, and the scent-marking rates increase prior to and during the mating season. In addition, males have a gular gland in the sternal region that is used more frequently as the mating system approaches. Males rub the sternal mark on trees/substrates and often overmark (or endorse) female anogenital scent-marks by first biting the area of the female's mark, then rubbing a sternal mark, followed by an anogenital mark (Powzyk, 1997; Patel, 2009).

Female dominance (or female *power*; see Lewis, 2020) has been recorded in all *Propithecus* species; the sexes are essentially monomorphic with some females being slightly larger than males (Richard, 1978; Pochron et al., 2003; Kappeler & Schaffler, 2008; Ramanamisata et al., 2014). *Propithecus* females often show authority with a bite and/or cuff (hand slap) toward a male to procure a desired food item, resting place, or sleeping spot.

Avahi socioecology can be generally summarized as involving members of small family groups engaging in lots of huddling, or at least sitting in close proximity to the other members of a family group. While there are good conceptualizations of the specialized folivorous feeding ecology and ranging behavior of *Avahi* (Thalmann, 2001, 2003; Thalmann & Geissmann, 2006; Norscia & Borgognini-Tarli, 2008)—that is, highly selective feeding on relatively rare large tree species—insight into the subtle details of *Avahi* social life awaits future detailed fieldwork.

PREDATION

Endemic predators of indriids and lemurids include the carnivoran fossa (*Cryptoprocta ferox*, family Eupleridae), which is capable of preying on adult *Propithecus* and young *Indri* (Goodman et al., 1993; Powzyk, 1997; Wright et al., 1997; Hart, 2000, 2007; Powzyk & Thalmann, 2003; Colquhoun, 2006). Wright (1998), in reviewing a decade of behavioral data on *P. edwardsi* from Ranomafana National Park, noted that Milne-Edwards sifakas give a distinctive call in response to fossa and use nocturnal sleeping sites that are higher up in the forest canopy than their diurnal resting sites. This choice of higher, more inaccessible nocturnal sleeping sites is consistent as a counter-maneuver to nocturnal hunting by the fossa (Wright et al., 1997). The other 9 extant species of the endemic carnivoran family Eupleridae, such as the widely distributed ring-tailed mongoose (*Galidia elegans*), and the introduced Indian civet (*Viverricula indica*), are all smaller than *C. ferox*. While these carnivores present a predation risk to the small nocturnal lemuriforms (Wright & Martin, 1995; Hart, 2000, 2007), none represent the predation risk posed by *C. ferox*. The fossa will include small nocturnal lemurs in its diet (Goodman et al., 1993; Rasoloarison et al., 1995; Ganzhorn & Kappeler, 1996; Goodman, 2003) but is also capable of subduing prey that weigh as much as 90% of its own body weight (Hawkins, 2003). In response to a terrestrial predator, notably the fossa, *Indri* give an exhaled "hoot";

if the predator increases its proximity to an indri group, members will issue a loud "honk-honk" call. Feral dogs and cats are also predators of lemuriforms (Gould & Sauther, 2007; Brockman et al., 2008).

Avian species that are predatory threats to lemuriforms include the Madagascar harrier-hawk (*Polyboroides radiatus*) and the Madagascar buzzard (*Buteo brachypterus*) (Goodman et al., 1993; Brockman, 2003). Both *P. radiatus* and *B. brachypterus* have been known to attack infant *Indri*, and both of these raptors are large enough to prey on *Eulemur* taxa. When an aerial threat is detected, adult *Indri* quickly sound a roaring bark alarm and often back down a tree to take cover. *Eulemur*, too, will give acoustically distinctive calls if a harrier-hawk is sighted circling overhead—for example, the black lemur, *E. macaco*, invariably gives a "scream-whistle" call when a harrier-hawk is sighted (Colquhoun, 2007). Even the silhouette of a large bird flying overhead can trigger these distinctive alarm calls. Besides *P. radiatus* and *B. brachypterus*, other raptor species that may elicit alarm calls from lemurs include the Madagascar cuckoo hawk (*Aviceda madagascariensis*), black kite (*Milvus migrans*), Madagascar fish eagle (*Haliaeetus vociferoides*), Madagascar serpent eagle (*Eutriorchis astur*), France's sparrowhawk (*Accipiter francesii*), Madagascar sparrowhawk (*A. madagascariensis*), and Henst's goshawk (*A. henstii*) (Hart, 2000). With a wingspan approaching 1 m, *A. henstii* is a known predator of lemuriforms, particularly smaller species and immature individuals of larger taxa. The same can be said for France's sparrowhawk (*A. francesii*); while not large enough (wingspan = ~50 cm) to take an adult *Eulemur*, *A. francesii* is a threat to young *Eulemur* individuals that are beginning to locomote on their own. For example, a France's sparrowhawk was observed swooping within striking distance, but not attacking, a pair of four-month-old black lemurs (distant from any adult members of their group) that were clambering and playing on a liana about 2 m off the ground (Colquhoun, pers.obs.).

The Madagascar ground boa (*Acrantophis madagascariensis*) is a potential predator of indriids and lemurids when they come to the ground to forage or to play. *E. macaco* will mob boa constrictors (Colquhoun, 1993). Burney (2002) provides an anecdotal account of an adult female Coquerel's sifaka (*P. coquereli*) almost falling prey to an approximately 3-meter-long Madagascar ground boa. The sifaka was already encoiled by the boa, but still alive, when Burney and colleagues investigated mobbing calls being given by the sifaka group. Burney's party succeeded in freeing the sifaka from the boa's coils without harming the constrictor. Once freed, they noticed the female sifaka's nipples were enlarged, and there were two other females with infants in the sifaka group. They also noticed

a slight bulge in the snake's esophagus leading them to conclude that the snake had taken the female's infant and caught her also when she tried to save the infant (Burney, 2002). More recently, Gardner and co-authors (2015) report a rare observation of a group of Coquerel's sifakas (*P. coquereli*) swarming a large ground boa that had struck and encoiled an adult female member of the group; the sifakas scratched and bit the snake around its head and, after about 20 minutes, the captured sifaka succeeded in squirming loose from the snake's coils, biting and injuring the snake's jaw in the process. The injured jaw resulted in the snake's death due to starvation two months later, providing a case study illustration of costs that can be incurred during predation attempts on a group-living primate (Gardner et al., 2015).

Cases of boa constrictor predation on *Hapalemur* species have also been reported. Rakotondravony et al. (1998) found a Madagascar tree boa (*Sanzinia madagascariensis* = *Boa mantidra*) in the process of suffocating an adult *H. griseus*. It took the meter-long tree boa about an hour to suffocate the gentle lemur and another hour to swallow it. Eppley and Ravelomanantsoa (2015) discovered predation of an adult female *H. meridionalis* from their study population when they tracked a radio-collar signal to a large (155 cm, 2.95 kg) male Dumeril's ground boa (*Acrantophis dumereli*). The snake defecated the still functioning radio-transmitting data-logger tag about two weeks after the predation was discovered. Data from the radio transmitter proved to be retrievable, showing when the adult female *H. meridionalis* likely was captured and constricted by the snake. *H. meridionalis* regularly come to the ground to feed on grasses (Eppley, 2015; Eppley et al., 2016a), which may be an inherent predation risk to gentle lemurs (Eppley & Ravelomanantsoa, 2015).

CONSERVATION STATUS

Like most other Lemuriformes, lemurid and indriid species have been of conservation concern for many decades (e.g., Richard & Sussman, 1975; Tattersall, 1982; Martin, 1995), primarily from loss of habitat due to ongoing extensive deforestation (Green & Sussman, 1990; Vieilledent et al., 2018), from growth of human communities (Salmona et al., 2014), from agriculture and mining (Estrada et al., 2017, 2018), and from hunting for food and the bushmeat trade (Golden, 2009; Jenkins et al., 2011; Borgerson et al., 2016). Forest fragmentation throughout Madagascar has created small, isolated lemur populations and has limited or interrupted gene flow between populations (Quéméré et al., 2010a,b; Nunziata et al., 2016). An illegal pet trade in wild lemurs is an additional threat to lemur populations (Reuter et al., 2016). For example, trade in wild-caught ring-tailed lemurs, among other lemuriform species, has contributed to a steep decline of *L. catta* populations (Gould & Sauther, 2016; LaFleur et al., 2016b, 2018, 2019; but see Murphy et al., 2017). Broadly speaking, of the 21 species of lemurids listed in the International Union for the Conservation of Nature (IUCN) Red List of Threatened Species, 8 have been declared Vulnerable, 7 are Endangered, and 6 are Critically Endangered; similarly, of the 19 species of indriids, 4 are Vulnerable, 4 are Endangered, and 11 are Critically Endangered (IUCN, 2020; see Table 3.2).

Taxa of the genera *Hapalemur* and *Prolemur* serve as a microcosm of the range of conservation concerns and challenges looming over the taxonomic families Lemuridae and Indriidae. The marsh vegetation that supports the Alaotran gentle lemur has decreased dramatically over the past 50 years (Mutschler & Feistner, 1995; Pidgeon, 1996; Mutschler et al., 2001). This is primarily due to encroachment of agriculture around Lake Alaotra (Ratsimbazafy et al., 2013), but this gentle lemur species also is suffering from hunting pressure (Feistner & Rakotoarinosy, 1993). In 1994, the total population size of *H. alaotrensis* was estimated to be approximately 7,500 individuals (Mutschler & Feistner, 1995). An identical follow-up survey conducted in 1999 estimated a 30% decline in the total population size of the species (Mutschler et al., 2001; Ratsimbazafy et al., 2013). Another species, *H. aureus*, has a patchy geographic distribution throughout their range and low population densities in the habitats that they occupy. Given this, the global population of the species may be no more than 1,500 individuals (Irwin et al., 2005).

Table 3.2   Current Conservation Status of Lemurid and Indriid Taxa

Species	IUCN Conservation Status	References[a]
Lemur catta	Endangered	LaFleur and Gould, 2020
Eulemur fulvus	Vulnerable	Irwin and King, 2020
E. rufus	Vulnerable	Razafindramanana et al., 2020a
E. rufifrons	Vulnerable	Johnson et al., 2020b
E. albifrons	Vulnerable	Borgerson et al., 2020a
E. sanfordi	Endangered	Chikhi et al., 2020
E. cinereiceps	Critically Endangered	Johnson et al., 2020a
E. collaris	Endangered	Donati et al., 2020b
E. coronatus	Endangered	Reuter et al., 2020
E. macaco	Endangered	Andriantsimanarilafy et al., 2020
E. flavifrons	Endangered	Volampeno et al., 2020
E. rubriventer	Vulnerable	Irwin et al., 2020a
E. mongoz	Endangered	Razafindramanana et al., 2020b
Varecia variegata	Critically Endangered	Louis et al., 2020c
V. rubra	Critically Endangered	Borgerson et al., 2020b
Hapalemur griseus	Vulnerable	Irwin et al., 2020b
H. occidentalis	Vulnerable	Eppley et al., 2020c
H. meridionalis	Vulnerable	Donati et al., 2020a
H. alaotrensis	Critically Endangered	Ralainasolo et al., 2020
H. aureus	Critically Endangered	Wright et al., 2020b
Prolemur simus	Critically Endangered	Ravaloharimanitra et al., 2020
Indri indri	Critically Endangered	King et al., 2020
Propithecus diadema	Critically Endangered	Irwin, 2020
P. candidus	Critically Endangered	Patel, 2020a
P. edwardsi	Endangered	Wright et al., 2020a
P. perrieri	Critically Endangered	Heriniaina et al., 2020
P. tattersalli	Critically Endangered	Semel et al., 2020
P. verreauxi	Critically Endangered	Louis et al, 2020b
P. deckenii	Critically Endangered	King and Rakotonirina, 2020
P. coronatus	Critically Endangered	Razafindramanana et al., 2020c
P. coquereli	Critically Endangered	Louis et al, 2020d
Avahi laniger	Vulnerable	Patel, 2020b
A. mooreorum	Endangered	Eppley et al., 2020b
A. peyrierasi	Vulnerable	Eppley and Patel, 2020a
A. betsileo	Endangered	Eppley and Patel, 2020b
A. ramanantsoavanai	Vulnerable	Eppley et al., 2020d
A. meridionalis	Endangered	Donati et al., 2020c
A. occidentalis	Vulnerable	Eppley et al., 2020e
A. unicolor	Critically Endangered	Louis et al., 2020e
A. cleesei	Critically Endangered	Louis et al., 2020a

[a] Individual species conservation assessments can be accessed directly through the IUCN Red List (IUCN, 2020).

P. simus occur at low population densities and are considered more territorial and terrestrial than *H. aureus* or *H. griseus* (Tan, 2000; Wright et al., 2008). The greatest immediate conservation threats to the continued existence of *P. simus* are habitat destruction and hunting (Meier & Rumpler, 1987; Wright et al., 2008, 2009; Delmore et al., 2009). However, significant concerns also exist over the possible future impacts of climate change on *P. simus*, and the implications this could have on maintaining genetic diversity in *P. simus* subpopulations (Eronen et al., 2017; Hawkins et al., 2018). Over the past decade, the overall population size of *P. simus* has been estimated at only 200–400 individuals scattered across at least 12 small subpopulations (Wright et al., 2008, 2009; Delmore et al., 2009; Gron, 2010). However, recent reassessment of *P. simus* distribution and population genetics has yielded a total population estimate of approximately 1,000 individuals (Dolch et al., 2004, 2008; Wright et al., 2008, 2009; Hawkins et al., 2018).

Conservation prospects for the genera *Lemur, Eulemur, Varecia, Indri, Propithecus,* and *Avahi* are also increasingly of concern. Descriptions of the entire Infraorder Lemuriformes as the most endangered group of mammals on the planet has become a widespread reflection of the deteriorating conservation status of almost all known lemur taxa (Schwitzer et al., 2014b). In 2020, the Madagascar Section of the IUCN/SSC Primate Specialist Group released a conservation status update classifying ~94% of all lemur species as Threatened (Vaughan, 2020; see also Table 3.2). In this context, "threatened" is a general label to reflect that ~94% of lemur taxa have been assessed as falling into one of the IUCN conservation threat categories of Vulnerable, Endangered, or Critically Endangered.

As an example, Perrier's sifaka (*P. perrieri*) has been alarmingly affected by current habitat loss from human activities and may now be limited to 500 animals total (Mayor & Lehman, 1999; Salmona et al., 2013). Banks et al. (2007) estimated that the effective population size (i.e., total pool of breeding age adults) of *P. perrieri* is not likely to exceed approximately 230 individuals. The IUCN Red List has classified *P. perrieri* as Critically Endangered, and the species also has been previously included in the "25 Most Endangered Primates in the World" list (Mittermeier et al., 2006).

The interconnected (and often synergistic) forces that are degrading the conservation statuses of lemuriform taxa are formidable, spanning spatiotemporal scales from immediate on the ground contexts, such as the impacts of cyclones or El Niño (Dunham et al., 2008; Lewis & Rakotondranaivo, 2011; Zhang et al., 2019), to long-term phenomena, such as habitat fragmentation, that could take decades to unfold (Cardillo et al., 2006; Irwin et

al., 2010; Volampano et al., 2015; Razafindratsima et al., 2018; Schüßler et al., 2018; Eppley et al., 2020a). Lemur researchers find themselves in the unenviable situation of being in an accelerating information loop. Over the last 25 years, there has been an information avalanche of new insights on lemuriform behavioral ecology—we have never known more about all aspects of lemuriform biology, from physiology to population genetics, to social behavior, to reproductive biology and socioecology, to feeding ecology, comparative ecology, synecology, and community ecology, to ethnoprimatology. Yet, at the same time that we gain valuable new data from field studies, there is the undeniable assessment that situations continue to worsen concerning lemuriform conservation, the future existence of Madagascar's lemurs, and the protection of their endangered habitats (Vaughn, 2020). In short, at a time when we now know more about lemuriform primates than ever before, there has never been a greater need to apply that multifaceted lemuriform biology body of knowledge to in situ lemuriform conservation (Schwitzer et al., 2013, 2014b).

Ganzhorn and colleagues (2001) sounded an early alarm on the critical conservation status of three habitat types among Madagascar's biodiverse biomes: the eastern littoral forest, the western dry deciduous forest, and the evergreen forests in the island's mountainous domain. But even the sober assessment by Ganzhorn et al. (2001) did not anticipate the rapid unfolding of the current challenges facing lemuriform conservation (Schwitzer et al., 2014b; Vaughn, 2020). The use of satellite mapping (and, where available, older aerial survey photography) of forest cover to monitor ongoing forest disturbance, fragmentation, and loss has advanced by leaps and bounds (e.g., compare Green & Sussman, 1990; Irwin et al., 2005; Vieilledent et al., 2018); at the same time, the alarming growth and rapid pace of these negative factors has been underscored (Schüßler et al., 2018; Tinsman, 2020). While a good understanding of the altitudinal distributions of lemur taxa has been gained (Goodman & Ganzhorn, 2004), and it is clear that lower elevation forest habitats harbor the greatest levels of lemuriform abundance and species diversity (Schwitzer et al., 2013, 2014b; Campera et al., 2020), these are exactly the same forest habitats and lemur populations that are under the greatest threat from anthropogenic pressures (Green & Sussman, 1990; Baden et al., 2019; Campera et al., 2020). Where the areas to conserve for maximum protection of lemuriform biodiversity cannot be effectively protected from anthropogenic pressures because of ongoing human activities, there are also additional present conditions and future outcomes on the horizon that even further complicate lemuriform conservation. We know, for example, that while lemurid and indriid taxa can persist in disturbed or degraded forest habitats, there are costs to this both in terms of decreased adult body condition (i.e., wasting) and compromised immature growth and development (i.e., stunting) (Irwin et al., 2019). The prospects of future altered climatic patterns (Graham et al., 2016; Kamilar & Tecot, 2016; Carvalho et al., 2019), and the effects of those climatic shifts on lemur habitats (e.g., Morelli et al., 2020; Stewart et al., 2020), will create latent extinction risks for lemuriform populations and species (Cardillo et al., 2006). Conversely, any future extinctions of large-seed dispersing lemur species would also create a domino effect of reduced recruitment of the large-seeded tree species and a concomitant effect on forest plant community composition and structure (Federman et al., 2016; Razafindratsima et al., 2018).

An illustration of the long-term problem of latent extinction risk to lemur populations comes from a Population Viability Analysis (PVA) recently conducted by Volampeno et al. (2015) on the blue-eyed black lemur (*E. flavifrons*) population resident in the Sahamalaza Peninsula National Park, in northwestern Madagascar. The Ankarafa Forest within the park harbors what is thought to be one of the largest remaining *E. flavifrons* populations (estimated at 228 individuals) in the very limited geographic distribution of the species (Mittermeier et al., 2008, 2010; Schwitzer et al., 2014a; Volampeno et al., 2015; Tinsman, 2020). Results of the multiple iteration (n = 100) PVA modeling for each of six different scenarios (in which population characteristics were varied), over simulation periods of 100 years, indicate that this population of *E. flavifrons* would become extirpated under all six modeled scenarios. Time to first extinction of *E. flavifrons* sub-populations ranged from just over 11 years under severe variable conditions, to just over 44 years under more moderate conditions. In all six scenarios, however, by the end of the 100-year simulation period, the lemur population had just 20–30% survival probability, i.e., the modeled population in each of the six scenarios was approaching functional extinction by 100 years (Volampeno et al., 2015).

For some lemuriform populations, however, the fast-changing conditions on the ground will lead to the extirpation of populations, either through extensive habitat loss, or from bushmeat hunting, or a combination of the two (Godfrey & Irwin, 2007) before latent extinction processes or climate change effects play out in years ahead. Ripple and co-authors (2016) present data showing that no other primate habitat nation has more endemic primate species that are threatened by bushmeat hunting than Madagascar. The illegal live capture of lemurs for the pet trade is an additional threat. In a multisite survey across 17 communities over northern and central Madagascar, Reuter et al. (2016) found the keeping of lemurs to be a widespread practice involving a variety of lemuriform taxa. Nonetheless, live capture of lemurs was most often by the individuals that kept the lemurs, sometimes just for brief periods of about one week, although some lemurs were kept for years. Reuter and colleagues (2016) concluded from their survey that, unlike trade in live primates in some other regions of the world, trade in pet lemurs was not a highly organized, networked operation. Still, Reuter et al. (2016) cautioned that even a decentralized pet trade could pose a threat to several endangered and critically endangered lemuriform species.

The extremely limited and fragmented geographic distributions of many lemuriform species will become more fragmented as a result of continuing anthropogenic practices and activities. Illustrating this emergency situation is an example that crystallizes the conservation situation of Madagascar's lemurs. In 1990, a previously unrecognized form of woolly lemur was identified in the Tsingy de Bemaraha National Park in western Madagascar; its distribution seemed very closely tied to the western edge of Bemaraha's limestone formations and the vegetation that grows among the crags adjacent to the foot of the limestone massif (Thalmann & Geissmann, 2000). This woolly lemur was sighted at two forest localities just inside the boundary of the national park, as well as at a third site, a very disturbed patch of forest, just outside the park boundaries. In 2005, that previously unidentified woolly lemur was assigned the taxonomic name *Avahi cleesei* (Thalmann & Geissmann, 2005), Cleese's woolly lemur, named in honor of British actor, comedian, and writer John Cleese of *Monty Python* fame, for his long-time service to lemur conservation. In the type description defining *A. cleesei*, Thalmann and Geissmann (2005) explicitly recommended that the newly named Cleese's woolly lemur go directly to the Endangered, or even Critically Endangered, category on the IUCN Red List. The type description also reported that the degraded forest patch where *A. cleesei* individuals were sighted in 1994 had been destroyed by 2003 (Thalmann & Geissmann, 2005). In short, at present we know little about *A. cleesei* other than it appears to have a patchy and highly localized distribution (Dammhahn et al., 2013), favoring sub-humid forest habitats. First brought to the attention of the primatological community in 1990, by 2020 *A. cleesei* was considered Critically Endangered, and now, just 30 years since *A. cleesei* was first noted, it may be slipping toward extinction before it has been studied in any detail. Most alarming is the fact that the decline of Cleese's woolly lemur is *not* an isolated case across the Infraorder Lemuriformes. We are facing dire straits concerning the future of lemur conservation.

There are applied research measures that can be taken to bolster the forest habitats of lemurs and foster conservation of these microendemic primate taxa (Wilmé et al., 2006). Targeted tree-planting, particularly the planting of lemur food tree species, is conceptualized as eventually regenerating forest ecology through lemur dispersal of the seeds of those trees (Steffens, 2020). Similarly, precise selection of pioneer tree species that could also serve as a food plant for lemuriforms would be another approach to not just maintaining forest cover but also regenerating habitat corridors between larger blocks of forest (Eppley et al., 2015b). Such reforestation projects operating at a local level have been incorporated into several lemur conservation projects in different regions across Madagascar (Schwitzer et al., 2013, 2014b). Indeed, developing community-based management of local conservation initiatives and protected areas is one of three key elements in the most recent IUCN Action Plan for lemur conservation (Schwitzer et al., 2013), together with promotion of responsible ecotourism, and continuing long-term collaborative field research in key lemur habitats (Schwitzer et al., 2014b). Recently, there has been growth in Madagascar's New Protected Area (NPA) system of community-managed protected areas. While many of these NPAs are small in area, and there are still questions about their effectiveness over the long term (Gardner et al., 2018), prior to political and public health crises, a number of NPAs had successfully become destinations for ecotourists (Schwitzer et al., 2013). There are also collaborative feedback mechanisms being applied to allow local communities to participate as partners in the management of protected areas, and for both protected area managers and researchers to work collaboratively (Schwitzer et al., 2014b; Pyhälä et al., 2019).

SUMMARY

Lemurids and indriids have been the foci of substantial research efforts dating back to the first lemur field studies, and this research represents a major body of comparative data in the larger history of field primatology. As well, lemurs are an important source of comparative data in ethnoprimatological research to better understand the dynamics of interactions between nonhuman primate populations and adjacent human populations.

Recent research attention on lemurid and indriid diets and feeding ecologies has illuminated how widespread cathemeral activity patterns are among the lemurids. Feeding ecology research on both families has elucidated both the degrees to which lemurids and indriids exhibit flexible feeding ecologies and, conversely, the extent to which certain species may be feeding ecology obligates. For example, a comparison of feeding ecology data on *Eulemur* and *Propithecus* taxa (Sato et al., 2016) yielded the somewhat surprising conclusion that *Eulemur* species are largely frugivore specialists, while *Propithecus* species (often previously considered to be folivore specialists) have, in comparison, broad selections of food species and good ability to employ food-switching tactics as local conditions demand.

Research into lemurid and indriid social structure and social organization has revealed not only broad patterns of female social dominance in many taxa, but also the extent to which some lemurid taxa exhibit flexibility in their social organization (e.g., fission-fusion within multi-male/multi-female social groups). Long-term field studies have clarified our understanding of dispersal patterns across these taxa. However, species representation in field studies of social structure and social organization is uneven, so what might appear to be interspecific comparative patterns need to be considered open for reinterpretation as data on less-studied species become available. In particular, details on the social structure and social organization dynamics of most *Avahi* species are poorly known. Therefore, our current understanding of lemurid and indriid social dynamics must be seen as preliminary and incomplete, pending additional data on less-studied species.

By its nature, the study of predator-prey dynamics is problematic (e.g., presence of a field observer could provide a buffer for potential prey species and affect the behavior of potential predatory species). Consequently, a good deal of the available data concerning predator pressure on lemurid and indriid species comes from opportunistic observation of independent predation attempts or events—a sort of compilation of case studies in predator-prey ecology. In particular, it is of note that many of those opportunistic observations of predation attempts involve *Propithecus* species. Whether the relatively large body sizes of sifaka species makes them preferred targets of potential predators (especially the fossa, but also large raptors such as Madagascar harrier-hawk) deserves future research attention (Karpanty & Goodman, 1999; Karpanty & Grella, 2001).

Finally, despite the appeal and general popularity that lemurs hold for people, and regardless of how fascinating lemur biology may be for primatologists, for that popularity and fascination to persist into the future, we will clearly need concerted, coordinated, and long-term conservation efforts to prevent lemur extinctions. Jones et al. (2019) have provided a major guidance and commentary document that outlines a productive way forward for achieving sustainable communities and addressing Madagascar's biodiversity crisis. The commentary proposes five positive steps that the government of Madagascar can take to effectively conserve biodiversity while providing benefits for people: (1) tackle and control environmental crime, e.g., poaching of rosewood and ebony trees from protected areas in the eastern rainforests, or capturing lemurs and other rare species for the pet trade; (2) invest in Madagascar's protected areas; (3) ensure that major infrastructure projects limit impacts on biodiversity; (4) strengthen tenure rights of local stakeholders over natural resources; and (5) address the growing crisis surrounding Madagascar's fuel wood demands and the pressures that expanding wood use put on protected areas. With international support, this five-step agenda for addressing Madagascar's biodiversity crisis, in order to also address the needs of her people, builds on previous valuable work. But, as Jones and co-authors (2019) warn, time would appear to be short to bring about a positive turning point for Madagascar's biodiversity, including the intriguing lemurids and indriids.

REFERENCES CITED—CHAPTER 3

Albrecht, G. H., Jenkins, P. D., & Godfrey, L. R. (1990). Ecogeographic size variation among the living and subfossil prosimians of Madagascar. *American Journal of Primatology*, 22, 1-50.

Andriaholinirina, V. N., Fausser, J.-L., & Rabarivola, J. C. (2003). Étude comparative de *Hapalemur simus* (Gray, 1870) de deux sites de la province autonome de Fianarantsoa, Madagascar: Forêt Secondaire de Park National de Ranomafana. *Lemur News*, 8, 9-13.

Andrianandrasana, H. T., Randriamahefasoa, J., Durbin, J., Lewis, R. E., & Ratsimbazafy, J. H. (2005). Participatory ecological monitoring of the Alaotra wetlands in Madagascar. *Biodiversity and Conservation*, 14 (11), 2757-2774.

Andriantsimanarilafy, R. R., Borgerson, C., Clarke, T., Colquhoun, I. C., Cotton, A., Donati, G., Eppley, T. M., Heriniaina, R., Irwin, M., Johnson, S., Mittermeier, R. A., Patel, E., Ralainasolo, F. B., Randrianasolo, H., Randriatahina, G., Ratsimbazafy, J., Ravaloharimanitra, M., Razafindramanana, J., Reuter, K. E.,... & Wright, P. (2020). *Eulemur macaco* (amended version of 2020 assessment). *The International Union for the Conservation of Nature Red List of Threatened Species* 2020: e.T8212A115563301. https://dx.doi.org/10.2305/IUCN.UK.2020-3.RLTS.T8212A182235113.en.

Ankel-Simons, F. (2007). *Primate anatomy: An introduction* (3rd ed.). Academic Press.

Baden, A., Wright, P., Louis, E., & Bradley, B. (2013). Communal nesting, kinship, and maternal success in a social primate. *Behavioral Ecology and Sociobiology*, 67(12), 1939-1950.

Baden, A. L., Mancini, A. N., Federman, S., Holmes, S. M., Johnson, S. E., Kamilar, J., Louis, E. E., Jr., & Bradley, B. J. (2019). Anthropogenic pressures drive population genetic structuring across a critically endangered lemur species range. *Scientific Reports*, 9, 16276.

Baden, A. L., Oliveras, J., & Gerber B. D. (2021). Sex-segregated range use by black-and-white ruffed lemurs (*Varecia variegata*) in Ranomafana National Park, Madagascar. *Folia Primatologica*, 92(1), 12-34.

Banks, M. A. (2013). *Determinants of abundance and the distribution of primates in northern Madagascar.* [Doctoral dissertation, Stony Brook University].

Banks, M. A., Ellis, E. R., Antonio, & Wright, P. C. (2007). Global population size of a critically endangered lemur, Perrier's sifaka. *Animal Conservation*, 10(2), 254-262.

Birkinshaw, C. R., & Colquhoun, I. C. (1998). Pollination of *Ravenala madagascariensis* and *Parkia madagascariensis* by *Eulemur macaco* in Madagascar. *Folia Primatologica*, 69 (5), 252-259.

Birkinshaw, C. R., & Colquhoun, I. C. (2003). Lemur food plants. In S. M. Goodman, & J. P. Benstead (Eds.), *The natural history of Madagascar* (pp. 1207-1220). University of Chicago Press.

Bonadonna, G., Torti, V., Randrianarison, R. M., Martinet, N., Gamba, M., & Giacoma, C. (2014). Behavioral correlates of extra-pair copulation in *Indri indri*. *Primates*, 55, 119-123.

Borgerson, C., Eppley, T. M., Donati, G., Colquhoun, I. C., Irwin, M., Johnson, S., Louis, E. E., Patel, E., Ralainasolo, F. B., Ravaloharimanitra, M., Razafindramanana, J., Reuter, K. E., Volampeno, S., Wright, P., & Zaonarivelo, J. (2020a). *Eulemur albifrons*. *The International Union for the Conservation of Nature Red List of Threatened Species* 2020: e.T8204A115561853.

Borgerson, C., Eppley, T. M., Patel, E., Johnson, S., Louis, E. E., & Razafindramanana, J. (2020b). *Varecia rubra*. *The International Union for the Conservation of Nature Red List of Threatened Species* 2020: e.T22920A115574598.

Borgerson, C., McKean, M. A., Sutherland, M. R., & Godfrey, L. R. (2016). Who hunts lemurs and why they hunt them. *Biological Conservation*, 197, 124-130.

Britt, A., Randriamandratonirina, N. J., Glasscock, K. D., & Iambana, B. R. (2002). Diet and feeding behaviour of *Indri indri* in a low-altitude rain forest. *Folia Primatologica*, 73(5), 225-239.

Brockman, D. K. (2003). *Polyboroides radiatus* predation attempts on *Propithecus verreauxi*. *Folia Primatologica*, 74, 71-74.

Brockman, D. K., Godfrey, L. R., Dollar, L. J., & Ratsirarson, J. (2008). Evidence of invasive *Felis silvestris* predation on *Propithecus verreauxi* at Beza Mahafaly Special Reserve, Madagascar. *International Journal of Primatology*, 29, 135-152.

Brown, J. H. (1995). *Macroecology*. University of Chicago Press.

Buettner-Janusch, J., Tattersall, I., & Sussman, R. W. (1975). History of study of the Malagasy lemurs, with notes on major museum collections. In I. Tattersall, & R.W. Sussman (Eds.), *Lemur biology* (pp. 3-11). Springer.

Burney, D. A. (2002). Sifaka predation by a large boa. *Folia Primatologica*, 73, 144-145.

Campera, M., Santini, L., Balestri, M., Nekaris, K. A. I., & Donati, G. (2020). Elevation gradients of lemur abundance emphasise the importance of Madagascar's lowland rainforest for the conservation of endemic taxa. *Mammal Review, 50*(1), 25-37.

Cardillo, M., Mace, G. M., Gittleman, J. L., & Purvis, A. (2006). Latent extinction risk and the future battlegrounds of mammal conservation. *Proceedings of the National Academy of Sciences, 103*(11), 4157-4161.

Carvalho, J. S., Graham, B., Rebelo, H., Bocksberger, G., Meyer, C. F. J., Wich, S., & Kühl, H. S. (2019). A global risk assessment of primates under climate and land use/cover scenarios. *Global Change Biology, 25*, 3163-3178.

Chikhi, L., Le Pors, B., Louis, E. E., Jr., Mittermeier, R. A., Ralainasolo, F. B., Ralison, J., Randrianambinina, B., Randriatahina, G., Rasamimananana, H., Rasoloharijaona, S., Ratelolahy, F., Ratsimbazafy, J., Reuter, K. E., Volampeno, S., & Zaonarivelo, J. (2020). *Eulemur sanfordi. The International Union for the Conservation of Nature Red List of Threatened Species* 2020: e.T8210A115562894.

Colquhoun, I. C. (1997). *A predictive socioecological study of the black lemur (Eulemur macaco macaco) in northwestern Madagascar.* [Doctoral dissertation, Washington University].

Colquhoun, I. C. (1998a). The lemur community of Ambato Massif: An example of the species richness of Madagascar's classified forests. *Lemur News, 3*, 11-14.

Colquhoun, I. C. (1998b). Cathemeral behaviour of *Eulemur macaco macaco* at Ambato Massif, Madagascar. *Folia Primatologica, 69*(S1), 22-34.

Colquhoun, I. C. (2005). Primates in the forest: Sakalava ethnoprimatology and synecological relations with black lemurs at Ambato Massif, Madagascar. In J. D. Paterson, & J. Wallis (Eds.), *Commensalism and conflict: The human-primate interface; Special topics in primatology* (Vol. 4, pp. 90-117). American Society of Primatologists.

Colquhoun, I. C. (2006). Predation and cathemerality: Comparing the impact of predators on the activity patterns of lemurids and ceboids. *Folia Primatologica, 77*(1-2), 143-165.

Colquhoun, I. C. (2007). Anti-predator strategies of cathemeral primates: Dealing with predators of the day and the night. In S. L. Gursky, & K. A. I. Nekaris (Eds.), *Primate anti-predator strategies. Developments in primatology: Progress and prospects* (pp. 146-172). Springer.

Colquhoun, I. C. (2011). A review and interspecific comparison of nocturnal and cathemeral strepsirrhine primate olfactory behavioural ecology. *International Journal of Zoology, 2011*(1), 1-11.

Curtis, D. (1997). *The mongoose lemur (Eulemur mongoz): A study in behaviour and ecology.* [Doctoral dissertation, University of Zurich].

Curtis, D. (2006). Cathemerality in lemurs. In L. Gould, & M. L. Sauther (Eds.), *Lemurs: Ecology and adaptation. Developments in primatology: Progress and prospects* (pp. 133-157). Springer.

Curtis, D., & Rasmussen, M. (2002). Cathemerality in lemurs. *Evolutionary Anthropology, 11*, 83-86.

Curtis, D., & Rasmussen, M. (2006). The evolution of cathemerality in primates and other mammals: A comparative and chronoecological approach. *Folia Primatologica, 77*(1-2), 178-193.

Dammhahn, M., Markolf, M., Lührs, M.-L., Thalmann, U., & Kappeler, P. M. (2013). Lemurs of the Beanka Forest, Melaky Region, western Madagascar. In S. M. Goodman, L. Gautier, & M. J. Raherilalao (Eds.), The Beanka Forest, Melaky Region, western Madagascar. *Malagasy Nature, 7*, 259-270.

de Wit, M. J. (2003). Madagascar: Heads it's a continent, tails it's an island. *Annual Review of Earth and Planetary Sciences, 31*, 213-248.

Delmore, K., Keller, M., Louis, E. E., Jr., & Johnson, S. (2009). Rapid primatological surveys of the Andringitra forest corridors: Direct observation of the greater bamboo lemur (*Prolemur simus*). *Lemur News, 14*, 49-52.

Dolch, R., Fiely, J. L., Ndriamiary, J.-N., Rafalimandimby, J., Randriamampionona, R., Engberg, S. E., & Louis, E. E., Jr. (2008). Confirmation of the greater bamboo lemur, *Prolemur simus*, north of the Torotorofotsy wetlands, eastern Madagascar. *Lemur News, 13*, 14-17.

Dolch, R., Hilgartner, R. D., Ndriamiary, J.-N., & Randriamahazo, H. (2004). The grandmother of all bamboo lemurs: Evidence for the occurrence of *Hapalemur simus* in fragmented rainforest surrounding the Torotorofotsy marshes, central eastern Madagascar. *Lemur News, 9*, 24-26.

Donati, G., Balestri, M., Campera, M., & Eppley, T. M. (2020a). *Hapalemur meridionalis. The International Union for the Conservation of Nature Red List of Threatened Species* 2020: e.T136384A115582831.

Donati, G., Balestri, M., Campera, M., Hyde Roberts, S., Račevska, E., Ramanamanjato, J.-B., & Ravoahangy, A. (2020b). *Eulemur collaris. The International Union for the Conservation of Nature Red List of Threatened Species* 2020: e.T8206A115562262.

Donati, G., Balestri, M., Campera, M., Norscia, I., & Ravoahangy, A. (2020c). *Avahi meridionalis. The International Union for the Conservation of Nature Red List of Threatened Species* 2020: e.T136369A115582568.

Donati, G., Lunardini, A., & Kappeler, P. M. (1999). Cathemeral activity of red-fronted brown lemurs (*Eulemur fulvus rufus*) in the Kirindy Forest/CFPF. In B. Rakotosamimanana, H. Rasamimanana, J. Ganzhorn, & S. Goodman (Eds.), *New directions in lemur studies* (pp. 119-137). Springer.

Donati, G., Santini, L., Razafindramanana, J., Boitani, L., & Borgognini-Tarli, S. M. (2013). (Un-)Expected nocturnal activity in "diurnal" *Lemur* catta supports cathemerality as one of the key adaptations of the lemurid radiation. *American Journal of Physical Anthropology, 150*, 99-106.

Dunham, A. E., Erhart, E. M., Overdorff, D. J., & Wright, P. C. (2008). Evaluating effects of deforestation, hunting, and El Niño events on a threatened lemur. *Biological Conservation, 141*, 287-297.

Eppley, T. M. (2015). *Ecological flexibility of the southern bamboo lemur (*Hapalemur meridionalis*) in southeast Madagascar.* [Doctoral dissertation, University of Hamburg].

Eppley, T. M., Borgerson, C., Sawyer, R. M., & Fenosoa, Z. S. E. (2020b). *Avahi mooreorum. The International Union for the Conservation of Nature Red List of Threatened Species* 2020: e.T16971566A115588141.

Eppley, T. M., & Donati, G. (2009). Grazing lemurs: Exhibition of terrestrial feeding by the southern gentle lemur, *Hapalemur meridionalis*, in the Mandena littoral forest, southeast Madagascar. *Lemur News*, 14, 16-20.

Eppley, T. M., Donati, G., & Ganzhorn, J. U. (2016a). Determinants of terrestrial feeding in an arboreal primate: The case of the southern bamboo lemur (*Hapalemur meridionalis*). *American Journal of Physical Anthropology*, 161(2), 328-342.

Eppley, T. M., Donati, G., Ramanamanjato, J.-B., Randriatafika, F., Andriamandimbiarisoa, L. N., Rabehevitra, D., Ravelomanantsoa, R., & Ganzhorn, J. U. (2015b). The use of an invasive species habitat by a small folivorous primate: Implications for lemur conservation in Madagascar. *PLoS ONE*, 10(11), e0140981.

Eppley, T. M., Ganzhorn, J. U., & Donati, G. (2015a). Cathemerality in a small, folivorous primate: Proximate control of diel activity in *Hapalemur meridionalis*. *Behavioral Ecology and Sociobiology*, 69, 991-1002.

Eppley, T. M., Ganzhorn, J. U., & Donati, G. (2016b). Latrine behaviour as a multimodal communitory signal station in wild lemurs: The case of *Hapalemur meridionalis*. *Animal Behavior*, 111, 57-67.

Eppley, T. M., & Patel, E. (2020a). *Avahi peyrierasi. The International Union for the Conservation of Nature Red List of Threatened Species* 2020: e.T136285A115581938.

Eppley, T. M., & Patel, E. (2020b). *Avahi betsileo. The International Union for the Conservation of Nature Red List of Threatened Species* 2020: e.T136767A115585814.

Eppley, T. M., Patel, E., Andriamisedra, T. R., Ranaivoarisoa, F. N., Peterson, C. R., Ratsimbazafy, J., & Louis, E. E. (2020d). *Avahi ramanantsoavanai. The International Union for the Conservation of Nature Red List of Threatened Species* 2020: e.T136434A115583017.

Eppley, T. M., Patel, E., Reuter, K. E., & Steffens, T. S. (2020e). *Avahi occidentalis. The International Union for the Conservation of Nature Red List of Threatened Species* 2020: e.T2435A115559730.

Eppley, T. M., & Ravelomanantsoa, R. (2015). Predation of an adult southern bamboo lemur *Hapalemur meridionalis* by a Dumeril's boa *Acrantophis dumerili. Lemur News*, 19, 2-3.

Eppley, T. M., Razafindramanana, J., Borgerson, C., Patel, E., & Louis, E. E. (2020c). *Hapalemur occidentalis. The International Union for the Conservation of Nature Red List of Threatened Species* 2020: e.T9678A115565375.

Eppley, T. M., Santini, L., Tinsman, J. C., & Donati, G. (2020a). Do functional traits offset the effects of fragmentation? The case of large-bodied diurnal lemur species. *American Journal of Primatology*, 82(4), ee23104.

Eppley, T. M., Tan, C. L., Arrigo-Nelson, S. J., Donati, G., Ballhorn, D. J., & Ganzhorn, J. U. (2017b). High energy or protein concentrations in food as possible offsets for cyanide consumption by specialized bamboo lemurs in Madagascar. *International Journal of Primatology*, 38(5), 881-899.

Eppley, T. M., Verjans, E., & Donati, G. (2011). Coping with low-quality diets: A first account of the feeding ecology of the southern gentle lemur, *Hapalemur meridionalis*, in the Mandena littoral forest, southeast Madagascar. *Primates*, 52(1), 7-13.

Eppley, T. M., Watzek, J., Dausmann, K. H., Ganzhorn, J., & Donati, G. (2017a). Huddling is more important than rest site selection for thermoregulation in southern bamboo lemurs. *Animal Behaviour*, 127, 153-161.

Erkert, H. G., & Kappeler, P. M. (2004). Arrived in the light: Diel and seasonal activity patterns in wild Verreaux's sifakas (*Propithecus v. verreauxi*; Primates: Indriidae). *Behavioral Ecology and Sociobiology*, 57, 174-186.

Eronen, J. T., Zohdy, S., Evans, A. R., Tecot, S. R., Wright, P. C., & Jernvall, J. (2017). Feeding ecology and morphology make a bamboo specialist vulnerable to climate change. *Current Biology*, 27, 1-6.

Estrada, A., Garber, P. A., Mittermeier, R. A., Wich, S., Gouveia, S., Dobrovolski, R., Nekaris, K. A. I., Nijman, V., Rylands, A. B., Maisels, F., Williamson, E. A., Bicca-Marques, J., Fuentes, A., Jerusalinsky, L., Johnson, S., Rodrigues de Melo, F., Oliveira, L., Schwitzer, C., Roos, C.,... & Setiawan, A. (2018). Primates in peril: The significance of Brazil, Madagascar, Indonesia and the Democratic Republic of the Congo for global primate conservation. *PeerJ*, 6, e4869.

Estrada, A., Garber, P. A., Rylands, A. B., Roos, C., Fernandez-Duque, E., Di Fiore, A., Nekaris, K. A. I., Nijman, V., Heymann, E. W., Lambert, J. E., Rovero, F., Barelli, C., Setchell, J. M., Gillespie, T. R., Mittermeier, R. A., Arregoitia, L. V., de Guinea, N., Gouveia, S., Dobrovolski, R.,... & Li, B. (2017). Impending extinction crisis of the world's primates: Why primates matter. *Science Advances*, 3(1), e1600946.

Everson, K. M., Jansa, S. A., Goodman, S. M., & Olson, L. E. (2020). Montane regions shape patterns of diversification in small mammals and reptiles from Madagascar's most evergreen forest. *Journal of Biogeography*, 47, 2059-2072.

Fausser, J.-L., Prosper, P., Donati, G., Ramanamanjato, J.-B., & Rumpler, Y. (2002). Phylogenic relationships between *Hapalemur* species and subspecies based on mitochondrial DNA sequences. *BMC Evolutionary Biology*, 2(1), 4.

Federman, S., Dornburg, A., Daly, D. C., Downie, A., Perry, G. H., Yoder, A. D., Sargis, E. J., Richard, A. F., Donoghue, M. J., & Baden, A. L. (2016). Implications of lemuriform extinctions for the Malagasy flora. *Proceedings of the National Academy of Sciences*, 113 (18), 5041-5046.

Feistner, A. T. C., & Rakotoarinosy, M. (1993). Conservation of the gentle lemur *Hapalemur griseus alaotrensis* at Lac

Alaotra, Madagascar: Local knowledge. *Dodo, Journal of the Wildlife Preservation Trusts, 29,* 54-65.

Fleagle, J. G. (2013). *Primate adaptation and evolution* (3rd ed.). Academic Press.

Freed, B. Z. (1996). *Co-occurrence among crowned lemurs (*Lemur coronatus*) and Sanford's lemurs (*Lemur fulvus sanfordi*) of Madagascar.* [Doctoral dissertation, Washington University].

Freed, B. Z. (2006). Polyspecific associations of crowned lemurs and Sanford's lemurs in Madagascar. In L. Gould, & M. L. Sauther (Eds.), *Lemurs: Ecology and adaptation. Developments in primatology: Progress and prospects* (pp. 111-131). Springer.

Gamba, M., Torti, V., Estienne, V., Randrianarison, R. M., Valente, D., Rovara, P., Bonadonna, G., Friard, O., & Giacoma, C. (2016). The indris have got rhythm! Timing and pitch variation of a primate song examined between sexes and age classes. *Frontiers in Neuroscience, 10,* 249.

Ganzhorn, J. U. (1988). Food partitioning among Malagasy primates. *Oecologia, 75*(3), 436-450.

Ganzhorn, J. U. (1995). Cyclones over Madagascar: Fate or fortune? *Ambio, 24*(2), 124-125.

Ganzhorn, J. U., Goodman, S. M., Nash, S., & Thalmann, U. (2006). Lemur biogeography. In J. G. Fleagle, & S. Lehman (Eds.), *Primate biogeography* (pp. 223-248). Springer.

Ganzhorn, J. U., & Kappeler, P. M. (1996). Lemurs of the Kirindy Forest. In J. U. Ganzhorn, & J. P. Sorg (Eds.), *Ecology and economy of a tropical dry forest in Madagascar* (pp. 257-274). Primate Report.

Ganzhorn, J. U., Lowry II, P. P., Schatz, G. E., & Sommer, S. (2001). The biodiversity of Madagascar: One of the world's hottest hotspots on its way out. *Oryx, 35*(4), 346-348.

Ganzhorn, J. U., Wilmé, L., & Mercier, J.-L. (2014). Explaining Madagascar's biodiversity. In I. R. Scales (Ed.), *Conservation and environmental management in Madagascar* (pp. 17-43). Routledge.

Garbutt, N. (1999). *Mammals of Madagascar.* Pica Press.

Gardner, C. J., Nicoll, M. E., Birkinshaw, C., Harris, A., Lewis, R. E., Rakotomalala, D., & Ratsifandrihamanana, A. N. (2018). The rapid expansion of Madagascar's protected area system. *Biological Conservation, 220,* 29-36.

Gardner, C. J., Radolalaina, P., Rajerison, M., & Greene, H. W. (2015). Cooperative rescue and predator fatality involving a group-living strepsirrhine, Coquerel's sifaka (*Propithecus coquereli*) and a Madagascar ground boa (*Acrantophis madagascariensis*). *Primates, 56,* 127-129.

Glander, K. E., & Powzyk, J. A. (1998). Morphometrics of wild *Indri indri* and *Propithecus diadema diadema*. *Folia Primatologica, 69*(S1), 399.

Glander, K. E., Wright, P. C., Daniels, P., & Merenlender, A. (1992). Morphometrics and testicle size of rainforest lemur species from southeastern Madagascar. *Journal of Human Evolution, 22,* 1-17.

Glander, K. E., Wright, P. C., Seigler, D. S., Randrianasolo, V., & Randrianasolo, B. (1989). Consumption of cyanogenic bamboo by a newly discovered species of bamboo lemur. *American Journal of Primatology, 19,* 119-124.

Godfrey, L. R., & Irwin, M. T. (2007). The evolution of extinction risk: Past and present anthropogenic impacts on the primate communities of Madagascar. *Folia Primatologica, 78,* 405-419.

Godfrey, L. R., Samonds, K. E., Baldwin, J. W., Sutherland, M. R., Kamilar, J. M., & Allfisher, K. L. (2020). Mid-Cenozoic climate change, extinction, and faunal turnover in Madagascar, and their bearing on the evolution of lemurs. *BMC Evolutionary Biology, 20,* 97.

Godfrey, L. R., Samonds, K. E., Jungers, W. L., Sutherland, M. R., & Irwin, M. T. (2004). Ontogenetic correlates of diet in Malagasy lemurs. *American Journal of Physical Anthropology, 123,* 250-276.

Golden, C. D. (2009). Bushmeat hunting and use in the Makira forest, north-eastern Madagascar: A conservation and livelihoods issue. *Oryx, 43,* 386-392.

Goodman, S. M. (2003). Predation on lemurs. In S. M. Goodman, & J. P. Benstead (Eds.), *The natural history of Madagascar* (pp. 1221-1228). University of Chicago Press.

Goodman, S. M., & Ganzhorn, J. U. (2004). Biogeography of lemurs in the humid forests of Madagascar: The role of elevational distribution and rivers. *Journal of Biogeography, 31* (1), 47-55.

Goodman, S. M., & Langrand, O. (1996). A high mountain population of the ring-tailed lemur (*Lemur catta*) on the Andringatra Massif, Madagascar. *Oryx, 30*(4), 259-268.

Goodman, S. M., O'Connor, S., & Langrand, O. (1993). A review of predation on lemurs: Implications for the evolution of social behavior in small nocturnal primates. In P. M. Kappeler, & J. U. Ganzhorn (Eds.), *Lemur social systems and their ecological basis* (pp. 51-66). Plenum Press.

Goodman, S. M., Rakotoarisoa, S. V., & Wilmé, L. (2006). The distribution and biogeography of the ring-tailed lemur (*Lemur catta*) in Madagascar. In A. Jolly, R. W. Sussman, N. Koyama, & H. Rasamimanana (Eds.), *Ring-tailed lemur biology:* Lemur catta *in Madagascar. Developments in primatology: Progress and prospects* (pp. 3-15). Springer.

Gould, L. (2006a). *Lemur catta* ecology: What we know and what we need to know. In L. Gould, & M. L. Sauther (Eds.), *Lemurs: Ecology and adaptation. Developments in primatology: Progress and prospects* (pp. 255-274). Springer.

Gould, L. (2006b). Male sociality and integration during the dispersal process in *Lemur catta*: A case study. In A. Jolly, R. W. Sussman, N. Koyama, & H. Rasamimanana (Eds.), *Ringtailed lemur biology:* Lemur catta *in Madagascar. Developments in primatology: Progress and prospects* (pp. 296-310). Springer.

Gould, L., & Sauther, M. L. (2007). Anti-predator strategies in a diurnal prosimian, the ring-tailed lemur (*Lemur catta*), at the Beza Mahafaly Special Reserve, Madagascar. In S. L. Gursky, & K. A. I. Nekaris (Eds.), *Primate anti-predator strategies. Developments in primatology: Progress and prospects* (pp. 275-288). Springer.

Gould, L., & Sauther, M. L. (2016). Going, going, gone...Is the iconic ring-tailed lemur (*Lemur catta*) headed for imminent extirpation? *Primate Conservation, 30,* 89-101.

Graham, T. L., Matthews, H. D., & Turner, S. E. (2016). A global-scale evaluation of primate exposure and vulnerability to climate change. *International Journal of Primatology, 37* (2), 158-174.

Grassi, C. (2001). *The behavioral ecology of* Hapalemur griseus griseus: *The influences of microhabitat and population density on this small-bodied prosimian folivore.* [Doctoral dissertation, University of Texas].

Grassi, C. (2002). Sex differences in feeding, height, and space use in *Hapalemur griseus. International Journal of Primatology, 23*(3), 677-693.

Grassi, C. (2006). Variability in habitat, diet, and social structure of *Hapalemur griseus* in Ranomafana National Park, Madagascar. *American Journal of Physical Anthropology,* 131(1), 50-63.

Green, G. M., & Sussman, R. W. (1990). Deforestation history of the eastern rain forests of Madagascar from satellite images. *Science, 248,* 212-215.

Gron, K. J. (2010). Primate factsheets: Greater bamboo lemur (*Prolemur simus*) taxonomy, morphology and ecology. *Primate Info Net.* https://primate.wisc.edu/primate-info-net /pin-factsheets/pin-factsheet-greater-bamboo-lemur/.

Groves, C. P. (2001). *Primate taxonomy.* Smithsonian Institution Press.

Groves, C. P. (2005). Order Primates. In D. E. Wilson, & D. M. Reeder (Eds.), *Mammal species of the world: A taxonomic and geographic reference, Vol.1* (3rd ed., pp. 111-184). Johns Hopkins University Press.

Hart, D. (2000). *Primates as prey: Ecological, morphological and behavioral relationships between primates and their predators.* [Doctoral dissertation, Washington University].

Hart, D. (2007). Predation on primates: A biogeographical analysis. In S. L. Gursky, & K. A. I. Nekaris (Eds.), *Primate anti-predator strategies* (pp. 27-59). Springer.

Hawkins, C. E. (2003). *Cryptoprocta ferox,* fossa, fosa. In S. M. Goodman, & J. P. Benstead (Eds.), *The natural history of Madagascar* (pp. 1360-1363). University of Chicago Press.

Hawkins, M. T. R., Culligan, R. R., Frasier, C. L., Dikow, R. B., Hagenson, R., Lei1, R., & Louis, E. E., Jr. (2018). Genome sequence and population declines in the critically endangered greater bamboo lemur (*Prolemur simus*) and implications for conservation. *BMC Genomics, 19,* 445.

Hekkala, E. R., Rakotondratsima, M., & Vasey, N. (2007). Habitat and distribution of the ruffed lemur, *Varecia,* north of the Bay of Antongil in northeastern Madagascar. *Primate Conservation, 22*(1), 89-95.

Heriniaina, R., Hosnah, H. B., & Zaonarivelo, J. (2020). *Propithecus perrieri. The International Union for the Conservation of Nature Red List of Threatened Species* 2020: e.T18361A115573556.

Herrera, J. P. (2016). Testing the adaptive radiation hypothesis for the lemurs of Madagascar. *Royal Society Open Science,* 4, 161014.

Hill, W. C. O. (1953). *Primates (comparative anatomy and taxonomy). I. Strepsirrhini.* Edinburgh University Press.

Irwin, M. T. (2006). Ecologically enigmatic lemurs: The sifakas of the eastern forests (*Propithecus candidus, P. diadema, P. edwardsi, P. perrieri,* and *P. tattersalli*). In L. Gould, & M. L. Sauther (Eds.), *Lemurs: Ecology and adaptation. Developments in primatology: Progress and prospects* (pp. 305-326). Springer.

Irwin, M. T. (2020). *Propithecus diadema. The International Union for the Conservation of Nature Red List of Threatened Species* 2020: e.T18358A115572884.

Irwin, M. T., Frasier, C. L., & Louis, E. E. (2020b). *Hapalemur griseus. The International Union for the Conservation of Nature Red List of Threatened Species* 2020: e.T9673A115564580.

Irwin, M. T., Johnson, S. E., & Wright, P. C. (2005). The state of lemur conservation in south-eastern Madagascar: Population and habitat assessments for diurnal and cathemeral lemurs using surveys, satellite imagery and GIS. *Oryx,* 39(2), 204-218.

Irwin, M. T., & King, T. (2020). *Eulemur fulvus. The International Union for the Conservation of Nature Red List of Threatened Species* 2020: e.T8207A115562499.

Irwin, M. T., King, T., Ravoloharimanitra, M., & Razafindramanana, J. (2020a). *Eulemur rubriventer. The International Union for the Conservation of Nature Red List of Threatened Species* 2020: e.T8203A115561650.

Irwin, M. T., Samonds, K. E., Raharison, J.-L., Junge, R. E., Mahefarisoa, K. L., Rasambainarivo, F., Godfrey, L. R., & Glander, K. E. (2019). Morphometric signals of population decline in diademed sifakas occupying degraded rainforest habitat in Madagascar. *Science Reports, 9,* 8776.

Irwin, M. T., Samonds, K. E., Raharison, J.-L., & Wright, P. C. (2004). Lemur latrines: Observations of latrine behavior in wild primates and possible ecological significance. *Journal of Mammalogy,* 85(3), 420-427.

Irwin, M. T., Wright, P. C., Birkinshaw, C., Fisher, B. L., Gardner, C. J., Glos, J., Goodman, S. M., Loiselle, P., Rabeson, P., Raharison, J.-L., Raherilalao, M. J., Rakotondravony, D., Raselimanana, A., Ratsimbazafy, J., Sparks, J. S., Lucienne Wilmé, L., & Ganzhorn, J. U. (2010). Patterns of species change in anthropogenically disturbed forests of Madagascar. *Biological Conservation, 143,* 2351-2362.

Janson, C., & Verdolin, J. (2005). Seasonality of primate births in relation to climate. In D. K. Brockman, & C. P. van Schaik (Eds.), *Seasonality in primates: Studies of living and extinct human and non-human primates* (pp. 307-350). Cambridge University Press.

Jenkins, R. K. B., Keane, A., Rakotoarivelo, A. R., Rakotomboavonjy, V., Randrianandrinina, F. H., Razafimanahaka, J. H., & Jones, J. P. G. (2011). Analysis of patterns of bushmeat consumption reveals extensive exploitation of protected species in eastern Madagascar. *PLoS ONE,* 6(12), e27570.

Johnson, S. (2006). Evolutionary divergence in the brown lemur species complex. In L. Gould, & M. L. Sauther (Eds.), *Lemurs: Ecology and adaptation. Developments in primatology: Progress and prospects* (pp. 187-210). Springer.

Johnson, S., Andriamisedra, T. R., Donohue, M. E., Ralainasolo, F. B., Birkinshaw, C., Ludovic, R., & Ratsimbazafy, J. (2020a). *Eulemur cinereiceps. The International Union for the Conservation of Nature Red List of Threatened Species* 2020: e.T8205A115562060.

Johnson, S., Narváez-Torres, P. R., Holmes, S. M., Wyman, T. M., Louis, E. E., & Wright, P. C. (2020b). *Eulemur rufifrons. The International Union for the Conservation of Nature Red List of Threatened Species* 2020: e.T136269A115581600.

Jolly, A. (1966). *Lemur behavior: A Madagascar field study.* University of Chicago Press.

Jolly, A. (2003). *Lemur catta*, ring-tailed lemur, *maky*. In S. M. Goodman, & J. P. Benstead (Eds.), *The natural history of Madagascar* (pp. 1329-1332). University of Chicago Press.

Jolly A. (2012). Berenty Reserve, Madagascar: A long time in a small space. In P. Kappeler, & D. Watts (Eds.), *Long-term field studies of primates* (pp. 21-44). Springer.

Jolly, A., Koyama, N., Rasamimanana, H., Crowley, H., & Williams, G. (2006). Berenty Reserve: A research site in southern Madagascar. In A. Jolly, R.W. Sussman, N. Koyama, & H. Rasamimanana (Eds.), *Ring-tailed lemur biology: Lemur catta in Madagascar. Developments in primatology: Progress and prospects* (pp. 32-42). Springer.

Jolly, A., & Sussman, R. W. (2006). Notes on the history of ecological studies of Malagasy lemurs. In L. Gould, & M. L. Sauther (Eds.), *Lemurs: Ecology and adaptation. Developments in primatology: Progress and prospects* (pp. 19-39). Springer.

Jones, J. P. G., Ratsimbazafy, J., Ratsifandrihamanana, A. N., Watson, J. E. M., Andrianandrasana, H. T., Cabeza, M., Cinner, J. E., Goodman, S. M., Hawkins, F., Mittermeier, R. A., Rabearisoa, A. L., Rakotonarivo, O. S., Razafimanahaka, J. H., Razafimpahanana, A. R., Wilmé, L., & Wright, P. C. (2019). Last chance for Madagascar's biodiversity. *Nature Sustainability, 2*, 350-352.

Jungers, W. L., Godfrey, L. R., Simons, E. L., & Chatrath, P. S. (1995). Subfossil *Indri indri* from the Ankarana Massif of northern Madagascar. *American Journal of Physical Anthropology, 97*(4), 357-366.

Kamilar, J. M., & Tecot, S. R. (2016). Anthropogenic and climatic effects on the distribution of *Eulemur* species: An ecological niche modeling approach. *International Journal of Primatology, 37*(1), 47-68.

Kappeler, P. M., Cuozzo, F. P., Fichtel, C., Ganzhorn, J. U., Gursky-Doyen, S., Irwin, M. T., Ichino, S., Lawler, R., Nekaris, K. A. I., Ramanamanjato, J.-B., Radespiel, U., Sauther, M. L., Wright, P. C., & Zimmermann, E. (2017). Long-term field studies of lemurs, lorises, and tarsiers. *Journal of Mammalogy, 98*(3), 661-669.

Kappeler, P. M., & Fichtel, C. (2012). A 15-year perspective on the social organization and life history of sifaka in Kirindy Forest. In P. M. Kappeler, & D. Watts (Eds.), *Long-term field studies of primates* (pp. 101-121). Springer.

Kappeler, P. M., & Schäffler, L. (2008). The lemur syndrome unresolved: Extreme male reproductive skew in sifakas (*Propithecus verreauxi*), a sexually monomorphic primate with female dominance. *International Journal of Primatology, 62*, 1007-1015.

Kappeler, P. M., van Schaik, C. P., & Watts, D. P. (2012). The values and challenges of long-term field studies. In P. Kappeler, & D. Watts (Eds.), *Long-term field studies of primates* (pp. 3-18). Springer.

Karpanty, S. M., & Goodman, S. M. (1999). Diet of the Madagascar harrier-hawk, *Polyboroides radiatus*, in southeastern Madagascar. *Journal of Raptor Research, 33*, 313-316.

Karpanty, S. M., & Grella, R. (2001). Lemur responses to diurnal raptor calls in Ranomafana National Park, Madagascar. *Folia Primatologica, 72*, 100-103.

King, T., Dolch, R., Randriahaingo, H. N. T., Randrianarimanana, L., & Ravaloharimanitra, M. (2020). *Indri indri. The International Union for the Conservation of Nature Red List of Threatened Species* 2020: e.T10826A115565566.

King, T., & Rakotonirina, L. (2020). *Propithecus deckenii. The International Union for the Conservation of Nature Red List of Threatened Species* 2020: e.T18357A115572684.

Kress, J. W., Schatz, G. E., Andrianifahanana, M., & Morland, H. S. (1994). Pollination of *Ravenala madagascariensis* (Strelitziaceae) by lemurs in Madagascar: Evidence of an archaic coevolutionary system? *American Journal of Botany, 81*, 542-551.

LaFleur, M., Clarke, T. A., Ratzimbazafy, J., & Reuter, K. (2016a). Ring-tailed lemur—*Lemur catta*, Linnaeus, 1758. In C. Schwitzer, R. A. Mittermeier, A. B. Rylands, F. Chiozza, E. A. Williamson, E. J. Macfie, J. Wallis, & A. Cotton (Eds.), *Primates in peril: The world's 25 most endangered primates 2016-2018* (pp. 35-37). Conservation International.

LaFleur, M., Clarke, T. A., Reuter, K. E., Schaefer, M. S., terHorst, C. (2019). Illegal trade of wild-captured *Lemur catta* within Madagascar. *Folia Primatologica, 90*, 199-214.

LaFleur, M., Clarke, T. A., Reuter, K., & Schaeffer, T. (2016b). Rapid decrease in populations of wild ring-tailed lemurs (*Lemur catta*) in Madagascar. *Folia Primatologica, 87*, 320-330.

LaFleur, M., & Gould, L. (2020). *Lemur catta. The International Union for the Conservation of Nature Red List of Threatened Species* 2020: e.T11496A115565760.

LaFleur, M., Gould, L., Sauther, M., Clarke, T., & Reuter, K. (2018). Restating the case for a sharp population decline in *Lemur catta. Folia Primatologica, 89*, 295-304.

LaFleur, M., Sauther, M., Cuozzo, F., Yamashita, N., Youssouf, I. A. J., & Bender, R. (2014). Cathemerality in wild ring-tailed lemurs (*Lemur catta*) in the spiny forest of Tsimanampetsotsa National Park: Camera trap data and preliminary behavioral observations. *Primates, 55*, 207-217.

Langer, P. (2003). Lactation, weaning period, food quality, and digestive tract differentiations in Eutheria. *Evolution, 57*, 1196-1215.

Lehman, S. M., & Mayor, M. (2004). Dietary patterns in Perrier's sifakas (*Propithecus diadema perrieri*): A preliminary study. *American Journal of Primatology, 62,* 115-122.

Lehman, S. M., Mayor, M., & Wright, P.C. (2005a). Ecogeographic size variations in sifakas: A test of the resource seasonality and resource quality hypotheses. *American Journal of Physical Anthropology, 126,* 318-328.

Lehman, S. M., Rajaonson, A., & Day, D. (2005b). Composition of the lemur community in the Vohibola III Classified Forest, SE Madagascar. *Lemur News, 10,* 16-19.

Lehman, S. M., Ratsimbazafy, J., Rajaonson, A., & Day, S. (2006). Ecological correlates to lemur community structure in southeast Madagascar. *International Journal of Primatology, 27*(4), 1023-1040.

Leigh, E. G., Jr., Hladik, A., Hladik, C. M., & Jolly, A. (2007). The biogeography of large islands, or how does the size of the ecological theater affect the evolutionary play? *Revue d'Écologie, 62,* 105-168.

Lewis, R. J. (2008). Social influences on group membership in *Propithecus verreauxi verreauxi. International Journal of Primatology, 29,* 1249-1270.

Lewis, R. J. (2020). Female power: A new framework for understanding "female dominance" in lemurs. *Folia Primatologica, 91,* 48-68.

Lewis, R. J., & Rakotondranaivo, F. (2011). The impact of Cyclone Fanele on sifaka body condition and reproduction in the tropical dry forest of western Madagascar. *Journal of Tropical Ecology, 27*(4), 429-432.

Loudon, J. E., Sauther, M. L., Fish, K. D., Hunter-Ishikawa, M., & Ibrahim, Y. J. (2006). One reserve, three primates: Applying a holistic approach to understand the interconnections among ring-tailed lemurs (*Lemur catta*), Verreaux's sifaka (*Propithecus verreauxi*), and humans (*Homo sapiens*) at Beza Mahafaly Special Reserve, Madagascar. *Ecological and Environmental Anthropology, 2,* 54-74.

Louis, E. E., Bailey, C. A., Sefczek, T. M., King, T., Radespiel, U., & Frasier, C. L. (2020d). *Propithecus coquereli. The International Union for the Conservation of Nature Red List of Threatened Species* 2020: e.T18355A115572275.

Louis, E. E., Bailey, C. A., Sefczek, T. M., Raharivololona, B., Schwitzer, C., & Wilmet, L. (2020e). *Avahi unicolor. The International Union for the Conservation of Nature Red List of Threatened Species* 2020: e.T41579A115579946.

Louis, E. E., Raharivololona, B., Schwitzer, C., & Wilmet, L. (2020a). *Avahi cleesei. The International Union for the Conservation of Nature Red List of Threatened Species* 2020: e.T136335A115582253.

Louis, E. E., Sefczek, T. M., Bailey, C. A., Raharivololona, B., Lewis, R., & Rakotomalala, E. J. (2020b). *Propithecus verreauxi. The International Union for the Conservation of Nature Red List of Threatened Species* 2020: e.T18354A115572044.

Louis, E. E., Sefczek, T. M., Raharivololona, B., King, T., Morelli, T. L., & Baden, A. (2020c). *Varecia variegata. The International Union for the Conservation of Nature Red List of Threatened Species* 2020: e.T22918A115574178.

MacArthur, R. H. (1972). *Geographical ecology: Patterns in the distribution of species.* Princeton University Press.

Markolf, M., & Kappeler, P. M. (2013). Phylogeographic analysis of the true lemurs (genus *Eulemur*) underlines the role of river catchments for the evolution of micro-endemism in Madagascar. *Frontiers in Zoology, 10,* 70.

Martin, R. D. (1995). Prosimians: From obscurity to extinction? In L. Alterman, G. A. Doyle, & M. K. Izard (Eds.), *Creatures of the dark: The nocturnal prosimians* (pp. 535-563). Plenum Press.

Martinez, B. (2008). Occurrence of bamboo lemurs, *Hapalemur griseus occidentalis*, in an agricultural landscape on the Masoala Peninsula. *Lemur News, 13,* 11-14.

Martinez, B. T., & Razafindratsima, O. H. (2014). Frugivory and seed dispersal patterns of the red-ruffed lemur, *Varecia rubra*, at a forest restoration site in Masoala National Park, Madagascar. *Folia Primatologica, 85,* 228-243.

Mayor, M. I., & Lehman, S. M. (1999). Conservation of Perrier's sifaka (*Propithecus diadema perrieri*) in Analamerena Special Reserve, Madagascar. *Lemur News, 4,* 21-23.

Meier, B., Albignac, R., Peyriéras, A., Rumpler, Y., & Wright, P. (1987). A new species of *Hapalemur* (Primates) from south east Madagascar. *Folia Primatologica, 48*(3-4), 211-215.

Meier, B., & Rumpler, Y. (1987). Preliminary survey of *Hapalemur simus* and of a new species of *Hapalemur* in eastern Betsileo, Madagascar. *Primate Conservation, 8,* 40-43.

Mertl-Millhollen, A. S., Blumenfeld-Jones, K., Raharison, S. M., Tsaramanana, D. R., & Rasamimanana, H. (2011). Tamarind tree seed dispersal by ring-tailed lemurs. *Primates, 52,* 391.

Milton, K. (1978). Behavioral adaptation to leaf-eating by the mantled howler monkey (*Alouatta palliata*). In G. G. Montgomery (Ed.), *The ecology of arboreal folivores* (pp. 535-549). Smithsonian Institution Press.

Mittermeier, R. A., Ganzhorn, J. U., Konstant, W., Glander, K., Tattersall, I., Groves, C., Rylands, A., Hapke, A., Ratsimbazafy, J., Mayor, M., Louis, E., Rumpler, Y., Schwitzer, C., & Rasoloarison, R. (2008). Lemur diversity in Madagascar. *International Journal of Primatology, 29*(6), 1607-1656.

Mittermeier, R. A., Konstant, W. R., Nicoll, M. E., & Langrand, O. (1992). *Lemurs of Madagascar: An action plan for their conservation 1993-1999.* Conservation International.

Mittermeier, R. A., Louis, E. E., Jr., Langrand, O., Schwitzer, C., Gauthier, C., Rylands, A. B., Rajaobelina, S., Ratsimbazafy, J., Rasoloarison, R., Hawkins, F., Roos, C., Richardson, M., & Kappeler, P. M. (2014). *Lemuriens de Madagascar.* D'Histoire Naturelle.

Mittermeier, R. A., Louis, E. E., Jr., Richardson, M., Schwitzer, C., Langrand, O., Rylans, A. B., Hawkins, F., Rajaobelina, S., Ratsimbazafy, J., Rasoloarison, R., Roos, C., Kappeler, P. M., & Mackinnon, J. (2010). *Lemurs of Madagascar* (3rd ed.). Conservation International.

Mittermeier, R. A., Valladares-Pádua, C., Rylands, A. B., Eudey, A. A., Butynski, T. M., Ganzhorn, J. U., Kormos, R., Aguiar, J. M., & Walker, S. (2006). Primates in peril: The world's 25 most endangered primates, 2004-2006. *Primate Conservation, 20,* 1-28.

Morelli, T. L., Smith, A. B., Mancini, N. A., Balko, E. A., Borgerson, C., Dolch, R., Farris, Z., Federman, S., Golden, C. D., Holmes, S. M., Irwin, M., Jacobs, R. L., Johnson, S., King, T., Lehman, S. M., Louis, E. E., Jr., Murphy, A., Randriahaingo, H. N. T., Randrianarimanana, H. L. L.,... & Baden, A. L. (2020). The fate of Madagascar's rainforest habitat. *Nature Climate Change, 10,* 89-96.

Moses, K. L., & Semple, S. (2011). Primary seed dispersal by the black-and-white ruffed lemur (*Varecia variegata*) in the Manombo forest, south-east Madagascar. *Journal of Tropical Ecology, 27,* 529-538.

Murphy, A. J., Ferguson, B., & Gardner, C. J. (2017). Recent estimates of ring-tailed lemur (*Lemur catta*) population declines are methodologically flawed and misleading. *International Journal of Primatology, 38,* 623-628.

Mutschler, T. (1998). *The Alaotran gentle lemur (*Hapalemur griseus alaotrensis*): A study in behavioural ecology.* [Doctoral dissertation, University of Zurich].

Mutschler, T. (1999). Folivory in a small-bodied lemur: The nutrition of the Alaotran gentle lemur (*Hapalemur griseus alaotrensis*). In B. Rakotosamimanana, H. Rasamimanana, J. U. Ganzhorn, & S. M. Goodman (Eds.), *New directions in lemur studies* (pp. 221-239). Plenum Press.

Mutschler, T., & Feistner, A. T. C. (1995). Conservation status and distribution of the Alaotran gentle lemur *Hapalemur griseus alaotrensis. Oryx, 29*(4), 267-274.

Mutschler, T., Nievergelt, C. M., & Feistner, A. T. C. (2000). Social organization of the Alaotran gentle lemur (*Hapalemur griseus alaotrensis*). *American Journal of Primatology, 50*(1), 9-24.

Mutschler T., Randrianarisoa A. J., & Feistner A. T. C. (2001). Population status of the Alaotran gentle lemur *Hapalemur griseus alaotrensis. Oryx, 35*(2), 152-157.

Mutschler, T., & Tan, C. L. (2003). *Hapalemur*, bamboo or gentle lemurs. In S. M. Goodman, & J. P. Benstead (Eds.), *The natural history of Madagascar* (pp. 1324-1329). University of Chicago Press.

Napier, J. R., & Walker, A. C. (1967). Vertical clinging and leaping—a newly recognized category of locomotor behaviour of primates. *Folia Primatologica, 6*(3), 204-219.

Nievergelt, C. M., Mutschler, T., & Feistner, A. T. C. (1998). Group encounters and territoriality in wild Alaotran gentle lemurs (*Hapalemur griseus alaotrensis*). *American Journal of Primatology, 46*(3), 251-258.

Nievergelt, C. M., Mutschler, T., Feistner, A. T. C., & Woodruff, D. S. (2002). Social system of the Alaotran gentle lemur (*Hapalemur griseus alaotrensis*): Genetic characterization of group composition and mating system. *American Journal of Primatology, 57*(4), 157-176.

Norscia, I., & Borgognini-Tarli, S. M. (2008). Ranging behavior and possible correlates of pair-living in southeastern *Avahis* (Madagascar). *International Journal of Primatology, 29,* 153-171.

Norscia, I., Carrai, V. C., Ceccanti, B., & Borgognini Tarli, S. M. (2005). Termite soil eating in Kirindy sifakas (Madagascar):

Proposing a new proximate factor. *Folia Primatologica, 76,* 119-122.

Nunziata, S. O., Wallenhorst, P., Barrett, M. A., Junge, R. E., Yoder, A. D., & Weisrock, D. W. (2016). Population and conservation genetics in an endangered lemur, *Indri indri*, across three forest reserves in Madagascar. *International Journal of Primatology, 37*(6), 688-702.

Ohba, M., Samonds, K. E., LaFleur, M., Ali, J. R., & Godfrey, L. R. (2016). Madagascar's climate at the K/P boundary and its impact on the island's biotic suite. *Palaeogeography, Palaeoclimatology, Palaeoecology, 441,* 688-695.

Oliver, L. K. (2017). *Coexistence of confamilial, folivorous indriids,* Propithecus diadema *and* Indri indri, *at Betampona Strict Nature Reserve, Madagascar.* [Doctoral dissertation, Washington University].

Olson, E. R., Marsh, R. A., Bovard, B. N., Randrianarimanana, H. L. L., Ravaloharimanitra, M., Ratsimbazafy, J. H., & King, T. (2013). Habitat preferences of the critically endangered greater bamboo lemur (*Prolemur simus*) and densities of one of its primary food sources, Madagascar giant bamboo (*Cathariostachys madagascariensis*), in sites with different degrees of anthropogenic and natural disturbance. *International Journal of Primatology, 34,* 486-499.

Overdorff, D. J., & Johnson, S. (2003). *Eulemur*, true lemurs. In S. M. Goodman, & J. P. Benstead (Eds.), *The natural history of Madagascar* (pp. 51-74). University of Chicago Press.

Overdorff, D. J., Strait, S. G., & Telo, A. (1997). Seasonal variation in activity and diet in a small-bodied folivorous primate, *Hapalemur griseus*, in southeastern Madagascar. *American Journal of Primatology, 43*(3), 211-223.

Oxnard, C. E., Crompton, R. H., & Liebermann, S. S. (1990). *Animal lifestyles and anatomies: The case of the prosimian primates.* University of Washington Press.

Parga, J. A., Sauther, M. L., Cuozzo, F. P., Jacky, I. A. Y., Gould, L., Sussman, R. W., Lawler, R. R., & Pastorini, J. (2015). Genetic evidence for male and female dispersal in wild *Lemur catta. Folia Primatologica, 86,* 66-75.

Pastorini, J., Forstner, M. R. J., & Martin, R. D. (2002). Phylogenetic relationships among Lemuridae (Primates): Evidence from mtDNA. *Journal of Human Evolution, 43,* 463-478.

Patel, E. R. (2006). Activity budget, ranging, and group size in silky sifakas (*Propithecus candidus*). *Lemur News, 11,* 42-45.

Patel, E. R. (2009). Silky sifaka, *Propithecus candidus*, 1871. In R. A. Mittermeier, J. Wallis, A. B. Rylands, J. U. Ganzhorn, J. F. Oates, E. A. Williamson, E. Palacios, E. W. Heymann, M. C. M. Kierulff, L. Yongcheng, J. Supriatna, C. Roos, S. Walker, L. Cortes-Ortiz, & C. Schwitzer (Eds.), *Primates in peril: The world's 25 most endangered primates 2008-2010* (pp. 23-26). Conservation International.

Patel, E. (2020a). *Propithecus candidus. The International Union for the Conservation of Nature Red List of Threatened Species* 2020: e.T18360A115573359.

Patel, E. (2020b). *Avahi laniger. The International Union for the Conservation of Nature Red List of Threatened Species* 2020: e.T2434A115559557.

Pebsworth, P. A., Huffman, M. A., Lambert, J. E., & Young, S. L. (2019). Geophagy among nonhuman primates: A systematic review of current knowledge and suggestions for future directions. *American Journal of Physical Anthropology,* 168(S67), 164-194.

Petter, J.-J. (1962). Recherche sur l'ecologie et l'ethologie des lemuriens Malgaches. *Memoirs du Museum National d'Histoire Naturelle—Serie A, Zoologie,* XXVII(1), 1-146.

Petter, J.-J., Albignac, R., & Rumpler, Y. (1977). *Faune de Madagascar 44: Mammifères lémuriens (Primates prosimiens).* Orstom.

Petter, J.-J., & Peyriéras, A. (1970), Observations éco-éthologiques sur les lémuriens malgaches du genre *Hapalemur. Terre et Vie,* 24(3), 356-382.

Petter, J.-J., & Peyriéras, A. (1975). Preliminary notes on the behaviour and ecology of *Hapalemur griseus.* In I. Tattersall, & R. W. Sussman (Eds.), *Lemur biology* (pp. 281-286). Plenum Press.

Pidgeon, M. (1996). An ecological survey of Lake Alaotra and selected wetlands of central and eastern Madagascar in analyzing the demise of Madagascar pochard *Aythya innotata. Work Group Birds Madagascar Region,* 6, 17-19.

Pochron, S. T., Fitzgerald. J., Gilbert, C. C., Lawrence, D., Grgas, M., Rakotonirina, G., Ratsimbazafy, R., Rakotosoa, R., & Wright, P. C. (2003). Patterns of female dominance in *Propithecus diadema edwardsi* of Ranomafana National Park, Madagascar. *American Journal of Primatology,* 61, 173-185.

Pollock, J. I. (1975). Field observations on *Indri*: A preliminary report. In I. Tattersall, & R. W. Sussman (Eds.), *Lemur biology* (pp. 287-311). Plenum Press.

Pollock, J. I. (1977). The ecology and sociology of feeding in *Indri indri.* In T. H. Clutton-Brock (Ed.), *Primate ecology: Studies of feeding and ranging behaviour in lemurs, monkeys, and apes* (pp. 37-69). Academic Press.

Pollock, J. I. (1979). Female dominance in *Indri indri. Folia Primatologica,* 31, 143-164.

Pollock, J. I. (1986). The song of the Indris (*Indri indri*; Primates: Lemuroidea): Natural history, form, and function. *International Journal of Primatology,* 7, 225-264.

Powzyk, J. A. (1997). *The socio-ecology of two sympatric indriids:* Propithecus diadema diadema *and* Indri indri, *a comparison of feeding strategies and their possible repercussions on species-specific behaviors.* [Doctoral dissertation, Duke University].

Powzyk, J. A., & Mowry, C. B. (2006). The feeding ecology and related adaptations of *Indri indri.* In L. Gould, & M. L. Sauther (Eds.), *Lemurs: Ecology and adaptation. Developments in primatology: Progress and prospects* (pp. 353-368). Springer.

Powzyk, J. A., & Thalmann, U. (2003). *Indri indri, Indri.* In S. M. Goodman, & J. P. Benstead (Eds.), *The natural history of Madagascar* (pp. 1342-1345). University of Chicago Press.

Pyhälä, A., Eklund, J., McBride, M. F., Rakotoarijaona, M. A., & Cabeza, M. (2019). Managers' perceptions of protected area outcomes in Madagascar highlight the need for species monitoring and knowledge transfer. *Conservation Science and Practice,* 1(2), e6.

Quéméré, E., Crouau-Roy, B., Rabarivola, C., Louis, E. E., Jr., & Chikhi, L. (2010a). Landscape genetics of an endangered lemur (*Propithecus tattersalli*) within its entire fragmented range. *Molecular Ecology,* 19, 1606-1621.

Quéméré, E., Louis, E. E., Jr., Ribéron, A., Chikhi, L., & Crouau-Roy, B. (2010b). Non-invasive conservation genetics of the critically endangered golden-crowned sifaka (*Propithecus tattersalli*): High diversity and significant genetic differentiation over a small range. *Conservation Genetics,* 11, 675-687.

Quinn, A., & Wilson, D. E. (2002). *Indri indri. Mammalian Species,* 694, 1-5.

Rabarivola, C., Prosper, P., Zaramody, A., Andriaholinirina, N., & Hauwy, M. (2007). Cytogenetics and taxonomy of the genus *Hapalemur. Lemur News,* 12, 46-49.

Rakotondravony, D., Goodman, S. M., & Soarimalala, V. (1998). Predation on *Hapalemur griseus griseus* by *Boa manditra* (Boidae) in the Littoral Forest of eastern Madagascar. *Folia Primatologica,* 69, 405-408.

Rakotondravony, D., & Razafindramahatra, L. V. (2004). Contribution à l'étude des populations de *Hapalemur aureus* dans le couloir forestier Ranomafana—Andringitra. *Lemur News,* 9, 28-32.

Rakotonirina, L. H. F., Rajaonson, A., Ratolojanahary, J. H., Missirli, J. M., & Fara, L. R. (2013). Southern range extensions for the critically endangered black-and-white ruffed lemur *Varecia variegata* and greater bamboo lemur *Prolemur simus. Primate Conservation,* 26(1), 49-55.

Rakotonirina, L., Rajaonson, A., Ratolojanahary, T., Rafalimandimby, J., Fanomezantsoa, P., Ramahefasoa, B., Rasolofoharivelo, T., Ravaloharimanitra, M., Ratsimbazafy, J., Dolch, R., & King, T. (2011). New distributional records and conservation implications for the critically endangered greater bamboo lemur *Prolemur simus. Folia Primatologica,* 82, 118-129.

Ralainasolo, F. B., Ratsimbazafy, J. H., & Stevens, N. J. (2008). Behavior and diet of the critically endangered *Eulemur cinereiceps* in Manombo forest, southeast Madagascar. *Madagascar Conservation & Development,* 3(1), 38-43.

Ralainasolo, F. B., Raveloarimalala, M. L., Randrianasolo, H., Reuter, K. E., Heriniaina, R., Clarke, T., Ravaloharimanitra, M., Volampeno, S., Donati, G., Razafindramanana, J., Andriantsimanarilafy, R. R., Randriatahina, G., Irwin, M., Eppley, T. M., & Borgerson, C. (2020). *Hapalemur alaotrensis. The International Union for the Conservation of Nature Red List of Threatened Species* 2020: e.T9676A115564982.

Ralainasolo, F. B., Waeber, P. O., Ratsimbazafy, J., Durbin, J., & Lewis, R. (2006). The Alaotra gentle lemur: Population estimation and subsequent implications. *Madagascar Conservation & Development,* 1(1), 9-10.

Ramanamisata, R., Pichon, C., Razafindraibe, H., Simmen, B. (2014). Social behavior and dominance of the crowned sifaka (*Propithecus coronatus*) in northwestern Madagascar. *Primate Conservation, 28*, 93-97.

Ramiadantsoa, T., Ovaskainen, O., Rybicki J., & Hanski, I. (2015). Large-scale habitat corridors for biodiversity conservation: A forest corridor in Madagascar. *PLoS ONE, 10*(7), e0132126.

Rasmussen, D. T. (1985). A comparative study of breeding seasonality and litter size in eleven taxa of captive lemurs (*Lemur* and *Varecia*). *International Journal of Primatology, 6*(5), 501-517.

Rasmussen, M. A. (1999). *Ecological influences on activity cycle in two cathemeral primates,* Eulemur mongoz *(mongoose lemur) and* Eulemur fulvus fulvus *(common brown lemur).* [Doctoral dissertation, Duke University].

Ratsimbazafy, J. (2006). Diet composition, foraging, and feeding behavior in relation to habitat disturbance: Implications for the adaptability of ruffed lemurs (*Varecia v. editorium*) in Manombo Forest, Madagascar. In L. Gould, & M. L. Sauther (Eds.), *Lemurs: Ecology and adaptation. Developments in primatology: Progress and prospects* (pp. 403-422). Springer.

Ratsimbazafy, J. H., Ralainasolo, F. B., Rendigs, A., Contreras, J. M., Andrianandrasana, H., Mandimbihasina, A. R., Nievergelt, C. M., Lewis, R., & Waeber, P. O. (2013). Gone in a puff of smoke? *Hapalemur alaotrensis* at great risk of extinction. *Lemur News, 17*, 14-18.

Ravaloharimanitra, M., King, T., Wright, P., Raharivololona, B., Ramaherison, R. P., Louis, E. E., Frasier, C. L., Dolch, R., Roullet, D., Razafindramanana, J., Volampeno, S., Randriahaingo, H. N. T., Randrianarimanana, L., Borgerson, C., & Mittermeier, R. A. (2020). *Prolemur simus. The International Union for the Conservation of Nature Red List of Threatened Species* 2020: e.T9674A115564770.

Ravaloharimanitra, M., Ratolojanahary, T., Rafalimandimby, J., Rajaonson, R., Rakotonirina, L., Rasolofoharivelo, T., Ndriamiary, J. N., Andriambololona, J., Nasoavina, C., Fanomezantsoa, P., Rakotoarisoa, J. C., Youssouf, Ratsimbazafy, J., Dolch, R., & King, T. (2011). Gathering local knowledge in Madagascar results in a major increase in the known range and number of sites for critically endangered greater bamboo lemurs (*Prolemur simus*). *International Journal of Primatology, 32*, 776-792.

Raxworthy, C. J., & Rankotondraparany. F. (1988). Mammals report. In N. Quansah, (Ed.), *Manogarivo Special Reserve (Madagascar) 1987/88 expedition report* (pp. 121-131). Madagascar Environmental Research Group.

Razafindramanana, J., Eppley, T. M., Rakotondrabe, R., Rakotoarisoa, A. A., Ravaloharimanitra, M., & King, T. (2020a). *Eulemur rufus. The International Union for the Conservation of Nature Red List of Threatened Species* 2020: e.T8209A115562696.

Razafindramanana, J., Eppley, T. M., Rakotondrabe, R., Roullet, D., Irwin, M., & King, T. (2020b). *Eulemur mongoz. The International Union for the Conservation of Nature Red List of Threatened Species* 2020: e.T8202A115561431.

Razafindramanana, J., Salmona, J., King, T., Roullet, D., Eppley, T. M., Sgarlata, G. M., & Schwitzer, C. (2020c). *Propithecus coronatus. The International Union for the Conservation of Nature Red List of Threatened Species* 2020: e.T18356A115572495.

Razafindratsima, O. H., Gentles, A., Drager, A. P., Razafimahaimodison, J.-C., Ralazampirenena, C. J., & Dunham, A. E. (2018). Consequences of lemur loss for above-ground carbon stocks in a Malagasy rainforest. *International Journal of Primatology, 39*, 415-426.

Reuter, K. E., Eppley, T. M., Hending, D., Pacifici, M., Semel, B., & Zaonarivelo, J. (2020). *Eulemur coronatus. The International Union for the Conservation of Nature Red List of Threatened Species* 2020: e.T8199A115561046.

Reuter, K. E., Gilles, H., Wills, A. R., & Sewall, B. J. (2016). Live capture and ownership of lemurs in Madagascar: Extent and conservation implications. *Oryx, 50*(2), 344-354.

Richard, A. F. (1978). *Behavioral variation: Case study of a Malagasy lemur.* Bucknell University Press.

Richard, A. F. (1985). Social boundaries in a Malagasy prosimian, the sifaka (*Propithecus verreauxi*). *International Journal of Primatology, 6*, 553-568.

Richard, A. F. (2003). *Propithecus*, sifakas. In S. M. Goodman, & J. P. Benstead (Eds.), *The natural history of Madagascar* (pp. 1345-1348). University of Chicago Press.

Richard, A. F., & Nicoll, M. E. (1987). Female social dominance and basal metabolism in a Malagasy primate, *Propithecus verreauxi. American Journal of Primatology,* l2, 309-314.

Richard, A. F., & Sussman, R. W. (1975). Future of the Malagasy lemurs: Conservation or extinction? In I. Tattersall, & R. W. Sussman (Eds.), *Lemur biology* (pp. 335-350). Plenum Press.

Rigamonti, M. M., Spezio, C., Poli, M. D., & Fazio, F. (2005). Laterality of manual function in foraging and positional behavior in wild indri (*Indri indri*). *American Journal of Primatology, 65*, 27-38.

Ripple, W. J., Abernethy, K., Betts, M. G., Chapron, G., Dirzo, R., Galetti, M., Levi, T., Lindsey, P. A., Macdonald, D. W., Machovina, B., Newsome, T. M., Peres, C. A., Wallach, A. D., Wolf, C., & Young, H. (2016). Bushmeat hunting and extinction risk to the world's mammals. *Royal Society Open Science, 3*(10), 160498.

Rowe, N., & Myers, M. (Eds.). (2016). *All the world's primates.* Pogonias Press.

Salmona, J., Jan, F., Rasolondraibe, E., Besolo, A., Ousseni, D. S., Beck, A., Zaranaina, R., Rakotoarisoa, H., Rabarivola, C. J., & Lounès Chikhi, L. (2014). Extensive survey of the endangered Coquerel's sifaka *Propithecus coquereli. Endangered Species Research, 25*, 175-183.

Salmona, J., Jan, F., Rasolondraibe, E., Zaranaina, R., Saïd Ousseni, D., Mohamed-Thani, I., Rakotonanahary, A., Ralantoharijaona, T., Kun-Rodrigues, C., Carreira, M., Wohlhauser, S., Ranirison, P., Zaonarivelo, J. R., Rabarivola, C. J., & Chikhi, L. (2013). Survey of the critically endangered Perrier's sifaka (*Propithecus perrieri*) across most of its distribution range. *Lemur News, 17*, 9-12.

Samonds, K. E., Godfrey, L. R., Ali, J. R., Goodman, S. M., Vences, M., Sutherland, M. R., Irwin, M. T., & Krause, D. W. (2012). Spatial and temporal arrival patterns of Madagascar's vertebrate fauna explained by distance, ocean currents, and ancestor type. *Proceedings of the National Academy of Sciences, 109*(14), 5352-5357.

Santini, L., Rojas, D., & Donati, G. (2015). Evolving through day and night: Origin and diversification of activity pattern in modern primates. *Behavioral Ecology, 26*(3), 789-796.

Sato, H. (2012). Frugivory and seed dispersal by brown lemurs in a Malagasy tropical dry forest. *Biotropica, 44*(4), 479-488.

Sato, H., Santini, L., Patel, E. R., Campera, M., Yamashita, N., Colquhoun, I. C., & Donati, G. (2016). Dietary flexibility and feeding strategies of *Eulemur*: A comparison with *Propithecus*. *International Journal of Primatology, 37*, 109-129.

Sauther, M. L. (1992). *Effect of reproductive state, social rank and group size on resource use among free-ranging ring-tailed lemurs (*Lemur catta*) of Madagascar*. [Doctoral dissertation, Washington University].

Sauther, M. L., & Sussman, R. W. (1993). A new interpretation of the organization and mating systems of the ring-tailed lemur (*Lemur catta*). In P. M. Kappeler, & J. U. Ganzhorn (Eds.), *Lemur social systems and their ecological basis* (pp. 111-121). Plenum Press.

Sauther, M. L., Sussman, R. W., & Gould, L. (1999). The socioecology of the ringtailed lemur: Thirty-five years of research. *Evolutionary Anthropology, 8*(4), 120-132.

Schmidt-Nielsen, K. (1975). Scaling in biology: The consequences of size. *Journal of Experimental Zoology, 194*(1), 287-307.

Schmidt-Nielsen, K. (1984). *Scaling: Why is animal size so important?* Cambridge University Press.

Schüßler, D., Radespiel, U., Ratsimbazafy, J. H., & Mantilla-Contrerasa, J. (2018). Lemurs in a dying forest: Factors influencing lemur diversity and distribution in forest remnants of north-eastern Madagascar. *Biological Conservation, 228*, 17-26.

Schwartz, J. H. (1974). Observations on the dentition of the Indriidae. *American Journal of Physical Anthropology, 41*(1), 107-114.

Schwitzer, C., Mittermeier, R. A., Davies, N., Johnson, S., Ratsimbazafy, J., Razafindramanana, J., Louis., E. E, Jr., Rajaobelina, S. (Eds.). (2013). *Lemurs of Madagascar: A strategy for their conservation 2013-2016*. IUCN SSC Primate Specialist Group.

Schwitzer, C., Mittermeier, R. A., Johnson, S. E., Donati, G., Irwin, M., Peacock, H., Ratsimbazafy, J., Razafindramanana,, J., Louis, E. E., Jr., Chikhi, L., Colquhoun, I. C., Tinsman, J., Dolch, R., LaFleur, M., Nash, S., Patel, E., Randrianambinina, B., Rasolofoharivelo, T., & Wright, P. C. (2014b). Averting lemur extinctions amid Madagascar's political crisis. *Science, 343*, 842-843.

Schwitzer, C., Randriatahina, G. H., & Volampeno, S. (2014a). Sclater's black lemur or blue-eyed black lemur *Eulemur flavifrons* (Gray, 1867). In C. Schwitzer, R. A. Mittermeier, A. B. Rylands, L. A. Taylor, F. Chiozza, E. A. Williamson,

J. Wallis, & F. E. Clark (Eds.), *Primates in peril: The world's 25 most endangered primates 2012-2014* (pp. 29-32). Conservation International.

Semel, B. P., Baden, A. L., Salisbury, R. L., McGee, E. M., Wright, P. C., & Arrigo-Nelson, J. (2019). Assessing the function of geophagy in a Malagasy rain forest lemur. *Biotropica, 51*(5), 769-780.

Semel, B., Semel, M., Salmona, J., & Heriniaina, R. (2020). *Propithecus tattersalli. The International Union for the Conservation of Nature Red List of Threatened Species* 2020: e.T18352A115571806.

Simons, E. L., & Rumpler, Y. (1988). *Eulemur*: A new generic name for species of lemur other than *Lemur catta*. *Comptes Rendus de l'Académie des Sciences, 307*, 547-551.

Steffens, K. J. E. (2020). Lemur food plants as options for forest restoration in Madagascar. *Restoration Ecology, 28*(6), 1517-1527.

Stewart, B. M., Turner, S. E., & Matthews, H. D. (2020). Climate change impacts on potential future ranges of non-human primate species. *Climatic Change, 162*, 2301-2318.

Sussman, R. W. (1991). Demography and social organization of free-ranging *Lemur catta* in the Beza Mahafaly Reserve, Madagascar. *American Journal of Physical Anthropology, 84*, 43-58.

Sussman, R. W. (1992). Male life history and intergroup mobility among ringtailed lemurs (*Lemur catta*). *International Journal of Primatology, 13*, 395-413.

Sussman, R. W. (2011). A brief history of of primate field studies. In C. J. Campbell, A. Fuentes, K. C. MacKinnon, S. K. Bearder, & R. M Stumpf (Eds.), *Primates in perspective* (2nd ed., pp. 6-11). Oxford University Press.

Sussman, R. W., & Ratsirarson, J. (2006). Beza Mahafaly Special Reserve: A research site in southwestern Madagascar. In A. Jolly, R. W. Sussman, N. Koyama, & H. Rasamimanana (Eds.), *Ring-tailed lemur biology:* Lemur catta *in Madagascar. Developments in primatology: Progress and prospects* (pp. 43-52). Springer.

Sussman, R. W., Richard, A. F., Ratsirarson, J., Sauther, M. L., Brockman, D. K., Gould, L., Lawler, R., & Cuozzo, F. P. (2012). Beza Mahafaly Special Reserve: Long-term research on lemurs in southwestern Madagascar. In P. M. Kappeler, & D. Watts (Eds.), *Long-term field studies of primates* (pp. 45-66). Springer.

Tan, C. L. (1999). Group composition, home range size, and diet of three sympatric bamboo lemur species (Genus *Hapalemur*) in Ranomafana National Park, Madagascar. *International Journal of Primatology, 20*(4), 547-566.

Tan, C. L. (2000). *Behavior and ecology of three sympatric bamboo lemur species (genus* Hapalemur*) in Ranomafana National Park, Madagascar*. [Doctoral dissertation, Stony Brook University].

Tan, C. L. (2006). Behavior and ecology of gentle lemurs (genus *Hapalemur*). In L. Gould, & M. L. Sauther (Eds.), *Lemurs: Ecology and adaptation. Developments in primatology: Progress and prospects* (pp. 369-381). Springer.

Tattersall, I. (1979). Patterns of activity in the Mayotte lemur, *Lemur fulvus mayottensis*. *Journal of Mammalogy, 60*(2), 314-323.

Tattersall, I. (1982). *The primates of Madagascar*. Columbia University Press.

Tattersall, I. (1987). Cathemeral activity in primates: A definition. *Folia Primatologica, 49*(3-4), 200-202.

Tattersall, I., & Sussman, R. W. (1975). Notes on the topography, climate and vegetation of Madagascar. In I. Tattersall, & R. W. Sussman (Eds.), *Lemur biology* (pp. 13-21). Plenum Press.

Tattersall, I., & Sussman, R. W. (2016). Little brown lemurs come of age: Summary and perspective. *International Journal of Primatology, 37*, 3-9.

Tecot, S. R. (2010). It's all in the timing: Birth seasonality and infant survival in *Eulemur rubriventer*. *International Journal of Primatology, 31*, 715-735.

Thalmann, U. (2001). Food resource characteristics in two nocturnal lemurs with different social behavior: *Avahi occidentalis* and *Lepilemur edwardsi*. *International Journal of Primatology, 22*, 287-324.

Thalmann, U. (2003). *Avahi*, woolly lemurs. In S. M. Goodman, & J. P. Benstead (Eds.), *The natural history of Madagascar* (pp. 1340-1342). University of Chicago Press.

Thalmann, U., & Geissmann, T. (2000). Distribution and geographic variation in the western woolly lemur (*Avahi occidentalis*) with description of a new species (*A. unicolor*). *International Journal of Primatology, 21*(6), 915-941.

Thalmann, U., & Geissmann, T. (2005). New species of woolly lemur *Avahi* (Primates: Lemuriformes) in Bemaraha (central western Madagascar). *American Journal of Primatology, 67*(3), 371-376.

Thalmann, U., & Geissmann, T. (2006). Conservation assessment of the recently described John Cleese's woolly lemur, *Avahi cleesei* (Lemuriformes, Indridae). *Primate Conservation, 21*, 45-49.

Thalmann, U., Geissmann, T., Simona, A., & Mutschler, T. (1993). The indris of Anjanaharibe-Sud, northeastern Madagascar. *International Journal of Primatology, 14*, 357-381.

Tinsman, J. (2020). *Geospatial and genomic tools for conserving the critically endangered blue-eyed black lemur (*Eulemur flavifrons*) and the sportive lemurs (genus* Lepilemur*)*. [Doctoral dissertation, Columbia University].

Vasey, N. (1997). *Community ecology and behavior of* Varecia variegata rubra *and* Lemur fulvus albifrons *on the Masoala Peninsula, Madagascar*. [Doctoral dissertation, Washington University].

Vasey, N. (2000). Niche separation in *Varecia variegata rubra* and *Eulemur fulvus albifrons*: I. Interspecific patterns. *American Journal of Physical Anthropology, 112*, 411-431.

Vasey, N. (2002). Niche separation in *Varecia variegata rubra* and *Eulemur fulvus albifrons*: II. Intraspecific patterns. *American Journal of Physical Anthropology, 118*, 169-183.

Vasey, N. (2003). *Varecia*, ruffed lemurs. In S. M. Goodman, & J. P. Benstead (Eds.), *The natural history of Madagascar* (pp. 1332-1336). University of Chicago Press.

Vasey, N. (2006). Impact of seasonality and reproduction on social structure, ranging patterns, and fission-fusion social organization in red ruffed lemurs. In L. Gould, & M. L. Sauther (Eds.), *Lemurs: Ecology and adaptation. Developments in primatology: Progress and prospects* (pp. 275-304). Springer.

Vasey, N., & Tattersall, I. (2002). Do ruffed lemurs form a hybrid zone? Distribution and discovery of *Varecia*, with systematic and conservation implications. *American Museum Novitates, 3376*, 1-26.

Vaughan, A. (2020). Almost all lemur species are now officially endangered. *New Scientist, 247* (3291), 15.

Vieilledent, G., Grinand, C., Rakotomalala, F. A., Ranaivosoa, R., Rakotoarijaona, J.-R., Allnutt, T. F., & Achard, F. (2018). Combining global tree cover loss data with historical national forest cover maps to look at six decades of deforestation and forest fragmentation in Madagascar. *Biological Conservation, 222*, 189-197.

Volampeno, M. S. N., Randriatahina, G. H., Kalle, R., Wilson, A.-L., & Downs, C. T. (2015). A preliminary population viability analysis of the critically endangered blue-eyed black lemur (*Eulemur flavifrons*). *African Journal of Ecology, 53*, 419-427.

Volampeno, S., Randriatahina, G., Schwitzer, C., & Seiler, M. (2020). *Eulemur flavifrons. The International Union for the Conservation of Nature Red List of Threatened Species* 2020: e.T8211A115563094.

Waeber, P. O., & Hemelrijk, C. K. (2003). Female dominance and social structure in Alaotran gentle lemurs. *Behaviour, 140*(10), 1235-1246.

Warren, R. (1997). Habitat use and support preference of two free-ranging saltatory lemurs (*Lepilemur edwardsi* and *Avahi occidentalis*). *Journal of Zoology, 241*(2), 325-341.

Warter, S., Randrianasolo, G., Dutrillaux, B., & Rumpler, Y. (1987). Cytogenetic study of a new subspecies of *Hapalemur griseus*. *Folia Primatologica, 48*(1-2), 50-55.

White, F. (1983). *The vegetation of Africa. A descriptive memoir to accompany the Unesco/AETFAT/ UNSO Vegetation Map of Africa*. United Nations Educational.

Wilmé, L., Goodman, S. M., & Ganzhorn, J. U. (2006). Biogeographic evolution of Madagascar's microendemic biota. *Science, 312*(5776), 1063-1065.

Wilson, D. E., & Hanlon, E. (2010). *Lemur catta* (Primates: Lemuridae). *Mammalian Species, 42* (854), 58-74.

Wright, P. C. (1986). Diet, ranging behavior and activity pattern of the gentle lemur (*Hapalemur griseus*) in Madagascar. *American Journal of Physical Anthropology, 69*, 283.

Wright, P. C. (1998). Impact of predation risk on the behaviour of *Propithecus diadema edwardsi* in the rain forest of Madagascar. *Behaviour, 135*(4), 483-512.

Wright, P. C. (1999). Lemur traits and Madagascar ecology: Coping with an island environment. *Yearbook of Physical Anthropology, 42*, 31-72.

Wright, P. C., Daniels, P., Meyers, D. M., Overdorff, D. J., & Rabesoa, J. (1987). A census and study of *Hapalemur* and *Propithecus* in southeastern Madagascar. *Primate Conservation, 8*, 84-88.

Wright P. C., Erhart, E., Tecot, S., Baden, A., Arrigo-Nelson, S., Herrera, J., Morelli, T. L., Blanco, M. B., Deppe, A. M., Atsalis, S., Johnson, S., Ratelolahy, F., Tan, C., & Zohdy, S. (2012). Long-term lemur research at Centre Valbio, Ranomafana National Park, Madagascar. In P. Kappeler, & D. Watts (Eds.), *Long-term field studies of primates* (pp. 67-100). Springer.

Wright, P., Hearthstone, E., Andrianoely, D., Donohue, M. E., & Otero-Jiménez, B. J. (2020a). *Propithecus edwardsi. The International Union for the Conservation of Nature Red List of Threatened Species* 2020: e.T18359A115573104.

Wright, P., Hearthstone, E., Donohue, M. E., Andrianoely, D., Otero-Jiménez, B. J., & Lauterbur M. E. (2020b). *Hapalemur aureus. The International Union for the Conservation of Nature Red List of Threatened Species* 2020: e.T9672A115564398.

Wright, P. C., Heckscher, S. K., & Dunham, A. E. (1997). Predation on Milne-Edward's sifaka (*Propithecus diadema edwardsi*) by the fossa (*Cryptoprocta ferox*) in the rain forest of southeastern Madagascar. *Folia Primatologica, 68*, 34-43.

Wright, P. C., Johnson, S. E., Irwin, M. T., Jacobs, R., Schlichting, P., Lehman, S., Louis, E. E., Jr., Arrigo-Nelson, S. J., Raharison, J.-L., Rafalirarison, R. R., Razafindratsita, V., Jonah Ratsimbazafy, J., Ratelolahy, F. J., Dolch, R., & Tan, C. L. (2008). The crisis of the critically endangered greater bamboo lemur (*Prolemur simus*). *Primate Conservation, 23*, 5-17.

Wright, P. C., Larney, E., Louis, E. E., Jr., Dolch, R., & Rafaliarison, R. R. (2009). Greater bamboo lemur *Prolemur simus* (Gray, 1871). In R. A. Mittermeier, J. Wallis, A. B. Rylands, J. U. Ganzhorn, J. F. Oates, E. A. Williamson, E. Palacios, E. W. Heymann, M. C. M. Kierulff, Y. Long, J. Supriatna, C. Roos, S. Walker, L. Cortés-Ortiz, & C. Schwitzer (Eds.), *Primates in peril: The world's 25 most endangered primates, 2008-2010* (pp. 11-14). Conservation International.

Wright P. C., & Martin L. B. (1995). Predation, pollination and torpor in two nocturnal prosimians: *Cheirogaleus major* and *Microcebus rufus* in the rain forest of Madagascar. In L. Alterman, G. A. Doyle, & M. K. Izard (Eds.), *Creatures of the dark: The nocturnal prosimians* (pp. 45-60). Plenum Press.

Wright, P. C., & Tan, C. L. (2016a). Golden bamboo lemur. In N. Rowe, & M. Myers (Eds.), *All the world's primates* (pp. 82-83). Pogonias Press.

Wright, P. C., & Tan, C. L. (2016b). Greater bamboo lemur. In N. Rowe, & M. Myers (Eds.), *All the world's primates* (p. 86). Pogonias Press.

Yamashita, N. (2002). Diets of two lemur species in different microhabitats in Beza Mahafaly Special Reserve, Madagascar. *International Journal of Primatology, 23*, 1025-1051.

Yamashita, N., Tan, C. L., Vinyard, C. J., & Williams, C. (2010). Semi-quantitative tests of cyanide in foods and excreta of three *Hapalemur* species in Madagascar. *American Journal of Primatology, 72*, 56-61.

Yamashita, N., Vinyard, C. J., & Tan, C. L. (2009). Food mechanical properties in three sympatric species of *Hapalemur* in Ranomafana National Park, Madagascar. *American Journal of Physical Anthropology, 139*(3), 368-381.

Yoder, A. D. (1997). Back to the future: A synthesis of strepsirrhine systematics. *Evolutionary Anthropology, 6*(1), 11-22.

Yoder, A. D., Cartmill, M., Ruvolo, M., Smith, K., & Vilgalys, R. (1996). Ancient single origin for Malagasy primates. *Proceedings of the National Academy of Sciences, 93*, 5122-5126.

Yoder, A. D., Irwin, J. A., Goodman, S. M., & Rakotoarisoa, S. V. (1999). *Lemur catta* from the Andringatra Massif are *Lemur catta. Lemur News, 4*, 32-33.

Yoder, A. D., & Nowak, M. D. (2006). Has vicariance or dispersal been the predominant biogeographic force in Madagascar? Only time will tell. *Annual Review of Ecology, Evolution, and Systematics, 37*(1), 405-431.

Young, A. L. Richard, A. F., & Aiello, L. C. (1990). Female dominance and maternal investment in strepsirrhine primates. *American Naturalist, 135*, 473-488.

Zhang, L., Ameca, E. I., Cowlishaw, G., Pettorelli, N., Foden, W., & Mace, G. M. (2019). Global assessment of primate vulnerability to extreme climatic events. *Nature Climate Change, 9*, 554-561.

CHAPTER 4

Tarsiers

MYRON SHEKELLE, SHARON L. GURSKY, ANGELA ACHORN, AND
IAN C. COLQUHOUN

THE TARSIERS ARE ENIGMATIC LITTLE PRIMATES (ADULT BODY weights range between 100–200 g). Indeed, their small size, together with their strictly nocturnal activity patterns, and their arboreal lifeways, have all contributed to the challenge of conducting quantitative observations on wild tarsiers. Compared to the depth of detailed field research conducted on diurnal anthropoid primates, long-term field studies of tarsiiform species were initiated approximately 20 years after the first long-term studies of various diurnal anthropoid species. As a result, research into tarsier behavioral ecology continues to produce new insights, including the identification of tarsier species that are new to science. Over the last 40 years, field research on tarsiers has clearly shown that previous conceptions of tarsier behavioral ecology, biogeography, phylogeny, and taxonomy were inaccurate and oversimplified. What is being revealed is considerable tarsiiform variability, both intraspecifically and interspecifically. It is not an overstatement to say that there still is a great deal to learn about wild tarsiers. This is increasingly an important matter, because without a clear understanding of the behavioral ecology of tarsiers, we will be hobbled in efforts to address the conservation biology issues facing their populations (specifically habitat fragmentation, habitat loss, hunting, and, in some regions, illegal capture for the pet trade).

TARSIER PHYLOGENY AND TAXONOMY

The phylogenetic and taxonomic positions of tarsiers within the Order Primates have been among the most vexing issues within primate biology (e.g., Pocock, 1918; Le Gros Clark, 1956; Simons, 1972; Szalay, 1975; Rosenberger & Szalay, 1980; Martin, 1993; Goodman et al., 1998; Yoder, 2003; Fleagle, 2013). Primate phylogeny and taxonomy can be broadly condensed into stating that the tarsiiforms, the strepsirrhine lemuriforms and lorisiforms, the platyrrhines (New World monkeys), and catarrhines (Old World monkeys and apes) constitute five unambiguous high-level taxonomic monophyletic groups that comprise the Order Primates. That is, the lemuriforms (lemurs) and lorisiforms (lorises and galagos) are sister groups of strepsirrhine primates, while the tarsiers are the sister group to haplorrhine primates (i.e., both the New World [platyrrhine] and Old World [catarrhine] lineages (Martin, 2012).

Historically, tarsiers have either been regarded as an evolutionary "grade" of primate intermediate between the more primitive lemuriforms and lorisiforms and the more derived anthropoid primates (e.g., Zuckerman, 1932; Hill, 1955, 1972; Le Gros Clark, 1956), or they have been classified with the lemurs, lorises, and galagos in the Suborder Prosimii, while the New World and Old World monkeys, apes, and humans were all classified together in the Suborder Anthropoidea (e.g., Simpson, 1945; Le Gros Clark, 1956; Rosenberger & Szalay, 1980). This latter classification was based on numerous traits, interpreted to be ancestral characters, that apparently linked the tarsiers with lemurs, lorises, and galagos—traits such as nocturnality, small body size, parenting strategies (e.g., infant parking and oral transport), and anatomical characteristics shared with lemuriforms and lorisiforms, including an unfused mandibular symphysis, a grooming claw on the second toe (plus, unique to tarsiers, a grooming claw on the third toe), multiple pairs of mammae, and a bicornuate uterus (e.g., Pocock, 1918; Cartmill & Kay, 1978; Rosenberger & Szalay, 1980; Schwartz & Tattersall, 1987; Fleagle, 2013). In contrast to this classification, other traits have been interpreted as shared derived characters that offer evidence that extant tarsiers are the sister taxon to the anthropoid primates, the two clades together forming the Suborder Haplorrhini (Aiello, 1986; Fleagle, 2013). Thus, a current classification is to taxonomically subdivide the Order Primates into the Suborders Strepsirrhini (Lorisiformes and Lemuriformes) and Haplorrhini (Tarsiiformes and Anthropoidea) (Fleagle, 2013). The upshot is that current phylogenetic and taxonomic interpretations, for the most part, see the tarsiers as much more closely aligned biologically to the anthropoid primate lineages (i.e., New World monkeys, Old World monkeys, and apes) than they are to either the lorisiform or lemuriform primate radiations (e.g., Pollock & Mullin, 1987; Rasmussen, 1994; Martin, 2012; but see Rosenberger & Preuschoft, 2012, for an alternative opinion).

Tarsier taxonomy has been in a period of revision recently. The number of recognized taxa is increasing and the correct nomenclature for some familiar taxa has been challenged (e.g., Groves, 1998). These changes have produced an increase in the number of recognized genera from one to three and an increase in the number of species from 3 to at least 14. The current taxonomic classification of the extant tarsiiforms is presented in Table 4.1.

Until recently, extant Tarsiiformes were all classified within a single genus, *Tarsius*, following Hill (1955), who recognized three species, *Tarsius syrichta*, *T. bancanus*, and *T. spectrum*, with multiple subspecies recognized within each species. Hill's species are allopatrically distributed within distinct biogeographic regions: (1) islands of the southern Philippines (that formed the Ice Age landmass known as Greater Mindano); (2) a subset of the islands that formed Ice Age Sundaland (the Malay peninsula on the Asian mainland, as well as the islands of Sumatra, Java, and Borneo); and (3) islands that formed Ice Age Sulawesi and additional surrounding island chains (Groves, 1976; see Figure 4.1). The tarsiiforms inhabiting these biogeographic regions are often referred to, respectively, as Philippine, Western, and Eastern tarsiers (Shekelle, 2008b; Driller et al., 2015). Subsequently, Groves and Shekelle (2010) formally reclassified the Philippine, Western, and Eastern (or Sulawesi) tarsiers into three separate genera: *Carlito*, *Cephalopachus*, and *Tarsius*, respectively (Merker et al., 2009; Shekelle et al., 2010; see also Figure 4.1). The rationale for this reclassification included previously unreported morphological variation, and several lines of evidence that were unknown to Hill, including genetic data that point to an ancient divergence of the tarsiiform lineage from other primate stock (Springer et al., 2012), plus field data showing large differences between taxa in vocalizations and social behavior.

Taxonomy of Eastern Tarsiers

Following the taxonomic revisions of Groves and Shekelle (2010), together with data on recently described species (Shekelle et al., 2017, 2019), the genus *Tarsius* is restricted to the Sulawesian tarsier species and consists of at least 12 species listed in Table 4.1. These species are,

Table 4.1 Current Taxonomy of the Tarsiiforms

Tarsier Geographic / Taxonomic Group [a]	Species / Subspecies Name	Reference
Philippine, or Central, group (*Carlito syrichta* ssp.)	*Carlito syrichta carbonarius*	Groves and Shekelle, 2010
	Carlito syrichta fraterculus	Groves and Shekelle, 2010
	Carlito syrichta syrichta	Groves and Shekelle, 2010
Western group (*Cephalopachus bancanus* ssp.)	*Cephalopachus bancanus bancanus*	Groves and Shekelle, 2010
	Cephalopachus bancanus borneanus	Groves and Shekelle, 2010
	Cephalopachus bancanus natunensis	Groves and Shekelle, 2010
	Cephalopachus bancanus saltator	Groves and Shekelle, 2010
Eastern group (*Tarsius* spp.)	*Tarsius niemitzi* (=*T. spectrum*) [b]	Groves et al., 2008; Shekelle et al., 2019
	Tarsius spectrumgurskyae (=*T. spectrum*) [b]	Shekelle et al., 2017, 2019
	Tarsius supriatnai (=*T. spectrum*) [b]	Shekelle et al., 2017, 2019
	Tarsius dentatus (=*T. dianae*) [b]	Groves and Shekelle, 2010; Shekelle et al., 2017, 2019
	Tarsius fuscus (=*T. spectrum*) [b]	Groves and Shekelle, 2010; Shekelle et al., 2017, 2019
	Tarsius pelengensis	Groves and Shekelle, 2010; Shekelle et al., 2019
	Tarsius pumilus	Groves and Shekelle, 2010; Shekelle et al., 2019
	Tarsius sangirensis (=*T. spectrum*) [b]	Groves and Shekelle, 2010; Shekelle et al., 2019
	Tarsius tarsier (=*T. spectrum*) [b]	Groves and Shekelle, 2010; Shekelle et al., 2017, 2019
	Tarsius tumparai (=*T. spectrum*) [b]	Groves and Shekelle, 2010; Shekelle et al., 2019
	Tarsius wallacei (=*T. spectrum*) [b]	Groves and Shekelle, 2010; Shekelle et al., 2019
	Tarsius lariang (=*T. spectrum*) [b]	Merker and Groves, 2006; Shekelle et al., 2019

[a] Groves et al. (2008) detail the taxonomic history of tarsiers, elucidating the distinctiveness of Western, Philippine, and Eastern tarsiers (see also Groves, 1998); Groves & Shekelle (2010) erected genus-level separation of Western (*Cephalopachus*), Philippine (*Carlito*), and Eastern (*Tarsius*) tarsiers.
[b] Groves et al. (2008) note inexact use of *T. spectrum* since the late 1700s; it is a junior synonym of *T. tarsier* and many other *Tarsius* taxa (see also Shekelle et al., 2019).

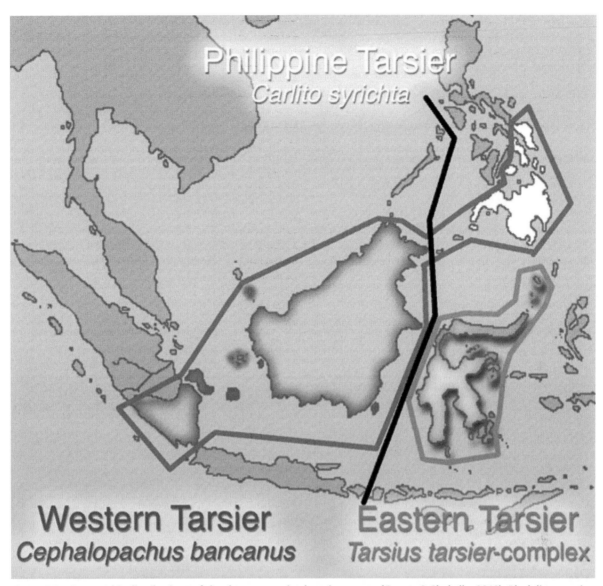

Figure 4.1 Geographic distributions of the three recognized tarsier genera (Groves & Shekelle, 2010). Black line running roughly north–south between the island of Borneo (to the west), and the island of Sulawesi and the Philippines (to the east), represents Wallace's Line.

(Adapted from Shekelle, 2008a: Fig. 1; Open Access source)

for the most part, hard to decipher taxonomically and, as with many other nocturnal taxa, their presence was first suspected by obvious differences in their vocalizations among geographically distinct populations (MacKinnon & MacKinnon, 1980; Niemitz et al., 1991; Nietsch & Niemitz, 1993; Nietsch, 1994, 1999; Shekelle, 1997; Nietsch & Kopp, 1998; Merker & Groves, 2006; Shekelle, 2008a,b; Burton & Nietsch, 2010; Merker et al., 2010). Subsequent to their identification during the course of field research on their calls, subtle morphological differences were also noted (Merker & Groves, 2006; Shekelle et al., 2019). Biogeographic and bioacoustic evidence also suggests that additional distinct species might still be recognized in the future (Shekelle et al., 2010, 2019), as does current genetic evidence. For instance, Sumampow et al. (2020)

recently studied tarsiers from three volcanic islands in Bunaken National Park, North Sulawesi, Indonesia, comparing genetic sequences for five nuclear genes to existing genotypes from tarsiers at sites on the main island of Sulawesi. Their results indicate that the three volcanic island populations have identical genetic sequences for all five genes and that they form a genetic clade distinct from tarsiers of Sulawesi proper.

Taxonomy of Western and Philippine Tarsiers

The geographic variation exhibited across the Western and Philippine tarsiers, and its taxonomic ramifications, are only imperfectly understood. Available museum specimens for comparative study may provide limited information, since most date to the era of opportunistic sampling in the wild, with little attention focused on the geographic distribution of biological variation. In the case of Philippine tarsiers, museum specimens are heavily skewed toward one region (i.e., Mindanao). Systematic field studies of variation among wild Western and Philippine tarsiers have been almost nonexistent thus far.

Despite scarce data, Hill (1955) accepted three subspecies of Philippine tarsier: *Carlito syrichta carbonarius* (from the island of Mindanao), *Ca. s. fraterculus* (from the island of Bohol), and *Ca. s. syrichta* (found in the remaining areas of the Philippine tarsier's geographic extent of occurrence). A number of subsequent authors, however, do not recognize these subspecies (e.g., Niemitz, 1984c; Groves, 2001).

Hill (1955) recognized four subspecies of Western tarsiers: *Cephalopachus bancanus borneanus* (from the island of Borneo), *Ce. b. saltator* (from Belitung, Indonesia), *Ce. b. natunensis* (from the South Natuna Island of Serasen, Indonesia), and the nominate taxon, *Ce. b. bancanus* from the island of Bangka, Indonesia (from whence the species gets its name) and, by default, everywhere else Western tarsiers are found—chiefly the southern regions of Sumatra, Indonesia. Niemitz (1984c) does not support recognition of all four subspecies.

Systematic phylogeographic field studies to definitively answer questions of taxonomy in *Carlito* and *Cephalopachus* are in their infancy for the former, and almost nonexistent at present for the latter. Merker et al. (2007, 2009) examined 12 microsatellite markers in Philippine tarsiers (*Carlito*) that hypothetically could be useful for studying phylogeographic and other questions related to population genetics and conservation genetics. More recently, Brown et al. (2014) reported evidence that Philippine tarsiers are partitioned into at least three genetic groups that might differ slightly from Hill's taxonomy, but they urge taxonomic caution until a more thorough study can be conducted. Detailed phylogeographic work has yet to be carried out for Western tarsiers (*Cephalopachus*). Morphology and bioacoustics can assist with systematic studies of taxonomy but, when possible, genetic evidence is paramount. Given the challenges of conducting high-resolution genetic work on tissue samples from tarsier habitat countries, it could be many years before full details are available regarding the taxonomy of extant tarsiers (but see Blanco et al., 2020, concerning the rapid biodiversity assessment work that can be done with the use of mobile genetics labs under remote field conditions). In the interim, it would be unfortunate, but not surprising, if some tarsier taxa were to go extinct before they have even been identified and described by science (Wright, 2003; Shekelle et al., 2019).

DISTRIBUTION AND BIOGEOGRAPHY

There are presently no known instances of sympatrically distributed tarsiers. Thus, to understand tarsier taxonomy, one must also understand their distribution and biogeography.

Genetic data show that the tarsier genera lineages are far older than has been generally appreciated, with the divergence of crown clade tarsiers dating to the middle Miocene or earlier when a species-rich variety of tarsier-like primates were broadly distributed across the northern hemisphere. Today, only three tarsier genera—all of which are distributed parapatrically or allopatrically—exist on a scattering of islands in insular Southeast Asia. The three genera are each endemic to a distinct biogeographic region: Western tarsiers (genus *Cephalopachus*) on Borneo and Sumatra, plus smaller adjacent islands; Philippine Islands tarsiers (genus *Carlito*); and Eastern tarsiers (genus *Tarsius*) on Sulawesi and adjacent islands.

The Eastern tarsiers are found throughout Sulawesi, on all of the landmasses that formed Ice Age Greater Sulawesi, as well as being distributed on numerous smaller, surrounding islands that were never linked to Sulawesi. This is strong evidence that dispersal by rafting must have occurred for these primates to establish themselves on islands beyond Sulawesi (Shekelle, 2014; Driller et al., 2015). Eastern tarsiers possess the most primitive morphology among tarsiiform taxa, based upon their relatively shortest legs, shortest hands, smallest eyes, furriest tail, and lack of a sitting pad on the tail composed of dermal skin. It is possible that, despite the Miocene events that drove tarsiiforms extinct on mainland Asia, tarsiers survived only on the proto-Sulawesi archipelago and dispersed to Sundaland and the Philippines from there (Groves, 1976; Shekelle, 2014; Driller et al., 2015).

The other two tarsier genera, *Cephalopachus* and *Carlito*, form a clade that split away from the Eastern tarsier (*Tarsius*) clade perhaps 10–20 mya. Generic separation between the Western and Philippine tarsier lineages has been questioned in the past, but recent partial mtDNA analysis by Zahadin and colleagues (2019) produced gene trees that showed clear clustering of each of the three tarsier genera.

The Western tarsiers (genus *Cephalopachus*) are found on a curiously restricted subset of Sundaland and are entirely absent from mainland Asia and other areas that were contiguous during glacial maxima of the late Pleistocene. Since the distribution of extant *Ce. bancanus* does not conform to areas of exposed land during the late Pleistocene ice ages, this is strong evidence that curious, and as yet unknown, biogeographic processes were at play in shaping the distribution of these primates (Shekelle, 2008a; cf. Merker et al., 2009). *Cephalopachus* possesses the most specialized morphology, based on their relatively longest legs, longest hands, largest eyes, and possession of a sitting pad on the tail composed of dermal skin with papillary ridges. It is possible that tarsiers arrived from Sulawesi, crossing Wallace's Line almost surely by rafting, around 10–20 mya. Groves and Shekelle (2010) recognize a single species of Western tarsier—*Ce. Bancanus*, found on several islands of the Sunda Shelf, including Borneo, the southeastern lowlands of Sumatra, and numerous smaller islands in between, notably Bangka and Belitung (Hill, 1955). A phylogeographic study of *Cephalopachus*, to uncover hidden taxonomic variation and to elucidate the mysterious biogeographic processes that produced this strange geographic distribution, would be highly desirable, but is logistically infeasible at the current time.

Philippine tarsiers (genus *Carlito*) are found on the islands of Ice Age Greater Mindanao, with geographic ranges that are completely consistent with the area of exposed land during glacial maxima of the Pleistocene. Overall, Philippine tarsiers possess an intermediate morphology to that of the Eastern and Western tarsiers in almost all ways save one—*Carlito* has the least furry tail. Otherwise, their leg length, hand length, eye size, and sitting pad are all intermediate to the Western and Eastern tarsier morphologies. It is possible that Philippine tarsiers arrived from Sundaland, re-crossing Wallace's Line, either by rafting or perhaps on a dry raft of land that extended from Sundaland to form part of Mindanao, around 5–10 mya. Groves and Shekelle (2010) recognize a single species of Philippine tarsier—*Carlito syrichta*. The extant Philippine tarsier, *Ca. syrichta*, is found on the Philippine islands of Samar, Leyte, Bohol, Mindanao, and several smaller adjacent islands (Hill, 1955).

Eastern tarsiers have a nearly continuous distribution on Sulawesi, including offshore islands that were connected to Sulawesi during the ice ages, and several surrounding island chains that were not (Hill, 1955; Shekelle, 2008a). With isolation on separate islands and millions of years to evolve, Shekelle and Leksono (2004) hypothesized that as many as 16 or more taxa may exist within the genus *Tarsius*, which collectively preserve a pattern of geographic distribution formed by their isolation on separate islands during the Miocene and Pliocene, and range fragmentation during the Pleistocene. Thus, while the taxonomy of *Tarsius* is the most thoroughly studied of the three extant tarsier genera, it is likely that the number of recognized taxa within this genus on Sulawesi will continue to increase, as several acoustic forms remain unnamed and several regions remain unsurveyed (e.g., Burton & Nietsch, 2010; Shekelle et al., 2010, 2017, 2019). The elevational range of Eastern tarsiers goes from sea level to over 2,200 masl.

TARSIER MORPHOLOGY

All living tarsiers are small, nocturnal, vertical clinging and leaping, faunivorous animals and, as such, they are anatomically and ecologically distinctive vis-à-vis other primate taxa. Among tarsiers, however, there exists previously underemphasized variability on this theme. Many anatomical and behavioral traits show a clinal distribution, with species among the Eastern tarsiers being the most generalized, Western tarsiers being the most specialized, and Philippine tarsiers being intermediate. Other differences among taxa include such things as body weight, intermembral indices, finger pad size, patterns of locomotor behavior, habitat selection, nesting sites, communication, and social and ranging behavior, among other differences.

For example, statistical differences in morphometrics exist among tarsier species, with Western tarsiers having the relatively longest legs, feet, and hands, and Eastern tarsiers the shortest legs, feet, and hands (Niemitz, 1977). This led Niemitz (1979b) to predict that Eastern tarsiers are less specialized for vertical clinging and leaping, come to the ground more often, and exhibit utilization of a broader range of habitats than do Western or Philippine tarsiers. Each of these predictions was later supported by field studies of Eastern tarsiers (MacKinnon & MacKinnon, 1980; Gursky, 1997).

For most species of tarsiers, fingers and toes exhibit distinct terminal digital pads that assist them in gripping smooth vertical surfaces, such as bamboo stalks. For example, a tarsier is able to walk head-first down a rain-slickened bamboo stalk. The relatively largest pads are found on Western tarsiers, and the smallest on *T. pumilus*, the Eastern pygmy, or montane, tarsier. In the latter species, the pad is almost nonexistent, and long, keeled, claw-like nails protrude beyond the ends of the fingers and toes. In fact, the most distinctive taxon of all tarsiers in terms of morphology and ecology is *T. pumilus*. In Lore Lindu National Park, central Sulawesi, *T. pumilus* is allopatric from lowland congeners and has an altitudinal distribution that lies between 2,000–2,300 masl (Grow et al., 2013). *T. pumilus* lives at very high altitudes in montane forests, where vertical surfaces are covered with moss. Musser and Dagosto (1987) speculate that the fingers and toes of this species are adapted for clinging to the moss-covered branches of the montane forest. As the common name suggests, the pygmy tarsier is the smallest species among the extant tarsiers.

The body of a tarsier measures only about 12–13 cm long, but the tail is often twice that length. Besides being used for support on vertical substrates, the tail is also used to control momentum and direction during leaps (Preuschoft et al., 1979; Niemitz, 1984b). The tail of

Eastern tarsiers is lightly haired along its length (as is the tail of all infant tarsiers) with a tuft of denser, longer hair along the distal third. The tails of Philippine and Western tarsiers are nearly naked except for the terminal tuft of hair, which is more prominent in Western tarsiers. The tail in these two species has developed a smooth sitting pad on the ventral surface, and the sitting pad of Western tarsiers is further characterized by well-developed papillary ridges, similar to the dermal ridges found on fingers and toes (Hill, 1955).

The huge eyes of tarsiers (Figure 4.2) lack a reflective *tapetum lucidum*, and unlike most nocturnal animals whose eyes reflect light vividly, the eyes of tarsiers only glow a dull orange when a light is shone on them. Western tarsiers have, relative to head and body size, the largest eyes among tarsiers, while Sulawesi tarsiers have the smallest.

TARSIER FIELD STUDIES

In comparison to the study of the larger diurnal primates, scientific study of tarsiers in their natural habitat is a relatively recent occurrence. This is partly due to the difficulty of observing a small-bodied (120 g) nocturnal primate that does not possess a *tapetum lucidum*, the "eye-shine" that researchers rely on to locate and follow most other nocturnal mammals in the dark. Until the 1970s, knowledge of tarsier behavior was limited to, and shaped by, numerous early reports from brief, casual observations made on tarsiers in their natural habitat and in captivity

Figure 4.2 Close-up of *Tarsius spectrumgurskyae*, endemic to northeastern Sulawesi.
(Photo credit: Meldy Tamenge, image licensed under Creative Commons Attribution-Share Alike 4.0 International license)

(Cuming, 1838; Le Gros Clark, 1924; Catchpole & Fulton, 1939; Cook, 1939; Lewis, 1939; Wharton, 1950; Hill et al., 1952; Ulmer, 1960, 1963; Harrison, 1962, 1963).

The 1970s saw the first intensive field studies of tarsier behavior. Fogden (1974) adapted methods used in the study of birds (e.g., mist nets for capture and marking animals with leg bands around the tarsus) to study *Ce. bancanus borneanus* near Kuching, Sarawak, Malaysia. Niemitz (1972, 1974, 1979a,b, 1984a–f) adopted Fogden's methods and added forest enclosures, to study semi-captive tarsiers in their native environment at the same locality where Fogden had worked. The caging enclosed natural vegetation in a boundary area between patches of secondary and primary forest.

Technological advancements produced miniaturized telemetry equipment which allowed for radio-tracking and all-night follows of animals. The first of these studies was by Crompton and Andau (1986, 1987). Thus, for the first two decades of systematic field studies of wild tarsiers, nearly all were conducted on one subspecies, *Cephalopachus bancanus borneanus*. These reports described animals that were largely solitary foragers. The one exception during this period was research by MacKinnon and MacKinnon (1980) who studied *Tarsius tarsier* (now *T. spectrumgurskyae*) at Tangkoko Nature Reserve, Indonesia, on Sulawesi's northeastern peninsula, finding surprisingly different social behavior from that of *Ce. b. borneanus*. They explicitly noted that the tarsier social groups they observed revealed gibbon-like small family groups, exhibiting, for example, coordinated female-male duetting call bouts (MacKinnon & MacKinnon, 1977, 1980). In their study of Eastern tarsiers, the MacKinnons found that tarsiers usually sleep at the same sleeping site each day, give loud calls at dawn, and often were approachable to within 2 m. It was possible for the MacKinnons to maintain contact with these tarsiers for long periods simply using flashlights.

Subsequently, behavioral and ecological studies of wild, free-ranging tarsiers shifted mostly to Sulawesi, with some work also conducted in the Philippines. Thus, together with the studies on *Ce. b. borneanus*, studies of free-ranging and semi-captive tarsiers have focused on 7 of 19 currently recognized taxa (Table 4.1). These studies represent the primary behavioral and ecological literature which compares the differences between the member species of the family Tarsiidae.

Radio collars have been successfully used in many tarsier field studies and have enabled researchers to conduct observations on individuals continuously throughout the night. In turn, better and more complete field data have resulted, expanding our knowledge of tarsier behavioral ecology.

TARSIER ECOLOGY

Locomotion and Habitat Preferences

Western tarsiers are found in a wide variety of lowland forest habitats. Niemitz (1979b) reports evidence of tarsiers in primary forest, shrub, and mangrove, in coastal areas, and even bordering plantations. Because of locations of capture, Fogden and earlier authors (e.g., LeGros Clark, 1924; Davis, 1962; Harrison, 1963) considered primary forest vegetation to be a marginal habitat for *Ce. bancanus*. However, Niemitz (1979b) believes that Bornean tarsiers are common in primary forest, but that they may have larger home ranges and, therefore, may occur at lower population densities in primary forest habitat. Crompton and Andau (1987) found no evidence that primary forest represents marginal habitat for tarsiers in Sabah, Indonesia, but they did find them to have larger home ranges than did Fogden or Niemitz. Eastern tarsiers are found in nearly all habitats of Sulawesi, except those with dense human populations, or agricultural areas where all potential nest sites have been cleared, or where insecticides or herbicides are used.

The longest field study of any Philippine tarsier population has been just eight months in duration (Neri-Arboleda, 2001). Consequently, knowledge about any of the Philippine tarsier populations is quite limited. Early observations of *Ca. syrichta* found coastal lowland regions of primary and secondary forests to be the habitat type where the species was most common (Fulton, 1939; Wharton, 1948). These early surveys found tarsiers on Leyte in a wide variety of habitats but absent from areas that lacked large trees. This is in direct contrast to the observations of Neri-Arboleda (2001) and Gursky (pers .obs.), who found that Philippine tarsiers were more abundant in young secondary forest than in old secondary forest where tree diameter at breast height (DBH) was substantially greater. Neri-Arboleda (2001) argues that the greater density of tarsiers in the younger secondary forest reflects the greater abundance of small-diameter vertical supports found in these younger forests. Smaller substrates can be beneficial in providing a more secure grip while foraging and may explain the preference of tarsiers for these substrates. Brown et al. (2014) collected genetic samples of 77 *Ca. syrichta* from 17 sites across the Philippine islands of Mindanao, Samar, Leyte, Bohol, and Dinagat, with all three recognized subspecies represented in their study sample. Their analyses uncovered three major *Ca. syrichta* evolutionary lineages, but these do not correspond to the current taxonomic arrangement of *Ca. syrichta* subspecies (Brown et al., 2014). Most recently, Bejar et al. (2020) describe preliminary observations on Philippine tarsier habitat use in an isolated forest fragment on Mindanao in order to add to the limited body of knowledge on *Ca. syrichta* ecology and habitat usage. They report that tarsiers show preference for small-diameter supports of less than 5 cm DBH, utilizing heights less than 7.5 m. According to Bejar et al. (2020), the pervasiveness of small-sized trees in the special protection zone is consistent with the preferred habitat characteristics of tarsiers, as observed in other tarsier populations (e.g., the Bohol-Samar-Leyte tarsiers; Gursky et al., 2011) and other tarsier species (Yustian, 2007; Shekelle & Salim, 2009; Merker et al., 2010) where sleeping and foraging trees are typically in the same size range as most of the associated trees. An exception is the pygmy tarsier (*T. pumulis*), which primarily uses large trees for sleeping (Grow et al., 2013).

The locomotor morphology of tarsiers is specialized for vertical clinging and leaping (Niemitz, 1977, 1979b, 1984b; Preuschoft et al., 1979). All species that have been studied use leaping between vertical supports as the major means of locomotion, though there is some subtle variation among species. Climbing is the second most common locomotor category used by tarsiers. For the most part, it consists of shinnying up and sliding down vertical supports, rather than climbing in the usual sense (Oxnard et al., 1990). Some walking and frog-like or bipedal hopping is done, usually while tarsiers are on the ground searching for insects. A "cantilever" movement, where the animal stretches out perpendicularly from an arboreal support while holding on with its feet, is used to manually capture and ingest flying insects (Niemitz, 1984b). Tarsiers leap around 60–70% of the time, but the amount of time they use other categories of locomotion differs between species. For example, *Ce. bancanus* relies on leaping and climbing almost exclusively (94% of locomotion), whereas *T. tarsier*, *T. dentatus*, and *Ca. syrichta* appear to be much less specialized, with comparatively greater reliance on hopping, walking, and cantilevering (Crompton & Andau, 1986; Tremble et al., 1993; Dagosto & Gebo, 1997). Western tarsiers use climbing mainly for short distance vertical displacements associated with foraging (Crompton & Andau, 1986). *T. dentatus* and *T. tarsier* use quadrupedal climbing, walking, and running in a wide variety of horizontal displacements, including somewhat rapid movement on the ground (Tremble et al., 1993).

Ce. bancanus and *Ca. syrichta* spend most of their time less than 2 m above the ground on small vertical supports of less than 5 cm in diameter. Although these tarsiers will range as high as to 6–10 m, this is rarely done (Niemitz, 1984b; Crompton & Andau, 1986; Dagosto & Gebo, 1998). Tarsiers from Sulawesi, however, have much greater vertical ranging patterns, sometimes nesting at heights of 10–20 m. *Ce. bancanus* in Sabah, for example, spend over 75% of their time below 1 m and only 5% above 3 m, whereas *T. dentatus* spend 23% of their time above 3 m. *T. tarsier* also spend more time higher in the forest than *Ce. bancanus*. This may be related to the fact that both Dian's tarsier (*T. dentatus*) and the spectral tarsier (see footnote to Table 4.1) were studied in forest habitat with greater canopy height than was the case in Western tarsier habitat (Niemitz, 1984e; Tremble et al., 1993), or there may be a real difference in habitat choice.

Different activities are performed at different heights. For example, Crompton and Andau (1986) found that *Ce. bancanus* rests and grooms significantly higher (around 2 m) than it travels and that it travels significantly higher (around 1 m) than it forages (mean = 0.66 m). Feeding has the lowest mean height, 0.53 m. *T. tarsier* also rested at higher levels than it traveled and traveled higher than it foraged and fed (MacKinnon & MacKinnon, 1980). In general, tarsiers use higher and larger supports when traveling *between* foraging sites than when foraging, and they use locomotor modes other than leaping less often (Crompton & Andau, 1986).

When resting, *T. dentatus* sits more than does *Ce. bancanus*, and this may be related to its more frequent use of oblique and horizontal branches. The latter species uses the specialized sitting pad on its tail and presses tightly against the vertical substrate while clinging. In this way, weight is distributed to the sitting pad and the clinging tarsier actually sits on its tail (Neimitz, 1984b). *T. dentatus* and *T. tarsier* also sit on their tails when using vertical supports. However, these species often sit on horizontal or oblique supports. Their tails do not possess a sitting pad, and the portion of the tail distal to the part that makes contact with the substrate will often loop over a branch or simply hang beneath the animal (Tremble et al., 1993; Shekelle, pers.comm.).

Ce. bancanus also shows less variation in the orientation of supports it uses; 90% of the time it is found on vertical or only slightly angled supports, whereas *T. dentatus* uses oblique and horizontal branches much more frequently (over 22%). Both *Ce. bancanus* and *T. dentatus* use small-diameter supports in similar proportions, with about 70% of their time being spent on supports smaller than 3 cm. The ground is used infrequently, mainly while capturing prey. On Sulawesi, tarsiers use a bipedal hop to move between trees in more open habitat, such as coconut groves or along dry stream beds.

The average length of tarsier leaps is slightly over 1 m; only on rare occasions do they leap farther than 3 m, although they can do so with ease (Niemitz, 1984b; Crompton & Andau, 1986). Niemitz (1984b) notes that the mean distance of a leap is nearly double the mean separation of 3–4 cm saplings in the pole forest where he

studied *Ce. bancanus*. While being chased by conspecifics, tarsiers can leap up to 5–6 m. *Ce. bancanus* has been observed to leap 30 times its head-to-body length (about 13 cm) without losing height (Niemitz, 1979b).

Activity Cycles and Sleeping Sites

Tarsiers are entirely nocturnal and begin their activity around sunset. Activity is continuous throughout the night, with bouts of leaping occurring at intervals of about one hour, alternating with periods of rest (Niemitz, 1979b). However, tarsiers also have peaks of activity in the early evening and again just before sunrise. Niemitz (1984e) found there was a minor peak in activity early and a major peak late in the evening in two *Ce. b. borneanus* individuals, whereas Crompton and Andau (1987) observed more travel in the early evening in the same subspecies. However, different species spend differing amounts of time in various activities. In Sulawesi, the most energetic periods for *T. tarsier* occur during the first half hour of activity and, again, during the final hour of activity, when groups re-formed and gave loud calls as they moved to their sleeping sites (MacKinnon & MacKinnon, 1980).

T. tarsier males spend approximately 48–62% of their night foraging, 26–30% traveling, 11–21% resting, and 5–11% socializing (Crompton & Andau, 1986; Tremble et al., 1993; Gursky, 1997). The only observable difference in activity patterns across *Tarsius* species is that *T. dentatus* spend nearly twice as much time resting in comparison with other species.

Regarding *Ca. syrichta*, Wojciechowski and colleagues (2019) report that during the non-mating season, solitary tarsiers spend the majority of their waking hours scanning, followed by resting, foraging, and traveling. During the mating season, scanning was still the most commonly reported activity for both sexes; however, time spent resting noticeably decreased, while time spent traveling and foraging increased. According to cross-species comparisons of activity budgets carried out by Wojciechowski et al. (2019), *Ca. syrichta* and *Ce. bancanus* spend, on average, a similar amount of time feeding (based on data from Crompton & Andau, 1986; Roberts & Kohn, 1993), while *T. spectrumgurskyae* females spend a greater amount of time feeding (data from Gursky, 1997, 2005a). However, when scanning, foraging, and feeding were combined, *Ca. syrichta* and *T. spectrumgurskyae* values did not differ significantly. Importantly, this research was conducted using two wild-caught individuals (one male, one female) that were transferred to outdoor enclosures for observation (Wojciechowski et al., 2019). No quantitative data on nightly activity patterns have been obtained for *T. pumilus*.

There is a large amount of variation among tarsier species with regard to sleeping site preferences. This has important consequences for determination of social groups and mating systems, since tarsier social groups are often identified, in practice, by which animals sleep together. Spectral tarsiers typically do not fashion nests but sleep in tree hollows, to which they regularly return each morning (MacKinnon & MacKinnon, 1980). Spectral tarsiers at Tangkoko seem to have a preference for nesting in large strangler figs (MacKinnon & MacKinnon, 1980; Gursky 1997, 2007a). The nest site, therefore, is more like a huge arboreal "cave" than simply a hole in a tree. While some groups have as many as three sleeping trees, the majority have only one sleeping site. The mean circumference of these strangler fig sleeping trees was 287 cm, ranging from 30–700 cm. The mean height of the sleeping trees was 20.2 m, ranging from 6–38 m. MacKinnon and MacKinnon (1980) also observed several *T. tarsier* groups utilizing vine tangles and grass platforms when groups were found in secondary forest and grassland areas. Social groups as large as 8 individuals (Gursky, 1997), or with as many as 6 adults may sleep together in a sleeping tree (MacKinnon & MacKinnon, 1980). Virtually nothing is known of tarsier behavior within the sleeping tree. Sleeping trees are highly stable and are used for several years without major interruption (Gursky, 2010b). At Tangkoko, park rangers have been bringing tourists to the same sleeping trees since the study by the MacKinnons in the late 1970s (Shekelle, pers.comm.).

Niemitz et al. (1991), in their description of *T. dentatus*, state that this species does not return to the same sleeping site each night. This description agrees with the MacKinnons' (1980) observation that tarsiers in the nearby Palu Valley do not use the same sleeping site each night. Tremble et al. (1993), however, found that their focal groups of Dian's tarsiers (*T. dentatus*) did return to the same sleeping site each night. Tremble et al. (1993) studied tarsiers at the same locality as Niemitz et al. (1991); six years following that study by Tremble and colleagues, the same sleeping sites were still being used (Shekelle, pers.comm.). At two sites (Kamarora & Posangke), Gursky (1997, 1998) also found that *T. dentatus* returned to the same nest site each morning. Sleeping sites for this tarsier include large tangles of lianas or other dense vegetation, tree cavities, and fallen logs (Tremble et al., 1993; Shekelle, pers.comm.). The type of sleeping site used might account for variation in whether or not groups return to their sleeping sites on consecutive nights. As many as six tarsiers used individual sleeping sites and the prominent duetting of *T. dentatus* indicates that more than two adults are often present (Shekelle, pers.comm.).

In the wild, Western tarsiers sleep singly unless it is a mother and her infant (Niemitz, 1979b; Crompton & Andau, 1987; Roberts & Kohn, 1993). Sleeping sites are likely chosen for shelter from the rain and for camouflage (Niemitz, 1979b; Crompton & Andau, 1987). Niemitz (1979b) found that tarsiers at his field site would sleep clinging to nearly vertical branches at elevations of 2 m and less. Crompton and Andau (1987) found tarsiers typically sleep in liana tangles and naturally formed mats of creepers situated 4–5.5 m above the ground on 50- to 90- degree supports but were occasionally as low as 2.5 m. Thus, the sleeping sites utilized by *Ce. bancanus* were low in the forest canopy. Crompton and Andau (1987) further note that *Ce. bancanus* sleeping sites tended to cluster at the edge of tarsier home ranges, in areas of overlap with neighbors of the opposite sex. This contrasts with *T. tarsier* whose sleeping sites tend to be located near the center of the group's home range. Roatch et al. (2011) evaluated the microhabitat of Bornean tarsiers. Through analysis of vegetation plots, they found tarsiers preferred tree heights of 3.6 m and tree diameters at breast height (DBH) of 6.2 cm. They also found that, similar to other tarsier species, Bornean tarsiers tend to frequent the forest understory and shrub layer below 5 m above the ground.

Until relatively recently, the only reports of sleeping behavior of wild *Ca. syrichta* were anecdotal accounts from early naturalists. Cuming (1838) stated that *Ca. syrichta* sleeps under the roots of trees, particularly bamboo. Cook (1939) reported that sleeping sites include grass, lianas, underbrush, hemp, with only one tarsier found to nest in a tree. Philippine tarsiers have also been observed to sleep in stands of bamboo (Catchpole & Fulton, 1939), and in hollow trees (Lewis, 1939). Among captive *Ca. s. syrichta*, males and females often sleep huddled together (Wright et al., 1985). However, Dagosto and Gebo (1997) report that *Ca. s. syrichta* in Leyte, Philippines, always sleep singly, never in groups except for mother-infant pairs, a pattern also observed by Neri-Arboleda (2001) with *Ca. s. fraterculus*, on Bohol Island. *Ca. s. syrichta* in Leyte utilized three or four different sleeping sites. These sleeping sites were primarily in large *Arctocarpus, Pterocarpus,* or *Ficus* trees, were located low to the ground (< 2.5 m), and were surrounded by very dense vegetation. The average tree height of sleeping sites was 7.5 m. *Ca. s. fraterculus* on the island of Bohol primarily utilized one sleeping site but also had several alternative sites. Males utilize an average of seven to eight sleeping sites, while females tend to limit their sleeping sites to three or four different locations (Dagosto & Gebo, 1997). According to Neri-Arboleda (2001), female tarsiers on Bohol were more likely to return to their favored sleeping site than were males. The sleeping sites of the Bohol *Ca. s. fraterculus* were highly variable, ranging from rock crevices near the ground (3%), to vine tangles (10%), dense screwpine (*Pandanus*) thickets (15%), tree holes (45%), and forked branches in trees (27%).

Grow and Gursky-Doyen (2010) report that pygmy tarsiers (*T. pumilus*) also sleep in groups. The pygmy tarsiers leave their sleeping tree each evening before dusk, returning to the same sleeping tree each morning. Unlike lowland tarsiers, pygmy tarsier groups include multiple adult males. Additionally, Grow and colleagues (2013) found that pygmy tarsier sleeping sites in Lore Lindu National Park, Sulawesi, Indonesia, range from 30–164 m from the main road. This is likely due to the fact that insects are larger and more abundant near forest edges.

Diet and Feeding Behavior

The feeding ecologies of tarsiers are highly specialized (Niemitz, 1979a, 1984f; MacKinnon & MacKinnon, 1980; Crompton & Andau, 1986; Jablonski & Crompton, 1994; Gursky, 1997, 2000b, 2007a; Neri-Arboleda, 2001). Tarsiers are the only primates with a diet composed entirely of animal prey (both invertebrates—mainly insects—and small vertebrates). In over 1,000 hours of observations of *Ce. bancanus* under semi-captive conditions (i.e., a forest-caged enclosure), no plant foods were eaten, even though natural vegetation was readily available (Niemitz, 1979b, 1984f). Further, although many observations of naturalistic foraging have been conducted (Davis, 1962; Harrison, 1963; Fogden, 1974; MacKinnon & MacKinnon, 1980; Niemitz, 1984f; Crompton & Andau, 1986), there are no reports of plant foods of any sort (i.e., leaves, leaf buds, petioles, flowers, nectar, fruit, seeds/nuts, or bark) being eaten by free-ranging tarsiers. However, Niemitz (1984f) occasionally observed *Ce. bancanus* bite a leaf but never observed them chewing or digesting leaves. The same pattern has also been observed in parked infant spectral tarsiers (Gursky, pers.obs.).

Observations by Niemitz (1979a,b, 1984a–f) on forest-caged *Ce. bancanus* indicated that their diet includes beetles (35%), ants (21%), locusts (16%), cicadas (10%), cockroaches (8%), and small arachnids and vertebrates (11%: spiders, birds, small fruit bats, and poisonous snakes).

Spectral tarsiers prefer moths (Lepidopterans), and crickets and cockroaches (Orthopterans). These two arthropod orders comprise more than 50% of the diet. Gursky observed that during the dry season, when these types of insects were less abundant, *T. tarsier* substantially increased their consumption of beetles (Coleoptera) and ants (Hymenoptera) (Gursky, 2000b). The mean size of insects captured by *T. tarsier* was 1.9 cm, ranging from 1–6 cm. The majority of the insects consumed were located on, or underneath, a leaf (43%), or were flying insects (35%). Smaller percentages of insects were captured on either branch surfaces (13%) or on the ground (9%). Tremble and colleagues (1993) found the dietary items of wild *T. dentatus* include moths, crickets, and a lizard.

Tarsiers spend most of their night foraging, aurally and visually searching for prey from small, vertical perches just above the ground (Niemitz, 1979b, 1984f; Crompton & Andau, 1986). They locate their prey primarily by sound and only secondarily by sight (Crompton, 1995). Tarsiers have voracious appetites, eating many large prey items in one night (Jablonski & Crompton, 1994). A small percentage of flying insects are captured by cantilevering out from an arboreal support and manually catching insects on the wing. In addition to insects, *Ce. bancanus* has also been observed feeding on snakes (even poisonous ones), many species of birds, a few species of bats (Niemitz, 1973, 1979b, 1984f; see also Harrison, 1963), as well as lizards (Fogden, 1974), "flying" frogs (species of frogs that have the ability to achieve gliding flight), and small land crabs (Jablonski & Crompton, 1994). Toads, snails, small rats, and shrews were not eaten by either *Ce. bancanus* or *T. tarsier* despite these potential prey being available.

The jaw muscles and teeth of tarsiers are highly specialized for acquisition and processing of relatively large animal prey. However, Jablonski and Crompton (1994) believe that these specializations are not optimal for a single function but are adaptive compromises between a number of competing pressures, which is consistent with the highly varied diets of tarsiers. The jaws and teeth allow tarsiers to efficiently bite and chew tough, chitinous insect exoskeletons and vertebrate bones, as well as soft, chewy insect internal organs. Often, when tarsiers catch vertebrates, they will eat everything, including bones, bird beaks, and feet. Jablonski and Crompton (1994) also argue that the tarsier diet and their digestive system entails compromise. Tarsiers have a simple gut (Hill, 1955), but it takes a long time for chitinous insect exoskeletons to pass through the digestive system. Jablonski and Crompton (1994) believe that tarsiers are able to digest insect exoskeletons by a process of hind-gut fermentation in the relatively large cecum they possess.

Considering the foraging pattern and diet of tarsiers, Niemitz (1979b, 1984a, 1985) suggests that tarsiers occupy a niche essentially similar to that of small forest understory owls (specifically, the Scops owls, genus *Otus*, family Strigidae). Tarsiers show striking convergence with owls both morphologically (e.g., large eyes and the ability to rotate their heads 180 degrees) and behaviorally. Niemitz (1985) lists 33 parallelisms, convergences, and analogies between tarsiers and owls, including the following four behavioral similarities: (1) owls and tarsiers feed on similar prey; (2) they locate their prey acoustically, rather than by smell or vision; (3) they are both completely noiseless during locomotion; and (4) they utilize ambush-type predation by moving about during the night just above ground level. Where the undergrowth is too dense to permit any flying species, Niemitz (1985) believes that tarsiers replace owls completely.

Home Range Use

Home range estimates are quite variable among numerous field studies. Without following the animals throughout the night, Niemitz (1979b, 1984a) estimated *Ce. bancanus* live in home ranges of 0.9–1.6 ha and have a population density of approximately 80 tarsiers/km². From trapping records, Fogden (1974) gave a home range size estimate of 2.5–3 ha for the same species. These are minimal estimates because they are centered mainly on trapping locations and/or sleeping sites. Crompton and Andau (1987), following animals throughout the night, found home ranges between 4.5–9.5 ha for females and 8.75–11.25 ha for males. Crompton and Andau (1987) report that, on a nightly basis, males utilized 50–75% of their total home ranges and females utilized 66–100% of their ranges. Given these results, population densities would be much lower than previously estimated, equaling approximately 15–20 tarsiers/km².

T. tarsier home ranges have been estimated to be about 1 ha (n = 8) (MacKinnon & MacKinnon, 1980), but this estimate was not derived from all-night follows. Using all-night follows, Gursky (1997) found female home ranges average 2.32 ha, and males use ranges of approximately 3.0 ha. The mean nightly path length (NPL) for males was 791 m (n = 5), while the mean NPL for females was 448 m (n = 8). Tarsier population density at Gursky's site was high: approximately 156 animals/km².

Tremble et al. (1993), also using radio-tracking, found *T. dentatus* have small home ranges, just 0.5–0.8 ha. However, Merker (2006) and Merker and Yustian (2008) found that the home range of Dian's tarsier was analogous to spectral tarsiers, averaging 1.1–1.8 ha. The nightly path length varied from 800–1,400 m. They noted there was substantial overlap between male and female ranges, and no statistical difference in range size usage between the sexes. However, the size of the home ranges varied significantly between habitat types, with the smallest ranges in slightly disturbed areas and the largest ranges in a heavily disturbed plantation (Merker, 2006). Thus, for these tarsiers, home range size is inversely correlated with habitat quality. In slightly and intermediately disturbed habitats, individuals covered slightly less than half of their total home range in a single night. In undisturbed and heavily disturbed habitats, Dian's tarsiers used 30% and 35%, respectively, of their ranges during their nightly forays.

According to Grow and Gursky-Doyen (2010), female pygmy tarsiers have a home range encompassing at least 1.2 ha. While being radio-tracked, the female traveled high in the canopy, out of visual range. The male range exceeded the capability of the radio tracking equipment and, therefore, must have been substantially larger than 2 ha.

In their short-term study, Dagosto and Gebo (1997) report home ranges of two male *Ca. syrichta* to be 0.6 and 1.7 ha. Dagosto and Gebo's (1998) preliminary study of tarsiers in Leyte suggests that the mean NPL was 301 m/night for two males. An eight-month study of *Ca. syrichta* was conducted by Neri-Arboleda (2001) who found that the home range was 6.86 ha for males (n = 4) and 2.76 ha for females (n = 6). Interestingly, Neri-Arboleda (2001) also had one home range of a subadult that averaged 13.4 ha, slightly larger than Crompton and Andau's (1987) figure of 11.25 ha for male *Ce. bancanus*. Neri-Arboleda (2001) noted that a plotting of home range area never plateaued, suggesting that the home range size would be even larger with additional data collection. The mean NPL for *Ca. syrichta* females was 1,118 m, ranging from 820–1,304 m (Neri-Arboleda, 2001). The mean NPL for males was 1,636 m. While these are comparable with those of *Ce. bancanus,* they are substantially greater than that observed for *T. tarsier* (Gursky, 1998) or the Leyte population of *Ca. syrichta.*

Thus, in some tarsier species and populations, individuals may have small home ranges, whereas the home ranges of individuals in other tarsier taxa or populations may be much larger. Some of these apparent differences may be due to variation in the field methodologies employed by various researchers, interspecific or intraspecific population levels, variability in behavioral ecology, differences in habitat (e.g., floristic diversity, forest structure, levels of anthropogenic disturbance), or a combination of these factors. When studied at the same site, by the same researcher using the same methods, however, the emergent pattern is that home range area is inversely correlated with habitat quality.

Home ranges appear to be stable over long periods. For example, Fogden (1974) trapped an adult male in the same area in 1965 and 1966. In 1972, the same animal (identified by the bird ring on its leg) was trapped by Niemitz (1979b) 30 m away from the original trap site. Gursky (2010a) also reports evidence of long-term site fidelity by spectral tarsiers. She observed that of the 29 individuals that were relocated after five years, 86% of them were relocated at their original sleeping tree, while

14% were located at a new sleeping tree. The median distance from the original to the new sleeping tree was about 620 m for adult males and 110 m for adult females.

Tarsiers appear to be territorial, with some marginal overlap between ranges and much variability among genera. Gursky (2007a) found that approximately 15% of spectral tarsier territories overlap. Crompton and Andau (1987) report that most activity in *Ce. bancanus* was concentrated around overlapping home range boundaries or near sleeping sites, which also were mostly near the boundaries. Further, activity was not concentrated in one area of the home range, but in two to four well-dispersed sections of the home range. Crompton and Andau (1987) believe that dispersed areas of clustered use within relatively large home ranges make it unlikely that home ranges could be "patrolled" each night, despite long night ranges by males.

Predation and Competition

Both Niemitz (1979b, 1984f) and MacKinnon and MacKinnon (1980) comment on the fact that tarsiers frequently appear incautious to potential predators. Large owls, snakes, civets, and slow lorises (*Nycticebus coucang*) elicited essentially no alarm response from tarsiers. In fact, as stated above, Niemitz (1973, 1979b, 1984f) reports an observation of a *Ce. bancanus* catching and eating a 30-cm-long banded Malaysian coral snake (*Calliophis intestinalis*), a poisonous snake with venom that is myotoxic but rarely fatal (Tan et al., 2019). It is thought that tarsiers have an unpleasant odor and, perhaps, the taste of the unwary tarsier may be its best defense against predation (Niemitz, 1979a,b; MacKinnon & MacKinnon, 1980). The quick, saltatory locomotion and dense habitat of *Ce. bancanus* makes it a challenge to most predators, particularly snakes. The MacKinnons observed alarm calls given by *T. tarsier* toward feral and domestic cats, and the MacKinnons' pet cat managed to catch and partially eat an adult male tarsier. Shekelle reports one observation of a monitor lizard (*Varanus*) attacking a tarsier, and two observations of attempted predation by unidentified snakes (unpub.data).

Gursky's (1997, 2007a,b) observations differ from those described above. She found that tarsiers regularly give alarm calls or show predator avoidance in several situations. Spectral tarsiers give alarm calls to the Malay palm civet (*Viverra tangalunga*), large birds of prey, monitor lizards (*Varanus*), and some snakes (Gursky 2005b, 2006, 2007a,b). In terms of anti-predator behavior, upon hearing the calls of predatory birds, tarsiers freeze and remain motionless for up to several minutes. One of the 18 individuals that Gursky captured and marked was found partially eaten. She concludes that predation pressure is high on *T. tarsier* at Tangkoko.

In addition to giving alarm calls when confronted by a potential predator, spectral tarsiers have also been observed mobbing their predators. Mobbing involves members of a group approaching and confronting a predator in a coordinated fashion. Among the types of predators that elicit group mobbing, snakes seem to be the most consistent recipients of this type of predator-directed behavior (Gursky, 2002a, 2003b). Between 1994–2004, she observed 19 bouts where snakes were mobbed by tarsiers. Gursky (2002a) recorded predation of *T. tarsier* by a reticulated python (*Python reticularis*) in Tangkoko Nature Reserve, Sulawesi, Indonesia. Despite six minutes of continuous mobbing by group members, including one individual actually biting the python, the snake never loosened its hold on the tarsier it had seized. She also observed a close encounter between an unidentified snake and a tarsier. The number of tarsiers participating in a mobbing bout varied from 3–10 individuals, with an average mobbing party consisting of 5 individuals (Gursky, 2007b). The durations of mobbing bouts ranged from 15–49 minutes, with a mean bout time of 33 minutes. Adult males, usually more than one, were more likely to participate in mobbing bouts, and mob for significantly longer, than adult females. Given that most spectral tarsier groups contain only a single adult male, this implies that adult males from neighboring groups may also participate in mobbing bouts (perhaps if a snake encounter occurs in the vicinity of tarsier home range boundaries).

Gursky (2005b, 2006, 2007b) observed mobbing in groups without immature offspring, as well as groups with offspring. Females carrying infants regularly participated in mobbing bouts (that is, they purposefully transported their infants toward the dangerous predator instead of away from it). However, while studying the response of infants to model snakes, Gursky (2003b) found that nursing infants—even those less than one week of age—alarm-called when exposed to a potential snake predator, despite having never seen one. The mother then responded to the infant by calling, which sometimes led to mobbing.

From observations of encounters between tarsier groups and sympatric snake species, Gursky (2006) reports that snakes are more likely to retreat from the tarsiers when there are many individuals involved in mobbing the snake. The tarsiers are more likely to retreat from a snake encounter when fewer individuals mob the snake. As the number of mobbers increases, the number of minutes that the predatory species is present decreases. That is, predators withdrew more quickly, the more intensely they were mobbed. Given the importance of crypticity for snakes, the removal of the surprise element is clearly negative for the snake's hunting success; thus, snakes should be more likely to remove quickly to a new area after the surprise element has been lost. Snakes spend less time in the area following mobbing calls than when the tarsiers only emit an alarm call or just ignore the snake.

Gursky (2003b) conducted an experiment to evaluate how spectral tarsier infants that were parked respond to the presence of a potential predator, exposing parked infants to falcon and civet wooden models and to rubber models of large snakes. Still, mothers continue to park their infants in the tree or a nearby tree. They do not avoid the tree or nearby area. The most obvious result from this experiment is that the infants move from their parked location following exposure to a potential predator. In addition to moving away, spectral tarsier infants also repeatedly give a series of alarm calls in response to the predator models. Infants emit an alarm call in response to the presentation of the predator model, regardless of their age. However, the type of alarm call they emit varies depending on the type of potential predator. The infants consistently emit a twittering alarm call in response to both the bird of prey models and avian vocalizations, whereas they emit a harsh loud call three times in rapid succession in response to the model snakes. Even when the infants are less than one week old, they emit these specific types of alarm calls in response to each potential predator type.

Among Philippine tarsiers, three cases of predation have been reported to date. Neri-Arboleda et al. (2002) report possible predation by a cat, and Řeháková-Petrů and colleagues (2012a) report predation on a female tarsier by a water monitor lizard, and on an infant tarsier by an unknown predator.

Numerous studies of nocturnal mammals, including many nocturnal primates, have consistently shown that many (but not all) species will restrict their foraging activity, restrict their movement, reduce their use of open space, reduce the duration of the activity period, or switch their activity to darker periods of the night in response to bright moonlight (Morrison, 1978; Wolfe & Summerlin, 1989). The only consistent exceptions to this pattern have been observed in four primates: the night monkeys (genus *Aotus*), the southern lesser galago (*Galago moholi*), the Mysore slender loris (*Loris tardigradus lydekkerianus*) (Trent et al., 1977; Erkert & Grober, 1986; Nash, 1986; Bearder et al., 2002), and tarsiers.

The lunar phobia exhibited by most nocturnal mammals is believed to be a form of predator avoidance (Morrison, 1978; Wolfe & Summerlin, 1989; Daly et al., 1992; Kotler et al., 1993; Bearder et al., 2002). This implies that *T. tarsier* and a few other nocturnal strepsirrhines are increasing their exposure to predators when they increase their activity during full moons, which raises intriguing

questions. However, Gursky observed that *T. tarsier* captured three times as many insects during full moons than during moonless (new moon) nights (Gursky, 2003a). These numbers do not simply reflect a bias in observation conditions, because this pattern was also observed in terms of the numbers of insects captured in the insect traps during the different moon phases (Gursky, 1997, 2000b, 2003a). Thus, for *T. tarsier* the benefits of foraging during full moons are substantial.

To deal with the potential increase in predation during full moons, *T. tarsier* modifies other aspects of its behavioral repertoire. Most interestingly, they increase the frequency that group members travel together. In particular, the frequency that any two group members are observed together increases substantially during full moons. It is believed that the more individuals there are in the group to scan for predators, the less time each individual will have to spend in vigilance; individuals in smaller groups often experience higher mortality (or lower breeding success) than members of large groups (Krebs & Davies, 1984; Dunbar, 1988; Kappeler & Ganzhorn, 1993; Clutton-Brock et al., 1999).

In summary, *T. tarsier*, like three other nocturnal primates, do not exhibit lunar phobia. Instead, they increase their activity during full moons and decrease their activity during new moons. The increased activity during full moons may increase their exposure to predators. To counter this putative increased risk of predation, *T. tarsier* increase the frequency that they travel in groups.

The most likely competitors of *Ce. bancanus*, and possibly other species of tarsiers, are owls, insectivorous bats, slow loris (*Nycticebus coucang*), and tree shrews. Owls may be competitors in open areas, but Niemitz (1985) does not believe they compete with tarsiers in dense vegetation. Gursky (1997) observed eight encounters between tarsiers and bats in which the tarsiers were challenged by bats over insect prey; in seven of the eight cases, the tarsier got the insect. Munds and co-authors (2013) proposed, albeit from limited data, that *Ce. b. borneanus* are ecologically impacted by sympatric slow lorises. Other diurnal sympatric primates, such as macaques, proboscis monkeys, gibbons, and orangutans are not thought to compete with tarsiers (Niemitz, 1984f).

Because tarsiers and slow lorises are the only nocturnal primates occurring in Southeast Asia, Harcourt (1999) hypothesized that the two taxa exhibit checkerboard distribution (meaning that only one or the other occurs) on 12 small (<12,000 km^2) islands in insular Southeast Asia and that this pattern is the result of competition between the two largely faunivorous primate species. Additionally, he suggested that slow lorises are able to survive on smaller islands than tarsiers are (Harcourt, 1999). Nijman and Nekaris (2010) reevaluated Harcourt's findings using an expanded data set including 49 islands with tarsiers or slow lorises. They found that tarsiers and slow lorises live on islands of similar size, and that they both inhabit an equal proportion of small, medium, and large islands. Notably, the authors did not find any evidence that tarsiers and slow lorises exhibit a checkerboard distribution (Nijman & Nekaris, 2010).

SOCIAL BEHAVIOR

Studies have raised many intriguing questions regarding tarsier social organization and mating systems. It is clear that tarsiers include some species that are gregarious in their social organization, while others are either mostly solitary or dispersed foragers. Regarding mating systems, wide variation exists among the genera and species. Even within species studied at the same site, evidence for mating systems have been interpreted in various ways.

Systematic behavioral studies of wild Western tarsiers have thus far been limited to studies of *Ce. b. borneanus* from Sarawak and Sabah, Malaysia, and of *Ce. b. saltator* on Belitung Island, Indonesia. From his data on trapping, Fogden (1974) suggested that the social structure of *Ce. bancanus* was solitary but social, similar to that of the west African dwarf galago (*Galagoides demidoff*) or the gray mouse lemur (*Microcebus murinus*) from western Madagascar. Fogden (1974) reported that he saw a male and female *Ce. bancanus* traveling together on eight occasions. In one situation, a male was seen within a few days with two different females. Niemitz (1979b, 1984a: p.121) believed *Ce. bancanus* typically live in pairs; however, he stated, "It is questionable whether Bornean tarsiers are strictly monogamous. Members of both sexes, however, live *synterritorially* for a long period developing strong pair bonds."

Crompton and Andau (1987) did not find this pattern in their radio-tracked animals. They found that individuals foraged and slept alone. In fact, the four individuals they studied were never seen together in 120 hours of following. Vocal (and probably olfactory) communication was common, but always at a distance. Calling appeared to be associated with range boundaries and areas of range overlap. Varying numbers of individuals, sometimes as many as five, often engaged simultaneously in these calling concerts. The range of one radio-collared female was visited by at least four males, and male ranges overlapped the ranges of several females. Crompton and Andau (1987) believe that the social structure of *Ce. bancanus* seems to be closer to the generalized dispersed polygyny pattern of many lorisines and galagines than to monogamy or pair-bonding, though distinguished from these by the extreme degree of solitariness and the lack of

contact at sleeping sites. In captivity, Wright et al. (1985) noted that the amount of contact between *Ce. bancanus* individuals is minimal compared to that for *Ca. syrichta*. Individuals of the latter species play, allogroom, urine-mark, genital-sniff, and call frequently throughout the night, while the Western tarsiers interact infrequently and rest alone for long periods.

Systematic behavioral studies of wild Philippine tarsiers are limited to studies of *Ca. s. fraterculus* on Bohol Island, and of *Ca. s. syrichta* from Leyte. According to Neri-Arboleda (2001), *Ca. syrichta* are solitary on Bohol. They forage and sleep solitarily, except for females carrying their infants. Neri-Arboleda (2001) never observed an individual in close proximity (< 10 m) with a conspecific during nocturnal activity. A single observation was made of a male and female having sleeping sites 5 m apart. However, Gursky (pers.obs.) did occasionally observe male and female individuals together. The Leyte *Ca. syrichta* are also solitary according to Dagosto et al. (2001); all the tarsiers were alone at their sleep sites, and three of the four males in the study were never within close proximity to another tarsier. However, Dagosto et al. (2001) did observe one male sharing a sleep site with another tarsier (sex unknown) on three out of eight nights of observation. At the sleeping site, the two tarsiers were never observed closer than about 1 m from each other.

Eastern tarsiers are gregarious and live in small, gibbon-like, family groups. Duetting is almost ubiquitous among these tarsiers, being absent only among *T. pumilus*. Only tarsiers at Tangkoko National Park, Indonesia (now classified as *T. spectrumgurskyae*), *T. dentatus, T. pumilus*, and one undescribed population from the island of Buton, Indonesia, have been the focus of systematic behavioral field studies. MacKinnon and MacKinnon (1980) believed *T. tarsier* to be "monogamous" (i.e., pair-bonded). The bonded pairs they reported on were highly stable, with only minor changes (essentially of a demographic nature, i.e., births, deaths, age changes, and immigration) occurring over the 15-month study period. Family members ranged closely together, maintaining some auditory and visual contact throughout the night. The MacKinnons reported that *T. tarsier* pair members relocate each other in the morning and give a characteristic duet song, the male and female contributing different components to the song. Similar duets are given during intergroup encounters. Pair members frequently come together during the night, sitting together, sometimes intertwining tails, and often scent-marking each other (MacKinnon & MacKinnon, 1980). During breeding season, copulations occur between pair-bonded adults.

T. tarsier may live in groups containing more than one adult of the same sex. In fact, one group studied by the MacKinnons (1980) contained three adult females and three adult males. Nietsch and Niemitz (1993) and Gursky (1997, 1999) found that spectral tarsiers often live in groups with more than one adult female. Nietsch and Niemitz (1993) observed one group with three adult females, and in 33 groups censused by Gursky (1997), group size ranged from two to eight individuals and five of the censused groups contained more than one female. Other groups contained one or more subadults. Gursky (2010a,b) also observed that approximately 19% of spectral tarsier groups are polygynous. She noted that polygynous groups are more likely to use strangling figs as their sleeping sites, as well as larger-diameter sleeping site trees than the monogamous groups. She suggests that tarsiers do not exhibit obligate monogamy (Kleiman, 1977) but, rather, that facultative polygyny in spectral tarsiers is limited by access to high-quality sleeping sites, in particular, mature *Ficus* trees with large DBH.

Shekelle (1997, 2003) drew attention to the variation in natural history of tarsiers distributed across northern and central Sulawesi. He has documented multiple distinct "acoustic forms" of tarsiers distributed across northern and central Sulawesi (Shekelle, 2008b). This populational variation was also examined by Nietsch (1999). Shekelle (2013) subsequently noted considerable flexibility in social group and sleeping group compositions, and sleeping site choice, among *T. sangerensis*—groups may contain multiple adult females, multiple adult males, or both. Some groups even contain multiple mating pairs and show evidence of incipient fission-fusion groupings and multilevel societies. Other extra adults beyond the basic pair were, in fact, adult offspring of a mating pair. With large strangler figs being a limited resource in the forest and also being correlated with tarsier group size, Shekelle (2013) observed that the quality of the home base (i.e., the preferred sleeping site) affects both tarsier social group size and composition.

Recent studies of *T. tarsier* have demonstrated that this species exhibits substantial amounts of gregariousness compared to other nocturnal strepsirrhine primates (Gursky 2002b, 2005a). For example, in *T. tarsier* groups, approximately 28% of their nightly forays outside the sleeping tree were spent at distances from one another of less than or equal to 10 m. Almost 40% of their night was spent at less than 20 m from one another, including 11% of their time being spent in actual physical contact with other members of the group (Gursky, 2002b).

Based on preliminary data, pygmy tarsiers also live in groups. Grow and Gursky-Doyen (2010) observed a group with at least four individuals. Based on the presence of a fourth group member and the group composition of other species of tarsier, *T. pumilus* probably exhibits a multi-male/multi-female group composition. Tremble et al. (1993) also observed *T. dentatus* in groups with more than one adult of the same sex living in the same home range (see also Niemitz, 1979b, 1984a).

On the basis of behavioral and genetic data, Driller et al. (2009) have concluded that the dominant mating system of *T. lariang* is monogamy. They found that 10 out of 11 social groups were monogamous. Nonetheless, paternity tests implied the presence of offspring resulting from extra-pair matings, so this may represent a situation of *social* monogamy, as opposed to reproductive monogamy.

Dispersal Patterns

The only data concerning dispersal patterns come from Gursky's work (2010a) on spectral tarsiers. She observed that both sexes are equally likely to disperse. The median dispersal distance for spectral tarsiers from their natal range was 437.5 m. The median dispersal distance for males was 829 m (ranging from 621–1,301 m), whereas the median dispersal distance for females was 342 m (ranging from 113–974 m). In other words, males typically disperse twice as far as females.

Partly as a result of the differences in dispersal distances, males and females form groups differently. Females regularly form their territory adjacent to their natal territory, but in Gursky's (2010a) study population, no dispersing males established territories adjacent to the territory of their parents. Comparing individuals that exhibit secondary dispersal with those that remain at their original sleeping site shows some interesting differences. Site fidelity of individuals relocated at their original sleeping tree five years later is highly related to the characteristics of the sleeping tree. Specifically, sleeping trees that were continuously used over that five-year period had a mean DBH of 314.71 cm, significantly larger than sleeping trees of individuals that exhibited secondary dispersal, which averaged a DBH of only 118.75 cm. Similarly, individuals with strong site fidelity resided in sleeping trees with a mean height of 27 m, compared to a sleeping tree height average of just 12 m for individuals that secondarily dispersed. There was also a noticeable difference in the social system between individuals that exhibited strong site fidelity and those that exhibited secondary dispersal. Eight of the 25 individuals that Gursky (2010a) observed residing at the same sleeping tree after five years were in polygynous groups and 17 were in monogamous groups. Four individuals that had moved to a new sleeping tree were in monogamous groups. Individuals in polygynous groups were significantly more likely to exhibit site fidelity than were individuals that were in monogamous groups. The mean sleeping tree height for individuals residing in a polygynous group was 24.8 m versus 19.3 m for individuals residing in a monogamous group. The mean DBH of sleeping trees for individuals residing in a polygynous group was 488 cm versus only 101 cm for individuals residing in a monogamous group. These results illustrate that both DBH and tree height are important variables affecting grouping patterns and sleeping site fidelity.

Acoustic and Olfactory Communication

Tarsiers are known to use a variety of acoustic and olfactory signals, and there is much variation among tarsier taxa in the extent to which they are employed. Generally, those studies which find extensive use of olfactory communication also find relatively less acoustic communication (e.g., Niemitz, 1984d), while those that find extensive acoustic communication report relatively less olfactory communication (e.g., MacKinnon & MacKinnon, 1980; Crompton & Andau, 1986, 1987). The most striking difference is between species in the genus *Tarsius* that live in family groups and, like gibbons, exhibit sexually dimorphic duetting, or family chorus behavior (MacKinnon & MacKinnon, 1980), and tarsiers of the *Cephalopachus-Carlito* clade (i.e., Western and Philippine) that do not duet and do not live in family groups. As with other nocturnal primate taxa (e.g., Bearder et al., 2013), vocalizations have been critically important for identifying taxonomically cryptic tarsier species (e.g., Nietsch & Kopp, 1998; Shekelle, 2008b).

Spectral tarsiers rely heavily on acoustic signals and are known to have at least 14 discrete calls (MacKinnon & MacKinnon, 1980; Nietsch & Kopp, 1998). Behavioral contexts have been established for several of these calls. Most prominent is the duet call which typically is sung in the morning, shortly before returning to the sleeping site. Parameters of the call have excellent characteristics for transmitting the location of the signaler (Niemitz, 1984d). Thus, the call is thought to function as a means for unifying the social group (MacKinnon & MacKinnon, 1980). The volume of the call, however, is far louder than is required for such a purpose, and the call is further thought to advertise a group's presence to neighboring groups.

Western tarsiers are solitary foragers and live in dispersed groups, where the largest sleeping group is a mother and a single infant. Western tarsiers are the least social of all tarsiers, and adult males do not tolerate the presence of infants in captivity. They are considered to be the least vocal of all tarsiers, and early accounts described Western tarsiers as typically silent (Le Gros Clark, 1924). Later, Harrison (1963) noted a distress call, and vocalizations associated with courtship and mating. Niemitz

(1984d) described four basic patterns for *Ce. bancanus* vocalizations including a short whistle, a "grate," a long call, and the clicks of infants. Male and female calls could not be distinguished. In contrast to this, Crompton and Andau (1986, 1987) described *Ce. bancanus* as exhibiting a variety of vocalizations and midnight-calling concerts. It is not clear if the differences are due to methodology (Crompton & Andau's study benefited from the development of tiny radio transmitters that, for the first time, allowed all-night follows of tarsiers) or possibly as yet unrecognized taxonomic variation between the tarsiers of Sarawak, where the earlier studies took place, and those of Sabah, where Crompton and Andau (1986, 1987) worked.

Philippine tarsiers are solitary foragers living in dispersed groups; largest sleeping groups are formed by females and their single dependent infants. Philippine tarsiers are intermediate between Western and Eastern tarsiers in their expression of sociality, although captive adult male Philippine tarsiers tolerate the presence of infants. Philippine tarsiers are thought also to be intermediate in their use of vocal communication between the other two genera. Wright and Simons (1989) studied the calls of a pair of captive *Ca. syrichta*, recording three different types of calls under three different behavioral conditions. Dagosto and Gebo (1997) noted wild Philippine tarsiers calling from the edges of their home ranges. Neri-Arboleda's (2001) study found that males and females vocalize from a distance (a high-pitched single whistle 3–5 seconds apart), and they also use a bird-like melodious trill. Although she never observed an actual scent-marking event, numerous scent-marks were discerned throughout the forest.

A mostly complete acoustic repertoire for Philippine tarsiers was assembled recently by Řeháková-Petrů and colleagues (2012b). They identified a total of eight different calls, five of which are used for communication among adults, two that are specific to mother-infant interactions, and one that is a distress call heard during capture. Frequency range of the calls varies from 9752 Hz to frequencies greater than 22 kHz. Calls were most commonly heard at sunset and less frequently at sunrise. Thus, as with so many other traits, Philippine tarsiers appear to be intermediate between Eastern and Western tarsiers, in the complexity of their acoustic communication.

Philippine tarsiers were the first primate species documented to both produce and hear pure ultrasound (Ramsier et al., 2012). More recently, Gursky (2015) reported that spectral tarsiers also emit multiple types of pure ultrasonic vocalizations. It is, therefore, reasonable to assume that all tarsiers use pure ultrasound to some degree.

High-frequency sounds can only travel short distances. Thus, their function is limited by the distance over which the sound can be discerned. Gursky (2019) hypothesized that tarsier ultrasonic vocalizations may function as a form of echolocation, allowing them to navigate through the forest; 42% of the ultrasonic calls (n = 257) recorded during this study were emitted within one second prior to locomoting. Additionally, two of the calls—trills and frequency-modulated calls—are never given except prior to, or during, locomotor bouts. These findings, at first approximation, suggest that tarsiers use these vocalizations as aids in navigating their habitat—for instance, to discern how far away a tree is from their current position (Gursky, 2019). Additional research, using simultaneous radio telemetry and video recording, may allow an increase of locomotor behavior samples, which can then be linked conclusively to the ultrasonic vocalizations. This will help determine whether ultrasonic calls are, in fact, used for navigation by spectral tarsiers (Gursky, 2019).

Compared to lowland Sulawesian and Philippine tarsiers, montane pygmy tarsiers engage in a form of communication that lacks scent-marking and lower-frequency vocalizations (Grow, 2019). In fact, Grow found that pygmy tarsiers communicate almost exclusively in pure ultrasound. Although research suggests that predation threats and habitat acoustics may influence the use of high-frequency vocalizations, this particular study found a relationship between vocal frequency and body mass in pygmy tarsiers, which may represent retained primitive traits rather than derived adaptations to a montane evergreen habitat (Grow, 2019).

Among Eastern tarsiers, duet calls have been used to identify taxonomically cryptic species. The practice began with observations by MacKinnon and MacKinnon (1980) who first noted that several distinct taxa in Sulawesi each had their own calling pattern; they predicted that there are more forms to discover in southern Sulawesi and offshore island groups. Several researchers followed up on this, such that some 15–16 acoustically distinct tarsier forms have been identified—each one a putative species (Niemitz, 1984d; Niemitz et al., 1991; Nietsch, 1993, 1994, 1999; Nietsch & Niemitz, 1993; Shekelle et al., 1997; Nietsch & Kopp, 1998; Shekelle, 2003, 2008a,b; Merker & Groves, 2006; Burton & Nietsch, 2010; Merker et al., 2010). On this basis, several new species have been identified, such that the number of species recognized within the *Tarsius* species complex has grown (Groves & Shekelle, 2010; Shekelle et al., in prep).

Niemitz (1979a) found olfactory communication among *Ce. bancanus* to be quite complex at his study site in Sarawak. There, tarsiers urinate on vertical substrates, and some tree trunks are used exclusively by a single individual. Females in estrus mark branches with their vulva. The epigastric gland on the abdomen is more highly developed in tarsier males, and secretions from this gland are rubbed on branches. The circumoral gland around the mouth is also rubbed on branches, as well as an individual's own tail tuft.

Olfactory communication among spectral tarsiers includes marking each other with the circumoral gland, with the epigastric gland, or with the anogenital region. Branches are marked with the epigastric gland as well as urine. Gursky (2010b) found that the primary function of scent-marking behavior by spectral tarsiers is territorial or group defense. She found that the majority of scent-marks were deposited along group boundaries, especially areas of overlap. Gursky (2010b) also observed that the number of scent-marks increased on nights when there was an intergroup encounter relative to nights when there was not. Lastly, scent-marking increased during the dry season when resources were more limited. MacKinnon and MacKinnon (1980) further note that spectral tarsiers communicate through close contact facial expressions, body posture, and intimate contact.

Reproduction and Infant Development

Except for the studies by Gursky (1997) and by Neri-Arboleda (2001), little is known of reproductive and mating behaviors among wild tarsiers. Thus, the majority of the available data on reproductive and mating behaviors are based on observations of captive tarsiers. Nothing in the literature concerning captive tarsiers indicates major differences among tarsier species. The estrous cycle of each of the species is typically 24 days (Catchpole & Fulton, 1939; Ulmer, 1963; Wright et al., 1986a; Permadi et al., 1994). Seasonality in tarsier births remains an open issue. Neither studies of captive births in *Ce. bancanus* and *Ca. syrichta* (Haring & Wright, 1989; Roberts & Kohn, 1993), nor a survey of wild *Ce. bancanus* placentae (Zuckerman, 1932), show evidence of seasonality. Nevertheless, seasonal birth peaks which follow local rainy seasons have been noted in *Ce. bancanus* near Kuching (Sarawak, East Malaysia) by Le Gros Clark (1924), Fogden (1974), and Niemitz (1977), with mating in October–December and births in January–March. Seasonal birth peaks have also been observed in *T. tarsier* at Tangkoko by MacKinnon and MacKinnon (1980) and Gursky (1997, 2000a, 2007a). Although Neri-Arboleda (2001) did not have data for an entire year on *Ca. syrichta*, the March–October data she did collect are suggestive of birth seasonality in this species.

In captive *Ce. bancanus* housed in pairs, male and female calling and scent-marking increased when the female was in proestrus and estrus. Males began initiating contact during proestrus, but were agonistically rebuffed by females. Once in estrus, females initiated contact. Courtship lasted for 60–90 minutes before a brief copulatory bout lasting from 60–90 seconds (Wright et al., 1986b).

The gestation period in *Ce. bancanus* is exceptionally long for such a small animal, lasting 178–180 days (Izard et al., 1985; Roberts, 1994; Gursky, 1997). Normally, females give birth to one infant. The infant is exceptionally large at birth relative to the mother, weighing from 20–33% of the mother's weight (Roberts, 1994; Gursky, 1999). Much of this weight is accounted for by the infant's brain, which is 60–70% of the adult brain size. In fact, compared to 26 other prosimian species, tarsiers have the largest relative neonatal size, the longest gestation length, and the largest relative neonatal brain size (Roberts, 1994).

Roberts (1994) found that *Ce. bancanus* had the slowest recorded fetal growth rate for any mammal, and that the relative postnatal growth rate to maturity was the lowest in a sample of 26 prosimian species. However, he also found that the relative rate of behavioral development, especially in foraging and locomotion, was extremely rapid. After about three weeks, neuromuscular coordination of the growing infant improves dramatically. By six weeks of age, the infant is no longer carried by the mother, and within two months young *Ce. bancanus* begin to catch their own prey.

For *Ca. syrichta*, weaning was determined when infants no longer shared a sleeping site with their mother. In captive tarsiers, lactation is extremely short, and much more variable, being measured from 49–120 days (Haring & Wright, 1989; Roberts & Kohn, 1993; Permadi et al., 1994; Roberts, 1994; Gursky, 1997). The infant is nutritionally independent by 80 days and by 100 days it is able to capture prey at frequencies similar to those of adults. These skills are achieved with no aid from either parent (Roberts, 1994). This rapid behavioral development is needed for such a specialized predator to attain early nutritional independence.

Among three pairs of captive spectral tarsiers, Hidayatik et al. (2018) report that the most frequent courtship behaviors were scent-marking and genital marking for females, and genital inspection for males. Copulations lasted between three–four minutes, and occurred only once for each pair during the nine-month observation period. Interestingly, while female tarsiers vocalized during or after copulation, Hidayatik and colleagues (2018) did not hear audible courtship calls by males. This pattern differs from what Wright et al. (1986b) reported for *Ce. bancanus* but resembles what Haring et al. (1985) reported for *Ca. syrichta*.

One of the most unusual reproductive behaviors of tarsiers concerns their parenting strategy. Tarsier mothers do not continually transport their infants on their bodies. Historically, tarsiers were thought to be rather nonchalant, with parents parking their infants through the entire night and coming back to get infants at dawn (Charles-Dominque, 1977; MacKinnon & MacKinnon, 1980; Niemitz, 1984a). However, a study of *T. tarsier* infants has demonstrated that parents are not negligent but are optimizing their behavior given numerous constraints. Specifically, Gursky (1997) found that although infants are parked throughout the territory, the mean length of each parking bout was a mere 27 minutes. She also found that the mean number of locations that the infant was parked averaged 11 per night and that the average distance between mothers and their parked infants was 4 m. In fact, *T. tarsier* mothers, more often than not, foraged in the same tree where their infant was parked. These results taken together indicate that *T. tarsier* are not inattentive parents. The mother caches the infant in one tree where she forages and then carries the infant (in her mouth) into another tree, parks it, and begins foraging again within the tree where she parked her infant.

When parked, infants of *T. tarsier* are often visited and played with by the family members of both sexes (MacKinnon & MacKinnon, 1980; Gursky, 1994, 1997, 2007a). Gursky (1997) observed subadult females retrieving fallen infants, allowing infants to take insect prey, and playing, grooming, and "babysitting" with infants. Adult males were seen playing and grooming with them as well.

CONSERVATION

Humans represent the greatest threat to tarsiers. Animals are opportunistically captured and eaten by traditional peoples (Sumampow et al., 2020; Gursky, pers.obs.). Another human threat comes from the pet trade; tarsiers are regularly captured and sold for pets, often dying within a few days. However, the largest threat to most tarsiers, by far, is loss of habitat due to diverse purposes such as industrial-scale agricultural plantations, timber, and other, usually short-term, economic exploitation (Neimitz, 1984f; MacKinnon, 1987). This is especially so in the Philippines (Myers, 1987).

Tarsier conservation status ranges from Near Threatened for the Philippine tarsier, *Carlito syrichta*, to Critically Endangered for the Siau Island tarsier, *Tarsius tumpara* (Table 4.2). The International Union for the Conservation of Nature (IUCN) Primate Specialist Group selected *T. pumilus*, the pygmy tarsier, as one of the world's 25 most endangered primates for the period 2012–2014 (Schwitzer et al., 2014), and *Ca. syrichta* was

Table 4.2 Current Conservation Status of Tarsiiform Taxa

Species	IUCN Conservation Status [a]
Philippine (Central) tarsier group:	
Carlito syrichta (3 subspecies)	Near Threatened
Western (Horsfield's or Bornean) tarsier group:	
Cephalopachus bancanus (4 subspecies)	Vulnerable
Eastern (Sulawesi) tarsier group—*Tarsius* spp.:	
T. niemitzi—Niemitz's tarsier	Not Assessed (Endangered?) [b]
T. spectrumgurskyae—Gursky's spectral tarsier	Vulnerable
T. supriatnai—Jatna's tarsier	Vulnerable
T. dentatus—Dian's tarsier (=*T. dianae*)	Vulnerable
T. fuscus—Makassar tarsier	Data Deficient (Vulnerable?) [b]
T. lariang—Lariang tarsier	Data Deficient (Decreasing)
T. pelengensis—Peleng Island tarsier	Endangered
T. pumilus—Mountain, or pygmy, tarsier	Data Deficient (Decreasing)
T. sangirensis—Sangihe Island tarsier	Endangered
T. tarsier—Spectral tarsier (=*T. spectrum*)	Vulnerable
T. tumpara—Siau Island tarsier	Critically Endangered
T. wallacei—Wallace's tarsier	Data Deficient (Decreasing)

[a] Conservation status taken from the IUCN Red List (2020), unless otherwise noted.
[b] Shekelle et al., 2019.

included as one of the world's 25 most endangered primates for the period 2014–2016 (Shekelle et al., 2015).

Several unresolved questions impede our ability to accurately estimate the threat to tarsiers at this time. The first is taxonomic uncertainty, and at the current rate of discovery of new tarsier species, it is probable that heretofore unknown species will disappear before they have been identified (Shekelle & Salim, 2009). A second issue is the relative merit of disturbed habitat for sustaining tarsiers—are these sources or sinks? (See *T. pelengensis*, Shekelle, 2020a.) A third issue is the altitude at which tarsier populations are sustainable; as lowland primary habitats are destroyed before those at high elevation, most of the remaining forests are possibly above the natural range of tarsiers (see *T. dentatus*, Shekelle, 2020b). All three of these issues affect Philippine tarsiers. Furthermore, habitat loss in the Philippines is critical, with many regions having almost no remaining primary forest.

Shekelle and Salim (2009) began using remote sensing to estimate threat status to *T. sangirensis* and *T. tumpara*. They note that both species are at high risk from a small extent of occurrence, a small area of occupancy, small population size, high risk of volcanism, high human density, fragmented populations, and lack of conservation areas for either species. *T. sangirensis* is presently categorized as Endangered and *T. tumpara* as Critically Endangered (IUCN, 2020).

Neri-Arboleda (2010) conducted a population viability analysis of *Ca. syrichta* using a metapopulation model called Analysis of the Likelihood of Extinction (ALEX). The minimum habitat area that could attain a less than 5% probability of extinction within 100 years is 60 ha. Therefore, 60 ha is considered the minimum viable habitat area for *Ca. syrichta*. Each tarsier has a home range of approximately 2.5 ha with little overlap, thus the minimum viable habitat of 60 ha is estimated to contain 24 female individuals. Neri-Arboleda (2010) also noted that slight increases in adult mortality made the population very unstable and caused substantial increases in the probability of extinction. This estimate of 60 ha also takes into consideration that the forested area is fully occupied, which is most often not the case. Finally, the estimate is also based on one eight-month study in the Philippines.

Density values are somewhat deceptive because tarsier population density is known to change substantially with both altitude and habitat type. For instance, regarding altitude, the density of both *T. tarsier* and *T. dentatus* was significantly lower at higher altitudes than at lower altitudes (Gursky, 1998). Initial interpretations of population densities seemed to show higher tarsier numbers in secondary disturbed forest patches than in primary forest habitat, although this may have been a unique or short-term situation. In addition, given numerous observations of anti-predator responses (Gursky, 2003a,b) and indications of predation on *T. tarsier*, it is fair to assume that natural mortality may be relatively high for these primates.

In regard to the fate of *Ce. bancanus* populations under future climatic conditions and what the effects of climate change might be on tarsier populations, Welman et al. (2017) found Western tarsiers to have a fairly narrow thermal neutral zone of between 25–30 °C, suggesting they are cold-sensitive (core body temperatures averaged ~31°C). Wellman and co-authors (2017) also suggest that *Ce. bancanus* may prove to be vulnerable to future higher temperatures resulting from global warming.

An analysis of information in Table 4.2 indicates that, even though detailed field data do not exist for some recently recognized taxa, all tarsier populations are considered to be decreasing. Available data indicate that tarsiers can exist in disturbed and degraded forest habitats. However, the overall pattern from population density data is that, compared to primary habitats, tarsiers persist in disturbed/degraded habitats only at lower population densities. Tarsier densities at some disturbed habitat sites are much lower than population densities calculated for prime tarsier habitat sites.

Most tarsier taxa appear to have maximum elevation occurrence limits well below 1,500 masl, with many taxa showing upper limits of elevational occurrence between 750–1,200 m. The mountain, or pygmy, tarsier, *T. pumilus* is an exception, occupying montane moss forest habitat in central Sulawesi from 1,800–2,200 masl (IUCN, 2020). It is also known that those lower elevation forests that are key habitats across the geographic distributions of tarsiiform taxa are the same forest habitats that are subject to the greatest anthropogenic pressures (Supriatna et al., 2020). Given that tarsiers in disturbed/degraded habitat occur at lower population densities than populations in primary habitats, the long-term persistence of tarsier populations in secondary habitats must be a concern. The presence today of tarsiers in secondary or degraded habitats does not ensure the long-term viability of those populations in the face of the combined effects of continuing human activity (particularly conversion of lower elevation forests for large-scale agricultural production) and, additionally, the ongoing fragmentation of remaining intact habitat (Supriatna et al., 2020) by humans and, in some regions, the capture of tarsiers for the pet trade. Deforestation together with habitat degradation and fragmentation of remaining habitat represent the greatest threats to the future survival of all tarsier species. At least four tarsier taxa on Sulawesi (*T. pelengensis, T. sangiriensis, T. tarsier,* and *T. tumpara*) do not occur at present in protected areas, and while larger protected areas of montane habitat have been established in central Sulawesi, lower elevation forests (which include the preferred habitats of multiple tarsier taxa) are not as well represented under existing protected areas systems. Forest conversion and habitat disturbance, in turn, put tarsier populations at risk of increased hunting pressures as well as the capture of animals for sale. These pressures are unlikely to diminish in the future, which will only drive the diminishing long-term viability of tarsier populations (Shekelle et al., 2017; Supriatna et al., 2020).

SUMMARY

All tarsiers are found in a variety of primary and secondary habitats largely between sea level and ~1,000–1,100 masl, with the exception of the highland-dwelling *T. pumilus* in central Sulawesi (Grow et al., 2013). The majority of species are of similar body size, with adult male weights on average ~120–140 g, and adult female body weights ~90% of males. All tarsiers have a diet that is unique among primates consisting of 100% live-caught animal prey; they are faunivores with no plant matter in their diet. While dietary preferences exist among genera and species, tarsier diets are quite generalized; they will eat almost anything they can catch and kill that is not too poisonous to consume.

All tarsiers are highly specialized for vertical clinging and leaping and are active mainly on small vertical supports in the undergrowth, usually less than 3 m above the ground. Variation exists among the genera to the extent of their specialization for vertical clinging and leaping, with Western tarsiers being most specialized, Eastern tarsiers least specialized, and Philippine tarsiers intermediate.

The most striking difference among tarsiers is reflected in their social organization; one clade (*Tarsius*) is gregarious and the other (*Carlito/Cephalopachus*) consists of solitary foragers. This indicates that sociality, or gregarious behavior, either evolved independently within tarsiers (separately from the development of sociality in strepsirrhine and anthropoid primates), or that gregarious behavior was primitive for Tarsiidae and has since been lost in the Central/Western tarsier clade. Yi et al. (2015) argue that the latter scenario is the most likely. Since all-night follows in the context of long-term field studies have only been conducted for a tiny percentage of known tarsier species, the range of behavioral variation which may be present in tarsiers over their entire geographical range is poorly understood at present.

In all known cases, females give birth to a single offspring, usually once per year. The infant is transported by mouth during the first three weeks or more, in a cache-and-carry behavior. After this, the infant becomes increasingly independent. The neonate is large in relation to the mother's body weight, and the brain of the infant accounts for 60–70% of its weight. Neonatal and postnatal growth are extremely slow, but behavioral development is very rapid. Infants move on their own within weeks and are as efficient as adults in capturing prey within 100 days. The combination of slow physical growth, large brain size, and rapid behavioral development is thought to be a compromise between the energetic/dietary requirements of the mother and the necessity for the infant to become competent in its highly specialized locomotor and predatory skills, and thus attain independence.

Efforts to understand tarsier conservation and threats to their future existence are impeded chiefly by three areas of uncertainty: (1) unrecognized cryptic taxonomic variation, (2) the extent to which secondary habitats are suitable for sustaining tarsier populations over the long term, and (3) the degree to which high-elevation montane forests are suitable for sustaining tarsier populations over the long term. Many tarsier taxa are surely among the most threatened primate species in the world. It is an unfortunate, but likely, prediction that some tarsier species will go extinct before they are known to science, owing to habitat destruction and degradation, pesticide use, nonnative predators (feral cats and dogs), and human exploitation for food and the pet trade.

REFERENCES CITED—CHAPTER 4

Aiello, L. C. (1986). The relationships of the tarsiiformes: A review of the case for the Haplorrhini. In B. Wood, L. Martin, & P. Andrews (Eds.), *Major topics in primate and human evolution* (pp. 47-65). Cambridge University Press.

Bearder, S. K., Butynski, T. M., & de Jong, Y. A. (2013). Vocal profiles for the galagos: A tool for identification. *Primate Conservation*, 27, 75.

Bearder, S. K., Nekaris, K. A. I., & Buzzell, A. (2002). Dangers in the night: Are some nocturnal primates afraid of the dark? In L. E. Miller (Ed.), *Eat or be eaten: Predator sensitive foraging among primates* (pp. 21-43). Cambridge University Press.

Bejar, S. G. F., Duya, M. R. M., Duya, M. V., Galindon, J. M. M., Pasion, B. O., & Ong, P. S. (2020). Living in small spaces: Forest fragment characterization and its use by Philippine tarsiers (*Tarsius syrichta* Linnaeus, 1758) in Mindanao Island, Philippines. *Primates*, 61, 529-542.

Blanco, M. B., Greene, L. K., Rasambainarivo, F., Toomey, E., Williams, R. C, Andrianandrasana, L., Larsen, P. A., & Yoder, A. D. (2020). Next-generation technologies applied to age-old challenges in Madagascar. *Conservation Genetics*, 21, 785-793.

Brown, R. M., Weghorst, J. A., Olson, K. V., Duya, M. R. M., Barley, A. J., Duya, M.V., Shekelle, M., Neri-Arboleda, I., Esselstyn, J. A., Dominy, N. J., Ong, P. S., Moritz, G. L., Luczon, A., Diesmos, M. L. L., Diesmos, A. C., & Siler, C. D. (2014). Conservation genetics of the Philippine tarsier: Cryptic genetic variation restructures conservation priorities for an Island Archipelago primate. *PLoS ONE*, 9(8), e104340.

Burton, J. A., & Nietsch, A. (2010). Geographical variation in duet songs of Sulawesi tarsiers: Evidence for new cryptic species in south and southeast Sulawesi. *International Journal of Primatology*, 31(6), 1123-1146.

Cartmill, M., & Kay, R. F. (1978). Craniodental morphology, tarsier affinities, and primate suborders (1978). In D. J. Chivers, & K. A. Joysey (Eds.), *Recent advances in primatology (Vol. 3—Evolution)* (pp. 205-214). Academic Press.

Catchpole, H. R., & Fulton, J. F. (1939). Tarsiers in captivity. *Nature*, 144, 514.

Charles-Dominique, P. (1977). *Ecology and behaviour of nocturnal primates: Prosimians of equatorial West Africa*. Columbia University Press.

Clutton-Brock, T. H., Gaynor, D., McIlrath, G. M., Maccoll, A. D. C., Kansky, R., Chadwick, P., Manser, M., Skinner, J. D., & Brotherton, P. N. M. (1999). Predation, group size and mortality in a cooperative mongoose, *Suricata suricatta*. *Journal of Animal Ecology*, 68, 672-683.

Cook, N. (1939). Notes on captive *Tarsius carbonarius*. *Journal of Mammalogy*, 20(2),173-178.

Crompton, R. H. (1995). "Visual predation," habitat structure, and the ancestral primate niche. In L. Alterman, G. A. Doyle, & K. Izard (Eds.), *Creatures of the dark: The nocturnal prosimians* (pp. 11-30). Plenum Press.

Crompton, R. H., & Andau, P. M. (1986). Locomotion and habitat utilization in free-ranging *Tarsius bancanus*: A preliminary report. *Primates*, 27, 337-355.

Crompton, R. H., & Andau, P.M. (1987). Ranging, activity, rhythms, and sociality in free-ranging *Tarsius bancanus*: A preliminary report. *International Journal of Primatology*, 8, 43-71.

Cuming, H. (1838). On the habit of some species of mammal from the Philippine Islands. *Proceedings of the Zoological Society of London*, 6, 67-68.

Dagosto, M., & Gebo, D. L. (1997). A preliminary study of the Philippine tarsier in Leyte. *Asian Primates*, 6, 5-8.

Dagosto, M., & Gebo, D. L. (1998). A preliminary study of the Philippine tarsier (*Tarsius syrichta*) in Leyte. *American Journal of Physical Anthropology*, 105(S26), 73.

Dagosto, M., Gebo, D. L., & Dolino, C. (2001). Positional behavior and social organization of the Philippine tarsier (*Tarsius syrichta*). *Primates*, 42(3), 233-243.

Daly, M., Behrends, P. R., Wilson, M. I., & Jacobs, L. F. (1992). Behavioural modulation of predation risk: Moonlight avoidance and crepuscular compensation in a nocturnal desert rodent, *Dipodomys merriami*. *Animal Behaviour*, 44(1), 1-9.

Davis, D. D., (1962). Mammals of the lowland rain-forest of North Borneo. *Bulletin of the Singapore National Museum*, 31, 1-129.

Driller, C., Merker, S., Perwitasari-Farajallah, D., Sinaga, W., Anggraeni, N., & Zischler, H. (2015). Stop and go—waves of tarsier dispersal mirror the genesis of Sulawesi Island. *PLoS ONE*, 10(11), e0141212.

Driller, C., Perwitasari-Farajallah, D., Zischler, H., & Merker, S. (2009). The social system of Lariang tarsiers (*Tarsius lariang*) as revealed by genetic analyses. *International Journal of Primatology*, 30(2), 267-281.

Dunbar, R. I. M. (1988). *Primate social systems*. Cornell University Press.

Erkert, H. G., & Grober, J. (1986). Direct modulation of activity and body temperature of owl monkeys (*Aotus lemurinus griseimembra*) by low light intensities. *Folia Primatologica*, 47, 171-188.

Fleagle, J. F. (2013). *Primate adaptation and evolution* (3rd ed.). Academic Press.

Fogden, M. P. L. (1974). A preliminary field-study of the western tarsier, *Tarsius bancanus* Horsfield. In R. D. Martin, G. A. Doyle, & A. C. Walker (Eds.), *Prosimian biology* (pp. 151-166). University of Pittsburgh Press.

Fulton, J. E. (1939). A trip to Bohol in quest of *Tarsius*. *Yale Journal of Biology and Medicine*, 11, 561-573.

Goodman, M., Porter, C. A., Czelusniak, J., Page, S. L., Schneider, H., Shoshani, J., Gunnell, G., & Groves, C. P. (1998). Toward a phylogenetic classification of primates based on DNA evidence complemented by fossil evidence. *Molecular Phylogenetics and Evolution*, 9(3), 585-598.

Groves, C. P. (1976). The origin of the mammalian fauna of Sulawesi (Celebes). *Zeitschrift für Säugetierkunde*, 41(4), 201-216.

Groves, C. P. (1998). Systematics of tarsiers and lorises. *Primates*, 39, 13-27.

Groves, C. P. (2001). *Primate taxonomy*. Smithsonian Books.

Groves, C. P., & Shekelle, M. (2010). The genera and species of Tarsiidae. *International Journal of Primatology*, 31, 1071-1082.

Groves, C., Shekelle, M., & Brandon-Jones, D. (2008). Taxonomic history of the tarsiers, evidence for the origins of Buffon's tarsier, and the fate of *Tarsius spectrum* Pallas, 1778. In M. Shekelle, C. Groves, I. Maryanto, H. Schulze, & H. Fitch-Snyder (Eds.), *Primates of the oriental night* (pp. 1-12). LIPI Press.

Grow, N. B. (2019). Cryptic communication in a montane nocturnal haplorrhine, *Tarsius pumilus. Folia Primatologica*, 90, 404-421.

Grow, N. B., & Gursky, S. (2010). Preliminary data on the behavior, ecology, and morphology of pygmy tarsiers (*Tarsius pumilus*). *International Journal of Primatology*, 31, 1174-1191.

Grow, N. B., Gursky, S., & Duma, Y. (2013). Altitude and forest edges influence the density and distribution of pygmy tarsiers (*Tarsius pumilus*). *American Journal of Primatology*, 75(5), 464-477.

Gursky, S. L. (1994). Infant care in spectral tarsier (*Tarsius spectrum*) Sulawesi, Indonesia. *International Journal of Primatology*, 15(6), 843-853.

Gursky, S. L. (1997). *Modeling maternal time budget: The impact of lactation and infant transport on the time budget of the spectral tarsier*, Tarsius spectrum. [Doctoral dissertation, Stony Brook University].

Gursky S. L. (1998). Conservation status of the spectral tarsier *Tarsius spectrum*: Population density and home range size. *Folia Primatologica*, 69(S1), 191-203.

Gursky, S. L. (1999). The Tarsiidae: Taxonomy, behavior and conservation status. In P. Dolhinow & A. Fuentes (Eds.), *The nonhuman primates* (pp. 140-145). Mayfield.

Gursky, S. L. (2000a). Sociality in the spectral tarsier, *Tarsius spectrum. American Journal of Primatology*, 51(1), 89-101.

Gursky, S. L. (2000b). Effect of seasonality on the behavior of an insectivorous primate, *Tarsius spectrum. International Journal of Primatology*, 21, 477-495.

Gursky S. L. (2002a). Predation on a wild spectral tarsier (*Tarsius spectrum*) by a snake. *Folia Primatologica*, 73, 60-62.

Gursky S. L. (2002b). Determinants of gregariousness in the spectral tarsier (Prosimian: *Tarsius spectrum*). *Journal of Zoology*, 256, 401-410.

Gursky S. L. (2003a). Lunar philia in a nocturnal primate. *International Journal of Primatology*, 24, 351-367.

Gursky S. L. (2003b). Predation experiments on infant spectral tarsiers (*Tarsius spectrum*). *Folia Primatologica*, 74(5-6), 272-284.

Gursky, S. L. (2005a). Associations between adult spectral tarsiers. *American Journal of Physical Anthropology*, 128, 74-83.

Gursky, S. L. (2005b). Predator mobbing in *Tarsius spectrum. International Journal of Primatology*, 26, 207-221.

Gursky, S. L. (2006). Function of snake mobbing in spectral tarsiers. *American Journal of Physical Anthropology*, 129(4), 601-608.

Gursky, S. L. (2007a). *The spectral tarsier (Primate field studies)*. Routledge.

Gursky S. L. (2007b) The response of spectral tarsiers toward avian and terrestrial predators. In S. L. Gursky, & K. A. I. Nekaris (Eds.), *Primate anti-predator strategies (Developments in primatology: Progress and prospects)* (pp. 241-252). Springer.

Gursky, S. L. (2010a). Dispersal patterns in *Tarsius spectrum. International Journal of Primatology*, 31, 117-131.

Gursky, S. L. (2010b). The function of scentmarking in spectral tarsiers. In S. Gursky, & J. Supriatna (Eds.), *Indonesian primates (Developments in primatology: Progress and prospects)* (pp. 359-369). Springer.

Gursky, S. L. (2015). Acoustic characterization of ultrasonic vocalizations by spectral tarsiers. *Folia Primatologica*, 86, 153-163.

Gursky, S. L. (2019). Echolocation in a nocturnal primate? *Folia Primatologica*, 90, 379-391.

Gursky, S., Salibay, C. C., & Cuevas, C. Z. (2011). Population survey of the Philippine tarsier (*Tarsius syrichta*) in Corella, Bohol. *Folia Primatologica*, 82, 189-196.

Harcourt, A. H. (1999). Biogeographic relationships of primates on south-east Asian islands. *Global Ecology and Biogeography*, 8, 55-61.

Haring, D. M., & Wright, P. C. (1989). Hand-raising an infant tarsier, *Tarsius syrichta. Zoo Biology*, 8, 265-274.

Haring, D. M., Wright, P. C., & Simons, E. L. (1985). Social behaviour of *Tarsius syrichta* and *Tarsius bancanus. American Journal of Physical Anthropology*, 66, 179.

Harrison, B. (1962). Getting to know about *Tarsius. Malaysian Nature Journal*, 16, 197-204.

Harrison, B. (1963). Trying to breed *Tarsius. Malaysian Nature Journal*, 17, 218-231.

Hidayatik, N., Yusuf, T. L., Agil, M., Iskandar, E., Sajuthi, D. (2018). Sexual behaviour of the spectral tarsier (*Tarsius spectrum*) in captivity. *Folia Primatologica*, 89, 157-164.

Hill, W. C. O. (1955). *Primates: Comparative anatomy and taxonomy. II. Haplorrhini: Tarsioidea.* Edinburgh University Press.

Hill, W. C. O. (1972). *Evolutionary biology of the primates.* Academic Press.

Hill, W. C. O., Porter, A, Southwick, M. D. (1952). The natural history, ectoparasites and pseudo-parasites of the tarsiers (*Tarsius carbonarius*), recently living in the society's menagerie. *Proceedings of the Zoological Society of London,* 122(1), 79-119.

IUCN. (2020). *The International Union for the Conservation of Nature (IUCN) Red List of Threatened Species.* Version 2020-3. www.iucnredlist.org.

Izard, M. K., Wright, P. C., & Simons, E. L. (1985). Gestation length in *Tarsius bancanus. American Journal of Primatology,* 9, 327-331.

Jablonski, N. G., & Crompton, R. H. (1994). Feeding behavior, mastication, and tooth wear in the western tarsier (*Tarsius bancanus*). *International Journal of Primatology,* 15, 29-59.

Kappeler, P. M., & Ganzhorn, J. U. (1993). *Lemur social systems and their ecological basis.* Plenum Press.

Kleiman, D. G. (1977). Monogamy in mammals. *Quarterly Review of Biology,* 52(1), 39-69.

Kotler, B., Brown, J., & Hasson, O. (1993). Factors affecting gerbil foraging, behavior and rates of owl predation. *Ecology,* 72, 2249-2260.

Krebs, J. R., & Davies, N. B. (Eds.). (1984). *Behavioural ecology: An evolutionary approach* (2nd ed.). Sinauer Associates, Inc.

Le Gros Clark, W. E. (1924). Notes on the living tarsier (*Tarsius spectrum*). *Proceedings of the Zoological Society of London,* 94(1), 217-223.

Le Gros Clark, W. E. (1956). *History of the primates: An introduction to the study of fossil man* (5th ed.). University of Chicago Press.

Lewis, G. C. (1939). Notes on a pair of tarsiers from Mindanao. *Journal of Mammalogy,* 20, 57-61.

MacKinnon, J., & MacKinnon, K. (1977). The formation of a new gibbon group. *Primates,* 18, 701-708.

MacKinnon, J., & MacKinnon, K. (1980). The behavior of wild spectral tarsiers. *International Journal of Primatology,* 1, 361-379.

MacKinnon, K. (1987). Conservation status of the primates of Malasia, with special reference to Indonesia. *Primate Conservation,* 8, 175-183.

Martin, R. D. (1993). Primate origins: Plugging the gaps. *Nature,* 363, 223-234.

Martin, R. D. (2012). Primer: Primates. *Current Biology,* 22(18), 785-790.

Merker, S. (2006). Habitat-specific ranging patterns of Dian's tarsiers (*Tarsius dianae*) as revealed by radiotracking. *American Journal of Primatology,* 68(2), 111-125.

Merker, S., Driller, C., Dahruddin, H., Wirdateti, Sinaga, W., Perwitasari-Farajallah, D., & Shekelle, M. (2010). *Tarsius wallacei*: A new tarsier species from Central Sulawesi occupies a discontinuous range. *International Journal of Primatology,* 31, 1107-1122.

Merker, S., Driller, C., Perwitasari-Farajallah, D., Pamungkasb, J., & Zischlera, H. (2009). Elucidating geological and biological processes underlying the diversification of Sulawesi tarsiers. *Proceedings of the National Academy of Sciences,* 106(21), 8459-8464.

Merker, S., Driller, C., Perwitasari-Farajallah, D., Zahner, R., & Zischler, H. (2007). Isolation and characterization of 12 microsatellite loci for population studies of Sulawesi tarsiers (*Tarsius* spp.). *Molecular Ecology Notes,* 7, 1216-1218.

Merker, S., & Groves, C. P. (2006). *Tarsius lariang*: A new primate species from western central Sulawesi. *International Journal of Primatology,* 27(2), 465-485.

Merker, S., & Yustian, I. (2008). Habitat use analysis of Dian's tarsiers (*Tarsier dianae*) in mixed species plantation in Sulawesi, Indonesia. *Primates,* 49, 161-164.

Morrison, D. (1978). Lunar phobia in a neotropical fruit bat, *Artibeus jamaicensis* (Chiroptera, Phyllostomidae). *Animal Behavior,* 26, 852-855.

Munds, R. A., Ali, R., Nijman, V., Nekaris, K. A. I., Goossens, B. (2013). Living together in the night: Abundance and habitat use of sympatric and allopatric populations of slow lorises and tarsiers. *Endangered Species Research,* 22, 269-277.

Musser, G. G., & Dagosto, M. (1987). The identity of *Tarsius pumilus*, a pygmy species endemic to the montane mossy forests of Central Sulawesi. *American Museum Novitates,* 2867, 1-53.

Myers, N. (1987). Trends in the destruction of rain forests. In C. W. Marsh & R. A. Mittermeier, (Eds.), *Primate conservation in the tropical rain forest* (pp. 3-22). Alan R. Liss.

Nash, L. T. (1986). Influence of moonlight level on travelling and calling patterns in two sympatric species of galago in Kenya. In D. M. Taub & F. A. King (Eds.), *Current perspectives in primate social dynamics* (pp. 357-367). Van Nostrand Reinhold Co.

Neri-Arboleda, I. (2001). *Ecology and behavior of Tarsius syrichta in Bohol, Philippines: implications for conservation.* [Master's thesis, University of Adelaide].

Neri-Arboleda, I. (2010). Strengths and weaknesses of a population viability analysis for Philippine tarsiers (*Tarsius syrichta*). *International Journal of Primatology,* 31, 1192-1207.

Neri-Arboleda, I., Stott, P., & Arboleda, N. P. (2002). Home ranges, spatial movements and habitat associations of the Philippine tarsier (*Tarsius syrichta*) in Corella, Bohol. *Journal of Zoology,* 257, 387-402.

Niemitz, C. (1972). Puzzle about *Tarsius. Sarawak Museum Journal,* 20, 329-337.

Niemitz, C. (1973). *Tarsius bancanus* (Horsfield's Tarsier) preying on snakes. *Laboratory Primates Newsletter,* 12, 18-19.

Niemitz, C. (1974). A contribution to the postnatal behavioral development of *Tarsius bancanus*, Horsfield, 1821, studied in two cases. *Folia Primatologica,* 21, 250-276.

Niemitz, C. (1977). Zur funktionsmorphologie und biometrie der gattung *Tarsius*, Storr, 1780. *Courier Forschungsinstitut Senckenberg,* 25, 1-161.

Niemitz, C. (1979a). Results of a field study on the western tarsier (*Tarsius bancanus borneanus* Horsfield, 1821) in Sarawak. *Sarawak Museum Journal*, 27, 171-228.

Niemitz, C. (1979b). Outline of the behavior of *Tarsius bancanus*. In G. A. Doyle, & R. D. Martin (Eds.), *The study of prosimian behavior* (pp. 631-660). Academic Press.

Niemitz, C. (1984a). An investigation and review of the territorial behaviour and social organization of the genus *Tarsius*. In C. Niemitz (Ed.), *The biology of tarsiers* (pp. 117-128). Gustav Fischer Verlag.

Niemitz, C. (1984b). Locomotion and posture of *Tarsius bancanus*. In C. Niemitz (Ed.), *The Biology of Tarsiers* (pp. 191-226). Gustav Fischer Verlag.

Niemitz, C. (1984c). *The biology of tarsiers*. Gustav Fischer Verlag.

Niemitz, C. (1984d). Vocal communication of two tarsier species (*Tarsius bancanus* and *Tarsius spectrum*). In C. Niemitz (Ed.), *The biology of tarsiers* (pp. 129-142). Gustav Fischer Verlag.

Niemitz, C. (1984e). Activity rhythms and the use of space in semi-wild Bornean tarsiers, with remarks on wild spectral tarsiers. In C. Niemitz (Ed.), *The biology of tarsiers* (pp. 85-116). Gustav Fischer Verlag.

Niemitz, C. (1984f). Synecological relationships and feeding behaviour of the genus *Tarsius*. In C. Niemitz (Ed.), *The biology of tarsiers* (pp. 59-76). Gustav Fischer Verlag.

Niemitz, C. (1985). Can a primate be an owl? Convergences in the same ecological niche. *Fortschritte der Zoologie*, 30, 667-670.

Niemitz, C., Nietsch, A., Warter, S., & Rumpler, Y. (1991). *Tarsius dianae*: A new primate species from Central Sulawesi (Indonesia). *Folia Primatologica*, 56, 105-116.

Nietsch, A. (1993). *Beiträge zur Biologie von Tarsius spectrum in Sulawesi*. [Doctoral dissertation, Free University Berlin].

Nietsch, A. (1994). A comparative study of vocal communication in Sulawesi tarsiers. *The XVth Biennial Congress of the International Primatological Society*, 320.

Nietsch, A. (1999). Duet vocalizations among different populations of Sulawesi tarsiers. *International Journal of Primatology*, 20(4), 567-583.

Nietsch, A., & Kopp., M.-L. (1998). Role of vocalization in species differentiation of Sulawesi tarsiers. *Folia Primatologica*, 69(S1), 71-78.

Nietsch, A., & Niemitz, C. (1993). Diversity of Sulawesi tarsiers. *Deutsche Gesellschaft für Saugetierkunde*, 67, 45-46.

Nijman, V., & Nekaris, A. (2010). Checkerboard patterns, interspecific competition, and extinction: Lessons from distribution patterns of tarsiers (*Tarsius*) and slow lorises (*Nycticebus*) in insular southeast Asia. *International Journal of Primatology*, 31, 1147-1160.

Oxnard, C. E., Crompton, R. H., & Lieberman, S. S. (1990). *Animal lifestyles and anatomies: The case of the prosimian primates*. University of Washington Press.

Permadi, D., Tumbelaka, L. I., & Yusef, T. L. (1994). Reproductive pattern of *Tarsius* spp. in the captive breeding. *International Journal of Primatology*, 17, 1059-1069.

Pocock, R. I. (1918). On the external characters of the lemurs and of *Tarsius*. *Proceedings of the Zoological Society of London*, 88(1-2), 19-53.

Pollock, J. I., & Mullin, R. J. (1987). Vitamin C biosynthesis in prosimians: Evidence for the anthropoid affinity of *Tarsius*. *American Journal of Physical Anthropology*, 73, 65-70.

Preuschoft, H., Fritz, M., & Niemitz, C. (1979). Biomechanics of the trunk in primates and problems of leaping in *Tarsius*. In M. E. Morbeck, H. Preuschoft, & N. Gomberg (Eds.), *Environment, behavior and morphology: Dynamic interactions in primates* (pp. 327-345). Gustav Fischer Verlag.

Ramsier, M. A., Cunningham, A. J., Moritz, G. L., Finneran, J. J., Williams, C. V., Ong, P. S., Gursky-Doyen, S. L., & Dominy, N. J. (2012). Primate communication in the pure ultrasound. *Biology Letters*, 8, 508-511.

Rasmussen, D. T. (1994). The different meanings of a tarsioid-anthropoid clade and a new model of anthropoid origin. In J. G. Fleagle, & R. F. Kay (Eds.), *Anthropoid origins* (pp. 335-360). Plenum Press.

Řeháková-Petrů, M., Peške, L., & Daněk, T. (2012a). Predation on a wild Philippine tarsier (*Tarsius syrichta*). *Acta Ethologica*, 15, 217-220.

Řeháková-Petrů, M., Policht, R., & Peške, L. (2012b). Acoustic repertoire of the Philippine tarsier (*Tarsius syrichta fraterculus*) and individual variation of long-distance calls. *International Journal of Zoology*, 2012, 602401.

Roatch, N., Munds, R. A., Ali, R., Nijman, V., Nekaris, K. A. I., Goossens, B. (2011). Bornean loris and tarsier (*Nycticebus menagensis* and *Tarsius bancanus borneanus*) abundance and micro-habitat divergences in a degraded forest in Sabah, Malaysian Borneo. *American Journal of Physical Anthropology*, 144, 254.

Roberts, M. (1994). Growth, development, and parental care in the western tarsier (*Tarsius bancanus*) in captivity: Evidence for a "slow" life-history and nonmonogamous mating system. *International Journal of Primatology*, 15, 1-28.

Roberts, M., & Kohn, F. (1993). Habitat use, foraging behavior and activity patterns in reproducing western tarsiers, *Tarsius bancanus* in captivity. *Zoo Biology*, 12, 217-232.

Rosenberger, A. L., & Preuschoft, H. (2012). Evolutionary morphology, cranial biomechanics and the origins of tarsiers and anthropoids. *Palaeobiodiversity and Palaeoenvironments*, 92(4), 507-525.

Rosenberger, A. L., & Szalay, F. S. (1980). On the tarsiform origins of anthropoidea. In R. L. Ciochon, & A. B. Chiarelli (Eds.), *Evolutionary biology of the New World monkeys and continental drift* (pp. 139-157). Plenum Press.

Schwartz, J. H., & Tattersall, I. (1987). Tarsiers, adapids and the integrity of strepsirrhini. *Journal of Human Evolution*, 16(1), 23-40.

Schwitzer, C., Mittermeier, R. A., Rylands, A. B., Taylor, L. A., Chiozza, F., Williamson, E. A., Wallis, J., & Clark, F. E. (Eds.). (2014). *Primates in peril: The world's 25 most endangered primates 2012-2014*. IUCN SSC Primate Specialist Group (PSG), International Primatological

Society (IPS), Conservation International (CI), and Bristol Zoological Society.

Shekelle, M. (1997). The natural history of the tarsiers of north and central Sulawesi. *Sulawesi Primate Newsletter,* 4(2), 4-11.

Shekelle, M. (2003). *Taxonomy and biogeography of eastern tarsiers.* [Doctoral dissertation, Washington University].

Shekelle, M. (2008a). Distribution of tarsier acoustic forms, north and central Sulawesi: With notes on primary taxonomy of Sulawesi's tarsiers. In M. Shekelle, I. Maryanto, C. Groves, H. Schulze, & H. Fitch-Snyder (Eds.), *Primates of the oriental night* (pp. 35-50). LIPI Press.

Shekelle, M. (2008b). The history and mystery of the mountain tarsier, *Tarsius pumilus. Primate Conservation,* 23, 121-124.

Shekelle, M. (2013). Observations of wild Sangihe Island tarsiers *Tarsius sangirensis. Asian Primates Journal,* 3(1), 18-23.

Shekelle, M. (2014). A case of primate rafting and island hopping: Long distance dispersal and successful colonization over open ocean in a volcanic archipelago. *American Journal of Physical Anthropology,* 153, 238.

Shekelle, M. (2020a). *Tarsius pelengensis. The IUCN Red List of Threatened Species* 2020: e.T21494A17977515. https://dx.doi.org/10.2305/IUCN.UK.2020-3.RLTS.T21494A17977515.en.

Shekelle, M. (2020b). *Tarsius dentatus. The IUCN Red List of Threatened Species* 2020: e.T21489A17977790. https://dx.doi.org/10.2305/IUCN.UK.2020-3.RLTS.T21489A17977790.en.

Shekelle, M., Groves, C. P., Maryanto, I., & Mittermeier, R. A. (2017). Two new tarsier species (Tarsiidae, primates) and the biogeography of Sulawesi, Indonesia. *Primate Conservation,* 31, 61-69.

Shekelle, M., Groves, C. P., Maryanto, I., Mittermeier, R. A., Salim, A., & Springer, M. S. (2019). A new tarsier species from the Togean Islands of central Sulawesi, Indonesia, with references to Wallacea and conservation on Sulawesi. *Primate Conservation,* 33, 65-73.

Shekelle, M., Gursky, S., Merker, S., & Ong, P. (2015). Philippine Tarsier *Carlito syrichta* (Linnaeus, 1758), Philippines. In C. Schwitzer, R. A. Mittermeier, A. B. Rylands, F.Chiozza, E. A. Williamson, J. Wallis, & A. Cotton (Eds.), *Primates in peril: The world's 25 most endangered primates 2014-2016* (pp. 43-44). IUCN SSC Primate Specialist Group (PSG), International Primatological Society (IPS), Conservation International (CI), and Bristol Zoological Society.

Shekelle, M., & Leksono, S. M. (2004). Conservation strategy in Sulawesi island using *Tarsius* as flagship species [Translated from Indonesian]. *Biota,* 9, 1-10.

Shekelle, M., Meier, R., Wahyu, I., Wirdateti, W., & Ting, N. (2010). Molecular phylogenetics and chronometrics of Tarsiidae based on 12S mtDNA haplotypes: Evidence for Miocene origins of crown tarsiers and numerous species within the Sulawesian clade. *International Journal of Primatology,* 31, 1083-1106.

Shekelle, M., Mukti, S., Ichwan, L., & Masala, Y. (1997). The natural history of the tarsiers of north and central Sulawesi. *Sulawesi Primate Newsletter,* 2, 4-11.

Shekelle, M., & Salim, A. (2009). An acute conservation threat to two tarsier species in the Sangihe Island chain, north Sulawesi, Indonesia. *Oryx,* 43(3), 419-426.

Simons, E. L. (1972). *Primate evolution: An introduction to man's place in nature.* Macmillan.

Simpson, G. G. (1945). The principles of classification and a classification of mammals. *Bulletin of the American Museum of Natural History,* 85, 1-350.

Springer, M. S., Meredith, R. W., Gatesy, J., Emerling, C. A., Park, J., Rabosky, D. L., Stadler T., Steiner, C., Ryder, O. A., Janečka, J. E., Fisher, C. A., Murphy, W. J. (2012). Macroevolutionary dynamics and historical biogeography of primate diversification inferred from a species supermatrix. *PLoS ONE,* 7(11), e49521.

Sumampow, T., Shekelle, M., Beier, P., Walker, F. M., & Hepp, C. M. (2020). Identifying genetic relationships among tarsier populations in the islands of Bunaken National Park and mainland Sulawesi. *PLoS ONE,* 15(3), e0230014.

Supriatna, J., Shekelle, M., Fuad, H., Winarni, N., Dwiyahreni, A., SriMariati, M., Margules, C., Prakoso, B., & Zakaria, Z. (2020). Deforestation on the Indonesian island of Sulawesi and the loss of primate habitat. *Global Ecology and Conservation,* 24, e01205.

Szalay, F. S. (1975). Phylogeny, adaptations, and dispersal of tarsiiform primates. In W. P. Luckett & F. S. Szalay (Eds.), *Phylogeny of primates: A multidisciplinary approach* (pp. 91-125). Plenum Press.

Tan, K. Y., Liew, J. L., Tan, N. H., Quah, E. S. H., Ismail, A. K., & Tan, C. H. (2019). Unlocking the secrets of banded coral snake (*Calliophis intestinalis, Malaysia*): A venom with proteome novelty, low toxicity and distinct antigenicity. *Journal of Proteomics,* 192, 246-257.

Tremble, M., Muskita, Y., & Supriatna, J. (1993). Field observations of *Tarsius dianae* at Lore Lindu Nation Park, Central Sulawesi, Indonesia. *Tropical Biodiversity,* 1(2), 67-76.

Trent, B. K., Tucker, M. E., Lockard, J. S. (1977). Activity changes with illumination in slow loris *Nycticebus coucang. Applied Animal Ethology,* 3(3), 281-286.

Ulmer, F. A. (1960). A longevity record for the Mindanao tarsier. *Journal of Mammalogy,* 41, 512.

Ulmer, F. A. (1963). Observations on the tarsier in captivity. *Zoological Gardens (Leipzig),* 27, 106-121.

Welman, S., Tuen, A. A., & Lovegrove, B. G. (2017). Searching for the haplorrhine heterotherm: Field and laboratory data of free-ranging tarsiers. *Frontiers in Physiology,* 8, 745.

Wharton, C. H. (1948). Seeking Mindanao's strangest creatures. *National Geographic,* 94, 389-399.

Wharton, C. H. (1950). The tarsier in captivity. *Journal of Mammalogy,* 31, 260-268.

Wojciechowski, F. J., Kaszycka, K. A., Wielbass, A. M., Řeháková, M. (2019). Activity patterns of captive Philippine tarsiers (*Tarsius syrichta*): Differences related to sex and social context. *Folia Primatologica,* 90(2), 109-123.

Wolfe, J., & Summerlin, C. (1989). The influence of lunar light on nocturnal activity of the old-field mouse. *Animal Behavior,* 37, 410-414.

Wright, P. C. (2003). Are tarsiers silently leaping into extinction? In P. C. Wright, E. L. Simons, & S. Gursky (Eds.), *Tarsiers: Past, present, and future* (pp. 296-308). Rutgers University Press.

Wright, P. C., Haring, D. M., & Simons, E. L. (1985). Social behavior of *Tarsius syrichta* and *Tarsius bancanus*. *American Association of Physical Anthropologists*, 66, 179.

Wright, P. C., Izard, M. K., & Simons, E. L. (1986a). Reproductive cycles in *Tarsius bancanus*. *American Journal of Primatology*, 11, 207-215.

Wright, P. C., & Simons, E. L. (1989). Calls of the Mindanao tarsier (*Tarsius syrichta*). *American Journal of Physical Anthropology*, 66, 236.

Wright, P. C., Toyama, L. M., & Simons, E. L. (1986b). Courtship and copulation in *Tarsius bancanus*. *Folia Primatologica*, 46(3), 142-148.

Yi, Y., Shekelle, M., Jang, Y., & Choe, J. C. (2015). Discriminating tarsier acoustic forms from Sulawesi's northern peninsula [poster]. *American Association of Physical Anthropologists*, 84, 25-29.

Yoder, A. D. (2003). The phylogenetic position of genus *Tarsius*: Which side are you on? In P. C. Wright, E. L. Simons, & S. Gursky (Eds.), *Tarsiers: Past, present, and future* (pp. 161-177). Rutgers University Press.

Yustian, I. (2007). *Ecology and conservation status of* Tarsius bancanus saltator *on Belitung Island, Indonesia.* [Doctoral dissertation, Universität Göttingen].

Zahidin, M. A., Jalil, N. A., Naharuddin, N. M., Rahman, M. R. A., Gani, M., & Abdullah, M. T. (2019). Partial mtDNA sequencing data of vulnerable *Cephalopachus bancanus* from the Malaysian Borneo. *Data in Brief Volume*, 25, 104133.

Zuckerman, S. (1932). *The social life of monkeys and apes.* Kegan Paul.

CHAPTER 5

Reflections on Prosimian Biology

ROBERT D. MARTIN

AFTER A GESTATION PERIOD OF SOME 50 YEARS, now is an appropriate time to take stock and review the development of a special subsection of primatology—prosimian biology. For my PhD thesis, completed in 1967, I studied the behavior and morphology of tree-shrews, at that time formally included in the Order Primates in George Gaylord Simpson's influential 1945 classification. Because they had been widely accepted as the most primitive living primates, I started out with the notion that studying tree-shrews could yield useful pointers to adaptations of ancestral primates. However, I discovered within a few months that common tree-shrews have highly unusual maternal behavior, with mothers giving birth to two to four altricial offspring in a separate nest visited only once every 48 hours for suckling. This minimal degree of maternal behavior among mammals contrasts starkly with the intensive care of (mostly single) precocial infants that is typical for primates. After detailed and wide-ranging research, my PhD thesis concluded that tree-shrews have little to tell us about the ancestral primate condition.

Having been sidetracked in this way, I decided to switch to genuine primates, focusing on a relatively primitive representative for my postdoctoral research. This is how I came to conduct a two-year project on mouse lemurs, mentored by Jean-Jacques Petter—a pioneer of prosimian biology—at the general ecology section of the National Museum of Natural History in Brunoy, France. Jean-Jacques advised me to visit Madagascar as part of my research in order to observe the natural behavior and ecology of lemurs in the wild. This I did for six months in 1968, starting with a three-month Oxford University expedition. Thus, began my lifelong career devoted to exploration of the origins and diversification of primates.

Fifty years ago, prosimians (lemurs, lorisiforms, and tarsiers) had been relatively little-studied in comparison to nonhuman simians (monkeys and apes). Regarded as generally primitive, prosimians (literally meaning *before the simians*) were widely labeled "lower primates," in contrast to the "higher primates" (a.k.a., simians). Accordingly, prosimians were generally held to be of limited interest for interpretation of human evolution. It is, indeed, true that extant lemurs, lorisiforms, and tarsiers

have generally retained a greater proportion of primitive characters than their simian counterparts. A key characteristic shared by most lemurs, all lorisiforms, and all tarsiers is restriction of activity to nighttime (*nocturnality*), whereas activity of simians (with the sole exception of owl monkeys) is predominantly confined to daylight hours (*diurnality*). Although some investigators disagree, many primatologists accept the inference that ancestral primates were nocturnal. It follows that this is a primitive feature, which is accompanied by a host of other features, such as greater reliance on olfactory, acoustic, tactile, and nocturnal visual communication.

But one major drawback of characterizing all prosimians as "lower primates" has been that they were seen by many as being more closely related to one another than to any of the "higher primates." Back in 1967, the crucial distinction between primitive and derived features for phylogenetic reconstructions was only dimly recognized in primatology. One major development over the past 50 years has been explicit recognition of the necessity to distinguish shared primitive, shared derived, and convergent characters when reconstructing relationships. This, combined with the steadily increasing availability of molecular data, has led to the far more reliable phylogenetic trees that we have today.

A primary casualty of the failure to distinguish between primitive and derived resemblances was delayed recognition of the fact that tarsiers are, in fact, more closely related to simians than to prosimians. This was indicated by certain early findings, notably the discovery of invasive placentation in tarsiers by the nineteenth-century pioneer Ambrosius Hubrecht. But most investigators steeped in skeletal morphology were reluctant to accept such evidence from "soft parts." The nocturnal habits of tarsiers were widely interpreted as a primitive feature shared with most other prosimians, and it took some time before it was generally recognized that nocturnal behavior is a secondary development in tarsiers. Following a long-running and sometimes acrimonious debate, it is now firmly established that the primate evolutionary tree has two main sister groups of extant species, one containing lemurs and lorisiforms (strepsirrhines) and the other containing tarsiers, monkeys, apes, and humans (haplorrhines).

Most mammals are either nocturnal (active between dusk and dawn) or diurnal (active between dawn and dusk). Intermediate conditions are relatively rare among small and medium-sized mammals, which include most primates. Presumably, adaptation to one of these two activity modes generally precludes overlap between them. However, an important discovery arising from field studies of lemurs was that members of the family Lemuridae (*Lemur*, *Eulemur*, *Hapalemur*, and *Prolemur*) can be active during nighttime as well as by day. A milestone paper published in 1982 by Ian Tattersall introduced the new term *cathemerality* for this unusual kind of activity pattern, which has since been studied in considerable detail in field studies of lemurs.

One particularly striking change applying across all three major groups of prosimians over the past 50 years has been cumulative recognition of additional species. In their influential *Handbook of Living Primates* (1967), John and Pru Napier recognized a grand total of 35 prosimian species: 21 lemurs, 11 lorisiforms, and 3 tarsiers. A conspicuous feature of all chapters included in this section of *The Natural History of Primates: A Systematic Survey of Ecology and Behavior* is a concerted effort to present an authoritative overview of current species and their current names (*taxonomy*). For nocturnal lemurs alone, Sylvia Atsalis recognizes a total of 68 species, Anna Nekaris et al. list 39 lorisiform species, while Myron Shekelle et al. tabulate a minimum of 14 tarsier species. Some of the additional species stem from novel discoveries of hitherto unknown populations during field studies, as exemplified by the golden bamboo lemur (*Hapalemur aureus*). One quite unexpected finding concerned the dwarf bushbabies studied by Pierre Charles-Dominique during his groundbreaking studies in Gabon. Dwarf bushbabies were long allocated to the single species *Galago demidovii*, but detailed study of vocalizations, subtle details of pelage patterns, and penile morphology revealed that there are actually two different species with overlapping distributions but ecological differences, now recognized as *Galagoides demidoff* and *Galagoides thomasi*. This discovery provided an explanation for some puzzling contrasts in behavior in the dwarf bushbabies observed by Charles-Dominique.

In most cases, however, recognition of new species has resulted from a combination of field observations with detailed investigation of specimens housed in collections. Repeatedly, it has emerged that previously recognized "species" in fact included members of two or more biologically distinct populations. In practice, this has often simply meant elevating previously recognized subspecies

to species rank. But with nocturnal prosimians there has also been a particular limitation, in that classical external distinguishing features of diurnal primates such as pelage pattern and facial appearance often show far less differentiation. Emphasizing the fact that nocturnal prosimians are active in a very different sensory environment, recognition of *cryptic* species has depended far more on special features of communication by night, especially vocalizations. Many new species have been recognized as a result of thorough revision, including more detailed examination of museum specimens and use of increasingly available molecular information. Moreover, detection of previously unrecognized differences has also led to increased allocation to separate genera.

In some cases, increases in numbers of species and genera for nocturnal prosimians have been spectacular. In Madagascar, for instance, since 1967 the number of mouse lemur species has grown from 2 species allocated to the single genus *Microcebus* to 26. One of the original species has now been subdivided into 24 *Microcebus* species, while the other has been raised to the rank of a separate genus, *Mirza*, containing 2 species. Equally astounding is the example of sportive lemurs, originally classified as a single species in the genus *Lepilemur* and now recognized as 26 separate species, albeit still allocated to a single genus. Species recognition for bushbabies in Africa has also undergone a major expansion. Starting with 6 species in the single genus *Galago* recognized in 1967, there are now 22 bushbaby species allocated to 6 genera. Also notable, if numerically less remarkable, is the case of the tarsiers. Originally allocated to 3 species in the single genus *Tarsius*, they are now subdivided into at least 14 species in 3 genera.

Steady accumulation of new information regarding prosimian biology has been reflected in a series of milestones in the form of particularly influential publications devoted specifically to prosimian biology. These increasingly reflected the fruits of expanding field studies of prosimians in Madagascar, Africa, Asia, and Southeast Asia. The new findings were incorporated into a broad-based literature also covering morphology, physiology, behavior, and reproduction along with phylogenetic studies taking all aspects into account. Here, special mention is needed for the pioneering contributions of my mentor Jean-Jacques Petter in his 1962 compendium *Recherches sur l'écologie et l'éthologie des lémuriens malgaches* and of his wife, Arlette Petter-Rousseaux, who opened up new avenues of research into the reproductive biology of lemurs. Soon afterward, in 1966, the appearance of Alison Jolly's monograph *Lemur Behavior: A Madagascan Field Study*—reporting results of her field study of ring-tailed lemurs and sifakas—paved the way for a now well-established tradition of long-term investigation using quantitative data collection covering ecological aspects as well as behavior.

My own field studies of mouse lemurs in Madagascar, conducted in 1968 and 1970, regrettably lasted only eight months in all. But I did try to follow the basic principles championed by Alison Jolly as far as possible. As it happened, my second study in 1970 coincided with an international conference on lemurs and conservation in the capital of Madagascar, Antananarivo. Through a connection with the London Missionary Society, I had managed to rent a large house for the duration of the conference. My wife, Anne-Elise, and I (actually at the tail-end of our honeymoon) shared that accommodation with several people engaged in the first wave of detailed investigation of prosimian biology: Pierre Charles-Dominique, Alison Jolly, Alison Richard, Bob Sussman, Alan Walker. Among other things, this provided a golden opportunity to get to know Bob and his wife Linda, launching one of our most treasured lifelong friendships.

Almost as a natural development from those encounters in Madagascar in 1970, I was soon deeply engaged in planning an international conference—together with Gerry Doyle and Alan Walker—that was held in London in 1972. The proceedings, including over 50 contributions, were published in an edited volume entitled *Prosimian Biology* in 1974. In the same year as the conference, 1972, I also published a much-cited review arising from my postdoctoral research, entitled "Adaptive Radiation and Behavior of the Malagasy Lemurs," in *Philosophical Transactions of the Royal Society of London*. The conference had several significant outcomes, but one of the most important for me personally was the connection established with Gerry Doyle, who later invited me to serve as external examiner for Simon Bearder's PhD thesis (1974) on the ecology and behavior of the thick-tailed bushbaby, now known as *Otolemur crassicaudatus*. Five years down the line, Gerry and I published a second co-edited volume entitled *The Study of Prosimian Behavior*. That same year (1979) saw the publication of Bob Sussman's seminal edited book *Primate Ecology: Problem Oriented Field Studies*. As its title indicates, that volume bore witness to an important transition in field studies, from basic fact-finding surveys to investigations specifically designed to test principles and hypotheses. In Bob's own case, his initial field research project examined the behavioral ecology of ring-tailed lemurs and brown lemurs at two sites in western and southeastern Madagascar. His primary focus was on the expected niche differences arising from potential competition between related species living in sympatry. A year later, in 1980, a similar theme was explored in contributions to *Nocturnal Malagasy Primates: Ecology, Physiology and Behavior*, a volume co-edited by Harold Cooper et al. This volume reported results from a comparative study of five nocturnal lemur species in seasonal dry forest in western Madagascar.

Following that initial period (1966–1980), a trickle of synthetic publications devoted in whole or in part to nocturnal prosimians gradually became a torrent that has flowed ever faster over the past 40 years. It is only possible to mention some of the highlights here. The first is undeniably Ian Tattersall's single-authored 1982 book *The Primates of Madagascar*, which laid a foundation for many subsequent developments. Another milestone was publication in 1995 of *Creatures of the Dark: The Nocturnal Prosimians*, edited by Lon Alterman, Gerald Doyle, and Kay Izard; containing proceedings from another major international conference, this brought together many key participants in the second wave of research into prosimian biology.

Special methods employed to study animals active by night gradually progressed from the early days. For lemurs and lorisiforms, a basic technique for following study subjects at night was using headlamps to generate eyeshine. This permits investigators to locate otherwise invisible individuals over substantial distances. The capacity to do so directly depends on the presence of a reflecting *tapetum lucidum* behind the retina, a shared derived feature of strepsirrhine primates. All tarsier species, by contrast, lack a reflecting tapetum. This is one of the features indicating that they differ distinctly from lemurs and lorisiforms and that their nocturnal habits are a secondary feature. But in practical terms it, unfortunately, also means that tarsiers do not show eyeshine. Particularly in their case, use of radio-telemetry with miniaturized transmitters (in any case invaluable in dense forest)—an approach pioneered by Pierre Charles-Dominique in his study of five nocturnal prosimians in Gabon—rapidly became a major asset for field studies. Fortunately for me, during my postdoctoral research with Jean-Jacques Petter in Brunoy, I met up with Pierre, who had returned from his field studies in Gabon to work at the same institution. He introduced me to many pioneering observations made during his field study of five lorisiform species in Gabon and we discussed at length the many striking similarities between bushbabies and mouse lemurs.

Radio-telemetry has increasingly benefited primate field studies, including my own contributions. In one early development, I teamed up with Simon Bearder to conduct a two-year investigation of the behavior and ecology of lesser bushbabies in South Africa. In fact, I had originally been awarded a research grant by the Royal Society of London to employ radio-tracking in a field study of lesser mouse lemurs in Madagascar. Unfortunately, this plan unexpectedly hit the buffers when the Malagasy government abruptly stopped issuing research permits to foreign nationals, a barrier that persisted for several years. Fortunately, through Gerry Doyle, I had become acquainted with Simon (and with his outstanding skills as a fieldworker) and had made a preliminary visit under his guidance to the field site where he had studied lesser bushbabies for his MSc thesis. So, I rejigged my plan, with Simon as the postdoctoral researcher funded by the grant, to investigate the population of bushbabies with which he was already very familiar. Using radio-telemetry as one of our key methods, we conducted the first behavioral study of a nocturnal primate species using quantitative sampling, incorporating established techniques from studies of diurnal monkeys and apes, such as continuous following throughout the activity period combined with point sampling for data collection.

Along the way, in addition to radio-telemetry, field primatologists have since adopted numerous other technological advances. Many have been particularly connected with the advent of a broader, three-pronged approach to field studies in which behavioral ecology has been joined by monitoring of hormones and genetic relationships. Here, special attention has been given to noninvasive approaches, notably using samples of urine or feces for hormone assays and analyses of extracted genetic material. At a pivotal stage, the volume *Field and Laboratory Methods in Primatology: A Practical Guide*, co-edited by Joanna Setchell and Deborah Curtis, admirably presented the wide range of techniques available. First published in 2003, this guide was so successful that an updated second edition followed in 2011. Another prosimian review volume deserving special mention is *Lemurs: Ecology and Adaptation* (2006), co-edited by Lisa Gould and Michelle Sauther. As a relatively recent update on field research on Madagascar's lemurs, it is a particularly valuable source of information.

With respect to field research on lemurs over recent decades, the achievements of Patricia Wright deserve special mention. In 1986, she set out to find the greater bamboo lemur (*Prolemur simus*), widely thought to be extinct. Not only was her search for that species in the eastern rainforest of Madagascar successful, but she also discovered a previously undocumented species there— the golden bamboo lemur (*Hapalemur aureus*). At Stony Brook University, Pat founded the Institute for the Conservation of Tropical Environments (ICTE), dedicated to research and science-based conservation in tropical habitats, with a particular focus on Madagascar. ICTE coordinates scientific research across Madagascar but has been most active in Ranomafana National Park, where it operates Centre ValBio, a now renowned, state-of-the-art research station.

One initial handicap in interpreting the behavior of nocturnal prosimians was the entrenched view that they are essentially primitive in all respects, including their social life. In most cases, in the course of general surveys or opportunistic studies, human observers encountered only single individuals at night, and nocturnal prosimians were generally described as "solitary." This label was taken as the polar opposite of "social" applied to diurnal primates, almost all of which live in recognizable groups of individuals that move around together. However, the first detailed behavioral studies of nocturnal prosimians revealed regular encounters at night between individuals with overlapping home ranges, often matched by shared daytime nesting sites. Using such indicators, extended social networks can be identified. On this basis, building on initial findings reported by Pierre Charles-Dominique and Marcel Hladik, Simon Bearder explicitly recognized that *all* primates show some form of social organization and that "solitary" life in nocturnal prosimians is actually the opposite of "gregarious" life in most diurnally active primates. Here, "solitary" life is not the opposite of "social."

Now that social networks of so many nocturnal prosimian species have been investigated with long-term, detailed field studies, it has emerged that they exhibit adult sex ratios corresponding to those in the main types of gregarious groups found in diurnal primates: (1) single male/single female units; (2) single male/multi-female groups; (3) multi-male/single female groups; (4) multi-male/multi-female groups. In 1993, Eleanor Sterling aptly advocated recognizing three components that emphasize the diversity of social patterns in nocturnal primates: *social system*, *spacing system*, and *mating system*. Subsequently, in 2000, Alexandra Müller and Urs Thalmann examined the diversity of social organization among nocturnal mammals generally, paying special attention to the spacing system afforded by home range overlap.

Recent genetic studies have revealed mismatches between social system and mating system in nocturnal prosimians resembling those reported for certain diurnal simians. For instance, Joanna Fietz et al. (2000) found with socially monogamous dwarf lemurs that offspring had been sired through extra-pair copulations in 44% of cases examined. For lesser mouse lemurs, Ute Radespiel et al. (2002) and Martine Perret et al. (2003) have even reported cases of mixed paternity for infants in the same litter. Similar results have been reported for lorisiforms, notably by Pimley et al. (2005) for pottos.

Field studies of prosimians have extended and enriched our knowledge of primate diversity in sometimes unexpected ways. They exhibit, for instance, a wide variety of dietary specializations, some of which are quite unusual. A prominent example is the consumption of plant *exudates* (predominantly gums), adding *gummivory* (also known as *exudativory*) to the triad of dietary categories—insectivory, frugivory, and folivory—traditionally recognized for diurnal primates. In fact, my own introduction to gummivory came in 1973, when Simon Bearder first showed me lesser bushbabies performing this behavior at his study site. As Anna Nekaris et al. note in their chapter on lorisiforms, consumption and digestion of gum taken together may represent the most fundamental feeding adaptation of bushbabies, given that the diets of all species include at least some gum. Indeed, exudate feeding has also been identified in pottos and various slow lorises, so it is common among lorisiforms generally. In fact, needle-clawed bushbabies (*Euoticus*) are gum-feeding specialists which have a needle-clawed counterpart in Madagascar, the fork-crowned lemurs (*Phaner*). The "needle claws" in both cases are flat nails with a central keel terminating in a sharp point, adapted for prolonged head-down clinging to broad tree trunks, a feeding posture typical of gummivores. Eventually, it became clear that certain simian primates—notably the patas monkey, *Erythrocebus*—also include gums in their diets. But this feeding specialization is particularly prevalent among nocturnal lorisiforms and lemurs. And it seems that the *toothscraper* at the anterior end of the lower jaw (another shared derived feature of strepsirrhine primates) originated as an adaptation for gummivory. As the alternative name *toothcomb* indicates, this unique structure (consisting of two bilaterally flattened canines flanking four similarly shaped incisors with their crowns in an almost horizontal array) is also commonly used for grooming the fur. But that is unlikely to be its primary function. Close-up field observations have revealed that the toothscraper plays an important part in collecting exudates from tree trunks. In fact, the chapter by Anna Nekaris et al. notes that, to trigger a flow of gum, slow lorises can gouge large holes in trunks in a few seconds. Reportedly, the crunching sound produced by the gouging of a slow loris can be heard at a distance of 10 m. Eventually, enough field studies had included observations of gummivory/exudativory in primates to elicit a dedicated volume published in 2010 entitled *The Evolution of Exudativory in Primates*, edited by Anne Burrows and Leanne Nash. Papers in that book revealed that this widespread feeding adaptation had been found in at least 69 primate species belonging to 12 families, accounting for almost a fifth of extant primate species.

An extremely important shift in emphasis in primate field studies that has taken place over recent decades has been mounting emphasis on conservation biology. Of course, initial awareness of threats to natural primate populations existed 50 years ago, but conservation efforts are now very much at the leading edge. I commented on this trend in a chapter entitled "Prosimians: From Obscurity to Extinction?" included in the 1995 conference volume *Creatures of the Dark*. As tree-living mammals, primates are particularly menaced by habitat degradation (notably deforestation), but hunting for bushmeat and capture for trade also contribute. At the outset, little thought was given to conservation of the relatively poorly documented nocturnal prosimians, widely regarded as obscure inhabitants of biomes containing larger and more visible diurnal mammal species. Although it was reported some decades ago that the bizarre nocturnal aye-aye had become very rare and possibly even extinct, that was seen as a special case. A recent overview of threatened primates by Alejandro Estrada et al. (2017) with a line-up of leading proponents of conservation—including Bob Sussman—reported that 60% of primate species are now threatened with extinction and that populations are declining in about 75% of them. The authoritative *Red List* issued by the International Union for the Conservation of Nature (IUCN), updated in 2020, revealed in particular that 103 of 107 recognized lemur species are at least to some degree threatened by extinction, being classified as either Vulnerable, Endangered, or Critically Endangered. Estrada et al. (2017) showed that in almost all prosimian families some nocturnal species are in the Critically Endangered category. Ironically, the aye-aye (the sole living representative of the family Daubentoniidae) is now classified as Endangered, representing an intermediate level of threat. As noted in Shekelle et al's chapter, certain tarsier species surely count among the world's most threatened primates. Indeed, some suspected additional tarsier species may well become extinct before their existence can be confirmed.

Looking back over the past 50 years, it is pleasing to see how far we have progressed with respect to the key issue of primate origins, notably regarding inference of adaptations that characterized ancestral primates. It has always been accepted that, undoubtedly, primates of modern aspect (*euprimates*) evolved in trees. In the process, they developed certain adaptations such as large, forward-facing eyes—permitting stereoscopic (3D) vision—and locomotion characterized by grasping feet and hands with digits bearing nails rather than claws. Several authors mistakenly linked the emphasis on vision to a reduced sense of smell (olfaction) and a shorter snout, but some nocturnal primates combine large eyes with an unreduced olfactory apparatus. The other features spawned the long-accepted *arboreal theory* of primate origins that was particularly advocated by Grafton Elliot Smith (1927) and Frederick Wood Jones (1929).

In the 1970s, however, Matt Cartmill pointed out that various non-primate mammals live in trees yet lack the features attributed to arboreal origins of primates. Tree-shrews—most of which are partially or exclusively arboreal—provide an object lesson here. They have small, sideward-facing eyes—permitting little in the way of stereoscopic vision—and their locomotion is characterized by claw-bearing digits on feet and hands that grapple rather than grasp. In 1972 and 1974, following instructive comparisons across mammals, Cartmill proposed the alternative *visual predation theory* to explain how primates originated. He proposed that forward-facing eyes together with grasping, nail-bearing hands and feet were adaptations for nocturnal foraging for insects and fruit on terminal branches. A third theory, first proposed by Bob Sussman in 1991, explicitly linked primate origins to the evolution of flowering plants in his *angiosperm theory*. Sussman interpreted early primates as omnivores that ate not only fruits, flowers, gums, and nectars of flowering plants but also insects feeding on them. Cartmill revisited the issue in a 1992 paper, and two decades later Sussman et al. re-examined primate origins in 2013.

After our first encounter in Madagascar 50 years ago, Bob Sussman and I would meet up from time to time and engage in lengthy, animated discussions. It often emerged that our ideas had progressed along parallel tracts in the meantime, so I eagerly awaited each new meeting. On one such occasion, my awakening inklings about a potential link between primates and flowering plants were dramatically confirmed when Bob explained his initial version of the angiosperm theory.

There has been much lively discussion of primate origins over the years, with a general tendency to see the three theories cited above as opposing alternatives. However, they all have merits and are certainly not mutually exclusive. Matt Cartmill was surely correct in pointing out that life in the trees does not in itself explain the origins of primate visual characteristics (witness the tree-shrews). On the other hand, Bob Sussman was doubtless on to something with his inspired interpretation that primate origins and the evolutionary radiation of the angiosperm plants were directly linked. One cross-connecting discovery—which could emerge only from steadily multiplying field studies of prosimians—was that early primates were probably adapted not just for life in the trees but for life in the *fine branch niche* that is typically present at the peripheries of trees and shrubs. The many striking similarities between the relatively primitive mouse lemurs and bushbabies that Pierre Charles-Dominique and I identified in a joint paper half a century ago provided clues to a set of features shared by the relatively

small-bodied, nocturnal ancestors that gave rise to all modern primates.

REFERENCES AND SUPPLEMENTAL SOURCES— CHAPTER 5

Alterman, L., Doyle, G. A., & Izard, M. K. (Eds.). (1995). *Creatures of the dark: The nocturnal prosimians*. Plenum Press.

Bearder, S. K., & Martin, R. D. (1980). The social organization of a nocturnal primate revealed by radio-tracking. In C. J. Amlaner, & D. W. Macdonald (Eds.), *A handbook on biotelemetry and radio tracking* (pp. 633-648). Pergamon Press.

Burrows, A. M., & Nash, L. T. (Eds.). (2010). *The evolution of exudativory in primates*. Springer.

Cartmill, M. (1972). Arboreal adaptations and the origin of the Order Primates. In R. H. Tuttle (Ed.), *The functional and evolutionary biology of primates* (pp. 97-122). Aldine-Atherton.

Cartmill, M. (1974). Rethinking primate origins. *Science*, 184, 436-443.

Cartmill, M. (1992). New views on primate origins. *Evolutionary Anthropology*, 1, 105-111.

Charles-Dominique, P. (1971). Éco-éthologie des prosimiens du Gabon. *Biologica Gabonica*, 7, 121-228.

Charles-Dominique, P. (1972). Écologie et vie sociale de *Galago demidovii* (Fischer 1808, Prosimii). *Zeitschrift für Tierpsychologie, Beiheft*, 9, 7-41.

Charles-Dominique, P. (1977). *Ecology and behaviour of nocturnal primates*. Columbia University Press.

Charles-Dominique, P., & Hladik, C. M. (1971). Le lépilémur du sud de Madagascar: Écologie, alimentation et vie sociale. *Terre et Vie*, 25, 3-66.

Charles-Dominique, P., & Martin, R. D. (1970). Evolution of lorises and lemurs. *Nature*, 227, 257-260.

Cooper, H., Hladik, A., Hladik, C., Pages, E., Pariente, G., Petter-Rousseaux, A., Petter, J., Schilling, A., & Charles-Dominique, P. (Eds.). (1980). *Nocturnal Malagasy primates: Ecology, physiology and behavior*. Academic Press.

Curtis, D. J., Zaramody, A., & Martin, R. D. (1999). Cathemerality in the mongoose lemur, *Eulemur mongoz*. *American Journal of Primatology*, 47, 279-298.

Doyle, G. A., & Martin, R. D. (Eds.). (1979). *The study of prosimian behavior*. Academic Press.

Driller, C., Merker, S., Perwitasari-Farajallah, D., Sinaga, W., Anggraeni, N., & Zischler, H. (2015). Stop and go—waves of tarsier dispersal mirror the genesis of Sulawesi Island. *PLoS ONE*, 10(11), e0141212.

Driller, C., Perwitasari-Farajallah, D., Zischler, H., & Merker, S. (2009). The social system of Lariang tarsiers (*Tarsius lariang*) as revealed by genetic analyses. *International Journal of Primatology*, 30, 267-281.

Elliot Smith, G. (1927). *Essays on the evolution of man*. 2nd edition. Oxford University Press.

Eppley, T. M., Watzek, J., Ganzhorn, J. U., & Donati, G. (2016). Predator avoidance and dietary fibre predict diurnality in the cathemeral folivore *Hapalemur meridionalis*. *Behavioral Ecolology and Sociobiology, 71*(1-4), 1-12.

Estrada, A., Garber, P. A., Rylands, A. B., Roos, C., Fernandez-Duque, E., Di Fiore, A., Nekaris, K. A. I., Nijman, V., Heymann, E. W., Lambert, J. E., Rovero, F., Barelli, C., Setchell, J. M., Gillespie, T. R., Mittermeier, R. A., Arregoitia, L. V., de Guinea, M., Gouveia, S., Dobrovolski, R.,... & Li, B. (2017). Impending extinction crisis of the world's primates: Why primates matter. *Science Advances, 3*, e1600946.

Fausser, J.-L., Prosper, P., Donati, G., Ramanamanjato, J.-B., & Rumpler, Y. (2002). Phylogenic relationships between *Hapalemur* species and subspecies based on mitochondrial DNA sequences. *BMC Evolutionary Biology, 2*, 4.

Gould, L., & Sauther, M. L. (Eds.). (2006). *Lemurs: Ecology and adaptation*. Springer.

Hidayatik, N., Agil, M., Heistermann, M., Iskandar, E., Yusuf, T. L., & Sajuthi, D. (2018). Assessing female reproductive status of spectral tarsier (*Tarsius tarsier*) using fecal steroid hormone metabolite analysis. *American Journal of Primatology, 80*(11), e22917.

Hidayatik, N., Yusuf, T. L., Agil, M., Iskandar, E., & Sajuthi, D. (2018). Sexual behaviour of the spectral tarsier (*Tarsius spectrum*) in captivity. *Folia Primatologica, 89*(2), 157-164.

Hubrecht, A. A. W. (1898). Über die Entwicklung der Placenta von *Tarsius* und *Tupaia*, nebst Bemerkungen über deren Bedeutung als haemopoeitische Organe. *Proceedings of the International Congress of Zoology, 4*, 345-411.

Jolly, A. (1966). *Lemur behavior: A Madagascan field study*. University of Chicago Press.

Martin, R. D. (1972). A preliminary field-study of the lesser mouse lemur (*Microcebus murinus*, J. F. Miller 1777). *Zeitschrift für Tierpsychologie, Beiheft, 9*, 43-89.

Martin, R. D. (1972). Adaptive radiation and behaviour of the Malagasy lemurs. *Philosophical Transactions of the Royal Society London B, 264*(862), 295-352.

Martin, R. D. (1990). *Primate origins and evolution: A phylogenetic reconstruction*. Princeton University Press.

Martin, R. D. (1995). Prosimians: From obscurity to extinction? In L. Alterman, G. A. Doyle, & M. K. Izard (Eds.), *Creatures of the dark: The nocturnal prosimians* (pp. 535-563). Plenum Press.

Martin, R. D. (2012). Primer: Primates. *Current Biology, 22*, 785-790.

Martin, R. D., Doyle, G. A., & Walker, A. C. (Eds.). (1974). *Prosimian biology*. Duckworth.

Merker, S. (2006). Habitat-specific ranging patterns of Dian's tarsiers (*Tarsius dianae*) as revealed by radiotracking. *Americam Journal of Primatology, 68*, 111-125.

Merker, S., Driller, C., Perwitasari-Farajallah, D., Pamungkas, J., & Zischler, H. (2009). Elucidating geological and biological processes underlying the diversification of Sulawesi tarsiers. *Proceedings of the National Academy of Sciences U.S.A., 106*, 8459-8464.

Müller, A. E., & Thalmann, U. (2000). Origin and evolution of primate social organisation: A reconstruction. *Biological Reviews, 75*, 405-435.

Mutschler, T., Nievergelt, C. M., & Feistner, A. T. C. (2000). Social organization of the Alaotran gentle lemur (*Hapalemur griseus alaotrensis*). *American Journal of Primatology, 50*, 9-24.

Napier, J. R., & Napier, P. H. (1967). *A handbook of living primates*. Academic Press.

Niemitz, C. (1984). *The biology of tarsiers*. Gustav Fischer Verlag.

Nievergelt, C. M., Mutschler, T., Feistner, A. T. C., & Woodruff, D. S. (2002). Social system of the Alaotran gentle lemur (*Hapalemur griseus alaotrensis*): Genetic characterization of group composition and mating system. *American Journal of Primatology, 57*, 157-176.

Perret, M., Gachot-Neveu, H., & Andrès, M. (2001). Genetic determination of paternity in captive grey mouse lemurs: Pre-copulatory sexual competition rather than sperm competition in a nocturnal prosimian? *Behaviour, 138*, 1047-1063.

Petter, J.-J. (1962). Recherches sur l'écologie et l'éthologie des lémuriens malgaches. *Mémoires du Muséum National d'Histoire Naturelle, série A (Zoologie), 27*, 1-146.

Petter, J.-J., Albignac, R., & Rumpler, Y. (1977). *Mammifères lémuriens (Primates Prosimiens)* (Vol. 44) of *Faune de Madagascar*. ORSTOM.

Petter-Rousseaux, A. (1964). Reproductive physiology and behavior of the Lemuroidea. In J. Buettner-Janusch (Ed.), *Evolutionary and genetic biology of the primates* (Vol. 2, pp. 91-132). Academic Press.

Pimley, E. R., Bearder, S. K., & Dixson, A. F. (2005). Examining the social organization of the Milne-Edwards's potto *Perodicticus potto edwardsi*. *American Journal of Primatology, 66*, 317-330.

Setchell, J. M., & Curtis, D. J. (Eds.). (2011). *Field and laboratory methods in primatology: A practical guide* (2nd Ed.). Cambridge University Press.

Sussman, R. W. (Ed.). (1979). *Primate ecology: Problem oriented field studies*. John Wiley.

Sussman, R. W. (1991). Primate origins and the evolution of angiosperms. *American Journal of Primatology, 23*, 209-223.

Sussman, R. W., Rasmussen, D. T., & Raven, P. H. (2013). Rethinking primate origins again. *American Journal of Primatology, 75*(2), 95-106.

Tan, C. L. (1999). Group composition, home range size, and diet of three sympatric bamboo lemur species (genus *Hapalemur*) in Ranomafana National Park, Madagascar. *International Journal of Primatology, 20*, 547-566.

Tattersall, I. (1982). *The primates of Madagascar*. Columbia University Press.

Tattersall, I. (1988). Cathemeral activity in primates: A definition. *Folia Primatologica, 49*, 200-202.

Wilmé, L., Goodman, S. M., & Ganzhorn, J. U. (2006). Biogeographic evolution of Madagascar's microendemic biota. *Science, 312*, 1063-1065.

Wood Jones, F. (1929). *Man's place among the mammals*. Edward Arnold.

PART II

NEOTROPICAL MONKEYS

CHAPTER 6

Tamarins, Callimicos, and Marmosets
The Evolutionary and Ecological Challenges of Body Size Reduction and Reproductive Twinning

Paul A. Garber and Robert W. Sussman

Tamarins, Goeldi's monkeys (hereafter referred to as callimicos), and marmosets form a monophyletic group (subfamily Callitrichinae) of small-bodied New World primates that fills a unique ecological role in the forests of Central and South America. They are the smallest of living anthropoids, with adult body mass ranging from 120 g in the pygmy marmoset (*Cebuella pygmaea*) to 650 g in the golden-headed lion tamarin (*Leontopithecus chrysomelas*).

TAXONOMY AND PHYLOGENY

At present, eight genera are recognized: *Saguinus* (large-bodied tamarins), *Leontocebus* (small-bodied tamarins, previously grouped as species and subspecies of *Saguinus fuscicollis* and *S. nigricollis*), *Leontopithecus* (lion tamarins), *Callimico* (callimico), *Callithrix* (Eastern Brazilian marmosets), *Mico* (Amazonian marmosets), *Callibella* (dwarf marmosets), and *Cebuella* (pygmy marmosets). Unfortunately, for most of the twentieth century, their taxonomic status and evolutionary relationships to each other and to other platyrrhines were misunderstood, and this has affected our understanding of their behavior and ecology. Traditionally, these primates were regarded either as a single family, the Callitrichidae, or as two distinct families, the Callitrichidae (which at the time included the genera *Saguinus*, *Callithrix*, *Cebuella*, and *Leontopithecus*) and the Callimiconidae (which included the single genus and single species *Callimico goeldii*; Hershkovitz, 1977). In this latter classification, callimicos were distinguished from tamarins and marmosets because females give birth to a single infant (all other callitrichines regularly twin), and individuals retain their third upper and lower molars (all other callitrichines have lost their third molars) that bear a reduced hypocone or fourth cusp. However, these traits represent the primitive platyrrhine condition and, therefore, their presence in callimicos cannot be used to argue against a close evolutionary relationship with tamarins and marmosets. Moreover, there is evidence to suggest that ancestral callimicos lost the ability to twin, and the production of a single offspring

represents a derived condition (Martin, 1992; Porter & Garber, 2004). Finally, based on recent morphological and biochemical information (Davis, 1996; Schneider & Rosenberger, 1996; Porter & Garber, 2004; Cortés-Ortiz, 2009), there is no justification in separating callimicos into a distinct family from tamarins and other marmosets. Rather, there is now a strong consensus that callimicos are best regarded as a sister taxon to the marmoset clade, and not a primitive or ancestral callitrichine. In fact, tamarins of the genus *Saguinus* appear to most closely represent the basal member of this radiation (Cortés-Ortiz, 2009; Buckner, 2015).

Assuming that all extant platyrrhines shared a last common ancestor some 35 mya, nuclear DNA analyses indicate that callitrichines split from a cebid ancestor (the Cebidae is the New World primate family that includes squirrel monkeys, capuchin monkeys, night monkeys, and all tamarins, callimicos, and marmosets) approximately 16 mya, with *Saguinus* diverging from the basal callitrichine stock during the middle Miocene, some 14.1 mya (Chaves et al., 1999; Opazo et al., 2006; Buckner et al., 2015). Over the next 2–4 million years, the lineage leading to lion tamarins (~13.4 mya) and the lineage leading to callimico diverged (~11.2 mya callimicos split from the remainder of the marmoset clade). Approximately 9.1 mya the large-bodied tamarins and small-bodied tamarins separated, and some 5.4 mya the Eastern Brazilian marmosets split from the lineage leading to the Amazonian marmosets (*Mico*), dwarf marmosets (*Callibella*), and pygmy marmosets (*Cebuella*) (Buckner et al., 2015). This latter group of marmosets appears to have shared a last common ancestor around 4.4 mya (Figure 6.1; Opazo et al., 2006; Buckner et al., 2015). In this chapter, we regard all eight tamarin, marmoset, and callimico genera as members of the subfamily Callitrichinae and the family Cebidae.

Callitrichines represent an extremely successful radiation that includes 12 species of large-bodied tamarins, 10 species of small-bodied tamarins, 4 species of lion tamarins, 1 species of callimico, 14 species of

Callitrichine Phylogeny and Taxonomy

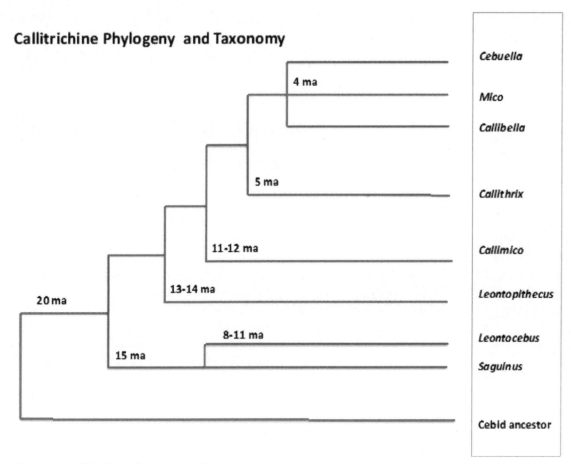

Figure 6.1 Callitrichine phylogeny and taxonomy.
(Adapted from Buckner et al., 2015; Cortés-Ortiz, 2009; Rylands et al., 2016)

Amazonian marmosets, 1 species of Amazonian dwarf marmoset, 2 species of Amazonian pygmy marmosets, and 6 species of eastern Brazilian marmosets (Rowe & Meyers, 2016; see Rylands et al., 2009, and Rylands & Mittermeier, 2009, for a list of all callitrichine species). Grouping all 50 species into a single taxonomic group serves to highlight the adaptive unity of their social, mating, and infant care systems, as well as their feeding ecology. For example, all species live in cohesive, small multi-male/multi-female social groups characterized by high reproductive output (in most species this involves reproductive twinning [the twins are dizygotic], lactational anestrus, and the potential for the same female to give birth twice during the year), cooperative infant care (principally provided by adult males), and the ability of the dominant female to behaviorally and/or hormonally impair or inhibit reproductive success in subordinate females (this may occur by reproductive suppression, stress-induced abortion, or infanticide by the dominant female; Digby et al., 2011). Tamarins and marmosets have the potential to produce four infants per year. No other group of anthropoids can produce more than a single infant in one year.

DISTINCTIVE TRAITS AND MISCONCEPTIONS OF CALLITRICHINE BEHAVIOR AND BIOLOGY

Although there remains some disagreement as how to best describe the range of variation in the mating behavior of callitrichines (in some wild groups of *Callithrix jacchus* and *Callithrix flaviceps* from two to four females have been observed to give birth within a one-month period; Hilario & Ferrari, 2010b; Digby et al., 2011), we argue that groups are best considered polyandrous (a single female mates with several males) or polygynous (multiple males and multiple females mate during a given reproductive period). In the case of callitrichines, it is important to distinguish between a mating system (the set of individuals who engage in sexual behavior) and a breeding system (the set of individuals who are gene donors) (Garber et al., 2016). Based on current evidence, we argue that the breeding system of callitrichines is best described as a nonmonogamous single female breeding system (see Garber et al., 2016). There are no groups that have been studied in the wild for which there is genetic evidence that the same adult male has sired the twin offspring of the same sovereign reproductive female over successive breeding periods. In the absence of genetic data, it is not possible to determine the degree to which one or more resident males or extragroup males sire a female's offspring. However, in at least three species of wild tamarins (*Saguinus geoffroyi*, *S. labiatus*, and *S. mystax*) there is genetic evidence that each of the twin offspring born into a given litter was sired by a different male (reviewed in Garber et al., 2016).

In addition, all species of callitrichines possess elongated, laterally compressed, and pointed claw-like nails. These nails are histologically identical to the nails of all other New World primates, but function as claws enabling these small primates to cling to large vertical trunks or sharply inclined supports (see Figure 6.2) that would otherwise be too large to be spanned by their tiny hands and feet (Garber, 1980b; Garber et al., 2009; Garber, 2011). Most species devote a substantial part of their day foraging and feeding on trunk resources such as plant exudates, insects and small vertebrates hidden in bark crevices and tree holes, or fungi that grow on the culms of bamboo or on decaying wood. Several callitrichines also cling to vertical trunks while scanning the ground for vertebrate and invertebrate prey or potential predators (Garber, 1992; Porter & Garber, 2004). The presence of claw-like nails (a derived digital morphology) *is*, in fact, one of the primary positional adaptations that allow callitrichines to exploit their unique ecological niche.

In the early 1980s, there were several common misconceptions regarding tamarins, marmosets, and callimicos. They were regarded as the ecological equivalent of tree squirrels (Eisenberg, 1977; Szalay & Delson, 1979), strictly monogamous (Leutenegger, 1980; Redican & Taub, 1981), and morphologically primitive (Hershkovitz, 1977). However, in a review of their ecology and social behavior, Sussman and Kinzey (1984) stressed that field studies did not support these earlier contentions. However, many of these misconceptions continued to persist through the mid- to late 1990s, and several continue today (Garber et al., 1996a).

Callitrichinae were traditionally divided into two functional groups based on the morphology of lower incisor and canine teeth. The first group included *Callithrix*, *Mico*, *Callibella*, and *Cebuella*, which exhibit the "short-tusked condition" in which the lower incisors are narrow, elongate, and reach the occlusal level of the canines (see Fleagle, 1999). These callitrichine taxa embed their claw-like nails into the tree trunks to adopt a stable posture and use their lower anterior dentition as a wedge to gouge holes into the bark to stimulate the flow of exudates (gums, saps, and resins). *Cebuella* and *Callithrix* (especially *C. jacchus* and *C. penicillata*) have the most functionally effective bark-gouging morphology and devote the most time gouging and feeding on exudates (Vinyard et al., 2009). Although *Cebuella* and *Callithrix*

Figure 6.2   A common marmoset (*Callithrix jacchus*) embedding its claw-like nails into tree bark while clinging to a vertical support.
(Photo credit: Paul A. Garber)

were once thought to be sister taxa or congeners (Barroso et al., 1997; Porter et al., 1997), biochemical and biogeographical evidence indicates that *Cebuella* shares a more recent common ancestor with Amazonian marmosets of the genus *Mico* than to *Callithrix*. The most parsimonious explanation is that this highly specialized, exudate-gouging, incisor-canine masticatory complex evolved independently in pygmy marmosets and the common ancestor of black-tufted marmosets (*Callithrix penicillata*) and common marmosets (*C. jacchus*). In callitrichines as well as in other lineages of exudate-feeding primates (*Phaner, Euoticus, Microcebus, Nycticebus*; see Burrows & Nash, 2010), this specialized gouging and digital morphology appear to serve a similar adaptive function. *Mico* (Verincini, 2009) and *Callibella* (Augiar & Lacher, 2009) also gouge holes in tree bark and consume exudates, but appear to do so less frequently than *Callithrix* and *Cebuella*.

The second group includes the large- and small-bodied tamarins, lion tamarins, and callimicos (*Saguinus, Leontocebus, Leontopithecus,* and *Callimico*) that exhibit a more typical anthropoid lower canine-incisor relationship in which the incisors are spatulate and shorter than the canine teeth (see Fleagle, 1999). This is called the long-tusked condition. Although tamarins, lion tamarins, and callimicos also consume plant exudates, they do not gouge holes in tree bark to stimulate exudate flow. Rather these primates consume exudates that are produced as a result of insect parasitism, weathering, and natural wounds to the trunk, secreted in the pods of legumes (principally of the genera *Parkia* and *Inga*), or exudates present on the stilted roots of palms of the genera *Iriartea* and *Euterpe* (Peres, 2000a; Garber & Porter, 2010). Occasionally, tamarins have been observed to parasitize exudate holes gouged by sympatric pygmy marmosets (Garber, pers.obs.). Thus, short-tusked and long-tusked are descriptive, not taxonomic terms. They are useful in distinguishing two broadly different adaptive groups, but not different phylogenetic clades. It must be noted, however, that there exists considerable variability in diet, dentition, gut morphology, home range, day range, group size, and insect and exudate-feeding behavior among callitrichines, and any distinction based solely on incisor/canine proportions masks important ecological and behavioral differences (Coimbra-Filho & Mittermeier, 1977; Ferrari & Lopes Ferrari, 1989; Ferrari & Martins, 1992; Ferrari et al., 1993; Porter et al., 2009; Vinyard et al., 2009; Abreu et al., 2016).

Another misconception is that small body mass represents the primitive condition for New World primates, and therefore callitrichines are among the most primitive of living platyrrhines (Hershkovitz, 1977). Although small body size is a hallmark of callitrichine evolution,

over the past 16 million years, individual tamarin and marmoset taxa have undergone at least six independent body-size reduction events. Body-size reduction appears to have initially occurred in the common ancestor of tamarins and marmosets after diverging from other cebids, and was likely associated with reproductive twinning, cooperative infant caregiving, and the potential for high reproductive output. However, secondary and independent body-size reduction events have occurred in different lineages, in different habitats, and during different time periods in *Leontocebus, Mico, Callithrix, Cebuella,* and *Callibella* (Buckner et al., 2015).

GEOGRAPHICAL DISTRIBUTION

Following the recent classification of Cortés-Ortiz (2009), Rylands and Mittermeier (2009), and Rylands et al. (2009), we distinguish five marmoset genera: *Callithrix, Mico, Callibella, Cebuella,* and *Callimico*. The genus *Callithrix* is distributed across three main habitat types in northeastern Brazil: the Caatinga, the Cerrado, and the Atlantic Forest (Rylands & Mittermeier, 2009: fig. 2.5). *Callithrix* species range in adult body mass from approximately 280–410 g (see Table 6.1) (Ferrari, 1993; Smith & Jungers, 1997; Garber et al., 2019). *Callibella*, or the dwarf marmoset, is represented by a single species, *Callibella humilis,* which is located in a highly restricted area of south-central Amazonia (Rylands et al., 2009; Augiar & Lacher, 2009). It is described as "slightly larger in size than *Cebuella* (adult head-body length = 160–170 mm; total length = 380–390 mm; weight = 150–185 g), but sharing more physical and behavioral characteristics with *Mico*" (van Roosmalen & van Roosmalen, 2003: p.7). Very little is known about this species in the wild, although it appears to inhabit riparian forests (Aguiar & Lacher, 2009).

The 14 species of *Mico* (formerly assigned to the Argentata group of marmosets of the genus *Callithrix*) weigh between 327–473 g, and are found mainly in "the southern Amazon Basin between the Madeira and the Tocantins rivers and as far south as northeastern Paraguay" (Digby et al., 2011: p.92). In the case of the pygmy marmoset, two species are currently recognized: *Cebuella pygmaea* inhabiting the northern Amazon and *C. niveiventris* inhabiting the southern Amazon (Boubli et al., 2018). These species appear to have diverged some 2.2 mya. *Cebuella* is the smallest living anthropoid, with nonpregnant adult females weighing about 120 g (range from 85–140 g, n = 63; Soini, 1993). Pygmy marmosets are found principally in riparian forests and are widely distributed throughout the upper Amazon region (see distribution maps in Rylands & Mittermeier, 2009, and Buckner et al., 2015).

Table 6.1 Body Mass in Wild-Trapped Adult Tamarins, Callimicos, and Marmosets

Species	Body Mass (g)
Saguinus geoffroyi	486-507
Saguinus labiatus	458-504
Saguinus midas	450-533
Saguinus mystax	512-594
Saguinus oedipus	410-420
Leontocebus fuscicollis	328-333
Leontocebus illigeri	292-296
Leontocebus nigrifrons	354-369
Leontocebus weddelli	315-412
Leontopithecus chrysomelas	586-653
Leontopithecus rosalia	598-620
Callimico goeldii	360-535
Mico argentatus	430
Mico emiliae	327
Mico humeralifer	-
Mico intermedius	473
Callibella humilis	150-175
Cebuella pygmaea	85-140
Callithrix geoffroyi	410
Callithrix jacchus	280-323
Callithrix penicillata	327

Callimico is represented by a single species, *C. goeldii* (Rylands et al., 2009). Although not well documented, callimicos are reported throughout the upper Amazon region of Colombia, possibly Ecuador, Peru, Bolivia, and Brazil (Hershkovitz, 1977; see maps in Rylands & Mittermeier, 2009). However, within this range, its distribution is patchy and discontiguous, and population densities are extremely low. Porter (2006) reports that in most areas surveyed, population density for callimicos averaged 11.96 individuals/km^2 with a range across sites of 1.6–36 individuals/km^2. Given callimicos' widespread but discontinuous distribution, we expect that the population may represent more than one species or subspecies.

All but one set of long-term field studies of callimicos are from a single site in northern Bolivia's Pando region (see Porter, 2001a,b,c, 2004, 2007; Porter & Garber, 2009, 2010; Porter et al., 2009). Based on limited data from the wild, callimicos are reported to weigh approximately 500 g (from 360–535) (Encarnacion & Heymann, 1997; Garber & Leigh, 2001). *Callimico* is distinct from most other marmosets by its exploitation of habitats containing areas of bamboo forest, and a strong reliance on fungi as a dietary staple (some species of lion tamarins, *Leontopithecus caissara*, and marmosets, *Callitrix jacchus* and *C. aurita*, also are reported to consume fungi) (reviewed in Porter & Garber, 2004, 2010; Prado & Valladares-Padua, 2004; Hilario & Ferrari, 2010a).

The 12 species of *Saguinus* differ in size and pelage characteristics, especially in the presence of mustaches, bare faces, and distinctive coloration. The large-bodied tamarins are distributed in Central American (Panama) and South America north of the Amazon, and west of the Madeira River south of the Rio Amazonas. *Saguinus* exhibits a more extensive distribution, especially in northern and eastern Amazonia than does *Leontocebus* (see Buckner et al., 2015, for maps indicating the evolution and distribution of large- and small-bodied tamarins). Mean adult body weights of the large-bodied tamarins range from ~410–594 g (see Table 6.1) (Garber, 1993a; Smith & Jungers, 1997). All species studied to date have a diet that includes insects, small vertebrates, ripe fruits, floral nectar, and exudates (Garber, 1993a,b; Garber & Porter, 2010; Digby et al., 2011).

The present-day distribution of *Leontocebus* or the small-bodied tamarins overlaps extensively with the large-bodied tamarins in several areas of the Amazon. In these areas of overlap, *Saguinus* and *Leontocebus* (sometimes along with callimicos) form mixed species troops (Garber, 1988a,b; Garber et al., 1993b; Heymann & Buchanan-Smith, 2000). The small-bodied tamarins are generally found throughout the western Amazon and as far south as the Pando region of northern Bolivia. Body mass in *Leontocebus* ranges from 292–412 g (Garber & Teaford, 1986; Garber et al., 2016). In addition to body mass, the small-bodied tamarins differ from the large-bodied tamarins in that they tend to travel, feed, and forage more commonly in the forest understory, more frequently leap between vertical trunks, consume a greater proportion of exudates, and use trunks as a foraging platform to scan the ground in search of prey (Garber, 1993a; Garber et al., 2012).

Four species of lion tamarins are currently recognized. Each is designated Endangered (*Leontopithecus rosalia*, *L. chrysomelas*, and *L. chrysopygus*) or Critically Endangered (*L. caissara*) by the International Union for the Conservation of Nature (IUCN) due to their low population sizes and limited geographical distribution (Seuanez et al., 2002). They represent the largest-bodied callitrichine species with body weights of wild adult golden lion tamarins (*Leontopithecus rosalia*) and the golden-headed lion tamarin (*L. chrysomelas*) ranging from 586–653 g (Dietz et al., 1994; Smith & Jungers, 1997; Oliveira et al., 2011). They appear to have increased in body mass over their evolutionary history. Lion tamarins are reported to have originated in the northern Atlantic forests of Brazil, and then dispersed into the southern Atlantic forest (Buckner et al., 2015). Today, lion tamarins are found in three restricted regions of eastern Brazil (IUCN, 2019). Rylands (1996) has argued that the genus *Leontopithecus* evolved in mature forest and exploits a diet principally of insects, small vertebrates, and ripe fruit. Lion tamarins have been described as micromanipulators (Peres, 1986) and forage for insects and small vertebrates in bromeliads, tree holes, and other areas used by prey to refuge. As indicated by Kierulff et al. (2002: p.159), although *Leontopithecus* may have evolved to exploit mature forest habitats, "most studies have been conducted in or near degraded areas," and therefore our understanding of lion tamarin behavioral ecology is limited.

HABITAT SELECTION
The importance of a mix of habitat types and access to exudate-producing trees within the home range of tamarin and especially marmosets, and fungi and bamboo forest in the case of callimicos, has been reported by a number of authors (Terborgh, 1983; Yoneda, 1984a,b; Snowden & Soini, 1988; Ferrari & Lopes Ferrari, 1989; Garber, 1993a; Rylands, 1996; Porter & Garber, 2004; Porter & Garber, 2010). In this regard, Rylands (1996) has argued that each major callitrichine radiation has specialized to exploit a related but alternative set of resources and habitat types. For example, he argues that whereas *Leontopithecus* has

adapted to exploit insects and ripe fruit in mature forest, *Cebuella* feeds on exudates year-round and is found principally in seasonally inundated riverine forest (Rylands, 1996). In contrast, *Callithrix* has adapted to exploit a year-round exudate-tree-gouging and insect-feeding niches in dry, savanna, Cerrado, Caatinga, and highly seasonal habitats in the Atlantic forests of Brazil (Digby & Barreto, 1996; Ferrari et al., 1996; Correa et al., 2000; Rylands et al., 2009). In fact, common marmosets are reported to exploit one of the driest habitats of any primate, the Caatinga, which is a thorn-scrub semidesert with annual rainfall as low as 350 mm, including years with no measurable rainfall (Garber et al., 2019). In the Caatinga, along with exudates, common marmosets commonly exploit fruits, flowers, and the flesh of thorny cacti of the genera *Pilosocereus* and *Cereus* (Abreu et al., 2016; Garber et al., 2019).

The ecology of Amazonian marmosets of the genus *Mico* is less well known, with long-term field data available for only two species, *M. intermedius* (Rylands, 1986a) and *M. argentatus* (Veracini, 2009). The Aripunana marmoset (*M. intermedius*) inhabits areas of secondary and edge vegetation in tropical moist forests and has a diet dominated by fruits, nectar, exudates, and animal prey. The silvery marmoset (*M. argentatus*) is reported to prefer secondary and edge forest and to exploit a diet of exudates, fruits, and insects throughout most of the year (Veracini, 2009). Nectar is consumed during the dry season (Veracini, 2009).

Callimico is distinguished from virtually all other marmosets in its reliance on fungi as a staple year-round resource, but also consumes insects, small vertebrates, exudates, and ripe fruits (see Figure 6.3) (Porter & Garber, 2004, 2010; Garber & Porter, 2010). Callimicos commonly inhabit the understory of the forest. Pook and Pook (1981) reported that 88% of their sightings of *Callimico* were at 5 m or lower. Porter (2001a, 2004) reported similar findings (80% of observations at a height of 5 m or lower). These monkeys also come to the ground to forage for insects and escape on the ground when frightened, and as in the case of the small-bodied tamarins, commonly leap between vertical trunks in the lowest levels of the forest understory (Garber & Leigh, 2001; Garber et al., 2005, 2009; Garber & Porter, 2009; Garber et al., 2012). *Callimico* have the largest home ranges reported for any marmoset taxa (100–150 ha). A thorny bamboo habitat provides food (fungi, insects, small vertebrates) and possibly refuge from predators. One of their main dietary staples, *Ascopolyporous* fungus, is found only in bamboo forest on the internodes of bamboo culms of the species *Guadua weberbaueri*. There are estimated to exist some 180,000 km^2 of aborescent bamboo-dominated

Figure 6.3 Goeldi's marmoset, *Callimico goeldii*.
(Photo credit: Malene Thyssen; GNU Free Documentation License, Creative Commons Attribution Share-Alike 2.5)

forests discontinuously distributed in southwestern Amazonia (Judziewicz et al., 1999; Griscom & Ashton, 2003, 2006). Individual stands of bamboo may extend across an area of only a few hectares or cover thousands of kilometers (Griscom & Ashton, 2006). This patchy distribution is influenced by a variety of factors including soil type (*Guadua* does well in areas of poorly drained soil), and the occurrence of large- and small-scale natural disturbances such as fires and tree-fall gaps (Judziewicz et al., 1999). Thus, although callimico exploit several different types of forests, arborescent bamboo-dominated forests are hypothesized to have played a major role in shaping their behavioral ecology and evolution (Porter et al., 2007).

Saguinus is found to exploit secondary and terra firma forest and consume insects, small vertebrates, fruits, floral nectar, and exudates. In Panama, *Saguinus geoffroyi* occur mainly in low, dense primary forests, gallery forests, and in abandoned, overgrown slash-and-burn agricultural fields (Dawson, 1977; Garber, 1980b). *S. labiatus*, *S. midas*, and *S. mystax* are found both in primary and secondary forests (Mittermeier & van Roosmalen, 1981; Norconk, 1986; Garber, 1988a; Peres, 1992; Garber,

1993a,b; Porter, 2001b). In extensive surveys of Suriname forests, Mittermeier and van Roosmalen (1981) found *S. midas* living in all forest types inhabited by other monkey species. They were most frequently seen in high rainforest, but of the eight species studied, *S. midas* was the only one to prefer edge habitats to non-edge habitats within the forest. In a survey in Guyana, *S. midas* also was found in a variety of habitats, but was more common in secondary forest near human habitation (Sussman & Phillips-Conroy, 1995). As indicated earlier, several species of *Leontocebus* are reported to form stable long-lasting polyspecific associations with other callitrichines, such as *S. imperator*, *S. labiatus*, *S. mystax*, *Callimico goeldii*, as well as *Callithrix emiliae* (Soini, 1987a; Garber, 1988a, 1993b; Rylands, 1993; Lopes & Ferrari, 1994; Porter, 2001a,b,c). In tamarin mixed-species groups, individuals of each species feed, forage, travel, rest, and cooperatively defend a single home range, although associations in some species are more cohesive than in others (e.g., *Leontocebus nigrifrons–Saguinus mystax* mixed-species associations are more cohesive than *L. weddelli–S. imperator* or *L. weddelli–S. labiatus* mixed-species associations). However, when forming a mixed-species group, diet, patterns of habitat utilization, and social organization of *Leontocebus* do not appear to differ compared to groups composed of a single species.

Lion tamarins utilize extremely large home ranges (40–321 ha) and feed opportunistically on plant exudates and floral nectar. However, lion tamarins may rely more on vertebrate prey than do many other callitrichines (Kierulff et al., 2002). In the case of the black-faced lion tamarin (*Leontopithecus caissara*) fungi is reported to account for 12.9% of total feeding time (Kierulff et al., 2002). Lion tamarins exploit highly seasonal as well as aseasonal tropical and subtropical mature rainforests (Coimbra-Filho, 1976; Rylands, 1993, 1996, but see Dietz et al., 1997; Kierulff et al., 2002). In the golden-headed lion tamarin (*L. chrysomelas*) and the golden lion tamarin (*L. rosalia*), epiphytic bromeliads represent an important foraging microhabitat to locate invertebrate and vertebrate prey. The forests occupied by all species, except the black lion tamarin (*L. chrysopygus*), are abundant in bromeliads. *L. chrysopygus* lives in forests that are more seasonal than those of other lion tamarins. They are reported to exploit a home range of between 113–277 ha and appear to feed on exudates (15.2% of feeding observations) more than the other lion tamarin species (Passos & Carvalho, 1991; Rylands, 1993, 1996; Kierulff et al., 2002). In contrast, *L. caissara* is reported to exploit fewer exudates (1.3%) and have the largest home range (321 ha) and greatest mean day range (2,235 m) reported for any lion tamarin species (Kierulff et al., 2002).

Marmosets of the genus *Callithrix* are distributed across eastern Brazil. *C. aurita*, *C. flaviceps*, *C. geoffroyi*, and *C. kuhlii* occupy both aseasonal evergreen forests and the more inland, seasonal, semi-deciduous forests of the Atlantic coast of Brazil. Based on DNA analysis, *C. aurita* appears to represent the most ancestral form (Buckner et al., 2015). *C. jacchus* and *C. penicillata*, which diverged from a common ancestor approximately one mya (Buckner et al., 2015), are found in patches of Atlantic forest, savanna-woodlands (Cerrado), and dry thorn-scrub and semidesert biome (Caatinga) of northeastern Brazil. Cacti are a common plant species in many Caatinga habitats, and common marmosets consume cactus fruits, cactus flowers, and cactus flesh (Figure 6.4) (Amora et al., 2013; Abreu et al., 2016). Unifying features of all marmoset species are their use of edge or secondary growth habitats (Faria, 1986; Rylands, 1987; Stallings, 1988; Ferrari & Mendes, 1991; Rylands & Faria, 1993; Ferrari et al., 1996), relatively small home range (see below), and their ability to gouge holes and exploit plant exudates when ripe fruits and other resources are unavailable. Some populations of *Callithrix kuhlii* vary from this pattern. Raboy et al. (2008) report that their study groups of

Figure 6.4   Ripe fruit of the cactus, *Cereus jamacaru*, consumed by common marmosets in a Caatinga habitat of northeastern Brazil.
(Photo credit: Paul A. Garber)

C. kuhlii consumed more fruit species and fewer exudate species, and exploited larger day ranges and home ranges than generally reported for other marmoset populations.

The natural habitat of *Cebuella* consists principally of riparian and floodplain forests (Ramirez et al., 1977; Izawa, 1979; Soini, 1982). In fact, Soini (1993) considers this species to be a habitat specialist, using mainly the edges and interior of seasonally inundated mature floodplain forests. When *Cebuella* has been reported in mature non-flooded forest (Hernandez-Camacho & Cooper, 1976), it is probably the result of humans having altered the original habitat (Soini, 1982, 1993). *Cebuella* often occupies edge habitat, however, whether they are in seasonally inundated or in mature forests (Izawa, 1976, 1979; Castro & Soini, 1977; Freese et al., 1982; Soini, 1982; Terborgh, 1983).

POSITIONAL BEHAVIOR (POSTURE AND LOCOMOTION)

It was once thought that tamarins and marmosets filled a niche similar to that occupied by squirrels. For example, Eisenberg (1977: p.13) stated: "In many respects, the New World callitrichids appear to occupy ecological niches which are exploited in the Paleotropics by rodents of the family Sciuridae." Szalay and Delson (1979: p.288), in an extensive review of primate evolution, concluded that, "The marmosets lead a squirrel-like life." It is the locomotor behavior, especially, that had been described as squirrel-like (e.g., Bates, 1864; Enders, 1935; Cruz Lima, 1945; Napier & Napier, 1967; Thorington, 1968; Hladik, 1970; Hershkovitz, 1972; Cartmill, 1974; Dawson, 1976). The idea that callitrichines filled niches generally occupied by tree squirrels in Africa and Asia seemed reasonable on first inspection since tree squirrels (sciurids), who migrated from North America, did not reach the tropical forests of Central and South America until approximately three to five mya (Patterson & Pascual, 1972; Webb, 1978), whereas callitrichines diverged from a cebine ancestor some 16 mya. Thus, for the majority of their evolutionary history, tamarins, callimicos, and marmosets occupied forests devoid of arboreal squirrels. In the absence of tree squirrel competitors, it was suggested that callitrichines opportunistically radiated into a vacant tree squirrel niche (Hershkovitz, 1972, 1977; Eisenberg, 1977). However, tamarins, marmosets, and callimicos are not squirrel-like in their diet, movements, or habitat selection. Detailed studies relating positional behavior and feeding ecology have been conducted on several species of calltrichines (Garber, 1980a,b, 1984a, 1991, 1993a, 1998, 2011; Garber & Sussman, 1984; Youlatos, 1999, 2009; Garber & Leigh, 2001; Garber et al., 2005, 2009a, 2012; Garber & Porter, 2009). For example, Garber and

Sussman (1984) collected quantitative data on *Saguinus geoffroyi* and *Sciurus granatensis* (the red-tailed squirrel) inhabiting the same forest in Panama. Later, Garber and colleagues studied the positional behavior and feeding ecology of *Saguinus mystax, Leontocebus nigrifrons, L. weddelli, S. labiatus,* and *Callimico goeldii* in Peru, Brazil, and Bolivia, as well as *Callimico goeldii, Callithrix jacchus,* and *Cebuella pygmaea* in captivity (see references above). The results indicated that tropical sciurids and callitrichines use arboreal supports in a very different manner. Whereas the red-tailed squirrels avoided thin flexible supports, 50% of travel, 66% of fruit feeding, and 85% of insect predation in the Panamanian tamarin took place on supports less than 3 cm in diameter. Unlike squirrels, tamarins did not transport plant foods obtained in the canopy from thin supports to larger horizontal boughs to feed; rather, they consumed fruits where they were acquired (Garber & Sussman, 1984). The Panamanian tamarins ranged through the canopy via a series of long acrobatic leaps that began and ended on thin, fragile, terminal branches. Their claw-like nails did not limit movement on small supports and allowed them to forage and feed on large tree trunks and sharply inclined branches to obtain resources such as bark-refuging insects, vertebrate prey, and plant exudates. Although the tamarins do employ their claw-like nails to ascend and cling to large vertical trunks, they commonly traveled and fed on thin terminal branches and behaved like other primates with grasping extremities, and not like squirrels.

In a related study, Garber and Leigh (2001) examined patterns of positional behavior in sympatric populations of *Saguinus labiatus, Leontocebus weddelli,* and *Callimico goeldii* in northwestern Brazil. They found that *Callimico* occupied lower levels of the forest understory than either *L. weddelli* or *S. labiatus.* Moreover, although leaping accounted for 38–45% of the positional repertoire for each species, in callimicos 55% were described as trunk-to-trunk leaping with relatively large noncompliant vertical supports serving as both take-off and landing platforms. Trunk-to-trunk leaping accounted for only 8.4% of leaps in *S. labiatus* and 20% of leaps in *L. weddelli* (Garber & Leigh, 2001). Moreover, in tamarins the vast majority of trunk-to-trunk leaps were estimated to be <1 m in distance, whereas in callimicos approximately 33% of trunk-to-trunk leaps exceeded 1 m and 13% exceeded 2 m in distance. In a related set of studies of positional behavior conducted in the Pando region of northern Bolivia, Garber and Porter (2009) found that the mean distance leaped between vertical trunks by *C. goeldii* was 1.8 m (range = 0.2-4.3), with 9% exceeding a horizontal distance of 3 m. In the case of *L. weddelli* (Garber et al., 2012), the mean distance leaped between vertical trunks

was 1.4 m and only 2% of leaps exceeded a horizontal distance of 3 m. Finally, in a series of experimental studies of trunk-to-trunk leaping in captive *Callimico goeldii, Cebuella pygmaea,* and *Callithrix jacchus* (Garber et al., 2005, 2009), data indicate that whereas common marmosets and pygmy marmosets lost height during trunk-to-trunk leaps of 1 m, callimicos gain height during these leaps. This is consistent with the fact that callimicos have elongated hindlimbs compared to all other callitrichine taxa and, along with some species of small-bodied tamarins, are the only callitrichine species reported to habitually employ trunk-to-trunk leaping during travel. Interestingly, unlike vertical clinging and leaping in prosimians, when callitrichines leap from trunk to trunk, they always land forelimbs first and not hindlimbs first (Garber et al., 2005, 2009).

Quantitative data on positional behavior in wild Amazonian or eastern Brazilian marmosets, except for the pygmy marmoset, *Cebuella pygmaea,* are extremely limited. Given their greater reliance on plant exudates, marmosets are more likely to frequently adopt vertical clinging postures and spend more time exploiting trunks than tamarins. *Cebuella* was initially described by Kinzey et al. (1975) as a vertical clinger and leaper, although it is clear that whereas they spend large amounts of time clinging to vertical trunks ("claw" clinging) and ascending vertical trunks during travel (scansorial locomotion or "claw" climbing), only 12.9% of travel involved trunk-to-trunk leaping (Youltaos, 2009). In fact, data collected by Youlatos (2009: p.284) indicate that "terminal leaping," which is described as leaping "with the body held mainly horizontal or inclined," accounted for almost twice as many leaps in pygmy marmosets than did trunk-to-trunk leaping. Like *Saguinus, Cebuella* utilized its claw-like nails primarily for clinging to large vertical supports while gouging and feeding and for scansorial locomotion (23.7%) during travel (Youlatos, 1999, 2009; Garber, 2011).

It is important to point out that in many callitrichines, vertical clinging is best thought of as a feeding/foraging posture that is separate from activities associated with leaping. The primary exceptions are *Callimico* and *Leontocebus,* both of which preferentially inhabit the understory, cling to trunks when feeding on exudates, fungi, and during prey foraging, but also engage in frequent trunk-to-trunk leaping. In *Saguinus,* leaping principally occurs in the canopy and on small, flexible supports, and not from one vertical trunk to another (Garber & Leigh, 2001). When trunk-to-trunk leaping occurs in callitrichines, it tends to involve travel in the lowest layers of the forest undercanopy (e.g., *Cebuella, Leontocebus, Callimico,* and possibly *Saguinus bicolor;* Egler, 1992).

DIET AND FEEDING ECOLOGY

Marmosets, tamarins, and callimicos feed on five primary types of food items: arthropods/small vertebrates, ripe fruits, floral nectar, plant exudates, and fungi; however, they utilize these items at different frequencies. Because of their nutritional content, seasonal availability, and distribution, each of these foods requires different foraging strategies and techniques of exploitation. The ranging patterns and positional behaviors required for obtaining these food types also differ. The callitrichines represent an adaptive array of trunk foragers (Garber, 1992). Trunks represent an important foraging microhabitat for several groups of insect, fungi, and exudate feeders such as woodpeckers, woodcreepers, certain phalangerid marsupials, and several primates. Given that the taxa most derived in terms of small body size, dental morphology, and digestive morphology (both *Callithrix jacchus* and *Cebuella* have enlarged and complex hindguts) are specialized exudate feeders, tree-gouging and obligate exudate feeding are unlikely to represent the ancestral condition for callitrichines. Rather, the initial callitrichine feeding niche may best be represented in terms of obligate ripe fruit, arthropod, and small vertebrate eating, as well as opportunistic exploitation of exudates and arthropod/small vertebrate predation on trunks (Sussman & Kinzey, 1984; Garber, 1992). This is the pattern most common in *Saguinus* and *Leontocebus*. In *Callibella*, *Callithrix*, and *Cebuella*, the evolution of derived dental and masticatory specializations and independent body-size reduction events may be associated with a dependence on exudates as a dietary staple.

Insect and Vertebrate Prey

The proportion of insects or leaves in the diet of primates is generally related to body size (Charles-Dominique, 1971; Kay, 1975). Most small primates are able to obtain the majority of their protein and lipid needs from arthropods and small vertebrates, whereas larger primates exploit leaves to fulfill some or most of their protein requirement. Marmosets, callimicos, and tamarins are the smallest living anthropoids, and a large proportion of their diet consists of animal prey (though this may be less so in some of the smaller, exudate-feeding marmosets like *Cebuella*). Approximately 30–80% of the feeding and foraging time is accounted for by arthropods (Dawson, 1976, 1979; Ramirez et al., 1977; Garber, 1980a,b, 1984b, 1993a,b; Terborgh, 1983; Snowdon & Soini, 1988; Stevenson & Rylands, 1988; Rylands & de Faria, 1993; Correa et al., 2000; Martin & Setz, 2000; Porter, 2001a,b; Kierulff et al., 2002; Knogge & Heymann, 2003; Porter et al., 2009).

In many tropical forests the leaf litter and structurally diverse vegetation of the understory includes the greatest diversity and abundance of insects (Janzen, 1973a,b). Many marmosets, callimicos, and tamarins forage for insects in the dense tangle of branches and vines in the shrub layer of the forest, glean insects off leaves in the mid-to-upper canopy, search bromeliad whorls, and as indicated previously, forage for arthropods on tree trunks. The small size of callitrichines enables them to move about relatively quietly on terminal branches in the canopy. In this regard, Garber (1993a) identified three patterns of arthropod foraging among *Saguinus* and *Leontocebus*: (1) active hunting on small supports, (2) visual scanning of exposed prey using a stealthy type of locomotion on medium-sized branches, and (3) microhabitat-specific foraging on large vertical supports or within bromeliad whorls and tank plants (this is common in lion tamarins) using visual or tactile scanning. Examples follow:

S. geoffroyi, using the first foraging pattern, hunts and captures insects by moving cautiously but rapidly on thin flexible supports in the dense vegetation some 1–5 m above the ground (Garber, 1980a, 1984b). Climbing, grasping, and jumping accounted for 81% of the positional repertoire (postural and locomotor behaviors) during foraging. Success is related to the ability of the monkey to move cryptically and with minimal disturbance in the understory. Prey are captured from leaf surfaces by striking rapidly with the forelimbs while maintaining a firm grip with the hindlimbs.

The second mode of foraging is seen in *S. mystax*, *S. labiatus*, *S. imperator*, and *S. midas* (Terborgh, 1983, 1985; Garber, 1988a, 1993b), and in several species of *Callithrix* (Rylands & Faria, 1993). In these callitrichines, visual scanning plays a primary role in detecting exposed and mobile prey. They search for insects in living and dead leaves, tree cavities, and on branches while moving slowly and close to the substrate. This involves a stealthy, stalk-and-pounce, foliage-gleaning method in the understory and middle levels of the forest. A large percentage of foraging time is spent in more sedentary postures such as walking, standing, and sitting.

The third insect-foraging technique is most common in the small-bodied tamarins (*Leontocebus*), *Callimico goeldii*, and possibly *Saguinus bicolor*. As well as searching for insects in the dense underbrush using the second technique described above, in all areas where the small-bodied tamarins have been studied, they also forage for trunk-refuging animal prey as well as use trunks to scan the ground and nearby vegetation for prey (Garber, 1992, 1993a).

The small-bodied tamarins commonly embed their elongated and laterally compressed claw-like nails into the tree bark, and manually explore knotholes, crevices, and other regions of the trunk. This is facilitated by their elongated forelimbs, fingers, and narrow hands (Bicca-Marques, 1999), analogous in certain ways to the hand morphology in *Leontopithecus*. Trunks also provide a scanning platform for locating terrestrial prey. The use of trunks to prey on insects ecologically distinguishes *Leontocebus* from most *Saguinus* and represents one of the mechanisms facilitating the formation of mixed-species troops and minimizing competition for animal prey. In fact, the small-bodied tamarins are reported to feed on a different spectrum of insect prey compared to sympatric large-bodied tamarins (their plant-based diet is virtually identical; Garber, 1993b). For example, of 35 prey species eaten by mixed-species troops of *Leontocebus nigrifrons* and *Saguinus mystax* in Peru, 25 were captured by the smaller tamarin species and 7 by the larger tamarin species. Only three arthropod species were shared by both tamarin species (Nickle & Heymann, 1996). Moreover, a study using high-throughput sequencing to identify insect DNA present in the feces of wild *L. weddelli* in Bolivia found that whereas orthopterans (grasshoppers, walking sticks, and crickets) and blattids (cockroaches) accounted for the greatest percent relative abundance of prey, other arthropods including spiders, springtails, ants, mantids, flies, wasps, moths, and crickets also were consumed (Mallott et al., 2015).

In the case of callimico (Porter, 2001a,b, 2007; Porter et al., 2007; Porter & Garber, 2009), arthropods were found to account for a greater percentage of feeding time compared to sympatric *L. weddelli* or *S. labiatus*. During a yearlong study, arthropods comprised 10–50% of monthly feeding/foraging time. As in many other tamarin and marmoset species, large othopterans were the most common prey item consumed (41% of all insect prey; Porter, 2001b). Insect foraging occurred principally in the understory on thin flexible branches, in leaf litter, and directly on the ground. Porter reports that callimico was never observed to search tree holes and crevices as occurs in small-bodied tamarins and lion tamarins.

The lion tamarins, *Leontopithecus rosalia*, *L. caissara*, *L. chrysopygus*, and *L. chrysomelas* have been observed to use tree trunks and vertical clinging postures to procure arthropods. However, these species are best described as manipulative foragers that exploit prey concealed under bark and dead leaves, among vine tangles and in bamboo thickets, in the leaf litter, and hidden inside epiphytic bromeliads and palm fronds (Peres, 1986; Rylands, 1986a, 1993; Garber, 1992; Dietz et al., 1997; Kierulff et al., 2002; Raboy & Dietz, 2004). Lion tamarins use their long and slender arms and elongated and webbed fingers to probe and extract largely nonmobile prey (Bicca-Marques, 1999). According to Kierulff et al. (2002: p.173), "*Leontopithecus rosalia* captures sedentary and cryptic prey, such as adult othopterans and larvae of Coleoptera and Lepidotpera." *L. chrysopygus*, forages in a very similar manner. At one study site (Morro do Diablo State Park, Brazil) tree cavities (41%), palm crowns (22%), bamboo (12%), and vines (11%) were the most common foraging substrates (Kierulff et al., 2002). In *L. caissara*, however, bromeliads, bark crevices, and palm crowns were the primary sites of insect/veterbrate prey foraging. In *L. chrysomelas*, although epiphytic bromeliads represented the most common prey-foraging microhabitat (average 76.5% of substrates), tree holes and bark crevices (19%), palm fronds (15%), and dried leaves (3%) also were searched (Raboy & Dietz, 2004).

Some species of *Callithrix* (*C. flaviceps*, *C. geoffroyi*, and *C. kuhlii*) and *Mico* (*M. intermedius*) take advantage of insects disturbed by swarming army ants and are reported to spend up to three hours at a time within 2 m of the ground stealing prey from the ants or picking off arthropods disturbed by the swarm (Stevenson & Rylands, 1988; Rylands & de Faria, 1993). Prey located near the ground are underutilized by many New World primates (Heltne et al., 1981), but as mentioned above, some species of marmosets and tamarins (especially *Leontocebus* and *Callithrix jacchus*), as well as callimicos, hunt for arthropod and vertebrate prey on or near the ground.

Orthoptera, especially large grasshoppers, are reported to be the most common prey item of callitrichines. Orthopterans constituted 70% of the volume of prey material in stomach contents of *Saguinus geoffroyi* collected by Dawson (1976), and represented about 60% of the insect-feeding bouts of *S. imperator* and *Leontocebus weddelli* observed by Terborgh (1983) in Peru. Izawa (1978) considered large grasshoppers to be the single most important food item of *L. nigricollis* in Colombia. The most frequently observed insect prey for *Cebuella* (Soini, 1982), *Callithrix flaviceps*, *C. kuhlii*, *Saguinus mystax*, *Leontocebus nigrifrons*, and *Mico intermedius* were grasshoppers, including both the families Acrididae and Tettigoniidae (Stevenson & Rylands, 1988; Ferrari, 1993; Nickle & Heymann, 1996). *Callimico* also has been observed eating large grasshoppers (Porter, 2001a,c). Similarly, Maier et al. (1982) reported that *Callithrix jacchus* consumed animal prey such as locusts, grasshoppers, beetles, and lizards. In addition, observations of group hunting to acquire relatively large lizards has been reported in *C. jacchus* inhabiting the Caatinga (Garber et al., 2019). Other prey items consumed by callitrichines include Lepidoptera (larval and adult moths and butterflies), Coleoptera (beetles), ants, spiders, snails, frogs, nesting birds, and bird eggs. Despite their small body size, tamarins, marmosets, and callimicos commonly feed on insects that are larger in body size than those consumed by many larger-bodied South American monkeys (Terborgh, 1983; Nickle & Heymann, 1996). Small group size (4–10 individuals) and small body size may provide advantages to callitrichine co-foragers in locating both sedentary and flushed prey.

However, it is unlikely that body size and group size alone provides the full explanation. For example, when tamarins forage in mixed-species associations, troop size can exceed 20 individuals and orthopterans continue to be the most common prey consumed (Nickle & Heymann, 1996). Moreover, Peres (1991b, 1992) has suggested that one of the advantages to tamarins in forming a mixed-species troop is that flushing or disturbing prey increases capture rates. He provides data that brown-mantled tamarins (*Leontocebus fuscicollis avilapiresi*) benefit from locating prey flushed by mustached tamarins (*S. mystax*). Peres (1992) estimated that 66–73% of prey biomass consumed by brown-mantled tamarins came from prey flushed by mustached tamarins. Squirrel monkeys (*Saimiri* spp.) also are small but forage in large groups of 40–70 monkeys. There is no evidence, however, that individuals benefit during insect foraging by the prey-flushing actions of other group members. That said, squirrel monkeys occasionally form mixed-species troops with tufted capuchins (genus *Sapajus*) and it has been suggested that during joint foraging, one or both taxa may benefit from increased opportunities to locate new feeding sites or capture flushed animal prey (Terborgh, 1983). Detailed knowledge of prey behavior and microhabitat preference along with information on callitrichine positional behavior, capture techniques, manual dexterity, and forager nutritional and energetic requirements are needed before species differences in predatory behavior among callitrichines are fully understood.

In some tamarin species, the timing and location of foraging activities may be strongly influenced by the activity pattern of its prey. Dawson (1979) suggests that the prey-searching activities of *Saguinus geoffroyi* coincides with the period of greatest vulnerability of large orthopteran prey. Because of their lower surface-to-volume ratio, larger insects warm at a slower rate than do smaller ones. The tamarins he studied tended to search for prey in the late morning when large insects began to emerge from their nocturnal resting places but were not fully mobile. Although data were not collected on temporal differences in hunting success, it is possible that factors such as temperature, humidity, and time of day, along with the exploration of particular microhabitats, affect insect capture success.

Seasonality also may influence the choice of large insects as prey (Pack et al., 1999). The abundance and diversity of insects, as well as fruits, flowers, plant exudates, and leaves can fluctuate significantly throughout the year, especially in response to temporal changes in rainfall. Janzen and Schoener (1968) and Janzen (1973a,b) found that large-bodied insects were better able to withstand the rigors of the dry season than were smaller forms. Large orthopterans, coleopterans and homopterans were the three insect families that remained relatively abundant during the dry season. Dawson (1976) found that orthopterans (78%) and coleopterans (9%), along with lizards (10%), made up 97% of the animal prey captured by *S. geoffroyi* during the dry season. In the case of *Callimico*, based on sweep samples, Porter (2001a,b) reported an increase in orthopteran availability in the months of December and January (rainy season), and a decrease in July (dry season).

Fruits, Flowers, and Nectar

Fruits, flowers, and nectar contain a high proportion of nonstructural carbohydrates and sugars. Certain oily fruits also may contain significant amounts of lipids. Although these resources may differ in seasonal availability and renewal rates, they are treated together in this discussion because they are exploited by callitrichines using similar foraging strategies. Marmosets, callimicos, and tamarins eat ripe fruits, but tamarins and lion tamarins appear to include more fruit in their diet than do many marmosets and callimicos.

Estimates of the proportion of fruit and floral nectar in the diet of *Saguinus* vary from about 40–86% (Hladik & Hladik, 1969; Dawson, 1976; Garber, 1980a,b, 1993a; Mittermeier & van Roosmalen, 1981; see Digby et al., 2011). In *Leontopithecus rosalia* (Peres, 1986; Dietz et al., 1997), *L. chrysomelas* (Rylands, 1993), *L. chrysopygus* (Valladares-Padua, 1993), and *L. caissara* (Kierulff et al., 2002), ripe fruits, floral nectar, and flowers account for 75–90% of feeding observations. The fruits and flowers utilized, as with the insects exploited, are generally dispersed and of low density. Most of the fruits are found scattered on vines and bushes or patchily distributed in the periphery of tree crowns (Fleagle & Mittermeier, 1980; Yoneda, 1981; Terborgh, 1983; Crandlemire-Sacco, 1986; Garber, 1993a; Dietz et al., 1997).

Additionally, many fruiting species consumed by tamarins are reported to ripen in a piecemeal fashion (Terborgh, 1983). Thus, only a small quantity of ripe fruit is available on the tree at any one time. In addition, many of the tree and liana species exploited by large- and small-bodied tamarins are characterized by intraspecific synchrony (Garber, 1988b, 1993a), with most or all trees of the same species bearing fruit or flowers at the same time. This adds an element of predictability to the tamarin resource base. In addition, although fruit production per tree per day was limited, trees often fruited over a period of several weeks. During the dry season, fruit-feeding continued but was supplemented with either exudates or floral nectar (Garber, 1988b, 1993b). Terborgh (1983) suggests that fruit production in tree species commonly utilized by tamarins was quite low and not normally sufficient to satisfy large groups of capuchins or squirrel monkeys. One indication of limited fruit production is the fact that a mixed-species tamarin troop will visit an average of 12–14 trees each day, in addition to exploiting animal prey in order to obtain a nutritionally balanced diet (Garber, 1993b). In both Panama (*Saguinus*) and Peru (*Saguinus* and *Leontocebus*), species of tamarins were observed to feed on many species of plants throughout the year, but they also showed a propensity to concentrate on a small set of favored species over the course of several days or weeks (Garber, 1993b; Porter & Garber, 2013). According to Terborgh (1983), in southern Peru, *S. imperator* and *L. weddelli* harvested identical plant resources while moving in mixed-species groups, even though the two species searched independently for insects in different microhabitats (see also Peres, 1996; Garber, 1993a). During the course of a year, *L. weddelli* and *S. imperator* relied on only five to seven plant species as major resources to supply the bulk of their carbohydrate and caloric needs (Terborgh, 1983).

All species of *Saguinus* and *Leontocebus* studied to date swallow large seeds (1–2 cm in length) covered by a thin aril. These seeds pass through their digestive tract unharmed (Terborgh, 1983; Garber, 1986, 1994b; Garber & Kitron, 1997, 2013), are typically voided singly, and are dispersed 150–300 m from the parent tree. On average, seeds swallowed and voided by both small- and large-bodied tamarins are larger in size than seeds swallowed by baboons, forest guenons, macaques, and even chimpanzees (Garber, 1986, 1994b). Garber and Kitron (1997, 2013) have proposed that seed swallowing in tamarins serves an important curative function in mechanically expelling spiny-headed worms (*Acanthocephala*) from their digestive tract. These parasites can attach to the gut lining and cause inflammation, lesions, and death. Exposure and re-exposure to these parasites may result from consuming orthopteran prey, which serve as intermediate hosts. Daily swallowing and voiding of seeds may serve to dislodge intestinal parasites prior to or during the initial stages of attachment (Garber & Kitron, 1997, 2013). In terms of their ecological impact on the forest, tamarins act as important seed dispersers for a number of plant species (Garber, 1986).

In contrast, seed swallowing is less common in marmosets. In addition, the amount of fruit eaten by different species of marmosets is inversely related to the proportion of exudates in their diet. Initial studies indicate that *Mico intermedius* and *Callithrix kuhlii* are highly frugivorous whereas *C. penicillata*, *C. jacchus*, and *C. flaviceps* are more highly exudativorous with a limited amount of fruit in their diet. *Mico argentatus* was somewhat intermediate (59% exudates and 36% fruit) (Digby et al., 2011). Rylands (1993) points out two factors that may affect the proportion of fruit in the diet of various marmoset species and populations. First, in all long-term studies, it appears that fruit-eating is highly seasonal and eaten where and when it is available. Second, given the renewal rates of many plant exudates, gouging holes in the morning requires that individuals return to the same site later that day to feed. Marmosets have a relatively small day range (approximately 1,000 m), and therefore after gouging, group members may remain in the general area of excavated exudates sites, limiting their access to more widely distributed resources such as ripe fruit. *Callithrix* marmosets are characterized by a more derived dental gouge (lower incisor/canine dental arrangement) than *Mico* (Amazonian marmosets), and therefore may rely more on year-round exudate feeding or more effectively exploit exudates when ripe fruits are in limited supply.

The pygmy marmoset is an extreme exudate-feeding specialist, and fruit and other plant parts form a more minor portion of its diet (Ramirez et al., 1977). Yepez et al. (2005) found that among four study groups, exudate feeding/gouging accounted for 20–34% of the total activity budget. The only activity that occurred more frequently than exudate feeding, was resting. As is reported for *Saguinus*, *Leontocebus*, and *Leontopithecus*, floral nectar, when available, also is consumed by *Cebuella* (Terborgh, 1983) and by *Mico humeralifer*, *Callithrix kuhlii*, *C. jacchus*, and *C. flaviceps* (Stevenson & Rylands, 1988; Ferrari & Strier, 1992). At the beginning of the dry season, 25% of feeding observations on *Mico humeralifer* were on flowers of the vine *Mendoncia aspera*. The floral nectar of *Symphonia globulifera* (Clusiacae) was an important food source for *Callithrix kuhlii* and *Mico argentatus* (Veracini, 2009), as it is for several species of *Saguinus*, *Leontocebus*, and *Leontopithecus* (Peres, 1986, 1989b; Garber, 1988b; Raboy & Dietz, 2004; Porter & Garber, 2009), and nectar of *Mabea fistulifera* was consumed by *C. flaviceps* during a period of relative fruit scarcity.

Ripe fruits account for approximately 25–30% of yearly feeding time for *Callimico* (Porter, 2001b,c; Porter et al., 2009). Porter argues that callimicos exploit a very small set of fruit species that are not generally abundant in the forest. Fruit feeding for callimico is seasonal with fruits accounting for 73% of total feeding time in the wet season month of February but only 3% of feeding time in the month of May (the beginning of the dry season; Porter 2001a,b). Callimicos tend to consume fruits located in the understory or low levels of the canopy. In the case of some large emergent trees, callimicos only consume fruits or pods that have fallen to the ground (Porter, 2001a,b; Porter et al., 2009). In the wet season, fruits commonly account for over 50% of callimico's total feeding time (Porter et al., 2009).

Exudates

Plant exudates is a general term that includes saps, gums (saps and gums are water soluble), and resins (resins are alcohol soluble) and is a resource fed upon by a number of primate species including humans, Old World monkeys, apes, and prosimians (Bearder & Martin, 1980; Nash, 1986; Porter et al., 2009; Power, 2010). Resins are generally considered toxic. Saps are typically composed of simple sugars and other nutrients and are easily digested. Gums, which are a form of dietary fiber, are composed of beta-linked complex polysaccharides and "require fermentation by gut microbes before nutrients are available to animals that feed on it" (Power, 2010: p.29). All species of callitrichines studied have been observed to feed on exudates (Garber, 1992; Kierulff et al., 2002; Porter et al., 2009; Garber & Porter, 2010). Callitrichines feed on three main types of exudates; trunk/branch exudates, pod exudates, and root exudates (Porter et al., 2009; Garber & Porter, 2010). In fact, callitrichines can be described as an adaptive radiation of exudate-feeding primates with derived behavioral and anatomical traits that enable the efficient exploitation of this resource (Garber, 1992).

Garber (1980b, 1984b) proposed that plant exudates are a nutritionally important resource for tamarins. During his study of the Panamanian tamarin, exudates accounted for 14% of total feeding time. The tamarins fed primarily on the exudates of two tree species, *Anacardium excelsum* (Figure 6.5) and *Spondias mombin*, both members of the Anacardeaceae, an important family of tropical exudate-producing trees (Howes, 1949). A nutritional analysis of *A. excelsum* exudate indicated the presence of water, protein, and several minerals, in particular, calcium. The calcium/phosphorus content of *A. excelsum* exudate ranged from 31:1 to 142:1. A high calcium/phosphorus ratio appears to characterize other plant exudates as well (Bearder & Martin, 1980; Garber, 1993a; Garber et al., 2009). Given that many insects consumed by the tamarins are characterized by a low calcium and high phosphorus content, Garber (1980a,b) proposed that in order to avoid an inadequate mineral balance, it may be necessary for insectivorous primates to include a resource

Figure 6.5 A Panamanian tamarin (*Saguinus geoffroyi*) clinging to the trunk of *Anacardium excelsum* while feeding on exudates. Note how the body of the tamarin blends into the dark and light color patterns of the trunk.
(Photo credit: Paul A. Garber)

in their diet that is high in calcium and low in phosphorus. Plant exudates may serve such a function for callitrichines. This has been suggested for exudate-feeding in galagos as well (Bearder & Martin, 1980). Given that callitrichine females produce twin infants, and have the potential to produce four offspring in a 12-month period, obtaining an appropriate balance of macronutrients and minerals is essential for fetal development and reproductive success.

The four species of *Leontopithecus* are reported to be opportunistic exudate-feeders and, like *Saguinus and Leontocebus*, consume exudates mainly during the dry season when fruit availability is low (Peres, 1986, 1989a,b; Passos & Carvalho, 1991; Rylands, 1993; Kierluff et al., 2002). *Leontopithecus rosalia* has been reported to chew bark to stimulate exudate flow (Peres, 1989a), although this needs to be confirmed by additional studies. In most cases this may not be gouging, but rather pulling or scraping hardened exudate from the bark. Furthermore, enlarged mandibular and maxillary incisors in *L. chrysomelas* may reflect some incisor harvesting adaptations, especially for wood-boring insects (Rosenberger & Coimbra-Filho, 1984), but gouging behavior to elicit exudate flow has not been observed in any other species of lion tamarin (Carvalho & Carvalho, 1989; Rylands, 1989, 1993).

Small- and large-bodied tamarins, *Callimico goeldii*, and *Leontopithecus chrysomelas* consume the liquid-like exudate found in the pods of emergent trees of the genus *Parkia*, in particular *P. nitida* and *P. velutina* (Rylands, 1993; Peres, 2000a,b; Raboy & Dietz, 2004; Porter et al., 2009; Garber & Porter, 2010). Pod exudates are water soluble and composed principally of nonstructural carbohydrates. In this regard they are more like nectar or sap than gum, and more easily broken down in the gut (Peres, 2000a; Power, 2010). Porter et al. (2009) report that in callimicos, pod exudates accounted for 19% of total feeding time during the dry season. Similarly, Garber and Porter (2010) report that pod exudates were an important dry season resource for mixed-species troops of *Leontocebus nigrifrons* and *Saguinus mystax* in Peru, accounting for 35.2% of total plant feeding time during the month of June. Callimicos also exploit exudates produced from natural injuries to the stilt roots of palm species such as *Iriartea deltoidea* and *Euterpe precatoria*. Root exudates were consumed principally in the dry season and are estimated to contain as much metabolizable energy as trunk exudates consumed by *Cebuella*, *Saguinus*, and *Leontocebus* (Porter et al., 2009).

Virtually nothing is known concerning the feeding ecology of the dwarf marmoset, *Callibella humilis*. However, based on jaw and skull morphology, Aguiar and Lacher (2010: p.373) predict that "*Callibella* is likewise specialized for intensive exudate-gouging, but perhaps to a lesser degree than *Cebuella*." For all marmoset species, plant exudates are not simply a dietary supplement but represent a major source of food. Species differences in the degree of exudate-feeding appears to be related to the degree of development of the dental gouge (lower incisors and canines) and the availability and distribution of ripe fruit (Maier et al., 1982; Natori, 1986, 1990; Natori & Shigehara, 1992; Kinzey, 1997b). Given the ability of marmosets to gouge holes to elicit the flow of exudates, this is a dependable food resource for *Mico*, *Callibella*, *Cebuella*, and *Callithrix*. The trees utilized for exudate feeding by marmosets are normally riddled with gouged holes, which are usually round or transversely elongated (Figure 6.6). It is important to point out that given callitrichine evolutionary relationships, many similarities in exudate feeding morphology between *Callithrix* and *Cebuella* are best understood as the result of parallel adaptations. Both common marmosets (*Callithrix jacchus*) and pygmy marmosets (*Cebuella pygmaea*) have a large

Figure 6.6   A common tree of the Caatinga biome of northeastern Brazil (*Prosopis juliflora*, Fabaceae) with exudate holes gouged by common marmosets.
(Photo credit: Paul A. Garber)

complex cecum and internal ribbon-like structures (tae-niae) which may facilitate more complete digestion of the complex carbohydrates contained in some plant exudates (Coimbra-Filho et al., 1980; Ferrari, 1992; Garber, 1992). Gut adaptations that characterize common marmosets may be related to the fact that their geographical distribution includes extremely dry savanna (Cerrado) and desert-like habitats (Caatinga) of low primary productivity. In these habitats, exudates may represent a dependable year-round food resource (Garber et al., 2019).

In *Cebuella*, the proportion of time spent feeding on insects and exudates shifts throughout the day. For example, exudate holes may be excavated just before nightfall, possibly to maximize exudate flow during the night, so that it can be harvested immediately in the morning (Ramirez et al., 1977). In the case of tamarins, Heymann and Smith (1999) found that exudate feeding was more extensive during the afternoon than in the morning. They argued that this resulted in an increase in the amount of time (overnight) exudates remained in the gut and facilitated increased microbial fermentation. In contrast, Porter et al. (2009) found that callimicos consumed pod exudates more frequently in the afternoon than expected. Given that pod exudates should be more like sap or nectar than gum in their nutritional properties, this is not consistent with the predictions of Heymann and Smith (1999). Thus, it appears that several factors, including the location of sleeping and exudate sites, the requirements of nutrient balancing, the availability of other food types, as well as species-specific patterns of exudate production in tropical trees, affect the timing of exudate feeding in callitrichines (Porter et al., 2009).

ACTIVITY PATTERNS

Terborgh (1983) found that the activity pattern of both *Leontocebus weddelli* and *Saguinus imperator* varied seasonally. At Manu National Park in southeastern Peru, the wet season corresponded to a period of high food availability and the birth of infants. Soon after birth, the infants were carried, mainly by adult males, adding 20% or more to the weight of the carrier. During this time, group travel was reduced, rest increased (60% of the time), and the group restricted its activities to a small portion of its home range. During the dry season, when floral nectar was a principal source of simple carbohydrates (e.g., in July), the tamarins visited hundreds of flowers to satisfy themselves, and the proportion of time spent foraging increased. When ripe fruits were available, they were harvested at a much higher rate and foraging and feeding time was reduced. Garber (1993b) and Tovar (1994) also observed seasonal differences in activity patterns, especially foraging and feeding behavior, in large and small

tamarin species. Overall, however, mixed-species groups of *S. mystax* and *L. nigrifrons* were active approximately 10 hours and 20 minutes per day (Garber, 1993b), with between 30–40% of each day's activity pattern associated with grooming and resting. Members of the mixed-species troop traveled almost 2 km per day and visited 12–14 feeding trees. Travel to feeding sites appeared to be goal directed, and often the tamarins traveled to the nearest tree of a target tree species (Garber, 1989). Overall, there is evidence that both small- and large-bodied tamarins retain detailed spatial information regarding the availability and distribution of a large number of potential feeding trees in their home range and employ route-based travel and landmark cues to efficiently navigate and monitor resources in large- and small-scale spaces (Garber, 1989; Garber & Hannon, 1993; Garber & Dolins, 1996; Garber, 2000; Porter & Garber, 2013; Garber & Porter, 2014).

Peres (1991b) found that mixed-species troops of *Saguinus mystax* and *Leontocebus fuscicollis avilapiresi* averaged 9 hours and 12 minutes of activity per day. Approximately 22% of the day was spent resting and engaging in intragroup social behavior, with the small tamarins devoting more time to resting than did the large tamarins. The small tamarins also traveled more in the early morning and late afternoon and rested more during midday (Peres, 1991b). Tamarins in this mixed-species troop engaged in 18.9 plant feeding bouts per day. In the case of mixed-species troops of *L. weddeli* and *S. imperator*, Terborgh (1983) found that resting accounted for 44% of the daily activity budget of the small tamarin species (Terborgh, 1983), which was considerably higher than for the large-bodied tamarin.

Lion tamarins also are reported to travel more during early morning and late afternoon and spend more time resting or stationary during midday (Kierulff et al., 2002). Plant-feeding peaks are reported to occur between 0700–0800 hours and at 1500 hours. Kierulff et al. (2002) report that the majority of insect foraging in *Leontopithecus chrysopygus* occurred between 900 hours and 1300 hours. Raboy and Dietz (2004) found that golden-headed tamarins (*L. chrysomelas*) devote 30–35% of their day to travel and rest only 5–10% of the day. The majority of resting occurred between 1100 hours and 1300 hours. *Callithrix* and *Cebuella*, like tamarins, generally arise shortly after dawn (Soini, 1988; Stevenson & Rylands, 1988). Activity cycles vary between species and populations; however, resting can account for a significant portion of the day in some populations. For example, in *Callithrix jacchus* resting can account for between 15–53% of daily activity (Stevenson & Rylands, 1988; Garber et al., 2019). Other *Callithrix* populations have been observed to spend approximately 30% or less of the day resting, and seasonal differences in activity patterns have been reported in a number of *Callithrix* studies (see Passamani, 1998; De la Fuente et al., 2014). As Digby and Barreto (1996) report, *C. jacchus* rested approximately 30% of the day when groups contained no dependent infants but rested almost 60% of the day when carrying infants. Thus, reproductive factors associated with infant care may play an important role in callitrichine activity budgets.

Although callitrichines are reported to live in small, cohesive social groups, the activities of individuals in a *Cebuella* group appear to be less coordinated (Ramirez et al., 1977; Soini, 1988). "While some individuals were eating exudate, others would hunt insects and others would rest in the crown of a tree, so that at any one time the troop members were dispersed, often over a large portion of their (very small) range" (Ramirez et al., 1977: p.99). Given that pygmy marmosets have home ranges that are often less than 0.5 ha (range = 0.1–1.2 ha; Yepez et al., 2005), dispersed individuals can easily maintain visual and/or vocal contact with the remainder of the group. *Cebuella* also is reported to devote between 40–45% of its activity budget to rest, which may be required to digest and process beta-linked complex structural carbohydrates found in many plant exudates (Yepez et al., 2005).

Sleeping trees used by Panamanian tamarins were generally higher than surrounding vegetation and/or lacked extensive vine connections with other trees. *Saguinus* and *Leontocebus* rarely return to the same sleeping tree on consecutive nights (Snowden & Soini, 1988; Garber, pers.obs.). *Leontocebus weddelli* are reported to frequently sleep in holes in trees as has been reported in *Leontopithecus* (Coimbra-Filho, 1977; Rylands, 1982; Peres, 1986; Carvalho & Carvalho, 1989; Kierulff et al., 2002), and because of this, lion tamarins may be limited to forests where tree holes are available (Rylands, 1993). Golden lion tamarins are reported to leave their sleeping trees 15–75 minutes after sunrise and return to their sleeping roost 30 minutes to 2 hours before sunset (Peres, 1986). The roosting habits of tamarins and marmosets may represent an anti-predator a strategy.

Like the small- and large-bodied tamarins, lion tamarins, callimicos, *Callithrix*, *Mico*, and *Cebuella* enter sleeping trees while it is still light. Sleeping sites are often dense, vine-covered vegetation (Rylands, 1981; Maier et al., 1982; Soini, 1988; Stevenson & Rylands, 1988) or, more rarely, tree holes (Moynihan, 1976a; Izawa, 1979; Rylands, 1982). *Callithrix jacchus* and *Cebuella* groups have been observed using the same sleeping sites continuously or repeatedly (Soini, 1988; Stevenson & Rylands, 1988), whereas *Mico humeralifer* and *Callithrix kuhlii* did not sleep in the same site on consecutive nights (Rylands, 1982).

In a comparative study of the behavior and ecology of mixed-species troops of *Callimico, Leontocebus weddelli,* and *Saguinus labiatus,* Porter (2007) reports that callimicos tended to leave their sleeping sites earlier than did tamarins and entered their sleeping sites a bit later. Callimico sleeping sites were principally tangles of vines and leaves (Porter, 2007). Callimicos, like lion tamarins, exploit very large home ranges. However, in contrast to *Leontopithecus, Leontocebus,* and *Saguinus*, day ranges in callimicos are generally smaller and more similar to *Callithrix* and *Mico*. Porter et al. (2007) and Porter and Garber (2010) report that despite exploiting a home range of 114–150 ha, day range for callimicos was only 925 m². Observations indicate that callimicos are either not territorial or rarely encounter neighboring groups (groups generally have discontiguous ranges with large areas with no groups). Daily range use appears to involve the intensive use of core areas over periods of days or weeks followed by movement to a new part of their range (Porter & Garber, 2010). Porter (2007) reports that callimicos devoted more of their activity budget to resting and less of their activity budget to traveling than did sympatric tamarins.

Finally, Porter and Garber (2010) describe two cases in which habituated groups of callimicos systematically lost membership over the course of one to two weeks and finally, with only the breeding female remaining, abandoned their range. In one case, a member of the habituated group returned to the area several years later (Porter & Garber, 2010). *Cebuella* is reported to abandon its range with the death or decrease in productivity of major exudate trees. In the case of callimicos, it remains unclear what set of factors may lead to home range abandonment.

PREDATION

Because of their small size, callitrichines are potentially vulnerable to a large number of predators. Terrestrial predators that can prey on callitrichines include the tayra (*Eira barbara*, a mustelid) and smaller felids such as the ocelot (*Felis pardalis*), the jaguarundi (*F. yagouroundi*), and the margay (*F. wiedii*) (Soini, 1988). The tayra has been observed carrying a dead tamarin (Moynihan, 1970), and a number of predation attempts by tayra have been reported (Galef et al., 1976; Izawa, 1978; Snowdon & Soini, 1988). Emmons (1987) identified the remains of *Leontocebus* in ocelot scats and Dietz (cited in Kinzey, 1997b) observed an ocelot kill a reintroduced lion tamarin. Dietz also witnessed lion tamarins being chased by capuchin monkeys. Capuchins are aggressive toward callitrichines in some areas where both taxa overlap, whereas in other areas capuchins and marmosets interact without aggression or fear (Rylands, 1982; Muskin, 1984).

Callitrichine species differ in their response to predators. Callimicos are extremely cryptic and often become silent, freeze, or descend to the ground when encountering a potential predator. Other callitrichines, including small-bodied tamarins, may mob terrestrial carnivores, and give calls similar to the vocalizations they use toward dogs and unfamiliar humans (Neyman, 1977; Izawa, 1978; Rylands, 1981; Stevenson & Rylands, 1988; Passamani, 1995). At least some species are reported to utter different alarm calls for terrestrial and aerial predators (Epple, 1975a; Neyman, 1977), and some react differently toward fecal scents of predatory and nonpredatory mammals (Caine & Weldon, 1989).

Large arboreal snakes also are predators of callitrichines. Both Heymann (1987) and Dietz (in Kinzey, 1997b) have observed snakes approach, capture, kill, and eat adult tamarins. In the first case, an anaconda preyed upon a mustached tamarin crossing low over a river, and in the second, a boa constrictor captured a lion tamarin. Bartecki and Heymann (1987) witnessed a group of *Leontocebus nigrifrons* mobbing a snake.

Birds of prey are undoubtedly the most important predators of callitrichines (Moynihan, 1970; Ferrari & Lopes Ferrari, 1990; Caine, 1993; Ferrari, 1993), and a limited number of attempted attacks and successful attacks have been reported (Dawson, 1976; Neyman, 1977; Izawa, 1978; Terborgh, 1983; Goldizen, 1987; Soini, 1988; Heymann, 1990a,b; Caine, 1993). Goldizen (1987) saw an average of one raptor attack per week for each mixed *Saguinus imperator–Leontocebus weddelli* troop she studied at Manu National Park in Peru, and Heymann (1990b) noted alarm events by mixed-species groups of *S. mystax–L. nigrifrons* every two to three hours, over half of which were directed toward avian predators. Izawa (1978) witnessed six raptor attacks on *L. nigricollis*, two of which were successful. Tamarins give specific, high-pitched alarm calls and then initiate escape behavior or become silent and immobile upon the appearance of most large flying objects including low-flying aircraft. Garber (pers.comm.) has observed mustached tamarins repeatedly chase large nonpredatory birds perched above them in exposed areas of the tree crown. Anti-predator behaviors such as freezing, dropping low in the canopy, alarm calls, rapid flight, crypticity, and the careful selection of sleeping sites are common. When moving to a sleeping tree, the animals move rapidly and silently and, if disturbed, soon after reaching the tree, will choose another sleeping site. Once the group has entered a nighttime sleeping tree, the monkeys huddle together in a dense tangle of leaves or vines, or in tree holes. There also is increased vigilance (Caine, 1987, 1993).

Being the smallest anthropoids, the pygmy marmosets are potential prey for an especially large number of predators. Moynihan (1976b) and Soini (1988) suggest that *Cebuella* anti-predator behaviors are designed to avoid attracting attention rather than distracting the predator. They do not mob predators as some tamarins do, and they move exceedingly slowly or in quick spurts and then freeze. Their coloration is highly cryptic and they almost always move in the dark, dense underbrush below the canopy.

It also has been suggested that in some callitrichine species, a particular group member may play an important role in sentinel behavior or predator vigilance, especially during feeding and traveling (Goldizen, 1987; Zullo & Caine, 1988; Caine, 1993). Vigilance and sentinel behavior, alarm calling, and mobbing are common cooperative behaviors within both single-species and mixed-species tamarin troops (Heymann, 1990a; Castro, 1991, 1993; Peres, 1993; Savage et al., 1996b; Buchanan-Smith & Hardie, 1997). Caine (1993) believes that predation is one of the major selective pressures leading to many aspects of the unique social system of the callitrichines. For example, not all adults in tamarin and marmoset groups breed (although many adults may be involved in copulatory behavior); thus, adding a small number of additional group members may not have immediate costs in increasing reproductive competition but will provide immediate benefits in increasing the number of infant caretakers and individuals engaged in predator vigilance. In addition, where two or three callitrichine species coexist (except for *Cebuella*), they often form mixed-species troops.

Callimicos are subject to many of the same predators as are other marmosets and tamarins (Pook & Pook, 1981). They are, perhaps, the most cryptic of the callitrichines. Callimicos travel close to, and occasionally even on, the ground; they move silently, and they may rest or remain stationary and silent for extended periods of time as an anti-predation strategy. Moynihan (1976a) states that even indigenous hunters often are unaware of groups of callimicos inhabiting local areas of forest. Their black coloration, dark brown eyes, and silent movement within the shadows of the forest understory, in tree falls, and on the ground make callimicos a challenge for scientific study. If the group is disturbed while in a tree, individuals climb down to a height of 1–4 m and move away using a series of trunk-to-trunk leaps, or flee on the ground. Groups have been observed to mob ground predators, such as snakes or tayras. Callimicos sleep and rest in dense vegetation and sometimes in bamboo thickets. They form polyspecific associations with small- and large-bodied tamarins, which may reduce the risk of predation (Pook & Pook, 1982; Porter, 2001a, 2007). One unusual aspect of callimico-tamarin mixed-species troops is that the same callimico group may form such an association with up to six different tamarin groups (Porter, 2007). Moreover, when callimicos are part of a mixed-species group, there is evidence of niche expansion (Porter & Garber, 2007). When alone, callimicos were found to travel farther distances, rest less, and exploit more fungi, less fruit, and fewer insects than when part of a mixed-species troop (Porter & Garber, 2007).

Several authors have argued that one of the advantages to primates of forming a mixed-species troop is enhanced predator vigilance and protection through increasing the number of eyes and ears focused on the environment. Additional evidence in support of a predator detection advantage is the fact that species that form polyspecific associations (polyspecific associations are the same as mixed-species troops) are characterized by vertical height separation, with different species primarily inhabiting different levels of the forest canopy. This may offer benefits in increasing the likelihood of detecting ground, arboreal, and aerial predators. In the case of tamarin mixed-species troops, *Leontocebus* inhabits the forest understory and the other tamarin species (i.e., *Saguinus mystax, S. imperator, S. labiatus*) inhabit the upper canopy. In the case of callimico-tamarin mixed-species troops, callimicos inhabit the lowest levels of the forest understory (<5 m in height, and on or near the ground), the small-bodied tamarins exploit the understory and low levels of the canopy, and the large-bodied tamarin species exploits the mid-upper canopy. In this way, the different tamarin species, potentially, can be alert for and alarm-call in the presence of different types of predators. However, in a review on predation risk and evidence of predator-sensitive foraging in single- and mixed-species tamarin troops, Garber and Bicca-Marques (2002) found no support for the contention that when part of a single-species group, tamarins were more predator sensitive than when the same individuals were part of a mixed-species troop. In addition, neither predator attacks, rates of alarm calling, nor successful predation events were higher in single-species tamarin groups compared to mixed-species troops. Thus, a shared vigilance model may not fully explain the primary benefit to some or all callitrichines of forming a mixed-species troop (Garber & Bicca-Marques, 2002).

SOCIAL STRUCTURE AND BREEDING SOVEREIGNTY

Marmosets, callimicos, and tamarins live in relatively small groups. Based on censuses of a number of populations, group size ranges from 1–19 individuals (Table 6.2). Long-term studies suggest that the smaller and larger number of tamarins are temporary phenomena: average group size is approximately 5–8 for *Saguinus, Leontocebus,* and *Cebuella;* 4–7 for *Leontopithecus;* 6–15 for *Mico;* 6–14 for *Callithrix;* 6–8 for *Callibella;* and 4–7 for *Callimico* (Digby et al., 2011). Despite the fact that only a single female in each group generally produces offspring, in the majority of cases groups contain multiple adult males and multiple adult females (Garber et al., 2016). Hilario and Ferrari (2010b) report that one group of *Callithrix flaviceps* contained between 2–5 adult males and as many as 6 adult females (4 of which gave birth within a 2-month period).

Despite recent field studies to the contrary, callitrichines have long been regarded by some as being strictly monogamous, socially monogamous, or pair-bonded (Kleiman, 1977b; Leutenegger, 1980; Redican & Taub, 1981; Evans, 1983; Evans & Poole, 1983). For example, Redican and Taub (1981: p.208) state that, "marmosets and tamarins are known to be organized in monogamous family units." Others have considered marmosets and tamarins to live in extended family groups in which a monogamous pair allows adult offspring to remain with the group after they mature (Eisenberg et al., 1972; Epple, 1975b; Moynihan, 1976a). Assumptions of monogamy as the modal breeding system for callitrichines were based, in large part, on the facts that sexual dimorphism in body mass is absent or minimal, that in captivity same-sex adults are often aggressive to each other, and that regardless of the number of adult female present in a captive group, only a single female breeds. This has resulted in the exclusive study of family groups in the laboratory, despite evidence from field studies that have contradicted this view of callitrichine social and breeding systems. In most captive breeding colonies of *Saguinus geoffroyi, S. oedipus, Leontocebus weddelli, Leontopithecus rosalia, Callithrix jacchus, Cebuella pygmaea,* and *Callimico goeldii* (Dummond, 1971; Rothe, 1975; Epple, 1975a, 1976; Hearn, 1977; Abbott & Hearn, 1978), individuals are housed as family groups with only one adult female and one adult male. Older offspring of both sexes are removed from the group as they reach sexual maturity. In captivity, subdominant adult females (*Callithrix jacchus*) do not exhibit hormonal changes indicative of an ovarian cycle (Hearn, 1977), although in some contexts daughters may not be suppressed to the same degree as unrelated females (Saltzman et al., 1997). Reproduction suppression may include both behavioral and hormonal mechanisms. Digby et al. (2011) reviewed patterns of reproductive suppression/inhibition in captive callitrichines. They indicate that in captive lion tamarins and callimicos, adult daughters living in family groups exhibit normal signs of ovulation. In captive *Leontopithecus*

Table 6.2 Group Size in Tamarins, Callimicos, and Marmosets

Species	Group Size		Reference
	(Range)	(Mean)	
Saguinus geoffroyi	5-10	7	Garber, 1980a
Saguinus imperator	2-8	4	Terborgh, 1983
Saguinus labiatus	-	6.6	Sussman and Kinzey, 1984
Saguinus labiatus	2-10	6.1	Puertas et al., 1995
Saguinus midas	-	3.4	Thorington, 1968
Saguinus midas	-	6	Mittermeier and van Roosmalen, 1981
Saguinus mystax	4-11	7	Garber et al., 1993a
Saguinus mystax	3-12	5.5	Heymann, 1998
Leontocebus nigricollis	-	6.3	Izawa, 1978
Leontocebus nigrifrons	-	5.7	Sussman and Kinzey, 1984
Leontocebus weddelli	2-10	-	Digby et al., 2011
Leontocebus weddelli	4-8	5.8	Goldizen, 1989
Leontopithecus caissara	4-7	7	Baker et al., 2002
Leontopithecus chrysomelas	5-8	6.7	Baker et al., 2002
Leontopithecus chrysomelas	3-9	5	Baker et al., 2002
Leontopithecus. chrysopygus	3-6	4.3	Baker et al., 2002
Leontopithecus. chrysopygus	2-7	3.6	Baker et al., 2002
Leontopithecus. chrysopygus	3-7	4.7	Baker et al., 2002
Leontopithecus rosalia	2-11	5.4	Baker et al., 2002
Callimico goeldii	-	4	Porter and Garber, 2010
Callimico goeldii	-	5.5	Porter and Garber, 2010
Callimico goeldii	7-9	-	Rehg, 2009
Mico intermedius	9-15	-	Digby et al., 2011
Mico humeralifer	4-13	10	Stevenson and Rylands, 1988
Mico argentatus	6-10	-	Veracini, 2009
Cebuella pygmaea	2-9	0.4	Soini, 1982
Callithrix aurita	6-11	-	Ferrari et al., 1996
Callithrix flaviceps	11-15	-	Ferrari et al., 1996
Callithrix geoffroyi			Digby et al., 2011
Callithrix jacchus [a]	2-16	8.3	Garber et al., 2019
Callithrix jacchus [b]			Garber et al., 2019
Callithrix kuhlii	2-6	4.2	Raboy et al., 2008
Callithrix penicillata	3-16	-	Digby et al., 2011
Callithrix penicillata	4-9	6.7	Stevenson and Rylands, 1988

[a] Common marmosets inhabiting the Atlantic Forest of northeastern Brazil.
[b] Common marmosets inhabiting a Caatinga biome in northeastern Brazil.

rosalia, by 16 months of age, both adult daughters and non-kin subordinate females exhibit normal ovarian cycles (French et al., 2002). In contrast, in *Saguinus*, adult daughters are usually anovulatory. In marmosets "captive *Callithrix* females are intermediate, with up to 50% or more of eldest daughters ovulating while living in their natal families" (Digby et al., 2011: p.101).

As a mating system, monogamy is defined as an exclusive breeding relationship between one male and one female that extends over several reproductive periods (Kleiman, 1977b). In order to demonstrate monogamy, genetic data on paternity are required. As a social system, monogamy is characterized as either a nuclear family or an extended family. In both cases, the majority of group members are expected to be closely related. Field data on the composition of callitrichine groups are not consistent with the expectations of a monogamous social organization, and groups containing a single adult male and a single adult female are extremely rare. For example, Dawson (1976, 1977), Neyman (1977), Izawa (1978), Rylands (1981), Garber et al. (1984), Soini (1988), Stevenson and Rylands (1988), Goldizen (1990), Baker et al. (1993), Garber (1993b), Digby (1994), Goldizen et al. (1996), Oliveira and Ferrari (2008), Porter (2001c), and Garber et al. (2016) have all found that wild groups commonly contain more than one adult female and more than one adult male. In the absence of data on paternity, however, it remains unclear how often a single male sires the dizygotic twin offspring, how often the twins each have different fathers, and how often a second female is impregnated but then aborts. As discussed below, in some groups of *Callithrix jacchus* (Digby 1994), *C. flaviceps* (Ferrari & Digby, 1996; Hilario & Ferrari, 2010b), *Callimico goeldii* (Porter & Garber, 2004), *Leontopithecus rosalia* (Dietz & Baker, 1993), *L. chrysomelas* (Raboy & Dietz, 2004), *Saguinus oedipus* (Savage et al., 1996a,b), and *Leontocebus weddelli* (Goldizen et al., 1996), two females have been observed to successfully give birth in the same group during the same breeding season. In the case of wild *Leontopithecus rosalia*, polygyny or groups with two simultaneously breeding females is reported in 10% of social groups and in most cases the two breeding females were mother and adult daughter (Dietz & Baker, 1993).

In addition, there is evidence from field studies in which individual animals were identified that more than one resident adult male may mate with the group's breeding female(s), and that extra-group copulations occur (Ferrari & Digby, 1996). Despite difficulties in determining paternity, multiple sires of the same litter have been confirmed in *Saguinus geoffroyi*, *S. mystax*, and *S. labiatus* (Huck et al., 2005; Suarez, 2007; Díaz-Muñoz, 2011). Given the possibility that two different breeding males can each sire one of the dizygotic twin offspring along with field data that, depending on the species, 5–35% of groups may contain two or more actively breeding females (reviewed in Garber et al., 2016), we argue that a concept of monogamy does not offer the best framework from which to understand callitrichine social and mating systems. Rather, we describe tamarins, marmosets, and callimicos as exhibiting a nonmonogamous single female breeding system in which females may mate with several adult males and there is no evidence of a pair-bond between a single male and a single female (Garber et al., 2016).

Several studies of callitrichines have identified the immigration and emigration of adults into and out of established groups. In Colombia, Neyman (1977) identified 22 adult cotton-top tamarins known to have entered or left groups, including both males and females, and all had dental wear indicating they were not young adults. Judging by nipple condition, at least two females had previously given birth. Savage et al. (1996a) report that their study groups of *S. oedipus* were somewhat more stable; nevertheless, immigration and emigration of adults of both sexes were common. Males tended to enter groups in the context of death or migration of one or more resident males. Adult females tended to enter groups that already contained a breeding female. The immigrating female generally did not usurp the breeding sovereignty of the resident breeding female. These authors state that "wild groups [of cotton-top tamarins] contained individuals of various ages and both sexes that were unrelated to the reproductively active animals in the group" (Savage et al., 1996a: p.94). Izawa (1978) also observed changes in group composition in *Leontocebus nigricollis* in Colombia, though his animals were not individually marked.

In a four-year study in Peru, Terborgh and Goldizen (1985) found that groups of marked *Leontocebus weddelli* only occasionally contained a single pair of adults. During a 37-month period when group compositions were known, a group had one adult female and one adult male 27% of the time, one adult female and more than one adult male 61% of the time, more than one adult of each sex 8% of the time, and adults of only one sex 4% of the time. Immigration and emigration of adult males and females occurred commonly, with adults transferring into or out of all seven intensively studied groups. In four two-male groups in which copulations were observed, both males copulated. In three of these groups, two males were observed mating at about equal frequencies with a single female. In the fourth instance, both adult males copulated with the two adult females in the group. Some of the observed copulations occurred in view of the group's other male without provoking any sign of aggression. In addition, one group was observed with two parous females and with two litters of young, suggesting that both females had reproduced within the group.

Goldizen et al. (1996) present 13 years of information on patterns of mating and reproduction in Weddelli's saddle-back tamarins (*Leontocebus weddelli*). Their data indicate that males and females tend to migrate/disappear from their natal group between 2.5–3.5 years of age. Most documented successful migrations involve transferring to neighboring groups. Males tend to migrate into groups already containing other adult females, whereas females tend to migrate into groups that did not currently contain a breeding female. Several adult females remained in their natal group and inherited the sovereign breeding position.

Mustached tamarins (*Saguinus mystax*) are characterized by a similar pattern of mating and migration. Garber et al. (1993a) monitored group size and migration patterns in *S. mystax* inhabiting an island in northeastern Peru. The tamarins were not native to the island; intact tamarin groups were trapped and relocated to this area in 1978–1980. In a follow-up study conducted some 10 years later, 91 individuals residing in 13 social groups were retrapped, examined, and marked. Group size averaged seven, and groups normally contained more than one adult male and one adult female. The greatest number of adult males in any group was three and the greatest number of adult females was four. Based on dental wear, there was evidence of what was called age-related female reproductive sovereignty. In all groups studied, the oldest female in the group was the breeding female. In groups containing nine or more individuals, two females were reproductively active (i.e., two pregnant, or one pregnant and one lactating); however, in none of these cases did both females produce offspring (Garber et al., 1993a).

It is likely that behavioral and physiological mechanisms similar to ovulatory suppression or stress-associated abortion may occur in subordinate females. Mechanisms of reproductive suppression and inhibition are known to vary between callitrichine species (e.g., ovulation in subordinate female *Leontopithecus rosalia* and *Callimico goeldii* does not appear to be physiologically suppressed), and also may vary depending on relatedness, familiarity (Abbott et al., 1993; Saltzman et al., 1997), or environmental factors (Savage et al., 1997). In golden lion tamarins, common marmosets, and occasionally in small-bodied tamarins, when two females in the same group are reproductively active, they are reported to be mother and daughter. However, French et al. (2003) report that in wild golden lion tamarins, adult daughters residing in their natal group may fail to ovulate and experience periods of ovulatory insufficiency. The fact that female lion tamarin reproductive biology is characterized by a set of traits that are distinct from other callitrichines ("shorter ovarian cycle, considerably shorter gestation, a higher rate of infertile postpartum ovulation, and a markedly seasonal pattern of gestation"; French et al., 2002: p.155), coupled with evidence of rapid male weight gain during the breeding period, is a reminder that despite the appearance of adaptive unity, individual callitrichine taxa have evolved alternative social and mating strategies. Other examples include evidence of seasonal changes in male testis size (*Saguinus mystax*) and significant differences in testis size among adult males residing in the same group (*Leontocebus nigrifrons* and *S. mystax*; Garber et al., 1996b).

In tamarins, it appears that patterns of migration take several forms. Although individuals are known to migrate singly, this may be more common in females than in males. In mustached tamarins and golden lion tamarins, for example, paired male migration is known to occur and, in at least some cases, these males were related (Garber et al., 1993a; Garber, 1994a; Baker et al., 2002). Also, in *Leontopithecus rosalia*, most successful immigration events occurred in the context of replacing a same-sex adult (Baker et al., 2002). That is, the immigration event either followed or closely preceded the death or emigration of a reproductively active group member. Moreover, in Weddell's saddleback tamarins, females are reported to migrate once, whereas secondary dispersal

appears to occur more frequently in adult males. Tamarins also become part of a new group through the process of group fissioning. In the wild, group size can increase rapidly, but groups larger than 13 or more animals are generally unstable. Garber (1988a) reports that his study group of 13 mustached tamarins split into groups of 8 and 5 immediately after the birth of a new set of twin infants. The group of 8 plus 2 infants remained in the home range, and the group of 5, which consisted of both adult males and females, migrated together out of the area. In common marmosets (*Callithrix jacchus*) inhabiting a Caatinga biome, groups never were found to exceed nine individuals, and once they reach that size they may disband or split into smaller groups (Garber et al., 2019).

Huck et al. (2005) examined kinship and paternity in eight groups of mustached tamarins in northeastern Peru. They found that although all or almost all adult male group members mated with the breeding female, and on occasion females engaged in extra-group copulations, a single resident male commonly sired the group's twin infants. In this population, individuals residing in smaller groups were less closely related to each other (R = 0.06) than individuals residing in intermediate sized (R = 0.49) or larger groups (R = 0.30). Overall, mean relatedness among group members was relatively high (R = 0.31); however, male group members were not more closely related to each other than were female group members. Similarly, a study by Garber et al. (2016) found that in *Leontocebus weddelli*, dispersal was bisexual and that "relatedness among adult females of the same groups averaged 0.31 and among adult males was 0.26" (Garber et al., 2016: p.298).

Kleiman, Dietz, Baker, and colleagues collected extensive data on migration and mating patterns of *Leontopithecus rosalia*. Their data on 17 habituated groups studied over a seven-year period (Dietz et al., 1994; Baker & Dietz, 1996) indicate that groups typically contain one reproductive female and one to two adult males (70% of groups contained more than one adult male, including adult male offspring of the breeding female, and 40% of groups contained at least two adult non-natal males; Baker et al., 2002). Approximately 10% of groups included two breeding females. In *Leontopithecus rosalia* (as is the case for *Saguinus mystax*; Garber et al., 1993a), there is evidence of paired male migration. And, although the absence of genetic information assigning kinship to individuals based on behavior and movement patterns is problematic, in many of these groups it is possible that some of these male pairs are related (father-son, full or half sibs). In two-male groups, both males may copulate with the breeding female, although generally there is a strong dominance hierarchy, suggesting that one male is more likely to sire the group's offspring.

Preliminary paternity data for golden lion tamarins provide evidence that extra-group breeding is more common than generally assumed, based solely on behavioral observations. Seuanez et al. (2002: p.131) state that "in at least 5 out of 152 cases (3 percent) in which genotype data for all resident males, resident females, and offspring are available... it appears that intragroup males could not have sired the offspring." These authors also have identified four cases in which group offspring did not share genetic information with their "presumed" mother, suggesting the possibility of adoption. Overall, Baker et al. (2002) suggest that in *L. rosalia*, groups are principally kin-based and contain one breeding female, one or more adult and potentially breeding male, plus unrelated adults and offspring. Additional behavioral and genetic data are required to more fully understand golden lion tamarin mating strategies and social organization. Less is known about the mating and social systems of the three other lion tamarin species.

At their study site, Baker and Dietz (1996) indicate that the number of immigrating individuals per group per year in *L. rosalia* averaged 0.48 (range 0–1.15). Although these authors argue that this is a low rate of immigration, an alternative explanation is possible. Golden lion tamarin groups contain approximately three adults, two or three of which contribute genes to the next litter. Field data indicate that a group's breeding female generally produces only two offspring per year (one twin litter). If on average there is a change in the adult composition of a group by 0.5 each year, then (0.5/3.5 adults) 14% of the potential gene pool of that group changes each year. If the immigrant becomes a breeding animal (male or female) then 50% of the gene pool may be changed. Thus, in species characterized by a limited number of breeding animals per group, even a small change in the number of resident adults can have an important impact on individual reproductive success and gene frequencies in the local population.

Long-term data on marmoset social and mating systems are available for *Callithrix jacchus, C. flaviceps,* and *Mico humeralifer.* Female marmosets tend to give birth to two sets of twins per year, and group sizes of 6–16 individuals are common. Based on observations of copulations among group members, the mating system of these marmoset species can be described as polyandrous or polygynous (i.e., groups with more than one breeding female). Polygynous groups of *Callithrix* have been studied by Digby (1994), Digby and Baretto (1996), Mendes Pontes and Monteiro da Cruz (1995), Nievergelt et al. (2000), Yamamoto et al. (2009), and Hilario and Ferrari (2010b).

In the case of *C. jacchus* observed by Digby and Baretto (1996) in the Atlantic Forest of northeastern Brazil, all three study groups contained two reproductively active females, although the dominant female typically had the highest reproductive success (Digby et al., 2011). In several cases, the second reproductively active female was the adult daughter of the breeding female. Similarly, Yamamoto et al. (2009) found that 3 of 14 (21%) common marmoset groups inhabiting the Atlantic Forest contained multiple breeding females. In *Callithrix aurita* and *C. flaviceps,* two or more females also have been observed giving birth simultaneously in the same group (Rylands, 1986b; Digby & Ferrari, 1994; Ferrari et al., 1996; Hilario & Ferrari, 2010b; Digby et al., 2011). However, studies of *C. jacchus* provide evidence of six cases of infanticide by adult female group members, and in five of those cases, the observed/suspected killer was the dominant female (Digby et al., 2011). In the case of *C. jacchus,* infanticide represents a type of female breeding competition, which is analogous to hormonal and behavioral reproductive suppression, and serves to limit female reproductive opportunities and breeding success.

In a genetic study of three *Callithrix jacchus* groups by Nievergelt et al. (2000), the social system was described as extended families and composed of closely related individuals (R > 0.5) as well as unrelated individuals (R = 0). However, they reported that the "relatedness between adult males from the same groups is significantly lower than between adult females" (Nievergelt et al., 2000: p.11). Genetic data also indicated that several groups contained currently breeding males who were recent immigrants into these groups. Although the behaviorally dominant male was assumed to sire most of the infants in his group, there was evidence of changes in adult-male group membership and female promiscuity, in the form of extra-group copulations with males from neighboring groups (24% of all copulations; Digby, 1999). This pattern is not consistent with the expectations of a monogamous mating pattern. In a second study of kinship in *C. jacchus,* Faulkes et al. (2003) examined 59 individuals from nine groups in two geographically separated populations. They found that adult group members were represented by several different haplotypes and genetically more "heterogeneous than the extended family model that has been suggested previously" (Faulkes et al., 2003: p.1107). In one group composed of nine individuals, Faulkes et al. (2009) identified five individuals from different matrilines (Faulkes et al., 2009). Overall, it appears that in many tamarin and marmoset species, groups are composed of both related and unrelated adults.

Pygmy marmosets exhibit a social system similar to that reported in other callitrichines. Group size in *Cebuella* averages 3–8 individuals, including 1.4 adult males and 1.4 adult females (Soini, 1987, 1988; Yepez et al., 2005). Adults of both sexes are often aggressive to same-sex conspecifics. This results in relatively early natal migration of subadult and young adult offspring. Two-thirds of groups studied by Soini (1988) contained only one fully adult male and one fully adult female, although groups with as many as three adult males have been documented (Garber, pers.comm.). Dominant males are reported to form a consort with a breeding female and guard them from resident group and extra-group males. Subordinate males have been observed to copulate with the breeding female, but as has been reported in lion tamarins, based on behavioral observations, it is assumed that the dominant male sires most offspring. The degree to which particular males or females form an exclusive long-term breeding relationship and the length of reproductive sovereignty remain unknown. Migrations of adult male and adult female pygmy marmosets are reported to be more common than in *Callithrix*.

Like *Callithrix*, female *Cebuella* commonly give birth to two twin litters per year. Although subadult daughters and unrelated males aid in infant care, pygmy marmosets are unusual in that distance traveled per day is extremely short (<100 m) and infants are often left unattended or "parked" while adults forage and excavate exudate holes (Soini, 1988). It is likely that the costs and benefits of cooperative infant care differ among callitrichine taxa, and this plays a major role in determining group size, composition, and social interactions. For example, in some groups of *Callithrix* and *Callimico,* mothers have been found to serve as primary caretakers or co-equal infant caretakers with males (Digby et al., 2011).

Callimico exhibits a pattern of mating similar in many ways to other callitrichines despite giving birth to a single infant. Long-term studies of callimicos in the wild in Bolivia by Porter and colleagues (2001a,b,c, 2004, 2007; Porter & Garber, 2004, 2009) indicate that callimicos live in groups of four to nine individuals. Given that twinning appears to be have been lost in *Callimico* after it diverged from a common ancestor with other marmosets, the appearance of two offspring and two nursing females in some wild groups suggests a breeding system in which multiple adult females may simultaneously produce offspring (Porter & Garber, 2004). As in the case of other callitrichines, adult male group members provide cooperative care of the young. In a detailed study of patterns of infant care in one wild group of callimicos, Porter and Garber (2009) reported that two males mated with the group's single adult female, and both males and the infant's mother were in close proximity to, carried, and shared food with the infant.

In reviewing callitrichine breeding systems, Garber et al. (1993a, 2016) have suggested that the ancestral mating system of tamarins, marmosets, and callimicos is small multi-male/multi-female groups. Polyandry in many species is likely to result from high levels of female reproductive competition and the ability of a dominant female to suppress/inhibit breeding in subordinate females. Not only does the dominant female attempt to control the reproductive opportunities of subordinate females but in doing so, she also controls male breeding opportunities by reducing the number of reproductively active females in the population. However, there is some indication that the breeding female may have a very limited reproductive tenure (2–4 years; Goldizen et al., 1996; Garber, 1997) and, therefore, waiting to inherit the reproductive position of the breeding female may be an effective reproductive strategy for female callitrichines. Cooperative caregiving (see discussion below) in callitrichines plays a major role in increasing the individual reproductive success of both males and females.

In many tamarin and marmoset species, although the breeding female is dominant to other group members, rates of within-group aggression are extremely low (Garber, 1997; Ferrari, 2009; Porter & Garber, 2009; Garber et al. 2009b; De la Fuente et al., 2019). In a recent paper, Yamamoto et al. (2009) presented two models to explain female mating competition in common marmosets: the Optimal Skew Model and the Incomplete Control Model. The Optimal Skew Model suggests that the dominant female controls the breeding opportunities of other group females and only allows subordinates to breed when it benefits the dominant (e.g., there are more than enough helpers in the group to care for her infants). The Incomplete Control Model suggests that under certain social and ecological conditions (e.g., large number of females in the group, habitat with high resource productivity), the dominant female is unable to control the reproductive activity of subordinate adult females. These authors concluded that field observations of common marmosets suggest that mating flexibility is most consistent with the inability of a dominant female to fully control the breeding activities of subordinate females, i.e., the Incomplete Control Model.

Reproductive Output

Many New World monkeys are characterized by a relatively slow reproductive output and have offspring every two or three years (Garber & Leigh, 1997). This occurs in squirrel monkeys, capuchin monkeys, spider monkeys, sakis, woolly monkeys, and woolly spider monkeys and is not closely associated with body size. Exceptions to this slow reproductive output are callitrichines, who can have two twin litters per year, and female titi monkeys (*Callicebus*) and night monkeys (*Aotus*) who are reported to give birth to a single offspring once every year (howler monkeys also are relatively fast breeders and have an interbirth interval of approximately 18 months). In the case of *Aotus* and *Callicebus*, the single adult male and single adult female in the group are reported to form a pair-bond and are considered to exhibit obligate monogamy or serial monogamy (Fernandez-Duque, 2011). In both taxa, extra-maternal infant care by the group's lone adult male reduces the cost of reproduction to the mother, allowing her to more rapidly redirect energy into her next reproductive event. In this regard, Garber and Leigh (1997) suggest that *Aotus* and *Callicebus* evolved monogamy independently, and in each taxa this social and breeding system serves to increase the reproductive output of pair-bonded males and females. Callitrichines also have evolved a specialized social and mating system designed to increase reproductive output (maximum of four offspring per year) and reduce the postnatal costs to the mother of infant care. Cooperative infant care in tamarins, callimicos, and marmosets represents a system in which adult males, adult females, kin, and non-kin help raise offspring born into their social group (Garber, 1997).

Cooperative infant care has been described in several avian and mammalian species and provides an explanation as to why adults, especially males, help raise offspring they may not have sired. Stacey (1979, 1982) describes a social system in acorn woodpeckers (*Melanerpes formicivordes*) analogous to that found in some callitrichines. In the woodpeckers, immigration and emigration by adults is common, and data collected on marked individuals indicate that groups are composed largely of unrelated adults of both sexes. In each group, only a single female is usually reproductively active, and she may mate with a number of males, all of whom subsequently provide care for the young. Stacey (1979) points out that the system, referred to as a communal or cooperative breeding system, is relatively rare but taxonomically widespread among birds and some social carnivores. This system is characterized by the presence, in addition to an adult male and female pair, "of one or more extra individuals, or helpers [Skutch, 1961] that participate in the care and feeding of young" (Stacey, 1979: p.53). The presence of helpers in these groups has been shown to have a significant positive effect on infant survivorship. In the case of some tamarins and marmoset species, there is similar evidence that groups with at least two adult male helpers are more successful in raising their young than groups with fewer adult males (Sussman & Garber, 1987; Koenig & Rothe, 1991; Koenig, 1995; Garber, 1997; Heymann, 2000).

One might ask why adults would care for unrelated offspring. Several explanations are possible. In the case of nonreproductive males, helping might provide a critical opportunity for practicing caregiving skills that are required to ensure the survivorship of their own young when they begin breeding. In the case of reproductively active males, infant caregiving may be a form of courtship behavior exhibited to increase their likelihood of mating with the breeding female. Given that callitrichine females may be lactating and ovulating at the same time (callitrichines are the only primate in which a lactating female successfully nursing twin infants resumes ovulation and conceives within days of parturition), the breeding female can directly assess male parenting skills during the period that she is most fertile and likely to conceive. Finally, since the twins are dizygotic and two eggs are fertilized, it is possible that two different males in the group sire offspring. Thus, male caregiving may involve paternal uncertainty (the breeding female will mate with several males), individual fitness benefits associated with caring for offspring, inclusive fitness gains associated with helping to raise kin's offspring, or a courtship strategy to increase individual fitness by forming a socio-sexual bond with a breeding female. However, any discussion of inclusive fitness must include some estimate of the cost to individual fitness of remaining in the present group as a nonbreeder relative to migrating and breeding successfully in another group. It is likely that for most callitrichine species, breeding opportunities in both their resident group and in neighboring groups play a prominent role in natal and secondary dispersal.

Reproductive Biology

Callitrichines are uniquely specialized among mammals in having chimeric dizygotic twins (except for *Callimico goeldii*, which appear to have lost twinning evolutionarily and give birth to a single offspring; Porter & Garber, 2004), which lie within a common chorion during embryonic development. During gestation the twins share fetal blood cells and are immunologically identical (Ross et al., 2007). Although tamarins and marmosets normally bear twins and might be described as obligate twinners (Jaquish, 1993), single births (19%) and triplets (30%) have been reported in captivity (see table in Hershkovitz, 1977: p.422). In captive common marmosets, quadruplets and even quintuplets have been reported (Tardif et al., 2003). Given that both maternal weight and maternal nutrition appear to play a determining role in ovulation number and litter size (Rutherford & Tardif, 2009), triplet, quadruplet, and quintuplet births are more frequent in captivity and rarely if ever occur in the natural habitat.

In captivity, the interbirth interval for *Callithrix* is commonly five to seven months (Stevenson & Rylands, 1988), indicating the potential for two sets of twins per year. *Cebuella* may have two breeding peaks per year (Christen, 1974), and in the wild there appears to be two birth peaks per year in both *Callithrix jacchus* and *Mico humeralifer* (Stevenson & Rylands, 1988; Garber et al., 2019). However, in natural populations of *Saguinus oedipus* in Colombia, almost 90% of births occurred between March–June, and the interbirth interval averages 11 months (332 ± 53.6 days; Savage et al., 1997). In *S. geoffroyi*, there is a birth peak at the beginning of the wet season between April–June (Moynihan, 1970; Dawson, 1976), and there may be a secondary minor birth peak from November–February (Dawson, 1976), although like many tamarins, *S. geoffroyi* commonly gives birth once per year. Birth seasonality is reported for other species as well. In addition, there is evidence of seasonal changes in male testes size in *S. mystax* and *Leontocebus nigrifrons* (Garber et al., 1996b; Garber, 1997). Whether this is influenced by abiotic factors of climate or day length or by pheromonal cues produced by females or other group males remains uncertain. In many multi-male groups of *S. mystax* and *L. nigrifrons*, one male tends to have significantly larger testes than other group males (Garber et al., 1996b). In *Leontopithecus rosalia*, Dietz et al. (1994) report that males experience up to a 10% increase in body weight (685 gm) in the late wet season (May) in order to more successfully compete for mating opportunities. By June, however, male body weight has dropped by 12% (mean body weight = 609 g). These authors suggest, "The seasonal variation in male body mass appears to be explained by male–male competition, not female choice" (Dietz et al., 1994: p.127).

The gestation period is 148 ± 5 days for *Callithrix jacchus* (Hearn, 1977), 135–142 days for *Cebuella* (Hershkovitz, 1977), 140–165 days for *Saguinus* spp. (Epple, 1970; Gengozian et al., 1977; Cross & Martin, 1981) (except for *Saguinus oedipus*, which has the longest gestation reported for any callitrichine at 184 days; Digby et al., 2011), 150 days for *Leontocebus*, 155 days for *Callimico* (Martin, 1992), and 125–132 days for *Leontopithecus* (Kleiman, 1977a; Digby et al., 2011). Thus, the largest and the smallest callitrichines have the shortest gestation lengths. Values for *Cebuella* and *Leontopithecus* are comparable with that of *Aotus*, which has a gestation length of 117–159 days (Fernandez-Duque, 2011).

It is usually assumed that callitrichine males carry infants most of the time, beginning immediately after birth, as do male *Aotus* (Wright, 1981) and *Callicebus* (Kinzey, 1981). However, in *Leontopithecus rosalia* and *Callimico*, the mother is the principal carrier for the first 2–3 weeks

postpartum in captivity and 10–15 days in the wild (Porter & Garber, 2009). In all species of callitrichines in the wild, multiple adult group members actively carry and provision the infants. Dixson and George (1982) provide a possible endocrinological basis for this behavior. They report that male *Callithrix jacchus* carrying twin offspring had plasma prolactin levels five times higher than those of males without infants. Prolactin has been implicated in the control of maternal behavior, and this is an indication that males exhibiting strong parental behavior also may have high prolactin levels (possibly stimulated, in part, by the infants attempting to nurse from their male carriers). In related studies, Mota and Sousa (2000) found increased prolactin levels in adult and subadult infant caretakers in captive common marmosets, and Prudom et al. (2008) found that the scent of infants can lower testosterone levels in male common marmosets who have experience caring for infants.

HOME RANGE

Both small- and large-bodied tamarins exploit home ranges that vary in size from approximately 10 ha to over 100 ha. Tamarins often emit long calls in the early morning and this may help to identify the current location of neighboring groups. Dawson (1976, 1979) radio-tracked two groups of *Saguinus geoffroyi* in Panama: one in a lowland area that had a home range of 26 ha and a second group that inhabited an upland area of 32 ha. In both habitats, the home ranges of neighboring groups overlapped. Thirteen percent of the home range of the lowland group was shared by four other groups. Intergroup interactions, including vocalization and occasional physical contact, occurred frequently in boundary areas. Neyman (1977) and Savage (1990) report that groups of *S. oedipus* in Colombia had home ranges that varied in area from 8–12.4 ha.

In mixed-species troops of *Saguinus mystax* and *Leontocebus nigrifrons* (home range of 40 ha), Garber (1988a) found that the location of inter-troop vocal battles and aggressive encounters changed seasonally and were directly associated with the locations of large productive feeding sites. As new trees began fruiting, neighboring troops would range into the same areas. The ability to defend large productive feeding trees from other mixed-species troops played an important role in tamarin feeding and ranging activities. In a related study on single-species groups of mustached tamarins, Garber et al. (1993b) demonstrated that the frequency and severity of intergroup encounters were related to three factors: the presence of productive feeding sites, the particular location in the home range (e.g., in the vicinity of boundary areas), and whether group females were in estrus. These three factors varied temporally, and this accounted for much of the variation in intergroup interactions. Finally, Peres (1991b) reported that mixed-species troops of *S. mystax* and *L. fuscicollis avilapiresi* in Brazil exploited home ranges of over 100 ha. In comparison, single-species groups of *L. illigeri* in Peru (Soini, 1987a) had a home range of approximately 16 ha, and single-species groups of *L. weddelli* (Crandlemire-Sacco, 1986) were reported to have a home range of approximately 33 ha. Additional data on tamarin home range size are presented in Table 6.3.

Home ranges of *Callithrix* (0.5–36 ha) are generally smaller than those of *Saguinus* (10–100 ha) and larger than those of *Cebuella* (0.1–1.5 ha). Both lion tamarins and callimicos exploit very large ranges, although range size can be quite variable (Ferrari & Digby, 1996; Yepez et al., 2005; Digby et al., 2011) (Table 6.3). *Callithrix* groups generally have overlapping ranges (in some groups of *C. jacchus*, home range overlap between neighboring groups may be 80–90%; Digby et al., 2011), and rather than defend strict territorial boundaries, individuals may defend mates or important feeding sites from neighboring groups (but see Ferrari et al., 1996, concerning one population of *C. aurita*). Stevenson and Rylands (1988) report 17%, 23%, and 52% overlap in each of three groups of *C. jacchus*, and 22% overlap in neighboring groups of *Mico humeralifer*. Ferrari (1988) reported 87.5% home range overlap in his study groups of *C. flaviceps*, and Digby (1994) reported 86% home range overlap in her study groups of *C. jacchus*. In this population, home range size was small (3.9 ha) and the density of animals was extremely high (Digby, 1994). In another study, *C. jacchus* aggressively defended its main exudate tree but not its entire range (Maier et al., 1982). Garber et al. (2019) report that common marmosets inhabiting the Caatinga and the Atlantic Forest tend to have similar-sized home ranges of between 4–5 ha.

In a study of *C. penicillata* inhabiting gallery forest, Lacher et al. (1981) observed overlapping ranges in three groups utilizing the same exudate trees. The groups moved along the stream up and down the gallery forest and seemed to maintain "sliding" home ranges. Marmosets frequently scent-mark exudate holes (Coimbra-Filho & Mittermeier, 1977; Ramirez et al., 1977; Lacher et al., 1981; Rylands, 1981). This behavior has been considered to have a territorial function (Coimbra-Filho, 1972), but the groups studied by Lacher et al. (1981) did not avoid holes marked previously by other groups. Marking may be a way of indicating the relative time that has elapsed since a particular exudate source was last utilized and may not have high territorial significance.

Cebuella live in very small non-overlapping home ranges. When exudate sites stop producing, they abandon their range and move to a new home range. Thus, the lifetime range of a *Cebuella* group is characterized by the exploitation of a series of successive small home ranges (Soini, 1982; Terborgh, 1983). In a study of exudate feeding in 13 groups of pygmy marmosets, Yepez et al. (2005) report that home ranges were 0.1–1.2 ha in areas with neighboring groups located an average of 200 m apart (range = 55–700 m). Interactions between groups of *Cebuella* have not been described. Home range area and daily path length for *Cebuella* is listed in Table 6.3. In pygmy marmosets there is no evidence of a direct relationship between group size and home range size (Yepez et al., 2005).

Most authors (e.g., Ramirez et al., 1977) agree that density and distribution of marmosets are directly related to density and distribution of exudate trees. However, Terborgh (1983) argued that the widely scattered distribution of *Cebuella* in southern Peru suggests that more than suitable exudate trees are needed for an adequate home range. In addition to locating feeding sites, these small primates must have suitable dense cover in which to rest and forage for insects and nighttime resting sites that are well protected by vines and foliage. Intensive and prolonged exploitation of particular exudate sites probably imposes a severe burden on the trees; however, Terborgh (1983) observed that most trees heal their wounds and survive for years after *Cebuella* have abandoned them. There also is a need for an alternative resource base (such as fruits, floral nectar, animal prey). Nevertheless, *Cebuella* appears to have adopted a highly specialized strategy for exploiting exudates (Soini, 1982). In part, this strategy involves taking advantage of a select set of exudate species (trees) present in their range. For example, Yepez et al. (2005) found that pygmy marmosets did not feed on the most common exudate species in their range, nor did they select exudate species randomly. Rather, individual groups appeared to have particular preferences for a certain or small number of exudate-producing tree species, and this differed between groups.

In the case of *Callimico*, groups are reported to live in widely separated and non-overlapping areas (Porter & Garber, 2004). It has been suggested that the ecology of *Callimico* is closely tied to areas of secondary forest, primary forest with dense understory, and bamboo forest (Izawa & Yoneda, 1981; Christen, 1999; Porter et al., 2007; Porter & Garber, 2010). Like *Cebuella*, callimicos rarely engage in territorial encounters with neighboring groups. In this regard, although callimicos may form mixed-species troops with two species of tamarins (Porter et al., 2007, report that *Leontocebus weddelli*, *Saguinus labiatus*, and *Callimico goeldii* formed a polyspecific association with one or both tamarin species during 81% of

Table 6.3 Home Range and Day Range in Tamarins, Callimicos, and Marmosets

Taxonomic Name	Home Range (ha)	Day Range (m)	Reference
Saguinus geoffroyi	26-32	2006	Sussman and Kinzey, 1984
Saguinus labiatus	30	-	Porter, 2007
Saguinus midas	31.1-42.5	-	Digby et al., 2011
Saguinus mystax	40	1946	Garber, 1989
Saguinus mystax	149	1150-2700	Peres, 1991
Saguinus niger	15-35	-	Digby et al., 2011
Saguinus tripartitus	16-21	500-2300	Digby et al., 2011
Leontocebus fuscicollis avilapiresi	149	1150-2700	Peres, 1991
Leontocebus nigricollis	30-50	1000	Sussman and Kinzey, 1984
Leontocebus nigrifrons	40	1849	Garber, 1989
Leontocebus weddelli	30	1718-1821	Porter, 2007; Porter et al., 2020
Leontopithecus caissara	321	2235	Kierulff et al., 2002
Leontopithecus chrysomelas	40	1410-2175	Kierulff et al., 2002
Leontopithecus chrysomelas	93-118	1684-2044	Kierulff et al., 2002
Leontopithecus chrysopygus	113-199	1725	Kierulff et al., 2002
Leontopithecus chrysopygus	277	1904	Kierulff et al., 2002
Leontopithecus rosalia	21-73	1339-1553	Kierulff et al., 2002
Leontopithecus rosalia	65-229	1745-1873	Kierulff et al., 2002
Callimico goeldii	114-150	950	Porter, 2006; Porter and Garber, 2010
Mico argentatus	4-35	-	Digby et al., 2011
Mico argentatus	15.5	1042 (630-1710)	Veracini, 2009
Mico humeralifer	28.3	770-2120	Stevenson and Rylands, 1988
Mico intermedius	22.1	772-2115	Digby et al., 2011
Cebuella pygmaea	0.8-1.3	-	Stevenson and Rylands, 1988
Cebuella pygmaea	2.8-3.0	-	Stevenson and Rylands, 1988
Callithrix aurita	26.5-35.3	959-986	Digby et al., 2011
Callithrix flaviceps	33.8-35.5	884-1222	Digby et al., 2011
Callithrix jacchus [a]	0.7-5.2	912-1243	Digby et al., 2011
Callithrix jacchus [b]	2.2-12.25	727-1007	Garber et al., 2019
Callithrix kuhlii	10-58	830-1120	Digby et al., 2011
Callithrix penicillata	3.5-10	1000	Digby et al., 2011

[a] Common marmosets inhabiting the Atlantic Forest of northeastern Brazil.
[b] Common marmosets inhabiting a Caatinga biome in northeastern Brazil.

observations), neighboring callimico groups rarely come into contact and exhibit minimal or no home range overlap. This may result in part from the fact that callimicos have large home ranges (100–150 ha) and low biomass, and exhibit a pattern of habitat utilization characterized by the intensive use of a small number of core areas and then a shift to new core areas located in other parts of their range (Porter et al., 2007). Core areas may be associated with particular habitat types and the temporal availability of fungi and exudates. Thus, unlike some *Saguinus* species that regularly visit the borders of their range, callimicos exhibit a more "marmoset"-like pattern characterized by a relatively short day range and the concentrated use of core areas.

Lion tamarins have the largest home ranges of any callitrichine taxa, ranging from 21–394 ha (Table 6.3). Home range overlap among neighboring groups of *Leontopithecus rosalia* has been reported to be up to 61% (Peres, 1986). Peres (1986) found that intergroup encounters at Poco d'Antas occurred once every 1.6–2.5 days. Average time engaged in each intergroup encounter was 1–1.5 hours and involved both long calls and face-to-face confrontations. At this site, lion tamarins tend to have relatively smaller home ranges and higher population density than reported at other sites (average home range = 46 ha at Pocos d'Antas, Brazil, with 12 individuals/km² compared to the União Biological Reserve, Brazil, in which *L. rosalia* are reported to have home ranges averaging 150 ha with 3.5 individuals/km²; Kierulff et al., 2002). Home range overlap among other lion tamarin species is reported to be between 7–14%, and home range area is reported to be significantly larger than that reported for *L. rosalia* (see Table 6.3 in this chapter, also see Table 7.1 in Digby et al., 2011, and Table 7.6 in

Kierulff et al., 2002). In the case of *L. chrysomleas*, groups are reported to defend home ranges of over 120 ha, but focus their daily and weekly activities within a small set of core areas within their range (Raboy & Dietz, 2004). For example, Raboy and Dietz (2004) report that 50% of all activities of *L. chrysomleas* occur in only 11% of their home range. Black lion tamarins (*L. chrysopygus*) exploit the largest home range of any callitrichine species (276–394 ha) but have a group size of only up to seven individuals (Digby et al., 2011).

Most recorded day ranges for tamarins are relatively long for such small animals. The average day range of *Saguinus geoffroyi* was 2,061 ± 402 m (Dawson, 1979), and a group used about 56% of its home range in any given day. Based on full-day follows in which the location of focal animals was recorded every two minutes, Garber (1993b) estimated day range in a mixed-species group of *S. mystax* and *Leontocebus nigrifrons* to average approximately 1,800 m per day. In other studies of *Saguinus* and *Leontocebus*, daily path length varies from 1,150–2,700 m. The average day range of *Callithrix* species is reported to vary from 800–1,300 m. Day range for *C. kuhlii* was about 1,000 m (Stevenson & Rylands, 1988). In *C. flaviceps* (Ferrari, 1988) and *C. jacchus,* day ranges vary from approximately 860–1,243 m (Digby, 1994; Garber et al., 2019). In the case of *Mico humeralifer*, mean day range was 1,469 ± 193 m over 12 months with a range of 773–2,115 m, and day range for *Mico argentatus* averaged 1,042 m (n = 83 days, range 630–1,710 m/day; Veracini, 2009).

In contrast, very small day ranges have been reported for some groups of *Callithrix jacchus* and for *Cebuella*. Stevenson reported average day ranges of 100 m and 200 m for each of two groups of *C. jacchus* she studied (Stevenson & Rylands, 1988). Terborgh (1983) observed a group of *Cebuella* to spend most of its time in just four or five trees, and the group studied by Ramirez et al. (1977) spent 80% of its daylight time in a core area comprising only one-third of its 0.3 ha home range. Similarly, Soini (1982) reported a daily path length of 280–300 m for pygmy marmosets.

In the case of *Callimico*, Porter et al. (2007) estimate an average day range of under 1,000 m, during a yearlong study in Bolivia. In contrast, Pook and Pook (1982) suggest a day range of 2,000 m. Among lion tamarins, day range varies from 950–2,400 m in *Leontopithecus rosalia*, from 1,000–3,400 m in *L. caissara*, from 1,400–2,200 m in *L. chrysomelas*, and from 1,350–3,100 m in *L. chrysopygus* (Digby et al., 2011). Thus, the ranging patterns of individual marmoset and tamarin species are variable across sites and species. In part, this is likely to reflect differences in the availability and distribution of preferred feeding and sleeping sites. In addition, some species appear to concentrate their activities to a relatively small core area within their range for several days or weeks and then travel larger than average distances to other parts of their range and exploit that area over some extended period of time. In other species, groups travel in broad arcs, visiting much of their home range over the course of a few weeks. Finally, Porter et al. (2007) and Porter and Garber (2010) examined data on group biomass (measure of group size times body mass divided by home range size), and day range across 18 callitrichine species (including all genera except *Callibella*) and did not find evidence of a strong relationship. Peres (2000b) reported a similar finding. Thus, across species, the precise set of factors that affect patterns of ranging and habitat utilization remains unclear.

CALLITRICHINE CONSERVATION

Callitrichines are widely distributed across Panama, the Amazon Basin, and as far east as the Atlantic forests of northern and southern Brazil. The primary conservation threats to tamarins, callimicos, and marmosets are habitat loss and capture for the legal and illegal pet trade (Shanee et al., 2013; Rowe & Jacobs, 2016). Given their small body size, human hunting for consumption is not a major threat to callitrichine populations. However, several species have a very circumscribed distribution, either the result of historical biogeography or due to severe habitat degradation. In either case, such species are at particular risk of extinction. For example, all four species of lion tamarins are listed as Endangered by the International Union for the Conservation of Nature (IUCN, 2019), as human activity over the past 500 years has degraded more than 90% of the Atlantic forests of Brazil. Similarly, cotton-top tamarins (*Saguinus oedipus*) and pied tamarins (*S. bicolor*) are listed as Critically Endangered and buffy-headed marmosets (*Callithrix flaviceps*) are listed as Endangered, in response to habitat loss (IUCN, 2019). Many other callitrichine species are considered Vulnerable, Near Threatened, or Least Concern. Alarmingly, recent changes in conservation laws in Brazil (the primate-richest country in the world with some 116 primate species), and the increasing conversion of tropical rainforest to monocultures, pastures for cattle grazing, mining camps, and fossil fuel exploration driven by global commodities trade is predicted to result in accelerated primate population decline and extinction by the end of the century, including many callitrichine species (Estrada et al., 2018, 2019).

SUMMARY OF CALLITRICHINE BEHAVIOR AND ECOLOGY

In the past, callitrichine primates have been characterized as primitive anthropoids and squirrel-like both in their locomotor behavior and general ecology. They also have been described as living in monogamous family groups and being strictly territorial. Field studies, however, have provided more complete and accurate pictures of callitrichine behavior and ecology and a better understanding of the morphological adaptations of this primate subfamily. Tamarins, callimicos, and marmosets exhibit a suite of highly derived morphological and behavioral traits and are no longer regarded as morphologically primitive for New World primates. In fact, they share a close evolutionary relationship with *Cebus*, *Saimiri*, and *Aotus*, and appear to have diverged from an ancestral cebine lineage some 16 mya. They definitely are not the ecological equivalents of tree squirrels.

The home range of most species of callitrichine include a mix of habitat types and commonly contain an abundance of edge or secondary forest. However, each major callitrichine taxon appears to have radiated into a dietary niche that varies in their dependence on fruits, arthropods, exudates, nectar, fungi, and small vertebrates. The evolution of claw-like nails on all digits excluding the hallux (which is present in all callitrichines and, therefore, reasonably assumed to have occurred soon after callitrichines split from cebid ancestors) appears principally to represent a postural adaptation that enables these small-bodied monkeys to cling to large vertical and sharply inclined supports when exploiting trunk resources such as exudates, insects, fungi, and small vertebrates and to use trunks as a perch to scan the ground for prey (Garber, 1992). *Callithrix* and *Cebuella* (also possibly *Callibella*) represent the most specialized exudate-feeding callitrichines, with derived morphological features of the dentition for gouging holes in tree bark to obtain exudates and of the cecum for digesting structural carbohydrates. Although callimicos also consume exudates, the majority of their feeding time involves the exploitation of fungi that grow on the vertical culms of bamboo and on decaying tree trunks.

Callitrichines live in small groups, generally ranging from 5 to 10 individuals. In the wild, the majority of groups do not conform in size and sex-age composition to a nuclear family. Although it is possible that, in some cases, groups may represent extended families with multiple adult males and multiple adult females, based on the evidence at present, callitrichines are best described as living in small multi-male/multi-female groups that commonly contain only one breeding female (Garber et al., 2016). The breeding female often mates with multiple adult males, and multiple adult group members cooperatively aid in the rearing of twin offspring (e.g., carrying, protecting, and provisioning infants). We argue that the modal mating system is polyandrous. This has been referred to as a "communal breeding system"; however, it is probably better described as a nonmonogamous single female breeding group with a communal infant-rearing system. Occasionally two females in a group produce twin offspring during the same breeding period, and these groups are polygynous. If monogamy is defined as an exclusive mating relationship between a single male and a single female that continues over several breeding periods, we see no preponderant behavioral or genetic evidence to support this as a common mating strategy in wild callitrichines (Garber et al., 2016). Furthermore, there is virtually no evidence in the wild that a single male and a single female form a "pair-bond," as has been described for *Aotus* or *Callicebus*. A recent study by De la Fuente et al. (2019) indicates that the within-group social hierarchy in wild common marmosets is "single female dominant," with the breeding female occupying the alpha rank and all other group members exhibiting a relatively equal rank below the breeding female.

Because Amazonian marmosets and eastern Brazilian marmosets are able to gouge into tree bark and thereby stimulate the flow of exudates year-round, they may be less dependent on the seasonal availability of other resources than are tamarins, lion tamarins, and callimicos. This allows them to have smaller home ranges and to move less throughout the day. It also may allow *Callithrix* and *Cebuella* to more regularly produce two sets of twins in a given year, and in the case of *Callithrix*, to occasionally have two simultaneously breeding females in a group. The presence of two breeding females has not been documented in *Cebuella*. Overall, the adaptive radiation of tamarins, callimicos, and marmosets is best understood by studying the interrelationships among phylogeny, diet, dental, gut, and positional morphology; reproductive biology; patterns of infant care; and the social and mating strategies that define this primate subfamily.

Acknowledgments: This chapter is dedicated to the living memory and outstanding scientific contributions of my friend, mentor, and colleague, the late Robert W. Sussman. For the past five decades, Bob Sussman was a leading figure in advancing our understanding of primate behavior, ecology, and conservation. I also thank Chrissie, Sara, and Jenni for inspiring me to protect the world's primates. I also could not have studied tamarins, callimicos, and marmosets without the friendship, expertise, and assistance of many field guides, in particular Eriberto Mermao and Walter Mermao, Edilio Nacimento, and my many colleagues and collaborators.

REFERENCES CITED—CHAPTER 6

Abbott, D. H., Barrett, J., & George, L. M. (1993). Comparative aspects of the social suppression of reproductions in female marmosets and tamarins. In A. B. Rylands (Ed.), *Marmosets and tamarins: Systematics, behaviour, and ecology* (pp. 152-163). Oxford University Press.

Abreu, F., De la Fuente, M. F. C., Schiel, N., & Souto, A. (2016). Feeding ecology and behavioral adjustments: Flexibility of a small Neotropical primate (*Callithrix jacchus*) to survive in a semiarid environment. *Mammal Research, 61*, 221-229.

Amora, T. D., Beltrão-Mendes, R., & Ferrari, S. F. (2013). Use of alternative plant resources by common marmosets (*Callithrix jacchus*) in the semi-arid Caatinga scrub forests of northeastern Brazil. *American Journal of Primatology, 75*, 333-341. DOI: 10.1002/ajp.22110.

Augiar, J. M., & Lacher, T. E., Jr. (2009). Cranial morphology of the dwarf marmoset *Callibella* in the context of callitrichid variability. In S. M Ford, L. M. Porter, & L. C. Davis (Eds.), *The smallest anthropoids: The marmosets/callimico radiation* (pp. 355-380). Springer.

Baker A. J., Bales, K., & Dietz, J. D. (2002). Mating system and group dynamics in lion tamarins. In D. G. Kleiman, & A. B. Rylands (Eds.), *Lion tamarins: Biology and conservation* (pp. 188-212). Smithsonian Institution Scholarly Press.

Baker, A. J., & Dietz, J. M. (1996). Polygyny and female reproductive success in golden lion tamarins, *Leontopithecus rosalia*. *Animal Behaviour, 46*, 1067-1078.

Baker, A. J., Dietz, J. M., & Kleiman, D. G. (1993). Behavioural evidence for monopolization of paternity in multi-male groups of golden lion tamarins. *Animal Behavior, 46*, 1091-1103.

Barroso, C. M. L., Schneider, H., Schneider, M. P. C., Sampaio, I., Harada, M. L., Czelusniak, J., & Goodman, M. (1997). Update on the phylogenetic systematics of New World monkeys: Further DNA evidence for placing marmoset (*Cebuella*) within the genus *Callithrix*. *International Journal of Primatology, 18*, 651-674.

Bartecki, U., & Heymann, E. W. (1987). Sightings of red uakaris, *Cacajao calvus rubicundus*, at Rio Blanco Peruvian Amazonia. *Primate Conservation, 8*, 34-36.

Bates, H. W. (1864). *The naturalist on the River Amazon* (2nd Ed.). John Murray Press.

Bearder, S. K., & Martin, R. D. (1980). Acacia gum and its use by bush-babies, *Galago senegalensis* (Primates: Lorisidae). *International Journal of Primatology, 1*, 103-128.

Bicca-Marques, J. C. (1999). Hand specialization, sympatry, and mixed-species associations in callitrichines. *Journal of Human Evolution, 36*, 349-378.

Boubli, J. P., da Silva, M. N. F., Rylands, A. B., Nash, S. D., Bertuol, F., Nunes, M., Mittermeier, R. A., Byrne, da Silva, F. E., Röhe, Sampaio, I., Schneider, H., Farias, I. P. & Hrbek, T. (2018). How many pygmy marmoset (*Cebuella* Gray, 1870) species are there? A taxonomic reappraisal based on new molecular evidence. *Molecular Phylogenetics and Evolution, 120*, 170-182.

Buchanan-Smith, H., & Hardie, S. M. (1997). Tamarin mixed-species groups: An evaluation of a combined captive and field approach. *Folia Primatologica, 68*, 272-286.

Buckner, J. C., Lynch Alfaro, J. W., Rylands, A. B., & Alfaro, M. E. (2015). Biogeography of the marmosets and tamarins (Callitrichidae). *Molecular Phylogenetics and Evolution, 82*, 413-425.

Burrows, A. M., & Nash, L. T. (Eds.). (2010). *The evolution of exudativory in primates*. Springer.

Caine, N. G. (1987). Vigilance, vocalizations, and cryptic behavior at retirement in captive groups of red-bellied tamarins (*Saguinas labiatus*). *American Journal of Primatology, 12*, 241-250.

Caine, N. G. (1993). Flexibility and co-operation as unifying themes in *Saguinus* social organization: The role of predation pressures. In A. B. Rylands (Ed.), *Marmosets and tamarins: Systematics, behavior, and ecology* (pp. 200-219). Oxford University Press.

Caine, N. G., & Weldon, P. J. (1989). Responses by red-bellied tamarins (*Saguinas labiatus*) to fecal scents of predatory and non-predatory Neotropical mammals. *Biotropica, 21*, 186-189.

Cartmill, M. (1974). Pads and claws in arboreal locomotion. In F. A. Jenkins (Ed.), *Primate locomotion* (pp. 45-83). Academic Press.

Carvalho, C. T., & Carvalho, C. F. (1989). A organizacao social dos suis-pretos (*Leontopithecus chrysopygus* Mikan) na reserva em Teodoro Sampaio, Sao Paulo (Primates, Callitrichidae). *Revista. Brasileira Zoologica, 6*, 707-717.

Castro, R. (1991). *Behavioral ecology of two coexisting tamarin species (*Saguinus fuscicollis nigrifrons *and* Saguinus mystax mystax, Callitrichidae, Primates*) in Amazonian Peru*. [Doctoral dissertation, Washington University].

Castro, R., & Soini, P. (1977). Field studies on *Saguinus mystax* and other callitrichids in Amazonian Peru. In D. G. Kleiman (Ed.), *The biology and conservation of the Callitrichidae* (pp. 73-78). Smithsonian Institution Press.

Charles-Dominique, P. (1971). Eco-ethologie des prosimiens du Gabon. *Biologica Gabonica, 7*, 121-228.

Chaves, R., Sampaio, I., Schneider, M., Schneider, H., Page, S., & Goodman, M. (1999). The place of *Callimico goeldii* in the Callitrichinae phylogenetic tree: Evidence from the von Willebrand Factor gene intron II sequences. *Molecular Phylogeny and Evolution, 13*, 392-404.

Christen, A. (1974). Fortpflanzungsbiologie und verhalten bei *Cebuella pygmaea* und *Tamarin tamarin*. *Fortschritte der Verhaltensforschung, 14*, 1-78.

Christen, A. (1999). Survey of Goeldi's monkeys (*Callimico goeldii*) in northern Bolivia. *Folia Primatologica, 70*, 107-111.

Coimbra-Filho, A. F. (1972). Aspectos inéditos do comportamento de sagüis do gênero *Callithrix* (Callitrichidae, Primates). *Revista Brasileira Biologia, 32*, 505-512.

Coimbra-Filho, A. F. (1976). *Leontopithecus rosalia chrysopygus* (Mikan, 1823) o mico-leao do Estado de Sao Paulo (Callitrichidae-Primates). *Silvicultura, Sao Paulo, 10*, 1-36.

Coimbra-Filho, A. F. (1977). Natural shelters of *Leontopithecus rosalia* and some ecological implications (Callitrichidae: Primates). In D. G. Kleiman (Ed.), *The biology and conservation of the Callitrichidae* (pp. 79-89). Smithsonian Institution Press.

Coimbra-Filho, A. F., & Mittermeier, R. A. (1977). Tree gouging, exudate-eating and the "short-tusked" condition in *Callithrix* and *Cebuella*. In D. G. Kleiman (Ed.), *The biology and conservation of the Callitrichidae* (pp. 105-115). Smithsonian Institution Press.

Coimbra-Filho, A.F., Rocha, N. C., & Pissinatti, A. (1980). Morfofisiologia do ceco e sua correlacao com o tipo odontologico em Callitrichidae (Platyrrhini, Primates). *Revista Basileira. Biologica*, 4, 141-147.

Correa, H. K. M., Coutinho, P. E. C., & Ferrari, S. F. (2000). Between-year differences in the feeding ecology of highland marmosets (*Callithrix autita* and *Callithrix flaviceps*) in southeastern Brazil. *Journal of Zoology*, 252, 421-427.

Cortés-Ortiz, L. (2009). Molecular phylogenetics of Callitrichidae with emphasis on marmosets and callimico. In S. M. Ford, L. M. Porter, & L. C. Davis (Eds.), *The smallest anthropoids: The marmosets/callimico radiation* (pp. 3-24). Springer.

Crandlemire-Sacco, J. L. (1986). *The ecology of saddle-backed tamarins, Saguinas fuscicollis, of southeastern Peru*. [Doctoral dissertation, University of Pittsburgh].

Cross, J. F., & Martin, R. D. (1981). Calculation of gestation period and other reproductive parameters for primates. *Dodo, Journal of the Jersey Wildlife Preserve Trust*, 18, 30-43.

Cruz Lima, E. D. (1945). *Mammals of Amazonia, (Vol. 1), General introduction and primates*. Belem do Par´a.

Davis, L. C. (1996). Functional and phylogenetic implications of ankle morphology in Goeldi's monkey (*Callimico goeldii*). In M. A. Norconk, A. L. Rosenberger, & P. A. Garber, (Eds.), *Adaptive radiations of Neotropical primates* (pp. 133-156). Plenum Press.

Dawson, G. A. (1976). *Behavioral ecology of the Panamanian tamarin, Saguinus oedipus (Callitrichidae: Primates)*. [Doctoral dissertation, Michigan State University].

Dawson, G. A. (1977). Composition and stability of social groups of the tamarin, *Saguinus oedipus geoffroyi*, in Panama: Ecological and behavioral implications. In D. G. Kleiman (Ed.), *The biology and conservation of the Callitrichidae* (pp. 23-37). Smithsonian Institution Press.

Dawson, G. A. (1979). The use of time and space by the Panamanian tamarin, *Saguinus oedipus*. *Folia Primatologica*, 31, 253-284.

De la Fuente, M. F., Schiel, N., Bicca-Marques, J. C., Caselli, C. B., Souto, A., & Garber, P. A. (2019). Balancing contest competition, scramble competition, and social tolerance at feeding sites in wild common marmosets (*Callithrix jacchus*). *American Journal of Primatology*, 81(4), e22964.

De la Fuente, M. F., Souto, A., Sampaio, M. B., & Schiel, N. (2014). Behavioral adjustments by a small Neotropical primate (*Callithrix jacchus*) in a semiarid Caatinga environment. *The Scientific World Journal*, 2014, 326524.

Díaz-Muñoz, S. (2011). Paternity and relatedness in a polyandrous nonhuman primate: Testing adaptive hypotheses of male reproductive cooperation. *Animal Behavior*, 82, 563-571.

Dietz, J. M., & Baker, A. J. (1993). Polygyny and female reproductive success in golden lion tamarins, *Leontopithecus rosalia*. *Animal Behaviour*, 46, 1067-1078.

Dietz, J. M., Baker, A. J., & Miglioretti, D. (1994). Seasonal variation in reproduction, juvenile growth, and adult body mass in golden lion tamatins (*Leontopithecus rosalia*), *American Journal of Primatology*, 34, 115-132.

Dietz, J. M., Peres, C. A., & Pinder, L. (1997). Foraging ecology and use of space in wild golden lion tamarins (*Leontopithecus rosalia*). *American Journal of Primatology*, 41, 289-303.

Digby, L. J. (1994). *Social organization and reproductive strategies in a wild population of common marmosets (Callithrix jacchus)*. [Doctoral dissertation, University of California, Davis].

Digby, L. J. (1999). Sexual behavior and extragroup copulations in a wild population of common marmosets (*Callithrix jacchus*). *Folia Primatologica*, 70, 136-145.

Digby, L. J., & Barreto, C. E. (1996). Activity and ranging patterns in common marmosets (*Callithrix jacchus*): Implications for reproductive strategies. In M. A. Norconk, A. L. Rosenberger, & P. A. Garber (Eds.), *Adaptive radiations of Neotropical primates* (pp. 173-185). Plenum Press.

Digby, L. J., & Ferrari, S. F. (1994). Multiple breeding females in free-ranging groups of *Callithrix jacchus*. *International Journal of Primatology*, 15, 389-397.

Digby, L. J., Ferrari, S. F., & Saltzman, W. (2011). Callitrichines: The role of competition in cooperatively breeding species. In C. J. Campbell, A. Fuentes, K. C. MacKinnon, S. K. Bearder, & R. M. Stumpf (Eds.), *Primates in perspective* (2nd ed., pp. 91-107). Oxford University Press.

Dixson, A. F., & George, L. (1982). Prolactin and parental behavior in a male New World primate. *Nature*, 299, 551-553.

Dummond, F. V. (1971). Comments on the minimum requirements in the husbandry of the golden marmoset (*Leontopithecus rosalia*). *Laboratory Primate Newsletter*, 10, 30-37.

Egler, S. G. (1992). Feeding ecology of *Saguinus bicolor bicolor* (Callitrichidae: Primates) in a relic forest in Manaus, Brazilian Amazonia. *Folia Primatologica*, 59, 61-76.

Eisenberg, J. F. (1977). Comparative ecology and reproduction of New World monkeys. In D. G. Kleiman (Ed.), *The biology and conservation of the Callitrichidae* (pp. 13-22). Smithsonian Institution Press.

Eisenberg, J. F., Muckenhirn, N. A., & Rudran, R. (1972). The relation between ecology and social structure in primates. *Science*, 176, 863-874.

Emmons, L. H. (1987). Comparative feeding ecology of felids in a Neotropical rainforest. *Behavioral Ecology and Sociobiology*, 20, 271-283.

Encarnacion, C. F., & Heymann, E. W. (1997). Body mass of wild *Callimico goeldii*. *Folia Primatologica*, 69, 368-371.

Enders, R. K. (1935). Mammalian life histories from Barro Colorado Island. *Bulletin of the Museum of Comparative Zoology, 78*, 385-502.

Epple, G. (1970). Maintenance, breeding and development of marmoset monkeys (Callitrichidae) in captivity. *Folia Primatologica, 12*, 56-76.

Epple, G. (1975a). The behavior of marmoset monkeys (Callitrichidae). In L. A. Rosenblum (Ed.), *Primate behavior (Vol. 4)* (pp. 195-239). Academic Press.

Epple, G. (1975b). Parental behavior in *Saguinus fuscicolis* (Callitrichidae). *Folia Primatologica, 24*, 221-238.

Epple, G. (1976). Chemical communication and reproductive processes in nonhuman primates. In R. L. Doty (Ed.), *Mammalian olfaction, reproductive processes, and behavior* (pp. 257-282). Academic Press.

Estrada, A., Garber, P. A., & Chaudhary A. (2019). Expanding global commodities trade and consumption place the world's primates at risk of extinction. *PeerJ, 7*, e7068. DOI: 10.7717/peerj.7068.

Estrada, A., Garber, P. A., Mittermeier, R. A., Wich, S., Gouvei, S., Dobrovolski, R., Nekaris, K. A. I., Nijman, V., Rylands, A. B., Maisels, F., Williamson, E. A., Bicca-Marques, J. C., Fuentes, A., Jerusalinsky, L., Johnson, S., Rodriguez de Melo, F., Oliveira, L., Schwitzer, C., Roos, C.,... & Setiawan, A. (2018). Primates in peril: The significance of Brazil, Madagascar, Indonesia and the Democratic Republic of the Congo for global primate conservation. *PeerJ, 6*, e4869. DOI: 10.7717/peerj.4869.

Evans, S. (1983). The pair-bond of the common marmoset, *Callithrix jacchus jacchus*: An experimental investigation. *Animal Behaviour, 31*, 651-658.

Evans, S., & Poole, T. B. (1983). Pair-bond formation and breeding success in the common marmoset *Callithrix jacchus jacchus*. *International Journal of Primatology, 4*, 83-97.

Faria, D. S. (1986). Tamnho, composicao de um grupo social e area de vivencia (home-range) do sagui (*Callithrix jacchus*) na mata ciliar do corrego Capetinga, Brasilia, DF. In M. T. Mello (Ed.), *A primatologica no Brasil-2* (pp. 87-105). Sociedade Brasileira de Primatologia.

Faulkes, C. G., Arruda, M. F., & Monteiro da Cruz, M. A. O. (2003). Matrilineal genetic structure within and among populations of the cooperatively breeding common marmoset, *Callithrix jacchus*. *Molecular Ecology, 12*, 1101-1108.

Faulkes, C. G., Arruda, M. F., & Monteiro da Cruz, M. A. O. (2009). Genetic structure within and among populations of the common marmoset, *Callithrix jacchus*: Implications for cooperative breeding. In S. M. Ford, L. M. Porter, & L. C. Davis (Eds.), *The smallest anthropoids: The marmosets/callimico radiation* (pp. 103-117). Springer.

Fernandez-Duque, E. (2011). Aotinae: Social monogamy in the only nocturnal anthropoid. In C. J. Campbell, A. Fuentes, K. C. MacKinnon, S. K. Bearder, & R. M. Stumpf (Eds.), *Primates in perspective* (2nd ed., pp. 140-154). Oxford University Press.

Ferrari, S. F. (1988). *The behaviour and ecology of the buffy-headed marmoset,* Callithrix flaviceps *(O. Thomas, 1903).* [Doctoral dissertation, University College].

Ferrari, S. F. (1992). The care of infants in a wild marmoset (*Callithrix flaviceps*) group. *American Journal of Primatology, 26*, 109-118.

Ferrari, S. F. (1993). Ecological differentiation in the Callitrichidae. In A. B. Rylands (Ed.), *Marmosets and tamarins: Systematics, behaviour, and ecology.* (pp. 314-328). Oxford University Press.

Ferrari, S. F. (2009). Social hierarchy and dispersal in free-ranging buffy-headed marmosets (*Callithrix flaviceps*). In S. M. Ford, L. M. Porter, & L. C. Davis (Eds.), *The smallest anthropoids: The marmosets/callimico radiation* (pp. 155-166). Springer.

Ferrari, S. F., Correa, H. K. M., & Coutinho, P. E. G. (1996). Ecology of the "southern" marmosets (*Callithrix aurita* and *Callithrix flaviceps*): How different, how similar? In M. A. Norconk, A. L. Rosenberger, & P. A. Garber (Eds.), *Adaptive radiations of Neotropical primates* (pp. 157-171). Plenum Press.

Ferrari, S. F., & Digby, L. J. (1996). Wild *Callithrix* groups: Stable extended families? *American Journal of Primatology, 38*, 19-28.

Ferrari, S. F., & Lopes Ferrari, M. A. (1989). A re-evaluation of the social organization of the Callitrichidae, with reference to ecological differences between genera. *Folia Primatologica, 52*(3-4), 132-147.

Ferrari, S. F., & Lopes Ferrari, M. A. (1990). Predator avoidance behavior in the buffy-headed marmoset (*Callithrix flaviceps*). *Primates, 31*, 323-338.

Ferrari, S. F., Lopes Ferrari, M. A., & Krause, A. (1993). Gut morphology of *Callithrix nigriceps* and *Saguinus labiatus* from western Brazilian Amazonia. *American Journal of Physical Anthropology, 90*, 487-493.

Ferrari, S. F., & Martins, E. (1992). Gummivory and gut morphology in two sympatric callitrichids (*Callithrix emiliae* and *Saguinus fuscicollis weddelli*) from western Brazilian Amazonia. *American Journal of Physical Anthropology, 88*, 97-103.

Ferrari, S. F., & Mendes, S. L. (1991). Buffy-headed marmosets 10 years on. *Oryx, 25*, 105-109.

Ferrari, S. F., & Strier, K. B. (1992). Exploitation of *Mabea fistulifera* nectar by marmosets (*Callithrix flaviceps*) and muriquis (*Brachyteles arachnoides*) in south-east Brazil. *Journal of Tropical Ecology, 8*, 225-239.

Fleagle, J. G. (1999). *Primate adaptation and evolution.* Academic Press.

Fleagle, J. G., & Mittermeier, R. A. (1980). Locomotor behavior, body size and comparative ecology of seven Surinam monkeys. *American Journal of Physical Anthropology, 52*, 301-314.

Freese, C. H., Heltne, P. G., Castro, N., & Whitesides, G. (1982). Patterns and determinants of monkey densities in Peru and Bolivia, with notes on distribution. *International Journal of Primatology, 3*, 53-90.

French, J. A., Bales, K. L., Baker, A. J., & Dietz, J. M. (2003). Endocrine monitoring of wild dominant and subordinate female *Leontopithecus rosalia*. *International Journal of Primatology, 24*, 281-1300.

French, J. A., de Vleeschouwer, K., Bales, K. L., & Heistermann, M. (2002). Lion tamarin reproductive biology. In D. G. Kleiman, & A. B. Rylands (Eds.), *Lion tamarins: Biology and conservation* (pp. 133-156). Smithsonian Institution Scholarly Press.

Galef, B. G., Mittermeier, R. A., & Bailey, R. C. (1976). Predation by the tayra (*Eira barbara*). *Journal of Mammalogy, 57*, 760-761.

Garber, P. A. (1980a). *Locomotor behavior and feeding ecology of the Panamanian tamarin* (Saguinus oedipus geoffroyi, *Callitrichidae, Primates).* [Doctoral dissertation, Washington University].

Garber, P. A. (1980b). Locomotor behavior and feeding ecology of the Panamanian tamarin (*Saguinus oedipus geoffroyi*, Callitrichidae, Primates). *International Journal of Primatology, 1*, 185-201.

Garber, P. A. (1984a). The ecology of the exudate feeding and the importance of plant exudates in the diet of the Panamanian tamarin. *International Journal of Primatology, 5*, 1- 15.

Garber, P. A. (1984b). Use of habitat and positional behavior in a Neotropical primate, *Saguinus oedipus*. In J. G. H. Cant, & P. S. Rodman (Eds.), *Adaptations for foraging in nonhuman primates: Contributions to an organismal biology of prosimians, monkeys and apes* (pp. 113-133). Columbia University Press.

Garber, P. A. (1986). The ecology of seed dispersal in two species of callitrichid primates (*Saguinus mystax* and *Saguinus fuscicollis*) in Amazonian Peru. *American Journal of Primatology, 10*, 155-170.

Garber, P. A. (1988a). Diet, foraging patterns, and resource defense in a mixed species troop of *Saguinus mystax* and *Saguinus fuscicollis* in Amazonian Peru. *Behavior, 105*, 18-34.

Garber, P. A. (1988b). Foraging decisions during nectar feeding by tamarin monkeys: *Saguinus mystax* and *Saguinus fuscicollis* in Amazonian Peru. *Biotropica, 20*, 100-106.

Garber, P. A. (1989). The role of spatial memory in primate foraging patterns: *Saguinus mystax* and *Saguinus fuscicollis*. *American Journal of Primatology, 19*, 203-216.

Garber, P. A. (1991). A comparative study of positional behavior in three species of tamarin monkeys. *Primates, 32*, 219-230.

Garber, P. A. (1992). Vertical clinging, small body size, and the evolution of feeding adaptations in the Callitrichidae. *American Journal of Physical Anthropology, 88*, 469-482.

Garber, P. A. (1993a). Feeding ecology and behaviour of the genus *Saguinas*. In A. B. Rylands (Ed.), *Marmosets and tamarins: Systematics, behavior, and ecology* (pp. 273-295). Oxford University Press.

Garber, P. A. (1993b). Seasonal patterns of diet and ranging in two species of tamarin monkeys: Stability versus variability. *International Journal of Primatology, 14*, 145-166.

Garber, P. A. (1994a). A phylogenetic approach to the study of tamarin and marmoset social systems. *American Journal of Primatology, 34*, 199-219.

Garber, P. A. (1994b). Aspects of fruit eating and seed dispersal in Panamanian (*Saguinus geoffroyi*) and moustached tamarins (*Saguinus mystax*). *Regional Conference Proceedings of the American Zoo and Aquarium Association*, 364-369.

Garber, P. A. (1997). One for all and breeding for one: Cooperation and competition as a tamarin reproductive strategy. *Evolutionary Anthropology, 5*, 187-199.

Garber, P. A. (1998). Within and between site variability in moustached tamarin (*Saguinus mystax*) positional behavior during food procurement. In A. Rosenberger, J. G. Fleagle, H. McHenry, & E. Strasser (Eds.), *Primate locomotion* (pp. 61-78). Plenum Press.

Garber, P. A. (2000). The ecology of group movement: Evidence for the use of spatial, temporal, and social information in some primate foragers. In S. Boinski, & P. A. Garber (Eds.), *On the move: How and why animals travel in groups* (pp. 261-298). University of Chicago Press.

Garber, P. A. (2011). Primate positional behavior and ecology. In C. J. Campbell, A. Fuentes, K. C. MacKinnon, S. Bearder, & R. Stumpf (Eds.), *Primates in perspective* (2nd Ed., pp. 548-563). Oxford University Press.

Garber, P. A., & Bicca-Marques, J. C. (2002). Evidence of predator sensitive foraging in small- and large-scale space in free-ranging tamarins (*Saguinus fuscicollis, Saguinus imperator,* and *Saguinus mystax*). In L. Miller (Ed.), *Eat or be eaten: Predator sensitive foraging in primates* (pp. 138-153). Cambridge University Press.

Garber, P. A, Bicca Marques, J. C., & Azevedo-Lopes, M. A. O. (2009b). Primate cognition: Integrating social and ecological information in decision-making. In P. A. Garber, A. Estrada, J. C. Bicca-Marques, E. W. Heymann, & K. B. Strier (Eds.), *South American primates: Comparative perspectives in the study of behavior, ecology, and conservation* (pp. 365-385). Springer.

Garber, P. A., Blomquist, G., & Anzenberger, G. (2005). Kinematic analysis of trunk-to-trunk leaping in *Callimico goeldii*. *International Journal of Primatology, 26*, 223-240.

Garber, P. A., Caselli, C. B., McKenney, A. C., Abreu, F. de, Fuente, M. F. De la, Araujo, A., Arruda, M. F., Souto, A., Schiel, N., & Bicca-Marques, J. C. (2019). Trait variation and trait stability in common marmosets (*Callithrix jacchus*) inhabiting ecologically distinct habitats in northeastern Brazil. *American Journal of Primatology, 81*(7), e23018. DOI: org/10.1002/ajp.23018.

Garber, P. A., & Dolins, F. L. (1996). Testing learning paradigms in the field: Evidence for use of spatial and perceptual information and rule-based foraging in wild moustached tamarins. In M. Norconk, A. L. Rosenberger, & P. A. Garber (Eds.), *Adaptive radiation of Neotropical primates* (pp. 201-216). Plenum Press.

Garber, P. A., Encarnacion, F., Moya, L., & Pruetz, J. D. (1993a). Demographic and reproductive patterns in moustached tamarin monkeys (*Saguinus mystax*): Implications of reconstructing platyrrhine mating systems. *American Journal of Primatology, 29*, 235-254.

Garber, P. A., & Hannon, B. (1993). Modeling monkeys: A comparison of computer generated and naturally occurring foraging patterns in two species of Neotropical primates. *International Journal of Primatology, 14*, 827-852.

Garber, P. A., & Kitron, U. (1997). Seed swallowing in tamarins: Evidence of a curative function or enhanced foraging efficiency. *International Journal of Primatology, 18*, 523-538.

Garber, P. A., & Kitron, U. (2013). Why do tamarins swallow such large seeds? A response to Heymann's commentary. *International Journal of Primatology, 34*, 450-454.

Garber, P. A., & Leigh, S. R. (1997). Ontogenetic variation in small-bodied New World primates: Implications for patterns of reproduction and infant care. *Folia Primatologica, 68*, 1-22.

Garber P. A., & Leigh, S. R. (2001). Patterns of positional behavior in mixed species troops of *Callimico goeldii*, *Saguinus labiatus*, and *Saguinus fuscicollis* in northwestern Brazil. *American Journal of Primatology, 54*, 17-31.

Garber, P. A., McKenney, A. C., & Mallott, E. K. (2012). The ecology of trunk-to-trunk leaping in *Saguinus fuscicollis*: Implications for understanding locomotor diversity in callitrichines. *Neotropical Primates, 19*(1), 1-7.

Garber, P. A., Moya, L., & Malaga, C. (1984). A preliminary field study of the moustached tamarin monkey (*Saguinas mystax*) in northeastern Peru: Questions concerned with the evolution of a communal breeding system. *Folia Primatologica, 42*, 17-32.

Garber, P. A., Moya, L., Pruetz, J. D., & Ique, C. (1996b). Social and seasonal influences on reproductive biology in male moustached tamarins (*Saguinus mystax*). *American Journal of Primatology, 38*, 29-46.

Garber, P. A., & Porter, L. M. (2009). Trunk-to-trunk leaping in wild *Callimico goeldii* in northern Bolivia. *Neotropical Primates, 16*, 9-14.

Garber, P. A., & Porter, L. M. (2010). The ecology of exudate feeding and exudate production in *Saguinus* and *Callimico*. In A. Burrows, & L. Nash (Eds) *The evolution of exudativory in primates* (pp. 89-108). Springer.

Garber, P. A., & Porter, L. M. (2014). Navigating in small-scale space: The role of landmarks and resource monitoring in understanding saddleback tamarin travel. *American Journal of Primatology, 76*, 447-459.

Garber, P. A., Porter, L. M., Spross, J., & Di Fiore, A. (2016). Tamarins: Insights into monogamous and non-monogamous single female social and breeding systems. *American Journal of Primatology, 78*, 298-314.

Garber, P. A., Pruetz, J. D., & Isaacson, J. (1993b). Patterns of range use, range defense, and intergroup spacing in moustached tamarin monkeys (*Saguinus mystax*). *Primates, 34*, 11-25.

Garber, P. A., Rosenberger, A. L., & Norconk, M. A. (1996a). Marmoset misconceptions. In M. A. Norconk, A. L. Rosenberger, & P. A. Garber (Eds.), *Adaptive radiations of Neotropical primates* (pp. 87-95). Plenum Press.

Garber P. A., Sallenave, A., Blomquist, G. E., & Anzenberger, G. (2009a). A comparative study of the kinematics of trunk-to-trunk leaping in *Callimico goeldii*, *Callithrix jacchus*, and *Cebuella pygmaea*. In S. Ford, L. M. Porter, & L. Davis (Eds.), *The smallest anthropoids: The marmoset/callimico radiation* (pp. 259-277). Springer.

Garber, P. A., & Sussman, R. W. (1984). Ecological distinctions in sympatric species of *Saguinus* and *Sciureus*. *American Journal of Physical Anthropology, 65*, 135-146.

Garber, P. A., & Teaford, M. F. (1986). Body weights in mixed species troops of *Saguinus mystax mystax* and *Saguinus fuscicollis nigrifrons* in Amazonian Peru. *American Journal of Physical Anthropology, 71*, 331-336.

Gengozian, N., Batson, J. S., & Smith, T. A. (1977). Breeding of tamarins (*Saguinus* ssp.) in the laboratory. In D. G. Kleiman (Ed.), *The biology and conservation of the Callitrichidae* (pp. 163-171). Smithsonian Institution Press.

Goldizen, A. W. (1987). Tamarins and marmosets: Communal care of offspring. In D. L. Cheney, R. M. Seyfarth, R. W. Wrangham, & T. T. Struhsaker (Eds.), *Primate societies* (pp. 34-43). University of Chicago Press.

Goldizen, A. W. (1989). Social relationships in a cooperatively polyandrous group of tamarins (*Saguinus fuscicollis*). *Behavioral Ecology and Sociobiology, 24*, 79-89.

Goldizen, A. W. (1990). A comparative perspective on the evolution of tamarin and marmoset social systems. *International Journal of Primatology, 11*, 63-83.

Goldizen, A. W., Mendelson, J., van Vlaardingen, M., & Terborgh, J. (1996). Saddle-back tamarin (*Saguinus fuscicollis*) reproductive strategies: Evidence from a thirteen-year study of a marked population. *American Journal of Primatology, 38*, 57-84.

Griscom, B., & Ashton P. (2003). Bamboo control of forest succession: *Guadua sarcocarpa* in southwestern Peru. *Forest Ecological Management, 175*, 445-454.

Griscom, B, & Ashton, P. (2006). A self-perpetuation bamboo disturbance cycle in a Neotropical forest. *Journal of Tropical Ecology, 22*, 587-597.

Hearn, J. P. (1977). The endocrinology of reproduction in the common marmoset *Callithrix jacchus*. In D. G. Kleiman (Ed.), *The biology and conservation of the Callitrichidae* (pp. 163-177). Smithsonian Institution Press.

Heltne, P. G., Wojcik, J. F., & Pook, A. G. (1981). Goeldi's monkey, genus *Callimico*. In A. F. Coimbra-Filho, & R. A. Mittermeier (Eds.), *Ecology and behavior of Neotropical primates, Vol. 1.* (pp. 169-209). Academia Brasileira de Ciências.

Hernandez-Camacho, J., & Cooper, R. W. (1976). The non-human primates of Colombia. In R. W. Thorington Jr., & P. G. Heltne (Eds.), *Neotropical primates: Field studies and conservation* (pp. 35-69). National Academy of Sciences.

Hershkovitz, P. (1972). The recent mammals of the Neotropical region: A zoogeographic and ecological review. In A. Keast, F. Erk, & B. Glass (Eds.), *Evolution, mammals and southern continents* (pp. 311-341). State University of New York Press.

Hershkovitz, P. (1977). *Living New World monkeys (Platyrrhini), Vol. 1.* University of Chicago Press.

Heymann, E. W. (1987). A field observation of predation on a moustached tamarin (*Saguinas mystax*) by an anaconda. *International Journal of Primatology*, 8, 193-195.

Heymann, E. W. (1990a). Interspecific relations in a mixed species troop of moustached tamarins, *Saguinas mystax*, and saddle-backed tamarins, *Saguinus fuscicollis* (Platyrrhini: Callitrichidae), at the Rio Blanco, Peruvian Amazon. *American Journal of Primatology*, 21, 115-127.

Heymann, E. W. (1990b). Reactions of wild tamarins *Saguinas mystax* and *Saguinus fuscicollis* to avian predators. *International Journal of Primatology*, 11, 327-337.

Heymann, E. W. (1998). Sex differences in olfactory communication in a primate, the moustached tamarin, *Saguinus mystax* (Callitrichinae). *Behavioral Ecology and Sociobiology*, 43, 37-45.

Heymann, E. W. (2000). The number of adult males in callitrichine groups and its implications for callitrichine social evolution. In P. M. Kappeler (Ed.), *Primate males: Causes and consequences of variation in group composition* (pp. 64-71). Cambridge University Press.

Heymann E. W., & Buchanan-Smith, H. M. (2000). The behavioural ecology of mixed species troops of callitrichine primates. *Biol. Rev.*, 75, 169-190.

Heymann, E., & Smith, A. (1999). When to feed on gums: Temporal patterns of gummivory in wild tamarins, *Saguinus mystax* and *Saguinus fuscicollis* (Callitrichinae). *Zoo Biology*, 18, 459-471.

Hilario, R. R. & Ferrari, S. F. (2010a). Feeding ecology of a group of buffy-headed marmosets (*Callithrix flaviceps*): Fungi as a preferred resource. *American Journal of Primatology*, 72, 515-552.

Hilario, R. R., & Ferrari, S. F. (2010b). Four breeding females in a free-ranging group of buffy-headed marmosets (*Callithrix flaviceps*). *Folia Primatologica*, 81(1), 31-40.

Hladik, A., & Hladik, C. M. (1969). Rapports trophiques entre végétation et primates dans la forêt de Barro Colorado (Panama). *La Terre et La Vie*, 23, 25-117.

Hladik, C. M. (1970). Les singes du Nouveau Monde. *Science Nature*, 102, 1-9.

Howes, F. N. (1949). *Vegetable gums and resins.* Chronica Botanica Co.

Huck, M., Lottker, P., Bohle, U.-R., & Heymann, E. W. (2005). Paternity and kinship patterns in polyandrous moustached tamarins (*Saguinus mystax*). *American Journal of Physical Anthropology*, 127, 449-464.

IUCN. (2019). *International Union for the Conservation of Nature Red List of Threatened Species.* Version 2018-2. http://www.iucnredlist.org.

Izawa, K. (1976). Group sizes and compositions of monkeys in the upper Amazon Basin. *Primates*, 17, 367-399.

Izawa, K. (1978). A field study of the ecology and behavior of the black-mantled tamarin (*Saguinus nigricollis*). *Primates*, 19, 241-274.

Izawa, K. (1979). Studies on peculiar distribution pattern of *Callimico. Kyoto University Overseas Research Reports of New World Monkeys*, 1, 1-19.

Izawa K., & Yoneda, M. (1981). Habitat utilization of nonhuman primates in a forest of the western Pando, Bolivia. *Kyoto University Overseas Research Reports of New World Monkeys*, 2, 13-22.

Janzen, D. H. (1973a). Sweep samples of tropical foliage insects: Description of study sites, with data on species abundances and size distribution. *Ecology*, 54, 659-686.

Janzen, D. H. (1973b). Sweep samples of tropical foliage insects: Effects of seasons, vegetation types, elevation, time of day and insularity. *Ecology*, 54, 687-708.

Janzen, D. H., & Schoener, T. W. (1968). Differences in insect abundance and diversity between wetter and drier sites during a tropical dry season. *Ecology*, 49, 98-110.

Jaquish, C. E. (1993). *Genetic, behavioral and social effects on fitness components in marmosets and tamarins (Family: Callitrichidae).* [Doctoral dissertation, Washington University].

Judziewicz, E., Clark, L., Londoño, X., & Stern, M. (1999). *American bamboos.* Smithsonian Institution Press.

Kay, R. F. (1975). The functional adaptations of primate molar teeth. *American Journal of Physical Anthropology*, 43, 195-216.

Kierulff, M. C. M., Raboy, B. E., Procopio de Oiveira, P., Miller, K., Passos, F. C., & Prado, F. (2002). Behavioral ecology of lion tamarins. In D. G. Kleiman, & A. B. Rylands (Eds.), *Lion tamarins: Biology and conservation* (pp. 157-187). Smithsonian Institution Scholarly Press.

Kinzey, W. G. (1981). The titi monkeys, genus *Callicebus*. In A. F. Coimbra-Filho, & R. A. Mittermeier (Eds.), *Ecology and behavior of Neotropical primates* (Vol. 1, pp. 241-276). Academia Brasileira de Ciências.

Kinzey, W. G. (1997b). Synopsis of New World primates (16 genera). In W. G. Kinzey (Ed.), *New World primates: Ecology, evolution and behavior* (pp. 169-324). Aldine de Gruyter.

Kinzey, W. G., Rosenberger, A. L., & Ramirez, M. (1975). Vertical clinging and leaping in a Neotropical anthropoid. *Nature*, 255, 327-328.

Kleiman, D. G. (1977a). Characteristics of reproduction and sociosexual interactions in pairs of lion tamarins (*Leontopithecus rosalia*) during the reproductive cycle. In D. G. Kleiman (Ed.), *The biology and conservation of the Callitrichidae* (pp. 181-190). Smithsonian Institution Press.

Kleiman, D. G. (1977b). Monogamy in mammals. *Quarterly Review of Biology*, 52, 39-69.

Knogge, C., & Heymann, E. W. (2003). Seed dispersal by sympatric tamarins, *Saguinus mystax* and *Saguinus fuscicollis*: Diversity and characteristics of plant species. *Folia Primatologica*, 74, 33-47.

Koenig, A. (1995). Group size, composition, and reproductive success in wild common marmosets (*Callithrix jacchus*). *American Journal of Primatology*, 35, 311-317.

Koenig, A., & Rothe, H. (1991). Social relationships and individual contributions to cooperative behavior on common marmosets. *Primates, 32,* 183-195.

Lacher, T. E., Jr., Bouchardet da Fonseca, G. A., Alves, C., Jr., & Magalhaes-Castro, B. (1981). Exudate-eating, scent marking and territoriality in wild population of marmosets. *Animal Behaviour, 29,* 306-307.

Leutenegger, W. (1980). Monogamy in callitrichids: A consequence of phyletic dwarfism? *International Journal of Primatology, 1,* 95-98.

Lopes, M. A., & Ferrari, S. F. (1994). Foraging behavior in tamarin group (*Saguinus fuscicollis weddelli*) and interactions with marmosets (*Callithrix emiliae*). *International Journal of Primatology, 15,* 373-388.

Maier, W., Alonso, C., & Langguth, A. (1982). Field observations on *Callithrix jacchus jacchus* L. *Zeitschrift für Säugetierkunde, 47,* 334-346.

Mallott, E. K., Malhi, R. S., & Garber, P. A. (2015). High-throughput sequencing of fecal DNA to identify insects consumed by wild Weddell's saddleback tamarins (Cebidae, Primates) in Bolivia. *American Journal of Physical Anthropology, 156,* 474-481.

Martin, R. D. (1992). Goeldi and the dwarfs: The evolutionary biology of the small New World monkeys. *Journal of Human Evolution, 22,* 367-393.

Martins, M. M., & Setz, E. Z. F. (2000). Diet of buffy tufted-ear marmosets (*Callithrix aurita*) in a forest fragment in southeastern Brazil. *International Journal of Primatology, 21,* 467-476.

Mendes Pontes, A. R., & Monteiro da Cruz, M. A. O. (1995). Home range, intergroup transfers, and reproductive status of common marmosets *Callithrix jacchus* in a forest fragment. *Primates, 36,* 335-347.

Mittermeier, R. A., & van Roosmalen, M. G. M. (1981). Preliminary observations on habitat utilization and diet in eight Surinam monkeys. *Folia Primatologica, 36,* 1-39.

Mota, M. T, & Sousa, M. B. C. (2000). Prolactin levels of fathers and helpers related to alloparental care in common marmosets, *Callithrix jacchus. Folia Primatologica, 71,* 22-26.

Moynihan, M. (1970). Some behavior patterns of platyrrhine monkeys, II. *Saguinus geoffroyi* and some other tamarins. *Smithsonian Contributions to Zoology, 28,* 1-77.

Moynihan, M. (1976a). *The New World primates.* Princeton University Press.

Moynihan, M. (1976b). Notes on the ecology and behavior of the pygmy marmoset (*Cebuella pygmaea*). In R. W. Thorington Jr., & P. G. Heltne (Eds.), *Neotropical primates: Field studies and conservation* (pp. 79-84). National Academy of Sciences.

Muskin, A. (1984). Preliminary observations on *Callithrix aurita* (Callitrichidae, Cebidae). In M. T. de Mello (Ed.), *A primatologia no Brasil* (pp. 79-82). Sociedade Brasileira de Primatologia.

Napier, J. R., & Napier, P. H. (1967). *A handbook of living primates.* Academic Press.

Nash, L. T. (1986) Dietary, behavioral, and morphological aspects of gummivory in primates. *Yearbook of Physical Anthropology, 29,* 113-137.

Natori, M. (1986). Interspecific relationships of *Callithrix* based on dental characters. *Primates, 27,* 321-336.

Natori, M. (1990). Numerical analysis of taxonomical status of *Callithrix kuhli* based on measurements of the postcanine dentition. *Primates, 31,* 555-562.

Natori, M., & Shigehara, M. (1992). Interspecific differences in lower dentition among eastern Brazilian marmosets. *Journal of Mammalogy, 73,* 668-671.

Neyman, P. F. (1977). Aspects of the ecology and social organization of free-ranging cotton-top tamarins *Saguinus oedipus* and the conservation status of the species. In D. G. Kleiman (Ed.), *The biology and conservation of the Callitrichidae* (pp. 39-71). Smithsonian Institution Press.

Nickle, D. A., & Heymann, E. W. (1996). Predation on orthoptera and other orders of insects by tamarin monkeys, *Saguinus mystax mystax* and *Saguinus fuscicollis nigrifrons* (Primates: Callitrichidae), in north-eastern Peru. *Journal of the Zoological Society of London, 239,* 799-819.

Nievergelt, C. M., Digby, L. J., Ramiakrishnan, U., & Woodruff, D. S. (2000). Genetic analysis of group composition and breeding system in a wild common marmoset (*Callithrix jacchus*) population. *International Journal of Primatology, 21,* 1-20.

Norconk, M. A. (1986). *Interaction between primate species in a near tropical forest: Mixed species troops* of Saguinus mystax *and* S. fuscicollis *(Callitrichidae).* [Doctoral dissertation, University of California, Los Angeles].

Oliveira, A., & Ferrari, S. F. (2008). Habitat exploitation by free-ranging *Saguinus niger* in eastern Amazonia. *International Journal of Primatology, 29,* 1499-1510.

Oliveira, L., Neves, L., Raboy, B., & Dietz, J. (2011). Abundance of jackfruit (*Artocarpus heterophyllus*) affects group characteristics and use of space by golden-headed lion tamarins (*Leontopithecus chrysomelas*) in Cabruca agroforest. *Environmental Management, 48,* 248-262.

Opazo. J., Wildman, E., Pyrchitko, T., Johnson, R., & Goodman, M. (2006). Phylogenetic relationships and divergence times among New World monkeys (Platyrrhini, Primates). *Molecular Phylogeny and Evolution, 40,* 274-280.

Pack, K. S., Henry, O., & Sabatier, D. (1999). The insectivorous-frugivorous diet of the golden-handed tamarin (*Saguinas midas midas*) in French Guiana. *Folia Primatologica, 70,* 1-7.

Passamani, M. (1995). Field observation of a group of Geoffroy's marmosets mobbing a margay cat. *Folia Primatologica, 64,* 163-166.

Passamani, M. (1998). Activity budget of Geoffroy's marmoset (*Callithrix geoffroyi*) in an Atlantic Forest in southeastern Brazil. *American Journal of Primatology, 46,* 333-340.

Passos, F. C., & Carvalho, C.T. (1991). Importancia de exudatos na alimentacao do mico-leao preto, *Leontopithecus chrysopygus* (Callitrichidae, Primates). *Resumos. XVIII*

Congresso Brasileiro de Zoologia. Salvasor, *Universidad Federal da Bahia*, 392.

Patterson, B., & Pascual, R. (1972). The fossil mammalian fauna of South America. In A. Keast, F. Erk, & B. Glass (Eds.), *Evolution, mammals and southern continents* (pp. 247-309). University of New York Press.

Peres, C. A. (1986). *Costs and benefits of territorial defense in golden lion tamarins,* Leontopithecus rosalia. [Doctoral dissertation, University of Florida].

Peres, C. A. (1989a). Exudate-eating by wild golden lion tamarins, *Leontopithecus rosalia. Biotropica,* 21, 287-288.

Peres, C. A. (1989b). Costs and benefits of territorial defense in wild golden lion tamarins, *Leontopithecus rosalia. Behavioral Ecology and Sociobiology,* 25, 287-288.

Peres, C. A. (1991). *Ecology of mixed-species groups of tamarins in Amazonian terra firme forests.* [Doctoral dissertation, Cambridge University].

Peres, C. A. (1992). Prey-capture benefits in mixed species group of Amazonian tamarins, *Saguinus fuscicollis* and *S. mystax. Behavioral Ecology and Sociobiology,* 31, 339-347.

Peres, C. A. (1993). Anti-predation benefits in a mixed-species group of Amazonian tamarins. *Folia Primatologica,* 61, 61-76.

Peres, C. A. (1996). Food patch structure and plant resource partitioning in interspecific associations of Amazonian tamarins. *International Journal of Primatology,* 17, 695-723.

Peres C. A. (2000a). Identifying keystone plant resources in tropical forests: The case of gums from *Parkia* pods. *Journal of Tropical Ecology,* 16, 287-317.

Peres, C. A. (2000b). Territorial defense and the ecology of group movements in small-bodied Neotropical primates. In S. Boinski, & P. A. Garber (Eds.), *On the move: How and why animals travel in groups* (pp. 100-124). Chicago University Press.

Pook, A. G., & Pook, G. (1981). A field study of the socioecology of the Goeldi's monkey (*Callimico goeldii*) in northern Bolivia. *Folia Primatolo*gica, 35, 288-312.

Pook, A. G., & Pook, G. (1982). Polyspecific association between *Saguinus fuscicollis, Saguinus labiatus, Callimico goeldii* and other primates in north-western Bolivia. *Folia Primatologica,* 35, 196-216.

Porter, C. A., Czelusniak, J., Schneider, H., Schneider, M. P. C., Sampaio, I., & Goodman, M. (1997). Sequences of primate e-globin gene: Implications for systematics of the marmosets and other New World primates. *Gene,* 205, 59-71.

Porter, L. M. (2001a). Benefits of polyspecific associations for the Goeldi's monkey (*Callimico goeldii*). *American Journal of Primatology,* 54, 143-158.

Porter, L. M. (2001b). *Callimico goeldii* and *Saguinus*: Dietary differences between sympatric callitrichines in northern Bolivia. *American Journal of Physical Anthropology,* 22, 961-992.

Porter, L. M. (2001c). Social organization, reproduction and rearing strategies of *Callimico goeldii*: New clues from the wild. *Folia Primatologica,* 72, 69-79.

Porter, L. M. (2004). Differences in forest utilization and activity patterns among three sympatric callitrichines: *Callimico goeldii, Saguinus fuscicollis* and *S. labiatus. American Journal of Physical Anthropology,* 124, 139-153.

Porter, L. M. (2006). Distribution and density of *Callimico goeldii* in northwestern Bolivia. *American Journal of Primatology,* 68, 235-243.

Porter, L. M. (2007). *The behavioral ecology of callimicos and tamarins in northwestern Bolivia.* Prentice Hall.

Porter, L. M., & Garber, P. A. (2004). Goeldi's monkeys: A primate paradox? *Evolutionary Anthropology,* 13, 104-115.

Porter, L. M., & Garber, P. A. (2007). Niche expansion in a cryptic primate (*Callimico goeldii*) during polyspecific associations. *American Journal of Primatology,* 69, 1-14.

Porter, L. M., & Garber, P. A. (2009). Social behavior of *Callimico*: Mating strategies and infant care. In S. Ford, L. Porter, & L. Davis (Eds.), *The smallest anthropoids: The marmoset/callimico radiation* (pp. 87-102). Springer.

Porter, L. M., & Garber, P. A. (2010). Mycophagy and its influence on habitat use and ranging patterns in *Callimico goeldii. American Journal of Physical Anthropology,* 142, 468-475.

Porter, L. M., & Garber, P. A. (2013). Foraging and spatial memory in Weddell's saddleback tamarins (*Saguinus fuscicollis weddelli*) when moving between distant and out-of-sight goal. *International Journal of Primatology,* 34, 30-48.

Porter, L. M., Garber, P. A., & Nacimento, E. (2009). Exudates as a fallback food for *Callimico goeldii. American Journal of Primatology,* 71, 120-129.

Porter, L. M., Sterr, S. M., & Garber, P. A. (2007). Habitat use, diet, and ranging patterns of *Callimico goeldii. International Journal of Primatology,* 28, 1035-1058.

Power, M. L. (2010). Nutritional and digestive challenges to being a gum-feeding primate. In A. Burrows, & L. Nash (Eds.), *The evolution of exudativory in primates* (pp. 25-44). Springer.

Prado, F., & Valladares-Padua, C. (2004). Feeding ecology of a group of black-headed lion tamarin, *Leontopithecus caissara* (Primates: Callithrichidae), in the Superagui National Park, Southern Brazil. *Primatologia no Brasil,* 8, 145-154.

Prudom, S. L., Broz, C. A., Schultz-Karken, N., Ferris, C. T., Snowdon, C., & Ziegler, T. E. (2008). Exposure to infant scent lowers serum testosterone in father common marmosets (*Callithrix jacchus*). *Biology Letters,* 4, 603-605.

Puertas P., Encarnación F., Aquino R., & Garcia J. (1995). Analisis poblacional del pichico pecho anaranjado, *Saguinus labiatus,* en el sur oriente Peruano. *Neotropical Primates,* 3, 4-7.

Raboy, B. E., Canale, G. R., & Dietz, J. M. (2008). Ecology of *Callithrix kuhlii* and a review of eastern Brazilian marmosets. *International Journal of Primatology,* 29, 449-467.

Raboy, B. E., & Dietz, J. (2004). Diet, foraging, and use of space in wild golden-headed lion tamarins. *American Journal of Primatology,* 63, 1-15.

Ramirez, M. F., Freese, C. H., & Revilla, J. (1977). Feeding ecology of the pygmy marmoset, *Cebuella pygmaea*, in northeastern Peru. In D. G. Kleiman (Ed.), *The biology and conservation of the Callitrichidae* (pp. 91-104). Smithsonian Institution Press.

Redican, W. K., & Taub, D. M. (1981). Male parental care in monkeys and apes. In M. E. Lamb (Ed.), *The role of the father in child development* (2nd Ed., pp. 203-258). John Wiley & Sons.

Rehg, J. A. (2009). Ranging patterns of *Callimico goeldii* (callimico) in a mixed species group. In S. Ford, L. Porter, & L. Davis (Eds.), *The smallest anthropoids: The marmoset/callimico radiation* (pp. 241-258). Springer.

Rosenberger, A. L., & Coinbra-Filho, A. F. (1984). Morphology, taxonomic status, and affinities of the lion tamarins *Leontopithecus* (Callitrichidae, Primates). *Folia Primatologica, 42,* 149-179.

Ross, C. N., French, J. A., & Orti, G. (2007). Germ-line chimerism and paternal care in marmosets (*Callithrix kuhlii*). *Proceedings of the National Academy of Sciences, 104,* 6278-6282.

Rothe, H. (1975). Some aspects of sexuality and reproduction in groups of captive marmosets (*Callithrix jacchus*). *Zeitschrift für Tierpsychologie, 37,* 255-273.

Rowe, N., & Jacobs, R. (2016). Tamarins and marmosets, subfamily Callitrichinae. In N. Rowe, & M. Myers (Eds.), *All the world's primates* (pp. 329-330). Pogonias Press.

Rowe, N., & Myers, M., Eds. (2016). *All the world's primates* (pp. 329-330). Pogonias Press.

Rutherford, J. N., & Tardif, S. D. (2009). Mother's little helper? The placenta and its role in intrauterine maternal investment in the common marmoset (*Callithrix jacchus*). In S. M. Ford, L. M. Porter, & L. C. Davis (Eds.), *The smallest anthropoids: The marmosets/callimico radiation* (pp. 301-329). Springer.

Rylands, A. B. (1981). Preliminary field observations on the marmoset, *Callithrix humeralifer intermedius* (Hershkovitz, 1977) at Dardanelos, Rio Aripuana, Mato Grasso. *Primates, 22,* 46-59.

Rylands, A. B. (1982). *The behavior and ecology of three species of marmosets and tamarins (Callitrichidae, Primates) in Brazil.* [Doctoral dissertation, University of Cambridge].

Rylands, A. B. (1986a). Ranging behavior and habitat preference of a wild marmoset group, *Callithrix humeralifer* (Callitrichidae, Primates). *Journal of the Zoological Society of London, 210*(4), 489-415.

Rylands, A. B. (1986b). Infant-carrying in a wild marmoset group, *Callithrix humeralifer:* Evidence for a polyandrous mating system. In M. T. de Mello (Ed.), *A primatologia no Brasil, II* (pp. 131-144). Sociedade Brasileira de Primatologia.

Rylands, A. B. (1987). Primate communities in Amazonian forests: Their habitats and food resources. *Experientia, 43,* 265-279.

Rylands, A. B. (1989). Sympatric Brazilian callitrichids: The black tufted-ear marmoset, *Callithrix kuhlii*, and the golden-headed lion tamarin, *Leontopithecus chrysomelus. Journal of Human Evolution, 18,* 679-695.

Rylands, A. B. (1993). The ecology of the tamarins, *Leontopithecus*: Some intrageneric differences and comparisons with other callitrichids. In A. B. Rylands (Ed.), *Marmosets and tamarins: Systematics, behavior, and ecology* (pp. 296-313). Oxford University Press.

Rylands, A. B. (1996). Habitat and the evolution of social and reproductive behavior in Callitrichidae. *American Journal of Primatology, 38,* 5-18.

Rylands, A. B., Coimbra-Filho, A. F., & Mittermeier, R. A. (2009). The systematics and distributions of the marmosets (*Callithrix, Callibella, Cebuella* and *Mico*) and Callimico (*Callimico*) (Callitrichidae: Primates). In S. M. Ford, L. M. Porter, & L. C. Davis (Eds.), *The smallest anthropoids: The marmosets/callimico radiation* (pp. 25-62). Springer.

Rylands, A. B., & de Faria, D. S. (1993). Habitats, feeding ecology and home range size in the genus *Callithrix*. In A. B. Rylands (Ed.), *Marmosets and tamarins: Systematics, behavior, and ecology* (pp. 262-272). Oxford University Press.

Rylands, A. B, Heymann, E. W., Lynch Alfara, J., Buckner, J. C., Roos C., Matauschek, C., Boubli, J. P., Sampaio, R., & Mittermeier, R. A. (2016). Taxonomic review of the New World tamarins (Primates, Callitrichidae). *Zoological Journal of the Linnean Society, 177,* 1003-1028. https://doi.org/10.1111/zoj.12386.

Rylands, A. B., & Mittermeier, R. A. (2009). The diversity of the New World primates (Platyrrhini: An annotated taxonomy). In P. A. Garber, A. Estrada, J. C. Bicca-Marques, E. W. Heymann, & K. B. Strier (Eds.), *South American primates: Comparative perspectives in the study of behavior, ecology, and conservation* (pp. 23-54). Springer.

Saltzman, W., Schultz-Darkin, N. J., & Abbott, D. H. (1997). Familial influences on ovulatory function in common marmosets (*Callithrix jacchus*). *American Journal of Primatology, 41,* 159-177.

Savage, A. (1990). *The reproductive biology of the cotton-top tamarin (*Saguinus oedipus oedipus*) in Colombia.* [Doctoral dissertation, University of Wisconsin].

Savage, A., Giraldo, L. H., Soto, L. H., & Snowdon, C. T. (1996a). Demography, group composition, and dispersal in wild cotton-top tamarin (*Saguinus oedipus*) groups. *American Journal of Primatology, 38,* 85-100.

Savage, A., Shideler, S. E., Soto, L. H., Causado, J., Ciraldo, L. H., Lasley, B. L., & Snowdon, C. T. (1997). Reproductive events of wild cotton-top tamarins (*Saguinus oedipus*) in Columbia. *American Journal of Primatology, 43,* 329-337.

Savage, A., Snowden, C. T., Giraldo, L. H., & Soto, L. H. (1996b). Parental care patterns and vigilance in wild cotton-top tamarins (*Saguinus oedipus*). In M. A. Norconk, A. L. Rosenberger, & P. A. Garber (Eds.), *Adaptive radiations of Neotropical primates* (pp. 187-199). Plenum Press.

Schneider, H., & Rosenberger, A. L. (1996). Molecules, morphology, and platyrrhine systematics. In M. A. Norconk,

A. L. Rosenberger, & P. A. Garber (Eds.), *Adaptive radiations of Neotropical primates* (pp. 3-19). Plenum Press.

Seuanez, H. N., Di Fiore, A., Moreira, M. A. M., Almeida, C. A. D. S., & Canavez, F. C. (2002). Genetics and evolution of lion tamarins. In D. G. Kleiman, & A. B. Rylands (Eds.), *Lion tamarins: Biology and conservation* (pp. 117-12). Smithsonian Institution Scholarly Press.

Shanee, S., Shanee, N., & Allgas-Marchena, N. (2013). Primate surveys in the Marañón-Huallaga Landscape, Northern Peru with notes on conservation. *Primate Conservation, 27*, 3-11.

Skutch, A. F. (1961). Helpers among birds. *Condor, 63*, 198-226.

Smith, R. J., & Jungers, W. L. (1997). Body mass in comparative primatology. *Journal of Human Evolution, 32*, 523-559.

Snowdon, C. T., & Soini, P. (1988). The tamarins, genus *Saguinus*. In R. A. Mittermeier, A. F. Coimbra-Filho, & G. A. B. da Fonseca (Eds.), *Ecology and behavior of Neotropical primates* (Vol. 2, pp. 223-298). World Wildlife Fund.

Soini, P. (1982). Ecology and population dynamics of the pygmy marmoset, *Cebuella pygmaea*. *Folia Primatologica, 39*, 1-21.

Soini, P. (1987). Ecology of the saddle-backed tamarin *Saguinus fuscicollis illigeri* on the Rio Pacaya, northeastern Peru. *Folia Primatologica, 49*, 11-32.

Soini, P. (1988). The pygmy marmoset, genus *Cebuella*. In R. A. Mittermeier, A. F. Coimbra-Filho, & G. A. B. da Fonseca (Eds.), *Ecology and behavior of Neotropical primates* (Vol. 2, pp. 79-129). World Wildlife Fund.

Soini, P. (1993). The ecology of the pygmy marmoset, *Cebuella pygmaea*: Some comparisons with two sympatric tamarins. In A. B. Rylands (Ed.), *Marmosets and tamarins: Systematics, behavior, and ecology* (pp. 257-261). Oxford University Press.

Stacey, P. B. (1979). Kinship, promiscuity, and communal breeding in the acorn woodpecker. *Behavioral Ecology and Sociobiology, 6*, 53-66.

Stacey, P. B. (1982). Female promiscuity and male reproductive success in social birds and mammals. *American Naturalist, 120*, 51-64.

Stallings, J. R. (1988). *Small mammal communities in an eastern Brazilian park*. [Master's thesis, University of Gainesville].

Stevenson, M. F., & Rylands, A. B. (1988). The marmoset monkeys, genus *Callithrix*. In R. A. Mittermeier, A. F. Coimbra-Filho, & G. A. B. da Fonseca (Eds.), *Ecology and behavior of Neotropical primates* (Vol. 2, pp. 131-222). World Wildlife Fund.

Suarez, S. S. (2007). *Paternity, relatedness and socio-reproductive behavior in a population of wild red-bellied tamarins* (Saguinus labiatus*)* [Doctoral dissertation, New York University].

Sussman, R. W., & Garber, P. A. (1987). New interpretation of the social organization and mating system of the Callitrichidae. *International Journal of Primatology, 8*, 73-92.

Sussman, R. W., & Kinzey, W. G. (1984). The ecological role of the Callitrichidae: A review. *American Journal of Physical Anthropology, 64*, 419-444.

Sussman, R. W., & Phillips-Conroy, J. E. (1995). A survey of the distribution and density of the primates of Guyana. *International Journal of Primatology, 16*, 761-791.

Szalay, F. S., & Delson, E. (1979). *Evolutionary history of the primates*. Academic Press.

Tardif, S. D., Smucny, D. A., Abbott, D. H., Mansfield, K., Schultz-Darken, N., & Yamomato, M. E. (2003). Reproduction in captive common marmosets (*Callithrix jacchus*). *Comparative Medicine, 53*, 364-368.

Terborgh, J. (1983). *Five New World primates: A study in comparative ecology*. Princeton University Press.

Terborgh, J. (1985). The ecology of Amazon primates. In G. T. Prance, & T. E. Lovejoy (Eds.), *Amazonia* (pp. 284-304). Pergamon Press.

Terborgh, J., & Goldizen, A. W. (1985). On the mating system of the cooperatively breeding saddle-back tamarin (*Saguinus fuscicollis*). *Behavioral Ecology and Sociobiology, 16*, 293-299.

Thorington, R. W., Jr. (1968). Observations of the tamarin, *Saguinus midas*. *Folia Primatologica, 9*, 95-98.

Tovar, N. V. (1994). Activity patterns of *Saguinus nigricollis hernandezi* at the Tinigua National Park, Colombia. *Field Studies of New World Monkeys, La Macarena, Columbia, 9*, 23-31.

Valladares-Padua, C. (1993). *The ecology, behavior and conservation of the black lion tamarin* (Leontopithecus chrysopygus, *Mikan, 1823*). [Doctoral dissertation, University of Florida].

van Roosmalen, M. G. M., & van Roosmalen, T. (2003). The description of a new marmoset genus, *Callibella* (Callitrichinae, Primates) including its molecular phylogenetic status. *Neotopical Primates, 11*, 1-10.

Veracini, C. (2009). Habitat use and ranging behavior of the silvery marmoset (*Mico argentatus*) at Caxiuana National Forest (Eastern Brazilian Amazonia). In S. M. Ford, L. M. Porter, & L. C. Davis (Eds.), *The smallest anthropoids: The marmosets/callimico radiation* (pp. 221-240). Springer.

Vinyard, C. J., Wall, C. E., Williams, S. H., Mork, A. L., Armfield, B. A., de Oliveira Melo, L. C., Valenca-Montenegro, M. M., Valle, Y. B. M., de Oliveira, M. A. B., Lucas, P. W., Schmidt, D., Taylor, A. B., & Hylander, W. L. (2009). The evolutionary morphology of tree gouging in marmosets. In S. M. Ford, L. M. Porter, & L. C. Davis (Eds.), *The smallest anthropoids: The marmosets/callimico radiation* (pp. 395-410). Springer.

Webb, S. D. (1978). A history of savanna vertebrates in the New World. Part II. South America and the great faunal interchange. *Annual Review of Ecology and Systematics, 9*, 393-426.

Wright, P. C. (1981). The night monkeys, genus *Aotus*. In A. F. Coimbra-Filho, & R. A. Mittermeier (Eds.), *Ecology and behavior of Neotropical primates* (Vol. 1, pp. 211-240). Academia Brasileira de Ciências.

Yamamoto, M. E., Arruda, M. F., Alencar, A. I., Sousa, M. B. C., & Araujo, A. (2009). Mating systems and female-female competition in the common marmoset, *Callithrix jacchus*.

In S. M. Ford, L. M. Porter, & L. C. Davis (Eds.), *The smallest anthropoids: The marmosets/callimico radiation* (pp. 119-133). Springer.

Yepez, P, de la Torre, S., & Snowdon, C. T. (2005). Interpopulation differences in exudate feeding of pygmy marmosets in Ecuadorian Amazonia. *American Journal of Primatology, 66,* 145-158.

Yoneda, M. (1981). Ecological studies of *Saguinus fuscicollis* and *Saguinus labiatus* with reference to habitat segregation and height preference. *Kyoto University Overseas Research Reports of New World Monkeys, 2,* 43-50.

Yoneda, M. (1984a). Comparative studies on vertical separation, foraging behavior and traveling mode of saddle-backed tamarins (*Saguinus fuscicollis*) and red chested moustached tamarins (*Saguinus labiatus*) in northern Bolivia. *Primates, 25,* 414-442.

Yoneda, M. (1984b). Ecological study of the saddle-backed tamarin (*Saguinus fuscicollis*) in Northern Bolivia. *Primates, 25,* 1-12.

Youlatos, D. (1999). Positional behavior of *Cebuella pygmaea* in Yasuni National Park, Ecuador. *Primates, 40,* 543-550.

Youlatos, D. (2009). Locomotion, postures, and habitat use in pygmy marmosets (*Cebuella pygmaea*). In S. M. Ford, L. M. Porter, & L. C. Davis (Eds.), *The smallest anthropoids: The marmosets/callimico radiation* (pp. 279-297). Springer.

Zullo, J., & Caine, N. (1988). The use of sentinels by captive red-bellied tamarins. *American Journal of Primatology, 14,* 455.

CHAPTER 7

Owl Monkeys
Under the Moonlight

ANNEKE M. DELUYCKER

OWL MONKEYS, ALSO KNOWN AS NIGHT MONKEYS OR douroucoulis, are found in the genus *Aotus* and are the only nocturnal taxa of anthropoids. They live in small groups composed of an adult female, adult male, and their offspring (see Figure 7.1). Owl monkeys are also characterized by strong male parental care of offspring. When first described, the genus only included a single species, *Aotus trivirgatus*. The most recent phenotypic, karyotypic, and molecular genetic data have led researchers to recognize between 11 and 12 species within the genus *Aotus* (Defler & Bueno, 2007; Menezes et al., 2010). Species of owl monkeys are characterized by gray-tan to brown fur coloration, with either gray or red fur on the sides and underbellies. In all species, there are three distinct black stripes on the face, two encircling each side of the face and one extending from the forehead down to the bridge of the nose. They are divided phenotypically, karyotypically, and geographically into two groups, the "grey-necked" group, found north of the Rio Amazonas, and the "red-necked" group, found south of the Rio Amazonas (Groves, 2001). Most species of *Aotus* are found in lowland tropical forest, but some species can be found in higher altitude montane forest (between 2,000–3,000 masl).

The first studies on owl monkeys were conducted in captivity or were short-term. Moynihan (1964) studied the behavioral patterns of *Aotus* in captivity, and Thorington et al. (1976) released a young captive male on Barro Colorado Island in Panama and radio-tracked it for nine days. After this latter study, the authors concluded that it probably would be impossible to conduct a study of *Aotus* without using radio-tracking. Wright, however, conducted a 2-month (400 hour) study of *Aotus* in a primary tropical forest in Peru (Wright, 1978), and a 20-month study in Peru and Paraguay (Wright, 1984, 1985, 1986, 1990, 1994). These studies were achieved by observation only; however, "individuals often could not be identified" (Wright, 1985: p.27), supporting the fact that direct observation studies of nocturnal, arboreal owl monkeys are extremely difficult, especially in tropical forests. Most species of *Aotus* are strictly nocturnal, but at least one species, the Azara's owl monkey (*Aotus*

azarae) inhabiting the Gran Chaco forests of Argentina, Paraguay, and Bolivia, regularly exhibits both nocturnal and diurnal activity patterns. This type of activity pattern, termed "cathemerality," can aid in making observations that are otherwise difficult or impossible in a nocturnal setting. Fernandez-Duque and Erkert (2006) suggest the best way to obtain reliable, quantitative data on activity movements at night is through the use of an accelerometer or data logger device. Juarez et al. (2011) give a review of the benefits and drawbacks of using radio collars for long-term free-ranging owl monkey studies.

Other studies on *Aotus* have been conducted in northern Argentina (Rathbun & Gache, 1980; Arditi, 1992), northeastern Peru (Aquino & Encarnación, 1986a, 1988; Aquino et al., 1990), Bolivia (Garcia & Braza, 1987, 1989, 1993), Ecuador (Fernandez-Duque et al., 2008a), and Colombia (Solano, 1995; Marín-Gómez, 2008). Since 1996, a long-term field site (The Owl Monkey Project) has been established in the Formosa Province of the Gran Chaco in Argentina, a region of gallery forest where studies of *A. azarae*, the only reported cathemeral owl monkey, are being conducted (Fernandez-Duque et al., 2001, 2002, 2006, 2008b, 2010; Fernandez-Duque & Huntington, 2002; Fernandez-Duque, 2003, 2009, 2016; Fernandez-Duque & Rotundo, 2003; Rotundo et al., 2005; Fernandez-Duque & Erkert, 2006; Wolovich et al., 2008a; Huck et al., 2011, 2014a,b; Huck & Fernandez-Duque, 2012, 2013, 2017; Fernandez-Duque & van der Heide, 2013; Corley et al., 2017a,b; Savagian & Fernandez-Duque, 2017).

DISTRIBUTION, HABITAT, AND LOCOMOTION

The genus *Aotus* has one of the most extensive geographical ranges of any Neotropical primate species (Figure 7.2), ranging from Panama to northern Argentina, and from lowland forests at sea level to almost 3,000 masl in the Andes Mountains (Rylands et al., 2000, 2006; Defler, 2004; Cornejo et al., 2008). It is a small monkey, ranging from 800–1,250 g in body mass, and there is no marked sexual dimorphism. Owl monkeys do not have prehensile tails; they are quadrupedal, moving on the tops of branches, but detailed analysis of their locomotion and

Figure 7.1 Family group of *Aotus vociferans* (Spix's night monkey); male, female, and offspring at Puerto Asís, Putumayo, Colombia.
(Photo credit: Martejas, Conservación del medio ambiente; https://gramho.com/media/2230175879850163010)

substrate use is not available. They are found in all strata of the forest, from near the ground to the top of the canopy (Moynihan, 1964; Durham, 1975; Wright, 1978), but rarely use the forest floor (Wright, 1981). Owl monkeys are found in a wide range of forest habitats, including primary, secondary, montane, and gallery forest (Rathbun & Gache, 1977; Aquino & Encarnación, 1994; Wright, 1994) as well as heavily altered environments, including forest remnants and fragments (Castaño et al., 2010; Shanee et al., 2013).

DIET

Aotus is a fruit eater that supplements its diet with flowers and nectar, young leaves, and insects. From stomach contents, Hladik et al. (1971) estimated the diet as follows: 65% fruit, 30% foliage (including shoots, young leaves, pith, buds, flowers, and sap), and 8% animal prey. Wright (1985, 1989) observed *Aotus* feeding almost exclusively on fruit during the wet season. Leaf eating and insect consumption are nearly impossible to quantify in a nocturnal *Aotus* species (Wright, 1985). In studies of cathemeral *A. azarae*, Wright (1985) in Paraguay and Arditi (1992) in Argentina reported owl monkey groups

consuming 40% leaves during the diurnal and crepuscular periods. However, Giménez (2004) reported *A. azarae* consuming only 15% leaves. The diet of one group of *A. miconax* in a Peruvian montane cloud forest consisted of 42% fruit, 5% flowers, 6% leaves, 25% buds, and 19% insects (Shanee et al., 2013). In captivity, *A. nancymaae* spent a significant amount of time engaged in insect-foraging and also consumed frogs, lizards, geckos, and a bird (Wolovich et al., 2010). Insect foraging most often occurs at dawn or dusk, consists of grabbing flying insects out of the air or from branches, and includes orthopterans, beetles, moths, and spiders as prey (Wright, 1989).

Many of the fruits consumed by *Aotus* at night are also utilized by sympatrically occurring species of larger-bodied monkeys during the day. During the drier months at Manú, Peru, *Aotus* ate mainly from large-crowned, superabundant sources, such as figs and *Brosimum*, and floral nectar. In fact, approximately 90% of their diet consisted of nectar in the month of July. In the harsh dry season of the Argentinean Chaco, groups of *A. azarae* consumed fruit from crucial fruiting species within core areas of exclusive use, including *Chrysophyllum gonocarpum*, *Guazuma ulmifolia*, and *Ficus* spp. (Fernandez-Duque & van der Heide, 2013).

Flowers may be important food for *Aotus* during resource-poor seasons. During the dry season, the diet of *Aotus* is very similar to that of *Lemur mongoz* in Madagascar. At Manú, owl monkeys feed on the flowers of *Combretum fruticosum* and *Quararibea cordata* (Janson et al., 1981). Much like the flowers utilized by *L. mongoz* in Madagascar and by marsupials in Australia, these flowers have characteristics adapted for nonflying mammal pollinators (Sussman & Tattersall, 1976). In the Chaco of Paraguay and Argentina, when fruit availability is low, *A. azarae* regularly feeds on flowers of *Tabebuia heptaphylla* (Wright, 1989). In addition, floral nectar consumption may provide an important food resource in fragmented habitats. In an agro-forestry area dominated by *Inga edulis* in Colombia, Marín-Gómez (2008) observed *A. lemurinus* eating the fruit aril and flower nectar from *Inga* trees.

ACTIVITY CYCLE

Aotus is the only species of anthropoid active at night. It descends from its sleeping tree 15 minutes after sunset and returns just before dawn (Thorington et al., 1976; Wright, 1978). Sometime during the middle of the night, it usually has one or two rest periods. The amount of rest during the evening depends upon the season, the availability of fruit, and upon the fullness of the moon. Owl monkeys are more active during the wet season when more fruit is available, and during nights of full moon

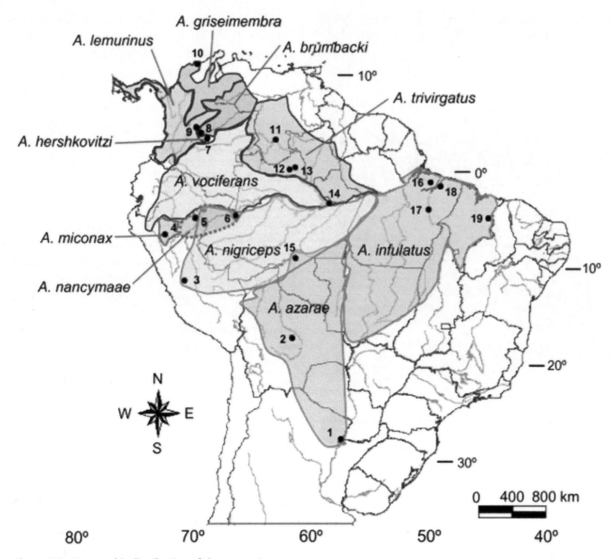

Figure 7.2 Geographic distribution of the genus *Aotus*.
(Map adapted from Menezes et al., 2010; image licensed under Creative Commons Attribution 2.0 Generic license)

when locomotor activity, feeding activity, and the number of intergroup agonistic encounters increases (Wright, 1981).

In the southernmost portion of its range, within forests of dry Chaco in Argentina and Paraguay, *A. azarae* has been observed to be active and foraging during the day (Rathbun & Gache, 1977). The Gran Chaco region is characterized by a mosaic of grasslands, dry forests, and gallery forests and has pronounced seasonality of rainfall, food resources, temperature, and day length (Fernandez-Duque et al., 2002). Wright (1985, 1989) believes that diurnal activity in these populations is related to competitive release (from larger, frugivorous monkeys) and an absence of diurnal predators. Indeed, only one other monkey lives in sympatry with *A. azarae* in the Argentinean Chaco: the black howler monkey (*Alouatta caraya*). Other hypotheses for cathemerality point to harsher, unstable environments or pronounced seasonality of food

resources (Overdorff & Rasmussen, 1995). However, Garcia and Braza (1987) and Fernandez-Duque (2003) suggest that thermoregulation, influenced by moonlight, may play a stronger role in influencing cathemerality in *A. azarae* in the Argentinean Chaco. For this species, nocturnal activity increased during nights of full moonlight, and diurnal activity was affected by changes in ambient temperatures, depending on moon phase. Specifically, diurnal activity was more pronounced during the colder winter months, presumably to take advantage of the warmer, brighter part of the day (Fernandez-Duque & Erkert, 2006). Avoiding activity on extremely cold, moonlight nights seems to be an efficient thermoregulatory strategy for this species (Fernandez-Duque, 2003).

In all studies conducted on species of *Aotus* in tropical forests, owl monkeys sleep during the daytime inside tree holes (Aquino & Encarnación, 1986a; Wright, 1989). In contrast, in the Argentinean Chaco, cathemeral-living owl monkeys (i.e., *A. azarae*) do not sleep inside tree holes, but instead choose dense vine tangles, dense foliage thickets, or open branches as sleeping sites (Rathbun & Gache, 1980; Wright, 1989; Garcia & Braza, 1993). Garcia and Braza (1993) argue that abundant lianas help protect against predators and sleeping on open branches confers a thermal advantage during sunny daytime hours. In a study by Savagian and Fernandez-Duque (2017), *A. azarae* group members slept more frequently under dense foliage during the wet, hot summer and preferred to sleep at sites with light foliage, which provides more direct sun exposure during the dry, colder winter. In tropical forests of Peru, *A. nigriceps* used 3–5 sleeping trees (Wright, 1978, 1983), returning to one of three sleeping sites, located within 30 m of each other, for a period of nine weeks (Wright, 1981). In Ecuador, *A. vociferans* showed sleeping site preferences and used only 5 sleeping trees, often staying in one tree over several days before changing to another (Fernandez-Duque et al., 2008a). However, in the Paraguayan Chaco, one group of *Aotus* used 42 different sleeping sites, 12 of which were used more than once—most were on open branches and a few were in vine tangles or hollow trees (Wright, 1985). In the Argentinean Chaco, five groups of *A. azarae* used 17 tree species, sleeping mostly in the top fifth of the canopy, midway between the trunk and crown exterior (Savagian & Fernandez-Duque, 2017). Aquino

and Encarnación (1986a) found that the primary factors in sleeping site selection were protection from predators, easy access, shelter from adverse weather, and sufficient space.

It appears that owl monkeys have become secondarily specialized for nocturnal activity (Martin, 1990), and evidence from the fossil record indicates that this shift to nocturnality evolved at least 12–15 mya (Setoguchi & Rosenberger, 1987). *Aotus* possesses certain characteristics associated with a nocturnal lifestyle, such as large eye orbits, small body size, and larger olfactory bulbs than diurnal New World primates (Martin, 1990). Among anthropoids, owl monkeys have the largest eyes relative to body size (see Figure 7.3). *Aotus* has also retained some diurnal visual characteristics, indicating the genus is derived from diurnal ancestors (Hershkovitz, 1983; Martin, 1990). These characteristics include the lack of a *tapetum lucidum,* the reflecting layer of cells behind the retina (Ogden, 1975), and the presence of a fovea in some individuals (Martin & Ross, 2005). *Aotus* does not possess color vision and is thus functionally monochromatic due to a defect in the opsin gene required to produce short-wavelength sensitive cone photopigment (Levenson et al., 2007). However, it is unclear whether or not this lack of S-cone pigment is related to a shift to nocturnality because many other nocturnal primate species retain the S opsin gene and have dichromacy (Kawamura, 2016).

Figure 7.3   Close-up of *Aotus nigriceps* showing the large eye orbits of these nocturnal primates.
(Photo credit: DuSantos—2008; image licensed under Creative Commons Attribution 2.0 Generic license)

PREDATION

Wright (1982, 1989) argues that one of the major advantages of a nocturnal activity pattern in *Aotus* may be predator avoidance. The owl monkey spends most of its time in the high canopy and travels quickly and noisily in the tall trees. It also feeds high in the canopy, often on exposed, conspicuous flowers during periods when the trees are leafless (Janson et al., 1981). When frightened, *Aotus* actively mobs the potential danger (Wright, 1989). During 1,000 hours of observation on *Aotus* at Manú in Peru, Wright (1982) saw no predatory attacks. The only nocturnal predatory birds at the Manú study site are owls that are too small to prey upon *Aotus*. Wright (1989) argues that diurnal predators (particularly large raptors) and diurnal competitors for resources, factors influencing nocturnality of *A. nigriceps* in Peru, are absent where cathemeral owl monkeys range. In the Argentinian Chaco study population, where *Aotus* is cathemeral, two predation attacks on groups of owl monkeys were attempted by a spectacled owl (*Pulsatrix perspicillata*) and a bicolored hawk (*Accipiter bicolor*) at dawn and dusk when the monkeys were most active; aggressive interactions with tayras (*Eira barbara*) were observed also (Fernandez-Duque, 2003, 2009).

Body size may play an important role in differences in predator avoidance strategies as well. The majority of nocturnal species of *Aotus* are small-bodied (range 700–900 g) and are able to utilize tree holes and cavities as sleeping sites. However, the larger-bodied Azara's owl monkeys (~1,254 g) in the Chaco may be too large for all members to simultaneously use tree holes, hence in all reports they are found to sleep in vine tangles or on branches (Wright, 1989; Fernandez-Duque, 2007). Another possibility is that reduced aerial predation pressure in the habitat of Azara's owl monkey may have allowed them to seek more open sleeping sites, while simultaneously affording them the advantage of increased exposure to sunlight, which can benefit this species during the cold winters when temperatures can drop below 0° C (Garcia & Braza, 1993; Erkert et al., 2012). Other cathemeral primates, such as brown lemurs (*Eulemur* spp.) and bamboo lemurs (*Hapalemur* spp.) also live in highly seasonal habitats in Madagascar and have few aerial predators (Tattersall, 2008). This indicates that similar selective pressures may have permitted the switch to cathemerality in Azara's owl monkey.

SOCIAL STRUCTURE, ORGANIZATION, AND REPRODUCTION

Species of *Aotus* live in family groups consisting of an adult male and adult female pair and their offspring—a total group size most often of two to five individuals with an average of four (Table 7.1) (Moynihan, 1964; Heltne, 1977; Rathbun & Gache, 1977; Green, 1978; Wright, 1978, 1986; Fernandez-Duque et al., 2008a). Owl monkeys give birth to singletons, and twins have only occurred on very rare occasions (Aquino et al., 1990; Huck et al., 2014b). Larger groups have been observed, but these seem to be unstable, merged associations of more than one group at superabundant food sources (Moynihan, 1964; Durham, 1975; Rathbun & Gache, 1977). Long-term studies on *A. azarae* suggest that a large percentage of the adults, up to 25–30%, temporarily live as "floaters," solitary individuals that move among territories of various social groups for several weeks or up to a year (Fernandez-Duque et al., 2006). The floaters are typically young adults seeking reproductive opportunities after dispersing from natal groups or older adults evicted from their groups by incoming floaters. Floaters try to maintain relatively close proximity to owl monkey groups, while still avoiding the group's main core ranges, which can be a way to avoid any unnecessary risks and agonistic encounters with group members (Huck & Fernandez-Duque, 2017). Studies have revealed that pair-living males and females are frequently replaced by adult intruders, coupled with occasionally intense intrasexual competition leading to reduced infant survivorship (Fernandez-Duque & Huck, 2013). In nearly all cases, newly formed pairs of *A. azarae* do not start to reproduce until after a one-year period (Fernandez-Duque, 2012). Recently, a genetic study has been conducted on the Azara's owl monkey; results indicated that no extra-pair copulations occurred in this population and, to date, it is the only primate species for which genetic monogamy is reported (Huck et al., 2014a).

Within the group, activity and movement are very cohesive and group members are seldom more than 10 m apart from one another (Wright, 1978). Both allogrooming and agonistic encounters between group members are extremely rare, and Wright (1981) could discern no dominance hierarchy within the group. Care of infants by siblings (allocare) is uncommon and has been recorded infrequently in captivity (Jantschke et al., 1998). Infants tend to have their fathers as nearest neighbors more frequently than their mothers (Gustison et al., 2009). For *A. azarae* in the Gran Chaco, an eviction of a resident male by a solitary male was observed, and the new male contributed to infant care soon after entering the social group (Fernandez-Duque et al., 2008b). During the three-day eviction process, infant care was shared by the siblings. Food sharing between adult pair-mates and between adults and offspring has been seen both in the wild (Rotundo et al., 2005; Wolovich et al., 2008a) and in captivity (Wolovich et al., 2006, 2008b). In wild *A. azarae*, males transferred food to offspring more often than

Table 7.1 Population Densities and Average Group Sizes of Various Aotus Species

Species	Study Site	Group[a] Density (grp/km²)	Individual Density (ind/km²)	Avg. Group Size	No. of Groups	Study Type[b] (Method); No. of Sites	Type of Habitat	References
A. azarae	Formosa, Argentina	10	29	2.9 (1-4)	25	C (LT); 1 site	Seasonal, lowland, riparian	Rathbun and Gache (1980)
A. azarae	Formosa, Argentina	5.5	12.8	2.3	12	C (LT); 1 site	Seasonal gallery forest	Zunino et al. (1985)
A. azarae	Agua Dulce, Paraguay	4.7	14.4	3.1	21	BS/C; 1 site	Seasonal forest	Stallings et al. (1989)
A. azarae	Teniente Enciso, Paraguay	3.3	8.9	2.7	6	BS/C; 1 site	Seasonal forest	Stallings et al. (1989)
A. azarae	Guaycolec, Formosa, Argentina	8	25	3.3	47	C (LT); 1 site	Seasonal gallery forest	Arditi and Placci (1990)
A. azarae	Guaycolec, Formosa, Argentina	16	64	4 (2-6)	11	BS (F); 1 site	Seasonal gallery forest	Fernandez-Duque et al. (2001)
A. lemurinus	Dapa, Valle de Cauca, Colombia	52.4	113	2.2 (1-5)	59	C (LT); 1 site	Cloud forest; degraded, successional forest	Hirche et al. (2017)
A. nancymaae	Rio Tahuayo, Peru	8.75	25	3.4 (2-5)	42	C (F); 1 site	Seasonally flooded lowland forest	Aquino and Encarnación (1986b)
A. nancymaae	Near Rio Amazonas, Peru	6.0 (3.9, 6.4, 6.6, 6.9); 11.3 (9.7, 10, 11.7, 11.8, 13.2)	24.4 (16, 26.2, 27, 28.3); 46.3 (39.8, 41, 48, 48.5, 54.2)	--	98	C (LT); 9 sites	Highland forest; lowland forest	Aquino and Encarnación (1988)
A. nancymaae	Rio Amazon, Loreto, Peru	1.6, 12	3.2, 24	1.9, 2	17, 43	C (LT); 8 sites	Flooded forest	Maldonado and Peck (2014)
A. nigriceps	Cocha Cashu, Manú NP, Peru	10	36-40	4.1 (2-5)	9	BS (F); 1 site	Primary tropical forest	Wright (1985)
A. nigriceps	Lower Rio Urubamba and Tambo, Peru	11.1	31.1	2.8 (2-4)	26	C (LT); 7 sites	Dense, semi-dense, open primary forest	Aquino et al. (2013)
A. vociferans	Near Rio Amazonas, Peru	2.4 (0.6, 2.2, 3.2, 3.5); 10 (5.9, 11.8, 12.5)	7.9 (2, 7.3, 10.6, 11.5); 33 (19.6, 38.9, 41.2)	--	33	C (LT); 7 sites	Highland forest; lowland forest	Aquino and Encarnación (1988)
A. vociferans	Rio Curaray, Peru	7.2	26	3.6 (2-5)	21	C (LT); 4 sites	High and low forest, palm-dominated terraces	Aquino et al. (2014)
A. vociferans	Rio Amazon, Amazonas, Colombia	7.9, 11.0, 13.3	25.9, 24, 44	3.5, 2, 3.3	46, 40, 46	C (LT); 8 sites	Terra firme forest	Maldonado and Peck (2014)
A. zonalis	Chagres NP, Panama	8.3 (7.1, 9.5)	17 (14.3, 19.7)	2.1	16	C (LT); 2 sites	Lowland semi-deciduous forest, secondary gallery forest	Svensson et al. (2010)

a Densities are reported as means, ranges, or exact values at various sites; averages are indicated in bold; ranges or exact values at sites are given in parentheses.
b "Study type" includes: C=Census, BS=Behavioral/Ecological Study. "Method" includes: LT=Line Transect, F=Observational Follows.

females did, which emphasizes the strong infant-male relationship (Wolovich et al., 2008a).

As in titi monkeys (family Pitheciidae, subfamily Callicebinae), *Aotus* infants are carried predominantly by the adult male. However, it has been observed that the adult female starts out being the main carrier of the infant during its first few weeks after birth (one week in *A. lemurinus*, Dixson & Fleming, 1981, and *A. azarae*, Rotundo et al., 2005; two weeks in *A. boliviensis*, Jantschke et al., 1998). The newborn is typically held initially in the ventral groin area on the adult's body (Dixson & Fleming, 1981). By three weeks of age, the infant spends about 75% to over 80% of its time on the back (dorsal side) of the adult male (Wright, 1984; Rotundo et al., 2005). By three months of age, the infant spends most of the time alone (71%), rides on the male for only 26% of the time, and spends 3% of its time on the female while nursing. By five months, the infant rides on the father only in stressful situations. Adults carrying infants usually travel in the first (leader) position during group movements (Rotundo et al., 2005).

Both males and females reach sexual maturity (full body mass and fully developed subcaudal gland) at approximately four years of age (Juárez et al., 2003; Fernandez-Duque, 2004) but may not reproduce until they are at least five or six years old (Fernandez-Duque, 2009; Huck et al., 2011). The mean age at first birth in captivity was found to be much earlier at three and a half years (Gozalo & Montoya, 1990). Hunter et al. (1979) report a gestation period of 133 days, but Wolovich et al. (2008b) report a shorter mean gestation length of 117 days (n = 4 individuals of *A. nancymaae*) and 121 days (n = 1 individual of *A. azarae*) for captive *Aotus*. Fernandez-Duque (2011) reports a gestation length of 121 days for *A azarae* in the wild (n = 1). The period of lactation is approximately six months (Dixson & Fleming, 1981; Wolovich et al., 2008b) and from 150–240 days for *A. azarae* in the wild (Rotundo et al., 2005). The interbirth period interval is about one year (Fernandez-Duque et al., 2002). In the Chaco, the mating season occurs during the winter months, characterized by low rainfall and colder temperatures (Feranandez-Duque, 2009). The birthing periods show strong seasonality, occurring between the months that correspond to a peak in rainfall, as well as peaks in fruits, flowers, new leaves, and higher insect abundance in the habitat (Fernandez-Duque et al., 2002). A birth peak during the rainy season has also been recorded for *Aotus* in Peru (Wright, 1986).

A 10-year study of the birthing, mating, and dispersal patterns of cathemeral owl monkeys (*A. azarae*) in the Argentinean Chaco revealed that both males and females disperse from their natal groups at similar ages (approximately 3 years old) but with some variation in age of dispersal; the youngest to disperse was 2.2 years old and the oldest dispersal age was 4.9 years old (Fernandez-Duque, 2009). Wright (1986) found a similar age of dispersal for *A. nigriceps* in Peru. In the Chaco, all individuals dispersed away from their natal groups and no individual inherited their birth territory, although in some cases the dispersal distance was not far from the natal territory. Agonistic interactions among group members are not common in Azara's owl monkey but increased during natal dispersal periods, with most agonistic interactions directed by adults toward subadults (Corley et al., 2017b). Non-natal social group members typically react aggressively toward a solitary individual that has previously dispersed from its natal group and is attempting to enter a new group. The solitary individuals may occasionally replace the resident same-sex adult through aggressive interactions, including chasing, grabbing, and/or biting (Fernandez-Duque, 2009). Dispersals occurred during all but one month during the year, but the majority of dispersals were concentrated during the spring/summer months characterized by high rainfall and warmer temperatures. The highest rate of dispersal of both sexes was seen at the onset of the birth season, when food is most abundant. Some subadults (2–3 years old) dispersed during or shortly after the dry season, possibly to avoid food competition (Fernandez-Duque, 2009; Huck et al., 2011). Dispersal events appear to be highly risky events, as individuals experience aggression from other groups. Occasionally, an individual may wander through other groups' territories for hours or days, giving hoot calls, before returning to its own natal group. Such pre-dispersal events may attract potential mates (Wright, 1989) or allow individuals to assess dispersal opportunities (Fernandez-Duque, 2009). Dispersal events in *A. azarae* may be influenced by changes in group composition (i.e., eviction of resident male or female), a peak in births, higher food abundance, and/or changes in composition of neighboring groups (Fernandez-Duque, 2009). A hormonal study of wild Azara's owl monkeys in the Argentinean Chaco by Corley et al. (2017a) has revealed that female owl monkeys start ovarian cycling while they are still residing within their natal groups (at 2–3 years of age), and hypothesized that the lack of reproduction within natal groups is not hormonally suppressed but, rather, is regulated by behavioral mechanisms (e.g., agonism) to avoid inbreeding.

Since *Aotus* species are nocturnal, it seems owl monkeys rely heavily on chemical communication, and it is likely that olfactory cues are used in social communication. Mated pairs use a variety of chemical communication signals, including scent-marking and urine-washing.

Urine-washing in *Aotus* may be used to mark travel paths or advertise territories. It may also be used to signal sexual receptivity, as one study revealed group members urine-washing around and separate from a mating/mounting event (Wolovich & Evans, 2007). *Aotus* possesses glands in the pectoral, face, and subcaudal regions. The subcaudal gland is unique in its position and the structure of hairs surrounding it (Hill et al., 1959). In captivity, *Aotus* performed both sternal and subcaudal scent-marking (Wolovich & Evans, 2007). Captive *Aotus* exhibit certain socio-sexual behaviors such as anogenital sniffing, urine drinking, and arching; mate scent-marking was performed more frequently by the male (Wolovich & Evans, 2007). In a captive study of *A. nancymaae*, males had a larger subcaudal gland than females, and males engaged in investigative behaviors, such as genital inspections, more frequently than females (Spence-Aizenberg et al., 2018). These findings differ greatly from other olfactory signaling patterns found in primate species where males invest greatly in infant care.

Although they do not perform vocal duets as do other socially monogamous primates, owl monkeys include a loud "hoot" call in their vocal repertoire (resembling the hoot calls of owls), which can be heard over long distances (Moynihan, 1964). Preliminary studies indicate some sexual differences in hoot calls; thus, these may play an important role in mate attraction (Depeine et al., 2008).

RANGING AND TRAVEL BEHAVIOR

Aotus groups maintain small, stable home ranges. Unlike some pair-living diurnal species (such as titi monkeys), *Aotus* groups do not maintain exclusive use of their home ranges, and there is extensive overlap of range boundaries with neighboring groups (Wright, 1978; Fernandez-Duque & van der Heide, 2013; Wartmann et al., 2014). *Aotus* intergroup encounters consist of arching, pilo-erection, and resonant whoop-calling (Wright, 1978). In a long-term (10-year) study on *A. azarae*, groups displayed substantial variation in size of home ranges and core areas. Significantly, range overlap of neighboring groups was 48%, but the groups occupied almost exclusive core areas with only minimal overlap (10%) (Fernandez-Duque, 2016). This suggests groups have core spaces with exclusive access to food, sleeping sites, and mates (Wartmann et al., 2014). Core areas provide continual food resources even during resource-poor seasons (Fernandez-Duque & van der Heide, 2013), and female distribution may be dictated by food availability in these frequently used core areas (Fernandez-Duque, 2016). Overlapping territories may allow neighboring groups to share limited resources on a time-sharing basis or may allow dispersing individuals to interact socially with others

for future mating opportunities (Fernandez-Duque & van der Heide, 2013). In a study in Ecuador, Fernandez-Duque et al. (2008a) found that a radio-collared group of *A. vociferans* seemed to maintain a territorial or exclusive use of space over an 18-month period during which the group ranged over a 6 ha area; no other owl monkey groups were observed to use the same area.

Aotus groups have short night ranges and small, stable home ranges. For example, *A. nigriceps* in Peru (Wright, 1994) inhabited a home range of 4–17 ha, and *A. vociferans* in Ecuador had a home range of approximately 6 ha (Fernandez-Duque et al., 2008a). For *A. nigriceps* in Peru, nightly ranges during the wet season averaged 829 m per night, and in the dry season travel averaged 252 m (Wright, 1978). This may indicate that owl monkeys spend more time on fewer resources during the dry season and spend more time resting (Wright, 1981). The travel pattern of *Aotus* has been described as goal directed, from one fruit (or flower) tree to the next (Wright, 1982).

Moonlight appears to have a strong effect on ranging behavior, with groups traveling farther during full-moon nights. From an 18-month study of *A. vociferans* in Yasuní, Ecuador, Fernandez-Duque et al. (2008a) reported a home range size of 6.3 ha and an average nightly path length of 645 m (range = 150–1,358 m). During full-moon nights, the group traveled an average of 795 m, but during new-moon nights, the group traveled an average of only 495 m.

CONSERVATION

Out of the 11 currently recognized species of owl monkeys, the International Union for the Conservation of Nature (IUCN) lists one species as Endangered (*A. miconax*), four are listed as Vulnerable (*A. brumbacki, A. griseimembra, A. lemurinus,* and *A. nancymaae*), three are Near Threatened (*A. nigriceps, A. vociferans, A. zonalis*), two are Least Concern (*A. azarae, A. trivirgatus*), and one is Data Deficient (*A. jorgehernandezi*) (IUCN, 2019). Owl monkeys are threatened by several factors, including habitat loss, habitat fragmentation, and habitat conversion to alternative uses. A study on Andean owl monkeys (*A. lemurinus*) in Colombia revealed that owl monkey groups used shade-grown coffee plantations throughout the year, thus providing support that shade coffee plantations can be used as complementary "refuge" areas for *Aotus* while simultaneously providing local farmers with economic incentives (Guzmán et al., 2016).

Other primary threats to owl monkeys include capture for the pet trade and for biomedical research. Owl monkeys are one of the few primates to show symptoms of malaria and so have been highly sought after as a primate model for malaria research (e.g., Herrera et al., 2002).

However, violations of national and international trade regulations were exposed in a study showing that illegal trade of two to three species of *Aotus* to a biomedical laboratory in the Brazil-Colombia-Peru tri-border zone has occurred. These illegal importations emphasize the importance of enforcement of trade regulations (Maldonado et al., 2009). All owl monkeys are listed under Appendix II of the Convention on International Trade in Endangered Species (CITES). Continuation of the current extensive trade in species of *Aotus* may be unsustainable and will potentially lead to severe population declines or extirpation from particular regions (Svensson et al., 2016). Working to create protected areas in increasingly fragmented habitat areas, as well as enforcing trade in these taxa, is necessary to maintain populations of owl monkeys throughout their geographic range.

SUMMARY

Owl monkeys are pair-living, arboreal primates and are the only nocturnal anthropoid. Genetic monogamy has been reported in at least one species, thus emphasizing the long-term resilience and permanency of the male and female pair's relationship. Owl monkey groups maintain an exclusive core ranging area with extensive overlap among home range areas occupied by other neighboring groups. Ranging is strongly influenced by moonlight. On full-moon nights, owl monkeys have longer average nightly ranges. One species of owl monkey (*A. azarae*) exhibits cathemerality and inhabits the harsher environment of the South American Gran Chaco region, characterized by colder winters with semi-deciduous, seasonally dry forests (pastures, palm savannas, galleries) typically producing less resources than tropical rainforests. Owl monkeys are mostly frugivorous, supplementing their diet with leaves, insects, and flowers. Both male and female owl monkeys care for infants, but there is stronger male parental care, with the male most frequently carrying the infant, as well as engaging in playing, grooming, and sharing food. Olfactory communication plays an important role in the social pair-bond, with males investigating female scent marks more frequently than the reverse. Both male and female subadults disperse from their natal territory between two and five years of age, with prospective "floaters" staying nearby but outside of the core area of new groups. Owl monkeys also communicate by hooting or whooping calls which may serve as long-distance contact calls, contact avoidance during intergroup encounters, and potential mate attraction.

REFERENCES CITED—CHAPTER 7

Aquino, R., Cornejo, F. M., & Heymann, E. W. (2013). Primate abundance and habitat preference on the lower Urubamba and Tambo Rivers, central-eastern Peruvian Amazonia. *Primates, 54,* 377-383.

Aquino, R., & Encarnación, F. (1986a). Characteristics and use of sleeping sites in *Aotus* (Cebidae: Primates) in the Amazon lowlands of Peru. *American Journal of Primatology, 11,* 319-331.

Aquino, R., & Encarnación, F. (1986b). Population structure of *Aotus nancymai* (Cebidae: Primates) in Peruvian Amazon lowland forest. *American Journal of Primatology, 11,* 1-7.

Aquino, R., & Encarnación, F. (1988). Population densities and geographic distribution of night monkeys (*Aotus nancymai* and *Aotus vociferans*) (Cebidae: Primates) in northeast Peru. *American Journal of Primatology, 14,* 375-381.

Aquino, R., & Encarnación, F. (1994). Owl monkey populations in Latin America: Field work and conservation. In J. F. Baer, R. E. Weller, & I. Kakoma (Eds.), *Aotus: The owl monkey* (pp. 59-95). Academic Press.

Aquino, R., López, L., García, G., & Heymann, E. (2014). Diversity, abundance, and habitats of the primates in the Río Curaray Basin, Peruvian Amazon. *Primate Conservation, 28,* 1-8.

Aquino, R., Puertas, P., & Encarnación, F. (1990). Supplemental notes on population parameters of northeastern Peruvian night monkeys, genus *Aotus* (Cebideae). *American Journal of Primatology, 21,* 215-221.

Arditi, S. (1992). Seasonal variations in activity and diet of *Aotus azarae* and *Alouatta caraya* in Formosa, Argentina. *Boletín Primatológico Latinoamericano, 3,* 11-30.

Arditi, S. I., & Placci, G. L. (1990). Hábitat y densidad de *Aotus azarae* y *Alouatta caraya* en Riacho Pilagá, Formosa. *Boletín Primatológico Latinoamericano, 2,* 29-47.

Castaño, J. H., Ramirez, D. C., & Botero, J. E. (2010). Ecology of nocturnal night monkey (*Aotus lemurinus*) in sub-Andean forest fragments of Colombia. In P. Pereira-Bengoa, P. R. Stevenson, M. L. Bueno, & F. Nassar-Montoya (Eds.), *Primatologia en Colombia: Avances al principio del milenio* (pp. 67-90). Fundación Universitaria San Martin.

Corley, M., Valeggia, C., & Fernandez-Duque, E. (2017a). Hormonal correlates of development and natal dispersal in wild female owl monkeys (*Aotus azarae*) of Argentina. *Hormones and Behavior, 96,* 42-51.

Corley, M. K., Siyang, X., & Fernandez-Duque, E. (2017b). The role of intragroup agonism in parent-offspring relationships and natal dispersal in monogamous owl monkeys (*Aotus azarae*) of Argentina. *American Journal of Primatology, 79,* 1-12.

Cornejo, F., Aquino, R., & Jimenez, C. (2008). Notes on the natural history, distribution, and conservation status of the Andean night monkey, *Aotus miconax* Thomas, 1927. *Primate Conservation, 23,* 1-4.

Defler, T. R. (2004). *Primates of Colombia.* Conservation International.

Defler, T. R., & Bueno, M. I. (2007). *Aotus* diversity and the species problem. *Primate Conservation, 22,* 55-70.

Depeine, C. D, Rotundo, M., Juarez, C. P., & Fernandez-Duque, E. (2008). Hoot calling in owl monkeys (*Aotus azarae*) of

Argentina: Sex differences and function. *American Journal of Primatology, 70*(S1), 69.

Dixson, A. F., & Fleming, D. (1981). Parental behavior and infant development in owl monkeys (*Aotus trivirgatus griseimembra*). *Journal of Zoology, 194,* 25-39.

Durham, N. M. (1975). Some ecological, distributional, and group behavioral features of Atelinae in southern Peru; with comments on interspecific relations. In R. A. Tuttle (Ed.), *Socioecology and psychology of primates* (pp. 87-101). Mouton.

Erkert, H. G., Fernandez-Duque, E., Rotundo, M. A., & Scheideler, A. (2012). Seasonal variation of temporal niche in wild owl monkeys (*Aotus azarai azarai*) of the Argentinean Chaco: A matter of masking? *Chronobiology International, 29*(6), 702-714.

Fernandez-Duque, E. (2003). Influences of moonlight, ambient temperature, and food availability on the diurnal and nocturnal activity of owl monkeys (*Aotus azarae*). *Behavioral Ecology and Sociobiology, 54*(5), 431-440.

Fernandez-Duque, E. (2004). High levels of intrasexual competition in sexually monomorphic owl monkeys (*Aotus azarai*). *Folia Primatologica, 75,* 260.

Fernandez-Duque, E. (2007). Aotinae: Social monogamy in the only nocturnal haplorrhines. In C. Campbell, A. Fuentes, K. MacKinnon, M. Panger, & S. K. Bearder (Eds.), *Primates in perspective* (pp. 139-154). Oxford University Press.

Fernandez-Duque, E. (2009). Natal dispersal in monogamous owl monkeys (*Aotus azarae*) of the Argentinian Chaco. *Behaviour, 146*(4-5), 583-606.

Fernandez-Duque, E. (2011). The Aotinae: Social monogamy in the only nocturnal haplorrhines. In C. J. Campbell, A. Fuentes, K. C. MacKinnon, S. Bearder, & R. Stumpf (Eds.), *Primates in perspective* (2nd ed., pp. 139-154). Oxford University Press.

Fernandez-Duque, E. (2012). Owl monkeys *Aotus* spp. in the wild and in captivity. *International Zoo Yearbook, 46,* 80-94.

Fernandez-Duque, E. (2016). Social monogamy in wild owl monkeys (*Aotus azarae*) of Argentina: The potential influences of resource distribution and ranging patterns. *American Journal of Primatology, 78,* 355-371.

Fernandez-Duque, E., de la Iglesia, H., & Erkert, H. G. (2010). Moonstruck primates: Owl monkeys (*Aotus*) need moonlight for nocturnal activity in their natural environment. *PLoS ONE, 5*(9), e12572. DOI: 10.1371/journal.pone.0012572.

Fernandez-Duque, E., Di Fiore, A., & Carrillo-Bilbao, G. (2008a). Behavior, ecology, and demography of *Aotus vociferans* in Yasuni National Park, Ecuador. *International Journal of Primatology, 29*(2), 421-431.

Fernandez-Duque, E., Di Fiore, A., Rotundo, M., & Juarez, C. (2006). Demographics and ranging of floaters in socially monogamous owl monkeys (*Aotus azarae*). *American Journal of Primatology, 68*(S1), 86.

Fernandez-Duque, E., & Erkert, H. G. (2006). Cathemerality and lunar periodicity of activity rhythms in owl monkeys

of the Argentinean Chaco. *Folia Primatologica, 77*(1-2), 123-138.

Fernandez-Duque, E., & Huck, M. (2013). Till death (or an intruder) do us part: Intra-sexual competition in a monogamous primate. *PLoS ONE, 8,* e53724.

Fernandez-Duque, E., & Huntington, C. (2002). Disappearances of individuals from social groups have implications for understanding natal dispersal in monogamous owl monkeys (*Aotus azarae*). *American Journal of Primatology, 57*(4), 219-225.

Fernandez-Duque, E., Juarez, C. P., & Di Fiore, A. (2008b). Adult male replacement and subsequent infant care by male and siblings in socially monogamous owl monkeys (*Aotus azarae*). *Primates, 49*(1), 81-84.

Fernandez-Duque, E., & Rotundo, M. (2003). Field methods for capturing and marking Azarai night monkeys. *International Journal of Primatology, 24,* 1113-1120.

Fernandez-Duque, E., Rotundo, M., & Ramirez-Llorens, P. (2002). Environmental determinants of birth seasonality in night monkeys (*Aotus azarae*) of the Argentinean Chaco. *International Journal of Primatology, 23*(3), 639-656.

Fernandez-Duque, E., Rotundo, M., & Sloan, C. (2001). Density and population structure of owl monkeys (*Aotus azarae*) in the Argentinian Chaco. *American Journal of Primatology, 53*(3), 99-108.

Fernandez-Duque, E., & van der Heide, G. (2013). Dry season resources and their relationship with owl monkey (*Aotus azarae*) feeding behavior, demography, and life history. *International Journal of Primatology, 34,* 752-769.

Garcia, J. E., & Braza, F. (1987). Activity rhythms and use of space of a group of *Aotus azarae*. *Primates, 28,* 337-342.

Garcia, J. E., & Braza, F. (1989). Densities comparisons using different analytic-methods in *Aotus azarae*. *Primate Report, 25,* 45-52.

Garcia, J. E., & Braza, F. (1993). Sleeping sites and lodge trees of the night monkey (*Aotus azarae*) in Bolivia. *International Journal of Primatology, 14,* 467-477.

Giménez, M. C. (2004). *Dieta y comportamiento de forrajeo en verano e invierno del mono mirikiná* (Aotus azarae azarae) *en bosques secos y húmedos del Chaco Argentino.* [Bachelor's thesis, University of Buenos Aires].

Gozalo, A., & Montoya, E. (1990). Reproduction of the owl monkey (*Aotus nancymai*) (Primates: Cebidae) in captivity. *American Journal of Primatology, 21,* 61-68.

Green, K. M. (1978). Primate censusing in northern Colombia: A comparison of two techniques. *Primates, 19,* 537-550.

Groves, C. (2001). *Primate taxonomy.* Smithsonian Institution.

Gustison, M. L., Snowdown, C. T., & Fernandez-Duque, E. (2009). Social monogamy: Proximity maintenance and foraging patterns in the wild Argentinian owl monkey (*Aotus azarae azarae*). *American Journal of Primatology, 71*(S1), 64.

Guzmán, A., Link, A., Castillo, J. A., & Botero, J. E. (2016). Agroecosystems and primate conservation: Shade coffee as potential habitat for the conservation of Andean night monkeys in the northern Andes. *Agriculture, Ecosystems and Environment,* 215, 57-67.

Heltne, P. G. (1977). *Census of* Aotus *in the north of Colombia.* Pan American Health Organization.

Herrera, S., Perlaza, B. L., Bonelo, A., & Arévalo-Herrera, M. (2002). *Aotus* monkeys: Their great value for anti-malaria vaccines and drug testing. *International Journal of Parasitology,* 32, 1625-1635.

Hershkovitz, P. (1983). Two new species of night monkeys, genus *Aotus* (Cebidae, primates): A preliminary report on *Aotus* taxonomy. *American Journal of Primatology,* 4, 209-243.

Hill, W. C. O., Appleyard, H. M., & Auber, L. (1959). The specialized area of skin glands in *Aotus* Humboldt (Simiae Platyrrhini). *Transactions of the Royal Society of Edinburgh,* 63(3), 535-551.

Hirche, A., Jimenez, A., Roncancio-Duque, N., & Ansorge, H. (2017). Population density of *Aotus* cf. *lemurinus* (Primates: Aotidae) in a subAndean forest patch on the eastern slopes of the western Andes, region of Dapa, Yumbo, Valle de Cauca, Colombia. *Primate Conservation,* 31, 1-7.

Hladik, C. M., Hladik, A., Bousset, J., Valdebouze, P., Viroben, G., & Delort-Laval, J. (1971). The diet of primates on the Barro-Colorado Island (Panama): Results of quantitative analyses. *Folia Primatologica,* 16, 85-122.

Huck, M. A., & Fernandez-Duque, E. (2012). Children of divorce: Effects of adult replacements on previous offspring in Argentinean owl monkeys. *Behavioral Ecology and Sociobiology,* 66, 505-517.

Huck, M. A., & Fernandez-Duque, E. (2013). When dads help: Male behavioral care during primate infant development. In K. B. H. Clancy, K. Hinde, & J. N. Rutherford (Eds.), *Building babies: Primate development in proximate and ultimate perspective* (pp. 361-385). Springer.

Huck, M., & Fernandez-Duque, E. (2017). The floater's dilemma: Use of space by wild solitary Azara's owl monkeys, *Aotus azarae,* in relation to group ranges. *Animal Behaviour,* 127, 33-41.

Huck, M. A., Fernandez-Duque, E., Babb, P., & Schurr, T. (2014a). Correlates of genetic monogamy in socially monogamous mammals: Insights from Azara's owl monkeys. *Proceedings of the Royal Society B,* 282, 20140195. http://dx.doi.org/10.1098/rspb.2014.0195.

Huck, M. A., Rotundo, M., & Fernandez-Duque, E. (2011). Growth, development and age categories in male and female wild monogamous owl monkeys (*Aotus azarai*) of Argentina. *International Journal of Primatology,* 32, 1133-1152.

Huck, M. A., van Lunenburg, M., Dávalos, V., Rotundo, M., Di Fiore, A., & Fernadez-Duque, E. (2014b). Double effort: Parental behaviour of wild Azara's owl monkeys in the face of twins. *American Journal of Primatology,* 66, 505-517.

Hunter, J., Martin, R. D., Dixson, A. F., & Rudder, B. C. C. (1979). Gestation and inter-birth intervals in the owl monkey (*Aotus trivirgatus griseimembra*). *Folia Primatologica,* 31, 165-175.

IUCN. (2019). *International Union for the Conservation of Nature Red List of Threatened Species.* Version 2019-2. https://www.iucnredlist.org.

Jacobs, G. H. (1977). Visual capacities of the owl monkey (*Aotus trivirgatus*)—I. Spectral sensitivity and color vision. *Vision Research,* 17, 811-820.

Jacobs, G. H., Deegan, J. F., Neitz, J., Crognale, M. A., & Neitz, M. (1993). Photopigments and color vision in the nocturnal monkey, *Aotus. Vision Research,* 33, 1773-1783.

Janson, C. H., Terborgh, J., & Emmons, L. H. (1981). Non-flying mammals as pollinating agents in the Amazonian forest. *Biotropica,* 13(2), 1-6.

Jantschke, B., Welker, C., & Klaiber-Schuh, A. (1998). Rearing without paternal help in the Bolivian owl monkey *Aotus azarae boliviensis*: A case study. *Folia Primatologica,* 69, 115-120.

Juárez, C. P., Rotundo, M. A., Berg, W., & Fernandez-Duque, E. (2011). Costs and benefits of radio-collaring on the behavior, demography and conservation of owl monkeys (*Aotus azarai*) in Formosa, Argentina. *International Journal of Primatology,* 32, 69-82.

Juárez, C. P., Rotundo, M. A., & Fernandez-Duque, E. (2003). Behavioural sex differences in the socially monogamous night monkeys of the Argentinean Chaco. *Revista de Ethologia,* 5 (Sl), 174.

Kawamura, S. (2016). Color vision diversity and significance in primates inferred from genetic and field studies. *Genes and Genomics,* 38, 779-791.

Levenson, D. H., Fernandez-Duque, E., Evans, S., & Jacobs, G. H. (2007). Mutational changes in S-cone opsin genes common to both nocturnal and cathemeral *Aotus* monkeys. *American Journal of Primatology,* 69, 7757-7765.

Maldonado, A. M., Nijman, V., & Bearder, S. K. (2009). Trade in night monkeys *Aotus* spp. in the Brazil-Colombia-Peru tri-border area: International wildlife trade regulations are ineffectively enforced. *Endangered Species Research,* 9, 143-149.

Maldonado, A. M., & Peck, M. R. (2014). Research and in situ conservation of owl monkeys enhances environmental law enforcement at the Colombian-Peruvian border. *American Journal of Primatology,* 76, 658-669.

Marín-Gómez, O. H. (2008). Nectar consumption by *Aotus lemurinus* and its possible role in flower pollination of *Inga edulis* (Fabales: Mimosoideae). *Neotropical Primates,* 15(1), 30-32.

Martin, R. D. (1990). *Primate Origins and Evolution.* Princeton University Press.

Martin, R. D., & Ross, C. F. (2005). The evolutionary and ecological context of primate vision. In J. Kremers (Ed.), *The primate visual system: A comparative approach* (pp. 1-36). Wiley.

Menezes, A. N., Bonvicino, C. R., & Seuanez, H. N. (2010). Identification, classification and evolution of owl monkeys (*Aotus,* Illiger 1811). *BMC Evolutionary Biology,* 10, 248-262.

Moynihan, M. (1964). Some behavior patterns of platyrrhine monkeys. The night monkey (*Aotus trivirgatus*). *Smithsonian Miscellaneous Collections,* 146, 1-84.

Ogden, T. E. (1975). The receptor mosaic of *Aotus trivirgatus*: Distribution of rods and cones. *Journal of Comparative Neurology,* 163, 165-183.

Overdorff, D. J., & Rasmussen, M. A. (1995). Determinants of nighttime activity in "diurnal" lemurid primates. In L. Alterman, G. A. Doyle, & M. K. Izard (Eds.), *Creatures of the dark* (pp. 61-74). Plenum Press.

Rathbun, G. B., & Gache, M. (1977). *The status of* Aotus trivirgatus *in Argentina*. Centro Argentino de Primates and National Institutes of Health Report.

Rathbun, G. B., & Gache, M. (1980). Ecological survey of the night monkey, *Aotus trivirgatus,* in Formosa Province, Argentina. *Primates, 21,* 211-219.

Rotundo, M., Fernandez-Duque, E., & Dixson, A. F. (2005). Infant development and parental care in free-ranging *Aotus azarae azarae* in Argentina. *International Journal of Primatology, 26*(6), 1459-1473.

Rylands, A. B., Groves, C. P., Mittermeier, R. A., Cortés-Ortíz, L., & Hines, J. J. H. (2006). Taxonomy and distributions of Mesoamerican primates. In A. Estrada, P. Garber, M. Pavelka, & L. Luecke (Eds.), *New perspective in the study of Mesoamerican primates. Distribution, ecology, behavior, and conservation* (pp. 29-79). Springer.

Rylands, A. B., Schneider, H., Langguth, A., Mittermeier, R. A., Groves, C. P., & Rodríguez-Luna, E. (2000). An assessment of the diversity of New World primates. *Neotropical Primates, 8,* 61-93.

Savagian, A., & Fernandez-Duque, E. (2017). Do predators and thermoregulation influence choice of sleeping sites and sleeping behavior in Azara's owl monkeys (*Aotus azarae azarae*) in northern Argentina? *International Journal of Primatology, 38,* 80-99.

Setoguchi, T., & Rosenberger, A. L. (1987). A fossil owl monkey from La Venta, Colombia. *Nature, 326,* 692-694.

Shanee, S., Allgas, N., & Shanee, N. (2013). Preliminary observations on the behavior and ecology of the Peruvian night monkey (*Aotus miconax*: Primates) in a remnant cloud forest patch, north eastern Peru. *Tropical Conservation Science, 6,* 138-148.

Solano, C. (1995). Activity patterns and habitat use of the owl monkey, *Aotus brumbacki* (Primate: Cebidae) at Tinigua National Park, Colombia. *Bulletin of the Ecological Society of America, 76,* 390.

Spence-Aizenberg, A., Williams, L. E., & Fernandez-Duque, E. (2018). Are olfactory traits in a pair-bonded primate under sexual selection? An evaluation of sexual dimorphism in *Aotus nancymaae. American Journal of Physical Anthropology, 166,* 884-894.

Stallings, J. R., West, L., Hahn, W., & Gamarra, I. (1989). Primates and their relation to habitat in the Paraguayan Chaco. In K. H. Redford, & J. F. Eisenberg (Eds.), *Advances in neotropical mammalogy* (pp. 425-442). Sandhill Crane Press.

Sussman, R. W., & Tattersall, I. (1976). Cycles of activity, group composition and diet of *Lemur mongoz mongoz* (Linnaeus, 1766) in Madagascar. *Folia Primatologica, 26,* 270-283.

Svensson, M. S., Samudio, R., Bearder, S. K., & Nekaris, K. A. I. (2010). Density estimates of Panamanian owl monkeys (*Aotus zonalis*) in three habitat types. *American Journal of Primatology, 72*(2), 187-192.

Svensson, M. S., Shanee, S., Shanee, N., Bannister, F. B., Cervera, L., Donati, G., Huck, M., Jerusalinsky, L., Juarez, C. P., Maldonado, A. M., Martinez Mollinedo, J., Méndez-Carvajal, P. G., Molina Argandoña, M. A., Mollo Vino, A. D., Nekaris, K. A. I., Peck, M., Rey-Goyeneche, J., Spaan, D., & Nijman, V. (2016). Disappearing in the night: An overview on trade and legislation of night monkeys in South and Central America. *Folia Primatologica, 87,* 332-348.

Tattersall, I. (2008). Avoiding commitment: Cathemerality among primates. *Biological Rhythm Research, 39*(3), 213-228.

Thorington, R. W., Muckenhirn, N. A., & Montgomery, G. G. (1976). Movements of a wild night monkey (*Aotus trivirgatus*). In R. W. Thorington, & P. G. Heltne (Eds.), *Neotropical primates: Field studies and conservation* (pp. 32-34). National Academy of Sciences.

Wartmann, F. M., Juárez, C. P., & Fernandez-Duque, E. (2014). Size, site fidelity, and overlap of home ranges and core areas in the socially monogamous owl monkey (*Aotus azarae*) of northern Argentina. *International Journal of Primatology, 35,* 919-939.

Wolovich, C. K., & Evans, S. (2007). Sociosexual behavior and chemical communication of *Aotus nancymaae. International Journal of Primatology, 28,* 1299-1313.

Wolovich, C. K., Evans, S., & French, J. (2008b). Dads do not pay for sex but do buy the milk: Food sharing and reproduction in owl monkeys (*Aotus* spp.). *Animal Behaviour, 75,* 1155-1163.

Wolovich, C. K., Feged, A., Evans, S., & Green, S. M. (2006). Social patterns of food sharing in monogamous owl monkeys. *American Journal of Primatology, 68,* 663-674.

Wolovich, C. K., Perea-Rodriguez, J. P., & Fernandez-Duque, E. (2008a). Food transfers to young and mates in wild owl monkeys (*Aotus azarae*). *American Journal of Primatology, 70*(3), 211-212.

Wolovich, C. K., Rivera, J., & Evans, S. (2010). Insect-foraging in captive owl monkeys (*Aotus nancymaae*). *Folia Primatologica, 81,* 63-72.

Wright, P. C. (1978). Home range, activity pattern, and agonistic encounters of a group of night monkeys (*Aotus trivirgatus*) in Peru. *Folia Primatologica, 29,* 43-55.

Wright, P. C. (1981). The night monkey, genus *Aotus*. In A. F. Coimbra-Filho, & R. A. Mittermeier (Eds.), *Ecology and behavior of neotropical monkeys* (Vol. 1, pp. 211-240). Academia Brasileira de Ciências.

Wright, P. C. (1982). Adaptive advantages of nocturnality in *Aotus. American Journal of Physical Anthropology, 57,* 242.

Wright, P. C. (1983). Day-active night monkeys (*Aotus trivirgatus*) in the Chaco of Paraguay. *American Journal of Physical Anthropology, 60,* 272.

Wright, P. C. (1984). Biparental care in *Aotus trivirgatus* and *Callicebus moloch*. In M. Small (Ed.), *Female primates: Studies by women primatologists* (pp. 59-75). A. R. Liss.

Wright, P. C. (1985). *The costs and benefits of nocturnality for* Aotus trivirgatus *(the night monkey)*. [Doctoral dissertation, City University of New York].

Wright, P. C. (1986). Ecological correlates of monogamy in *Aotus* and *Callicebus*. In J. Else, & P. Lee (Eds.), *Primate ecology and conservation* (pp. 159-167). Cambridge University Press.

Wright, P. C. (1989). The nocturnal primate niche in the New World. *Journal of Human Evolution, 18,* 635-658.

Wright, P. C. (1990). Patterns of paternal care in primates. *International Journal of Primatology, 11,* 89-102.

Wright, P. C. (1994). The behavior and ecology of the owl monkey. In J. F. Baer, R. E. Weller, & I. Kakoma (Eds.), *Aotus: The owl monkey* (pp. 97-112). Academic Press.

Zunino, G. E., Galliari, C. A., & Colilas, O. J. (1985). Distribución y conservación del mirikiná (*Aotus azarae*) en Argentina: Resultados preliminares. In M. T. de Mello (Ed.), *A primatologia no Brasil* (Vol. 2, pp. 97-112). Sociedade Brasileira de Primatologia.

CHAPTER 8

Titi Monkeys

Tail-Twining in the Trees

ANNEKE M. DELUYCKER

TITI MONKEYS ARE SMALL-BODIED (1–2 KG) PLATYR-RHINE PRIMATES OF the family Pitheciidae. They are diurnal-living primates that display a pair-bonded, pair-living social structure and organization with strong, intensive paternal offspring care. Males are the primary carriers of infants shortly after birth and engage in the majority of social interactions with offspring (see Figure 8.1).

Mason (1966, 1968) carried out the first socio-ecological field study of titi monkeys in central Colombia. Later field research concentrating on titi monkeys was conducted in Peru and Brazil by Kinzey and his students and colleagues (Kinzey et al., 1977; Kinzey, 1977a,b; Easley, 1982; Kinzey & Wright, 1982; Kinzey & Becker, 1983; Wright, 1984, 1985, 1986; Easley & Kinzey, 1986), as well as studies in Colombia by Robinson (1977, 1979a,b, 1981) and Defler (1983). More recently, a number of longer-term behavioral ecology studies (≥6 months) have been completed or are currently being conducted on species of titi monkeys. These include studies in Brazil (Müller, 1995a,b, 1996; Heiduck, 1997, 1998, 2002; Price & Piedade, 2001a,b; Cäsar & Young, 2008; Souza-Alves, 2010; Caselli & Setz, 2011; Souza-Alves et al., 2011a,b; de Santana, 2012; Souza-Alves, 2013; Nagy-Reis & Setz, 2017), Colombia (Polanco-Ochoa, 1992; Polanco-Ochoa & Cadena, 1993; Polanco-Ochoa et al., 1994; Palacios et al., 1997; Palacios & Rodriguez, 2013; Acero-Murcia et al., 2018), Ecuador (Youlatos & Pozo Rivera, 1999; Carrillo-Bilbao et al., 2005; Fernandez-Duque et al., 2013; Spence-Aizenberg et al., 2016; Van Belle et al., 2016), and Peru (Bossuyt, 2002; Nadjafzadeh, 2005; DeLuycker, 2007, 2012, 2013, 2014; Lawrence, 2007; Nadjafzadeh & Heymann, 2008). A number of researchers have documented new species (bringing the total number of recognized titi monkey species to 35) and are updating geographic distributions and taxonomic clarifications (Kobayashi & Langguth, 1999; van Roosmalen et al., 2002; Rowe & Martinez, 2003; DeLuycker, 2006; Felton et al., 2006; Wallace et al., 2006; Martinez & Wallace, 2007; Aquino et al., 2008; Boveda-Penalba et al., 2009; Röhe & Silva Jr., 2009; Vermeer, 2009; Defler et al., 2010; Vermeer et al., 2011; Gualda-Barros et al., 2012; Sampaio et al., 2012; Dalponte et al., 2014; Gusmão & Costa, 2014; Vermeer

& Tello-Alvarado, 2015; Printes et al., 2018; Boubli et al., 2019; Gusmão et al., 2019; Rocha et al., 2019). Emerging as one of the most species-rich and widespread primate groups, a recent classification by Bryne et al. (2016) recognizes three genera: *Callicebus*, *Cheracebus*, and *Plecturocebus*. Previously considered monogeneric, since its initial taxonomic description (*Callicebus*, Thomas, 1903), new molecular evidence (Byrne et al., 2016), combined with biogeographic and morphological data, has given insights into further relationships among lineages and the biogeographic history of this diverse clade (Byrne et al., 2018; Carneiro et al., 2018).

HABITAT AND LOCOMOTION

Titi monkeys have an extensive range across varied habitat types in South America (Figure 8.2). The geographic distribution of titi monkeys extends from the Andean foothills in Colombia, Ecuador, Peru, and Bolivia, and south of the Río Orinoco in Venezuela to the dry forests of northeast Paraguay, occupying a large portion of the Brazilian Amazon Basin. Their distribution also includes parts of the Atlantic coastal forests and dry forest regions of Brazil (van Roosmalen et al., 2002). Members of the genus *Cheracebus* (*torquatus* group) are found in the Orinoco, Negro, and upper Amazon Basins in Brazil, Venezuela, Colombia, Ecuador, and Peru. Members of the genus *Plecturocebus* (*donacophilus* and *moloch* groups) inhabit the greater part of the southern and western Amazon Basin and dry Chaco forest regions of Brazil, Ecuador, Peru, Bolivia, and Paraguay, as well as one species inhabiting a small portion of the upper Caquetá Basin in Colombia. Members of the genus *Callicebus* (*personatus* group) are geographically separated from all other callicebines, and inhabit the Atlantic coastal forests and dry Caatinga forests of eastern Brazil.

Titi monkeys inhabit a wide range of habitat types including primary forest, secondary forest, gallery forest, deciduous forest, semi-deciduous forest, moist restinga broadleaf forest, dry Chaco forest, sub-humid Pantanal riverine forests, transitional Cerrado and Chiquitano forests, seasonal Caatinga forest, flooded forest, swamp forest, terra firma, Andean premontane forest, and

Figure 8.1 Male San Martín titi monkey (*Plecturocebus oenanthe*) carrying infant.
(Photo credit: Anneke DeLuycker)

fragmented forest (Hershkovitz, 1990; van Roosmalen et al., 2002; de Freitas et al., 2011; Rumiz, 2012).

In the southern part of their range, some species of the genus *Cheracebus* have overlapping geographic ranges with some species of the genus *Plecturocebus*; however, species of titi monkey are rarely found in sympatry. In one comparative field study, Kinzey and Gentry (1979) argued that where they are found in sympatry, *Plecturocebus toppini* (formerly *P. brunneus*) and *Cheracebus lucifer* occupy very different habitats and may be "habitat isolates" as they did not see individuals of the different species occupying the same habitat areas. In Kinzey's (1981) study, *P. toppini* seemed to prefer areas of dense canopy forest, but also used low forest with abundant lianas and bamboo thickets, whereas *C. lucifer* seemed to prefer well-stratified, tall forests of upland or flooded forests (varillal) of many different soil types. However, in other studies, researchers have observed *C. lucifer* in disturbed areas (Johns, 1991; Defler, 1994, 2003). *P. ornatus* is seen to inhabit areas of low, bushy forest vegetation which is often vine-abundant, with poorly drained

soils that sometimes become partially flooded (Mason, 1968; Moynihan, 1976; Polanco-Ochoa & Cadena, 1993; Defler, 1994). Several species of titis are commonly found living in disturbed, degraded, fragmented, or secondary forest sites (e.g., Polanco-Ochoa, 1992; DeLuycker, 2007; Souza-Alves et al., 2011a; Acero-Murcia et al., 2018); however, other studies seem to suggest species of titi monkeys prefer primary forest and avoid disturbed areas (e.g., Heiduck, 2002; Kulp & Heymann, 2015). The extent to which titi monkeys are tolerant of, and able to thrive in, altered or disturbed habitats remains unclear and warrants further study.

A number of titi monkey species have been observed to perform most daily activities, such as feeding and foraging, in the lower levels of the forest strata, often below 10 m; this includes *P. toppini* (Wright, 1985; Buchanan-Smith et al., 2000), *Callicebus nigrifrons* (Trevelin et al., 2007), *P. oenanthe* (DeLuycker, 2007), and *P. ornatus* (Polanco-Ochoa & Cadena, 1993). *Cheracebus lucifer* spends most of its time in the middle strata of the forest (15–25 m high) but feeds mostly in the emergent layer

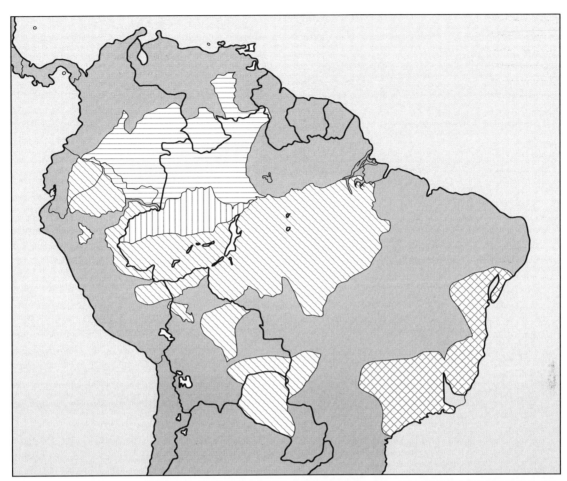

Figure 8.2 Geographic distribution of the three titi monkey genera. *Cheracebus* (horizontal lines) found in NW Brazil, S Colombia, and S Venezuela; *Callicebus* (cross-hatching) inhabit Atlantic coastal region of E Brazil; *Plecturocebus* (diagonal lines) found in central and W Brazilian Amazonia, N Paraguay, S, central, and N Bolivia, E and N Peru, and E Ecuador; region of sympatry between species of *Cheracebus* and *Plecturocebus* (vertical lines). (Redrawn from Byrne et al., 2016)

(Easley, 1982; Lawler et al., 2006), using mostly medium and terminal branches, and most locomotion is quadrupedal walking (Kinzey et al., 1977; Easley, 1982). Feeding is done above-branch quadrupedally or while sitting. When in the understory, feeding on fruits or insects, titi monkeys often use a vertical clinging-and-leaping postural stance, although this represents a very small proportion of time (Lawler et al., 2006; DeLuycker, 2007). While feeding, *P. oenanthe* sometimes use a suspensory posture (DeLuycker, 2007). *P. toppini* spend more time in the lower forest strata, using mostly short-distance leaping within this habitat type, and utilizing branches of smaller diameter than *C. lucifer* (Kinzey, 1981; Wright, 1985; Crandlemire-Sacco, 1988; Lawler et al., 2006). *P. oenanthe* use mostly smaller, horizontal branches and spend the majority of their time in the understory and lower canopy of secondary forest (DeLuycker, 2007). Titi monkeys have been observed to engage in a variety of activities on the ground, including foraging for insect prey, feeding on soil, resting, playing, moving between open

areas or gaps, retrieving fallen infants, or escaping/hiding from predators (Souza-Alvez et al., 2019b). The titi monkey tail is not prehensile but rather used mainly for balance and in social tactile affiliative communication in a characteristic tail-twining posture (see Figure 8.3).

DIET

Titi monkeys are mainly frugivorous, consuming between 40–80% ripe fruit, but species vary in the amount of the protein source in the diet represented as leaves, insects, or seeds. These differences may be related to a number of other behavioral, morphological, or ecological differences among the species. Animal prey, mainly insects, make up a significant proportion of the diet of almost all species of titi monkeys with the exception of *Callicebus personatus* and *Callicebus melanochir*, both inhabiting eastern Atlantic Forests of Brazil (DeLuycker, 2012; Heymann & Nadjafzadeh, 2013). Titi monkeys capture both solitary and social insects, whether exposed or concealed in rolled leaves (Kinzey, 1977a; Easley, 1982; Wright, seeds) can be important resources, especially in fragmented habitats or as a dependable resource available year-round in habitats with annual fluctuations in food availability (Terborgh, 1983; DeLuycker, 2007; Caselli & Setz, 2011; Souza-Alves et al., 2011a; Nagy-Reis & Setz, 2017). In a few studies, titi monkeys relied on fruits from hemiparasitic epiphytes of the mistletoe families (including Loranthaceae and Viscaceae) during the entire year, and these items belonged to the top 10 fruit species eaten (DeLuycker, 2007; Acero-Murcia et al., 2018). Other food items, consumed to a lesser degree, include flowers, pith/meristem, petioles, soil, and bark (Kinzey, 1981). Titi monkeys, in general, display a great amount of dietary flexibility in food type choice and seem to rely on specific fruit species that are available year-round.

Seed eating seems to be an important dietary resource for some species of titis. For example, *Cheracebus lugens* in Amazonian Colombia include a high proportion of seeds in the diet over the entire year, and some researchers consider them to be specialized seed predators based on dietary, behavioral, and morphological characteristics related to a seed-eating niche (Palacios et al., 1997; Palacios & Rodriguez, 2013). Some species of *Callicebus* in the Atlantic Forest of Brazil rely heavily on seeds as a dietary component. *C. melanochir* obtain most of their dietary protein from young seeds (22–26%) and consume fewer leaves (14–17%) (Müller, 1996; Heiduck, 1997). In a study of *C. nigrifrons*, the group obtained most of its protein from seeds (22%) and included fewer leaves (19%) and invertebrates (10%) (Nagy-Reis & Setz, 2017). In another study of *C. nigrifrons* during a bamboo masting year, one-third of the diet was made up of seeds from one species of bamboo, *Merostachys fischeriana* (dos Santos et al., 2012). *C. coimbrai* supplement their fruit diet with mostly leaves (20%) but also eat insects and seeds during the dry season (Souza-Alves et al., 2011a). *C. personatus* do not seem to supplement their fruit diet with either insects or seeds, only leaves (18–26%), although these data are from short-term studies (Kinzey & Becker, 1983; Price & Piedade, 2001a).

From available data, it seems that titi monkeys tend to concentrate on a small number of fruit species over various seasons. In a 12-month period, both Kinzey (1977a) and Easley (1982) found that *Cheracebus lucifer* ate 55–57 species of fruit; however, only 3–6 species made up over 50% of the diet depending on the season. For *Plecturocebus toppini* at Los Amigos in Peru, 2–4 species of fruit made up more than 50% of the diet, depending on the season (Lawrence, 2007). Only 1–2 species of fruit made up more than 50% of the diet for *P. oenanthe* in the Alto Mayo, Peru (DeLuycker, 2007). However, in the diet of *Callicebus melanochir* in Brazil, 10 species of fruit made

Figure 8.3 Titi monkeys (*Plecturocebus cupreus*) with their tails intertwined.
(Photo credit: Steven G. Johnson; https://commons.wikimedia.org/wiki/)

1985; Nadjafzadeh & Heymann, 2008; DeLuycker, 2012). In Amazonian Peru, *Cheracebus lucifer* obtained most of its protein from insects (16–20%) and ate few leaves (4–9%) (Kinzey, 1977a; Easley, 1982). This was similar to *P. oenanthe* in a premontane forest in Andean Peru, which also obtained most of its protein from insects (22%) and less from leaves and other vegetative material (11–16%) (DeLuycker, 2012). *P. toppini* in southern Peru and *P. discolor* in Ecuador obtained most of their protein from both leaves (31–39%) and insects (11–20%) (Wright, 1986; Youlatos & Pozo-Rivera, 1999; Lawrence, 2007). Alternatively, *P. caquetensis* living in a forest fragment in Colombia supplemented its low fruit diet with mostly leaves (31%) and seeds (21%) and only ate arthropods 3% of the time (Acero-Murcia et al., 2018). Titi monkeys tend to prefer young leaves. In addition, vegetative food items obtained from lianas/vines (leaves, tendrils, buds, up more than 50% of the diet (Müller, 1996), which may be due to its greater reliance on seeds. Overall, most species of titi monkeys utilize a large number of fruit species, while a small number of fruit species make up a greater proportion of the diet. Easley (1982) hypothesized that at Estación Biológica Callicebus (EBC) in Amazonian Peru, one species of palm, *Jessenia polycarpa*, was the most commonly eaten fruit and may be the most important fruit staple in the diet of *Cheracebus lucifer*, thereby making the distribution and phenology of these palm trees important determinants of the species' ranging behavior. The palm, *J. polycarpa*, is relatively abundant, available throughout the entire year, and evenly distributed within the home range of groups of *C. lucifer*, while the rest of the utilized food resources are scarce, unpredictable, and only fruit for very short periods. Easley argued that at EBC the territorial behavior of the titis may be related to the distribution of this dependable and stable resource.

Titi monkeys typically rely on either seeds, young leaves, and/or insects as fallback foods or resources that become important when preferred foods are scarce; thus, their consumption of these items increases during times of food scarcity or decreased food availability. These foods also provide protein and lipids (Kinzey & Norconk, 1993; Lambert, 2011), but it is unclear whether nutritional quality, distribution, or abundance of these sources is determining what species of titi monkeys prefer among various habitats. In several studies, titi monkeys adjusted feeding behavior in response to the availability of food items, similar to other frugivorous primates living in seasonal environments (e.g., Heiduck, 1997; Palacios et al., 1997; DeLuycker, 2007; Lawrence, 2007; Souza-Alves et al., 2011a; Nagy-Reis & Setz, 2017). Titi monkeys also seem to alter their foraging behavior based on fruit availability. During the dry season (when fruit abundance is low), titi monkeys switch to eating fallback foods, decrease their time spent traveling, decrease their travel rate, and decrease their travel distance, while broadening the diversity of plant species used (DeLuycker, 2007; Nagy-Reis & Setz, 2017). Titi monkeys display a flexible foraging strategy, and follow a pattern of energy-minimization, especially in seasonal habitats.

ACTIVITY CYCLE AND SLEEPING SITE USE

Titi monkey groups are active during the daytime period for approximately 10–11.5 hours and spend between 31–63% of the day resting and between 13–38% of the day feeding (Kinzey, 1978; Easley, 1982; Kinzey & Becker, 1983; Polanco-Ochoa et al., 1994; Müller, 1996; DeLuycker, 2007). The daily activity patterns of various species of titi monkeys may reflect their dietary propensities. For example, *Plecturocebus toppini* and

Cheracebus lucifer have significant differences in distribution of activities throughout the day. *C. lucifer* feed more or less continuously with short rests between feeding but without a long rest break. Fruit feeding, which peaks in early morning, mid-morning, mid-afternoon, and late afternoon, is interspersed with insect and leaf eating throughout the day (Easley, 1982; Kinzey & Becker, 1983). Two hours before sunset, *C. lucifer* begin resting and grooming in their night sleeping tree. The monkeys sleep in the canopy, about 25 m high, on open, large horizontal branches and choose different sleeping sites each night (Kinzey, 1981; Easley, 1982). *P. toppini* have an activity pattern more similar to folivorous primates, i.e., with two major feeding peaks, one in the morning and one in the late afternoon, and a long afternoon resting period. After the late afternoon feeding period, which is spent mostly on leaves, the animals move to their sleeping site for the night (Kinzey, 1977b). Both *Callicebus personatus* (in Brazil), which also supplement their mostly fruit diet with leaves, and *P. ornatus* (in Colombia) displayed peaks of feeding and resting similar to those described for *P. toppini* (Kinzey & Becker, 1983; Polanco-Ochoa et al., 1994). The sleeping sites of *P. toppini* tend to consist of vine tangles far below the top of the canopy, about 15 m high (Kinzey, 1981; Wright, 1985), and often the group returns to the same sleeping sites (Mason, 1968; Kinzey, 1981) but never uses the same sleeping site on two consecutive nights. *Cherocebus lucifer* used 44 different sleep trees in a six-month study (Kinzey et al., 1977), and *P. toppini* used 26 different sleep trees in a yearlong study (Wright, 1985). Similar to *P. toppini*, *P. ornatus* in Colombia sleep in the canopy of trees 15–20 m tall within a tangle of vines (Polanco-Ochoa & Cadena, 1993).

In the Alto Mayo, Peru, *P. oenanthe* fed continuously throughout the day with multiple short peaks of feeding activity (DeLuycker, 2007) similar to that seen in *C. lucifer*, a species which also spends a large proportion of time hunting for insect prey. *P. oenanthe* used only five sleeping sites in a 3-ha home range, and the most frequently used sleeping sites were a stand of bamboo and a tall tree with relatively open branches (Figure 8.4). These titis also used sleeping trees with thick vine clusters, but much less frequently. The use of a small number of sleeping sites has also been seen in *Callicebus melanochir* in Brazil, that use only five sleeping sites in a 22-ha area and prefer tall trees with dense foliage for their sleeping trees (Heiduck, 2002), and *Callicebus coimbrai* in Brazil, that use only three sleeping sites in a 14.4-ha area, all of which were the same tree species (Souza-Alves et al., 2011b). During nighttime sleeping and day resting, group members huddled together with their tails entwined.

Figure 8.4 Family group of San Martín titi monkeys (*Plecturocebus oenanthe*) at sleeping site tree, with adult male and female engaged in tail-twining.
(Photo credit: Anneke DeLuycker)

PREDATION

When threatened, titi monkeys typically flee into dense vegetation or vine thickets and remain cryptic. Observations of predation on titi monkeys are rare. Wright (1985) observed *P. toppini* being attacked multiple times by predatory birds (*Spizaetus ornatus*, *Morphnus guianensis*, *Accipiter bicolor*, and other *Accipiter* spp.) and once by an ocelot (*Felis pardalis*) in 900 hours of observation, although these attacks were all unsuccessful. In Peru, Bossuyt (2002) observed two attempted predations by raptors on juvenile *P. toppini*. In Ecuador, de Luna et al. (2010) report predation attempts on *P. discolor* by tayra (*Eira barbara*) and harpy eagles (*Harpia harpyia*). DeLuycker (2007) observed an unsuccessful attack on an adult female *P. oenanthe* in Peru by two roadside hawks (*Buteo magnirostris*). C. Freese (pers.comm. in Kinzey et al., 1977) reported sighting a large predatory bird catching a titi monkey. Both Cisneros-Heredia et al. (2005) and Lawrence (2007) observed an arboreal boa constrictor preying on a titi monkey (though the boa let the monkey escape). Both Sampaio and Ferrari (2005) in Brazil and Lawrence (2007) in Peru saw an adult capuchin monkey (*Sapajus apella*) killing and eating a young titi monkey.

Titi monkeys are generally very cryptic in coloration, and while moving and resting during the day, group members synchronize activities (Easley, 1982; Terborgh,

1983; Wright, 1985; Crandlemire-Sacco, 1988; DeLuycker, 2007). They prefer to rest during the day in dense vegetation in the mid-canopy (Terborgh, 1983). Titi monkeys most likely choose their nightly sleeping sites for predator protection as well as for social reasons. The open, large branches preferred by *Cheracebus lucifer*, *Callicebus personatus*, and *Plecturocebus oenanthe* allow the monkeys to see approaching predators. Wright (1985) observed that only 8 of 26 sleeping trees of *P. toppini* provided cryptic sleeping sites (e.g., vine tangles). She argued that, at least at Manú, Peru, diurnal activity by titi monkeys stresses predator avoidance, whereas the night sleeping sites are relatively unprotected because nocturnal, arboreal predators at this location are not as great of a threat. A study in Peru by Adams and Kitchen (2018) found that *P. toppini* produced specific loud and repetitive alarm calls that deterred potential predators such as ocelots.

SOCIAL STRUCTURE AND ORGANIZATION

Titi monkeys exhibit a pair-bonded social structure. They live in groups, composed of an adult female, an adult male, and their offspring, associating within a common home range. Offspring care is performed mostly by the male and includes carrying/transport, grooming, cleaning/inspection, playing, rescue, and protection. Group size ranges from two to seven individuals (see Table 8.1).

Table 8.1 Group Size, Composition, Home/Day Range Size, and Population Density Among Titi Monkey Species

Taxa and Site [a]	Avg. Group Size (N)	Group Size Range	Group Composition Adult M:F	Group Composition No. of Offspring [b]	Home Range [c] (range)	Day Range [d] (range)	Group Density [e] (G/km²)	Individual Density [e] (Ind/km²)	No. Groups Studied	References
Callicebus:										
C. personatus										
Linhares, Brazil	4.5 (2)	4-5	1:1	2	11.5	1000 (800-1300)	-	-	2	Price and Piedade, 2001b
Triângulo Mineiro, Brazil	-	-	-	-	33.8	1270	-	-	1	Neri and Rylands, 1997; Neri, 1997
C. melanochir										
Lemos Maia, Brazil	4.5 (2)	3-6	1:1	1-4	23.8	1000	4	17	2	Müller, 1995b, 1996
Lemos Maia, Brazil	4 (1)	3-4	1:1	1-2	22	1020 (610-1370)	-	-	1	Heiduck, 1997, 1998
C. nigrifrons										
Cantareira Park, Brazil	2.2	1-5	-	-	-	-	-	12	39	Trevelin et al., 2007
Ribeirão Cachoeira, Brazil	-	3-5	1:1	-	17	1222 (631-1412)	-	-	1	Nagy-Reis and Setz, 2017
C. coimbrai										
Fazenda Trapsa, Brazil	3.6	2-5	1:1	1-3	-	-	4.3	14.6	15	Chagas and Ferrari, 2011
Cheracebus:										
C. lucifer										
Estacion Biologica Callicebus, Peru	3 (1)	2-3	1:1	1	29 (14-30)	819 (500-1400)	-	-	1	Kinzey, 1977a
Estación Biológica Callicebus, Peru	3.9 (15)		1:1	0-3	18.5	820	-	20	1	Easley, 1982
C. lugens										
Rio Tomo, Colombia	4.0 (5)		-		14.2 (9-22)		-	32	5	Defler, 1983
Caparú, Colombia	4.0 (1)	3-5	1:1	1-3	22.3	-	5	-	1	Palacios et al., 1997
Caparú, Colombia	-		1:1	1-2	16-20	-	5.1	18	7	Defler, 2003
Plecturocebus:										
P. toppini										
Estación Biológica Cocha Cashu, Peru	4.2 (6)	2-5	1:1	1-3	6.9 (6-8)	552 (150-1450); 670	-	20-26	2	Wright, 1985, 1986
Estación Biológica Cocha Cashu, Peru	4.0	2-7	-		11.5 (6-18)		-	-	7	Bossuyt, 2002
Estación Biológica Los Amigos, Peru	4.3 (7)	2-6	1:1	0-4	2.5 (1.9-3.2)	-	-	41-51	7	Lawrence, 2007
P. ornatus										
Socay, Colombia	3.4 (9)	2-4	1:1	1-2	0.3-0.5	570 (315-870)	-	-	9	Mason, 1966, 1968
Meta, Colombia	3.0 (3)	3	1:1	2	3.3-4.2	860 (578-1152)	-	57	3	Robinson, 1977, 1979a
La Macarena, Colombia	3.5 (2)	2-5	1:1	0-3	9.7		-	8	2	Polanco-Ochoa and Cadena, 1993

(continued)

Table 8.1 (Continued)

Taxa and Site [a]	Avg. Group Size (N)	Group Size Range	Adult M:F	No. of Offspring [b]	Home Range [c] (range)	Day Range [d] (range)	Group Density [e] (G/km²)	Individual Density [e] (Ind/km²)	No. Groups Studied	References
P. cupreus										
Estación Biológica Quebrada Blanco, Peru	4.5 (2)	4-5	1:1	2-3	-	-	-	-	2	Nadjafzadeh, 2005
Estación Biológica Quebrada Blanco, Peru	4.0 (2)	3-5	1:1	1-3	9.0 (6.7-11.4)	-	-	-	2	Kulp and Heymann, 2015
P. discolor										
Tiputini Biodiversity Station, Ecuador	5.7 (7)	2-7	1:1	0-5	-	-	-	-	7	Van Belle et al., 2016
Tiputini Biodiversity Station, Ecuador	3.5 (4)	-	-	-	6.1	-	13.6 / 16.4[b]	47.6 / 57.4[b]	4	Dacier et al., 2011
P. oenanthe										
Moyobamba, Peru	5 (1)	4-5	1:1	2-3	1.6-2.6	663 (478-877)	-	-	1	DeLuycker, 2006, 2007

[a] Data are from long-term studies (>6 Months).
[b] "Offspring" includes subadult, juvenile, and infant age classes.
[c] Home Range is in hectares.
[d] Day Range is in meters.
[e] Two estimates of population density were derived; the first number is from using the point-count playback method and the second from using home range size estimates.
- Indicates no data.

Titi monkeys tend to synchronize group activity; aggressive interactions among conspecifics are extremely rare, and there appear to be no dominance hierarchies within the group (Kinzey, 1981). Grooming is an important social activity for titis, occurring especially during day resting periods and (at least in *Cheracbus lucifer*, *Callicebus personatus*, and *Plecturocebus oenanthe*) also at night rest sites before sleeping (Kinzey & Wright, 1982; Kinzey & Becker, 1983; DeLuycker, 2007).

From several long-term studies (with reliable identification of adults), it seems that the adult pair is relatively permanent or at least prolonged over multiple breeding seasons. However, there is evidence of possible extra-pair paternity (Van Belle et al., 2016), and the pair may not "mate for life." In a long-term study, Lawrence (2007) witnessed the dissolution of a pair-bond. The pair-bond in titi monkeys has been characterized by species-typical attributes such as frequent grooming, tail-twining, small intra-pair distances, huddling, close-following, and displays during encounters with other groups (Mason, 1966, 1968; Kinzey, 1981; Robinson, 1981; Wright, 1985). In several captive studies, males and females show signs of distress when separated and strongly prefer pair mates over strangers (Mendoza & Mason, 1986; Anzenberger, 1988; Fernandez-Duque et al., 1997). The contribution to, and maintenance of, the pair-bond seems to vary among species of titi monkeys. For *P. discolor* in Ecuador, pair mates contributed equally to grooming bouts, and the male contributed more to maintaining proximity (Spence-Aizenberg et al., 2016). Pair mates also contributed equally in grooming bouts for *P. toppini* in Peru, as well as the maintenance of proximity (Lawrence, 2007). For *P. oenanthe* in Peru, the male contributed more to grooming bouts (more often and at higher rates) than females, and the male initiated and maintained proximity and close contact more often than the female (DeLuycker, 2007).

Titi monkeys give birth to one offspring at a time. Twin births have rarely been observed; only three twin births have been reported in the wild (Knogge & Heymann, 1995; Lawrence, 2007; de Santana et al., 2014) and two in captivity (Valeggia et al., 1999). In all three cases in the wild, only one of the twins survived to weaning, suggesting a potential energetic cost to some aspect of infant care in titis (de Santana et al., 2014). There is one observed case of adoption of an infant in a group of *Callicebus nigrifrons* (Cäsar & Young, 2008), and the two paired adults successfully reared both infants. At Manú, Peru, most births of *P. toppini* were concentrated at the beginning of the rainy season (Wright, 1984). In Ecuador, most births occurred during the rainy season and mean interbirth interval averaged 14.2 months (Van Belle et al., 2016). In captivity, births occur year-round (Valeggia et al., 1999); the mean interbirth interval for captive titis is 318±19 days or 11.9 months (n=60) (Valeggia et al., 1999) and 13.1 months recorded in another study (Jantschke et al., 1995). The length of the gestation period ranges from 124–136 days (Jantschke et al., 1995). The length of the ovarian cycle is 17–21 days (Sassenrath et al., 1980; Valeggia et al., 1999).

Within 24–48 hours of birth, the adult male titi begins to carry the infant (Wright, 1984, 1990; Jantschke et al., 1995; DeLuycker, 2014). The male carries the infant the majority (95–98%) of the time, with the infant only moving onto the female to nurse, and the male engages in more grooming, cleaning/inspection, and play with infants than the female (DeLuycker, 2014; Spence-Aizenberg et al., 2016). In one study of P. oenanthe in Peru, a daytime birthing event was witnessed, and the male remained in contact with or in close proximity (≤ 1 m) to the female during the entire event (DeLuycker, 2014). The male made physical contact with the infant almost immediately after birth and started carrying the infant within 24 hours. Based on observations of activity patterns and foraging rates, DeLuycker (2014) hypothesizes that the immediate, extensive care given by the male serves to establish his bond with the infant as well as reinforce his caretaking duties, which additionally frees the female from such tasks and allows her more time to forage. The prolonged and intense tactile bond between the male (putative father) and infant is clearly evident in titi monkeys. Indeed, in captive studies, it has been shown that male titi fathers experience higher levels of prolactin (a hormone connected with paternal care) than non-fathers (Schradin et al., 2003). In addition, an infant exhibits an increase in the adrenocortical stress response when separated from its father, even while in the presence of the mother (Hoffman et al., 1995). This may be the only physiological evidence of an attachment between a mammalian infant and the putative father where the determining factor is the source of contact (from the father) and not nourishment (from the mother). Infants begin to move on their own between 2.5–3.5 months but are still partially carried by the male until 4–6 months of age (DeLuycker, 2014). Furthermore, adult male care of the young extends into juvenile stages of development, and juveniles tend to maintain proximity to adults more than the reverse (Kinzey, 1981; Jantschke et al., 1995; DeLuycker, 2007; Fernandez-Duque et al., 2013). Occasional food sharing or food transfer by the adult male to the offspring has been reported in a few species of titi monkeys (Kinzey, 1977a; Starin, 1978; Wright, 1985; Spence-Aizenberg et al., 2016); however, in at least one study (with reliable individual identification), no food sharing was observed (e.g., DeLuycker, 2007). Weaning occurs between 4–6 months. Sexual maturity is probably attained between 3–4 years of age (Kinzey, 1981). In captivity females gave birth to their first infant at about 3.7 years of age (Valeggia et al., 1999). Subadults leave the natal group at around 3–4 years of age, and their departure is abrupt and without any precipitous agonistic activities; males tend to disperse later than females (Kinzey, 1981; Bossuyt, 2002). Subadults may remain in their natal group until after sexual maturity, potentially to assist in sibling infant caretaking duties (allocare), and more than one adult has been seen in several groups of titi monkey species (Defler, 1983; Price & Piedade, 2001b; Bicca-Marques et al., 2002; Felton et al., 2006). Instances of allocare of infants (e.g., carrying and grooming) by juveniles or subadults in the group has been seen for brief periods of time in some species, but most social interactions among offspring occur in the form of playing (Bossuyt, 2002; DeLuycker, 2014; de Santana et al., 2014). Bossuyt (2002) suggested that due to delayed dispersal of subadults (3.5–4 years of age) in these groups, allocare may provide aid to siblings that are at greater risk of predation.

Another commonality among species of titi monkeys is that males and females perform loud calls in duets (occasionally joined by or performed solo by subadults); these coordinated loud calls are exchanged between neighboring groups. The bioacoustic structure and organization of titi monkey calls have been investigated in a few species (P. discolor: Moynihan, 1966; P. ornatus: Robinson, 1979b; P. cupreus: Müller & Anzenberger, 2002; and Callicebus nigrifrons: Caselli et al., 2014). All studies found that titi monkey calls (i.e., syllables) are made up of phrases which are then combined into longer sequences, with males and females emitting the same syllables and phrase types but at different times. These calls can project over 1 km in distance and have been described as an "operatic turkey gobble" (Kinzey, 1981).

The function of these calls appears to be different among species. Male P. ornatus give loud calls almost every morning to identify the locality of the group. If the group is near a boundary or close to another calling group, the female joins the male in calling. The resulting "duet" stimulates the two groups to approach the boundary and to interact (Robinson, 1981). During 160 days, Robinson (1981) observed 121 interactions between P. ornatus groups. The interactions involve elaborate and extended displays in which calling, chasing, and rushing occur but rarely do animals make contact (Mason, 1968; Robinson, 1979a). Thus, P. ornatus loud calls, and especially duets, can be considered distance-decreasing signals between groups that are maintaining defined territories. In

Cheracebus lucifer and *Callicebus personatus*, on the other hand, neighboring groups do not approach one another after exchanging loud calls, and for *C. lucifer*, no ritualized boundary encounters were observed in over 1,000 hours of observation (Kinzey & Becker, 1983; Kinzey & Robinson, 1983; Price & Piedade, 2001b). Solo male calls and duets are given either spontaneously or in response to neighboring groups. However, calls are given from well within the home range and groups do not respond by moving toward one another and, thus, do not interact at boundaries. Loud calls in these two species help groups to maintain distance between one another and ensure avoidance of intergroup interactions (Kinzey & Becker, 1983; Kinzey & Robinson, 1983). These patterns of intergroup spacing appear to be related to the ranging patterns of the species. Wright (2013) found that *P. toppini* at Manú gave significantly more dawn loud calls when fruit was abundant, suggesting a resource defense function, and that loud calls were not associated with mating. In a study on *Callicebus nigrifrons*, Caselli et al. (2014) found support for the hypothesis that loud calls during intergroup communication are cooperative displays used by the mated pair to regulate access to important food resources such as fruits. In studies of reactions of mated individuals to unfamiliar conspecifics, researchers have found that duetting was always performed in the presence of another and never when the pair was alone, supporting the suggestion that duetting serves to maintain the social integrity of the mated pair (Anzenberger et al., 1986; Anzenberger, 1988). In a field study on *P. toppini* in Peru, Lawrence (2007) found from playback experiments that both males and females initiated and participated in duet calling. Additional evidence related to resource availability supports the suggestion that the pair-bond may function as cooperative resource defense and potentially mate defense, but the function of duetting in relation to the pair-bond still warrants further study.

RANGING BEHAVIOR

There tends to be very little overlap (<15%) of conspecific groups' home ranges in most species of titi monkeys thus far studied (Mason, 1968; Kinzey et al., 1977; Terborgh, 1983; Wright, 1985; Müller, 1995b; Heiduck, 2002; Van Belle et al., 2016). Titi monkeys display active territorial defense by engaging in almost daily intergroup interactions (Mason, 1968; Robinson, 1981; Lawrence, 2007) and/or use frequent loud calls at range boundaries to mediate intergroup spacing (Robinson, 1981; Terborgh, 1983). Species of *Cheracebus* and *Plecturocebus* have relatively shorter day range lengths and smaller home ranges than species of *Callicebus* (see Table 8.1). The day range lengths of groups of *P. ornatus*, *P. toppini*, *P. oenanthe*, and

Cheracebus lucifer are similar—averaging 570 m (Mason, 1968) to 860 m (Polanco-Ochoa & Cadena, 1993) for *P. ornatus*, 611 m for *P. toppini* (Wright, 1985), 663 m for *P. oenanthe* (DeLuycker, 2007), and 820 m for *C. lucifer* (Kinzey et al., 1977; Easley, 1982). However, the home ranges of *P. ornatus* (0.3–0.5 ha: Mason, 1968; 5–14 ha: Polanco-Ochoa & Cadena, 1993), *P. toppini* (6–8 ha: Wright, 1986), and *P. oenanthe* (2–3 ha: DeLuycker, 2007) are on average smaller than those of *C. lucifer* (19–29 ha: Kinzey et al., 1977; Easley, 1982).

Groups of *Callicebus personatus*, *C. melanochir*, and *C. nigrifrons* (all species inhabiting the Atlantic Forests of Brazil) have generally larger home ranges of 11–34 ha and larger average day ranges of 1,000–1,270 m (see Table 8.1), although there is considerable overlap depending on the season. Interestingly, Price and Piedade (2001b) found that neighboring groups of *C. personatus* in Linhares Forest Reserve in Brazil showed more than 20% overlap, and they appeared not to have exclusive access to resources within a large part of the areas used. Furthermore, intergroup encounters were extremely rare, no aggression or displacement was seen if groups did encounter each other, and groups did not show tendency to call within boundary areas, which suggests that at this site, *C. personatus* does not defend or maintain exclusive use of space.

Very few primate species show this territorial pattern of intergroup spacing. Generally, it is assumed that animals divide space in this manner if they use evenly distributed, predictable, and self-renewing resources and, thus, are able to utilize their home range more or less evenly (Pyke et al., 1977). In Peru, Easley (1982) suggested that *Cheracebus lucifer* in Peru establishes and maintains boundaries on the basis of the number of *J. polycarpa* palm trees necessary to provide adequate yearlong production. Easley's study group of *C. lucifer* made a complete circuit of its home range every 6–10 days, traveling through most quadrants of its range on each circuit. The group did not show strong seasonal differences in home range use. Thus, the group could constantly monitor both seasonally available fruit resources and *J. polycarpa* trees in fruit. This same group also gradually shifted its boundaries over a seven-year period (1974–1980), while its home range size remained relatively constant from year to year. Although able to provide only circumstantial evidence, Easley and Kinzey (1986) suggest that this shift of range boundaries is related to the establishment of new territories for offspring. Since offspring of pair-bonded groups necessarily leave the natal group, an adaptive advantage may be gained by providing some offspring with a "known" area. Observations have been made of titi monkey adult offspring forming overlapping or adjacent home ranges with their natal home ranges (Bicca-Marques et al., 2002).

CONSERVATION

Out of the currently recognized species of titi monkey, four are considered by the International Union for the Conservation of Nature (IUCN, 2019) as Critically Endangered (*Callicebus barbarabrownae*, *Plecturocebus caquetensis*, *P. oenanthe*, and *P. olallae*). In addition, two species are considered Endangered (*Callicebus coimbrai* and *P. modestus*). Four species are listed as Vulnerable (*Callicebus melanochir*, *Callicebus personatus*, *Cheracebus medemi*, and *P. ornatus*) and two species as Near Threatened (*Callicebus nigrifrons* and *P. pallescens*), with the remaining 23 species listed as Least Concern or Data Deficient.

Habitat loss, habitat fragmentation, and habitat conversion are the major threats facing this taxon. Destruction of forest and conversion to agriculture and cattle ranching are common in many regions of these primates' tropical habitats. For example, the critically endangered San Martín titi monkey (*P. oenanthe*), living in the foothills of the Andes Mountains in northeastern Peru, has experienced extreme habitat loss (DeLuycker, 2006). The rate of deforestation in the region of San Martín is increasing as demand for coffee plantations, rice fields, and cattle-grazing lands are proliferating due in large part to a major agrarian program established by the government of Peru. This contributed to a large influx of migrants into the area, and the region now experiences one of the fastest immigration rates in the country (DeLuycker, 2006). *P. oenanthe* is found in an already restricted geographic range, and estimates suggest the current range of this species is less than 14,000 km², with at least 34% of lowland forest habitat lost and 95% of remaining forest fragments likely too small to support viable populations (Schaffer-Smith et al., 2016).

Forest loss and fragmentation are particularly deleterious for species of titi monkeys. Boyle (2016) found that titis experienced an 18% modification of land cover within their geographic ranges from 2000–2012. Habitat loss and fragmentation are particularly critical in the Atlantic Forest regions of Brazil, which has lost nearly 90% of its original forest cover (Ribeiro et al., 2009). The five species of *Callicebus* are threatened with severe population declines due to predicted degradation of the amount, quality, and connectivity of their habitats, which are expected to worsen due to effects of climate change (Gouveia et al., 2016).

Other threats to titi monkeys include destruction of gallery and riparian forests as a result of land clearance, dredging, and associated activities involved in illegal gold mining and illicit crop-growing activity (Wagner et al., 2009). A study has shown that anthropogenic noise pollution produced from mining activity in the Atlantic Forest in Brazil may constrain the ability of titi monkeys in their long-distance vocal communication (Duarte et al., 2017). Finally, threats such as disease, logging, hunting for the bushmeat trade, and live capture for the pet trade are common and pose a threat for some species in several areas.

Based on current studies and predictive models, conservation solutions to protect titi monkeys should include increased establishment of protected areas (e.g., Natural Protected Areas, Regional Conservation Areas, Private Conservation Areas). In addition, solutions should also include regeneration of lost habitat, establishment of appropriate habitat corridors to connect fragmented areas, development of educational programs within local communities, regulation and enforcement of land use and ownership laws, and formulation of sustainable methods for agrarian initiatives—all imperative for the conservation of these species.

SUMMARY

Titi monkeys are small-bodied (1–2 kg), pair-bonded, pair-living, diurnal primates exhibiting strong male parental care. Paternal care begins shortly after birth of the infant and consists of the adult male carrying the infant the majority of the time, as well as grooming, play, rescue, and protection. Infants are transferred from the father to the mother only for brief nursing bouts. Titi monkeys rely heavily on tactile contact; social interactions include long grooming bouts among conspecifics. Titis also engage in tail-twining, in which they braid their tails around each other's tails as they sit closely together during resting periods and in their sleeping trees. Tactile communication is important in maintenance of the pair-bond and cohesion of the family group. Other activities that reinforce the pair-bond include close contact, close following, scent-marking investigation, and coordinated duet calls, all of which serve as long-distance maintenance of defined home ranges. Titis are generally territorial, with only minimal overlap of home ranges among neighboring groups. Defense of territories may be related to maintaining access to abundant, constant, and predictable fruiting tree resources. Titi monkeys are mainly frugivorous, but during the dry, low-resource season, they may rely on resources that are available year-round, including lianas/vines, seeds, insects, epiphytic plants, and young leaves. The majority of titi monkey species seem to rely on a small proportion of fruit species depending on the season. The amount of the protein source incorporated in their diet (represented by leaves, insects, or young seeds) varies. The major anthropogenic threats facing these species include habitat loss, with resulting deleterious effects of fragmentation and conversion of habitat to human use.

REFERENCES CITED—CHAPTER 8

Acero-Murcia, A., Almario, L. J., García, J., Defler, T. R., & López, R. (2018). Diet of the Caquetá titi (*Plecturocebus caquetensis*) in a disturbed forest fragment in Caquetá, Colombia. *Primate Conservation, 32*, 31-47.

Adams, D. B., & Kitchen, D. M. (2018). Experimental evidence that titi and saki alarm calls deter an ambush predator. *Animal Behaviour, 145*, 141-147.

Anzenberger, G. (1988). The pairbond in the titi monkey (*Callicebus moloch*): Instrinsic versus extrinsic contributions of the pairmates. *Folia Primatologica, 50*, 188-203.

Anzenberger, G, Mendoza, S. P., & Mason, W. A. (1986). Comparative studies of social behavior in *Callicebus* and *Saimiri*: Behavioral and physiological responses of established pairs to unfamiliar pairs. *American Journal of Physical Anthropology, 11*, 37-51.

Aquino, R., Terrones, W., Cornejo, F., & Heymann, E. W. (2008). Geographic distribution and possible taxonomic distinction of *Callicebus torquatus* populations in Peruvian Amazonia. *American Journal of Primatology, 70*, 1181-1186.

Bicca-Marques, J. C., Garber, P. A., & Azevedo-Lopes, M. A. O. (2002). Evidence of three resident adult male group members in a species of monogamous primate, the red titi monkey (*Callicebus cupreus*). *Mammalia, 66*(1), 138-142.

Bossuyt, F. J. (2002). Natal dispersal of titi monkeys (*Callicebus moloch*) at Cocha Cashu, Manú National Park, Peru. *American Journal of Physical Anthropology, 117*(S34), 47.

Boubli, J. P., Byrne, H., da Silva, M. N. F., Silva, J., Araujo, R. C., Bertuol, F., Goncalves, J., de Melo, F. R., Rylands, A. B., Mittermeier, R. A., Silva, F. E., Nash, S. D., Canale, G., Alencar, R. D., Rossi, R. V., Carneiro, J., Sampaio, I., Farias, I. P., Schneider, H., & Hrbeck, T. (2019). On a new species of titi monkey (Primates: *Plecturocebus* Byrne et al., 2016), from Alta Floresta, southern Amazon, Brazil. *Molecular Phylogenetics and Evolution, 132*, 117-137.

Boveda-Penalba, A. J., Vermeer, J., Rodrigo, F., & Guerra-Vasquez, F. (2009). Preliminary report on the distribution of *Callicebus oenanthe* on the eastern feet of the Andes. *International Journal of Primatology, 30*, 467-480.

Boyle, S. (2016). Pitheciids in fragmented habitats: Land cover change and its implications for conservation. *American Journal of Primatology, 78*, 534-549.

Buchanan-Smith, H. M., Hardie, S. M., Caceres, C., & Prescott, M. J. (2000). Distribution and forest utilization of *Saguinus* and other primates of the Pando Department, northern Bolivia. *International Journal of Primatology, 21*, 353-379.

Byrne, H., Alfaro, J. W., Sampaio, I., Farias, I., Schneider, H., Hrbek, T., & Boubli, J. (2018). Titi monkey biogeography: Parallel Pleistocene spread by *Plecturocebus* and *Cheracebus* into a post-Pebas western Amazon. *Zoologica Scripta*, 1-19.

Byrne, H., Rylands, A. B., Carneiro, J. C., Alfaro, J. W., Bertuol, F., da Silva, M. N. F., Messias, M., Groves, C., Mittermeier, R., Farias, I., Hrbek, T., Schneider, H., Sampaio, I., & Boubli, J. (2016). Phylogenetic relationships of the New World titi monkeys (*Callicebus*): First appraisal of taxonomy based on molecular evidence. *Frontiers in Zoology, 13*(10), 1-25.

Carneiro, J., Sampaio, I., Júnior, J. D., Farias, I., Hrbek, T., Pissinatti, A., Silva, R., Junior, A., Boubli, J., Ferrari, S. F., & Schneider, H. (2018). Phylogeny, molecular dating and zoogeographic history of the titi monkeys (*Callicebus*, Pitheciidae) of eastern Brazil. *Molecular Phylogenetics and Evolution, 124*, 10-15.

Carrillo-Bilbao, G., Di Fiore, A., & Fernandez-Duque, E. (2005). Dieta, forrajeo y presupuesto de tiempo en cotoncillos (*Callicebus discolor*) del Parque Nacional Yasuní en la Amazonia Ecuatoriana. *Neotropical Primates, 13*, 7-11.

Cäsar, C., & Young, R. J. (2008). A case of adoption in a wild group of black-fronted titi monkeys (*Callicebus nigrifrons*). *Primates, 49*, 146-148.

Caselli, C. B., Mennill, D. J., Bicca-Marques, J. C., & Setz, E. Z. F. (2014). Vocal behavior of black-fronted titi monkeys (*Callicebus nigrifrons*): Acoustic properties and behavioral contexts of loud calls. *American Journal of Primatology, 76*, 788-800.

Caselli, C. B., & Setz, E. Z. F. (2011). Feeding ecology and activity pattern of black-fronted titi monkeys (*Callicebus nigrifrons*) in a semideciduous tropical forest of southern Brazil. *Primates, 52*, 351-359.

Chagas, R. R. D., & Ferrari, S. F. (2011). Population parameters of the endangered titi monkey *Callicebus coimbrai* Kobayashi and Langguth 1999, in the fragmented landscape of southern Sergipe, Brazil. *Brazilian Journal of Biology, 71*, 569-575.

Cisneros-Heredia, D. F., Leon-Reyes, A., & Seger, S. (2005). Boa constrictor predation on a titi monkey, *Callicebus discolor*. *Neotropical Primates, 13*, 11-12.

Crandlemire-Sacco, J. (1988). An ecological comparison of two sympatric primates: *Saguinus fuscicollis* and *Callicebus moloch* of Amazonian Peru. *Primates, 29*, 465-475.

Dacier, A., de Luna, A. G., Fernandez-Duque, E., & Di Fiore, A. (2011). Estimating population density of Amazonian titi monkeys (*Callicebus discolor*) via playback point counts. *Biotropica, 43*, 135-140.

Dalponte, J. C., Silva, F. E., & Silva-Júnior, J. S. (2014). New species of titi monkey, genus *Callicebus* Thomas, 1903 (Primates, pitheciidae), from southern Amazonia, Brazil. *Pap Avulsos Zoo, São Paulo, 54*(32), 457-472.

Defler, T. R. (1983). Some population characteristics of *Callicebus torquatus lugens* (Humboldt, 1812) (Primates: Cebidae) in eastern Colombia. *Lozania, 38*, 1-9.

Defler, T. R. (1994). *Callicebus torquatus* is not a white-sand specialist. *American Journal of Primatology, 33*, 149-154.

Defler, T. R. (2003). Density of species and spatial organization of a primate community: Caparu Biological Station, Department of Vaupes, Colombia. In V. Pereira-Bengoa, F. Nassar-Montoya, & A. Savage (Eds.), *Primatologia del Nuevo Mundo: Biologia, medicina, manejo y conservacion* (pp. 23-39). Centro de Primatologia Arguatos.

Defler, T. R, Bueno, M. L., & García J. (2010). *Callicebus caquetensis*: A new and critically endangered titi monkey from southern Caquetá, Colombia. *Primate Conservation, 25*, 1-9.

de Freitas, E. B., De-Carvalho, C. B., & Ferrari, S. F. (2011). Abundance of *Callicebus barbarabrownae* (Hershkovitz 1990), (Primates: Pitheciidae) and other nonvolant mammals in a fragment of arboreal Caatinga in northeastern Brazil. *Mammalia, 75*, 339-343.

de Luna, A. G., Sanmiguel, R., Di Fiore, A., & Fernandez-Duque, E. (2010). Predation and predation attempts on red titi monkeys (*Callicebus discolor*) and equatorial sakis (*Pithecia aequatorialis*) in Amazonian Ecuador. *Folia Primatologica, 81*, 86-95.

DeLuycker, A. M. (2006). Preliminary report and conservation status of the Río Mayo titi monkey, *Callicebus oenanthe* Thomas, 1924, in the Alto Mayo Valley, northeastern Peru. *Primate Conservation, 21*, 33-39.

DeLuycker, A. M. (2007). *The ecology and behavior of the Rio Mayo titi monkey (*Callicebus oenanthe*) in the Alto Mayo, northern Peru.* [Doctoral dissertation, Washington University].

DeLuycker, A. M. (2012). Insect prey foraging strategies in *Callicebus oenanthe* in northern Peru. *American Journal of Primatolology, 74*, 450-461.

DeLuycker, A. M. (2013). Diet and foraging ecology of the critically endangered Andean titi monkey (*Callicebus oenanthe*) in northern Peru. *American Journal of Primatolology, 75*, 93.

DeLuycker, A. M. (2014). Observations of a daytime birthing event in wild titi monkeys (*Callicebus oenanthe*): Implications of the male parental role. *Primates, 55*, 59-67.

de Santana, M. M. (2012). *Ecologia comportamental de um grupo de guigó-de-coimbra (*Callicebus coimbrai* Kobayashi and Langguth 1999) no leste de Sergipe.* [Master's thesis, Universidade Federal de Sergipe].

de Santana, M. M., Souza-Alves, J. P., & Ferrari, S. (2014). Twinning in titis (*Callicebus coimbrai*): Stretching the limits of biparental infant caregiving? *Neotropical Primates, 21*, 190-192.

dos Santos, G. P., Galvão, C., & Young, R. J. (2012). The diet of wild black-fronted titi monkeys *Callicebus nigrifrons* during a bamboo masting year. *Primates, 53*, 265-272.

Duarte, M. H. L., Kaizer, M. C., Young, R. L., Rodriguez, M., & Sousa-Lima, R. S. (2017). Mining noise affects loud call structures and emission patterns of wild black-fronted titi monkeys. *Primates, 59*, 89-97.

Easley, S. P. (1982). *Ecology and behavior of* Callcebus torquatus, *Cebidae, primates.* [Doctoral dissertation, Washington University].

Easley, S. P., & Kinzey, W. G. (1986). Territorial shift in the yellow-handed titi monkey (*Callicebus torquatus*). *American Journal of Primatolology, 11*, 307-318.

Felton, A., Felton, A. M., Wallace, R. B., & Gómez, H. (2006). Identification, behavioral observations, and notes on the distribution of the titi monkeys *Callicebus modestus*

Lönnbert, 1939 and *Callicebus olallae*, Lönnberg, 1939. *Primate Conservation, 20*, 41-46.

Fernandez-Duque, E., Di Fiore, A., & de Luna, A. G. (2013). Pair-mate relationships and parenting in equatorial saki monkeys (*Pithecia aequatorialis*) and red titi monkeys (*Callicebus discolor*) of Ecuador. In L. M. Veiga, A. A. Barnett, S. F. Ferrari, & M. A. Norconk (Eds.), *Evolutionary biology and conservation of titis, sakis, and uacaris* (pp. 295-302). Cambridge University Press.

Fernandez-Duque, E., Mason, W. A., & Mendoza, S. P. (1997). Effects of duration of separation on responses to mates and strangers in the monogamous titi monkey (*Callicebus moloch*). *American Journal of Primatology, 43*, 225-237.

Gouveia, S. F., Souza-Alves, J. P., Rattis, L., Dobrovolski, R., Jerusalinsky, L., Beltrão-Mendes, R., & Ferrari, S. (2016). Climate and land use changes will degrade the configuration of the landscape for titi monkeys in Eastern Brazil. *Global Change Biology, 22*, 2003-2012.

Gualda-Barros, J., Nascimento, F. O., & Amaral, M. K. (2012). A new species of *Callicebus* Thomas 1903 (Primates, Pitheciidae) from the states of Mato Grosso and Pará, Brazil. *Pap Avulsos Zoologia (São Paulo), 52*(23), 261-279.

Gusmão, A. C., & Costa, T. M. (2014). Registro de *Callicebus cinerascens* (Spix, 1823) no Vale do Guaporé, Rondônia, Brasil. *Neotropical Primates, 21*, 211-214.

Gusmão, A. C., Messias, M. R., Carneiro, J. C., Shneider, H., de Alencar, T. B., Calouro, A. M., Dalponte, J. C., Mattos, F., Ferrari, S. F., Buss, G., Azevedo, R. B., Júnior, S., Nash, S. D., Rylands, A. B., & Barnett, A. A. (2019). A new species of titi monkey, *Plecturocebus* Byrne et al., 2016 (Primates, Pitheciidae) from southwestern Amazonia, Brazil. *Primate Conservation, 33*, 21-35.

Heiduck, S. (1997). Food choice in masked titi monkeys (*Callicebus personatus melanochir*): Selectivity or opportunism? *International Journal of Primatology, 18*(4), 487-503.

Heiduck, S. (1998). *Nahrungsstrategien Schwarzköpfiger Springaffen (Callicebus personatus melanochir).* Cuvillier.

Heiduck, S. (2002). The use of disturbed and undisturbed forest by masked titi monkeys *Callicebus personatus melanochir* is proportional to food availability. *Oryx, 36*, 133-139.

Hershkovitz, P. (1990). Titis, New World monkeys of the genus *Callicebus* (Cebidae, Primates). *Fieldiana Zoology, 55*, 1-109.

Heymann, E. W., & Nadjafzadeh, M. N. (2013). Insectivory and prey foraging techniques in *Callicebus*—a case study of *Callicebus cupreus* and a comparison to other pitheciids. In L. M. Veiga, A. A. Barnett, S. F. Ferrari, & M. A. Norconk (Eds.), *Evolutionary biology and conservation of titis, sakis and uakaris* (pp. 215-224). Cambridge University Press.

Hoffman, K. A., Mendoza, S. P., Hennessy, M. B., & Mason, W. A. (1995). Responses of infant titi monkeys, *Callicebus moloch*, to removal of one or both parents: Evidence for paternal attachment. *Developmental Psychobiology, 28*, 399-407.

IUCN. (2019). *The International Union for the Conservation of Nature Red List of Threatened Species.* Version 2020-1. https://www.iucnredlist.org.

Jantschke, B., Welker, C., & Klaiber-Schuh, A. (1995). Notes on breeding of the titi monkey *Callicebus cupreus*. *Folia Primatologica*, 65(4), 210-213.

Johns, A. D. (1991). Forest disturbance and Amazonian primates. In H. O. Box (Ed.), *Primate responses to environmental change* (pp. 115-135). Chapman & Hall.

Kinzey, W. G. (1977a). Diet and feeding behaviour of *Callicebus torquatus*. In T. H. Clutton-Brock (Ed.), *Primate ecology: Studies of feeding and ranging behaviour in lemurs, monkeys, and apes* (pp. 127-151). Academic Press.

Kinzey, W. G. (1977b). Positional behavior and ecology in *Callicebus torquatus*. *Yearbook of Physical Anthropology*, 20, 468-480.

Kinzey, W. G. (1978). Feeding behaviour and molar features in two species of titi monkey. In D. J. Chivers, & J. Herbert (Eds.), *Recent advances in primatology* (*Vol. 1: Behaviour*, pp. 372-375). Academic Press.

Kinzey, W. G. (1981). The titi monkeys, genus *Callicebus*. In A. F. Coimbra-Fihlo, & R. A. Mittermeier (Eds.), *Ecology and behavior of Neotropical primates* (pp. 241-276). Brasilian Academy of Sciences.

Kinzey, W. G., & Becker, M. (1983). Activity pattern of the masked titi monkey, *Callicebus personatus*. *Primates*, 24, 337-343.

Kinzey, W. G., & Gentry, A. H. (1979). Habitat utilization in two species of *Callicebus*. In R. W Sussman (Ed.), *Primate ecology: Problem-oriented field studies* (pp. 89-100). John Wiley & Sons.

Kinzey, W. G., & Norconk, M. A. (1993). Physical and chemical properties of fruits and seeds eaten by *Pithecia* and *Chiropotes* in Surinam and Venezuela. *International Journal of Primatology*, 14, 2017-2227.

Kinzey, W. G., & Robinson, J. G. (1983). Intergroup loud calls, range size, and spacing in *Callicebus torquatus*. *American Journal of Physical Primatology*, 60, 539-544.

Kinzey, W. G., Rosenberger, A. L., Heisler, P. S., Prowse, D. L., & Trilling, J. S. (1977). A preliminary field investigation of the yellow handed titi monkey, *Callicebus torquatus torquatus*, in northern Peru. *Primates*, 18, 159-181.

Kinzey, W. G., & Wright, P. C. (1982). Grooming behavior in the titi monkey (*Callicebus torquatus*). *American Journal of Primatology*, 3(1-4), 267-275.

Knogge, C., & Heymann, E. W. (1995). Field observation of twinning in the dusky titi monkey, *Callicebus cupreus*. *Folia Primatologica*, 65(2), 118-120.

Kobayashi, S., & Langguth, A. L. (1999). A new species of titi monkey, *Callicebus* Thomas, from northeastern Brazil (Primates, Cebidae). *Revista Brasileira Zoologia*, 16(2), 531-551.

Kulp, J., & Heymann, E. W. (2015). Ranging, activity budget, and diet composition of red titi monkeys (*Callicebus cupreus*) in a primary forest and forest edge. *Primates*, 56, 273-278.

Lambert, J. E. (2011). Primate nutritional ecology: Feeding biology and diet at ecological and evolutionary scales. In C. J. Campbell, A. Fuentes, K. C. MacKinnon, S. K. Bearder, &

R. M. Stumpf (Eds.), *Primates in perspective* (pp. 482-495). Oxford University Press.

Lawler, R. R., Ford, S. M., Wright, P. C., & Easley, S. P. (2006). The locomotor behavior of *Callicebus brunneus* and *Callicebus torquatus*. *Folia Primatologica*, 77, 228-239.

Lawrence, J. M. (2007). *Understanding the pair bond in brown titi monkeys* (Callicebus brunneus): *Male and female reproductive interests*. [Doctoral dissertation, Columbia University].

Martinez, J., & Wallace, R. B. (2007). Further notes on the distribution of endemic Bolivian titi monkeys, *Callicebus modestus* and *Callicebus olallae*. *Neotropical Primates*, 14, 47-54.

Mason, W. A. (1966). Social organization of the South American monkey, *Callicebus moloch*, a preliminary report. *Tulane Studies in Zoology*, 13, 23-28.

Mason, W. A. (1968). Use of space by *Callicebus* groups. In P. C. Lee (Ed.), *Primates: Studies in adaptation and variability* (pp. 200-216). Holt, Rinehart and Winston.

Mendoza, S., & Mason, W. (1986). Contrasting responses to intruders and to involuntary separation by monogamous and polygynous New World monkeys. *Physiology and Behavior*, 38, 795-801.

Moynihan, M. (1966). Communication in the titi monkey, *Callicebus*. *Journal of Zoology*, 150, 77-127.

Moynihan, M. (1976). *The New World primates*. Princeton University Press.

Müller, A. E., & Anzenberger, G. (2002). Duetting in the titi monkey *Callicebus cupreus*: Structure, pair specificity and development of duets. *Folia Primatologica*, 73, 104-115.

Müller, K. (1995a). *Ecology and feeding behavior of masked titi monkeys* (Callicebus personatus melanochir, Cebidae, primates) *in the Atlantic rainforest of eastern Brazil*. [Doctoral dissertation, University of Berlin].

Müller, K. (1995b). Ranging in masked titi monkeys (*Callicebus personatus*) in Brazil. *Folia Primatologica*, 65, 224-228.

Müller, K. (1996). Diet and feeding ecology of masked titis (*Callicebus personatus*). In M. A. Norconk, A. L. Rosenberger, & P. A. Garber (Eds.), *Adaptive radiations of Neotropical primates* (pp. 383-401). Plenum Press.

Nadjafzadeh, M. N. (2005). *Strategies and techniques of prey acquisition of copper red titi monkeys,* Callicebus cupreus, *compared to the sympatric tamarin* Saguinus mystax *and* Saguinus fuscicollis *in northeastern Peru*. [Master's thesis, University of Bochum].

Nadjafzadeh, M. N., & Heymann, E. W. (2008). Prey foraging of red titi monkeys, *Callicebus cupreus*, in comparison to sympatric tamarins, *Saguinus mystax* and *Saguinus fuscicollis*. *American Journal of Physical Anthropology*, 135, 56-63.

Nagy-Reis, M. B., & Setz, E. Z. (2017). Foraging strategies of black-fronted titi monkeys (*Callicebus nigrifrons*) in relation to food availability in a seasonal tropical forest. *Primates*, 58, 149-158.

Neri, F. M. (1997). *Manejo de* Callicebus personatus, *Geoffroy 1812, resgatados: Uma tentative de reintrodução e estudos ecológicos de um grupo silvestre na Reserva do Patrimônio Natural Galheiro.* [Master's thesis, Universidade Federal de Minas Gerais].

Neri, F. M., & Rylands, A. B. (1997). Tamanho da área de uso e distâncias percorridas por sauás, *Callicebus personatus*, região do Triângulo Mineiro. In *Programa e resumos. VIII Congresso Brasileiro de Primatologia, V Reunião Latino-Americano de Primatologia, João Pessoa, Paraíba, 10–15 de agosto de 1997* (p. 42). Sociedade Brasileira de Primatologia.

Palacios, E., & Rodriguez, A. (2013). Seed eating by *Callicebus lugens* at Caparu Biological Station, on the lower Apaporis River, Colombian Amazonia. In L. M. Veiga, A. A. Barnett, S. F. Ferrari, & M. A. Norconk (Eds.), *Evolutionary biology and conservation of titis, sakis and uakaris* (pp. 225-231). Cambridge University Press.

Palacios, E., Rodriguez, A., & Defler, T. R. (1997). Diet of a group of *Callicebus torquatus lugens* (Humboldt, 1812) during the annual resource bottleneck in Amazonian Columbia. *International Journal of Primatology,* 18(4), 503-522.

Polanco-Ochoa, R. (1992). *Aspectos etológicos y ecológicos de* Callicebus cupreus ornatus *Gray 1870 (Primates: Cebidae) en el Parque Nacional Natural Tinigua, La Macarena, Meta, Colombia.* [Tesis de grado, Universidad Nacional de Colombia].

Polanco-Ochoa, R., & Cadena, A. (1993). Use of space by *Callicebus cupreus ornatus* (Primates; Cebidae) in Macarena, Colombia. *Field Studies of New World Monkeys, La Macarena, Colombia,* 8, 19-32.

Polanco-Ochoa, R., Garcia, J. E., & Cadena, A. (1994). Utilización del tiempo y patrones de actividad de *Callicebus cupreus* (Primates: Cebidae) en La Macarena, Colombia. *Trianea,* 5(5), 305-322.

Price, E. C., & Piedade, H. M. (2001a). Diet of northern masked titi monkeys (*Callicebus personatus*). *Folia Primatologica,* 72, 335-338.

Price, E. C., & Piedade, H. M. (2001b). Ranging behavior and intraspecific relationships of masked titi monkeys (*Callicebus personatus personatus*). *American Journal of Primatology,* 53, 8-92.

Printes, R. C., Buss, G., Azevedo, R., Ravetta, A. L., & Silva, G. N. (2018). Update on the geographic distributions of two titi monkeys, *Plecturocebus hoffmannsi* (Thomas 1908) and *P. baptista* (Lönnberg, 1939), in two protected areas in the Brazilian Amazon. *Primate Conservation,* 32, 1-8.

Pyke, G. H., Pulliam, H. R., & Charnov, E. L. (1977). Optimal foraging: A selective review of theory and tests. *Quarterly Review of Biology,* 52, 137-154.

Ribeiro, M. C., Metzger, J. P., Martensen, A. C., Ponzoni, F. Z., & Hirota, M. M. (2009). The Brazilian Atlantic Forest: How much is left and how is the remaining forest distributed? Implications for conservation. *Biological Conservation,* 142, 1141-1153.

Robinson, J. G. (1977). *Vocal regulation of spacing in the titi monkey,* Callicebus moloch. [Doctoral dissertation, University of North Carolina].

Robinson, J. G. (1979a). Vocal regulation of use of space by groups of titi monkeys *Callicebus moloch. Behavioral Ecology and Sociobiology,* 5, 1-15.

Robinson, J. G. (1979b). An analysis of the organization of vocal communication in the titi monkey *Callicebus moloch. Zeitschrift fur Tierpsychologie,* 49, 381-405.

Robinson, J. G. (1981). Vocal regulation of inter- and intra-group spacing during boundary encounters in the titi monkey, *Callicebus moloch. Primates,* 22, 161-172.

Rocha. A., Barnett, A. P. A., & Sprionello, W. R. (2019). Extension of the geographic distribution of *Plecturocebus baptista* (Pitheciidae, primates) and a possible hybrid zone with *Plecturocebus hoffmansi*: Evolutionary and conservation implications. *Acta Amazonica,* 49, 330-333.

Röhe, F., & Silva-Júnior, J. S. (2009). Confirmation of *Callicebus dubius* (Pitheciidae) distribution and evidence of invasion into the geographic range of *Callicebus stephennashi. Neotropical Primates,* 16, 71-73.

Rowe, N., & Martinez, W. (2003). *Callicebus* sightings in Bolivia, Peru and Ecuador. *Neotropical Primates,* 11, 32-35.

Rumiz, D. I. (2012). Distribution, habitat and status of the white-coated titi monkey (*Callicebus pallescens*) in the Chaco-Chiquitano forests of Santa Cruz, Bolivia. *Neotropical Primates,* 19, 8-15.

Sampaio, D. T., & Ferrari, S. F. (2005). Predation of an infant titi monkey (*Callicebus moloch*) by a tufted capuchin (*Cebus apella*). *Folia Primatologica,* 76, 113-115.

Sampaio, R., Dalponte, J. C., Rocha, E. C., Hack, R. O. E., Gusmão, A. C., Aguiar, K. M. O., Kuniy, A. A., & Silva-Júnior, J. S. (2012). Novos registros com uma extensão da distribuição geográfica de *Callicebus cinerascens* (Spix, 1823). *Mastozoologica Neotropical,* 19, 159-164.

Sassenrath, E. N., Mason, W. A., Fitzgerald, R. C., & Kenney, M. D. (1980). Comparative endocrine correlates of reproductive states in *Callicebus* (titi) and *Saimiri* (squirrel) monkeys. *Anthropologia Contemporanea,* 3, 265.

Schaffer-Smith, D., Swenson, J. J., & Bóveda-Penalba, A. J. (2016). Rapid conservation assessment for endangered species using connectivity models. *Environmental Conservation,* 43(3), 221-230.

Schradin, C., Reeder, D. M., Mendoza, S. P., & Anzenberger, G. (2003). Prolactin and paternal care: Comparison of three species of monogamous New World monkeys (*Callicebus cupreus, Callithrix jacchus,* and *Callimico goeldii*). *Journal of Comparative Psychology,* 117, 166-175.

Souza-Alves, J. P. (2010). *Ecologia alimentar de un grupo de guigó-de-Coimbra-Filho (*Callicebus coimbrai *Kobayashi & Langguth, 1999): Perspectivas para a conservação da espécie na paisagem fragmentada do sul de Sergipe.* [Master's thesis, Universidade Federal de Sergipe].

Souza-Alves, J. P. (2013). *Ecology and life-history of Coimbra-Filho's titi monkeys (*Callicebus coimbrai*) in the Brazilian Atlantic Forest.* [Doctoral dissertation, Universidade Federal de Paraíba].

Souza-Alves, J. P., Caselli, C. B., Gestich, C. C., & Nagy-Reis, M. B. (2019a). Should I store, or should I sync? The breeding strategy of two small Neotropical primates under predictable resource availability. *Primates, 60,* 113-118.

Souza-Alves, J. P, Fontes, I. P., Chagas, R. R. D., & Ferrari, S. F. (2011a). Seasonal versatility in the feeding ecology of a group of titis (*Callicebus coimbrai*) in the northern Brazilian Atlantic Forest. *American Journal of Primatology, 73,* 1-11.

Souza-Alves, J. P, Fontes, I. P., & Ferrari, S. F. (2011b). Use of sleeping sites by a titi group (*Callicebus coimbrai*) in the Brazilian Atlantic Forest. *Primates, 52,* 155-161.

Souza-Alves, J. P., Mourthe, I., Hilário, R. R., Bicca-Marques, J. C., Rehg, J., Gestich, C. C., Acero-Murcia, A. C., Adret, P., Aquino, R., Berthet, M., Bowler, M., Calouro, A. M., Canale, G. R., Cardoso, N., Caselli, C. B., Casar, C., Chagas, R. R. D., Clyvia, A., Corsini, C. F.,... & Barnett, A. A. (2019b). Terrestrial behavior in titi monkeys (*Callicebus, Cheracebus,* and *Plecturocebus*): Potential correlates, patterns, and differences between genera. *International Journal of Primatology, 40,* 553-572.

Spence-Aizenberg, A., Di Fiore, A., & Fernandez-Duque, E. (2016). Social monogamy, male-female relationships, and biparental care in wild titi monkeys (*Callicebus discolor*). *Primates, 57,* 103-112.

Starin, E. D. (1978). Food transfer by wild titi monkeys (*Callicebus torquatus torquatus*). *Folia Primatologica, 30,* 145-151.

Terborgh, J. (1983). *Five New World primates: A study in comparative ecology.* Princeton University Press.

Thomas, O. (1903). Notes on South American monkeys, bats, carnivores, and rodents, with descriptions of new species. *Annals and Magazine of Natural History, including Zoology, Botany, and Geology, 12,* 455-464.

Trevelin, L. C., Port-Carvlho, M., Silveira, M., & Morell, E. (2007). Abundance, habitat use and diet of *Callicebus nigrifrons* Spix (Primates, Pitheciidae) in Cantareira State Park, São Paulo, Brazil. *Revista Brasileira de Zoologia,* 24, 1071-1077.

Valeggia, C. R., Mendoza, S. P., Fernandez-Duque, E., Mason, W. A., & Lasley, B. (1999). Reproductive biology of female titi monkeys (*Callicebus moloch*) in captivity. *American Journal of Primatology,* 47(3), 183-195.

Van Belle, S., Fernandez-Duque, E., & Di Fiore, A. (2016). Demography and life history of wild red titi monkeys (*Callicebus discolor*) and equatorial sakis (*Pithecia aequatorialis*) in Amazonian Ecuador: A 12-year study. *American Journal of Primatology, 78,* 204-215.

van Roosmalen, M. G. M., van Roosmalen, T., & Mittermeier, R. A. (2002). A taxonomic review of the titi monkeys, genus *Callicebus* Thomas, 1903, with the description of two new species, *Callicebus bernhardi* and *Callicebus stephennashi,* from Brazilian Amazonia. *Neotropical Primates, 10,* 1-52.

Vermeer, J. (2009). On the identification of *Callicebus cupreus* and *Callicebus brunneus. Neotropical Primates, 16,* 69-71.

Vermeer, J., & Tello-Alvarado, J. C. (2015). The distribution and taxonomy of titi monkeys (*Callicebus*) in central and southern Peru, with the description of a new species. *Primate Conservation, 29,* 9-29.

Vermeer, J., Tello-Alvarado, J. C., Moreno-Moreno, S., & Guerra-Vásquez, F. (2011). Extension of the geographical range of white-browed titi monkeys (*Callicebus discolor*) and evidence for sympatry with San Martín titi monkeys (*Callicebus oenanthe*). *International Journal of Primatology, 32,* 924-930.

Wagner, M., Castro, F., & Stevenson, P. R. (2009). Habitat characterization and population status of the dusky titi (*Callicebus ornatus*) in fragmented forests, Meta, Colombia. *Neotropical Primates, 16,* 18-24.

Wallace, R. B., Gómez, H., Felton, A., & Felton, A. M. (2006). On a new species of titi monkey, genus *Callicebus* Thomas (Primates, Pitheciidae), from western Bolivia with preliminary notes on distribution and abundance. *Primate Conservation, 20,* 29-39.

Wright, P. C. (1984). Biparental care in *Aotus trivirgatus* and *Callicebus moloch.* In M. Small (Ed.), *Female primates: Studies by women primatologists* (pp. 59-75). A R. Liss.

Wright, P. C. (1985). *The costs and benefits of nocturnality for* Aotus trivirgatus *(the night monkey).* [Doctoral dissertation, City University of New York].

Wright, P. C. (1986). Ecological correlates of monogamy in *Aotus* and *Callicebus.* In J. Else, & P. Lee (Eds.), *Primate ecology and conservation* (pp. 159-167). Cambridge University Press.

Wright, P. C. (1990). Patterns of parental care in primates. *International Journal of Primatology, 11,* 89-102.

Wright, P. C. (2013). *Callicebus* in Manu National Park: Territory, resources, scent marking and vocalizations. In L. M. Veiga, A. A. Barnett, S. F. Ferrari, & M. A. Norconk (Eds.), *Evolutionary biology and conservation of titis, sakis and uakaris* (pp. 232-239). Cambridge University Press.

Youlatos, D., & Pozo-Rivera, W. (1999). Preliminary observations on the songo songo (dusky titi monkey, *Callicebus moloch*) of northeastern Ecuador. *Neotropical Primates,* 7(2), 45.

CHAPTER 9

Sakis, Bearded Sakis, and Uakaris

Platyrrhine Seed Predators

CHRISTOPHER A. SHAFFER AND PATRICIA T. ORMOND

THE PITHECIINAE ARE ONE OF TWO SUBFAMILIES—THE OTHER being Callicebinae—in the family Pitheciidae. The Pitheciinae includes three genera, the sakis (genus *Pithecia*: Figure 9.1), bearded sakis or cuixius (genus *Chiropotes*: Figure 9.2), and uakaris (genus *Cacajao*: Figure 9.3). The pitheciines are distinguished by their dental adaptations for seed eating (Kay et al., 2008). In fact, the defining feature of the subfamily is their ability to rely on seeds for a higher percentage of their diet than any other group of primates (Norconk, 2011). Kinzey (1992) and Norconk et al. (2009) have characterized them as "sclerocarpic" (*hard fruit*) foragers due to their preference for thick-husked, "mechanically protected" fruit. Their highly specialized craniodental morphology allows the pitheciines to extract the soft, nutritious seeds from fruits with tough and hard exocarp. Yet despite their shared morphological adaptations, pitheciines vary considerably in their ecology and behavior.

The defining morphological characteristics of the Pitheciinae are their highly specialized dental and cranial adaptations for seed eating (Figure 9.4). Sakis, bearded sakis, and uakaris have highly procumbent incisors, which they use to both plane hard pericarp from large fruits (Figure 9.5) and scrape mesocarp from the interior of fruit (Kinzey, 1992; Ledogar, 2009; Norconk & Veres, 2011; Ledogar et al., 2013). Their widely splayed, robust canines increase their gape width, allowing them to puncture fruit that is larger than their dental arcade. Their seed crushing adaptations include flat molars and very robust mandibles with expanded attachment for the masseter. Further, their premolars are molariform and their molars have crenulated enamel, allowing for increased surface area and ability to hold seeds during mastication (Ledogar, 2009; Norconk & Veres, 2011; Ledogar et al., 2013). Bearded sakis and uakaris also often have sagittal crests, with expanded temporalis muscles that allow them to produce considerable bite force.

All of the pitheciines are highly frugivorous, with fruit making up between 60% to over 95% of their diets (Soini, 1986; Van Roosmalen et al., 1988; Ayres, 1989; Kinzey & Norconk, 1993; Peres, 1993; Setz, 1993; Norconk, 1996; Homburg, 1997; Aquino, 1998; Boubli, 1999; Defler,

1999; Peetz, 2001; Barnett et al., 2005; Veiga, 2006; Veiga & Ferrari, 2006; Bowler, 2007; Norconk, 2007; Palminteri et al., 2011; Shaffer, 2013a; Boyle et al., 2016). Consistent with their specialized morphology, seeds make up a large portion of fruit consumed (40–87.2%). The fruits of their preferred plant families Lecythidaceae, Chrysobalanaceae, Fabaceae, and Sapotaceae often have tough exocarp but relatively soft seeds (Kinzey & Norconk, 1990, 1993; Boubli, 1999; Peetz, 2001; Barnett et al., 2005; Bowler, 2007; Norconk, 2007; Boubli & De Lima, 2009). This extremely high reliance on seeds makes them virtually unique in the primate world, and Norconk (2011) argued that the saki/uakaris are the most specialized primate seed predators based on dietary and morphological traits. Bearded sakis and uakaris appear to consume a higher percentage of seeds than any other primates (Norconk, 2011; Norconk & Veres, 2011; Shaffer, 2013a). Along with, and likely because of, this high reliance on seeds, the pitheciines generally consume a large number of different plant species, with most species making up less than 1% of annual feeding time. The total number of plant species consumed by pitheciines over the course of a year exceeded 130 in several studies (Bowler & Bodmer, 2009; Palminteri et al., 2011; Shaffer, 2013a; Boyle et al., 2016). They show a preference for slowly maturing fruit and seeds available in dry season. This may account for their ability to live in oligotrophic species-poor forests or flooded forests (Boubli et al., 2008; Shaffer et al., 2019). The ability to consume a diet composed primarily of seeds appears somewhat to insulate the pitheciines from seasonal shortages of food resources (Ayres, 1989; Kinzey & Norconk, 1993; Peetz, 2001; Bowler, 2007; Norconk, 2011; Norconk & Veres, 2011; Boyle et al., 2016). Seeds have a much longer temporal availability than ripe fruit and many pitheciines are able to consume resources from the same tree species (or even individual trees) for four months (see Table 9.1) (Kinzey & Norconk, 1993; Norconk & Veres, 2011; Shaffer, 2014). Combined with their ability to exploit a wide variety of different fruit types, this increased temporal availability may make the pitheciines relatively immune from the negative effects of seasonal resource scarcity (Norconk & Veres, 2011; Shaffer, 2013a).

Figure 9.1 *Pithecia pithecia* **pair showing sexual dichromatism.**
(Photo credit: R. W. Sussman)

The seed eating niche currently occupied by *Chiropotes*, *Pithecia*, and *Cacajao* appears to have a long history of occupation in South America. One of the first shared-derived adaptations to appear in the Neotropical primate fossil record is specialized anterior dentition associated with seed eating (Kay & Takai, 2006). The first evidence of this adaptation appears in the middle Miocene (12–15 mya) in Colombia and Argentina. Fossils from these areas show remarkable similarities to the modern saki/uakaris and suggest that the titi monkey clade had separated from pitheciines before 15 mya (Kay & Takai, 2006; Kay et al., 2008). The fossil record from the middle Miocene also contains a group of extinct species called Soriacebidae that had anatomical adaptations consistent with seed eating. However, they are distinct from pitheciines and may have occupied the seed predator niche independently from, but concurrent with, the pitheciines.

Despite similarities stemming from their adaptations to this seed-eating niche, pitheciines vary tremendously in their ranging behavior and activity budgets (Defler, 2001; Bowler, 2007; Gregory, 2011; Norconk, 2011; Shaffer, 2013d; Palminteri et al., 2015; Van Belle et al., 2018). *Pithecia* are generally found in smaller groups, and have smaller home and day ranges, and much shorter activity periods than *Chiropotes* and *Cacajao*. *Cacajao* is characterized by extremely large group size and day and home ranges as extensive as any platyrrhine. *Chiropotes* appear to fit somewhere in between *Pithecia* and *Cacajao* in terms of behavioral ecology, with large groups, large home ranges, and long day ranges. Recent research from a diverse range of study sites shows that the genus is highly variable in its use of space and grouping patterns (Boyle et al., 2009; Boyle & Smith, 2010; Shaffer, 2013a,b; Gregory & Norconk, 2014).

Figure 9.2 **Bearded saki (***Chiropotes satanas***) male showing temporal swellings and prominent beard.**
(Photo credit: Stephen Ferrari)

Until the beginning of the twenty-first century, the pitheciines were considered the least studied Neotropical primates (Mittermeier, 2013). Bearded sakis, sakis, and uakaris are notoriously difficult to study for many reasons (Pinto et al., 2013). All of the pitheciines are extremely fast moving (Walker, 1996; Walker & Ayers, 1996; Vie et al., 2001; Boyle et al., 2009; Norconk, 2011). *Pithecia* have small group sizes and show cryptic behavior unless habituated. *Chiropotes* have large home and day ranges and spend most of their time in the upper parts of the canopy. *Cacajao* have extremely large ranges and are found in some of the most inhospitable habitats for researchers. Nevertheless, several long-term studies in the past decade have dramatically increased our understanding of the behavioral ecology of these animals. An edited volume dedicated to the Pitheciidae was published in 2013 (Veiga et al., 2013) and a special issue of the *International Journal of Primatology* on the latest knowledge of pitheciid behavioral ecology followed three years later (Barnett et al., 2016). While much remains to be learned, long-term study sites representing a variety of habitat types have been established for all three genera throughout Amazonia and the Guiana Shield: in Peru

Figure 9.3 Male *Cacajao calvus calvus*.
(Photo credit: R. W. Sussman)

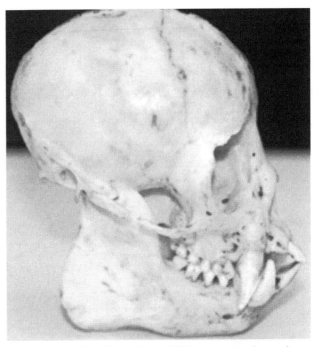

Figure 9.4 Skull and dentition of *Chiropotes sagulatus*, showing adaptations for seed eating. Notice the robust mandible with expanded attachment for the masseter muscle, flaring canines, and procumbent incisors.
(Photo credit: Jessica Joganic)

(Bowler, 2007; Palminteri et al., 2011), in Brazil (Veiga & Silva, 2005; Barnett et al., 2008; Boyle, 2008, 2014), in Venezuela (Norconk, 1996; Peetz, 2001), in Guyana (Shaffer, 2012), and in Suriname (Gregory, 2011; Thompson, 2011).

GEOGRAPHIC DISTRIBUTION, ECOLOGY, AND BODY SIZE

The shared-derived dental specializations and corresponding high reliance on seed eating of the sakis/uakaris put them in a very unique ecological niche (Kay & Takai, 2006; Kay et al., 2008; Boubli & De Lima, 2009). The geographic distribution and behavioral ecology of sakis/uakaris can best be understood in relation to the niche they occupy. Sclerocarpic seed predation makes the pitheciines particularly well adapted to the forests of the Guiana Shield and Central Amazonia, and two of the genera (*Chiropotes* and *Cacajao*) are concentrated there. Many species of plants that produce dry fruits with mechanically protected seeds are highly abundant in the ancient Guiana Shield forests (Mori, 1989; ter Steege, 1993; ter Steege et al., 2006). Fabacae, Lecythidaceae, and Chrysobalanaceae are much more abundant and species diverse in the forests of the Guiana Shield than they are in Western Amazonia (ter Steege, 1993; ter Steege et al., 2006). These families are characterized by dry dehiscent and indehiscent fruits and pods and feature prominently in the diets of all three genera of pitheciines at every site

they have been studied (Norconk, 2011; Ayres & Prance, 2013; Boyle et al., 2016). These families are much less well represented in Western Amazonia, where families that are rare in the Guiana Shield (e.g., Moraceae and Bombacaceae) are dominant. Seven of the top 10 genera in the Guiana Shield are legumes compared to none of the top 10 in Western Amazonia (ter Steege et al., 2006). The 10 most abundant Guiana Shield genera (*Carapa, Lecythis, Aldina, Pentaclethra, Alexa, Dicorynia, Eperua, Catostemma, Mora,* and *Dicymbe*) are some of the most important plants in the diet of the pitheciines, especially uakaris and bearded sakis (Norconk, 2011; Boyle et al., 2016). Many Lecythidaceae species appear to have originated in the Guiana Shield and only recently expanded their range to parts of the Amazon Basin (Mori, 1989).

Another characteristic of the forests of the Guiana Shield and Central Amazonia that distinguishes them from Western Amazonian forests, and likely to be of extreme importance to the Pitheciinae, is that they are composed of trees with much larger seeds (ter Steege et al., 2000, 2006). A study by ter Steege et al. (2006) found that legumes had a 20% higher average seed mass than non-legumes in the Guiana Shield but were similar in size to non-legumes in Western Amazonia. Larger seeds are particularly common in the seasonally flooded, blackwater *igapo* habitats that make up much of the range of *Cacajao*. Several studies have shown that pitheciines,

Figure 9.5 *Pouteria speciosa* (Sapotaceae) fruit with the exocarp removed by a bearded saki. Sakis use their specialized procumbent incisors to scrape off the firm, adherent mesocarp, but were not observed eating the seed of this plant species.

especially bearded sakis and uakaris, prefer larger seeded species (Ayres, 1986; Norconk, 1996; Bowler & Bodmer, 2009; Norconk & Veres, 2011; Shaffer, 2013a; Boyle et al., 2016). Thus, large seeds and the presence of certain plant families appear to determine habitat suitability for the pitheciines, especially bearded sakis and uakaris.

In a study of the geographic distribution and abundance of the pitheciines, Stevenson (2001) found that fruit abundance alone did not predict pitheciine biomass or the number of pitheciine species present at 30 Neotropical field sites. Instead, the abundance of pitheciines was strongly related to the abundance of trees in the genus *Eschweilera* (family Lecythidaceae).

The genus *Pithecia* is widely distributed throughout Amazonia and the Guiana Shield (Figure 9.6) (Peres, 1993; Ford, 1994; Marroig & Cheverud, 2005; Norconk, 2011). Hershkovitz (1987) separated the genus into two groups based on geographic distribution and pelage color: the Amazonian (*P. monachus*) group and the Guianan

(*P. pithecia*) group. Marroig and Cheverud (2005) used cranial morphology to classify all six Amazonian sakis as distinct species and the two Guianan sakis as subspecies of *P. pithecia*. A more recent and comprehensive taxonomic revision by Marsh (2014) recognizes 16 species of *Pithecia*. Sakis are the least robust and least derived of the pitheciines, with a less highly specialized craniodental morphology than *Chiropotes* and *Cacajao*.

White-faced or Guianan sakis (*P. pithecia*) are found north of the Amazon River and occupy a variety of habitat types throughout Venezuela and the Guianas. They range in body size from 1.4–1.9 kg, with males slightly larger than females (Ford, 1994). White-faced sakis show dramatic sexual dichromatism, with males having black body pelage and white, or yellowish, fur around their faces (see Figure 9.1). In contrast, females are a gray-brown and lack the striking facial coloration of the males. Western Amazonian sakis are found south of the Amazon River in Brazil and throughout this region of Amazonia (Norconk et al., 2013). They are larger than white-faced sakis, with body size ranging from 2 kg in *P. irrorata* to over 3 kg in *P. albicans*. Although lacking the extreme sexual dichromatism seen in *P. pithecia*, Western Amazonian saki males show a range of facial and pelage coloration that distinguishes them from gray-brown females (Norconk, 2011).

Chiropotes is intermediate in size between *Pithecia* and *Cacajao*, with body size ranging from 2.5–3.7 kg. Males are slightly larger than females, but sexual dichromotism is absent. They are distinguished by their characteristic beards and bulbous temporal swellings, both of which are more prominent in males. Bearded sakis are found both north and south of the Amazon River but are restricted to areas east of the Caroni River in Venezuela (Figure 9.7) (Hershkovitz, 1985; Ferrari & Lopes, 1996; Walker, 1996). Nevertheless, bearded sakis show a somewhat patchy distribution in Eastern Amazonia, possibly due to riverine and habitat barriers (De Granville, 1982;

Table 9.1 Monthly Percentages of Feeding Time on Top 10 Plant Species Consumed Annually by Guianan Bearded Sakis (*Chiropotes sagulatus*)

Plant Species[a]	Jan.	Feb.	Mar.	Apr.	May	Jun.	Sept.	Oct.	Nov.	Dec.
Manilkara bidentate	46.22	38.54	5.75	-	-	-	-	-	-	-
Swartzia leiocalycina	0.89	0.13	0.70	12.80	37.64	42.55	-	-	-	-
Unknown (Sheu)	-	-	16.89	28.31	5.03	6.61	-	-	-	-
Prieurella spp.	7.19	17.79	13.61	4.55	-	-	-	-	-	-
Geissospermum sericeum	-	0.29	5.00	19.66	10.31	7.89	-	-	-	-
Pouteria cuspidate	5.36	3.78	3.01	4.38	3.36	2.70	18.30	11.41	6.52	1.70
Goupia glabra	-	-	-	-	-	-	-	7.57	38.10	18.47
Licania densiflora	-	0.54	0.43	1.26	-	0.30	1.20	13.78	5.36	26.17
Eschweilera sagotiana	7.07	9.88	1.92	0.51	0.16	0.47	-	0.86	0.22	-
Brosimum parinarioides	-	-	0.36	-	-	4.19	26.63	13.62	9.72	3.66

[a] Data from Shaffer (2013a).
- Indicates no data.

Bongers et al., 2001; Norconk, 2011). *Chiropotes* and *P. pithecia* are sympatric throughout much of their geographic ranges, although *P. pithecia* is more widely distributed (perhaps because of their more flexible habitat requirements). Hershkovitz (1985) identified two species of *Chiropotes*: *C. satanas* with three subspecies, and *C. albinasus*. Bonvicino et al. (2003) revised this taxonomy using genetic data and pelage color. They recommended elevating all of the subspecies of *C. satanas* to species status and classifying the most westerly group as a fifth species. This is currently the most accepted taxonomy of the genus, although the systematics of the western bearded sakis may require further revision.

Uakaris (*Cacajao* spp.) show the most restricted geographic range of the pitheciines (Figure 9.8) and are almost entirely allopatric with bearded sakis (*Chiropotes* spp.) (Hershkovitz, 1985; Ayres, 1986; Barnett & Brandon-Jones, 1997; Boubli, 1999; Bowler, 2007). They are similar in size to bearded sakis, with males weighing approximately 3.5 kg and females weighing 3 kg (Hershkovitz, 1985). Uakaris have a very distinctive appearance, with a bare face, prominent temporal swellings, and the shortest tail of any platyrrhine. Hershkovitz (1987) identified two species of uakaris: bald-headed uakaris (*C. calvus*) with four subspecies, and black-headed uakaris (*C. melanocephalus*) with two subspecies. Uakaris are also the most restricted of the pitheciines in their habitat requirements.

The geographic distribution of bearded sakis and uakaris may be limited to areas with a high abundance of Lecythidaceae species, with the western migration of Lecythidaceae genera setting the extent of uakari distribution into Western Amazonia (Ayres, 1986; Ayres & Prance, 2013). As species of Lecythidaceae slowly migrated toward western Amazonia, the geographic range of *Cacajao* likely followed (Ayres & Prance, 2013). The lack of overlap in the ranges of *Chiropotes* and *Cacajao* probably reflects competitive exclusion, given their extremely similar feeding ecology, locomotion, use of vertical space, and ranging behavior (Ayres, 1986; Boubli et al., 2008; Boubli & De Lima, 2009). In contrast, *Pithecia* coexists with either *Chiropotes* or *Cacajao* in virtually all areas. Dietary flexibility and a less specialized dental morphology may make *Pithecia* less limited by the availability of

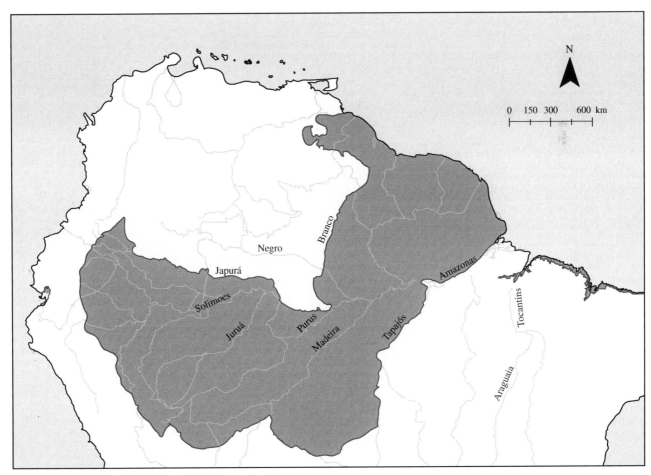

Figure 9.6 Geographic range of the genus *Pithecia*.
(Modified from De Sousa e Silva Júnior et al., 2013)

legumes and Lecythidaceae species. The ability of *Pithecia* to incorporate a high percentage of leaves in their diet may contribute to their more flexible habitat requirements. In addition, the flexibility in use of vertical space exhibited by *Pithecia* may have allowed them to disperse across seasonally dry habitats and xeric, scrubby forests. These environments appear to have prevented the migration of bearded sakis into French Guiana. Interestingly, in the absence of sympatric *Chiropotes* or *Cacajao* species, western Amazonian *P. albicans* show larger body sizes, larger home ranges, and a preference for higher-canopy travel than other *Pithecia* species (Peres, 1993; Norconk, 2011). Norconk (2011) suggested this may represent competitive release and is consistent with the Pitheciinae radiating out of the Guiana Shield and occupying western Amazonia more recently.

All of the pitheciines have bushy tails; sakis and bearded sakis have long tails and uakaris have tails measuring only one-third of their head-to-body length (Norconk, 2011). These bushy tails play an important role in the social behavior of bearded sakis and uakaris (Fontaine, 1981; Fernandes, 1993; Walker & Ayers, 1996; Peetz, 2001; Defler, 2003; Shaffer, 2013c; Gregory & Bowler,

2016). Tail wagging, often accompanied by vocalizations, occurs during predator sightings and after alarm calls, as well as during intergroup encounters and the coalescence of groups after fissioning. Bearded sakis often wag their long tails arched over their heads. It appears that bushy tails may have evolved as a medium-distance visual cue for intraspecies communication in sakis and uakaris (Norconk, 2007).

SAKIS

Habitat and Locomotion

Members of the genus *Pithecia* are quite flexible in their habitat requirements, especially compared to *Cacajao* and *Chiropotes*. They are found in a variety of habitat types including tropical dry forests, highland and lowland wet forests, and gallery forests (Norconk, 2011). In addition, they tolerate seasonal environments well, and their dispersal into French Guiana suggests savanna habitats are not barriers. This habitat flexibility compared to the other pitchiines can be explained by their broader diet, including their ability to regularly ingest leaves, and their locomotion (moving through low- and

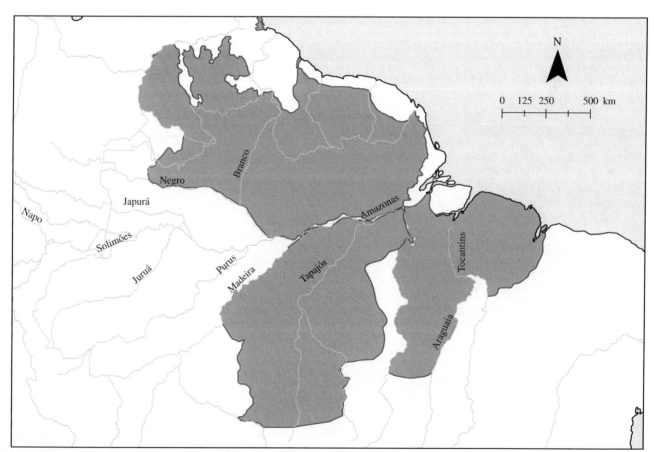

Figure 9.7 Geographic range of the genus *Chiropotes*.
(Modified from De Sousa e Silva Júnior et al., 2013)

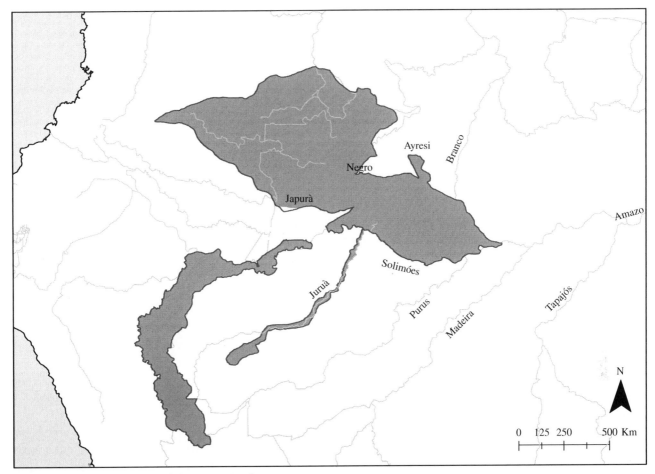

Figure 9.8 Geographic range of the genus *Cacajao*.
(Modified from De Sousa e Silva Júnior et al., 2013)

middle-canopy levels as well as upper forest strata compared to the upper forest level preferred by *Chiropotes* and *Cacajao*).

Pithecia have traditionally been thought to cope well with forest fragmentation and not be limited to undisturbed, closed canopy forest. Many of the studies of white-faced sakis come from fragmented and island habitats. However, Norconk (2011) reported that live births ceased after the tenth year of her study in a population of *P. pithecia* on Lago Guri Island in Venezuela. Similarly, Lenz (2013) suggested that a Brazilian population of white-faced sakis was declining in a fragmented landscape.

Sakis are arboreal quadrupeds but spend a significant amount of their time clinging and leaping on vertical supports (Walker, 1998). They spend most of their time in the lower levels of the canopy and understory and can spend several hours a day on the ground. Jumping and leaping are their primary locomotive modes of travel. They can travel very quickly in quick spurts, so much so that a white-faced saki is known as a "flying jack" or "breezy monkey" in Guyana. They can also be very cryptic after periods of rapid travel, remaining motionless for long periods of time in dense vegetation. *Pithecia* use larger supports than do *Chiropotes* or *Cacajao* and are much more likely to use vertical supports. They often assume a vertical body orientation during feeding and resting.

There are some differences between southern Amazonian *Pithecia* and Guianan *Pithecia* in terms of locomotion and microhabitat preferences that may be related to the sympatry of *P. pithecia* with bearded sakis and uakaris (Peres, 1993; Walker, 1996; Davis & Walker-Pacheco, 2013). The larger Amazonian sakis appear to prefer upper levels of the forest and exhibit less vertical clinging and leaping. Peres (1993) suggested that Amazonian sakis, especially *P. albicans*, may have evolved larger body sizes, larger home ranges, and a preference for upper-canopy levels in the absence of competition with sympatric bearded sakis and uakaris.

Diet and Ranging Behavior

Like all of the pitheciines, *Pithecia* rely heavily on seeds; the percentage of time eating seeds varies from around 40% to over 80% in different studies (Kinzey & Norconk, 1993; Cunningham & Janson, 2007; Frisoli, 2009). Sakis generally show a preference for immature seeds. While able to eat the large seeds ingested by *Cacajao* and *Chiropotes*, *Pithecia* are also able to extract tiny seeds of multiloculated fruit using their specialized canines (Norconk, 2011). Preferred plant families for *Pithecia* include Sapotaceae, Lecythidaceae, and Fabaceae (Boyle et al., 2016). Most of the remaining diet of *Pithecia* is made up of ripe fruit, which becomes their most important food source when widely available. *Pithecia* appear to be more selective than *Chiropotes* in their preference for fruit in only one stage of maturity, generally exploiting different tree species for flowers, ripe fruit, and seeds (Norconk, 1996).

Leaves are also an important part of the diet of *Pithecia*, making up 10% of feeding time in several studies (Norconk et al., 1998, 2002; Boyle et al., 2016). Norconk et al. (1998) found that Venezuelan white-faced sakis consumed leaves daily throughout the year. Flowers and insects, while occupying a relatively small percentage of saki feeding time, may be important seasonal resources in many habitats. Wasps and grasshoppers eaten by white-faced sakis contain the highest crude protein in their diets. *Pithecia*, while also adapted for sclerocarpic foraging, appear to be more limited in their ability to exploit mechanically protected seeds (Norconk & Veres, 2011). Dietary and digestive studies suggest that *Pithecia* may be *less* limited than the other Pitheciinae in their ability to exploit chemically protected seeds (Milton, 1984; Norconk et al., 2002, 2009). The gut transit time of *Pithecia* is much longer than that of *Chiropotes* or *Cacajao* and may be related to digestive fermentation of fibrous seeds and/ or leaves. While bearded sakis and uakaris have transit times of about five hours (among the shortest of any platyrrhine), *P. pithecia* and *P. monachus* have transit times in excess of 15 hours (Milton, 1984; Norconk et al., 2002). In fact, the only platyrrhine genus with a longer transit time is the highly folivorous *Alouatta*. This dramatic difference between pitheciine species suggests that *Pithecia* may be adapted to exploiting resources that are high in secondary compounds, like leaves, and possibly chemically protected seeds. Consistent with these adaptations, Norconk et al. (2009) found that *Pithecia* consume a diet high in lipids and also high in dietary fiber. The estimated energy value of their diet was much lower than that of *Cacajao* and *Chiropotes*. Similarly, Norconk et al. (2002) reported that the diet of *P. pithecia* contains acid-detergent fiber levels twice as high, and non-detergent fiber levels 60% higher, than the diet of *Chiropotes sagulatus*, despite the similar percentage of seeds in both diets. However, little data exist on the specific digestive morphology or biomicrobiome dynamics of any of the pitheciines.

While still large, *Pithecia* home and day ranges are much smaller than *Cacajao* and *Chiropotes*. Estimates of home range vary from 10 ha in some *P. pithecia* populations, to over 200 ha in *P. albicans,* but are generally around 50 ha in continuous forest habitats (Peres, 1993; Setz, 1993; Homburg, 1998; Norconk et al, 2003; Norconk, 2006, 2007; Palminteri et al., 2015; Van Belle et al., 2018). Day ranges vary from 500–2,800 m, averaging about 1–2 km (Setz, 1993; Vie et al., 2001; Cunningham, 2003; Norconk, 2006; Palminteri et al., 2015; Van Belle et al., 2018). The smaller ranges of *Pithecia* compared to other pitheciines may be related to both smaller group sizes and their ability to consume seeds and leaves with higher percentages of fiber and secondary compounds (Shaffer, 2013b; Palminteri et al., 2015). However, their relatively long day ranges compared to the sizes of their home ranges suggest that day range length may also be related to territoriality, with selection for the ability to regularly monitor the boundaries of their home ranges (Palminteri et al., 2015; Van Belle et al., 2018).

Activity, Social Structure, and Social Organization

Pithecia are characterized by short daily activity periods (8–10 hours), leading them to be called "the lemurs of the New World" by some researchers (Setz, 1993; Vie et al., 2001; Norconk, 2011). They are unusual among primates because they settle in for the night long before sunset with daily activity usually ending around 16:30. *Pithecia* divide their time equally among travel, feeding, and rest. Cycles of activity, especially time spent traveling and resting, vary seasonally at some sites (Setz, 1993; Norconk et al., 1998; Norconk, 2011).

Sakis were long thought to form pair-bonded groups with one male, one female, and dependent offspring (Buchanan et al., 1981; Hershkovitz, 1985; Poyas & Bartlett, 2009). This is consistent with census data from several populations, showing 2–5 individuals per group (Mittermeier, 1977; Oliveira et al., 1985; Soini, 1986; Peres, 1993; Kessler, 1998; Ferrari et al., 1999; Heymann et al., 2002). Several other surveys (Lehman et al., 2001; Norconk, 2006) show larger group sizes, ranging up to 12 individuals. Norconk (2011) and Thompson (2015) have suggested that slow dispersal of young adults may account for the "extra" adults seen in pair-bonded groups. Norconk (2011: p.128) has characterized their group sizes as "small but flexible and responsive to variables such as population density, food distribution and sympatry with *Chiropotes* and *Cacajao*." Thompson (2015) suggested that variable group sizes in white-faced sakis represent a trade-off between successful resource defense (favoring increased group size and extra-pair individuals) and sole reproductive access to mates (favoring pair-bonding).

Consistent with a pair-bonded social structure, sakis exhibit territorial behaviors in field studies and in captivity (Norconk et al., 2003; Norconk, 2006; Di Fiore et al., 2007; Thompson & Norconk, 2011; Thompson, 2015; Van Belle et al., 2018). These behaviors include aggressive chasing and threats, as well as specific vocalizations limited to intergroup encounters. Between-group aggression appears to be related to both resource defense and mate guarding (Thompson, 2015). While adult female sakis are the primary caregivers, adult males show some level of infant care. White-faced saki males have been observed playing and sharing food with older infants and even carrying infants in some populations. Scent marking, although relatively rare compared to callitrichids, may be an important form of communication in *Pithecia* and appears to be related to sexual behavior and intergroup communication (Kinzey, 1986; Brumloop et al., 1994; Setz & Gaspar, 1997; Gleason, 1998).

Predation

Due to their small size, *Pithecia* are prey for a number of species, including eagles, tayra, ocelots, jaguars, and large snakes (Adams & Erhart, 2009). They avoid predation through their cryptic habits, jumping quickly into dense foliage and remaining extremely quiet for long periods of time (Adams & Erhart, 2009). Human hunting of sakis is relatively rare. While they are occasionally hunted for food, they are not as important a food source as the larger Neotropical primates (Peres, 2000). Nevertheless, they are sometimes captured for pets and are often hunted for their bushy tails (Mittermeier, 1991; Shaffer, 2017a). These are used as dusters, both traditionally by indigenous people, and as modern curios (Mittermeier, 1991; Shaffer, 2017a).

BEARDED SAKIS

Habitat and Locomotion

Chiropotes have traditionally been thought to be strict in their habitat requirements, restricted to terra firma closed canopy, undisturbed lowland rainforest (Mittermeier & van Roosmalen, 1981; Hershkovitz, 1985, 1987). Research from several sites throughout eastern Amazonia is now showing that they are much more flexible in habitat requirements than previously thought. Bearded saki populations have been studied in fragmented habitats, island environments, dry forests, and seasonally inundated forests (Peetz, 2001; Veiga & Silva, 2005; Pinto, 2008; Gregory, 2011; Shaffer, 2012; Boyle, 2014; Shaffer et al., 2019). Several studies have specifically addressed the ability of *Chiropotes* to subsist in forest fragments of various sizes, some as small as 19 ha (Kinzey & Norconk, 1993; Peetz, 2001; Port-Carvalho & Ferrari, 2004; Norconk, 2006; Veiga, 2006; Boyle, 2008; Silva & Ferrari, 2009; Boyle et al., 2012). Research has shown that their incredibly diverse

diet and flexibility in ranging and grouping patterns allow bearded sakis to do relatively well in disturbed environments. The minimum size of the fragment in which a bearded saki group can persist, however, appears to be relatively large (i.e., 100 ha) (Boyle et al., 2012).

In all habitats, bearded sakis spend most of their time in the upper parts of the canopy. They are arboreal quadrupeds but are also adept leapers (Walker, 1996, 2005). Unlike *Pithecia*, *Chiropotes* leap from a pronograde position and land on the terminal branches of a neighboring tree rather than on larger vertical supports (Walker, 2005). Although rare, *Chiropotes* do exhibit suspensory postures, especially during feeding and play (Walker, 2005; Shaffer, 2012). During these movements, *Chiropotes* hang from their hindlimbs and use their non-prehensile tails to support themselves.

Diet and Ranging Behavior

Bearded sakis are reported to be highly granivorous in all studies, with seeds making up as much as 75% of the diet (van Roosmalen et al., 1988; Norconk, 1996; Peetz, 2001; Boyle, 2008; Gregory & Norconk, 2013; Shaffer, 2013c; Boyle et al., 2016). Seeds from the Sapotaceae and Lecythidaceae families are generally preferred, although *Chiropotes* have a tremendously diverse diet, but most plant species make up less than 1% of the annual diet. Bearded sakis also consume fruit pulp, both unripe and ripe. Preferred fruit resources are often from slowly maturing species, and *Chiropotes* will frequently utilize the same plant species in a variety of states of maturity. This pattern of feeding allows them to avoid competition with sympatric frugivores. For example, Guyanese bearded sakis (*C. sagulatus*) will often feed from the same individual trees for three months, two months before any of the other seven primate species that share the resource (Shaffer, 2013a).

Insects form a small part of annual diet but can be very important seasonally (Ayres & Nessimian, 1982; Mittermeier et al., 1983; Frazao, 1991; Peetz, 2001; Veiga & Ferrari, 2006; Shaffer, 2013c). For example, in both Venezuela and Guyana, preying on insects makes up over 10% (and up to 40%) of *C. sagulatus* feeding time in some dry season months (Peetz, 2001; Shaffer, 2013c). Bearded sakis target social insects like caterpillars that can be found in high concentrations during certain parts of the year. Flowers are an important part of the bearded saki diet in some studies but are rarely eaten in others (Norconk, 1996; Peetz, 2001; Veiga, 2006). Although they do so infrequently, bearded sakis also consume leaves and pith.

Along with *Cacajao*, *Chiropotes* range as far as any other Neotropical primate (Shaffer, 2013b). Home and day range estimates are confounded by the fact that most studies of *Chiropotes* have been conducted on islands or in fragmented habitats (Setz, 1993; Bobadilla & Ferrari, 2000; Peetz, 2001; Boyle, 2008; Boyle et al., 2009). Nevertheless, day ranges average 3–4 km but can be as long as 7 km. Home ranges in continuous habitats range from 250 to over 1,000 ha (van Roosmalen et al., 1981; Boyle, 2008; Pinto, 2008; Gregory, 2011; Shaffer, 2013b). Bearded sakis show seasonal differences in ranging behavior in some studies. Peetz (2001) reported a decrease in daily path lengths during the dry season for *C. sagulatus* in Venezuela. Gregory (2011) found that traveling by bearded sakis in Suriname remained relatively consistent throughout the year, although feeding party size decreased in the long dry season. Similarly, daily path lengths for *C. sagulatus* in Guyana varied relatively little seasonally (Shaffer, 2013b), although they were shortest in the dry season when fruit was most scarce. It is unclear why bearded sakis (and uakaris, see below) range so far, although a likely explanation is the need to minimize the effects of the toxins from any one type of seed (Kinzey & Norconk, 1993; Norconk, 1996; Shaffer, 2013b). Shaffer (2013b) found that *C. sagulatus* path lengths were more strongly correlated with the percentage of seeds in the diet than any other variable, including group size. However, this hypothesis remains to be tested with empirical data on the secondary compounds of bearded saki foods.

Activity, Social Structure, and Social Organization

Like most primates, bearded sakis are active from sunrise to sunset. While some members of the group are often active before and after these times, the majority of individuals are resting and the group does not travel. Compared to most other Neotropical primates, bearded sakis spend much of their day traveling and feeding, resting only about 20% of the time (Peetz, 2001; Veiga, 2006; Shaffer, 2013d). Feeding shows typical primate bimodal patterns, with peaks in the morning and late evening. The hottest hours of the day 11:00–13:00 are generally spent resting and engaging in social behavior, but feeding and travel are not infrequent. *Chiropotes* are almost always on the move and feeding bouts in individual trees average only 10 minutes (Gregory, 2011; Shaffer, 2014). Travel is generally rapid, and animals frequently grab and eat food items as they travel.

While social behavior is relatively rarely observed in bearded sakis, they do show high levels of male affiliation, with males commonly huddling and grooming each other (Peetz, 2001; Veiga & Silva, 2005; Veiga, 2006; Shaffer, 2013d; Gregory & Bowler, 2016). One of the most common bearded saki social behavior consists of a ritualized

pattern of lining up in groups of two to eight animals and synchronizing the wagging of tails, shown in Figure 9.9 (Peetz, 2001; Shaffer, 2013d; Gregory & Norconk, 2014; Gregory & Bowler, 2016). These groups almost always consist of adult males. In addition, males engage in an affiliative behavior of "piling up," during which individuals lie on top of one another in a pile of two to five individuals (Shaffer, 2013d; Gregory & Norconk, 2014; Gregory & Bowler, 2016). Veiga (2006) found that 90% of male interactions were affiliative, and Shaffer (2013d) observed a similar pattern. Researchers in Guyana and Venezuela also have found that bearded saki males frequently play with and groom infants and juveniles (Peetz, 2001; Shaffer, 2013d).

Bearded sakis live in large, multi-male/multi-female groups ranging in size from 8 individuals (in fragmented habitats) to over 65 (van Roosmalen et al., 1981; Ayres, 1989; Norconk & Kinzey, 1994; Ferrari & Lopes, 1996; Ferrari et al., 1999; Peetz, 2001; Boyle, 2008; Silva & Ferrari, 2009; Gregory, 2011; Shaffer, 2013d). Groups in continuous forest number between 35–65 individuals at most sites (Boyle, 2008; Pinto, 2008; Gregory, 2011; Shaffer, 2013d). *Chiropotes* are highly flexible in their social organization, with groups at many sites frequently breaking up into subgroups. However, the specific pattern of this subgrouping varies from site to site (Setz, 1993; Norconk, 2011; Shaffer, 2013a). Some researchers have reported relatively cohesive groups traveling between food patches but group fissioning taking place upon entering patches (Ayres, 1989; Norconk & Kinzey, 1994), while others have suggested a more regular pattern of fission and fusion similar to that exhibited by primates with high fission-fusion dynamics like *Ateles* (Veiga et al., 2006). Shaffer (2013d) found that subgrouping was common among a group of 65 *C. sagulatus* individuals in Guyana, and that subgroup size was strongly correlated with the size of food patches. When fruit became scarce, especially during the long dry season, subgroup sizes were regularly below 20. Similar patterns have been observed in Brazil (Veiga, 2006; Pinto, 2008) and Suriname (Gregory & Norconk, 2014). Therefore, while large group sizes are characteristic of bearded sakis in continuous forest habitats, their flexible grouping behavior appears to allow them to reduce intragroup feeding competition during periods of resource scarcity.

Relatively little is known about intergroup interactions among bearded sakis, as encounters between two or more groups are rarely observed (Shaffer, 2013d; Gregory & Bowler, 2016). In addition, because of their large home and day ranges, *Chiropotes* groups rarely encounter one another. Nevertheless, bearded saki home ranges appear to be highly overlapping and encounters do occur (Shaffer, 2013d). During intergroup encounters, all animals of the group alarm call and most move rapidly away from the neighboring group. Shaffer (2013d) observed aggressive interactions between adult males during these encounters, with individuals suffering relatively severe falls (although injuries appeared to be minimal). Consistent with their high levels of male affiliation, males from the same group cooperated in the aggressive intergroup encounters.

Ayres (1981) hypothesized that bearded saki groups were composed of a number of male/female units. Unfortunately, the composition of subgroups and their consistency throughout the year remains unclear since no researchers have thus far been able to identify individual animals (Gregory & Bowler, 2016).

Predation

While little quantitative data are available, predation on bearded sakis may be quite high in some areas (Ferrari et al., 2004; Martins et al., 2005; Shaffer, 2012; Barnett et al., 2017). For example, out of 44 all-day follows of Guianan bearded sakis (*C. sagulatus*) in a partially flooded forest in Guyana, Shaffer (2012) observed four predatory attacks (3 eagles and 1 tayra). Four predatory attacks (all raptors) were observed on red-nosed bearded sakis

(*C. albinasus*) during 217 hours of observation in Igapó forests in Brazil (Barnett et al., 2017). Predators include eagles, especially the harpy eagle, tayras, large snakes, and possibly ocelots and jaguars. In a study of harpy eagles in Guyana, Rettig et al. (1978) found that remains of bearded sakis and white-faced sakis were more common than the remains of any other primate, excluding *Cebus*. Bearded sakis respond to arboreal predator threats by jumping to lower levels of the canopy while loudly alarm calling (Shaffer, 2012; Barnett et al., 2017). They will also mob predators, with four to eight individuals jumping toward the predator (Barnett et al., 2017). It is possible that a high level of predation may be an important driver for the evolution of large group sizes in bearded sakis.

While they are not as commonly targeted as larger-bodied species, human hunting of bearded sakis does occur, both for food and for their tails (Peres, 2000; Norconk, 2011; Shaffer et al., 2017a,b). Hunting may be a significant pressure in some areas. For example, Guianan bearded sakis (*C. sagulatus*) are an important source of bushmeat for the Waiwai in southern Guyana, due to their large group size and substantial population density in the area (Shaffer et al., 2017a). They are the second

Figure 9.9 Two bearded saki individuals engaged in "lining up," a ritualized affiliative behavior between males.
(Illustration credit: Michelle Bezanson)

most frequently harvested primate for the Waiwai, after the much larger spider monkey (*Ateles paniscus*; Shaffer et al., 2017b). While individual bearded sakis are taken opportunistically throughout the year, the Waiwai intensively harvest bearded sakis during the late wet season when they are considered to be at their fattest. Like *Pithecia*, *Chiropotes* are also targeted for their bushy tails (Figure 9.10).

UAKARIS

Habitat and Locomotion

Cacajao are the most specialized of the pithciines in their habitat requirements, primarily inhabiting seasonally flooded areas along black-water rivers (*igapo*) and white-water rivers (*varzea*) (Barnett et al., 2013). While long characterized as flooded forest specialists, more recent studies have shown that some uakari populations spend almost all of their time in terra firma habitats (Heymann & Aquino, 2010). Nevertheless, most populations of *Cacajao* have *igapo* and *varzea* forests as their primary environments (Barnett et al., 2013). These unique forests are found along tributaries of the Rio Negro, Amazon, and Orinoco rivers and are flooded for up to nine months of the year. They support fewer primate species than terra firma forests, and uakaris appear to be the only primate that travels far into flooded areas.

Igapo are high in tannins, with low pH and low nutrient concentrations, resulting in lower plant species diversity (Ayres & Prance, 2013). They are characterized by plants with large-seeded fruits. *Varzea* habitats have relatively nutrient-rich soils due to suspended sediments washed down from the Andes Mountains. Many genera of the plant family Lecythidaceae, which are preferred by uakari, are abundant in *varzea* forests (Ayres, 1989; Ayres & Prance, 2013). In both habitats, seeds provide reliable and readily available resources where other common primate foods, such as ripe fruit, are less abundant.

Like bearded sakis, uakaris are above-branch quadrupeds that frequently leap during travel (Walker & Ayres, 1996). They spend most of their time in the upper levels of the canopy but forage seasonally in lower strata. In some sites, they may feed on the ground during parts of the year, especially during the dry season in flooded forests. For example, Ayres (1986) estimated that *Cacajao* spent 20–40% of foraging time on the ground during the dry season. After floodwaters retreat in these habitats, water-dispersed seeds become abundant on the forest floor.

Diet and Ranging Behavior

Seeds are the most important food item in the uakari diet, taking up between 46–67% of feeding time (Ayres, 1989; Aquino, 1998; Boubli, 1999; Defler, 1999; Defler, 2003; Barnett et al., 2005; Bowler et al., 2009; Boyle et al., 2016). Uakaris show a preference for large seeds, which tend to be highly abundant in *igapo* habitats (Boubli, 1999; Boubli et al., 2008). These large seeds are generally higher in lipids and protein than smaller seeds and lower in condensed tannins (Ayres, 1986; Norconk & Conklin-Brittain, 2004; Norconk, 2011). Like the other pitheciines, *Cacajao* are highly reliant on the Lecythidaceae, Fabaceae, and Sapotaceae plant families (Boyle et al., 2016). The Euphorbiaceae are also very important in some habitats (Boyle et al., 2016). Most of the rest of their diet is made up of ripe fruit. Flowers, leaves, and insects are less important but may be significant seasonally.

Like bearded sakis, *Cacajao* range over extremely large areas, with home and day ranges as large as any other Neotropical primate, and larger than for any other arboreal primate of their size (Defler, 2001; Barnett et al., 2005; Bowler, 2007). The black uakari day range averages 2,300 m, and home range is estimated at a minimum of 1,053 ha (Boubli, 1997). Ranging patterns show a great deal of seasonal variation, with uakaris frequently traveling into flooded areas when river levels rise.

Figure 9.10 A ceremonial duster made from the bushy tail of the bearded saki by the Waiwai Amerindians of Guyana.

Activity, Social Structure, and Social Organization

Uakaris show similar activity patterns to bearded sakis. They are active from dusk until dawn and spend most of the day traveling and feeding, resting rarely (Ayres, 1986; Aquino & Encarnacion, 1999; Boubli, 1999; Defler, 2001; Barnett et al., 2005). Unlike *Pithecia*, *Cacajao* shows little seasonal variation in activity.

The social structure and social organization of *Cacajao* is poorly known. Uakaris form large, multi-male groups of about 20–40 individuals (and possibly up to 200 individuals during some times of the year) (Ayres, 1986, 1989; Setz, 1993; Boubli, 1999; Bennett et al., 2001; Defler, 2001; Bowler & Bodmer, 2011). These large groups appear to represent multilevel societies composed of smaller multi-male groups in the core, with multi-female groups and bachelor groups on the periphery. The sex ratio in large groups is approximately 1:1. The specific composition of these groups, however, and the reason for the variation in size across studies is unclear. Research has shown that *Cacajao* groups frequently break into subgroups, especially during feeding. These subgroups may remain apart for several days. In a few studies, groups were observed to be cohesive (Boubli, 1999). Patterns of subgrouping and the size of feeding parties vary seasonally in some populations (Defler, 2003), while remaining relatively stable in others (Boubli, 1999). *Cacajao* males engage in affiliative interactions, although they do not exhibit the ritualized huddling behavior of bearded sakis (Bezerra et al., 2011; Bowler et al., 2012; Barnett et al., 2013; Gregory & Bowler, 2016). Males exhibit coalitionary behavior, using affiliative interactions with core males to aggressively exclude periphery males (Bowler et al., 2012; Gregory & Bowler, 2016). Male philopatry is strongly suspected, but dispersal patterns are relatively unstudied. Almost nothing is known about intergroup interactions in uakaris.

Predation

Like the other pitheciines, uakaris may be preyed upon by a variety of avian, mammalian, and reptilian predators. While not hunted as frequently as the larger atelines, Peres (2000) found that *Cacajao* were a more important source of food for humans across Amazonia than the other pitheciines. Ucayali bald uakaris (*C. calvus ucayalii*) and black-headed uakaris (*C. ouakary*) are intensively hunted by some indigenous groups while the human-like appearance of bald-headed uakaris (*C. calvus*) make them taboo in other areas (Barnett et al., 2013).

CONSERVATION

The conservation outlook for the pitheciines as a whole is less pessimistic only when compared to the dire state of most primate species, as their habitats include some of the least disturbed and least densely populated areas of South America (Norconk, 2011; Boyle et al., 2016). The species that live in the Guiana Shield (i.e., *Pithecia pithecia*, *Chiropotes sagulatus*) are at lower risk, as Guyana, Suriname, and French Guiana retain approximately 80% of their original forest (Boyle et al., 2016). Nevertheless, many pitheciine populations are affected by large-scale habitat disturbance and hunting, and several species are very pressured by anthropogenic activities. In general, the species inhabiting Central Amazonia are the most at risk, due to large-scale forest fragmentation. The International Union for the Conservation of Nature Red List warns that the population levels of 21 species of pitheciines are decreasing and only 3 are considered stable; furthermore, 7 species are listed as Vulnerable (IUCN, 2020). Bearded sakis are often found in fragmented forests, although their wide-ranging behavior means that viable fragments may need to be an order of magnitude larger than for most other primates (Boyle, 2014; Norconk & Conklin-Britain, 2015; Boyle et al., 2016). The black-bearded saki is the most threatened of the pitheciine species due to its restricted geographic range in the densely populated southeastern Brazilian Amazon (Boyle, 2014). Currently classified as Endangered by IUCN Red List criteria (Veiga et al., 2008; IUCN, 2020; Port-Carvalho et al., 2021), deforestation for highway construction, logging, and cattle ranching makes the future of this primate extremely bleak. Due to their restricted geographic range, uakaris are the most vulnerable pitheciine genus. While human population densities are relatively low in most areas where uakaris are found, many species have very small distributions and human hunting is more intense than for the other pitheciines (Boyle et al., 2016).

SUMMARY

The pitheciines are a unique group of primates, united and distinguished from other platyrrhines by their anatomical specializations for seed predation. They are the most granivorous of all primates, and virtually every aspect of their behavioral ecology is defined by their sclerocarpic seed-eating niche. Nevertheless, they vary widely in their habitat requirements, ranging patterns, and social structure. While there has been a tremendous increase in our understanding of pitheciines in the past decade, many questions about their behavioral ecology remain unanswered. This is especially true for social behavior, social organization, and mating patterns. For example, it remains unclear why bearded sakis and uakaris form such large groups, and why these groups range over such extensive areas (Shaffer, 2013d; Gregory & Bowler,

2016). In addition, because no studies of bearded sakis and uakaris have yet been able to identify individuals, the composition of their subgroups (i.e., if they are composed of related males) is unknown. Finally, studies of the variation in the social organization of *Pithecia*, both geographically and at the same sites across seasons, are still in their infancy, but in the future can provide essential data on the selective pressures favoring pair-living in primates (Thompson, 2015). With long-term study sites of all genera established throughout Amazonia and the Guianas, researchers may soon be able to shed light on these important questions.

REFERENCES CITED—CHAPTER 9

Adams, D. B., & Erhart, E. (2009). *A preliminary study on vocal communication in the Gray's bald-faced saki monkey, Pithecia irrorate.* [Master's thesis, Texas State University].

Aquino, R. (1998). Some observations on the ecology of *Cacajao calvus ucayalii* in the Peruvian Amazon. *Primate Conservation, 18*, 21-24.

Aquino, R., & Encarnación, F. (1999). Observaciones preliminares sobre la dieta de *Cacajao calvus ucayalii* en el nor-oriente Peruano. *Neotropical Primates, 7*, 1-5.

Ayres, J. M. (1981). *Observations on the ecology and behavior of cuxius (*Chiropotes albinasus *and* Chiropotes satanas, *Cebidae: Primates).* [Master's thesis, Instituto Nacional de Pesquisas da Amazônia].

Ayres, J. M. (1989). Comparative feeding ecology of the uacari and bearded saki, *Cacajao* and *Chiropotes. Journal of Human Evolution,18*(7), 21-24.

Ayres, J. M. C. (1986). *Uakaris and Amazonian flooded forest.* [Doctoral dissertation, University of Cambridge].

Ayres, J. M., & Nessimian, J. L. (1982). Evidence for insectivory in *Chiropotes satanas. Primates, 23*, 458-459.

Ayers, J. M., & Prance, D. T. (2013). On the distribution of pitheciine monkeys and lecythidaceae trees in Amazonia. In L. M. Veiga, A. A. Barnett, S. F. Ferrari, & M. A. Norconk (Eds.), *Evolutionary biology and conservation of titis, sakis and uacaris* (pp. 127-139). Cambridge University Press.

Barnett A. A., Bezerra, B., Ross, C., & MacLarnon, A. (2008). Hard fruits and black waters: The conservation ecology of the golden-backed uacari, *Cacajao melanocephalus ouakary*, an extreme diet- and habitat-specialist. Presentation No 692, [abstract]. *XXIInd Congress of the International Primatological Society*, 248-250.

Barnett, A. A., Bowler, M., Bezerra, B. M., & Defler, T. R. (2013). Ecology and behavior of uacaris (genus *Cacajao*). In L. M. Veiga, A. A. Barnett, S. F. Ferrari, & M. A. Norconk (Eds.), *Evolutionary biology and conservation of titis, sakis and uacaris* (pp. 151-172). Cambridge University Press.

Barnett, A. A., Boyle, S. A., & Thompson, C. L. (2016). Pitheciid research comes of age: Past puzzles, current progress, and future priorities. *American Journal of Primatology, 78*(5), 487-492.

Barnett, A. A., & Brandon-Jones, D. (1997). The ecology, biogeography and conservation of the uakaris, *Cacajao* (Pitheciinae). *Folia Primatologica, 68*, 223-235.

Barnett, A. A., Castilho, C., Shapley, R. L., & Anicacio, A. (2005). Diet, habitat selection and natural history of *Cacajao melanocephalus ouakary* in Jau National Park, Brazil. *International Journal of Primatology, 26*(4), 949-969.

Barnett, A. A., Silla, J. M., de Oliveira, T., Boyle, S. A., Bezerra, B. M., Spironello, W. R., Setz, E. Z., de Silva, R. F., de Albuquerque, T. S., Todd, L. M., & Pinto, L. P. (2017). Run, hide, or fight: Anti-predation strategies in endangered red-nosed cuxiú (*Chiropotes albinasus*, Pitheciidae) in southeastern Amazonia. *Primates, 58*(2), 353-360.

Bennett, C. L., Leonard, S., & Carter, S. (2001). Abundance, diversity, and patterns of distribution of primates on the Tapiche River in Amazonian Peru. *American Journal of Primatology, 54*(2), 119-126.

Bezerra, B. M., Barnett, A. A., Souto, A., & Jones, G. (2011). Ethnogram and natural history of golden-backed uakaris (*Cacajao melanocephalus*). *International Journal of Primatology, 32*, 46-68.

Bobadilla, U. L., & Ferrari, S. F. (2000). Habitat use by *Chiropotes satanas utahicki* and synoptic platyrrhines in eastern Amazonia. *American Journal of Primatology, 50*, 215-224.

Bongers, F., Charles-Dominique, P., Forget, P. M., & Thery, M. (2001). *Nouragues: Dynamics and plant-animal interactions in a neotropical rainforest.* Dordrecht: Kluwer Academic Publishers.

Bonvicino, C. R., Boubli, J. P., Otazú, I. B., Almeida, F. C., Nascimento, F. F., Coura, J. R., & Seuánez, H. N. (2003). Morphologic, karyotypic, and molecular evidence of a new form of *Chiropotes* (Primates, pitheciinae). *American Journal of Primatology, 61*, 123-133.

Boubli, J. P. (1997). *Ecology of the black uacari monkey* Cacajao melanocephalus melanocephalus *in the Pico de Neblina National Park, Brazil.* [PhD dissertation, University of California—Berkeley].

Boubli, J. P. (1999). Feeding ecology of black-headed uacaris (*Cacajao melanocephalus melanocephalus*) in Pico da Neblina National Park, Brazil. *International Journal of Primatology, 20*, 719-749.

Boubli, J. P., da Silva, M. N. F., Amado, M. V., Hrbek, T., Pontual, F. B., & Farias, I. P. (2008). A taxonomic reassessment of *Cacajao melanocephalus* Humboldt (1811), with the description of two new species. *International Journal of Primatology, 29*, 723-741.

Boubli, J. P., & De Lima, M. G. (2009). Modeling the geographical distribution and fundamental niches of *Cacajao* spp. and *Chiropotes israelita* in Northwestern Amazonia via a maximum entropy algorithim. *International Journal of Primatology, 30*, 217-228.

Bowler, M. (2007). *The ecology and conservation of the red uacari monkey on the Yavari River, Peru.* [Doctoral dissertation, University of Kent].

Bowler, M., & Bodmer, R. (2009). Social behavior in fission-fusion groups of red uacari monkeys (*Cacajao calvus ucayalii*). *American Journal of Primatology*, 71(12), 976-987.

Bowler, M., & Bodmer, R. (2011). Diet and food choice in Peruvian red uakaris (*Cacajao calvus ucayalii*): Selective or opportunistic seed predation? *International Journal of Primatology*, 32(5), 1109-1122.

Bowler, M., Knogge, C., Heymann, E. W., & Zinner, D. (2012). Multilevel societies in New World primates? Flexibility may characterize the organization of Peruvian red uakaris (*Cacajao calvus ucayalii*). *International Journal of Primatology*, 33(5), 1110-1124.

Bowler, M., Murrieta, J. N., Recharte, M., & Puertas, P. E. (2009). Peruvian red uakari monkeys (*Cacajao calvus ucayalii*) in the Pacaya-Samiria National Reserve—A range extension across a major river barrier. *Neotropical Primates*, 16(1), 34-37.

Boyle, S. A., (2008). *The effects of forest fragmentation on primates in the Brazilian Amazon.* [Doctoral dissertation, Arizona State University].

Boyle, S. A. (2014). Pitheciids in fragmented habitats: Land cover change and its implications for conservation. *American Journal of Primatology*, 78(5), 534-549.

Boyle, S. A., Lourenço, W. C., da Silva, L. R., & Smith, A. T. (2009). Travel and spatial patterns change when *Chiropotes satanas chiropotes* inhabit forest fragments. *International Journal of Primatology*, 30, 515-531.

Boyle, S. A., & Smith, A. T. (2010). Behavioral modifications in northern bearded saki monkeys (*Chiropotes satanas chiropotes*) in forest fragments of central Amazonia. *Primates*, 51(1), 43-51.

Boyle, S. A., Thompson, C. L., Deluycker, A., Alvarez, S. J., Alvim, T. H. G., Aquino, R., Bezerra, B. M., Boubli, J. P., Bowler, M., Caselli, C. B., Chagas, R. R. D., Ferrari, S. F., Fontes, I. P., Gregory, T., Haugaasen, T., Heiduck, S., Hores, R., Lehman, S., de Menlo, F. R.,... & Barnett, A. A. (2016). Geographic comparison of plant genera used in frugivory among the pitheciids *Cacajao*, *Callicebus*, *Chiropotes*, and *Pithecia*. *American Journal of Primatology*, 78(5), 493-506.

Boyle, S. A., Zartman, C. E., Spironello, W. R., & Smith, A.T. (2012). Implications of habitat fragmentation on the diet of bearded saki monkeys in central Amazonian forest. *Journal of Mammalogy*, 93(4), 959-976.

Brumloop, A., Homburg, I., Peetz, A., & Riehl, R. (1994). Gular scent glands in adult white-faced saki, *Pithecia pithecia pithecia*, and field observations on scent-marking behaviour. *Folia Primatologica*, 63, 212-215.

Buchanan, D. B., Mittermeier, R. A., & Van Roosmalen, M. G. M. (1981). The saki monkeys, genus *Pithecia*. In A. F. Coimbra-Filho, & R. A. Mittermeier (Eds.), *Ecology and behavior of neotropical primates* (pp. 391-418). Academia Brasileira de Ciências.

Cunningham, E. P. (2003). *The use of memory in* Pithecia pithecia*'s foraging strategy.* [Doctoral dissertation, City University of New York]. Available from: University Microfilms, Ann Arbor, 64-03:973.

Cunningham, E., & Janson, C. (2007). Integrating information about location and value of resources by white-faced saki monkeys (*Pithecia pithecia*). *Animal Cognition*, 10(3), 293-304.

Davis, L. C., & Walker-Pacheco, S. E. (2013). Functional morphology and positional behavior in the pitheciini. In L. M. Veiga, A. A. Barnett, S. F. Ferrari, & M. A. Norconk (Eds.), *Evolutionary biology and conservation of titis, sakis and uacaris* (pp. 84-96). Cambridge University Press.

Defler, T. R. (1999). Fission-fusion in the black-headed uacari (*Cacajao melanocephalus*) in eastern Colombia. *Neotropical Primates*, 7(1), 5-8.

Defler, T. R. (2001). *Cacajao melanocephalus ouakary* densities on the lower Apaporis River, Colombian Amazon. *Primate Report*, 61, 31-36.

Defler, T. R. (2003). Densidad de especies y organización espacial de una comunidad de Primates: Estación Biológica Caparú, Departamento del Vaupés, Colombia. In F. Nassar, & V. Pereira (Eds.), *Primatología del Nuevo Mundo* (pp. 21-37). Centro de Primatología Araguatos Ltd.

De Granville, J. J. (1982). Rain forest and xeric flora refuges in French Guiana. In G. T. Prance (Ed.), *Biological diversification in the tropics* (pp. 159-181). Columbia University Press.

De Sousa e Silva Júnior, J., Figueiredo-Ready, W., & Ferrari, S. (2013). Taxonomy and geographic distribution of the Pitheciidae. In L. M. Veiga, A. A. Barnett, S. F. Ferrari, & M. A. Norconk (Eds.), *Evolutionary biology and conservation of titis, sakis and uacaris* (pp. 31-42). Cambridge University Press.

Di Fiore, A., Fernandez-Duque, E., & Hurst, D. (2007). Adult male replacement in socially monogamous equatorial saki monkeys (*Pithecia aequatorialis*). *Folia Primatologica*, 78(2), 88-98.

Fernandes, M. F. B. (1993). Tail-wagging as a tension relief mechanism in pithecines. *Folia Primatologica*, 61(1), 52-56.

Ferrari, S. F., Emidio-Silva, C., Lopes, M. A., & Bobadilla, U. L. (1999). Bearded sakis in south-eastern Amazonia—back from the brink? *Oryx*, 33(4), 346-351.

Ferrari, S. F., & Lopes, M. A. (1996). Primate populations in eastern Amazonia. In M. A. Norconk, A. L. Rosenberger, & P. A. Garber (Eds.), *Adaptive radiations of Neotropical primates* (pp. 53-67). Plenum Press.

Ferrari, S. F., Pereira, W. L. A., Santos, R. R., & Veiga, L. M. (2004). Fatal attack of a boa constrictor on a bearded saki (*Chiropotes satanas utahicki*). *Folia Primatologica*, 75(2), 111-113.

Fontaine, R. (1981). The uakaris, genus *Cacajao*. In A. F. Coimbra-Filho, & R. A. Mittermeier (Eds.), *Ecology and behavior of Neotropical primates* (pp. 443-493). Academia Brasileira de Ciencias.

Ford, S. M. (1994). Evolution of sexual dimorphism in body weight in platyrrhines. *American Journal of Primatology*, 34(2), 221-244.

Frazão, E. (1991). Insectivory in free-ranging bearded saki (*Chiropotes satanas chiropotes*). *Primates*, 32, 245-247.

Frisoli, L. K. (2009). *A behavioral investigation of the* Pithecia *niche in northeastern Peru.* [Master's thesis, Winthrop University].

Gleason, T. M. (1998). The ecology of olfactory communication in Venezuelan white-faced sakis. *American Journal of Primatology, 45,* 183.

Gregory, T. (2011). *Socioecology of the Guianan bearded saki,* Chiropotes sagulatus. [Doctoral dissertation, Kent State University].

Gregory, T., & Bowler, M. (2016). Male-male affiliation and cooperation characterize the social behavior of the large-bodied pitheciids, *Chiropotes* and *Cacajao*: A Review. *American Journal of Primatology, 78*(5), 550-560.

Gregory, T., & Norconk, M. A. (2013). Comparative socioecology of sympatric, free-ranging white-faced and bearded saki monkeys in Suriname: Preliminary data. In L. M. Veiga, A. A. Barnett, S. F. Ferrari, & M. A. Norconk (Eds.), *Evolutionary biology and conservation of titis, sakis and uacaris* (pp. 285-294). Cambridge University Press.

Gregory, T., & Norconk, M. A. (2014). Bearded saki socioecology: Affiliative male-male interactions in large, free-ranging groups in Suriname. *Behaviour, 151*(4), 493-533.

Hershkovitz, P. (1985). A preliminary taxonomic review of the South American bearded saki monkey's genus *Chiropotes* (Cebidae, Platyrrhini), with the description of a new subspecies. *Fieldiana, 27,* 1-46.

Hershkovitz, P. (1987). The taxonomy of the South American sakis, genus *Pithecia* (Cebidae, platyrhinni): A preliminary report and critical review with the description of a new species and a new subspecies. *American Journal of Primatology, 12*(4), 387-468.

Heymann, E. W., & Aquino, R. (2010). Peruvian red uakaris (*Cacajao calvus ucayalii*) are not flooded-forest specialists. *International Journal of Primatology, 31*(5), 751-758.

Heymann, E. W., Encarnacion, C. F., & Canaquin-Y., J. E. (2002). Primates of the Río Curaray, Northern Peruvian Amazon. *International Journal of Primatology, 23*(1), 191-201.

Homburg, I. (1997). *Okologie and sozialverhalten einer gruppe von weissgesicht-sakis (*Pithecia pithecia pithecia *Linnaeus 1766) im Estado Bolıvar, Venezuela.* [Doctoral dissertation, Universitat Bielefeld].

Homburg, I. (1998). *Ökologie und sozial verhalten von weissegesicht-sakis. Eine freilandstudie in Venezuela.* [Doctoral dissertation, University of Göttingen].

IUCN. (2020). *The International Union for the Conservation of Nature Red List of Threatened Species.* www.iucnredlist.org.

Kay, R. F., & Takai, M. (2006). Pitheciidae and other platyrrhine seed predators: The dual occupation of the seed predator niche during platyrrhine evolution. *XXI International Congress of Primatology,* Entebbe Uganda.

Kay, R. F., Vizcaino, S. F., Bargo, M. S., Perry, J. M. G., Prevosti, F. J., & Fernicola, J. C. (2008). Two new fossil vertebrate localities in the Santa Cruz Formation (late early–early middle Miocene, Argentina), ~51 degrees South l Latitude. *Journal of South American Earth Sciences, 25*(2), 187-195.

Kessler, P. (1998). Primate densities in the natural reserve of Nouragues, French Guiana. *Neotropical Primates, 6*(2), 45-46.

Kinzey, W. G. (1986). New World primate field studies: What's in it for anthropology? *Annual Review of Anthropology, 15*(1), 121-148.

Kinzey, W. G. (1992). Dietary and dental adaptations in the Pitheciinae. *American Journal of Physical Anthropology, 88*(4), 499-514.

Kinzey, W. G., & Norconk, M. A. (1990). Hardness as a basis of fruit choice in two sympatric primates. *American Journal of Physical Anthropology, 81*(1), 5-15.

Kinzey, W. G., & Norconk, M. A. (1993). Physical and chemical properties of fruit and seeds eaten by *Pithecia* and *Chiropotes* in Surinam and Venezuela. *International Journal of Primatology, 14*(2), 207-227.

Ledogar, J. A. (2009). Postcanine occlusal loading and relative dental arcade width in pitheciine primates. [Abstracts of AAPA poster and podium presentations]. *American Journal of Physical Anthropology, 138*(S48), 173.

Ledogar, J. A., Winchester, J. M., St. Clair, E. M., & Boyer, D. M. (2013). Diet and dental topography in pitheciine seed predators. *American Journal of Physical Anthropology, 150*(1), 107-121.

Lehman, S. M., Prince, W., & Mayor, M. (2001). Variations in group size in white-faced sakis (*Pithecia pithecia*): Evidence for monogamy or seasonal congregations? *Neotropical Primates, 9,* 96-101.

Lenz, B. B. (2013). *The effects of cattle ranching on a primate community in the central Amazon.* [Doctoral dissertation, Tulane University].

Marroig, G., & Cheverud, J. M. (2005). Size as a line of least evolutionary resistance: Diet and adaptive morphological radiation in New World monkeys. *Evolution, 59*(5), 1128-1142.

Marsh, L. K. (2014). A taxonomic revision of the saki monkeys, *Pithecia* Desmarest, 1804. *Neotropical Primates, 21*(1), 1-165.

Martins, S. S., de Lima, E. M., & Silva, J. S. (2005). Predation of a bearded saki (*Chiropotes utahicki*) by a harpy eagle (*Harpia harpyja*). *Neotropical Primates, 13*(1), 7-10.

Milton, K. (1984). The role of food-processing factors in primate food choice. In P. S. Rodman, & J. G. H. Cant (Eds.), *Adaptations for foraging in nonhuman primates* (pp. 249-279). Columbia University Press.

Mittermeier, R. A. (1977). *Distribution, synecology, and conservation of Surinam monkeys.* [Doctoral dissertation, Harvard University].

Mittermeier, R. A. (1991). Hunting and its effects on wild primate populations in Suriname. In J. G. Robinson, & K. H. Redford (Eds.), *Neotropical wildlife use and conservation* (pp. 93-106). University of Chicago Press.

Mittermeier, R. A. (2013). In a nutshell: A brief introduction to early explorations in the pitheciids. In L. M. Veiga, A. A. Barnett, S. F. Ferrari, & M. A. Norconk (Eds.), *Evolutionary biology and conservation of titis, sakis and uacaris* (pp. xvi-xxx). Cambridge University Press.

Mittermeier, R. A., Konstant, W. R., Ginsberg, H., Van Roosmalen, M. G. M., & Cordeiro da Silva, E., Jr. (1983). Further evidence of insect consumption in the bearded saki monkey, *Chiropotes satanas chiropotes*. *Primates*, 24, 602-605.

Mittermeier, R., & Van Roosmalen, M. G. M. (1981). Preliminary observations on habitat utilization and diet in eight Surinam monkeys. *Folia Primatologica*, 36(1-2), 1-39.

Mori, S. A. (1989). Diversity of Lecythidaceae in the Guianas. In L. B. Holm-Nielsen, I. C. Nielsen, & H. Balslex (Eds.), *Tropical rainforests: Botanical dynamics, speciation and diversity* (pp. 319-332). Academic Press.

Norconk, M. A. (1996). Seasonal variation in the diets of white-faced and bearded sakis (*Pithecia pithecia* and *Chiropotes satanas*) in Guri Lake, Venezuela. In M. A. Norconk, A. L. Rosenberger, & P. A. Garber (Eds.), *Adaptive radiations of Neotropical primates* (pp. 403-423). Plenum Publishers.

Norconk, M. A. (2006). Long-term study of group dynamics and female reproduction in Venezuelan *Pithecia pithecia*. *International Journal of Primatology*, 27(3), 653-674.

Norconk, M. A. (2007). Sakis, uakaris, and titi monkeys: Behavioral diversity in a radiation of primate seed predators. In C. J. Campbell, A. Fuentes, K. C. MacKinnon, M. Panger, & S. K. Bearder (Eds.), *Primates in perspective*, (1st ed., pp. 123-138). Oxford University Press.

Norconk, M. A. (2011). Sakis, uakaris, and titi monkeys: Behavioral diversity in a radiation of primate seed predators. In C. J. Campbell, A. Fuentes, K. C. MacKinnon, M. Panger, & S. K. Bearder (Eds.), *Primates in perspective* (2nd ed., pp. 123-138). Oxford University Press.

Norconk, M. A., & Conklin-Brittain, N. L. (2004). Variation on frugivory: The diet of Venezuelan white-faced sakis. *International Journal of Primatology*, 25(1), 1-26.

Norconk, M. A., & Conklin-Brittain, N. L. (2015). Bearded saki feeding strategies on an island in Lago Guri, Venezuela. *American Journal of Primatology*, 78(5), 507-522.

Norconk, M. A., Grafton, B. W., & Conklin-Brittain, N. L. (1998). Seed dispersal by neotropical seed predators. *American Journal of Primatology*, 45(1), 103-126.

Norconk, M. A., Grafton, B. W., & McGraw, W. S. (2013). Morphological and ecological adaptations to seed predation—a primate-wide perspective. In L. M. Veiga, A. A. Barnett, S. F. Ferrari, & M. A. Norconk (Eds.), *Evolutionary biology and conservation of titis, sakis and uacaris* (pp. 55-71). Cambridge University Press.

Norconk, M. A., & Kinzey, W. G. (1994). Challenge of Neotropical frugivory: Travel patterns of spider monkeys and bearded sakis. *American Journal of Primatology*, 34(2), 171-183.

Norconk, M. A., Oftedal, O. T., Power, M. L., Jakubasz, M., & Savage, A. (2002). Digesta passage and fiber digestibility in captive white-faced sakis (*Pithecia pithecia*). *American Journal of Primatology*, 58(1), 23-34.

Norconk, M. A., Raghanti, M. A., Martin, S. K., Grafton, B. W., Gregory, L. T., & DiDijn, B. P. E. (2003). Primates of Brownsberg Natuurpark, Suriname, with particular attention to the pitheciins. *Neotropical Primates*, 11(2), 94-100.

Norconk, M. A., & Veres, M. (2011). Physical properties of fruit and seeds ingested by primate seed predators with emphasis on sakis and bearded sakis. *Anatomical Record*, 294(12), 2092-2111.

Norconk, M. A., Wright, B. W., Conklin-Brittain, N. L., & Vinyard, C. J. (2009). Mechanical and nutritional properties of food as factors in platyrrhine dietary adaptations. In P. A. Garber, A. Estrada, J. C. Bicca-Marques, E. W. Heymann, & K. B. Strier, (Eds.), *South American primates: Comparative perspectives in the study of behavior, ecology, and conservation* (pp. 279-319). Springer.

Oliveira, J. M. S., Lima, M. G., Bonvincino, C., Ayres, J. M., & Fleagle, J. G. (1985). Preliminary notes on the ecology and behavior of the Guianan saki (*Pithecia pithecia*, Linnaeus 1766; Cebidae, Primate). *Acta Amazonica*, 15(1-2), 249-263.

Palminteri, S., Powell, G., Endo, W., & Kirby, C. (2011). Usefulness of species range polygons for predicting local primate occurrences in southeastern Peru. *American Journal of Primatology*, 73(1), 53-61.

Palminteri, S., Powell, G. V. N., & Peres, C. A. (2015). Determinants of spatial behavior of a tropical forest seed predator: The roles of optimal foraging, dietary diversification, and home range defense. *American Journal of Primatology*, 78(5), 523-533.

Peetz, A. (2001). *Ecology and social organization of the bearded saki* Chiropotes satanas chiropotes *(Primates: Pitheciinae) in Venezuela*. Ecotropical Monographs.

Peres, C. A. (1993). Notes on the ecology of buffy saki monkeys (*Pithecia albicans*, Gray 1860): A canopy seed-predator. *American Journal of Primatology*, 31(2), 129-140.

Peres, C. A. (2000). Effects of subsistence hunting on vertebrate community structure in Amazonian forests. *Conservation Biology*, 14(1), 240-253.

Pinto, L. P. (2008). *Food ecology of the red-nosed cuxiu* Chiropotes albinasus *(Primates: Pithecidae) in the Tapajós National Forest, Pará*. [Doctoral dissertation, State University of Campinas].

Pinto, L. P., Barnett, A. A., Bezerra, B. M., & Boubli, J. P. (2013). Why we know so little: The challenges of fieldwork on the pitheciids. In L. M. Veiga, A. A. Barnett, S. F. Ferrari, & M. A. Norconk (Eds.), *Evolutionary biology and conservation of titis, sakis and uacaris* (pp. 145-150). Cambridge University Press.

Port-Carvalho, M., & Ferrari, S. F. (2004). Occurrence and diet of the black bearded saki (*Chiropotes satanas satanas*) in the fragmented landscape of western Maranhão, Brazil. *Neotropical Primates*, 12(1), 17-21.

Port-Carvalho, M., Muniz, C. C., Fialho, M. S., Alonso, A. C., Jerusalinsky, L. & Veiga, L. M. 2021. *Chiropotes satanas* (amended version of 2020 assessment). *The IUCN Red List of Threatened Species* 2021: e.T39956A191704509. https://dx.doi.org/10.2305/IUCN.UK.2021-1.RLTS.T39956A191704509.e.

Poyas, A. L., & Bartlett, T. Q. (2009). Pair bonding in socially monogamous primates: A comparative study of the white-cheeked gibbon (*Nomascus leucogenys*) and the white-faced saki (*Pithecia pithecia*). *American Journal of Physical Anthropology, 138*(S48), 214.

Rettig, N. (1978). Breeding behavior of the harpy eagle (*Harpia harpyja*). *The Auk, 95*(4), 629-643.

Setz, E. Z. F. (1993). *Ecologia alimentar de um grupo de parauacus* (Pithecia pithecia chrysocephala) *em um fragmento floral na Amazonia Central.* [Doctoral dissertation, Universidade Estadual de Campinas].

Setz, E. Z. F., & Gaspar, D. A. (1997). Scent-marking behaviour in free-ranging golden-faced saki monkeys, *Pithecia pithecia chrysocephala*: Sex differences and context. *Journal of Zoology, 241*(3), 603-611.

Shaffer, C. A. (2012). *Ranging behavior, group cohesiveness, and patch use in northern bearded sakis* (Chiropotes sagulatus) *in Guyana.* [Doctoral dissertation, Washington University].

Shaffer, C. A. (2013a). GIS analysis of patch use and group cohesiveness of bearded sakis (*Chiropotes sagulatus*) in the Upper Essequibo Conservation Concession, Guyana. *American Journal of Physical Anthropology, 150*(2), 235-246.

Shaffer, C. A. (2013b). Ecological correlates of ranging behavior in bearded sakis (*Chiropotes sagulatus*) in a continuous forest in Guyana. *International Journal of Primatology, 34,* 515-532.

Shaffer, C. A. (2013c). Feeding ecology of northern bearded sakis (*Chiropotes sagulatus*) in Guyana. *American Journal of Primatology, 75*(6), 568-580.

Shaffer, C. A. (2013d). Activity patterns, intergroup encounters, and male affiliation in free-ranging bearded sakis (*Chiropotes sagulatus*). *International Journal of Primatology, 34*(6), 1190-1208.

Shaffer, C. A., Marawanaru, E., & Yukuma, C. (2017a). An ethnoprimatological approach to assessing the sustainability of primate subsistence hunting of indigenous Waiwai in the Konashen community owned conservation concession, Guyana. In K. M. Dore, E. P. Riley, & A. Fuentes (Eds.), *Ethnoprimatology: A practical guide to research on the human-nonhuman primate interface* (pp. 232-250). Cambridge University Press.

Shaffer, C. A., Milstein, M. S., Yukuma, C., & Marawanaru, E. (2017b). Sustainability and comanagement of subsistence hunting in an indigenous reserve in Guyana. *Conservation Biology, 31*(5), 1119-1131.

Shaffer, C. A., Wright, B. A., & Wright, K. (2019). Primate community structure at three flooded forest sites in Guyana: Ecology and conservation. In A. A. Barnett, I. Matsuda, & N. Katarzyna (Eds.), *Primates in flooded habitats* (pp. 226-235). Cambridge University Press.

Silva, S. S. B., & Ferrari, S. F. (2009). Behavior patterns of southern bearded sakis (*Chiropotes satanas*) in the fragmented landscape of eastern Brazilian Amazonia. *American Journal of Primatology, 71*(1), 1-7.

Soini, P. (1986). A synecological study of a primate community in the Pacaya-Samiria National Reserve, Peru. *Primate Conservation, 7,* 63-71.

Stevenson, P. R. (2001). The relationship between fruit production and primate abundance in Neotropical communities. *Biological Journal of the Linnean Society, 72*(1), 161-178.

ter Steege, H. (1993). *Patterns of tropical rain forest in Guyana.* Tropenbos International.

ter Steege, H., Castellanos, H., Van Andel, T., Duivenvoorden, J., De Oliveira, A. A., Ek, R., & Shaffer, C. A. (2014). Spatial foraging in free ranging bearded sakis: Traveling salesmen or Lilwah, R., Maas, P., & Mori, S. (2000). An analysis of the floristic composition and diversity of Amazonian forests including those of the Guiana Shield. *Journal of Tropical Ecology, 16*(6), 801-828.

ter Steege, H., Pitman, N. C. A., Phillips, O. L., Chave, J., Sabatier, D., Duque, A., Molino, J., Prevost, M., Spichiger, R., Castellanos, H., Von Hilderbrand, P., & Vasquez, R. (2006). Continental-scale patterns of canopy tree composition and function across Amazonia. *Nature, 443,* 444-447.

Thompson, C. L. (2011). *Sex, aggression, and affiliation: The social system of white-faced saki monkeys* (Pithecia pithecia). [Doctoral dissertation, Kent State University].

Thompson, C. L. (2015). To pair or not to pair: Sources of social variability with white-faced saki monkeys (*Pithecia pithecia*) as a case study. *American Journal of Primatology, 78*(5), 561-572.

Thompson, C. L., & Norconk, M. (2011). Within-group social bonds in white-faced saki monkeys (*Pithecia pithecia*) display male-female pair preferences. *American Journal of Primatology, 73*(10), 1051-1061.

Van Belle, S., Porter, A., Fernandez-Duque, E., & Di Fiore, A. (2018). Ranging behavior and potential for territoriality in equatorial sakis (*Pithecia aequatorialis*) in Amazonian Ecuador. *American Journal of Physical Anthropology, 167*(4), 701-712.

van Roosmalen, M. G. M., Mittermeier, R. A., & Fleagle, J. G. (1988). Diet of the northern bearded saki (*Chiropotes satanas chiropotes*): A Neotropical seed predator. *American Journal of Primatology, 14*(1), 11-35.

van Roosmalen, M. G. M., Mittermeier, R. A., & Milton, K. (1981). The bearded sakis, genus *Chiropotes*. In A. F. Coimbra-Filho, & R. A. Mittermeier (Eds.), *Ecology and behavior of Neotropical primates* (Vol. 1., pp. 419-441). Brazilian Academy of Sciences.

Veiga, L. M. (2006). Ecologia e comportamento do cuxiú-preto (*Chiropotes satanas*) na paisagem fragmentada da Amazônia oriental. [Doctoral dissertation, Universidade Federal do Pará].

Veiga, L. M., Barnett, A. A., Ferrari, S. F., & Norconk, M. A. (Eds.). (2013). *Evolutionary biology and conservation of titis, sakis and uacaris.* Cambridge University Press.

Veiga, L. M., & Ferrari, S. F. (2006). Predation on arthropods by southern bearded sakis (*Chiropotes satanas*) in eastern Brazilian Amazonia. *American Journal of Primatology, 68*(2), 209-215.

Veiga, L. M., Pinto, L. P., & Ferrari, S. F. (2006). Fission-fusion sociality in bearded sakis (*Chiropotes albinasus* and *Chiropotes satanas*) in Brazilian Amazonia. *Proceedings of the XXIst Congress of the International Primatology Society, Uganda,* 709.

Veiga, L. M., Silva, J. S., Jr., Ferrari, S. F., & Rylands, A. B. (2008). *Chiropotes satanas. The IUCN Red List of Threatened Species* 2008: e.T39956A10297662. http://dx.doi.org/10.2305/IUCN.UK.2008.RLTS.T39956A10297662.en.

Veiga, L. M., & Silva, S. S. B. (2005). Relatives or just good friends? Affiliative relationships among male bearded sakis (*Chiropotes satanas*) in Tucurui, Brazil. In J. C. Bicca-Marques (Ed.), *Programa e Livro de Resumos: XI Congresso Brasileiro de Primatologia* (p. 174). Sociedade Brasileira de Primatologia.

Vie, V., Mau, N. V., Pomarede, C. D., Schwartz, J. L., Laprade, R., Frutos, R., Rang, C., Masson, L., Heitz, F., & Le Grimellec, C. (2001). Lipid-induced pore formation of the *Bacillus thuringiensis* Cry1Aa Insecticidal Toxin. *Journal of Membrane Biology,* 180, 195-203.

Walker, S. E. (1996). The evolution of positional behavior in the saki-uakaris (*Pithecia, Chiropotes,* and *Cacajao*). In M. A. Norconk, A. L. Rosenberger, & P. A. Garber (Eds.), *Adaptive radiations of neotropical primates* (pp. 335-367). Plenum Press.

Walker, S. E. (1998). Fine-grained differences within positional categories. In E. Strasser, J. G. Fleagle, A. L. Rosenberger, & H. M. McHenry (Eds.), *Primate locomotion* (pp. 31-43). Springer.

Walker, S. E. (2005). Leaping behavior of *Pithecia pithecia* and *Chiropotes satanas* in eastern Venezuela. *American Journal of Primatology,* 66(4), 369-387.

Walker, S. E., & Ayers, J. M. (1996). Positional behavior of the white uacari (*Cacajao calvus calvus*). *American Journal of Physical Anthropology,* 101(2), 161-172.

CHAPTER 10

Ecological Niches and Behavioral Strategies of Capuchins and Squirrel Monkeys

KATHERINE C. MACKINNON AND MICHELLE BEZANSON

THIS CHAPTER PROVIDES AN OVERVIEW OF THE ECOLOGY and behavior of the family Cebidae, which includes two subfamilies: Cebinae (the capuchin monkeys, with two genera: *Cebus* and *Sapajus*) and Saimiriinae (the squirrel monkeys, with one genus: *Saimiri*) (Rylands & Mittermeier, 2009; Lynch Alfaro et al., 2012a; Schneider & Sampaio, 2015). All genera are well-known and well-studied New World primates both in the wild and in captive settings. Here we follow the taxonomic classifications of Rylands et al. (2000, 2006), Defler (2004), Groves (2001a, 2005), Fragaszy et al. (2004a), Rylands and Mittermeier (2009), Lynch Alfaro et al. (2012a), and Schneider and Sampaio (2015) in terms of the species and subspecies names (see Table 10.1; and Groves, 2001b, 2004, for a general discussion of primate taxonomy). For many foci of the discussion, namely, ecological niches and aspects of behavior, we will highlight both differences and commonalities of the three genera to underscore adaptive strategies.

Typically, Cebidae species within each genus fill similar ecological roles and, generally, can be considered ecological equivalents (Kinzey, 1997; Lynch Alfaro et al., 2011). Small-bodied and energetic, capuchins and squirrel monkeys range in size from ~2 kg (*Saimiri*) to 4 kg (some of the larger *Cebus* and *Sapajus*). The *Saimiri* tail is long and non-prehensile, and helps with balance during locomotion and feeding. While the *Cebus* and *Sapajus* tails are long and prehensile, they differ from the prehensile tails of atelines (e.g., howler monkeys, spider monkeys, woolly monkeys, and muriquis) in several aspects of anatomy; ateline tails have a hairless underside on the distal quarter with a fleshy muscular pad and dermatoglyphs, all lacking in capuchins (Rosenberger, 1983; Lemelin, 1995; Bergeson, 1998; Organ et al., 2009, 2011; Organ, 2010). Normally, a single Cebidae infant is born and carried by the mother (and other group members) from birth. The Cebidae dental formula is 2:1:3:3/2:1:3:3, with the presence of a third premolar differentiating all New World monkeys from Old World species. All digits have flat nails (unlike the Callitrichidae, which have sharply curved, claw-like nails). Cebidae occur throughout the tropical forested regions of Central and South America and, while many species have a wide distribution, several species and subspecies are quite restricted in geographic range and, thus, are more of a conservation concern.

TAXONOMY

Until about 15 years ago, *Cebus* (Erxleben, 1777) was the sole genus recognized for capuchins, and consisted of four species: *C. apella, C. albifrons, C. olivaceus,* and *C. capucinus,* with ~30 subspecies generally recognized (Ford & Hobbs, 1996), but there was much debate over these classifications (see Groves 1993; Rylands et al.,1995). The four *Cebus* species were traditionally split into two groups based on the presence or absence of tufted hair patterns on the head: *C. apella* in the tufted group, and *C. albifrons, C. olivaceus,* and *C. capucinus* in the non-tufted group (Hershkovitz, 1949, based on Elliot, 1913). At the beginning of this century, Groves (2001a) and Rylands et al. (2000) argued for *C. nigritus, C. libidinosus,* and *C. xanthosternos* as distinct species, not subspecies of *C. apella* (see Fragaszy et al., 2004a). The Ka'apor capuchin (*C. kaapori*), first described by Queiroz in 1992, was initially considered a subspecies of *C. olivaceus* but was later elevated to species status (Ferrari & Queiroz, 1994; Silva & Cerqueira, 1998; Carvalho et al., 1999; and see Groves, 2001a). Throughout a series of compelling research articles published in a 2012 special issue of the *American Journal of Primatology*, Lynch Alfaro and colleagues argued convincingly for a second capuchin genus to be reinstated: *Sapajus* (Kerr, 1792), based upon extensive and multifaceted sets of data from biogeographical, genetic, morphological, phylogenetic, and behavioral evidence (Ruiz-Garcia et al., 2010, 2012; Lynch Alfaro et al., 2011, 2012a; Boubli et al., 2012).

There is one recognized genus for squirrel monkeys, *Saimiri,* which now includes eight species with variable subspecies (following Rylands et al., 2013; see also Cropp & Boinski, 2000; Rylands et al., 2000, 2013; Groves, 2001a, 2005; Lynch Alfaro et al., 2015). Current data from morphological, behavioral, and molecular studies all support this taxonomic organization, although it had been disputed when based solely on morphological data (Ayres, 1985; Boinski & Cropp, 1999; Groves,

Table 10.1 Currently Recognized Species of Cebidae

Taxon [a]	Common Name	Male Body Mass (kg) [b]	N	Female Body Mass (kg) [b]	N
Infraorder: Platyrrhini	Platyrrhines/New World monkeys				
Superfamily: Ceboidea/Ateloidea					
Family: Cebidae	Cebids				
Subfamly: Cebinae	Cebines				
Saimiri [c]					
Saimiri oerstedii	Central American squirrel monkey	0.897	11	0.680	7
Saimiri boliviensis	Bolivian squirrel monkey	0.911	17	0.711	19
Saimiri vanzolinii	Vanzolini's squirrel monkey	0.959	9	0.650	4
Saimiri sciureus	Common squirrel monkey	0.779	40	0.662	40
Saimiri ustus	Golden-backed squirrel monkey	0.910	-	0.795	-
Saimiri collinsi	Collins' squirrel monkey	-	-	-	-
Saimiri macrodon	Ecuadorian squirrel monkey	-	-	-	-
Saimiri cassiquiarensis	Humboldt's squirrel monkey	-	-	-	-
Cebus					
Cebus capucinus (ssp. *capucinus*)	Colombian white-faced/white-headed capuchin	3.68	16	2.54	10
Cebus imitator (*C. capucinus imitator*)	Panamanian/Central American white-faced capuchin	3.18	26	2.29	15
Cebus albifrons	Humboldt's white-fronted capuchin	3.29	28	2.52	10
Cebus cuscinus	Shock-headed capuchin	-	-	-	-
Cebus unicolor	Spix's white-fronted capuchin	-	-	-	-
Cebus olivaceus	Wedged-capped/weeper capuchin	3.29	28	2.52	10
Cebus kaapori	Ka'apor tufted capuchin	3.65	-	-	-
Sapajus					
Sapajus apella	Brown capuchin/black-capped capuchin	3.65	51	2.52	38
Sapajus macrocephalus	Large-headed capuchin	3.1	-	2.4	-
Sapajus libidinosus	Bearded capuchin	3.5	-	2.1	-
Sapajus cay	Azara's/hooded capuchin	3.0-3.1	-	-	-
Sapajus flavius	Blond capuchin	2.9	4	2.15	4
Sapajus robustus	Crested capuchin/robust tufted capuchin	3.8	10	2.5	10
Sapajus xanthosternos	Buff-headed/yellow-breasted capuchin	3.0-3.5	-	-	-
Sapajus nigritus	Black capuchin/black-horned capuchin	2.0-3.3	-	-	-

[a] Cebid taxonomy from Rylands and Mittermeier (2009), Mittermeier and Rylands (2013), Rylands et al. (2013), Schneider and Sampaio (2015), Lynch Alfaro et al. (2012a,b, 2015).
[b] Body masses from Smith and Jungers (1997), Fragaszy et al. (2004a, 2016), Ford and Davis (1992).
[c] Eight *Saimiri* species are recognized, but many data for *S. collinsi*, *S. macrodon*, *S. cassiquiarensis* are still subsumed under previously expansive *S. sciureus*.

2001a, 2005; Chiou et al., 2011; Rylands et al., 2013; Lynch Alfaro et al., 2015). For many years *Saimiri* included only one species—*S. sciureus* in South America. The smaller population of squirrel monkeys in Central America was thought by some to have been introduced during pre-Colombian times (Hershkovitz, 1969; c.f. Hershkovitz, 1984; Costello et al., 1993; Cropp & Boinski, 2000), but genetic data now support *S. oerstedii* as a distinct species (see Boinski, 1999; Cropp & Boinski, 2000; Rylands et al., 2013). As Jack (2011) points out, the historical practice of lumping all South American squirrel monkeys together as *S. sciureus* can make trying to decipher earlier reports confusing. For example, most of our current knowledge about *S. boliviensis* comes from Mitchell's (1990, 1994) study at Manú, Peru, where the species used to be referred to as *S. sciureus* (see Boinski, 1999). The species name of *sciureus* is now only used for those populations in the Guiana Shield area (Guyana, Suriname, French Guiana, northern Brazil) and the northeastern Amazonian Basin (Venezuela, Colombia, and Ecuador) (Boinski, 1999; Boinski & Cropp, 1999;

Rylands et al., 2013). Recent taxonomic adjustments can also make parsing out specific information on "newer" species difficult. For example, studies are now published on *S. collinisi* (e.g., Stone & Silva Jr., 2016; Stone, 2018; Stone & Ruivo, 2020), but work for that species was published under *S. sciureus* as recently as 2014.

DISTRIBUTION

Capuchin monkeys are found throughout Central and South America, from Honduras to Argentina (Kinzey, 1997; Eisenberg & Redford, 1989) as shown in Figures 10.1 and 10.2. *C. capucinus* (with subspecies *capucinus* and *imitator*) is the only cebid species indigenous to Central America, ranging from northern Honduras to the northwestern coast of Ecuador (Rowe, 1996; Kinzey, 1997); the rest of the *Cebus* and *Sapajus* species are found only in South America (see Lynch Alfaro et al., 2011). Most of what we know about this genus comes from long-term field studies at the following sites: *C. capucinus imitator* in Santa Rosa National Park and Lomas Barbudal National Reserve, Costa Rica; *C. olivaceus* in Hato

Figure 10.1 Geographic distribution of *Cebus* species based on spatial data from IUCN Red List (IUCN, 2020).
(Map derived from Martins-Junior et al. (2018: p.700); image licensed under terms of Creative Commons Attribution License-type CC-BY, permitting unrestricted use provided original article properly cited)

Figure 10.2 Geographic distribution of *Sapajus* species based on spatial data from IUCN Red List (IUCN, 2020).
(Map derived from Martins-Junior et al. (2018: p.700); image licensed under terms of Creative Commons Attribution License-type CC-BY, permitting unrestricted use provided original article properly cited)

Masaguaral, Venezuela; *S. apella* in Iguazu National Park, Argentina, Manú National Park, Peru, and Raleighvallen-Voltzberg in the Central Suriname Reserve, Suriname; and *S. libidinosus* at several sites in Brazil (Fragaszy et al., 2004a). *C. albifrons* has been the subject of shorter-term studies in Colombia (Defler, 1979a,b, 1982), Peru (Terborgh & Janson, 1986), Trinidad (Phillips, 1998; Phillips & Abercrombie, 2003), and Ecuador (Jack & Campos, 2012; Campos & Jack, 2013). Historically, substantial decades-long field studies of wild New World monkeys across all taxa have lagged behind the better-known examples of Old World monkey and ape research. Nonetheless, that is changing (e.g., see Norconk et al., 1996; Strier, 1999, 2010; Defler, 2004; Estrada et al., 2006; Strier & Boubli, 2006; Campbell, 2008; Strier & Mendes, 2009; Gama Campillo, 2011; Kappeler & Watts, 2012; Kowalewski et al., 2015; Kalbitzer & Jack, 2018; Urbani et al., 2018). For Cebidae, some studies of *C. capucinus imitator* in Costa Rica now span over three decades (e.g., see Fedigan & Jack, 2012; Kalbitzer et al., 2017; Kalbitzer & Jack, 2018), and ongoing work on several other capuchin

species is yielding much-needed information, e.g., *Sapajus libidinosus* (Liu et al., 2009; Spagnoletti et al., 2011; Duarte et al., 2012; Izar et al., 2012; Massaro et al., 2012, 2016; Falótico & Ottoni, 2013; Fragaszy et al., 2013, 2016; Verderane et al., 2013; De Moraes et al., 2014; Eshchar et al., 2016), *Sapajus nigritus* (Garber et al., 2011, 2012; Izar et al., 2012; Scarry, 2013, 2017), and *Sapajus xanthosternos* (Canale et al., 2013; Flesher, 2015; Beltrão-Mendes & Ferrari, 2019).

Saimiri species are found in two distinct geographical areas (Figure 10.3). *S. oerstedii* occurs only in Central America across a small tract of lowland forest from the central Pacific coast of Costa Rica to the Pacific coast of western Panama. The other species inhabit the lowland forests of the Amazon Basin from Guyana to Paraguay (Boinski et al., 2002). Extensive studies of wild *Saimiri* have been conducted at only a handful of sites: *S. oerstedii* in Corcovado, Costa Rica (Boinski, 1987a,b,c, 1988a, 1994; Boinski et al., 1998), *S. boliviensis* in Manú, Peru (Mitchell, 1990, 1994; Boinski, 1999, 1994), *S. sciureus* in Raleighvallen-Voltzberg, Suriname (Boinski, 1999;

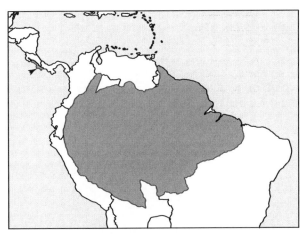

Figure 10.3 Geographic distribution of *Saimiri* species.
(Map derived from Carneiro et al. (2016); image licensed under terms of Creative Commons Attribution License-type CC-BY, permitting unrestricted use provided original article properly cited)

Boinski et al., 2002), and more recently, *S. collinsi* in eastern Brazilian Amazonia (Stone, 2006, 2007a,b,c, 2008, 2014, 2018; Ruivo & Stone, 2014, 2020; Stone & Silva Jr., 2016; Ruivo et al., 2017; Mercês et al., 2018).

GENERAL ECOLOGY AND BEHAVIOR
Habitat Types and Locomotion
Members of the Cebidae are found in a wide range of habitats throughout the Neotropics. Capuchins (Figure 10.4) inhabit primary rainforest, highly seasonal and deciduous dry forest, swampy lowland areas, and mosaic habitats with extensive anthropogenic disturbances; they do very well in areas of secondary growth and can easily cross open gaps on the ground or along fence lines between forest patches.

Among capuchins, *Sapajus apella* can be found in sympatry with *Cebus albifrons* or *C. olivaceus* in parts of their range. Capuchins and squirrel monkeys can also be sympatric, but *Saimiri* tends to occur less frequently in primary forest—they seem to thrive in more disturbed and secondary areas of forest regrowth characterized by low shrubby areas, tangled lianas, and stands of bamboo (Sussman & Phillips-Conroy, 1995; Boinski et al., 2002; Stone, 2007a). Like capuchins, squirrel monkeys are quite adaptable and can be found in differing habitat types (particularly in South America) due to their diets and locomotor styles. Where they are sympatric, capuchins and squirrel monkeys have been observed in mixed-species associations (Terborgh, 1983; Podolsky, 1990; Lehman et al., 2006; Haugaasen & Peres, 2009; Levi et al., 2013).

In a classic study 40 years ago, Mittermeier and van Roosmalan (1981) described and quantified the use of different habitat types by eight primate species in Suriname. The species studied were *Saguinus midas*, *Pithecia pithecia*, *Chiropotes satanas*, *Saimiri sciureus*, *Alouatta seniculus* (now *A. macconnelli*), *Ateles paniscus*, *Cebus* (now *Sapajus*) *apella*, and *C. nigrivittatus* (now *C. olivaceus*). All of the species, except *Saimiri sciureus*, were most often seen in high levels of the canopy. However, this may reflect varying degrees of habituation as many subsequent studies have shown a greater use of different forest levels for many of these species, e.g., *Pithecia* is frequently documented as occupying mid to low levels of the canopy when well habituated (Gleason & Norconk 2002; Norconk 2006, 2011; Barnett et al., 2012; Thompson 2013), and some upper-canopy specialists like *Chiropotes* have even been seen on the ground for brief moments at some study sites (Barnett et al., 2012). In the 1981 study, *Sapajus* (formerly *Cebus*) *apella* was the least specialized in habitat choice, followed closely by *Saimiri sciureus*. *C. olivaceus* utilized the upper canopy of primary rainforest the most, as did *Ateles paniscus*. Of the eight Suriname species, only *Saimiri sciureus* and *Sapajus apella* used forest edge areas extensively, although *Pithecia pithecia* and *Alouatta seniculus* (now *A. macconnelli*) were seen in edges over 15% of the time. *Saimiri sciureus* used the understory more than either *Sapajus apella* or *Cebus olivaceus*, and the two capuchins species were observed to spend more time in the middle- and upper-canopy levels (Mittermeier & van Roosmalen, 1981). In the 40 years since that study, there has been an explosion of Cebidae research, and multiple reports of variable habitat use have been published by many researchers across dozens of sites (e.g., Fedigan et al., 1985, 1988, 1998; Boinski, 1987c, 1989; Chapman, 1987, 1988; Norconk et al., 1996; Garber & Paciulli, 1997; Garber & Rehg, 1999; Fedigan & Jack, 2001, 2012; Boinski et al., 2003; Fragaszy et al., 2004a, 2013; MacKinnon, 2006; Stone, 2007a,b,c, 2014, 2018; Bezanson, 2009; Lynch Alfaro et al., 2011, 2012a,b, 2015; Emidio & Ferreira, 2012; De Oliveira et al., 2014; Flesher, 2015; Stone & Silva Jr., 2016; Ruivo et al., 2017; Beltrão-Mendes & Ferrari, 2019; Ruivo & Stone, 2020).

While no platyrrhine species can be classified as terrestrial, capuchins will routinely visit the ground to explore and forage, particularly in deciduous forest habitats during the dry season when visibility is greater. For example, in Santa Rosa National Park, Costa Rica, *C. capucinus imitator* visits the ground during the dry season to sift through leaf litter, break open old fallen seed pods and desiccated fruits for invertebrates/insect larvae, and to eat the fallen acorns of lowland oak trees (*Quercus oleoides*) (MacKinnon, 2006). In northern Argentina and dry forest areas of Brazil, *Sapajus* will spend time on or

Figure 10.4   Capuchin monkey (*Cebus imitator*), Guanacaste, Costa Rica.
(Photo credit: David M. Jensen; Creative Commons Attribution-Share Alike 3.0 Unported license)

near the ground while group members break open palm nuts with heavy rocks using large, terrestrial "anvil" rock surfaces in the landscape (Fragaszy et al., 2004b, 2013; Liu et al., 2009; Spagnoletti et al., 2011; Massaro et al., 2012; Verderane et al., 2013; Visalberghi et al., 2013; Eshchar et al., 2016).

Traditionally, capuchins and squirrel monkeys have been described as arboreal quadrupeds (Napier & Napier, 1967; Hershkovitz, 1977). However, more detailed examinations of capuchin locomotor behavior are providing a complex picture of their extensive positional repertoire (see Table 10.2). Thus far, positional behavior in white-faced capuchins has been specifically considered in 10 published studies (Oppenheimer, 1968; Fontaine, 1985, 1990; Gebo, 1992; Bergeson, 1996, 1998; Garber & Rehg, 1999; Bezanson, 2009, 2012; Bezanson & Morbeck, 2013). Fontaine (1985) reported on locomotion, posture, and tail use in five species on Barro Colorado Island, Panama, and captive groups housed at Monkey Jungle in Florida.

Fontaine's (1985) examination in Panama focused on *Alouatta*, *Cebus*, and *Ateles* and described *Cebus* as distinguishable from *Alouatta* and *Ateles* as a non-suspensory New World primate that emphasized quadrupedal climbing and leaping. Bergeson (1996, 1998) specifically addressed gap crossing and all categories of suspensory behavior. While white-faced capuchins engaged in a smaller proportion of hindlimb and tail-assisted suspensory postures (2.7% of total feeding and foraging samples) than *Alouatta* (12.7%) or *Ateles* (33.3%), Bergeson (1998) concluded that suspensory postures were important for foraging on all resource classes. Garber and Rehg (1999) and Bezanson (2009) found high proportions of tail-assisted suspensory postures in capuchins at La Suerte Biological Field Station in Costa Rica.

There appears to be some locomotor variation among capuchin species. *Sapajus apella* has been observed to engage in more quadrupedal locomotion and less leaping and climbing than does *Cebus capucinus* (Gebo, 1992;

Table 10.2 Proportion of Time Spent in Most Common Positional Modes (*Saimiri, Cebus,* and *Sapajus*)

Species References	*Saimiri boliviensis* Fontaine (1990)	*Saimiri oerstedii* Boinski (1989)	*C. capucinus imitator* Bergeson (1996)	*C. capucinus imitator* Garber and Rehg (1999)	*C. capucinus imitator* Bezanson (2009)	*Cebus olivaceus* Youlatos (1998)	*Cebus olivaceus* Wright (2005)	*Sapajus apella* Youlatos (1998)	*Sapajus apella* Wright (2005)
Locomotor Modes:									
Quadrupedal Walk	73.58	21.36	12.2	13.6	22.4	31.6	30.0	24.2	42.0
Quadrupedal Run	-	-	0.2	<1.0	-	-	11.0	-	3.0
Climb	4.26	1.08	3.0	22.5	2.8	39.5	11.0	39.4	11.0
Leap	20.63	1.82	0.1	1.3	1.5	10.5	22.0	11.3	17.0
Postural Modes:									
Stationary [a]	-	70.39	-	-	-	-	-	-	-
Sit/Squat/Recline	68.7	-	39.2	24.5	40.0	35.6	-	74.9	-
Quad. Stand/Crouch	21.86	-	25.1	12.6	20.2	34.0	-	11.3	-
Tail Only Suspension	-	-	2.7	3.7	2.0	17.2	-	9.5	-
Tail and Hindlimb Suspension	-	-	-	8.9	0.7	-	-	-	-
Hindlimb Suspension	0.78	-	-	-	-	-	-	-	-
Horizontal Tripod / Inverted Biped	-	-	7.0	12.5	2.8	-	-	-	-
Vertical Tripod	-	-	7.7	-	1.0	-	-	-	-

[a] Pools standing, sitting, and crouching.

Youlatos, 1998; Wright, 2005, 2007). *Saimiri* locomotion is heavily characterized by quadrupedal walking and running (Fleagle & Mittermeier, 1980; Fleagle et al., 1981; Boinski, 1989). Fleagle et al. (1981) found that 55% of observed locomotor bouts by squirrel monkeys were walking or running, 45% were leaping, and only 3% involved climbing. Squirrel monkeys are fast and agile, and are able to maneuver through tangled branches with great speed and dexterity (Figure 10.5). Their long non-prehensile tails are mainly used for balance while traveling and resting; they occasionally use suspensory postures, gripping with hands and back feet but usually only during feeding bouts when they are trying to reach insects and fruits. Overall, the positional behavior of *Saimiri* is highly flexible and reflects seasonal adaptability in foraging strategies (Boinski, 1989; Fontaine, 1990).

Capuchins have a prehensile tail which evolved independently from the larger-bodied Atelinae (spider, howler, woolly monkeys, and muriquis) (Rosenberger, 1983). Overall, the tail is used in approximately 35% of locomotor and postural behaviors (e.g., sitting or resting on a branch) and in 60+% of foraging and feeding observations (Bergeson, 1996, 1998; Bezanson, 2009, 2012). Suspension of the entire body mass by the tail is relatively less common when compared to atelines (Bezanson, 2012). Tail suspension is also more common in younger individuals. For example, in a study on the ontogeny of positional behavior, Bezanson (2012) found that tail suspension (when the tail was observed to bear equal or greater mass than any other limb) peaked in young juveniles (9.9% of all behaviors) and decreased in older individuals (older juveniles 5.7%, adults 2.6%). Capuchins often use an inverted bipedal suspensory position (horizontal tripod, see Table 10.2) when feeding on fruits, exploring terminal branches, or processing palm nuts near the tree trunk. Such a posture also allows capuchins to use their free hands to explore other nearby substrates, such as dead branches and clumps of lianas, as they search for invertebrates. With the sturdy support of their prehensile tail and two back feet, they can more efficiently exploit various substrates for embedded and hidden foods (Bergeson, 1996; Garber & Rehg, 1999; Bezanson, 2009, 2012). Independent infants and juveniles may hang for a few seconds by the tail alone, particularly during play bouts, but then soon drop down to a lower branch when it releases (Bezanson, 2012).

Diet and Activity Budgets

Most primates are omnivorous to varying degrees, and their diets include a combination of fruits, insects, and various reproductive parts of plants (e.g., flowers, petioles) (Sussman, 1978, 1991; Harding, 1981; Terborgh,

Figure 10.5 Squirrel monkey (*Saimiri* sp.).
(Photo credit: Kongkham6211; Creative Commons Attribution-Share Alike 4.0 International license)

1983; Chivers & Wood 1984; Cords, 1986; Chapman, 1987; Martin, 1990; Chapman et al., 2003; Di Fiore, 2003; Stone, 2007a; Bogart & Pruetz, 2011; Fleagle, 2013; Rothman et al., 2014). Generally speaking, with an increase in body size, the proportion of insect material in the diet decreases and proportion of plant foods increases (Clutton-Brock & Harvey, 1980; Kay & Covert, 1984; Janson & Boinski, 1992; Leigh, 1994; Lambert, 1998; Rosenberger et al., 2011). For most mammals, insects are easy to digest but require skillful proficiency in order to find and catch. Plant material such as leaves, on the other hand, can be harder to digest and lower in value nutritionally, yet this type of food is easier to locate and eat. Thus, we would expect primates that specialize on insects to be relatively small-bodied and to spend a great deal of time foraging and less time resting. Capuchins and squirrel monkeys conform to these expectations and spend >80% of their overall activity budgets traveling and feeding and <20% in social activities and resting (Freese & Oppenheimer, 1981; Terborgh, 1983; Robinson, 1986; Fragaszy, 1990a; Mitchell, 1990; Rose, 1994; Bergeson, 1996; Stevenson et al., 2000; MacKinnon, 2006; Matthews, 2009; Izar et al., 2012; De Oliveira et al., 2014).

Similarly, but on a smaller scale, daily activity budgets can be influenced and constrained by dietary strategies (see Table 10.3). Insect food comes in small, moveable "packets" of readily digestible protein and carbohydrates,

so primate species that depend on this food type have to alternate their energetic bouts of searching and capturing such food with rest periods (Altmann, 2009; Mosdossy et al., 2015; and see Melin et al., 2007 for effects of color vision on insect capture in capuchins). Usually the daily feeding patterns of capuchins and squirrel monkeys unfold in the following general way: Immediately after the group wakes up in the early morning, ripe fruit is consumed if there is some available near the sleeping trees. Insects can also be eaten during these early hours, but there is more of a shift to insect foraging once the group starts to travel away from the sleeping area. Foraging for insects continues into the late morning, with capuchins concentrating especially on embedded larvae. There might be short rest periods at various times in midday (anywhere from ~1000–1400 in the afternoon), although this depends on ambient temperature and food availability. If it is a relatively cool day and/or the monkeys haven't eaten enough, rest periods will be short or nonexistent as they continue to forage. Later in the afternoon there might be another bout of feeding on fruits somewhere else in the group's home range—or they may circle back and end up near the sleeping trees from the previous night (Presotto et al., 2018). Capuchins and squirrel monkeys often continue to travel and forage until dusk. More detailed descriptions of daily activity cycles in Cebidae species can be found in Baldwin and Baldwin

Table 10.3 Diet of adult *Saimiri*, *Cebus*, and *Sapajus*

Species	Fruit	Plant Parts	Seeds	Flowers	Invertebrates	Vertebrates	Other	References and Site
Saimiri sciureus	21-23%				79-97%			Mitchell, 1990; Boinski et al., 2002; Stone, 2007a (Eastern Amazonia)
Saimiri oerstedii	10%				90%			Mitchell, 1990; Boinski et al., 2002; Stone, 2007a (Eastern Amazonia)
Cebus capucinus	65%	15%			20%			Freese and Oppenheimer, 1981 (Barro Colorado Island, Panama)
Cebus capucinus	40.1%	0.9%	19.7%	2.5%	35.1%	0.9%		Rose, 2000 (Santa Rosa, CR)
Cebus capucinus	43.4%				49.2%	6.5%		Bergeson, 1996 (Santa Rosa, CR, and La Selva, CR)
Cebus capucinus	62.1%	1.1%	10.6%		21.3%	2.0%		MacKinnon, 2006 (Santa Rosa, CR)
Cebus kaapori	74.1%		10.2%		12.6% [a]	<1.0%	<3.0%	De Oliveira et al., 2009 (Goianésia do Pará, Pará State, Brazil)
Sapajus apella	53.9%	6.3%	16.0%	11.1%			15.4%	Galetti and Pedroni, 1994 (Santa Genebra Reserve, Campinas São Paolo, Brazil)

[a] Arthropods only.

(1971, 1972, 1981), Freese and Oppenheimer (1981), Terborgh (1983), Janson (1990a,b), Zimbler-DeLorenzo and Stone (2011), Jack (2011), and Fragaszy et al. (2004a). Seasonal patterns of rainfall, temperature, and daylight length, along with seasonality of births (in *Saimiri*) and changes in availability and quantity of food and water resources, can all influence foraging patterns and group behavior throughout the year (Robinson, 1984; Fedigan & Jack, 2001, 2012; Campos & Fedigan, 2009; Zimbler-DeLorenzo & Stone, 2011; Nowak & Lee, 2013). Even so, variations in the general foraging patterns of these frugivore-insectivore primates seem to result in only minor adjustments to cycles necessitated by small body size and dietary adaptations (Terborgh, 1983; Robinson, 1986; Janson & Boinski, 1992; Phillips, 1995; Stone, 2007a,c; Melin et al., 2014; Mosdossy et al., 2015; but see Campos & Fedigan 2009).

Diet and Foraging Behaviors

Capuchins are extremely active animals and spend most of their day foraging, traveling, and engaged in social behaviors. They are generally classified as either frugivore-insectivores or omnivorous feeders, and their diet typically consists of 60–80% fruit, 20–30% animal material (invertebrate and vertebrate), and 10% plant material—with much variation among species, study sites, and geographical regions (e.g., Oppenheimer, 1968; Freese, 1977; Chapman, 1987, 1988; Mitchell, 1989; Chapman & Fedigan, 1990; Fedigan, 1990; Janson & Boinski, 1992; Rose, 1994; Phillips, 1995; Panger et al., 2002; Fragaszy et al., 2004a; MacKinnon, 2006; Jack, 2011; McKinney, 2011; Lynch Alfaro et al., 2012a; de Oliveira et al., 2014).

Unlike squirrel monkeys, who quickly search substrates and then keep moving, capuchins can be meticulously diligent foragers, often spending over 30 minutes at one location depending on targeted food items, season, and overall food availability. They use a diversified set of foraging behaviors and seem to leave no substrate uninvestigated; they routinely tear off tree bark, break open branches, rummage through terrestrial and arboreal bromeliad stem bases, crumble apart large arboreal termite nests, investigate the dead bases of palm fronds, push around dead branches and sift through leaf litter while on the ground, explore tangled clumps of green vines and woody lianas, and put their fingers, hands, and entire arms into tree holes and crevices (Thorington, 1968; Freese & Oppenheimer, 1981; Robinson, 1986; Janson & Boinski, 1992; Panger et al., 2002; Rose et al., 2003; MacKinnon, 2006; Lynch Alfaro et al., 2012b; Canale et al., 2013; Nowak & Lee, 2013). Although they specialize in eating fruit and invertebrates, they are able to process a remarkably diverse array of food sources; in addition to leaves, flowers, petioles, seeds, and nuts, they also eat vertebrate prey such as birds, eggs, lizards, frogs, bats, squirrels, tree rats, mice, coati pups (Izawa, 1978; Newcomer & De Farcy, 1985; Fedigan, 1990; Galetti, 1990; Perry & Rose, 1994; Rose, 1994, 1997; Panger et al., 2002; Resende et al., 2003; Rose et al., 2003; Fragaszy et al., 2004a; MacKinnon, 2006; Milano & Cunha et al., 2006; Monteiro-Filho, 2009; Izar et al., 2012; Canale et al., 2013) and even, rarely, other primates (see Sampaio & Ferrari, 2005; Carretero-Pinzón et al., 2008). The combination of extractive foraging skills and an opportunistic feeding strategy allow capuchins to successfully utilize multiple habitats, such as extremely seasonal tropical dry forest or primary and secondary rainforest at varying elevations, and more marginal anthropogenic areas.

While *Saimiri* also consume fruits, they have a consistently higher proportion of insects in their diets compared to capuchins. There are seasonal fluctuations, but they generally spend upward of 80% of their foraging time looking for and ingesting insect prey (Mittermeier & van Roosmalan, 1981; Terborgh, 1983; Boinski, 1988a; Janson & Boinski, 1992; Zimbler-DeLorenzo & Stone, 2011). However, Pinheiro's et al. (2013) study reports a marked difference in food preferences, with plant material making up ~70% of the diet in two *Saimiri* social groups in eastern Amazonia. Unlike the heavy use of manipulative, extractive foraging techniques by capuchins, squirrel monkeys search substrates quickly and then move on; they usually spend only ~two to four minutes in any one tree while insect foraging (Thorington, 1968; Janson & Boinski, 1992; Zimbler-DeLorenzo & Stone, 2011, but see Pinheiro et al., 2013, for differing activity budgets and dietary profiles). Like capuchins, squirrel monkeys consume caterpillars, some of which are protected by irritating urdicating hairs and stinging spines. To get around these mechanical barriers, squirrel monkeys, like capuchins, employ creative methods of food processing, including wrapping the caterpillar in a leaf and vigorously rubbing it directly on a branch before consuming it (Boinski & Fragaszy, 1989; Panger et al., 2002; Pinheiro et al., 2018; Stone, 2018).

In terms of the fruit component of their diets, squirrel monkeys eat a diverse group of plant species that produce small berry-like fruits, such as wild figs (genus *Ficus*). Seasonal variation affects the amounts and types of fruits ingested at different localities (Zimbler-DeLorenzo & Stone, 2011; Pinheiro et al., 2013; Stone & Ruivo, 2020). For example, at one Costa Rican site, *S. oerstedi* ate fruit from 45 species of 20 plant families, and at a Peruvian site, *S. sciureus* ate over 150 species from 42 families (see Mitchell et al., 1991). When in season, figs are a preferred fruit for many animals, and at the Manú study site in Peru, figs accounted for >50% of fruit eating by primates in all months except one (Terborgh, 1983, 1986). In some habitats with distinct dry seasons, large trees bearing a lot of fruit are rare during these months, and the asynchronically fruiting *Ficus* trees can become an even more important food source for primates—sometimes making up as much as 77% of the diet and over 90% of the fruit species eaten. Such clustered food sources with large mastings of fruit influence day ranges and foraging behaviors so much that 80% of *Saimiri* food, for several days in a row, might come from only two to three large, widely scattered trees. During these times, *Saimiri* are often found in foraging associations with *Cebus* and *Sapajus* where they occur sympatrically (see maps, Figures 10.1, 10.2, and 10.3). Additionally, synchronization of weaning time with peak fruit availability has been documented in *S. collinsi* in eastern Amazonia (Stone & Ruivo, 2020).

When capuchins are found in marginal habitats and/or areas of marked seasonal fluctuations, they exhibit a high degree of behavioral plasticity in diet (Panger et al., 2002; McKinney, 2011; De Oliveira et al., 2014). For example, they can become almost entirely frugivorous or insectivorous seasonally (Chapman, 1987). In tropical dry forests, where fruits are scarce during the months without rainfall, they can survive almost entirely on invertebrate larvae, insect prey, and parts of epiphytic bromeliads and orchids (e.g., pith from stem bases, bromeliad fruits, etc.) (Brown & Zunino, 1990; MacKinnon, 2006), and/or they can become seed predators on genera like *Luehea* and *Slonea* (Galetti & Pedroni, 1994; O'Malley & Fedigan, 2005a) as well as on cashew nuts, genus *Anacardium* (Visalberghi et al., 2016), and larger palm nuts from *Attalea*, *Astrocarium*, and *Orbignya* species (Spagnoletti et al., 2012; for additional discussion, see Altmann 2009 for a review of omnivorous primates and fallback foods). This dietary flexibility, coupled with extended geographic distribution has allowed capuchins to be successful in a wide range of habitat types from extremely seasonal dry forest, to primary rainforest, to mosaic landscapes marked by anthropogenic disturbances, all at varying elevations.

Tool Use

Captive and field studies describe complex and sophisticated behaviors such as tool use, social transmission, complicated social networks, and the maintenance of behavioral traditions in *Cebus* and *Sapajus* (Fedigan, 1993; Perry et al., 2003a,b; Fragaszy et al., 2004a,b; O'Malley & Fedigan, 2005a,b; Ottoni & Izar, 2008; Lynch Alfaro et al., 2012b; MacKinnon & Fuentes, 2011; Falótico & Ottoni, 2013; Humle & Newton-Fisher, 2013; MacKinnon, 2013). Perhaps the capuchin species best known for tool use are *Sapajus apella* and *S. libidinosus* (see Ottoni & Izar, 2008; Mendes et al., 2015). Reports going back more than 30 years describe (1) how capuchins repeatedly pound palm nuts against a hard surface or pound on the nuts with selectively chosen rocks—often upon anvil surfaces used by many generations of monkeys (Izawa & Mizuno, 1977; Struhsaker & Leland, 1977; Anderson, 1990; Boinski et al., 2000; Ottoni & Mannu, 2001; Canale et al., 2009; Izar et al., 2012; Massaro et al., 2012; Spagnoletti et al., 2012; Fragaszy et al., 2013; Visalberghi et al., 2013; Mendes et al., 2015); (2) how marine oysters are opened by the use of stones and other oysters (shells), and how terrestrial crabs are extracted from their shells (Parker & Gibson, 1977; Fernandes, 1991; Westergaard & Suomi, 1995; Port-Carvalho et al., 2004); (3) and how sticks are used to clean wounds, as well as being incorporated into displays and threats against a potentially dangerous object (Cooper & Harlow, 1961; Westergaard & Fragaszy, 1987;

Richie & Fragaszy, 1988; Boinski, 1988b; Visalberghi, 1990).

Among the non-tufted capuchins, researchers have suggested that while wild *Cebus capucinus* are able to learn an association, they do not understand cause/effect relationships in object manipulation and, thus, cannot use tools (Visalberghi & Limongelli, 1994; Garber & Brown, 2004; and see Panger, 1998, for "object" vs. "tool" use definitions). However, at Santa Rosa and other sites in Costa Rica, *C. capucinus* pound, rub, and tap food sources against various substrates (Chevalier-Skolnikoff, 1990; Panger et al., 2002; O'Malley & Fedigan, 2005a,b; MacKinnon, 2006). They wrap some food items in leaves before rubbing them against a branch, especially those prey that have mechanical or chemical defenses, e.g., *Automeris* and *Ecpantheria* caterpillars and *Sloanea terniflora* fruit (O'Malley & Fedigan, 2005a). Capuchins will also bite off one end of a caterpillar and hold it away from themselves for a few seconds while the green innards drip out and then consume the casing of the body (O'Malley & Fedigan, 2005a,b). As many species of caterpillars consume host plants that are high in secondary compounds, this processing technique may reduce the bitter taste or even harmful effects of plant poisons. Capuchins, often in association with raptor species (e.g., roadside hawks, *Rupornis magnirostris*; double-toothed kites, *Harpagus bidentatus*), also follow army ants to catch flushed-out insects (Panger et al., 2002; Perry et al., 2003b; Fragaszy et al., 2004a). Such behaviors observed in wild study groups suggest that capuchins do indeed understand cause-and-effect relationships. Additionally, Barrett et al. (2018) documented the first habitual use of stone tools as extractive aids in foraging by *C. capucinus imitator* in Panama. This population on one island in Coiba National Park relies on hammerstone and anvil tool use to expose hidden and protected food items such as hermit crabs, marine snails, terrestrial crabs, embedded seeds, and other items (Barrett et al., 2018). These behaviors have persisted at this location since at least 2004 and expand our understanding of stone tool use potential in the non-tufted capuchins. Such findings require us to abandon the assumption that only the tufted capuchin species are capable of using stone tools when foraging. By contrast, there have been no observations of stone tool use by squirrel monkeys.

Predation

Birds of prey are the most common predators of New World monkeys, and the smaller species (especially callitrichids) are exposed to a greater amount of danger than larger ones (Di Fiore, 2002; Hart & Sussman, 2005; Ferrari, 2009; Barnett et al., 2011; Miller & Treves, 2011;

Barnett, 2015; and see McGraw & Berger, 2013, for review). Terborgh (1983) suggested that aerial predators constitute the only serious daytime threat to Neotropical arboreal monkeys, but there are many published reports over the past 35+ years since then of lethal attacks by other predators, particularly felids and boa constrictors (e.g., Chapman, 1986; Isbell, 1994, 2006, 2009; Hart & Sussman, 2005; Quintino & Bicca-Marques, 2013; and see Miller, 2002, and Miller & Treves, 2011, for review). Night sleeping sites may be selected to minimize the chances of surprise attacks by nocturnal predators. For example, DiBitetti et al. (2000) found that predation avoidance and general safety from falling are predominant factors driving sleeping site preferences for capuchin monkeys in Argentina's Iguazu National Park (see Colquhoun, 2006, for anti-predation cathemerality adaptations in lemurids and ceboids). However, capuchins show much variation across species and sites, and several other hypotheses (e.g., forest structure and demographic characteristics) have received support for sleeping site preferences (Zhang, 1995; Holmes et al., 2011; Smith et al., 2018).

Like most primates, squirrel monkeys give alarm calls when medium to large birds fly overhead, and there has been direct observation of raptor predation on these monkeys (e.g., Mitchell, 1990; Boinski et al., 2003; Stone, 2007b,c; Lopes et al., 2015). Juveniles, in particular, modify their behavior in certain forest strata and seasons to avoid predation (Stone, 2007c). In one study, groups were subject to a diurnal predator attack every six to seven days; further, five confirmed predations by raptors were observed in Costa Rica and four in Peru (Mitchell et al., 1991). Squirrel monkeys mob and vocalize at dogs, snakes, tayras, and cats (Freese & Oppenheimer, 1981; Westoll et al., 2003; Stone, 2007b,c), and in addition to raptors, are preyed upon by tayras (Galef et al., 1976), ocelots (Mitchell, 1990), and larger cats like jaguars (Emmons, 1987). Squirrel monkeys likely have increased protection from predators by forming mixed-species groups with capuchins and by living in their own large social groups (sometimes 40+ individuals), such that the chance of any one individual being taken is statistically reduced (Baldwin & Baldwin, 1971; Baldwin, 1985; Pinheiro et al., 2011; Levi et al., 2013).

Capuchins have a well-developed alarm call repertoire, with differing vocalizations that reflect location and perceived urgency for protection from aerial predators (e.g., raptors), terrestrial predators (e.g., cats), and snakes (Digweed et al., 2005, 2007; Fichtel et al., 2005; Wheeler, 2008, 2010; Coss et al., 2019). Adult males, particularly the alpha male, spend a great deal of time being vigilant and monitoring the environment; they seem to be more focused on detecting other neighboring groups or individuals in the area, but predation detection also increases concomitantly with the heightened alertness (Rose, 1994; Rose & Fedigan, 1995; Hirsch, 2002; Jack 2011). The alarm call of capuchins is a distinct raspy bark and is immediately responded to by conspecifics, as well as by other animals nearby (Fichtel et al., 2005; Wheeler, 2010; Coss et al., 2019).

Capuchins are larger than squirrel monkeys and are probably preyed upon by fewer raptor species, but independent infants and small juveniles are still vulnerable. Terborgh (1983) reported two unsuccessful attacks on capuchins by harpy eagles, and in a study in Guyana, *Sapajus* was the second most common prey in nest remains and by far the most common monkey species in these eagles' diets, others being *Alouatta*, *Pithecia*, and *Chiropotes* (Fowler & Cope, 1964; Rettig, 1978; Eason, 1989). An average of one capuchin was caught every 25 days, and there was no selection by age of the monkeys, suggesting these are opportunistic captures by the birds. Suscke et al. (2017) describe several encounters between different individual harpy eagles and a group of yellow-breasted capuchin monkeys (*Sapajus xanthosternos*) in the Atlantic Forest of Brazil. The interactions lasted an average of 58 minutes, and the capuchins used specific behavioral strategies against the harpy eagle that were not used in reaction to other aerial predators. Suscke et al. (2017) did not observe any successful predation events, but they note that after one of those encounters an infant capuchin disappeared from the group. Capuchins also give alarm calls when seeing jaguaroundis, jaguars, ocelots, margays, dogs, boa constrictors, rattlesnakes, large colubrid snakes (e.g., indigo snakes), king vultures, and caimans (Freese & Oppenheimer, 1981; Digweed et al., 2005; Fichtel et al., 2005; Wheeler, 2008, 2010). Chapman (1986) observed a capuchin being swallowed by a boa constrictor, and Tórrez et al. (2012) describe a fatal attack on a capuchin by a jaguar on Barro Colorado Island, Panama. While predation events are rare to see, there is a growing list of anecdotal reports from field researchers. For an overview of how predation has shaped nonhuman primate and human sociality, see Hart and Sussman (2005), and see Isbell (2009) for the evolution of primate vision vis-à-vis predator detection.

Capuchin and Squirrel Monkey Mixed-Species Associations

In areas where they are sympatric, polyspecific associations of capuchin and squirrel monkeys are quite common. These associations can last from a few hours to several days, and usually there are very low levels of agonism. In earlier discussions, researchers hypothesized that capuchins were the ones actively maintaining these associations and were the main benefactors (Baldwin & Baldwin, 1971; Klein & Klein, 1973; Freese, 1977). Terborgh (1983), however, found that *Saimiri* were attracted to and followed capuchins. For example, during slower traveling and foraging periods, squirrel monkeys were often out in front of the mixed group, but if they were not eventually followed by the capuchins, they circled back around. Capuchins always led the mixed-group movement to specific fruit trees, and if they stopped at a fruit source not normally used by *Saimiri* (e.g., for palm nuts), then the squirrel monkeys would continue on, independently foraging for a while, but returning frequently to monitor the capuchin group. Capuchins, however, have not been observed waiting for or checking on *Saimiri* in such a manner. Interestingly, in Manú, Peru, squirrel monkeys appear to actively search for capuchin groups, and *Saimiri* cover much more area (twice as far per hour) when traveling alone than when in mixed groups; the squirrel monkeys immediately travel toward and join a capuchin group when one is detected. The opposite effect was true for capuchins in the Manú study; when they were foraging with *Saimiri*, they traveled up to 40% more than when by themselves. Similarly, in a short-term study at Manú, Podolsky (1990) found that as *Saimiri* traveled through *Sapajus* home ranges, the capuchins sometimes led squirrel monkeys to large fruiting trees. He also found that feeding rates in large fruit trees were modified by the presence of the other species: positively for *Sapajus*, but negatively for *Saimiri*. More recently, Levi et al. (2013) found that interspecific interactions between two species of capuchins and one species of squirrel monkey in Guyana were as important as habitat and fruit availability in determining the distribution and abundance of these species. These studies suggest that the effects of interspecific associations on resource use differ for capuchins and squirrel monkeys and are related to factors such as group- and body-size differences, as well as the size of the food resources themselves.

Social Organization and Group Size

Most capuchin species live in multi-male/multi-female social groups of 4–30 individuals (Freese & Oppenheimer, 1981; Fragaszy et al., 2004a; Jack, 2011). *C. capucinus* has an average group size of 15–17 (Fedigan et al., 1985, 1996, 2004a; Chapman, 1990; Fragaszy, 1990a; Fedigan & Jack, 2012; Perry, 2012), and *S. apella* occur in smaller groups of about 12 individuals (Robinson & Janson, 1987; Jack, 2011; Lynch Alfaro et al., 2012a; but see Bezerra et al., 2014, for group estimates of up to 46 members for the critically endangered *Sapajus flavius*). There is usually an alpha male and female in the group with variable levels of intragroup agonism across species (Freese, 1978; Fedigan, 1993; Perry, 1997; Manson, 1999; Fragaszy et al., 2004a; Kalbitzer et al., 2017; see Jack, 2011; Jack & Fedigan, 2018).

Since 1983, Fedigan and colleagues have conducted over 10 censuses in Santa Rosa National Park, Costa Rica, to document population changes in three species of primates—*Cebus capucinus*, *Alouatta palliata*, and *Ateles geoffroyi*) (Fedigan & Jack, 2012; Kalbitzer & Jack, 2018). In 2007, 39 capuchin groups were counted (594 individuals); in 2003, 49 groups were counted (655 individuals); in 1999, 31 groups (521 individuals) were documented with an average group size of 17 animals; and in 1983, 20 groups were counted (226 individuals) (Fedigan & Jack, 2012). The capuchin numbers have continued to increase, most likely as a result of their behavioral plasticity, omnivorous diet, and ability to occupy early-regenerating habitats (compared to the other two primate species at this location). Among all of these censuses, the ratio of adult males in the *Cebus* groups increased, due to a male-biased birth sex ratio, and adult male immigration from surrounding areas into the protected Santa Rosa groups (Fedigan et al., 1996; Fedigan & Jack, 2001, 2004, 2012).

Capuchins are generally a male-transfer species but even after an individual leaves his natal group, occasional encounters may occur between the male and members of his former group. He may leave and return many times or may be followed by a younger juvenile to a neighboring group (Jack, 2003; Jack & Fedigan, 2004a,b,c; Jack et al., 2012; and see Jack & Fedigan, 2018, on alpha male capuchins as keystone individuals). Parallel dispersal occurs in *C. capucinus* and lasts through multiple migration events (Jack & Fedigan, 2004a,b). Females may occasionally transfer, although it is rare by comparison (Jack & Fedigan, 2009). Capuchin intergroup encounters are common in some areas, with variable outcomes (Perry, 1996; Fedigan & Jack, 2004; Crofoot, 2007, 2013; Crofoot et al., 2008, 2011; Crofoot & Wrangham, 2010; Crofoot & Gilby, 2012; Scarry, 2013, 2017; Schoof & Jack, 2013), and not all individuals in these groups are strangers to each other. Adult females may have dispersed offspring and siblings in neighboring groups, males may encounter mothers and siblings, and juveniles may encounter siblings. Adult males have been documented making "visits" to a neighboring group for short periods of time. Male reunion displays have been described, not only among males within a group who are separated for brief amounts of time (captive data for *S. apella* from Phillips et al., 1994; Matheson et al., 1996; Phillips & Shauver Goodchild, 2005; wild data for *C. capucinus* from Jack & Fedigan, 2006; Schoof & Jack, 2013; wild data for *S. apella* from Lynch Alfaro, 2007, 2008; wild data for *S. nigritus* from Scarry, 2013,

2017), but also among males in different groups during intergroup encounters (Schoof & Jack, 2013).

Saimiri live in multi-male/multi-female social groups with group size varying greatly; the number of animals per group appears to correlate with habitat condition and resource availability. Groups of 10–300+ animals have been reported, but those over 100 represent temporarily combined groups (Baldwin & Baldwin, 1981; Terborgh, 1983; Mitchell et al., 1991; Zimbler-DeLorenzo & Stone, 2011). Most commonly, groups average between 20–75 individuals (Robinson & Janson, 1987; Zimbler-DeLorenzo & Stone, 2011), which still represent the largest groupings of any Neotropical primate. Perhaps not surprisingly, groups with many individuals are more common in large, undisturbed forests, and smaller groups can be found in patchier, disturbed areas, or where squirrel monkeys are hunted (Baldwin & Baldwin, 1981; Zimbler-DeLorenzo & Stone, 2011).

When squirrel monkeys are foraging for insects, they can be widely dispersed, using vocalizations to maintain contact. During these times they are in constant motion, flushing out insects as they move through the canopy; small raptors and other birds often follow groups of foraging squirrel monkeys, swooping in to capturing disturbed insect prey (Thorington, 1968; Boinski & Scott, 1988; Terborgh, 1990; Zimbler-DeLorenzo & Stone, 2011). At other times of the day, particularly when feeding on a clumped food source like fruit, members of a squirrel monkey group can be in very close proximity to one another (Boinski, 1987c, 1989, 1994, 1996; Stone, 2007a).

Squirrel monkeys in different geographical areas display a remarkable flexibility in social organization and dispersal patterns (see Boinski, 1999; Boinski et al., 2005; Kauffman et al., 2005; Jack, 2011; Stone, 2017). For example, Terborgh (1983) and Mitchell (1990; Mitchell et al., 1991) found that in Peru, *S. sciureus* females are philopatric, have a dominance hierarchy, and form kin-based coalitions. Young males transfer out of their natal group at approximately four years of age, sometimes join a bachelor group, and then enter into another multi-male/multi-female social group. Adult males have a dominance hierarchy and their interactions with one another are agonistic. Males tend to transfer groups every one to two years. Adult females harass and peripheralize males and are dominant to them (Boinski et al., 2005).

Saimiri oerstedii act very differently in Costa Rica. In those groups, females have no clear dominance hierarchies, are much more tolerant of one another, and are the dispersing sex upon maturity (Boinski 1987a,b,c, 1999; Mitchell et al., 1991; Boinski & Mitchell, 1994; Boinski et al., 2005). Since males are philopatric in this region, growing up in their natal group they have close relationships with members of their birth cohort but not with the newer females or infants. Occasionally, such male cohorts might leave their group and work together to displace the adult males in another group. In contrast to the Peruvian populations, there is little agonism within groups, and males are highly vigilant for predators (Boinski, 1994; Boinski et al., 2003). During the short mating season, adult males cooperate in the mobbing of females, holding them down and sniffing their genitals (Boinski, 1988a, 1999). Strong bonds form between philopatric members of a social group, and this is likely the reason for many of the differences between the Peruvian and Costa Rican *Saimiri* (Boinski, 1994; Boinski & Mitchell, 1994; Boinski et al., 2002). Additionally, the fact that *S. sciureus* in the Peru study display female hierarchies and more agonism might be tied to resource distribution; food patches are defendable and of high quality (compared to the more even distribution at the Costa Rican sites), so there is more direct competition among group members (Mitchell et al., 1991; Boinski, 1996, 1999; Boinski et al., 2002).

In eastern Brazilian Amazonia, Stone (2014) found that "fatted" males spend significantly more time near females and less time alone than the unfatted males. Fatted males also spend less time foraging and more time engaged in sociosexual activities, suggesting a trade-off between general health maintenance and reproductive behaviors. This extra weight (up to 20% of their entire body weight) is actually retained water that settles in between the skin and muscle tissue as a result of circulating androgens. Stone (2014) also found that the 2-month mating season accounted for 62% of all male-male agonism observed over a 12-month period. However, unlike the Peruvian squirrel monkeys, males did not coerce females to mate, and females often rejected males. This suggests that male fattening in *Saimiri* is a product of both intra- and intersexual selection (Jack, 2011).

Social Behavior

Capuchins lead rich social lives and show higher levels of affiliative behaviors compared to agonistic interactions. In most studies of capuchins, relatively few aggressive interactions have been reported among group members when *stable hierarchies* are present (Fedigan, 1993; Hall & Fedigan, 1997; Perry, 1997, 1998, 2012; Fragaszy et al., 2004a; Perry et al., 2003a,b; Jack & Fedigan, 2004a,b,c, 2006; Fedigan & Bergstrom, 2010; Crofoot et al., 2011; Izar et al., 2012; Mendonca-Furtado et al., 2014). Dyadic and triadic interactions occur between and within the sexes. Male capuchins, in particular, show extensive variation in their relationships with each other, ranging from despotic to highly cooperative and affiliative (Perry, 1998, 2012; Lynch Alfaro, 2005; Jack, 2011;

Izar et al., 2012; Mendonca-Furtado et al., 2014). Extensive visual monitoring and communication (postural, vocal, tactile, visual), grooming, solicited assistance during conflicts, and long bouts of being in contact and/or proximity with others in the group are some of the daily components of capuchin repertoires (Gros Louis, 2002; Perry, 2012). As female capuchins are philopatric, they form close and long-lasting bonds with one another, and affiliative patterns are strongly influenced by matrilineal kinship (Fedigan, 1993; Izawa, 1994; Manson, 1999; Perry et al., 2003a,b, 2008; Izar et al., 2012; Perry, 2012; Verderane et al., 2013).

C. capucinus males in the same group have low rates of agonistic interactions and will support each other during intergroup conflicts. Dominance rank is based on complex social strategies including dyadic and triadic interactions, and females might actively solicit and mate with all males in a group throughout the year (Fedigan, 1993; Perry, 1997; Carnegie et al., 2005, 2006, 2011a,b). Similar social patterns exist in *C. albifrons* (Defler, 1979b; Janson, 1986; Matthews, 2009). Thus, for these two capuchin species the alpha male is not the sole (potentially) reproducing male in the group; however, in a review of 20+ years of *C. capucinus* life history data from Santa Rosa National Park in Costa Rica, Jack and Fedigan (2006) found that the rank of alpha did afford those males significantly higher reproductive success, at least during their tenure at the top (see Muniz et al., 2010, for similar findings at another site).

As noted earlier, male capuchins are quite vigilant and will scan the tree line for several minutes at a time from high up in the canopy. They seem to focus on neighboring groups and other males, rather than on potential predators (van Schaik & van Noordwijk, 1989; Rose, 1994; Rose & Fedigan, 1995; Hirsch, 2002; Jack, 2011). Fedigan (1993) suggests that males are more preoccupied with factors external to the group, while females focus more on internal group dynamics (females usually do not take part in intergroup interactions or in territory defense directly, but some will, on occasion, participate, and they have the scars to prove it). Coordinated male takeovers do occur, where incoming males challenge the resident alpha, and are sometimes associated with infant deaths; such dramatic shifts in the social group can affect subsequent conceptions and births (e.g., see Fedigan, 1993, for *C. capucinus*). The alpha male position is one that affords group centrality in all capuchin species. He is the subject of intense grooming and monitoring by other members, and he usually has a calm temperament. Infants and juveniles play on or near the alpha (and other males), and while there is usually not direct caregiving, there are a range of immature-adult male social interactions within a group that contribute to the survival of younger animals (e.g., see Pereira & Fairbanks, 2002; MacKinnon, 2011; Sherrow & MacKinnon, 2011, for review).

Interestingly, although they live in larger social groups, squirrel monkeys have lower rates of social interactions than capuchins (Baldwin & Baldwin, 1971, 1972; Boinski et al., 2002; Jack, 2011; Stone, 2017). They do not groom much, but do have a broad range of greeting and postural displays (e.g., sniffing, huddling, genital displays, urine-washing) used in affiliative and dominance interactions and during the mating season (Mitchell et al., 1991; Ruivo & Stone, 2014; Ruivo et al., 2017; and see Jack, 2011, and Zimbler-DeLorenzo & Stone, 2011, for reviews), as well as allomothering interactions and play behaviors (Williams et al., 1994; Stone, 2008).

Mating Systems and Reproduction

Capuchin mating patterns vary widely; some species are characterized by a promiscuous mating system where all adult females mate with all adult males in a group at varying times throughout the year, while in other species the alpha male does most or all of the mating (Janson, 1984; Robinson & Janson, 1987; Fedigan, 1993; Carosi et al., 2005; Jack & Fedigan, 2006; Fedigan et al., 2008; Izar et al., 2012; Mendonça-Furtado et al., 2014). For example, in *C. capucinus,* all adult males might mate with the adult females in a group over the course of a year, but as noted earlier, there is strong evidence for differential reproductive success with higher-ranking males siring more offspring (Jack & Fedigan, 2006), as well as reproductive seasonality in female capuchins (Carnegie et al., 2005, 2011a,b; Fedigan et al., 2008). Estrus is difficult to determine in wild *C. capucinus* based on behavioral observations alone as there are often post-conceptive, nonreproductive matings with the full array of sexual behaviors and cues (Manson et al., 1997; Carnegie et al., 2006; see Campbell, 2011, for a discussion on problems using the term "estrus"). This contrasts sharply with *C. olivaceus* and *S. apella*, where generally only the alpha male mates with the receptive females of a group (Janson, 1984; Phillips et al., 1994; Carosi et al., 2005; Lynch Alfaro, 2005). Noninvasive methods for collecting urine and fecal samples from known individuals in habituated study groups, along with newer hormone extraction techniques that can be done in the field, have resulted in an increasing number of studies that relate behavioral observations to hormonal profiles in wild living primates (Carnegie et al., 2006, 2011a,b).

Unlike the year-round mating in capuchins, the breeding and birth seasons in *Saimiri* are restricted to just a couple months (Baldwin & Baldwin, 1981; Boinski, 1987a,b; Zimbler-DeLorenzo & Stone, 2011; Stone,

2014). Adult males only produce sperm during the mating season, and in some populations take on a "fatted" appearance in the upper arms and shoulders as described earlier. Most females give birth every year, and Boinski (1987a,b) found that in *S. oerstedii* female ovulation was synchronized, and each female was receptive for only one or two days each mating season. This system is the most seasonally restricted reproductive cycle of any New World monkey and is similar to many prosimians. *S. sciureus* and *S. collinsii* females, on the other hand, do not have such restricted synchrony and generally give birth every other year (Mitchell, 1990; Zimbler-DeLorenzo & Stone, 2011; Stone, 2014; Stone & Ruivo, 2020).

Development and Life History

Compared to other Neotropical monkeys, and controlling for body size, capuchins have a long period of development and overall life span (Fleagle & Samonds, 1975; Mitchell, 1989; Fragaszy & Adams-Curtis, 1997; Fragaszy & Bard, 1997; Janson & van Schaik, 2002; Fragaszy et al., 2004a, 2016; MacKinnon, 2011, 2013). Although life span in wild individuals is usually unknown from most field studies (but see Fedigan & Jack, 2012), capuchins in captivity have lived over 40 years (Fragaszy & Bard, 1997; Fragaszy et al., 2004a). *C. capucinus* females first give birth around age 6 or 7 (Fedigan & Rose, 1995; Carnegie et al., 2005, 2011a; Fedigan & Jack, 2012; Perry, 2012), while males at ages 7–10 years are still considered subadult but are on the threshold of adult status and are already engaged in sexual mountings with adult females (Jack et al., 2012). Generally, all capuchin species have a gestation length of approximately 160 days, which is considered average for primates of their weight (Harvey & Clutton-Brock, 1985; Fragaszy et al., 2004a); twinning is very rare. At Santa Rosa National Park in Costa Rica, the average interbirth interval for *C. capucinus* females is 2.25 years when the previous infant survives but only 1.05 years if the previous infant dies younger than 12 months of age (Fedigan & Rose, 1995; Fedigan et al., 1996; Carnegie et al., 2005, 2011a,b; Fedigan et al., 2008).

Infant capuchins crawl around on their mothers during the first week of life and might briefly crawl off her for up to 5 minutes at 1–2 months old (Fragaszy, 1990b; MacKinnon, 2002). By the third month, infants make longer forays off their mothers, travel greater distances from her, and crawl onto other individuals in the social group. All group members approach, stare at, sniff, touch, and inspect the infant, but certain individuals are allowed longer and closer contact, depending on their relationship with the infant's mother (Fragaszy, 1990b; Fragaszy et al., 1991; MacKinnon, 2002, 2011). Once an infant is continually crawling off its mother for short bouts at 8–10 weeks of age, its social world changes. It can decide whom to approach and when to leave, in addition to being the recipient of invitations for play and other affiliative contacts (Byrne & Suomi, 1995; MacKinnon, 2011). From this point on, the young capuchin lives in an increasingly complex social environment. From 4 months on, capuchin infants show a dramatic increase in locomotor independence and travel on their own for longer periods of time (MacKinnon, 2002; Bezanson, 2009, 2012). Youngsters might still be carried during rapid group movement, or across substantial gaps between trees and in areas that are difficult to traverse. When they are carried during this stage, it is usually by other adults and older juveniles. Weaning is gradual and generally occurs between 12–18 months of age (Fragaszy et al., 1991). As juveniles, capuchins spend more time foraging, play is common (Smith, 1978; Fontaine, 1994; and Palagi, 2011, 2018, for functions of play in primates), strong bonds among peers develop, and preferential relationships with certain older animals are formed. Older capuchin juveniles continue to have high rates of play behaviors (particularly "chase play" and rough-and-tumble "wrestle play") but might also show an increase in conflict with adult members. Older juvenile females engage in allomothering, while older juvenile males might take longer forays away from the central core of the social group, preparing to eventually transfer (Sherrow & MacKinnon, 2011).

Saimiri show a faster rate of development in comparison to capuchins, although they still have a relatively slow life history for a small-bodied primate (Ross, 1991). Generally, females start to give birth around 2.5 years of age, and gestation length is around 150 days on average (Lorenz et al., 1973; Kerber et al., 1977; Zimbler-DeLorenzo & Stone, 2011). Age at first reproduction also varies among species and between study sites. For example, wild female S. boliviensis have the latest age on average, at 3.5 years (Mitchell, 1990), while males start to breed even later, around 5–6 years of age (Scollay, 1980; Mitchell, 1990). Contrast this with wild female S. oerstedii who first breed at 2.5 years of age (Boinski, 1986) and males around 4 years (Robinson & Janson, 1987). Further differences are found in wild male S. sciureus first breeding at 4.5 years (Robinson & Janson, 1987), and females reproducing a bit earlier in captivity, between 3.5–4 years (Taub, 1980).

Infants leave their mothers to explore at ~5–7 weeks of age, become more independent at 2–4 months, and are usually weaned around 6 months of age (Baldwin & Baldwin, 1981; Fragaszy et al., 1991; Zimbler-DeLorenzo & Stone, 2011), but again we see variation in these benchmarks (Mitchell, 1990; Stone, 2007b,c, 2008; Zimbler-DeLorenzo & Stone, 2011). For example, in S. sciureus

infants are weaned around 6 months (Stone, 2004), but there is high variability between species (and between captive and field studies). In field studies, wild S. oerstedii are weaned at 4 months (Boinski & Fragaszy, 1989); however, S. boliviensis infants continue to nurse for up to 18 months (Mitchell, 1990). While the infant stage of development is somewhat rapid compared to capuchins, the juvenile period is marked by an extended period of relatively slow growth for several years (Ross, 1991; Stone, 2007b). Males are fully adult at 4–5.5 years, and females at approximately 2.5 years (Baldwin & Baldwin, 1981; Zimbler-DeLorenzo & Stone, 2011). Wild S. oerstedii breed every year (Boinski, 1987a) but wild S. boliviensis typically breed every 2 years (Mitchell, 1990). S. sciureus females may give birth yearly, although not all females in a social group do so every year (Stone, 2004). Observed maximum life span in captivity is 35 years for both males and females, although the average for wild individuals is around 16 years (Zimbler-DeLorenzo & Dobson, 2011). It should be noted that while squirrel monkeys have been used extensively in captive research, there is much variation in socioecology and life history traits between and among Saimiri species, as well as between wild populations of the same species studied at different locales (see Zimbler-DeLorenzo & Stone, 2011).

CONSERVATION

The biggest threats to populations of capuchin and squirrel monkeys are habitat loss and fragmentation (Schwarzkopf & Rylands, 1989; Gilbert, 2003; Boyle, 2008; Boyle & Smith, 2010; Estrada et al., 2012; Boyle et al., 2013). Agricultural practices, logging, infrastructure development (e.g., roads), and flooding for hydroelectric power generation are some of the main reasons for the increasing loss of intact forests where these primates live (Estrada et al., 2017, 2018, 2019). The development of agribusinesses for banana, pineapple, palm oil, coconut, and teak plantations are drivers of extensive habitat destruction and fragmentation in Central and South America (McKinney, 2011; Estrada et al., 2012). Even though small-bodied, these primates need healthy forests to thrive (e.g., see Paim et al., 2019, regarding Saimiri vanzolinii in Amazonia). For example, the minimum contiguous habitat required to sustain a group of tufted capuchins is estimated at around 100 ha, but the ideal area is likely 1,000 ha or even up to 23,000 ha, depending on location, forest type, and proximity to high levels of human activity (Gilbert & Setz, 2001; Spironello, 2001; Canale et al., 2013). The hunting of capuchin and squirrel monkeys for food, and human-wildlife conflicts when they are viewed as crop pests, are also threats to populations of some species (Hill, 2017, 2018; see da Silva et al., 2016, for the critically

Table 10.4 IUCN Red List Conservation Status of Cebidae Species/Subspecies

Cebid Taxa	IUCN Red List Status [a]
Saimiri	
Saimiri oerstedii - Central American squirrel monkey	Vulnerable
Saimiri oerstedii oerstdeii - Black-crowned Central American squirrel monkey	Endangered
Saimiri oerstedii citrinellus - Grey-crowned Central American squirrel monkey	Endangered
Saimiri boliviensis - Bolivian squirrel monkey	Least Concern
Saimiri boliviensis boliviensis - Bolivian squirrel monkey	Least Concern
Saimiri boliviensis peruviensis - Peruvian squirrel monkey	Least Concern
Saimiri vanzolinii - Vanzolini's squirrel monkey	Vulnerable
Saimiri sciureus - Common squirrel monkey	Least Concern
Saimiri sciureus albigena - Colombian squirrel monkey	Near Threatened
Saimiri sciureus cassiquiarensis - Humboldt's squirrel monkey	Least Concern
Saimiri sciureus macrodon - Ecuadorian squirrel monkey	Least Concern
Saimiri sciureus sciureus - Guianan squirrel monkey	Least Concern
Saimiri sciureus ustus - Golden-backed squirrel monkey	Near Threatened
Cebus	
Cebus capucinus - White-faced/white headed capuchin	Least Concern
Cebus capucinus capucinus - Colombian white-faced/white headed capuchin	Least Concern
Cebus capucinus curtus - Gorgona white-faced/white headed capuchin	Vulnerable
Cebus capucinus imitator - Panamanian white-faced/white headed capuchin	Least Concern
Cebus capucinus limitaneus - Honduran white-faced/white headed capuchin	Least Concern
Cebus albifrons - Humboldt's white-fronted capuchin	Least Concern
Cebus albifrons aequatorialis - Ecuadorian white-fronted capuchin	Critically Endangered
Cebus albifrons albifrons - White-fronted capuchin	Least Concern
Cebus albifrons cesarae - Rio Cesar white-fronted capuchin	Data Deficient
Cebus albifrons cuscinus - Shock-headed capuchin	Near Threatened
Cebus albifrons malitiosis - Santa Marta white-fronted capuchin	Endangered
Cebus albifrons trinitatus - Trinidad white-fronted capuchin	Critically Endangered
Cebus albifrons versicolor - Varied white-fronted capuchin	Endangered
Cebus albifrons leucocephalus - Sierra de Perijá white-fronted capuchin	Vulnerable
Cebus olivaceus - Weeper capuchin	Least Concern
Cebus olivaceus apiculatus - Hershkovitz's weeper/wedge-capped Capuchin	Least Concern
Cebus olivaceus brunneus - Brown weeper capuchin	Least Concern
Cebus olivaceus nigrivittatus - Gray weeper/wedge-capped capuchin	Least Concern
Cebus olivaceus olivaceus - Guianan weeper/wedge-capped capuchin	Least Concern
Cebus olivaceus castaneus - Chestnut capuchin	Least Concern
Cebus kaapori - Ka'apor capuchin	Critically Endangered
Sapajus	
Sapajus apella - Brown capuchin/black-capped capuchin	Least Concern
Sapajus apella apella - Guianan brown capuchin	Least Concern
Sapajus apella margaritae - Margarita Island capuchin	Critically Endangered
Sapajus macrocephalus - Large-headed capuchin	Least Concern
Sapajus libidinosus - Bearded capuchin	Near Threatened
Sapajus cay - Azara's capuchin/ hooded capuchin	Least Concern
Sapajus xanthosternos - Buff-headed capuchin/yellow-breasted capuchin	Critically Endangered
Sapajus nigritus - Black-horned capuchin	Critically Endangered
Sapajus flavius - Blond capuchin	Critically Endangered
Sapajus robustus - Crested capuchin/robust tufted capuchin	Endangered

[a] IUCN (2020).

endangered *Sapajus xanthosternos*; Gonzalez-Kirchner & Sainz de la Maza, 1998; Fragaszy et al., 2004a; Estrada, 2006; McKinney, 2011; Spagnoletti et al., 2017).

Generally speaking, if a species/subspecies is listed under the International Union for the Conservation of Nature (IUCN) Red List in the Least Concern category, it has a widespread distribution, there are few recognized major threats that might drive a significant population decline (at the present), and it displays an adaptability to a wide variety of habitat types. Table 10.4 lists capuchin and squirrel monkey species/subspecies that are at the least risk in terms of conservation status. For now, at least, these populations are surviving in the wild; however, there are many capuchins and squirrel monkeys (especially following more recent taxonomic reorganizations) that need more immediate conservation efforts and are listed by IUCN as Vulnerable, Endangered, and Critically Endangered (IUCN, 2020). For example, *Cebus kaapori* is one of the most threatened of all the Amazonian primates and has been recently recognized as one of the most threatened primates in the world (Ferrari & Queiroz, 1994; Lopes & Ferrari, 1996; Mittermeier et al., 2013).

With continuing (and increasing) habitat loss due to human activity and the concomitant anthropogenic effects of accelerated climate change, infectious disease outbreaks, and an ever-increasing demand for resources due to human overpopulation, approximately 75% of nonhuman primates are now in decline, and >60% of species are threatened with extinction (Estrada et al., 2017, 2018, 2019). Estrada et al. (2019) argue for a "greening" of trade practices; a reduced consumption of products using oil seed (e.g., palm oil, among others); less use of tropical timber, fossil fuels, metals, minerals, etc., from tropical habitats; and a global shift away from a reliance on meat products from the large-scale, industrialized factory farm complexes. The negative impacts of severe income inequality, and unsustainable worldwide demands for products and lifeways that benefit the more affluent, lie at the foundation of problems that need to be changed so that we have a chance to save some of the world's most enigmatic and ecologically important primate species, including cebids, from extinction.

SUMMARY

Capuchins and squirrel monkeys are a well-known and well-studied group of New World primates. Found throughout the tropical and subtropical ecosystems of Central and South America, these highly social, very active arboreal monkeys have successfully adapted to myriad habitats.

Taxonomic changes have recently split the genus *Cebus* (capuchins) into two recognized genera: *Cebus* and *Sapajus*, as well as adding to the number of species included in the genus *Saimiri* (squirrel monkeys). Highlighting both differences and commonalities of the three cebid genera underscores the adaptive strategies of this charismatic group of monkeys and their remarkable ecological and behavioral variations.

REFERENCES CITED—CHAPTER 10

Altmann, S. A. (2009). Fallback foods, eclectic omnivores, and the packaging problem. *American Journal of Physical Anthropology, 140*, 615-629.

Anderson, J. R. (1990). Use of objects as hammers to open nuts by capuchin monkeys. *Folia Primatologica, 54*, 138-145.

Ayres, J. M. (1985). On a new species of squirrel monkey, genus *Saimiri,* from Brazilian Amazonia (Primates, Cebidae). *Papéis Avulsos de Zoologia, 36*, 147-164.

Baldwin, J. D. (1985). The behavior of squirrel monkeys (*Saimiri*) in natural environments. In L. A. Rosenblum, & C. L. Coe (Eds.), *Handbook of squirrel monkey research* (pp. 35-53). Springer.

Baldwin, J. D., & Baldwin, J. I. (1971). Squirrel monkeys (*Saimiri*) in natural habitats in Panama, Colombia, Brazil, and Peru. *Primates, 12*, 45-61.

Baldwin, J. D., & Baldwin, J. I. (1972). The ecology and behavior of squirrel monkeys (*Saimiri oerstedi*) in a natural forest in western Panama. *Folia Primatologica, 18*, 161-184.

Baldwin, J. D., & Baldwin, J. I. (1981). The squirrel monkeys, genus *Saimiri*. In A. F. Coimbra-Filho, & R. A. Mittermeier (Eds.), *Ecology and behavior of Neotropical primates* (Vol. 1, pp. 277-330). Academia Brasileira de Ciencias.

Barnett, A. A. (2015). Primate predation by black hawk-eagle (*Spizaetus tyrannus*) in Brazilian Amazonia. *Journal of Raptor Research, 49*(1), 105-107.

Barnett, A. A., Boyle, S. A., Norconk, M. M., Palminteri, S., Santos, R. R., Veiga, L. M., Alvim, T. H.G., Bowler, M., Chism, J., Di Fiore, A., Fernandez-Duque, E., Guimarães, A. C. P., Harrison-Levine, A., Haugaasen, T., Lehman, S., MacKinnon, K. C., De Melo, F. R., Moreira, L. S., Moura, V. S.,... & Ferrari, S. F. (2012). Terrestrial activity in pitheciines (*Cacajao, Chiropotes,* and *Pithecia*). *American Journal of Primatology, 74*, 1106-1127.

Barnett, A. A., Deveny, A., Valsko, J., Spironello, W., & Ross, C. (2011). Predation on *Cacajao m. ouakary* and *Cebus albifrons* (Primates: Platyrrhini) by harpy eagles. *Mammalia, 75*, 169-172.

Barrett, B. J., Monteza-Moreno, C. M., Dogandžic, T., Zwyns, N., Ibáñez, A., & Crofoot, M. C. (2018). Habitual stone-tool-aided extractive foraging in white-faced capuchins, *Cebus capucinus. Royal Society Open Science, 5*, 181002.

Beltrão-Mendes, R., & Ferrari, S. F. (2019). Mangrove forests as a key habitat for the conservation of the critically endangered yellow-breasted capuchin, *Sapajus xanthosternos*, in the

Brazilian northeast. In K. Nowak, A. A. Barnett, & I. Matsuda (Eds.), *Primates in flooded forests: Ecology and conservation* (pp. 68-76). Cambridge University Press.

Bergeson, D. J. (1996). *The positional behavior and prehensile tail use of* Alouatta palliata, Ateles geoffroyi, *and* Cebus capucinus. [Doctoral dissertation, Washington University].

Bergeson, D. J. (1998). Patterns of suspensory feeding in *Alouatta palliata, Ateles geoffroyi*, and *Cebus capucinus*. In E. Strasser, J. Fleagle, A. Rosenberger, & H. McHenry (Eds.), *Primate locomotion: Recent advances* (pp. 45-60). Plenum Press.

Bezanson, M. (2006). Leap, bridge or ride? Ontogenetic influences on gap crossing in *Cebus* and *Alouatta*. In A. Estrada, P. A. Garber, M. Pavelka, & L. Luecke (Eds.), *New perspectives in the study of Mesoamerican primates* (pp. 333-348). Springer.

Bezanson, M. (2009). Life history and locomotion in *Cebus capucinus* and *Alouatta palliata*. *American Journal of Physical Anthropology*, 140, 508-517.

Bezanson, M. (2012). The ontogeny of prehensile-tail use in *Cebus capucinus* and *Alouatta palliata*. *American Journal of Primatology*, 74, 770-782.

Bezanson, M., & Morbeck, M. E. (2013). Future adults or old children? Integrating life history frameworks for understanding primate locomotor patterns. In K. Clancy, K. Hinde, & J. Rutherford (Eds.), *Building babies: Primate development in proximate and ultimate perspective* (pp. 435-458). Springer.

Bezerra, B. M., Bastos, M., Souto, A., Keasey, M. P., Eason, P., Schiel, N., Jones, G. (2014). Camera trap observations of nonhabituated critically endangered wild blonde capuchins, *Sapajus flavius* (formerly *Cebus flavius*). *International Journal of Primatology*, 35, 895-907.

Bogart, S. L., & Pruetz, J. D. (2011). Insectivory of savanna chimpanzees (*Pan troglodytes versus*) at Fongoli, Senegal. *American Journal of Physical Anthropology*, 145, 11-20.

Boinski, S. (1986). *The ecology of squirrel monkeys in Costa Rica*. [Doctoral dissertation, University of Texas at Austin].

Boinski, S. (1987a). Mating patterns in squirrel monkeys (*Saimiri oerstedii*): Implications for seasonal sexual dimorphism. *Behavior Ecology and Sociobiology*, 21, 13-21.

Boinski, S. (1987b). Birth synchrony in squirrel monkeys (*Saimiri oerstedii*): A strategy to reduce neonatal predation. *Behavior Ecology and Sociobiology*, 21, 393-400.

Boinski, S. (1987c). Habitat use by squirrel monkeys (*Saimiri oerstedii*) in Costa Rica. *Folia Primatologica*, 49, 151-167.

Boinski, S. (1988a). Sex differences in the foraging behavior of squirrel monkeys: Ecological implications. *Behavior Ecology and Sociobiology*, 21, 171-186.

Boinski, S. (1988b). Use of a club by a wild white-faced capuchin (*Cebus capucinus*) to attack a venomous snake (*Bothrops asper*). *American Journal of Primatology*, 14, 177-179.

Boinski, S. (1989). The positional behavior and substrate use of squirrel monkeys: Ecological implications. *Journal of Human Evolution*, 18, 659-677.

Boinski, S. (1994). Affiliative patterns among male Costa Rican squirrel monkeys. *Behaviour*, 130, 191-209.

Boinski, S. (1996). Vocal coordination of troop movement in squirrel monkeys (*Saimiri oerstedi* and *S. sciureus*) and white-faced capuchins (*Cebus capucinus*). In M. A. Norconk, A. L. Rosenberger, & P. A. Garber (Eds.), *Adaptive radiations of Neotropical primates* (pp. 251-269). Springer.

Boinski, S. (1999). The social organizations of squirrel monkeys: Implications for ecological models of social evolution. *Evolutionary Anthropology*, 8(3), 101-112.

Boinski, S., & Cropp, S. (1999). Disparate data sets resolve squirrel monkey (*Saimiri*) taxonomy: Implications for behavioral ecology and biomedical usage. *International Journal of Primatology*, 20, 237-256.

Boinski, S., Ehmke, E., Kauffman, L., Schet, S., & Vreedzaam, A. (2005). Dispersal patterns among three species of squirrel monkeys (*Saimiri oerstedii, S. boliviensis* and *S. sciureus*): II. Within-species and local variation. *Behaviour*, 142(5), 633-677.

Boinski, S., & Fragszy, D. M. (1989). The ontogeny of foraging in squirrel monkeys, *Saimiri oerstedii*. *Animal Behaviour*, 37(3), 415-428.

Boinski, S., Jack, K., Lamarsh, C., & Coltrane, J. (1998). Squirrel monkeys in Costa Rica: Drifting to extinction. *Oryx*, 32(1), 45-58.

Boinski, S., Kauffman, L., Westoll, A., Stickler, C. M., Cropp, S., & Ehmke, E. (2003). Are vigilance, risk from avian predators and group size consequences of habitat structure? A comparison of three species of squirrel monkey (*Saimiri oerstedii, S. boliviensis*, and *S. sciureus*). *Behaviour*, 140(11-12), 1421-1467.

Boinski, S., & Mitchell, C. L. (1994). Male residence and association patterns in Costa Rican squirrel monkeys (*Saimiri oerstedi*). *American Journal of Primatology*, 34, 157-169.

Boinski, S., Quatrone. R. P., & Swartz, H. (2000). Substrate and tool use by brown capuchins in Suriname: Ecological contexts and cognitive bases. *American Anthropologist*, 102, 741-761.

Boinski, S., & Scott, P. E. (1988). Association of birds with monkeys in Costa Rica. *Biotropica*, 20(2), 136-143.

Boinski, S., Sughrue, K., Lara, S., Quatrone, R., Henry, M., & Cropp, S. (2002). An expanded test of the ecological model of primate social evolution: Competitive regimes and female bonding in three species of squirrel monkeys (*Saimiri oerstedii, S. boliviensis*, and *S. sciureus*). *Behaviour*, 139, 227-261.

Boubli, J. P., Rylands, A. B., Farias, I., Alfaro, M. E., & Lynch Alfaro, J. W. (2012). *Cebus* phylogenetic relationships: A preliminary reassessment of the diversity of the untufted capuchin monkeys. *American Journal of Primatology*, 74(4), 381-393.

Boyle, S. A. (2008). Human impacts on primate conservation in central Amazonia. *Tropical Conservation Science*, 1, 6-17.

Boyle, S. A., Lenz, B. B., Gilbert, K. A., Spionello, W. R., Gomez, M. S., Sertz, E. Z. F., Marajo dos Reis, A., da Silva,

O. F., Keuroghlian, A., & Pinto, F. (2013). Primates of the Biological Dynamics of Forest Fragments project: A history. In L. Marsh, & C. Chapman (Eds.), *Primates in fragments. Developments in primatology: Progress and prospects* (pp. 57-74). Springer.

Boyle, S. A., & Smith, A. T. (2010). Can landscape and species characteristics predict primate presence in forest fragments in the Brazilian Amazon? *Biological Conservation, 143,* 1134-1143.

Brown, A. D., & Zunino, G. E. (1990). Dietary variability in *Cebus apella* in extreme habitats: Evidence for adaptability. *Folia Primatologica, 54,* 187-195.

Byrne, G., & Suomi, S. J. (1995). Development of activity patterns, social interactions, and exploratory behavior in infant tufted capuchins (*Cebus apella*). *American Journal of Primatology, 35,* 255-270.

Campbell, C. J. (Ed.). (2008). *Spider monkeys: Behavior, ecology, and evolution of the genus* Ateles. Cambridge University Press.

Campbell, C. J. (2011). Primate sexuality and reproduction. In C. J. Campbell, A. F. Fuentes, K. C. MacKinnon, R. Stumpf, & S. Bearder (Eds.), *Primates in perspective* (2nd Ed., pp. 464- 475). Oxford University Press.

Campos, F. A., & Fedigan, L. M. (2009). Behavioral adaptations to heat stress and water scarcity in white-faced capuchins (*Cebus capucinus*) in Santa Rosa National Park, Costa Rica. *American Journal of Physical Anthropology, 138,* 101-111.

Campos, F. A., & Jack, K. M. (2013). A potential distribution model and conservation plan to the critically endangered Ecuadorian capuchin, *Cebus albifrons aequatorialis. International Journal of Primatology, 34,* 899-916.

Canale, G. R., Guidorizzi, C. E., Kierulff, M. C. M., & Gatto, C. A. F. R. (2009). First record of tool use by wild populations of the yellow-breasted capuchin monkey (*Cebus xanthosternos*) and new records for the bearded capuchin (*Cebus libidinosus*). *American Journal of Primatology, 71,* 366-372.

Canale, G. R., Kierulff, M. C. M., & Chivers, D. J. (2013). A critically endangered capuchin monkey (*Sapajus xanthosternos*) living in a highly fragmented hotspot. In L. Marsh, & C. Chapman (Eds.), *Primates in fragments. Developments in primatology: Progress and prospects* (pp. 299-311). Springer.

Carnegie, S. D., Fedigan, L. M., & Melin, A. D. (2011a). Reproductive seasonality in female capuchins (*Cebus capucinus*) in Santa Rosa (Area de Conservación Guanacaste), Costa Rica. *International Journal of Primatology, 32,* 1076-1090.

Carnegie, S. D., Fedigan, L. M., & Ziegler, T. E. (2005). Behavioral indicators of ovarian phase in white-faced capuchins (*Cebus capucinus*). *American Journal of Primatology, 67,* 51-68.

Carnegie, S. D., Fedigan, L. M., & Ziegler, T. E. (2006). Post-conceptive mating in white-faced capuchins, *Cebus capucinus*: Hormonal and sociosexual patterns of cycling, noncycling, and pregnant females. In A. Estrada, P. A. Garber, M. S. M. Pavelka, & L. Luecke (Eds.), *New perspectives in the study of Mesoamerican primates. Developments in primatology: Progress and prospects* (pp. 387-409). Springer.

Carnegie, S. D., Fedigan, L. M., & Ziegler, T. E. (2011b). Social and environmental factors affecting fecal glucocorticoids in wild, female white-faced capuchins (*Cebus capucinus*). *American Journal of Primatology, 73,* 861-869.

Carneiro, J., da Silva Rodrigues-Filho, L.F., Schneider, H., & Sampaio, I. (2016). Molecular data highlight hybridization in squirrel monkeys (*Saimiri*, Cebidae). *Genetics and Molecular Biology, 39,* 539-546.

Carosi, M., Linn, G. S., & Visalberghi, E. (2005). The sexual behavior and breeding system of tufted capuchin monkeys (*Cebus apella*). *Advances in the Study of Behavior, 35,* 105-149.

Carretero-Pinzón, X., Defler, T. R., & Ferrari, S. F. (2008). Observation of black-capped capuchins (*Cebus apella*) feeding on an owl monkey (*Aotus brumbacki*) in the Colombian llanos. *Neotropical Primates, 15(2),* 62-63.

Carvalho, O., Jr., de Pinto, A. C. B., & Galetti, M. (1999). New observations on *Cebus kaapori* Queiroz, 1992, in eastern Brazilian Amazonia. *Neotropical Primates, 7,* 41-43.

Chapman, C. A. (1986). Boa constrictor predation and group response in white-faced *Cebus* monkeys. *Biotropica, 18(2),* 171-172.

Chapman, C. A. (1987). Flexibility in diets of three species of Costa Rican primates. *Folia Primatologica, 29,* 90-105.

Chapman, C. A. (1988). Patterns of foraging and range use by three species of Neotropical primates. *Primates, 29,* 177-194.

Chapman, C. A. (1990). Ecological constraints on group size in three species of Neotropical primates. *Folia Primatologica, 55,* 1-9.

Chapman, C. A., Chapman, L. J., Rode, K. D., Hauck, E. M., & McDowell, L. R. (2003). Variation in the nutritional value of primate foods: Among trees, time periods, and areas. *International Journal of Primatology, 24,* 317-333.

Chapman, C. A., & Fedigan, L. M. (1990). Dietary differences between neighboring *Cebus capucinus* groups: Local traditions, food availability or responses to food profitability? *Folia Primatologica, 54,* 177-186.

Chevalier-Skolnikoff, S. (1990). Tool use by wild *Cebus* monkeys at Santa Rosa National Park, Costa Rica. *Primates, 31,* 375-383.

Chiou, K. L., Pozzia, L., Lynch Alfaro, J. W., & Di Fiore, A. (2011). Pleistocene diversification of living squirrel monkeys (*Saimiri* spp.) inferred from complete mitochondrial genome sequences. *Molecular Phylogenetics and Evolution, 59,* 736-745.

Chivers, D. J., & Wood, B. A. (Eds.). (1984). *Food acquisition and processing in primates.* Plenum Press.

Clutton-Brock, T. H., & Harvey, P. H. (1980). Primates, brains and ecology. *Journal of Zoology, 190,* 309-323.

Colquhoun, I. C. (2006). Predation and cathemerality: Comparing the impact of predators on the activity patterns of lemurids and ceboids. *Folia Primatologica, 77,* 143-165.

Cooper, L. R., & Harlow, H. F. (1961). Note on a *Cebus* monkey's use of a stick as a weapon. *Psychological Reports, 8,* 418.

Cords, M. (1986). Interspecific and intraspecific variation in the diet of two forest guenons, *Cercopithecus ascanius* and *C. mitis*. *Journal of Animal Ecology, 55,* 811-827.

Coss, R. G., Cavanaugh, C., & Brennan, W. (2019). Development of snake-directed antipredator behavior by wild white-faced capuchin monkeys: III. The signaling properties of alarm-call tonality. *American Journal of Primatology, 81,* e22950.

Costello, R. K., Dickinson, C., Rosenberger, A. L., Boinski, S., & Szalay, F. S. (1993). *Squirrel monkey (genus* Saimiri) *taxonomy: A multidisciplinary study of the biology of the species*. Plenum Press.

Crofoot, M. C. (2007). Mating and feeding competition in white-faced capuchins (*Cebus capucinus*): The importance of short- and long-term strategies. *Behaviour, 144*(12), 1473-1495.

Crofoot, M. C. (2013). The cost of defeat: Capuchin groups travel further, faster and later after losing conflicts with neighbors. *American Journal of Physical Anthropology, 152,* 79-85.

Crofoot, M. C., & Gilby, I. C. (2012). Cheating monkeys undermine group strength in enemy territory. *Proceedings of the National Academy of Sciences, 109*(2), 501-505.

Crofoot, M. C., Gilby, I. C., Wikelski, M. C., & Kays, R. W. (2008). Interaction location outweighs the competitive advantage of numerical superiority in *Cebus capucinus* intergroup contests. *Proceedings of the National Academy of Sciences, 105*(2), 577-581.

Crofoot, M. C., Rubenstein, D. I., Maiya, A. S., & Berger-Wolf, T. Y. (2011). Aggression, grooming and group-level cooperation in white-faced capuchins (*Cebus capucinus*): Insights from social networks. *American Journal of Primatology, 73,* 821-833.

Crofoot, M. C., & Wrangham, R. W. (2010). Intergroup aggression in primates and humans: The case for a unified theory. In P. Kappeler, & J. Silk (Eds.), *Mind the gap* (pp. 171-195). Springer.

Cropp, S., & Boinski, S. (2000). The Central American squirrel monkey (*Saimiri oerstedii*): Introduced hybrid or endemic species? *Molecular Phylogenetics and Evolution, 16*(3), 350-365.

Cunha, A. A., Vieira, M. V., & Grelle, C. E. V. (2006). Preliminary observations on habitat, support use and diet in two non-native primates in an urban Atlantic forest fragment: The capuchin monkey (*Cebus* sp.) and the common marmoset (*Callithrix jacchus*) in the Tijuca forest, Rio de Janeiro. *Urban Ecosystems, 9,* 351-359.

da Silva, F. A., Canale, G. R., Kierulff, M. C. M., Duarte, G. T., Paglia, A. P., & Bernardo, C. S. S. (2016). Hunting, pet trade, and forest size effects on population viability of a critically endangered Neotropical primate, *Sapajus xanthosternos* (Wied-Neuwied, 1826). *American Journal of Primatology, 78,* 950-960.

Defler, T. R. (1979a). On the ecology and behavior of *Cebus albifrons* in eastern Colombia: 1. Ecology. *Primates, 20,* 475-490.

Defler, T. R. (1979b). On the ecology and behavior of *Cebus albifrons* in eastern Colombia: II. Behavior. *Primates, 20,* 491-502.

Defler, T. R. (1982). A comparison of intergroup behavior in *Cebus albifrons* and *C. apella*. *Primates, 23,* 385-392.

Defler, T. R. (2004). *Primates of Colombia*. Conservation International.

De Moraes, B. L. C., da Silva Souto, A., & Schiel, N. (2014). Adaptability in stone tool use by wild capuchin monkeys (*Sapajus libidinosus*). *American Journal of Primatology, 76,* 967-977.

De Oliveira, S. G., Lynch Alfaro, J. W., & Veiga, L. M. (2014). Activity budget, diet, and habitat use in critically endangered Ka'apor capuchin monkey (*Cebus kaapori*) in Pará State, Brazil: A preliminary comparison to other capuchin monkeys. *American Journal of Primatology, 76,* 919-931.

Di Bitetti, M. S., Vidal, E. M. L., Baldovino, M. C., & Benesovsky, V. (2000). Sleeping site preferences in tufted capuchin monkeys (*Cebus apella nigritus*). *American Journal of Primatology, 50,* 257-274.

Di Fiore, A. (2002). Predator sensitive foraging in ateline primates. In L. Miller (Ed.), *Eat or be eaten: Predator sensitive foraging among primates* (pp. 242-267). Cambridge University Press.

Di Fiore, A. (2003). Ranging behavior and foraging ecology of lowland woolly monkeys (*Lagothrix lagotricha poeppigii*) in Yasuni National Park Ecuador. *American Journal of Primatology, 59,* 47-66.

Digweed, S. M., Fedigan, L. M., & Rendall, D. (2007). Who cares who calls? Selective responses to the lost calls of socially dominant group members in the white-faced capuchin (*Cebus Capucinus*). *American Journal of Primatology, 69,* 829-835.

Digweed, S. M., Rendall, D., & Fedigan, L. M. (2005). Variable specificity in the anti-predator vocalizations and behaviour of the white-faced capuchin, *Cebus capucinus*. *Behaviour, 142*(8), 997-1021.

Duarte, M., Hanna, J., Sanches, E., Liu, Q., & Fragszy, D. (2012). Kinematics of bipedal locomotion while carrying a load in the arms in bearded capuchin monkeys (*Sapajus libidinosus*). *Journal of Human Evolution, 63*(6), 851-858.

Eason, P. (1989). Harpy eagle attempts predation on adult howler monkey. *The Condor, 91,* 469-470.

Eisenberg, J. F., & Redford, K. H. (1989). *Mammals of the Neotropics (Vol.1). The northern Neotropics*. University of Chicago Press.

Emidio, R. A., & Ferreira, R. G. (2012). Energetic payoff of tool use for capuchin monkeys in the Caatinga: Variation by season and habitat type. *American Journal of Primatology, 74,* 332-343.

Emmons, L. H. (1987). Comparative feeding ecology of felids in a Neotropical rainforest. *Behavior Ecology and Sociobiology, 20,* 271-283.

Eshchar. Y., Izar, P., Visalberghi, E., Resende, B., & Fragaszy, D. (2016). When and where to practice: Social influences on the development of nut-cracking in bearded capuchins (*Sapajus libidinosus*). *Animal Cognition, 19*(3), 605-618.

Estrada, A. (2006). Human and non-human primate co-existence in the Neotropics: A preliminary view of some agricultural practices as a complement for primate conservation. *Ecological and Environment Anthropology, 2*(2), 17-29.

Estrada, A., Garber, P. A., & Chaudhary, A. (2019). Expanding global commodities trade and consumption place the world's primates at risk of extinction. *PeerJ, 7,* e7068. https://doi.org /10.7717/peerj.7068.

Estrada, A., Garber, P. A., Mittermeier, R. A., Wich, S., Gouveia, S., Dobrovolski, R., Nekaris, K. A. I., Nijman, V., Rylands, A. B., Maisels, F., Williamson, E. A., Bicca-Marques, J., Fuentes, A., Jerusalinsky, L., Johnson, S., Rodrigues de Melo, F., Oliveira, L., Schwitzer, C., Roos, C.,... & Setiawan, A. (2018). Primates in peril: The significance of Brazil, Madagascar, Indonesia and the Democratic Republic of the Congo for global primate conservation. *PeerJ, 6,* e4869. https://doi.org/10.7717/peerj.4869.

Estrada, A., Garber, P. A., Pavelka, M. M., & Luecke, L. (Eds.). (2006). *New perspectives in the study of Mesoamerican primates: Distribution, ecology, behavior and conservation.* Springer.

Estrada, A., Garber, P. A., Rylands, A. B., Roos, C., Fernandez-Duque, E., Di Fiore, A., Nekaris, K. A. I., Nijman, V., Heymann, E. W., Lambert, J. E., Rovero, F., Barelli, C., Setchell, J. M., Gillespie, T. R., Mittermeier, R. A., Arregoitia, L. V., de Guinea, M., Gouveia, S., Dobrovolski, R.,... & Li, B. (2017). Impending extinction crisis of the world's primates: Why primates matter. *Science Advances, 3*(1), e1600946.

Estrada, A., Raboy, B. E., & Oliveira, L. C. (2012). Agroecosystems and primate conservation in the tropics: A review. *American Journal of Primatology, 74,* 696-711.

Falótico, T., & Ottoni, E. B. (2013). Stone throwing as a sexual display in wild female bearded capuchin monkeys, *Sapajus libidinosus. PLoS ONE, 8*(11), e79535. https://doi.org/10.1371 /journal.pone.0079535.

Fedigan, L. M. (1990). Vertebrate predation in *Cebus capucinus*: Meat eating in a Neotropical monkey. *Folia Primatologica, 54,* 196-205.

Fedigan, L. M. (1993). Sex differences in intersexual relations in adult white-faced capuchins (*Cebus capucinus*). *International Journal of Primatology, 14,* 853-877.

Fedigan, L. M. (2003). Impact of male takeovers on infant deaths, births, and conception in *Cebus capucinus* at Santa Rosa, Costa Rica. *International Journal of Primatology, 24,* 723-741.

Fedigan, L. M., & Bergstrom, M. (2010). Dominance among female white-faced capuchin monkeys (*Cebus capucinus*): Hierarchical linearity, nepotism, strength and stability. *Behaviour, 147*(7), 899-931.

Fedigan, L. M., Carnegie, S. D., & Jack, K. M. (2008). Predictors of reproductive success in female white-faced capuchins (*Cebus capucinus*). *American Journal of Physical Anthropology, 137,* 82-90.

Fedigan, L. M., Fedigan, L., Chapman, C. A., & Glander, K. E. (1985). A census of *Alouatta palliata* and *Cebus capucinus* monkeys in Santa Rosa National Park, Costa Rica. *Brenesia, 23,* 309-322.

Fedigan, L. M., Fedigan, L., Chapman, C. A., & Glander, K. E. (1988). Spider monkey home ranges: A comparison of radio telemetry and direct observation. *American Journal of Primatology, 16,* 19-29.

Fedigan, L. M., & Jack, K. (2001). Neotropical primates in a regenerating Costa Rican dry forest: A comparison of howler and capuchin population patterns. *International Journal of Primatology, 22,* 689-713.

Fedigan, L. M., & Jack. K. (2004). The demographic and reproductive context of male replacements in *Cebus capucinus. Behaviour, 141,* 755-775.

Fedigan, L. M., & Jack. K. (2012). Tracking Neotropical monkeys in Santa Rosa: Lessons from a regenerating Costa Rican dry forest. In P. M. Kappeler, & D. P. Watts (Eds.), *Long-term field studies of primates* (pp. 165-184). Springer.

Fedigan, L. M., & Rose, L. (1995). Interbirth interval variation in three sympatric species of Neotropical monkey. *American Journal of Primatology, 37,* 9-24.

Fedigan, L. M., Rose, L. M., & Avila, R. M. (1996). See how they grow: Tracking capuchin monkey populations in a regenerating Costa Rican dry forest. In M. Norconk, A. Rosenberger, & P. Garber (Eds.), *Adaptive radiations of Neotropical primates* (pp. 289-307). Plenum Press.

Fedigan, L. M., Rose, L. M., & Avila, R. M. (1998). Growth of mantled howler groups in a regenerating Costa Rican dry forest. *International Journal of Primatology, 19,* 405-432.

Fernandes, M. E. B. (1991). Tool use and predation of oysters (*Crassostrea rhizophorae*) by the tufted capuchin, *Cebus apella apella*, in brackish water mangrove swamp. *Primates, 32,* 529-531.

Ferrari, S. F. (2009) Predation risk and antipredator strategies. In P. A. Garber, A. Estrada, J. C. Bicca-Marques, E. W. Heymann, & K. B. Strier (Eds.), *South American primates* (pp. 251-278). Springer.

Ferrari, S. F., & Queiroz, H. (1994). Two new Brazilian primates discovered, endangered. *Oryx, 28*(1), 31-36.

Fichtel, C., Perry, S., & Gros-Louis, J. (2005). Alarm calls of white-faced capuchin monkeys: An acoustic analysis. *Animal Behaviour, 70,* 165-176.

Fleagle, J. G. (2013). *Primate adaptation and evolution* (3rd Ed.). Academic Press.

Fleagle, J. G., & Mittermeier, R. A. (1980). Locomotor behavior, body size, and comparative ecology of seven Surinam monkeys. *American Journal of Physical Anthropology, 52,* 301-314.

Fleagle, J. G., Mittermeier, R. A., & Skopec, A. L. (1981). Differential habitat use by *Cebus apella* and *Saimiri sciureus* in central Surinam. *Primates, 22,* 361-367.

Fleagle, J. G., & Samonds, K. W. (1975). Physical growth of *Cebus* monkeys (*Cebus albifrons*) during the first year of life. *Growth, 39*(1), 35-52.

Flesher, K. M. (2015). The distribution, habitat use, and conservation status of three Atlantic Forest monkeys (*Sapajus xanthosternos, Callicebus melanochir, Callithrix, Callithrix* sp.) in an Agroforestry/forest mosaic in southern Bahia, Brazil. *International Journal of Primatology, 36,* 1172-1197.

Fontaine, R. (1985). *Positional behavior in six cebid species.* [Doctoral dissertation, University of Georgia].

Fontaine, R. (1990). Positional behavior in *Saimiri boliviensis* and *Ateles geoffroyi. American Journal of Physical Anthropology, 82,* 485-508.

Fontaine, R. P. (1994). Play as physical flexibility training in five ceboid primates. *Journal of Comparative Psychology, 108,* 203-212.

Ford, S. M., & Davis, L. (1992). Systematics and body size: Implications for feeding adaptations in New World monkeys. *American Journal of Physical Anthropology, 88,* 415-468.

Ford, S. M., & Hobbs, D. G. (1996). Species definition and differentiation as seen in the postcranial skeleton of *Cebus.* In M. A. Norconk, A. L. Rosenberger, & P. A. Garber (Eds.), *Adaptive radiations of Neotropical primates* (pp. 229-249). Springer.

Fowler, J. M., & Cope, J. B. (1964). Notes on the harpy eagle in British Guiana. *Auk, 81,* 257-273.

Fragaszy, D. M. (1990a). Sex and age differences in the organization of behavior in wedge-capped capuchins, *Cebus olivaceus. Behavioral Ecology, 1,* 81-94.

Fragaszy, D. M. (1990b). Early behavioral development in capuchins (*Cebus*). *Folia Primatologica, 54,* 119-128.

Fragaszy, D. M., & Adams-Curtis, L. (1997). Developmental changes in manipulation in tufted capuchins (*Cebus apella*) from birth through two years and their relation to foraging and weaning. *Journal of Comparative Psychology, 111,* 201-211.

Fragaszy, D. M., Baer, J., & Adams-Curtis, L. (1991). Behavioral development and maternal care in tufted capuchins (*Cebus apella*) and squirrel monkeys (*Saimiri sciureus*) from birth through seven months. *Developmental Psychobiology, 24,* 375-393.

Fragaszy, D. M., & Bard, K. (1997). Comparison of development and life history in *Pan* and *Cebus. International Journal of Primatology, 18,* 683-701.

Fragaszy, D. M., Izar, P., Liu, Q., Eshchar, Y., Young, L. A., & Visalberghi, E. (2016). Body mass in wild bearded capuchins, (*Sapajus libidinosus*): Ontogeny and sexual dimorphism. *American Journal of Primatology, 8*(4), 473-484.

Fragaszy, D. M., Izar, P., Visalberghi, E., Ottoni, E. B., & de Oliveira, M. G. (2004b). Wild capuchin monkeys (*Cebus libidinosus*) use anvils and stone pounding tools. *American Journal of Primatology, 64,* 359-366.

Fragaszy, D. M., Liu, Q., Wright, B. W., Allen, A., Brown, C. W., & Visalberghi, E. (2013). Wild bearded capuchin monkeys (*Sapajus libidinosus*) strategically place nuts in a stable position during nut-cracking. *PLoS ONE, 8*(2), e56182.

Fragaszy, D. M., Visalberghi, E., & Fedigan, L. M. (2004a). *The complete capuchin: The biology of the genus* Cebus. Cambridge University Press.

Freese, C. H. (1977). Food habits of white-faced capuchins *Cebus capucinus* L. (Primates: Cebidae) in Santa Rosa National Park, Costa Rica. *Brenesia, 10,* 43-56.

Freese, C. H. (1978). The behavior of white-faced capuchins (*Cebus capucinus*) at a dry-season waterhole. *Primates, 19,* 275-286.

Freese, C. H., & Oppenheimer, J. R. (1981). The capuchin monkey: Genus *Cebus* sp. In A. Coibra-Filho, & R. Mittermeier (Eds.), *Ecology and behavior of Neotropical primates.* (Vol. 1., pp. 331-390). Academia Brasileira de Ciencias.

Galef, B. G., Mittermeier, J. R. A., Bailey, R. C. (1976). Predation by the tayra (*Eira barbara*). *Journal of Mammalogy, 57,* 760-761.

Galetti, M. (1990). Predation on the squirrel, *Sciurus aestuans,* by capuchin monkeys, *Cebus apella. Mammalia, 54*(1), 152-154.

Galetti, M., & Pedroni, F. (1994). Seasonal diet of capuchin monkeys (*Cebus apella*) in a semideciduous forest in south-east Brazil. *Journal of Tropical Ecology, 10*(1), 27-39.

Gama Campillo, L. M., Pozo Montuy, G., Contreras Sánchez, W. M., & Arriaga Weiss, S. L. (Eds.). (2011). *Perspectivas en primatología Mexicana.* Universidad Juárez Autónoma de Tabasco.

Garber, P. A., & Brown, E. (2004). Wild capuchins (*Cebus capucinus*) fail to use tools. *American Journal of Primatology, 62,* 165-170.

Garber, P. A., Gomes, D. F., & Bicca Marques, J. C. (2011). A field study of problem-solving using tools in capuchins (*Sapajus nigritus,* formerly *Cebus nigritus*). *American Journal of Primatology, 73,* 1-15.

Garber, P. A., Gomes, D. F., & Bicca-Marques, J. C. (2012). Experimental field study of problem-solving using tools in free-ranging capuchins (*Sapajus nigritus,* formerly *Cebus nigritus*). *American Journal of Primatology, 74,* 344-358.

Garber, P. A., & Paciulli, L. (1997). Experimental field study of spatial memory and learning in wild capuchins (*Cebus capucinus*). *Folia Primatologica, 68,* 236-253.

Garber, P. A., & Rehg, J. A. (1999). The ecological role of the prehensile tail in *Cebus capucinus. American Journal of Physical Anthropology, 110,* 325-339.

Gebo, D. L. (1992). Locomotor and postural behavior in *Alouatta palliata* and *Cebus capucinus. American Journal of Primatology, 26,* 277-290.

Gilbert, K. A. (2003). Primates and fragmentation of the Amazon forest. In L. K. Marsh (Ed.), *Primates in fragments: Ecology and conservation.* (pp. 145-157). Springer.

Gilbert, K. A., & Setz, E. A. (2001). Primates in a fragmented landscape: Six species in central Amazonia. In R. O. Bierregaard Jr., C. Gascon, T. E. Lovejoy, & R. Mesquita (Eds.), *Lessons from Amazonia: The ecology and conservation of a fragmented forest.* (pp. 262-270). Yale University Press.

Gleason, T. M., & Norconk, M. A. (2002). Predation risk and anti-predator adaptations in white-faced sakis, *Pithecia pithecia.* In L. Miller (Ed.), *Eat or be eaten: Predation sensitive foraging in primates* (pp. 169-184). Cambridge University Press.

Gonzalez-Kirchner, J. P., & Sainz de la Maza, M. (1998). Primates hunting by Guaymi Amerindians in Costa Rica. *Human Evolution, 13,* 15-19.

Gros Louis, J. (2002). Contexts and behavioral correlates of trill vocalizations in wild white-faced capuchin monkeys. *American Journal of Primatology, 57,* 189-202.

Groves, C. P. (1993). Order primates. In D. E. Wilson, & D. M. Reeder (Eds.), *Mammal species of the world: A taxonomic and geographic reference* (2nd Ed., pp. 243-277). Smithsonian Institution Press.

Groves, C. P. (2001a). *Primate taxonomy.* Smithsonian Institution Press.

Groves, C. P. (2001b). Why taxonomic stability is a bad idea, or why are there so few species of primates (or are there?). *Evolutionary Anthropology, 10,* 192-198.

Groves, C. P. (2004). The what, why and how of primate taxonomy. *International Journal of Primatology, 25,* 1105-1126.

Groves, C. P. (2005). Order primates. In D. E. Wilson, & D. M. Reeder (Eds.), *Mammal species of the world: A taxonomic and geographic reference* (3rd. Ed., pp. 111-184). Johns Hopkins University Press.

Hall, C. L., & Fedigan, L. M. (1997). Spatial benefits afforded by high ranked males in white-faced capuchins. *Animal Behaviour, 43,* 1069-1082.

Harding, R. S. O. (1981). An order of omnivores: Nonhuman primate diets in the world. In R. S. Harding, & G. Teleki (Eds.), *Omnivorous primates* (pp. 191-214). Columbia University Press.

Hart, D., & Sussman, R. W. (2005). *Man the hunted: Primates, predators, and human evolution.* Westview Press.

Harvey, P. H., & Clutton-Brock, T. H. (1985). Life history variation in primates. *Evolution, 39*(3), 559-581.

Haugaasen, T., & Peres, C. A. (2009). Interspecific primate associations in Amazonian flooded and unflooded forests. *Primates, 50,* 239-251.

Hershkovitz, P. (1949). Mammals of northern Colombia. Preliminary report No. 4: Monkeys (Primates), with taxonomic revisions of some forms. *Proceedings of the United States National Museum, 98,* 323-427.

Hershkovitz, P. (1969). The evolution of mammals on southern continents. *Quarterly Review of Biology, 44*(1), 1-70.

Hershkovitz, P. (1977). *Living New World monkeys (Platyrrhini), Vol. 1.* University of Chicago Press.

Hershkovitz, P. (1984). Taxonomy of squirrel monkeys, genus *Saimiri* (Cebidae, Platyrrhini): A preliminary report with description of a hitherto unnamed form. *American Journal of Primatology, 4,* 209-243.

Hill, C. M. (2017). Primate crop feeding behavior, crop protection, and conservation. *International Journal of Primatology, 38,* 385-400.

Hill, C. M. (2018). Crop foraging, crop losses, and crop raiding. *Annual Review of Anthropology, 47,* 377-394.

Hirsch, B. T. (2002). Social monitoring and vigilance behavior in brown capuchin monkeys (*Cebus apella*). *Behavior Ecology and Sociobiology, 52,* 458-464.

Holmes, T. D., Bergstrom, M. L., & Fedigan, L. M. (2011). Sleeping site selection by white-faced capuchins (*Cebus capucinus*) in the Area de Conservación Guanacaste, Costa Rica. *Ecological and Environmental Anthropology, 6,* 1-9.

Humle, T., & Newton-Fisher, N. E. (2013). Culture in nonhuman primates: Definitions and evidence. In R. Ellen, S. J. Lycett, & S. E. Johns (Eds.), *Understanding cultural transmission in anthropology: A critical synthesis* (pp. 80-101). Berghahn Books.

IUCN. (2020). *International Union for the Conservation of Nature Red List of Threatened Species.* www.iucnredlist.org.

Isbell, L. A. (1994). Predation on primates: Ecological patterns and evolutionary consequences. *Evolutionary Anthropology, 3,* 61-71.

Isbell, L. A. (2006). Snakes as agents of evolutionary change in primate brains. *Journal of Human Evolution, 51*(1), 1-35.

Isbell, L. A. (2009). *The fruit, the tree, and the serpent: Why we see so well.* Harvard University Press.

Izar, P., Verderane, M. P., Peternelli-dos-Santos, L., Mendonça-Furtado, O., Presotto, A., Visalberghi, E., & Fragaszy, D. (2012). Flexible and conservative features of social systems in tufted capuchin monkeys: Comparing the socioecology of *Sapajus libidinosus* and *Sapajus nigritus. American Journal of Primatology, 74,* 315-331.

Izawa, K. (1978). Frog-eating behavior of wild black-capped capuchin (*Cebus apella*). *Primates, 19,* 633-642.

Izawa, K. (1994). Social changes within a group of wild black-capped capuchins IV. *Field Studies of New World Monkeys, La Macarena, Columbia, 9,* 15-21.

Izawa, K., & Mizuno, A. (1977). Palm-fruit cracking behavior of wild black-capped capuchin (*Cebus apella*). *Primates, 18,* 773-792.

Jack, K. M. (2003). Explaining variation in affiliative relationships among male white-faced capuchins (*Cebus capucinus*). *Folia Primatologica, 74*(1), 1-16.

Jack, K. M. (2011). The cebines: Toward an explanation of variable social structure. In C. J. Campbell, A. Fuentes, K. C. MacKinnon, S. K. Bearder, & R. M. Stumpf (Eds.), *Primates in perspective.* (2nd Ed., pp. 108-122). Oxford University Press.

Jack, K. M., & Campos, F. A. (2012). Distribution, abundance and spatial ecology of the critically endangered Ecuadorian capuchin (*Cebus albifrons aequatroialis*). *Tropical Conservation Science*, 5(2), 173-191.

Jack, K. M., & Fedigan, L. M. (2004a). Male dispersal patterns in white-faced capuchins, *Cebus capucinus*. Part 1: Patterns and causes of natal emigration. *Animal Behaviour, 67*(4), 761-769.

Jack, K. M., & Fedigan, L. M. (2004b). Male dispersal patterns in white-faced capuchins, *Cebus capucinus*. Part 1: Patterns and causes of secondary dispersal. *Animal Behaviour, 67*(4), 771-782.

Jack, K. M., & Fedigan, L. M. (2004c). The demographic and reproductive context of male replacements in *Cebus capucinus*. *Behaviour, 141*(6), 755-775.

Jack, K. M., & Fedigan, L. M. (2006). Why be alpha male? Dominance and reproductive success in wild white-faced capuchins (*Cebus capucinus*). In A. Estrada, P. A. Garber, M. S. M. Pavelka, & L. Luecke (Eds.), *New perspectives in the study of Mesoamerican primates: Developments in primatology: Progress and prospects* (pp. 367-386). Springer.

Jack, K. M., & Fedigan, L. M. (2009). Female dispersal in a female-philopatric species, *Cebus capucinus*. *Behaviour, 146*(4-5), 471-497.

Jack, K. M., & Fedigan, L. M. (2018). Alpha male capuchins (*Cebus capucinus imitator*) as keystone individuals. In U. Kalbitzer, & K. Jack (Eds.), *Primate life histories, Sex roles, and adaptability* (pp. 99-115). Springer.

Jack, K. M., Sheller, C., & Fedigan, L. M. (2012). Social factors influencing natal dispersal in male white-faced capuchins (*Cebus capucinus*). *American Journal of Primatology, 74*, 359-365.

Janson, C. H. (1984). Female choice and mating system of the brown capuchin monkey *Cebus apella* (Primates: Cebidae). *Zeitschrift für Tierpsychologie, 65*, 177-200.

Janson, C. H. (1986). Capuchin counterpoint. *Natural History, 95*, 44-52.

Janson, C. H. (1990a). Social correlates of individual spatial choice in foraging groups of brown capuchin monkeys, *Cebus apella. Animal Behaviour, 40*(5), 910-921.

Janson, C. H. (1990b). Ecological consequences of individual spatial choice in foraging groups of brown capuchin monkeys, *Cebus apella. Animal Behaviour, 40*(5), 922-934.

Janson, C. H., & Boinski, S. (1992). Morphological and behavioral adaptations for foraging in generalist primates: The case of the cebines. *American Journal of Physical Anthropology, 88*, 483-498.

Janson, C. H., & van Schaik, C. P. (2002). Ecological risk aversion in juvenile primates: Slow and steady wins the race. In M. E. Pereira, & L. A. Fairbanks (Eds.), *Juvenile primates: Life history, development and behavior* (pp. 57-74). University of Chicago Press.

Kalbitzer, U., Bergstrom, M. L., Carnegie, S. D., Wikberg, E. C., Kawamura, S., Campos, F. A., Jack, K. M., & Fedigan, L. M. (2017). Female sociality and sexual conflict shape offspring survival in a Neotropical primate. *Proceedings of the National Academy of Sciences, 114*(8), 1892-1897.

Kalbitzer, U., & Jack, K. M. (Eds.). (2018). *Primate life histories, sex roles, and adaptability.* Springer.

Kappeler, P. M., & Watts, D. P. (Eds.). (2012). *Long-term field studies of primates.* Springer.

Kauffman. L., Ehmke, E., Schet, S., Vreedzaam, A., & Boinski, S. (2005). Dispersal patterns among three species of squirrel monkeys (*Saimiri oerstedii, S. boliviensis* and *S. sciureus*): I. Divergent costs and benefits. *Behaviour, 142*(5), 525-632.

Kay, R. F., & Covert, H. H. (1984). Anatomy and behavior of extinct primates. In D. J. Chivers, B. A. Wood, & A. Bilsborough (Eds.), *Food acquisition and processing in primates* (pp. 346-388). Plenum Press.

Kerber, W. T., Conaway, C. H., & Smith, D. M. (1977). The duration of gestation in the squirrel monkey (*Saimiri sciureus*). *Laboratory Animal Science, 5*(1), 700-702.

Kinzey, W. G. (1997). *New world primates: Ecology, evolution, and behavior.* Aldine de Gruyter.

Klein, L. L., & Klein, D. J. (1973). Observations on two types of Neotropical primate intertaxa associations. *American Journal of Physical Anthropology, 38*, 649-653.

Kowalewski, M., Garber, P., Cortés-Ortiz, L., Urbani, B., & Youlatos, D. (Eds.). (2015). *Howler monkeys. Developments in primatology: Progress and prospects.* Springer.

Lambert, J. E. (1998). Primate digestion: Interactions among anatomy, physiology, and feeding ecology. *Evolutionary Anthropology, 7*, 8-20.

Lehman, S. M., Sussman, R.W., Phillips-Conroy, J., & Prince, W. (2006). Ecological biogeography of primates in Guyana. In S. M. Lehman, & J. G. Fleagle (Eds.), *Primate biogeography: Progress and prospects* (pp. 105-130). Springer.

Leigh, S. R. (1994). Ontogenetic correlates of diet in anthropoid primates. *American Journal of Physical Anthropology, 94*, 499-522.

Lemelin, P. (1995). Comparative and functional myology of the prehensile tail in New World monkeys. *Journal of Morphology, 224*, 351-368.

Levi, T., Silvius, K., Oliveira, L., Cummings, A., & Fragoso, J. (2013). Competition and facilitation in the capuchin–squirrel monkey relationship. *Biotropica, 45*(5), 636-643.

Liu, Q., Simpson, K., Izar, P., Ottoni, E., Visalberghi, E., & Fragaszy, D. (2009). Kinematics and energetics of nut-cracking in wild capuchin monkeys (*Cebus libidinosus*) in Piauí, Brazil. *American Journal of Physical Anthropology, 138*, 210-220.

Lopes, G. P., & Ferrari, S. F. (1996). Preliminary observations on the Ka'apor capuchin *Cebus kaapori* Queiroz 1992 from eastern Brazilian Amazonia. *Biological Conservation, 76*(3), 321-324.

Lopes, G. P., Guimaraes, D. P., & Jaskulski, A. (2015). Predation of *Saimiri cassiquiarensis* (Lesson, 1840) (Primates: Cebidae) by *Spizaetus ornatus* (Daudin, 1800) (Accipitriformes: Accipitridae) in the Brazilian Amazon. *Atualidades Ornitológicas, 186*, 20.

Lorenz, R., Anderson, C. O., & Mason, W. A. (1973). Notes on reproduction in captive squirrel monkeys (*Saimiri sciureus*). *Folia Primatologica, 19*, 286-292.

Lynch Alfaro, J. W. (2005). Male mating strategies and reproductive constraints in a group of wild tufted capuchin monkeys (*Cebus apella nigritus*). *American Journal of Primatology, 67,* 313-328.

Lynch Alfaro, J. W. (2007). Subgroup patterns in a group of wild *Cebus apella nigritus*. *International Journal of Primatology, 28,* 271-289.

Lynch Alfaro, J. W. (2008). Scream-embrace displays in wild black-horned capuchin monkeys. *American Journal of Primatology, 70,* 551-559.

Lynch Alfaro, J. W., Alves, S. L., Silva Junior, J., Ravetta, A., & Messias, M. (2019). *Saimiri ustus. The International Union for the Conservation of Nature Red List of Threatened Species.* www.iucnredlist.org.

Lynch Alfaro, J. W., Boubli, J. P., Olson, L. E., Di Fiore, A., Wilson, B., Gutierrez-Espeleta, G. A., Chiou, K. L., Schulte, M., Neitzel, S., Ross, V., Schwochow, D., Nguyen, M. T. T., Farias, I., Janson, C. H., & Alfaro, M. E. (2011). Explosive Pleistocene range expansion leads to widespread Amazonian sympatry between robust and gracile capuchin monkeys. *Journal of Biogeography, 39,* 272-288.

Lynch Alfaro, J. W., Boubli, J. P., Paim, F. P., Ribas, C. C., da Silva, M. N. F., Messias, M. R., Rohe, F., Merces, M. P., Silva Junior, J. S., Silva, C. R., Pinho, G. M., Koshkarian, G., Nguyen, M. T. T., Harada, M. L., Rabelo, R. M., Queiroz, H. L., Alfaro, M. E., & Farias, I. P. (2015). Biogeography of squirrel monkeys (genus *Saimiri*): South-Central Amazon origin and rapid pan-Amazonian diversification of a lowland primate. *Molecular Phylogenetics and Evolution, 82(b),* 436-454.

Lynch Alfaro, J. W., De Sousa, E., Silva Junior, J., & Rylands, A. B. (2012a). How different are robust and gracile capuchin monkeys? An argument for the use of *Sapajus* and *Cebus*. *American Journal of Primatology, 74(4),* 273-286.

Lynch Alfaro, J. W., Matthews, L., Boyette, A. H., Macfarlan, S. J., Phillips, K. A., Falótico, T., Ottoni, E., Verderane, M., Izar, P., Schulte, M., Melin, A., Fedigan, L., Janson, C., & Alfaro, M. E. (2012b). Anointing variation across wild capuchin populations: A review of material preferences, bout frequency and anointing sociality in *Cebus* and *Sapajus*. *American Journal of Primatology, 73,* 1-16.

MacKinnon, K. C. (2002). *Social development of wild white-faced capuchin monkeys* (Cebus capucinus*) in Costa Rica: An examination of social interactions between immatures and adult males.* [Doctoral dissertation, University of California].

MacKinnon, K. C. (2006). Food choice by juvenile capuchin monkeys (*Cebus capucinus*) in a tropical dry forest. In A. Estrada, P. A. Garber, M. Pavelka, & L. Luecke (Eds.), *New perspectives in the study of Mesoamerican primates: Distribution, ecology, behavior, and conservation* (pp. 349-365). Springer.

MacKinnon, K. C. (2011). Social beginnings: The tapestry of infant and adult interactions. In C. J. Campbell, A. Fuentes, K. C. MacKinnon, R. Stumpf, & S. K. Bearder (Eds.), *Primates in perspective* (2nd Ed., pp. 440-455). Oxford University Press.

MacKinnon, K. C. (2013). Ontogeny of social behavior in the genus *Cebus* and the application of an integrative framework for examining plasticity and complexity in evolution. In K. Clancy, K. Hinde, & J. Rutherford (Eds.), *Building babies: Primate development in proximate and ultimate perspective* (pp. 387-408). Springer.

MacKinnon, K. C., & Fuentes, A. (2011). Primates, niche construction, and social complexity: The roles of social cooperation and altruism. In R. W. Sussman, & R. C. Cloninger (Eds.), *Origins of altruism and cooperation* (pp. 121-143). Springer.

Manson, J. H. (1999). Infant handling in wild *Cebus capucinus*: Testing bonds between females? *Animal Behaviour, 57,* 911-921.

Manson, J. H., Perry, S., & Parish, A. R. (1997). Nonconceptive sexual behavior in bonobos and capuchins. *International Journal of Primatology, 18,* 767-786.

Martin, R. D. (1990). *Primate origins and evolution: A phylogenetic reconstruction.* Chapman and Hall.

Martins-Junior, A.M.G., Carneiro, J., Sampaio, I., Ferrari, S. F., & Schneider, H. (2018). Phylogenetic relationships among capuchin (Cebidae, Platyrrhini) lineages: An old event of sympatry explains the current distribution of *Cebus* and *Sapajus*. *Genetics and Molecular Biology, 41,* 699-712.

Massaro, L., Liu, Q., Visalberghi, E., & Fragaszy, D. (2012). Wild bearded capuchins (*Sapajus libidinosus*) select hammer tools on the basis of both stone mass and distance from the anvil. *Animal Cognition, 15,* 1065-1074.

Massaro, L., Massa, F., Simpson, K., Fragaszy, D., & Visalberghi, E. (2016). The strategic role of the tail in maintaining balance while carrying a load bipedally in wild capuchins (*Sapajus libidinosus*): A pilot study. *Primates, 57,* 231-239.

Matheson, M. D., Johnson. J. S., & Feuerstein, J. (1996). Male reunion displays in tufted capuchin monkeys (*Cebus apella*). *American Journal of Primatology, 40,* 183-188.

Matthews, L. J. (2009). Activity patterns, home range size, and intergroup encounters in *Cebus albifrons* support existing models of capuchin socioecology. *International Journal of Primatology, 30,* 709-728.

McGraw, W. S., & Berger, L. R. (2013). Raptors and primate evolution. *Evolutionary Anthropology, 22,* 280-293.

McKinney, T. (2011). The effects of provisioning and crop-raiding on the diet and foraging activities of human-commensal white-faced capuchins (*Cebus capucinus*). *American Journal of Primatology, 73(5),* 439-448.

Melin, A. D., Fedigan, L. M., Hiramatsu, C., Sendall, C., & Kawamura, S. (2007). Effects of colour vision phenotype on insect capture by a free-ranging population of white-faced capuchins (*Cebus capucinus*). *Animal Behaviour, 73,* 205-214.

Melin, A. D., Young, H. C., Mosdossy, K. N., & Fedigan, L. M. (2014). Seasonality, extractive foraging and the evolution of primate sensorimotor intelligence. *Journal of Human Evolution, 71*, 77-86.

Mendes, F. D. C., Cardoso, R. M., Ottoni, E. B., Izar, P., Villar, D. N. A., & Marquezan, R. F. (2015). Diversity of nutcracking tool sites used by *Sapajus libidinosus* in Brazilian Cerrado. *American Journal of Primatology, 77*, 535-546.

Mendonca-Furtado, O., Edaes, M., Palme, R., Rodrigues, A., Siqueira, J., & Izar P. (2014). Does hierarchy stability influence testosterone and cortisol levels of bearded capuchin monkeys (*Sapajus libidinosus*) adult males? A comparison between two wild groups. *Behavioural Processes, 109*(A), 79-88.

Mercês, P. M., de Paula, W., & Júnior, J. S. S. (2018). New records of *Saimiri collinsii* Osgood, 1916 (Cebidae, Primates), with comments on habitat use and conservation. *Mammalia, 82*(5), 516-520.

Milano, M. Z., & Monteiro-Filho, E. L. A. (2009). Predation on small mammals by capuchin monkeys, *Cebus cay. Neotropical Primates, 16*(2), 78-80.

Miller, L. E. (Ed.). (2002). *Eat or be eaten. Predator sensitive foraging among primates.* Cambridge University Press.

Miller, L. E., & Treves, A. (2011). Predation on primates: Past studies, current challenges, and directions for the future. In C. J. Campbell, A. Fuentes, K. C. MacKinnon, S. K. Bearder, & R. M. Stumpf (Eds.), *Primates in Perspective* (2nd Ed., pp. 535-548). Oxford University Press.

Mitchell, B. J. (1989). *Resources, group behavior, and infant development in white-faced capuchin monkeys,* Cebus capucinus. [Doctoral dissertation, University of California].

Mitchell, C. L. (1990). *The ecological basis for female social dominance: A behavioral study of the squirrel monkey.* [Doctoral dissertation, Princeton University].

Mitchell, C. L. (1994). Migration alliances and coalitions among adult male South American squirrel monkeys (*Saimiri sciureus*). *Behaviour, 130*, 169-190.

Mitchell, C. L., Boinski, S., & van Schaik, C. P. (1991). Competitive regimes and female bonding in two species of squirrel monkeys (*Saimiri oerstedi* and *S. sciureus*). *Behavior Ecology and Sociobiology, 28*, 55-60.

Mittermeier, R. A., Rylands, A. B., & Wilson, D. E. (Eds.). (2013). *Handbook of the mammals of the world. Vol. 3: Primates.* Lynx Edicions.

Mittermeier, R. A., & van Roosmalen, M. G. M. (1981). Preliminary observations on habitat utilization and diet in eight Surinam monkeys. *Folia Primatologica, 36*, 1-39.

Mosdossy, K. N., Melin, A. D., & Fedigan, L. M. (2015). Quantifying seasonal fallback on invertebrates, pith, and bromeliad leaves by white-faced capuchin monkeys (*Cebus capucinus*) in a tropical dry forest. *American Journal of Physical Anthropology, 158*, 67-77.

Muniz, L., Perry, S., Manson, J. H., Gilkenson, H., Gros-Louis, J., & Vigilant, L. (2010). Male dominance and reproductive success in wild white-faced capuchins (*Cebus capucinus*) at Lomas Barbudal, Costa Rica. *American Journal of Primatology, 72*, 1118-1130.

Napier, J. R., & Napier, P. (1967). *A handbook of living primates.* Academic Press.

Newcomer, M. W., & De Farcy, D. D. (1985). White-faced capuchin (*Cebus capucinus*) predation on a nestling coati (*Nasua narica*). *Journal of Mammalogy, 66*(1), 185-186.

Norconk, M. A. (2006). Long-term study of group dynamics and female reproduction in Venezuelan *Pithecia pithecia. International Journal of Primatology, 27*(3), 653-674.

Norconk, M. A. (2011). Sakis, uakaris, and titi monkeys: Behavioral diversity in a radiation of primate seed predators. In C. J. Campbell, A. Fuentes, K. C. MacKinnon, S. K. Bearder, & R. M. Stumpf (Eds.), *Primates in perspective* (pp. 122-139). Oxford University Press.

Norconk, M. A., Rosenberger, A. L., & Garber, P. A. (Eds.). (1996). *Adaptive radiations of Neotropical primates.* Plenum Press.

Nowak, K., & Lee, P. C. (2013). "Specialist" primates can be flexible in response to habitat alteration. In L. Marsh, & C. Chapman (Eds.), *Primates in fragments. Developments in primatology: Progress and prospects* (pp. 199-211). Springer.

O'Malley, R. C., & Fedigan, L. M. (2005a). Variability in food-processing behavior among white-faced capuchins (*Cebus capucinus*) in Santa Rosa National Park, Costa Rica. *American Journal of Physical Anthropology, 128*, 63-73.

O'Malley, R. C., & Fedigan, L. M. (2005b). Evaluating social influences on food-processing behavior in white-faced capuchins (*Cebus capucinus*). *American Journal of Physical Anthropology, 127*, 481-491.

Oppenheimer, J. R. (1968). *Behavior and ecology of the white-faced monkey,* Cebus capucinus, *on Barro Colorado Island.* [Doctoral dissertation, University of Illinois].

Organ, J. M. (2010). Structure and function of platyrrhine caudal vertebrae. *Anatomical Record, 293*, 730-745.

Organ, J. M., Muchlinski, M. N., & Deane, A. S. (2011). Mechanoreceptivity of prehensile skin varies between ateline and cebine primates. *Anatomical Record, 294*, 2064-2072.

Organ, J. M., Teaford, M. F., & Taylor, A. B. (2009). Functional correlates of fiber architecture of the lateral caudal musculature in prehensile and nonprehensile tails of the Platyrrhini (Primates) and the Procyonidae (Carnivora). *Anatomical Record, 292*, 827-841.

Ottoni, E. B., & Izar, P. (2008). Capuchin monkey tool use: Overview and implications. *Evolutionary Anthropology, 17*, 171-178.

Ottoni, E. B., & Mannu, M. (2001). Semifree-ranging tufted capuchins (*Cebus apella*) spontaneously use tools to crack open nuts. *International Journal of Primatology, 22*, 347-358.

Paim, F. P., El Bizri, H. R., Paglia, A. P., & Queiroz, H. L. (2019). Long-term population monitoring of the threatened and endemic black-headed squirrel monkey (*Saimiri vanzolinii*) shows the importance of protected areas for primate conservation in Amazonia. *American Journal of Primatology, 81*, e22988.

Palagi, E. (2011). *Playing at every age: Modalities and potential functions in non-human primates.* In A. D. Pellegrini (Ed.), *The Oxford handbook of the development of play* (pp. 70-82). Oxford University Press.

Palagi, E. (2018). Not just for fun! Social play as a springboard for adult social competence in human and non-human primates. *Behavior Ecology and Sociobiology, 72,* 90.

Panger, M. (1998). Object-use in free-ranging white-faced capuchins (*Cebus capucinus*) in Costa Rica. *American Journal of Physical Anthropology, 106,* 311-321.

Panger, M., Perry, S., Rose, L. M., Gros-Louis, J., Vogel, E., MacKinnon, K. C., & Baker, M. (2002). Cross-site differences in foraging behavior of white-faced capuchins (*Cebus capucinus*). *American Journal of Physical Anthropology,* 119, 52-66.

Parker, S., & Gibson, K. (1977). Object manipulation, tool use and sensorimotor intelligence as feeding adaptations in *Cebus* monkeys and great apes. *Journal of Human Evolution, 6*(7), 623-641.

Pereira, M. E., & Fairbanks, L. A. (Eds.). (2002). *Juvenile primates: Life history, development, and behavior.* University of Chicago Press.

Perry, S. (1996). Intergroup encounters in wild white-faced capuchin monkeys (*Cebus capucinus*). *American Journal of Primatology, 17,* 309-330.

Perry, S. (1997). Male-female social relationships in wild white-faced capuchins (*Cebus capucinus*). *Behaviour, 135,* 477-510.

Perry, S. (1998). Male-male social relationships in wild white-faced capuchin monkeys (*Cebus capucinus*). *Behaviour, 135,* 139-172.

Perry, S. (2012). The behavior of wild white-faced capuchins: Demography, life history, social relationships, and communication. *Advances in the Study of Behavior, 44,* 135-181.

Perry, S., Baker, M., Fedigan, L. M., Gros-Louis, J., Jack, K., MacKinnon, K. C., Manson, J. H., Panger, M., Pyle, K., & Rose, L. M. (2003a). Social conventions in wild white-faced capuchins: Evidence for traditions in a Neotropical primate. *Current Anthropology, 44*(2), 241-268.

Perry, S., Manson, J. H., Muniz, L., Gros-Louis, J., & Vigilant, L. (2008). Kin-biased social behaviour in wild adult female white-faced capuchins, *Cebus capucinus. Animal Behaviour, 76*(1), 187-199.

Perry, S., Panger, M., Rose, L., Baker, M., Gros-Louis, J., Jack, K., MacKinnon, K. C., Manson, J. H., Fedigan, L., & Pyle, K. (2003b). Traditions in wild white-faced capuchin monkeys. In D. M. Fragaszy, & S. Perry (Eds.), *The biology of traditions: Models and evidence* (pp. 391-425). Cambridge University Press.

Perry, S., & Rose, L. M. (1994). Begging and transfer of coati meat by white-faced capuchin monkeys, *Cebus capucinus. Primates, 35,* 409-415.

Phillips, K. (1995). Resource patch size and flexible foraging in white-faced capuchins (*Cebus capucinus*). *International Journal of Primatology, 16,* 509-519.

Phillips, K. A. (1998). Tool use in wild capuchin monkeys (*Cebus albifrons trinitatis*). *American Journal of Primatology, 46,* 259-261.

Phillips, K. A., & Abercrombie, C. L. (2003). Distribution and conservation status of the primates of Trinidad. *Primate Conservation, 19,* 19-22.

Phillips, K. A., Bernstein, I. S., Dettmer, E. L., Devermann, E. L., & Powers, M. (1994). Sexual behavior in brown capuchins (*Cebus apella*). *International Journal of Primatology, 15,* 907-917.

Phillips, K. A., & Shauver Goodchild, L. M. (2005). Reunion displays in captive male brown capuchins (*Cebus apella*). *Primates, 46,* 121-125.

Pinheiro, T., Ferrari, S. F., & Lopes, M. A. (2011). Polyspecific associations between squirrel monkeys (*Saimiri sciureus*) and other primates in eastern Amazonia. *American Journal of Primatology, 73,* 1145-1151.

Pinheiro, T., Ferrari, S. F., & Lopes, M. A. (2013). Activity budget, diet, and use of space by two groups of squirrel monkeys (*Saimiri sciureus*) in eastern Amazonia. *Primates, 54,* 310-308.

Pinheiro, T., Guevara, R., & Lopes, M. A. (2018). Behavioral changes of free-living squirrel monkeys (*Saimiri collinsi*) in an urban park. *Primate Conservation, 32,* 89-94.

Podolsky, R. D. (1990). Effects of mixed-species association on resource use by *Saimiri sciureus* and *Cebus apella. American Journal of Primatology, 21,* 147-158.

Port-Carvalho, M., Ferrari, S. F., & Magalhaes, C. (2004). Predation of crabs by tufted capuchins (*Cebus apella*) in eastern Amazonia. *Folia Primatologica, 75,* 154-156.

Presotto, A., Verderane, M. P., Biondi, L., Mendonca-Furtado, O., Spagnoletti, N., Madden, M., & Izar, P. (2018). Intersection as key locations for bearded capuchin monkeys (*Sapajus libidinosus*) traveling within a route network. *Animal Cognition, 21,* 393-405.

Queiroz, H. L. (1992). A new species of capuchin monkey, genus *Cebus* Erxleben 1977 (Cebidae, Primates), from eastern Brazilian Amazonia. *Goeldiana Zoologia, 15,* 1-13.

Quintino, E. P., & Bicca-Marques, J. C. (2013). Predation of *Alouatta puruensis* by boa constrictor. *Primates, 54,* 325-330.

Resende, B. D., Greco, V. L. G., Ottoni, E. B., & Izar, P. (2003). Some observations on the predation of small mammals by tufted capuchin monkeys (*Cebus apella*). *Neotropical Primates, 11*(2), 103-104.

Rettig, N. L. (1978). Breeding behavior of the harpy eagle (*Harpya harpyja*). *Auk, 95,* 629-643.

Ritchie, B. G., & Fragaszy, D. M. (1988). Capuchin monkey (*Cebus apella*) grooms her infant's wound with tools. *American Journal of Primatology, 16,* 345-348.

Robinson, J. G. (1984). Diurnal variation in foraging and diet in the wedge-capped capuchin *Cebus olivaceus. Folia Primatologica, 43,* 216-228.

Robinson, J. G. (1986). Seasonal variation in use of time and space by the wedge-capped capuchin monkey *Cebus olivaceus*: Implications for foraging theory. *Smithsonian Contributions to Zoology, 431,* 1-60.

Robinson, J. G., & Janson, C. H. (1987). Capuchins, squirrel monkeys and atelines: Socioecological convergence. In B. Smuts, D. Cheney, R. Seyfarth, R. Wrangham, & T. Struhsaker (Eds.), *Primate societies* (pp. 69-82). University of Chicago Press.

Rose, L. M. (1994). Sex differences in diet and foraging behavior in white-faced capuchins (*Cebus capucinus*). *International Journal of Primatology, 15*, 95-114.

Rose, L. M. (1997). Vertebrate predation and food-sharing in *Cebus* and *Pan*. *International Journal of Primatology, 18*, 727-765.

Rose, L. M., & Fedigan, L. M. (1995). Vigilance in white-faced capuchins, *Cebus capucinus*, in Costa Rica. *Animal Behaviour, 49*, 63-70.

Rose, L. M., Perry, S., Panger, M., Jack, K., Manson, J., Gros-Louis, J., MacKinnon, K.C., & Vogel, E. (2003). Interspecific interactions between *Cebus capucinus* and other species at three Costa Rican sites. *International Journal of Primatology, 4*(4), 759-796.

Rosenberger, A. L. (1983). Tale of tails: Parallelism and prehensility. *American Journal of Physical Anthropology, 60*, 103-107.

Rosenberger, A. L., Halenar, L., & Cooke, S. B. (2011). The making of platyrrhine semifolivores: Models for the evolution of folivory in primates. *Anatomical Record, 294*, 2112-2130.

Rosenberger, A. L., & Strier K. B. (1989). Adaptive radiation of the ateline primates. *Journal of Human Evolution, 18*, 717-750.

Ross, C. (1991). Life history patterns of New World monkeys. *International Journal of Primatology, 12*, 481-502.

Rothman, J. M., Raubenheimer, D., Bryer, M. A. H., Takahashi, M., & Gilbert, C. C. (2014). Nutritional contributions of insects to primate diets: Implications for primate evolution. *Journal of Human Evolution, 71*, 59-69.

Rowe, N. (1996). *The pictorial guide to the living primates*. Pogonias Press.

Ruivo, L. P., & Stone, A. I. (2014). Jealous of mom? Interactions between infants and adult males during the mating season in wild squirrel monkeys (*Saimiri collinsi*) in Pará, Brazil. *Neotropical Primates, 21*, 165-170.

Ruivo, L. P., & Stone, A. I. (2020). Synchronization of weaning time with peak fruit availability in squirrel monkeys (*Saimiri collinsi*) living in Amazonian Brazil. *American Journal of Primatology, 82*, e23139.

Ruivo, L. P., Stone, A. I., & Fienup, M. (2017). Reproductive status affects the feeding ecology and social association patterns of female squirrel monkeys (*Saimiri collinsi*) in an Amazonian rainforest. *American Journal of Primatology, 79*(6), e22657. DOI: 10.1002/ajp22657.

Ruiz-García, M., Castillo, M. I., Ledezma, A., Leguizamon, N., Sanchez, R., Chinchilla, M., & Gutierrez-Espeleta, G. A. (2012). Molecular systematics and phylogeography of *Cebus capucinus* (Cebidae, Primates) in Colombia and Costa Rica by means of the mitochondrial COII gene. *American Journal of Primatology, 74*(4), 366-380.

Ruiz-García, M., Castillo, M. I., Vásquez, C., Rodriguez, K., Pinedo, M., Shostell, J., & Leguizamon, N. (2010). Molecular phylogenetics and phylogeography of the white-fronted capuchin (*Cebus albifrons*; Cebidae, Primates) by means of mtCOII gene sequences. *Molecular Phylogenetics and Evolution, 57*, 1049-1061.

Rylands, A. B., Groves, C. P., Mittermeier, R. A., Cortés-Ortiz, L. & Hines, J. J. (2006). Taxonomy and distributions of Mesoamerican primates. In A. Estrada, P. A. Garber, M. Pavelka, & L. Luecke (Eds.), *New perspectives in the study of Mesoamerican primates* (pp. 29-79). Springer.

Rylands, A. B., & Mittermeier, R. A. (2009). The diversity of the New World primates (Platyrrhini): An annotated taxonomy. In P. A. Garber, A. Estrada, J. C. Bicca-Marques, E. W. Heymann, & K. B. Strier (Eds.), *South American primates* (pp. 23-54). Springer.

Rylands, A. B., Mittermeier, R. A., Bezerra, B. M., Palm, F. P., & Queiroz, H. L. (2013). Family Cebidae (squirrel monkeys and capuchins). In R. Mittermeier, A. Rylands, & D. Wildon (Eds.) *Handbook of the mammals of the world* (Vol. 3, pp. 390-413). Lynx Edicions.

Rylands, A. B., Mittermeier, R. A., & Rodriguez-Luna, E. (1995). A species list for the New World primates (Platyrrhini): Distribution by country, endemism, and conservation status according to the Mace-Lande System, *Neotropical Primates, 3*, 113-160.

Rylands, A. B., Schneider, H., Langguth, A., Mittermeier, R. A., Groves, C. P., & Rodríguez-Luna, E. (2000). An assessment of the diversity of New World primates. *Neotropical Primates, 8*, 61-93.

Sampaio, D. T., & Ferrari, S. F. (2005). Predation of an infant titi monkey (*Callicebus moloch*) by tufted capuchins (*Cebus apella*). *Folia Primatologica, 76*, 113-115.

Scarry, C. J. (2013). Between-group contest competition among tufted capuchin monkeys, *Sapajus nigritus*, and the role of male resource defence. *Animal Behaviour, 85*(5), 931-939.

Scarry, C. J. (2017). Male resource defence during intergroup aggression among tufted capuchin monkeys. *Animal Behaviour, 123*, 169-178.

Schneider, H., & Sampaio, I. (2015). The systematics and evolution of New World primates: A review. *Molecular Phylogenetics and Evolution, 82*(B), 348-357.

Schoof, V. A. M., & Jack, K. M. (2013). The association of intergroup encounters, dominance status, and fecal androgen and glucocorticoid profiles in wild male white-faced capuchins (*Cebus capucinus*). *American Journal of Primatology, 75*, 107-115.

Schwarzkopf, L., & Rylands, A. B. (1989). Primate species richness in relation to habitat structure in Amazonian rainforest fragments. *Biological Conservation, 48*, 1-12.

Scollay, P. A. (1980). Cross-sectional morphometric data on a population of semifree-ranging squirrel monkeys, *Saimiri sciureus* (Iquitos). *American Journal of Physical Anthropology, 53*, 309-316.

Sherrow, H. M., & MacKinnon, K. C. (2011). Juvenile and adolescent primates: The application of life history theory.

In C. J. Campbell, A. Fuentes, K. C. MacKinnon, R. Stumpf, & S. K. Bearder (Eds.), *Primates in perspective* (2nd Ed., pp. 455-464). Oxford University Press.

Silva Junior, J. S., & Cerqueira, R. (1998). New data and a historical sketch on the geographical distribution of the Ka'apor capuchin, *Cebus kaapori* Queiroz, 1992. *Neotropical Primates, 6*, 118-121.

Smith, E. O. (1978). *Social play in primates.* Academic Press.

Smith, R. L., Hayes, S. E., Smith, P., & Dickens, J. K. (2018). Sleeping site preferences in *Sapajus cay* Illiger 1815 (Primates: Cebidae) in a disturbed fragment of the Upper Paraná Atlantic Forest, Rancho Laguna Blanca, eastern Paraguay. *Primates, 59*, 79-88.

Smith, R. J., & Jungers, W. L. (1997). Body mass in comparative primatology. *Journal of Human Evolution, 32*(6), 523-559.

Spagnoletti, N., Cardoso, T. C. M., Fragaszy D., & Izar, P. (2017). Coexistence between humans and capuchins (*Sapajus libidinosus*): Comparing observational data with farmers' perceptions of crop losses. *International Journal of Primatology, 38*, 243-262.

Spagnoletti, N., Visalberghi, E., Ottoni, E., Izar, P., & Fragaszy, D. (2011). Stone tool use by adult wild bearded capuchin monkeys (*Cebus libidinosus*). Frequency, efficiency and tool selectivity. *Journal of Human Evolution, 61*(1), 97-107.

Spagnoletti, N., Visalberghi, E., Verderane, M. P., Ottoni, E., Izar, P., & Fragaszy, D. (2012). Stone tool use in wild bearded capuchin monkeys, *Cebus libidinosus.* Is it a strategy to overcome food scarcity? *Animal Behaviour, 83*(5), 1285-1294.

Spironello, W. R. (2001). The brown capuchin monkey (*Cebus apella*): Ecology and home range requirements in Central Amazonia. In R. O. Bierregaard Jr., C. Gascon, T. E. Lovejoy, & R. Mesquita (Eds.), *Lessons from Amazonia: The ecology and conservation of a fragmented forest* (pp. 271-283). Yale University Press.

Stevenson, P. R., Quiñones, M. J., & Ahumada, J. A. (2000). Influence of fruit availability on ecological overlap among four Neotropical primates at Tinigua National Park, Colombia. *Biotropica, 32*, 533-544.

Stone, A. I. (2004). *Juvenile feeding ecology and life history in a Neotropical primate, the squirrel monkey.* [Doctoral dissertation, University of Illinois at Urbana–Champaign].

Stone, A. I. (2006). Foraging ontogeny is not linked to delayed maturation in squirrel monkeys (*Saimiri sciureus*). *Ethology, 112*, 105-115.

Stone, A. I. (2007a). Responses of squirrel monkeys to seasonal changes in food availability in an eastern Amazonian forest. *American Journal of Primatology, 68*, 1-16.

Stone, A. I. (2007b). Ecological risk aversion and foraging behaviors of juvenile squirrel monkeys (*Saimiri sciureus*). *Ethology, 113*, 782-792.

Stone, A. I. (2007c). Age and seasonal effects on predator-sensitive foraging in squirrel monkeys (*Saimiri sciureus*): A field experiment. *American Journal of Primatology, 69*, 127-141.

Stone, A. I. (2008). Seasonal effects on play behavior in immature *Saimiri sciureus* in eastern Amazonia. *International Journal of Primatology, 29*, 195-205.

Stone, A. I. (2014). Is fatter sexier? Reproductive strategies in male squirrel monkeys, *Saimiri sciureus. International Journal of Primatology, 35*, 628-642.

Stone, A. I. (2017). The cebids. In A. Fuentes (Ed.), *International encyclopedia of primatology* (pp. 400-403). Wiley-Blackwell.

Stone, A. I. (2018). The foraging ecology of male and female squirrel monkeys (*Saimiri collinsi*) in eastern Amazonia, Brazil. In B. Urbani, M. Kowalewski, R. T. G. Cunha, S. de la Torre, & L. Cortes-Ortiz (Eds.), *La primatologia en Latinonamerica 2* (pp. 229-237). Ediciones IVIC.

Stone, A. I., & Ruivo, L. V. P (2020). Synchronization of weaning time with peak fruit availability in squirrel monkeys (*Saimiri collinsi*) living in Amazonian Brazil. *American Journal of Primatology, 82*, e23139.

Stone, A. I., & Silva Junior, J. S. (2016). Collins' squirrel monkey. In N. Rowe, & M. Myers (Eds.), *All the world's primates* (pp. 307-308). Pogonias Press.

Strier, K. B. (1999). *Faces in the forest: The endangered muriqui monkeys of Brazil.* Harvard University Press.

Strier, K. B. (2010). Long-term field studies: Positive impacts and unintended consequences. *American Journal of Primatology, 72*, 772-778.

Strier, K. B., & Boubli, J. P. (2006). A history of long-term research and conservation of northern muriquis (*Brachyteles hypoxanthus*) at the Estação Biológica de Caratinga/RPPN-FMA. *Primate Conservation, 20*, 53-63.

Strier, K. B., & Mendes, S. L. (2009). Long-term field studies of South American primates. In P. A. Garber, A. Estrada, J. C. Bicca-Marques, E. W. Heymann, & K. B. Strier (Eds.), *South American primates* (pp. 139-155). Springer.

Struhsaker, T. T., & Leland, L. (1977). Palm-nut smashing by *Cebus a. apella* in Columbia. *Biotropica, 9*, 124-126.

Suscke, P., Verderane, M., de Oliveira, R. S., Delval, I., Fernández-Bolaños, M., & Izar, P. (2017). Predatory threat of harpy eagles for yellow-breasted capuchin monkeys in the Atlantic Forest. *Primates, 58*, 141-147.

Sussman, R. W. (1978). Morpho-physiological analysis of diets: Species-specific dietary patterns in primates and human dietary adaptations. In W. G. Kinzey (Ed.), *Evolution of human behavior: Primate models* (pp. 151-182). State University of New York.

Sussman, R. W. (1991). Primate origins and the evolution of angiosperms. *American Journal of Primatology, 23*, 209-223.

Sussman, R. W., & Phillips-Conroy, J. E. (1995). A survey of the distribution and density of the primates of Guyana. *International Journal of Primatology, 16*, 761-791.

Taub, D. (1980). Age at first pregnancy and reproductive outcome among colony-born squirrel monkeys (*Saimiri sciureus*, Brazilian). *Folia Primatologica, 33*, 262-272.

Terborgh, J. (1983). *Five New World primates: A study in comparative ecology.* Princeton University Press.

Terborgh, J. (1986). Community aspects of frugivory in tropical forests. In A. Estrada, & T. H. Fleming (Eds.), *Frugivores and seed dispersal: Tasks for vegetation science* (pp. 371-384). Springer.

Terborgh, J. (1990). Mixed flocks and polyspecific associations: Costs and benefits of mixed groups to birds and monkeys. *American Journal of Primatology, 21,* 87-100.

Terborgh, J., & Janson, C. H. (1986). The socioecology of primate groups. *Annual Review of Ecology and Systematics, 17,* 111-136.

Thompson, C. L. (2013). Non-monogamous copulations and potential within-group mating competition in white-faced saki monkeys (*Pithecia pithecia*). *American Journal of Primatology, 75*(8), 817-824.

Thorington, R. W., Jr. (1968). Observations of squirrel monkeys in a Colombian forest. In L. A. Rosenblum, & R. W. Cooper (Eds.), *The squirrel monkey* (pp. 69-85). Academic Press.

Tórrez, L., Robles, N., González, A., & Crofoot, M. C. (2012). Risky business? Lethal attack by a jaguar sheds light on the costs of predator mobbing for capuchins (*Cebus capucinus*). *International Journal of Primatology, 33,* 440-446.

Urbani, B., Kowalewski, M., Cunha, R. T. G., de la Torre, S., & Cortes-Ortiz, L. (Eds.). (2018). *La primatologia en Latinonamerica* 2. Ediciones IVIC.

van Schaik, C. P., & van Noordwijk, M. A. (1989). The special role of male *Cebus* monkeys in predation avoidance and its effect on group composition. *Behavior Ecology and Sociobiology, 24,* 265-276.

Verderane, M. P., Izar. P., Visalberghi, E., & Fragaszy, D. (2013). Socioecology of wild bearded capuchin monkeys (*Sapajus libidinosus*): An analysis of social relationships among female primates that use tools in feeding. *Behaviour, 150*(6), 659-689.

Visalberghi, E. (1990). Tool use in *Cebus. Folia Primatologica, 54,* 146-154.

Visalberghi, E., Albani, A., Ventricelli, M., Izar, P., Schino, G., & Fragaszy, D. (2016). Factors affecting cashew processing by wild bearded capuchin monkeys (*Sapajus libidinosus,* Kerr 1792). *American Journal of Primatology, 78,* 799-815.

Visalberghi, E., Haslam, M., Spagnoletti, N., & Fragaszy, D. (2013). Use of stone hammer tools and anvils by bearded capuchin monkeys over time and space: Construction of an archeological record of tool use. *Journal of Archaeological Science, 40*(8), 3222-3232.

Visalberghi, E., & Limongelli, L. (1994). Lack of comprehension of cause-effect relations in tool-using capuchin monkeys (*Cebus apella*). *Journal of Comparative Psychology, 108*(1), 15-22.

Westergaard, G. C., & Fragaszy, D. M. (1987). The manufacture and use of tools by capuchin monkeys (*Cebus apella*). *Journal of Comparative Psychology, 101*(2), 159-168.

Westergaard, G. C., & Suomi, S. J. (1995). The stone tools of capuchins (*Cebus apella*). *International Journal of Primatology, 16,* 1017-1024.

Westoll, A., Boinski, S., Stickler, C., Cropp, S., Ehmke, E., & Kauffman, L. (2003). Are vigilance, risk from avian predators and group size consequences of habitat structure? A comparison of three species of squirrel monkey (*Saimiri oerstedii, S. boliviensis,* and *S. sciureus*). *Behaviour, 140*(11-12), 1421-1467.

Wheeler, B. C. (2008). Selfish or altruistic? An analysis of alarm call function in wild capuchin monkeys, *Cebus apella nigritus. Animal Behaviour, 76*(5), 1465-1475.

Wheeler. B. C. (2010). Production and perception of situationally variable alarm calls in wild tufted capuchin monkeys (*Cebus apella nigritus*). *Behavior Ecology and Sociobiology, 64,* 989-1000.

Williams, L., Gibson, S., McDaniel, M., Bazzel, J., Barnes, S., & Abee, C. (1994). Allomaternal interactions in the Bolivian squirrel monkey (*Saimiri boliviensis boliviensis*). *American Journal of Primatology, 34,* 145-156.

Wright, K. A. (2005). *Interspecific and ontogenetic variation in locomotor behavior and habitat use, and postcranial morphology in* Cebus apella *and* Cebus olivaceous [Doctoral dissertation, Northwestern University].

Wright, K. A. (2007). The relationship between locomotor behavior and limb morphology in brown (*Cebus apella*) and weeper (*Cebus olivaceous*) capuchins. *American Journal of Primatology, 69,* 736-756.

Youlatos, D. (1998). Positional behavior of two sympatric Guianan capuchin monkeys, the brown capuchin (*Cebus apella*) and the wedge-capped capuchin (*Cebus olivaceous*). *Mammalia, 62,* 351-365.

Zhang, S.-Y. (1995). Sleeping habits of brown capuchin monkeys (*Cebus apella*) in French Guiana. *American Journal of Primatology, 36,* 327-335.

Zimbler-DeLorenzo, H. S., & Dobson, F. S. (2011). Demography of squirrel monkeys (*Saimiri sciureus*) in captive environments and its effect on population growth. *American Journal of Primatology, 73,* 1041-1050.

Zimbler-DeLorenzo, H. S., & Stone, A. I. (2011). Integration of field and captive studies for understanding the behavioral ecology of the squirrel monkey (*Saimiri* sp.). *American Journal of Primatology, 73*(7), 607-622.

CHAPTER 11

Howler Monkeys, Spider Monkeys, Woolly Monkeys, and Muriquis

Behavioral Ecology of the Largest Primates of the Americas

Michelle Bezanson, Karen B. Strier, and Robert W. Sussman

In this chapter we provide a behavioral and ecological overview of the largest living nonhuman primates in the Americas. The Atelidae, which include howler monkeys, spider monkeys, woolly monkeys, and muriquis, are divided into two subfamilies: Atelinae and Alouattinae (taxonomic classifications follow Rylands & Mittermeier, 2009). Three genera are included in the subfamily Atelinae: *Ateles*, the spider monkeys; *Lagothrix*, the woolly monkeys; and *Brachyteles*, the muriquis (previously also known as woolly spider monkeys). There is only one genus in the Alouattinae subfamily: *Alouatta*, the howler monkeys. Members of the Atelidae have a number of fascinating anatomical adaptations that relate to large-bodied movement in an arboreal canopy. For example, all Atelidae have prehensile tails that differ from *Cebus* in bony anatomy, the presence of a friction pad, and muscular arrangement allowing greater total angular excursion in flexion and extension (German, 1982; Lemelin, 1995; Organ, 2010). Both *Ateles* and *Brachyteles* have lost most or all of the thumb (pollex) and the hands are elongated and used as hooks as they move both within and between trees using arm-swinging locomotor modes. In contrast to *Alouatta*, all three Atelinae genera have an erect or orthograde posture and their locomotor morphology allows more suspensory locomotion and feeding postures below branches.

The genus *Alouatta* (the howler monkeys) currently includes 14 species: *A. arctoidea, A. belzebul, A. caraya, A. discolor, A. guariba, A. juara, A. macconelli, A. nigerrima, A. palliata, A. pigra, A. puruensis, A. sara, A. seniculus,* and *A. ululata*. Howlers weigh between 3–>12 kg (Table 11.1), depending on species and sex. *Alouatta* is adapted behaviorally, and to some extent morphologically, for a folivorous diet. Howler monkeys are distributed from southern Mexico to northern Argentina and as far east as the coastal forests of Brazil (Figure 11.1). They have the largest geographical range of the platyrrhine genera recognized by experts today (Rylands & Mittermeier, 2009).

Alouatta holds a special role in the history of primatology because mantled howler monkeys were the subject of the first systematic field study of a primate species. Clarence Ray Carpenter (1934) studied *A. palliata* on Barro Colorado Island (BCI) in the early 1930s. Carpenter's study set new standards of scientific methodology for field research on primate behavior. In 1935, Carpenter conducted the first study of *Ateles*, also at BCI. Since the 1970s, many studies have been conducted on atelids, especially howler monkeys (see Di Fiore et al., 2011, for a detailed bibliography). In fact, between 1990–1995, approximately 700 articles and abstracts were published on howler monkeys alone (Kinzey, 1997).

Ateles weighs between about 6–10 kg (see Table 11.1). They are very slender with extremely long arms and legs, giving them a spidery appearance as indicated by their common name. There are seven species of *Ateles* recognized: *A. belzebuth, A. chamek, A. fusciceps, A. geoffroyi, A. hybridus, A. marginatus,* and *A. paniscus*. Spider monkeys have a wide geographical range extending from southern Mexico, through Central America and the Amazon, and as far south as northeastern Peru, eastern Bolivia, and northern Brazil (Figure 11.2).

Woolly monkeys have thick fur and relatively robust limbs when compared to other members of the Atelidae. They weigh approximately 5–10 kg (see Table 11.1). Five species are currently recognized, *Lagothrix cana, L. flavicada, L. lagotricha, L. lugens,* and *L. poeppigii*. They are found in the upper Amazon Basin of Brazil, Peru, and Ecuador, and the Andean headwaters of the Orinoco in Colombia and Venezuela (Ramirez, 1988) (Figure 11.3).

Brachyteles, or the muriqui, weigh 7–10.2 kg (Lemos de Sá & Glander, 1993). They are intermediate between *Ateles* and *Lagothrix* in build and pelage. Muriquis are found only in the Atlantic coastal forests of southeastern Brazil (Figure 11.4). Two species of muriqui are now recognized: *B. arachnoides*, the southern muriqui; and *B. hypoxanthus*, the northern muriqui (Chaves et al., 2019). The northern muriqui shows little, if any, sexual dimorphism in canine length, and some individuals of this species have retained vestigial thumbs (Lemos de Sá & Glander, 1993). However, southern muriquis may be more

Table 11.1 Taxonomy, Common Names, and Average Body Weights of Atelidae [a]

Scientific Name	Common Name	Male Wt. (kg)	N	Female Wt. (kg)	N
Infraorder: Platyrrhini	Platyrrhines, New World monkeys				
Superfamily: Ceboidea/Ateloidea					
Family: Atelidae	Atelids				
Subfamily: Alouattinae	Alouattins				
Genus: *Alouatta*	Howling monkeys, howlers				
Alouatta arctoidea	Ursine howler	-		-	
Alouatta belzebul	Red-handed howler	7.2	27	5.5	26
Alouatta caraya	S. Am. black, or black and gold, howler	6.8	≥19	4.6	13
Alouatta discolor	Spix's red-handed howler	-		-	
Alouatta guariba	Brown howler/red and black howler	6.1	4	4.5	3
Alouatta juara	Juruá red howler	-		-	
Alouatta macconnelli	Guianan red howler	7.6	3	5	2
Alouatta nigerrima	Black howler	-		-	
Alouatta palliata	Mantled howler	7.1	≥56	5.3	≥67
Alouatta pigra	Central American black howler	11.4	2	6.4	4
Alouatta puruensis	Purú's red howler	-		-	
Alouatta sara	Bolivian red howler	-		-	
Alouatta seniculus	Red howler	7.2	61	5.6	61
Alouatta ululata	Maranhão red howler	-		-	
Subfamily: Atelinae	Atelins				
Genus: *Ateles*	Spider monkeys				
Ateles belzebuth	White-bellied spider	8.5	12	8.1	15
Ateles chamek	Black-faced spider	-		-	
Ateles fusciceps	Brown-headed spider	-		-	
Ateles geoffroyi	Geoffroy's, black-handed	8.2	56	7.7	≥101
Ateles hybridus	Variegated, or brown spider	8.3	2	9.1	7
Ateles marginatus	White-whiskered spider	-		-	
Ateles paniscus	Red-faced black spider	9.1	20	8.4	42
Genus: *Lagothrix*	Woolly monkeys				
Lagothrix cana	Geoffroy's woolly	9.5	3	7.7	1
Lagothrix lagotricha	Humboldt's woolly/brown woolly	9	3	5.8	6
Lagothrix lugens	Colombian woolly	-		6	1
Lagothrix poeppigii	Poeppig's woolly	7.1	6	4.5	9
Lagothrix flavicauda	Peruvian yellow-tailed woolly	-		-	
Genus: *Brachyteles*	Muriqui, woolly spider				
Brachyteles arachnoides	Southern muriqui	10.2	1	8.5	1
Brachyteles hypoxanthus	Northern muriqui	9.4	3	8.3	3

[a] Adapted from Di Fiore et al. (2011), original references included therein; taxonomic information from Rylands (2000), Groves (2001), Rylands et al. (2006), Rylands and Mittermeier (2009), Di Fiore et al. (2015).

sexually dimorphic in body size and canine size, and lack thumbs. Southern muriquis also lack the distinct patterns of dyspigmentation on their faces and genitalia that northern muriquis display. In both species, adult males have large testes and females have a pendulous clitoris (Nishimura et al., 1988).

The geographic distribution of muriquis—the Atlantic coastal forests of southeastern Brazil—lies in an area with the largest human concentration in the country (Lemos de Sá & Strier, 1992, 1993). Only 12 populations of northern muriqui are known to exist in the states of Minas Gerais and Espirito Santos (Strier et al., 2017) and, with fewer than 1,000 individuals, it is classified as

Critically Endangered by the International Union for the Conservation of Nature (IUCN, 2019). The southern muriqui is found mainly in protected tracts of forest in the states of Rio de Janeiro and São Paulo, but some of the small remaining forest fragments in São Paulo and Parana still support small populations of this species; the southern muriqui is also classified as Critically Endangered (IUCN, 2019).

HABITAT AND RANGING BEHAVIOR
Howler Monkeys

Alouatta, the howler monkeys (Figure 11.5), are found in a variety of habitats including primary forests,

Figure 11.1 Geographic distribution of the genus *Alouatta* (in black) based on spatial data from the IUCN Red List (IUCN, 2020).

(Map by Miguel Rangel Jr., subject to Attribution-Share Alike Creative Commons License)

Figure 11.2 Geographic distribution of the genus *Ateles* (in black) based on spatial data from the IUCN Red List (IUCN, 2020).

(Map by Jack Hynes; public domain document)

secondary forests, edge habitats, forests fragments, mangroves, marginal habitats, and agroecosystems (Fedigan et al., 1998; Arroyo-Rodriguez & Dias, 2010; Estrada et al., 2012; Chaves & Bicca-Marques, 2017). In the Andes, howler monkeys are found in cloud forests up to 3,200 m. Mantled howling monkeys in Costa Rica are described as crop raiders in areas of severe anthropogenic habitat modification (McKinney et al., 2015). Chaves and Bicca-Marques (2017) also found crop raiding in howlers but interpret the behavior as preference for larger fruits and more easily acquired fruits rather than a result of deforestation. They explain the conservation value of cultivated fruits for monkeys in areas where there is no conflict between landowners and nonhuman primates.

Across species, howler home ranges are 2–182 ha, and howlers typically travel fewer than 3,000 m per day (Di Fiore et al., 2011). *Alouatta* groups do not return to the same sleeping tree on successive nights but sleep in convenient trees near evening feeding sites. Alternating between feeding on fruit and leaves necessitates their travel between feeding trees each day.

Although all howler populations consume some fruits, their ability to rely heavily on leaves means they can subsist on resources that are densely and evenly distributed but that also provide little energy per item ingested and are difficult to digest. Together these factors enable and necessitate howlers to minimize energy expenditure both in general activity and in movement and travel. Milton (1980) found mantled howler monkey travel was directly followed by feeding 90% of the time. In her study at BCI, the mean distance traveled was 443 m, and most daily travel was between 300–600 m. Milton's study groups used one or two pivotal trees each day. These trees were usually primary fruit resources on which the group focused its daily activity. The group would leave and return to the pivotal tree, balancing its diet with leaves and searching for nearby ripening fruit resources. When switching between these resources, the group was cohesive, and group members traveled in direct and single line fashion to a new pivotal tree. Usually, a male led the group in travel, and while traveling, the group traversed at least one and sometimes many fig trees. Other observers have not seen this pattern of resource utilization; earlier studies describe howlers as using very small areas for a limited period of time, rarely returning to the same tree in the same day, and then making one long continuous movement to a new part of their home range (Carpenter, 1935; Richard, 1970; Smith, 1977). In some studies, the

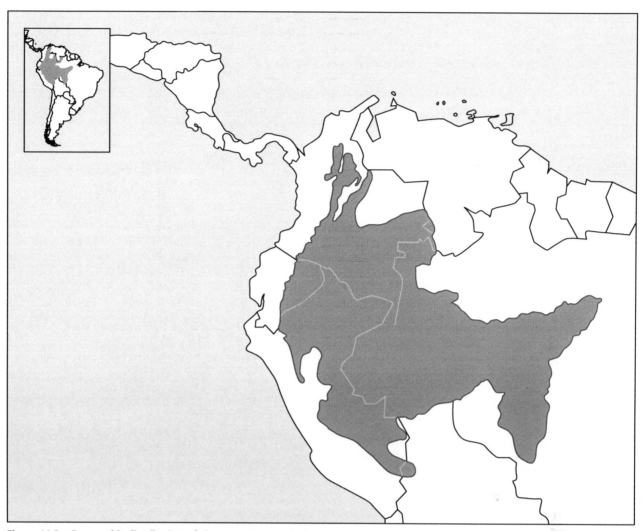

Figure 11.3 Geographic distribution of the genus *Lagothrix* (in gray).
(Redrawn from Di Fiore et al., 2015)

home range was traversed in one to two weeks (Richard, 1970; Baldwin & Baldwin, 1972; Gaulin & Gaulin, 1982), while other studies have described subgrouping patterns in *Alouatta* that resulted in different parts of the home range being used concurrently (Wang & Milton, 2003; Bezanson et al., 2008; McKinney, 2019).

These different patterns of range use may be dependent upon the distribution of resources or social preferences. There is considerable variation in home range size, both inter- and intraspecifically. Nonetheless, the average time spent traveling per day by howler groups is consistently very small relative to that of other atelids. Travel is further minimized by intergroup spacing mechanisms. Home ranges of howler groups overlap extensively and sometimes groups have only minimal areas of exclusive use within their ranges (Carpenter, 1935; Smith, 1977; Glander, 1978; Milton, 1980; Neville et al., 1988, updated in Di Fiore et al., 2011). Home range size is inversely related to howler population density (Crockett & Eisenberg,

1987). Spacing between groups is hypothesized to be maintained by dawn choruses and by loud calls given by adult males before and after travel to a new site (Carpenter, 1934; Baldwin & Baldwin, 1972; Whitehead, 1987; Chiarello, 1995b; Briseño-Jaramillo, 2017). When groups do come face-to-face, they usually either ignore or avoid each other, or else enter into a ritualized vocal battle after which one or both groups move in different directions. Thus, howlers may avoid excessive movement by (1) moving directly to new feeding sites, (2) not having preferred sleeping sites to which they return each night, and (3) using loud calls to maintain spacing between groups.

Spider Monkeys

In Amazonia, *Ateles* (Figure 11.6 and Figure 11.7) occur most frequently in mature, high forest in unflooded areas, although they also can be found in remnant/degraded forests, mangrove forests, edge areas, deciduous forests (Hernandez-Camacho & Cooper, 1976;

Figure 11.4 Geographic distribution of the genus *Brachyteles* (in gray) based on spatial data from the IUCN Red List (IUCN, 2020). Southern muriqui (*B. arachnoides*) on left; northern muriqui (*B. hypoxanthus*) on right.
(Map by Chermundy, subject to Attribution-Share Alike Creative Commons License)

Figure 11.5 *Alouatta palliata* in Costa Rica, La Suerte Biological Research Station.
(Photo credit: Michelle Bezanson)

Mittermeier & Coimbra-Filho, 1977; Rosenberger et al., 2009). Spider monkeys have large ranges due to their frugivorous diet, and in Santa Rosa National Park, Costa Rica, Freese (1976) observed *A. geoffroyi* to have a more extensive distribution than sympatric *Alouatta* along with a wider utilization of forest types. *Ateles* is reported to have home ranges that vary from 95 to over 900 ha (reviewed in Di Fiore et al., 2011).

Ateles are quite variable in their daily ranging patterns and in the sizes of foraging parties. Some groups remain together throughout the day (Durham, 1975; Coehlo Jr. et al., 1976; Wallace, 2008), other groups split into subgroups and reform at night using the same sleeping trees (Carpenter, 1935; Coehlo et al., 1976), and still others remain dispersed and even sleep 400–800 m apart (Klein & Klein, 1975). In Peru, day ranges vary from 465–4,070 m (avg. = 1,977 m), with males traveling more and feeding less than females (Symington, 1988a,b). However, average day ranges of >2,300 m are also reported (Di Fiore et al., 2011). Day ranges of groups and subgroups are directly dependent upon the density and dispersion of fruit

sources (Richard, 1970; Klein & Klein, 1977; Terborgh, 1983; Symington, 1988a; Di Fiore et al., 2008; Wallace, 2008; Rodrigues, 2017).

Spider monkeys are multiple central place foragers in that they choose one of a number of central places for sleeping sites each night (Chapman, 1989b). Thus, subgroups reduce travel costs by selecting sleeping sites close to current feeding areas. Group home ranges overlap 10–30% and relations between groups may be antagonistic, especially between adult males (Klein & Klein, 1975; Symington, 1988b; Chapman, 1990b). Adult males are described as using core areas that are larger than those of adult females (90 ha versus 50 ha in Costa Rica, for example), while females with infants have the smallest core areas (Fedigan et al., 1988). The core areas of individuals overlap, and male ranges tend to overlap those of several females. However, core areas of all adult females in the group overlap those of neighboring females with whom they often travel (Fedigan et al., 1988). Adult males may travel together and are frequently sighted near

Figure 11.6 *Ateles geoffroyi* in Costa Rica, La Suerte Biological Research Station.
(Photo credit: Shawn Hanna)

group boundaries, whereas females with infants often travel alone or in small groups and avoid these boundaries (Chapman, 1990b). In general, the social system and ranging pattern of spider monkeys are extremely similar to that of chimpanzees (Symington, 1990; Chapman et al., 1995).

Woolly Monkeys

Lagothrix (Figure 11.8) occur between sea level and 2,500 m altitude and can be found in gallery forest, palm forest, flooded and unflooded rainforest, swamp forest, and cloud forest (Di Fiore et al., 2011). Woolly monkeys are particularly vulnerable to forest disturbance and are not found in secondary forest areas (Ramirez, 1988). They have been observed in high, non-flooded forest to low-lying forest types (Peres, 1996). The yellow-tailed woolly monkey, *Lagothrix flavicauda*, is restricted to a very small area of mountain rainforest in northern Peru and is one of the rarest species in South America (De-Luycker & Heymann, 2007).

Lagothrix home ranges are relatively large with a population home range averaging 580 ha in highland Colombian woolly monkeys (García-Toro et al., 2019). Estimates of *Lagothrix* home ranges vary from about 100 to over 1,000 km² (reviewed in Di Fiore et al., 2011). These are the largest home ranges reported for strictly arboreal monkeys. However, given the large size of the individuals and of groups, this is not surprising (Peres, 1996). All observers indicate that home range boundaries of woolly monkeys overlap, but whereas some retain little if any areas of exclusive use, others overlap by 50% or less (Di Fiore et al., 2011). Likewise, although some groups do not defend home ranges against conspecifics (Ramirez, 1988; Nishimura, 1990; Peres, 1996), at other times intergroup encounters can be hostile, with males playing an active role. There is high variation across studies in average day ranges (540–2,880 m) and in whether groups are cohesive or divide into subgroups of various types during travel (reviewed in Di Fiore et al., 2011).

Figure 11.7 *Ateles geoffroyi* in Costa Rica, La Suerte Biological Research Station. The common name "spider monkey" originates from the unique silhouette of a leaping *Ateles*. (Photo credit: Shawn Hanna)

Muriquis

Brachyteles (Figure 11.9) exist both in the evergreen forests along the eastern coastal slopes of southeastern Brazil, where annual rainfall exceeds 2,000 mm, and to the west of these slopes in drier semi-deciduous forests. These latter forests have a pronounced dry season which lasts up to six months (Strier et al., 2001). The muriqui is found in primary forest habitats as well as in tall secondary forest and even in severely disturbed remnant forests (Milton, 1984a; Fonseca, 1985; Lemos de Sá & Strier, 1992). Strier (1992b) has suggested that muriquis may be found at higher densities in mixed habitats containing both primary and secondary vegetation. These habitats offer a greater variety of potential foods for the eclectic muriqui (Pinto et al., 1993; de Moraes et al., 1998). Muriquis typically spend most of their time in the mid- and upper canopy (Nishimura et al., 1988), but they have also been observed to spend some of their time on the ground (Mourthé et al., 2007; Tabacow et al., 2009).

Muriquis have relatively large home ranges, from 70–>300 ha that may increase with group size when densities permit (Lima et al., 2019). A group of seven southern muriquis at Fazenda Barreiro Rico (FBR), São Paulo had the smallest home range (Milton, 1984b), whereas home ranges of one group of northern muriquis in a 1,000-ha protected reserve near Caratinga, Minas Gerais (Reserva Particular Patrimonio Natural—Feliciano Miguel Abdala, RPPN-FMA) increased from 168 to 184 ha as the group increased from 26 to 34 members (Strier, 1993). There was also a 46% overlap of the study group range with that of an adjacent group (Strier, 1987a). By the time this study group had increased to some 60 members, its home range had increased to >300 ha (Dias & Strier, 2003). Muriqui population growth in this forest fragment is now thought to constrain increases in the group's home range size, leading to high overlap between adjacent groups (Lima et al., 2019) and increasing use of the forest floor (Strier & Ives, 2012).

Adjacent groups exchange long calls and often avoid one another by moving away from each other. Aggressive intergroup encounters occur in areas of overlap, usually over a particular large fruit patch, and both males and females participate (Strier, 1992b,c). These encounters are usually brief and involve vocal exchanges, chases, and touching. However, sometimes the two groups feed in adjacent trees, exchange "neigh" calls but do not threaten one another (Strier, 1992b,c). Muriquis do not defend territories (Strier, 1990b), but recent data suggest that they may defend core areas located near the borders of group home ranges (Lima et al., 2019).

Muriqui groups usually sleep at night in the mid-canopy. The whole group generally sleeps in neighboring trees, although sometimes a small group may settle in a single tree. There are no regularly used sleeping sites, but animals settle near late afternoon feeding sites. Long calls often are exchanged prior to settling for the night and serve to reduce distance between group members (Nishimura et al., 1988).

Day range length appears to vary with group size and with food distribution and availability. The small group of southern muriquis at Fazenda Barreiro Rico (FBR), São Paulo, had an average day range of 630±128 m (range 350–1,400 m) (Milton, 1984a), whereas northern muriqui day ranges at the RPPN-FMA averaged 1,283±642 m (range 141–3,403 m) when the group had 23–27 members (Strier, 1987a) and did not increase significantly despite a twofold (Dias & Strier, 2003) and ultimately a sixfold increase in group size (Lima et al., 2019). The larger group at RPPN-FMA depended more on fruit and had a relatively greater number of large fruit patches

Figure 11.8 *Lagothrix lagotricha poeppigii* in Ecuador, Tiputini Biodiversity Station.
(Photo credit: Kelsey Ellis)

Figure 11.9 Northern muriqui (*Brachyteles hypoxanthus*) mother and infant, Projeto Muriqui de Caratinga, RPPN Feliciano Miguel Abdala.
(Photo credit: Pablo Fernicola, https://creativecommons.org/licenses/by/4.0/)

available than did the group at Fazenda BR (Strier, 1989, 1991a). At RPPN-FMA, day ranges were longer during the wet months, when more fruit was available than during dry months. Because of their large size, muriquis must cover relatively long distances to obtain sufficient fruit, and their mode of suspensory locomotion enables them to do this. Strier (1987a) described a pattern of "camping out" at large fruit patches by the northern muriquis in her study group, which Talebi and Lee (2010) found also occured in their study group of southern muriquis at the continuous forest of Parque Estadual de Carlos Botelho.

POSITIONAL BEHAVIOR: LOCOMOTION AND POSTURE

Howler Monkeys

Alouatta have been described as slow, deliberate quadrupeds with relatively long stride lengths that engage in a variety of below-branch suspensory postures (Richard, 1970; Mendel, 1976; Fleagle & Mittermeier, 1980; Cant, 1986; Gebo, 1992; Bicca-Marques & Calegaro-Marques, 1993, 1995; Youlatos, 1993, 1998; Bergeson, 1996; Bezanson, 2009; Guillot, 2009; Youlatos & Gasc, 2011; Youlatos & Guillot, 2015). The genus *Alouatta* has been observed to walk quadrupedally 30–80% of all travel observations, climb 7–33%, and leap 1.5–4% of observations (Fleagle & Mittermeier, 1980; Bezanson, 2009; Guillot, 2009). Howlers rarely locomote by tail-arm suspension (Cant, 1986; Bergeson, 1996; Bezanson, 2012). In a comparison of sympatric atelines, Guillot (2009) found *Alouatta* to be more cautious with a more limited positional repertoire than *Ateles* or *Lagothrix*.

Alouatta primarily engage in above-branch sitting postures and below-branch suspensory postures while feeding (Table 11.2). Howlers have also been observed

climbing on relatively small supports, during which they use their prehensile tails in balance and support (Fleagle & Mittermeier, 1980; Bergeson, 1996; Bezanson, 2012). Adult howlers have been observed to use their prehensile tails as mass-bearing fifth limbs in 25–35% of feeding/foraging contexts (Johnson & Shapiro, 1998; Lawler & Stamps, 2002). Howler monkeys use their tails in suspensory postures while feeding in a horizontal tripod or inverted bipedal posture near the trunk of the tree. Howlers rarely come to the ground to cross open areas or to feed, but ground behavior has been observed in some habitats (e.g., Neville et al., 1988).

Spider Monkeys

Ateles, like *Alouatta*, run and walk quadrupedally along the tops of large branches. Among small terminal branches, where howlers move carefully and slowly, spiders use their arms to swing with great speed and agility. This fast, suspensory locomotion is a feature of positional behavior unique to hylobatids and spider monkeys and may be an adaptation in these primates to exploit widely dispersed and ephemeral food patches (Cant, 1986; Strier, 1992a; Youlatos, 2008). Suspensory locomotion has been observed to occur in 23–38% of locomotor modes (Mittermeier 1978; Cant et al., 2003). The arms and hands of spider monkeys are adapted for this agile, suspensory mode of locomotion. The arms are long in relation to body length and the hands are somewhat like suspensory hooks, with reduced or absent thumbs and long permanently curved fingers.

Swinging by the arms below branches is referred to as *brachiation*; however, spider monkeys do not brachiate in the true sense of the term like gibbons; rather they engage in arm-swinging suspensory postures that

Table 11.2 Percentage of Time Spent in Major Feeding Postures by *Alouatta, Ateles, Lagothrix,* and *Brachyteles*

Feeding Posture	Alouatta caraya [a]	Alouatta palliata [b]	Alouatta seniculus [c]	Ateles geoffroyi [d]	Ateles paniscus [e]	Lagothrix lagotricha [f]	Brachyteles hypoxanthus [g]
Sit/squat/recline	61.4	56.5	40.1	25.7	34.9	34.1	38.2
Stand/crouch	9.4	1.6	28.7	17.9	7.6	32.3	0.8
Tail-only suspension	–	17.8	–	–	18.8	–	0.5
Tail-hindlimb suspend	–	2.7	–	–	16.0	–	38.0 [h]
Tail-forelimb suspend	4.1	0	–	33.3	8.2	34.4	1.0
Horizontal tripod/ inverted biped	24.2 [i]	5.2	30.4	13.1	7.9	–	–
Tail-forelimb and -hindlimb	–	–	–	–	–	–	21.4

[a] Prates and Bicca-Marques (1995).
[b] Bezanson (2009).
[c] Guillot (2009).
[d] Bergeson (1996).
[e] Youlatos (2002).
[f] Defler (1999).
[g] Iurck et al. (2013).
[h] Tail-hindlimb suspension and horizontal tripod is combined.
[i] Horizontal tripod/inverted biped termed "bridge."

are often assisted by the prehensile tail (Mittermeier & Fleagle, 1976; Mittermeier, 1978; Bergeson, 1996; Youlatos, 2008). They usually use their tails as an additional limb while progressing through the branches and do not propel themselves solely with their arms (Richard, 1970; Youlatos, 2002, 2008). Spider monkeys also use their tails in postural modes; and in a behavior to help young individuals cross gaps, a female bridges a gap in the canopy by holding branches and then allowing juveniles to cross over her body.

Fleagle and Mittermeier (1980) found *A. paniscus* in Surinam to divide their travel fairly evenly between bimanual locomotion (38.6%), climbing (31.1%), and quadrupedal walking and running (25.4%). Leaping (4.2%) and bipedalism (0.7%) accounted for a small proportion of travel. Using different categories, Cant (1986) observed *A. geoffroyi* in Guatemala to use quadrupedalism 52% of the time, tail-arm suspension 25%, and postures involving appreciable tail, but not arm, suspension 23% of the time. Leaping was only seen in 1% of the observations. Thus, the tail was used in a major way approximately 50% of the time in this study, similar to more recent findings on other species of spider monkeys conducted elsewhere (reviewed in Youlatos, 2008) (see Table 11.2). In Costa Rica, Bergeson (1996) observed *A. geoffroyi* to engage in tail-assisted suspensory postures 33.3% of feeding and foraging observations. Spider monkeys also used different positional behaviors in different locations of the canopy. These postures did not conform to previous predictions. It is generally assumed that suspension is used more while feeding in the periphery of trees. However, Bergeson (1996, 1998) found that inverted bipedal suspensory/horizontal tripod postures were used more frequently close to the trunk, sitting was more common on medium supports in the middle of the canopy, and quadrupedal standing was used most in the terminal branches. He also observed high-tension prehensile tail use to be more common when spider monkeys fed on leaves than on fruit, although suspension near the trunk also allowed access to vine stems and fruits. These findings were similar to those on howler monkeys, though, overall, spiders used their tails more than howlers (70% versus 50% of all posture and locomotion).

Woolly Monkeys

Lagothrix are robustly built and in many ways resemble howler monkeys. Although similar to *Ateles* and *Alouatta*, in their above-branch quadrupedalism, climbing, and suspensory feeding postures (Defler, 1999; Cant et al., 2003), *Lagothrix* differ in frequency of these locomotor modes, engaging in almost twice as much quarupedalism as *Ateles*. Orthograde suspensory locomotion, using

hand-over-hand movements or assistance of the tail, occurs rarely (about 6% of the time) (Defler, 1999). The tail is used frequently for suspensory feeding postures and in pronograde forelimb swing locomotion (Cant et al., 2003). Woolly monkeys leap less frequently than spider monkeys and use larger substrates during travel. They are usually slow but can move rapidly if need be (Fooden, 1963; Moynihan, 1976). *Lagothrix* locomotion thus overlaps and is considered intermediate with that of sympatric *Ateles* and *Alouatta* (Guillot, 2009).

Muriquis

Brachyteles primarily use suspensory locomotion and climbing and can move very quickly through the canopy. In this large primate, as in *Ateles*, rapid suspensory locomotion may be critical for moving between dispersed patches of high-quality foods like fruit and nectar (Strier, 1987a; Rosenberger & Strier, 1989). *Brachyteles'* fingers are long and curved to hook over the tops of branches (Lemos de Sá & Glander, 1993). During rapid travel they follow each other on well-tested branches (Strier, 1992b). Quadrupedal walking and running are used only on large, mainly horizontal supports. Muriquis also employ leaping on a regular basis. During feeding, they sit, stand, and use suspensory postures. Most feeding takes place on small, terminal branches and involves below-branch suspension. The prehensile tail is used in many locomotor and postural contexts (Nishimura et al., 1988; Iurck et al., 2013; Table 11.2).

DIET
Howler Monkeys
Although there is substantial intra-generic variation in diet, *Alouatta* typically consume the largest proportion of leaves of the atelids. Depending on the season, between 25–70% of their diet is immature and mature leaves (Table 11.3). In general, younger leaves are higher in protein and water while lower in less-digestible fiber (Lambert, 2011). The morphology of the intestinal tract (Martin et al., 1985), molars (Kay & Hylander, 1980), salivary glands (Milton, 1987), and the gut histology (Hladik, 1967) of the howler monkey are adapted for a largely folivorous diet. However, these morphological features are not as highly specialized as those seen in the Colobinae or Indriidae. Milton (1978, 1984a, 1998) has argued that *Alouatta* are more "behavioral folivore" than "anatomical folivore." This means that howlers use behavioral strategies, such as long periods of rest and selective foraging with a focus on particular plant parts. However, they do have some anatomical strategies for dealing with a leafy diet, including an enlarged hindgut and relatively slow

Table 11.3 Percentage of Feeding Time Invested in Different Dietary Items by *Alouatta* (Howler Monkeys)

Species	Mature Leaves	Immature Leaves	Fruit	Flowers	Other	Site and References
A. belzebul	5.0	19.8	55.6	5.7	13.9	Paranaita, Brazil (Pinto and Setz, 2009)
A. caraya	41	4	24	17	10	Isla Brasilera, Argentina (Kowalewski, 2007)
A. palliata	19	50	13	18	0	La Pacifica, Costa Rica (Glander, 1978)
A. palliata	10	38	42	10	0	BCI, Panama (Milton, 1980)
A. palliata	27.0	27.8	34.8	7.9	1.6	Finca La Luz Nicaragua (Williams-Guillen, 2003)
A. pigra	7.9	37.2	40.8	10.6	3.4	Community Baboon Sanctuary (Silver et al., 1998)
A. seniculus	7.5	44.5	42.3	5.4	0.1	Finca Merenberg, Colombia (Gaulin and Gaulin, 1982)

food passage rates, which could explain why more than 30% of their energy is derived from fermentation end products (Milton, 1980, 1987a, 1998; Dias & Rangel-Negrín, 2015).

Due to the anatomical difficulties with digestion, howlers eat very small amounts of a great variety of species and concentrate on a few species for bulk (similar in many ways to the diet of sifakas, *Propithecus* spp.). For example, howlers at some research sites are reported to eat more than 100 different plant species (Milton, 1980; Stoner, 1996; Williams-Guillen, 2003; Dias & Rangel-Negrín, 2015), and a small number of more staple plant species (12–15) account for a large proportion (up to 80%) of the diet. Howlers are very selective in the leafy portion of their diet, usually choosing species which are not abundant in the forest and often selecting very few individual trees of particular species (e.g., Hladik & Hladik, 1969; Smith, 1977; Glander, 1978; Milton, 1980; Chapman, 1988a; Chiarello, 1994; Estrada et al., 1999).

The remainder of the howler diet is made up of fruit, flowers, and buds (between about 40–70%). Howlers are reported to eat both mature and immature fruits (Glander, 1978; Milton, 1978, 1980; Estrada et al., 1984; Julliot & Sabatier, 1993; Williams-Guillen, 2003; Righini et al., 2017; reviewed in Dias and Rangel-Negrín, 2015). At some sites, fruits comprise a smaller proportion of the howler diet, with very few species accounting for a great percentage of the fruits eaten (e.g., Milton, 1980; Estrada, 1984; Chapman, 1988a). In many forests, figs (*Ficus* ssp.) are an important component of the diet, sometimes accounting for up to 40% of time spent feeding. Figs are patchily distributed in the forest but are usually available throughout the year. In an extensive review of 116 published studies on howler diet, Dias and Rangel-Negrín (2015) describe the howler diet as varying inter- and intraspecifically by forest type, season, rainfall, group size, and habitat size.

The seeds of fruits eaten by howlers are not destroyed and are often more viable than controls in germination

experiments. Furthermore, the slow food passage rate results in seeds being dropped away from the parent plant. These factors indicate that howlers may be important in dispersing many of the fruit trees that they feed upon (Hladik & Hladik, 1969; Howe, 1980; Estrada & Coates-Estrada, 1984; Estrada et al., 1984; Chapman, 1989a; Arroyo-Rodríguez et al., 2014; Anzures-Dadda et al., 2016). In fact, Howe (1980) found that 70% of the seeds removed from a commonly used fruit tree in Panama were taken by howlers. Furthermore, it appears that some insects are ingested along with figs and other fruit. Howlers may be aiding the successful dispersal of these plants by "cleaning" them of insect parasites while gaining protein and other nutritional benefits not found in plant foods (Bravo & Zunino, 1998).

The relatively indigestible diet necessitates balancing nutrients by feeding on a mix of food types each day (Hladik & Hladik, 1969; Smith, 1977; Milton, 1980; Gaulin & Gaulin, 1982; Williams-Guillen, 2003). However, in a four-year study in Costa Rica, Chapman (1987) found that although howlers ate a mixture of leaves, fruits, and flowers in most months, they fed almost exclusively on flowers in one month and on leaves in two months of the study. He also found that the diet was not consistent from year to year. This dietary flexibility is a critical factor to survival in various habitats (Arroyo-Rodriguez & Dias, 2010).

Protein content appears to be an important factor in food choices of howler monkeys (but see Righini et al., 2017). They require approximately 15% of their daily diet in dry weight to be protein. Howlers have also been observed to come to the ground for water, salt, or dirt (Glander, 1978; Terborgh, 1983; Neville et al., 1988; Gilbert & Stouffer, 1989; Izawa & Lozano, 1990; Izawa et al., 1990). Dirt eating, or geophagy, has been observed in other leaf-eating primates, but its significance is still poorly understood (i.e., Oates, 1994). The resources ingested may serve as antacids, to help absorb toxins or to supply mineral supplements (Kay & Davies, 1994).

Spider Monkeys

In contrast to *Alouatta*, *Ateles* eats few leaves (Table 11.4). Approximately 80% of its diet, on average, consists of fruits of which almost all are ripe (Hladik & Hladik, 1969; Klein & Klein, 1977; Milton, 1981; Chapman, 1988a; Symington, 1988a; reviewed in Di Fiore et al., 2011). The remainder is made up of young leaves and buds, flowers, and dead or decaying wood. The wood is often obtained from the same sources over many months. Although some figs are eaten, these fruits are much less important to spider monkeys than to howlers. At some sites, spider monkeys eat palm fruits throughout the year and, depending on the season, these fruits account for 20–60% of the diet (Klein & Klein, 1975; Nunes, 1988).

The distribution of fruit resources is one of the determinants of *Ateles* social structure. Most fruits eaten by *Ateles* occur at low densities and are widely distributed, patchy, and ephemeral (Silva-Lopez et al., 1987; Chapman, 1988b, 1990a; Di Fiore et al., 2011). The number of individuals able to feed together is dependent upon food patch size and density. Although on some days spider monkeys may visit a large number of plants for fruit, the annual diet is not extremely diverse and two to three plant species often account for over half of the yearly diet (Hladik & Hladik, 1969; Cant, 1977; Klein & Klein, 1977; White, 1986; Chapman, 1988a).

When spider monkeys consume fruit, seeds are often swallowed whole and a major volume of the feces contains undamaged seeds. These seeds are defecated intact and are viable after defecation. Thus, similar to howler monkeys, spider monkeys are also important seed dispersal agents, especially for plants with relatively large seeds (Hladik & Hladik, 1969; Klein & Klein, 1975, 1977; Muskin & Fischgrund, 1981; Howe, 1983; Fleming et al., 1985; Chapman, 1989a; Stevenson et al., 2002; Di Fiore et al., 2011). Mittermeier and van Roosmalen (1981) found that *Ateles* were dispersal agents for at least 81.3% of the fruit species upon which they fed.

Woolly Monkeys

Like *Ateles*, *Lagothrix* at many sites is primarily frugivorous, with fruits accounting for up to 70–80% of the annual diet in some habitats (Soini, 1986; Ramirez, 1988; Nishimura, 1990; Peres, 1994) (Table 11.4). Seeds, leaves, flowers, bark, twigs, insects, and occasional vertebrates also are included in the diet. In some populations, however, insects can account for more than 25% of the diet at times (Stevenson, 2006; Fonseca et al., 2019). Because of the extremely worn incisors and canines of most mature museum specimens he examined, Fooden (1963) speculated that woolly monkeys ate hard-shelled fruits. However, others have theorized that their dental structure was secondarily adapted for leaf-eating (Kay, 1973; Eaglan, 1984; Rosenberger & Strier, 1989). In fact, during seasons when fruit is scarce, both of these foods can become important. In some months, seeds account for 25–35% of the diet and young leaves close to 50% (Soini, 1986; Peres, 1994). In a field site in Brazil, woolly monkeys selected primarily large food patches, regardless of feeding party size (Peres, 1996). Among the atelines, *Lagothrix* and *Ateles* have the most similar diets (Di Fiore et al., 2011). It would appear that reliance of *Lagothrix* on leaves, seeds, or insects during some seasons would lead to food niche separation between these two genera, especially at sites where they occur sympatrically.

Muriquis

Muriquis, like howlers, may have a highly folivorous diet, with leaves accounting for 22–67% of their annual diet and reaching 80–90% in some populations in dry months (Milton, 1984a). The annual diet of northern *Brachyteles* in the small forest at the RPPN Feliciano Miguel Abdala (RPPN-FMA) in Minas Gerais, Brazil, consists of 51% leaves, 32% fruit and seeds, 11% flowers and floral nectar, and 6% other plant material (Strier, 1991a). When available, fruit makes up over half of the diet, and for southern muriquis living in continuous

Table 11.4 Percentage of Feeding Time Invested in Different Dietary Items by *Ateles*, *Lagothrix*, and *Brachyteles*

Species	Leaves	Fruit	Flowers	Insects	Other	References
Ateles belzebuth	9	87	1	0.7	3	Dew (2005)
A. geoffroyi	15.5	76	6.4		2	Cant (1990)
A. geoffroyi	17.2	82.2	1.0		0.9	Campbell (2000)
A. paniscus	10.7	85.7	2.9		0.7	Wallace (2005)
Lagothrix lagotricha	11.4	78.9	0.1	4.9	4.7	Defler and Defler (1996)
L. flavicauda	23.3	46.3	1.8	19.1	9.4 (moss and buds)	Shanee (2014)
Brachyteles arachnoides	24.5	68.3	4.9		2.2	de Carvalho Jr. et al. (2004)
B. hypoxanthus	51	32	11		6	Strier (1991a)

forest at the Parque Estadual Carlos Botelho (PECB) in São Paulo, fruit accounts for an average of 75% of the diet (Talebi et al., 2005). Leaves probably are eaten for bulk and protein and not simply because of their relative abundance. Immature leaves and ripe fruit are preferred. Muriquis are probably important dispersers for many of the fruit species they eat (Strier, 1992b; Martins, 2008). During periods of fruit scarcity and flower abundance, floral nectar becomes a valuable resource and the muriquis might serve as pollinator for some of these plants (Milton, 1984a; Ferrari & Strier, 1992). Muriquis are able to include more fruit in their diets than sympatric howler monkeys because, with their rapid locomotion, they can find it more easily (Strier, 1992a). Muriquis are also dominant to sympatric howler and capuchin monkeys, which often wait for muriquis to finish eating before entering a fruiting tree (Strier, 1992b; Dias & Strier, 2000).

At the RPPN-FMA, northern muriquis ate 63 species of plant from 57 genera during a one-year study period, and both northern and southern muriquis were highly selective and chose a number of rare plant species (Milton, 1984a; Strier, 1991a). Muriquis are able to tolerate high levels of tannins, presumably due to the relatively rapid rate of food passage through the digestive tract (Milton, 1984a,e; Strier, 1992b). Besides large body size, the dentition, jaw morphology, and digestive system of the muriquis show adaptations for folivory (Gaulin, 1979; Rosenberger & Strier, 1989). Muriquis obtain most of their water from the plants they consume, although in the dry season they drink from tree cavities and standing water (Strier, 1992b).

Activity Patterns

Primates that depend mainly on plant materials for food can eat large quantities at one sitting but must have concomitant long periods of rest for digestion (Clutton-Brock & Harvey, 1977). Both *Ateles* and *Alouatta* fit this pattern. In *Ateles*, long rest periods follow periods of intensive feeding when the monkeys rapidly ingest large quantities of food (about 10% of the animal's weight). There may be three to five periods of alternating activity and rest, and rest becomes more frequent and longer during the midday (Richard, 1970; Klein & Klein, 1977). *Lagothrix* are reported to rest less and be more active than *Ateles*, which may reflect their larger group sizes or slower rate of travel (reviewed in Di Fiore et al., 2011).

Ramirez (1988) observed *Lagothrix* to remain active throughout the day, but most travel and feeding occurred in the first hours of the morning and last two hours of the day. The animals mainly rested in midday. The pattern is similar in *Alouatta*, eating during several sessions throughout the day, separated by resting periods of two or more hours (Gaulin & Gaulin, 1982). Peaks of feeding occur in the early morning and late afternoon, and resting peaks at midday. Howlers rest up to 50–80% of the day (Smith, 1977; Gaulin & Gaulin, 1982; Neville et al., 1988; Bergeson, 1996; Estrada et al., 1999; Bezanson et al., 2008), and rest more than spider monkeys throughout the day (Richard, 1970; Klein & Klein, 1977; Gaulin & Gaulin 1982; Crockett, 1987).

The activity pattern of *Brachyteles* varies from site to site and is dependent upon seasonality, habitat, group size, and food quality and distribution (reviewed in Di Fiore et al., 2011). For example, the small group of highly folivorous muriquis at Fazenda Barreiro Rico in São Paulo spent more time resting and feeding and less time traveling than a larger group of more frugivorous northern muriquis at the RPPN-FMA or than southern muriquis in a large, continuous forest (Milton, 1984a; Strier, 1987a; Talebi & Lee, 2010). The northern muriquis at the RPPN-FMA shifted their activity times between seasons, resting during the hot mid-afternoon in summer and remaining inactive until midday during winter. However, the actual amount of time they spent feeding (19%) and traveling (29%) remained constant throughout the year (Strier, 1987a). Although, generally, the activity budgets of northern males and females were similar, a lactating female spent a greater proportion of her time feeding than other adults (Strier, 1987a).

Predation

In primates, it is relatively rare to observe predation events. Perhaps due to their slower-moving, energy-maximizing behavioral repertoire, researchers have observed several predation events in howler monkeys. Peres (1990a) and Sherman (1991) each have observed a harpy eagle kill an adult male *Alouatta seniculus,* and in Peru, Terborgh (1983) and Eason (1989) have observed unsuccessful attacks on howlers by these eagles. Carpenter (1934) observed a possible attack by an ocelot on a young howler, and Peetz et al. (1992) observed a jaguar killing a howler (see also, Cuarón, 1997, and other accounts in Ferrari, 2009). Bianchi and Mendes (2007) collected 60 ocelot fecal samples and found howler (*A. guariba,* n = 12 bones from 7 individuals), muriqui (n = 1 individual), and capuchin (n = 1 individual) bones. Chapman (1986) reports an unsuccessful attack of a howler by a boa constrictor. As described above, howlers often come to the ground for water and may be vulnerable at these times.

In the Andes, *Lagothrix* have been observed to be hunted by eagles (Lehman, 1959), and in Columbia, two species of eagles are known as "Aguilas Churuqueres" (woolly monkey hawks) because of their reputation as *Lagothrix* predators (Hernandez-Camacho & Cooper, 1976). Jaguars may also hunt *Lagothrix* (Ramirez, 1988). The woolly monkey is approximately the same size and weight as the spider monkey (see Table 11.1), and spider monkeys are vulnerable to many of the same raptor predators (Shimooka et al., 2008; Ferrari, 2009). Emmons (1987) reported remains of spider monkeys in the feces of jaguars, and puma have also been confirmed as spider monkey predators in large tracts of forest, whereas ocelots prey on both spider and howler monkeys in small forest fragments (reviewed in Ferrari, 2009). Alarm calls are given by spider monkeys at sightings of large eagles (Janson & van Schaik, 1993). Generally, however, these large monkeys do not show specific adaptations for predator avoidance and even during rest periods, they sprawl conspicuously about in the trees.

Jaguars, ocelots, small carnivores, feral dogs, harpy eagles, and other avian predators may present a threat to younger muriquis (Nishimura et al., 1988; Olmos, 1994; Bianchi & Mendes, 2007). Alarm calls are given to potential predators by muriquis. Historically, humans have hunted muriquis for food. Although hunting is still a major problem in many areas, habitat destruction is thought to be the primary cause of population decline for the northern muriqui (Milton, 1986; Mittermeier et al., 1989; Strier, 1992b).

SOCIAL STRUCTURE AND SOCIAL ORGANIZATION
Howler Monkeys

Howler monkeys are traditionally described as living in cohesive multi-male/multi-female groups. Average group sizes of *Alouatta palliata* range from 9–22 animals, whereas other howler monkey species have smaller average groups of 3–10 individuals (summarized in Di Fiore et al., 2011). It appears that howlers exhibit great plasticity in grouping patterns and subgrouping/fissioning has been observed at several research sites (Gaulin et al., 1980; Leighton & Leighton, 1982; Chapman, 1988a; Fedigan et al., 1998; Stevenson et al., 1998; Bezanson et al., 2008). Groups often contain 1–4 adult males and have a sex ratio of approximately 1 male to 2–4 females. However, in *A. palliata*, groups contain approximately 3 adult males and 9 adult females (Di Fiore et al., 2011). Recently formed groups are smaller than established ones (Crockett & Eisenberg, 1987), and study sites with high population density have larger groups with more females and lower male:female ratios than do low density sites (Crockett, 1996; Fedigan et al., 1998). The home ranges overlap extensively, and in some groups, no part of the home range is used exclusively (Carpenter, 1965; Baldwin & Baldwin, 1972; Smith, 1977; Milton, 1980; Sekulic, 1982a,b).

Howlers may avoid contact between groups throughout the day with choruses of loud howling. The hyoid bone and larynx are adapted for eliciting these loud vocalizations. It has also been hypothesized that loud calling may be multifunctional and serve as a mechanism to avoid feeding competition and defend resources (Van Belle et al., 2014). Intergroup interactions are rare, usually peaceful, and seem to function to maintain group integrity and immediate control of a resource rather than absolute space. In fact, two groups may intermittently share a food tree during a single feeding bout, with one group resting nearby, waiting its turn, while the other is feeding (Glander, 1975). Kowalewski (2007) found that male *A. caraya* generally cooperate during intergroup encounters and behave similarly to *Brachyteles* and *Lagothrix* (Strier, 1992a; Nishimura, 2003). Thus, howlers are not territorial (Neville et al., 1988, but see Crockett & Eisenberg, 1987). This pattern of interaction between groups is reminiscent of that seen between groups of *Eulemur fulvus* in southwestern Madagascar.

In howler monkeys, both male and female juveniles emigrate from their natal group. This can result in groups consisting of unrelated individuals unless same-sexed relatives transfer together. In addition, at Hacienda La Pacifica, Costa Rica, secondary transfer has been documented from 1983–2005 by Clarke and Glander (2010). They found that 35% of adult male and 29% of adult female La Pacifica howlers were "potential secondary transfers" (Clarke & Glander, 2010: p.245). Overt social interaction is often difficult to observe within howler groups. Extremely little allogrooming occurs in *A. palliata* and *A. pigra*, but it is common in *A. caraya*, *A. seniculus*, and *A. fusca* (Neville et al., 1988; Chiarello, 1995a; Sánchez-Villagra et al., 1998). In most early studies, intergroup aggression was rarely observed (Altmann, 1959; Bernstein, 1964; Richard, 1970; Neville, 1972; Smith, 1977; Jones, 1980). Smith (1977) noted that adults at BCI spent less than 1% of their time in even the slightest response to one another, and Crockett (1984) observed only two serious fights in over 1,500 hours of observation. Relatively recent studies show a similar pattern in *A. caraya*, *A. guariba*, *A. palliata*, and *A. pigra* where aggressive interactions are extremely rare and affiliative behaviors are much more common (Pavelka & Knopff, 2002; Dias & Rodriguez, 2006; Kowalewski, 2007). However, long-term field studies of individually identified mantled howler monkeys (*A. palliata*) in Costa Rica (e.g., Glander, 1975, 1980, 1992; Clarke & Glander, 1984; Clarke et al., 1986) and red howlers (*A. seniculus*) in Venezuela (e.g., Rudran, 1979;

Crockett, 1984; Crockett & Eisenberg, 1987; Eisenberg, 1991; Crockett & Pope, 1993) have revealed that dominance interactions are extremely important to howler social organization.

Glander (1975, 1992) captured all of the monkeys in his group at Hacienda La Pacifica using a tranquilizer gun, and each individual was marked. In this way, he was able to trace the life histories of individuals for over 30 years. In his detailed study of mantled howler social behavior, Glander observed frequent aggressive behavior within groups. For example, he recorded approximately 400 dominance interactions in 172 days of observation (Glander, 1975), and resident alpha males were violently attacked by three extra-group males (Glander, 1992). Scars and wounds, seemingly from bites, have often been seen on howlers (Carpenter, 1934; Chivers, 1969; Baldwin & Baldwin, 1972; Otis et al., 1981; Crockett, 1984), but rarely were attacks witnessed (see, however, Glander, 1992). Glander believes that fights are often missed because they are of short duration and totally silent.

Among red howler monkeys, Crockett and Eisenberg (1987) hypothesized that most groups may be functionally uni-male, in that only the alpha male copulates during peak estrus, and paternity exclusion analysis in one study indicated that only one male sired the group's infants (Pope, 1990). However, this does not appear to be the case in all howlers as recent studies indicate that *A. caraya*, *A. guariba*, *A. palliata*, *A. pigra*, and *A. seniculus* females mate with many different adult males and engage in extra-group matings as well (Agoramoorthy & Hsu, 2000; Wang & Milton, 2003; Fialho & Setz, 2007; Van Belle et al., 2009; Kowalewski & Garber, 2010). In red howlers, male-male competition has been described as severe and males have died of injuries during fights (Clarke, 1983; Crockett & Sekulic, 1984; Izawa & Lozano, 1994). Males are reported to invade and take over groups, sometimes forming temporary coalitions, and this is sometimes associated with infant deaths or disappearances (Rudran, 1979; Clarke, 1983; Crockett & Sekulic, 1984; Izawa & Lozano 1994). This pattern is very similar to that reported in black howlers (*A. pigra*) and black and gold howlers (*A. caraya*) (Van Belle et al., 2010) and in Hanuman langurs of India. However, howler social interactions appear to be quite variable and male-male competition is rare at other field sites (Wang & Milton, 2003; Kowlewski, 2007; Bezanson et al., 2008).

For mantled howlers, Fedigan et al. (1998) found that group size and sex ratio were correlated with population density. At high-density sites, mantled howler groups were larger and contained relatively fewer adult males. They suggest that in this species the number of male "positions" within groups may be strictly limited because of

higher levels of competition between males. If this is the case, in high-density habitats there may be few opportunities for males to form new groups and it may be more difficult for them to enter established ones. In regenerating forests with low howler population density, males may have increased dispersal opportunities, and there may be some relaxation of factors causing male mortality. There is evidence for this at Santa Rosa National Park, Costa Rica (Fedigan et al., 1998). Moreover, similar differences in the population densities, groups sizes, and group compositions of *A. pigra* have been reported based on comparisons of census data taken in continuous versus fragmented forests (Van Belle & Estrada, 2006).

Thus, population density and other demographic factors appear to influence a number of aspects of howler social behavior. Such factors as group size, migration patterns, and intragroup relatedness, in turn, are related to variations in social behavior between populations (Clarke et al., 1998; Di Fiore et al., 2011). Although howlers have been the subject of a great deal of research, much remains to be examined to determine the relationships that exist between social structure and ecology in these monkeys. Even though there are relatively long-term data from many research sites and across many species of the genus *Alouatta*, Clarke and Glander (2010) point out that many years of data over several generations can change our ideas about variation in social behavior.

Generally, there is no seasonality in most species of howler breeding or births, though births in some populations occur in clusters during the dry season (Glander, 1980; Clarke & Glander, 1984; Crockett & Eisenberg, 1987; Crockett & Rudran, 1987a,b; Neville et al., 1988; Fedigan et al., 1998; Strier et al., 2001; Kowalewski & Zunino, 2004). Glander (1980) collected statistics on reproductive cycles and development in *A. palliata*: The female estrus cycle is between 11–24 days, with a mean of 16 days (n = 23). The female is receptive for 2–4 days during each cycle. Gestation lasts for 180–194 days, with a mean of 186 (n = 4). These data are similar to those reported for other species (Neville et al., 1988; Di Fiore et al., 2011). The mean time for 16 complete interbirth intervals in 7 females was 22.5 months (Glander, 1980), though shorter intervals of 16–17 months have been reported in other populations (Milton, 1982; Neville et al., 1988; Crockett & Pope, 1993). The infant is carried ventrally by the mother until about 3 weeks and then on her back. Infants begin to move independently with short excursions off their mother by 7–8 weeks. A young howler is still carried during difficult crossings up to 6 months of age. After this time, the infant moves independently and by age 2, juveniles match adults in locomotor frequencies

(Bezanson & Morbeck, 2013). Infants are weaned by about 10–14 months of age (Crockett & Pope, 1993).

Sexual maturity is reached at about 3.5 years for females and 3.5–7 years for males (Crockett & Pope, 1993; Di Fiore et al., 2011). First births have been reported at between 4–7 years of age (median = 5 years). Males first become fathers at 6–8 years (median = 7 years) (Pope, 1990). As stated above, juveniles normally are forced to emigrate (Clarke & Glander, 1984; Glander, 1992). Full weight (4–5 kg) is reached in the female at 5–6 years and in the male (5–6 kg) by 7–8 years (Crockett & Pope, 1993). In Glander's (1980) study population, a 16-year-old female and a male estimated to be 21 years old were still reproductively active.

Spider Monkeys

The social organization of *Ateles*, first described by Raymond Carpenter in 1935, is complex. The basic social structure is "fission-fusion," which is characterized by the existence of multi-male/multi-female social groups, which fragment into smaller, widely dispersed, foraging subgroups of varying size and composition. Social groups contain approximately 15–40+ members (Klein & Klein, 1975, 1977; van Roosmalen, 1985; Chapman, 1988a; Symington, 1988b; reviewed in Shimooka et al., 2008, and Di Fiore et al., 2011).

Adjacent groups maintain separate home ranges, although range overlap varies from none (van Roosmalen, 1985) to 10–30% overlap (Klein & Klein, 1975; Symington, 1988b). Intergroup relations have been characterized as agonistic, particularly among males, and most of this agonism occurs near home range boundaries (Klein, 1972, 1974; Symington, 1987a,b). However, females also are known to visit adjacent groups and participate in friendly intergroup interactions. These visits can last from several hours to an entire day, with some overnight stays (van Roosmalen & Klein, 1988). Females migrate between groups, whereas males do not (van Roosmalen & Klein, 1988; Symington, 1988b, 1990; updated in Di Fiore et al., 2011).

The larger group rarely, if ever, aggregates in its entirety. Rather, small temporary subgroups are formed from the larger unit. Subgroups range from 1–35, with a modal size of 2–5 individuals documented in several studies (i.e., Klein & Klein, 1975, 1977; Cant, 1977; van Roosmalen, 1985; van Roosmalen & Klein, 1988; Symington 1988a,b, 1990; Ahumada, 1989; Chapman, 1989a, 1990a). Klein and Klein (1975, 1977) were the first to do a detailed study of spider monkey (*A. belzebuth*) social organization in Colombia. They observed that the median size of subgroups was 3.5 independently locomoting animals. The modal subgroup size was 2 animals (21%

of the sightings), and subgroups of 4 animals comprised 16% of the sightings. Isolated animals and groups of 8 or more each made up 15% of the total. Although precise numbers differ from study to study, the patterns are quite similar from site to site (Chapman et al., 1995).

Age and sexual composition of subgroups of spider monkeys are variable and consist of all possible permutations. Based on Klein and Klein (1977) male/female subgroups were the most common, accounting for 51% of the sample. These groups often contained from 1–5 males and, in some instances, males outnumbered females. Subgroups containing only females accounted for 45% of the observed subgroups and ranged in size from 2–11 animals. Entirely male subgroups consisted of 2–4 males. Subgroup stability was assumed to be similar in all three types, lasting from 15 minutes to over a day. Individual animals also moved alone and often remained separated from other animals for 1–3 days. Solitary females were sighted most frequently, accounting for 66.7% of the isolated individuals; females with infants accounted for 10.7%, and solitary males for 22.7% of the cases. Juveniles and infants were not observed outside of a subgroup.

Although generally similar, the age/sex composition of subgroups varies from site to site (reviewed in Shimooka et al., 2008; Di Fiore et al., 2011). At Santa Rosa, Costa Rica, for example, Chapman found that the most common subgroups were those of adult females with or without infants (48%). Solitary females, with or without dependent infants, accounted for 33% of the sightings. Mixed-sex groups made up 16%, and adult males accounted for only 2% of the subgroups sited (Chapman, 1990b; Chapman et al., 1995). Symington (1987a, 1990) examined associations between dyads, noting the amount of time specific individuals spent together in the same subgroup. She found that certain males had much higher levels of association than did either female-female or male-female dyads. The highest frequency of grooming occurred between males, and females groomed each other the least frequently. Fedigan and Baxter (1984) described a similar pattern among *Ateles*, but in their study males were most affiliative both with other males and with adult females. The pattern of male-biased affiliations has been found in other populations of spider monkeys (e.g., Slater et al., 2009) and in other primate species in which males are philopatric and remain in close proximity to male relatives throughout their life.

Spider monkeys use food calls to manipulate foraging subgroups, and this may decrease feeding competition between individuals (Chapman & Lefebvre, 1990). Furthermore, subgroups typically forage in small, localized areas, following roughly a circular path and returning to one of a number of regularly used sleeping sites in the evening (Chapman, 1989b). Chapman (1989b) characterizes spider monkeys as multiple central place foragers because they reduce travel costs by sleeping in trees closest to the feeding area being used and then change to a new sleeping site once local resources are exhausted. Thus, the fission-fusion social structure of *Ateles* allows the adjustment of subgroup size to seasonal fluctuations in the size of resource patches.

Within subgroups of *Ateles*, females may associate with one or several males and reproduction can occur throughout most of the year (Carpenter, 1935; Klein, 1971; Symington, 1987b). Estrus females may actively choose their mating partners (Robinson & Janson, 1987), and they may mate promiscuously with multiple males (Campbell, 2006; Campbell & Gibson, 2008). The gestation period is 7.5 months (Eisenberg, 1973), and births are typically single although occasional cases of twin births have been reported (reviewed in Campbell & Gibson, 2008). The interbirth interval averages 35 months (range = 25–42 months) (van Roosmalen & Klein, 1988; Symington, 1988b). In the wild, infants move from the mother's belly to her back at around 4–6 months. Spider monkey infants are almost totally dependent upon their mothers for 10 months. After this, they gradually become more independent. Even after young are moving independently, they are watched closely by adult females, who use their own bodies to form bridges in the canopy for their young to cross. Carpenter (1935) describes one instance of a female spider monkey allowing 5 immature individuals to cross over her back. By 24–36 months, juveniles are no longer carried by their mothers, though they remain close to her until they are up to 50 months of age (van Roosmalen, 1985). Spider monkeys are considered preadults between 50–65 months of age (Eisenberg, 1976; van Roosmalen & Klein, 1988), but they do not reproduce in the wild until they reach 7–8 years of age (Symington, 1987b).

Woolly Monkeys

Lagothrix social structure and organization is similar to that of *Ateles* in many, but not all, respects. Reports of group size are quite variable, ranging from 12–49 animals (reviewed in Di Fiore, et al., 2011). Group size and grouping patterns appear to be related to habitat. Some groups have been described as multi-male and cohesive (Izawa, 1976; Nishimura, 1990), although group members are often spread out and may, therefore, appear to resemble the fluid, fission-fusion associations of *Ateles* (Di Fiore et al., 2011; Ellis & Di Fiore, 2019). Ramirez (1988) found that groups split into subgroups, usually of 2–6 individuals who traveled and fed together, and Defler (1996) studied

a group that commonly traveled in two subgroups separated by 100–200 m but kept in vocal contact. Stevenson (1998) reported that in his study population, woolly monkeys did not form cohesive subgroups except for mothers and infants. Proximity between group members varied and depended upon age/sex class, activity, and individual identity.

Little is known about reproduction or infant development in natural populations. However, mating is reported to be promiscuous within the group (Nishimura et al., 1992; Stevenson, 1998; reviewed in Di Fiore et al., 2011). The gestation period has been reported at approximately 210–225 days (Napier & Napier, 1967; Williams, 1974; Mack & Kafka, 1978). Woolly monkeys achieve locomotor independence at two years old and become socially independent at three years (Schmitt & Di Fiore, 2014). Young females and perhaps some males disperse from their natal group (Di Fiore, 2011; Schmitt & Di Fiore, 2014). There is a lag between dispersal, which occurs at about six years of age, and a female's first reproduction, which occurs around nine years of age. Similar to *Ateles*, birth intervals are about three years (reviewed in Di Fiore et al., 2011).

Muriquis

Brachyteles has a multi-male/multi-female social structure and a promiscuous, polygamous mating system. Groups range in size from 13–80 individuals (reviewed in Di Fiore et al., 2011), and in the growing population of northern muriquis at Reserva Particular do Patrimônio Natural–Feliciano Miguel Abdala (RPPN-FMA), groups of more than 100 individuals have been documented (Strier et al., 2006). Initially, two types of social organization were described from studies of different populations. The southern muriquis at Fazenda Barreiro Rico (FBR) lived in fluid, fission-fusion groups similar to those of spider monkeys. Milton (1984b) observed small groups of 3–5 females, their young offspring, and 1–2 preadults to occupy discrete portions of the forest, whereas adult and preadult males moved throughout the study area. Males moved alone or in groups of up to 8 individuals and joined associations with one or more females for periods of a few minutes to over a week. However, the only permanent associations at FBR were between mothers and offspring. Although southern muriquis in the continuous forest at the RPPN-FMA in Brazil live in larger groups of up to 39 individuals (Talebi et al., 2005), they nonetheless form small subgroups similar to those at FBR (de Moraes et al., 1998).

At the RPPN-FMA, however, the 22–27 members of one study group maintained a cohesive social organization with group members traveling and feeding together for the first six years of the study. This was similar to the cohesive social organization that Lemos de Sá (1988) found during a yearlong study of one group of about 20 northern muriquis in an even smaller (40 ha) forest than that at the RPPN-FMA. By 1988, however, the main study group at the RPPN-FMA began to split into smaller temporary subgroups on occasion (Strier et al., 1993). Over the years, the size of this and other muriqui groups in this protected forest has continued to grow, and their grouping patterns have become increasingly fluid (Dias & Strier, 2003).

As is common in other atelines, muriqui males are philopatric (Strier, 1990a) and preadult females typically disperse between five to seven years of age (Strier et al., 2006). At the RPPN-FMA, the group members normally remain close to one another and they spend more time than expected closer to individuals of their own sex. Furthermore, muriqui males spend more time in close association than do any other primate species studied to date (Strier, 1990a). They also exhibit extremely low rates of aggression (Milton, 1985a,b; Mendes, 1987; Strier, 1990a, 1992b), although lethal intragroup aggression in southern muriquis at Parque Estadual Carlos Botelho (PECB) has been observed (Talebi et al., 2009). Few, if any, agonistic encounters occurring over access to resources or sexual partners have been observed among northern muriquis during more than 25 years of research. There are no noticeable dominance hierarchies and females are codominant to males. Muriquis do not mutual or allogroom, but they do embrace one another. Embraces probably relieve tension and help reinforce social bonds (Strier, 1992b).

In northern muriquis, male relationships are affiliative and are based on a patrilineal kinship system, which Strier (1994a) has described as a brotherhood. Females, on the other hand, transfer between groups (Strier, 1990a, 1993), a characteristic common among many Neotropical monkeys and among all of the atelines (Strier, 1994a,b). This period of transfer is stressful, and immigrating females are sometimes chased by resident females. Recent immigrants may remain peripheral to their new groups for several months (Strier, 1993), or they may rapidly integrate into their new group (Printes & Strier, 1999). Females may also move between groups prior to settling in the one in which they remain once they begin to reproduce (Strier et al., 2006). Males begin to develop strong male-male relationships by the time they are preadults, the same age at which females begin to migrate (Strier, 1993). Among northern muriquis, adult males maintain proximity to other group members more than females do (Strier et al., 2002).

Strier (1992a) has suggested that the muriqui, which modifies its grouping associations in response to patchy food resources and yet remains within calling proximity, may represent an intermediate social organization on a continuum between the cohesive and fission-fusion social systems of other Atelidae. Thus, in habitats where large patches of preferred fruit are scarce, muriquis might be expected to form small, fluid associations, whereas at sites with more abundant large food patches, cohesive groups would be expected. Alternatively, increases in group size may lead to a greater tendency for fissioning, which would have the effect of reducing feeding competition (Strier, 1987a,b, 1989). De Moraes et al. (1998) tested these predictions by comparing the size of food patches and feeding subgroups of southern muriquis at PECB and northern muriquis at what is now the RPPN-FMA. Contrary to expectations, at PECB, where food patches were significantly larger than those at RPPN-FMA, feeding parties were smaller. De Moraes et al. (1998) speculated that higher densities of muriquis and other sympatric primates at the RPPN-FMA might make large associations more advantageous to these muriquis than to those living at lower population densities at PECB. In addition, the disturbed forest at RPPN-FMA might offer a greater diversity of alternative foods, such as leaves, for larger and more cohesive groups.

Muriquis of both sexes routinely mate with multiple partners (Milton, 1985a,b; Strier, 1986, 1991b; Possamai et al., 2007). Although individual males differ in their success in mating, this was not related to male-male competition or hierarchical interactions; rather, female choice appeared to be the most significant component (Strier, 1994a; Possamai et al., 2005, 2007). Females actively initiate sexual inspections and copulations with particular males, and easily avoid unwanted attention from others. Mothers also avoid mating with sexually active sons (Strier, 1997). In a 60-month period, all sexually active males and females within the focal study group were observed to copulate, and 13% of the 527 observed copulations were performed by extra-group males. Fourteen of 17 females participated in these extra-group copulations. However, in subsequent years, no extra-group copulations involving any of the females were observed (Possamai et al., 2005, 2007). It was not evident why certain males were more successful in mating than others. Strier (1997) hypothesized that familiarity and established social bonds might be important factors in female muriqui mate choice, and Possamai et al. (2007) found that some females did mate preferentially with their closest social associates.

Females do not usually cycle throughout the year and cycling among group females is asynchnonic (Strier & Ziegler, 1994). The average ovulatory cycle lasts about 21 days, and gestation length is roughly 7.2 months (Strier & Ziegler, 1997). Births occur throughout the year, but there is a peak during the dry season in May–September (Strier, 1991b, 1996) that is not found in sympatric brown howler monkeys at the RPPN-FMA (Strier et al., 2001). During the first year, infants are fully dependent upon their mother (Strier, 1991b, 1993). They begin to move onto mothers' backs at 6 months and, at this time, may be left alone for short periods as the mothers feed. At around one year of age, infants begin to travel and feed on their own. Weaning occurs between 18–24 months. Strier (1993, 1996) considers individuals of 5–7 years, while they are still smaller than adults, to have well-developed genitalia. This also coincides with the age at which males produce visible ejaculate during the copulations (Possamai et al., 2005) and when females emigrate from their natal groups (Strier et al., 2006). Females usually undergo a delay between dispersal and the onset of ovarian cycling and their first births (Strier & Ziegler, 2000), and most dispersing females in the RPPN-FMA population have given birth to their first infants at 9–10 years of age (Strier et al., 2006). Mothers with surviving infants give birth at approximately three-year intervals (n = 34), whereas similar to other primates, those who lost unweaned infants have shorter interbirth intervals (14–28 months, n = 3) (Strier, 1999).

CONSERVATION

The Atelidae, like all primates, are experiencing population declines that are largely the result of anthropogenic activities in their forested habitats (Estrada et al., 2017, 2019). In Table 11.5, we list Atelidae species, their geography, their conservation status, and major threats reported by the International Union for the Conservation of Nature. *Ateles hybridus, Lagothrix lugens, L. falvicauda, Bryachyteles arachnoides*, and *B. hypoxanthus* are listed in the highest risk category of Critically Endangered, which means they are soon facing extinction in the wild (IUCN, 2020).

Humans hunt all of the atelid species, though the amount of hunting pressure depends upon the species and the area. Atelids are all hunted for food, but the meat of *Ateles* and *Lagothrix* is the most prized (Marsh et al., 1987; Peres, 1990 a,b). Correspondingly, these two genera are becoming increasingly rare in many areas of their ranges (Di Fiore et al., 2011). In addition to being hunted for food, these monkeys are hunted for bait, for the pet trade, for curios (e.g., stuffed monkeys, necklaces, etc.), and to a lesser extent for research. *Alouatta* is hunted in some regions for medicinal purposes where drinking from its specialized hyoid bone is thought to cure goiters or stuttering, or to ease labor pains. In Peru, a medication made from the hair of howler monkeys is thought to relieve coughing (Mittermeier & Coimbra-Filho, 1977). The density of these species in various regions is directly related to hunting and other human pressures (Mittermeier, 1987; Redford & Robinson, 1987; Redford, 1992; Sussman & Phillips-Conroy, 1995; Di Fiore et al., 2011). While hunting is now considered a serious threat to many primate species, the larger conservation issue relates to forest loss associated with industrial agriculture, pastureland for cattle, logging, mining, and fossil fuel extraction (Estrada et al., 2019). Primatologists are in the early stages of studying the effects of climate change on large-bodied primate species, and the next 10–30 years will be critical to prevent the extinction of many plants and animals (Estrada et al., 2019).

SUMMARY

The atelids include one of the most widely occurring genera (e.g., *Alouatta*) and one of the most restricted genera (*Brachyteles*) of all primates in the Americas. Their diets are among the most generalized of all primates, but the intraspecific variation in the percentage of leaves versus fruits in the diets of species such as southern muriquis can exceed that of interspecific and intergeneric differences—an intriguing finding that is especially evident in comparisons of sympatric genera at different sites (Strier, 1992a). Nonetheless, it is difficult to evaluate the degree to which we are documenting recent responses to altered environmental conditions versus evolutionary adaptations that include the capacity for highly variable diets.

Similar intraspecific and intrageneric variation characterizes the social patterns of atelids. From the uni- to multi-male howler monkey troops, to the cohesive and fluid fission-fusion dynamics of the atelines, flexibility in group size and grouping associations is a key atelid trait. Although similar levels of social flexibility are also found in other primates, the ability to adjust to fluctuations in demographic conditions may be particularly important in primates with slow life histories, such as those of the atelids. As their habitats are increasingly fragmented and their populations increasingly isolated, both the social and ecological flexibility of the atelids may be pushed to its limits.

Acknowledgments: We thank all of the people living near the primate groups that we have studied, and who have facilitated and/or contributed to our research. We thank Donna Hart, Ian Colquhoun, and Linda Sussman for resurrecting and editing this volume. Bob Sussman invited us to contribute this chapter because he generously wanted to involve more experts. We thank him for inviting us; and we thank him for inspiring us, inspiring current primatologists, and inspiring future primatologists.

Table 11.5 Atelidae Species: Geographic Distribution, IUCN Status/Population Trends, and Conservation Threats

Species	Geographic Distribution	IUCN [a] Conservation Status	Conservation Threats
Alouatta arctoidea	Venezuela, Colombia	Least Concern, declining	Crops; livestock farming and ranching; hunting and trapping
Alouatta belzebul	Brazil	Least Concern, declining	Housing and urban area development; crops; livestock farming and ranching; roads and railroads; hunting and trapping; logging and wood harvesting
Alouatta caraya	Argentina, Bolivia, Brazil, Paraguay, Uruguay	Least Concern, declining	Housing and urban area development; crops; livestock farming and ranching; hunting and trapping; logging and wood harvesting
Alouatta discolor	Brazil	Vulnerable, declining	Housing and urban area development; crops; livestock farming and ranching; roads and railroads; hunting and trapping; logging and wood harvesting
Alouatta guariba	Brazil, Argentina	Least Concern, declining	Housing and urban area development; crops; livestock farming and ranching; utilities and service lines; hunting and trapping; logging and wood harvesting; endemic disease
Alouatta juara	Colombia, Ecuador, Venezuela, Brazil, Peru	Least Concern, declining	Housing and urban area development; hunting and trapping; logging and wood harvesting
Alouatta macconnelli	Brazil, Colombia, Venezuela	Least Concern, trend unknown	Livestock farming and ranching; mining and quarrying; hunting and trapping; logging and wood harvesting
Alouatta nigerrima	Brazil	Least Concern, declining	Housing and urban area development; crops; livestock farming and ranching; hunting and trapping
Alouatta palliata	Mexico, Guatemala, Belize, Honduras, Nicaragua, Costa Rica, Panama, Colombia, Ecuador, Peru	Least Concern, trend unknown	Housing and urban area development; crops; livestock farming and ranching; hunting and trapping; logging and wood harvesting
Alouatta pigra	Belize, Guatemala, Mexico	Endangered, declining	Housing and urban area development; crops; livestock farming and ranching; hunting and trapping; logging and wood harvesting; storms and flooding
Alouatta puruensis	Bolivia, Brazil, Peru	Least Concern, trend unknown	Housing and urban area development; crops; livestock farming and ranching; hunting and trapping; logging and wood harvesting
Alouatta sara	Bolivia, Peru	Least Concern, declining	Housing and urban area development; crops; hunting and trapping; logging and wood harvesting
Alouatta seniculus	Brazil, Ecuador, Colombia, Peru	Least Concern, declining	Housing and urban area development; hunting and trapping; logging and wood harvesting
Alouatta ululata	Brazil	Endangered, declining	Housing and urban area development; crops; livestock farming and ranching; roads and railroads; hunting and trapping; logging and wood harvesting
Ateles belzebuth	Bolivia, Peru	Endangered, declining	Housing and urban area development; crops; livestock farming and ranching; mining and quarrying; hunting and trapping; logging and wood harvesting; dams and water management/use
Ateles chamek	Brazil, Bolivia, Peru	Endangered, declining	Housing and urban area development; crops; livestock farming and ranching; roads and railroads; hunting and trapping; logging and wood harvesting
Ateles fusciceps	Colombia, Ecuador, Panama	Endangered, declining	Housing and urban area development; crops, livestock farming and ranching; shipping lanes; hunting and trapping; logging and wood harvesting
Ateles geoffroyi	Mexico, Guatemala, Belize, Honduras, Costa Rica, Panama, El Salvador	Endangered, declining	Crops; livestock farming and ranching; hunting and trapping; logging and wood harvesting; fire and fire suppression
Ateles hybridus	Colombia, Venezuela	Critically Endangered, declining	Housing and urban area development; commercial and industrial areas; crops; livestock farming and ranching; roads and railroads; hunting and trapping; logging and wood harvesting
Ateles marginatus	Brazil	Endangered, declining	Housing and urban area development; crops; livestock farming and ranching; roads and railroads; hunting and trapping; logging and wood harvesting
Ateles paniscus	Brazil, Guiana, French Guiana, Suriname	Vulnerable, declining	Crops; livestock farming and ranching; hunting and trapping; logging and wood harvesting

(continued)

Table 11.5 (Continued)

Species	Geographic Distribution	IUCN [a] Conservation Status	Conservation Threats
Lagothrix cana	Peru, Brazil, Bolivia	Endangered, declining	Housing and urban area development; crops; livestock farming and ranching; mining and quarrying; roads and railroads; hunting and trapping; logging and wood harvesting; dams and water management/use
Lagothrix lagotricha	Colombia, Ecuador, Brazil	Vulnerable, declining	Housing and urban area development; crops; livestock farming and ranching; hunting and trapping
Lagothrix lugens	Colombia	Critically Endangered, declining	Housing and urban area development; crops; livestock farming and ranching; roads and railroads; hunting and trapping
Lagothrix poeppigii	Peru, Ecuador, Brazil	Vulnerable, declining	Housing and urban area development; crops; livestock farming and ranching; hunting and trapping
Lagothrix flavicauda	Peru	Critically Endangered, declining	Housing and urban area development; crops; livestock farming and ranching; mining and quarrying; roads and railroads; hunting and trapping; logging and wood harvesting
Brachyteles arachnoides	Brazil	Critically Endangered, declining	Tourism; crops; livestock farming and ranching; wood and pulp plantations; mining and quarrying; hunting and trapping; logging and wood harvesting; dams and water management/use
Brachyteles hypoxanthus	Brazil	Critically Endangered, declining	Crops; livestock farming and ranching; wood and pulp plantations; hunting and trapping; logging and wood harvesting; recreational activities; fire and fire suppression

[a] IUCN (2020).

REFERENCES CITED—CHAPTER 11

Agoramoorthy, G., & Hsu, M. J. (2000). Extragroup copulation among wild red howler monkeys in Venezuela. *Folia Primatologica*, 71(3), 147-151.

Ahumada, J. A. (1989). Behavior and social structure of free-ranging spider monkeys (*Ateles belzebuth*) in La Macarena. *Field Studies of New World Monkeys La Macarena Colombia*, 2, 7-31.

Altmann, S. A. (1959). Field observations on a howling monkey society. *Journal of Mammalogy*, 40(3), 317-330.

Anapol, F., & Lee, S. (1994). Morphological adaptation to diet in platyrrhine primates. *American Journal of Physical Anthropology*, 94(2), 239-261.

Anzures-Dadda, A., Manson, R. H., Andresen, E., & Martínez, M. L. (2016). Possible implications of seed dispersal by the howler monkey for the early recruitment of a legume tree in small rain-forest fragments in Mexico. *Journal of Tropical Ecology*, 32(4), 340-343.

Arroyo-Rodríguez, V., Andresen, E., Bravo, S. P., & Stevenson, P. R. (2014). Seed dispersal by howler monkeys: Current knowledge, conservation implications, and future directions. In M. M. Kowalewski, P. A. Garber, L. Cortés, B. Urbani, & D. Youlatos (Eds.), *Howler monkeys, behavior, ecology, and conservation* (pp. 111-139). Springer.

Arroyo-Rodriguez, V., & Dias, P. A. (2010). Effects of habitat fragmentation and disturbance on howler monkeys: A review. *American Journal of Primatology*, 72(1), 1-16.

Baldwin, J. D., & Baldwin, J. I. (1972). Population density and use of space by howling monkeys (*Alouatta villosa*) in southwestern Panama. *Primates*, 13(4), 371-379.

Bergeson, D. J. (1996). *The positional behavior and prehensile-tail use of* Alouatta palliata, Ateles geoffroyi, *and* Cebus capucinus. [Doctoral dissertation, Washington University].

Bergeson, D. J. (1998). Patterns of suspensory feeding in *Alouatta palliata, Ateles geoffroyi*, and *Cebus capucinus*. In E. Strasser, J. Fleagle, A. Rosenberger, & A. McHenry (Eds.), *Primate locomotion: Recent advances* (pp. 45-60). Plenum Press.

Bernstein, I. (1964). A field study of the activities of howler monkeys. *Animal Behaviour*, 12(1), 84-91.

Bezanson, M. (2009). Life history and locomotion in *Cebus capucinus* and *Alouatta palliata*. *American Journal of Physical Anthropology*, 140(3), 508-517.

Bezanson, M. (2012). The ontogeny of prehensile-tail use in *Cebus capucinus* and *Alouatta palliata*. *American Journal of Primatology*, 74(8), 770-782.

Bezanson M., Garber, P. A., Murphy, J. T., & Premo, L. S. (2008). Patterns of subgrouping and spatial affiliation in a community of mantled howling monkeys (*Alouatta palliata*). *American Journal of Primatology*, 70(3), 282-293.

Bezanson, M., & Morbeck, M. E. (2013). Future adults or old children? Integrating life history frameworks for understanding primate locomotor patterns. In K. Clancy, K. Hinde, & J. Rutherford (Eds.), *Building babies: Primate development in proximate and ultimate perspective* (pp. 435-458). Springer.

Bianchi, R. C., & Mendes, S. L. (2007). Ocelot (*Leopardus pardalis*) predation on primates in Caratinga Biological Station, southeast Brazil. *American Journal of Primatology*, 69(10), 1173-1178.

Bicca-Marques, J. C., & Calegaro-Marques, C. (1993). Feeding postures in the black howler monkey, *Alouatta caraya*. *Folia Primatologica*, 60(3), 169-172.

Bicca-Marques, J. C., & Calegaro-Marques, C. (1995). Locomotion of black howlers in a habitat with discontinuous canopy. *Folia Primatologica*, 64(1-2), 55-61.

Bravo, S. P., & Zunino, G. E. (1998). Effects of black howler monkey (*Alouatta caraya*) seed ingestion on insect larvae. *American Journal of Primatology*, 45(4), 411-415.

Briseño-Jaramillo, M., Biquand, V., Estrada, A., & Lemasson, A. (2017). Vocal repertoire of free-ranging black howler monkeys (*Alouatta pigra*): Call types, contexts, and sex-related contributions. *American Journal of Primatology*, 79(5), e22630.

Campbell, C. J. (2000). Fur rubbing behavior in free-ranging black-handed spider monkeys (*Ateles geoffroyi*) in Panama. *American Journal of Primatology*, 51(3), 205-208.

Campbell, C. J. (2006). Copulation in free-ranging black-handed spider monkeys (*Ateles geoffroyi*). *American Journal of Primatology*, 68(5), 507-511.

Campbell, C. J., & Gibson, N. (2008). Spider monkey reproduction and sexual behavior. In C. J. Campbell (Ed.), *Spider monkeys: Behavior, ecology and evolution of the genus Ateles* (pp. 266-287). Cambridge University Press.

Cant, J. G. H. (1977). *Ecology, locomotion and social organization of spider monkeys* (Ateles geoffroyi). [Doctoral dissertation, University of California, Davis].

Cant, J. G. H. (1986). Locomotion and feeding postures of spider and howling monkeys: Field study and evolutionary interpretation. *Folia Primatologica*, 46(1), 1-4.

Cant, J. G. H. (1990). Feeding ecology of spider monkeys (*Ateles geoffroyi*) at Tikal, Guatemala. *Human Evolution*, 5(3), 269.

Cant, J. G. H. (1992). Positional behavior and body size of arboreal primates: A theoretical framework for field studies and an illustration of its application. *American Journal of Physical Anthropology*, 88(3), 273-283.

Cant, J. G. H, Youlatos, D., & Rose, M. D. (2003). Suspensory locomotion of *Lagothrix lagotricha* and *Ateles Belzebuth* in Yasuni National Park, Ecuador. *Journal of Human Evolution*, 44(6), 685-699.

Carpenter, C. R. (1934). *A field study of the behavior and social relations of howling monkeys.* Johns Hopkins University Press.

Carpenter, C. R. (1935). Behavior of red spider monkeys (*Ateles geoffroyi*) in Panama. *Journal of Mammalogy*, 16(3), 171-180.

Carpenter, C. R. (1965). The howlers of Barro Colorado Island. In I. DeVore (Ed.), *Primate behavior: Field studies of monkeys and apes* (pp. 250-291). Holt, Rinehart, and Winston.

Chapman, C. A. (1986). Boa constrictor predation and group response in white-faced *Cebus* monkeys. *Biotropica*, 18(2), 171-172.

Chapman, C. A. (1987). Flexibility in diets of three species of Costa Rican primates. *Folia Primatologica*, 40(2), 90-105.

Chapman, C. A. (1988a). Patterns of foraging and range use by three species of Neotropical primates. *Primates,* 29(2), 177-194.

Chapman, C. A. (1988b). Patch use and patch depletion by the spider and howling monkeys of Santa Rosa National Park, Costa Rica. *Behaviour*, 105(1-2), 99-116.

Chapman, C. A. (1989a). Primate seed dispersal: The fate of dispersed seeds. *Biotropica,* 21(2), 148-154.

Chapman, C. A. (1989b). Spider monkey sleeping sites: Use and availability. *American Journal of Primatology*, 18(1), 53-60.

Chapman, C. A. (1990a). Ecological constraints on group size in three species of neotropical primates. *Folia Primatologica*, 55(1), 1-9.

Chapman, C. A. (1990b). Association patterns of spider monkeys: The influence of ecology and sex on social organization. *Behavioral Ecology and Sociobiology*, 26(6), 409-414.

Chapman, C. A., & Lefebvre, L. (1990). Manipulating foraging group size: Spider monkey food calls at fruiting trees. *Animal Behaviour*, 39(5), 891-896.

Chapman, C. A., Wrangham, R. W., & Chapman, L. J. (1995). Ecological constraints on group size: An analysis of spider monkey and chimpanzee subgroups. *Behavioral Ecology and Sociobiology*, 36(1), 59-70.

Chaves, Ó. M., & Bicca-Marques, J. C. (2017). Crop feeding by brown howlers (*Alouatta guariba clamitans*) in forest fragments: The conservation value of cultivated species. *International Journal of Primatology*, 38(2), 263-281.

Chaves, P. B., Magnus, T., Jerusalinsky, L., Talebi, M., Strier, K. B., Breves, B., Tabacow, F., Teixeira, R. H. F., Moreira, L., Hack, R. O. E., Milagres, A., Pissinatti, A., de Melo, F. R., Pessutti, C., Mendes, S. L., Margarido, T. C., Fagundes, V., Di Fiore, A., & Bonatto, S. L. (2019). Phylogeographic evidence for two species of muriqui (genus *Brachyteles*). *American Journal of Primatology*, 81(12), e23066.

Chiarello, A. G. (1994). Diet of the brown howler monkey *Alouatta fusca* in a semi-deciduous forest fragment of southeastern Brazil. *Primates*, 35(1), 25-34.

Chiarello, A. G. (1995a). Grooming in brown howler monkeys, *Alouatta fusca. American Journal of Primatology*, 35(1), 73-81.

Chiarello, A. G. (1995b). Role of loud calls in brown howlers, *Alouatta fusca. American Journal of Primatology*, 36(3), 213-222.

Chivers, D. J. (1969). On the daily behavior and spacing of howling monkey groups. *Folia Primatologica*, 10(1-2), 48-102.

Clarke, M. R. (1983). Infant-killing and infant disappearance following male takeovers in a group of free-ranging howling monkeys (*Alouatta palliata*) in Costa Rica. *American Journal of Primatology*, 5(3), 241-247.

Clarke, M. R., & Glander, K. E. (1984). Female reproductive success in a group of free-ranging howling monkeys (*Alouatta palliata*) in Costa Rica. In M. F. Small (Ed.), *Female primates: Studies by women primatologists* (pp. 111-126). Alan R. Liss.

Clarke, M. R., & Glander, K. E. (2010). Secondary transfer of adult mantled howlers (*Alouatta palliata*) on Hacienda La Pacifica, Costa Rica: 1975-2009. *Primates*, 51(3), 241-249.

Clarke, M. R., Glander, K. E., & Zucker E. L. (1998). Infant-nonmother interactions of free-ranging mantled howlers (*Alouatta palliata*) in Costa Rica. *International Journal of Primatology, 19*(3), 451-472.

Clarke, M. R., & Zucker, E. L. (1994). Survey of the howling monkey population at La Pacifica: A seven-year follow-up. *International Journal of Primatology, 15*(1), 61-74.

Clarke, M. R., Zucker, E. L., & Scott, N. J. (1986). Population trends of the mantled howler groups of La Pacifica, Guanacaste, Costa Rica. *American Journal of Primatology, 11*(1), 79-88.

Clutton-Brock, T. H., & Harvey, P. H. (1977). Species differences in feeding and ranging behavior in primates. In T. H. Clutton-Brock (Ed.), *Primate ecology: Studies of feeding and ranging behavior in lemurs, monkeys, and apes* (pp. 557-584). Academic Press.

Coelho, A. M., Jr., Coelho, L., Bramblett, C., Bramblett, S., & Quick, L. (1976). Ecology, population characteristics, and sympatric association in primates: A socio-bioenergetic analysis of howler and spider monkeys in Tikal, Guatemala. *Yearbook of Physical Anthropology, 20*, 96-135.

Crockett, C. M. (1984). Emigration by female red howler monkeys and the case for female competition. In M. F. Small (Ed.), *Female primates: Studies by women primatologists* (pp. 159-173). Alan R. Liss.

Crockett, C. M. (1987). Diet, dimorphism, and demography: Perspectives from howlers to hominids. In W. G. Kinzey (Ed.), *The evolution of human behavior: Primate models* (pp. 115-135). SUNY Press.

Crockett, C. M. (1996). The relation between red howler monkey (*Alouatta seniculus*) troop size and population growth in two habitats. In M. A. Norconk, A. L. Rosenberger, & P. A. Garber (Eds.), *Adaptive radiations of Neotropical primates* (pp. 489-501). Plenum Press.

Crockett, C. M., & Eisenberg, J. F. (1987). Howlers: Variation in group size and demography. In B. B. Smuts, D. L. Cheney, R. M. Seyfarth, R. W. Wrangham, & T. T. Struhsaker (Eds.), *Primate societies* (pp. 54-68). University of Chicago Press.

Crockett, C. M., & Pope, T. (1993). Consequences of sex differences in dispersal for juvenile red howler monkeys. In M. E. Pereira, & L. A. Fairbanks (Eds.), *Juvenile primates: Life history, development, and behavior* (pp. 104-118). Oxford University Press.

Crockett, C. M., & Rudran, R. (1987a). Red howler monkey birth data. I: Seasonal variation. *American Journal of Primatology, 13*(4), 347-368.

Crockett, C. M., & Rudran, R. (1987b). Red howler monkey birth data. II: Interannual, habitat, and sex comparisons. *American Journal of Primatology, 13*(4), 369-384.

Crockett, C. M., & Sekulic, R. (1984). Infanticide in red howler monkeys (*Alouatta seniculus*). In G. Hausfater, & S. B. Hrdy (Eds.), *Infanticide: Comparative and evolutionary perspectives* (pp. 173-192). Transaction Publishers.

Cuaron, A. D. (1997). Conspecific aggression and predation costs for a solitary mantled howler monkey. *Folia Primatologica, 68*(2), 100-105.

de Carvalho O., Jr., Ferrari, S. F., & Strier, K. B. (2004). Diet of a muriqui group (*Brachyteles arachnoides*) in continuous primary forest. *Primates, 45*(3), 201-204.

Defler T. R. (1996). Aspects of the ranging pattern in a group of wild woolly monkeys (*Lagothrix lagotricha*). *American Journal of Primatology, 38*(4), 289-302.

Defler T. R. (1999). Locomotion and posture in *Lagothrix lagotricha*. *Folia Primatologica, 70*(6), 313-327.

Defler, T. R., & Defler S. B. (1996). Diet of a group of *Lagothrix lagothricha lagothricha* in southeastern Colombia. *International Journal of Primatology, 17*(2), 161-190.

DeLuycker, A. M., & Heymann, E. W. (2007). Peruvian yellow-tailed woolly monkey, *Oreonax flavicauda* (Humboldt, 1912). In R. A. Mittermeier, J. Wallis, A. B. Rylands, J. U. Ganzhorn, J. F. Oates, E. A. Williamson, E. Palacios, E. W. Heyman, M. C. M. Kierulff, L. Yongcheng, J. Supriatna, C. Roos, S. Walker, L. Cortes-Ortiz, & C. Schwitzer (Eds.), *Primates in peril: The world's 25 most endangered primates 2006–2008* (pp. 20-21). Bioone.

de Moraes, P. L. R., de Carvalho, O., & Strier, K. B. (1998). Population variation in patch and party size in muriqui (*Brachyteles arachnoides*). *International Journal of Primatology, 19*(2), 325-337.

Dew, J. L. (2005). Foraging, food choice, and food processing by sympatric ripe-fruit specialists: *Lagothrix lagotricha poeppigii* and *Ateles belzebuth belzebuth*. *International Journal of Primatology, 26*(5), 1107-1135.

Di Fiore, A., Chaves, P. B., Cornejo, F. M., Schmitt, C. A., Shanee, S., Cortes-Ortiz, L., Fagundes, V., Roos, C., & Pacheco, V. (2015). The rise and fall of a genus: Complete mtDNA genomes shed light on the phylogenetic position of yellow-tailed woolly monkeys, *Lagothrix lagotricha*, and on the evolutionary history of the family Atelidae (Primates: Platyrrhini). *Molecular Phylogenetics and Evolution, 82*(B), 495-510.

Di Fiore, A., Link, A., & Campbell, C. J. (2011). The atelines: Behavioral and socioecological diversity in a New World monkey radiation. In C. J. Campbell, A. Fuentes, K. C. MacKinnon, S. K. Bearder, & R. M. Stumpf (Eds.), *Primates in perspective* (2nd Ed., pp. 155-188). Oxford University Press.

Di Fiore, A., Link, A., & Dew, J. L. (2008). Diets of wild spider monkeys. In C. J. Campbell (Ed.), *Spider monkeys: Behavior, ecology and evolution of the genus* Ateles (pp. 81-137). Cambridge University Press.

Dias, L. G., & Rodriguez, L. E. (2006). Seasonal changes in male associative behavior and subgrouping of *Alouatta palliata* on an island. *International Journal of Primatology, 27*(6), 1635-1651.

Dias, L. G., & Strier, K. B. (2000). Agonistic encounters between muriquis, *Brachyteles arachnoides hypoxanthus* (Primates: Cebidae), and other animals at the Estação Biológica de Caratinga, Minas Gerais, Brazil. *Neotropical Primates, 8*(4), 138-141.

Dias, L. G., & Strier, K. B. (2003). Effects of group size on ranging patterns in *Brachyteles arachnoides hypoxanthus*. *International Journal of Primatology, 24*(2), 209-221.

Dias, P. A. D., & Rangel-Negrín, A. (2015). Diets of howler monkeys. In M. M. Kowalewski, P. A. Garber, L. Cortés, B. Urbani, & D. Youlatos (Eds.), *Howler monkeys: Behavior, ecology, and conservation. Developments in primatology: Progress and prospects* (pp. 21-56). Springer.

Eaglen, R. H. (1984). Incisor size and diet revisited: The view from a platyrrhine perspective. *American Journal of Physical Anthropology, 64*(3), 263-275.

Eason, P. (1989). Harpy eagle attempts predation on adult howler monkey. *Condor, 91*(2), 469-470.

Eisenberg, J. F. (1973). Reproduction in two species of spider monkeys, *Ateles fuciceps* and *Ateles geoffroyi*. *Journal of Mammalogy, 54*(4), 955-957.

Eisenberg, J. F. (1976). *Communication mechanisms and social integration in the black spider monkey,* Ateles fusciceps robustus, *and related species.* Smithsonian Institution Press.

Eisenberg, J. F. (1991). Mammalian social organization and the case of *Alouatta*. In J. G. Robinson, & L. Tiger (Eds.), *Man and beast revisited* (pp. 127-138). Smithsonian Institution Press.

Ellis, K., & Di Fiore, A. (2019). Variation in space use and social cohesion within and between four groups of woolly monkeys (*Lagothrix lagotricha poeppigii*) in relation to fruit availability and mating opportunities at the Tiputini Biodiversity Station, Ecuador. In R. Reyna Hurtado, & C. Chapman (Eds.), *Movement ecology of Neotropical forest mammals* (pp. 141-171). Springer.

Emmons, L. E. (1987). Comparative feeding ecology of felids in Neotropical rain forests. *Behavioral Ecology and Sociobiology, 20*(4), 271-283.

Estrada, A. (1984). Resource use by howler monkeys (*Alouatta palliata*) in the rain forest of Los Tuxtlas, Vera Cruz, Mexico. *International Journal of Primatology, 5*(2), 105-131.

Estrada, A., & Coates-Estrada, R. (1984). Fruit-eating and seed dispersal by howling monkeys (*Alouatta palliata*) in the tropical rain forest of Los Tuxtlas, Mexico. *American Journal of Primatology, 6*(2), 77-91.

Estrada, A., Coates-Estrada, R., & Vazquez-Yanes, C. (1984). Observation on fruiting and dispersers of *Cecropia obtusifolia* at Los Tuxtlas, Mexico. *Biotropica, 16*(4), 315-318.

Estrada, A., Garber, P. A., & Chaudhary, A. (2019). Expanding global commodities trade and consumption place the world's primates at risk of extinction. *PeerJ, 7*, e7068.

Estrada, A., Garber, P. A., Rylands, A. B., Roos, C., Fernandez-Duque, E., Di Fiore, A., Nekaris, K. A. I., Nijman, V., Heymann, E. W., Lambert, J. E., Rovero, F., Barelli, C., Setchell, J. M., Gillespie, T. R., Mittermeier, R. A., Arregoitia, L. V., de Guinea, M., Gouveia, S., Dobrovolski, R.,... & Li, B. (2017). Impending extinction crisis of the world's primates: Why primates matter. *Science Advances, 3*(1), e1600946.

Estrada, A., Raboy, B. E., & Oliveira, L. C. (2012). Agroecosystems and primate conservation in the tropics: A review. *American Journal of Primatology, 74*(8), 696-711.

Estrada, A., Saúl, J. S., Ortíz Martínez, T., & Coates-Estrada, R. (1999). Feeding and general activity patterns of a howler monkey (*Alouatta palliata*) troop living in a forest fragment at Los Tuxtlas, Mexico. *American Journal of Primatology, 48*(3), 167-183.

Fedigan, L. M., & Baxter, M. F. (1984). Sex differences and social organization of free-ranging spider monkeys (*Ateles geoffroyi*). *Primates, 25*(3), 279-294.

Fedigan, L. M., Fedigan, L., Chapman, C., & Glander, K. E. (1988). Spider monkey home ranges: A comparison of radio telemetry and direct observation. *American Journal of Primatology, 16*(1), 19-29.

Fedigan L. M., Rose L. M., & Avila, R. M. (1998). Growth of mantled howler groups in a regenerating Costa Rican dry forest. *International Journal of Primatology, 19*(3), 405-433.

Ferrari, S. F. (2009). Predation risks and antipredator strategies. In P.A. Garber, A. Estrada, J. C. Bicca-Marques, E. Heymann, & K. B. Strier (Eds.), *South American primates: Comparative perspectives in the study of behavior, ecology, and conservation. Developments in primatology: Progress and prospects* (pp. 251-277). Springer.

Ferrari, S. F., & Strier, K. B. (1992). Exploitation of *Mabea fistulifera* nectar by marmosets (*Callithrix flaviceps*) and muriquis (*Brachyteles arachnoides*) in south-east Brazil. *Journal of Tropical Ecology, 8*(3), 225-239.

Fialho, M. S., & Setz, E. Z. F. (2007). Extragroup copulations among brown howler monkeys in southern Brazil. *Neotropical Primates, 14*(1), 28-30.

Fleagle, J. G., & Mittermeier, R. A. (1980). Locomotor behavior, body size, and comparative ecology of seven Surinam monkeys. *American Journal of Physical Anthropology, 52*(3), 301-314.

Fleming, T. H., Williams, C. F., Bonacorso, F. J., & Herbst, L. H. (1985). Phenology, seed dispersal, and colonization in *Muntingia calabura*, a Neotropical pioneer tree. *American Journal of Botany, 72*(3), 383-391.

Fonseca, G. B. A. (1985). Observations on the ecology of the muriqui (*Brachyteles arachnoides*, E. Geoffroyi 1806): Implications for its conservation. *Primate Conservation, 5*, 48-52.

Fonseca, M. L., Cruz, D. M., Rojas, D. C. A., Crespo, J. P., & Stevenson, P. R. (2019). Influence of arthropod and fruit abundance on the dietary composition of highland Colombian woolly monkeys (*Lagothrix lagotricha lugens*). *Folia Primatologica, 90*(4), 240-257.

Fontaine, R. P. (1985). *Positional behavior in six cebid species.* [Unpublished doctoral dissertation, University of Georgia].

Fooden, J. (1963). A revision of the woolly monkeys (genus *Lagothrix*). *Journal of Mammalogy, 44*(2), 213-247.

Freese, C. H. (1976). Censusing *Alouatta palliata, Ateles geoffroyi*, and *Cebus capucinus* in the Costa Rican dry forest. In R. W. Thorington, & P. G. Hetne (Eds.), *Neotropical primates: Field studies and conservation* (pp. 4-9). National Academy of Science.

García-Toro, L. C., Link, A., Páez-Crespo, E. J., & Stevenson, P. R. (2019). Home range and daily traveled distances of highland Colombian woolly monkeys (*Lagothrix lagothricha lugens*): Comparing spatial data from GPS collars and direct follows. In R. Reyna Hurtado, & C. Chapman (Eds.),

Movement ecology of Neotropical forest mammals (pp. 173-193). Springer.

Gaulin, S. J. C. (1979). A Jarmon/Bell model of primate feeding niches. *Human Ecology*, 7(1), 1-17.

Gaulin, S. J. C., & Gaulin, C. K. (1982). Behavioral ecology of *Alouatta seniculus* in Andean cloud forest. *International Journal of Primatology*, 3(1), 1-32.

Gaulin, S. J. C., Knight, D. H., & Gaulin, C. K. (1980). Local variance in *Alouatta* group size and food availability on Barro Colorado Island. *Biotropica*, 12, 137-143.

Gebo, D. L. (1992). Locomotor and postural behavior in *Alouatta palliata* and *Cebus capucinus*. *American Journal of Primatology*, 26(4), 277-290.

German, R. Z. (1982). The functional morphology of caudal vertebrae in New World monkeys. *American Journal of Physical Anthropology*, 58(4), 453-460.

Glander, K. E. (1975). Habitat description and resource utilization: A preliminary report on mantled howler monkey ecology. In R. H. Tuttle (Ed.), *Socioecology and psychology of primates* (pp. 37-57). Mouton Press.

Glander, K. E. (1978). Howling monkey feeding behavior and plant secondary compounds: A study of strategies. In G. G. Montgomery (Ed.), *The ecology of arboreal folivores* (pp. 561-574). Smithsonian Press.

Glander, K. E. (1980). Reproduction and population growth in free-ranging mantled howling monkeys. *American Journal of Physical Anthropology*, 53(1), 25-36.

Glander, K. E. (1992). Dispersal patterns in Costa Rican mantled howling monkeys. *International Journal of Primatology*, 13(4), 415-435.

Groves, C. P. (2001). *Primate taxonomy*. Smithsonian Institution Press. *International Journal of Primatology*, 13(4), 415-436.

Guillot, D. M. (2009). *Positional behavior in wild ateline primates: A comparative analysis of* Alouatta seniculus, Lagothrix poeppigii *and* Ateles belzebuth. [Doctoral dissertation, Boston University].

Hernández-Camacho, J., & Cooper, R. W. (1976). The nonhuman primates of Colombia. In R. W. Thorington, & P. G. Hetne (Eds.), *Neotropical primates: Field studies and conservation* (pp. 35-69). National Academy of Science.

Hladik, A., & Hladik, C. M. (1969). Rapports trophiques entre végétation et primates dans la forêt de Barro Colorado (Panama). *Terre et Vie*, 1, 25-117.

Howe, H. F. (1980). Monkey dispersal and waste of a neotropical fruit. *Ecology*, 61(4), 944-959.

Howe, H. F. (1983). Annual variation in a Neotropical seed-dispersal system. In S. L. Sutton, T. C. Whitmore, & A. C. Chadwick (Eds.), *Tropical rain forest: Ecology and management* (pp. 211-227). Blackwell.

IUCN. (2019). *The International Union for the Conservation of Nature Red List of Threatened Species*. Version 2020-1. https://www.iucnredlist.org.

IUCN. (2020). *The International Union for the Conservation of Nature Red List of Threatened Species*. Version 2020-3. https://www.iucnredlist.org.

Iurck, M. F., Nowak, M. G., Costa, L. C., Mendes, S. L., Ford, S. M., & Strier, K. B. (2013). Feeding and resting postures of wild northern muriquis (*Brachyteles hypoxanthus*). *American Journal of Primatology*, 75(1), 74-87.

Izawa, K. (1976). Group sizes and compositions of monkeys in the upper Amazon Basin. *Primates*, 17(3), 367-399.

Izawa, K., Kimura, K., & Ohnishi, Y. (1990). Chemical properties of soils eaten by red howler monkeys (*Alouatta seniculus*) in La Macarena, Colombia. II. *Field Studies of New World Monkeys*, 4, 27-37.

Izawa, K., & Lozano, H. M. (1990). Frequency of soil eating by a group of wild howler monkeys (*Alouatta seniculus*) in La Macarena, Colombia. *Field Studies of New World Monkeys*, 4, 47-56.

Izawa, K., & Lozano, H. M. (1994). Social changes within a group of red howler monkeys in La Macarena, Colombia. V. *Field Studies of New World Monkeys*, 9, 15-21.

Janson, D. H., & van Schaik, C. P. (1993). Ecological risk aversion in juvenile primates: Slow and steady wins the race. In M. E. Pereira, & L. A. Fairbanks (Eds.), *Juvenile primates: Life history, development, and behavior* (pp. 57-76). Oxford University Press.

Johnson, S. E., & Shapiro, L. J. (1998). Positional behavior vertebral morphology in atelines and cebines. *American Journal of Physical Anthropology*, 105(3), 333-354.

Jones, C. B. (1980). The function of status in the mantled howler monkey, *Alouatta palliata*, Gray: Intraspecific competition for group membership in a folivorous Neotropical primate. *Primates*, 21(3), 389-405.

Julliot, C., & Sabatier, D. (1993). Diet of the red howler monkey (*Alouatta seniculus*) in French Guiana. *International Journal of Primatology*, 14(4), 527-550.

Kay, R. F. (1973). *Mastication, molar tooth structure, and diet in primates*. [Doctoral Dissertation, Yale University].

Kay, R. N., & Davies, A. G. (1994). Digestive physiology. In A. G. Davies, & J. F. Oates (Eds.), *Colobine monkeys: Their ecology, behaviour and evolution* (pp. 229-249). Cambridge University Press.

Kinzey, W. G. (Ed.). (1997). *New world primates: Ecology, evolution, and behavior*. Aldine.

Klein, L. L. (1971). Observations on copulation and seasonal reproduction of two species of spider monkeys, *Ateles belzebuth* and *Ateles geoffroyi*. *Folia Primatologica*, 15(3-4), 233-248.

Klein, L. L. (1972). *The ecology and social organization of the spider monkey (*Ateles belzebuth*)*. [Doctoral dissertation, University of California, Berkeley].

Klein, L. L. (1974). Agonistic behavior in Neotropical primates. In R. Holloway (Ed.), *Primate aggression, territoriality, and xenophobia* (pp. 77-122). Academic Press.

Klein, L. L., & Klein, D. J. (1975). Social and ecological contrasts between four taxa of Neotropical primate. In R. A. Tuttle (Ed.), *Socioecology and psychology of primates* (pp. 153-181). Academic Press.

Klein, L. L., & Klein, D. J. (1977). Feeding behavior of the Colombian spider monkey (*Ateles belzebuth*). In T. H.

Clutton-Brock (Ed.), *Primate ecology: Studies of feeding and ranging behaviour in lemurs, monkeys, and apes* (pp. 153-181). Academic Press.

Kowalewski, M. M. (2007). *Patterns of affiliation and co-operation in howler monkeys: An alternative model to explain social organization in non-human primates.* [Doctoral dissertation, University of Illinois at Urbana Champaign].

Kowalewski, M. M., & Garber, P. A. (2010). Mating promiscuity and reproductive tactics in female black and gold howler monkeys (*Alouatta caraya*) inhabiting an island on the Parana River, Argentina. *American Journal of Primatology*, 72, 734-748.

Kowalewski, M. M., & Zunino, G. E. (2004). Birth seasonality in *Alouatta caraya*. *International Journal of Primatology*, 25(2), 383-400.

Lambert, J. E. (2011). Primate nutritional ecology: Feeding biology and diet at ecological and evolutionary scales. In C. J. Campbell, A. Fuentes, K. C. MacKinnon, S. K. Bearder, & R. M. Stumpf (Eds.), *Primates in perspective* (pp. 512-534). Oxford University Press.

Lawler, R. R., & Stamps C. (2002). The relationship between tail use and positional behavior in *Alouatta palliata*. *Primates*, 43(2), 147-152.

Lehman, F. C. (1959). Contribuciones al studio de la faunne de Colombia. XIV. Nuevas observaciones sobre *Oroasetus isidori* (Des Murs). *Novedades Colombianas*, 1(4), 169-195.

Leighton, M., & Leighton, D. R. (1982). The relationship of size of feeding aggregate to size of food patch: Howler monkeys (*Alouatta palliata*) feeding in *Trichilia cipo* fruit trees on Barro Colorado Island. *Biotropica*, 14, 81-90.

Lemelin, P. (1995). Comparative and functional myology of the prehensile tail in New World monkeys. *Journal of Morphology*, 224(3), 351-368.

Lemos de Sá, R. M. (1988). *Situação de uma população mono-carveiro (Primates, Cebidae) arȩch, em fragment da mata Atlântica (MG), e implçáo para sua conservação.* [Master's thesis, Universidade de Brasil].

Lemos de Sá, R. M., & Glander, K. E. (1993). Capture techniques and morphometrics for the woolly spider, muriqui (*Brachyteles arachnoides*, E. Geoffroyi, 1806). *American Journal of Primatology*, 29(2), 145-153.

Lemos de Sá, R. M., Pope, T. R., Struhsaker, T. T., & Glander, K. E. (1993). Sexual dimorphism in canine length of woolly spider monkeys (*Brachyteles arachnoides*, E. Geoffroyi, 1806). *International Journal of Primatology*, 14(5), 755-763.

Lemos de Sá, R. M., & Strier, K. B. (1992). A preliminary comparison of forest structure and use by two isolated groups of woolly spider monkeys, *Brachyteles arachnoides*. *Biotropica*, 24(3), 455-459.

Lima, M., Mendes, S. L., & Strier, K. B. (2019). Habitat use in a population of the northern muriqui (*Brachyteles hypoxanthus*). *International Journal of Primatology*, 40, 470-495.

Mack, D. S., & Kafka, H. (1978). Breeding and rearing of woolly monkeys at the National Zoological Park, Washington. *International Zoo Yearbook*, 18, 117-122.

Marsh, C. W., Johns, A. D., & Ayres, J. M. (1987). Effects of habitat disturbance on rain forest primates. In C. W. Marsh, & R. A. Mittermeier (Eds.), *Primate conservation in the tropical rain forest* (pp. 83-107). Alan R. Liss.

Martin, R. D., Chivers, D. J., MacLarnon, A. M., & Hladik, C. M. (1985). Gastrointestinal allometry in primates and other mammals. In W. L. Jungers (Ed.), *Size and scaling in primate biology* (pp. 61-89). Plenum Press.

Martins, M. M. (2008). Fruit diet of *Alouatta guariba* and *Brachyteles arachnoides* in southeastern Brazil: Comparison of fruit type, color, and seed size. *Primates*, 49(1), 1-8.

McKinney, T. (2019). Ecological and behavioural flexibility of mantled howlers (*Alouatta palliata*) in response to anthropogenic habitat disturbance. *Folia Primatologica*, 90, 456-469.

McKinney, T. Westin, J. L., & Serio-Silva, J. C. (2015). Anthropogenic habitat modification, tourist interactions and crop-raiding in howler monkeys. In M. M. Kowalewski, P. A. Garber, L. Cortes-Ortiz, B. Urbani, & D. Youlatos (Eds.), *Howler monkeys* (pp. 281-311). Springer.

Mendel, F. (1976). Postural and locomotor behavior of *Alouatta palliata* on various substrates. *Folia Primatologica*, 26(1), 36-53.

Milton, K. (1978). Behavioral adaptations to leaf eating. In G. G. Montgomery (Ed.), *The ecology of arboreal folivores* (pp. 535-550). Smithsonian Press.

Milton, K. (1980). *The foraging strategy of the howler monkeys: A study in primate economics.* Columbia University Press.

Milton, K. (1981). Food choice and digestive strategies of two sympatric primate species. *American Naturalist*, 117(4), 496-505.

Milton, K. (1982). Dietary quality and population regulation in a howler monkey population. In E. G. Leigh, A. S. Rand, & D. M. Windsor (Eds.), *The ecology of a tropical forest* (pp. 273-289). Smithsonian Institution Press.

Milton, K. (1984a). Habitat, diet, and activity patterns of free-ranging woolly spider monkeys (*Brachyteles arachnoides* E. Geoffrey 1806). *International Journal of Primatology*, 5(5), 491-514.

Milton, K. (1984b). Diet and social structure of free-ranging woolly spider monkeys. *American Journal of Physical Anthropology*, 63(2), 195.

Milton, K. (1985a). Multi-male mating and absence of canine dimorphism in woolly spider monkeys (*Brachyteles arachnoides*). *American Journal of Physical Anthropology*, 68(4), 519-523.

Milton, K. (1985b). Mating patterns of woolly spider monkeys, *Brachyteles arachnoides*: Implications for female choice. *Behavioral Ecology and Sociobiology*, 17(1), 53-59.

Milton, K. (1986). Ecological background and conservation priorities for woolly spider monkeys (*Brachyteles arachnoides*). In K. Benirschke (Ed.), *Primates: The road to self-sustaining populations* (pp. 241-250). Springer.

Milton, K. (1987). Behavior and ecology of the woolly spider monkey, *Brachyteles arachnoides*. *International Journal of Primatology*, 8(5), 422.

Milton, K. (1998). Physiological ecology of howlers (*Alouatta*): Energetics and digestive considerations and comparison with the Colobinae. *International Journal of Primatology*, 19(3), 513-548.

Mittermeier, R. A. (1978). Locomotion and posture in *Ateles geoffroyi* and *Ateles paniscus*. *Folia Primatologica*, 30(3), 161-193.

Mittermeier, R. A. (1987). Effects of hunting on rain forest primates. In C. W. Marsh, & R. A. Mittermeier (Eds.), *Primate conservation in the tropical rain forest* (pp. 109-146). Alan R. Liss.

Mittermeier, R. A., & Coimbra-Filho, A. F. (1977). Primate conservation in Brazilian Amazonia. *Primate Conservation*, 117-166.

Mittermeier, R. A., & Fleagle, J. G. (1976). The locomotor and postural behavior repertoires of *Ateles geoffroyi* and *Colobus guereza*, and a reevaluation of the locomotor category semibrachiation. *American Journal of Physical Anthropology*, 45(2), 235-256.

Mittermeier, R. A., Kinzey, W. G., & Mast, R. B. (1989). Neotropical primate conservation. *Journal of Human Evolution*, 18(7), 597-610.

Mittermeier, R. A., & van Roosmalen, M. G. M. (1981). Preliminary observations on habitat utilization and diet in eight Suriname monkeys. *Folia Primatologica*, 36(1-2), 1-39.

Mourthé, I. M., Guedes, D., Fidelis, J., Boubli, J. P., Mendes, S. L., & Strier, K. B. (2007). Ground use by northern muriquis (*Brachyteles hypoxanthus*). *American Journal of Primatology*, 69(6), 706-712.

Moynihan, M. (1976). *The New World primates*. Princeton University Press.

Muskin, A., & Fischgrund, A. J. (1981). Seed dispersal of *Stemmadenia* (Apocynaceae) and sexually dimorphic feeding strategies by *Ateles* in Tikal, Guatemala. *Biotropica*, 13(2), 78-80.

Napier, J. R., & Napier, P. H. (1967). *A handbook of living primates*. Academic Press.

Neville, M. K. (1972). Social relations within troops of red howler monkeys (*Alouatta seniculus*) *Folia Primatologica*, 18(1-2), 47-77.

Neville, M. K., Glander, K. E., Braza, F., & Rylands, A. B. (1988). The howling monkeys, genus *Alouatta*. In R. A. Mittermeier, A. B. Rylands, A. F. Coimbra-Filho, & G. A. B. Fonseca (Eds.), *Ecology and behavior of Neotropical primates* (Vol. 2, pp. 349-453). World Wildlife Fund.

Nishimura, A. (1990). A sociological and behavioral study of woolly monkeys, *Lagothrix lagotricha*, in the Upper Amazon. *Science and Engineering Reviews of Doshisha University*, 31(2), 87-121.

Nishimura, A. (2003). Reproductive parameters of wild female *Lagothrix lagotricha*. *International Journal of Primatology*, 24(4), 707-772.

Nishimura, A., Fonseca, G. A. B., Mittermeier, R. A., Young, A. L., Strier, K. B., & Valle, C. M. C. (1988). The muriqui, genus *Brachyteles*. In R. A. Mittermeier, A. B. Rylands, A. Coimbra-Filho, & G. A. B. Fonseca (Eds.), *Ecology and behavior of Neotropical primates* (Vol. 2, pp. 577-610). World Wildlife Fund.

Nishimura, A., Wilches, A. V., & Estrada, C. (1992). Mating behaviors of woolly monkeys (*Lagothrix lagotricha*) at La Macarena, Colombia (III): Reproductive parameters viewed from a long-term field study. *Field Studies of New World Monkeys: La Macarena Colombia*, 7, 1-7.

Nunes, A. (1988). Diet and feeding ecology of *Ateles belzebuth belzebuth* at Maraca Ecological Station, Roraima, Brazil. *Folia Primatologica*, 69(2), 61-76.

Olmos, F. (1994). Jaguar predation on muriqui (*Brachyteles arachnoides*). *Neotropical Primates*, 2, 16.

Organ, J. M. (2010). Structure and function of platyrrhine caudal vertebrae. *Anatomical Record*, 293(4), 730-745.

Otis, J. S., Froehlich, J. W., & Thorington, R. W. (1981). Seasonal and age-related differential mortality by sex in the mantled howler monkey, *Alouatta palliata*. *International Journal of Primatology*, 2(3), 197-205.

Pavelka, M. S. S., & Knopff, K. H. (2002). Diet and activity in black howler monkeys (*Alouatta pigra*) in southern Belize: Does degree of frugivory influence activity level? *Primates*, 45(2), 105-111.

Peetz, A., Norconk, M. A., & Kinzey, W. G. (1992). Predation by jaguar on howler monkeys (*Alouatta seniculus*) in Venezuela. *American Journal of Primatology*, 28(3), 223-228.

Peres, C. A. (1990a). Effects of hunting on western Amazonian primate communities. *Biological Conservation*, 54(1), 47-59.

Peres, C. A. (1990b). A harpy eagle successfully captures an adult male red howler monkey. *Wilson Bulletin*, 102, 560-561.

Peres, C. A. (1994). Diet and feeding ecology of gray woolly monkeys (*Lagothrix lagotricha cana*) in central Amazonia: Comparisons with other atelines. *International Journal of Primatology*, 15, 333-372.

Peres, C. A. (1996). Use of space, foraging group size, and spatial group structure in gray woolly monkeys (*Lagothrix lagotricha cana*) at Urucu, Brazil: A review of the Atelinae. In M. A. Norconk, A. L. Rosenberger, & P. A. Garber (Eds.), *Adaptive radiations in Neotropical primates* (pp. 467-488). Plenum Press.

Peres, C. A. (1997). Effects of habitat quality and hunting pressure on arboreal folivore densities in Neotropical forests: A case study of howler monkeys (*Alouatta* species). *Folia Primatologica*, 68(3-5), 199-222.

Pinto, L. P., Costa, C. M. R., Strier, K. B., & de Fonseca, G. A. B. (1993). Habitat, density, and group size of primates in a Brazilian tropical forest. *Folia Primatologica*, 61(3), 135-143.

Pinto, L. P., & Setz, E. Z. F. (2009). Diet of *Alouatta belzebul discolor* in an Amazonian rain forest of northern Mato Grosso State, Brazil. *International Journal of Primatology*, 25(6), 1197-1211.

Pope, T. R. (1990). The reproductive consequences of male cooperation in the red howler monkey: Paternity exclusion in multi-male and single-male troops using genetic markers. *Behavioral Ecology and Sociobiology*, 27(6), 439-446.

Possamai, C. B., Young, R. J., Mendes, S. L., & Strier, K. B. (2007). Socio-sexual behavior of female northern muriquis (*Brachyteles hypoxanthus*). *American Journal of Primatology*, 69 (7), 766-776.

Possamai, C. B., Young, R. J., Oliveira, R. C. F., Mendes, S. L., & Strier, K. B. (2005). Age-related variation in copulations of male northern muriquis (*Brachyteles hypoxanthus*). *Folia Primatologica*, 76(1), 33-36.

Prates H. S., & Bicca Marques J. (2008). Age-sex analysis of activity budget, diet, and positional behavior in *Alouatta caraya* in an orchard forest. *International Journal of Primatology*, 29(3), 703-715.

Printes, R. C., & Strier, K. B. (1999). Behavioral correlates of dispersal in female muriquis (*Brachyteles arachnoides*). *International Journal of Primatology*, 20(6), 941-960.

Ramirez, M. (1988). The woolly monkey, genus *Lagothrix*. In R. A. Mittermeier, A. B. Rylands, A. Coimbra-Filho, & G. A. B. Fonseca (Eds.), *Ecology and behavior of Neotropical primates* (Vol. 2, pp. 539-575). World Wildlife Fund.

Redford, K. H. (1992). The empty forest. *BioScience*, 42(6), 412-422.

Redford, K. H., & Robinson, J. G. (1987). The game of choice: Patterns of Indian and colonist hunting in the Neotropics. *American Anthropologist*, 89(3), 650-667.

Richard, A. (1970). A comparative study of the activity patterns and behavior of *Alouatta villosa* and *Ateles geoffroyi*. *Folia Primatologica*, 12(4), 241-263.

Righini, N., Garber, P. A., & Rothman, J. M. (2017). The effects of plant nutritional chemistry on food selection of Mexican black howler monkeys (*Alouatta pigra*): The role of lipids. *American Journal of Primatology*, 79(4), 1-15.

Robinson, J. G., & Janson, C. H. (1987). Capuchins, squirrel monkeys, and atelines: Socioecological convergence with Old World primates. In B. B. Smuts, D. L. Cheney, R. M. Seyfarth, R. W. Wrangham, & T. T. Struhsaker (Eds.), *Primate societies* (pp. 69-82). University of Chicago Press.

Rodrigues, M. A. (2017). Female spider monkeys (*Ateles geoffroyi*) cope with anthropogenic disturbance through fission–fusion dynamics. *International Journal of Primatology*, 38(5), 838-855.

Rosenberger, A. L., & Matthews, L. J. (2008). *Oreonax*—not a genus. *Neotropical Primates*, 15(1), 8-12.

Rosenberger, A. L., & Strier, K. B. (1989). Adaptive radiation of the ateline primates. *Journal of Human Evolution*, 18(7), 717-750.

Rosenberger, A. L., Tejedor, M. F., Cooke, S., Helenar, L., & Pekkar, S. (2009). Platyrrhine ecophylogenetics, past and present. In P. A. Garber, A. Estrada, J. C. Bicca Marques, E. W. Heymann, & K. B. Strier (Eds.), *South American primates: Comparative perspectives in the study of behavior, ecology, and conservation* (pp. 69-113). Springer.

Rudran, R. (1979). The demography and social mobility of a red howler (*Alouatta seniculus*) population in Venezuela. In J. Eisenberg (Ed.), *Vertebrate ecology in the northern Neotropics* (pp. 107-126). Smithsonian Institution.

Rylands, A. B. (2000). An assessment of the diversity of New World primates. *Neotropical Primates*, 8, 61-93.

Rylands, A. B., Groves, C. P., Mittermeier, R. A., Cortés-Ortiz, L., & Hines, J. J. H. (2006). Taxonomy and distributions of Mesoamerican primates. In A. Estrada, P. A. Garber, M. Pavelka, & L. Luecke (Eds.), *New perspectives in the study of Mesoamerican primates: Distribution, ecology, behavior, and conservation* (pp. 29-79). Springer.

Rylands, A. B., & Mittermeier, R. A. (2009). The diversity of the New World primates (Platyrrhini): An annotated taxonomy. In P. A. Garber, A. Estrada, J.-C. Bicca-Marques, E. Heymann, & K. B. Strier (Eds.), *South American primates: Comparative perspectives in the study of behavior, ecology, and conservation* (pp. 23-54). Springer.

Sanchez-Villagra, M. R., Pope, T. R., & Salas, V. (1998). Relations of intergroup variation in allogrooming to group social structure and ectoparasite loads in red howlers (*Alouatta seniculus*). *International Journal of Primatology*, 19(3), 473-491.

Schmitt, C. A., & Di Fiore, A. (2014). Life history, behavior, and development of wild immature lowland woolly monkeys (*Lagothrix poeppigii*) in Amazonian Ecuador. In T. Defler, & P. R. Stevenson (Eds.), *The woolly monkey: Behavior, ecology, systematics and captive research* (pp. 113-146). Springer.

Sekulic, R. (1982a). Behavior and ranging patterns of a solitary female red howler (*Alouatta seniculus*). *Folia Primatologica*, 38(3-4), 217-232.

Sekulic, R. (1982b). Daily and seasonal patterns of roaring and spacing in four red howler (*Alouatta seniculus*) troops. *Folia Primatologica*, 39(102), 22-48.

Shanee, S. (2014). Yellow-tailed woolly monkey (*Lagothrix flavicauda*) proximal spacing and forest strata use in La Esperanza, Peru. *Primates*, 55(4), 515-523.

Sherman, P. T. (1991). Harpy eagle predation on a red howler monkey. *Folia Primatologica*, 56(1), 53-56.

Shimooka, Y., Campbell, C. J., Di Fiore, A., Felton, A. M., Izawa, K., Link, A., Nishimura, A., Ramos-Fernandez, G., & Wallace, R. B. (2008). Demography and group composition of *Ateles*. In C. J. Campbell (Ed.), *Spider monkeys: Behavior, ecology and evolution of the genus* Ateles (pp. 328-348). Cambridge University Press.

Silva-Lopez, G., Jimenez-Huerta, J., & Benitez-Rodriguez, J. (1987). Monkey populations in disturbed areas: A study on *Ateles* and *Alouatta* at Sierra de Santa Martha, Veracruz, Mexico. *American Journal of Primatology*, 12(3), 355-356.

Silver, S. C., Ostro, L. E. T., Yeager C. P., & Horwich, R. (1998). Feeding ecology of the black howler monkey (*Alouatta pigra*) in northern Belize. *American Journal of Primatology*, 45(3), 263-279.

Slater, K. Y., Schaffner, C. M., & Aureli, F. (2009). Sex differences in the social behavior of wild spider monkeys (*Ateles geoffroyi yucatanensis*). *American Journal of Primatology*, 71(1), 21-29.

Smith, C. C. (1977). Feeding behaviour and social organization in howling monkeys. In T. H. Clutton-Brock (Ed.), *Primate ecology: Studies of feeding and ranging behaviour in lemurs, monkeys, and apes* (pp. 97-126). Academic Press.

Soini, P. (1986). A synecological study of a primate community in the Pacaya-Samiria National Reserve, Peru. *Primate Conservation*, 7, 63-71.

Stevenson, P. R. (1998). Proximal spacing between individuals in a group of woolly monkeys (*Lagothrix lagotricha*) in Tinigua National Park, Colombia. *International Journal of Primatology*, 19(6), 299-311.

Stevenson, P. R. (2006). Activity and ranging patterns of Colombian woolly monkeys in north-western Amazonia. *Primates*, 47(3), 239-247.

Stevenson, P. R., Castellanos, M. C., Pizarro, J. C., & Garavito, M. (2002). Effects of seed dispersal by three ateline monkey species on seed germination at Tinigua National Park, Colombia. *International Journal of Primatology*, 23(6), 1187-1204.

Stevenson, P. R., Quiñones, M. J., & Ahumanda, J. A. (1998). Effects of fruit patch availability on feeding subgroup size and spacing patterns in four primate species at Tingua National Park, Colombia. *International Journal of Primatology*, 19(2), 313–324.

Stoner, K. E. (1996). Habitat selection and seasonal patterns of activity and foraging of mantled howler monkeys (*Alouatta palliata*) in northeastern Costa Rica. *International Journal of Primatology*, 17, 1-30.

Strier, K. B. (1987a). Demographic patterns of one group of free-ranging woolly spider monkeys. *Primate Conservation*, 8, 73-74.

Strier, K. B. (1987b). Activity budgets of woolly spider monkeys, or muriquis (*Brachyteles arachnoides*). *American Journal of Primatology*, 13(4), 385-396.

Strier, K. B. (1989). Effects of patch size on feeding associations in muriquis (*Brachyteles arachnoides*). *Folia Primatologica*, 52(1-2), 70-77.

Strier, K. B. (1990a). New World primates, new frontiers: Insights from the woolly spider monkey, or muriqui (*Brachyteles arachnoides*). *International Journal of Primatology*, 11(1), 7-19.

Strier, K. B. (1990b). Demography, ecology and conservation: An example from southeastern Brazil. *American Journal of Physical Anthropology*, 81(2), 302-303.

Strier, K. B. (1991a). Diet in one group of woolly spider monkeys, or muriquis (*Brachyteles arachnoides*). *American Journal of Primatology*, 23(2), 113-126.

Strier, K. B. (1991b). Demography and conservation of an endangered primate, *Brachyteles arachnoides*. *Conservation Biology*, 5(2), 214-218.

Strier, K. B. (1992a). Atelinae adaptations: Behavioral strategies and ecological constraints. *American Journal of Physical Anthropology*, 88(4), 515-524.

Strier, K. B. (1992b). *Faces in the forest: The endangered muriqui monkeys of Brazil*. Oxford University Press.

Strier, K. B. (1992c). Causes and consequences of nonaggression in the woolly spider monkey, or muriqui (*Brachyteles arachnoides*). In J. Silverberg, & J. Patrick Gray (Eds.), *Aggression and peacefulness in humans and other primates* (pp. 100-116). Oxford University Press.

Strier, K. B. (1993). Growing up in a patrifocal society: Sex differences in the spatial relations of immature muriquis (*Brachyteles arachnoides*). In M. E. Pereira, & L. A. Fairbanks (Eds.), *Juveniles: Comparative socioecology* (pp. 138-147). Oxford University Press.

Strier, K. B. (1994a). Brotherhoods among atelins. *Behaviour*, 130, 151-167.

Strier, K. B. (1994b). Myth of the typical primate. *Yearbook of Physical Anthropology*, 37, 233-271.

Strier, K. B. (1996). Reproductive ecology of female muriquis. In M. Norconk, A. Rosenberger, & P. Garber (Eds.), *Adaptive radiations of New World primates* (pp. 511-532). Plenum Press.

Strier, K. B. (1997). Mate preferences in wild muriqui monkeys (*Brachyteles arachnoides*): Reproductive and social correlates. *Folia Primatologica*, 68, 120-133.

Strier, K. B. (1999). Predicting primate responses to "stochastic" demographic events. *Primates*, 40, 131-142.

Strier, K. B., Boubli, J. P., Possamai, C. B., & Mendes, S. L. (2006). Population demography of northern muriquis (*Brachyteles hypoxanthus*) at the Estação Biológica de Caratinga/Reserva Particular do Patrimônio Natural-Feliciano Miguel Abdala, Minas Gerais, Brazil. *American Journal of Physical Anthropology*, 130(2), 227-237.

Strier, K. B., Dib, L. T., & Figueira, J. E. C. (2002). Social dynamics of male muriquis (*Brachyteles arachnoides hypoxanthus*). *Behaviour*, 139 (2-3), 315-342.

Strier, K. B., & Ives, A. R. (2012). Unexpected demography in the recovery of an endangered primate population. *PLoS ONE*, 7(9), e44407.

Strier, K. B., Mendes, F. D. C., Rímoli, J., & Rímoli, A. O. (1993). Demography and social structure in one group of muriquis (*Brachyteles arachnoides*). *International Journal of Primatology*, 14, 513-526.

Strier, K. B., Mendes, S. L., & Santos, R. R. (2001). Timing of births in sympatric brown howler monkeys (*Alouatta fusca clamitans*) and northern muriquis (*Brachyteles arachnoides hypoxanthus*). *American Journal of Primatology*, 55(2), 87-100.

Strier, K. B., Possamai, C. B., Tabacow, F. P., Pissinatti, A., Lanna, A. M., Rodrigues de Melo, M., Moreira, L., Talebi, M., Breves, P., Mendes, S. L., & Jerusalinsky, L. (2017). Demographic monitoring of wild muriqui populations: Criteria for defining priority areas and monitoring intensity. *PLoS ONE*, 12(12), e0188922.

Strier, K. B., & Ziegler, T. E. (1994). Insights into ovarian function in wild muriqui monkeys (*Brachyteles arachnoides*). *American Journal of Primatology*, 32(1), 31-40.

Strier, K. B., & Ziegler, T. E. (1997). Behavioral and endocrine characteristics of the reproductive cycle in wild muriqui monkeys, *Brachyteles arachnoides*. *American Journal of Primatology*, 42, 299-310.

Strier, K. B., & Ziegler, T. E. (2000). Lack of pubertal influences on female dispersal in muriqui monkeys (*Brachyteles arachnoides*). *Animal Behaviour*, 59, 849-860.

Sussman, R. W., & Phillips-Conroy, J. (1995). A survey on the distribution and density of the primates of Guyana. *International Journal of Primatology, 16*(5), 761-792.

Symington, M. M. (1987a). Sex ratio and maternal rank in wild spider monkeys: When daughters disperse. *Behavioral Ecology and Sociobiology, 20*(6), 421-425.

Symington, M. M. (1987b). Predation and party size in the black spider monkey, *Ateles paniscus chamek*. *International Journal of Primatology, 8*(5), 534.

Symington, M. M. (1988a). Food competition and foraging party size in the black spider monkey (*Ateles paniscus chamek*). *Behavior, 105*(1-2), 117-134.

Symington, M. M. (1988b). Demography, ranging patterns, and activity budgets of black spider monkeys (*Ateles paniscus chamek*) in the Manu National Park, Peru. *American Journal of Primatology, 15*(1), 45-57.

Symington, M. M. (1990). Fission-fusion social organization in *Ateles* and *Pan*. *International Journal of Primatology, 11*(1), 47-60.

Tabacow, F. P., Mendes, S. L., & Strier, K. B. (2009). Spread of a terrestrial tradition in an arboreal primate. *American Anthropologist, 111*(2), 238-249.

Talebi, M., Bastos, A., & Lee, P. C. (2005). Diet of southern muriquis in continuous Brazilian Atlantic forest. *International Journal of Primatology, 26*(5), 1175-1187.

Talebi, M. G., Beltrao-Mendes, R., & Lee, P. C. (2009). Intra-community coalitionary lethal attack of an adult male southern muriqui (*Brachyteles arachnoides*). *American Journal of Primatology, 71*(10), 860-867.

Talebi, M. G., & Lee, P. C. (2010). Activity patterns of *Brachyteles arachnoides* in the largest remaining fragment of Brazilian Atlantic Forest. *International Journal of Primatology, 31*(4), 571-583.

Terborgh, J. (1983). *Five New World primates: A study in comparative biology*. Princeton University Press.

Van Belle, S., & Estrada, A. (2006). Demographic features of *Alouatta pigra* populations in extensive and fragmented forests. In A. Estrada, P. A. Garber, M. Pavelka, & L. Luecke (Eds.), *New perspectives in the study of Mesoamerican primates* (pp. 121-142). Springer.

Van Belle, S., Estrada, A., & Garber, P. A. (2014). The function of loud calls in black howler monkeys (*Alouatta pigra*): Food, mate, or infant defense? *American Journal of Primatology, 76*(12), 1196-1206.

Van Belle, S., Estrada, A., Ziegler, T. E., & Strier, K. B. (2009). Sexual behavior across ovarian cycles in wild black howler monkeys (*Alouatta pigra*). *American Journal of Primatology, 71*(2), 153-164.

Van Belle, S., Kulp, A. E., Thiessen-Bock, R., Garcia, M., & Estrada, A. (2010). Observed infanticides following a male immigration event in black howler monkeys, *Alouatta pigra*, at Palenque National Park, Mexico. *Primates, 51*(4), 279-284.

van Roosmalen, M. G. (1985). Habitat preferences, diet, feeding strategy, and social organization of the black spider monkey (*Ateles paniscus paniscus*, Linnaeus, 1758) in Suriname. *Acta Amazonia, 15*, 1-238.

van Roosmalen, M. G., & Klein, L. L. (1988). The spider monkeys, genus *Ateles*. In A. B. Gustavo, R. A. Mittermeier, A. B. Rylands, A. Coimbra-Filho, & G. A. Fonseca (Eds.), *Ecology and behavior of Neotropical primates* (Vol. 2, pp. 455-539). World Wildlife Fund.

Wallace, R. B. (2005). Seasonal variations in diet and foraging behavior of *Ateles chamek* in a southern Amazonian tropical forest. *International Journal of Primatology, 26*(5), 1053-1075.

Wallace, R. B. (2008). Factors influencing spider monkey habitat use and ranging patterns. In C. J. Campbell (Ed.), *Spider monkeys: Behavior, ecology and evolution of the genus* Ateles (pp. 138-154). Cambridge University Press.

Wang, E., & Milton, K. (2003). Intragroup social relationships of male *Alouatta palliata* on Barro Colorado Island, Republic of Panama. *International Journal of Primatology, 24*(6), 1227-1244.

White, F. (1986). Census and preliminary observation of the black-faced spider monkey (*Ateles paniscus chamek*) in Manu National Park, Peru. *American Journal of Primatology, 11*(2), 125-132.

Whitehead, J. M. (1987). Vocally mediated reciprocity between neighbouring groups of mantled howling monkeys, *Alouatta palliata palliata*. *Animal Behaviour, 35*(6), 1615-1627.

Williams, L. (1974). *Monkeys and the social instinct: An inter-living study from the woolly monkey sanctuary*. Monkey Sanctuary Publications.

Williams-Guillen, K. (2003). *The behavioral ecology of mantled howling monkeys (*Alouatta palliata*) living in a Nicaraguan shade coffee plantation*. [Doctoral dissertation, New York University].

Youlatos, D. (1993). Passages within a discontinuous canopy: Bridging in the red howler monkey (*Alouatta seniculus*). *Folia Primatologica, 61*, 144-147.

Youlatos, D. (1998). Seasonal variation in the positional behavior of red howling monkeys (*Alouatta seniculus*). *Primates, 39*(4), 449-457.

Youlatos, D. (2002). Positional behavior of black spider monkeys (*Ateles paniscus*) in French Guiana. *International Journal of Primatology, 23*(5), 1071-1093.

Youlatos, D. (2008). Locomotion and positional behavior of spider monkeys. In C. J. Campbell (Ed.), *Spider monkeys: Behavior, ecology and evolution of the genus* Ateles (pp. 185-219). Cambridge University Press.

Youlatos, D., & Gasc, J. P. (2011). Gait and kinematics of arboreal quadrupedal walking of free-ranging red howlers (*Alouatta seniculus*) in French Guiana. In K. D'Aout, & E. E. Vereecke (Eds.), *Primate locomotion: Linking field and laboratory research* (pp. 271-287). Springer.

Youlatos, D., & Guillot, D. (2015). Howler monkey positional behavior. In M. M. Kowalewski, P. A. Garber, L. Cortés, B. Urbani, & D. Youlatos (Eds.), *Howler monkeys: Behavior, ecology, and conservation* (pp. 191-218). Springer.

CHAPTER 12

Primate Field Research in the Neotropics

Early History, Systematics, and Conservation

ANTHONY B. RYLANDS AND RUSSELL A. MITTERMEIER

CARPENTER'S (1934, 1935) PIONEER FIELD STUDIES OF THE howler monkeys (*Alouatta palliata*) and spider monkeys (*Ateles geoffroyi*) on the island of Barro Colorado, Panama, were an auspicious start to field primatology in South and Central America. Beyond expeditions for museum collections, and excepting a census of *Alouatta palliata* on Barro Colorado in 1951 (Collias & Southwick, 1952), however, no further field studies were carried out until the 1960s and early 1970s. In the 1970s, they were inspired very largely by the desire to count primates rather than investigate their behavior. The surge in the use of primates in biomedical research in the 1950s and 1960s, and the expediency for the United States of obtaining them from Central and South America, resulted in the trading and death of thousands upon thousands of monkeys, especially squirrel monkeys, obtained from Leticia and Barranquilla in Colombia and Iquitos in Peru (Ávila Pires, 1976). Besides the welfare aspects, this raised major conservation concerns (Heltne, 1967a,b) and highlighted the need to understand the scale of the slaughter, the offtake, and the stocks of this "commodity." The Pan American Health Organization (PAHO) sponsored primate population surveys in Central America, Colombia, Peru, Bolivia, and Guyana to address the stocks part (for example, Muckenhirn et al., 1975; Struhsaker et al., 1975; Freese, 1976; Neville et al., 1976; Freese et al., 1982).

Despite the considerable investment in these censuses, survey techniques were yet in their infancy, and there was little to no understanding of the habitat preferences, the taxonomy (see below), and the geographic distributions of the primates. With the rapid local depletions of primate numbers, and the high mortality incurred during capture, maintenance, and transport, there was increasing difficulty in supplying the laboratories, and one option tested was that of ranching them on islands. A trader in Leticia, Mike Tsalickis, released 5,690 squirrel monkeys onto a 400-ha fluvial island, Santa Sofia, upstream from Leticia, between 1967–1970, and in 1971 estimated a population of 20,698, which seemingly resolved the supply problem. Independent censuses in 1972, however, counted less than 1,000 (Bailey et al., 1974), revealing the bogus claims of the supplier and dashing hopes

of ranching on a large scale, at least for squirrel monkeys (Mittermeier et al., 1977). Attempts to regulate this damaging trade worldwide (Southwick et al., 1970) led to the adoption, in 1979, by the United Nations World Health Organization (WHO) of a "Policy Statement on Use of Primates for Biomedical Purposes" prepared by the IUCN SSC Primate Specialist Group (Mittermeier et al., 1982).

The primate trade and interest in behavior and ecology mobilized researchers to initiate the first field studies in the 1960s and first half of the 1970s. Examples include: *Aotus zonalis* in Panama (Moynihan, 1964); *Alouatta caraya* in Argentina (Pope, 1966); *Alouatta palliata* and *Ateles geoffroyi* in Panama and Guatemala (Chivers, 1969; Baldwin & Baldwin, 1972a; Mittermeier, 1973; Coelho et al., 1976); *Plecturocebus ornatus* in a small forest patch in Colombia (Mason, 1968); *Saimiri oerstedii* and *Saguinus geoffroyi* in Panama (Baldwin & Baldwin, 1972b, 1976; Dawson, 1978); *Saimiri* in Amazonia (Baldwin & Baldwin, 1971); *Cebus* in Central America and Venezuela (Oppenheimer, 1967; Oppenheimer & Oppenheimer, 1973); *Ateles chamek* and *Cheracebus* (Durham, 1975; Kinzey, 1975) and the demographic and behavioral studies of the Proyecto Peruano de Primatologia (Encarnación et al., 1990) in Peru; *Cebuella pygmaea* and *Saguinus oedipus* (Moynihan, 1976; Neyman, 1978) in Colombia; the primate community at the Serrania de La Macarena, Colombia, especially *Ateles belzebuth* (Klein, 1975; Klein & Klein, 1976) and studies of the primates in the Caquetá Basin in Colombia (Izawa, 1975, 1978; Nishimura & Izawa, 1975); and *Alouatta seniculus* in Venezuela and Trinidad (Neville, 1972). In the 1970s, there was also a burgeoning interest, particularly in the United Kingdom and Germany, in *Callithrix jacchus*, which, twinning healthily in captivity, was proving a fruitful model for research in reproductive physiology (Rothe et al., 1978). In 1976, Miranda Stevenson carried out the first field study of this genus in the Tapacurá Ecological Station in Pernambuco, Northeast Brazil (Stevenson & Rylands, 1988), engendering long-term field research initiatives there and also in Natal and João Pessoa.

The nascent status of Neotropical primate field research by the mid-1970s is quite well portrayed in Thorington and Heltne's 1976 book, *Neotropical Primates: Field Studies and Conservation.* To add to that, however, there had been some activity in the Brazilian Atlantic Forest, with the establishment of the Tijuca Biological Bank, Rio de Janeiro, to breed lion tamarins, *Leontopithecus*, for research and reintroduction in 1972–1974, and the revelatory observations on the ecology and behavior of the species by Adelmar F. Coimbra-Filho (Coimbra-Filho, 1970; Coimbra-Filho & Mittermeier, 1972, 1973), as well as the discovery that marmosets (*Callithrix*) gouge tree trunks to eat gum (Coimbra-Filho, 1972; Coimbra-Filho & Mittermeier, 1976). In 1979, Coimbra-Filho inaugurated the Rio de Janeiro Primate Center (CPRJ), the first of its kind in Brazil for research on and the conservation of Brazilian primates (Rylands et al., 2002). Thorington and Heltne (1976: p.1) noted that, to their knowledge, "there are simply no data available on the status of Brazilian populations of primates in the Amazon," and also noted the "absence of material dealing with any of the species of the Pitheciinae." They were quite right then, but Mittermeier and Coimbra-Filho (1977; also Mittermeier et al., 1978) subsequently published Mittermeier's report on a four-month expedition he carried out in 1973 up the Rio Solimões and its tributaries in Brazil to retrace the steps of the English explorer-naturalists Henry Walter Bates and Alfred Russel Wallace and search for the three then-recognized uakaris, *Cacajao*, and the white-nosed saki, *Chiropotes albinasus*. He found the uakaris, becoming the first outsider to see them in the wild (Mittermeier & Coimbra-Filho, 1977), but decided against carrying out his doctoral research on them because of the challenging logistics. He instead went to Suriname to carry out a synecological study of the eight Suriname primates, including *Chiropotes sagulatus* and *Pithecia pithecia* (Fleagle & Mittermeier, 1980; Mittermeier & van Roosmalen, 1981). He was later joined by van Roosmalen, who made a detailed study of *Ateles paniscus* (1985).

In the early 1970s, Colombia, Peru, and Brazil banned the export of wild primates, and an agitated PAHO set out to establish primate breeding centers in the three countries. The Brazilian Ministry of Health had determined that one of these be at the National Institute for Amazon Research (INPA), recently expanded to a large campus in the Manaus suburbs, with the oxymoronic mission to research the means to "open-up" the Brazilian Amazon without wrecking it—this following the failure of the experiment to colonize the recently constructed Transamazon Highway (Goodland & Irwin, 1975). The director, bee geneticist Warwick Kerr, was wisely uncomfortable with this imposition, and, to buy time, in 1976, he serendipitously employed a recently graduated Rylands to begin the task of building it. Securing assurances that after

one year Rylands be allowed to dedicate himself to field research, in 1977, he went to INPA's Humboldt Pioneer Nucleus on the Rio Aripuanã in the northern Mato Grosso, there to begin his study of *Mico intermedius* (Rylands, 1979). He was joined that year by José Márcio Ayres, who was to study *Chiropotes albinasus* (Ayres, 1981). Ayres went on to do his doctoral research on *Cacajao calvus*—the study that Mittermeier had decided against 10 years earlier (Ayres, 1986). The site of PAHO's desired Brazilian National Primate Center was moved to Belém, Pará, in 1978 (Muniz & Kingston, 1985).

While much of the attention in the early days focused on Amazonia and northern South America, in 1977 Mittermeier, then director of the World Wildlife Fund–US Primate Program, initiated a decade-long survey (1979–1989) of the Brazilian Atlantic Forest's 26 primates (21 of them endemic). Aguirre (1971) had carried out a remarkable range-wide survey of the muriqui, *Brachyteles*, but, besides Coimbra-Filho's studies of lion tamarins and marmosets, very little was known of the primates in the region, by then reduced to only a fragmented 8% of its original extent. This survey put the Atlantic Forest on the map. A highly significant result was what is now a 38-year study at the Fazenda Montes Claros in Caratinga, Minas Gerais, of the muriqui, *B. hypoxanthus*, led by Karen Strier. To date, it has trained more than 75 Brazilian primatologists, and Caratinga now has the largest remaining population of this species (Strier, 1992). It also led to the creation, in 1989, of a conservation and ecology graduate program at the Universidade Federal de Minas Gerais, supported by the U.S. Fish and Wildlife Service. From 1983 to 1989, Milton Thiago de Mello, professor at the University of Brasília, organized annual two-month primate field courses for Latin American countries and, after a gap, abbreviated versions continue to be held to this day.

The rest is (more recent) history. In 1988, participants of the XII International Primatological Congress, organized by Thiago de Mello in Brasília, were enthusiastically surprised by the abundance and quality of primate field research in Mexico, Central America, Brazil, and other South American countries.

Returning to the 1960s and the exodus of wild primates for biomedical research, Hershkovitz (1965) reproached the laboratories for their nescience and disregard as to the correct identification or even the origin of the monkeys they were receiving. He pointed to a "lack of comprehensive and authoritative taxonomic revisions of the large majority of living primate species" (p.1157) and to the potentially damaging consequences for the validity of their experimental results. Cooper (1968) made a brave attempt for the squirrel monkeys, as did

Thorington in 1985, but it was Hershkovitz himself who dedicated much of his life to address this lack for the Neotropical monkeys. Napier and Napier (1967) listed 64 species, and 160 species and subspecies of New World primates. Thirty-two of them are today synonyms, so only 128 of today's taxonomic tally—175 species and 216 taxa—were then recognized. Reasons for the increase in the number of taxa are numerous, but 46 were newly described as from 1966—16 by Hershkovitz through taxonomic revisions, but the remainder largely through field research, which has contributed so strongly, not just from the discovery of new distinct phenotypes, but from the associated and crucial understanding of the ranges of the species comprising each genus.

The beta taxonomy of the Neotropical monkeys also underwent a major upheaval in the late 1970s. Since the classification of Mivart (1865), most taxonomies had placed the platyrrhines in two, variously named, families: the Callitrichidae (marmosets and tamarins) and the Cebidae (the remainder), as in the 1960s. Hershkovitz (1977), like Dollman (1933), was puzzled by *Callimico*, and placed it in a third monotypic family, Callimiconidae. Based on morphological studies, and contesting the gradistic approach of Hershkovitz (1977), Rosenberger (1980, 1981) argued cogently for a rearrangement that placed the marmosets and tamarins in a redefined Cebidae, with them and just the cebines—squirrel monkeys and capuchins. Pitheciines and atelines, he placed in the Atelidae. Robert Sussman adopted this taxonomy in his book *Primate Ecology and Social Structure: New World Monkeys* (2000), but treated the tail-twining *Aotus* and *Callicebus* as the Aotinae subfamily of the Cebidae, while Rosenberger had them firmly placed in the Atelidae with the pitheciines. Rosenberger later (2011, 2020) placed his pitheciines as a separate family. Rosenberger's classification has been consistently confirmed by molecular genetic studies, except in the placement of the night, or owl, monkeys. Genetic analyses mysteriously but consistently place *Aotus* in the Cebidae, while morphology and behavior place the genus with the titis in the Pitheciidae (Schneider & Rosenberger, 1996). But for the place of *Aotus*, Rosenberger's beta taxonomy is now universally accepted, and the one that prevails in this book.

The corollary of Rosenberger's (1980, 1981) revision was his ecological approach to his evolutionary interpretations of form and function in the morphology characterizing his three families, the revised Cebidae, the Pitheciidae, and the Atelidae, that were divided into subfamilies Callitrichinae, Cebinae, Homunculinae (owl monkeys and titis), Pitheciinae, and Atelinae (including the tribe Alouattini). These gave rise to four so-called adaptive zones (2011) in what Rosenberger refers to as

an ecophylogenetic model of platyrrhine evolution. The key elements in this model are (1) body size and locomotion, and (2) food, its acquisition and processing. They intricately influence or are influenced by habitat selection in the broad- (forest types) and the meso-sense (forest strata), and besides, have repercussions on their social organization and reproduction (Rosenberger, 2020). Mittermeier's synecological study of the Surinamese primates was an early source providing needed clues to this, as were the findings of John Terborgh and Charles Janson studying five sympatric primates at Cocha Cashu, Manu, Peru (Terborgh, 1983; Terborgh & Janson, 1986), and the Colombia-Japan long-term partnership for primate research program from 1986 at the Centro de Investigaciones Ecológicas Macarena in the Colombian Amazon (Nishimura et al., 1995). Rylands (1987) attempted a synthesis of body size, habitat, diet, and lifestyle of the Amazonian primates overall, indicating adaptive zones (but not calling them that) in an effort to understand how things are divided up in communities typically of 10–12, but reaching as many as 15, sympatric primates in the upper reaches of the basin. The primate chapters in this book summarize our greatly increased understanding of Rosenberger's adaptive zones and the ecological roles these primates play in the maintenance of Neotropical forest communities. More than that, the findings of field studies of the Neotropical primates, short- and long-term, revealed the fragility of what Strier (1994) refers to as the cercopithecine model of a "typical primate," established during the preeminence of Old World primate studies in the 1960s to 1980s (Strier, 1994). New World primates were found to be distinct ("atypical") in such aspects as dispersal, hierarchical social structures mediated by aggressive competition, and female choice, mating systems, and infant rearing (for example, polyandry, female reproductive suppression, and cooperative breeding that so remarkably characterize the callitrichids, and paternal care in the pair-living titi and owl monkeys) (Strier, 1994).

The early days of field primatology in the Old World were largely anthropological and ethological pursuits—at first descriptive as to how the primates live in their natural environments, but later emphasizing the need for systematic data to allow for comparative studies in evolutionary and ecological theory, considering especially their relatedness to humans, and pursuing such themes as variation in diet, ranging behavior, social organization, and mating systems, and the correlates of living in different habitats, on the ground or in the trees (Strier, 1994). Field primatology focused first on the relatively easily observable semiterrestrial baboons, macaques, gray langurs, and chimpanzees. The Neotropics, despite Carpenter's pioneering studies, were left behind in those early days.

It took the devastating primate trade to snap it out of its dormancy, and conservation has, right from start, always been a strong underlying motive for primate field studies there. Field research quickly went beyond estimating stocks for biomedical research, because it was in the late 1970s that Amazonia was all of a sudden being "opened-up," and people began to take proper notice of the devastation, rampant since the earliest days of colonization, of the forests elsewhere, i.e., in northern Colombia, western Ecuador, Central America, southern Mexico, northern Argentina, and eastern Brazil. Although Neotropical field primatology hit the market, so to speak, when ecological and evolutionary correlates of behavior were in vogue, there has always been an underlying urgency to find and study primates to contribute to and inform conservation measures (Strier, 1997).

Particularly noteworthy in the Neotropics, and indicative of today's strong community of primatologists and their focus on conservation, is the proliferation of national primate societies and even a Latin America–wide society. This all began with the Brazilian Primatological Society, founded in 1979, followed by the Latin American Primatological Society in 1986 (it languished but was revived in 2013), and today there are also national societies for Argentina, Bolivia, Colombia, Ecuador, Mexico, and Peru. These societies cover a wide range of research topics, but they all have a strong focus on conservation. The IUCN SSC Primate Specialist Group in turn has always had an especially strong representation in the Neotropics, now with four regional sections: Brazil and the Guianas, the Tropical Andes, Mesoamerica, and, most recently, the Southern Cone. This community has not only greatly impacted primatology and primate conservation, but many of its members have taken on leadership roles in national, regional, and global conservation issues in general.

REFERENCES CITED—CHAPTER 12

Aguirre, A. C. (1971). O mono Brachyteles arachnoides (É. Geoffroy). Situação atual da espécie no Brasil. Academia Brasileira de Ciências, Rio de Janeiro, 1-51.

Ávila-Pires, F. D. (1976). Nonhuman primates of the American continent. First Inter-American Conference on Conservation and Utilization of American Nonhuman Primates in Biomedical Research, 3-7.

Ayres, J. M. C. (1981). Observações sobre a ecologia e o comportamento dos cuxiús (Chiropotes albinasus e Chiropotes satanas, Cebidae: Primates). [Master's thesis, Fundação Universidade do Amazonas].

Ayres, J. M. C. (1986). Uakaris and Amazonian flooded forests. [Doctoral dissertation, Cambridge University].

Bailey, R. C., Baker, R. S., Brown, D. S., von Hildebrand, P., Mittermeier, R. A., Sponsel, L. E., & Wolf, K. (1974). Progress of a breeding project for non–human primates in Colombia. Nature, 248, 453-455.

Baldwin, J. D., & Baldwin, J. I. (1971). Squirrel monkeys (Saimiri) in natural habitats in Panama, Colombia, Brazil and Peru. Primates, 12, 45-61.

Baldwin, J. D., & Baldwin, J. I. (1972a). Population density and use of space in howling monkeys (Alouatta villosa) in southwestern Panama. Primates, 13, 371-379.

Baldwin, J. D., & Baldwin, J. I. (1972b). The ecology and behavior of squirrel monkeys (Saimiri oerstedii) in a natural forest in western Panama. Folia Primatologica, 18, 161-184.

Baldwin, J. D., & Baldwin, J. I. (1976). Primate populations in Chiriqui, Panama. In R. W. Thorington Jr., & P. G. Heltne (Eds.), Neotropical primates: Field studies and conservation (pp. 20-31). National Academy of Sciences.

Carpenter, C. R. (1934). A field study of the behavioral and social relations of howling monkeys (Alouatta palliata). Comparative Psychology Monographs, 10(2), 168.

Carpenter, C. R. (1935). Behavior of red spider monkeys in Panama. Journal of Mammalogy, 16(3), 171-180.

Chivers, D. J. (1969). On the daily behavior and spacing of howling monkey groups. Folia Primatologica, 10, 48-102.

Coelho, A. M., Jr., Coelho, L., Bramblett, C., Bramblett, S., & Quick, L. (1976). Ecology, population characteristics, and sympatric associations in primates: A socioenergetic analysis of howler and spider monkeys in Tikal, Guatemala. Yearbook of Physical Anthropology, 20, 96-135.

Coimbra-Filho, A. F. (1970). Consideracões gerais e situação atual dos micos-leões escuros, Leontideus chrysomelas (Kühl, 1820) e Leontideus chrysopygus (Mikan, 1823) (Callithricidae, primates). Revista Brasileira de Biologia, 30, 249-268.

Coimbra-Filho, A. F. (1972). Aspectos inéditos do comportamento de sagüis do gênero Callithrix (Callithricidae, primates). Revista Brasileira de Biologia, 32, 505-512.

Coimbra-Filho, A. F., & Mittermeier, R. A. (1972). Taxonomy of the genus Leontopithecus Lesson, 1840. In D. Bridgewater (Ed.), Saving the lion marmoset, proceedings of the Wild Animal Propagation Trust Golden Lion Marmoset Conference (pp. 7-22). Wild Animal Propagation Trust.

Coimbra-Filho, A. F., & Mittermeier, R. A. (1973). Distribution and ecology of the genus Leontopithecus Lesson, 1840 in Brazil. Primates, 14, 47-66.

Coimbra-Filho, A. F., & Mittermeier, R. A. (1976). Exudate-eating and tree-gouging in marmosets. Nature, 262, 630.

Collias, N., & Southwick, C. (1952). A field study of population density and social organization in howling monkeys. Proceedings of the American Philosophical Society, 96(2), 143-156.

Cooper, R. W. (1968). Squirrel monkey taxonomy and supply. In L. A. Rosenblum, & R. W. Cooper (Eds.), The squirrel monkey (pp. 1-29). Academic Press.

Dawson, G. (1978). Troop size, composition and instability in the Panamanian tamarin (S. oedipus geoffroyi: Ecological and behavioral implications. In D. G. Kleiman (Ed.), The *biology and conservation of the Callitrichidae* (pp. 23-38). Smithsonian Institution Press.

Dollman, G. (1933). *Primates*. Series 3. Trustees of the British Museum.

Durham, N. M. (1975). Some ecological, distributional, and group behavioral features of Atelinae in southern Peru: With comments on interspecific relations. In R. H. Tuttle (Ed.), *Socioecology and psychology of primates* (pp. 87-101). Mouton.

Encarnación, F., Moya, L., Moro. J., & Málaga, C. (1990). Misíon, y objetivos. In Proyecto Peruano de Primatología (Ed.), *La primatologia en el Perú: Investigaciones primatológicas (1973-1985)* (pp. 3-14). Dirección General de Forestal y Fauna, Unidade Agraria Departamental XXII-Loreto, 1990.

Fleagle, J. G., & Mittermeier, R. A. (1980). Locomotor behavior, body size, and comparative ecology of seven Surinam monkeys. *American Journal of Physical Anthropology, 52,* 301-314.

Freese, C. H. (1976). Censusing *Alouatta palliata, Ateles geoffroyi,* and *Cebus capucinus* in the Costa Rican dry forest. In R. W. Thorington Jr., & P. G. Heltne (Eds.), *Neotropical primates: Field studies and conservation* (pp. 4-9). National Academy of Sciences.

Freese, C. H., Heltne, P. G., Castro, R. N., & Whitesides, G. (1982). Patterns and determinants of monkey densities in Peru and Bolivia, with notes on distributions. *International Journal of Primatology, 3,* 53-90.

Goodland, H. S., & Irwin, R. J. A. (1975). *Amazon jungle: Green hell to red desert? An ecological discussion of the environmental impact of the highway construction program in the Amazon Basin.* Elsevier.

Heltne, P. G. (1967a). Animals from the Amazon Basin. *Science, 157,* 134.

Heltne, P. G. (1967b). Latin America: Call for conservation. *Science, 158,* 717.

Hershkovitz, P. (1965). Primate research and systematics. *Science, 147,* 1156-1157.

Hershkovitz, P. (1977). Living New World monkeys (Platyrrhini) with an introduction to primates, Volume 1. Chicago University Press.

Izawa, K. (1975). Foods and feeding behavior of monkeys in the upper Amazon Basin. *Primates, 16,* 295-316.

Izawa, K. (1978). A field study of the ecology and behavior of the black-mantle tamarin (*Saguinus nigricollis*). *Primates, 19,* 241-274.

Kinzey, W. G. (1975). The ecology of locomotion in *Callicebus torquatus. American Journal of Physical Anthropology, 42,* 312.

Klein, L. (1975). Social and ecological contrasts between four taxa of Neotropical primates (*Ateles belzebuth, Alouatta seniculus, Saimiri sciureus, Cebus apella*). In R. H. Tuttle (Ed.), *Socioecology and psychology of primates* (pp. 59-85). Mouton.

Klein, L. L., & Klein, D. J. (1976). Neotropical primates: Aspects of habitat usage, population density, and regional distribution in La Macarena, Colombia. In R. W. Thorington Jr., and P. G. Heltne (Eds.), *Neotropical primates: Field studies and conservation* (pp. 70-78). National Academy of Sciences.

Mason, W. A. (1968). Use of space by Callicebus groups. In P. C. Jay (Ed.), *Primates: Studies in adaptation and variability* (pp. 200-216). Holt, Rinehart and Winston.

Mittermeier, R. A. (1973). Group activity and population dynamics of the howler monkeys on Barro Colorado Island. *Primates, 14,* 1-19.

Mittermeier, R. A., Bailey, R. C., & Coimbra-Filho, A. F. (1978). Conservation status of the Callitrichidae in Brazilian Amazonia, Surinam, and French Guiana. In D. G. Kleiman (Ed.), *The biology and conservation of the Callitrichidae* (pp. 127-146). Smithsonian Institution Press.

Mittermeier, R. A., Bailey, R. C., Sponsel, L. E., & Wolf, E. (1977). Primate ranching: Results of an experiment. *Oryx, 13,* 449-453.

Mittermeier, R. A., & Coimbra-Filho, A. F. (1977). Primate conservation in Brazilian Amazonia. In H. S. H. Prince Rainier III, & G. H. Bourne (Eds.), *Primate conservation* (pp. 117-166). Academic Press.

Mittermeier, R. A., Scott, R. F., Chivers, D. J., & Else, J. G. (1982). World Health Organization (WHO) and Ecosystem Conservation Group (ECG) adopt Primate Specialist Group's policy statement on use of primates for biomedical purposes. *IUCN/SSC Primate Specialist Group Newsletter* (2), 2-7.

Mittermeier, R. A., & van Roosmalen, M. G. M. (1981). Preliminary observations on habitat utilization and diet in eight Surinam monkeys. *Folia Primatologica, 36*(1-2), 1-39.

Mivart, S. G. (1865). Contributions towards a more complete knowledge of the axial skeleton in the primates. *Proceedings of the Zoological Society of London,* 542-592.

Moynihan, M. (1964). Some behavior patterns of platyrrhine monkeys: 1. The night monkey (*Aotus trivirgatus*). *Smithsonian Miscellaneous Collections, 146*(5), 84.

Moynihan, M. (1976). Notes on the ecology and behavior of the pygmy marmoset (*Cebuella pygmaea*) in Amazonian Colombia. In R. W. Thorington Jr., & P. G. Heltne (Eds.), *Neotropical primates: Field studies and conservation* (pp. 79-84). National Academy of Sciences.

Muckenhirn, N. A., Mortensen, B. K., Vessey, S., Fraser, C. E. O., & Singh, B. (1975). *Report on a primate survey in Guyana.* Pan American Health Organization.

Muniz, J. A. P. C., & Kingston, W. R. (1985). The Brazilian National Primate Center. *Primate Conservation,* (6), 39-41.

Napier, J. R., & Napier, P. H. (1967). *A handbook of living primates.* Academic Press.

Neville, M. K. (1972). The population structure of red howler monkeys (*Alouatta seniculus*) in Trinidad and Venezuela. *Folia Primatologica, 17,* 56-86.

Neville, M., Castro, N., Marmol, A., & Revilla, J. (1976). Censusing primate populations in the Reserved Area of the Pacaya and Samiria Rivers, Department Loreto, Peru. *Primates, 17,* 151-181.

Neyman, P. (1978). The ecology and social behavior of the cottontop tamarin in Colombia. In D. G. Kleiman (Ed.), *The biology and conservation of the Callitrichidae* (pp. 39-72). Smithsonian Institution Press.

Nishimura, A., & Izawa, K. (1975). The group characteristics of woolly monkeys (*Lagothrix lagotricha*) in the upper Amazonian Basin. In S. Kondo, M. Kawai, & A. Ehara (Eds.), *Contemporary primatology* (pp. 351-357). S. Karger.

Nishimura, A., Izawa, K., & Kimura, K. (1995). Long-term studies of primates at La Macarena, Colombia. *Primate Conservation* (16), 7-14.

Oppenheimer, J. R. (1967). The diet of *Cebus capucinus* and the effect of *Cebus* on the vegetation. *Bulletin of the Ecological Society of America*, 48, 138.

Oppenheimer, J. R., & Oppenheimer, E. C. (1973). Preliminary observations of *Cebus nigrivittatus* (Primates: Cebidae) on the Venezuelan llanos. *Folia Primatologica*, 19, 409-436.

Pope, B. L. (1966). Population characteristics of howler monkeys (*Alouatta caraya*) in northern Argentina. *American Journal of Physical Anthropology*, 24, 361-370.

Rosenberger, A. L. (1980). Gradistic views and adaptive radiation of platyrrhine primates. *Journal of Morphology and Anthropology*, 71(2), 157-163.

Rosenberger, A. L. (1981). Systematics: The higher taxa. In A. F. Coimbra-Filho, & R. A. Mittermeier (Eds.), *The ecology and behavior of neotropical primates* (Vol. 1, pp. 9-27). Academia Brasileira de Ciências.

Rosenberger, A. L. (2011). Evolutionary morphology, platyrrhine evolution and systematics. *Anatomical Record*, 294, 1955-1974.

Rosenberger, A. L. (2020). *New World monkeys: The evolutionary odyssey*. Princeton University Press.

Rothe, H., Wolters, H.-J., & Hearn, J. P. (Eds.). (1978). *Biology and behaviour of marmosets*. Eigenverlag H. Rothe.

Rylands, A. B. (1979). Observações preliminares sobre o sagüi, *Callithrix humeralifer intermedius* Hershkovitz, 1977, em Dardanelos, Rio Aripuanã, Mata Grosso. *Acta Amazonica*, 9(3), 589-602.

Rylands, A. B. (1987). Primate communities in Amazonian forests: Their habitats and food resources. *Experientia*, 43(3), 265-279.

Rylands, A. B., Mallinson, J. J. C., Kleiman, D. G., Coimbra-Filho, A. F., Mittermeier, R. A., Câmara, I. de G., Valladares-Padua, C., & Bampi, M. I. (2002). A history of lion tamarin conservation and research. In D. G. Kleiman, & A. B. Rylands (Eds.), *Lion tamarins: Biology and conservation* (pp. 3-41). Smithsonian Institution Press.

Schneider, H., & Rosenberger, A. L. (1996). Molecules, morphology, and platyrrhine systematics. In M. A. Norconk, A. L. Rosenberger, & P. A. Garber (Eds.), *Adaptive radiations of the Neotropical primates* (pp. 3-19). Plenum Press.

Southwick, C. H., Siddiqi, M. R., & Siddiqi, M. F. (1970). Primate populations and biomedical research. *Science*, 170, 1051-1054.

Stevenson, M. F., & Rylands, A. B. (1988). The marmosets, genus *Callithrix*. In R. A. Mittermeier, A. B. Rylands, A. F. Coimbra-Filho, & G. A. B. da Fonseca (Eds.), *Ecology and behavior of Neotropical primates* (Vol. 2, pp. 349-453). World Wildlife Fund.

Strier, K. B. (1992). *Faces in the forest: The endangered muriqui monkeys of Brazil*. Oxford University Press.

Strier, K. B. (1994). Myth of the typical primate. *Yearbook of Physical Anthropology*, 37, 233-271.

Strier, K. B. (1997). Behavorial ecology and conservation biology of primates and other animals. *Advances in the Study of Behavior*, 26, 101-158.

Struhsaker, T. T., Glander, K., Chirivi, H., & Scott, N. J. (1975). A survey of primates and their habitats in northern Colombia. *Primate Censusing Studies in Peru and Colombia*, 43-79.

Sussman, R. W. (2000). *Primate ecology and social structure. Volume 2: New World monkeys*. Pearson Custom Publishing.

Terborgh, J. (1983). *Five New World primates: A study in comparative ecology*. Princeton University Press.

Terborgh, J., & Janson, C. H. (1986). The socioecology of primate groups. *Annual Review of Ecology and Systematics*, 17, 111-135.

Thorington, R. W., Jr. (1985). The taxonomy and distribution of squirrel monkeys (*Saimiri*). In C. L. Coe, & L. A. Rosenblum (Eds.), *Handbook of squirrel monkey research* (pp. 1-33). Springer.

Thorington, R. W., Jr., & Heltne, P. G. (Eds.). (1976). *Neotropical primates: Field studies and conservation*. National Academy of Sciences.

van Roosmalen, M. G. M. (1985). Habitat preferences, diet, feeding strategy and social organization of the black spider monkey (*Ateles paniscus paniscus* Linnaeus, 1758) in Surinam. *Acta Amazonica*, 15(S3-4), 238.

PART III

AFRICAN AND ASIAN MONKEYS

CHAPTER 13

The African Colobines

Behavioral Flexibility and Conservation in a Changing World

Eva C. Wikberg, Elizabeth Kelley, Robert W. Sussman, and Nelson Ting

The African colobines (Tribe Colobini, or colobus monkeys) are a group of arboreal, medium-sized primates; their modern distribution is confined to the forested regions of the African tropics. They have nearly absent thumbs and specialized digestive systems adapted to a folivorous and granivorous diet. Despite the specialized physiological and morphological adaptations of this clade, the African colobines as a whole show a much greater range of variation in their behavior and ecology than previously thought, which has been revealed by longitudinal studies and between-site comparisons across Africa. However, while some African colobine populations show remarkable flexibility in their behavior, others seem to be much more limited. Although the source of this variation in behavioral flexibility is not completely understood, the implications have become increasingly obvious—the populations that show an aptitude for adaptability are better able to cope with currently changing environments, while the others are becoming increasingly threatened. The majority of African colobines are threatened by extinction, and all species are in states of population decline (Table 13.1). Five taxa, *Piliocolobus waldroni* (Miss Waldron's red colobus), *P. epieni* (Niger Delta red colobus), *P. pennantii* (Pennant's red colobus), *P. rufomitratus* (Tana River red colobus), and *Colobus vellerosus* (white-thighed black-and-white colobus) have been listed among the world's 25 most endangered primates (Mittermeier et al., 2007, 2009; Schwitzer et al., 2019). In fact, *P. waldroni* is likely already extinct due to extensive levels of human hunting and habitat loss. In response to such threats, colobus behaviors such as grouping patterns, foraging behaviors, and habitat and strata use have changed over time (e.g., Galat-Luong & Galat, 2005). The question is whether their degree of behavioral flexibility and current conservation efforts are sufficient for these populations to persist in landscapes with increasing human populations, proximity, and/or activities.

In Greek, the word *kolobus* means mutilated (Oates & Davies, 1994). The "mutilated," or nearly absent thumb of African colobines is a defining feature, as shown in Figure 13.1, and the basis for the name of this group of primates (Oates & Davies, 1994). Early literature actually described colobus monkeys as brachiators or semi-brachiators because of their reduced thumbs (Straus, 1949; Napier & Napier, 1967; Stern & Oxnard, 1973). Although the red colobus has enhanced arm mobility, possibly associated with more suspensory foraging postures, compared to some other cercopithecines (Dunham et al., 2016), field observations show that colobines are basically quadrupedal, above-branch walkers (Kingdon, 1971; Rose, 1978; Gebo et al., 1994; Gebo & Chapman, 1995; McGraw, 1998a,b).

Colobines first appear in the fossil record by the late Miocene, around 12.5 mya in East Africa (Rossie et al., 2013). The early period of their evolution saw an extensive distribution that included Eurasia, a diet that was less folivorous than today, and locomotion that was semiterrestrial (Delson, 1994; Gilbert et al., 2010; Nakatsukasa et al., 2010; Rossie et al., 2013). The fossil record of colobines in Africa is particularly diverse and extends from the late Miocene into the Pleistocene, with the early members displaying a short thumb and thus possessing a close relationship to the living forms. This is concordant with dates inferred using molecular data, which show that modern colobus monkeys diversified over 7.5 mya (Ting, 2008b). Thus, despite many of the African fossil colobines being morphologically unlike any of the species living today (Fleagle, 1988; Delson, 1994), the majority of extinct African colobines likely represent one large shared evolutionary lineage with the modern colobus monkeys that was even more diverse in the past. While these broad strokes of colobus monkey evolutionary history are starting to become clearer with more recent molecular research and fossil discoveries, much more needs to be done to understand the details of how the fossils are related to one another, how exactly they are related to the living forms, and how the living species have diverged from one another in relation to the biogeographic history of Africa.

The digestive specialization that differentiates the Colobinae from other primates is the presence of a multichambered stomach which enables these primates to digest foods that are toxic for other primate groups. Specifically, colobine monkeys have large sacculated

Table 13.1 African Colobine Species, Habitat Preferences, and IUCN Conservation Status

Scientific Name	Common Name	Habitat Preferences	IUCN Red List Category [a]
Procolobus verus	Olive colobus	Forest: dry, moist lowland, swamp. Forest: heavily degraded	Vulnerable
Piliocolobus badius	Upper Guinea red colobus	Forest: dry, moist lowland, swamp	Endangered
P. bouvieri	Bouvier's red colobus	Forest: moist lowland, swamp	Endangered
P. epieni	Niger Delta red colobus	Forest: moist lowland, swamp	Critically Endangered
P. foai	Foa's red colobus	Forest: moist submontane and montane	Endangered
P. gordonorum	Udzungwa red colobus	Forest: moist lowland, moist montane	Vulnerable
P. kirkii	Zanzibar red colobus	Forest: moist lowland, mangrove, swamp. Swamp: moist	Endangered
P. oustaleti	Oustalet's red colobus	Forest: dry, moist lowland, swamp. Savanna: dry	Vulnerable
P. langi	Lang's red colobus	Forest: moist lowland	Endangered
P. lulindicus	Ulindi red colobus	Forest: moist lowland	Endangered
P. parmentieri	Lomami red colobus	Forest: moist lowland	Endangered
P. pennantii	Pennant's red colobus	Forest: moist lowland, swamp	Critically Endangered
P. preussi	Preuss's red colobus	Forest: moist lowland, moist montane	Critically Endangered
P. semlikiensis	Semliki red colobus	Forest: moist lowland	Vulnerable
P. temminckii	Temminck's red colobus	Forest: dry, moist lowland, mangrove, swamp Savanna: dry, moist	Endangered
P. tephrosceles	Ashy red colobus	Forest: dry, moist lowland, swamp. Savanna: dry	Endangered
P. rufomitratus	Tana River red colobus	Forest: moist lowland, swamp	Critically Endangered
P. tholloni	Tshuapa red colobus	Forest: moist lowland	Vulnerable
P. waldroni	Miss Waldron's red colobus	Forest: moist lowland, swamp. Forest: heavily degraded	Critically Endangered
Colobus angolensis	Angolan colobus	Forest: moist lowland, moist montane	Vulnerable
C. guereza	Guereza	Forest: moist lowland and montane. Forest: heavily degraded	Least Concern
C. polykomos	King colobus	Forest: dry, moist lowland	Endangered
C. satanas	Black colobus	Forest: moist lowland, swamp, moist montane	Vulnerable
C. vellerosus	White-thighed, or ursine, black-and-white colobus	Forest: moist lowland Savanna: moist	Critically Endangered

[a] IUCN (2020).

Figure 13.1 *Colobus* hand.
(Photo credit: Keith Thompson)

stomachs that are typically divided into four (*Pilicolobus* and *Procolobus*) or three chambers (*Colobus*) (Matsuda et al., 2019). Most alloenzymatic dietary carbohydrates are fermented in the fore-gut (Lambert & Fellner, 2012). These fermentation chambers contain bacteria and are adaptations for a folivorous diet, a system analogous in fundamental aspects to the digestive system of ruminants (Bauchop, 1977; McKey et al., 1981). However, because most colobines weigh less than 10 kg, thus making fore-gut microbial fermentation and volatile fatty acid production (VFA) less efficient, fore-gut fermentation among this group appears to be an effective strategy to minimize the negative effects of toxins and not as an effective means of obtaining energy (Dasilva, 1992).

Colobines also have enlarged salivary glands that secrete large amounts of saliva to further aid in the digestion of foods that are high in tannins and other compounds. Premolars and molars that have high, sharp cusps and deep lateral notches are thought to fold and slice leaves effectively and may thus be an adaptation for eating leaves (Oates & Davies, 1994). However, their molars may be equally well suited for processing certain types of seeds, particularly ones with thin and flexible coverings (Lucas & Teaford, 1994).

TAXONOMY, MORPHOLOGY, AND DISTRIBUTION

The extant African colobines consist of three major groups: the olive colobus, the red colobus, and the black-and-white colobus. The taxonomy and classification of these monkeys has changed substantially over the years, and there is still no clear consensus regarding the number of genera and species that should be recognized. Here, we follow the classification in Zinner and colleagues (2013), as shown in Table 13.1, except for the red colobus populations inhabiting the eastern Democratic Republic of Congo (DRC), where we recognize four (instead of three) taxa following Colyn (1991) due to their highly unstable taxonomy. Zinner et al. (2013) recognize each of the three living colobus monkey groups as distinct genera: *Procolobus* (olive colobus), *Piliocolobus* (red colobus), and *Colobus* (black-and-white colobus). The olive colobus and red colobus are more closely related to one another, which is reflected in their genetics (Ting, 2008a,b) and shared anatomical characteristics, such as a four-chambered stomach, a small larynx, separate ischial callosities, and the presence of a perineal organ in males and sexual swellings in females (Pocock, 1935; Hill, 1952). We recognize 1 species of olive colobus, 18 species of red colobus, and 5 species of black-and-white colobus. It is important to note that the number of red colobus species recognized, compared to the number of black-and-white colobus species, is not necessarily reflective of a large difference in

diversity between these groups, but rather a difference in the manner in which species are diagnosed. While the red colobus monkeys have been split into numerous species with few subspecies, there exist numerous subspecies of black-and-white colobus that could potentially be elevated to species status if treated like the red colobus taxa.

Olive Colobus

The olive colobus, *Procolobus verus*, is a monotypic species. *P. verus* is the smallest of all of the Colobinae with an average adult body weight of 4.2 kg for females and 4.7 kg for males (Delson et al., 2000). This colobine is small, slender, has a reddish-brown to gray coat, and adult males often have a short sagittal crest on the crown of the skull (Oates et al., 1994; Figure 13.2).

The olive colobus is restricted to the coastal forests of West Africa from Sierra Leone to Nigeria, although the distribution is discontinuous in the countries of Togo, Benin, and Nigeria (Oates et al., 1994; Groves, 2007; Figure 13.3). Surveys in the country of Benin indicate

Figure 13.2   Male olive colobus (*Procolobus verus*).
(Photo credit: W. Scott McGraw)

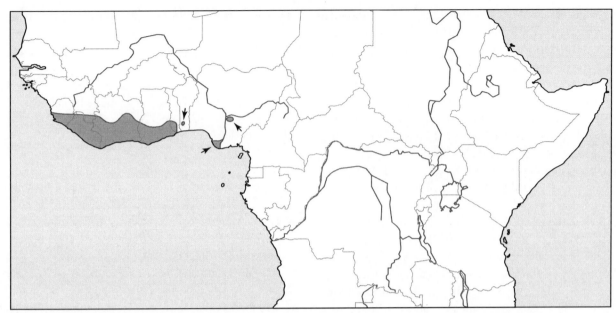

Figure 13.3   Distribution of olive colobus monkeys (*Procolobus verus*).
(From Ting, 2008a)

that this species has a confirmed geographic distribution of 25,403 km^2 (Djego-Djossou et al., 2014). Although the olive colobus is one of the first of the Colobinae to have been researched in the wild (e.g., Booth, 1957), certain aspects of its natural history are not as well known as that of some long-term study populations of red colobus and black-and-white colobus.

Red Colobus

The second major group is the red colobus monkey, genus *Piliocolobus*. This group of primates has coats with distinctive color patterns that vary in shades of red, brown, black, and white (Figure 13.4). Body weights vary greatly, with only some of the species being sexually dimorphic. For example, while Zanzibar red colobus weigh as little as 5.5–6 kg, ashy red colobus males in Kibale National Park, Uganda, weigh around 10.5 kg (Oates et al., 1994; Delson et al., 2000), but females in Kibale tend to be smaller and weigh around 7 kg (Oates et al., 1994). In contrast, female Temminck's red colobus monkeys can be larger or the same size as males at the Abuku Nature Reserve in The Gambia, while the males tend to have slightly larger canines (Starin, 1994).

Red colobus monkeys can be found in forested areas throughout equatorial Africa from Senegal to Zanzibar (Table 13.1). However, their modern natural distribution is patchy, with large biogeographic gaps in areas where one might expect them to occur absent of human disturbance, such as the forests of Rio Muni (Equatorial Guinea), southeastern Cameroon, and Gabon (Oates et al., 1994; Figure 13.5).

The classification of the red colobus monkeys is unstable with no consensus among taxonomists. Consequently, the genus name for this group and the species names of certain red colobus populations vary in the literature depending on the publication year and author(s) of the publication. For example, the name of one of the best-studied colobine populations, the red colobus of Kibale National Park in Uganda, has been referred to as *Colobus badius tephrosceles* (Struhsaker, 1975), *Procolobus badius tephrosceles* (Chapman & Chapman, 2002), *Procolobus rufomitratus tephrosceles* (Simons et al., 2017), or *Piliocolobus tephrosceles* (Simons et al., 2019). Although the genus and species names have changed throughout the years, the subspecies names have remained relatively stable (e.g., *tephrosceles*).

Black-and-White Colobus

The third major group of living colobus monkeys is the genus *Colobus*, the black-and-white colobus—which, as the name implies, has distinctive long, flowing black-and-white fur (Figure 13.6). One exception to the color scheme is *C. satanas*, which has solid black pelage. This primate group is found throughout Equatorial Africa where suitable forest exists. *Colobus* is divided into five species *C. polykomos*, *C. satanas*, *C. vellerosus*, *C. angolensis*, and *C. guereza* (Groves, 2007). The first three species are found in West and Central Africa whereas *C. angolensis* and *C. guereza* are widely distributed in Central and East Africa with overlapping ranges in the northeastern Republic of Congo (see Figure 13.7).

Figure 13.4 Ashy red colobus (*Piliocolobus tephrosceles*).
(Photo credit: Nelson Ting)

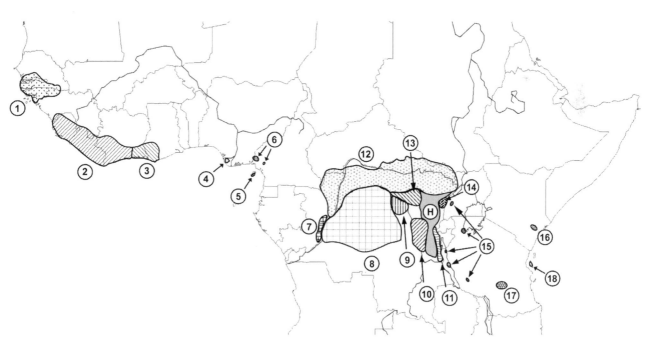

Figure 13.5 Distribution of red colobus (*Piliocolobus*) species. Area marked with an H refers to a putative zone of hybridization between adjacent taxa. 1=*P. temminckii*, 2=*P. badius*, 3=*P. waldroni*, 4=*P. epieni*, 5=*P. pennantii*, 6=*P. preussi*, 7=*P. bouvieri*, 8=*P. tholloni*, 9=*P. parmentieri*, 10=*P. lulindicus*, 11=*P. foai*, 12=*P. oustaleti*, 13=*P. langi*, 14=*P. semlikiensis*, 15=*P. tephrosceles*, 16=*P. rufomitratus*, 17=*P. gordonorum*, 18=*P. kirkii*.
(From Ting, 2008a)

Figure 13.6 Adult female white-thighed black-and-white colobus (*Colobus vellerosus*) in a teak tree.
(Photo credit: Eva Wikberg)

ECOLOGY

Many studies have focused on the African colobines' ecology; most of them report that forest is the primary habitat for these monkeys, that they rely on a folivorous diet, and that they have small home ranges. While these primates have clear adaptations for an arboreal lifestyle and are adept climbers and routine leapers while in the trees, many colobus monkey populations are known to come to the ground and move terrestrially when necessary.

Population Densities and Habitat Preference

The reported population density is low for olive colobus, but ranges from very low to very high for red colobus, and ranges from very low to high for black-and-white colobus (see Table 13.2). Many studies report

increasing population densities with indicators of habitat quality, such as soil fertility (which may reflect nutritional content of the foliage: Thomas, 1991; but see Rothman et al., 2015), basal area of food trees (Mbora & Meikle, 2004; Anderson et al., 2007), and tree species richness or diversity (Mammides et al., 2009; Kankam & Sicotte 2013). Protein-to-fiber ratios in African colobine habitat appear to be a particularly important demarcation of species abundance for this taxonomic group since 62% of African colobine biomass variance across nine sites appears to be explained by this factor (Chapman & Chapman, 2002).

Population status is also affected by protection from habitat degradation and hunting, as well as forest fragment size, and distance to other forest fragments. For example, *Piliocolobus gordonorum* (Araldi et al., 2014) and *Colobus vellerosus* (Saj et al., 2006; Kankam et al., 2010; Kankam & Sicotte, 2013) populations are affected by these factors, but see Mammides et al. (2009) for a lack of relationship between disturbance and group density for *C. guereza*. However, some populations may continue to exist at relatively high densities in disturbed habitats by locomoting on the ground and altering their feeding behavior, such as spending more time feeding and feeding on a greater diversity of plants (*P. tephrosceles*: Milich et al., 2014a), feeding more on lianas (*C. vellerosus*: Wong et al., 2006; *C. angolensis*: Dunham, 2017), and utilizing planted and/or exotic trees (e.g., *C. vellerosus* foraging on planted teak trees as in Figure 13.8; *C. angolensis*: Dunham & Lambert, 2016). Conversely, high densities can

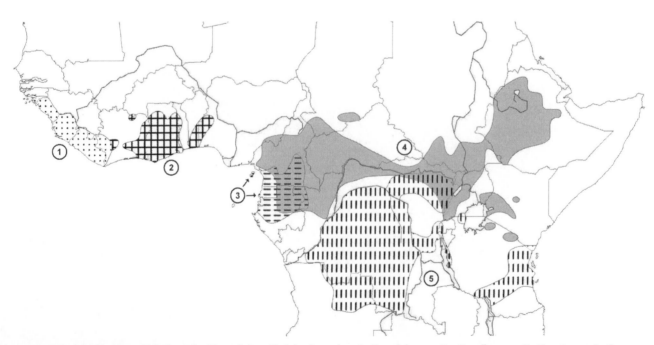

Figure 13.7 Distribution of black-and-white colobus (*Colobus*) species. 1 = *C. polykomos*, 2 = *C. vellerosus*, 3 = *C. satanas*, 4 = *C. guereza*, 5 = *C. angolensis*. (From Ting, 2008a)

Table 13.2 Demographic Composition of African Colobines at Different Sites [a]

Species	Site	Forest Type	Density (ind/km²)	Group Size [b]	N	♂ [b]	♀ [b]	References
Procolobus. verus	Taï Nat. Park, Côte d'Ivoire	Lowland moist primary	14	7.1 (2-12)	10	1.5 (1-3)	2.6 (1-6)	Korstjens (2001), Korstjens and Schippers (2003)
P. verus	Tiwai, Sierra Leone	Lowland moist secondary	11	8.5 (3-11)	1	2.0 (1-2)	3.5 (1-5)	Oates (1988), Oates and Whitesides (1990)
Piliocolobus badius	Taï Nat. Park, Côte d'Ivoire	Lowland moist primary	158	52.3 (41-64)	4	10.5 (6-15)	18.3 (14-22)	Korstjens (2001)
P. badius	Tiwai, Sierra Leone	Lowland moist secondary	66	33.0	1	7	14-22	Davies et al. (1999)
P. epieni	Gbanraun, Nigeria	Lowland marsh	119	30-80 [c]	4	6-7	26	Werre (2000)
P. gordonorum	Udzungwa Mountains National Park and Scarp Forest Reserve, Tanzania	Montane moist, mixed evergreen, semi-deciduous, dry	-	7-83	39	1-6	4-29	Struhsaker et al. (2004)
P. gordonorum	Kilombero Valley, Tanzania	Semi-deciduous, dry, deciduous, flood plain	-	8-50	15	1-8	4-26	Struhsaker et al. (2004)
P. kirkii	Jozani (forest), Zanzibar	Secondary, flood plain	176	31.1 (23-36)	3	3.9	12	Siex (2003)
P. kirkii	Jozani (shamba), Zanzibar	Regenerating, cultivated areas	784	37.5 (20-65)	4	4.7	14.9	Siex (2003)
P. oustaleti	Badane, Central African Republic	Seasonally flooded	-	11.7 3-18	3	1+	1-2+	Galat-Luong and Galat (1979)
P. pennantii	Gbanraun, Nigeria	Lowland disturbed marsh	-	46.0 (15-80) [c]	4	7	26	Werre (2000)
P. preussi	Korup, Cameroon	Lowland moist primary	-	47+ (24-80+) [c]	7	-	-	Struhsaker (1975)
P. rufomitratus	Tana River, Kenya	Lowland disturbed gallery	33-253	11.2 (4-24)	9	1.1 (1-2)	5.6 (2-11)	Marsh (1978, 1979, 1981c), Decker (1994)
P. temminckii	Abuko Nature Reserve, The Gambia	Gallery, savanna woodland	124	25.9 (18-32)	2	2.1 (2-3)	10.8 (8-14)	Starin (1991)
P. temminckii	Fathala, Senegal	Open dense riverine	105-433	25.2 (14-62)	6	5.5 (4-13)	10.7 (5-27)	Gatinot (1975, cited in Milich et al., 2014), Gatinot (1977)
P. temminckii	Cantanhez National Park, Guinea-Bissau	Fragmented forest savanna mosaic, riverine	-	27+	1	3	10	Minhós et al. (2015)
P. tephrosceles	Gombe, Tanzania	Disturbed mixed	-	82	1	11	24	Clutton-Brock (1975a)
P. tephrosceles	Kibale, Uganda	Medium-altitude moist mixed	160-300	28.44 (9-130)	20	2-10	3-25	Struhsaker (1975, 2000), Thomas (1991), Siex and Struhsaker (1999), Miyamoto et al. (2013), Gogarten et al. (2015), Chapman et al. (2010, calculated in Milich et al., 2014)
P. tephrosceles	Mahale Mountains, Tanzania	Montane, medium-altitude semi-deciduous, gallery, woodland, bamboo	-	(10-50)	-	-	-	Nishida (1972)

(continued)

Table 13.2 (Continued)

Species	Site	Forest Type	Density (ind/km²)	Group Size [b]	N	♂ [b]	♀ [b]	References
P. tholloni	Salonga, DRC	Lowland moist primary	-	60+	1	-	-	Maisels et al. (1994)
Colobus angolensis	Diani, Kenya	Lowland fragmented dry	-	6.5 (3-11)	25	-	-	Kanga (2001), Kanga and Heidi (1999, 2000)
C. angolensis	Ituri, DRC	Lowland moist primary	3	13.9 (6-20)	8	2-5+	6-8+	Bocian (1997)
C. angolensis	Lake Nabugabo	Lowland moist evergreen	-	10.33 (4-23)	12 core units	2.6 (1-8)	3.8 (1-6)	Stead and Teichroeb (2019)
C. angolensis	N.E. Tanzania (Usambara Mts.)	Medium-altitude moist	-	4.9 (2-9)	10	1.3 (1-2)	1.6 (1-3)	Groves (1973)
C. angolensis	Nyungwe, Rwanda	Montane moist mixed	-	>300	2	-	-	Fimbel et al. (2001), Fashing et al. (2007)
C. angolensis	Salonga, DRC	Lowland moist primary	-	3-7	5	-	-	Maisels et al. (1994)
C. angolensis	Bole, Ethiopia	Lowland gallery	-	7 (3-10)	12	1.6 (1-4)	2.1 (1-3)	Dunbar (1987)
C. angolensis	Budongo, Uganda	Medium-altitude mixed	20-49	6.9 (2-13)	25	1.1 (1-2)	3.6 (1-6)	Suzuki (1979), Plumtre (unpub. data in Fashing, 2011)
C. angolensis	Chobe, Uganda	Gallery	-	12 (12)	1	2	3	Leskes and Acheson (1971, cited in Fashing, 2011)
C. angolensis	Ituri, DRC	Lowland moist mixed	17	7.8 (5-11)	2	1.5 (1-2)	2.5 (2-3)	Bocia (1997)
C. angolensis	Kakamega, Kenya	Medium-altitude moist secondary	150	12.8 (5-23)	5	2.4 (1-6)	3.8 (3-5)	Fashing (2001a,c), Fashing and Cords (2000)
C. angolensis	Kibale, Uganda	Medium-altitude moist secondary	100	7.84-11.4 (9-15)	7	1.4 (1-4)	3.4 (3-4)	Oates (1977a,c, 1994), Gogarten et al. (2015)
C. angolensis	Kibale surrounding fragments, Uganda	Lowland moist	-	6.2 (3-11)	29	1.2 (1-2)	1.8 (1-4)	Onderdonk and Chapman (2000)
C. angolensis	Kyambura Gorge, Uganda	Lowland gallery	-	8 (3-13)	24	1.1 (1-2)	4.7 (2-7)	Krüger et al. (1998)
C. angolensis	Lake Naivasha, Kenya	Medium-altitude gallery	-	19	1	2	3	Rose (1978)
C. angolensis	Lake Shalla, Ethiopia	Lowland gallery	-	7.8 (6-10)	6	(1)	(1-3)	Dunba (1987)
C. angolensis	Limuru, Kenya	Highland fragmented	-	9.8 (6-15)	2	3.1 (2-5)	3.3 (2-5)	Schenkel and Schenkel-Hulliger (1967, cited in Fashing, 2011)
C. polykomos	Cantanhez National Park, Guinea-Bissau	Fragmented forest savanna mosaic, riverine	-	10	1	1	4	Minhos et al. (2015)
C. polykomos	Taï Nat. Park, Côte d'Ivoire	Lowland moist primary	47	16.2 (14-19)	10	1.2 (1-2)	4.7 (4-6)	Korstjens (2001)
C. polykomos	Tiwai, Sierra Leone	Lowland moist secondary	67	11	1	3	4	Dasilva (1989, 1994), Oates (1994), Whitesides et al. (1988)
C. satanas	Douala-Edea, Cameroon	Lowland moist primary coastal	38	15 (11-16)	1	2.3 (1-3)	6 (6)	McKey (1978), McKey and Waterman (1982)
C. satanas	Forêt des Abeilles, Gabon	Lowland moist primary	8	17 (7-25)	3	>1-6	-	Fleury and Gautier-Hion (1999), Brugière et al. (2002)

C. satanas	Lopé, Gabon	Lowland moist mixed	11	10	1	2	5	Harrison and Hladik (1986), Harrison (pers .comm., cited in Oates, 1994), White (1994)
C. vellerosus	Bia, Ghana	Lowland moist rain	-	16	2	3 (2-4)	6.5 (6-7)	Olson (1980, cited in Fashing, 2011)
C. vellerosus	Boabeng-Fiema, Ghana	Lowland dry semi-deciduous	66-143	14.3-15.10 (4-38)	19	2.9 (1-5)	5.3 (3-11)	Fargey (1992), Saj et al. (2005), Wong and Sicotte (2006), Kankam and Sicotte (2013)
C. vellerosus	Boabeng-Fiema surrounding fragments, Ghana	Lowland dry semi-deciduous	13-72	10.33 (3-17)	10	-	-	Wong and Sicotte (2006), Kankam and Sicotte (2013)
C. vellerosus	Dinaoudi Sacred Grove, Côte d'Ivoire	Woodland	1000 [d]	-	-	-	-	Gonedelé Bi et al. (2010)
C. vellerosus	Kikélé Sacred Forest, Benin	Dry semi-deciduous	-	18	1	Multiple	Multiple	Djègo-Djossou et al. (2015)
C. vellerosus	Soko Sacred Grove, Côte d'Ivoire	Woodland	-	8	1 [e]	-	-	Gonedelé Bi et al. (2010)
C. vellerosus	Tanoé Forest, Côte d'Ivoire	Swamp	-	2	2	-	-	Gonedelé Bi et al. (2010)

[a] Adapted from Fashing (2011).
[b] Reported value is the mean with the range of values in parentheses. N=the largest number of groups that group size is based on.
[c] Approximate numbers.
[d] 30 individuals in a 3-ha fragment.
[e] Present in 2000 but likely locally extinct in 2003.
- indicates no data.

also be indicative of an unstable population that is temporarily compressed through habitat loss, which may be the cause of high densities of *P. gordonorum* in the small, isolated Magombera Forest Reserve in the Kilombero Valley, Tanzania (Araldi et al., 2014) and *P. kirkii* on Zanzibar (Siex & Struhsaker, 1999; Nowak & Lee, 2013a). The highest reported population densities are *P. kirkii* at 784 individuals/km^2 in the shamba (garden) habitat on Zanzibar (Siex, 2003) and *C. vellerosus* occurring at a density of 1,000 individuals/km^2 in the small Dinaoudi Sacred Grove in Côte d'Ivoire (Gonedelé Bi et al., 2010). Such high-density populations may cause damage to food trees, particularly during the lean season (Siex, 2003).

Olive colobus monkeys (*Procolobus verus*) tend to be restricted to low elevation riparian and disturbed habitats with dense lower- to mid-canopy vegetation (Booth, 1956, 1957; Oates, 1981; Oates et al., 1994; McGraw, 1998a,b; Fashing, 2011; Saj & Sicotte, 2013). The red colobus and black-and-white colobus tend to reside in the canopy where substrate boughs are sturdier, while the olive colobus is primarily found in the forest understory (McGraw, 1998a,b). The home range overlap between groups of olive colobus and black-and-white colobus is 73% on Tiwai Island, Sierra Leone, where all three major groups are represented (Davies et al., 1999). The olive colobus often form mixed-species associations with other forest dwelling African monkeys (e.g., *Cercopithecus diana*) throughout its range in Sierra Leone (Oates, 1994).

Overall, red colobus prefer high canopy and are found at highest densities in relatively pristine lowland rainforests (Davies, 1994; Oates, 1994; Chapman et al., 2000; Struhsaker et al., 2004; Fashing, 2011; Araldi et al., 2014). Most red colobus species seem to be particularly vulnerable to the effects of logging, habitat degradation, and forest fragmentation as they are often the first primates to decline in the face of such conditions (Struhsaker, 2005). However, there are certain red colobus populations that have shown some tolerance to disturbed forest. For example, the highly seasonal Fathala Forest in Saloum Delta National Park, Senegal, drastically declined, both in terms of forest coverage and tree species diversity, over the course of 30 years (Galat-Luong & Galat, 2005). *Piliocolobus temminckii* adjusted to these changes by adopting higher levels of frugivory, terrestriality, a tendency to form polyspecific associations with green monkeys, a tendency to frequent more open habitats, and the use of mangrove swamps for refuge and foraging (Galat-Luong & Galat, 2005). Likewise, within previously logged compartments of Kibale in Uganda, population densities of *P. tephrosceles* at first declined but then stabilized as the monkeys learned to deal with the effects of deforestation and associated food loss through behavioral strategies;

these included diversification of diet and increased foraging time (Milich et al., 2014a), eventually leading to an increasing population size within the park (Chapman et al., 2018). However, it is important to note that this primate has virtually disappeared from heavily degraded forest fragments outside the park (Chapman et al., 2013), demonstrating some limits to the level of disturbance it can tolerate. On Zanzibar Island (*P. kirkii*) and in the Udzungwa Mountains (*P. gordonorum*) in Tanzania, red colobus populations exist outside protected areas that have persisted in disturbed low plant diversity habitats by crop raiding in the surrounding agricultural matrix (Nowak & Lee, 2013b). Thus, while the red colobus monkeys prefer the canopy of mature lowland rainforest and generally fare poorly in disturbed habitats, some species are capable of surviving in moderately degraded and/or seasonal forests. Why this variation exists and the long-term conservation implications are unclear.

Black-and-white colobus populations frequent many different kinds of habitat—bamboo, lowland swamp and coastal forests, moist savanna, woodland, dry thicket, highly seasonal gallery forests, secondary roadside forests, and montane mature forests (Kingdon, 1971; Dorst & Dandelot, 1972; Oates, 1994; Oates & Davies, 1994). They come to the ground and move terrestrially between trees when necessary (Sabater Pi, 1973; Dunbar & Dunbar, 1974; Rose, 1978; Oates, 1994). *Colobus guereza*, in particular, prefer lower strata and secondary forest (Rose, 1978; Thomas, 1991) and tend to show highest densities in relatively young secondary forests, in narrow strips of riparian forests, in small relic forests, and in logged forests (Struhsaker & Oates, 1975; Plumptre & Reynolds, 1994). The ability of *C. guereza* to utilize all levels of forest strata and adapt to different ecological conditions appears to be essential in their capacity to exploit changing and degraded habitat. In fact, this species can show increases in population size within heavily fragmented forests where logging has occurred (Davies, 1994; Oates, 1994; Chapman et al., 2000, 2010; Gillespie & Chapman, 2008). *C. angolensis* is also able to inhabit a variety of forest types, although it does not fare as well in disturbed environments (Marshall et al., 2005; Rovero & Struhsaker, 2007; Rovero et al., 2012; Barelli et al., 2014). However, Dunham's (2017) study indicates that *C. angolensis* may cope with habitat changes by foraging more on exotic plant species and lianas. Likewise, *C. vellerosus* is able to inhabit small, disturbed forest patches (Wong & Sicotte, 2006; Kankam et al., 2010; Kankam & Sicotte, 2013), but their ranging patterns and diet differ across forest fragments of different sizes and levels of disturbance (Wong et al., 2006) with unknown consequences on individual health and fitness.

There is generally a difference in the behavioral flexibility of red colobus and black-and-white colobus. At Kibale, this is most obvious, as the red colobus (*P. tephrosceles*) do not fare as well in logged forest and forest fragments, while the black-and-white colobus (*C. guereza*) are capable of thriving in such habitat (Chapman et al., 2018). Red colobus monkeys (*P. pennantii*) are also less flexible in habitat use than the sympatric black-and-white colobus (*C. satanas*) at Bioko Island, Equatorial Guinea. At this site, *P. pennantii* only inhabits the primary lowland rainforest and montane forest, where it frequents the middle canopy strata and never descends to the ground (Gonzalez-Kirchner, 1997). In comparison, although *C. satanas* is also most prevalent in mid-canopy primary rainforest, *C. satanas* inhabits all canopy levels, and can be found in numerous habitats that include high altitudinal sites over 2,500 masl in mountain forest, secondary forest, and alpine grassland (Gonzalez-Kirchner, 1997). Likewise, compared to the red colobus (*P. gordonorum*) in the Udzungwa Mountains of Tanzania, sympatric black-and-white colobus (*C. angolensis*) appear to be relatively more common at higher elevations and affected by forest degradation to a slightly lesser degree (Marshall et al., 2005; Rovero & Struhsaker, 2007; Rovero et al., 2012; Barelli et al., 2014). It is important to note that while these differences exist among red colobus and black-and-white colobus species, both groups tend to fare worse in disturbed forests when compared to their cercopithecine cousins, especially in the presence of human hunting.

Diet and Foraging Behavior

Much of what is known about the diet of the olive colobus monkey (*Procolobus verus*) is based on an early study by Oates (1988) on Tiwai Island, Sierra Leone, and two comparative studies that involved all three colobus genera (*Procolobus verus, Piliocolobus badius, Colobus polykomos*) in sympatry at Tiwai (Davies et al., 1999) and at Taï National Park in Côte d'Ivoire (Korstjens, 2001). Taï consists of primary rainforest, while Tiwai is made up of mature secondary forest and regenerating farmland (see Table 13.2). Furthermore, leopards have been extirpated and chimpanzees occur at low abundance at Tiwai (McGraw, pers.comm.). Such contrasting habitat types and predation pressures are likely to lead to considerable differences in their behaviors. Oates (1988) found that young leaves may compose as much as 85% of the olive colobus diet in the late wet season at Tiwai, and much of the consumed foliage comes from climber plants such as *Acacia pennata*. In the dry season, however, olive colobus monkeys will spend about one-quarter of their feeding time on both ripe and unripe small and medium-sized seeds (Oates, 1988). The olive colobus is also selective in the leaves it consumes and will tend to ignore the most common trees in the forest (Oates, 1988). Specifically, Oates (1988) found that they will consume only young leaves that are low in tannin and fiber content and will largely ignore the blades of mature leaves. The comparative study at Tiwai found that olive colobus incorporated fewer fruit, flowers, and/or seeds compared to red colobus and black-and-white colobus, and all three groups consumed seeds at different stages of ripeness and pod development with very little overlap in the plant species consumed among the groups (Davies et al., 1999). The olive colobus at Taï also consumed more leaves and less fruits, seeds, and flowers than the red colobus and black-and-white colobus, but unlike the Tiwai study, the olive colobus at Taï relied on a diet consisting almost exclusively of young leaves (Korstjens, 2001; Korstjens et al., 2007, Table 13.3). It is possible that the olive colobus monkeys at Tiwai incorporate less young leaves and more mature leaves and fruit in their diet because they inhabit secondary forest and farmland rather than primary forest.

The red colobus feeds mainly on leaf parts, especially on young leaves and leaf buds, but its diet is diverse in both species consumed and plant parts consumed (see Table 13.3). *Piliocolobus tephrosceles* at Kibale in Uganda and Gombe in Tanzania are highly selective, do not feed on the most abundant species of plants (Struhsaker, 1975; Clutton-Brock, 1975a; Struhsaker, 2010), and the species of plants consumed do not seem to be based on proximity, protein-to-fiber ratios, or even secondary compound levels (Chapman & Chapman, 2002). Although a red colobus group's diet can be relatively stable across years, dietary diversity increased over time according to Struhsaker (1975, 1978). Furthermore, the diets between groups of this species can vary greatly (Chapman et al., 2002; Struhsaker, 2010). For example, Struhsaker (2010) calculated only a 17.3% dietary overlap in the top food items consumed by two groups spaced just 10 km apart. There are also marked differences between groups in time spent feeding on different plant parts, such as young leaves (48.8–87.0%) and flowers (2.0–22.7%) (Chapman et al., 2002). Yet despite this variation in diet among neighboring groups, nutritional analyses indicate that all of the groups still tend to receive the same nutrient concentrations, which suggests that the animals are selective in the foods they consume to obtain optimal protein-to-fiber ratios (Ryan et al., 2012). Some of this variation is linked to habitat disturbance. As an example, the pattern of selectivity is relaxed in circumstances where the forest is logged and preferred food items are absent (Milich et al., 2014a). *P. tephrosceles* expands its diet and foraging

time in logged habitats to compensate for loss of resources (Milich et al., 2014a).

In contrast to *P. tephrosceles* dietary differences across habitat types at Kibale (Wasserman & Chapman, 2003), the proportion of plant species incorporated in the *P. kirkii* diet was similar across the Jozani forest and shamba habitats in Zanzibar and changed very little over time in the forest habitat, although the number of species in the diet decreased dramatically (Mturi, 1993; Siex, 2003). In large protected areas of forest comparable to Kibale, *P. badius* at Taï in Côte d'Ivoire and *P. tholloni* at Salonga National Park in the DRC consume a similarly high number of plant species (Maisels et al., 1994; Wachter et al., 1997; Korstjens et al., 2007). However, *P. tholloni* at Salonga had a strong preference for legumes from the Caesalpinioideae family, yet such preference for legumes does not occur at other sites where legumes are abundant (Maisels et al., 1994; Davies et al., 1999). Conversely, red colobus incorporated fewer numbers of tree species in their diet in highly seasonal forests with lower diversity at both the eastern and western extremes of the red colobus range (*P. temminckii* in the Fathala forest in Senegal: Gatinot, 1978; *P. kirkii* on Zanzibar Island: Siex, 2003). The Tana River red colobus in Kenya (*P. rufomitratus*) occupies a highly seasonal forest, much of which has been cleared for farmland. This forest has few tree species and only 22 species are consumed by the red colobus there (Marsh, 1979a; Marsh, 1981a,b,c). Behavioral and phytochemical research at Tana River indicates that *P. rufomitratus* avoid mature leaves that are highly fibrous and prefer young leaves that are high in nitrogen, water, and protein/acid-detergent fibers ratios (Mowry et al., 1996). However, when their preferred foods are not available, *P. rufomitratus* selectively consume the mature leaves that are relatively high in protein (Mowry et al., 1996).

Of the black-and-white colobus species, two have a predominantly folivorous diet, but with some between-site differences in their degree of frugivory, while seeds are an important food source for the other three species (see Table 13.3). Leaves composed 78.5–94% of the total diet for eight *C. guereza* groups with overlapping ranges at Kibale (Harris & Chapman, 2007). Although there was variation in relation to what plant species they consumed, all groups fed heavily on the leaves of *Celtis durandii* (Harris & Chapman, 2007). This one tree species accounted for over 70% of their diet in a three-month study (Clutton-Brock, 1975a) and over 50% of the yearly total (Oates, 1977b). Interestingly, *C. guereza* is the only primate at Kibale to consume mature leaves of this species (Struhsaker, 1975), and their diet includes more mature leaves than the sympatric red colobus diet (Struhsaker, 1978). Oates (1977b) found a direct

Table 13.3 Percent (%) Plant Part Consumed by African Colobines at Different Sites (9-24 Months of Data) [a]

Species	Site	% YL	%ML	% UL	% FL	% FR	% SD	% OT	# Spp. in Diet	References
Procolobus verus	Taï, Côte d'Ivoire	83	1	1	4	8	–	3	–	Korstjens et al. (2007)
P. verus	Tiwai, Sierra Leone	59	11	4	7	19	14	0	50+	Oates (1988)
Piliocolobus badius	Taï, Côte d'Ivoire	24-46	4-7	0	20-30	29-37	–	1-2	153	Wachter et al. (1997), Korstjens et al. (2007)
P. badius	Tiwai, Sierra Leone	32	20	0	16	31	25	1	51	Davies et al. (1999)
P. kirkii	Jozani-forest, Zanzibar	50-51	7-9	4-7	5-8	26-32	2	0-6	21-63	Mturi (1993)
P. kirkii	Jozani-shamba, Zanzibar	55	7	6	6	5	0	22	27	Siex (2003)
P. pennantii	Gbanraun, Nigeria	56	10	0	9	16	12	9	19+	Werre (2000)
P. preussi	Korup, Cameroon	89	0	0	10	1	–	0	17	Usongo and Amubode (2001)
P. rufomitratus	Tana River (Mchelelo), Kenya	52-61	2-11	0-1	6-13	22-25	1	2-4	22-28	Marsh (1981b)
P. rufomitratus	Tana River (Baomo), Kenya	46	1	0	27	26	–	1	26	Decker (1994)
P. temminckii	Abuko, Gambia	35	12	0	9	42	3	3	89+	Starin (1991)
P. temminckii	Fathala, Senegal	42	5	0	9	36	19	9	39	Gatinot (1978), Oates (1994)
P. tephrosceles	Gombe, Tanzania	35	44	0	7	11	–	3	58+	Clutton-Brock (1975a)
P. tephrosceles	Kibale (Kanyawara), Uganda	33-70	10-25	7-8	5-15	1-17	2-10	57	–	Struhsaker (1978, 2010) Wasserman and Chapman (2003)
P. tephrosceles	Kibale surrounding fragments, Uganda	60	22	2	2	7	–	7	–	
P. tephrosceles	Kibale (logged), Uganda	79	7	6	2	6	–	1	–	Wasserman and Chapman (2003)
P. tholloni	Salonga, DRC	54	6	–	1	38	31	0	84	Maisels et al. (1994)
Colobus angolensis	Diani, Kenya	58	13	0	14	14	10	0	76	Dunham (2017)
C. angolensis	Ituri, DRC	26	2	22	7	28	22	15	37	Bocian (1997)
C. angolensis	Lake Nabugabo, Uganda	65	3	0	<1	31	<1	1	22+	Arseneau-Robar et al. (2020)
C. angolensis	Nyungwe, Rwanda	25-30	7-40	1-7	1-5	17-23	–	6-37	45+- 59+	Fimberl et al. (2001), Vedder and Fashing (2002)
C. guereza	Budongo (unlogged), Uganda	–	–	63	6	29	12	2	–	Plumptre (unpub. data in Fashing, 2011)
C. guereza	Budongo (unlogged), Uganda	–	–	51	7	40	11	2	–	Plumptre (unpub. data in Fashing, 2011)
C. guereza	Ituri, DRC	30	4	24	3	25	22	15	31	Bocian (1997)
C. guereza	Kakamega, Kenya	24	7	23	1	39	1	8	37+	Fashing (2001b)
C. guereza	Kibale (Kanyawara), Uganda	65-81	5-13	1-3	1-2	5-15	–	2-8	43	Oates (1977a, 1994)
C. guereza	Kibale surrounding fragments, Uganda	65	14	1	6	12	–	2	–	Wasserman and Chapman (2003)
C. guereza	Kibale (logged), Uganda	78	5	3	3	10	–	0	0	Wasserman and Chapman (2003)
C. polykomos	Taï, Côte d'Ivoire	28	20	0	3	48	–	1	–	Korstjens et al. (2007)
C. polykomos	Tiwai, Sierra Leone	30	26	2	3	35	32	3	46+	Dasilva (1989)
C. satanas	Douala-Edea, Cameroon [b]	21	18	–	3	53	53	5	84+	McKey et al. (1981)
C. satanas	Forêt des Abeilles, Gabon [b]	35	3	0	12	50	41	0	109	Gautier-Hion et al. (1997), Fleury and Gautier-Hion (1999)
C. satanas	Lopé, Gabon [b]	23	3	0	5	64	60	4	65	Harris (unpub. data in Fashing, 2011)
C. vellerosus	Boabeng-Fiema, Ghana	35-53	25-36	5	1-6	2-11	4-9	0	34-42	Saj and Sicotte (2007a), Teichroeb and Sicotte (2009)
C. vellerosus	Kikélé Sacred Forest, Benin	–	–	53	3	33	3	9	35	Djègo-Djossou et al. (2015)

[a] YL=Young Leaves, ML=Mature Leaves, UL=Unidentified Leaves, FL=Flowers, FR=Fruit (with or without seeds), SD=Seeds, OT=Other (from Fashing, 2011, Table 13.2, with the exception of more recently published *C. vellerosus* studies—data rounded to the nearest whole number).
[b] Total exceeds 100% in some cases due to fruits and seeds being combined into one category.
– Indicates no data.

correlation between *C. durandii* young leaf availability and the proportion of young leaves in *C. guereza* diets. There was also an increase in the diversity of *C. guereza* diet when *C. durandii* young leaves were unavailable (Clutton-Brock, 1975a; Oates, 1977b). Chemical analyses showed that young leaves in comparison to mature leaves of *C. durandii* contained twice as much protein and 25% less tannin and 75% as much lignin, both of which are digestion-inhibitors (Oates et al., 1977). A small proportion of the diet of *C. guereza* in Kibale is made up of water plants and soil, but the animals go to great lengths to obtain these resources (Oates et al., 1977). The water plants have higher levels of sodium, iron, manganese, and zinc than dry-land plants eaten by this colobus species (Oates et al., 1977). The clay eaten by the monkeys contains considerably more magnesium, iron, and copper than neighboring soils, but may also be used to absorb plant toxins or to adjust the pH of the fore-stomach (Oates, 1978). These dietary preferences may be related to nutritional deficiencies and high toxic levels of the specialized, leaf-dominated diet of *C. guereza*, which contains digestion-inhibitors such as tannin and lignin. There are some interesting between-site differences in the *C. guereza* diet, possibly due to differences in fruit availability (Harris & Chapman, 2007). At Kibale, fruit consumption ranges from 1.6–17.7% of the total diet (Harris & Chapman, 2007). In contrast, *C. guereza* in the Kakamega Forest of Kenya spend almost as much time feeding on whole fruits as they do on young leaves (Fashing, 2001a); this study also notes that a majority of these fruits are from Moraceae, a tree family that is uncommon at the other sites where *C. guereza* have been studied.

C. vellerosus at Boabeng-Fiema in Ghana has a diet most like that of *C. guereza* in Kibale with 25–36% of feeding time spent consuming mature leaves, 35–53% consuming young leaves, and 5% consuming leaves that could not be classified (Saj & Sicotte, 2007a; Teichroeb & Sicotte, 2009). The Boabeng-Fiema colobus spend 1–11% of their time consuming fruits, flowers, and seeds, and they also occasionally feed on bark and soil (Saj & Sicotte, 2007a; Teichroeb & Sicotte, 2009). Their diet varies seasonally with up to 100% of the diet consisting of mature leaves when other food items are unavailable, while they forage more on young leaves, fruits, flowers, and seeds during the dry season (Saj & Sicotte, 2007a; Teichroeb & Sicotte, 2018). Across study years, the young leaf composition remained the same in one group while it increased threefold in another (Saj & Sicotte, 2007a; Teichroeb & Sicotte, 2018). Although the causes of these longitudinal changes remain unknown, they might be linked to changes in availability of young leaves in certain parts of this heterogeneous forest, because the earlier study

reported that the monkeys consume different plant parts according to their availability (Saj & Sicotte, 2007a). *C. vellerosus* in smaller forest fragments spend 50% of their time feeding on species that groups in the Boabeng-Fiema forest do not eat (Wong et al., 2006). The groups in the smaller forest fragments also feed more on lianas, spend up to twice as much time foraging on fruits, flowers, and seeds, and even incorporate insects in their diet that the groups in the Boabeng-Fiema forest have not been observed to eat (Wong et al., 2006). Similar to *C. vellerosus* in these small forest fragments in Ghana, *C. vellerosus* in the Kikélé Sacred Forest, Benin, also spend a large proportion of their time feeding on fruits (Djègo-Djossou et al., 2015).

In contrast to *C. guereza* and *C. vellerosus*, *C. satanas* at Doula-Edéa, Cameroon, spend over half of their time consuming seeds while discarding most fruit flesh (McKey et al., 1981). Through studies on this species in Lopé, Gabon, it appears that the diet of *C. satanas* is relatively similar across sites, with seeds constituting the majority of the diet, followed secondly by young leaves (Fleury & Gautier-Hion, 1999). The seeds eaten contain the same chemical defenses found in the mature leaves that are ignored by this species, but nutrient concentration is higher in the seeds (McKey et al., 1981). In Gabon, *C. satanas* sometimes feed in mixed-species associations with frugivorous primates such as mangabeys and guenons. When in these mixed-species groups, *C. satanas* consume more seeds and fruit than when they are feeding alone (Gautier-Hion et al., 1996).

C. polykomos have similar dietary patterns as *C. satanas*. At Tiwai Island in Sierra Leone, *C. polykomos* consumes a preponderance of seeds (about one-third of total diet) (Davies et al., 1999). DaSilva (1994) found that unripe seeds at this site were the most important food source for the species throughout the year. In fact, DaSilva (1994) observed that *C. polykomos* would start consuming seeds as soon as the seeds reached full size and would deplete all seeds before they were fully ripe, except for the most common species. At Tai, *Pentaclethra macrophylla* seeds and pods made up 27% of the food items *C. polykomos* consumed during a four-year period (McGraw et al., 2016), and it accounts for a large proportion of their diet during certain months (Korstjens, 2001; Korstjens et al., 2005).

Among *C. angolensis* in Salonga National Park, DRC, seeds seem to be preferred over leaves (Maisels et al., 1994; Table 13.3). Much like the area that *C. satanas* inhabits in Douala-Edea, Cameroon (McKey et al., 1981), and *C. polykomos* on Tiwai Island in Sierra Leone (Davies et al., 1999), the soil quality at Salonga is poor, and Maisels et al. (1994) hypothesize that *C. angolensis* seed consumption is high at this site because the soils are ideal for leguminous trees, mainly Caesalpinoideae and Papilionoideae (see DaSilva, 1994). Seeds are nutrient-rich in protein and lipids, but are also high in secondary compounds such as alkaloids and condensed tannins (Oates et al., 1990; DaSilva, 1992; Maisels et al., 1994). In contrast, *C. angolensis* in the Diani Forest, Kenya, forage mostly on young leaves while seeds only make up 6–15% of the diet in different groups (Dunham, 2017). The diet of *C. angolensis* at Lake Nabugabo, Uganda, consists mostly of fruit and young leaves (Arseneau-Robar et al., 2020). Fruits are consumed based on availability, and when fruits are scarce, they consume more young leaves (Arseneau-Robar et al., 2020). These colobus monkeys might not have to fall back on mature leaves as there is a relatively high abundance of young leaves year-round (Arseneau-Robar et al., 2020).

To summarize, the African colobines show a surprisingly large degree of dietary variation within and between species and populations considering their morphological and physiological adaptions for folivory and granivory. Some of the variation of a species' diet over time and space (Table 13.3) may be linked to the relative levels of toxins, digestion inhibitors, and nutrient content across different geographic localities (Hladik, 1978; McKey et al., 1981; Moreno-Black & Bent, 1982) as well as the availability of specific tree species and plant parts (e.g., Fashing, 2001a). These dietary studies underscore the point that field studies need to be replicated both across sites and across time for a full understanding of a species dietary flexibility (Chapman et al., 2002). The ability to alter their diets in terms of what plant species or plant parts they focus on, in combination with increasing their time foraging, may allow colobines to occupy a variety of habitat types and continue to exist in human-modified habitats. Unfortunately, we lack an understanding of how differences in diet across sites impact individual health and fitness and, thus, the trajectory for the majority of populations. An exception comes from the well-studied populations at Kibale National Forest, Uganda, where the red colobus (*Piliocolobus tephrosceles*) are more susceptible than the black-and-white colobus (*C. guereza*) to a majority of the parasites found along edges of forest fragments (Gillespie & Chapman, 2008). The parasite *Oesophagostomum* is particularly harmful to the red colobus and spread through feces; infected individuals have died from this parasite through dehydration caused by gastrointestinal complications (Gillespie et al., 2005; Chapman et al., 2006a). The reason may be that *P. tephrosceles* tend to be nutritionally stressed within the forest edges, which compromises their immune response, while *C. guereza* thrive in this habitat (Chapman et al., 2006a,b; Gillespie

& Chapman, 2008). This difference in nutrition may make *P. tephrosceles* more susceptible to parasite infections and explain why *C. guereza* fare better in altered and changing habitats than *P. tephrosceles* (Chapman et al., 2006b; Gillespie & Chapman, 2008).

Ranging Behavior

African colobines tend to have small home ranges under 100 ha with day ranges that are rarely longer than 2,000 m (Fashing, 2011). The exceptions are among the black-and-white colobus, where *C. satanas* and *C. angolensis*, specifically, can range over exceptionally large areas (see Table 13.4). While the largest home ranges occur in some *Colobus* groups, *Procolobus verus*, the olive colobus, tend to have the longest day ranges where the three genera are sympatric (Korstjens, 2001). A primary factor in *P. verus* long day ranges is due to following the more mobile *Cercopithecus diana* in polyspecific associations (Korstjens et al., 2007). However, the high-quality diet of *P. verus* may be a contributing factor to why this species is able to maintain longer daily travel distances than other African colobines (Korstjens et al., 2007). Home range overlap is reported to be 14% for olive colobus (Fashing, 2011).

The home ranges of red colobus groups vary from 5–114 ha, with a home range overlap between 6–99% (Fashing, 2011), although some of the between-study variation may be due to different methodologies (Fashing, 2011). The mean day range of the red colobus at Kibale is about half the length of daily travel of sympatric cercopithecine monkeys (Struhsaker, 1980). However, there is considerable variation in daily path length within and across sites, ranging from 0–1,900 m (Table 13.4). Based on his research at Gombe, Clutton-Brock (1975b) suggested that the ranging behavior of *P. tephrosceles* is closely related to variation in food availability and distribution. He found that (1) the amount of time spent in different quadrats throughout the year was related to availability of three tree species on which the animals fed most, (2) the number of nights spent in three different sleeping areas was related to the availability of preferred foods in or around the area, and (3) the animals distributed their time more evenly across their range in months during which more widely dispersed resources (e.g., flowers, fruits, and shoots) were available. Use of these resources correlated with peaks in day range length, monthly ranging area, evenness of quadrant utilization, and rate of movement throughout the range (Clutton-Brock, 1975b). Marsh's (1981c) and Decker's (1994) subsequent studies at Tana River, Kenya, on *P. rufomitratus* support Clutton-Brock's (1975b) findings that the ranging distances of red colobus groups correlate with diet distribution and availability (Decker, 1994). Yet, while

Table 13.4 Ranging Behaviors of African Colobines at Different Sites [a]

Species	Site	No. of Groups	Home Range (ha) [b]	Daily Path Length (m)	References
Procolobus verus	Taï, Côte d'Ivoire	2	56	482-2105	Korstjens (2001)
P. verus	Tiwai, Sierra Leone	1	28	-	Oates and Whitesides (1990)
Piliocolobus temminckii	Abuko, Gambia	1	34	-	Starin (1991)
P. temminckii	Fathala, Senegal	1	20	-	Gatinot (1977)
P. badius	Taï, Côte d'Ivoire	2	58-65	300-1532	Holenweg et al. (1996), Höner et al. (1997), Korstjens (2001)
P. badius	Tiwai, Sierra Leone	1	53	-	Davies et al. (1999)
P. epieni	Gbanraun	1	73	450-1900	Werre (2000)
P. kirkii	Jozani (forest), Zanzibar	3	25-60	30-1532	Mturi (1991, 1993, cited in Fashing, 2011)
P. kirkii	Jozani (shamba), Zanzibar	4	13	0-690	Siex (2003)
P. kirkii	Uzi, Zanzibar	3	5	-	Nowak (2008)
P. gordonorum	Udzungwa, Tanzania	4	11	411-1718	Home range based on behavior-corrected polygons by Steel (2012)
P. tephrosceles	Gombe, Tanzania	1	114	-	Clutton-Brock (1975b)
P. tephrosceles	Kibale, Uganda	1-2	22-35	223-1185	Struhsaker (1975), Chapman and Pavelka (2005)
P. rufomitratus	Tana River (Mchelelo)	1	9-12	200-1100	Marsh (1978, 1979a,b, 1981c)
P. rufomitratus	Tana River (Baomo S.)	1	13	180-780	Decker (1994)
Colobus angolensis	Ituri, DRC	1	371	312-1914	Bocian (1997)
C. angolensis	Nyungwe, Rwanda	1	2440	-	Fashing et al. (2007)
C. angolensis	Diani Forest, Kenya	3	8.7	-	Dunham (2017)
C. guereza	Bole, Ethiopia	7	1.7-2.5	75-500	Dunbar and Dunbar (1974)
C. guereza	Budongo, Uganda	25	14	-	Marler (1969), Suzuki (1979)
C. guereza	Budongo, Uganda	6	10-33	-	Plumptre (unpub. data in Fashing, 2011)
C. guereza	Ituri, DRC	1	100	-	Bocian (1997)
C. guereza	Kakamega, Kenya	5	18	166-1360	Fashing (2001a,c)
C. guereza	Kibale, Uganda	2	10-15	288-1004	Oates (1977b,c), Chapman and Pavelka (2005)
C. polykomos	Taï, Côte d'Ivoire	2	77	241-1341	Korstjens (2001)
C. polykomos	Tiwai, Sierra Leone	1	24	350-1410	Dasilva (1989), Oates (1994)
C. satanas	Douala-Edea, Cameroon	1	60	<100-800	McKey (1978), McKey and Waterman (1982)
C. satanas	Forêt des Abeilles, Gabon	1	573	20-1983	Fleury and Gautier-Hion (1999)
C. satanas	Lopé, Gabon	1	184	40-1100	Harrison (1986 and unpub. data, cited in Fashing, 2011)
C. vellerosus	Bia, Ghana	1	48	75-752	Olson (1986)
C. vellerosus	Boabeng-Fiema, Ghana	4	11	364	Teichroeb and Sicotte (2009)

[a] Adapted from Fashing (2011, Table 13.3).
[b] Mean home range size if reported, otherwise data ranges are given.

Clutton-Brock (1975b) found that day ranges were shortest when new leaves and fruit were least available, Marsh (1981c) found that ranging distances were longest during that period, and Decker (1994) found that daily path lengths were most variable at this time. These differences may be attributed to variances in available vegetation between the two sites. Marsh (1981c) notes that at Gombe, mature leaves are abundant and easily obtainable when young leaves are scarce; however, at Tana River, food sources are widely dispersed.

For three species of black-and-white colobus (*Colobus guereza*, *C. vellerosus*, and *C. polykomos*), most populations have small home ranges of 8–28 ha (Table 13.4), travel a few hundred meters every day, and have a degree of home range overlap of 20–70% (Fashing, 2011). But there are some exceptions. In contrast to most other populations of black-and-white colobus that show home range defense (Fashing, 2011), *C. guereza* at Bole Valley, Ethiopia, rarely defend the boundaries of their very small home ranges of 2 ha (Table 13.4), which appears to be fixed and stable among the groups (Dunbar & Dunbar,

1974). *C. guereza* at Bole are exceptionally sedentary relative to their conspecifics, and it is possible that this extreme energy-conserving lifestyle is due to the population density being at the habitat-carrying capacity (Dunbar & Dunbar, 1974; Table 13.3). When they did range farther during the study, increased mobility did not appear to be associated with primary food item availability; instead it was associated with less commonly consumed but nutritiously important foods (Dunbar & Dunbar, 1974). Conversely, at Ituri Forest, DRC, where population density is very low, there is little home range overlap and home ranges are as large as 100 ha (Fashing, 2001b, 2011; Table 13.4). Habitat types and sizes may also explain within-species variation in ranging patterns (see Table 13.2).

For example, the *C. vellerosus* in Boabeng-Fiema, Ghana, that occupy a 192-ha fragment of mosaic habitat (Fargey, 1992), have smaller home ranges and less variable daily path lengths (Saj & Sicotte, 2007b; Teichroeb & Sicotte, 2009, 2018) than *C. vellerosus* in Bia National Park (Ghana), which consists of 30,000 ha of less disturbed, moist semi-deciduous forest (Olson, 1986). A similar pattern in terms of home range size is reported for *C. polykomos*. In the primary rainforests of Taï National Park, Côte d'Ivoire, *C. polykomos* groups have three times larger home ranges as those in the more human-modified habitat at Tiwai, Sierra Leone, although the daily path lengths were similar at these two sites (DaSilva, 1989; Oates, 1994; Korstjens, 2001).

Table 13.5 Activity Budget Comparisons of African Colobines at Different Sites

Species	Site	%Rest	%Feed	%Move	%Social	%Other	References
Procolobus verus	Taï, Côte d'Ivoire	35	39	19	7	0	McGraw (1998)
P. verus	Tiwai, Sierra Leone	40	27	25	–	8	Oates (1994)
Piliocolobus badius	Abuko, Gambia	52	21	13	13	0	Starin (1991)
P. badius	Taï, Côte d'Ivoire	30	45	19	6	–	McGraw (1998)
P. badius	Tiwai, Sierra Leone	55	37	5	<3 [c]	0	Davies (pers.comm, cited in Oates, 1994)
P. kirkii	Jozani (forest), Zanzibar	47	29	12	7	5	Siex (2003)
P. kirkii	Jozani (shamba), Zanzibar	44	29	6	15	7	Siex (2003)
P. pennantii	Gbanraun, Nigeria	33	37	25	6	0	Werre (2000)
P. rufomitratus	Tana River Mchelelo, Kenya [a]	55	30	7	8	0	Marsh (1978)
P. rufomitratus	Tana River Mchelelo, Kenya [b]	48	29	22	2	0	Decker (1994)
P. rufomitratus	Tana River Maomo S., Kenya [b]	50	23	24	3	0	Decker (1994)
P. tephrosceles	Gombe, Tanzania	54	25	8	<9 [d]	–	Clutton-Brock (1974)
P. tephrosceles	Kibale, Uganda	34	45	9 [e]	<8 [d]	0	Struhsaker (1975) [e]
P. tephrosceles	Kibale, Uganda	30-40	30-53	6-12	2-8	0-3	Struhsaker (2010) [f]
P. tephrosceles	Kibale, Uganda	25-30	40-51	16-29	5-10	–	Snaith and Chapman (2008)
P. tephrosceles	Kibale (old-growth), Uganda	41	35	18	16	–	Milich et al. (2014)
P. tephrosceles	Kibale (logged), Uganda	36	46	14	4	–	Milich et al. (2014)
Colobus angolensis	Ituri, DRC	43	27	24	5	1	Bocian (1997)
C. angolensis	Lake Nabugabo, Uganda	40	28	25	7	1	Arseneau-Robar et al. (2020)
C. angolensis	Nyungwe, Rwanda	43	27	24	5	1	Fashing et al. (2007)
C. angolensis	S.E. Kenya	64	22	3	4	–	Wijtten et al. (2012)
C. guereza	Ituri, DRC	52	19	22	5	0	Bocian (1997)
C. guereza	Kakamega, Kenya	63	26	2	7	2	Fashing (2001a)
C. guereza	Kibale, Uganda	57	20	5	11	7	Oates (1977b)
C. polykomos	Taï, Côte d'Ivoire	34	46	15	5	0	McGraw (1998), Korstjens and Dunbar (2007)
		55	31	13	6	5	
C. polykomos	Tiwai, Sierra Leone	61	28	9	1	1	Dasilva (1992)
C. satanas	Duoala-Edea, Cameroon	60	23	4	14	0	McKey and Waterman (1982)
C. vellerosus	Boabeng-Fiema, Ghana	59	24	15	3	2	Teichroeb et al. (2003)
C. vellerosus	Boabeng-Fiema (surrounding fragments), Ghana	69	22	7	3	0	Wong and Sicotte (2007)
C. vellerosus	Kikélé, Ghana	57	26	13	3	1	Djègo-Djossou et al. (2015)

[a] Study years: 1973-1974.
[b] Study years: 1986-1988.
[c] Includes behaviors classified as "other."
[d] Includes self-cleaning, grooming, and playing.
[e] Omitted the category "clinging" in which only infants engage.
[f] Ranges represent values for the monthly activity budgets of two groups over six months.
– Indicates no data.

Most populations of *C. satanas* and *C. angolensis* have much larger home ranges than the other black-and-white colobus species, but with variable daily path lengths. The *C. satanas* population in Douala-Edea, Cameroon, has much smaller home range size than other populations of this species, and is largely sedentary when seeds are available, as the group obtains a large part of their food from four or five plants of a single species (McKey, 1978). Yet when mature leaves are the primary food item, *C. satanas* at Douala-Edea will travel much farther (McKey, 1978). In contrast, leaf consumption by *C. satanas* at Makandé in Gabon was associated with shorter day ranges (McKey & Waterman, 1982). Notably, groups of *C. satanas* at Makandé were regularly observed in new areas of their range, had no favored core areas, and ranged very little in the areas used the previous year (Fleury & Gautier-Hion, 1999). Therefore, this population has been characterized as seminomadic, a behavior that is highly atypical among arboreal forest monkeys in Africa, specifically (Fleury & Gautier-Hion, 1999), and among primates in general. Fleury and Gautier-Hion (1999) suggest that this seminomadic ranging behavior is an adaptation to the relatively seasonal and homogenous environment that characterizes this site. While the ecological correlates associated with this behavior merit further investigation, it is an example of extreme ecological and behavioral flexibility.

While seminomadic ranging behavior has been observed among *C. satanas*, a group of *C. angolensis* has been documented migrating 13 km from its former home range in the Nyungwe Forest of Rwanda (Fashing et al., 2007). This group of *C. angolensis* is also atypical among *Colobus* in that it has a home range of around 2,440 ha, an estimated movement rate of 141 m/hour, and a group size of around 300 individuals (Fashing et al., 2007). Fashing et al. (2007) hypothesize that while the abundance of mature leaves and lichens are associated with the unusually large group size of *C. angolensis* at this site, food patch depletion and renewal time may necessitate these exceptional ranging behaviors. Although not as extreme, *C. angolensis* elsewhere are also characterized as having high travel rates and relatively low resting rates (DaSilva, 1992; Fashing et al., 2007). Thus, the interplay between habitat, diet, and ranging within and among *Colobus* species is complex. However, most of the African colobines have small home ranges and short daily path lengths in comparison to sympatric cercopithecines.

Predation

Predation on African colobines is best discussed in the context of predator ecology and predator-prey counterstrategies. Although there are many potential predators of African colobines, such as dogs, golden cats, pythons, and spotted hyenas (Hart et al., 1996; Galat-Luong & Galat, 2005), there are three major nonhuman predators on colobus monkeys: leopards (*Panthera pardus*), crowned hawk-eagles (*Stephanoaetus coronatus*), and chimpanzees (*Pan troglodytes*). In Taï National Park, where all three African colobine genera live sympatrically, all three of these predators are also present, and the risk of predation by each predator differs among the genera. While the greatest predation risk for olive colobus (*Procolobus verus*) is from crowned hawk-eagles, leopards and chimpanzees are both major predators on red colobus (*Piliocolobus badius*), but leopards have the highest predation rate on black-and-white colobus (*Colobus polykomos*) (Schultz et al., 2004). However, an analysis of bones found in or underneath crowned hawk-eagle nests at Taï suggests a slightly different pattern since red colobus make up the largest proportion of prey, followed by olive colobus, and then black-and-white colobus (McGraw et al., 2006)

Although *Procolobus verus* are regular prey of crowned hawk-eagles, predation risk for this species is mitigated through polyspecific associations with *Cercopithecus diana* (Schultz & Noë, 2002). Lone and isolated monkeys may be particularly at risk from predation by crowned hawk-eagles; solitary adult *Colobus guereza* males are regularly targeted by this aerial predator in Kibale (Skorupa, 1989). In fact, the "roaring" predator alarm calls by *C. guereza* may actually attract the eagles' attention in these instances (Davies, 1994). Nevertheless, there was no difference in abundance of male versus female remains found in association with the eagle nests at Taï (McGraw et al., 2006). Red colobus monkeys are also subject to predation by crowned hawk-eagles, but they do not give loud calls (Davies, 1994). Instead, males will form coalitions against crowned hawk-eagles, a behavior that appears to be effective in deterring this aerial predator (Struhsaker & Leakey, 1990).

Since leopards are ineffective hunters above 10 m high in the rainforest, in general, they are more of a major predator threat to colobus populations in savanna habitats (Struhsaker, 1975). However, both *Piliocolobus badius* and *C. polykomos* are regular prey of leopards in Taï, where leopards hunt the monkeys primarily on the ground, during the day, and in portions of the rainforest where the monkeys occur in relatively high densities (Zuberbühler & Jenny, 2002). Remains of black-and-white colobus have also been reported in leopard scat and food at Mahale, Tanzania (Nishida, 1972), and at Ituri Forest, DRC (Hart et al., 1996), although some of these findings may reflect leopard scavenging on eagle kills, rather than that of direct colobus predation (see Oates, 1994; Hart et al., 1996). Local extirpation of leopards and other large predators may be associated with rapid increases in colobus population numbers.

The third major predator of colobus monkeys, the chimpanzee, is atypical as a predator since this gregarious frugivorous primate is not dependent on meat as a primary food source (Teelen, 2008). In the early years of primate research, observations of colobus predation by chimpanzees were interpreted as opportunistic (Goodall, 1965; Struhsaker, 1975). Similarly, although antagonistic interactions were observed between the two species (Struhsaker, 1975), red colobus monkeys were not thought to be targeted any more than other primates or ungulates (Takahata et al., 1984). In the more recent literature, views of chimpanzee predation on red colobus monkeys have changed. Red colobus monkeys have now been found to be the primary prey items for chimpanzees at four sites—Mahale, Gombe, Ngogo (Kibale), and Taï. Red colobus monkeys at these sites can comprise as much as 91% of the chimpanzee populations' total prey diet (Uehara, 1997), and multiple kills can occur within one hunting bout (Stanford et al., 1994; Watts & Mitani, 2002). Researchers disagree as to whether these high predation rates are reflective of recently disrupted ecosystems such as high population densities and habitat loss (Fourrier et al., 2008; Lwanga et al., 2011), or whether earlier studies reflected unusually low predation rates because of the effects provisioning bananas had on the dynamics of the overall primate community (in Gombe specifically) (Stanford et al., 1994). However, it is notable that early studies across sites never noted chimpanzee predation as a major threat (e.g., Nishida, 1972). Although mixed-species associations (Bshary & Noë, 1997; Noë & Bshary, 1997; Wachter et al., 1997) and some behavioral responses occur when encountering chimpanzee parties (Stanford, 2005), red colobus monkeys do not seem to have developed a very effective counterstrategy to chimpanzee predation (Zuberbühler et al., 1999; Stanford, 2005). There is no consistent pattern in the sex and age class of targeted colobus prey by chimpanzees across sites (Stanford et al., 1994; Uehara, 1997; Teelen, 2008), which may support the hypothesis that high-intensity hunting of red colobus, in particular, by chimpanzees is relatively recent, unstable, and unsustainable (Fourrier et al., 2008; Teelen, 2008; Lwanga et al., 2011). At Ngogo, where chimpanzee predation has been estimated to remove annually as much as 53% of the red colobus population (Teelen, 2008), chimpanzee hunting of red colobus monkeys has recently declined as red colobus have become very rare, thus causing encounters between the two species to decline (Watts & Amsler, 2013). It remains to be seen whether chimpanzee hunting trends at this site reflect a cyclical pattern characteristic of sustainable hunting, or whether chimpanzees will inevitably be responsible for the extirpation of red colobus at Ngogo and other sites (Watts & Amsler, 2013), which indicates an aberrant situation not compatible with long-term evolutionary adaptations.

Lastly, humans have been for centuries the biggest threat to colobus monkeys. In the past, *Colobus* were targeted for their long, flowing fur (Booth, 1956). Today, hunting for meat consumption is a greater threat, and it plays a major role in the critically endangered status of *Piliocolobus pennantii*, *P. preussi*, and *P. epieni*, the probable extinction of *P. waldroni*, and the highly threatened status of red colobus monkeys in West and Central West Africa (Gonzalez-Kirchner, 1997; Oates et al., 2000; Struhsaker, 2005; Mittermeier et al., 2007). *Colobus* are also hunted in some areas as a preferred resource, such as *C. satanas* on Bioko Island, Equatorial Guinea (Gonzalez-Kirchner & Sainz de la Maza, 1993), and have been extirpated in certain areas because of hunting. *C. vellerosus*, in particular, is critically endangered and has been extirpated in many national parks and reserves, possibly linked to illegal hunting; most remaining groups of this species are found only in isolated forest patches throughout their range (Gonedelé et al., 2010).

Activity Budget

Colobus monkeys would be expected to spend a great deal of time resting because of the relative indigestibility of their foods. However, just as there is variability in the amount of digestion-inhibiting materials consumed by different species and different populations, there is a corresponding variability in activity budgets (Table 13.5).

Available data on olive colobus activity at Tiwai Island, Sierra Leone, and Taï, Côte d'Ivoire suggest they spend a relatively high percentage of time moving (19–25%) in comparison to other African colobines. This may be because their activity budget is dependent on other primate taxa that compose their mixed-species associations (Oates, 1994). For example, olive colobus monkeys will typically end resting and feeding sessions to follow *Cercopithecus diana* as they begin their travel bouts (Oates, 1994). As far as other activities, olive colobus spend 35–40% of their time resting, 27–39% feeding, and 7–8% in social or other activities (Table 13.5). Red colobus spend 30–55% of their time resting, 21–45% feeding, 3–22% moving, and 2–15% in social activities, while black-and-white colobus spend 34–69% resting, 19–46% feeding, 2–25% moving, and 1–7% in social activities. Generally, there is a large range of variation in the way time is allocated within each genus.

Some of the variations in activity budgets have been linked to environmental factors. For example, the exceptionally high percentage of time spent resting and low percentage spent moving in some populations of black-and-white colobus may be linked to habitats with low food abundance, small fragment sizes, and/or behavioral thermoregulation to save energy (*Colobus angolensis*: Wijtten et al., 2012; *C. guereza*: Fashing et al., 2007; *C. polykomos*: Dasilva, 1989, 1992; *C. vellerosus*: Djègo-Djossou et al., 2015). However, cross-site comparisons are often complicated because group size is a major determinant of activity budgets. Indeed, feeding time increases with group size across species (Korstjens & Dunbar, 2007), across groups within the same population (*Piliocolobus tephrosceles*: Milich et al., 2014a; *C. vellerosus*: Teichroeb & Sicotte, 2018), and within groups over time (*P. tephrosceles*: Gogarten et al., 2014). Therefore, it is not surprising that, among known populations, one of the most active is a group of 300 *C. angolensis* in the Nyungwe Forest, Rwanda (Fashing et al., 2007). Similarly, the time spent socializing may be influenced by group composition. For example, the time spent in social activity was higher in the *C. vellerosus* group that contained more females, as they are more avid groomers than males (Teichroeb et al., 2003), and females spend more time grooming other females in groups with more infants (Wikberg et al., 2015). Time spent socializing may also be linked to population density, since *P. kirkii* in the shamba (garden) environment engaged in social grooming, play, and aggression significantly more often than *P. kirkii* in the lower-density forest population (Siex, 2003). A similar trend with increasing affiliation and aggression has also been documented over time in the rapidly increasing population of *C. vellerosus* at Boabeng-Fiema, Ghana (Wikberg & Sicotte, unpub. data).

SOCIALITY

Many studies have described basic information about African colobine grouping patterns, although there have been relatively few longitudinal studies that focus on more detailed analyses of individual social behaviors. This is partly because it is difficult to observe social behaviors of arboreal primates and to identify individual colobus monkeys. Colobines also display more subtle social behaviors at lower rates than many other cercopithecines. Despite these challenges, the integration of molecular techniques and observational data have highlighted many new aspects of colobine social behavior.

Grouping Patterns

Cohesion varies, from groups that are spatially very unified in black-and-white colobus, to groups more spatially diffuse in red colobus, to fission-fusion societies in the olive colobus. Some red colobus groups have also been observed to exhibit fission-fusion societies, such as *Piliocolobus gordonorum* in a recently degraded forest of the Udzungwa Mountains, Tanzania, *P. kirkii* in the Jozani forest of Zanzibar, and *P. tephrosceles* in logged areas of Kibale National Forest, Uganda—all of which may be a response to food distribution and abundance (Starin, 1994; Siex, 2003; Struhsaker et al., 2004; Marshall et al., 2005; Nowak & Lee, 2011).

Booth (1957) first observed small, multi-male olive colobus groups consisting of 6–20 animals, but later studies documented that smaller groups with only one or two adult males and several females are more common (Table 13.2; Oates, 1988; Oates & Whitesides, 1990; Korstjens, 2001; Korstjens & Schippers, 2003). Red colobus groups often contain at least twice as many females as males but vary markedly in size within and between sites. Red colobus groups are sometimes reported to be smaller in lower-quality habitats (*P. gordonorum*: Struhsaker et al., 2004; but see Siex, 2003, for lack of *P. kirkii* group size difference between two habitats), smaller at higher altitudes (*P. tephrosceles*: Nishida, 1972), increase in size over time with food abundance (*P. tephrosceles*: Gogarten et al., 2015), and decrease in size with habitat loss (*P. temminckii*: Galat-Luong & Galat, 2005).

Black-and-white colobus groups contain one or multiple males and females (Table 13.2). With the exception of an early study of *Colobus satanas* (Malbrant & Maclatchy, 1949, cited in Sabeter Pi, 1973) and one population of *C. angolensis,* black-and-white colobus tend to form smaller groups than red colobus. This is the case even when they are sympatric with relatively similar diets such as *C. guereza* and *P. tephrosceles* at Kibale (Oates, 1977c; Chapman & Pavelka, 2005; Harris et al., 2010). Small forest patches may contain particularly small groups (*C. guereza*: Oates, 1977c; Dunbar, 1987; von Hippel, 1996; Krüger et al., 1998; Onderdonk & Chapman, 2000; possibly also *C. vellerosus*: Wong & Sicotte, 2006; Gonedelé Bi et al., 2010). In contrast to the other species of black-and-white colobus, small groups of *C. angolensis* form aggregations of as many as hundreds of individuals (Oates, 1994). Stead and Teichroeb (2019) describe the multilevel structure of *C. angolensis* at Lake Nabugabo, Uganda, as: (1) cohesive core units including one to several males and females; (2) several core units tolerating each other but rarely showing direct interactions that form a clan; and (3) several clans making up a band. In contrast to the more frequent fissions and fusions at Lake Nabugabo, *C. angolensis* form more stable large aggregations at the high-altitude site of the Nyungwe Forest in Rwanda, which may be related to the abundance of relatively nutritious mature leaf foliage found there (Fimbel et al., 2001; Chapman & Pavelka, 2005).

Dispersal Patterns

The African colobines show a range of dispersal patterns, often with considerable within-population variation in male and/or female dispersal. Olive colobus disperse frequently in relation to other primate species, perhaps because the costs associated with dispersal are mitigated by dispersing in parallel with conspecifics and by associating with *Cercopithecus diana* before joining a new group (Korstjens & Schippers, 2003). Although both sexes disperse, female dispersal occurs more often than male dispersal (Korstjens & Schippers, 2003). Females may reduce feeding competition by dispersing to smaller uni-male groups, while males may reduce mate competition by dispersing from groups with unfavorable male:female ratios (Korstjens & Schippers, 2003).

Most red colobus populations show at least occasional male and female dispersal, although dispersal patterns appear to vary across populations in terms of which sex is the most frequent to disperse. In the population of *Piliocolobus tephrosceles* at Kibale, all females and half of the males dispersed (Struhsaker & Pope, 1991). As young males approach sub-adulthood, they are frequently harassed by adult males, and Struhsaker (1975) believes that it is in this manner that a young male is either admitted into the adult-male subgroup or expelled from the group. Miyamoto and colleagues (2013) suggest that female, but not male, dispersal increases with group size, based on their study of patterns of relatedness values in two groups. As expected for a population with female dispersal and male philopatry, a group of *P. temminckii* in the Cantanhez National Park, Guinea-Bissau, consisted mostly of males with high relatedness values while this was not the case for females (Minhos et al., 2015). Both male and female *P. temminckii* at Abuko, The Gambia, transfer among groups at relatively similar frequencies; females typically dispersing with other age mates (Starin, 1994). Indications of male dispersal were observed during a five-month study of eight small, all-male groups and eight solitary male *P. rufomitratus* in the Tana River Delta, Kenya (Decker, 1994). In the shamba population of *P. kirkii* in Zanzibar, dispersal is male-biased although both sexes may disperse (Siex, 2003). At Taï National Park, male and female *P. badius* were also

observed associating with groups of *Colobus polykomos* (Korstjens et al., 2002); thus, similar to the olive colobus, some red colobus may mitigate the costs of dispersal by transferring together with conspecifics or associate with another species.

While early work suggested that dispersal was male-biased in the black-and-white colobus, recent studies based on longitudinal observation and genetic data have revealed variable dispersal and residency patterns in this taxon. Both observational and genetic data show male-biased dispersal in *C. guereza* at Kibale (Oates, 1977c; Harris et al., 2009) and *C. vellerosus* at Boabeng-Fiema, Ghana (Teichroeb et al., 2011; Wikberg et al., 2012). Males sometimes transfer between groups together (i.e., engage in parallel dispersal) and, as a result, adult males may reside with close adult male kin in their new group (Oates, 1977c; Harris et al., 2009; Teichroeb et al., 2011; Wikberg et al., 2012). The primary benefit of engaging in parallel dispersal may be to gain a numerical advantage and more easily overcome the resistance of the resident male(s) (Teichroeb et al., 2011). Harris and colleagues (2009) report that although no successful female transfer between *C. guereza* groups was recorded during 5.5 years, genetic data revealed several closely related female-female dyads residing in different groups, possibly because group fission events or female dispersal took place before the start of the study. A combination of observational and genetic data from *C. vellerosus* indicate that about half of the females dispersed while the other half have remained in their natal groups (Wikberg et al., 2012). It is possible that females disperse to reduce female-female competition as younger females are sometimes evicted by older females, and females tend to emigrate from large groups and immigrate into small groups (Teichroeb et al., 2009). However, larger groups also suffer from more frequent male takeovers and infanticide (Teichroeb et al., 2012), and females often disperse after male takeovers, perhaps to find a higher-quality male and to reduce the future risk of infanticide (Sicotte et al., 2017). Females often engage in parallel dispersal, which allows familiar females to remain together (Teichroeb et al., 2009; Wikberg et al., 2012). This could be a way to reduce the social costs associated with dispersal, as familiarity via long-term co-residency shapes their social relationships (Wikberg et al., 2014a,b, 2015). Because both male and female dispersal have been observed during relatively short studies of *C. polykomos* and *C. satanas* (Oates, 1994; Korstjens et al., 2002), female dispersal may be more common in these species than in *C. guereza* and *C. vellerosus*. *C. angolensis* males and females have been observed to disperse between core units, but all cases of male dispersal occurred within bands while some females dispersed between bands (Stead & Teichroeb, 2019). Thus, this may be the only black-and-white colobus species with female-biased dispersal, at least in terms of dispersal distance.

Intragroup Interactions
Male and Female Agonism

Males typically show higher frequencies of displacements than females, leading to detectable male dominance hierarchies in at least some African colobines (e.g., Struhsaker 1975, 2010; Newton & Dunbar, 1994; Teichroeb & Sicotte, 2008b; Wikberg et al., 2013). In *P. tephrosceles*, for example, low rates of submissive behaviors preclude the detection of linear dominance hierarchies among females (Tombak et al., 2019), while a male hierarchy is detectable based on displacements (Struhsaker, 1975, 2010). Male *C. vellerosus* can also be ranked into a hierarchy based on displacements, avoids, and pant-grunts (Teichroeb & Sicotte, 2008b). In contrast, observers have been unable to detect a linear female hierarchy in some groups of *C. vellerosus*, while in other groups, female dominance hierarchies are as strongly expressed as other primate populations classified as despotic (Wikberg et al., 2013). Both sexes form linear dominance hierarchies in *C. polykomos* (Newton & Dunbar, 1994; Korstjens et al., 2002). Female *C. polykomos* show higher rates of within-group agonistic interactions compared to the sympatric *Piliocolobus badius* (Korstjens et al., 2002) and the more closely related *C. vellerosus* (Wikberg et al., 2013) and *C. guereza* (Harris, 2005). Higher rates of agonism and expression of dominance hierarchies are likely linked to higher levels of competition when foraging on contestable resources (e.g., Korstjens et al., 2002; Wikberg et al., 2013).

Coalitionary aggression is rare or absent in most populations of African colobines (female *P. badius* and *C. polykomos*: Korstjens et al., 2007; female *P. tephrosceles*: Tombak et al., 2019; female *C. guereza*: Dunbar, 2018; female and male *C. vellerosus*: Saj et al., 2007; Teichroeb et al., 2013; Wikberg et al., 2014a). Females in at least some groups of *C. vellerosus* form coalitions during intergroup encounters or during male attacks of infants and females (Saj et al., 2007; Wikberg et al., 2014a; Wikberg, unpub. data). Furthermore, females in this population appear to show an increase in agonistic interactions (Wikberg et al., 2013) and kin-biased coalitions over time (Wikberg et al., 2014a), perhaps linked to increased population densities and feeding competition. A similar pattern has been observed in *P. kirkii* in Zanzibar. Contact agonism is much more frequent in the densely inhabited shamba habitat compared to the forest habitat (Siex, 2003; Nowak & Lee, 2011). Siex (2003) suggests that the increase in social interactions, compared to an earlier study (Mturi, 1993), may be due to more than a ninefold increase in population density leading to reduced per capita food abundance.

Male Affiliation

Olive colobus males tolerate but rarely groom each other (Oates, 1994; Korstjens & Noë, 2004; McGraw & Zuberbühler, 2008). The affiliation pattern among male red colobus differs between study populations (Struhsaker, 2010). For example, adult male *P. tephrosceles* at Kibale and *P. epieni* in Gbanraun, Nigeria, form a very close subgroup that rest together, facilitate group movement and cohesion, and frequently groom each other (Struhsaker, 1975; Struhsaker & Leland, 1979; Werre, 2000; Struhsaker et al., 2004; Struhsaker, 2010). In contrast, male *P. temminckii* did not groom each other often (Starin, 1991). Male *P. kirkii* were never seen grooming each other and only rarely groomed other group members (Siex, 2003), which may be linked to frequent male dispersal (Struhsaker, 2010).

C. polykomos at Cantanhez Forests National Park, Guinea-Bissau, show less grooming in male-male than male-female and female-female dyads (Minhos et al., 2015). Similarly, male *C. vellerosus* rarely groom, including closely related male-male dyads and parallel male immigrants (Teichroeb et al., 2013). As a matter of fact, parallel immigrant males often cooperate to take over the group from the previous resident males, but once successful, expel each other one by one (Teichroeb et al., 2011). However, males were in one-meter proximity more often in groups with only one cohort of male immigrants compared to groups that consisted of males that had entered the group at different times (Wikberg et al., 2012). This indicates that parallel immigrant males that remain together after successful immigration may form somewhat more tolerant relationships than with other males. In contrast, male *C. angolensis* groom each other and rest together relatively frequently (Bocian, 1997).

Female Affiliation

In olive colobus, females rarely affiliate with each other (Korstjens et al., 2007) and grooming bouts occur most often between adult males and adult females (Oates, 1994). Similarly, red colobus females form stronger bonds with males than females (*Piliocolobus badius*: Korstjens et al., 2002; *P. epieni*: Werre, 2000; *P. kirkii*: Siex, 2003; *P. tephrosceles*: Struhsaker, 1975). When *P. tephrosceles* females at Kibale groom each other, they do so in a reciprocal fashion, and grooming bonds do not predict their reproductive success (Tombak et al., 2019). In contrast, *P. temminckii* show stronger female-female than male-male or male-female bonds (Starin, 1994; Minhos et al., 2015). Minhos and colleagues (2015) conclude that this pattern is not driven by kinship as their study group was mostly made up of unrelated females and related males. Starin (1994) suggests that females may be able to maintain strong bonds with familiar females by engaging in parallel dispersal.

For black-and-white colobus, female-female dyads form the strongest grooming relationships in *C. guereza* (Oates, 1977c; Harris, 2005; Minhós et al., 2015) and *C. vellerosus* (Teichroeb et al., 2003; Saj et al., 2007; Wikberg et al., 2012). There is no significant difference in time females spend grooming male and female partners, if partner availability is taken into account in *C. polykomos* (Korstjens et al., 2002). Female *C. angolensis* at Ituri, DRC, form the core of the social groups, and their bonds are described as strong with reciprocal grooming (Bocian, 1997). In contrast, male-female dyads form stronger grooming bonds than male-male and female-female dyads in *C. angolensis* at Nabugabo, Uganda (Arseneau-Robar et al., 2018). Although females spend relatively low percentages of time grooming, they often form differentiated social relationships and certain aspects of the grooming networks are remarkably varied even within populations. For example, female grooming network centralization (i.e., the degree of inequity in incoming and outgoing grooming ties) varies as much across groups of *C. vellerosus* as it does across 40 primate species (Wikberg et al., 2015). This variation in network centralization is linked to the stability of female group membership, and long-term resident females are central players in the grooming network (Wikberg et al., 2014a, 2015). In groups without immigrant females, female *C. vellerosus* show kin bias in affiliation and infant handling (Bădescu et al., 2014; Wikberg et al., 2014a,b). In contrast, female kin do not preferentially affiliate with each other if they have different immigration statuses, possibly because they do not recognize each other as kin or because they will not gain any fitness benefits from doing so (Wikberg et al., 2014a,b). Similarly, relatedness values do not predict affiliation in *C. polykomos* (Minhós et al., 2015), although it is unclear if this can be explained by differences in dispersal status.

Female-female bonds tend to be weaker in African colobine populations that show more frequent female dispersal. In at least some groups, females form more unequal grooming networks in the presence of recent female immigrants, because these recent female immigrants are less likely to receive grooming from resident females. Thus, some of the differences in bond strength between African colobine populations, groups, and dyads can at least partly be explained by their dispersal pattern.

Intergroup Interactions

Olive colobus vocalize, threat, and chase opposing groups (Korstjens & Noë, 2004). Red colobus intergroup encounters are often tolerant but can include low-intensity aggression, such as branch shaking and vocalizations displayed by *Piliocolobus kirkii* in the Zanzibar forest groups (Siex, 2003), or occasional physical fights observed in *P. tephrosceles* (Struhsaker, 2010). The high-density shamba subpopulation of *P. kirkii* shows higher rates of intergroup aggression than the forest groups (Siex, 2003). During between-group encounters, black-and-white colobus occasionally engage in friendly behaviors (i.e., play, copulations, touches, and grooming) but more frequently show no reaction to each other or show aggressive stiff-leg displays, loud call displays, jump displays, chasing, or wrestling (Oates, 1977c; Fleury & Gautier-Hion, 1999; Sicotte & MacIntosh, 2004; Korstjens et al., 2005; Harris, 2010; Teichroeb & Sicotte, 2018; Wikberg et al., 2020).

In most colobus populations, males are the main participants in intergroup encounters (Fashing, 2011). Korstjens and colleagues (2007) describe how olive colobus females disappear out of view while the males participate in an intergroup encounter. These males sometimes display together and chase opposing groups away (Korstjens & Noë, 2004; McGraw & Zuberbühler, 2008). The occurrence of male aggression is not related to the presence of receptive females even though extra-group copulations occur (Korstjens & Noë, 2004).

It is typically male red colobus that engage in intergroup display behaviors (Siex, 2003; Struhsaker, 2010), although a few sites show contrasting patterns (Starin, 1994; Korstjens et al., 2002; Siex, 2003). For example, both male and female *P. kirkii* participate in intergroup agonism in the densely populated shamba habitat, which Siex (2003) suggests is linked to high population numbers. Female *P. temminckii* at Abuko, The Gambia, are particularly aggressive toward male non-group members, occasionally resulting in fatalities; Starin (1994) suggests this may be due to the lack of sexual size dimorphism in the population.

Black-and-white colobus males engage in intergroup aggression more often than females (Oates, 1977c; von Hippel, 1996; Fashing, 2001c; Sicotte & MacIntosh, 2004; Korstjens et al., 2005; Sicotte et al., 2007; Harris, 2010; Teichroeb & Sicotte, 2018). Male *C. angolensis* group members cooperate to chase away other males. Male *C. vellerosus* occasionally form coalitions to deter extra-group males; these coalitions are more likely to occur between high-ranking males even though they do not form particularly strong affiliative bonds (Teichroeb et al., 2013). The benefits of these coalitions are unclear, because groups with fewer—rather than more—males and groups with a larger alpha male are more likely to win the encounter

(*C. guereza*: Harris, 2010; *C. vellerosus*: Teichroeb & Sicotte, 2018). Resident males are likely to defend access to females during these encounters, and they are particularly aggressive when encountering extra-group and potentially immigrant males (Fashing 2001c; Sicotte & MacIntosh, 2004; Harris, 2010; Teichroeb et al., 2011).

Female *C. polykomos* sometimes join their resident males in chasing away intruding males (Bocian, 1997). Female *C. guereza* and *C. vellerosus* also occasionally initiate roaring bouts, jump displays, and chases of other groups, either with or without their resident male(s) (Oates, 1977c; Fashing, 2001c; Sicotte & MacIntosh, 2004; Sicotte et al., 2007; Harris, 2010; Teichroeb & Sicotte, 2018). *C. guereza* and *C. vellerosus* individuals are more likely to engage in aggressive intergroup interactions if the encounter occurs within the core of their home range or in important feeding areas, suggesting that groups defend access to food (Fashing, 2001c; Harris, 2006; Wikberg & Sicotte, unpub.data). Female intergroup aggression occurs more frequently in *C. polykomos,* and at least some females participate in the majority of aggressive encounters (Korstjens et al., 2002). Female *C. polykomos*, but not males, show a seasonal pattern of intergroup aggression peaking when important food items are available (Korstjens et al., 2005).

There is remarkable variation across and within species of African colobines in grouping patterns, dispersal, and social interactions. For example, some species like *P. kirkii* and *C. vellerosus* show considerable variation in frequencies and types of social within-group and between-group interactions. Unfortunately, the long-term consequences of this variation remain unknown for most populations. This is an important topic for future studies because it can help researchers understand whether increased interaction rates in high-density populations are adaptive and sustainable long term. If this is not the case, elevated interaction rates could lead to a future population collapse; ideally, management strategies should be put in place to return behaviors to baseline levels before the population collapses (Berger-Tal et al., 2011). Using behaviors as indicators of future populations changes is an intriguing concept, but primatologists have yet to see if using these early warning signals actually provides enough time to put effective conservation strategies in place.

REPRODUCTION AND INFANT DEVELOPMENT
Reproductive Strategies

Olive colobus monkeys at Tiwai, Sierra Leone, and Taï, Côte d'Ivoire, are seasonal breeders, with mating concentrated in the wet season and a birth peak occurring during the dry season (Oates, 1994; Korstjens & Noë, 2004; Table 13.6). During the mating season, female olive colobus exhibit highly visible sexual swellings, and male testes size increases notably (Korstjens & Schippers, 2003; Korstjens & Noë, 2004). Red colobus females also exhibit exaggerated sexual swellings, and the duration of maximal genital tumescence is shorter in logged areas, possibly due to decreased food abundance (Milich et al., 2014b). Mating and births in red colobus populations are aseasonal (Struhsaker, 1975) but with birth peaks that are particularly noticeable for red colobus living in seasonal habitats (Oates, 1994), such as *P. temminckii* at Abuko, The Gambia (Starin, 1994). Black-and-white colobus lack exaggerated sexual swellings, but females in some species may exhibit slightly raised perineal regions (Nunn, 1999). Copulations among *C. guereza* are not distributed seasonally (Oates, 1977c), although some female *C. guereza* and *C. vellerosus* tend to cycle and elicit copulations around the same time (Harris & Monfort, 2006; Teichroeb & Sicotte, 2010). This could be linked to the occurrence of infant deaths after a new male takes over the alpha position, leading to estrus synchrony of females that lost their infants around the same time (Sicotte et al., 2007; Teichroeb & Sicotte, 2008a; Teichroeb et al., 2012).

Red colobus reach adulthood when they are two to five years old, while black-and-white colobus reach sexual maturity when they are four to seven years of age (see Table 13.6). However, some of the apparent variation between studies may be an artifact of small sample sizes, and in the case of males, the difficulties of defining adulthood. For example, male *C. vellerosus* copulate with females as subadults (i.e., 3–7 years of age) but do not gain full body size until they are about seven years old (Wikberg, pers.obs.).

African colobines have a reported cycle length of 24–28 days, a gestation period of about 4–6 months, and an interbirth interval that ranges up to 4.5 years. Interbirth intervals are longer when the infant survives to nutritional independence (Harris & Monford, 2006; Vayro et al., 2016; but see Korstjens & Noë, 2004). Some of the dissimilarity in these life history variables can be explained by body size and/or degree of folivory (Zimmerman & Radespiel, 2013). For example, the somewhat shorter interbirth intervals in *C. vellerosus* compared to other *Colobus* is expected based on their slightly smaller body size (Vayro et al., 2016). However, Harris and Montfort (2006) highlight that *Colobus* gestation is shorter than other similar-sized primates for unknown reasons.

Group membership and alpha status are likely important factors in determining male mating and reproductive success in olive colobus (Korstjens & Schippers,

Table 13.6 African Colobine Reproductive Parameters[a]

Species	Study Site	Birth Pattern	Age at Sexual Maturity (Years)[b]	Cycle Length (Days)	Gestation Length (Days)	Birth Interval (Months)	Nursing (Months)	References
Procolobus verus	Tiwai, Sierra Leone	Seasonal	–	–	180	–	–	Oates (1994)
P. verus	Taï, Ivory Coast	Seasonal	–	28	116	20[c]	12	Korstjens and Schippers (2003), Korstjens and Noë (2004), Korstjens (pers.comm., cited in Fashing, 2011)
Piliocolobus temminckii	Abuko, Gambia	Peak	M: 2.3 F: 2.9	–	174	29.4	–	Starin (1988, 1991)
P. badius	Tiwai, Sierra Leone	Peak	–	–	–	–	–	Davies (pers.comm., cited in Oates, 1994)
P. kirkii	Jozani (forest), Zanzibar	Peak	–	–	–	26-55[c]	–	Siex and Struhsaker (1999), Siex (2003)
P. kirkii	Jozani (shamba), Zanzibar	Peak	–	–	–	32-35[c]	–	Siex and Struhsaker (1999), Siex (2003)
P. pennantii	Gbanraun, Nigeria	Peak	–	–	–	–	–	Werre (2000)
P. rufomitratus	Tana River Kenya	Peak	–	–	–	–	–	Marsh (1979a)
P. tephrosceles	Kibale, Uganda	Peak	M: 2.9-4.8 F: 3.2-3.8	–	–	24.4 27.5[d]	–	Struhsaker (1975), Struhsaker and Leland (1979, 1987), Struhsaker and Pope (1991)
Colobus angolensis	Ituri, DRC	–	–	–	–	25.2	–	Bocian (1997)
C. angolensis	Nyungwe, Rwanda	Aseasonal	–	–	–	–	–	Fimbel et al. (2001)
C. guereza	Kibale, Uganda	Aseasonal	M:6 F: 4	24[e]	158[e]	21.5-25.2	11-12.8	Struhsaker and Leland (1987), Harris and Monfort (2006), Fashing (2011), Chapman et al. (1990), van Schaik and Isler (2012)
C. guereza	Kakamega, Kenya	Aseasonal	–	–	–	17[c]	–	Fashing (2002), Fashing (unpub.data, cited in Fashing, 2011)
C. polykomos	Tiwai, Sierra Leone	Seasonal	–	–	–	24	12	Dasilva (1989, cited in Oates, 1994), van Schaik and Isler (2012)
C. polykomos	Taï, Ivory Coast	Aseasonal	–	–	–	–	–	Korstjens (2001)
C. vellerosus	Boabeng-Fiema, Ghana	Aseasonal	M: 7[f] F: 5.8	24[e]	168.5[e]	16.5	14.6	Vayro et al. (2016), Crotty (2016)

[a] Compiled from Fashing (2011) and Crotty (2016).
[b] M=Males; F=Females.
[c] Numbers are calculated as the number of female months divided by the number of infants born.
[d] Birth intervals following infant death are excluded.
[e] Length is based on hormonal data; all other numbers are approximate lengths based on observational data.
[f] Estimated based on gaining full adult size but males copulate before this age.
– Indicates no data.

2003), red colobus (Struhsaker & Leland, 1979), and black-and-white colobus (*C. guereza*: Fashing, 2001c; *C. vellerosus*: Teichroeb & Sicotte, 2010; Wikberg, unpub. genetic data). Although extra-group copulations occur (*Procolobus verus*: Korstjens & Schippers, 2003; *C. guereza*: Fashing, 2001c; *C. vellerosus*: Sicotte & MacIntosh, 2004), no extra-group paternity has been detected (Wikberg, unpub.data) from the scant genetic data available from most of these populations. At least in olive colobus, males with no female group members obtain mating partners by visiting females in polyspecific associations and by females with swellings visiting with the apparent purpose of mating (Korstjens & Noë, 2004). Where red colobus matings have been observed, the dominant male of the group does most of the mounting and copulation, although he is frequently harassed by young juveniles during copulation (Struhsaker & Leland, 1979). Alpha male *C. vellerosus* sire all infants in uni-male groups, but they do not completely monopolize reproduction in multi-male groups (Wikberg, unpub.data).

Infant Development and Handling

Olive colobus infants are darker in color than adults but lack the conspicuous coat color of many other African colobine infants (Oates, 1994). Red colobus infants are born black and gray and do not assume adult coloration until about 3.5 months (Struhsaker & Oates, 1975; Struhsaker, 2010). Black-and-white colobus infants, with the exception of *C. satanas*, show the most conspicuous natal color of all African colobines and are born white (Oates, 1994). Infants gradually transition from white to a gray coat (Figure 13.8); this transition is faster for males than females and for infants in multi-male groups compared to uni-male groups of *C. vellerosus* (Bădescu et al., 2016). Most infants gain the adult black-and-white coloration when they are 3-4 months old, but some infants only take 2 months to transition into an adult coat (Oates, 1977c; Bădescu et al., 2016).

Colobus infants nurse for approximately 12 months (Table 13.6). Infant *C. vellerosus* may not nurse as long in unstable multi-male groups as they would in multi-male groups with stable male group membership, which may be a strategy that mothers in unstable groups employ to reduce the risk of new immigrant males attacking their infants (Teichroeb & Sicotte, 2008a; Vayro, 2017).

Handling of infant olive colobus is limited to the mother who carries her infant in her mouth when traveling, a behavior that is atypical among primates generally and haplorrhines (anthropoids) specifically (Booth, 1957; Oates, 1994; Oates et al., 1994). Newborn red colobus monkeys interact very little with other monkeys in their group (Struhsaker & Oates, 1975; Struhsaker, 2010), and females at Kibale do not appear to exchange grooming for access to infants (Tombak et al., 2019). However, females with infants tolerate the presence of other females, and some infant handling by non-mothers (i.e., allomothering) has been observed in *Piliocolobus kirkii* in Zanzibar (Siex, 2003; Struhsaker, 2010; Nowak & Lee, 2011). In contrast to most red colobus, young black-and-white colobus infants are handled by other group members, especially by nulliparous females and maternal female kin (Oates, 1977c; Brent et al., 2008; Bădescu et al., 2014). Although infant handling by non-mothers is

Figure 13.8   Black-and-white colobus infant color transition; white (left) transitioning to gray (right).
(Photo credits: Andrea Donaldson (*Colobus angolensis*, left photo); Eva Wikberg (*C. vellerosus*, right photo)

widespread in the primate order, it is only in a few species, such as *C. guereza* and *C. vellerosus*, that infants of such young age (under one week) are taken from their mothers (Oates, 1977c; Brent et al., 2008). Group members are less interested in other females' infants after the infant transition to adult coat color (Oates, 1977c; Brent et al., 2008; Bădescu et al., 2014). Young infant male *C. vellerosus* tend to receive the most interest and handling by non-mothers (Bădescu et al., 2014). Adult male red colobus and black-and-white colobus do not interact much with infants (Struhsaker & Leland, 1979; Struhsaker, 2010), with the exception of *C. angolensis* (Fashing, 2011). Infants begin to participate in play groups with other infants when they are a few months old (Struhsaker & Oates, 1975; Oates, 1977c) and can become independent of their mothers as early as 5.5–7 months (Oates, 1977c; Saj & Sicotte, 2005; Teichroeb & Sicotte, 2008a).

At Kibale, 32% of *P. tephrosceles* individuals die before they reach two years of age (Struhsaker & Pope, 1991). At Gombe, Tanzania, the probability of dying in any six-month period during the first two years of life ranges from 0.14 to 0.31 (Fourrier et al., 2008); predation by chimpanzees is a major source of mortality at Gombe (Fourrier et al., 2008). Of the *C. vellerosus* infants at Boabeng-Fiema, Ghana, 54% die within their first year (Teichroeb & Sicotte, 2008a). One leading cause that explains 29% of infant deaths in this black-and-white colobus population, is attacks by extra-group males, new immigrant males, and/or new alpha males (Sicotte & MacIntosh, 2004; Saj & Sicotte, 2005; Sicotte et al., 2007; Teichroeb & Sicotte, 2008a; Teichroeb et al., 2012). The majority of infants do not survive these attacks, although they are defended by both adult male and female group members (Saj & Sicotte, 2005; Saj et al., 2007; Teichroeb & Sicotte, 2008a). A few cases of males killing infants have been observed in *C. guereza* and *P. tephrosceles*, although it appears to be a very rare occurrence in red colobus (Struhsaker & Leland, 1985; Onderdonk, 2000; Harris & Monfort, 2003; Chapman & Pavelka, 2005). It is not surprising that it is more common to observe black-and-white colobus males killing infants than red colobus, because the former show more frequent male dispersal and groups contain fewer co-residing male kin (refer to discussion of dispersal patterns). Males are not expected to kill infants sired by closely related males as this would likely be detrimental to their inclusive fitness (Hamilton, 1964; but see Schoof et al., 2015). One olive colobus infant that was conceived before the mother immigrated into her current group died of unknown causes, but male olive colobus have not been observed to kill infants (Korstjens & Noë, 2004).

BEHAVIORAL FLEXIBILITY AND CONSERVATION

Over the years, cross-site comparisons and longitudinal studies have revealed an astonishing amount of African colobine behavioral and ecological variation between species, within species, and even within a population. What might the implications of this variation be? It is clear that some colobus taxa have proven to be much more ecologically flexible, diverse, and even resilient than what was imagined prior to extensive field research. As a result, it is now not surprising that certain populations have been able to persist in disturbed habitats, at least to a certain degree. This is clearly demonstrated in their conservation status (see Table 13.1), especially in relation to the two biggest colobus monkey extinction threats—forest degradation and hunting. Generally, the species that are able to adapt to forest degradation are faring better than others. Conversely, those that are not able to adapt are more threatened, but none of the colobus species fare well in the face of high levels of forest disturbance, particularly when combined with human hunting.

Colobus guereza may be the most well-known of the colobus monkeys due to its beautiful pelage, its presence in zoos, studies conducted across different sites, and the longitudinal research at Kibale National Park, which included some of the first detailed African colobine behavioral research by J. F. Oates in the 1970s. However, it is now apparent that the ability of this species to exist in heavily disturbed forest is exceptional for a colobus monkey, which is one of the reasons why it is the only African colobine species listed as Least Concern by the International Union for the Conservation of Nature (IUCN) and not currently threatened with extinction, although it is cited as decreasing in numbers (IUCN, 2020). Likewise, the ability of the olive colobus (*Procolobus verus*) to persist in highly seasonal and heavily degraded forest may be the reason it is the only West African colobine species that is not endangered; however, it is listed by IUCN as Vulnerable and decreasing (IUCN, 2020). Sadly, except for *C. guereza*, all colobus species are listed by IUCN (2020) as Vulnerable, Endangered, or Critically Endangered. Of particular concern are the red colobus, one of the most endangered primate taxa in the world (Struhsaker, 2005). While these monkeys are capable of reaching high levels of local abundance, thus somewhat masking their threatened status, some of the species have extremely limited distributions. The Niger Delta red colobus (*Piliocolobus epieni*), Tana River red colobus (*P. rufomitratus*), Preuss's red colobus (*P. preussi*), and Pennant's red colobus (*P. pennantii*) are all in extremely precarious states, and their persistence into the future will require better protection.

It is not a coincidence that most of the critically endangered red colobus species reside in West Africa. While behavioral flexibility has aided some colobus monkey populations occupying moderately degraded forest, this flexibility shows its greatest limits in aiding species survival in the face of human hunting. In this region, severe habitat loss is compounded by high levels of unsustainable human hunting, and colobus monkeys in general are known to be easy prey due to their large group sizes, loud vocalizations, and poor anti-predator strategies, especially when compared to their cercopithecine and ape cousins. These factors have driven Miss Waldron's red colobus (*P. waldroni*) to likely extinction, as the only evidence of its continued existence for the last several decades has been a tail, a skin, a photo of a carcass provided to scientists conducting surveys in the region, and reported vocalizations (McGraw & Oates, 2002; McGraw, 2005). Likewise, the white-thighed black-and-white colobus (*C. vellerosus*) is critically endangered due to similar factors, with the only secure population residing in the Boabeng-Fiema Monkey Sanctuary in Ghana. This population was on the brink of local extinction when the communities of Boabeng and Fiema initiated a conservation program in the 1970s, and today the habitat that the villagers set aside for the project supports an increasing and expanding population (e.g., Fargey, 1992; Saj et al., 2006; Wong & Sicotte, 2006; Kankam et al., 2010) that seems to be adapting well to a modestly sized forest. This is a good example, along with Temminck's red colobus (*P. temminckii*), Zanzibar red colobus (*P. kirkii*), and ashy red colobus (*P. tephrosceles*), of how behavioral flexibility can help species adapt to changing conditions as long as their forests are well protected from human hunting and continued degradation.

Ultimately, much needs to be done to secure the future of colobus monkeys in Africa. Continued surveys are required to monitor population trends and establish new baselines for future comparisons. Outreach efforts are necessary to educate foreign and local communities about the status of these monkeys and the threats that they face, especially as exploitation of forest resources becomes increasingly commercial in a globalized marketplace. Long-term research is required to determine if the behavioral variation we are seeing in various colobus species is indeed a sign of adaptation or worrying signals that will lead to declining individual health and eventual population collapse in the future. Lastly, successful colobus conservation will not happen without local community involvement and an increase in the political will of local and national governments. Without these actions, the survival of the majority of colobus monkey species across Africa looks increasingly bleak.

SUMMARY

For a group of specialized arboreal folivores, the African colobines are much more diverse and complex than one might first assume. It is now clear that initial field studies, particularly at Kibale National Forest in Uganda, likely skewed perceptions of these animals, causing assumptions of cohesiveness regarding their patterns of arboreality, folivory, and sociality. In reality, the behavioral ecology of the Kibale populations of red colobus and black-and-white colobus are not necessarily representative of all colobus monkeys; further, longitudinal studies now show that those early cross-sectional results are not generalizable over time. This variation across time and locations highlights that there are few fixed or species-typical patterns, but we nevertheless attempt to summarize the most common patterns here in terms of their ecology, activity budget, sociality, reproduction, and infant development.

The African colobines occur in different habitat types in the forested regions of the African tropics (Tables 13.1, 13.2). Population density is relatively low for most populations, but they can occur at high densities in high-quality habitats or in areas with high levels of habitat loss, although this latter case is likely unsustainable and due to population compression. Indeed, most populations of African colobines are in decline, and the population status is tightly linked to habitat degradation, habitat size, distance to other forest fragments, and human hunting. While some populations appear to cope with close proximity to humans and human-modified habitats by locomoting on the ground and altering their feeding behavior, the limits to these adaptations and long-term consequences are unclear.

In most populations, individuals consume large amounts of leaves or seeds (Table 13.3), which is expected based on their morphological and physiological adaptions for folivory and granivory. Despite these adaptations, some populations of red colobus and black-and-white colobus incorporate a large amount of fruit in their diet. Some of the variation in a species' diet over time and space may be linked to availability and nutritional content of specific food items.

Most African colobines have small home ranges (<100 ha) and short daily path lengths (<2,000 m) in comparison to sympatric cercopithecines, and may deal with high population densities and/or low food abundance either by increasing or decreasing ranging and home range overlap (Table 13.4). A few populations of black-and-white colobus range over larger areas, and some are even considered seminomadic or may occasionally migrate. Many colobus monkeys spend a high proportion of time resting (Table 13.5), which is likely linked to the relative indigestibility of their foods and their energy-conserving lifestyle. The variation in activity budgets across and within populations may be due to the occurrence of mixed-species associations, population density, group size, diet, feeding competition, and/or food availability.

African colobines live in multi-female/uni-male or multi-male/multi-female groups, although all-male bands and even multitiered societies occur. Group cohesion ranges from spatially very cohesive in black-and-white colobus groups, to more spatially diffuse in red colobus groups, to fission-fusion societies in the olive colobus. Most African colobine populations show at least occasional dispersal by both sexes, although many show either a male or a female bias in distance or frequency of dispersal. Dispersal patterns are often, but not perfectly, associated with social relationships, as the more philopatric sex tends to form stronger same-sex bonds. Although detectable dominance hierarchies have been reported from a few populations, displacements and coalitionary aggression are relatively rare compared to many cercopithecines. It is more common for aggression to be directed to new or potential immigrants and opposing groups, and males are often the main participants in these encounters.

Olive colobus monkeys are seasonal breeders while red colobus and black-and-white colobus are aseasonal breeders but with a birth peak in some seasonal habitats (Table 13.6). Infants typically have a conspicuous coat pattern distinct from adults, and they transition to the adult coat around 3 months of age and nurse for approximately 12 months. Infant development may be sped up during high-infanticide conditions, but this is based on one black-and-white colobus population in which infanticide is a major source of infant mortality. However, access to food and other factors also likely affect speed of infant development. Chimpanzee predation is a leading source of infant red colobus mortality in some populations, which Fourrier et al. (2008) argue is linked to disrupted ecosystems.

The variation we now see in diet, ranging, activity budget, group cohesion, group size, dispersal patterns, and rates of social interactions within colobus monkey populations may allow them to persist under a range of environmental conditions. For example, grouping patterns appear to be linked to food abundance in some cases, which leaves open the possibility that the African colobines can, at least to some extent, respond to environmental changes leading to lowered food abundance by reducing group cohesion or group size. Also, despite dispersal being a relatively phylogenetically constrained trait in other primates (Di Fiore & Rendall, 1994), there is a remarkable degree of variation in dispersal patterns within and between African colobine populations that could improve their ability to persist in dynamic environments. The ability to recolonize forest patches relatively quickly after hunting is eliminated (Boabeng-Fiema black-and-white colobus: Wong & Sicotte, 2006; Kankam et al., 2010) or logging has ceased (Kibale red colobus: Chapman et al., 2018) could partly be due to both sexes showing at least occasional dispersal. If dispersal and other behavioral traits are not constrained by phylogeny, colobus monkeys may cope with changing environments by modifying their behaviors. However, the consequences of changing various aspects of behavior need to be furthered assessed in order to understand whether these changes are adaptive and sustainable in the long-term, or if they will eventually lead to a population collapse (Berger-Tal et al., 2011). This is particularly pertinent given that nearly all colobus monkey species are threatened with extinction, a situation that will not improve unless researchers cooperate with local people, conservationists, and government authorities to incorporate new knowledge into more effective management plans.

These beautiful and interesting monkeys remain among the most endangered primates in the world, and there is still much unknown about their taxonomy, physiology, ecology, sociality, and reproduction. Continued monitoring of long-term study populations, as well as the initiation of new longitudinal studies will help us determine the causes and consequences of the variation that exists within and between species of African colobines.

REFERENCES CITED—CHAPTER 13

Anderson, J., Cowlishaw, G., & Rowcliffe, J. M. (2007). Effects of forest fragmentation on the abundance of *Colobus angolensis palliatus* in Kenya's coastal forests. *International Journal of Primatology, 28*, 637-655.

Araldi, A., Barelli, C., Hodges, K., & Rovero, F. (2014). Density estimation of the endangered Udzungwa red colobus (*Procolobus gordonorum*) and other arboreal primates in the Udzungwa Mountains using systematic distance sampling. *International Journal of Primatology, 35*, 941-956.

Arseneau-Robar, T. J. M., Changasi, A. H., Turner, E., & Teichroeb, J. A. (2020). Diet and activity budget in *Colobus angolensis ruwenzorii* at Nabugabo, Uganda: Are they energy maximizers? *Folia Primatologica*, 1-20.

Arseneau-Robar, T. J. M., Joyce, M. M., Stead, S. M., & Teichroeb. J. A. (2018). Proximity and grooming patterns reveal opposite-sex bonding in Rwenzori Angolan colobus monkeys (*Colobus angolensis ruwenzorii*). *Primates*, 59(3), 267-279.

Bădescu, J., Sicotte, P., Ting, N., & Wikberg, E. C. (2014). Female parity, maternal kinship, infant age and sex influence natal attraction and infant handling in a wild colobine (*Colobus vellerosus*). *American Journal of Primatology*, 77(4), 376-387. DOI: 10.1002/ajp.22353.

Bădescu, I., Wikberg, E. C., MacDonald, L. J., Fox, S. A., Vayro, J., & Sicotte, P. (2016). Infanticide pressure accelerates infant development in a wild primate. *Animal Behaviour*, 114, 231-239.

Barelli, C., Gallardo Palacios, J. F., & Rovero, F. (2014). Variation in primate abundance along an elevational gradient in the Udzungwa Mountains of Tanzania. In N. Grow, S. Gursky-Doyen, & A. Krzton (Eds.), *High altitude primates. Developments in primatology: Progress and prospects.* Springer.

Bauchop, T. (1977). Foregut fermentation. In R. T. J. Clarke, & T. Bauchop (Eds.), *Microbial ecology of the gut* (pp. 223-250). Academic Press.

Berger-Tal, O., Polak, T., Oron, A., Lubin, Y., Kotler, B. P., & Saltz, D. (2011). Integrating animal behavior and conservation biology: A conceptual framework. *Behavioral Ecology*, 22, 236-239.

Bocian, C. M. (1997). *Niche separation of black-and-white colobus monkeys* (Colobus angolensis *and* C. guereza) *in the Ituri Forest.* [Doctoral dissertation, City University of New York].

Booth, A. H. (1956). The distribution of primates in the Gold Coast. *Journal of the West African Science Association*, 2, 122-133.

Booth, A. H. (1957). Observations on the natural history of the olive colobus monkey, *Procolobus verus* (van Beneden). *Proceedings of the Zoological Society of London*, 129, 421-430.

Brent, L. J. N., Teichroeb, J. A., & Sicotte, P. (2008). Preliminary assessment of natal attraction and infant handling in wild *Colobus vellerosus*. *American Journal of Primatology*, 70, 101-105.

Brugiere, D., Gautier, J. P., Moungazi, A., & Gautier-Hion, A. (2002). Primate diet and biomass in relation to vegetation composition and fruiting phenology in a rain forest in Gabon. *International Journal of Primatology*, 23, 999-1024.

Bshary, R., & Noë, R. (1997). Red colobus and diana monkeys provide mutual protection against predators. *Animal Behaviour*, 54, 1461-1474.

Chapman, C. A., Balcomb, S. R., Gillespie, T. R., Skorupa, J. P., & Struhsaker, T. T. (2000). Long-term effects of logging on African primate communities: A 28-year comparison from Kibale National Park, Uganda. *Conservation Biology*, 14, 207-217.

Chapman, C. A., Bortolamiol, S., Matsuda, I., Omeja, P. A., Paim, F. P., Reyna-Hurtado, R., Sengupta, R., & Valenta, K. (2018). Primate population dynamics: Variation in abundance over space and time. *Biodiversity and Conservation*, 27(5), 1221-1238.

Chapman, C. A., & Chapman, L. J. (2002). Foraging challenges of red colobus monkeys: Influence of nutrients and secondary compounds. *Comparative Biochemistry and Physiology*, 133(3), 861-875.

Chapman, C. A., Chapman, L. J., & Gillespie, T. R. (2002). Scale issues in the study of primate foraging: Red colobus of Kibale National Park. *American Journal of Primatology*, 117, 349-363.

Chapman, C. A., Ghai, R. R., Reyna-Hurtado, R., Jacob, A. L., Koojo, S. M., Rothman, J. M., Twinomugisha, D., Wasserman, M. D., & Goldberg, T. L. (2013). Going, going, gone: A 15-year history of the decline in abundance of primates in forest fragments. In L. K. Marsh, & C. A. Chapman (Eds.), *Primates in fragments: Complexity and resilience* (pp. 89-104). Springer.

Chapman, C. A., & Pavelka, M. S. M. (2005). Group size in folivorous primates: Ecological constraints and the possible influence of social factors. *Primates*, 46, 1-9.

Chapman, C. A., Speirs, M. L., Gillespie, T. R., Holland, T., & Austad, K. M. (2006a). Life on the edge: Gastrointestinal parasites from the forest edge and interior primate groups. *American Journal of Primatology*, 68, 397-409.

Chapman, C. A., Struhsaker, T. T., Skorupa, J. P., Snaith, T. V., & Rothman, J. M. (2010). Understanding long-term primate community dynamics: Implications of forest change. *Ecological Applications*, 20, 179-191.

Chapman, C. A., Walker, S., & Lefebvre, L. (1990). Reproductive strategies of primates: The influence of body size and diet on litter size. *Primates*, 31, 1-13.

Chapman, C. A., Wasserman, M. D., Gillespie, T. R., Spiers, M. L., Lawes, M. J., Saj, T. L., & Ziegler, T. E. (2006b). Do food availability, parasitism, and stress have synergistic effects on red colobus populations living in forest fragments? *American Journal of Primatology*, 131, 525-534.

Clutton-Brock, T. H. (1974). Activity patterns of red colobus (*Colobus badius tephrosceles*). *Folia Primatologica*, 21, 161-187.

Clutton-Brock, T. H. (1975a). Feeding behaviour of red colobus and black-and-white colobus in East Africa. *Folia Primatologica*, 23, 706-722.

Clutton-Brock, T. H. (1975b). Ranging behaviour of red colobus (*Colobus badius tephrosceles*) in the Gombe National Park. *Animal Behaviour*, 23, 706-722.

Colyn, M. (1991) L'importance zoogeographique du basin du fleuve Zaire pour la speciation: Le cas des primates simians. *Belgique Annales Sciences Zoologiques, Musee Royal de l'Afrique Central Tervuren*, 264, 1-250.

Crotty, A. M. M. (2016). *Influences of lactation length and the timing of weaning events in Colobus vellerosus.* [Master's thesis, University of Calgary].

Dasilva, G. L. (1989). *The ecology of the western black and white colobus* (Colobus polykomos polykomos *Zimmerman 1780*) *on a riverine island in south-eastern Sierra Leone.* [Doctoral dissertation, University of Oxford].

Dasilva, G. L. (1992). The western black-and-white colobus as a low-energy strategist: Activity budgets, energy expenditure and energy intake. *Journal of Animal Ecology, 61*, 79-91.

Dasilva, G. L. (1994). Diet of *Colobus polykomos* on Tiwai Island: Selection of food in relation to its seasonal abundance and nutritional quality. *International Journal of Primatology, 15*, 655-679.

Davies, A. G. (1994). Colobine populations. In A. G. Davies, & J. F. Oates (Eds.), *Colobine monkeys: Their ecology, behaviour and evolution* (pp. 285-310). Cambridge University Press.

Davies, A. G., Oates, J. F., & Dasilva, G. L. (1999). Patterns of frugivory in three West African colobine monkeys. *International Journal of Primatology, 20*, 327-357.

Decker, B. S. (1994). Effects of habitat disturbance on the behavioral ecology and demographics of the Tana River red colobus (*Colobus badius rufomitratus*). *International Journal of Primatology, 15*, 703-737.

Delson, E. 1994. Evolutionary history of the colobine monkeys in paleoenvironmental perspective. In A. G. Davies, & J. F. Oates (Eds.), *Colobine monkeys: Their ecology, behaviour and evolution* (pp. 11-43). Cambridge University Press.

Delson, E., Terranova, C. J., Jungers, W. L., Sargis, E. J., & Jablonski, N. G. (2000). Body mass in Cercopithecidae (Primates, Mammalia): Estimation and scaling in extinct and extant taxa. *Anthropological Papers of the American Museum of Natural History, 83*.

Di Fiore, A., & Rendall, D. (1994). Evolution of social organization: A reappraisal for primates by using phylogenetic methods. *Proceedings of the National Academy of Sciences of the United States of America, 91*, 9941-9945.

Djego-Djossou, S., Djego, J. G., Mensah, G. A., Huynen, M.-C., & Sinsin, B. (2014). Distribution du colobe vert olive, *Procolobus verus*, au Bénin et menaces pesant sur sa conservation. *African Primates, 9*, 23-34.

Djego-Djossou, S., Koné, I., Fandohan, A. B., Djègo, J. G., Huynen, M. C., & Sinsin, B. (2015). Habitat use by white-thighed colobus in the Kikélé Sacred Forest: Activity budget, feeding ecology and selection of sleeping trees. *Primate Conservation, 29*, 97-105.

Dorst, J., & Dandelot, P. (1972). *A field guide to the larger mammals of Africa.* Collins.

Dunbar, R. I. M. (1987). Habitat quality, population dynamics, and group composition in colobus monkeys (*Colobus guereza*). *International Journal of Primatology, 8*, 299-329.

Dunbar, R. I. M. (2018). Social structure as a strategy to mitigate the costs of group living: A comparison of gelada and guereza monkeys. *Animal Behaviour, 136*, 53-64.

Dunbar, R. I. M., & Dunbar, E. P. (1974). Ecology and population dynamics of *Colobus guereza* in Ethiopia. *Folia Primatologica, 21*, 36-60.

Dunham, N. T. (2017). Feeding ecology and dietary flexibility of *Colobus angolensis palliates* in relation to habitat disturbance. *International Journal of Primatology, 38*, 553-571.

Dunham, N. T., Kane, E. E., & McGraw, W. S. (2017). Humeral correlates of forelimb elevation in four West African cercopithecid monkeys. *American Journal of Physical Anthropology, 162*, 337-349.

Dunham, N. T., & Lambert, A. (2016). The role of leaf toughness on foraging efficiency in Angola black and white colobus monkeys (*Colobus angolensis palliatus*). *American Journal of Physical Anthropology, 161*(2), 343-354.

Fargey, P. J. (1992). Boabeng-Fiema monkey sanctuary—an example of traditional conservation in Ghana. *Oryx, 26*, 151-156.

Fashing, P. J. (2001a). Feeding ecology of guereza in the Kakamega Forest, Kenya: The importance of Moraceae fruit in their diet. *International Journal of Primatology, 22*, 579-609.

Fashing, P. J.(2001b). Activity and ranging patterns of guerezas in the Kakamega Forest: Intergroup variation and implications for intragroup feeding competition. *International Journal of Primatology, 22*, 549-577.

Fashing, P. J. (2001c). Male and female strategies during intergroup encounters in guereza (*Colobus guereza*): Evidence for resource defense mediated through males and a comparison with other primates. *Behavioral Ecology and Sociobiology, 50*, 219-230.

Fashing, P. J. (2002). Population status of black and white colobus monkeys (*Colobus guereza*) in the Kakamega Forest, Kenya: Are they really on the decline? *African Zoology, 37*, 119-126.

Fashing, P. J. (2011). African colobine monkeys: Their behavior, ecology, and conservation. In C. J. Campbell, A. Fuentes, K. C. MacKinnon, S. K. Bearder, & R. M. Stumpf (Eds.), *Primates in perspective* (2nd Ed., pp. 203-229). Oxford University Press.

Fashing, P. J., & Cords, M. (2000). Diurnal primate density and biomass in the Kakamega Forest: An evaluation of census methods and a comparison with other forests. *American Journal of Primatology, 50*, 139-152.

Fashing, P. J., Mulindahabi, F., Gakima, J.-B., Masozera, M., Mununura, I., Plumptre, A. J., & Nguyen, N. (2007). Activity and ranging patterns of *Colobus angolensis ruwenzorii* in Nyungwe Forest, Rwanda: Possible costs of large group size. *International Journal of Primatology, 28*, 529-550.

Fimbel, C., Vedder, A., Dierenfeld, E., & Mulindahabi, F. (2001). An ecological basis for large group size in *Colobus angolensis* in the Nyungwe Forest, Rwanda. *African Journal of Ecology, 39*, 83-92.

Fleagle, J. G. (1988). *Primate adaptations and evolution.* Academic Press.

Fleury, M. C., & Gautier-Hion, A. (1999). Seminomadic ranging in a population of black colobus (*Colobus satanas*) in Gabon and its ecological correlates. *International Journal of Primatology, 20*, 491-509.

Fourrier, M., Sussman, R. W., Kippen, R., & Childs, G. (2008). Demographic modeling of a predator-prey system and its implication for the Gombe population of *Procolobus rufomitratus tephrosceles*. *International Journal of Primatology, 29*, 497-508.

Galat, G., & Galat-Luong, A. (1979). Quelques observations sur l'ecologie de *Colobus pennanti oustaleti* en Empire Centrafrican. *Mammalia, 43,* 309-312.

Galat-Luong, A., & Galat, G. (2005). Conservation and survival adaptations of temminck's red colobus (*Procolobus badius temminckii*), in Senegal. *International Journal of Primatology, 26,* 585-603.

Gatinot, B. L. (1977). Le règime alimentaire du colobe bai au Sènègal. *Mammalia, 41,* 373-402.

Gatinot, B. L. (1978). Characteristics of the diet of the West African red colobus. In D. J. Chivers, & J. Herbert (Eds.), *Recent advances in primatology* (Vol. 1, pp. 253-255). Academic Press.

Gautier-Hion, A., Gautier, J.-P., & Moungazi, A. (1997). Do black colobus in mixed-species groups benefit from increased foraging efficiency? *Comptes Rendus, 320,* 67-71.

Gebo, D. L., & Chapman, C. A. (1995). Habitat, annual, and seasonal effects on positional behavior in red colobus monkeys. *American Journal of Physical Anthropology, 96,* 73-82.

Gebo, D. L., Chapman, C. A., Chapman, L. J., & Lambert, J. (1994). Locomotor response to predator threat in red colobus monkeys. *Primates, 35,* 219-223.

Gilbert, C. C., Goble, E. D., & Hill, A. (2010). Miocene Cercopithecoidea from the Tugen Hills, Kenya. *Journal of Human Evolution, 59,* 465-483.

Gillespie, T. R., & Chapman, C. A. (2008). Forest fragmentation, the decline of an endangered primate, and changes in host-parasite interactions relative to an unfragmented forest. *American Journal of Primatology, 70,* 222-230.

Gillespie, T. R., Chapman, C. A., & Greiner, E. C. (2005). Effects of logging on gastrointestinal parasite infections and infection risk in African primates. *Journal of Applied Ecology, 42,* 699-707.

Gogarten, J. F., Bonnell, R. T., Brown, L. M., Campenni, M., Wasserman, M. D., & Chapman, C. A. (2014). Increasing group size alters behavior of a folivorous primate. *International Journal of Primatology, 35,* 590-608.

Gogarten, J. F, Jacob, A. L., Ghai, R. R., Rothman, J. M., Twinomugisha, D., Wasserman, M. D., & Chapman, C. A. (2015). Group size dynamics over 15+ years in an African forest primate community. *Biotropica, 47,* 101-112.

Gonedelé, S., Bitty, A., Gnangbé, F., Bené, J. C., Koné, I., & Zinner, D. (2010). Conservation status of Geoffroy's pied colobus monkey *Colobus vellerosus* Geoffroy 1834 has dramatically declined in Côte D'Ivoire. *African Primates, 7,* 19-26.

Gonzalez-Kirchner, J. P. (1997). Behavioural ecology of two sympatric colobines on Bioko Island, Equatorial Guinea. *Folia Zoologica, 46,* 97-104.

Gonzalez-Kirchner, J. P., & Sainz de la Maza, M. (1993). Primates as a food source for the native human population of Equatorial Guinea. *Tropical Biodiversity, 1,* 163-168.

Goodall, J. (1965). Chimpanzees of the Gombe Stream Reserve. In I. DeVore (Ed.), *Primate behavior: Field studies of monkeys and apes* (pp. 425-473). Holt, Rinehart, and Winston.

Groves, C. P. (1973). Notes on the ecology and behaviour of the Angola colobus (*Colobus angolensis* P. L. Sclater 1860) in N.E. Tanzania. *Folia Primatologica, 20,* 12-26.

Groves, C. P. (2007). The taxonomic diversity of the Colobinae of Africa. *Journal of Anthropological Sciences, 85,* 7-34.

Hamilton, W. J. (1964). The genetical evolution of social behavior I and II. *Journal of Theoretical Biology, 7,* 1-52.

Harris, T. R. (2005). *Roaring, intergroup aggression, and feeding competition in black and white colobus monkeys* (Colobus guereza) *at Kanyawara, Kibale National Park, Uganda.* [Doctoral dissertation, Yale University].

Harris, T. R. (2006). Between-group contest competition for food in a highly folivorous population of black and white colobus monkeys (*Colobus guereza*). *Behavioral Ecology and Sociobiology, 61*(2), 317-329.

Harris, T. R. (2010). Multiple resource values and fighting ability measures influence intergroup conflict in guerezas (*Colobus guereza*). *Animal Behaviour, 79*(1), 89-98.

Harris, T. R., Caillaud, D., Chapman, C. A., & Vigilant, L. (2009). Neither genetic nor observational data alone are sufficient for understanding sex-biased dispersal in a social-group-living species. *Molecular Ecology, 18,* 1777-1790.

Harris, T. R., & Chapman, C. A. (2007). Variation in diet and ranging of black-and-white colobus monkeys in Kibale National Park, Uganda. *Primates, 48,* 208-221.

Harris, T. R., Chapman, C. A., & Monfort, S. L. (2010). Small folivorous primate groups exhibit behavioral and physiological effects of food scarcity. *Behavioral* Ecology, 21, 46-56.

Harris, T. R., & Monfort, S. L. (2003). Behavioral and endocrine dynamics associated with infanticide in a black and white colobus monkey (*Colobus guereza*). *American Journal of Primatology, 61,* 135-142.

Harris, T. R., & Monfort, S. L. (2006). Mating behavior and endocrine profiles of wild black-and-white colobus monkeys (*Colobus guereza*): Toward an understanding of their life history and mating system. *American Journal of Primatology, 68,* 383-396.

Harrison, M. J. S. (1986). Feeding ecology of black colobus, *Colobus satanas,* in central Gabon. In J. G. Else, & P. C. Lee (Eds.), *Primate ecology and conservation* (pp. 31-37). Cambridge University Press.

Harrison, M. J. S., Hladik, C. M. (1986). Un primate granivore: Le colobe noir dans la forêt du Gabon: Potentialité d'évolution du comportement alimentaire. *Terre et Vie, 41,* 281-298.

Hart, J. A., Katembo, M., & Punga, K. (1996). Diet, prey selection and ecological relations of leopard and golden cat in the Ituri Forest, Zaire. *African Journal of Ecology, 34,* 364-379.

Hill, W. C. O. (1952). On the external and visceral anatomy of the olive colobus monkey (*Procolobus verus*). *Proceedings of the Zoological Society of London, 122,* 127e186.

Hladik, A. (1978). Phenology of leaf production in the rain forest of Gabon: Distribution and composition of food for folivores. In G. G. Montgomery (Ed.), *The ecology of arboreal folivores* (pp. 51-71). Smithsonian Institution Press.

Holenweg, A. K., Noë, R., & Schabel, M. (1996). Waser's gas model applied to associations between red colobus and diana monkeys in the Taï National Park, Ivory Coast. *Folia Primatologica, 67,* 125-136.

Höner, O. P., Leumann, L., & Noë, R. (1997). Dyadic associations of red colobus and diana monkey groups in the Taï National Park, Ivory Coast. *Primates, 38,* 281-291.

IUCN. (2020). *International Union for the Conservation of Nature Red List of Threatened Species 2020-3.* www .iucnredlist.org.

Kanga, E. M. (2001). Survey of black and white colobus monkeys (*Colobus angolensis palliates*) in Shimba Hills National Reserve and Maluganji Sanctuary, Kenya. *American Journal of Primatology, 25,* 8-9.

Kanga, E. M., & Heidi, C. M. (1999/2000). Survey of the Angolan black-and-white colobus monkeys *Colobus angolensis palliates* in the Diani Forests, Kenya. *African Primates, 4,* 50-54.

Kankam, B. O., Saj, T. L., & Sicotte, P. (2010). How to measure "success" in community-based conservation projects: The case of the Boabeng-Fiema Monkey Sanctuary in Ghana. In K. P. Puplampu, & W. J. Tettey (Eds.), *The political economy of development in Ghana: Critiques of theory and practice* (pp. 115-141). Woeli Publishers.

Kankam, B. O., & Sicotte, P. (2013). The effect of forest fragment characteristics on abundance of *Colobus vellerosus* in the forest-savanna transition zone of Ghana. *Folia Primatologica, 84,* 74-86.

Kingdon, J. (1971). *East African mammals, Vol. 1.* Academic Press.

Korstjens, A. H. (2001). *The mob, the secret sorority, and the phantoms: An analysis of the socio-ecological strategies of the three colobines at Taï.* [Doctoral dissertation, University of Utrecht].

Korstjens, A. H., Bergmann, K., Deffernez, C., Krebs, M., Nijssen, E. C., van Oirschot, B. A. M., Paukert, C., & Schippers, E. Ph. (2007). How small-scale differences in food competition lead to different social systems in three closely sympatric colobines. In W. S. McGraw, K. Zuberbühler, & R. Noë (Eds.), *Monkeys of the Taï Forest: An African primate community* (pp. 72-108). Cambridge University Press.

Korstjens, A. H., & Dunbar, R. I. M. (2007). Time constraints limit group sizes and distribution in red and black-and-white colobus monkeys. *International Journal of Primatology, 28,* 551-575.

Korstjens, A. H., Nijssen, E. C., & Noë, R. (2005). Intergroup relationships in western black-and-white colobus, *Colobus polykomos polykomos. International Journal of Primatology, 26,* 1267-1289.

Korstjens, A. H., & Noë, R. (2004). Mating system of an exceptional primate, the olive colobus (*Procolobus verus*). *American Journal of Primatology, 62,* 261-273.

Korstjens, A. H., & Schippers, E. Ph. (2003). Dispersal patterns among olive colobus in Taï National Park. *International Journal of Primatology, 24,* 515-539.

Korstjens, A. H., Sterck, E. H. M., & Noë, R. (2002). How adaptive or phylogenetically inert is primate social behaviour? A test with two sympatric colobines. *Behavior, 139,* 203-225.

Krüger, O., Affeldt, E., Brackmann, M., & Milhahn, K. (1998). Group size and composition of *Colobus guereza* in Kyambura Gorge, southwest Uganda, in relation to chimpanzee activity. *International Journal of Primatology, 19,* 287-297.

Lambert, J. E., & Fellner, V. (2012). In vitro fermentation of dietary carbohydrates consumed by African apes and monkeys: Preliminary results for interpreting microbial and digestive strategy. *International Journal of Primatology, 33,* 263-281.

Leskes, A., & Acheson, N. H. (1971). Social organization of a free-ranging troop of black and white colobus monkeys (*Colobus abyssinicus*). In H. Kummer (Ed.), *Proceedings of the Third International Congress of Primatology, Zürich 1970, Vol. 3.* Karger.

Lwanga, J.S., Struhsaker, T. T., Struhsaker, P. J., Butynski, T. M., & Mitani, J. C. (2011). Primate population dynamics over 32.9 years at Ngogo, Kibale National Park. *American Journal of Primatology, 73,* 997-1011.

Maisels, F., Gautier-Hion, A., & Gautier, J.-P. (1994). Diets of two sympatric colobines in Zaire: More evidence on seed-eating in forests on poor soils. *International Journal of Primatology, 15,* 681-701.

Malbrant, R., & Maclatchy, A. (1949). Faune de l'équateur African Francais. In *Encyclopedie biologique, mammifères, Vol. 2.* Paul Le Chevaliers.

Mammides, C., Cords, M., & Peters, M. K. (2009). Effects of habitat disturbance and food supply on population densities of three primate species in the Kakamega Forest, Kenya. *African Journal of Ecology, 47,* 87-96.

Marler, P. (1969). *Colobus guereza*: Territoriality and group composition. *Science, 163,* 93-95.

Marsh, C. W. (1978). *Ecology and social organization of the Tana River red colobus* (Colobus badius rufomitratus*).* [Doctoral dissertation, University of Bristol].

Marsh, C. W. (1979a). Comparative aspects of social organization in the Tana River red colobus, *Colobus badius rufomitratus. Zeitschrift für Tierpsychologie, 51,* 337-362.

Marsh, C. W. (1979b). Female transference and mate choice among Tana River red colobus. *Nature, 281,* 568-569.

Marsh, C. W. (1981a). Diet choice among red colobus (*Colobus badius rufomitratus*) on the Tana River, Kenya. *Folia Primatologica, 35,* 147-178.

Marsh, C. W. (1981b). Time budget of Tana River red colobus. *Folia Primatologica, 35,* 30-50.

Marsh, C. W. (1981c). Ranging behaviour and its relation to diet selection in Tana River red colobus (*Colobus badius rufomitratus*). *Journal of Zoology, 195,* 473-492.

Marshall, A. R., Topp-Jørgensen, J. E., Brink, H., & Fanning, E. (2005). Monkey abundance and social structure in two high-elevation forest reserves in the Udzungwa Mountains of Tanzania. *International Journal of Primatology, 26,* 127-145.

Matsuda, I., Chapman, C. A., & Clauss, M. (2019). Colobine forestomach anatomy and diet. *Journal of Morphology, 280,* 1608-1616.

Mbora, D. N. M., & Meikle, D. B. (2004). Forest fragmentation and the distribution, abundance and conservation of the Tana River red colobus (*Procolobus rufomitratus*). *Biological Conservation, 118,* 67-77.

McGraw, W. S. (1998a). Comparative locomotion and habitat use of six monkeys in the Tai Forest, Ivory Coast. *American Journal of Physical Anthropology, 105,* 493-510.

McGraw, W. S. (1998b). Posture and support use of Old World monkeys (Cercipithecidae): The influence of foraging strategies, activity patterns, and the spatial distribution of preferred food items. *American Journal of Primatology, 46,* 229-250.

McGraw, W. S. (2005). Update on the search for Miss Waldron's red colobus monkey. *International Journal of Primatology, 26,* 605-619.

McGraw, W. S., Cooke, C., & Shultz, S. (2006). Primate remains from African crowned eagle (*Stephanoaetus coronatus*) nests in Ivory Coast's Tai Forest: Implications for primate predation and early hominid taphonomy in South Africa. *American Journal of Physical Anthropology, 131,* 151-165.

McGraw, W. S., & Oates, J. F. (2002). Evidence for a surviving population of Miss Waldron's red colobus. *Oryx, 36,* 223-226.

McGraw, W. S., van Casteren, A., Kane, E., Geissler, E., Burrows, B., & Daegling, D. J. (2016). Feeding and oral processing behaviors of two colobine monkeys in Tai Forest, Ivory Coast. *Journal of Human Evolution, 98,* 90-102.

McGraw, W. S., & Zuberbühler, K. (2008). Socioecology, predation, and cognition in a community of West African monkeys. *Evolutionary Anthropology, 17,* 254.

McKey, D. B. (1978). Soils, vegetation, and seed-eating by black colobus monkeys. In G. G. Montgomery (Ed.), *The Ecology of Arboreal Folivores* (pp. 423-437). Smithsonian Institution Press.

McKey, D. B., Gartlan, J. S., Waterman, P. G., & Choo, G. M. (1981). Food selection by black colobus monkeys (*Colobus satanas*) in relation to plant chemistry. *Biological Journal of the Linnean Society, 16,* 115-146.

McKey, D. B., & Waterman, P. G. (1982). Ranging behaviour of a group of black colobus (*Colobus satanas*) in the Douala-Edea Reserve, Cameroon. *Folia Primatologica, 39,* 264-304.

Milich, K. A., Bahr, J. M., Stumpf, R. M., & Chapman, C. A. (2014b). Timing is everything: Expanding the cost of sexual attraction hypothesis. *Animal Behaviour, 88,* 219-224.

Milich, K. A., Stumpf, R. M., Chambers, J. M., & Chapman, C. A. (2014a). Female red colobus monkeys maintain their densities through flexible feeding strategies in logged forests in Kibale National Park, Uganda. *American Journal of Physical Anthropology, 154,* 52-60.

Minhós, T., Sousa, C., Vicente, L. M., & Bruford, M. W. (2015). Kinship and intragroup social dynamics in two sympatric African colobus species. *International Journal of Primatology, 36*(4), 871-886.

Mittermeier, R. A., Ratsimbazafy, J., Rylands, A. B., Williamson, L., Oates, J. F., Mbora, D., Ganzhorn, J. U., Rodríguez-Luna, E., Palacios, E., Heymann, E. W., Cecília, M., Kierulff, M., Yongcheng, L., Supriatna, J., Roos, C., Walker, S., & Aguiar, J. M. (2007). Primates in peril: The world's 25 most endangered primates, 2006-2008. *Primate Conservation, 22,* 1-40.

Mittermeier, R. A., Wallis, J., Rylands, A. B., Ganzhorn, J. U., Oates, J. F., Williamson, E. A., Palacios, E., Heymann, E. W., Cecília, M., Kierulff, M., Yongcheng, L., Supriatna, J., Roos, C., Walker, S., Cortés-Ortiz, L., & Schwitzer, C. (2009). Primates in peril: The world's 25 most endangered primates 2008-2010. *Primate Conservation, 24,* 1-57.

Miyamoto, M. M., Allen, J. M., Gogarten, J. F., & Chapman, C. A. (2013). Microsatellite DNA suggests that group size affects sex-biased dispersal patterns in red colobus monkeys. *American Journal of Primatology, 75,* 478-490.

Moreno-Black, G. S., & Bent, E. F. (1982). Secondary compounds in the diet of *Colobus angolensis*. *African Journal of Ecology, 20,* 29-36.

Mowry, C. B., Decker, B. S., & Shure, D. J. (1996). The role of phytochemistry in dietary choices of Tana River red colobus monkeys (*Procolobus badius rufomitratus*). *International Journal of Primatology, 17,* 63-84.

Mturi, F. A. (1991). *The feeding ecology and behavior of the red colobus monkey* (Colobus badius kirkii*).* [Doctoral dissertation, University of Dar es Salaam].

Mturi, F. A. (1993). Ecology of the Zanzibar red colobus monkey, *Colobus badius kirkii* (Gray, 1968), in comparison with other red colobines. In J. C. Lovett, & S. K. Wasser (Eds.), *Biogeography and ecology of the rain forests of eastern Africa* (pp. 243-266). Cambridge University Press.

Nakatsukasa, M., Mbua, E., Sawada, Y., Sakai, T., Nakaya, H., Yano, W., & Kunimatsu, Y. (2010). Earliest colobine skeletons from Nakali, Kenya. *American Journal of Physical Anthropology, 143,* 365-382.

Napier, J. R., & Napier, P. H. (1967). *A handbook of living primates.* Academic Press.

Newton, P. N., & Dunbar, R. I. M. (1994). Colobine monkey society. In A. G. Davies, & J. F. Oates (Eds.), *Colobine monkeys: Their ecology, behaviour and evolution* (pp. 311-346). Cambridge University Press.

Nishida, T. (1972). A note on the ecology of the red colobus (*Colobus badius tephrosceles*) living in the Mahale Mountains. *Primates, 13,* 57-64.

Noë, R., & Bshary, R. (1997). The formation of red colobus-diana monkey associations under predation pressure from chimpanzees. *Proceedings of the Royal Society of London, 264,* 253-259.

Nowak, K. (2008). Frequent water drinking by Zanzibar red colobus (*Procolobus kirkii*) in a mangrove forest refuge. *American Journal of Primatology*, 70(11), 1081-1092.

Nowak, K., & Lee, P. C. (2011). Demographic structure of Zanzibar red colobus populations in unprotected coral rag and mangrove forests. *International Journal of Primatology*, 32, 24-45.

Nowak, K., & Lee, P. C. (2013a). Status of Zanzibar red colobus and Sykes's monkeys in two coastal forests in 2005. *Primate Conservation*, 27, 65-73.

Nowak, K., & Lee, P. C. (2013b). "Specialist" primates can be flexible in response to habitat alteration. In L. K. Marsh, & C. A. Chapman (Eds.), *Primates in fragments* (pp. 199-211). Springer.

Nunn, C. L. (1999). The evolution of exaggerated sexual swellings in primates and the graded-signal hypothesis. *Animal Behaviour*, 58, 229-246.

Oates, J. F. (1977a). The guereza and man. How man has affected the distribution and abundance of *Colobus guereza* and other black colobus monkeys. In H. S. H. Prince Rainier III, & G. H. Bourne (Eds.), *Primate conservation* (pp. 419-467). Academic Press.

Oates, J. F. (1977b). The guereza and its food. In T. H. Clutton-Brock (Ed.), *Primate ecology: Studies of feeding and ranging behaviour in lemurs, monkeys and apes* (pp. 275-321). Academic Press.

Oates, J. F. (1977c). The social life of the black-and-white colobus monkey, *Colobus guereza*. *Zeitschrift für Tierpsychologie*, 45, 1-60.

Oates, J. F. (1978). Water-plant and soil consumption by guereza monkeys (*Colobus guereza*): A relationship with minerals and toxins in the diet? *Biotropica*, 10, 241-253.

Oates, J. F. (1981). Mapping the distribution of West African rain-forest monkeys: Issues, methods, and preliminary results. *Annals of the New York Academy of Sciences*, 376, 53-64.

Oates, J. F. (1988). The diet of the olive colobus monkey, *Procolobus verus*. *International Journal of Primatology*, 9, 457-478.

Oates, J. F. (1994). The natural history of African colobines. In A. G. Davies, & J. F. Oates (Eds.), *Colobine monkeys: Their ecology, behaviour and evolution* (pp. 75-128). Cambridge University Press.

Oates, J. F., Abedi-Lartey, M., McGraw, W. S., Struhsaker, T. T., & Whitesides, G. H. (2000). Extinction of a West African red colobus monkey. *Conservation Biology*, 14, 1526-1532.

Oates, J. F., & Davies, A. G. (1994). What are the colobines? In A. G. Davies, & J. F. Oates (Eds.), *Colobine monkeys: Their ecology, behaviour and evolution* (pp. 1-9). Cambridge University Press.

Oates, J. F., Davies, A. G., & Delson, E. (1994). The diversity of living colobines. In A. G. Davies, & J. F. Oates (Eds.), *Colobine monkeys: Their ecology, behaviour and evolution* (pp. 45-73). Cambridge University Press.

Oates, J. F., Swain, T., & Zantovska, J. (1977). Secondary compounds and food selection by colobus monkeys. *Biochemical Systematics and Ecology*, 5, 317-321.

Oates, J. F., & Whitesides, G. H. (1990). Association between olive colobus (*Procolobus verus*), diana guenons (*Cercopithecus diana*), and other forest monkeys in Sierra Leone. *American Journal of Primatology*, 21, 129-146.

Oates, J. F., Whitesides, G. H., Davies, A. G., Waterman, P. G., Green, S. M., Dasilva, G. L., & Mole, S. (1990). Determinants of variation in tropical forest primate biomass: New evidence from West Africa. *Ecology*, 71(1), 328-343.

Olson, D. K. (1986). Determining range size for arboreal monkeys: Methods, assumptions, and accuracy. In D. M. Taub, & F. A. King (Eds.), *Current perspectives in primate social dynamics* (pp. 212-227). Van Nostrand Reinhold Company.

Onderdonk, D. A. (2000). Infanticide of a newborn black-and-white colobus monkey (*Colobus guereza*) in Kibale National Park, Uganda. *Primates*, 41, 209-212.

Onderdonk, D. A., & Chapman, C. A. (2000). Coping with forest fragmentation: The primates of Kibale National Park, Uganda. *International Journal of Primatology*, 21, 587-611.

Plumptre, A. J., & Reynolds, V. (1994). The effect of selective logging on the primate populations in the Budongo Forest Reserve, Uganda. *Journal of Applied Ecology*, 31, 631-641.

Pocock, R. I. (1935). The external characters of a female red colobus monkey (*Procolobus badius waldroni*). *Proceedings of the Zoological Society of London*, 105, 939-944.

Rose, M. D. (1978). Feeding and associated positional behavior of black-and-white colobus monkeys (*Colobus guereza*). In G. G. Montgomery (Ed.), *The ecology of arboreal folivores* (pp. 253-262). Smithsonian Institution Press.

Rossie, J. B., Gilbert, C. C., & Hill, A. (2013). Early cercopithecid monkeys from the Tugen Hills, Kenya. *Proceedings of the National Academy of Sciences*, 110(15), 5818-5822.

Rothman, J. M., Chapman, C. A., Struhsaker, T. T., Raubenheimer, D., Twinomugisha, D., & Waterman, P. G. (2015). Long-term declines in nutritional quality of tropical leaves. *Ecology*, 96, 873-878.

Rovero, F., Mtui, A. S., Kitegile, A. S., Nielsen, M. R. (2012). Hunting or habitat degradation? Decline of primate populations in Udzungwa Mountains, Tanzania: An analysis of threats. *Biological Conservation*, 146, 89-96.

Rovero, F., & Struhsaker, T. T. (2007). Vegetative predictors of primate abundance: Utility and limitations of a fine-scale analysis. *American Journal of Primatology*, 69, 1242-1257.

Ryan, A. M., Chapman, C. A., & Rothman, J. M. (2012). How do differences in species and part consumption affect diet nutrient concentrations? A test with red colobus monkeys in Kibale National Park, Uganda. *African Journal of Ecology*, 51, 1-10.

Sabater Pi, J. (1973). Contribution to the ecology of *Colobus polykomos satanas* (Waterhouse, 1838) of Rio Muni, Republic of Equatorial Guinea. *Folia Primatologica*, 19, 193-207.

Saj, T. L., Marteinson, S., Chapman, C., & Sicotte, P. (2007). Controversy over the application of current socioecological models to folivorous primates: *Colobus vellerosus* fits the predictions. *American Journal of Physical Anthropology, 133,* 994-1003.

Saj, T. L., Mather, C., & Sicotte, P. (2006). Traditional taboos in biological conservation: The case of *Colobus vellerosus* at the Boabeng Fiema Monkey Sanctuary, central Ghana. *Social Science Information, 45,* 285-310.

Saj, T. L., & Sicotte, P. (2005). Male takeover in *Colobus vellerosus* at Boabeng-Fiema Monkey Sanctuary, central Gabon. *Primates, 46,* 211-214.

Saj, T. L., & Sicotte, P. (2007a). Predicting the competitive regime of female *Colobus vellerosus* from the distribution of food resources. *International Journal of Primatology, 28,* 315-336.

Saj, T. L., &. Sicotte, P. (2007b). Scramble competition among *Colobus vellerosus* at Boabeng-Fiema, Ghana. *International Journal of Primatology, 28,* 337-355.

Saj, T. L., & Sicotte, P. (2013). Species profile for *Colobus vellerosus*. In J. Kingdon, D. Happold, & T. Butynski (Eds.), *Mammals of Africa* (pp. 109-111). Bloomsbury Publishing.

Saj, T. L., Teichroeb, S. A., & Sicotte, P. (2005). The population status and habitat quality of the ursine colobus (*Colobus vellerosus*) at Boabeng-Fiema, Ghana. In J. D. Paterson, & J. Wallis (Eds.), *Commensalism and conflict: The human-primate interface* (pp. 350-375). American Society of Primatologists Publishing.

Schenkel, R., & Schenkel-Hulliger, L. (1967). On the sociology of free-ranging colobus (*Colobus guereza caudatus* Thomas 1885). In D. Starck, R. Schneider, & H. J. Kuhn (Eds.), *Progress in primatology* (pp. 185-194). Karger.

Schoof, V. A. M., Wikberg, E. C., Fedigan, L. M., Jack, H. M., Ziegler, T. E., & Kawamura, S. (2015). Infanticides during periods of social stability: Paternity, infant age, and maternal resumption of cycling in white-faced capuchins (*Cebus capucinus*). *Neotropical Primates, 21,* 192-196.

Schultz, S., & Noë, R. (2002). The consequences of crowned eagle central-place foraging on predation risk in monkeys. *Proceedings of the Royal Society of London, 269,* 1797-1802.

Schultz, S., Noë, R., McGraw, W. S., & Dunbar, R. I. M. (2004). A community-level evaluation of the impact of prey behavioural and ecological characteristics on predator diet composition. *Proceedings of the Royal Society of London, 271,* 725-732.

Schwitzer, C., Mittermeier, R. A., Rylands, A. B., Chiozza, F., Williamson, E. A., Byler, D., Wich, S., Humle, T., Johnson, C., Mynott, H., & McCabe, G. (Eds.). (2019). *Primates in peril: The world's 25 most endangered primates 2018-2020.* SSC Primate Specialist Conservation International.

Sicotte, P., & MacIntosh, A. J. (2004). Inter-group encounters and male incursions in *Colobus vellerosus* in central Ghana. *Behaviour, 141,* 533-553.

Sicotte, P., Teichroeb, J. A., & Saj, T. L. (2007). Aspects of male competition in *Colobus vellerosus*: Preliminary data on male and female loud calling, and infant deaths after a takeover. *International Journal of Primatology, 28,* 627-636.

Sicotte, P., Teichroeb, J. A., Vayro, J. V., Fox, S., & Wikberg, E. C. (2017). Female dispersal post-takeover is related to male quality in *Colobus vellerosus*. *American Journal of Primatology, 79,* e22436.

Siex, K. S. (2003). *Effects of population compression on the demography, ecology, and behavior of the Zanzibar red colobus* (Procolobus kirkii). [Doctoral dissertation, Duke University].

Siex, K. S., & Struhsaker, T. T. (1999). Ecology of the Zanzibar red colobus monkey: Demographic variability and habitat stability. *International Journal of Primatology, 20,* 163-192.

Simons, N. D., Eick, G. N., Ruiz-Lopez, M. J., Hyeroba, D., Omeja, P. A., Weny, G., Zheng, H., Shankar, A., Frost, S. D. W., Jones, J. H., Chapman, C. A., Switzer, W. M., Goldberg, T. L., Sterner, K. N., & Ting, N. (2019). Genome-wide patterns of gene expression in a wild primate indicate species-specific mechanisms associated with tolerance to natural simian immunodeficiency virus infection. *Genome Biology and Evolution, 11*(6), 1630-1643.

Simons, N. D., Eick, G. N., Ruiz-Lopez, M. J., Omeja, P. A., Chapman, C. A., Goldberg, T. L., Ting, N., & Sterner, K. N. (2017). Cis-regulatory evolution in a wild primate: Infection-associated genetic variation drives differential expression of MHC-DQA 1 in vitro. *Molecular Ecology, 26*(17), 4523-4535.

Skorupa, J. P. (1989). Crowned eagles *Strephanoaetus coronatus* in rainforest: Observations on breeding chronology and diet at a nest in Uganda. *Ibis, 131,* 294-298.

Snaith, T. V., & Chapman, C. A. (2008). Red colobus monkeys display alternative behavioral responses to the costs of scramble competition. *Behavioral Ecology, 19,* 1289-1296.

Stanford, C. B. (2005). The influence of chimpanzee predation on group size and anti-predator behaviour in red colobus monkeys. *Animal Behaviour, 49,* 577-587.

Stanford, C. B., Wallis, J., Matama, H., & Goodall, J. (1994). Patterns of predation by chimpanzees on red colobus monkeys in Gombe National Park, 1982-1991. *American Journal of Physical Anthropology, 94,* 213-228.

Starin, E. D. (1988). Gestation and birth-related behaviors in Temminck's red colobus. *Folia Primatologica, 51,* 161-164.

Starin, E. D. (1991). *Socioecology of the red colobus in the Gambia with particular reference to female-male differences and transfer patterns.* [Doctoral dissertation, City University of New York].

Starin, E. D. (1994). Philopatry and affiliation among red colobus. *Behaviour, 130,* 253-270.

Stead, S. M., & Teichroeb, J. A. (2019). A multi-level society comprised of one-male and multi-male core units in an African colobine (*Colobus angolensis ruwenzorii*). *PLoS ONE, 14,* e0217666.

Steel, R. M. (2012). *The effects of habitat parameters on the behavior, ecology, and conservation of the Udzungwa* red colobus monkeys (Procolobus gordonorum). [Doctoral dissertation, Duke University].

Stern, J. T., & Oxnard, C. E. (1973). Primate locomotion: Some links with evolution and morphology. *Primatologia, Handbook of Primatology,* 4, 1-93.

Straus, W. L., Jr. (1949). Riddle of man's ancestry. *Quarterly Review of Biology,* 24, 200-223.

Struhsaker, T. T. (1975). *The red colobus monkey.* University of Chicago Press.

Struhsaker, T. T. (1978). Food habits of five monkey species in the Kibale Forest, Uganda. In D. J. Chivers, & J. Herbert (Eds.), *Recent advances in primatology, Vol. 1: Behaviour* (pp. 225-248). Academic Press.

Struhsaker, T. T. (1980). Comparison of the behavior and ecology of red colobus and redtail monkeys in the Kibale Forest, Uganda. *African Journal of Ecology,* 18, 33-51.

Struhsaker, T. T. (2000). The effects of predation and habitat quality on the socioecology of African monkeys: Lessons from the Islands of Bioko and Zanzibar. In P. F. Whitehead, & C. J. Jolly (Eds.), *Old World monkeys* (pp. 393-430). Cambridge University Press.

Struhsaker, T. T. (2005). Conservation of red colobus and their habitats. *International Journal of Primatology,* 26(3), 525-538.

Struhsaker, T. T. (2010). *The red colobus monkeys: Variation in demography, behavior, and ecology of endangered species.* Oxford University Press.

Struhsaker, T. T., & Leakey, M. (1990). Prey selectivity by crowned-hawk eagles on monkeys in the Kibale Forest, Uganda. *Behavioral Ecology and Sociobiology,* 26, 435-443.

Struhsaker, T. T., & Leland, L. (1979). Socioecology of five sympatric monkey species in the Kibale Forest, Uganda. *Advances in the Study of Behavior,* 9, 159-228.

Struhsaker, T. T., & Leland, L. (1985). Infanticide in a patrilineal society of red colobus monkeys. *Zeitschrift für Tierpsychologie,* 69, 89-132.

Struhsaker, T. T., & Leland, L. (1987). Colobines: Infanticide by adult males. In B. B. Smuts, D. L. Cheney, R. M. Seyfarth, R. W. Wrangham, & T. T. Struhsaker (Eds.), *Primate societies* (pp. 83-97). University of Chicago Press.

Struhsaker, T. T., Marshall, A. R., Detwiler, K., Siex, K., Ehardt, C., Lisbjerg, D. D., & Butynski, T. M. (2004). Demographic variation among Udzungwa red colobus in relation to gross ecological and sociological parameters. *International Journal of Primatology,* 25, 615-658.

Struhsaker, T. T., & Oates, J. F. (1975). Comparison of the behavior and ecology of red colobus and black-and-white colobus monkeys in Uganda: A summary. In R. H. Tuttle (Ed.), *Socio-ecology and psychology of primates* (pp. 103-123). The Hague.

Struhsaker, T. T., & Pope, T. R. (1991). Mating system and reproductive success: A comparison of two African forest monkeys (*Colobus badius* and *Cercopithecus ascanius*). *Behaviour,* 117, 182-205.

Suzuki, A. (1979). The variation and adaptation of social groups of chimpanzees and black and white colobus monkeys. In I. S. Bernstein, & E. O. Smith (Eds.), *Primate ecology and human origins* (pp. 153-173). Garland STPM Press.

Takahata, Y., Hasegawa, T., & Nishida, T. (1984). Chimpanzee predation in the Mahale Mountains from August 1979 to May 1982. *International Journal of Primatology,* 5, 213-233.

Teelen, S. (2008). Influence of chimpanzee predation on the red colobus population at Ngogo, Kibale National Park, Uganda. *Primates,* 49, 41-49.

Teichroeb, J. A., Saj, T. L., Paterson, J. D., & Sicotte, P. (2003). Effect of group size on activity budgets of *Colobus vellerosus* in Ghana. *International Journal of Primatology,* 24, 743-758.

Teichroeb, J. A., & Sicotte. P. (2008a). Infanticide in ursine colobus monkeys (*Colobus vellerosus*) in Ghana: New cases and a test of the existing hypothesis. *Behaviour,* 145, 727-755.

Teichroeb, J. A., & Sicotte, P. (2008b). Social correlates of fecal testosterone in male ursine colobus monkeys (*Colobus vellerosus*): The effect of male competition in aseasonal breeders. *Hormones and Behavior,* 54, 417-423.

Teichroeb, J. A., & Sicotte, P. (2009). Test of the ecological-constraints model on ursine colobus monkeys (*Colobus vellerosus*) in Ghana. *American Journal of Primatology,* 71, 49-59.

Teichroeb, J. A., & Sicotte, P. (2010). The function of male agonistic displays in ursine colobus monkeys (*Colobus vellerosus*): Male competition, female mate choice or sexual coercion? *Ethology,* 116, 366-380.

Teichroeb, J. A., & Sicotte, P. (2018). Cascading competition: The seasonal strength of scramble influences between-group contest in a folivorous primate. *Behavioral Ecology and Sociobiology,* 72(1), 1-16.

Teichroeb, J. A., Wikberg, E. C., Bădescu, I., MacDonald, L. J., & Sicotte, P. (2012). Infanticide risk and male quality influence optimal group composition for *Colobus vellerosus.* *Behavioral Ecology,* 6, 1348-1359.

Teichroeb, J. A., Wikberg, E. C., & Sicotte, P. (2009). Female dispersal patterns in six groups of ursine colobus (*Colobus vellerosus*): Infanticide avoidance is important. *Behaviour,* 146, 551-582.

Teichroeb, J. A., Wikberg, E. C., & Sicotte, P. (2011). Dispersal in male ursine colobus monkeys (*Colobus vellerosus*): Influence of age, rank and contact with other groups on dispersal decisions. *Behaviour,* 148, 765-793.

Teichroeb, J. A., Wikberg, E. C., Ting, N., & Sicotte, P. (2013). Factors influencing male affiliation and coalitions in a species with male dispersal and intense male-male competition, *Colobus vellerosus.* *Behaviour,* 151, 1045-1066.

Thomas, S. C. (1991). Population densities and patterns of habitat use among anthropoid primates of the Ituri Forest, Zaire. *Biotropica,* 23, 68-83.

Ting, N. (2008a). *Molecular systematics of red colobus monkeys (Procolobus [Piliocolobus]): Understanding the evolution of an endangered primate.* [Doctoral dissertation, City University of New York].

Ting, N. (2008b). Mitochondrial relationships and divergence dates of the African colobines: Evidence of Miocene origins for the living colobus monkeys. *Journal of Human Evolution,* 55(2), 312-325.

Tombak, K. J., Wikberg, E. C., Rubenstein, D. I., & Chapman, C. A. (2019). Reciprocity and rotating social favour among females in egalitarian primate societies. *Animal Behaviour,* 157, 189-200.

Uehara, S. (1997). Predation on mammals by the chimpanzee (*Pan troglodytes*). *Primates,* 38, 193-214.

Usongo, L. I., & Amubode, F. O. (2001). Nutritional ecology of Preuss's red colobus monkey (*Colobus badius preeussi* Rahm 1970) in Korup National Park, Cameroon. *African Journal of Ecology,* 39, 121-125.

van Schaik, C. P., & Isler, K. (2012). Life-history evolution. In J. C. Mitani, J. Call, P. M. Kappeler, R. A. Palombit, & J. B. Silk (Eds.), *The evolution of primate societies* (pp. 220- 244). University of Chicago Press.

Vayro, J. (2017*). Social and hormonal correlates of life history characteristics and mating patterns in female* Colobus vellerosus. [Doctoral dissertation, University of Calgary].

Vayro, J., Fedigan, L. M., Ziegler, T., Crotty, A., Ataman, R., Clendenning, R., Potvin-Rosselet, E., Wikberg, E. C., & Sicotte, P. (2016). Hormonal correlates of life-history characteristics in wild female *Colobus vellerosus. Primates,* 57, 509-519.

Vedder, A., & Fashing, P. J. (2002). Diet overlap and polyspecific associations of red colobus and diana monkeys in the Taï National Park, Ivory Coast. *Ethology,* 103, 514-526.

von Hippel, F. A. (1996). Interactions between overlapping multi-male groups of black-and-white colobus monkeys (*Colobus guereza*) in the Kakamega, Forest, Kenya. *American Journal of Primatology,* 38, 193-209.

Wachter, B., Schabel, M., & Noë, R. (1997). Diet overlap and polyspecific associations of red colobus and diana monkeys in the Taï National Park, Ivory Coast. *Ethology,* 103, 514-526.

Wasserman, M. D., & Chapman, C. A. (2003). Determinants of colobine monkey abundance: The importance of food energy, protein and fibre content. *Journal of Animal Ecology,* 72, 650-659.

Watts, D. P., & Amsler, S. J. (2013). Chimpanzee-red colobus encounter rates show a red colobus population decline associated with predation by chimpanzees at Ngogo. *American Journal of Primatology,* 75, 927-937.

Watts, D. P., & Mitani, J. C. (2002). Hunting behavior of chimpanzees at Ngogo, Kibale National Park, Uganda. *International Journal of Primatology,* 23, 1-28.

Werre, J. L. R. (2000). *Ecology and behavior of the Niger Delta red colobus* (Procolobus badius epieni*).* [Doctoral dissertation, City University of New York].

White, L. J. T. (1994). Biomass of rain forest mammals in the Lopé Reserve, Gabon. *Journal of Animal Ecology,* 63, 499-512.

Whitesides, G. H., Oates, J. F., Green, S. M., & Kluberdanz, R. P. (1998). Estimating primate densities from transects in a West African rain forest: A comparison of techniques. *Journal of Animal Ecology,* 57, 345-367.

Wijtten, Z., Hankinson, E., Pellissier, T., Nuttall, M., & Lemarkat, R. (2012). Activity budgets of Peters' Angola black-and-white colobus (*Colobus angolensis palliatus*) in an East African coastal forest. *African Primates,* 7, 203-210.

Wikberg, E. C., Christie, D., Sicotte, P., & Ting, N. (2020). Interactions between social groups of colobus monkeys (*Colobus vellerosus*) explain similarities in their gut microbiomes. *Animal Behaviour,* 163, 17-31.

Wikberg, E. C., Sicotte, P., Campos, F. A., & Ting, N. (2012). Between-group variation in female dispersal, kin composition of groups, and proximity patterns in black-and-white colobus monkeys (*Colobus vellerosus*). *PLoS ONE,* e48740.

Wikberg, E. C., Teichroeb, J. A., Bădescu, I., & Sicotte, P. (2013). Individualistic female dominance hierarchies with varying strength in a highly folivorous population of black-and-white colobus. *Behaviour,* 150, 295-320.

Wikberg, E. C., Ting, N., & Sicotte, P. (2014a). Kinship and similarity in residency status structure of female social networks in black-and-white colobus monkeys (*Colobus vellerosus*). *American Journal of Physical Anthropology,* 153, 365-376.

Wikberg, E. C., Ting, N., & Sicotte, P. (2014b). Familiarity is more important than phenotypic similarity in shaping social relationships in a facultative female dispersed primate, *Colobus vellerosus. Behavioural Processes,* 106, 27-35.

Wikberg, E. C., Ting, N., & Sicotte, P. (2015). Demographic factors are associated with between-group variation in the grooming networks of female colobus monkeys (*Colobus vellerosus*). *International Journal of Primatology,* 36, 124-142.

Wong, S. N. P., Saj, T. L., & Sicotte, P. (2006). Comparison of habitat quality and diet of *Colobus vellerosus* in forest fragments in Ghana. *Primates,* 47, 365-373.

Wong, S. N. P., & Sicotte, P. (2006). Population size and density of *Colobus vellerosus* at the Boabeng-Fiema Monkey Sanctuary and surrounding forest fragments in Ghana. *American Journal of Primatology,* 68, 465-476.

Wong, S. N. P., & Sicotte, P. (2007). Activity budget and ranging patterns of *Colobus vellerosus* in forest fragments in central Ghana. *Folia Primatologica,* 78, 245-254.

Zimmermann, E., & Radespiel, U. (2013). Primate life histories. In W. Henke, & I. Tattersall (Eds.), *Handbook of paleoanthropology* (pp. 1163-1205). Springer.

Zinner, D., Fickenscher, G. H., Roos, C., Anandam, M. V., Bennett, E. L., Davenport, T. R. B, Davies, N. J., Detwiler, K. M., Engelhardt, A., Eudey, A. A., Gadsby, E. L., Groves, C. P., Healy, A., Karanth, K. P., Molur, S., Nadler, T., Richardson, M. C., Riley, E. P., Rylands, A. B.,... & Whittaker, D. J. (2013). Family Cercopithecidae (Old World monkeys). In R. A. Mittermeier, A. B. Rylands, & D. E. Wilson (Eds.), *Handbook of the mammals of the world: 3. Primates* (pp. 550-754). Lynx.

Zuberbühler, K., & Jenny, D. (2002). Leopard predation and primate evolution. *Journal of Human Evolution, 43*, 873-886.

Zuberbühler, K., Jenny, D., & Bshary, R. (1999). The predation deterrence function of primate alarm calls. *Ethology, 105*, 477-490.

CHAPTER 14

Ecology and Behavior of the Arboreal African Guenons

Joanna E. Lambert, Robert W. Sussman, and Martha M. Lyke

The guenons (tribe Cercopithecini, subfamily Cercopithecinae) are a species-rich group of catarrhine monkeys found only in sub-Saharan Africa (see Table 14.1). These monkeys are among the most colorful primates; many species have brightly colored pelage and appear to be, as noted by Glenn and Cords (2002: p.xiii) "... multicolored experiments in body design" (Figure 14.1). Classification and systematics of the guenons have not been without contention and depending upon which taxonomy one uses, there are anywhere from 24–35 total guenon species (and over 70 subspecies) divided into 8 superspecies groups, making them the largest and most diverse group of primates in the Paleotropics (Leakey, 1988; Groves, 2001; Grubb et al., 2003; Jaffe & Isbell, 2011; Cords, 2012). Despite their known diversity, Colyn and Deleporte (2002) predicted that more species would be discovered. Indeed, since this proclamation, Hart et al. (2012) have described a new guenon species—*Cercopithecus lomamiensis*—discovered in the Lomami Basin, central Democratic Republic of Congo.

The common term "guenon" has been used strictly to refer only to the genus *Cercopithecus* and has also been used more broadly to include the closely related genera of *Allenopithecus*, *Chlorocebus*, *Miopithecus*, and *Erythrocebus* (Glenn & Cords, 2002; Butynski, 2002a). Here, we will employ the broader term and cover *Cercopithecus*, *Allenopithecus*, and *Miopithecus* (see Table 14.1 on taxonomy). The evolutionary radiation of the guenons was rapid and relatively recent—probably within the last one million years or so (Leakey, 1988; Ruvolo, 1988; Xing et al., 2007). Fossil guenons are extremely rare, with none older than three million years (but see Gilbert et al., 2014), and most information for the evolution of this genus derives from molecular and morphological analyses (Leakey, 1988; Tosi et al., 2003; Cords, 2012). Recent analysis (Kamilar et al., 2009) based on biogeography, Y-chromosome, X-choromosome, and mtDNA data provide strong support of allopatric speciation, with most divergence events a function of population shifts as forests expanded and retreated. This supports earlier models in which high species diversity and endemism—including cercopithecines—have been correlated with Pleistocene refugia (Hamilton, 1988; Kingdon, 1988; Colyn et al., 1991).

It seems likely that the guenon radiation derived from a form very much like the Central African guenon *Allenopithecus nigroviridis* (Allen's swamp monkey), a species found in swampy habitats in the Congo Basin (Kingdon, 1971, 1990; Ruvolo, 1988). Kingdon (1971: p.204) believes this species "represents in little changed condition a surviving population of the ancestral type of guenon monkey." This observation is supported by recent genetic analysis that divides the guenons into four distinct clades: (1) a basal lineage comprising *Allenopithecus*; (2) *Miopithecus*; (3) an arboreal lineage comprising all *Cercopithecus* species except *C. lhoesti*, and; (4) a terrestrial clade comprising *Chlorocebus* (cf. vervets, *Cercopithicus aethiops*), *Erythrocebus*, and *Cercopithecus lhoesti*.

It is in fact highly likely that speciation is ongoing within this clade (Xing et al., 2007; Detwiler, 2019). This proposition is supported by data demonstrating hybridization among multiple *Cercopithecus* species. There are documented hybrid crosses of *Cercopithecus erythrotis* and *C. cephys*, *C. campbelli campbelli* and *C. c. lowri*, *C. mitis stuhlmanni* and *C. m. klobi*, *C. nictitans* and *C. cephus*, *C. nictitans* and *C. petaurista* (in captivity), and the well-described hybridization between *C. mitis* and *C. ascanius* (Struhsaker et al., 1988; Detwiler, 2002; Detwiler et al., 2005). Hybridization between these two species is especially common in Gombe National Park, Tanzania, in a region that has been suggested to be a hybrid zone (Detwiler, 2002, 2019). Extreme species richness and speciation itself may have occurred via hybridization, in which populations of divergent taxa were contracted into refugia that then underwent interbreeding, resulting in new species (Detwiler, 2002; Detwiler et al., 2005; Kamilar et al., 2009).

Guenons can be found in a great range of habitat types throughout the continent, spanning the extremes of rainfall, altitude, and latitude (Table 14.1). With the exception of the more terrestrial forms (i.e., *Chlorocebus* [*Cercopithecus*] *aethiops* and *Erythrocebus*), most guenons are forest obligates, found in varying forest types, as described by Butynski (2002a):

Table 14.1 Taxonomy and Distribution of Arboreal Guenons [a]

Scientific Name	Common Name	Distribution	Habitat
Allenopithecus nigroviridis	Allen's swamp monkey	Central African Republic (CAR), Republic of Congo, Democratic Republic of Congo (DRC); Cameroon (uncertain)	Primary lowland riverine swamp forest
Miopithecus talapoin	Angolan talapoin/ Southern talapoin	Angola; DRC; Republic of Congo	Inundated primary riverine forest
Miopithecus ogouensis	Gabon talapoin/ Northern talapoin	Angola; Cameroon; Republic of Congo; Equatorial Guinea; Gabon	Inundated primary riverine forest
Cercopithecus ascanius	Redtail monkey	Angola; Burundi; CAR; DRC; Kenya; Rwanda; South Sudan; Tanzania; Uganda; Zambia	Moist deciduous forest; swamp forest; disturbed/secondary tropical moist forest
Cercopithecus cephus	Mustached guenon	Angola; Cameroon; CAR; Republic of Congo; DRC; Equatorial Guinea; Gabon	Primary tropical rainforest
Cercopithecus erythrogaster	White-throated guenon	Benin; Nigeria; Togo	Lowland rainforest
Cercopithecus erythrotis	Red-eared guenon	Cameroon; Nigeria; Equatorial Guinea (Bioko)	Lowland tropical forest
Cercopithecus petaurista	Lesser spot-nosed guenon	Côte d'Ivoire; Ghana; Guinea; Liberia; Sierra Leone; Togo; Guinea-Bissau	Primary rain forest; tropical evergreen seasonal lowland forest
Cercopithecus sclateri	Sclater's guenon	Nigeria	Moist tropical lowland forest
Cercopithecus diana	Diana monkey	Côte d'Ivoire; Guinea; Liberia; Sierra Leone	Tropical moist forest; tropical evergreen seasonal lowland forest; disturbed/secondary tropical moist forest
Cercopithecus roloway	Roloway monkey	Côte d'Ivoire; Ghana	Tropical moist forest; tropical evergreen seasonal lowland forest; disturbed/secondary tropical moist forest
Cercopithecus dryas	Dryas guenon	DRC	Tropical rainforest
Cercopithecus hamlyni	Owl-faced monkey	DRC; Rwanda	Bamboo forest; tropical montane forest; lowland forest
Cercopithecus lhoesti	L'Hoest's monkey	Burundi; DRC; Rwanda; Uganda	Tropical montane forest; tropical moist evergreen forest; mature lowland rainforest
Cercopithecus preussi	Preuss's monkey	Cameroon; Equatorial Guinea (Bioko); Nigeria	Primary tropical rainforest; disturbed/regenerating tropical rainforest
Cercopithecus solatus	Sun-tailed monkey	Gabon	Primary tropical rainforest
Cercopithecus albogularis	Sykes monkey	Kenya; Somalia; Tanzania	Lowland semi-deciduous tropical forest; coastal tropical forest; mangrove forest; disturbed/regenerating tropical forest
Cercopithecus doggetti	Silver monkey	Burundi; Tanzania; Rwanda; Uganda; DRC	Tropical montane forest; tropical rainforest
Cercopithecus kandti	Golden monkey	Rwanda; Uganda; DRC	Tropical montane forest; tropical rainforest
Cercopithecus mitis	Blue monkey	Angola; Burundi; Republic of Congo, DRC; Eswatini; Ethiopia; Kenya; Malawi; Mozambique; Rwanda; Somalia; South Africa; South Sudan; Tanzania; Uganda; Zambia; Zimbabwe	Most, semi-deciduous forest; moist evergreen rainforest; tropical montane forest
Cercopithecus nictitans	Putty-nosed/ Greater spot-nosed monkey	Cameroon; CAR; Republic of Congo; DRC; Côte d'Ivoire; Equatorial Guinea; Gabon; Liberia; Nigeria; Angola (uncertain)	Primary tropical rainforest
Cercopithecus campbelli	Campbell's guenon	Gambia; Guinea; Guinea-Bissau; Liberia; Senegal; Sierra Leone	Primary, secondary, swamp forest; tropical evergreen seasonal lowland forest
Cercopithecus denti	Dent's guenon	DRC; Rwanda; CAR; Uganda	Lowland tropical forest; submontane tropical forest
Cercopithecus lowei	Lowe's guenon	Côte d'Ivoire; Ghana	Tropical lowland rainforest
Cercopithecus mona	Mona monkey	Benin; Cameroon; Ghana; Nigeria; Togo	Mangrove swamp forest

Table 14.1 (Continued)

Scientific Name	Common Name	Distribution	Habitat
Cercopithecus pogonias	Crowned guenon/ Crested Mona monkey	Angola; Cameroon; CAR; Republic of Congo; DRC; Equatorial Guinea (Mainland, Bioko); Gabon	Primary tropical rainforest
Cercopithecus wolfi	Wolf's guenon	DRC	Primary, secondary lowland forest; swamp forest
Cercopithecus neglectus	De Brazza's monkey	Angola; Cameroon; CAR; Republic of Congo; DRC; Equatorial Guinea; Ethiopia; Gabon; Kenya; South Sudan; Uganda	Swampy, flooded areas of riverine forest; montane forest

[a] Adapted from IUCN (2019); Jaffe and Isbell (2011).

Guenons live in a wide variety of habitats. These range from woodlands (patas monkey, vervet monkey) to mangrove forests at sea level (northern talapoin monkey, *Miopithecus ogouensis*), to swamp forests (Allen's swamp monkey, *Allenopithecus nigroviridis*), through lowland, mid-altitude, and montane forests to bamboo forest at >3300 m (golden monkey, *Cercopithecus mitis kandti*; owl-faced monkey, *C. hamlyni*), to alpine moorland at 4500 m (Bale Mountains grivet monkey, *C. aethiops djamdjamensis*). The guenons also occur along the arid edges of the Kalahari and Sahara Deserts (green monkey, vervet monkey, patas monkey) as well as in the wettest places in Africa, such as Mt. Cameroon and Bioko Island, where mean annual rainfall is ca.1000 cm (Preuss's monkey, *Cercopithecus preussi*; red-eared monkey, *C. erythrotis*). (p.3)

Some of the most diverse communities of guenons, where multiple species live sympatrically, are found in the forests of Central and West Africa. These forests are intersected by numerous rivers and separated by at least one major savanna during arid periods, making them ideal for isolation and speciation of many groups of Central and West African mammals (Booth, 1957; Moreau, 1969; Kingdon, 1971; Hamilton, 1988; Oates, 1988; Chapman et al., 1999; Colyn & Deleporte, 2002). In the Ituri Forest of the Democratic Republic of Congo, for example, six species of *Cercopithecus* live sympatrically (Chapman et al., 1999) and in Taï National Park, Côte d'Ivoire, four species live sympatrically (Kane & McGraw, 2017)

FEEDING BIOLOGY

Diet

Guenons, like all cercopithecines, are "eclectic omnivores" (*sensu* Altmann, 1998) that consume a diversity of food types. Virtually all plant portions are consumed by one guenon species or another, including fruit, seeds, and leaves of all stages of maturity (unripe fruit, ripe fruit, immature/mature seeds, leaf buds, young/mature leaves), gum/sap, petioles, bark, and flowers/flower buds. Guenons also consume fungus and lichen, as well as both invertebrates and small vertebrate prey. In Uganda, for example, *Cercopithecus mitis* are known to hunt and kill galagos (*Galagoides demidoff*, *Galago senegalensis*) and *Cercopithecus ascanius* (redtail monkeys) have been documented to hunt and consume green pigeons (*Treron calva*) (Butynski, 1982; Furuichi, 2006). Additionally, *C. mitis* are reported to consume bats opportunistically at sites in both Kenya and Tanzania (Tapanes et al., 2016).

Figure 14.1 *Cercopithecus albogularis* from Zanzibar, illustrating the striking pelage for which guenons are famous. (Photo credit: Thomas T. Struhsaker)

Although all guenons are omnivores, consuming a diversity of foods from more than one trophic level, most are highly frugivorous and will consume ripe (and unripe) fruit when available. As summarized by Jaffe and Isbell (2011), averaging across all sites and research, the diet of forest-dwelling guenons in general comprises anywhere between 24.5–91% fruit. For example, in the Taï National Park, Côte d'Ivoire, C. diana, C. campbelli, and C. petaurista have an annual diet comprising almost 80% fruit (Buzzard, 2006a,b). When consuming fruit, guenons rely extensively on their cheek pouches—an important feeding adaptation found only in cercopithecine monkeys (Lambert & Whitham, 2001; Lambert, 2005; Buzzard, 2006c). In Kibale National Park, Uganda, cheek pouches are used most commonly by Cercopithecus ascanius (and by the gray-cheeked mangabey, Lophocebus albigena) when feeding on fruit and when the number of feeding conspecifics increased in a fruit patch (Lambert, 2005). Similarly, in Kakamega Forest, Kenya, C. mitis uses cheek pouches most often when feeding on fruits and increases their use when conspecifics were nearby (Smith et al., 2008).

Guenons can also consume a high percentage of nonreproductive plant parts, especially leafy material, primarily from trees but also from terrestrial herbaceous vegetation (THV) (e.g., C. mitis, Butynski, 1990; C. petaurista, Buzzard, 2006 a,b). In the Budongo Forest, Uganda, C. ascanius has been reported to have a diet at some times of the year that can be over 70% leaves (Sheppard, 2000). While the most terrestrial Cercopithecus species, C. lhoesti, consumes a wide variety of species and food types (e.g., fruit, leaves, shoots, mushrooms, lichen, gums, insects, cultivated crops, and insects), a major component of the diet is THV (Kaplin et al., 1998; Peignot et al., 1999; Kaplin & Moermond, 2000). In Rwanda, Kaplin and Moermond (2000) found that C. lhoesti feeds on fruit only around 25% of the time and THV 35% of the time. They also found an inverse relationship between invertebrate consumption and THV consumption in C. lhoesti diet compared with that of C. mitis, the blue monkey, suggesting that THV is an important source of protein for the former (Kaplin et al., 1998; Kaplin & Moermond, 2000).

High levels of folivory are almost certainly facilitated by the absolutely and relatively long digestive passage times observed among guenons (Lambert, 1998, 2001; Blaine & Lambert, 2012). Long digestive retention times enable high levels of bacterial fermentation of the fibrous portions of plant material; leaves and other nonreproductive plant parts are particularly high in these fractions. Thus, having lengthy digestive retention times can facilitate consumption of leaves and other nonreproductive plant parts and may be characteristic of the Cercopithecus genus, and perhaps of the Cercopithecinae in general. As an example, Maisels (1993) found that C. pogonias, C. erythrotis, and Lophocebus albigena had mean digestive retention times (MRT) of approximately 27, 26.7, and 38 hours, respectively. Caton (1999) reported an MRT of 32 hours for C. neglectus, while Chlorocebus (Cercopithecus) aethiops had an MRT of 31.5 hours (Clemens & Maloiy, 1981), and Papio anubis had an MRT of 37 hours (Clemens & Phillips, 1980). Moreover, in an analysis regressing transit times (MRT not available for most species) as a function of body size using all reported primate transit times, Lambert (1998) and Blaine and Lambert (2012) have found that Cercopithecinae are significantly further above the regression line than any other primate taxon.

Insectivory is also an important, but difficult to assess, aspect of guenon diets. In Kibale National Park, Uganda, C. mitis has been reported to have a diet comprising anywhere from 35.9–55% insects and other arthropods (Struhsaker, 1978; Butynski, 1982; Lambert, 2002; Jaffe & Isbell, 2011; Bryer et al., 2013). Gautier-Hion (1988) reported that Miopithecus talapoin has a diet of over 30% insects in Makokou, Gabon. When hunting insects, guenons use a variety of capture behaviors, including gleaning them from leaves, tree trunks, and branches, and grabbing at flying insects (Cords, 1986; Lambert, 2002; Jaffe & Isbell, 2011; Lambert, pers.obs.). The fact that most insects are rapidly moving and in tree canopies makes insectivory difficult to observe and quantify. However, recent molecular methods have been successful in the identification of arthropod prey from primate fecal samples, including guenons. Using a metagenomic approach, Lyke and colleagues (2019) identified DNA from 68 arthropod families present in C. mitis and C. ascanius fecal samples from Kibale National Park, Uganda. Prior to this study, few arthropod prey taxa were known to be consumed by the two groups (but see Bryer et al., 2015), despite the fact that arthropods consistently comprise a substantial portion (20–55%) of their diets (Butynski, 1990; Chapman et al., 2002; Lambert, 2002).

Adding to the complexity of primate insectivory is the fact that many fruits are heavily infested with insects, making it almost impossible to determine whether some fruit are consumed for the fruit pulp, the insect larvae, or both. For example, figs (genus Ficus) are consumed by all guenon species, and each fig species has its own (coevolved), unique, "live-in" wasp species that serves as a species-specific pollinator (Waser, 1977). It is very likely that the aseasonal fruiting behavior of Ficus is the result of the need to provide a continuous supply of figs for these wasps. Thus, figs are a year-round food supply for many animals and, in some cases, fruit-eating may be viewed as a form of insectivory (Waser, 1977). Indeed, fig wasp (*Agaonidae* spp.) DNA was present in fecal samples of *C. mitis* and *C. ascanius* collected during both the wet and dry seasons in Kibale National Park (Lyke et al., 2019). Just as in the New World, large, aseasonal fig trees are an important food source for many African forest-living frugivores, including *Cercopithecus* monkeys. Haddow (1952) described the activity at a fig tree in fruit:

> A tree of this kind presents an amazing picture, being full of monkeys and birds all feeding together... an immense fig in fruit was visited during the course of a single afternoon by three species of monkeys, *C. albigena*, *C. mitis* and *C. ascanius*, by chimpanzees, *P. troglodytes*, by two species of squirrels, by numerous hornbills belonging to no less than six species, by two species of Plantain-eater and one species of Lourie, and by a very large flock of Violet-backed Starlings. In the evening Green Pigeons, and Love-Birds, were also present, and at twilight fruit-bats began to arrive. At times during the afternoon as many as thirty or forty monkeys (not all of the same species) would be present in the tree at the same time, together with twenty to thirty hornbills, a few plantain-eaters and very numerous starlings, the noise and general activity being considerable. Under such circumstances the members of a monkey band will be found feeding very actively, all at one time stuffing themselves with fruit, frequently in company with other species. (p.9)

Feeding Flexibility

Guenons exhibit a very high level of feeding flexibility at all scales (intra- and interindividual, intra- and intergroup, intra- and interpopulation, intra- and interspecific) and at all units of measure (behavioral bout, daily, monthly, seasonally, annually). Given the cyclicity of climate in Africa both yearly and historically, it is likely that there has been strong selective pressures for dietary flexibility (see relevant papers in Fleagle, 1999; Lambert, 2002). In Kibale National Park, Rudran (1978b) recorded wide inter-monthly variation in the diet of blue monkeys (*C. mitis*): fruit range 17.3–75.5%, leaf range 2.5–57.8%, and seed range 0–11.7%. Cords (1986, 1987) has documented that although the diets of sympatric *C. mitis* and *C. ascanius* at Kakemega Forest, Kenya, are similar to the diet of these same two species in Kibale National Park, Uganda, important differences were found as well (Figure 14.2). In some cases, variation between populations of the same species can be greater than that observed between species (Chapman et al., 2002). For example, the diet of *C. pogonias* at one site in Gabon is more similar to that of *C. mona* in the Congo, than to populations of *C. pogonias* at two other sites in Gabon (Maisels & Gautier-Hion, 1994; Brugierè et al., 2002; Chapman et al., 2002). Chapman et al. (2002) report a range of 26.1–91.1% for fruit eating, 0.7–35.9% for insect eating, and 3.0–46.6% for leaf eating among populations of blue monkeys. They also found similar degrees of variation in the diets of redtail monkeys (*C. ascanius*), with fruit accounting for 13.2–61.3%, insects 14.5–31.2%, and leaves 6.5–73.6% of the diet of various redtail monkey groups or populations.

Despite the enormous variability in the diet of the guenons, several authors (e.g., Struhsaker, 1978; Gautier-Hion, 1988; Lawes, 1991; Lambert, 2002; Buzzard, 2006a) have found high dietary overlap among sympatric guenon species. Lambert (2002) compared the diets of three cercopithecines (*C. ascanius*, *C. mitis*, and *Lophocebus albigena*) and one colobine (*Colobus badius*) in Kibale, Uganda, and found that the three cercopithecines had 60% overlap in their plant diets, and 54% of the total plant species were consumed by three or more monkey species. She also found considerable overlap in the top five commonly eaten foods among the cercopithecines. Similarly, Buzzard (2006a) has found extensive overlap among sympatric *C. campbelli*, *C. petaurista*, and *C. diana* in the Taï National Park, Côte d'Ivoire, although there was a range in this dietary overlap from month to month. *C. campbelli* and *C. diana* had the highest mean overlap at 73% (range: 46–97%); *C. petaurista* and *C. diana* shared 65% of their diet (range: 54–77%); and *C. petaurista* and *C. campbelli* shared 59% (range: 50–67%). Lambert (2002) has argued that the ability to switch foods as an immediate response to the presence of competitors is an important coexistence mechanism for omnivorous species with high dietary overlap.

ECOLOGICAL ROLES
Seed Dispersal

Guenons are important seed dispersers and play a critical role in the regeneration of forests. Parent plants can benefit from having their seeds removed and dispersed by primates and other animals because, in this fashion, seeds escape the disproportionately high, density-dependent mortality that characterize the environment beneath the parent's crown. Most primates swallow the seeds of fruit they consume and defecate them some distance from the parent tree. However, as noted earlier, guenons—like all cercopithecine monkeys—have cheek pouches, which facilitate a distinct form of fruit processing and seed-handling, seed-cleaning, and seed-spitting. In addition to seed-spitting, guenons also disperse both large and small seeds by defecation and can also act sometimes as seed predators (Corlett & Lucas,

Figure 14.2 One of the most studied arboreal guenon species, *Cercopithecus ascanius*, in a fruiting *Ficus natalensis* tree, Kibale National Park, Uganda.
(Photo credit: Alain Houle)

1990; Rowell & Mitchell, 1991; Lambert, 1999; Kaplin & Moermond, 2000; Kaplin & Lambert, 2002; Fairgrieve & Muhumuza, 2003). For example, Lambert (2011) found that *Cercopithecus* species disperse more seeds than any other primate taxon in Kibale National Park and can, in a single day, move the seeds of as many as 33,840 fruits/km² in comparison to the total ape removal (by swallowing and defecating) of 1,398 fruits/km². The cercopithecine processed seeds are effectively moved into safe sites (Kaplin & Lambert, 2002; Lambert, 2011, 2002). In the Nyungwe National Park, Rwanda, Gross-Camp et al. (2009) found that cercopithecines' patterns were most predictive of effective seed dispersal. Nyungwe is home to nine cercopithecine species, seven of which belong to the genus *Cercopithecus*. The authors suggested that *Cercopithecus* had the highest potential for seed dispersal because they spend so much time in focal trees that they tend to overlap with the greatest number of fruiting tree species across all frugivore assemblages. In the Dja Reserve, Cameroon, Poulsen et al. (2002) demonstrated that seed-dispersing frugivores could be clustered into dietary guilds and subguilds. Of these, the subguild comprising three *Cercopithecus* species emerged as having the greatest potential for dispersing seeds, indicating their importance as seed dispersers relative to other frugivores

such as hornbills. In Nigeria, Chapman et al. (2010) found that *C. nictitans* disperse seeds across a wide range of habitats and have become a highly important disperser of seeds as other, larger-bodied taxa have been locally extirpated. Additionally, in South Africa, Linden et al. (2015) found that *C. mitis* were important seed dispersers in an Afromontane forest that was also lacking other larger-bodied, forest-dwelling mammals.

Although *Cercopithecus* monkeys can and do swallow seeds whole and defecate them, they also often clean pulp off small seeds and, over 80% of the time, spit them out close to the parent tree. Seed cleaning appears to mitigate effects of fungal pathogens and reduce seed mortality. This is especially important in tree species living in clumped groves, where seeds can have a high success rate even if dropped near parent trees. For example, the impact of *Cercopithecus* seed-processing and -handling can dramatically improve germination of a tree species like *Strychnos mitis* in Kibale National Park (Lambert, 2001).

As noted by Kaplin and Lambert (2002), *Cercopithecus* species may be the primary dispersers of some tree species in some habitats. For example, data from several studies of frugivores in Nyungwe National Park, Rwanda, suggest that two forest tree species with relatively large seeds are not visited by turacos, understory birds, or chimpanzees, but are regularly visited by *Cercopithecus* monkeys (Kaplin & Lambert, 2002). Similar findings come from studies conducted elsewhere. In Malawi, tree censuses over a 16-month period demonstrated that of 134 fruiting tree species, fruits of 5 species were visited by only *Cercopithecus* species (Dowsett-Lemaire, 1988). In the Budongo Forest, Uganda, *Cercopithecus* monkeys are the only frugivores known to visit and disperse the seeds of 6 important fruiting tree species (Plumptre & Reynolds, 1994; Kaplin & Lambert, 2002).

Predator Avoidance

While seed dispersal is an ecological role in which benefits are incurred by all interacting species, predation is clearly an interaction in which one species (the predator) benefits, and the other (the prey) does not. Guenons are consumed as prey species throughout Africa by both terrestrial and aerial predators, including chimpanzees (*Pan troglodytes*), leopards (*Panthera pardus*), golden cats (*Profelis aurata*), domesticated dogs (*Canis familiaris*), African rock pythons (*Python sebae*), monitor lizards (*Varanus niloticus*), and a variety of raptors (Gautier-Hion et al., 1983; Struhsaker & Leakey, 1990; Wahome et al., 1993; Mitani et al., 2001; Shultz et al., 2004; Jaffe & Isbell, 2011; Cords, 2012). Humans prey upon guenons as well; indeed, as discussed later, commercial "bushmeat" hunting is an exceptionally serious problem threatening virtually all guenon species.

The African crowned hawk-eagle (*Stephanoaetus coronatus*) is perhaps the most commonly identified guenon predator and indeed has been called a "primate specialist" predator (Hart, 2000). In Tanzania, *C. mitis* comprised nearly 90% of the prey remains at a crowned eagle roost. In Kibale National Park, Uganda, 83–88% of this eagle species' prey are monkeys of three genera: *Colobus, Cercopithecus,* and *Lophocebus* (Skorupa, 1989; Struhsaker & Leakey, 1990; Leland & Struhsaker, 1993; Sanders et al., 2003). Mitani et al. (2001) documented that cercopithecines are taken most frequently, and *C. ascanius* composed the majority of primate prey (68%), which may explain why *C. ascanius* females have been reported to scan the environment 96 minutes per day (Treves, 1998, 1999). Cercopithecines are also killed by leopards, and *C. mitis* can be preyed upon by these feline predators as the monkeys drop to the ground to escape from crowned hawk-eagles (Haddow, 1952; Sugiyama, 1968; Teleki, 1973; Hart et al., 1996). In the Taï National Park, Côte d'Ivoire, monkeys were the second favorite prey of leopards, and guenons were found in 19.5% of leopard scats (Kingdon, 1971).

Alarm calls are similar among guenons and are generally responded to by other guenon species (Struhsaker,

1970; Gautier-Hion, 1988). When in polyspecific associations, males of different species will combine alarm calls and mob predators (Gautier-Hion & Gautier, 1974). There is some degree of convergence in the warning calls of *C. ascanius* and *C. mitis,* with each reacting to the other (Marler, 1973). *C. mitis* react dramatically to crowned hawk-eagles and alert one another to these raptors' presence with loud, chirping vocalizations (Cords, 1987; Cordeiro, 1992). Upon sighting an eagle, group members immediately increase spatial cohesion and rapidly descend to understory trees; adult males give a loud call and other group members continue to chirp until after the bird departs (Marler, 1973; Cordeiro, 1992; Zuberbühler, 2002). Males will also charge and displace perched eagles (Cords, 1987; Cordeiro, 1992). *C. ascanius* give bird-like chirps, shake vegetation, and emit explosive warning calls. When alarmed, *C. ascanius* hurl themselves into the undergrowth and, thereby, create an effective warning to nearby monkeys (Kingdon, 1971). They will also hide in dense vegetation or descend to the ground and flee.

In contrast to *C. mitis* and *C. ascanius, C. neglectus* live in small groups that avoid interaction with conspecifics, and much of their behavior is adapted to avoid the attention of predators (Gautier-Hion & Gautier, 1974). These monkeys are characterized by their caution and capacity for concealment. They move slowly and silently and avoid jumping. Vocal exchanges are rare and discreet; in fact, *C. neglectus* often use olfactory marking for communication, a behavior not found in other *Cercopithecus* species. If a predator is encountered, the *C. neglectus* monkeys will attempt to escape silently on the ground or freeze and remain silent for up to several hours. If noticed by a predator, adult males will face the adversary and attempt to gain its attention by giving loud calls and jumping about loudly in the branches as the remainder of the group disperses and hides (Gautier-Hion & Gautier, 1974). The male is constantly vigilant, and his role in predator defense closely resembles that of the male of the savanna-dwelling *Erythrocebus patas*. In contrast, in playback experiments with wild *C. diana*, the females consistently alarm-called before the male, and males only gave predator-specific calls after females had given them (Stephan & Zuberbühler, 2016).

The diminutive *Miopithecus talapoin* live in multimale groups that are much larger than the other guenons, and because of their small size, they often inhabit small branches of the dense underbrush in inundated forests (Gautier-Hion, 1973). Both of these factors are probably efficacious in guarding against diurnal predators. *M. talapoin* have a complex series of mobbing choruses that are used as a system of alert and rapid collective defense. At night, sleeping sites also are chosen for predator defense.

Usually sleeping sites are on supports overhanging the water. Older males sleep on the outer and upper branches of the sleeping site, cutting off access to the ends of branches and to lower levels. Animals sleep in dense foliage or suspended "in air" on thin lianas that transmit the slightest vibrations; if disturbed during the night, they plunge into the water and disappear by swimming underwater. Nocturnal predators are mainly arboreal carnivores, pythons, and cobras. One unfortunate *M. talapoin* was found in the intestinal tract of a python (Gautier-Hion, 1973).

Birth, Growth, and Development

As with diet and social structure, most details regarding life history in guenons stem from research on *C. mitis* and *C. ascanius* (Figure 14.2). In both species, there is some seasonality of mating and birthing, although both events can occur in most months (Rudran, 1978b; Cords, 1987; Butynski, 1988). Gestation has been estimated as 162–190 days (Pazol et al., 2002). Births are single and interbirth intervals are approximately two years in *C. mitis* and between one and two years in *C. ascanius* (Cords, 1987, 2002), though there is a great deal of variability among females in both species. Recently, Kane (2017) reported that *C. diana* in Taï National Park, Côte d'Ivoire, also have approximately two-year interbirth intervals, but more distinct mating (March–May) and birthing (October–December) seasons. In all arboreal guenons, infants are carried ventrally by their mothers (Gautier-Hion, 1971). Female *C. mitis* are highly protective of their infants. Even after two months of age, the infant remains close to the mother, although by about that time it may be groomed by other adult females (Rudran, 1978b; Struhsaker & Leland, 1979). Allomothering occurs in some guenon species (Bourliere et al., 1970; Gautier-Hion, 1971; Cords, 1987). At thirteen weeks, *C. mitis* infants spend about 30% of their time more than 2 feet from their mothers. Infants are weaned at around five to six months of age, at the time of the new birth season (Haddow, 1952; Bourliere et al., 1970; Gautier-Hion, 1971). Juvenile status is reached at approximately seven months. Males reach maturity at six-plus years and females at five to six years (Cords, 1987). Among *C. ascanius*, group members other than the mother have physical contact with the infant in the first month and sometimes within the first week of birth (Struhsaker & Leland, 1979).

Few data exist on guenon development, but, as with many arboreal species, infants may develop at a relatively slow rate when compared to closely related terrestrial primate species (Chalmers, 1972; Cords, 2012). However, Cords (2002) has found that the rate at which *C. mitis* infants attained independence resembled that of similarly sized terrestrial species but was faster than the infants of the few other arboreal species that have been studied. This suggests that infant development in *C. mitis* reflects the risks of intra-group aggression and predation more than arboreality. In male guenons, adulthood is often accompanied by a very rapid growth phase and changes in vocal and behavioral repertoire. Social maturity may be inhibited in some males by the presence of a resident adult male and, thus, social maturity of a subadult male may not depend only upon age (Gautier-Hion & Gautier, 1976).

SOCIAL BEHAVIOR

Social Structure and Organization

Just as the guenons exhibit extreme diversity in habitat and diet, they also exhibit high diversity in their social structure and organization, ranging from monogamy to uni-male/multi-female groups to multi-male/multi-female groups (Glenn & Cords, 2002; Cords, 2012). Most information regarding social structure—and its variance—stems from the two best-studied forest guenon species: *C. mitis* and *C. ascanius*. On average, these two species live in one-male groups, although group size at different sites is variable, ranging from 10–50 individuals, and larger groups have been observed to fission (Struhsaker & Leland, 1979; Cords & Rowell, 1986; Windfelder & Lwanga, 2002). Most groups contain one adult "resident" male, although a small percentage of social groups will have more than one resident male (Aldrich-Blake, 1970; Rudran, 1978b; Cords, 1984, 2000a; Treves, 1998). Females remain in their natal group, whereas subadult males migrate upon their maturation. The tenure of resident males varies considerably in both species, and extra-group males either remain solitary or in small groups of two to five members (Galat-Luong, 1975; Struhsaker, 1977; Cords, 1984; Henzi & Lawes, 1988). These males are cryptic and have been difficult for researchers to follow and habituate (Cords, 2000a).

Within-group interindividual interactions among guenons are often subtle, especially when compared to other cercopithecines such as baboons. The social core of a *C. ascanius* group comprises several philopatric females that mainly interact with their own offspring and other females. Males and females interact infrequently outside of the mating season. Dominance hierarchies are not strikingly apparent among group females and juveniles (Cords, 1987). For example, at Kakamega, Kenya, the hierarchy among female *C. mitis* is linear, but dominance does not confer advantages in feeding or reproductive success (Foerster et al., 2011), and grooming (along with other social relationships) among females was widely and evenly spread throughout the group

(Cords, 2000b). All females groom with at least half of the others, and dyadic grooming tends to be reciprocal. Grooming patterns among adults are not kin-related and do not correlate with any agonistic dominance. Patterns of grooming among immature females correlate with the development of long-term social relationships, especially with female peers (Cords, 2000b), and mothers groom female infants more than male infants (Förster & Cords, 2002). Although two groupings of "higher-" and "lower"-ranking females can form, these groupings are typically vague, loose fitting, and irregular (Rowell et al., 1991). Moreover, vocalizations are highly flexible as well. Bouchet et al. (2012), for example, have found that in *C. neglectus*, vocalizations are similar to those of other guenons and that both phylogeny and sexual dimorphism play important roles in vocal communication. The sexual distinction, however, between adult male and female vocal repertoires is more flexible than previously reported; indeed, both sexes appear to be able to produce calls of the other sex. It may be that vocalizations are constrained by social roles (Bouchet et al., 2012).

There can be considerable variation in the cohesion of *C. mitis* groups from site to site. For example, *C. mitis* at Kibale live in cohesive, but widely dispersed, groups with a median spread of 50 m and a maximum of 120 m between members (Rudran, 1978a). At Budongo Forest in Uganda, where population density tends to be higher and home ranges much smaller, groups will break into subgroups that remain dispersed even throughout the night (Aldrich-Blake, 1970). In high-density populations, *C. mitis* appear to be territorial (Butynski, 1990). Although females do not seem to form agonistic alliances within the group, they do perform agonistic displays against females of neighboring groups. Males usually are not active in these disputes. Similarly, social groups of *Miopithecus talapoin* in Cameroon are often divided into subgroups in which individuals are typically within 100–200 m of each other, but subgroups were often out of sight of one another (Rowell, 1973). Constant contact calls are exchanged during group movement (Gautier-Hion, 1973, 1988).

C. hamlyni can be argued to be the most cryptic of all forest guenons. These monkeys are reported to live in one-male groups of less than 10 individuals. The relatively large and impressive males are very protective. When the group is disturbed, resident males approach the threat while other group members hide in the dense vegetation (Kingdon, 1990). Similarly, very little has been published on the social behavior of the elusive *C. lhoesti*. L'hoest's monkeys live in cohesive, one-male groups of 10–29 individuals (Kingdon, 1990; Kaplin & Moermond, 2000). Males tend to stay on the periphery and solitary males

often are seen in the vicinity of larger groups (Kingdon, 1990).

The social structure and organization of *C. neglectus* also differs from the above species (Gautier-Hion & Gautier, 1974; Gautier-Hion, 1988). These monkeys live in smaller groups of three to four individuals and have small overlapping home ranges, and sometimes groups intermingle peacefully. As in other guenons, *C. neglectus* sometimes maintain distance by the exchange of male loud calls and cohesion calls by females and young, though cohesion calls are rarely given. Most of the time, *C. neglectus* monkeys remain silent and their white pelage around the hindlimbs and tail is probably used as a visual signal. Group members maintain high interindividual distance, between 10–30 m apart 52% of the time.

Miopithecus talapoin also live in social groups that differ from those of most forest-dwelling guenons. Talapoin live in multi-male groups containing between 40–150 individuals, with crop-raiding groups being larger than those not dependent upon human agriculture (Gautier-Hion, 1971, 1988; Rowell, 1973). The male-to-female ratio is 1:2 on average. Home ranges are quite large (as much as 120 ha) and may not overlap; no intergroup interactions have been observed. Day ranges average 2,323 m. Gautier-Hion (1973) describes groups as cohesive, with adult males taking on the role of defender and giving calls that rally the group after alarm situations.

Most forest guenons live in polyspecific groups more frequently than monospecific ones (Struhsaker, 1969, 1981; Gartlan & Struhsaker, 1972; Gautier-Hion & Gautier, 1974; Gautier-Hion, 1978, 1980, 1988; Gautier-Hion et al., 1983; Cords, 1987, 1990, 2000). Indeed, Gautier-Hion (1978: p.278) has noted that "the difficulty is to find one consistently monospecific group to follow." Most of the polyspecific groups are stable and of long duration. For example, in Gabon, *C. nictitans* and *C. pogonias* groups remain together for 95–100% of the time. The two groups coordinate travel patterns and activities, also sharing sleeping sites and fruit sources. Although more independent, *C. cephus* groups can spend between 40–50% of their time in bi-specific or tri-specific associations (Gautier-Hion & Gautier, 1974; Gautier-Hion, 1978, 1980, 1988; Gautier-Hion et al., 1983). In Cameroon, Howard (1977) observed one stable tri-specific association including *C. nictitans*, *C. erythrotis*, and *C. mona*. *C. mona* was seen in a monospecific group in only 9 of 166 sightings, and while feeding in certain areas, *C. erythrotis* was virtually never encountered alone. The significance of these polyspecific associations will be discussed below.

Among most guenon species, intra-group cohesion and intergroup spacing of monospecific groups are maintained through loud calls exchanged by adult males.

These loud calls are similar from species to species and thus serve to identify the location of conspecific groups. When living in polyspecific associations, group coordination is facilitated by continuous contact calls given by the females and juveniles and through loud calls exchanged by males. Each of the groups living together may sleep in separate places or become separated after an alarm or an interspecific dispute. The male calls help the members of polyspecific groups to regroup after these events, though males of different species may play different roles in intergroup coordination. For example, in *C. pogonias* and *C. cephus* mixed groups, *C. pogonias* males call first and adopt the role of a leader male, while *C. cephus* males are always followers (Gautier-Hion, 1988).

While spacing between conspecific groups is maintained by the exchange of male loud calls, in the case of agonistic interactions between groups, females and juveniles become more vocal (Cords 1987, 2012; Brown, 2013). Such interactions are suggestive of territorial behavior. However, whether guenons are—or are not—territorial has been somewhat equivocal, and it is clear that there is variance among species and habitats (Gautier & Gautier-Hion, 1969; Struhsaker, 1970; Marler, 1973; Struhsaker & Leland, 1979; Cords, 1987; Gautier-Hion, 1988). For example, Brown (2013) has found that while *C. ascanius* does not exhibit boundary defense, they do defend discrete feeding patches. Moreover, interpretation as to whether a behavior is territorial or not also relates to the definition of "territorial" itself. Given the definition discussed by Sussman (1999), at least some groups and some species are not territorial either in monospecific or polyspecific groups. Howard (1977) studied group spacing between *C. erythrotis* and *C. mona* groups at his study site in Cameroon (one group of *C. nictitans* also was located within the site). Although the two groups of *C. erythrotis* had overlapping home ranges, they were never observed to engage in agonistic interactions. In his 13-month study, Howard observed 17 agonistic interactions among 4 groups of *C. mona*, at least 7 of which involved physical contact. Again, range boundaries of *C. mona* groups overlapped and these encounters seemed to function to maintain group autonomy and identity rather than any specific boundary. Interspecific interactions at the group level were never agonistic, despite displacements between individuals of different species. Given the high density of the population at this study site, Howard (1977) was struck by how few agonistic interactions occurred.

Mating Systems

Guenons were once believed to be exemplars of a one-male group structure with a "harem" polygynous mating system (Cords, 2012). In this type of system, competition

between males for breeding females should be intense, with male mating success relating to the ability of that male to maintain exclusivity to breeding females and to the length of tenure as a resident male (Cowlishaw & Dunbar, 2000). Some data support this interpretation of guenon group structure and organization. As noted previously, most groups include only one male. Furthermore, resident males can be intolerant of other males, who are often threatened and chased. A number of these resident males monopolize groups of females and, most likely, their reproduction over at least 12-month periods (Cords, 1984). Extra-group males of some species may form loose affiliations, traveling and feeding together, and sometimes displacing resident males. It is of note that infant killing has been observed during some group takeovers among guenons (van Schaik, 2000).

Although the above "harem" polygynous mating system is implicated from a number of studies, data from other studies suggest a more complex system (Cords, 1986, 2012; Enstam & Isbell, 2007). In several long-term studies in which individuals are known and recognizable (including extra-group males) and followed for long periods, this expected exclusive mating system has not been observed (Cords, 1984, 2000a; Henzi & Lawes, 1988). In fact, this clade probably illustrates the potential for confusion among group structure, mating systems, and genetic consequences of mating behavior better than any other (Cords, 2000a). Moreover, in three study sites where the reproductive behavior of guenons has been intensively studied, promiscuous mating behavior has been observed. At Kakamega, Kenya, and Kibale, Uganda, *C. mitis* and *C. ascanius* monkeys are sympatric (Cords 1986, 2000a, 2002; Butynski, 1988), and at Cape Vidal, South Africa, *C. mitis* is allopatric (Henzi & Lawes, 1987, 1988). In one group of Kakamega *C. mitis*, the long-term (4-year) resident male was joined by 18 different males (Cords, 1986). At Cape Vidal, up to 25 males joined the *C. mitis* group. In the Kakamega group, the nonresident males remained with the group from 1–44 days (median = 11.5 days). Several of these transient males left and reentered the group multiple times; when not in the group, they were seen alone, with other males or, in some cases, in neighboring groups. During the middle of the mating season, there were typically 8–11 males in the group per day. On days when more than one male was seen in the group, there was a mean number of 5.9 males ± 2.8 (n = 52). The resident male was present throughout the time.

During this influx of males, 13 of 15 females became sexually receptive, whereas none was receptive with only one resident male. Receptive periods lasted from 1–20 days (median = 11 days); the number of receptive females varied from 1–8 per day (mean = 3.8 ± 2.1, n = 52 days).

On 75% of the 52 days, there were 2 or more receptive females at the same time. The frequency of matings and the presence of more than one male closely paralleled the number of receptive females seen each day, with up to as many as 12 matings in a day (Cords, 2000a). By 2001, 6 groups had been monitored at Kakamega since 1979 (Cords, 2002); through these years, 23% of the breeding seasons (n = 98) were characterized by multi-male influxes. Every group monitored for more than 5 years had at least 2 multi-male influx years. Over a 21-year period, male influxes coincided with the number of receptive females in a group on both a seasonal and daily basis (Cords, 2002). Mating was initiated by following, approaching, and sexual solicitation; among 519 such interactions, 44% (229) were initiated by females and 56% by males. However, in those interactions that ended in copulation, females normally had control over when and with whom they copulated. In 31 of 35 copulations in which the interaction was observed in sufficient detail, it was the female who solicited the male (Cords et al., 1986). It is interesting to note that although the resident male remained agonistically dominant to the other males, and indeed was deferred to by them, he was not seen to copulate during the mating period. He showed relatively little interest in the females and vice versa. However, a resident male of a neighboring troop joined the group for 3 days and did copulate with the resident females.

Some females show little or no sexual interest in the long-term resident male if he is still in the group. They are, however, quite interested in newcomers. Female *C. mitis*, for example, sometimes literally go out of their way to mate with newcomers; indeed, among *C. mitis* females at Kakamega, a further 5–20% of a group's observed copulations were "sneaked" outside multi-male influxes (Cords & Rowell, 1986). Females also often engage in sexual behavior when conception is unlikely or impossible (Pazol et al., 2002). Nonresident males roam and can be successful at mating in different groups both within and between years. Furthermore, resident males can also roam and become nonresidents or be replaced altogether (Cords, 1984). At the Cape Vidal site, residency lasted an average of only 1.6 years in six groups over 2.5 years of study (Macleod et al., 2002).

Nonresident males do not necessarily attempt to take over groups in order to become the resident male. In fact, most appear to have a cavalier attitude to group life (Henzi & Lawes, 1988). They do not challenge and often are not challenged by the resident male, and yet they choose to remain temporary residents and to leave the group after relatively short periods, even in the absence of serious aggression (Tsingalia & Rowell, 1984; Cords, 1986; Henzi & Lawes, 1987, 1988). This may be due to

reduced motivation to challenge the resident male. Henzi and Lawes (1988) propose that nonresident males use two strategies: *hovering*, in which males remain in areas where there are several groups with a large number of females and invade those groups during the mating season, and *wandering*, in which these males travel over a larger range to locate areas with large numbers of receptive females. In any case, it appears that extra-group males have equivalent or even better lifetime fitness values than resident males. There may be factors other than an attempt to maximize mating opportunities that lie behind attempting to become a resident male (Cords 1986; Henzi & Lawes, 1988; see Macleod et al., 2002, for further discussion). Thus, as summarized by Cords (1987, 2012), the picture that emerges from these studies is that guenon mating systems are tremendously variable.

Polyspecific Associations

As noted earlier, a common and salient feature characterizing most guenon species is their propensity for forming polyspecific associations in which social groups of two or more species aggregate (see Table 14.2) (Struhsaker, 1981; Gautier-Hion et al., 1983; Cords, 1987, 1990; Gautier-Hion, 1988; Bryer et al., 2013). The vast majority of *Cercopithecus* species form polyspecific associations with other *Cercopithecus* species as well as with *Lophocebus, Cercocebus, Colobus,* and *Procolobus*; exceptions are *C. neglectus*, which never forms polyspecific associations, and *C. lhoesti*, which only rarely does (Struhsaker, 1981; Gautier-Hion, 1988; Jaffe & Isbell, 2011). When, why, and for how long polyspecific associations take place is certainly multifactorial, with variables such as food distribution and predation risk being particularly important (Bryer et al., 2013). Three categories of polyspecific associations can be identified based on the likelihood that an association will form and the duration of that association: *chance* associations, *short-term* but common associations, and *long-term* associations.

Chance polyspecific associations occur when two social groups of different species happen to come together in the same place during their daily activities. These types of associations are most common in areas with high population densities and/or between monkey species using mutual feeding sites (Waser, 1980, 1982). Short-term polyspecific associations are those that occur for relatively short periods of time but are observed more frequently—and are of longer duration—than would be expected by chance. In these associations, one or both species is likely gaining some advantage from the interaction (e.g., as in *Cebus* and *Saimiri*; Sussman, 1999). In West Africa, for example, guenon and colobus species commonly join groups of *C. diana* because this species is

Table 14.2 Guenon Polyspecific Associations [a]

Species	Polyspecific Associations Formed With:
Allenopithecus nigroviridis	*Cercopithecus ascanius, Cercopithecus wolfi*
Cercopithecus ascanius	*Cercopithecus mitis, Cercopithecus wolfi, Allenopithecus nigroviridis, Procolobus badius, Colobus guereza, Colobus angolensis, Lophocebus albigena*
Cercopithecus cephus	*Cercopithecus nictitans, Cercopithecus pogonias, Cercocebus torquatus, Lophocebus albigena*
Cercopithecus erythrogaster	*Cercopithecus mona, Cercocebus torquatus*
Cercopithecus erythrotis	*Lophocebus albigena*
Cercopithecus petaurista	*Cercopithecus campbelli, Cercopithecus diana, Procolobus verus*
Cercopithecus diana	*Cercopithecus campbelli, Cercopithecus petaurista, Cercopithecus nictitans, Procolobus verus, Procolobus badius, Procolobus polykomos*
Cercopithecus mitis	*Cercopithecus ascanius, Procolobus badius, Colobus guereza, Colobus polykomos, Lophocebus albigena*
Cercopithecus nictitans	*Cercopithecus pogonias, Cercopithecus diana, Cercopithecus cephus, Lophocebus albigena*
Cercopithecus campbelli	*Cercopithecus petaurista, Cercopithecus diana, Procolobus verus*
Cercopithecus mona	*Cercopithecus erythrogaster, Cercocebus torquatus*
Cercopithecus pogonias	*Cercopithecus cephus, Cercopithecus nictitans*
Cercopithecus wolfi	*Cercopithecus ascanius, Allenopithecus nigroviridis, Colobus polykomos, Lophocebus aterrimus*

[a] Data from: Struhsaker (1981), Waser (1982), Gautier-Hion et al. (1983), Glenn and Cords (2002), Jaffe and Isbell (2011). Only associations that have been described/published are included; data insufficient for all other species.

particularly vigilant toward prey (Cords, 1990; Höner et al., 1997; Noë & Bshary, 1997). Conversely, social groups of *Procolobus verus* frequently may join groups of other species because they are particularly vulnerable to predation (Noë & Bshary, 1997).

Long-term polyspecific associations were described by Sussman (1999) and are particularly well known among sympatric *Saguinus* species. These stable associations are also very common among the arboreal guenons (Gautier-Hion & Gautier, 1974; Howard, 1977; Gautier-Hion, 1978; Struhsaker, 1981; Gautier-Hion et al., 1983; Gautier-Hion, 1988; Cords, 1990; Kane & McGraw, 2017). The functioning of these relatively permanent polyspecific associations may be quite different from those of short-term and passing endurance. Guenon species engaged in a stable and fairly permanent association typically have compatible and complementary lifestyles, occupy overlapping forest strata, have similar home and day ranges, and have similar daily activity patterns. These species also tend to be primarily frugivorous, with inter-specific differences relating more to insect and vegetative plant portions of their diets. Vocal repertoires are homologous (Gautier-Hion et al., 1981; Gautier-Hion, 1988). As stated by Gautier-Hion (1988: p.461): "Thus the proverb 'birds of a feather flock together' seems to apply better to guenons' polyspecific associations than the principle of competitive exclusion (Gause, 1934)."

Struhsaker (1981) has argued that many primate polyspecific associations cannot be explained by chance alone, particularly as the "chance" models fail to take into account the duration of associations. Moreover, there are discernable and predictable patterns of who associates with whom, and which species initiates the association.

For example, in Kibale, among *C. ascanius, C. mitis*, and *Lophocebus albigena*, less abundant species tend to join more abundant species, species with smaller social groups tend to join species with larger social groups, species with larger home ranges join species with smaller home ranges, and larger-bodied species tend to join smaller-bodied species (which may be related to the former tendency) (Struhsaker, 1981).

If polyspecific associations are not simply chance phenomena, then an adaptive reason for the existence of these mixed groups is expected, with at least one species benefiting and no aspects of parasitism without some other mitigating benefit. An early, and unique, explanation for polyspecific associations in Kibale was forwarded by Freeland (1977), who suggested that such associations reduce the probability that individuals will be bitten by disease-carrying flies; this argument was based on a correlation between polyspecific associations and biting fly activity. However, later investigation failed to substantiate such a correlation (Struhsaker, 1981). More common explanations for polyspecific associations in African forest monkeys are those related to increased efficiency of food procurement and reduction of predator pressure (discussed in detail in Sussman, 1999).

In Kibale, *C. ascanius* have home ranges that are much smaller than those of *C. mitis* and *L. albigena* and presumably also have more intensive, fine-grain knowledge of food availability within these smaller areas. *C. mitis* and *L. albigena* may employ *C. ascanius* to locate these resources (Struhsaker, 1981). However, *C. ascanius* appear to also gain a foraging advantage by this association. Indeed, *C. ascanius* have been seen to follow, approach, and even line up behind *L. albigena* as they open and feed upon certain foods such as the extremely large ("soccer ball–sized") fruit of *Monodora myristica*, which they themselves are unable to physically open and access. *L. albigena* often leave partially eaten fruit still attached to the tree, allowing access to remaining mesocarp by *C. ascanius* (Struhsaker, 1981).

In West Africa, some mixed-species groups of *Cercopithecus* monkeys are highly integrated units (Gautier-Hion, 1980, 1988). The species in these mixed groups often share abundant fruit resources but avoid competition by reducing dietary overlap between species, as well as age and sex groups, when food is scarce. For example, within the mixed-species groups of guenons in Gabon, during periods of scarcity, *C. nictitans* males and females and *C. cephus* males decrease intake of fruit and animal matter and increase their intake of leaves. *C. pogonias* males and females and *C. cephus* females do not change their diets appreciably. In addition, *C. nictitans* males use the highest forest strata (20–30 m); *C. pogonias* males and *C. nictitans* females are found slightly lower in the canopy (20–25 m); *C. cephus* males and *C. pogonias* females usually forage 15–20 m high; and the lowest levels are utilized by *C. cephus* females. Classes of individuals colonizing the same forest level show the least overlap in their diets. Gautier-Hion (1980: pp.259 & 261) interprets these patterns: "By means of these differences not only the ecological separation between species is increased but there is a more complete exploitation of the milieu... polyspecific associations between guenons in West African rainforest cannot be accounted for by a simple grouping at large food sources. The fact that they do not disband as food is scarce but avoid competition by reducing overlap between species diets accounts for the strength of the bond which unites them."

Even in the absence of a foraging benefit to both species (i.e., mutualism), at the least a permanent polyspecific association between two or more species must not hinder the foraging efficiency of the species involved (i.e., commensalism). Indeed, because mixed groups associate when not foraging, other benefits are almost certainly involved, namely, predator protection (Gautier-Hion et al., 1983; Cords, 1987, 1990; Gautier-Hion, 1988). More individuals result in more vigilance and a larger area under surveillance, yielding improved detection and communication of the presence of a predator. Moreover, more animals (especially more males) available to mob a predator results in overall greater levels of predator deterrence. This may be very important for arboreal guenons that live in one-male groups in which the male is the most vigilant animal and the one communicating the presence of danger. Wolters and Zuberbühler (2003) reported on polyspecific associations between *C. campbelli* and *C.*

diana, and found that when these two species are in association, they spend less of their daily activity in vigilance and scanning for predators. The time otherwise spent on predator detection was used by the animals for foraging and feeding.

Furthermore, by fulfilling different roles, associated species can increase this advantage. Among the guenons of Gabon, depending on the source and nature of the danger, the probability of detection is not the same for each species. Being lower in the canopy, *C. cephus* have the major role in responding to danger from the ground, while adult male *C. pogonias* are the most vigilant polyspecific group member with regard to raptors such as *Stephanoaetus coronatus*. *C. nictitans* individuals position themselves higher in trees and hence are most likely to detect predators first; *C. nictitans* males react later and actively drive eagles away (Gautier-Hion et al., 1983; Gautier-Hion, 1988).

Gautier-Hion (1988) has suggested that the roles played within the social organization of polyspecific associations enable the different species to exploit a greater diversity of habitats and microhabitats. In Gabon, when alone, *C. cephus* groups are usually found in the densest understory, although this is the least resource-rich (especially fruit) microhabitat. The increased protection afforded by association with *C. pogonias* and *C. nictitans* enables them to enter more open, taller, and resource-rich forest (Gautier-Hion et al., 1981, 1983; Gautier-Hion, 1988). *C. cephus* is a relatively small and cryptic guenon species, and hence vulnerable to predation. Thus, its association in mixed groups makes it less vulnerable to danger from the most deadly predator, the crowned hawk-eagle. However, *C. cephus* normally has a smaller home range and may have a better knowledge of the resources in the area and may be more vigilant for terrestrial predators such as leopards and humans (Gautier-Hion, 1983). In fact, if the *C. cephus* group becomes separated from the others during intra-species disputes, *C. pogonias* and *C. nictitans* might wait for up to an hour with the *C. pogonias* male giving loud calls to rally *C. cephus*.

CONSERVATION

Throughout Africa, primate populations are being decimated by logging, habitat conversion, and a voracious illegal commercial bushmeat trade that is heavily impacting *Cercopithecus* monkeys (Wilkie et al., 1992; Oates, 1999; Struhsaker, 1999; Cowlishaw & Dunbar, 2000; Oates et al., 2000; Chapman & Peres, 2001; Ukizintambara & Thebaud, 2002; Butynski, 2002b; Mittermeier et al., 2007; Easton et al., 2011; Covey & McGraw, 2014). Indeed, 26% of *Cercopithecus* species are threatened with extinction, some critically so (Oates, 1999; Butynski,

2004; Mittermeier et al., 2007; Struhsaker, 2010). *C. diana, C. erythrogaster, C. erythrotis, C. hamlyni, C. lhoesti,* and *C. lowei* have been listed as Vulnerable by the International Union for the Conservation of Nature (IUCN, 2019), which, by definition, suggests that these species face a high risk of extinction. Other species are more severely threatened: *C. preussi, C. dryas,* and *C. sclateri* face a very high risk of extinction and are thereby classified as Endangered. *C. diana roloway* is classified as Critically Endangered and is one of the top 25 most-endangered primate species in the world (McGraw et al., 2017), though they were recently sighted in Ghana for the first time in 10 years (Osei et al., 2015). Virtually all other species are either undergoing population decline or have not been studied well enough to determine the stability of their populations in the wild (Jaffe & Isbell, 2011; Schwitzer et al., 2017). In Rwanda, for example, there remains only one, very poorly known population of *C. hamlyni* (Easton et al., 2011).

Causes of population decline in guenons across the continent are exceedingly complex, with multiple (and not mutually exclusive) contributing factors associated with hunting (primarily commercial) and habitat conversion and loss, all of which can be attributed to human population growth and resource use. Of all continents, Africa has by far the highest human population growth, approximately 2.6% per annum (World Resources Institute, 1998; Ukizintambara & Thebaud, 2002; Butynski, 2002b). In a meta-analysis of extinction risks in guenons, Ukizintambara and Thebaud (2002) have argued that all *Cercopithecus* species—including the most terrestrial and savanna-adapted species (i.e., *Chlorocebus* [*Cercopithecus*] *aethiops*)—are invariably associated with woodland and forest habitats; they are thus severely impacted by anthropogenic conversion of forests. In turn, forest conversion is directly linked to human population growth and resource use. Moreover, Ukizintambara and Thebaud (2002) found that most guenon species are distributed in countries that are undergoing the greatest human population growth rate (e.g., Cameroon).

Habitat conversion and loss characterizes the African continent. Greater than two-thirds of original forest habitat has been impacted by anthropogenic activity, and it has been predicted that some particularly biodiverse countries (e.g., Côte d'Ivoire and Nigeria) may lose all their moist forest in the early decades of the twenty-first century (World Resources Institute, 1998; Butynski, 2004). The two most important causes of habitat loss are logging and the clearing of forest for horticulture. Butynski (2002b) reports that 61% of undisturbed forest is threatened by logging, and 13% is threatened by human requirement for agricultural purposes. Thus, almost 75%

of all remaining forest (which is only one-third of the original cover) is threatened. Chapman and Peres (2001) have calculated that countries that are home to primates, on average, lose 125,140 km^2 of forest per year; for scale, they note that this is an area greater than the state of Mississippi (122,335km^2) or just smaller than Greece (131,985 km^2). From this calculation, they conclude that forest lost each year supports upward of 32 million primates (biomass = 123,000 tons).

However, while habitat loss has profoundly impacted the African continent, it has been suggested that hunting may be a more immediate and important threat to *Cercopithecus* species (Butynski, 2002b). Guenons are highly prized for their meat, and commercial trade of monkeys (as well as other mammals, birds, and reptiles) has exploded in Africa, particularly in Central and West Africa. For example, among the Fang of Guinea, Cameroon, and Gabon, *Cercopithecus* monkeys are some of the most hunted and trapped of all forest fauna (Pi & Groves, 1972), and in Côte d'Ivoire, all monkeys are considered very palatable and are hunted or trapped year-round (Bourliere et al., 1970). In Equatorial Guinea, Fa et al. (1995) reported market survey data collected over a year in two urban areas with a combined human population of 107,000. These data indicated the sale of 4,222 primate carcasses. In another bushmeat market located on the border of Liberia and Côte d'Ivoire, Covey and McGraw (2014) report that *C. petaurista* and *C. diana* are the two most abundant species of an estimated 9,500 dead primates sold annually. There is no evidence that illegal hunting is abating in any country in tropical Africa, and if present trends persist, some species are certain to become extinct in the near future (Butynski, 2002b).

Clearly, the issue is urgent and new methods to conserve primates are required (Baker et al., 2011). Lambert (2011) evaluated the genus *Cercopithecus* for its potential utility as an "umbrella species" in conservation tactics. An umbrella species is defined as a *species whose conservation confers protection to a large number of naturally co-occurring species and the important mutualisms among them* (after Roberge & Angelstam, 2004). Choosing which species have the potential to best serve as umbrella species in conservation can be difficult. Lambert (2011) used an index incorporating variables of species richness, species rarity, and species sensitivity to anthropogenic impacts in order to evaluate which primate species in African forests best meet the requirements as an umbrella. *Cercopithecus* emerged as the taxon best fitting the criteria. The fact that guenons are also important seed dispersers means that if guenons are targeted as a taxon for conservation, important mutualisms will be protected as well, with the cascading effects of plants also being protected. However, as she notes, unless immediate and dramatic conservation action is undertaken at all scales of governance, using a *Cercopithecus* umbrella tactic may become obsolete.

SUMMARY

The Old World monkey group known as the guenons, specifically, the genera *Cercopithecus, Allenopithecus,* and *Miopithecus,* range throughout sub-Saharan Africa. Guenons comprise a large and diverse clade that often live sympatrically with other closely related species. Indeed, it is common for three or more guenon species to reside in the same habitat and have overlapping home ranges. Most guenons have colorful pelage or markings that distinguish the different species. Guenons inhabit a broad range of habitat types, from lowland swamps to dense rainforests, though most are arboreal forest obligates.

Guenons not only live sympatrically with other guenons but with many other primate species, and frequently travel and forage in polyspecific associations. Polyspecific associations provide increased vigilance and predator detection, with guenons and other primate species recognizing each other's alarm calls and responding aptly. While these associations also increase food encounter rates, they also increase competition. Thus, sympatric species in polyspecific associations often forage in different strata of the canopy to reduce feeding competition. Guenons vary in their social and mating structures, though one-male/multi-female groups are common, and females generally remain in their natal groups while males disperse.

As with most cercopithecines, guenons are omnivorous, incorporating a broad range of both plant and animal food resources. Guenons consume many plant parts (e.g., fruits, leaves, flowers, stems) and types (e.g., lichen, herbaceous vegetation) as well as small invertebrate (e.g., spiders, insects) and vertebrate prey (e.g., bats, birds). Guenons exhibit flexible feeding strategies, including resource switching, which minimizes feeding competition. Despite this dietary variation, most guenons are highly frugivorous and consume ripe fruits when available. As such, guenons are important seed-dispersal agents and help shape the forests they inhabit. Their roles as key seed dispersers in many habitats make them candidates for increased conservation-based protections.

REFERENCES CITED—CHAPTER 14

Aldrich-Blake, F. (1970). Problems of social structure in forest monkeys. In J. H. Crook (Ed.), *Social behaviour of birds and mammals* (pp. 79-101). Academic Press.

Altmann, S. A. (1998). *Foraging for survival: Yearling baboons in Africa.* University of Chicago Press.

Baker, L. R., Arnold, T. W., Olubode, O. S., & Garshelis, D. L. (2011). Considerations for using occupancy surveys to monitor forest primates: A case study with Sclater's monkey (*Cercopithecus sclateri*). *Population Ecology, 53*(4), 549-561.

Blaine, K. P., & Lambert, J. E. (2012). Digestive retention times for Allen's swamp monkey and L'Hoest's monkey: Data with implications for the evolution of cercopithecine digestive strategy. *Integrative Zoology, 7*(2), 183-191.

Booth, A. (1957). Observations on the natural history of the olive colobus monkey, *Procolobus verus* (van Beneden). *Proceedings of the Zoological Society of London, 129*(3), 421-430.

Bouchet, H., Blois-Heulin, C., & Lemasson, A. (2012). Age- and sex-specific patterns of vocal behavior in De Brazza's monkeys (*Cercopithecus neglectus*). *American Journal of Primatology, 74*(1), 12-28.

Bourliere, F., Hunkeler, C., & Bertrand, M. (1970). Ecology and behaviour of Lowe's guenon (*Cercopithecus campbelli lowei*) in the Ivory Coast. In J. R. Napier, & P. H. Napier (Eds.), *Old World monkeys: Evolution, systematics and behaviour* (pp. 297-350). Academic Press.

Brown, M. (2013). Food and range defence in group-living primates. *Animal Behaviour, 85*(4), 807-816.

Brugiere, D., Gautier, J.-P., Moungazi, A., & Gautier-Hion, A. (2002). Primate diet and biomass in relation to vegetation composition and fruiting phenology in a rain forest in Gabon. *International Journal of Primatology, 23*(5), 999-1024.

Bryer, M. A., Chapman, C. A., Raubenheimer, D., Lambert, J. E., & Rothman, J. M. (2015). Macronutrient and energy contributions of insects to the diet of a frugivorous monkey (*Cercopithecus ascanius*). *International Journal of Primatology, 36*(4), 839-854.

Bryer, M. A., Chapman, C. A., & Rothman, J. M. (2013). Diet and polyspecific associations affect spatial patterns among redtail monkeys (*Cercopithecus ascanius*). *Behaviour, 150,* 277-293.

Butynski, T. M. (1982). Blue monkey (*Cercopithecus mitis stuhlmanni*) predation on galagos. *Primates, 23*(4), 563-566.

Butynski, T. M. (1988). Guenon birth seasons and correlates with rainfall and food. In A. Gautier-Hion, F. Bourliére, J. P. Gautier, & J. Kingdon (Eds.), *A primate radiation: Evolutionary biology of the African guenons* (pp. 284-322). Cambridge University Press.

Butynski, T. M. (1990). Comparative ecology of blue monkeys (*Cercopithecus mitis*) in high- and low-density subpopulations. *Ecological Monographs, 60*(1), 1-26.

Butynski, T. M. (2002a). The guenons: An overview of diversity and taxonomy. In M. E. Glenn, & M. Cords (Eds.), *The guenons: Diversity and adaptation in African monkeys* (pp. 3-13). Springer.

Butynski, T. M. (2002b). Conservation of the guenons: An overview of status, threats, and recommendations. In M. E. Glenn, & M. Cords (Eds.), *The guenons: Diversity and adaptation in African monkeys* (pp. 411-424). Springer.

Buzzard, P. J. (2006a). Cheek pouch use in relation to interspecific competition and predator risk for three guenon monkeys (*Cercopithecus* spp.). *Primates*, 47(4), 336-341.

Buzzard, P. J. (2006b). Ecological partitioning of *Cercopithecus campbelli*, *C. petaurista*, and *C. diana* in the Taï Forest. *International Journal of Primatology*, 27(2), 529-558.

Buzzard, P. J. (2006c). Ranging patterns in relation to seasonality and frugivory among *Cercopithecus campbelli*, *C. petaurista*, and *C. diana* in the Taï Forest. *International Journal of Primatology*, 27(2), 559-573.

Caton, J. M. (1999). Digestive strategy of the Asian colobine genus *Trachypithecus*. *Primates*, 40(2), 311-325.

Chalmers, N. (1972). Comparative aspects of early infant development in some captive cercopithecines. In F. Poirier (Ed.), *Primate socialization* (pp. 63-82). Random House.

Chapman, C. A., Chapman, L. J., Cords, M., Gathua, J. M., Gautier-Hion, A., Lambert, J. E., Rode, K., Tutin, C. E., & White, L. J. (2002). Variation in the diets of *Cercopithecus* species: Differences within forests, among forests, and across species. In M. E. Glenn, & M. Cords (Eds.), *The guenons: Diversity and adaptation in African monkeys* (pp. 325-350). Springer.

Chapman, C. A., Gautier-Hion, A., Oates, J. F., & Onderdonk, D. A. (1999). African primate communities: Determinants of structure and threats to survival. In J. G. Fleagle, C. H. Janson, & K. Reed (Eds.), *Primate communities* (pp. 1-37). Cambridge University Press.

Chapman, H. M., Goldson, S. L., & Beck, J. (2010). Postdispersal removal and germination of seed dispersed by *Cercopithecus nictitans* in a West African Montane Forest. *Folia Primatologica*, 81(1), 41-50.

Chapman, C. A., & Peres, C. A. (2001). Primate conservation in the new millennium: The role of scientists. *Evolutionary Anthropology*, 10(1), 16-33.

Clemens, E., & Maloiy, G. (1981). Organic acid concentrations and digesta movement in the gastrointestinal tract of the bushbaby (*Galago crassicaudatus*) and vervet monkey (*Cercopithecidae pygerythrus*). *Journal of Zoology*, 193(4), 487-497.

Clemens, E., & Phillips, B. (1980). Organic acid production and digesta movement in the gastrointestinal tract of the baboon and Sykes monkey. *Comparative Biochemistry and Physiology*, 66(3), 529-532.

Colyn, M., & Deleporte, P. (2002). Biogeographic analysis of Central African forest guenons. In M. E. Glenn, & M. Cords (Eds.), *The guenons: Diversity and adaptation in African monkeys* (pp. 61-78). Springer.

Colyn, M., Gautier-Hion, A., & van den Audenaerde, T. (1991). *Cercopithecus dryas* Schwarz 1932 and *C. salongo* Thys van den Audenaerde 1977 are the same species with an age-related coat pattern. *Folia Primatologica*, 56(3), 167-170.

Cordeiro, N. J. (1992). Behaviour of blue monkeys (*Cercopithecus mitis*) in the presence of crowned eagles (*Stephanoaetus coronatus*). *Folia Primatologica*, 59(4), 203-206.

Cords, M. (1984). Mating patterns and social structure in redtail monkeys (*Cercopithecus ascanius*). *Zeitschrift für Tierpsychologie*, 64(3-4), 313-329.

Cords, M. (1986). Interspecific and intraspecific variation in diet of two forest guenons, *Cercopithecus ascanius* and *C. mitis*. *Journal of Animal Ecology*, 55(3), 811-827.

Cords, M. (1987). Mixed-species association of *Cercopithecus* monkeys in the Kakamega Forest, Kenya. *University of California Publications in Zoology*, 1, 1-109.

Cords, M. (1990). Mixed-species association of East African guenons: General patterns or specific examples? *American Journal of Primatology*, 21(2), 101-114.

Cords, M. (2000a). Agonistic and affiliative relationships in a blue monkey group. In C. J. Jolly, & P. F. Whitehead (Ed.), *Old World monkeys* (pp. 453-479). Cambridge University Press.

Cords, M. (2000b). The number of males in guenon groups. In P. M. Kappeler (Ed.), *Primate males: Causes and consequences of variation in group composition* (pp. 84-96). Cambridge University Press.

Cords, M. (2002). Foraging and safety in adult female blue monkeys in the Kakamega Forest. In L. E. Miller (Ed.), *Eat or be eaten: Predator sensitive foraging among primates* (pp. 205-221). Cambridge University Press.

Cords, M. (2012). The behavior, ecology and social evolution of cercopithecine monkeys. In J. C. Mitani, J. Call, P. M. Kappeler, R. A. Palombit, & J. B. Silk (Eds.), *The evolution of primate societies* (pp. 91-112). University of Chicago Press.

Cords, M., & Rowell, T. (1986). Group fission in blue monkeys of the Kakamega Forest, Kenya. *Folia Primatologica*, 46(2), 70-82.

Corlett, R., & Lucas, P. (1990). Alternative seed-handling strategies in primates: Seed-spitting by long-tailed macaques (*Macaca fascicularis*). *Oecologia*, 82(2), 166-171.

Covey, R., & McGraw, W. S. (2014). Monkeys in a West African bushmeat market: Implications for cercopithecid conservation in eastern Liberia. *Tropical Conservation Science*, 7(1), 115-125.

Cowlishaw, G., & Dunbar, R. I. (2000). *Primate conservation biology*. University of Chicago Press.

Detwiler, K. M. (2002). Hybridization between red-tailed monkeys (*Cercopithecus ascanius*) and blue monkeys (*C. mitis*) in east African forests. In M. E. Glenn, & M. Cords (Eds.), *The guenons: Diversity and adaptation in African monkeys* (pp. 79-97). Springer.

Detwiler, K. M. (2019). Mitochondrial DNA analyses of *Cercopithecus* monkeys reveal a localized hybrid origin for *C. mitis doggetti* in Gombe National Park, Tanzania. *International Journal of Primatology*, 40(1), 28-52.

Detwiler, K. M., Burrell, A. S., & Jolly, C. J. (2005). Conservation implications of hybridization in African cercopithecine monkeys. *International Journal of Primatology*, 26(3), 661-684.

Dowsett-Lemaire, F. (1988). Fruit choice and seed dissemination by birds and mammals in the evergreen forests of upland Malawi. *Revue d'écologie*, 43, 251-286.

Easton, J., Chao, N., Mulindahabi, F., Ntare, N., Rugyerinyange, L., & Ndikubwimana, I. (2011). Status and conservation of the only population of the vulnerable owl-faced monkey *Cercopithecus hamlyni* in Rwanda. *Oryx*, 45(3), 435-438.

Enstam, K. L., & Isbell, L. A. (2007). The guenons (genus *Cercopithecus*) and their allies: Behavioral ecology of polyspecific associations. In C. J. Campbell, A. Fuentes, K. C. McKinnon, M. Panger, & S. K. Bearder (Eds.), *Primates in perspective* (pp. 252-274). Oxford University Press.

Fa, J. E., Juste, J., Del Val, J. P., & Castroviejo, J. (1995). Impact of market hunting on mammal species in equatorial Guinea. *Conservation Biology*, 9(5), 1107-1115.

Fairgrieve, C., & Muhumuza, G. (2003). Feeding ecology and dietary differences between blue monkey (*Cercopithecus mitis stuhlmanni* Matschie) groups in logged and unlogged forest, Budongo Forest Reserve, Uganda. *African Journal of Ecology*, 41(2), 141-149.

Fleagle, J. (1999). *Primate adaptation and evolution*. Academic Press.

Foerster, S., Cords, M., & Monfort, S. L. (2011). Social behavior, foraging strategies, and fecal glucocorticoids in female blue monkeys (*Cercopithecus mitis*): Potential fitness benefits of high rank in a forest guenon. *American Journal of Primatology*, 73(9), 870-882.

Förster, S., & Cords, M. (2002). Development of mother-infant relationships and infant behavior in wild blue monkeys (*Cercopithecus mitis stuhlmanni*). In M. E. Glenn, & M. Cords (Eds.), *The guenons: Diversity and adaptation in African monkeys* (pp. 245-272). Springer.

Freeland, W. (1977). Blood-sucking flies and primate polyspecific associations. *Nature*, 269(5631), 801.

Furuichi, T. (2006). Red-tailed monkeys (*Cercopithecus ascanius*) hunt green pigeons (*Treron calva*) in the Kalinzu Forest in Uganda. *Primates*, 47(2), 174-176.

Galat-Luong, A. (1975). Notes preliminaires sur l'écologie de *Cercopithecus ascanius schmidti* dans les environs de Bangui (RCA). *La Terre et la Vie*, 24, 288-297.

Gartlan, S. J., & Struhsaker, T. T. (1972). Polyspecific associations and niche separation of rain-forest anthropoids in Cameroon, West Africa. *Journal of Zoology*, 168, 221-266.

Gause, G. F. (1934). *The struggle for existence*. Williams and Wilkins.

Gautier, J.-P., & Gautier-Hion, A. (1969). Les associations polyspécifiques chez les Cercopithecidae du Gabon. *La Terre et la Vie*, 23, 164-201.

Gautier-Hion, A. (1971). L'écologie du talapoin du Gabon. *La Terre et la Vie*, 4, 427-490.

Gautier-Hion, A. (1973). Social and ecological features of talapoin monkey-comparison with sympatric cercopithecines. In R. P. Michael, & J. H. Crook (Eds.), *Comparative ecology and behavior of primates* (pp. 147-170). Academic Press.

Gautier-Hion, A. (1978). Food-niches and co-existence in sympatric primates in Gabon. In D. J. Chivers, & J. Herbert (Eds.), *Recent advances in primatology* (pp. 269-286). Academic Press.

Gautier-Hion, A. (1980). Seasonal variations of diet related to species and sex in a community of *Cercopithecus* monkeys. *Journal of Animal Ecology*, 237-269.

Gautier-Hion, A. (1988). The diet and dietary habits of forest guenons. In A. Gautier-Hion, F. Bourliére, J. P. Gautier, & J. Kingdon (Eds.), *A primate radiation: Evolutionary biology of the African guenons* (pp. 257-283). Oxford University Press.

Gautier-Hion, A., & Gautier, J.-P. (1974). Les associations polyspécifiques de cercopithèques du Plateau de M'passa (Gabon) (Part 1 of 4). *Folia Primatologica*, 22(2-3), 134-144.

Gautier-Hion, A., & Gautier, J.-P. (1976). Growth, sexual maturity, social maturity and reproduction in African forest cercopithecines [author's translation]. *Folia Primatologica*, 26(3), 165-184.

Gautier-Hion, A., Gautier, J.-P., & Quris, R. (1981). Forest structure and fruit availability as complementary factors influencing habitat use by a troop of monkeys (*Cercopithecus cephus*). *Revue d'écologie*, 35, 511-536.

Gautier-Hion, A., Quris, R., & Gautier, J.-P. (1983). Monospecific vs polyspecific life: A comparative study of foraging and antipredatory tactics in a community of *Cercopithecus* monkeys. *Behavioral Ecology and Sociobiology*, 12(4), 325-335.

Gilbert, C. C., Bibi, F., Hill, A., & Beech, M. J. (2014). Early guenon from the late Miocene Baynunah Formation, Abu Dhabi, with implications for cercopithecoid biogeography and evolution. *Proceedings of the National Academy of Sciences*, 111(28), 10119-10124.

Glenn, M. E., Cords, M. (2002). Preface. In M. E. Glenn, & M. Cords (Eds.), *The guenons: Diversity and adaptation in African monkeys* (pp. xiii-xvi). Springer.

Glenn, M. E., Matsuda, R., Bensen, K. J. (2002). Unique behavior of the mona monkey (*Cercopithecus mona*): All-male groups and copulation calls. In M. E. Glenn, & M. Cords (Eds.), *The guenons: Diversity and adaptation in African monkeys* (pp. 133-145). Springer.

Gross-Camp, N. D., Mulindahabi, F., & Kaplin, B. A. (2009). Comparing the dispersal of large-seeded tree species by frugivore assemblages in tropical Montane Forest in Africa. *Biotropica*, 41(4), 442-451.

Groves, C. P. (2001). *Primate taxonomy*. Smithsonian Institution Press.

Grubb, P., Butynski, T. M., Oates, J. F., Bearder, S. K., Disotell, T. R., Groves, C. P., & Struhsaker, T. T. (2003). Assessment of the diversity of African primates. *International Journal of Primatology*, 24(6), 1301-1357.

Haddow, A. (1952). Field and laboratory studies on an African monkey, *Cercopithecus ascanius schmidti* Matschie. *Proceedings of the Zoological Society of London*, 297-394.

Hamilton, A. (1988). Guenon evolution and forest history. In A. Gautier-Hion, F. Bourliére, J. P. Gautier, & J. Kingdon (Eds.), *A primate radiation: Evolutionary biology of the African guenons* (pp. 13-34). Oxford University Press.

Hart, D. L. (2000). *Primates as prey: Ecological, morphological and behavioral relationships between primate species and their predators*. [Doctoral dissertation, Washington University].

Hart, J., Katembo, M., & Punga, K. (1996). Diet, prey selection and ecological relations of leopard and golden cat in the Ituri Forest, Zaire. *African Journal of Ecology, 34*(4), 364-379.

Hart, J. A., Detwiler, K. M., Gilbert, C. C., Burrell, A. S., Fuller, J. L., Emetshu, M., Hart, T. B., Vosper, A., Sargis, E. J., & Tosi, A. J. (2012). Lesula: A new species of *Cercopithecus* monkey endemic to the Democratic Republic of Congo and implications for conservation of Congo's Central Basin. *PLoS ONE, 7*(9), e44271.

Henzi, S., & Lawes, M. (1987). Breeding season influxes and the behaviour of adult male samango monkeys (*Cercopithecus mitis albogularis*). *Folia Primatologica, 48*(3-4), 125-136.

Henzi, S., & Lawes, M. (1988). Strategic responses of male samango monkeys (*Cercopithecus mitis*) to a decline in the number of receptive females. *International Journal of Primatology, 9*(5), 479-495.

Höner, O. P., Leumann, L., & Noë, R. (1997). Dyadic associations of red colobus and Diana monkey groups in the Taï National Park, Ivory Coast. *Primates, 38*(3), 281-291.

Howard, R. (1977). *Niche separation among three sympatric species of* Cercopithecus *monkeys*. [Doctoral dissertation, University of Texas at Austin].

IUCN. (2019). *The International Union for the Conservation of Nature Red List of Threatened Species*. Version 2019-3. https://www.iucnredlist.org.

Jaffe, K. E., & Isbell, L. A. (2011). The guenons: Polyspecific associations in socioecological perspective. In C. J. Campbell, A. Fuentes, K. C. McKinnon, M. Panger, & S. K. Bearder (Eds.), *Primates in perspective* (pp. 277-300). Oxford University Press.

Kamilar, J. M., Martin, S. K., & Tosi, A. J. (2009). Combining biogeographic and phylogenetic data to examine primate speciation: An example using cercopithecine monkeys. *Biotropica 41*(4), 514-519.

Kane, E. E. (2017). *Socioecology, stress, and reproduction among female diana monkeys (*Cercopithecus diana*) in Côte d'Ivoire's Taï National Park*. [Doctoral dissertation, Ohio State University].

Kane, E. E., & McGraw, W. S. (2017). Dietary variation in diana monkeys (*Cercopithecus diana*): The effects of polyspecific associations. *Folia Primatologica, 88*(6), 455-482.

Kaplin, B. A., & Lambert, J. E. (2002). Effectiveness of seed dispersal by *Cercopithecus* monkeys: Implications for seed input into degraded areas. In D. J. Levey, W. R. Silva, & M. Galetti (Eds.), *Seed dispersal and frugivory: Ecology, evolution and conservation* (pp. 351-364). CABI Publishing.

Kaplin, B., & Moermond, T. (2000). Foraging ecology of the mountain monkey (*Cercopithecus l'hoesti*): Implications for its evolutionary history and use of disturbed forest. *American Journal of Primatology, 50*(4), 227-246.

Kaplin, B., Munyaligoga, V., & Moermond, T. (1998). The influence of temporal changes in fruit availability on diet composition and seed handling in blue monkeys (*Cercopithecus mitis doggetti*). *Biotropica, 30*(1), 56-71.

Kingdon, J. (1971). *Atlas of East African mammals. Vol. 1*. Academic Press.

Kingdon, J. (1988). What are face patterns and do they contribute to reproductive isolation in guenons? In A. Gautier-Hion, F. Bourliére, J. P. Gautier, & J. Kingdon (Eds.), *A primate radiation: Evolutionary biology of the African guenons* (pp. 227-245). Oxford University Press.

Kingdon, J. (1990). *Island Africa: The evolution of Africa's rare animals and plants*. Collins.

Lambert, J. E. (1998). Primate frugivory in Kibale National Park, Uganda, and its implications for human use of forest resources. *African Journal of Ecology, 36*(3), 234-240.

Lambert, J. E. (1999). Seed handling in chimpanzees (*Pan troglodytes*) and redtail monkeys (*Cercopithecus ascanius*): Implications for understanding hominoid and cercopithecine fruit-processing strategies and seed dispersal. *American Journal of Physical Anthropology, 109*(3), 365-386.

Lambert, J. E. (2001). Red-tailed guenons (*Cercopithecus ascanius*) and *Strychnos mitis*: Evidence for plant benefits beyond seed dispersal. *International Journal of Primatology, 22*(2), 189-201.

Lambert, J. E. (2002). Resource switching and species coexistence in guenons: A community analysis of dietary flexibility. In M. E. Glenn, & M. Cords (Eds.), *The guenons: Diversity and adaptation in African monkeys* (pp. 309-323). Springer.

Lambert, J. E. (2005). Competition, predation, and the evolutionary significance of the cercopithecine cheek pouch: The case of *Cercopithecus* and *Lophocebus*. *American Journal of Physical Anthropology, 126*(2), 183-192.

Lambert, J. E. (2011). Primate seed dispersers as umbrella species: A case study from Kibale National Park, Uganda, with implications for Afrotropical forest conservation. *American Journal of Primatology, 73*(1), 9-24.

Lambert, J. E., & Whitham, J. C. (2001). Cheek pouch use in *Papio cynocephalus*. *Folia Primatologica, 72*(2), 89.

Lawes, M. (1991). Diet of samango monkeys (*Cercopithecus mitis erythrarchus*) in the Cape Vidal dune forest, South Africa. *Journal of Zoology, 224*(1), 149-173.

Leakey, M. (1988). Fossil evidence for the evolution of the guenons. In A. Gautier-Hion, F. Bourliére, J. P. Gautier, & J. Kingdon (Eds.), *A primate radiation: Evolutionary biology of the African guenons* (pp. 7-12). Oxford University Press.

Leland, L., & Struhsaker, T. T. (1993). Teamwork tactics: Kibale Forest's monkeys and eagles each depend on strategic cooperation for survival. *Natural History, 102*, 42-48.

Linden, B., Linden, J., Fischer, F., & Linsenmair, K. E. (2015). Seed dispersal by South Africa's only forest-dwelling guenon, the samango monkey (*Cercopithecus mitis*). *African Journal of Wildlife Research, 45*(1), 88-100.

Lyke, M. M., Di Fiore, A., Fierer, N., Madden, A. A, & Lambert, J. E. (2019). Metagenomic analyses reveal previously unrecognized variation in the diets of sympatric Old World monkey species. *PLoS ONE, 14*(6), e0218245.

Macleod, M. C., Ross, C., & Lawes, M. J. (2002). Costs and benefits of alternative mating strategies in samango monkey males. In M. E. Glenn, & M. Cords (Eds.), *The guenons: Diversity and adaptation in African monkeys* (pp. 203-216). Springer.

Maisels, F. (1993). Gut passage rate in guenons and mangabeys: Another indicator of a flexible feeding niche? *Folia Primatologica,* 61(1), 35-37.

Maisels, F., & Gautier-Hion, A. (1994). Why are Caesalpinioideae so important for monkeys in hydromorphic rainforests of the Zaire basin? In J. I. Sprent, & D. McKey (Eds.), *Advances in legume systematics 5: The nitrogen factor* (pp. 189-204). Royal Botanical Gardens.

Marler, P. (1973). A comparison of vocalizations of red-tailed monkeys and blue monkeys, *Cercopithecus ascanius* and *C. mitis,* in Uganda 1. *Zeitschrift für Tierpsychologie,* 33(3-4), 223-247.

McGraw, S. W., Oates, J. F., & Dempsey, A. (2017). *Cercopithecus diana roloway* (Schreber 1774). In C. Schwitzer, R. A. Mittermeier, A. B. Rylands, F. Chiozza, E. A. Williamson, E. J. Macfie, J. Wallis, & A. Cotton (Eds.), *Primates in peril: The world's 25 most endangered primates 2016–2018* (pp. 40-43). Primate Specialists Group.

Mitani, J. C., Sanders, W. J., Lwanga, J. S., & Windfelder, T. L. (2001). Predatory behavior of crowned hawk-eagles (*Stephanoaetus coronatus*) in Kibale National Park, Uganda. *Behavioral Ecology and Sociobiology,* 49(2-3), 187-195.

Mittermeier, R. A., Ratsimbazafy, J., Rylands, A. B., Williamson, L., Oates, J. F., Mbora, D., Ganzhorn, J. U., Rodríguez-Luna, E., Palacios, E., & Heymann, E. W. (2007). Primates in peril: The world's 25 most endangered primates, 2006–2008. *Primate Conservation,* 22(1), 1-41.

Moreau, R. (1969). Climatic changes and the distribution of forest vertebrates in West Africa. *Journal of Zoology,* 158(1), 39-61.

Noë, R., & Bshary, R. (1997). The formation of red colobus–diana monkey associations under predation pressure from chimpanzees. *Proceedings of the Royal Society of London,* 264(1379), 253-259.

Oates, J. (1988). The distribution of *Cercopithecus* monkeys in West African forest. In A. Gautier-Hion, F. Bourliére, J. P. Gautier, & J. Kingdon (Eds.), *A primate radiation: Evolutionary biology of the African guenons* (pp. 79-103). Cambridge University Press.

Oates, J. F. (1999). *Myth and reality in the rain forest: How conservation strategies are failing in West Africa.* University of California Press.

Oates, J. F., Abedi-Lartey, M., McGraw, W. S., Struhsaker, T. T., & Whitesides, G. H. (2000). Extinction of a West African red colobus monkey. *Conservation Biology,* 14(5), 1526-1532.

Osei, D., Horwich, R. H., & Pittman, J. M. (2015). First sightings of the roloway monkey (*Cercopithecus diana roloway*) in Ghana in ten years and the status of other endangered primates in southwestern Ghana. *African Primates,* 10, 25-40.

Pazol, K., Carlson, A. A., & Ziegler, T. E. (2002). Female reproductive endocrinology in wild blue monkeys: A preliminary assessment and discussion of potential adaptive functions. In M. E. Glenn, & M. Cords (Eds.), *The guenons: Diversity and adaptation in African monkeys* (pp. 217-232). Springer.

Peignot, P., Fontaine, B., & Wickings, E. (1999). Habitat exploitation, diet and some data on reproductive behaviour in a semi-free-ranging colony of *Cercopithecus l'hoesti solatus,* a guenon species recently discovered in Gabon. *Folia Primatologica,* 70(1), 29-36.

Pi, J. S., & Groves, C. (1972). The importance of higher primates in the diet of the Fang of Rio Muni. *Man,* 7(2), 239-243.

Plumptre, A. J., & Reynolds, V. (1994). The effect of selective logging on the primate populations in the Budongo Forest Reserve, Uganda. *Journal of Applied Ecology,* 631- 641.

Poulsen, J. R., Clark, C. J., Connor, E. F., & Smith, T. B. (2002). Differential resource use by primates and hornbills: Implications for seed dispersal. *Ecology,* 83(1), 228-240.

Roberge, J. M., & Angelstam, P. (2004). Usefulness of the umbrella species concept as a conservation tool. *Conservation Biology,* 18(1), 76-85.

Rowell, T. E. (1973). Social organization of wild talapoin monkeys. *American Journal of Physical Anthropology,* 38(2), 593-597.

Rowell, T. E., & Mitchell, B. J. (1991). Comparison of seed dispersal by guenons in Kenya and capuchins in Panama. *Journal of Tropical Ecology,* 7(2), 269-274.

Rowell, T. E., Wilson, C., & Cords, M. (1991). Reciprocity and partner preference in grooming of female blue monkeys. *International Journal of Primatology,* 12(4), 319-336.

Rudran, R. (1978a). Socio ecology of the blue monkeys (*Cercopithecus mitis stuhlmanni*) of the Kibale Forest, Uganda. *Smithsonian Contributions to Zoology,* 249, 1-99.

Rudran, R. (1978b). Intergroup dietary comparisons and folivorous tendencies of two groups of blue monkeys (*Cercopithecus mitis stuhlmanni*). In G. G. Montgomery (Ed.), *The ecology of arboreal folivores* (pp. 483-503). Smithsonian Press.

Ruvolo, M. (1988). Genetic evolution in the African guenons. In A. Gautier-Hion, F. Bourliere, J. P. Gautier, & J. Kingdon (Eds.), *A primate radiation: Evolutionary biology of the African guenons* (pp. 127-149). Cambridge University Press.

Sanders, W. J., Trapani, J., & Mitani, J. C. (2003). Taphonomic aspects of crowned hawk-eagle predation on monkeys. *Journal of Human Evolution,* 44(1), 87-105.

Schwitzer, C., Mittermeier, R. A., Rylands, A. B., Chiozza, F., Williamson, E. A., Macfie, E. J., Wallis, J. & Cotton, A. (Eds.). (2017). *Primates in peril: The world's 25 most endangered primates 2016–2018.* Primate Specialists Group.

Sheppard, D. J. (2000). *Ecology of the Budongo Forest redtail: Patterns of habitat use and population density in primary and regenerating forest sites.* [Master's thesis, University of Calgary].

Shultz, S., Noë, R., McGraw, W. S., & Dunbar, R. (2004). A community-level evaluation of the impact of prey behavioural and ecological characteristics on predator diet composition. *Proceedings of the Royal Society of London,* 271(1540), 725-732.

Skorupa, J. P. (1989). Crowned eagles *Strephanoaetus coronatus* in rainforest: Observations on breeding chronology and diet at a nest in Uganda. *Ibis,* 131(2), 294-298.

Smith, L. W., Link, A., & Cords, M. (2008). Cheek pouch use, predation risk, and feeding competition in blue monkeys (*Cercopithecus mitis stuhlmanni*). *American Journal of Physical Anthropology,* 137(3), 334-341.

Stephan, C., & Zuberbühler, K. (2016). Persistent females and compliant males coordinate alarm calling in diana monkeys. *Current Biology,* 26(21), 2907-2912.

Struhsaker, T. T. (1969). Correlates of ecology and social organization among African cercopithecines. *Folia Primatologica,* 11(1-2), 80-118.

Struhsaker, T. T. (1970). Phylogenetic implications of some vocalizations of *Cercopithecus* monkeys. In J. R. Napier, & P. H. Napier (Eds.), *Old World monkeys* (pp. 367-444). Academic Press.

Struhsaker, T. T. (1977). Infanticide and social organization in the redtail monkey (*Ceraopithecus ascanius schmidti*) in the Kibale Forest, Uganda. *Zeitschrift für Tierpsychologie,* 45(1), 75-84.

Struhsaker, T. T. (1978). Food habits of five monkey species in the Kibale Forest, Uganda. *Recent Advances in Primatology,* 1, 225-248.

Struhsaker, T. T. (1981). Polyspecific associations among tropical rain-forest primates. *Zeitschrift für Tierpsychologie,* 57(3-4), 268-304.

Struhsaker, T. T. (1999). Primate communities in Africa: The consequence of long-term evolution or the artifact of recent hunting. In J. G. Fleagle, C. Janson, & K. E. Reed (Eds.), *Primate communities* (pp. 289-294). Cambridge University Press.

Struhsaker, T. T. (2010). *The red colobus monkeys: Variation in demography, behavior, and ecology of endangered species.* Oxford University Press.

Struhsaker, T. T., Butynski, T., & Lwanga, J. (1988). Hybridization between redtail (*Cercopithecus ascanius schmidti*) and blue (*C. mitis stuhlmanni*) monkeys in the Kibale Forest, Uganda. In A. Gautier-Hion, F. Boueliere, J.-P. Gautier, & J. Kingdon (Eds.), *A primate radiation: Evolutionary biology of the African guenons* (pp. 477-497). Cambridge University Press.

Struhsaker, T. T., & Leakey, M. (1990). Prey selectivity by crowned hawk-eagles on monkeys in the Kibale Forest, Uganda. *Behavioral Ecology and Sociobiology,* 26(6), 435-443.

Struhsaker, T. T., & Leland, L. (1979). Socioecology of five sympatric monkey species in the Kibale Forest, Uganda. *Advances in the Study of Behavior,* 9, 159-228.

Sugiyama, Y. (1968). Social organization of chimpanzees in the Budongo Forest, Uganda. *Primates,* 9(3), 225-258.

Sussman, R. W. (1999). *Primate ecology and social structure.* Pearson Custom Publishing.

Tapanes, E., Detwiler, K. M., & Cords, M. (2016). Bat predation by *Cercopithecus* monkeys: Implications for zoonotic disease transmission. *Ecohealth,* 13(2), 405-409.

Teleki, G. (1973). *The predatory behavior of wild chimpanzees.* Bucknell University Press.

Tosi, A. J., Disotell, T. R., Morales, J. C., & Melnick, D. J. (2003). Cercopithecine Y-chromosome data provide a test of competing morphological evolutionary hypotheses. *Molecular Phylogenetics and Evolution,* 27(3), 510-521.

Treves, A. (1998). The influence of group size and neighbors on vigilance in two species of arboreal monkeys. *Behaviour,* 453-481.

Treves, A. (1999). Has predation shaped the social systems of arboreal primates? *International Journal of Primatology,* 20(1), 35-67.

Tsingalia, H., & Rowell, T. E. (1984). The behaviour of adult male blue monkeys. *Zeitschrift für Tierpsychologie,* 64(3-4), 253-268.

Ukizintambara, T., & Thébaud, C. (2002). Assessing extinction risk in *Cercopithecus* monkeys. In M. E. Glenn, & M. Cords (Eds.), *The guenons: Diversity and adaptation in African monkeys* (pp. 393-409). Springer.

van Schaik, C. P. (2000). Infanticide by male primates: The sexual selection hypothesis revisited. In C. P. van Schaik, & C. H. Janson (Eds.), *Infanticide by males and its implications* (pp. 27-60). Cambridge University Press.

Wahome, J., Rowell, T., & Tsingalia, H. (1993). The natural history of de Brazza's monkey in Kenya. *International Journal of Primatology,* 14(3), 445-466.

Waser, P. (1977). Feeding, ranging and group size in the mangabey (*Cercocebus albigena*). In T. H. Clutton-Brock (Ed.), *Primate ecology: Studies of feeding and ranging behavior in lemurs, monkeys, and apes* (pp. 183-222). Academic Press.

Waser, P. M. (1980). Polyspecific associations of *Cercocebus albigena*: Geographic variation and ecological correlates. *Folia Primatologica,* 33(1-2), 57-76.

Waser, P. M. (1982). Primate polyspecific associations: Do they occur by chance? *Animal Behaviour,* 30(1), 1-8.

Wilkie, D. S., Sidle, J. G., & Boundzanga, G. C. (1992). Mechanized logging, market hunting, and a bank loan in Congo. *Conservation Biology,* 6(4), 570-580.

Windfelder, T. L., & Lwanga, J. S. (2002). Group fission in red-tailed monkeys (*Cercopithecus ascanius*) in Kibale National Park, Uganda. In M. E. Glenn, & M. Cords (Eds.), *The guenons: Diversity and adaptation in African monkeys* (pp. 147-159). Springer.

Wolters, S., & Zuberbühler, K. (2003). Mixed-species associations of diana and Campbell's monkeys: The costs and benefits of a forest phenomenon. *Behaviour,* 140(3), 371-385.

World Resources Institute. (1998). *World resources 1998-1999.* Oxford University Press.

Xing, J., Wang, H., Zhang, Y., Ray, D. A., Tosi, A. J., Disotell, T. R., & Batzer, M. A. (2007). A mobile element-based evolutionary history of guenons (tribe Cercopithecini). *BMCBiology,* 5(1), 5.

Zuberbühler. K. (2002). Effects of natural and sexual selection on the evolution of guenon loud calls. In M. E. Glenn, & M. Cords (Eds.), *The guenons: Diversity and adaptation in African monkeys* (pp. 289-306). Springer.

CHAPTER 15

Behavioral Ecology of Patas Monkeys and Vervets

Jill D. Pruetz, Robert W. Sussman, and Jennifer D. Cramer

Vervets (*Cercopithecus aethiops*, F. Cuvier, 1821 or *Chlorocebus aethiops*, Gray, 1870) and patas monkeys (*Erythrocebus patas*, Schreber, 1774) are semiterrestrial, savanna-dwelling primates that share a number of traits in common and exploit similar habitats. Along with other Old World monkeys, such as Allen's swamp monkey (*Allenopithecus nigroviridus*) and the talapoin (*Miopithecus talapoin*), they are usually considered separate from the forest guenons (i.e., *Cercopithecus* monkeys *sensu* Butynski, 2002), their closest living relatives in the Order Primates. However, these two species also exhibit important differences from one another, with patas monkeys being more similar to forest guenons than vervets in aspects of their behavior, but exhibiting derived features of anatomy and morphology, which have caused them to be classified in a separate genus from their closest living relatives.

TAXONOMIC CLASSIFICATION AND GEOGRAPHICAL DISTRIBUTION

Vervets and patas monkeys are members of the tribe Cercopithecini (subfamily Cercopithecinae, family Cercopithecidae), which consists of the African guenons and their closest relatives. The Cercopithicini is a fairly recent radiation and the most speciose tribe of Old World monkeys, probably evolving within the last million years and characterized by a relatively depauperate fossil record (Leakey, 1988). Vervets and patas monkeys are closely related species (Gautier, 1988; Martin & MacLarnon, 1988; Ruvolo, 1988; van der Kuyl et al., 1995; Disotell, 2000) that have been reported to hybridize in captivity (see Lernould, 1988). They are considered more closely related to each other than to other guenons based on chromosomal banding pattern evidence (Disotell, 1996), morphological traits (Strasser & Delson, 1987), and protein electrophoretic data (Ruvolo, 1988).

While vervets have traditionally been classified as belonging to the genus *Cercopithecus* and patas monkeys to the genus *Erythrocebus*, taxonomic classification varies. Groves (2001) assigned vervets to the genus *Chlorocebus*, rather than *Cercopithecus*, raising six subspecies to species status. Others (Nowak, 1999; Isbell & Jafee, 2013) refer to the group as a superspecies of *Cercopithecus aethiops*, with the various subspecies designated as vervets in East Africa (*C. a. pygerythrus*: southern Ethiopia to Angola and South Africa), grivets (*C. a. aethiops*: Sudan and Ethiopia), tantalus monkeys in western Central Africa (*C. a. tantalus*: Volta River, Ghana to Uganda), Bale mountain monkeys in Ethiopia (*C. a. djamdjamensis*), Malbrouck monkeys in Angola and southern Congo (*C. a. cynosuros*), and green monkeys in West Africa (*C. a. sabeus*: Senegal to the Volta River, Uganda). Recent genetic analyses of the vervet monkeys have given insight into their historic population genetics (Jasinska et al., 2013; Warren et al., 2015; Svardal et al., 2017). Warren and colleagues (2015) used the newly published *Chlorocebus* reference genome to estimate dates of divergence between populations. The most ancient split from the ancestor *Chlorocebus* population was the *sabaeus* group, around 531 kya, with other populations diverging until the *cynosuros* group last diverged around 129 kya. Svardal and colleagues (2017) found evidence for population divergence within *Chlorocebus* based on geographic distance; even after populations split, there was still a high level of gene flow across ancient populations in East and West Africa, and significant gene flow across populations more recently in East and South Africa.

Patas monkeys have traditionally been classified separately from other guenon species based on morphological characters such as their long limb length. Some taxonomic classifications include patas monkeys as a member of the genus *Cercopithecus* rather than the sole representative of the genus *Erythrocebus* (Dutrillaux et al., 1988; Gautier, 1988; Kingdon, 1988; Lernould, 1988; van der Kuyl et al., 1995), but most authors consider patas monkeys as a separate genus. Patas monkeys have been classified into two to three subspecies (*E. patas patas* in West Africa, *E. p. baumstarki* in Tanzania, and *E. p. pyrrhonotus* elsewhere in East Africa) (Isbell, 2013).

Both patas monkeys and vervets are found in sub-Saharan Africa from the west to the east coasts, although the patas monkeys' distribution does not reach the east coastal area of Africa, and it is found in a much narrower north–south geographical band than are vervets (see Figure 15.1) (IUCN, 2008a,b). The vervet is the most widely distributed and abundant African monkey after the savanna baboon (*Papio* spp.) and can be found from Senegal in far West Africa to Somalia on the eastern coast and from Sudan south to the tip of South Africa (Kingdon, 1997). Patas monkeys are found in the northern savannas of Africa but are not found in the southern savanna regions that vervets occupy (Tappen, 1960; Kingdon, 1997). Patas monkeys appear to be able to range much farther into the arid zones of the north than vervets because they are less dependent upon water sources and can use shorter trees for sleeping sites. In some areas of the patas monkey range, there is less than 500 mm

of rainfall per year (Nakagawa, 2007; Isbell et al., 2009), and the dry season lasts for up to eight months (Tappen, 1960). Patas monkeys are considered to occur at low densities compared to sympatric primate species (Enstam et al., 2002). Tanzania, their southernmost range in East Africa, has less than 1,000 individuals of the subspecies *E. p. baumstarki*, thought to inhabit *Acacia* woodlands in the northwest (De Jong et al., 2009).

CONSERVATION

Patas monkeys are considered a Near-Threatened species and decreasing in terms of conservation status by the International Union for Conservation of Nature

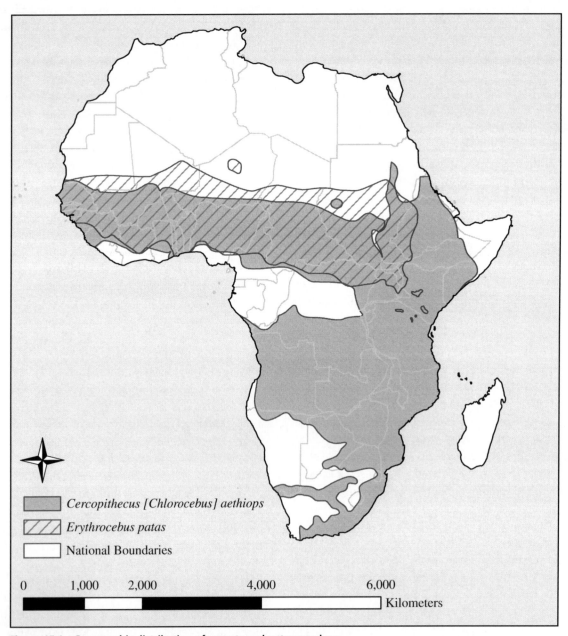

Figure 15.1 Georgraphic distribution of vervets and patas monkeys.
(From IUCN, 2021)

(IUCN, 2020). One subspecies (or species depending on the taxonomic authorities consulted), *baumstarki*, is listed as Critically Endangered (IUCN, 2020). Vervets are also considered to be decreasing by IUCN. While the Bale Mountain vervet in Ethiopia (*Cercopithecus aethiops djamdjamensis*) is listed as Vulnerable, however, the rest of the vervet subspecies are still considered Stable/Least Concern (IUCN, 2020). While neither patas monkeys nor vervets are under *immediate* threat of extinction, plans for maintaining their status and halting their decline, as well as educating the public and shaping policy, are important as the interaction between humans and these two species becomes increasingly contentious, particularly in agricultural, residential, and commercial areas. Patas monkeys and vervets are often referred to as pests because of their crop-raiding behavior and cohabitation with humans in much of their ranges. This close proximity with humans has led some researchers to apply an ethnoprimatological lens to study these relationships, as well as to identify solutions to mitigate human-primate conflict (see review in Turner et al., 2019). For example, the population trend for patas monkeys shows obvious and concerning decline, and in East Africa, specifically, there has been significant decrease in geographic range, as well as increasing gaps between populations (Kingdon et al., 2008). Where not protected, commercial trapping for medical research has been a major threat to vervet populations in the past. In addition, humans often hunt them with domestic dogs in agricultural areas where they are considered pests (Fedigan & Fedigan, 1988; Cheney & Seyfarth, 1990; Pruetz, pers.obs.).

ANATOMY AND MORPHOLOGY

Anatomically and morphologically, vervets and patas monkeys exhibit many similarities. Each has been classified as semiterrestrial (Kingdon, 1988) with hand and foot adaptations geared toward terrestrial locomotion (Kingdon, 1988; Strasser, 1992). Each species exhibits typical cercopithecid traits, such as ischial callosities and bilophodont molars, which allow more efficient food processing than simple molars. They also exhibit the characteristic cercopithecine cheek pouches for food storage and transport and where some digestion also occurs via a special enzyme (Lambert, 2005).

The two species differ in body size and weight, although there is some overlap. Both vervets and patas monkeys exhibit sexual dimorphism in body size and canine tooth size, but patas monkeys are among the most dimorphic of primates. Vervets are small, olive-green- or gray-backed monkeys with a white belly; their black faces are ringed to various degrees with white hairs. Females weigh from 2.5–5.3 kg (head and body

length is 40–61 cm), and males weigh approximately 7 kg (Haltenorth & Diller, 1988). Although their coloration is not as striking as some of their close relatives, the forest guenons, the various subspecies of vervets can be distinguished based on pelage differences, especially by markings on their heads and faces (Hill, 1966) as shown in Figure 15.2 and Figure 15.3. The patas male is almost twice as large as the female (head and body length is 50–60 cm) and is twice as heavy, weighing about 12 kg vs. 6 kg for the female (Hall, 1965; Sly et al., 1983) (see Figure 15.4). Patas monkeys have prominent white moustaches, earning them the nickname of "military monkey" or "Hussar monkey" in reference to uniforms of British Empire army officers. West African patas monkeys have black noses as adults, while East African patas monkeys have white noses at maturity. The patas monkey's red coat is an unusual color for the Cercopithecini (Haltenorth & Diller, 1988); East African patas are darker red then their West African counterparts.

Patas monkeys have traditionally been classified separately from other guenon species based on morphological characters such as their long limb length (Hurov, 1987; Strasser, 1992). The slender build is thought to be an adaptation to widely scattered savanna food resources (Isbell et

Figure 15.2 Juvenile vervet monkey, Ethiopia.
(Photo credit: Mary Willis)

Figure 15.3   Vervet monkey, South Africa.
(Photo credit: Jennifer Cramer)

Figure 15.4   Male patas monkey, Ghana.
(Photo credit: Edward Miafe)

al., 1998b). Patas monkeys are distinct among the terrestrial cercopithecines in that their tarsal bones are very elongated and exhibit a high crural index (tibial length/femoral length x 100) relative to body mass (Strasser, 1992). These specializations are related to cursorial, terrestrial locomotion. However, Chism and Rowell (1988) note that, although often assumed to be ecologically and behaviorally separate from other guenons because of such cursorial adaptations, patas monkeys are in fact more similar to other guenons than to the savanna baboons (*Papio hamadryas*). Gartlan and Brain (1968) describe vervets as being intermediate between the more arboreal guenons, such as the blue monkey (*Cercopithecus mitis*), and the "dry savanna country" patas monkeys. Compared to *C. mitis*, vervet monkeys differ in magnitude of about 5% in upper arm length but do not differ from patas monkeys in this respect (Jolly, 1964, cited in Gartlan & Brain, 1968). Vervets and patas monkeys differ in respect to lower arm length, with vervets being more similar to other guenons than to patas monkeys (Jolly, 1964, in Gartlan & Brain, 1968). The ratio of foot to lower leg length in vervets was intermediate between, and did not differ significantly from, arboreal guenons or patas monkeys (Jolly, 1964, in Gartlan & Brain, 1968).

Other interesting anatomical and morphological characteristics characterize vervets and patas monkeys. Vervets exhibit a relatively long (30 hours) gut-retention time for their body size (Lambert, 2002), suggesting they could be effective seed dispersers for certain fruiting plant species. Besides humans, patas monkeys are the only primate capable of producing copious amounts of sweat via a relatively large proportion of eccrine glands, thought to be an adaption to their arid environment (Mahoney, 1980). Finally, both vervet and patas monkey males, like some other guenons, are characterized by having bright blue scrota, a white perineal region, and red penis. Struhsaker (1967a) described the "red, white and blue" display of adult male vervets as a behavior characteristic of male dominance. In this display, the dominant individual walks with tail raised, explicitly displaying the colorful arrangement. Visual communication through scrotal coloration is particularly important between males (Brain, 1965; Struhsaker, 1967a; Durham, 1969; Henzi, 1985; Gerald, 1999, 2001) and may be important in intersexual communication (Gerald et al., 2010). The genitalia are less conspicuous in patas monkeys because of longer hair covering the region; their typical displays consist of bouncing off small tree trunks with all four limbs, which, with East African whistling thorn trees (*Acacia drepanolobium*), makes for an impressive auditory as well as visual display. While genital color may change with health in male vervets (Isbell, 1995), it does not appear to indicate changes in reproductive condition.

GROWTH, DEVELOPMENT, AND LIFE HISTORY

Both patas monkeys and vervets are characterized by relatively quick growth and development compared to other cercopithecines, even within the tribe Ceropithecini (Struhsaker, 1971; Chalmers, 1972; Butynski, 1988; Fedigan & Fedigan, 1988). Age at first birth is usually around three years in patas monkeys and can occur this early in vervets as well (but is variable), with vervet sexual maturity being reached at three to five years (Whitten & Turner, 2008). Like other anthropoid primates, vervet and patas monkey births are single and infants are carried clinging to their mother's belly. Infants become independent relatively early in these species. Such fast life history patterns are conducive to rapid population recovery and colonization of new areas under good conditions.

The estrus cycle in vervets lasts 30–33 days and gestation is around 160 days (Cheney et al., 1988; Ely et al., 1989). Infants are weaned beginning as early as 12 weeks, and weaning is usually complete by 8.5 months (Lee, 1984; Horrocks, 1986). Age at first birth can be as early as 2.5 years, although Isbell et al. (2009) found that mean age at first reproduction for Segera, Laikipia, Kenya, vervets was 5 years. At Amboseli, Kenya, female age at first birth in more productive habitats was 4.4 years as opposed to 5.4 years in a less productive habitat (Cheney et al., 1988). Females conceived in consecutive years at Segera, as well as other study sites (Isbell et al., 2009). Females normally reach full adult weight by 4–5 years. Males begin to reproduce at around 5 years and reach full size at age 6. Vervets can live for over 10 and up to 17 years of age (Cheney & Seyfarth, 1990).

Both patas monkeys and vervets exhibit a synchronized mating and birth season. In vervets, approximately 80% of births occur within a 2–3 month period (Fedigan & Fedigan, 1988). Birth seasons correspond with the rainy season at most sites (Butynski, 1988; Nakagawa, 2000). Vervets can exhibit rapid population growth rates and, even in marginal habitats, populations can be maintained or recover quickly when conditions improve (but see below regarding Amboseli vervets). At Isbell's Segera site, female vervet monkeys, but not patas, showed increased birth rates in response to increased rainfall (Isbell et al., 2009). At this site, mean interbirth interval for vervets was 13.3 months (Isbell et al., 2009), which was shorter than the mean interbirth intervals at Amboseli, ranging 13.8–21.3 months (Cheney et al., 1988).

In patas monkeys, estrus lasts 30–32 days (Rowell & Hartwell, 1978; Loy, 1981) and gestation is approximately 170 days (Gartlan, 1974; Loy, 1981; Sly et al., 1983). The birth season corresponds with the beginning of the dry season (Struhsaker & Gartlan, 1970; Gartlan, 1974; Chism & Rowell, 1986), and conception and weaning correspond to periods of high food availability, at least in Kenya (Chism et al., 1984), which appears to differ from the pattern seen in other guenons and in vervets (Butynski, 1988). However, Nakagawa (2000) examined the diet of vervets and patas monkeys in Cameroon in detail and concluded that patas monkeys were better able to extract energy during the dry season when their food availability was highest, so that the patas monkey pattern is indeed similar to that of other guenons who expend the greatest amount of energy via lactation during periods of greatest food availability (Butynski, 1988).

For patas monkeys, growth and development is even faster than in vervets. At 12 weeks, an infant spends 45% of the time independent of an adult. Four- to five-month-old infants rarely nurse and almost always move independently. At one year, the infant feeds itself and keeps up with the group on its own (Chism, 1986, 1999b). Primiparous females can conceive at 2.5 years, give birth at 3 years of age, and then annually bear young (Chism & Rowell, 1988; Isbell et al., 2009). At the Segera site, mean age at first reproduction was 3 years in a sample of 24 females (Isbell et al., 2009). Males reach adulthood at around 5–6 years (Rowell, 1977; Chism, 1999b). Mean interbirth interval for female patas monkeys studied at three different sites (two in Kenya and one in Cameroon) is 13.3 months, ranging from 11.8–14.4 months (Chism et al., 1984; Nakagawa, 2003; Isbell et al., 2009). Patas monkeys have among the fastest rates of maturation and of reproductive turnover of any Old World monkey (Chism et al., 1984; Nakagawa, 2003; Enstam & Isbell, 2007; Isbell et al., 2009). Isbell et al. (2009) interpret this as a response to the uncharacteristically high adult female mortality rate in this species relative to the typically higher infant mortality rate typifying most mammals. Of 85 infant patas born at the Segera site, 26% did not survive their first year, but the 29.8% infant mortality rate was lower than the 32.9% mortality rate of adult females there (Isbell et al., 2009). Similarly, the annual mortality rate of adult females in Cameroon averaged 26% (Nakagawa, 2003). Patas monkeys have been recorded to live for up to 24 years (Nowak, 1999), although none of the females born into Isbell's study groups in Laikipia, Kenya, lived more than 8 years, and the oldest females in the groups were estimated to be approximately 9 years of age (Isbell, 2013). In Cameroon, Nakagawa et al. (2003) recorded the oldest female living to about 17 years of age.

Alloparenting is common in both vervets and patas monkeys. Vervet infants, especially newborns, are handled, groomed, and carried extensively, sometimes as early as 4 hours after birth, and such attention continues throughout the first 12 weeks of life. As in other primates, adult females, and especially juvenile and subadult females, are attracted to infants (Struhsaker, 1971; Lancaster, 1972; Lee, 1984). Adult males rarely carry, and are normally kept away from infants by females, but non-related juvenile males sometimes care for them (Fedigan & Fedigan, 1988). In patas monkeys, alloparenting is also very common during the first few months, beginning as early as 12 hours after birth. Infants are often taken for long periods, and females who share each other's infants are usually those who have strong grooming and affiliative relationships. The infants also develop close relationships with these females (Chism, 1986, 1999a). Often, however, the allomother appears to mishandle the infant. Isbell et al. (2009) report that 18% of infants that died (n = 4) did so as a result of kidnapping and mishandling by adult females during one season of relatively lower reproduction at the Segera site. According to Sterck et al. (1997), infanticidal behavior is predicted to be characteristic of dispersal-egalitarian species like the patas monkey. Enstam et al. (2002) reported a possible infanticidal attack by a male patas monkey in their study group after he overthrew the resident male. Allomaternal care may be an important component of the patas monkey's ability to reproduce rapidly, allowing increased foraging efficiency of mothers, and has been interpreted as a likely adaptation to their harsh and highly seasonal environment (Rowell, 1977; Chism et al., 1984; Chism, 1986, 1999b).

FIELD STUDIES OF VERVETS AND PATAS MONKEYS

Vervets have been studied extensively at a number of sites, but patas monkeys have been studied extensively only at three sites across Africa (Kala Maloue, Cameroon, and Mutara and Laikipia, Kenya). A large number of vervet studies, however, have been short-term in nature. Long-term studies of vervets in their natural habitats for at least one year have been conducted in Kenya (Amboseli: Struhsaker, 1967b; Klein, 1978; Cheney & Seyfarth, 1983. Laikipia: Isbell & Pruetz, 1998; Isbell et al., 1998ab, 2009; Pruetz, 1999, 2009; Pruetz & Isbell, 2000; Enstam & Isbell, 2002; Isbell & Enstam, 2002. Samburu: Whitten 1983), in Cameroon (Kavanagh, 1980; Nakagawa, 1991, 2000, 2007), in Senegal (Harrison, 1983), and in South Africa (Henzi & Lucas, 1980; Barrett, 2005, 2006). Research has also been conducted on introduced, free-ranging vervets in Barbados (Fedigan et al., 1984; Chapman, 1985, 1987) and St. Kitts (Dore, 2013).

The longest-running field studies of wild vervets were conducted in East Africa. One of the best-known studies of multiple groups of vervets stems from the Amboseli, Kenya site. Vervets at Amboseli were studied by T. Struhsaker in the mid-1960s, by D. Klein in 1975, and by D. Cheney and R. Seyfarth and colleagues from 1977–1988.

The population drastically declined at the site, dropping from 215 individuals in 11 groups in 1977 to only 35 monkeys in three groups in 1988 (Cheney & Seyfarth, 1990). Ultimately, habitat destruction was deduced to be the cause of the decline as the water table rose and fever trees (*Acacia xanthophloea*) died off. The effect was compounded by relatively greater consumption of the remaining fever trees by the largely restricted Amboseli elephant population, leaving fewer trees for vervet food and refuge, which then also resulted in increased leopard predation (Isbell et al., 1990). Amboseli may also represent one of the harshest environments in which vervets have been studied to date, receiving less than 300 mm of rainfall per year.

Patas monkeys have been studied long-term at only a few sites. In Uganda, Hall (1965) and Struhsaker and Gartlan (1970) conducted the first studies of patas monkeys. A project conducted by a Japanese team at Kala Maloue National Park in northern Cameroon included vervets as well as patas monkeys. Monkeys at this site are provisioned occasionally and were studied intermittently from 1984–1989 (Nakagawa, 1989, 1991, 1992, 2000, 2007; Ohsawa et al., 1993; Ohsawa, 2003). Two studies of unprovisioned patas monkeys on Laikipia Plateau, Kenya, have been conducted, with one lasting two years at Mutara (Rowell, 1977, 1988; Chism et al., 1983, 1984; Chism & Rowell, 1986, 1988; Chism & Wood, 1994) and the longest continuing for 11 years at Segera (Isbell & Pruetz, 1998; Isbell et al., 1998a,b, 2009; Pruetz, 1999; Pruetz & Isbell, 2000; Carlson & Isbell, 2001; Enstam & Isbell, 2002; Enstam et al., 2002; Isbell & Enstam, 2002). A number of studies have also been conducted on introduced, free-ranging patas monkeys in Puerto Rico (Kaplan & Zucker, 1980; Gonzalez-Martinez, 1998). Much of what we know about patas monkeys comes from only two sites (if one considers the Laikipia Plateau in Kenya to be a single site); it should be noted that this unique and "simple" habitat (Pruetz & Isbell, 2000) may, in fact, be atypical of patas monkeys over much of their geographical range, save for their current distribution in East Africa.

HABITAT SELECTION AND RANGING BEHAVIOR

As mentioned previously, both vervets and patas monkeys are found across sub-Saharan Africa, particularly in savanna ecosystems. Savanna ecosystems include shrubland, woodland, and other types of wooded habitat dominated by a grass understory, in addition to the open grassland habitats (e.g., Serengeti Plains) that many envision when using the term "savanna." Neither patas monkeys nor vervets spend much time in treeless savannas and neither is found in rainforests, although vervets readily colonize forest edges and more closed canopy environments. Similarly, neither monkey species is found in desert regions. Vervets are described as a species that almost always includes riverine or gallery forest in their range, save for the Amboseli, Kenya, population, which ranges throughout a swampy area (Enstam & Isbell, 2007). Vervets are an edge species, using ecotones in most of the areas in which they occur. They are also able to use disturbed habitats and to adapt to human tourist areas, urban parks, and gardens, and are often considered pests. In areas where population densities fall too low, groups can become extinct (Hauser et al., 1986), but with edge areas and disturbed forests expanding in Africa, vervets are generally a very successful colonizing species (Fedigan & Fedigan, 1988). However, the northern limit of the species range is contracting due to deforestation (Kingdon, 1997).

Vervets are limited by the need for water and sleeping trees (Fedigan & Fedigan, 1988), while patas monkeys are not as environmentally restricted, and this is reflected in patas geographical distribution. Patas monkeys occupy *Acacia* woodland, woodland-savanna, and more open areas than vervets. Additionally, patas monkeys do not utilize riverine vegetation or forest and are even observed to actively avoid these areas within their home range in Laikipia, Kenya (Chism & Rowell, 1988; Pruetz, pers.obs.). However, at the Laikipia site, patas monkeys were observed to drink almost daily, in contrast to reports from a short-term study in Uganda (Hall, 1965).

Vervets rarely range far from trees, remaining relatively close to riverine or other forested areas even when using more open environments. These semiterrestrial monkeys spend about 20% of the time on the ground, almost 30% between 1–4 m high, and 35% of their time above 13 m (Rose, 1979). In a study of *C. aethiops* in Ethiopia, Dunbar and Dunbar (1974) found that they spend almost 60% of the time in the trees or bushes. Vervets primarily come to the ground to cross open areas and feed mainly in tall trees. A group in Ngel Nyaki Forest Reserve, Nigeria, was observed to cross grassland areas only to reach forest and forest edges (Agmen et al., 2009). In this montane forest of Nigeria, Agmen et al. (2009) found that their study group spent 72% of their time feeding along the forest edge, 20% of the time feeding in forest fragments, and only 8% feeding in grassland. In response to a terrestrial predator alarm call, vervets in Laikipia, Kenya, were observed to flee more than 100 m from a whistling thorn *Acacia* shrub woodland back to a taller, fever-tree riverine woodland (Pruetz, pers.obs.). Although patas monkeys climb into trees to feed, rest, or survey their environment, trees are widely scattered in their habitat so they seldom move from one tree to another arboreally, at least in East Africa where almost all travel takes place on the ground (Hall, 1965; Chism & Rowell, 1988; Isbell, 1998).

The quality of habitat, in terms of the abundance and distribution of foods important in the diet, affects vervet ranging behavior and habitat use at a number of sites. Home range size and daily path length varies in vervets, sometimes—but not always—corresponding with group size (Table 15.1) (see Enstam & Isbell, 2007). Harrison (1983) conducted a detailed, quantitative study of vervet ranging behavior in Senegal and found that range use was directly related to habitat structure, i.e., availability and distribution of important species of fruit and flowers, sleeping sites, competing neighboring groups, and (depending on the season) distribution of standing water. During daily ranging, the group did not recross its own path but used new areas for foraging on consecutive days (Harrison, 1983). Areas that were visited generally contained especially good food resources, sleeping trees, or water. The group spent half of its time in less than 10% of its range. Water availability did not influence ranging during the wet season (Harrison, 1983). Struhsaker (1967b) also found a relationship between ranging behavior and what he termed optimal habitat. In Amboseli, optimal habitat included associations of fever trees and the shrubs *Azima tetracantha* and *Salvadora*

persica. In three of four groups studied, monkeys spent 95% of their time in this habitat (Struhsaker, 1967b). In Senegal, vervets spend more time than expected in gallery forest, closed woodland, and scrub vegetation, and as much time as would be expected in open woodland (the most common habitat and the one most used) (Harrison, 1983).

Vervets living in poor quality habitats generally move farther and more irregularly over their total home range and have relatively larger home ranges (DeMoor & Steffens, 1972; Kavanagh, 1981). At Blydeberg Conservancy, South Africa, where food was considered abundant for vervets throughout the year, the study group's ranging appeared to be linked to the availability of fruiting plants and, especially, to water availability (Barrett et al., 2006). At Loskop Dam, South Africa, vervets living in higher-quality habitats with more diverse food resources optimized their diet by using more high-energy foods than vervets living in lower-quality habitats with less food diversity (Barrett et al., 2016). In Amboseli, Struhsaker (1967b) also found that three groups spent a great deal of time in relatively few quadrats (75% of time spent in eight 10,000 m² quadrats). However, a fourth group, living in a less optimal habitat, had to move more and use more of its total range; that group spent 75% of its time in 20 quadrats.

Table 15.1 Ranging and Grouping Behavior of Patas Monkeys and Vervets

	Group Size	Home Range Size	Daily Path Length	Site
Patas	45	4000 ha	3188 m	Segera, Kenya
	(1) 16, (2) 47	(1) 2340 ha, (2) 3200 ha	(1) 3830 m, (2) 4220 m	Mutara, Kenya
	11	440 ha	5168 m	Kala Maloue, Cameroon
	31	5200 ha	2250 m	Murchison Falls, Uganda
	21 average (N = 12)			Waza Reserve, Cameroon
Average [a]	27.9	3103 ha	3658 m	
Vervets	(1) 55-60, (2) 46-50	15.2 ha (N = 2)	(1) 956 m, (2) 898 m (mean = 928 m)	Odobullo Forest, Ethiopia
	33	77 ha	865 m	Blydeberg, South Africa
	(1) 7.3, (2) 16.4, (3) 17.2, (4) 49.9 (mean = 22.7)	(1) 34 ha, (2) 96 ha, (3) 19 ha, (4) 34 ha (mean = 45.75 ha)	1188 m	Amboseli, Kenya
	(1) 8, (2) 27 (mean = 17.5)	(1) 10 ha, (2) 40 ha (mean = 25 ha)	(1) 1025 m, (2) 1632 m (mean = 1328.5 m)	Segera, Kenya
		14 ha		Bakossi, Cameroon
	18.5	56 ha		Kala Maloue, Cameroon
		103 ha		Baffle Noir, Cameroon
		178 ha		Assirik, Senegal
			1500-2000 m	Ngel Nyaki, Nigeria
	23	(1) 99 ha avg wet season (N = 2), (2) 145 ha avg dry season (N = 2)		Loskop Dam Nature Reserve, South Africa
	(1) 26, (2) 36 (mean = 31)			Samara Private Game Reserve, South Africa
Average [a]	24.3	63.6 ha	1077 m	

[a] Obtained by averaging values for each site and then averaging across sites for the overall mean.
References: Agmen et al. (2009), Barrett (2005), Barrett et al. (2016), Cheney and Seyfarth (1987), Enstam and Isbell (2007), Hall (1965), Harrison (1983), Hauser et al. (1986), Isbell et al. (1998b), Jaffe and Isbell (2010), Mekonnen et al. (2010), Nakagawa (2007), Pasternak et al. (2013).

Patas monkeys are found in the driest grassland areas of Africa and have the largest home ranges relative to body size of any primate. Isbell (1998) maintains that the large patas monkey range is determined by widely dispersed food resources, availability of water, and avoidance of potential predators (by avoiding riverine areas). Patas monkeys in East Africa have been reported to inhabit home ranges of 2,340–5,200 ha (n = 4, average home range size = 3,685 ha) (Enstam & Isbell, 2007) but have a much smaller home range in Cameroon (260–440 ha) (Nakagawa, 2007). For the above East African patas monkey study groups, daily path length ranged from 2,250–4,220 m (3,372 m on average) (see Table 15.1). In comparison with sympatric vervets, patas monkeys at Segera travel three times farther per unit time, travel twice the distance between food sites, and spend one-third as much time at each food site (Isbell et al., 1998b). In addition to ranging widely, patas monkeys also disperse while moving and feeding. Chism and Rowell (1988) found that their study groups of 16–41 and 47–74 individuals exhibited average group spreads of 96 m and 179 m, on average.

Female patas monkeys have fixed home ranges, whereas males are part of a population within an area larger than any one female group. Within this area, at any one time, a male may be found as a solitary animal, as a member of an all-male group, or as a resident male in a female group. Adult females typically initiate group movement, lead progressions, and choose sleeping sites (Chism et al., 1984; Chism & Rowell, 1988; Chism, 1999b). Day ranges vary from day to day and between seasons. In some seasons, when resources are abundant, groups might spend several hours in a particular area. At other times, they move steadily throughout the day in search of food or water. The pattern of ranging indicated to Hall (1965) that the group concentrated on feeding in a given area for several days and then moved on to another area and repeated the process. Isbell and colleagues describe the patas monkeys they studied at Segera as having a more circuitous ranging pattern than sympatric vervets (Enstam & Isbell, 2007).

INTERGROUP AND INTERSPECIFIC RELATIONS

Both patas monkeys and vervets differ from most other guenons in that they do not frequently form polyspecific associations (Enstam & Isbell, 2007). Unlike savanna baboons, they have not been observed to regularly associate with other mammals, such as ungulates (e.g., bushbucks, *Tragelaphus sylvaticus*). Immigrating male vervets, however, have been seen to form temporary associations with baboon troops. Even though baboons prey on vervets (DeVore & Washburn, 1963; Struhsaker, 1967a; Altmann & Altmann, 1970), their interspecies relationship

toward one another is complex. The young of both species frequently play together. Both species respond to each other's alarm calls. They sometimes sleep together in the same trees and drink simultaneously at the same waterholes. Groups of the two species often are close together or intermingle and elicit very little reaction to one another (Struhsaker, 1967b; Altmann & Altmann, 1970).

Both patas monkey and vervet groups exhibit aggressive behavior toward neighboring groups, although their behaviors can vary. Vervets can be described as territorial at some sites, but patas monkeys do not defend specific areas of their ranges. At most study sites, vervet groups have been observed to defend some portion of the home range against other groups (Hall & Gartlan, 1965; Struhsaker, 1967a, 1976; Gartlan & Brain, 1968; Kavanagh, 1981; Harrison, 1983; Fedigan & Fedigan, 1988; Cheney & Seyfarth, 1990; Isbell et al., 1999; Enstam & Isbell, 2007). At Harrison's (1983) Assirik, Senegal, site, group ranges overlapped more during the seasons of greater food abundance, also corresponding with a greater number of intergroup encounters. DeMoor and Steffens (1972) observed that, under stressful conditions of food shortage in the dry season, groups were more tolerant of one another and ranges overlapped in areas where food was more available. Struhsaker (1967b) found vervet groups to be mutually tolerant near certain small waterholes, but Wrangham (1981) did not find this to be the case during drought conditions. Similarly, McDougall et al. (2010) found intense intergroup interactions among vervets around water sources during a drought in South Africa.

Intergroup interactions between vervets frequently occur near the boundaries of groups' home ranges. These encounters may involve most or all members of the group or, at other times, mainly dominant females or only the most dominant male or males engaging in ritualized encounters. These are characterized by "chuttering" vocalizations, jumping displays and, sometimes, penile displays, but physical contact between protagonists is rarely seen (Struhsaker, 1967c; Dunbar, 1974; Galat & Galat-Luong, 1976). Small groups may exhibit deceptive behavior by emitting predator alarm calls, which effectively dispel both their own and the other group's members. In a detailed study of intergroup encounters at Amboseli, Cheney (1981) reported that about half of the encounters involved aggressive chases, hits, or bites. The other half consisted solely of vocalizations, mainly between high-ranking females. Males rarely took part in these exchanges. In the former type of encounter, adult males and females focused a large proportion of aggression toward the other group's adult males, but adult males appeared to avoid threatening females from other groups. A small proportion of interactions involved friendly exchanges—either play among juveniles or grooming. Aggressive interactions occurred more frequently between groups that had previously exchanged males, but females were less aggressive in these situations. Male transfer between groups tended to be consistent, with males of one group transferring into the same new group.

Given their complex system of aggressive encounters, friendly interactions, and male transfer, the outcome of intergroup interchange among vervets is probably influenced by the past experiences of individuals with each other, throughout the population (Fedigan & Fedigan, 1988; Cheney & Seyfarth, 1990). Cheney (1981) believes that patterns of aggression may influence the distribution of male movement and that the outcome of intergroup encounters is affected by the number, age, relative rank, and genetic relatedness of animals, especially males, within and across groups. Vervet populations have, therefore, been interpreted as neighborhoods or communities of interacting, familiar individuals above the level of the group (Cheney & Seyfarth, 1990).

While patas monkeys engage in aggressive intergroup encounters, they do not defend their home ranges as vervets do (Enstam & Isbell, 2007). At Hall's (1965) site in Uganda, interactions between groups were very rare, and groups remained widely dispersed. In 627 observation hours spread over 12 months, Hall recorded only two intergroup encounters. However, at other sites groups have had broadly overlapping home ranges, encounter each other frequently, and engage in agonistic interactions (Strushaker & Gartlan, 1970; Chism et al., 1984; Chism & Rowell, 1988; Chism, 1999b). During these encounters, females and juveniles initiate and actively participate, while males act as passive bystanders except during conception periods (Chism, 1999a). In some areas, during the dry season, intergroup interactions occur around scarce water sources (Struhsaker & Gartlan, 1970; Gartlan & Gartlan, 1973; Gartlan, 1974). Even though these waterholes attracted many groups at the same time, friendly behaviors were never observed between groups, and all-male groups would never approach a waterhole when a heterosexual group was in the vicinity. Heterosexual groups always had priority to these scarce water sources (Gartlan, 1974). Isbell et al. (2009) recorded a seemingly peaceful mingling of two patas monkey groups, one of which had recently lost a resident male, for about a week at Segera, Kenya, although intergroup encounters were normally aggressive at the site.

ACTIVITY

Throughout the day vervet groups remain relatively cohesive and overall activity is approximately synchronized. Feeding generally occurs in two peaks during the day, but at some sites it tends to increase gradually throughout the day. Vervets are adaptable to seasonal variation in food availability and can adjust activity times, such as foraging and rest, to counter the effects of temperature extremes and food availability (Harrison, 1985; Fedigan & Fedigan, 1988). They have been observed to spend 20–45% of their time feeding, depending on conditions (Dunbar & Dunbar, 1974; Kavanagh, 1978; Wrangham & Waterman, 1981; Hamilton, 1985; Saj et al., 1999). For example, a study of diet and feeding in Nigeria reported that individuals fed for approximately 41% of the time and that 51% of feeding time was spent on fruit (Agmen et al., 2009). In Ethiopia, feeding accounted for 65.7% of vervets' activity budget, followed by moving (14.4%), resting (10.7%), social (7.1%), and other behaviors (2.4%) (Mekonnen et al., 2010). Crop-raiding vervets spend less time feeding than populations dependent on wild resources, and time spent feeding is inversely related to time resting, whereas time in social behavior remains fairly constant across populations (Dunbar & Sharman, 1984; Saj et al., 1999).

Like many primates, vervets often exhibit a midday resting period, and both day and night-time resting sites are chosen with safety in mind. Day resting sites are usually in clumps of small trees, in thickened scrub, or in night roosting trees (Moreno-Black & Maples, 1977). At night, vervets sleep in tall trees, and groups have favorite sleeping sites. Struhsaker (1967c) observed groups dividing into sleeping subgroups that sometimes slept up to 750 m apart. In the morning these subgroups regrouped and, from sunrise until the group moved away from the sleeping area, major social activity and sunning predominated. At most study sites, major group progressions occur twice daily: once in the morning when groups move away from their night sleeping sites and again as the groups move *to* the sleeping sites (Struhsaker, 1967a; DeMoor & Steffens, 1972; Dunbar & Dunbar, 1974).

Most vervet groups sleep at night in forested areas and usually have preferential sleeping trees or tree groves but also may avoid using the same sleeping sites sequentially (e.g., Agmen et al., 2009). Struhsaker (1967b) found that only about 34% of 92 potential sleeping groves were used as sleeping sites at Amboseli, Kenya. *Acacia*, the fever tree, was the most preferred tree and, to a lesser extent, *A. tortilis* (umbrella *Acacia*). Harrison (1983) found his group in Senegal slept in gallery forest and closed woodland 90% and 10% of the time, respectively. DeMoor and Steffens (1972), Dunbar (1974), and Kavanagh (1981) also reported this pattern of preferred sleeping sites in forest of dense vegetation, with daily ranging away from these sites. However, vervet groups living in poor habitats often do not have regular patterns of daily ranging and do not have preferred sleeping sites (DeMoor & Steffens, 1972; Kavanagh, 1981).

The patas monkey group studied by Isbell and colleagues at Segera, Kenya, spent less time feeding (11%) and more time moving (39%) than other monkey species (Isbell, 1998). Isbell (1998) suggests that the patas monkeys are able to exploit high-quality, but widely scattered foods usually reserved for smaller animals, due to anatomical adaptations allowing them to cover long distances efficiently.

Unlike vervets, patas monkeys rarely use the same sleeping site from one night to the next. Instead patas monkeys disperse into a number of small trees. Patas monkeys' choices of sleeping sites and night resting behavior appear to be adaptations to avoid nocturnal predation. In East Africa, group members sleep in widely scattered, small trees within their range. The group splits up, sleeping over a wide area in the open savanna (Hall, 1965; Chism et al., 1983; Chism, 1999a). Groups choose different sites each night and remain silent once they are settled (Hall, 1965; Chism et al., 1983; Chism & Rowell, 1988). Only one adult and one juvenile, or mothers and infants, sleep together at night. The group is very widely spread, sometimes reaching at least 250,000 m². Patas monkey group composition and habitat make it particularly vulnerable to nocturnal predation, and their widely dispersed sleeping patterns are hypothesized to be techniques for avoiding predators (Chism et al., 1983; Chism & Rowell, 1988; Isbell et al., 2009).

In Kenya, during the birth season, patas groups moved less and over shorter distances during the morning than during the afternoon. This corresponded with the timing of diurnal births (Chism et al., 1983). Resting during the day usually took place in a large shade tree and group members either all rested in one tree or sometimes in a cluster of two or three trees. The day-rest periods typically lasted between one and three hours. In Cameroon, patas group cohesion increased, and female distance to neighbors decreased, during the birth season (Muroyama, 2017). In Uganda, social behavior occurred at the beginning of the day. However, the patas had two peaks of feeding broken by a resting period during the hottest portion of the day. On some days, patas fed for several hours in a particular area, whereas on others they moved steadily throughout the day, only stopping to rest (Hall, 1965). Unlike other gummivore/insectivores, patas spend little time searching for insect prey, because they focus on social insects found in *Acacia* thorns, at least in East Africa (Isbell, 1998).

DIET

Like other savanna-dwelling primates, both vervets and patas monkeys have been described as omnivorous; each species includes fruit and other plant parts as well as vertebrate and invertebrate prey in their diet. However, vervets typically include a large proportion of fruit in the diet, as well as flowers at some sites. In fact, in East Africa they have been characterized as florivorous because of the high percentage of flowers in the diet (Klein, 1978; Wrangham & Waterman, 1981; Whitten, 1983). Both species rely to a large degree on *Acacia* species in their diet at a number of sites where they have been studied (patas monkeys: Chism & Wood, 1994; Isbell, 1998; Pruetz & Isbell, 2000; vervets: Struhsaker, 1967b; Kavanagh, 1978; Wrangham & Waterman, 1981; Whitten 1983; Lee, 1984; Isbell et al., 1998b; Nakagawa, 2000). Neither species feeds on large vertebrate species, although male patas monkeys in particular could easily hunt small to medium-sized mammals. Immature patas monkeys, as well as an adult male in Kenya, were observed to curiously examine, follow, and touch but ultimately abandon an African hare, although they feed on large lizards (*Agama agama*) and nestling birds (Pruetz, pers.obs.).

Like some other savanna primates, patas monkeys and vervets exhibit a relatively narrow diet in terms of plant species and parts eaten (Pruetz, 2006). In Senegal, Galat and Galat-Luong (1977) found *C. aethiops* fed on 17 species of plants. Struhsaker (1967b) observed Amboseli vervets consumed 24 plant species over the course of a year. In Cameroon, Kavanagh (1978) found that vervets fed on 41 plant species at Baffle Noir but only 26 at the Kala Maloue site. At the montane site of Ngel Nyaki Forest Reserve in Nigeria, Agmen et al. (2009) found that their vervet study group consumed at least 28 plant species over the course of 11 months, with 16 of those being fruit. Only 11 plant species were included in the diet of Bale Mountain vervets in an Ethiopian bamboo forest (Mekonnen et al., 2010). On average, vervets at these various study sites fed on less than 30 different plant species. In a 17-month study of patas monkeys in Laikipia, Kenya, Isbell (1998) observed patas monkeys feeding on only 13 plant species.

Vervets and patas monkeys depend to a large degree on *Acacia* tree foods at a number of sites where they have been studied. At the relatively poor (in terms of plant species diversity) Amboseli site, two species of *Acacia* (umbrella and fever tree) accounted for over 50% of the vervet diet (Wrangham & Waterman, 1981; Lee, 1984). In a thornbush, riverine forest in Senegal, the vervet diet consisted mainly of fruit, with other plant parts, and also *Acacia* gums, which contributed significantly to the diet (Galat & Galat-Luong, 1978). In the Laikipia Plateau, Kenya, vervets fed mainly on food items (such as swollen thorns, animals within swollen thorns, inflorescences, exudates, but mainly new growth and seeds) associated with the dominant tree species, the whistling-thorn acacia (*Acacia drepanolobium*); over 40% of the diet was made up of whistling-thorn and other *Acacia* species. This plant is defended by thorns and by protective biting ants (Isbell, 1998; Pruetz, 1999). At a more northern site in Kenya, Whitten (1983) found that the vervets' main food was *Acacia* flowers. At Baffle Noir and Kala Maloue, Cameroon *Acacia* species were also important diet items (Kavanagh, 1978). In East Africa, patas monkeys depend even more on nourishment from *Acacia* than either vervets or patas monkeys studied elsewhere. Whistling-thorn *Acacia* accounted for 83% of the diet of patas monkeys at the Segera, Laikipia, site (Isbell, 1998). Gums and swollen thorns of this tree made up 37% and 36% of the overall diet, respectively. In their study of Mutara, Laikipia, patas in Kenya, Olson and Chism (1984) reported that more than 20% of all feeding activity of these monkeys was on the gum of three species of *Acacia*.

The diet of vervets has been studied more extensively and in a wider variety of habitats than patas monkeys, and variation can be seen both between and within sites (see Table 15.2). Galat and Galat-Luong (1978) studied vervet groups in a thornbush, riverine forest and in a mangrove in Senegal. In both study sites, miscellaneous plant material (leaves, shoots, grass blades, buds, flowers, fresh thorns, rhizomes, and gums) accounted for 38.7% of the diet. At the Ngel Nyaki site in Nigeria, which was characterized as degraded forest scrubland and riparian forest fragments surrounded by grassland, over 63% of vervet plant diet consisted of fruits (Agmen et al., 2009). *Ficus* was the most heavily consumed plant species (Agmen et al., 2009). Kavanagh (1978) studied vervet diets in detail at three woodland savanna sites in Cameroon; at all sites, the majority of the diet consisted of fruit, flowers, foliage, or invertebrates. The most important food

items at two sites (Baffle Noir and Kala Maloue) were the flowers, fruits, and gums of the African nettle tree (*Celtis integrifolia*) and invertebrates (Kavanagh, 1978). The proportions of dietary components were different at the two sites. At Baffle Noir, the vervets ate fruit, flowers, and invertebrates more or less evenly, whereas at Kala Maloue they spent over half their time eating fruit. Harrison (1984) studied the diet of vervets at Mt. Assirik, Senegal, and found that they ate mainly fruit and flowers and maximized the number of fruit species eaten when fruit was abundant. Other plant parts only became major food items when fruits and flowers were unavailable (Harrison, 1984).

The large amount of variation in the number of species eaten by vervets relates to the plant diversity at different sites. At the two sites with low plant species diversity (Amboseli in Kenya and Senegal River in Senegal) gums and leaves were major dietary components due to low plant food availability. Plant parts other than fruit and flowers were supplemented in the diet when these reproductive parts were unavailable (Fedigan & Fedigan, 1988). An exception to the typical vervet omnivorous diet is the Bale Mountain monkey. In Ethiopia, Mekennon et al. (2010) found bamboo alone (mostly in the form of immature leaves) accounted for about three-fourths of this monkey's diet.

Savanna-dwelling primates live in highly seasonal environments, so it is not surprising that there are often significant differences in wet versus dry season diets and feeding behavior. In a two-year study, the diet of vervets in a poor habitat in Senegal changed significantly between seasons, with fruit constituting 8% of the diet during the rains but 42% in the dry season, and leaves (mainly grasses) accounting for 55% of the diet in the wet season and

Table 15.2 Diet of Patas Monkeys and Vervets

	% Fruit	% Flowers	% Leaves	% Exudate	% Animal Prey	% Seeds	Site [a]
Patas:							
	See "Seeds"	7.0	2.0	35.0	38.0	10.0	Segera, Laikipia, Kenya
	–	–	–	20.0	–	–	Mutara, Laikipia, Kenya
	0	65.0	6.0	12.0	7.0	0	Kala Maloue, Cameroon
Average:	See "Seeds"	36.0	4.0	20.6	22.5	5.0	
Vervets:							
	37.0	–	27.0	–	13.0	–	Senegal
	42.0	–	32.0	14.0	11.0	–	Senegal (forest)
	49.2	5.3	20.5	–	25.1 (insects)	–	Ngel Nyaki, Nigeria
	11.1	14.3	26.6	30.0	7.7	2.6	Amboseli, Kenya
	5.8	44.7	0	–	–	19.6	Samburu, Kenya
	10.0	8.0	2.0	37.0	23.0	8.0	Segera, Laikipia, Kenya
	9.6	3.1	85.1	–	2.3	–	Odobullu Forest, Ethiopia
Average:	23.5	15.1	27.6	27	13.7	10.1	

[a] *Sources*: Agmen et al. (2009), Chism (1984), Chism and Wood (1994), Galat and Galat-Luong (1978), Harrison (1984), Isbell (1998), Isbell et al. (1998b), Kavanagh (1978), Klein (1978), Lee (1984), Mekonnen et al. (2010), Nakagawa (2000), Pruetz (2006), Pruetz and Isbell (2000), Struhsaker (1967b), Whitten (1983); Wrangham and Waterman (1981).

only 24% (mainly *Acacia*) during the dry period (Galat & Galat-Luong, 1977). At Kavanagh's (1978) woodland study sites in Cameroon, diet fluctuated seasonally with food availability. Harrison (1984) found that Assirik, Senegal, vervets exhibited seasonal changes in plant species chosen, number of species eaten, and particular plant parts—all directly related to availability. Harrison (1984) compared the diets of vervet populations at six different study sites: a great deal of seasonal variability existed, but the same food parts were eaten at all sites.

Vervets almost always include some animal foods in their diet, but this varies among the sites where they have been studied. At one extreme, in the mangrove site in Senegal, Galat and Galat-Luong (1978) found that fiddler crabs accounted for almost 48.4% of animal food in the diet. During a drought in Senegal, also characterized by a population explosion of Nile rats, a competitor for the vervets' food, monkeys moved from including no animal prey in the diet to including Nile rats, eggs, and other vertebrates (Galat & Galat-Luong, 1977). Similarly, during a period of food shortage, vervets in South Africa have been observed feeding intensively on the eggs of egrets and weaver birds (Skinner & Skinner, 1974). In Amboseli, Kenya, Struhsaker (1967b) observed vervets feeding on a number of invertebrates as well as, on two occasions, vertebrates. At Kavanagh's (1978) three woodland study sites in Cameroon, vervets concentrated on termites and caterpillars when the swarming of the former and heavy infestations of the latter occurred; as previously mentioned, invertebrates could be considered one of the top two foods of the monkeys at two sites (Baffle Noir and Kala Maloue) where more detailed study was carried out. Harrison's (1984) comparison of vervets at his Assirik, Senegal, site with other populations caused him to conclude that invertebrates (but few vertebrates) are generally eaten regularly by vervets throughout the year regardless of plant food availability.

Where they have been studied, patas monkeys include less fruit in their diet than vervets and appear to focus on *Acacia* parts, with a large proportion of insect prey as well as gums in their diet. Patas monkeys have an atypical diet for their body size, which includes a significant amount of animal prey and gums (Chism & Rowell, 1988; Isbell, 1998). The patas monkeys that Isbell (1998) studied fed primarily on gums and insects, violating Kay's threshold, which states that primates less than 500 g tend to be insectivorous and gummivorous, while larger primates include more leaves and fruits in the diet (Kay, 1984). Isbell (1998) estimates that 30–40% of the diet of patas monkeys at Segera consisted of animal prey, mainly social ants. Mature swollen acacia thorns are opened to obtain the eggs, larvae, and pupae of obligate acacia ants. In addition, the monkeys ate agama lizards, geckos, chameleons, and a variety of invertebrates, including grasshoppers and caterpillars (about 5% of all food consumed). Olson and Chism (1984) reported that at Mutara, Kenya patas monkeys also ate substantial quantities of arthropods. Besides prey items and whistling-thorn *Acacia*, other plant parts made up less than 20% of the Segera monkeys' diet (Isbell, 1998). A minute proportion of the diet consisted of grasses. However, Hall (1965) observed patas monkeys in Uganda fed mainly on grasses, berries, fleshy fruits, pods, and seeds and to a lesser extent on mushrooms, ants, and grasshoppers. Lizards, birds' eggs, and red mud also were eaten occasionally. Some food was obtained from trees, but most was eaten while the monkeys were on the ground. Besides grasses, the most common food at Hall's study site was pods and fruit pulp of the tamarind tree (*Tamarindus indica*).

DRINKING

Savanna dwelling primates are probably most restricted in terms of ranging, and likely other activities, by the availability of water. Unlike arboreal, forest-living primates, the savanna monkeys do not obtain most of their water from plants and must depend upon standing water, especially during the dry season. During a period of drought at their South African site, in a semi-arid Nama Karoo biome, McDougall et al. (2010) found that two study groups of vervets moved outside of their regular territories to drink. One study group moved more than 750 m, bypassing the ranges of four other vervet groups, to arrive at a sufficient source of water. Vervets at this site also dig for water in dry streambeds, as has been observed for other savanna-dwelling primates such as chimpanzees (*Pan troglodytes verus*: Pruetz & Bertolani, 2009) and baboons (*Papio hamadryas*: Galat-Luong et al., 2006). At Isbell's Segera, Kenya, site an unhabituated vervet group was observed moving hundreds of meters across open savanna-woodland to drink at a cattle trough during the peak of the dry season (Pruetz, pers.obs.), but members of the smaller study group relied to some extent on plant parts for water, chewing then spitting out plants like *Sarcostemma*. Struhsaker (1967b) found permanent waterholes within the ranges of all vervet groups studied at Amboseli. Over the whole year, the vervets visited water sources about once every two days on average. However, during the dry season (June–October) water is restricted to a few permanent waterholes and the vervets visit them much more frequently (once every 10 hours). In fact, the home ranges of vervets are more restricted during this season because the monkeys concentrate their activity in the vicinity of permanent water. Wrangham (1981) found that access to water during drought was critical to the survival of some individuals in the Amboseli study groups. Similarly, at the South African site, McDougall et al. (2010) found that the 14 high-ranking females in vervet groups had priority of access to water during the drought. Since certain groups supplanted others from drinking, they noted that not all individuals of a group were able to drink and, therefore, low-ranking individuals likely suffered at this time (McDougall et al., 2010). However, there were no recorded cases of mortality associated with the drought, and the authors surmised that the monkeys either obtained additional water from food plants or otherwise adjusted physiologically to the severe conditions (McDougall et al., 2010).

The patas monkey is generally presumed to be less dependent upon water than other savanna primates, such as baboons, but their behavior varies where they have been studied. At Hall's study site in Uganda, patas monkeys were seen to go to standing water to drink on only four occasions. On many hot, dry days they ranged close to rain pools, wallows, or the Emi River without going to drink. During the dry season, they were seen to drink from pools, puddles, or water-filled holes in trees. The behavior of single groups of patas monkeys visiting waterholes in Uganda was different than when multiple groups were present. After waiting for the resident male to survey the area for several minutes, other individuals then followed him to the waterhole and remained there for less than 30 seconds (Hall, 1965). Struhsaker and Gartlan (1970) observed a concentration of patas monkey groups around water sources in the Waza Reserve, Cameroon. In this area, rainfall is about 650 mm/year, and virtually no rain falls from November–April. During the peak of the dry season in April, patas monkey groups were seen to concentrate around the few remaining water supplies; sometimes as many as two to three one-male groups and some all-male groups congregated near the waterholes, often remaining for long periods. Many social interactions occurred at these times, and one morning as many as 100 patas remained near a waterhole for five hours (Struhsaker & Gartlan, 1970). At the Segera site in Kenya, Isbell et al. (1998b) found her study group always eventually moved toward cattle troughs or ephemeral pools of water, or a river located at one end of the home range. Other groups' ranges included regular but scattered sources of water. Isbell (1998) suggests patas monkeys may minimize depleting local resources around waterholes by moving between widely scattered water sources rather than radiating from them when foraging.

PREDATION

The vervet is the smallest African savanna monkey and is, therefore, subject to greater predation pressure than other diurnal savanna-dwelling primates. At Amboseli, vervets were subject to at least 16 potential predators (Struhsaker, 1967b), to 11 predator species at study sites in Ethiopia (Dunbar & Dunbar, 1974), and to at least 12 at the Segera site in Kenya (Isbell et al., 1998a). The Blydeberg Conservancy site in South Africa is home to at least 10 potential vervet predators (Barrett, 2005). Major predators of vervets include leopards, serval cats, crowned hawk-eagles, martial hawk-eagles, pythons, and baboons (Struhsaker 1967b; Cheney & Wrangham, 1987; Cheney & Seyfarth, 1990; Isbell, 1990; Isbell et al., 1998a). Verreaux's eagle-owl may be an important predator during crepuscular and nocturnal hours and was met by alarm calls at the Segera site. Spotted hyenas, jackals, hunting dogs, small African felids, lions, and cheetahs are also possible predators and elicit low-intensity responses from vervets (Struhsaker, 1967b). In Senegal, vervets react to chimpanzees (*Pan troglodytes verus*) as they do to humans, refraining from giving alarm calls but hiding in vegetation. Vervets account for one of the most common vertebrate prey of chimpanzees in the area (Pruetz, unpub. data). Chimpanzees less frequently prey upon patas monkeys at Fongoli, Senegal (Pruetz & Marshack, 2009).

At the Segera, Laikipia, site, vervets and patas monkeys shared the same potential predators, which also included domestic dogs, tawny eagles, and striped hyenas, as well as those mentioned previously (Isbell et al., 1998a). Over the course of the study by Isbell and colleagues, vervets changed their reaction to domestic dogs, indicating some event occurred that caused the monkeys to later consider dogs a potential predator. Formerly, dogs were met with "strange human chutters," a low-intensity-level alarm call, but later dogs were met with terrestrial predator or "leopard alarm" calls (Pruetz, pers.obs.).

The main cause of mortality in vervets is thought to be predation (Enstam & Isbell, 2002). At Amboseli, between 1963–1987, 45–70% of vervet mortality was attributed to predation on healthy animals (Cheney & Wrangham, 1987; Cheney & Seyfarth, 1990; Isbell, 1990) and 70% of infant mortality was thought to be also due to predation (Cheney et al., 1988). Vervets' main defense against predators is arboreal escape, although this varies according to the predator context. Vervets retreat to trees in the face of danger from terrestrial predators. At night, vervet groups divide up and sleep in trees in small subgroups, and they visit waterholes at times when other vulnerable species are likely to be there (Struhsaker, 1967a). In order to escape potential predators, vervets have developed an elaborate set of warning vocalizations. In the face of major predators, mainly males and high-ranking adults give alarm calls, initially as an alarm and subsequently as a threat while mobbing the predator. As in other *Cercopithecus* species, vervet males spend more time in vigilance than other age-sex classes (Gartlan, 1968; Fedigan & Fedigan, 1988). Struhsaker (1967b) noticed that different warning vocalizations and responses were elicited by different types of predators. For example, vervet males have a specific alarm call for leopards, the most dangerous terrestrial predator. Seyfarth (1980) and Cheney and Seyfarth (1990) confirmed these observations using playbacks of recorded alarm calls. They found that vervets ran into trees for "leopard alarms" (terrestrial predator alarms), looked up for "eagle alarms" (avian predator alarms), and looked down for "snake alarms." Young vervets learned, with age and experience, which alarm calls were appropriate to which class of predator. For instance, young vervets have been observed to alarm-call at ostriches but are then ignored by other group members who might initially react to the terrestrial predator alarm call by escaping into the trees (Pruetz, pers.obs.). The normal response of vervets to humans is to move silently and rapidly into dense vegetation, but their reaction to humans can be quite variable given the circumstances. Recently, encounters with drones have elicited new alarm calls from vervet monkeys in West Africa (Wegdell et al., 2019).

Patas monkeys, the fastest of all primates and able to reach speeds of at least 55 km/h (about 35 mph) traditionally have been thought to escape danger by running on the ground (Tappen, quoted in Hall, 1965). However, over a four-year period in Laikipia, Isbell et al. (1998a) observed patas monkeys responding to predators by climbing trees as often as running away on the ground. Climbing into trees was often an initial response, followed by fleeing the area. The potential predators of the patas include leopards, jackals, lions, cheetahs, and domestic dogs (Chism & Rowell, 1988; Isbell et al., 1998a). Chism et al. (1983) list 12 species of mammals and 4 species of large raptorial birds that are potential predators. At the nearby Segera site, a total of 12 potential predators of patas monkeys were identified (Pruetz, 1999). Vigilance is the most important feature of the patas monkeys, though being cryptic, agile, and able to flee rapidly are also important predator defense mechanisms (Hall, 1965; Chism & Rowell, 1988; Isbell & Enstam, 2002). Before entering new or wooded areas and before approaching waterholes or night-sleeping sites, the animals are cautious and thoroughly scan the area. In dry, treeless areas, the male may run ahead 200–600 m while the rest of the group remains silent (Hall, 1965). However, the behavior of males varied at Isbell's Segera site, with some resident males exhibiting defensive behavior toward potential predators but other males refraining from such behavior as other group members take up this role (see below). Most of the time, patas monkeys move quietly and vocalize infrequently; larger,

more conspicuous males often minimize their visibility by remaining low in trees and moving at the periphery of the group (Chism & Rowell, 1988). When a predator is encountered, however, patas monkeys have been observed to "escort" (a mild form of mobbing) animals as dangerous as an adult leopard more than one kilometer to beyond the boundaries of the monkeys' home range (Isbell et al., 1998a; Pruetz, pers.obs.). Adult females and juveniles led the group, while the adult male remained at the rear (reactions by adult males vary, however). Based on their reaction to humans, Hall (1965) assumed that if a patas monkey group detects a predator, the male attempts diversionary tactics. While in a tree, his size and white color of his thighs make him conspicuous (more so in East Africa) and he noisily bounces on the branches, shaking them vigorously. He then may approach the predator, trying to gain its attention or to encourage a chase. As the male engages in these diversionary tactics, the females and young remain in their places often lying flat in the grass where their coat color blends in with the dry grass. Similarly, young boys in Senegal out hunting monkeys with dogs report that adult male patas monkeys may provide a diversion, only to reverse and attack dogs on more than one occasion, in some cases inflicting wounds to the dogs that were eventually fatal (P. Stirling, pers.comm.). Hall thought that the running speed of the patas is probably a major consideration in the diversionary tactic of the male and in any last resort attempts by other group members to escape a predator.

Chism and Rowell (1988) believe that vigilance and crypticity are as important as high-speed locomotion in avoiding predators. In Kenya, patas monkeys prefer to remain near woodland margins, which provide good visibility across open grassland and ready access to the safety of trees, and they depend on trees and rocks as a refuge from predators (Chism & Rowell, 1988). In fact, Isbell et al. (1998a) found that patas monkeys responded to mammalian predators in a similar manner as do vervets, with both species climbing trees as often as they ran. Thus, the cursorial adaptations of patas monkeys may be as important for covering long distances to exploit their widely scattered resources as they are for high-speed escape from predators (Chism & Rowell, 1988; Isbell, 1998; Isbell et al., 1998a; Isbell & Enstam, 2002). However, where Hall conducted his study, there were few trees and mainly grassland; therefore, high-speed locomotion is likely also an important component of the defense mechanisms of these monkeys.

Struhsaker and Gartlan (1970), during a severe dry period in Cameroon, observed interactions between patas monkeys and jackals at waterholes. In three encounters, jackals attacked young patas, but in all three cases, adult males chased the jackals away and the young monkeys were unhurt. In one of the above instances, three adult males chased a jackal and two of these males did not belong to the same group as the youngster. In Laikipia, Kenya, one resident adult male patas monkey made a practice of chasing black-backed jackals while the next year's resident male ignored them as did the other group members (Pruetz, pers.obs.). Nevertheless, during the latter male's tenure, a black-backed jackal moving through a patas monkey group grabbed an infant in its jaws and ran before dropping it, seemingly unharmed, after the group gave chase (Isbell et al., 1998a).

Another characteristic attributed to predator pressure is the tendency for patas monkeys to give birth during the day, usually in the morning when the group is less mobile. Chism et al. (1983) presented evidence that this reduces vulnerability of the females to nighttime predation. They also observed that patas were most likely to encounter predation at specific times during the day, and the timing of diurnal births corresponded with hours that the female was least likely to encounter predators or lose contact with her group. Given the large home ranges, long day ranges, unpredictable sleeping sites, and combined vigilance of group members, a female separated from the group would have little chance of survival. Groups reduce mobility during the birth season, and this seems related to the presence of numerous pregnant females and newborn infants. These factors may correspond to birth synchrony and to the unusually short birth season among patas monkeys (Chism et al., 1983).

SOCIAL STRUCTURE AND BEHAVIOR
Models of primate socioecology maintain that predator pressure and food availability are two of the most important factors influencing primate social structure and behavior (Wrangham, 1980; van Schaik, 1983; Sterck et al., 1997; Isbell & Young, 2002). Vervets and patas monkeys correspond to some, but not all, of the predictions of these models. Vervets generally live in larger groups than other *Cercopithecus* monkeys, and typically in multi-male groups, rather than in the one-male groups or units (OMU) that characterize other monkeys of the Cercopithecini, including patas monkeys. Such differences are hypothesized to be adaptations to the greater predator pressure of these small monkeys' open habitat. For example, arboreal *Cercopithecus* monkeys often move in mixed-species groups, adding more vigilant males but still occupying relatively small home ranges. Vervets have larger home ranges, but greater predator pressure, than other guenons. Increasing the number of adult males within a group is hypothesized to add a certain degree of safety to group members without any obvious overriding

deficits (Gartlan, 1968; Goss-Custard et al., 1972; Busse, 1976; Wrangham, 1980). Isbell et al. (2004), however, propose the limited dispersal hypothesis to explain this multi-male system. Environmental factors like habitat configuration with a small number of adjacent groups and the high cost of dispersal limit male dispersal opportunities. In addition, males are more likely to join a nearby group, where a genetically related male has already dispersed, further reducing infanticide risk.

Upper limits on vervet group size are likely determined by availability of food and water, and smaller group size is likewise apparently constrained by predator pressure. Fitting with predictions according to socioecological models (Wrangham, 1980; van Schaik, 1983), larger vervet groups at Amboseli inhabited areas with more fever trees, an important food at this site (Cheney & Seyfarth, 1987). Similarly, when groups fall to a minimum number of females, such as at the Amboseli site (Hauser et al., 1986) or at the Segera site (Jaffe & Isbell, 2010), they are unable to sustain themselves. The females then join other larger groups and the original group becomes extinct (Hauser et al., 1986). Such fusion in apparent response to predation pressure has been observed only 10 times in vervets (Hauser et al., 1986; Isbell et al., 1991; Jaffe & Isbell, 2010).

Vervets vary somewhat across sites in terms of social grouping behavior, tolerance, and cohesiveness. Larger groups are generally found in more open habitats where predator pressure is higher. The sex ratio usually ranges from 1:1 to 1:4 adult males to adult females (Melnick & Pearl, 1987). Males transfer between groups, with young males often leaving their natal group accompanied by brothers or peers and older males migrating alone. Females remain within their natal group, and membership is stable except under extreme conditions (Struhsaker, 1967a; Cheney & Seyfarth, 1990). Group members remain closely spaced (10–50 m) (Isbell & Enstam, 2002), especially under conditions of poor visibility, when crossing open spaces, and during intergroup encounters. In some areas, however, large groups split into subgroups, using different portions of the group range for extended periods of time (Galat & Galat-Luong, 1976; Kavanagh, 1981). In Amboseli, although subgroups are not formed during the day, sleeping subgroups form and roosting sites can be as far as 750 m apart. Struhsaker (1967a) hypothesizes that this allows the monkeys to remain inconspicuous to nocturnal predators.

Dominance hierarchies and subgroup alliances are important factors structuring social organization within vervet groups (Struhsaker, 1967a; Seyfarth, 1980; Wrangham, 1981; Whitten, 1983; Fedigan & Fedigan, 1988; Cheney & Seyfarth, 1990; Pruetz, 1999). Generally, the dominance hierarchy of females is linear, relatively stable, and matrilineally inherited (Cheney & Seyfarth, 1990), whereas the male hierarchy is linear but not stable, and males may change dominance rank many times during their lifetime. Dominant animals have priority of access to food, space, and grooming relationships (dominant animals received more and groomed less than subordinates), and are winners in agonistic encounters. Males are dominant over females, but female choice may have a great deal to do with male dominance hierarchies (Hector & Raleigh, 1992). Further, under certain circumstances, female hierarchies do not conform to certain popular models of primate social organization, in that competition over food is only one of many factors influencing female relationships (Pruetz, 1999). Adult females frequently lead group progressions and certain matrilines are dominant over others. Interestingly, McDougall et al. (2010) found that a vervet group that left its territory to find water during a drought in South Africa followed a male that had immigrated into the group the day before.

Although the dominance hierarchy of female vervets is often not statistically linear, it may serve to ensure priority of access for high-ranking individuals during times of resource stress (Pruetz, 2009). Dominant individuals have been observed to exhibit significantly greater reproductive success in terms of lower infant mortality at Samburu, Kenya (Whitten, 1983), and dominant females reproduce more frequently (Turner et al., 1987). Whitten (1983) linked this rank-related difference to the clumped nature of the monkeys' main food in Samburu, *Acacia* flowers. During a drought in Amboseli, Wrangham (1981) found that high-ranking females outcompeted lower-ranking individuals for access to water and that three of the latter, but only one of the former, died during this time period.

Besides dominance relationships, subgroup alliances of attracted individuals (most often matrilineal kin) provide mechanisms for intragroup organization. These alliances form for defensive coalitions, as grooming partners, and as sleeping subgroups and are important when the group grows too large and splits into two independent groups. In many cases, defensive coalitions between less dominant individuals against more dominant ones are able to neutralize or reverse dominance relations, although individual dominance positions are not affected. At Mawana Game Reserve, South Africa, Arseneau-Robar et al. (2017) found that females were more likely to be involved in aggressive encounters with other groups if important, long-term resources were involved, if they were higher in rank, if they did not have an infant, and if they had male support during the aggressive encounter. These findings suggest females make complex decisions about involvement in intergroup aggression, depending on the unique circumstances and context each time.

In a study of grooming relationships among adult females of three groups of vervets, Seyfarth (1980; Cheney & Seyfarth, 1983) found that high rates of grooming were correlated with high rates of alliance formation and/or proximity. Furthermore, alliances permitted females to drive away otherwise dominant males. Seventy percent of all observed female-female alliances were formed against males. This allowed females to consistently defeat males in competitive interactions (Seyfarth, 1980; Cheney, 1983; Cheney & Seyfarth, 1990). Close association over a lifetime allows opportunities for individuals to build strong alliances and is a primary mechanism for "kin recognition."

Vervet males migrate from their natal group, and there is no evidence that vervets recognize paternal kin. Alliances among male vervets are rare (Henzi, 1985; Cheney & Seyfarth, 1990). Cheney (1981) found that, although high-ranking adults are more aggressive than those of lower rank, high-ranking males did not account for the largest proportion of copulations. In two groups at Amboseli, dominance rank and copulation frequencies were not correlated. Recently deposed dominant males, and those males who had been in the group for a long time, regardless of rank, copulated frequently. Additionally, there was no positive correlation between female rank and reproductive success (Cheney et al., 1981). Infant survival and life span failed to correlate with high rank among vervet females (Whitten, 1983; Cheney & Seyfarth, 1990).

Patas groups range in size from 11–74 animals and average around 28 individuals (see Table 15.4). When first studied, patas monkeys were thought to live in one-male, polygynous groups. Although they are often seen in one-male groups, male membership in these groups is not stable nor is mating polygynous. In Kenya, Chism and Rowell (1986, 1988) found that the population contained several permanent groups of females with fixed home ranges. Within this large area, there was a relatively stable population of males. These males lived as solitaries, as members of all male associations, or as resident males in a female group; over a 4-year period, several recognized males changed from one of these conditions to another repeatedly. Male tenure in female groups averaged 3 months, with the longest lasting less than a year (Chism, 1999b). Chism and Rowell (1986, 1988) focused their study on two of the female groups within the study area. Several recognized males joined and mated with females on more than one occasion. During the short mating season, and some birth seasons, several males accompanied the groups. Many studies now have confirmed this social organization among patas monkeys (Harding & Olson,

1986; Oshawa et al., 1993; Chism & Rogers, 1997; Chism, 1999b; Carlson & Isbell, 2001). In one-male groups, adult sex ratio ranges from 1:3 to 1:19 adult males to females. At Chism and Rowell's Mutara study site, the two study groups ranged in size from 15–41 animals (group M1) and from 42–74 (group M2). The majority of the time one male accompanied each group, although female groups were seen without any adult males for brief periods, as was also observed at the nearby Segera site. During the 4-year study in Kenya, 6 different males were resident in M1 group and 11 in M2. Of 18 periods of male residence, 14 were for less than 2 months, the longest being for 9 months. Five males were resident at one time or another in both groups, and during the peak of mating season, several males often were present in one or both groups.

Resident male patas monkeys have been described as peripheral to the social group in terms of both spatial and social behavior, save for the conceptive season (Chism et al., 1984). The adult females remain in their natal group and make up a compact central core. Females interact mainly with close relatives, and interactions are generally relaxed (Chism 1999a). The male does not interfere in squabbles among the females, and females normally initiate and lead group progressions (Chism et al., 1984; Chism & Rowell, 1988). Female patas monkeys have been described as having more individualistic and egalitarian dominance relations than vervets (Table 15.3) (Rowell & Olsen, 1983; Isbell & Pruetz, 1998; Pruetz, 1999; Chism, 1999b; Pruetz & Isbell, 2000; but see Nakagawa, 2007). Females can be ranked into a linear pattern, but their hierarchy is not statistically linear at most sites like that of vervets (Pruetz, 2009). Although Nakagawa (2007) maintains that the clumped foods used by patas monkeys in Cameroon explains their higher rate of contest competition, no statistical analyses were done to establish linearity. Isbell and Young (2002: p.189) suggest that "latency to detection" of social hierarchies can be a measure of the importance of dominance in a species. However, Pruetz (2009) maintains that a statistically linear pattern is not likely to be recorded in patas monkeys given their rapid maturation and a dynamic turnover of adult females due to high mortality (at least in the Segera study group). Such a combination leads to an unstable adult female group membership from year to year, and a stable hierarchy would not be expected given the lack of maternal inheritance of status. Although social dominance may be more important as an organizing feature for vervets, we do not know as much about its potential importance for the less-studied patas monkey.

Although female patas group membership is stable from year to year, all-male groups do not have long-term stability (Gartlan, 1974; Chism & Rowell, 1986). All-male groups range in size from two to eight animals, averaging three to four (Gartlan & Gartlan, 1973; Gartlan, 1974; Chism & Rowell, 1986). They contain subadult, old, and prime adult males. Younger males leave their natal group between two and three+ years of age and rarely transfer alone (Chism & Rowell, 1986; Chism, 1999a). They then spend some time in all-male groups before attempting to join a female group (Chism & Rowell, 1986; Chism & Rogers, 1997). As stated above, males spend their entire lives frequently changing their social situation.

For patas monkeys, there is a discrete mating season, and conceptions usually occur during a two-month period. During this time, influxes of several adult and subadult males into heterosexual groups have been observed, and many of these males actively mate with receptive adult females (Carlson & Isbell, 2001). These males, however, are permanent residents of the study area. During the mating season, the female has a choice of males with which to mate and may mate with more than one. Paternity is not simply, or necessarily, an outcome of male competition and does not relate to dominance rank. In fact, the mating system is best described as a form of promiscuous polygyny (Chism & Rowell, 1986; Harding & Olson, 1986; Chism & Rogers, 1997; Chism, 1999a). The animals

Table 15.3 Female Vervet and Patas Monkey Social Relationships

Socioecological Variables [a]	Vervet Monkeys	Patas Monkeys	References [b]
Social category	Resident-nepotistic (tolerant?)	Resident-egalitarian	1, 2, 3, 5, 6, but see 7
Female-resident	Yes	Yes	1, 2, 3
Female relationships	Nepotistic	Individualistic?	3, 5, but see 7
Within-group contests	Strong	Weak (Kenya); Strong (Cameroon)	3
Between-group contests	Yes	Yes	1, 3, 6
Dominance hierarchy	Stable, linear, despotic	Not stable or linear for long periods (>15 months) (Kenya); linear pattern (Cameroon)	2, 3, 5, but see 7
Support by relatives	Common	Rare?	3
Food abundance	Limited	Limited	6
Spatial food distribution	Clumped	Dispersed (Kenya); clumped (Cameroon)	4, 6, 7

[a] Socioecological variable classifications modified from Isbell (1991), Wrangham (1980), van Schaik (1989), Sterck et al. (1997).
[b] References: 1—Chism and Rowell (1988), 2—Whitten (1983), 3—Cheney and Seyfarth (1990), 4—Isbell et al. (1998b), 5—Isbell and Pruetz (1998), 6—Isbell (1991), 7—Nakagawa (2007).

in the Kenya study area must be seen as an interbreeding population, or deme, larger than a single group, and the group must be seen mainly as an adaptation for foraging and predator protection and not necessarily or primarily as a reproductive unit. In fact, a particular male's tenure in the female group often ends during the mating season (Chism & Rogers, 1997) and, thus, presence in the group outside of the mating season does not guarantee access to receptive females. Obviously, advantages to a male that is attached to a female group outside of mating season are great—providing the protection of group membership, establishment of social relationships, and knowledge gained from resident females of water sources or predator presence and behavior.

SUMMARY

In general, patas monkeys and vervets share a number of traits, behavioral as well as anatomical. Both species are characterized as having overlapping home ranges with conspecific groups, aggressive intergroup interactions involving adult females, distinct mating and birthing seasons, female-resident social groups, and seasonally restricted food sources (Table 15.4). Patas monkeys exhibit

many traits similar to those of forest-living guenons, such as a mating system that can be described as ranging from female defense polygyny (i.e., one male monopolizes reproductive access to females) to promiscuity with multimale influxes (Cords, 1988; Carlson & Isbell, 2001), male intolerance of other adult males, female defense of home ranges, and a lack of male interaction with nonreceptive females (Rowell, 1988). Of these specific aspects of guenon society, vervets are characterized only by female defense of home ranges. Differences between the two species include size of day and home ranges, size of within-group spread, body size, and number of adult males resident in social groups. Vervets and patas monkeys are characterized by marked differences in female dominance relationships. Vervets exhibit a linear and stable female dominance hierarchy at most sites (Wrangham, 1981; Whitten, 1983; Cheney & Seyfarth, 1987; Isbell & Pruetz, 1998), while dominance among female patas monkeys is more egalitarian and individualistic (Isbell & Pruetz, 1998; but see Nakagawa, 1992, 2007). Whereas patas monkeys are characteristically described as savanna or savanna-woodland dwelling primates (Hall, 1965; Struhsaker & Gartlan, 1970; Chism & Rowell, 1988; Isbell

Table 15.4 Patas Monkey and Vervet Social Structure, Behavior, and Morphology

	Erythrocebus patas	*Cercopithecus aethiops*
Social Structure [a]		
Similarities:	Male dispersal	Male dispersal
	Overlapping home ranges	Overlapping home ranges
	Mean group size: 28	Mean group size: 23
Differences:	Unstable female dominance hierarchy	Stable female dominance hierarchy
	Linear pattern female dominance hierarchy	Statistically linear female dominance hierarchy
	One-male social group with multi-male influxes	Multi-male group
	Home ranges not defended	Home ranges defended
Behavior [a]		
Similarities:	Distinct mating season	Distinct mating season
	Distinct birth season	Distinct birth season
	Omnivorous diet	Omnivorous diet
	Acacias important in diet	*Acacias* important in diet
	Seasonal food resources	Seasonal food resources
	Semiterrestrial	Semiterrestrial
	Females aggressive in intergroup interactions	Females aggressive in intergroup interactions
Differences:	Larger day range (3,648 m mean)	Smaller day range (1,077 m mean)
	Larger home range (3,103 ha mean)	Smaller home range (64 ha mean)
	Various sleeping sites	Regularly used sleeping sites
	Inhabits patchy woodland habitats	Inhabits continuous canopy woodland habitats
Morphology [a]		
Differences:	Larger body size	Smaller body size
	(12 kg adult male, 7 kg adult female)	(5.6 kg adult female, 7 kg adult male)
	Longer limb length	Shorter limb length (relative to body size)
	Females mature earlier (2.5 years)	Females mature later (3.5 years)
	Sexual dimorphism: males ~45% larger than females	Sexual dimorphism: males ~20% larger than females

[a] Modified from Pruetz (2009). *Sources*: Nakagawa (1992), Fedigan and Fedigan (1988) review, Chism and Rowell (1988), Harrison (1983), Hall (1965), Struhsaker and Gartlan (1970), Loy and Harnois (1988), Loy et al. (1993), Kaplan and Zucker (1980), Whitten (1983), Struhsaker (1967a,b,c), Gartlan and Brain (1968), Henzi and Lucas (1980), Chism et al. (1984), Chism and Rowell (1986), Harrison (1984), Kavanagh (1978), Wrangham and Waterman (1981), Lee (1984), Struhsaker (1969), Harrison (1985), Turner et al. (1994), Tembo (1994), Hall and Gartlan (1965), Nakagawa (2000, 2007).

et al., 1998a), vervets are described as inhabiting riverine or gallery forest woodlands (Struhsaker, 1967b; Gartlan & Brain, 1968; Chism & Rowell, 1988; Pickford & Senut, 1988).

REFERENCES CITED—CHAPTER 15

Agmen, F. L., Chapman, H., & Bawuro, M. (2009). Seed dispersal by tantalus monkeys (*Chlorocebus tantalus tantalus*) in a Nigerian montane forest. *African Journal of Ecology*, 48(4), 1123-1128.

Altmann, S. A., & Altmann, J. (1970). *Baboon ecology*. University of Chicago Press.

Arseneau-Robar, T. J. M., Taucher, A. L., Schnider, A. B., van Schaik, C. P., & Willems, E. P. (2017). Intra- and inter-individual differences in the costs and benefits of intergroup aggression in female vervet monkeys. *Animal Behaviour*, 123, 129-137.

Barrett, A. S. (2005). *Foraging ecology of the vervet monkey (Chlorocebus aethiops) in mixed lowveld bushveld and sour lowveld bushveld of the Blydeberg Conservancy, Northern Province, South Africa*. [Magister's dissertation, University of South Africa].

Barrett, A. S., Barrett, L., Henzi, P., & Brown, L. R. (2016). Resource selection on woody plant species by vervet monkeys (*Chlorocebus pygerythrus*) in mixed–broad leaf savanna. *African Journal of Wildlife Research*, 46(1), 14-22.

Barrett, A. S., Brown, L. R., Barrett, L., & Henzi, S. P. (2006). Phytosociology and plant community utilization by vervet monkey of the Blydesberg Conservancy, Limpopo Province. *Koedoe*, 49, 49-68.

Brain, C. K. (1965). Observations on the behavior of vervet monkeys (*Cercopithecus aethiops*). *Zoologica Africana*, 1, 13-27.

Busse, C. D. (1976). Do chimpanzees hunt cooperatively? *Nature*, 112, 767-770.

Butynski, T. (1988). Guenon birth seasons and correlates with rainfall and food. In A. Gautier-Hion, F. Bourliere, J. Gautier, & J. Kingdon (Eds.), *A primate radiation: Evolutionary biology of the African guenons* (pp. 284-322). Cambridge University Press.

Carlson, A. A. & Isbell, L. A. (2001). Causes and consequences of single-male and multimale mating in free-ranging patas monkeys, *Erythrocebus patas*. *Animal Behaviour*, 62, 1047-1058.

Chalmers, N. R. (1972). Comparative aspects of early infant development in some captive cercopithecines. In F. E. Poirier (Ed.), *Primate socialization* (pp. 63-82). Random House.

Chapman, C. (1985). The influence of habitat on behaviour in a group of St. Kitts green monkeys. *Journal of Zoology*, 206, 311-320.

Chapman, C. (1987). Selection of secondary growth areas by vervet monkeys (*Cercopithecus aethiops*). *American Journal of Primatology*, 12(2), 217-221.

Cheney, D. L. (1981). Intergroup encounters among free-ranging vervet monkeys. *Folia Primatologica*, 35, 124-146.

Cheney, D. L. (1983). Extrafamilial alliances among vervet monkeys. In R. A. Hinde (Ed.), *Primate social relationships: An integrated approach* (pp. 103-111). Sineauer Associates.

Cheney, D. L., Lee, P. C., & Seyfarth, R. M. (1981). Behavioral correlates of non-random mortality among free-ranging female vervet monkeys. *Behavioral Ecology and Sociobiology*, 9, 153-161.

Cheney, D. L., & Seyfarth, R. M. (1983). Non-random dispersal in free-ranging vervet monkeys: Social and genetic consequences. *American Naturalist*, 122(3), 392-412.

Cheney, D. L., & Seyfarth, R. M. (1987). The influence of intergroup competition on the survival and reproduction of female vervet monkeys. *Behavioral Ecology and Sociobiology*, 21, 375-386.

Cheney, D. L., & Seyfarth, R. M. (1990). *How monkeys see the world*. University of Chicago Press.

Cheney, D. L., Seyfarth, R. M., Andelman, S. J., & Lee, P. C. (1988). Reproductive success in vervet monkeys. In T. H. Clutton-Brock (Ed.), *Reproductive success* (pp. 384-402). University of Chicago Press.

Cheney, D. L., & Wrangham, R. W. (1987). Predation. In B. B. Smuts, D. L. Cheney, R. M. Seyfarth, R. W. Wrangham, & T. T. Struhsaker (Eds.), *Primate societies* (pp. 227-239). University of Chicago Press.

Chism, J. (1984). Life history patterns of female patas monkeys. In M. Small (Ed.), *Female primates: Studies by women primatologists* (pp. 175-190). Alan L. Liss.

Chism, J. (1986). Development and mother-infant relationships among captive patas monkeys. *International Journal of Primatology*, 7, 49-81.

Chism, J. (1999a). Intergroup encounters in wild patas monkeys (*Erythrocebus patas*) in Kenya. *American Journal of Primatology*, 49(1), 43.

Chism, J. (1999b). Decoding patas social organization. In P. Dolhinow, & A. Fuentes (Eds.), *The nonhuman primates* (pp. 86-92). Mayfield Publishers.

Chism, J., Olson, D., & Rowell, T. (1983). Diurnal births and perinatal behavior among wild patas monkeys: Evidence of an adaptive pattern. *International Journal of Primatology*, 4, 167-184.

Chism, J., & Rogers, W. (1997). Male competition, mating success and female choice in a seasonally breeding primate (*Erythrocebus patas*). *Ethology*, 103(2), 109-126.

Chism, J., & Rowell, T. (1986). Mating and residence patterns of male patas monkeys. *Ethology*, 72, 31-39.

Chism, J., & Rowell, T. (1988). The natural history of patas monkeys. In A. Gautier-Hion, F. Bourliere, J. Gautier, & J. Kingdon (Eds.), *A primate radiation: Evolutionary biology of the African guenons* (pp. 412-438). Cambridge University Press.

Chism, J., Rowell, T., & Olson, D. K. (1984). Life history patterns of female patas monkeys. In M. Small (Ed.), *Female primates: Studies by women primatologists* (pp. 175-192). Alan R. Liss.

Chism, J., & Wood, C. S. (1994). Diet and feeding behavior of patas monkeys (*Erythrocebus patas*) in Kenya. *American Journal of Physical Anthropology*, 37(S18), 67.

Cords, M. (1988). Mating systems of forest guenons: A preliminary view. In A. Gautier-Hion, F. Bourliere, J. Gautier, & J. Kingdon (Eds.), *A primate radiation: Evolutionary biology of the African guenons* (pp. 323-339). Cambridge University Press.

Cords, M. (2000). The number of males in guenon groups. In P. Kappeler (Ed.), *Primate males: Causes and consequences of variation in group composition* (pp. 84-96). Cambridge University Press.

De Jong, Y. A., Butynski, T. M., Isbell, L. A., & Lewis, C. (2009). Decline in the geographical range of the southern patas monkey *Erythrocebus patas baumstarki* in Tanzania. *Oryx, 43,* 267-274.

DeMoor, P. P., & Steffens, F. E. (1972). The movements of vervet monkeys (*Cercopithecus aethiops*) within their ranges as revealed by radio-tracking. *Journal of Animal Ecology, 41,* 677-687.

DeVore, I., & Washburn, S. (1963). Baboon ecology and human evolution. In F. D. Howell, & F. Bourliere (Eds.), *African ecology and human evolution* (pp. 335-367). Routledge.

Disotell, T. (1996). The phylogeny of Old World monkeys. *Evolutionary Anthropology, 5*(1), 18-24.

Disotell, T. (2000). Molecular systematics of the Cercopithecidae. In P. F. Whitehead, & C. J. Jolly (Eds.), *Old World monkeys* (pp. 29-56). Cambridge University Press.

Dore, K. M. (2013). *An anthropological investigation of the dynamic human-vervet monkey (*Chlorocebus aethiops sabaeus*) interface in St. Kitts.* [Doctoral dissertation, University of Wisconsin-Milwaukee].

Dunbar, R. I. M. (1974). Observations on the ecology and social organization of the green monkey, *Cercopithecus sabaeus,* in Senegal. *Primates, 15,* 341-350.

Dunbar, R. I. M., & Dunbar, E. P. (1974). Ecological relations and niche separation between sympatric terrestrial primates in Ethiopia. *Folia Primatologica, 21,* 36-60.

Dunbar, R. I. M., & Sharman, M. (1984). Is social grooming altruistic? *Zeitschrift für Tierpsychologie, 64*(2), 163-173.

Durham, N. M. (1969). Sex differences in visual threat displays of West African vervets. *Primates, 10,* 91-95.

Dutrillaux, B., Muleris, M., & Couturier, J. (1988). Chromosomal evolution of Cercopithecinae. In A. Gautier-Hion, F. Bourliere, J. Gautier, & J. Kingdon (Eds.), *A primate radiation: Evolutionary biology of the African guenons* (pp. 150-159). Cambridge University Press.

Ely, R. M., Tarara, R. P., Worthman, C. M., & Else, J. G. (1989). Reproduction in the vervet monkey (*Cercopithecus aethiops*): III. The menstrual cycle. *American Journal of Primatology, 17*(1), 1-10.

Enstam, K. L., & Isbell, L. A. (2002). Comparison of responses to alarm calls by patas (*Erythrocebus patas*) and vervet (*Cercopithecus aethiops*) monkeys in relation to habitat structure. *American Journal of Physical Anthropology, 119,* 3-14.

Enstam, K. L., & Isbell, L. A. (2007). The guenons (genus *Cercopithecus*) and their allies: Behavioral ecology of polyspecific associations. In C. J. Campbell, A. Fuentes, K. C. MacKinnon, M. A. Panger, & S. K. Bearder (Eds.), *Primates in perspective* (pp. 252-274). Oxford University Press.

Enstam, K. L., Isbell, L. A., & De Maar, T. W. (2002). Male demography, female mating behavior, and infanticide in wild patas monkeys (*Erythrocebus patas*). *International Journal of Primatology, 23,* 85-104.

Fedigan, L., & Fedigan, L. M. (1988). *Cercopithecus aethiops*: A review of field studies. In A. Gautier-Hion, F. Bourliere, J. Gautier, & J. Kingdon (Eds.), *A primate radiation: Evolutionary biology of the African guenons* (pp. 389-411). Cambridge University Press.

Fedigan, L. M., Fedigan, L., Chapman, C., & McGuire, M. T. (1984). A demographic model of colonization by a population of St. Kitts vervets. *Folia Primatologica, 42*(3-4), 194-202.

Galat, G., & Galat-Luong, A. (1976). La colonization de la mangrove par *Cercopithecus aethiops sabaeus* au Senegal. *Revue d'écologie, 30,* 3-30.

Galat, G., & Galat-Luong, A. (1977). Demographie et regime alimentaire d'une troupe de *Cercopithecus aethiops sabaeus* en habitat marginal au nord Senegal. *La Terre et la Vie, 31,* 557-577.

Galat, G., & Galat-Luong, A. (1978). Diet of green monkeys in Senegal. In D. J. Chivers, & J. Herbert (Eds.), *Recent advances in primatology* (Vol. 1, pp. 257-258). Academic Press.

Galat-Luong, A., Galat, G., & Hagell, S. (2006). The social and ecological flexibility of Guinea baboons: Implications for Guinea baboon social organization and male strategies. In L. Swedell, & S. R. Leigh (Eds.), *Reproduction and fitness in baboons: Behavioral, ecological, and life history perspectives* (pp. 105-121). Springer.

Gartlan, J. S. (1968). Structure and function in primate society. *Folia Primatologica, 8,* 89-120.

Gartlan, J. S. (1974). Adaptive aspects of social structure in *Erythrocebus patas.* In S. Kondo, M. Kawai, A. Ehara, & S. Kawamura (Eds.), *Proceedings from the symposia of the fifth congress of the International Primatological Society* (pp. 161-171). Japan Science Press.

Gartlan, J. S., & Brain, C. K. (1968). Ecology and social variability in *Cercopithecus aethiops* and *C. mitis.* In P. C. Jay (Ed.), *Primates; Studies in adaptation and variability* (pp. 253-292). Holt, Rinehart & Winston.

Gartlan, J. S., & Gartlan, S. C. (1973). Some observations on the exclusively male groups of *Erythrocebus patas. Annales de la Faculte des Sciences du Cameroun, 12,* 121-144.

Gautier, J. (1988). Interspecific affinities among guenons as deduced from vocalizations. In A. Gautier-Hion, F. Bouliere, J. Gautier, & J. Kingdon (Eds.), *A primate radiation: Evolutionary biology of the African guenons* (pp. 194-226). Cambridge University Press.

Gerald, M. S. (1999). *Scrotal color in vervet monkeys (*Cercopithecus aethiops sabaeus*): The signal functions and potential proximate mechanisms of color variation.* [Doctoral dissertation, University of California at Los Angeles].

Gerald, M. S. (2001). Primate colour predicts social status and aggressive outcome. *Animal Behaviour, 61,* 559-566.

Gerald, M. S., Ayala J., Ruiz-Lambides A., Waitt C., & Weiss A. (2010). Do females pay attention to secondary sexual coloration in vervet monkeys (*Chlorocebus aethiops*)? *Naturwissenschaften, 97*(1), 89-96.

Gonzalez-Martinez, J. (1998). The ecology of the introduced patas monkey (*Erythrocebus patas*) population of southwestern Puerto Rico. *American Journal of Primatology, 45,* 351-365.

Goss-Custard, J. D., Dunbar, R. I. M., & Aldrich-Blake, F. P. G. (1972). Survival, mating and rearing strategies in the evolution of primate social structure. *Folia Primatologica, 17,* 1-19.

Groves, C. (2001). *Primate taxonomy.* Smithsonian Institution Press.

Hall, K. R. L. (1965). Ecology and behavior of baboons, patas, and vervet monkeys in Uganda. In H. Vagtborg (Ed.), *The baboon in medical research* (pp. 43-61). University of Texas Press.

Hall, K. R. L., & Gartlan, J. S. (1965). Ecology and behaviour of the vervet monkey, *Cercopithecus aethiops,* Lolui Island, Lake Victoria. *Proceedings of the Zoological Society of London, 145,* 37-57.

Haltenorth, T., & Diller, H. (1988). *A Field guide to the mammals of Africa including Madagascar.* Collins Sons & Co., Ltd.

Hamilton, W. J. (1985). Demographic consequences of a food and water shortage to desert chacma baboons, *Papio ursinus. International Journal of Primatology, 6,* 451-462.

Harding, R. S. O., & Olson, D. K. (1986). Patterns of mating among male patas monkeys (*Erythrocebus patas*) in Kenya. *American Journal of Primatology, 11*(4), 343-358.

Harrison, M. J. S. (1983). Age and sex differences in the diet and feeding strategies of the green monkey, *Cercopithecus sabaeus. Animal Behaviour, 31,* 969-977.

Harrison, M. J. S. (1984). Optimal foraging strategies in the diet of the green monkey, *Cercopithecus sabaeus,* at Mt. Assirik, Senegal. *International Journal of Primatology, 5*(5), 435-471.

Harrison, M. J. S. (1985). Time budget of the green monkey, *Cercopithecus sabaeus:* Some optimal strategies. *International Journal of Primatology, 6*(4), 351-376.

Hauser, M., Cheney, D., & Seyfarth, R. (1986). Group extinction and fusion in free-ranging vervet monkeys. *American Journal of Primatology, 11*(1), 63-77.

Hector, A. K., & Raleigh, M. J. (1992). The effects of temporary removal of the alpha male on the behavior of subordinate male vervet monkeys. *American Journal of Primatology, 26*(2), 77-87.

Henzi S. P. (1985). Genital signaling and the coexistence of male vervet monkeys (*Cercopithecus aethiops pygerythrus*). *Folia Primatologica, 45,* 129-147.

Henzi, S. P., & Lucas, J. W. (1980). Observations on the intertroop movement of adult vervet monkeys (*Cercopithecus aethiops*). *Folia Primatologica, 43,* 189-197.

Herzog, N. M., Parker, C. H., Keefe, E. R., Coxworth, J., Barrett, A., & Hawkes, K. (2014). Fire and home range expansion: A behavioral response to burning among savanna dwelling vervet monkeys (*Chlorocebus aethiops*). *American Journal of Physical Anthropology, 154,* 554-560.

Hill, W. C. O. (1966). *Primate comparative anatomy and taxonomy. VI: Catarrhini, Cercopithecoidea, Cercopithecinae.* Edinburgh University Press.

Horrocks, J. A. (1986). Life-history characteristics of a wild population of vervets (*Cercopithecus aethiops sabaeus*) in Barbados, West Indies. *International Journal of Primatology, 7,* 31-47.

Hurov, J. R. (1987). Terrestrial locomotion and back anatomy in vervets (*Cercopithecus aethiops*) and patas monkeys (*Erythrocebus patas*). *American Journal of Primatology, 13,* 297-311.

IUCN. (2008a). *Chlorocebus* spp.: Geographic range. *The International Union for the Conservation of Nature Red List of Threatened Species,* Version 2019-1. http:/www.iucnredlist.org.

IUCN. (2008b). *Erythrocebus patas*: Geographic range. *The International Union for the Conservation of Nature Red List of Threatened Species,* Version 2019-1. http:/www.iucnredlist.org.

IUCN. (2020). *Erythrocebus patas* and *Chlorocebus* spp. *The International Union for the Conservation of Nature Red List of Threatened Species,* Version 2020-1. http:/www.iucnredlist.org.

IUCN. (2021). *The International Union for the Conservation of Nature Red List of Threatened Species,* Version 2020-3. http:/www.iucnredlist.org.

Isbell, L. A. (1990). Sudden short-term increase in mortality of vervet monkeys (*Cercopithecus aethiops*) due to leopard predation in Amboseli National Park, Kenya. *American Journal of Primatology, 21*(1), 41-52.

Isbell, L. A. (1991). Contest and scramble competition: Patterns of female aggression and ranging behavior among primates. *Behavioral Ecology, 2,* 143-155.

Isbell, L. A. (1995). Seasonal and social correlates of changes in hair, skin, and scrotal condition in vervet monkeys (*Cercopithecus aethiops*) of Amboseli National Park, Kenya. *American Journal of Primatology, 36,* 61-70.

Isbell, L. A. (1998). Diet for a small primate: Insectivory and gummivory in the (large) patas monkey (*Erythrocebus patas pyrrhonotus*). *American Journal of Primatology, 45,* 381-398.

Isbell, L. A. (2013). *Erythrocebus patas* patas monkey (Hussar monkey, nisnas). In T. Butynski, J. Kingdon, & J. Kalina (Eds.), *Mammals of Africa: Vol. 2, Primates* (pp. 257-264). Bloomsbury Publications.

Isbell, L. A., Cheney, D. L., & Seyfarth, R. M. (1990). Costs and benefits of home range shifts among vervet monkeys (*Cercopithecus aethiops*) in Amboseli National Park, Kenya. *Behavioral Ecology and Sociobiology, 27,* 351-358.

Isbell, L. A., Cheney, D. L., & Seyfarth, R. M. (1991). Group fusions and minimum group sizes in vervet monkeys (*Cercopithecus aethiops*). *American Journal of Primatology, 25*(1), 57-65.

Isbell, L. A., Cheney, D. L., & Seyfarth, R. M. (2004). Why vervet monkeys (*Cercopithecus aethiops*) live in multimale groups.

In M. E. Glenn, & M. Cords (Eds.), *The guenons: Diversity and adaptation in African monkeys* (pp. 173-187). Springer.

Isbell, L. A., & Enstam, K. L. (2002). Predator (in)sensitive foraging in sympatric female vervets (*Cercopithecus aethiops*) and patas monkeys (*Erythrocebus patas*): A test of ecological models of group dispersion. In L. E. Miller (Ed.), *Eat or be eaten: Predator sensitive foraging among primates* (pp. 154-168). Cambridge University Press.

Isbell, L. A., & Jaffe, K. E. (2013). *Chlorocebus pygerythrus* vervet monkey. In T. Butynski, J. Kingdon, & J. Kalina, *Mammals of Africa: Vol. 2, Primates* (pp. 277-283). Bloomsbury Publications.

Isbell, L. A., & Pruetz, J. D. (1998). Differences between patas monkeys (*Erythrocebus patas*) and vervet monkeys (*Cercopithecus aethiops*) in agonistic interactions between adult females. *International Journal of Primatology, 19*(5), 837-855.

Isbell, L. A., Pruetz, J. D., Lewis, M., & Young, T. P. (1999). Rank differences in ecological behavior: A comparative study of patas monkeys (*Erythrocebus patas*) and vervets (*Cercopithecus aethiops*). *International Journal of Primatology, 20*(2), 257-272.

Isbell, L. A., Pruetz, J. D., & Young, T. P. (1998b). Movements of vervets (*Cercopithecus aethiops*) and patas monkeys (*Erythrocebus patas*) as estimators of food resource size, density, and distribution. *Behavioral Ecology and Sociobiology, 42*, 123-133.

Isbell, L. A., Pruetz, J. D., Young, T. P., & Lewis, M. (1998a). Locomotory activity differences between sympatric patas (*Erythrocebus patas*) and vervet monkeys (*Cercopithecus aethiops*): Implications for the evolution of long hindlimb length in *Homo*. *American Journal of Physical Anthropology, 105*(2), 199-207.

Isbell, L. A., & Young, T. P. (2002). Ecological models of female social relationships in primates: Similarities, disparities, and some directions for future clarity. *Behaviour, 139*, 177-202.

Isbell, L. A., Young, T. P., Jaffe, K. E., Chancellor, R. L., & Carlson, A. A. (2009). Demography and life histories of sympatric patas monkeys, *Erythrocebus patas*, and vervets, *Cercopithecus aethiops*, in Laikipia, Kenya. *International Journal of Primatology, 30*, 103-124.

Jaffe, K. E., & Isbell, L. A. (2010). Changes in ranging and agonistic behavior of vervet monkeys (*Cercopithecus aethiops*) after predator-induced group fusion. *American Journal of Primatology, 72*(7), 634-644.

Jasinska, A. J., Schmitt, C. A., Service, S. K., Cantor, R. M., Dewar, K., Jentsch, J. D., Kaplan, J. R., Turner, T. R., Warren, W. C., Weinstock, G. M., Woods, R. P., & Freimer, N. B. (2013). Systems biology of the vervet monkey. *ILAR Journal, 54*(2), 122-143.

Kaplan, J. R., & Zucker, E. (1980). Social organization in a group of free-ranging patas monkeys. *Folia Primatologica, 34*, 196-213.

Kavanagh, M. (1978). The diet and feeding behavior of *Cercopithecus aethiops tantalus*. *Folia Primatologica, 30*, 30-63.

Kavanagh, M. (1980). Invasion of the forest by an African savannah monkey: Behavioral adaptations. *Behavior, 73*, 238-260.

Kavanagh, M. (1981). Variable territoriality among tantalus monkeys in Cameroon. *Folia Primatologica, 36*, 76-98.

Kay, R. F. (1984). On the use of anatomical features to infer foraging behavior in extinct primates. In J. Cant, & P. Rodman (Eds.), *Adaptations for foraging in nonhuman primates* (pp. 21-53). Columbia University Press.

Kingdon, J. (1988). Comparative morphology of hands and feet in the genus *Cercopithecus*. In A. Gautier-Hion, F. Bourliere, J. Gautier, & J. Kingdon (Eds.), *A primate radiation: Evolutionary biology of the African guenons* (pp. 184-193). Cambridge University Press.

Kingdon, J. (1997). *Kingdon field guide to the mammals of Africa*. Princeton University Press.

Kingdon, J., Butynski, T. M., & De Jong, Y. (2008). *Erythrocebus patas*. The IUCN Red List of Threatened Species 2008: e.T8073A12884516. http://dx.doi.org/10.2305/IUCN.UK .2008.RLTS.T8073A12884516.en.

Klein, D. F. (1978). *The diet and reproductive cycle of a population of vervet monkeys* (Cercopithecus aethiops*)*. [Doctoral dissertation, New York University].

Lambert, J. E. (2002). Resource switching and species coexistence in guenons: A community analysis of dietary flexibility. In M. E. Glenn, & M. Cords (Eds.), *The guenons: Diversity and adaptation in African monkeys* (pp. 309-323). Springer.

Lambert, J. E. (2005). Competition, predation, and the evolutionary significance of the cercopithecine cheek pouch: The case of *Cercopithecus* and *Lophocebus*. *American Journal of Physical Anthropology, 126*(2), 183-192.

Lancaster, J. B. (1972). Play-mothering: The relations between juvenile females and young infants among free-ranging vervets. In F. E. Poirier (Ed.), *Primate socialization* (pp. 83-104). Random House.

Leakey, M. (1988). Fossil evidence for the evolution of the guenons. In A. Gautier-Hion, F. Bourliere, J. Gautier, & J. Kingdon (Eds.), *A primate radiation: Evolutionary biology of the African guenons* (pp. 7-12). Cambridge University Press.

Lee, P. C. (1984). Ecological constraints on the social development of vervet monkeys. *Behaviour, 91*, 245-262.

Lernould, J. (1988). Classification and geographical distribution of guenons: A review. In A. Gautier-Hion, F. Bourliere, J. Gautier, & J. Kingdon (Eds.), *A primate radiation: Evolutionary biology of the African guenons* (pp. 54-78). Cambridge University Press.

Loy, J. (1981). The reproductive and heterosexual behaviours of adult patas monkeys in captivity. *Animal Behaviour, 29*, 714-726.

Loy, J. Argo, B., Nestell, G., Vallett, S., & Wanamaker, G. (1993). A reanalysis of patas monkeys' "grimace and gecker" display and a discussion of their lack of formal dominance. *International Journal of Primatology, 14*(6), 879-893.

Loy, J., & Harnois, M. (1988). An assessment of dominance and kinship among patas monkeys. *Primates, 29*, 331-342.

Mahoney, S. A. (1980). Cost of locomotion and heat balance during rest and running from 9 to 55 C in a patas monkey. *Journal of Applied Physiology, 49*, 789-800.

Martin, R. D., & MacLarnon, A. M. (1988). Quantitative comparisons on the skull and teeth in guenons. In A. Gautier-Hion, F. Bourliere, J. Gautier, & J. Kingdon (Eds.), *A primate radiation: Evolutionary biology of the African guenons* (pp. 160-183). Cambridge University Press.

McDougall, M., Forshaw, N., Barrett, L., & Henzi, S. P. (2010). Leaving home: Responses to water depletion by vervet monkeys. *Journal of Arid Environments, 74*, 924-927.

McFarland, R., Barrett, L., Boner, R., Freeman, N. J., & Henzi, S. P. (2014). Behavioral flexibility of vervet monkeys in response to climatic and social variability. *American Journal of Physical Anthropology, 154*, 357-364.

Mekonnen, A., Bekele, A., Fashing, P. J., Hemson, G., & Atickem, A. (2010). Diet, activity patterns, and ranging ecology of the bale monkey (*Chlorocebus djamdjamensis*) in Odobullu Forest, Ethiopia. *International Journal of Primatology, 31*, 339-362.

Melnick, D. J., & Pearl, M. C. (1987). Cercopithecines in multimale groups: Genetic diversity and population structure. In B. B. Smuts, D. L. Cheney, R. M. Seyfarth, R. W. Wrangham, & T. T. Struhsaker (Eds.), *Primate societies* (pp. 121-134). University of Chicago Press.

Moreno-Black, G., & Maples, W. R. (1977). Differential habitat utilization of four Cercopithecidae in a Kenyan forest. *Folia Primatologica, 27*, 85-107.

Muroyama, Y. (2017). Variations in within-group inter-individual distances between birth and non-birth seasons in wild female patas monkeys. *Primates, 58*(1), 115-119.

Nakagawa, N. (1989). Activity budget and diet of patas monkeys in Kala Maloue National Park, Cameroon: A preliminary report. *Primates, 30*, 27-34.

Nakagawa, N. (1991). Comparative feeding ecology of patas monkeys and tantalus monkeys in Kala Maloue National Park, Cameroon (I): Patterns of range use. In A. Ehara, T. Kimura, O. Takenaka, & M. Iwamoto (Eds.), *Primatology Today* (pp. 119-122). Elsevier Science Publishers.

Nakagawa, N. (1992). Distribution of affiliative behaviors among adult females within a group of wild patas monkeys in a nonmating, nonbirth season. *International Journal of Primatology, 13*(1), 73-96.

Nakagawa, N. (2000). Foraging energetics in patas monkeys (*Erythrocebus patas*) and tantalus monkeys (*Cercopithecus aethiops*): Implications for reproductive seasonality. *American Journal of Primatology, 52*, 169-185.

Nakagawa, N. (2003). Difference in food selection between patas monkeys (*Erythrocebus patas patas*) and tantalus monkeys (*Cercopithecus aethiops tantalus*) in Kala Maloue National Park, Cameroon, in relation to nutrient content. *Primates, 41*, 3-11.

Nakagawa, N. (2007). Despotic wild patas monkeys (*Erythrocebus patas*) in Kala Maloue, Cameroon. *American Journal of Primatology, 69*, 1-13.

Nakagawa, N., Ohsawa, H., & Muroyama, Y. (2003). Life-history parameters of a wild, West African group of patas monkeys (*Erythrocebus patas*). *Primates, 44*, 281-290.

Nowak, R. M. (1999). *Walker's primates of the world.* Johns Hopkins University Press.

Ohsawa, H. (2003). Long-term study of the social dynamics of patas monkeys (*Erythrocebus patas*): Group male supplanting and changes to the multi-male situation. *Primates, 44*(2), 99-107.

Ohsawa, H., Inoue, M., & Takenaka, O. (1993). Mating strategy and reproductive success of male patas monkeys (*Erythrocebus patas*). *Primates, 34*, 533-544.

Olson, D., & Chism, J. (1984). Gum feeding as a dietary adaptation of wild patas monkeys (*Erythrocebus patas*) in Kenya. *International Journal of Primatology, 5*, 370.

Pasternak, G., Brown, L. R., Kienzle, S., Fuller, A., Barrett, L., & Henzi, S. P. (2013). Population ecology of vervet monkeys in a high latitude, semi-arid riparian woodland. *Koedoe, 55*(1), 1-9.

Pickford, M., & Senut. B. (1988). Habitat and locomotion in Miocene cercopithecoids. In A. Gautier-Hion, F. Bourliere, J. Gautier, & J. Kingdon (Eds.), *A primate radiation: Evolutionary biology of the African guenons* (pp. 35-53). Cambridge University Press.

Pruetz, J. D. (1999). *Socioecology of adult female vervet and patas monkeys in Laikipia, Kenya.* [Doctoral dissertation, University of Illinois at Urbana].

Pruetz, J. D. (2006). Feeding ecology of savanna chimpanzees at Fongoli, Senegal. In C. Boesch, G. Hohmann, & M. Robbins (Eds.), *The feeding ecology of great apes and other primates* (pp. 161-182). Cambridge University Press.

Pruetz, J. D. (2009). *The socioecology of adult female patas monkeys and vervets in Kenya.* Pearson Prentice Hall.

Pruetz, J. D., & Bertolani, P. (2009). Chimpanzee (*Pan troglodytes verus*) behavioral responses to stresses associated with living in a savanna-mosaic environment: Implications for hominin adaptations to open habitats. *PaleoAnthropology, 2009*, 252-262.

Pruetz, J. D., & Isbell, L. A. (2000). Ecological correlates of competitive interactions in female vervets (*Cercopithecus aethiops*) and patas monkeys (*Erythrocebus patas*) in simple habitats. *Behavioral Ecology Sociobiology, 49*, 38-47.

Pruetz, J. D., & Marshack, J. L. (2009). Predation on patas monkeys by savanna chimpanzees at Fongoli, Senegal. *PAN Africa News, 16*, 15-17.

Rose, M. D. (1979). Positional behavior of natural populations: Some quantitative results of a field study of *Colobus guereza* and *Cercopithecus aethiops*. In M. E. Morbeck, H. Preuschoft, & N. Gomberg (Eds.), *Environment, behavior, and morphology: Dynamic interactions in primates* (pp. 75-93). Fischer.

Rowell, T. E. (1977). Variation in age at puberty in monkeys. *Folia Primatologica, 27*, 284-296.

Rowell, T. E. (1988). The social system of guenons, compared with baboons, macaques and mangabeys. In A. Gautier-Hion, F. Bourliere, J. Gautier, & J. Kingdon (Eds.), *A primate radiation: Evolutionary biology of the African guenons* (pp. 439-451). Cambridge University Press.

Rowell, T. E., & Hartwell, K. M. (1978). The interaction of behavior and reproductive cycles in patas monkeys. *Behavioral Biology, 24*, 141-167.

Rowell, T. E., & Olson, D. K. (1983). Alternative mechanisms of social organization in monkeys. *Behaviour, 86*(1-2), 31-54.

Ruvolo, M. (1988). Genetic evolution in the African guenons. In A. Gautier-Hion, F. Bourliere, J. Gautier, & J. Kingdon (Eds.), *A primate radiation: Evolutionary biology of the African guenons* (pp. 127-139). Cambridge University Press.

Saj, T., Sicotte, P., & Paterson, J. D. (1999). Influence of human food consumption on the time budget of vervets. *International Journal of Primatology, 20*, 977-994.

Seyfarth, R. M. (1980). The distribution of grooming and related behaviors among adult female vervet monkeys. *Animal Behaviour, 28*, 798-813.

Skinner, J. D., & Skinner, C. P. (1974). Predation on cattle egret (*Bulbulcus ibis*) and masked weaver (*Ploceus valatus*) by vervet monkey (*Cercopithecus aethiops*). *South African Journal of Science, 70*, 157-158.

Sly, D. L., Harbaugh, S. W., London, W. T., & Rice, J. M. (1983). Reproductive performance of a laboratory breeding colony of patas monkeys (*Erythrocebus patas*). *American Journal of Primatology, 4*, 23-32.

Sterck, E. H. M., Watts, D. P., & van Schaik, C. P. (1997). The evolution of female social relationships in nonhuman primates. *Behavioral Ecology and Sociobiology, 41*, 291-309.

Strasser, E. (1992). Hindlimb proportions, allometry, and biomechanics in Old World monkeys (Primates, Cercopithecidae). *American Journal of Physical Anthropology, 87*, 187-213.

Strasser, E., & Delson, E. (1987). Cladistic analysis of cercopithecid relationships. *Journal of Human Evolution, 16*, 81-99.

Struhsaker, T. T. (1967a). Social structure among vervet monkeys (*Cercopithecus aethiops*). *Behaviour, 29*, 83-121.

Struhsaker, T. T. (1967b). Ecology of vervet monkeys (*Cercopithecus aethiops*) in the Maasai-Amboseli Game Reserve, Kenya. *Ecology, 48*, 891-904.

Struhsaker, T. T. (1967c). Behavior of vervet monkeys (*Cercopithecus aethiops*). *University of California Publications in Zoology, 82*, 1-64.

Struhsaker, T. T. (1969). Correlates of ecology and social organization among African cercopithecines. *Folia Primatologica, 11*, 80-118.

Struhsaker, T. T. (1971). Social behavior of mother and infant vervet monkeys (*Cercopithecus aethiops*). *Animal Behaviour, 19*, 233-250.

Struhsaker, T. T. (1976). A further decline in numbers of Amboseli vervet monkeys. *Biotropica, 8*, 211-214.

Struhsaker, T. T., & Gartlan, J. S. (1970). Observations on the behaviour and ecology of the patas monkey (*Erythrocebus patas*) in the Waza Reserve, Cameroon. *Journal of Zoology, 161*(1), 49-63.

Svardal, H., Jasinska, A. J., Apetrei, C., Coppola, G., Huang, Y., Schmitt, C. A., Jacquelin, B., Ramensky, V., Muller-Trutman, M., Antonio, M., Weinstock, G., Grobler, J. P., Dewar, K., Wilson, R. K., Turner, T. R., Warren, W. C., Freimer, N. B., & Nordborg, M. (2017). Ancient hybridization and strong adaptation to viruses across African vervet monkey populations. *Nature Genetics, 49*(12), 1705.

Tappen, N. C. (1960). Problems of distribution and adaptation of the African monkeys. *Current Anthropology, 1*, 91-120.

Tembo, A. (1994). Notes and records: Population characteristics of the vervet monkey in the Mosi-Oa-Tunya National Park, Zambia. *African Journal of Ecology, 32*, 72-74.

Turner, T. R., Anapol, F., & Jolly, C. J. (1994). Body weight of adult vervet monkeys (*Cercopithecus aethiops*) at four sites in Kenya. *Folia Primatologica, 63*, 177-179.

Turner, T. R., Schmitt, C. A., & Cramer, J. D. (2019). *Savanna monkeys: The genus Chlorocebus*. Cambridge University Press.

Turner, T. R., Whitten, P. L., Jolly, C. J., & Else, J. G. (1987). Pregnancy outcome in free-ranging vervet monkeys (*Cercopithecus aethiops*). *American Journal of Primatology, 12* (2), 197-203.

van der Kuyl, A. C., Kuiken, C. L., Dekker, J. T., & Goudsmit, J. (1995). Phylogeny of African monkeys based upon mitochondrial 12S rRNA sequences. *Journal of Molecular Evolution, 40*(2), 173-180.

van Schaik, C. P. (1983). Why are diurnal primates living in groups? *Behaviour, 87*(1-2), 120-144.

van Schaik, C. P. (1989). The ecology of social relationships among female primates. In V. Standen, & R. A. Foley (Eds.), *Comparative socioecology: The behavioural ecology of humans and other mammals* (pp. 195-218). Blackwell.

Warren, W. C., Jasinska, A. J., García-Pérez, R., Svardal, H., Tomlinson, C., Rocchi, M., Archidiacono, N., Capozzi, O., Minx, P., Montague, M. J., Kyung, K., Hillier, L. W., Kremitzki, M., Graves, T., Chiang, C., Hughes, J., Tran, N., Huang, Y., Ramensky, V.,... & Freimer, N. B. (2015). The genome of the vervet (*Chlorocebus aethiops sabaeus*). *Genome Research, 25*(12), 1921-1933.

Wegdell, F., Hammerschmidt, K., & Fischer, J. (2019). Conserved alarm calls but rapid auditory learning in monkey responses to novel flying objects. *Nature Ecology & Evolution, 3*, 1039-1042.

Whitten, P. L. (1983). Diet and dominance among female vervet monkeys (*Cercopithecus aethiops*). *American Journal of Primatology, 5*, 139-159.

Whitten, P. L., & Turner, T. R. (2008). Ecological and reproductive variance in serum leptin in wild vervet monkeys. *American Journal of Physical Anthropology, 137* (4), 441-448.

Wrangham, R. W. (1980). An ecological model of female-bonded primate groups. *Behaviour, 75*, 262-299.

Wrangham, R. W. (1981). Drinking competition in vervet monkeys. *Animal Behaviour, 29*, 904-910.

Wrangham, R. W., & Waterman, P. G. (1981). Feeding behaviour of vervet monkeys on *Acacia tortilis* and *Acacia xanthophloea*: With special reference to reproductive strategies and tannin production. *Journal of Animal Ecology, 50*(3), 715-731.

CHAPTER 16

Mangabeys, Mandrills, Drills, Baboons, and Geladas

PIA NYSTROM

MANGABEYS, MANDRILLS, DRILLS, BABOONS, AND GELA-DAS BELONG TO the subfamily Cercopithecinae, part of the cheek-pouched Afro-Asian monkey group, and to the African branch of the tribe Papionini (Jolly, 2007). Table 16.1 is a listing of the species belonging to the subtribe Papionina, the primates in this chapter. Untangling the phylogenetic relationships between members of this subtribe has required extensive research effort. Homoplasy is extensive in the papionins (Lockwood & Fleagle, 1999; Collard & Wood, 2001), a circumstance that has masked their accurate phylogeny in the past.

Historically, it was assumed that mandrills and drills were more closely related to baboons (genus *Papio*), and that all the mangabeys made up a monophyletic group (Szalay & Delson, 1979). However, an accumulation of morphological and biomolecular evidence has shown that drills and mandrills are sufficiently distinct from baboons to be placed within their own genus, *Mandrillus*. There are two subspecies of drills: the mainland drill *Mandrillus leucophaeus leucophaeus* and *M. l. poensis*, endemic to the island of Bioko, Equatorial Guinea. The mandrills, on the other hand, are presently considered to be a single species, but Tefler et al. (2003) have suggested that there are two distinct genetic subpopulations located north and south of the Ogooué River in Gabon. The distinctness of the two populations is further supported by the fact that each carries a unique simian type of immune deficiency virus (Souquiere et al., 2001). However, whether or not this diversity is sufficient to warrant subspecies status has yet to be confirmed.

It is not possible to discuss "mangabeys" as one taxonomic entity since the term refers not to two different genera but to members from two distinct groups. The terrestrial mangabeys (*Cercocebus*) are more closely related to mandrills and drills, while the arboreal mangabeys (*Lophocebus*) are more closely related to baboons and possibly geladas (Page & Goodman, 2001; Fleagle & McGraw, 2002). There are traits that clearly distinguish *Lophocebus* mangabeys from *Cercocebus* mangabeys and which unite *Lophocebus* with *Papio* and *Cercocebus* with *Mandrillus*. An example are the morphological traits associated with a unique foraging scheme shared by mandrills, drills, and terrestrial mangabeys (i.e., *Cercocebus*) to the exclusion of *Lophocebus* (Fleagle & McGraw, 1999). *Cercocebus* is more terrestrial, while *Lophocebus* is more arboreal in its habits. *Lophocebus* is slightly smaller in size and shows less sexual dimorphism compared with *Cercocebus* (Fleagle, 2013). Shared features—ranging from environmental adaptations, to slim long-legged physical appearance, to the "whoop-gobble" loud call—are due to evolutionary convergence.

A new species, the kipunji, discovered in remote upland forests of southern Tanzania (Jones et al., 2005), is considered sufficiently different both morphologically and genetically to warrant placement in its own genus, *Rungwecebus* (Davenport et al., 2006). Mitochondrial evidence suggests it originated from a *Papio* + *Lophocebus* hybridization that has led to a distinct and long-surviving taxon (Burrell et al., 2009).

The baboons (genus *Papio*) are widespread throughout sub-Saharan Africa, and there is much discussion surrounding the classification and taxonomy of the genus. This controversy may never be satisfactorily resolved because, as stated by Jolly (1993), they are in an intermediate stage of speciation and thus no "correct" classification can be determined. The idea that there was no clear ecological niche separation between the different baboon taxa has been used in support of a single species (*Papio hamadryas*) with a number of subspecies (e.g., Jolly, 1993; Kamilar, 2006). The presence of viable hybrid zones at many contact areas gives support to the single species classification (e.g., Newman et al., 2004). However, each baboon taxon is highly distinctive morphologically (i.e., body size, hair coloration, and dental morphology), which has led many to support a multispecies classification (Grubb et al., 2003; Jolly, 2007) as posed in Table 16.1.

The gelada (*Theropithecus gelada*) is a species endemic to Ethiopia (Gippoliti, 2010), and it has been suggested there may be two subspecies (Kingdon, 1997) based on reports of a small population of geladas in the highlands of the southern Rift Valley region that differ from the northern Rift Valley populations in size and hair color (Mori & Belay, 1990). This population may warrant a

Table 16.1 Habitat and Conservation Status of Mangabeys, Kipunji, Mandrills, Drills, and Baboons

Genus	Species	Species Authority	Common Name	Habitat [a]	IUCN Red List [b]
Cercocebus	agilis	Milne-Edwards, 1886	Agile mangabey	tropical swamp forest; terrestrial	Least Concern
Cercocebus	atys	Audebert, 1797	Sooty mangabey	subtropical dry and moist lowland forest; terrestrial	Near Threatened
Cercocebus	chrysogaster	Lydekker, 1900	Golden-bellied mangabey	subtropical moist lowland forest; terrestrial	Data Deficient
Cercocebus	galeritus	Peters, 1879	Tana River mangabey	subtropical moist lowland riverine and gallery forest; terrestrial	Endangered
Cercocebus	sanjei	Mittermeier, 1986	Sanje mangabey	submontane and montane forest; terrestrial	Endangered
Cercocebus	torquatus	Kerr, 1792	Red-capped mangabey	tropical lowland forest, mangrove, swamp; terrestrial	Endangered
Cercocebus	lunulatus	Temminck, 1853	White-naped mangabey	gallery forest, secondary forest, woodland savanna; terrestrial	Endangered
Lophocebus	albigena	Gray, 1850	Grey-cheeked mangabey	subtropical moist lowland forest; arboreal	Vulnerable
Lophocebus	aterrimus	Oudemans, 1890	Black mangabey	primary and secondary moist forest; arboreal	Vulnerable
Rungwecebus	kipunji	Ehardt, Butynski, Jones, and Davenport, 2005	Kipunji	montane and upper montane forest; arboreal	Endangered
Mandrillus	leucophaeus	Cuvier, 1807	Drill	lowland, submontane rainforest; terrestrial	Endangered
Mandrillus	sphinx	Linnaeus, 1758	Mandrill	evergreen rainforest; terrestrial	Vulnerable
Papio	anubis	Lesson, 1827	Anubis baboon	subtropical moist lowland forest, savanna, steppe; terrestrial	Least Concern
Papio	cynocephalus	Linnaeus, 1766	Yellow baboon	Miombo woodland, savanna; steppe, coastal littoral; terrestrial	Least Concern
Papio	hamadryas	Linnaeus, 1758	Hamadryas baboon	subtropical dry grassland, semidesert; terrestrial	Least Concern
Papio	kindae	Lönnberg, 1919	Kinda baboon	Miombo woodland; terrestrial	Least Concern
Papio	papio	Desmarest, 1820	Guinea baboon	subtropical dry grassland; terrestrial	Near Threatened
Papio	ursinus	Kerr, 1792	Chacma baboon	woodland, savanna, steppe, subdesert, montane; terrestrial	Least Concern
Theropithecus	gelada gelada	Rüppell, 1835	Northern (Western) gelada	subalpine grassland; terrestrial	Vulnerable
Theropithecus	gelada obscurus	Heuglin, 1863	Southern (Eastern) gelada	subalpine grassland; terrestrial	Least Concern

[a] Data derived from Kingdon (1997).
[b] IUCN (2020).

subspecies distinction due to clear biomolecular differences, since it is estimated that it has been isolated from the northern population for at least 350,000 years (Belay & Shotake, 1998).

GEOGRAPHIC DISTRIBUTION

The African papionins are widely distributed throughout sub-Saharan Africa, occupying a wide range of habitats. Drills, mandrills, and both the terrestrial and arboreal mangabeys are mainly found in forested habitats (mostly rainforest), while baboons tend to be more commonly found in savanna grassland and woodland, although some populations are found in densely forested habitats. Papionins are also found in high-altitude subalpine and in arid, semidesert habitats. The gelada is restricted to alpine grass meadows of Ethiopia.

Both *Cercocebus* and *Lophocebus* mangabeys are widely distributed in a central band throughout sub-Saharan Africa (Figures 16.1 and 16.2). They are, however, absent south of the Republic of the Congo and Democratic Republic of Congo (DRC), and the distribution in East Africa is highly fragmented. In general, the different species

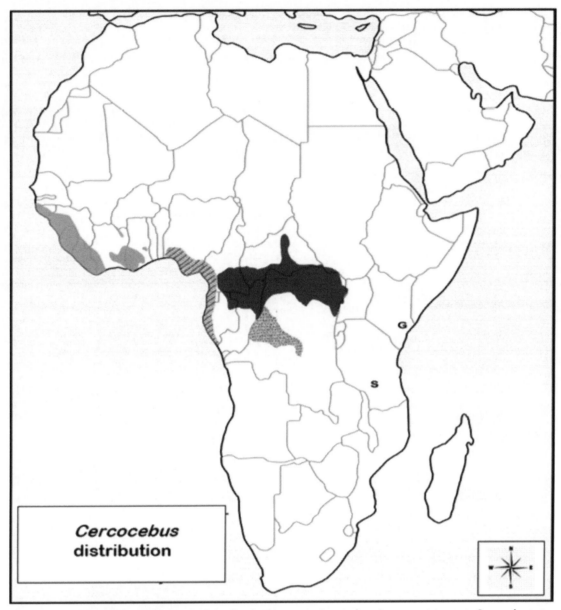

Figure 16.1 Geographic distributions of species in the genus *Cercocebus*. From west to east: *Cercocebus atys* (gray), *C. torquatus* (diagonal lines), *C. agilis* (black), *C. chrysogaster* (stippling). In East Africa, two *Cercocebus* species have very localized distributions: *C. galeritus* (G) is endemic to Kenya, occurring only in the gallery forests along the Tana River; and *C. sanjei* (S) is found only in the Eastern Arc Mountains of south-central Tanzania. (Map credit: Colin Merrony, Department of Archaeology, University of Sheffield)

are allopatrically distributed, although occasionally a *Lophocebus* species can be found living in sympatry with a *Cercocebus* species. Their ranges often overlap with other primates, such as mandrills, drills, guenons, colobines, and apes (Horn, 1987; Chapman & Chapman, 2000; Brugiere et al., 2002).

Members of the *Mandrillus* genus are found in the rainforest of western Central Africa, ranging from southeast Nigeria south to the Republic of the Congo (Figure 16.3) (Oates & Butynski, 2008a,b). Throughout their ranges, the drills and mandrills are allopatric, although historically there may have been some overlap in their

distributions along the Sanaga River in Cameroon (McGraw & Fleagle, 2006). Presently, it is estimated that the largest population density of mandrills is in Gabon. The gelada, *Theropithecus gelada*, is endemic to the highlands of Ethiopia and has a limited range (Figure 16.3). Geladas inhabit highland grassland plateau and escarpments at altitudes ranging from 2,000–4,500 masl (Gippoliti, 2008).

Papio comprises six distinct species (*P. anubis, P. cynocephalus, P. hamadryas, P. kindae, P. papio,* and *P. ursinus*). It is the most widely spread genus of all the African papionins, found throughout most of sub-Saharan Africa, with the exception of the west-central tropical

Figure 16.2 Geographic distributions of two species in the genus *Lophocebus*: *L. albigena* (light gray) and *L. aterrimus* (dark gray). The kipunji, genus *Rungwecebus* (X), is endemic to the Eastern Arc Mountains of south-central Tanzania.

(Map credit: Colin Merrony, Department of Archaeology, University of Sheffield)

rainforest region of the Republic of the Congo and DRC and a large part of Botswana (Figure 16.4). Throughout their range, the different baboon species are allopatric, but at adjacent range borders some hybridization and gene flow occur (Jolly & Phillips-Conroy, 2003; Tung et al., 2008; Jolly et al., 2011; Martinez et al., 2019). The anubis (*P. anubis*) is the most widely distributed of the baboon species, ranging from Guinea and Sierra Leone in the west across to Ethiopia in the east and south to Tanzania. The yellow baboon (*P. cynocephalus*) is found along the eastern coastal region from Kenya in the north to Mozambique in the south. Throughout its range it inhabits savanna-type habitats. The hamadryas baboon (*P. hamadryas*) is restricted to the eastern part of the Horn of Africa, including southeastern Ethiopia, eastern Eritrea, Djibouti, and part of northern Somalia. There are

also viable populations present in Yemen and in southern Saudi Arabia (Hapke et al., 2001). This species is able to survive in arid, almost desert habitats. The Kinda baboon (*P. kindae*) was previously considered a subspecies of *P. cynocephalus* but has now been accepted as a distinct species (Mittermeier et al., 2013). It is distributed inland from the west coast of Africa to Angola, DRC, and Zambia. The Guinea baboon (*P. papio*) is restricted to a limited area of West Africa, including mainly The Gambia, Guinea, Guinea-Bissau, and Senegal (Oates et al., 2008a). Over the last few decades, much of the Guinea baboon's habitat has been reduced due to human encroachment. However, with the presence of national parks some of the population is ensured limited protection (Galat-Luong et al., 2006; Oates et al., 2008a). The chacma (*P. ursinus*) is the most southern baboon—present from Angola and

Figure 16.3 Geographic distributions of species in the genus *Mandrillus*. The drill (*M. leucophaeus*) and the mandrill (*M. sphinx*) both have limited distributions in West-Central Africa: drills (black) and mandrills (light cross-hatching). The gelada (*Theropithecus gelada*) is endemic to East Africa in the highlands of Ethiopia (light gray).
(Map credit: Colin Merrony, Department of Archaeology, University of Sheffield)

Namibia in the west, to Zimbabwe and Mozambique in the east, and south throughout South Africa (Hoffmann & Hilton-Taylor, 2008).

EVOLUTIONARY HISTORY

Even though fossil discoveries may still be the most relied upon source of evidence when reconstructing primate evolutionary phylogeny, biomolecular analyses, mostly relying on mitochondrial DNA, are becoming equally important. This is especially true when the fossil record is sparse or nonexistent. Based on biomolecular evidence, the colobine and cercopithecine lineages diverged around 18 mya. Within the cercopithecine lineage, the papionins diverged 4–5 million years later, with macaques and baboons going their separate ways 8–11 mya (Zinner et al., 2009; Perelman et al., 2011). The divergence of the *Mandrillus* + *Cercocebus*

clade and the *Theropithecus* + *Papio* + *Lophocebus* clade is estimated to have occurred 6–8 mya, and 3–4 million years later the lineage leading to *Papio* diverged from the *Theropithecus* lineage (Perelman et al., 2011; Pugh & Gilbert, 2018). *Papio* baboons did not appear until 1.6–2.5 mya (Zinner et al., 2013; Rogers et al., 2019).

Support for biomolecular data is not always forthcoming, as the fossil record of the papionins is quite poor prior to the Plio-Pleistocene. During the Plio-Pleistocene, however, it is the papionin primates that dominate the fossil record (Jablonski, 2002). A well-represented lineage of papionins, the *Theropithecus* genus, had an extensive radiation in the early Pliocene; the extant *Theropithecus gelada* is but a relic population of what once was a widely distributed and common genus (Jablonski, 2002). There were at least six different species of *Theropithecus* ranging throughout

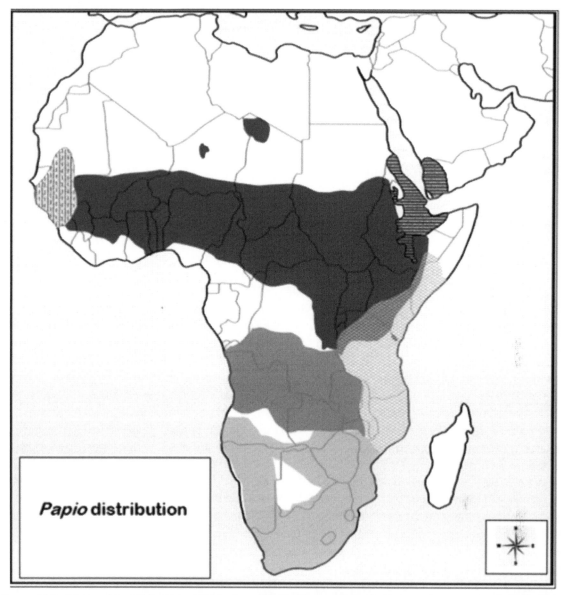

Figure 16.4 Geographic distributions of the six species in the genus *Papio* that occur over most of sub-Saharan Africa. From west to east: Guinea baboon, *P. papio* (vertical lines and dashes); anubis, or olive, baboon, *P. anubis* (dark gray); hamadryas baboon, *P. hamadryas* (horizontal lines on dark background). From north to south: yellow baboon, *P. cynocephalus* (cross-hatching) [*Note*: both hamadryas and yellow baboons exhibit some areas of distributional overlap with the anubis baboon.]; Kinda baboon, *P. kindae* (medium gray); chacma baboon, *P. ursinus* (light gray). (Map credit: Colin Merrony, Department of Archaeology, University of Sheffield)

Africa, southern Europe, and India (Rook et al., 2004; Alba et al., 2016). All members of the genus showed terrestrial adaptation; the early species were found in more wooded habitats and had dental specializations suggesting a more folivorous diet. All the fossil *Theropithecus* were substantially larger than the extant gelada. For comparison, the extinct *T. oswaldi* had an estimated body mass of ~100 kg, while another extinct species, *T. brumpti*, was twice as large as present-day geladas (Jablonski, 2002).

There is less evidence of mangabeys in the fossil record, and the potential evidence of either *Cercocebus* or *Lophocebus* tends to be quite fragmentary. Fossil remains

with affinity to *Cercocebus*, *Procercocebus* and *Soromandrillus* have been recovered from both South and East Africa and may date as far back as 3 mya (Gilbert, 2007, 2013; Pugh & Gilbert, 2018). Dental and cranial remains attributed to *Lophocebus* have been recovered from western Kenya and have been dated to the Plio-Pleistocene eras (Harrison & Harris, 1996).

LOPHOCEBUS MANGABEYS AND BABOONS
Morphology and Ecology: Lophocebus albigena
The gray-cheeked mangabeys are medium- to large-sized monkeys with long legs, a slender build, and long tails.

They have hollow cheekbones, strong jaws, and thick enamel on their molar teeth (McGraw et al., 2012). Their bodies are covered with black-brown hair; adult males have a mantle of lighter long hair over their shoulders. Infants have glossy black hair and, when newborn, they have pale pink skin on their face. The gray-cheeked mangabeys have a wide distribution, and different populations can be recognized by the variation in the color scheme of their capes. Adults have a crest of hair on top of the head, although it is not as prominent as seen in the black mangabeys (*L. aterrimus*). While there is no sexual dichromatism between males and females, there is some size dimorphism—adult males weigh 7.6–11 kg and adult females weigh 5.4–6.4 kg (Rowe 1996; Olupot & Waser, 2005).

Of all mangabey species, *Lophocebus albigena* has been the most extensively studied with most research conducted in Uganda (Olupot & Waser, 2001a; O'Driscoll et al., 2006; Arlet et al., 2014). Fewer studies have focused on the more western populations, although there have been several studies in Gabon (Tutin et al., 1997a; Brugiere et al., 2002) and in Cameroon (Poulsen et al., 2001a,b). Throughout its range, the gray-cheeked mangabey is found in primary and secondary moist evergreen to semideciduous forests. Annual rainfall ranges around 1,600–1,900 mm; it is bimodally distributed, which results in distinct dry and wet seasons, especially in the western regions (Poulsen et al., 2001a; Brugiere et al., 2002; Olupot & Waser, 2005). Temperature remains fairly constant throughout the year with a daily minimum of $16.2 \pm 0.4°C$ and maximum of $23.3 \pm 0.6°C$ (Butynski, 1990). Even though the gray-cheeked mangabeys prefer rainforest habitats, they can be quite flexible in their environmental adaptation and frequent different habitat types including grasslands, swamps, and tree plantations (Olupot et al., 1994; Tutin et al., 1997b). They reside in both logged and unlogged forest habitats. However, Olupot (2000) found that males captured in logged forests had a lower body mass compared to those trapped in unlogged forests, which suggests that they do not fare so well in disturbed habitats. They are most frequently encountered in dense canopy forest with tall trees and fairly open ground (Brugiere et al., 2005; Chapman et al., 2010). At Lopé Reserve in Gabon, Tutin et al. (1997a) found gray-cheeked mangabeys inhabit relatively small forest fragments and they move between different fragments. Such behavior has not been observed at Kibale National Park in Uganda where mangabeys tend to be absent from forest fragments (Onderdonk & Chapman, 2000).

The gray-cheeked mangabeys have large home ranges (300–600 ha), and solitary males may wander over even greater areas (Poulsen et al., 2001a; Janmaat et al., 2006; O'Driscoll et al., 2006). Different groups appear to maintain extensive overlap in their home ranges without displaying conflicts. It is possible that groups keep away from each other when moving through the forest to avoid confrontations (Olupot & Waser, 2005). Via playback experiments in Uganda, Brown (2014) and Brown and Waser (2018) determined that gray-cheeked mangabeys employ the long-distance calls of neighbors to maintain spatial buffers between groups. The groups have stable home ranges over years unless a group has fissioned and produced two daughter groups (Janmaat et al., 2009).

Gray-cheeked mangabeys often form polyspecific associations, especially with forest guenons and red colobus. In Kibale, the mangabeys associate with other primate species 45% of the time. Most often they are found together with red-tailed guenons (*Cercopithecus ascanius*) and red colobus (*Procolobus tephrosceles*), both of which prefer to feed on leafy materials (Chapman & Chapman, 2000). However, these polyspecific associations tend to be short-lived and center on feeding sites when there is an abundance of a preferred food source (Chapman & Chapman, 2000; Brugiere et al., 2002).

Arboreal feeding and foraging take place mainly at the middle- and upper-canopy levels and very rarely in the understory. These mangabeys seldom venture to the ground except during periods when preferred foods are scarce; at those times they may frequent swamps and even raid plantations (Olupot & Waser, 2001a; Poulsen et al., 2001a). The gray-cheeked mangabey is often referred to as an arboreal frugivore (Poulsen et al., 2001a; Fleagle, 2013). However, when exploring the content of their overall diet, it may be more accurate to classify them as a mixture of frugivore, granivore (seed-eater), and insectivore (Tutin et al., 1997b). When fruits are available, they feed on a few species, with figs being a highly prized source (Onderdonk & Chapman, 2000). When fruits are scarce, diet diversity increases and they include bark, flowers, leaves, pith, insects, and other animal food. In central Gabon, in a Ceasalpiniaceae (legume family)–dominated habitat, mangabeys get more than 50% of their food from legumes (Brugiere et al., 2002). At this location, seeds make up the bulk of the diet (57%), while fruit is only consumed when abundant (26%).

The crowned hawk-eagle (*Stephanoaetus coronatus*) poses the most significant predation threat to gray-cheeked mangabeys (Mitani et al., 2001; Arlet & Isbell, 2009). In Kibale male gray-cheeked mangabeys are preferred targets when compared to other age-sex groups (Struhsaker & Leakey, 1990). This is, in part, because males expose themselves more than females or immatures who tend to drop into lower levels of the forest when a raptor is sighted (Arlet & Isbell, 2009). While all adult males may engage in alarm calling, only the alpha male will actively mount a counterattack on the raptors. Crowned hawk-eagles tend to hunt in pairs; one eagle will approach and threaten the primates first, and so draw out and distract the alpha male, while the second bird will swoop down to strike their prey, kill it, and dismember and carry pieces back to the nest (Struhsaker & Leakey, 1990). In other species, such as the red colobus, all adult males engage in aggressive defense, so there is no such direct selection on the alpha male, rather all age and sex classes are preyed upon (Struhsaker & Leakey, 1990).

Other predators include chimpanzees, leopards and other felids, and humans. In Kibale, the mangabeys are not exposed to human hunting pressure and leopards are rare (Struhsaker & Leakey, 1990). Even though chimpanzees are sympatric, there has not been any recorded instance of chimpanzees preying on mangabeys (Olupot & Waser, 2001a). In Gabon, Cameroon, and Republic of the Congo, the hunting pressure for bushmeat is more severe, and it is not uncommon for mangabeys to end up in the bushmeat trade (Oates et al., 2008b).

Gray-cheeked mangabeys spend between 40–55% of the day feeding and foraging. When fruits are available, more time is devoted to feeding (Olupot et al., 1998; Poulsen et al., 2001a). Since they have large home ranges and long day ranges (O'Driscoll et al., 2006), a large portion of their daily activity budget goes toward travel (27%). Resting and social interactions, which mostly take place in the afternoons, are allocated less time (10% and 7%, respectively) (Poulsen et al., 2001a). Compared with other group members, adult males tend to spend less time feeding and foraging (31%) and more time resting (26%,) while they engage in social activities <5% of the time (Olupot & Waser, 2001a).

Social Structure and Organization: Lophocebus albigena

Gray-cheeked mangabeys live in relatively small groups with an average of 14 individuals (median range 5–25) (Olupot & Waser, 2001a; Brugiere et al., 2002; Arlet & Isbell, 2009). Most commonly they live in multimale/multi-female groups, although in smaller groups there may be only a single adult male present (Olupot & Waser, 2001a,b; Lambert, 2005). The ratio of adult males to females is around 1:2. However, the position of males is a complex affair since males may reside within social groups or live solitary lives. Furthermore, males residing within social groups may hold a central position or be peripheral. Central males appear to be more closely engaged with social activities in the group while peripheral males, sometimes found hundreds of meters away, pay less attention to group activities.

Males leave their natal group when they near maturity, as determined by emitting the first loud or whoop-gobble call (Janmaat et al., 2009), while females remain in their natal group throughout their lives (Arlet & Isbell, 2009). Males always emigrate alone, and dispersing males will remain solitary for a period of days or years before seeking access to another group (Olupot & Waser, 2005). In Kibale, Olupot and Waser (2005) found that adult males remain in a new group on average for 19 months, with 50% leaving before 3 years. Gaining residency in a new group usually entails a period of conflict between the resident males and the migrating individual. Nevertheless, if the migrating male is persistent, the resistance shown by group males soon wanes. Even after a male has gained residency in a group, he may go on social visits to neighboring groups. Such visits may last anywhere from a few minutes to days, but most last for less than a day as a resident male usually returns to his own group before nightfall (Olupot & Waser, 2005; Arlet et al., 2008). The underlying reason for this diverse male residence pattern is not well understood. Yet, it is clear that antagonism by group members does not lead a male to migrate because the level of aggression within the group remains the same after the migratory event as before (Olupot & Waser, 2001b, 2005). More likely it is part of male reproductive tactics (Arlet et al., 2008). The availability of estrous or cycling females is very likely the reason males move about. Males that transfer secondarily do so into groups that have more estrous or cycling females present (Olupot & Waser, 2005).

Gray-cheeked mangabey males have linear dominance hierarchies. However, dominance interactions are not very prominent, and it is not always possible to determine a male dominance hierarchy because aggression is rarely observed (Olupot & Waser, 2001b, 2005). This is corroborated when males are captured to be radio collared, as they show no signs of scars or wounds (Olupot & Waser, 2001a). Males found in a central position always appear better integrated within the social network of the group. The alpha male is more often than not found in a central position, and he is recognized as being the only individual to launch attacks on crowned hawk-eagles (Struhsaker & Leakey, 1990; Arlet & Isbell, 2009).

The activity budgets of solitary males are not very different from resident group males. The only difference that has been observed is that solitary males scan their surroundings significantly more often. In addition, solitary males tend to be very quiet. They rarely vocalize, and when they do, it is soft grunts rather than the characteristic whoop-gobble call given by adult resident males (Olupot & Waser, 2001a). Solitary males mostly associate with red colobus monkeys, which resident males have never been seen to do. Furthermore, solitary males are occasionally observed being groomed by red-tailed guenons and red colobus, although mostly they engage in autogrooming. Interestingly, when solitary males encounter each other they behave antagonistically toward each other (Olupot & Waser, 2001a).

All gray-cheeked mangabey females remain in the social group into which they are born, although when the number of individuals in a group reaches a certain size, the group may fission (Janmaat et al., 2009). When this happens, it is unknown if the fission occurs along familial lines or not. Adult females form distinct linear dominance hierarchies (Lambert, 2005; Chancellor & Isbell, 2009). While studying female interactions, Chancellor and Isbell (2009) found that agonistic interactions between females are uncommon and mostly entail approach-avoidance interactions (44%) or supplants (35%), and aggression involving physical contact is rare (4%). More than half (55%) of the agonistic interactions involve food, but it is tree bark being contested not fruit. High-ranking females monopolize trees with good sources of bark and remain feeding for long periods of time. Why bark is a contestable food source can only be guessed at, but it has been suggested that it is high in nutrients, including proteins and soluble sugars (Rogers et al., 1994). Low-ranking females, in turn, spend more time feeding on insects, which are good sources of protein.

Females cycle throughout the year and no discrete mating or birthing seasons have been noted (Arlet et al., 2008, 2015). Like other papionins, female gray-cheeked mangabeys display inflated perineal skin during the period when ovulation occurs. An average cycle length is 47 days but may range shorter or longer (Wallis, 1983). A male indicates his interest in an estrous female by performing so-called head-flags, where he repeatedly tosses his head over his shoulder (Wallis, 1983). Males and females indicate preference for each other by being nearest neighbors (Olupot & Waser, 2001b). Males that hold the nearest position to a female (and, therefore, are her preferred partner) have higher copulation rates. High-ranking males mate more often and tend to do so when the females are most likely to conceive (Arlet et al., 2008); they also are more likely to display aggression and chase other males away (Olupot & Waser, 2001b). Nevertheless, a female may be fickle and leave the consorting male in order to copulate with other males in the group. Females initiate copulations almost as frequently as males (42% and 58%, respectively). Copulations are over within a few seconds, and there are usually little if any interactions between the pair afterward. If there *is* a grooming bout, it is very brief. Even though copulations can be conspicuous, especially due to the female's distinctive post-copulation

call, other group members pay scant attention to the copulating pair (Wallis, 1983; Olupot & Waser, 2001b).

Morphology and Ecology: Lophocebus aterrimus

The most distinctive feature of the black mangabey is the very prominent crest of hair on top of the head and rather excessive cheek tufts that curl downward (Figure 16.5). The whole body is covered by black hair, except the cheek tufts, which are light gray and sometimes speckled with white. Albinism has been reported as common in this species (Horn, 1987; Eppley et al. 2010).

Very little is known about the life of the black mangabey as there have been few observations of this species in the wild and few systematic data collected (Horn, 1987; Gautier-Hion et al., 1999), although McGraw (1994) reported on habitat use and frequencies of polyspecific grouping for *L. aterrimus* in Lomako Forest, DRC. Therefore, limited understanding of its ecology exists; nor is there much known on how it may differ from the gray-cheeked mangabey. The two *Lophocebus* species are separated by the Congo River, with the black mangabeys living to the south in the heart of the Congo Basin. They are mainly found in tropical rainforest habitats, although

Figure 16.5 Black-crested mangabey (*Lophocebus aterrimus*) with dark face and prominent crest on top of head. (Photo credit: kevinscharer.com/arkive.org)

they may visit swamp forests occasionally. Horn (1987) observed several groups intermittently over a period of years in secondary forest habitats that showed strong evidence of human activity. The habitats could not have been too unproductive as they held seven different primate species including several guenon species (*Cercopithecus ascanius*, *C. wolfi*, *C. neglectus*, and *Allenopithecus nigroviridis*), bonobos (*Pan paniscus*), red colobus (*Piliocolobus tholloni*), and black-and-white colobus (*Colobus angolensis*).

Compared with the gray-cheeked mangabey of Uganda, the two focal groups of black mangabeys that Horn studied had relatively small home ranges (48 ha and 70 ha). There is extensive overlap in the home ranges for the groups (60–75%), but no territorial defense was ever recorded. The different groups even utilize the same sleeping sites, although not at the same time. The groups use all levels of the forest but tend to forage and travel in the middle levels (10–25 m) while they rest and socialize at higher levels (25–40 m). They seldom come down to lower levels or to the ground. Horn (1987) recorded activities of the two study groups: During the day the groups frequently split into smaller foraging parties, each of which contain at least one adult male. All group members seem to synchronize their activities. Feeding and foraging account for 36% of the time, but they actually spend more time resting (38%). Very little time is spent traveling (1.5%), and this usually takes place early in the day when the group sets out from their sleeping site to the primary feeding sites. They spend the rest of the time (24%) engaging in social interactions. They feed mainly on fruits (57%) and nuts and seeds (31%). Flowers (4.2%), leaf flush (3.3%), and insects (1.6%) are less important sources. They are seldom seen drinking but when they do, it is from standing pools in hollows of tree branches.

The black mangabey is extensively hunted throughout its range, and as a result tends to be very shy of people (Horn, 1987; Eppley et al., 2010). This hunting pressure is posing a severe threat to their survival (Hart et al., 2008b). In addition, they contend with natural predation from crowned hawk-eagles; although Horn (1987) observed eagle attacks, he never lost a member of his study groups. Bonobos (*Pan paniscus*), with whom the mangabeys share part of their habitat range, occasionally hunt and eat monkeys, including one immature black mangabey recorded by Surbeck et al. (2009). Whenever a predator is observed, a staccato bark alarm call is given and protection is sought in dense foliage.

Black mangabeys are highly vocal primates, and they have various calls that are frequently emitted during active periods. Their calls appear to be the same as those of gray-cheeked mangabeys. Waser (1982) compared recordings of staccato barks and whoop-gooble calls and found them indistinguishable from gray-cheeked mangabeys. The whoop-gobble calls are always given by adult males and tend to be more commonly given in the early part of the day (Horn, 1987). Horn frequently observed polyspecific associations; usually the mangabeys are found comingled with Wolf's guenons (*C. wolfi*, 57%) and red-tailed guenons (*C. Ascanius*, 20%), or both guenon species (16%). Guenons and mangabeys form polyspecific associations most of the time when feeding on fruit, as it appears there is no difference in preference. However, none of the guenons feed on the same seeds and nuts as mangabeys.

Social Structure and Organization: Lophocebus aterrimus

Very little is known of the black mangabey (*Lophocebus aterrimus*) social organization. Like other mangabeys, the black mangabey lives in multi-male/multi-female groups comprising 14–20 individuals (Horn, 1987). Horn (1987) never observed solitary males, but males from his two study groups occasionally visit neighboring groups when there are estrous females available.

Morphology and Ecology: Papio

Baboons (*Papio anubis*, *P. cynocephalus*, *P. hamadryas*, *P. kindae*, *P. papio*, and *P. ursinus*) are large, terrestrially adapted monkeys. They have relatively long limbs, with arms and legs of almost equal length resulting in an intermembral index close to 100 (Fleagle, 2013). Pronounced sexual dimorphism is evident in all six species. Male body mass can be twice as large as females (males 16–35 kg, females 8–18 kg; see Swedell, 2011). Chacma (*P. ursinus*) males are larger than the other species, while the hamadryas and Kinda are the lightest. There is less species variation in female body mass, although chacma females are the largest. Baboons have long muzzles and prominent brow ridges; relative to other cercopithecines, they have short tails, and their fingers are relatively short and stubby. They have sizable ischial callosities and females display prominent sexual swelling during estrus. In general, it is the males that display the most obvious color schemes, while females tend to have more uniformly drab hair. Within each species there is much local variability (i.e., they are polytypic), which suggests extensive gene flow between adjacent populations and taxa (Jolly, 2003). The different species can easily be differentiated based on their external appearance, specifically in hair coloration but also general body build and tail carriage. Overlaying these color differences is also a north–south divide in several characteristics. The northern species—anubis or olive, hamadryas, and Guinea (*P. papio*)—have long wavy hair over their shoulders and neck and prominent cheek tufts. The southern species—chacma (*P. ursinus*) and yellow (*P. cynocephalus*)—have shorter straight hair over the shoulders, no cheek tufts, but patches of lighter color on the muzzle below the eyes. Kinda baboons share most of the features of the southern species.

The Guinea baboon has reddish-brown hair; facial skin and paracallosal skin (lateral to the ischial callosities) are dark pink. The tail is relatively long and smooth without a kink at the base. The anubis baboon has olive-brown hair and dark brown facial skin, which is surrounded by lighter-colored cheek tufts. Their paracallosal skin is also dark in color, and the tail has a clear kink. Hamadryas have light brown hair, and adult males have light gray to white mantles; tails are long and smooth without a kink and end with a lion-like tuft. The facial skin and paracallosal skin (which is large in males) are pale to bright pink. Chacmas are large and generally have dark brown hair with much variability among populations in different locations. Their facial skin is dark, almost black, with long narrow muzzles. The yellow baboons are the most slender and long-limbed of the genus. The facial skin and paracallosal skin are dark brown. Hair on the upper body is yellowish-brown, while the underparts are pale creamy white and there is no mane. The Kinda baboon is the smallest of the genus (with males weighing no more than 16 kg) and show the lowest level of sexual dimorphism in body mass (Petersdorf et al., 2019). In all species the infants, up to about three months, have black hair and pink skin, which progressively turn into the adult coloration by the first year.

Baboons are ecologically flexible and are found in a great range of habitat types—lowland rainforest, woodland, bushland, and savanna grassland. They are able to thrive in a wide range of environments, from arid, almost desert areas (chacma, Guinea, and hamadryas) to rainforest (anubis), and at high altitudes (chacma). However, more often than not, baboons are found in open woodland and grassland where they often come into conflict with humans.

Even though the Guinea baboons have restricted distribution (see Figure 16.4), they populate a variety of habitats due to wide latitude distribution and altitude variations, from sea-level mangrove forests in Senegal to mountain foothills in Guinea (>1,000 m). Environments include Sahelian steppe (Mauritania), scrub savanna (Senegal, Mali), mosaic woodlands (Senegal), and secondary high forest (Guinea). In most of their range, they experience distinct dry and rainy seasons. Rainfall ranges from <200 mm in the more arid regions to >1,400 mm

in the southern forest habitats. Mean daily temperature ranges from around 20°C to over 30°C but can on occasion reach 50°C (Galat-Luong et al., 2006). Niokolo Koba, Senegal is the site of a long-term study of Guinea baboons in the wild (Sharman, 1981; Galat-Luong et al., 2006). At this location the baboons spend most of their time in scrub-woodland savanna but use the tall gallery forest trees as sleeping sites.

The anubis baboon is found in a great variety of ecological settings due to its extensive distribution, which includes 25 countries (see Figure 16.4) and diverse climatic conditions. Anubis inhabit rainforests, woodlands, savanna, and arid thorn scrub. At Comoé National Park in Côte d'Ivoire, where the anubis range in mixed forest and savanna grassland, annual rainfall reaches >1,000 mm and mean temperature is over 26°C (Kunz & Linsenmair, 2008). In Gashaka-Gumpti National Park in eastern Nigeria, the anubis live in closed forest and, while the mean annual temperature is the same as Comoé, annual rainfall is twice as high (Higham et al., 2009). In East Africa, anubis baboons are found in dry woodland and grassland. At Gilgil, Kenya, and Awash National Park, Ethiopia, their habitat consists mainly of open grassland with scattered thorn bush interspersed sporadically by *Acacia* trees. Rainfall, which is distributed biannually, ranges between 600–760 mm and mean maximum temperature is 25.5°C (Harding, 1976; Nystrom, 1992).

The chacma baboon is also found in a wide range of habitat types, including open woodland and grassland savanna. In addition, they inhabit areas that are climatically extreme, including deserts and high mountains. The western chacmas live in the Namib Desert in Namibia where they forage along ephemeral rivers passing through deep canyons (Cowlishaw, 1999). Trees and shrubs are present along a river's edge where the baboons are able to tap into subsurface water during long dry periods. Several groups have been studied at the southern end of the Kuiseb River (Hamilton et al., 1976; Hamilton, 1985), which is the most arid habitat recorded for nonhuman primates. Here they experience unrelenting high temperatures which stay more or less the same throughout the year (peak daily temperature 29.5±2.3°C) with a mean annual rainfall of just 27.2 mm. Vegetation is sparse and shade can be hard to find. The baboons attempt to keep their body temperature down by digging deep hollows in the sand of riverbeds and laying on their stomachs (Brain, 1990; Brain & Mitchell, 1999). Water is scarce and the baboons may go without drinking for months, surviving on fluids they get from food (Brain & Mitchell, 1999). In South Africa the chacma baboon is found throughout arid thornveld and mountainous habitat where there can be dramatic changes in daily temperatures; day temperatures reach over 30°C, while temperatures drop below freezing at night in the winter (e.g., van Doorn et al., 2010). In the Drakensberg Mountains chacma baboons are found in montane (1,280–1,829 m) and subalpine (1,829–2,865 m) areas (e.g., Henzi et al., 1992). At lower altitudes there are more bushes and trees present, but they are typically arid montane species (e.g., Proteacae and Ericacae). Due to the lack of trees, the baboons use cliff faces as sleeping sites. Annual rainfall may reach 2,000 mm, but most rain falls during the summer months leaving the land parched during the dry season with little but plant underground storage organs and seeds for the baboons to feed on.

Throughout their range the hamadryas baboon is found in relatively arid habitats compared with other baboon species. Both in the southwestern Saudi Arabian Peninsula and throughout most of the Horn of Africa, the environment consists of semidesert where vegetation is sparse and adapted to xeric conditions, i.e., mostly open, rocky areas where *Acacia* thorn scrub grows (Biquand et al., 1992). Since trees, especially tall trees, are rare, the baboons seek out vertical rock faces for sleeping sites. Water is scarce and standing water sources may be miles apart (Swedell, 2006). During the dry season, the baboons have been observed digging deep holes into ephemeral riverbeds to gain access to subterranean water. Within northern Ethiopia and Eritrea, hamadryas are found in more diverse environments, including moist woodland habitats at both low and high altitudes (Zinner et al., 2001).

At Filoha in Awash National Park, Ethiopia, the hamadryas have access to richer and more predictable resources. Filoha is a small wetland ecosystem consisting of marshland and swamps interspersed by hot springs where the baboons always have access to water. The area supports dense patches of doum palm trees, an important food source but also used as sleeping sites (Schreier & Swedell, 2008). Away from the wetland area, however, there is typical arid thorn scrub habitat dominated by *Acacia* and *Grewia*. Throughout the hamadryas range, the rainfall is bimodally distributed, although the main rains fall during July and August. Erer Gota, in eastern Ethiopia, is another site of hamadryas studies (see Kummer, 1995). During the dry season, arid thorn scrub territory is bare, but with the arrival of the rains, grasses and herbs become abundant. Throughout the hamadryas range, rainfall seldom exceeds 600 mm annually and in some areas is significantly less. Temperature can vary seasonally but being semidesert, the days are hot (mid- to upper 30s°C) and nighttime temperature can dip toward freezing during the dry season. It is apparent that hamadryas have adapted to foraging far and wide to gain sufficient food for survival (Kummer, 1995; Swedell, 2006).

The yellow baboon is most commonly found in Miombo (*Brachystegia*) woodland and grassland and, of all the baboon species, is probably the most adapted to savanna environments. In Amboseli National Park, Kenya, located just north of Mt. Kilimanjaro, study of the yellow baboons began in 1963 (Altmann & Altmann, 1970) and continues to the present. Rainfall is bimodal with most precipitation occurring in March–April. However, rainfall can be unpredictable and annual amounts vary between 250–550 mm. Temperature is typical of savanna areas (daily high is ~30°C and low is ~22°C). Another site where yellow baboons have been studied over many years is Mikumi National Park, in southeastern Tanzania (Norton et al., 1987; Kleindorfer & Wasser, 2004). Mikumi is located on a plain surrounded by hills and, like Amboseli, is mainly short grass savanna with scattered bushes and woodland on the hills.

Baboon home range size is not species specific; rather it is dictated by the productivity of the habitat. Baboons living in forested areas where resources are expected to be more abundant and more seasonally constant tend to have relatively small home ranges, while those living in more arid savanna habitats have more extensive home ranges. For example, the home ranges of forest-dwelling anubis baboons in Gombe, Tanzania, are 3.9-5.2 km² (Ransom, 1981) and in Uganda the mean is 4.7 km² (Rowell, 1966). In contrast, baboons in arid habitats have significantly larger home ranges, e.g., anubis at Gilgil wander over 20–31 km² (Smuts, 1985), Guinea baboons at Niokolo-Koba have home ranges of 32.5 km² (Sharman, 1981), and yellow baboons at Amboseli use 16–24.1 km² (Stacey, 1986). Home ranges of hamadryas baboons in Ethiopia are calculated at 28 km² (prior to GPS data) in Erer Gota (Sigg & Stolba, 1981) and estimated to be over 40 km² in Awash National Park (Swedell, 2006; Schreier, 2009).

Home ranges of neighboring groups show some overlap, although when neighboring groups encounter each other it is seldom with equanimity. Core areas, which mainly coincide with resource-rich areas (such as sleeping sites or preferred food resources), may be defended. Day ranges vary seasonally and often from day to day, measuring from just a few kilometers to 14 km. Hamadryas tend to have longer day ranges than other baboons, sometimes close to 20 km long (Kummer, 1968). The length of day ranges is thought to reflect the productivity of the habitat, but it can also indicate how widely water sources are distributed.

Water is an important limiting resource for baboons. Unlike arboreal, forest-living primates, baboons do not obtain most of their water from plants, especially not during the dry season. They depend upon standing water, either from permanent water sources, such as rivers and lakes, or from ephemeral sources such as pools of rainwater, which are abundantly available during the rainy season. At Moremi Game Reserve, part of the Okavango Delta located in Botswana, chacma baboons face seasonal inundations. During these floods, their habitat turns into islands; the baboons must cross long stretches of water during their daily foraging, which causes them much stress (Cheney & Seyfarth, 2007).

Baboons tend to live in large groups; on average groups number 30–60 individuals, although occasionally there may be over a hundred members. Yet, habitat productivity is the important determinant for group size. This is apparent at Kuiseb River in the Namib Desert, Namibia, where group sizes are small, ranging from 12–35 individuals (Hamilton et al., 1976). Likewise, in the Drakensberg Mountains of South Africa, groups tend to be small, averaging around 20 individuals (Henzi & Lycett, 1995). A further example of group size variability comes from Comoé National Park, Côte d'Ivoire (possibly the limit of anubis geographic distribution), where group size only numbers 15.3±11.1 individuals (Kunz & Linsenmair, 2008).

Baboons do not usually form polyspecific associations, although they may feed on grassland plains together with a range of different ungulate species. It is common for baboons to chase away another primate species (such as vervets) if encountered at a feeding site. Yellow baboons along the Tana River in Kenya commonly supplant mangabeys from feeding sites (Wahungu, 1998, 2001).

Baboons are eclectic omnivores that include a wide variety of plant species, invertebrates, and vertebrates in their diet (e.g., Norton et al., 1987; Kunz & Linsenmair, 2008; Swedell et al., 2008). Their diet may show both seasonal and year-to-year variations (e.g., Norton et al., 1987; van Doorn et al., 2010). Even though baboons may sample from a wide range of plant species, the core diet tends to come from a more limited list of species that is determined by locality (Post, 1982; Kunz & Linsenmair, 2008; Swedell et al., 2008). At Filoha, Awash National Park, Swedell et al. (2008) found that hamadryas baboons spend 47% of their time feeding on fruits from the doum palm (*Hyphaene thebaica*), which are an important staple food for most months of the year (range 2–78% of their diet). For the Tana River yellow baboons, the top 10 plant species make up 62% of their diet; fruit and seeds from doum and wild date palms are the most frequently utilized species, aside from a range of grasses (Bentley-Condit, 2009). At Comoé National Park, Côte d'Ivoire, the anubis baboons sample 79 out of 84 potential plant foods, but the top 10 plant species account for 60% of their diet (Kunz & Linsenmair, 2008). The use of various vegetative parts (varying by season) of specific plants, such as *Acacia*, tamarind, and grasses, is common in the diet of baboons. *Acacia* trees and bushes (including seeds, seed pods, flowers, gums, cambium, and leaf flush) are especially important food sources for baboons living in woodland savanna habitats. During the dry season, corms (swollen underground stem bases) and rhizomes become very important foods, sometimes making up as much as 90% of the diet.

Baboons include both invertebrates and vertebrates in their diet. These tend to make up no more than a small percentage of their annual diet but can add a considerable amount of protein at certain times of the year. Baboons include a wide range of insects in their diet, such as grubs found in dung or in the soil, and they feed extensively on social insects, such as ants and termites (Kunz & Linsenmair, 2008; Swedell et al., 2008). Baboons also include small vertebrates in their diet, e.g., lizards, birds (including chicks and eggs), hares, small antelopes (mainly neonates), vervet monkeys, galagos, and fruit bats. Since baboons frequently encounter human domesticates, some individuals have become proficient at catching young domestic stock, mainly sheep and goats (Strum, 1981). In areas near the ocean or near rivers, baboons feed on shellfish, such as mussels, limpets, and crabs; chacma baboons living along the coast in the Cape Peninsula habitually include marine foods, such as mussels and limpets in their diet (Lewis & O'Riain, 2017).

Primates living in open thorn scrub and grasslands have more potential predators than do forest-living ones. Baboons, despite being fairly large, may be attacked and killed by cheetahs, chimpanzees, crocodiles, dogs, eagles, hyenas, leopards, lions, and pythons (Hart & Sussman, 2005; Isbell et al., 2018). Of these, leopards, lions, and hyenas pose the largest threat to baboons. However, predation rates on baboons in most areas are low, which may be directly linked to their social structure and organization.

How baboons spend their days (i.e., their activity budget) has been studied extensively in most baboon species (see Table 16.2). Climatic conditions, habitat productivity, and competition from other species for resources clearly affect their behavior. All baboons show marked seasonal, as well as daily, variations in their activity budgets. For most species and populations, feeding and foraging occupy the majority of their days. Resting and traveling also account for a significant portion of their time, but some of the day is always spent in social interaction.

The majority of intragroup social interactions in many baboon populations take place in the mornings before groups depart from their sleeping sites and again in the evenings upon return to those sites. However, these early morning and late afternoon social sessions are reduced

Table 16.2 Mean Annual Activity Budgets for Selected Baboon Species

Papio Species	Habitat	Feed/Forage	Travel	Rest	Social	References
Anubis (*P. anubis*)	grassland	43.5	16.5	31.5	8.5	Forthman Quick, 1986
Anubis	grassland	45.7	32.0	12.5	8.7	Eley et al., 1989
Anubis	bushland	36.2	24.2	23.3	15.7	Aldrich-Blake et al., 1971
Anubis	bushland	32.5	25.0	31.3	12.0	Nagel, 1973
Anubis	bushland	28.2	21.0	41.6	9.2	Nystrom, 1992 [a]
Anubis	forest	41.8	25.6	18.8	13.9	Kunz and Linsenmair, 2008
Hamadryas (*P. hamadryas*)	bushland	30.9	25.3	26.3	17.2	Nagel, 1973
Hamadryas	bushland	29.0	27.6	23.5	19.9	Schreier, 2009 [a]
Yellow (*P. cynocephalus*)	grassland	44.0	26.0	23.0	6.0	Post, 1981
Yellow	grassland	35.0	24.0	32.0	9.3	Stacey, 1986
Yellow	grassland	46.5	26.0	8.9	18.6	Bronikowski and Altmann, 1996
Chacma (*P. ursinus*)	bushland	42.7	27.1	13.3	15.1	Hill et al., 2003
Chacma	bushland	34.0	32.0	24.0	10.0	Davidge, 1978
Guinea (*P. papio*)	bushland	21-39	22-58	19-26	2-13	Galat-Luong et al., 2006

[a] Denotes activity budget of adult males only.

in some localities at times of low food availability, and groups may leave their sleeping sites and go directly to feeding sites, moving more and feeding longer each day at the expense of social activity and resting. During dry seasons, baboons often take a siesta during the hottest part of the day, resting under the shade of bushes and trees, and often long periods of social interactions (e.g., play and grooming) take place before they move on. In the rainy season, this midday rest period may be replaced by a number of short rest periods while animals drink at the numerous water sources available.

Activity budgets are sensitive to many ecological factors, such as habitat productivity, which are determined largely by climatic conditions (e.g., Dunbar, 1992; Hill, 2006) and group size, which may change over time. At Gilgil and Amboseli, Kenya, baboons spend between 44–48% of the time feeding and foraging, 21–26% traveling, 22–23% resting, and 6–9% socializing (Post, 1981; Forthman Quick, 1986; Bronikowski & Altmann, 1996). Bronikowski and Altmann (1996) recorded exceptions to this routine; when yellow baboon groups are raiding agricultural crops, dramatic differences occur in resting time (8.9%) and social time (18.6%).

Chacma baboons have been studied over a number of years at the De Hoop Nature Reserve located in the coastal region of Western Cape Province, South Africa. This coastal population, which includes crustaceans and mollusks in their diet, spend 42.7% of their time feeding and foraging and travel 27.1% of the time. They spend only 13.3% of time resting, although grooming makes up 15.1% of their activity budget (Hill et al., 2003). In this more temperate Mediterranean-type environment, the baboons experience distinctive seasonal variations and their activity budget is significantly influenced by factors such as day length and temperature (Hill et al., 2003; Hill, 2006).

Social Structure and Organization: Papio

All baboon species live in large multi-male/multi-female social groups. However, how these social groups are structured differs; anubis, chacma, Kinda, and yellow baboons live in classic multi-male/multi-female social groups, while both the hamadryas and Guinea baboons deviate from this pattern in their own special ways. The typical "savanna" baboon social structure entails groups with approximately twice as many adult females as adult males, along with immature individuals (see review in Swedell, 2011). In forested habitats, however, there consistently appears to be a more equal distribution of males to females (Higham et al., 2009). Most males leave their natal group as subadults or young adults and may disperse a number of times throughout their lifetime, while females remain in their natal group permanently (Pusey & Packer, 1987). Related females form strong social clusters within the group called matrilines, and adult males often form special affiliative relationships with specific females that may last months or years (Smuts, 1985).

Hamadryas have a multilevel social structure in which the basic social element is the *one male unit* (OMU) comprising an adult male leader, a number of adult females, and their immature offspring (Swedell, 2006). An OMU retains social cohesiveness due to the adult male's possessive behavior directed toward the females. The intensity of these possessive behaviors shows regional variability, with OMU leader males from Saudi Arabia being much less intense in herding their females compared with Ethiopian hamadryas (Kummer, 1995). OMUs can also contain follower males, which most often are subadult or young adult males. Some OMUs spend time together while foraging and traveling, these are referred to as *clans*, and males within clans are kin (Städele et al., 2015). *Bands*, which have temporal stability, comprise a number of OMUs and clans as well as solitary "free-floating" males (Pines et al., 2011). Several bands may share sleeping sites; these aggregations are referred to as *troops*, but they do not appear to have a social function and do not extend beyond the sleeping area. An OMU may contain one to nine adult females and immatures younger than two to three years of age. Juveniles older than three years begin their divergent trajectories toward adult social life; males join up with solitary subadult males while females are recruited by other adult or subadult males to form initial units, precursors to OMUs.

Based on data from a long-term intermittent study of Guinea baboons at Niokolo Koba National Park, Senegal, Galat-Luong et al. (2006) concludes there is clear evidence of multilevel social structure in Guinea baboons akin to that seen in hamadryas, indicating greater homology than previously thought (Jolly, 2020). Fischer et al. (2017) confirms that Guinea baboons live in nested multilevel societies that are similar to hamadryas baboons, and Jolly (2020) presents an argument that male philopatry, the crucial factor shared by hamadryas and Guinea baboons, evolved from a common ancestor. Like hamadryas, Guinea baboon males are philopatric and form lasting social and coalition bonds with each other. Females may form bonds with a single male at a time but, as with hamadryas, these cross-sex bonds are not permanent as both taxa also have female-based dispersal (Kopp et al., 2015). The collection of genetic relatedness data, combined with spatial and social interaction analyses carried out at Niokolo Koba National Park, has resulted in much new information on the social organization of the Guinea baboon. Strong male-male affiliation, observable through ritualized greetings, and minimal aggression toward females has been documented (Patzelt et al., 2011, 2014). Units composed of primary males with females, along with sporadic secondary males, form the core of Guinea baboon society; unlike hamadryas society, however, there is less overt control over females (Goffe et al., 2016).

Anubis, chacma, and yellow male baboons leave their natal group, and this dispersal is more often than not a solitary affair. Males tend to emigrate for the first time as older subadults or young adults (7–9 years), while secondary or breeding dispersals may occur a number of times in a male's lifetime (Alberts & Altmann, 1995). The reason for migrating is most often related to a male's mating potential; if his chances are low, it behooves the male to seek better fortunes elsewhere, but gaining entrance into a new social group is not easy. A prime adult male may enter a new group in an obvious manner and begin to challenge top-ranking males for alpha position. However, more often than not, the potential émigré will take a peripheral position and watch the interactions

taking place between group members. Next, he may solicit friendly interactions with adult females because acceptance by females, especially dominant females, may hold the key to being accepted into the group. No matter the initial strategy, all males will need to assert their position within the male dominance hierarchy through direct interactions with group males.

In hamadryas society, males stay within their own clan or band, while females move across bands and even between troops (Swedell et al., 2011; Städele et al., 2015). Females do not voluntarily disperse but are forcefully abducted by males (Swedell et al., 2011). Females are moved out of their natal OMU as young juveniles or later "stolen" from other OMU leaders. In the former case, males accomplish this feat by paying much affiliative and protective attention to the juveniles. Adult females receive much more aggressive attention—so-called herding behavior—to convince the female not to stray from the male's side.

Both male and female baboons (other than hamadryas and Guinea) form linear dominance hierarchies. A female's dominance position is acquired based on maternal position and, therefore, related females tend to be close in rank, resulting in a matrilineal dominance hierarchy. In general, a female's dominance position remains stable throughout her life, although she may lose rank if for some reason members of her matriline perish (Cheney & Seyfarth, 2007). A male, in contrast, has to compete for his position within the social group, and his position will be contested by other males almost continuously. Subadult males begin their ascendency by challenging the adult females and, as young adults, they begin to compete among themselves for positions within the male dominance hierarchy. Rank is highly correlated with physical strength, and males are at their physical prime around 10–12 years of age. It is at this time they have the greatest potential to acquire their highest rank and, as they grow older, they begin to lose rank. It is generally thought that dominance rank confers advantages such as priority of access to scarce resources, e.g., prime feeding sites and receptive females. However, males can use a range of alternative tactics to circumvent dominance positions by forming coalitions with other males against dominant rivals. Such coalition partnership may be temporary or more permanent. Males also form special relationships with females that may provide an opportunity for a lower-ranking male to gain mating access. Swedell (2006) did not observe any overt competition by females to gain access to or groom the OMU leader. Rather she suggests that hamadryas females—and males—lack dominance ranking altogether, which makes them distinctly different from anubis, chacma, or yellow baboons.

Olive, yellow, Kinda, and chacma male baboons are at best uncomfortable around each other; they tend to keep a fair distance between each other and mostly come into contact when affirming their dominance relationships or when seeking coalition support against another male. Male hamadryas and Guinea baboons, alternatively, continuously test each other's positions within the dominance hierarchy using a series of ritualized behaviors, often referred to as greetings (*P. hamadryas*: Colmenares et al., 2000; *P. papio*: Dal Pesco & Fischer, 2018). These interactions may be the only time adult males come into physical contact with each other without being involved in aggressive battles. In addition to stylized posturing, lip smacking, and vocalizations, as two males approach each other, greetings may also entail touching each other's genitals. Hamadryas males in possession of females do not overtly show their relationships to other males but use these ritualized greetings at a distance. Philopatry provides the basis for cooperation among kin (Städele et al., 2015) and, unlike other baboon species, hamadryas and Guinea male baboons are philopatric and maintain social relationships with one another. As is often the case, patrilocal males show more affiliation toward each other, as seen in Guinea and hamadryas males (Swedell, 2006; Fischer et al., 2017).

In most baboons, the majority of affiliative interactions that take place within a baboon group occur between adult females or between mothers and their offspring. Females often form strong social bonds with related females, the strength of which appears to benefit the offspring's survival (Silk et al., 2009). Females use grooming as a means to gain and maintain both male and female friends. Grooming taking place between females tends to be more or less reciprocal, but females invest more effort into grooming males than they receive in return. The only exception may be during consortships when the consorting male seeks and maintains proximity to the female and spends an increasing amount of time grooming her and possibly also displaying possessive behaviors toward her (Smuts, 1985). Hamadryas males differ since they spend as much attention on their OMU females, irrespective of the females' reproductive condition, as do other baboon males when in consort. If a hamadryas male perceives his OMU as threatened, he will increase the rate of attention to his females even further (Swedell, 2006). This kind of affiliative attention to females is not evident in Guinea baboons (Fischer et al., 2017).

The affiliative attention females pay to adult males and the establishment of friendships with specific males is considered a female strategy to gain protection against harassment from other group members, a protection that also includes the female's offspring (Smuts, 1985).

In some chacma baboon populations, females establish friendships with one or two males in the later stages of pregnancy, friendships that are maintained while the offspring is dependent on nursing. It has been suggested that this is the females' strategy to circumvent potential infanticide (Cheney et al., 2004; Nguyen et al., 2009). From a male's point of view, such a friendship may be an alternative strategy to increase his reproductive potential since a female friend may be more inclined to mate with her male friend (Noë & Sluijter, 1990).

Grooming is not the only social cement in baboon groups; baboons are highly vocal primates and on a daily basis much intragroup and intergroup communication takes place. The vocalizations used fall into three broad categories: alarm calls, grunts, and geckering. Alarm calls, especially the loud call barks also referred to as wa-hoo calls, are mostly given by adult males in response to predators but can be given by all group members at times of distress. For example, if group members have been separated from each other during daily foraging, wa-hoo calls can seek and reestablish contact (Rendall et al., 2000). They are also given if a neighboring group has been found trespassing on a group's home range, especially its core area. Males also may give wa-hoo calls during dominance rank competition (Kitchen et al., 2005). Other calls that reflect distress are a range of "scream" calls that are given by any group member when threatened by a higher-ranking individual. Distressed infants or young juveniles, especially when having tantrums, emit a distinctive geckering vocalization. However, more commonly vocalizations are variants of soft grunts given between group members throughout the day. These intragroup communications tend to be more subtle and reliant on social context, are usually affiliative in intent, and can carry much information to the recipient as well as those within audible range. They may be given when initiating travel, as greetings, and as part of reconciliation (Cheney & Seyfarth, 1997; Rendall et al., 1998). Baboons also communicate using a range of visual signaling such as facial expressions, including lip smacking and other body postures, which may denote a threat or friendly intent. Adult males may send signals of threat or intimidation by yawning and exposing their well-developed canine teeth (Figure 16.6) or by raising their eyebrows and exposing white eyelids. Milder threats include ground-slapping or head-bobbing. In addition to screaming, individuals who are threatened or in distress may hold their tail straight up in the air and display a "fear grin."

Each newborn infant baboon receives considerable attention from the rest of the social group, especially from members of the mother's matriline and her friends (Silk et al., 2003). Even though both males and females

Figure 16.6 Male *Papio anubis* with teeth bared.
(Photo credit: Anup Shah, arkive)

contribute to the raising of offspring, it is the mother who will for the first three to six months provide the majority of care and protection (Figure 16.7). A mother's rank significantly influences how an immature is treated; offspring of high-ranking females receive less antagonism from other group members than do offspring of low-ranking females. Males, especially the female's male friends, tend to provide indirect care and protection through vigilance against predators or by breaking up conflict situations. Males may also function as babysitters while the mother is feeding and foraging. Additionally, baboon males have been observed to use young infants (those still with black hair) as shields against other adult males in so-called agonistic buffering interactions, which are interpreted as elements of dominance hierarchy (Strum, 1984). The infants used in these instances are usually willing participants and their mothers do not seem to show any concern.

The weaning process is usually complete by the time the infant is 12 months old. As infants and young juveniles become more independent of their mother, they join peer groups and engage in play behavior. However, juvenile females spend less time within peer groups as they grow older and begin to stay closer to their female kin group by engaging in grooming bouts. Interest in infant care is also evident in older juvenile females as they attempt to engage in allomothering whenever the

Figure 16.7 *Papio anubis* female grooming offspring.
(Photo credit: Shimelis Beyene)

opportunity presents itself (Altmann, 1980). Juvenile males spend much less time with the female kin group; their interest is directed toward adult male interactions, although both male and female juveniles are attracted to adult males and seek proximity and engage in grooming. Adult males appear to be quite tolerant of juveniles and allow them to feed close by, thus providing improved access to resources. Subadult males have a much more peripheral position in their natal group than female peers, and they pay close attention to what is going on in neighboring groups. During intergroup encounters, subadult males are often seen to mingle and interact with members of another group.

Females reach sexual maturity and begin displaying sexual swelling around four to five years of age. The length of the reproductive cycle in adult female baboons is varied, ranging between 30–40 days (see Swedell, 2011) and can be influenced by many factors (e.g., age, health, or nutrition). When females display estrus swellings, which peak around ovulation, they become reproductively receptive and males compete to gain mating access, although adult males prefer parous females. Most copulation takes place within consortships that may last for hours or days (Smuts, 1985; Weingrill et al., 2000). Peak swelling appears to be most attractive to males and, more often than by chance, it is top-ranking males that are mating with the female during this time. Most estrous females tend to mate with a number of males; this behavior has been explained as a way for the female to confuse paternity, gaining more males willing to protect what they may consider their own offspring and so decrease possibility of infanticide. However, we may be underestimating what baboons know about paternity. A study on yellow baboons at Amboseli National Park in Kenya has shown that fathers are more protective and affiliative to their biological offspring than they are to nonrelated offspring (Buchan et al., 2003; Charpentier et al., 2008). It is not yet understood how males may realize their paternity, but this study throws doubt on the notion that females engage in mating with multiple males primarily to confuse paternity.

There are alternative reproductive tactics used by both males and females to express their preference in mates. Males may form friendships with specific females (Smuts, 1985); these friendships entail close proximity and frequent affiliation between the two, usually in the form of grooming. During foraging, especially if the group is subdividing into smaller units, the friends will travel together. The male will come to the female's aid during conflict situations, and his protection will also include the female's previous offspring. These friendships are on a superficial level similar to the hamadryas OMU, except that male friends do not show possessive behaviors toward their female friends. Hamadryas males constantly pay attention to the whereabouts of their own females, and if a female strays away from the male leader, he will quickly herd her back. Well-trained females will only need a look from the male to resume a close position. This attentiveness on the part of the hamadryas male is independent of the reproductive condition or age of the female—to him a female is always attractive (Swedell, 2006). In other baboon species, females are usually left to their own devices; male attention is aroused mainly when the female is in estrus, especially at peak estrus. In addition, adult males appear to prefer females that have already given birth (Alberts et al., 2006; Fischer et al., 2017).

Gestation length in baboons is around 6 months (175–190 days) (Barrett et al., 2006). A female will give birth to her first offspring when she is 6–7 years old; for the rest of her life she may produce a new offspring every 1.8–2 years (e.g., Cheney et al., 2004; Higham et al., 2009). Interbirth intervals may be reduced under certain circumstances, such as if food resources are ample and predictable or if the female has lost her infant. In the latter case, a female resumes cycling again within days or weeks.

KIPUNJIS

Morphology and Ecology: Rungwecebus

The kipunji (*Rungwecebus kipunji*), a distinct taxon which evolved from *Papio + Lophocebus* hybridization (Burrell et al., 2009), was first recognized by scientists in the mid-1980s in areas around Mount Rungwe-Livingston montane forest and the Udzungwa Mountains in southwestern Tanzania, see Figure 16.2 (Jones et al., 2005; Davenport et al., 2006). These two locations are 350 km apart, and the presence of kipunjis in both suggest that these habitats were once contiguous (Jones et al., 2005).

The kipunji is a medium-sized monkey with long dense hair and a long tail. Adult males often have longer hair over their shoulders and some have tails that end with a slight tuft. Some of the adults also sport a broad crest of erect hair on top of the head and distinctive cheek tufts which have a graceful downward curve. Body hair is light-medium brown while part of the ventrum and tail are cream colored. Facial skin is dark brown and lacks contrasting color on the eyelids. The muzzle is slightly elongated and appears more distinct due to deep indentations (suborbital fossa) below the eyes (Davenport et al., 2006; Ehardt & Butynski, 2006a). There is no sexual dichromatism among the adults, but there are some subtle differences in coloration between the two locations where the species is found (Jones et al., 2005). Adult males are estimated to weigh 10–16 kg and are only slightly larger than females (Ehardt & Butynski, 2006a).

The kipunji is critically endangered, and it is estimated that the population size does not exceed 1,200 individuals. The majority of the population (93%) is found in areas around Mount Rungwe-Livingston montane forest while the rest is found in the Ndundulu Forest, part of the Udzungwa Mountains, Tanzania (Bracebridge et al., 2011, 2013). The two locations where kipunjis are found differ significantly from each other; the former comprises highly degraded forests, while the latter comprises pristine forest. The Ndundulu Forest is located in the northwest of the Udzungwa Mountains at an elevation of 1,300–1,750 masl. It is a submontane evergreen moist forest with canopy trees reaching 30–40 m, although canopy height decreases with altitude (Ehardt & Butynski, 2006a; Rovero & de Luca, 2007). The forest is mostly closed canopy and has great biodiversity despite the fact that sporadic selective logging took place in the northern part until the mid-1990s (Topp-Jørgensen et al., 2009). A large part of the Ndundulu Forest is located within the Udzungwa Mountain National Park, but locations where the kipunji is found are outside the park boundary. Nevertheless, the habitat is primary forest with little or no evidence of human activity (Bracebridge et al., 2011). Rainfall may reach 2,200 mm annually and is bimodally distributed with the bulk falling in March–May and a smaller peak in November–December. Due to the high elevation, temperature tends to be cool (13-17°C) with below-freezing temperatures at higher elevations (Jones et al., 2005).

Mount Rungwe Forest Reserve and Livingston Forest have a narrow, but not continuous, corridor of forest between them, thus the kipunji live in more or less isolated subpopulations (Bracebridge et al., 2011). Rungwe-Livingston is a montane and upper montane forest, with bamboo and plateau grassland at higher elevations. Annual rainfall is >2,000 mm and can be close to 3,000 mm at higher elevations. Just as at the Udzungwa Mountains, rainfall is bimodally distributed with distinct dry and wet seasons, and temperatures may fall below freezing (Ehardt & Butynski, 2006a). The forest is highly degraded with significant broken canopy and secondary forest, thick undergrowth, and few tall trees (Jones et al., 2005; Ehardt & Butynski, 2006a).

In southern Ndundulu Forest, Jones (2006) estimated there are 3–6 groups living in an area of 7.2 km². He found that groups range in size between 15–25 individuals, which brings the population to a total of 60–150 individuals. In Mt. Rungwe and the Livingstone Forest, 16 and 18 groups, respectively, have been located in over 70 km² (Davenport et al., 2006). Here the groups are larger, with a combined total of 1,042 individuals (Davenport & Jones, 2008; De Luca et al., 2009). Based on monitoring

of four study groups, mean daily travel is close to 1.3 km, and the home range is estimated at 306 ha, with extensive overlap between groups (De Luca et al., 2009).

Kipunji are arboreal and mainly found in mid- and upper-canopy levels. Even though they travel and flee in the canopy, they have also been observed foraging on the ground. An extensive dietary study undertaken in the Mt. Rungwe and Livingston Forests, over a period of 45 months, included 34 kipunji groups (Davenport et al., 2010). This study found that the kipunji is a diet generalist, with clear seasonal patterns in plant species consumed, as well as plant parts consumed. Overall, 122 plant species from 60 families are included in their diet. A plant species that appears to be especially important is the wild poplar, *Macaranga capensis* var. *capensis* (Euphorbia) from which they feed on young and mature leaves, leaf stalks, pith, and flowers, as well as bark. During the rainy season (November–April), they rely on fewer plant species (58 species), while during the dry season (May–June), 80 different species are utilized. Of the plant parts consumed, mature leaves make up the highest proportion (22%), followed by unripe fruit and seeds (14%), and ripe fruits (13%). Unripe fruits are consumed equally throughout the year, while ripe fruits are more commonly consumed during the rainy season. The kipunjis also eat young leaves (12%), bark (11%), stalks (9%), flowers (9%), pith (7%), and insects (2%). The remaining portion of the diet is made up of a range of plant and plant parts, including seed pods, rhizomes, shoots, tubers, epiphytes, herbs, climbers, mosses, fungi, and lichens (Davenport et al., 2010).

Hunting and habitat degradation are the greatest threats to the survival of the kipunji (Rovero et al., 2012; Bracebridge et al., 2013). There are few natural predators that threaten these mangabeys; only the leopard may be of concern, since it has been recorded to exist in forests both to the north and south of Ndudulu Forest. In other areas of the Udzungwa Mountains National Park, leopards and crowned hawk-eagles have been reported to be common and predatory on the kipunji (Topp-Jorgensen et al., 2009). In Rungwe-Livingston, kipunjis living close to human habitation are known to raid crops and occasionally are trapped and killed (Davenport et al., 2006).

Social Structure and Organization: Rungwecebus
Little is known about kipunji social groupings except that they live in multi-male/multi-female groups. As with other primates that are heavily hunted, they are extremely vigilant, shy, quiet, and afraid of humans. When encountered in the Nduldulu Forest, Tanzania, kipunji groups scatter and take hours to get back together.

They form polyspecific associations with other diurnal species, most commonly Sykes guenon (*Cercopithcus mitis albogularis*), which includes sleeping in neighboring trees (Ehardt & Butynski, 2006a). The kipunji moves about quietly and rarely vocalizes. When males alarm-call, they use a baboon-like loud, low-pitched "honk-bark" (Ehardt & Butynski, 2006a). This type of vocalization reflects their *Papio + Lophocebus* ancestry (Burrell et al., 2009). Kipunji also appear to use vocalizations to inform neighboring groups of their locations, possibly to reduce chances of encounters (De Luca et al., 2009).

GELADAS

Morphology and Ecology: Theropithecus

Fairly typical of papionins, prime adult male geladas are close to twice as large as females (males 15–22 kg and females 10–15 kg). Males are fully grown when they reach eight to nine years, while females give birth to their first offspring at five years. Even though both sexes have luxurious thick reddish-brown hair, males have longer hair with a mane and expanded cheek tufts (Figure 16.8). Gelada tails are well covered with hair and end in a lion-like tuft. Hands and feet are distinctly darker in coloration, as is the skin on the face. The area surrounding the eyes is lighter, providing a highly distinctive contrast when flashing the eyelids. Newborn infants have pink faces and ears and black hair; these differences, however, change and by three months of age infants have adult coloration. When threatening opponents (or humans), geladas retract and "flip"

their upper lip exposing the very bright pink gums, which in contrast to the dark face results in a startling effect.

Both male and female geladas have an hourglass-shaped bare patch on the chest extending toward the throat which ranges in color from bright red to pale pink. In males the chest color appears to be correlated with social status. Bergman et al. (2009) found that core unit leader males have a deeper red color compared with follower males. In addition, leader males with a larger number of adult females in their core unit have the most intense red chest patches. Dunbar and Dunbar (1975) noted that when a male loses the leadership of his core unit, his chest patch turns pale pink overnight and remains so for the rest of his life. In females the edges of this area, as well as the paracallosal skin, display nodule-like vesicles. These vesicles are affected by hormonal changes and during estrus become bright red, swollen, and filled with fluid. Since geladas spend most of their time sitting, the chest patch may serve as a means to display reproductive state in females. Males inspect and appear to sniff the anogenital and chest regions of females within their unit, especially if they display signs of estrus, suggesting that the vesicles carry olfactory information (Roberts et al., 2017).

Geladas are well adapted to a terrestrial life with limbs equal in length; the intermembral index is 100 (Fleagle, 2013). They have a thumb that is relatively long compared with the other fingers, and they have a highly developed ability to oppose their fingers. Their manual dexterity is not so different from humans (Napier, 1981). To accommodate

Figure 16.8   Male *Theropithecus gelada*, dry season Simien Mountains National Park, Ethiopia.
(Photo credit: Wikimedia Commons)

their grazing habits, geladas have a relatively long cecum and colon (Iwamoto, 1993b). To gain the maximum nutrients from their grass diet, geladas grind food into small-sized particles—but because they lack an efficient hind-gut fermentation system, they have to increase the volume consumed, which may explain the extensive amount of time devoted to feeding (Dunbar & Bose, 1991).

Geladas are confined to an area of volcanic uplift in Ethiopia which is dissected by numerous steep gorges through which rivers, such as the Blue Nile, flow. The species is dependent on the cliff faces of gorges for refuge and sleeping sites. Geladas have been studied extensively within the Simien Mountains (Beehner & Bergman, 2008; le Roux et al., 2011) as well as locations at lower altitude such as Bole Valley (e.g., Dunbar, 1992) and Guassa Plateau (Fashing & Nguyen, 2009). Populations located at the southern Rift Valley escarpment have also been observed (e.g., Mori & Belay, 1990; Belay & Shotake, 1998). Habitat comprises typical Afro-alpine flora, and the vegetation consists mainly of grassland interspersed with giant lobelias (*Lobelia rhyncopetalum*) and stands of heather (*Erica arborea*). There are some bushes and trees present, especially on the escarpments, but in general it is very open habitat. At some localities, geladas are found above the tree line (3,500 m), but the highest population density is found between 2,000–3,000 m. The climate is cool and temperate; at higher altitudes it is fairly wet with an annual rainfall as high as 1,500 mm. Even though there is precipitation during most months of the year, there are distinct dry and wet seasons. The presence of a dry season is more evident at lower altitudes where rainfall is reduced and restricted to the June–September wet season. Average daily temperature is 7–10°C, but during the day the temperature may reach 20–25°C. At night the temperature drops dramatically, and during the dry season (October–February) can dip below freezing. Hail storms are not uncommon in the rainy season, sometimes so heavy that hail leaves a white blanket over the ground. At lower altitudes, such as the Bole Valley, the habitat is more temperate with open grassland and scattered bushes and trees (e.g., *Acacia, Rosa, Carissa,* and *Solanum*); rainfall is around 1,100 mm and the average daily temperature is 15.9°C (Iwamoto, 1993a). At higher elevations the gelada is mostly in undisturbed habitat, but at lower altitude much of the land is farmed. This often results in conflicts as geladas are known to raid cultivated areas, especially barley fields (Yihune et al., 2008).

The geladas are unique among primates by being highly graminivorous, with over 80% of their diet made up of grasses (Iwamoto, 1979; Fashing & Nguyen, 2009). In the rainy season they feed almost exclusively on grass (leaves, stems, flowers, seeds, and rhizomes). During the dry season, when there is less grass available, herbs, fruits, and underground storage organs (e.g., lily bulbs) are utilized. In the cool and wet highlands, grasses and other vegetation resist decay longer, thus green grass is available for a longer portion of the year, especially along gorges and escarpments (Iwamoto, 1993a).

Living at high altitude places extra strain on the gelada in that they require more energy to stay warm, especially where the ambient temperature frequently drops below freezing. Dunbar and Bose (1991) investigated how geladas can survive on a grass diet in an environment that can be less than hospitable. They compared the efficiency of nutrient extraction from grass by geladas to other grazing and browsing animals. The gastrointestinal tract of the geladas is less efficient at extracting nutrients because their ability to ferment grass is low (Dunbar & Bose, 1991; Iwamoto, 1993b). Instead, the gelada has a distinctive dental morphology with hypsodont molars, i.e., high crowned teeth with long shearing blades. Hypsodont molars are highly efficient at slicing grass blades into fine particles that can be transformed into energy more readily (Jablonski, 1994). In addition, as the cold dry season approaches, geladas increase food intake, and to maximize protein intake they eat more herbs. This change in diet and volume of food consumed appears sufficient to get them through the "winter" season.

Despite being specialized grass feeders, geladas opportunistically take advantage of swarming insects. When desert locust swarmed the Guassa Plateau for a few days during June 2009, the geladas fed on almost nothing else for at least two days (Fashing et al., 2010). Gelada groups seldom forage more than 1–2 km away from a cliff edge. Day ranges vary on a daily as well as seasonal basis and can span 0.5–2.2 km (Iwamoto & Dunbar, 1983). The larger the group, the farther they must travel to gain sufficient food for all. Even though home range size is related to group size, it is usually fairly small (>1–3.5 km²) (Iwamoto & Dunbar, 1983). Since grass resources tend to be evenly distributed within their habitat, there is no need to defend unique home ranges and the range area of different bands overlaps extensively. Geladas are found in very high densities (an average of 72.5 ± 11.62 individuals/km²) which is three to four times higher than anubis baboons (Dunbar & Dunbar, 1975).

Geladas devote most of the day to feeding. The time spent feeding increases with altitude, which most likely is a reflection of thermoregulation requirements (Iwamoto & Dunbar, 1983). The fact that they need to devote so much time to feeding is not due to lack of food resources but to the geladas' particular feeding technique, which leads to a relatively low hourly food intake compared to graminivorous ungulates (Iwamoto, 1979), although it is high overall compared to other primates (Dunbar & Dunbar, 1974). The extensive time spent feeding is also necessary because of the low nutritional value and low digestibility of their main food items (Iwamoto, 1979).

Highly adapted to life on the ground, geladas seldom use trees even if they are present. When feeding, the geladas shuffle along sitting on their haunches picking off grass blades using pinching movements of the thumb and tip of second digit with high efficiency and speed, especially since they use both hands simultaneously. Additionally, their short and stubby fingers are very efficient at digging out roots and tubers (Dunbar, 1984). Geladas have been observed drinking two to three times daily from streams and rainwater puddles, although more and longer drinking sessions are required in the peak of the dry season (Kawai & Iwamoto, 1979).

Iwamoto and Dunbar (1983) provide detailed activity budgets from three different locations, and at all three sites geladas spend most the day feeding (36–62%). Females spend slightly more time feeding compared with adult males, which may be reflective of the reproductive demands on the females (Dunbar et al., 2002). Since grasses tend to be uniformly distributed, there is not much need to search for food, so only 15–20% of the day is spent moving. Shuffling while feeding accounted for most of the travel time in a day, which may be a way of conserving energy rather than standing up each time there is need to change position. The daily amount of time spent resting ranges between 5–26% and is inversely correlated with time spent feeding. Like most gregarious primate species, geladas are more likely to reduce resting time if they need to devote more time to feeding, rather than reducing time spent in social activities. The geladas spend 16–21% of their daily activity engaged in social interactions (Dunbar, 1984).

In the Simien Mountains there are few dangers present aside from humans and their dogs. However, in the highlands south of the Rift Valley, the geladas may be exposed to wild predators such as leopards and hyenas (Iwamoto et al., 1996). The geladas' main anti-predator behavior is to retreat rapidly to cliffs where they are safe from most potential threats. Occasionally, adult males mob and chase off predators. This has been observed repeatedly in cases of attacks by domestic dogs and even leopards (Iwamoto et al., 1996).

Social Structure and Organization: Theropithecus

Like Guinea and hamadryas baboons, the gelada has a multilevel social structure (Dunbar & Dunbar, 1975). The core unit in gelada society comprises an adult male,

a number of adult females (1–12), and their offspring of different ages. Gelada females within a core unit are natal and usually closely related to each other, while the male is unrelated to the females. The leader male may be the only adult male member of the core unit, although one to two follower males may be present, often former leader males (Bergman et al., 2009). These extra males are not reproductively active within the core unit except in very rare instances (Snyder-Mackler et al., 2012). The size of core units is highly variable; there can be as many as 50 individuals, but more often the number is below 20. When the number of females present in a core unit goes beyond 5, there is an increased risk for the leader to be challenged. Adult unattached males seek to dispose such leaders or at least take some of the females away to form a new core unit (Mori, 1979).

Within a shared home range, a number of core units aggregate into bands. Bands can include close to 300 individuals. Some core units join together and coordinate their activities more frequently than others, and Kawai et al. (1983) referred to these groupings as "teams." It is hypothesised that these teams are made up of a number of core units that have fissioned. Thus, the females may be closely related to each other and prefer to be within close proximity. Shotake and Nozawa (1984) have shown that members of bands show closer genetic affinity than between bands. Sometimes several bands or parts of bands join together at feeding sites forming what is referred to as herds. These are temporary and may number over 1,000 individuals. To accommodate all the extra males, male offspring tend to leave their natal core unit as juveniles or subadults. Most of them join up with other males in all-male groups (AMG). These groups are usually fairly small (2–15 individuals) ranging in age from older juveniles to physically mature males. Not all extra males, however, join an AMG; a few remain unattached and move freely about the band or even between bands. These unattached males can move so freely due to the complete lack of antagonism shown toward core unit leaders. In contrast, the older males in AMGs tend to continuously test core unit leaders' ability to maintain control over their females (Dunbar, 1993).

The only way for a gelada male to gain reproductive success is to take over the leadership of a core unit. As a leader he will have unchallenged mating access to females and, therefore, paternity certainty will be high. Once a male loses control of his core unit, he will not get a second chance, and his reproductive career is over. Defeated males may join an AMG where they will engage in social interactions with juveniles. Alternatively, they may stay on as follower males in their former core unit. The reason for staying may be to protect their young offspring against the incumbent leader who might commit infanticide. These defeated leaders tend to engage in intensive social interactions with infants and juveniles and come to their support during conflicts (Dunbar, 1984).

In most species where the group structure is based on a single adult male leader, infant killings are reported. Faced with this possibility, females introduce countermeasures to prevent such events from happening. Even though Dunbar and Dunbar (1975) did not observe any infant killings, they note that when a new male takes over a core unit, there is an increase in spontaneous abortions and interbirth intervals are reduced by six months. Beehner and Bergman (2008) also reported infanticide, most presumably related to the takeover of a core unit by a new male.

The gelada core unit differs significantly from the hamadryas one-male unit because gelada females are philopatric and, thus, females within core units are closely related. The majority of social interactions take place within the core unit (Tinsley Johnson et al., 2014). Female geladas form linear dominance hierarchies, and female rank is maternally inherited (le Roux et al., 2011). Females form the strongest social bond with kin, especially those of similar rank (Tinsley Johnson et al., 2014). Females with relatives or friends within the core unit tend to pay very little attention to the male leader, and may even ignore his request for grooming. Females spend most time grooming each other, and they will come to each other's aid during a fracas. The strongest affiliative dyads are formed between sisters or a mother and daughter (Dunbar & Dunbar, 1975). Females spend close to 20% of the day in social activity, mainly involving a female relative, followed by immature offspring, with the male leader receiving the least attention.

Males tend to have preferred female partners within the core unit. More often than not this female is low ranking or has no close relative within the core unit. He will lavish attention on this female and come to her aid when she is harassed, and this relationship may continue even after the leader has been disposed. The success of a core unit leader is based very much on the attention he pays to his females. Interestingly, the longer a male has been leader of a core unit, the less attention he pays to his females. However, it is recognized that stability within a core unit is not dependent on the male but rather on the strength of the female-female bonds (Dunbar & Dunbar, 1975).

Even though males reach sexual maturity around 5 years of age, they do not attain leadership of a reproductive unit until they are 8–10 years old. Females reach sexual maturity early and produce first offspring around 6 years of age with unusually long interbirth intervals

averaging 2.36 years (Roberts et al., 2017). Even though geladas breed throughout the year, there is a tendency for more births to occur during the November–March dry season (Dunbar, 1984). Infants are fully dependent on their mothers up to 3 months of age. At this point they have lost most of their infant coloration, and mothers do not let them nurse while they themselves are feeding; however, complete weaning does not take place until the infant is around 9 months old. Once infants are weaned, they join with peers and form play groups. If there is a deposed leader present, juveniles may spend much time interacting with him. Juvenile males remain within these peer groups until they join an AMG, while juvenile females have a closer relationship with their mothers and aunts.

CERCOCEBUS MANGABEYS, DRILLS, AND MANDRILLS

Morphology and Ecology: Cercocebus torquatus

Red-capped mangabeys are medium to large monkeys with only slight sexual dimorphism—females weigh 5–8 kg and males 7–12.5 kg (Maisels et al., 2007). Body hair on the back is slate gray, but the ventrum is light in color. Long dark tails grade into a white tip. The crown of the head has a dark red cap which is offset sharply by white-gray cheek tufts that extend down below the chin. Prominent white eyelids are clearly visible even when the eyes are wide open (Figure 16.9). Infants have lighter skin and hair than adults.

Red-capped mangabeys are found in the lowland Atlantic coastal forest biome, a closed-canopy lowland moist forest (Matthews & Matthews, 2002; Baya & Storch, 2010). According to Gartlan and Struhsaker (1972), the red-capped mangabey preferred habitat is mangrove, gallery, and swamp forest. However, throughout their range they are also found farther inland in hilly, steeply sloping closed canopy and gallery forest (Waltert et al., 2002; Maisels et al., 2007). The Atlantic coastal forest receives a high annual volume of rain (2,800 mm to >5,000 mm). The rainfall is bimodally distributed with most falling during the summer months; temperature varies little throughout the year and the monthly mean maximum is around 30°C (Astaras et al., 2011).

The red-capped mangabeys are semiterrestrial and can be observed from the forest floor to mid-canopy (40 m), although they spend most of their time at or near the ground (Mitani, 1991; Maisels et al., 2007). Red-capped mangabey groups are fairly conspicuous because they are highly vocal (Maisels et al., 2007). From a captive study it has been found that females are much more vocal than males. Females also have a wider range of vocalizations, most of which are food and contact calls, while males give

Figure 16.9   Collared or red-capped mangabey (*Cercocebus torquatus*) with typical conspicuous white eyelids. (Photo credit: Rufus46, Wikimedia Commons, cc-by-sa 3.0)

mainly threat calls (Bouchet et al., 2010). Groups tend to move and forage widely along the forest floor and underbrush, using vocalizations to maintain contact. They also run along the ground or quickly ascend into trees when escaping from threats (Mitani, 1991; Maisels et al., 2007) and often rest and socialize while on the forest floor (Mitani, 1989).

This species is frequently found in polyspecific associations. Most often red-capped mangabeys are together with crowned, putty-nosed, and moustached guenons. There appears to be little competition or interactions, since *C. torquatus* spend more time close to the ground while the guenons spend more time in the middle and upper canopy. Where red-capped mangabeys are sympatric with either mandrills or drills, they are occasionally found foraging together in large groups (Waltert et al., 2002; Maisels et al., 2007). Where their distribution ranges overlap with gray-cheeked mangabeys (*L. albigena*), they may form temporary mixed groups when fruits are abundant; at other times the two species go their separate ways (Mitani, 1991).

There has been little study of the home range size of red-capped mangabeys. At Campo Animal Reserve in southwestern Cameroon, Mitani (1991) estimated the home ranges to be fairly large (248 ha). He also noted extensive home range overlap between different groups, but neighboring groups appear to avoid each other rather than contest space. There are no accounts of predation on red-capped mangabeys, although crowned hawk-eagles and leopards may be potential predators (Mitani, 1989).

Humans are the most severe threat to the mangabeys as they are hunted for bushmeat throughout most of their range (Astaras et al., 2011).

There are no published records of activity budgets for red-capped mangabeys. They appear to spend most of the day traveling and foraging, moving from one sleeping site to the next, and they often subgroup when foraging (Dolado et al., 2016). Most resting and social interactions take place before ascending the sleeping trees in the late afternoon. Fruit appears to be the preferred food source. At Campo Animal Reserve, Mitani (1989) found that fruit was an important food source at all seasons (60%) and together with seeds made up 80% of the diet. At Campo, the top three plant species comprise more than 60% of food intake. However, fruits tend to be mostly available during the rainy season; at other times there is a shift to feeding on leaves and flowers, with insects also consumed. During dry seasons, when fruits are scarcer and less clumped, red-capped mangabeys forage over wider areas, including more diverse habitats.

Social Structure and Organization: Cercocebus torquatus

There have been no detailed studies on the social behavior of the red-capped mangabeys. It is known that they live in multi-male/multi-female groups, and group size ranges between 20–33 individuals (Matthews & Matthews, 2002; Waltert et al., 2002). The sex ratio appears to be slightly more adult females than males (1:1.25; Mitani, 1989).

Morphology and Ecology: Cercocebus atys

The upper bodies of sooty mangabeys (*Cercocebus atys*) are brown with contrasting cream-colored chests and bellies. The top of the head has short hair with prominent cheek whiskers. The face is light pink to gray, which makes the white eyelids less distinct, while the muzzle is darker than the rest of the face. Males weigh between 7–12 kg and females 4–7 kg (Swedell, 2011). Their loud calls are slightly different compared to other mangabey species because they add a bark at the end of the whoop-gobble calls (Gust, 1994).

Sooty mangabeys have a patchy distribution and are not well studied in their natural habitats, except in the Taï National Park, Côte d'Ivoire, where they are mainly found in tropical moist forest habitats (Range & Noë, 2002). The Taï National Park experiences distinct dry and wet seasons; mean annual rainfall is around 1,875 mm, mainly during the wet seasons, which are interspersed by two dry periods (December–January and July–August) (McGraw, 1998). There is a limited annual fluctuation in temperature with a mean of 24°C (Range & Noë, 2002).

The Taï Forest sooty mangabeys live in large groups of around 100 individuals (McGraw & Bshary, 2002), but in more western locations group sizes are smaller (a group size of 15 has been reported from Sierra Leone) (Harding, 1984). These are terrestrial monkeys that spend most of their foraging and traveling time on the forest floor (McGraw, 1998; Range & Noë, 2002) and are seldom seen in the higher layers of the canopy (McGraw, 1998). In the Taï Forest the mangabeys have large home ranges estimated to cover 6–7 km² (McGraw, 1998). There is extensive overlap in home ranges of adjacent groups; however, during intergroup encounters, females act aggressively toward other groups while males do not engage in such behavior (Range & Noë, 2002).

The biggest predator threats to sooty mangabeys come from crowned hawk-eagles, leopards, chimpanzees, and humans. Even though primates may not be the preferred prey for crowned hawk-eagles, a study in the Taï Forest showed that sooty mangabeys are killed at a higher rate than expected based on their population density (McGraw et al., 2006). The eagles do not appear to be discriminating, as all sex or age groups are approximately equally represented in the remains at nest sites.

In the Taï Forest, sooty mangabeys are occasionally found in polyspecific associations, especially with red colobus (*Procolobus badius*) and Diana monkeys (*Cercopithecus diana*). The mangabeys are effective sentinels for ground predators as they tend to alarm-call first, suggesting that they are better than the other primate species at detecting predators from a distance (McGraw & Bshary, 2002). It is not clear if there is any advantage to the sooty mangabeys when in these polyspecific associations, although there most likely is little or no disadvantage since virtually no dietary competition takes place between these species.

The diet of sooty mangabeys in the Taï Forest is mainly fruits and seeds (68%) and invertebrates (26%) which they find on the forest floor (Range & Noë, 2002). There appears to be little variation in the diet among age and sex classes. An important food source is seeds from *Sacoglottis gabonensis*, which during certain times of the year, form a major part of their diet (McGraw et al., 2011). Using camera trap recordings, sooty mangabeys have frequently been observed to scavenge chimpanzee nut-cracking sites for leftover nuts (van Pinxteren et al., 2018). The mangabeys spend most of their day feeding (28–38%) and foraging (24–45%) along the forest floor (McGraw, 1998; Range & Noë, 2002). At Taï, during the day they spend 19% of the time resting and 8% engaged in social interactions. The remaining time (10%) is spent traveling, which tends to occur mostly during the early part of the day as they set out for their target feeding place.

Social Structure and Organization: Cercocebus atys

Sooty mangabeys live in multi-male/multi-female social groups (Range & Noë, 2002). Females remain in their natal group throughout life, while males may leave upon reaching sexual maturity. Adult males, as is seen in other mangabey species, may be full-time or part-time residents in social groups, or they may roam around alone (Range & Noë, 2002). Part-time residents may remain in a group for weeks and then be absent for weeks. It is not known if the visiting males are familiar to the females, but it is clear that these part-time resident males are often found in the group center where they freely interact and mate with the females (Range, 2005). During group encounters, which may occur once or twice a week, nonresident males may enter the group and harass females and juveniles.

It is clear from both captive studies and observations in their natural habitat that sooty mangabeys form linear dominance hierarchies (Gust & Gordon, 1994; Range & Noë, 2002). However, while captive studies showed no evidence of a matrilineal social system (Ehardt, 1988a,b), Range (2006: p.511) found a major difference during her study of wild sooty mangabeys: "juvenile females' dominance ranks remained stable over time and were highly correlated with the dominance ranks of their mothers." Social behavior of wild juvenile sooty mangabeys does not differ from social behavior observed in juveniles of many matrilineal primate species (Range, 2006). Coalitions are mainly composed of juvenile females and other animals of similar rank, including mothers and siblings. In addition, juvenile females exhibit preferential association and grooming with close-ranking juvenile and adult females (Range, 2006).

Range (2006) also observed the social behavior of male juveniles, whose dominance ranks correlate with the rank of their mothers initially but decrease in correlation as the males age. Juvenile males show the same preference as juvenile females in their affiliation with adult females, but when they associate with other juvenile males, they prefer peers.

Range and Noë (2002) studied female social relationships in a group of wild sooty mangabeys and found that females do not associate randomly. Many of the females have one specific female grooming partner, while most females restrict grooming to a few partners. Much of the grooming takes place between females of similar rank, with some reflection of kinship. Female coalitions are observed infrequently; most coalitions are between alpha and beta females, and usually it is the alpha female who supports the second-ranking female. Range and Noë (2002) found that female foraging efficiency is correlated with rank and conclude that females associate with a limited number of females as expected when there is contest competition over access to resources. During male rank reversals, females increase protectiveness toward infants in order (1) to prevent the deposed leader from carrying infants away if he is threatened by the new alpha male or (2) to prevent the new alpha male from threatening infants (Busse & Gordon, 1983). Proximity is the most frequent affiliative interaction females engage in and most is directed at their youngest offspring (Ehardt, 1988a,b). Adult females also perform significantly more grooming than they receive (both in frequency and duration). Curiously, infants and juveniles receive less grooming than expected, while adult males receive more than expected (Ehardt, 1988b).

Sooty mangabeys breed throughout the year (Gust et al., 1990). Alpha males attempt to monopolize peak estrous females and may account for more than 50% of mountings during a female's peak swelling. However, peak estrous females may sneak copulations with other males as well (Gordon et al., 1991; Range, 2005). Like other mangabey females, the sootys show sexual swellings which begin to appear when they are close to 3 years old (Ehardt, 1988b; Gust et al. 1990). They also show post-conception swellings which tend to peak around 50 days after conception, and these swellings appear very similar to estrus swellings. Males mount females at about the same frequency during the two different swelling episodes (Gordon et al., 1991). In captivity females produce their first offspring around 4.5–5 years of age (Ehardt, 1988a,b; Gust et al., 1990).

Morphology and Ecology: Cercocebus lunulatus

The white-naped mangabey has a brownish-gray coat and can be distinguished from *C. atys* by its lighter color and pure white underparts (Abelló et al., 2018). *C. lunulatus* inhabits primary and secondary forest, gallery forest, and swamp forest, including mangrove and mosaic habitats, within the Upper Guinean rainforest of West Africa (Oates et al., 2016). It is a terrestrial primate that can utilize the forest canopy as well as woodland savanna (Abelló et al., 2018). This mangabey species is tolerant of a wide range of habitats and, in the absence of hunting, tolerates some degree of environmental degradation (Abelló et al., 2018).

Distribution is limited and patchy since the species is not in abundance anywhere in its range. Nine subpopulations have been confirmed, but the white-naped mangabey is now absent from reserves and national parks in which historical records exist (Oates, 2006). Camera trap studies have attempted to confirm the presence of white-naped mangabey populations based on reported rare sightings (Nolan et al., 2019). The species is severely threatened by logging, mining, and hunting. Mortality also arises from persecution by agriculturalists due to crop raiding (Dempsey et al., 2020). *C. lunulatus* is undoubtedly one of the most endangered African primates (Nolan et al., 2019).

Social Structure and Organization: Cercocebus lunulatus

To date, there have been no studies of social patterns in wild groups of white-naped mangabeys. Research on captive groups has yielded information on allogrooming (Alea et al., 1999; Pérez Pérez & Veà Baró, 1999), problem solving (Albiach-Serrano et al., 2007), and social compatibility in an artificially constructed surplus male group (Fàbregas & Guillén-Salazar, 2007). When four male white-naped mangabeys were housed together at Valencia Zoo, they exhibited low-intensity facial threats at high rates, but physical aggression was rare; affiliative behaviors included positive approaches and social grooming (Fàbregas & Guillén-Salazar, 2007).

Morphology and Ecology: Cercocebus agilis *and* C. chryogaster

There is very little information available on the agile mangabey (*Cercocebus agilis*) or the golden-bellied mangabey (*C. chryogaster*). The agile mangabey is a medium-sized monkey with clear sexual dimorphism; females on average are 60% of male mass (males 7–12 kg and females 5–7 kg). There is no evidence of sexual dichromatism, however. Agile mangabeys have short speckled brownish hair, while the throat, chest, belly, and inside of legs and arms are off-white, and the back has a dark band running down the center. The face has pale eyelids. The tail is fairly short and stubby with a lighter tip (Shah, 2003). There are both dark and light morphs within this species, but the darker form is more common (Groves, 2001); both color variants have been found to exist in the same groups (Shah, 2003). Infants have similar hair color to adults, although their faces are pale pink at birth.

The agile mangabey is usually reported present in primary and secondary periodically flooded swamp forests (Hart et al., 2008a; Devreese et al., 2013). It has been studied at the Mondika Research Station, located in the Dzanga-Ndoki National Park, Central African Republic (CAR), and at Bai Hokou, CAR. The habitat at Mondika comprises three types of vegetation: mixed forest, riverine gallery forest, and swamp forest. The area experiences a distinct dry and wet season, with the dry season usually occurring between December and February. The forest receives high rainfall, ranging between 1,000–2,000 mm per year. There is little variation in temperature throughout the year with mean monthly maximum temperature being just below 30°C while the minimum is slightly above 20°C (Lilly et al., 2002).

Despite reports of agile mangabeys inhabiting periodically flooded forest, at Mondika the mangabeys never visit inundated forest habitats but range consistently in terra firma forest (Shah, 2001, 2003). At the Mondika study site, *Lophocebus albigena* and *C. agilis* have overlapping home ranges, and both species show a preference for mixed-forest over riverine forest. *L. albigena* will occasionally visit gallery forest to feed on insects, and *C. agilis* will seasonally search for mushrooms there (Shah, 2001). However, just 100 km to the east, in Mbaéré-Bodingué National Park, CAR, the agile mangabeys have been sighted *only* in seasonally inundated forests, while the gray-cheeked mangabeys are never seen in these locations (Brugiere et al. 2005).

Agile mangabeys are semiterrestrial and spend a fair amount of time on the ground (~15–20%), with males using the ground more often than females, although both sexes in general tend to stay fairly close (0–10 m) to the ground (Shah, 2003). They prefer to travel on the ground especially when foraging. Their diet is highly omnivorous, including a diverse array of seeds and nuts, shoots of herbs and grasses, mushrooms, insects, bird eggs, and vertebrates (Shah, 2003). They prefer ripe fruit but also eat unripe or very mature fruits when encountered on the forest floor. Agile mangabeys have been observed to eat

from elephant dung, which may explain why this species has been found to host a wider range of intestinal parasites than other sympatric primates (Lilly et al., 2002). In the dry season, when fruit is rare, flexible subgroups travel far and wide in search of food. Since 2004 the agile mangabeys at Mondika have been observed to frequently hunt and eat meat (Knights et al., 2008). The hunts appear to be opportunistic and only adult males participate. They target mainly infant blue duikers, but other species are also captured and eaten. Of the 99 instances of hunting observed, 76% were successful (Knights et al., 2008). Even though hunting and eating of meat is most common during the dry season, meat eating occurs throughout the year.

The study groups at Mondika range over 400 ha (Shah, 2001), while at a different study site in Gabon, Quris (1975) reported home ranges half that size. Foraging groups comprise 20+ individuals with subgroups often joining together (Shah, 2001), but Knights et al. (2008) reported study groups ranging in size from 100–200+ individuals just a few years after Shah's (2003) seminal study. Likewise, at Bai Hokou group size remained consistently over 100 individuals with observations of temporary associations of several groups—so-called super groupings (Devereese et al., 2013). It is possible that food availability provided the opportunity for larger bands of agile mangabeys to aggregate. Predation was never observed at Mondika although potential predators include leopards, pythons, and eagles. Since the research station is very remote, hunting is not a conservation issue presently.

The golden-bellied mangabey (*C. chryogaster*) has a more robust body build, quite unlike other mangabeys (Kingdon, 1997). The body is covered with rich, dark olive hair on the back, top of head, and down the outsides of arms and legs. The chest is off-white in color and grades into bright orange on the stomach. Faces are dark brown, which makes the pale eyelids stand out. They have prominent, almost white cheek tufts swept back toward the ears. What further sets this species apart is their brightly colored rear; the naked skin is deep blue-purple, akin to drills (Kingdon, 1997). Furthermore, they do not hold their tails arched over their backs, as is typical of mangabeys, rather they carry their short stout tails arched backward away from the body.

The golden-bellied mangabeys have not been studied in their natural habitat. They are found south of the Congo River in the lowland tropical rainforests of the central Congo Basin. Recent observations suggest a much-reduced distribution pattern, potentially due to hunting pressure, and sightings are predominantly in swamp forest habitats (Inogwabini & Thompson, 2013). Nothing is known about their preferred habitat or their diet, although possibly they may be fairly similar to the agile mangabeys north of the Congo River. From sporadic sightings it is known that they are semiterrestrial and forest-living (Oates, 1986).

Social Structure and Organization: Cercocebus agilis *and* C. chrysogaster

The agile mangabey (*C. agilis*) is often found in small groups that are usually led by a single adult male, but solitary males frequently are observed peripheral to these groups (Shah, 2003). However, aggregations of more than 100 individuals at Mondika and at Bai Hokou suggest a more fluid group membership (Knights et al., 2008; Devreese et al., 2013). Nothing else is presently known about either the agile or golden-bellied mangabey (*C. chrysogaster*) social structure or organization. Walker et al. (2004) reported some reproductive parameters from a captive group of golden-bellied mangabeys: Two females began cycling at 2.5 and 2.6 years, although captive animals are known to be precocious in this regard. Gestation lasted between 5.8–5.9 months, which is equivalent to what is recorded for other mangabey species (Kinnaird, 1990).

Morphology and Ecology: Cercocebus galeritus

The Tana mangabey has the general mangabey body of slender limbs and light build, with a long tail usually held arched over the body. They have dark faces with contrasting white eyelids. Hair is speckled gray with an almost black face. Sexual dimorphism exists (males weigh 8–10 kg, females 5–6.5 kg), but no noticeable sexual dichromatism. Infants have grayish hair and pink faces which take on adult coloration around three months of age.

The Tana mangabey inhabits a limited area along the Tana River in eastern Kenya. The river originates in the foothills of Mt. Kenya, and throughout most of its course passes through arid, almost semidesert habitats. However, in the lower reaches of the river the habitat changes and there is a 60-mile stretch of forest. It is within this gallery forest that the Tana mangabeys live. The forest is naturally fragmented due to the meandering river, and as a result there are many oxbow lakes and swamp areas (Moinde-Fockler et al., 2007). The gallery forest, which in some locations may be up to 5 miles wide, is dependent on seasonal flooding and groundwater levels remaining fixed (Wieczkowski, 2004). Lowland evergreen forests are close to the river; the trees form a closed canopy and most are 10–20 m tall, with some reaching 30 m in height, and there is a fairly dense understory (Wieczkowski & Kinnaird, 2008). Moving away from the river, the habitat becomes increasingly arid, grading from deciduous woodland, to wooded grassland, to grassland (Muoria et al., 2003). Rainfall is unpredictable and usually does not exceed 500 mm annually (Karere et al., 2004). Most of the rainfall is bimodally distributed with peaks in March–April and November–December. Daily temperatures vary between a minimum of 17–25°C and a maximum of 30–38°C, with January and February being the hottest and driest part of the year (Karere et al., 2004). It is estimated that 70–80 forest patches are still in existence, and mangabeys are found in approximately half of them (Muoria et al., 2003; Moinde-Fockler et al., 2007). The plant community is fairly diverse and shows strong temporal variation in fruit and flower phenology. Nevertheless, the environment sustains 7 sympatric nonhuman primate species—the Tana mangabey, the Tana River red colobus (*Piliocolobus rufomitratus*), Sykes monkey (*Cercopithecus mitis*), yellow baboon (*Papio cynocephalus*), vervet monkey (*Chlorocebus aethiops*), thick-tailed galago (*Otolemur crassicaudatus*), and lesser galago (*Galago senegalensis*) (Wahungu, 1998; Wieczkowski & Kinnaird, 2008).

Butynski et al. (2000) estimated an average group size of 35.1 ± 4.6 based on census data from 12 groups. Several groups may form temporary conspecific associations comprising 50–60 individuals. Groups contain between 1–6 adult males and usually twice as many adult females.

The home range depends to a large extent on the forest patch size but can vary between 17–70 ha (Kinnaird, 1992a). When more than one group live in a forest patch, there is much overlap in the home ranges.

The Tana mangabeys can be considered semiterrestrial as they forage at all levels in the forest including the ground. They are omnivorous and consume a wide range of food sources, including fruits, seeds, insects, leaves, flowers, tree gum, honey, and occasional small animals like frogs, and lizards (Homewood, 1998). However, their primary food sources are seeds and fruit, with palm fruits and seeds (primarily *Phoenix reclinata*) being very important in their diet (Kinnaird, 1992c). Seeds and fruit from *P. reclinata* may make up as much as 62% of their diet in some months and, on an annual basis, account for up to 26% of their total diet (Kinnaird, 1992c; Wahungu, 1998). The Tana mangabeys rely heavily on these resources because palm fruit and seeds are out of synchrony with other plant species. Figs (*Ficus sycamorus*) are also a critical food source, especially because they are available throughout the year (Kinnaird, 1992b,c). Both these tree species grow along the riverbanks; however, while palm trees are common, fig trees are more scarce (Wahungu, 1998).

Yellow baboons are potential competitors with *C. galeritus* at a Tana River site in Kenya. A detailed study by Wahungu (1998) found that habitat and diet overlap of the two species is low in the dry season (August–October) but increases in the wet season (November–January) and reaches its peak in February directly after rains. Even though this is the time of highest fruit production, fruit abundance does not correlate with habitat overlap or diet overlap. The presence of baboons does, however, have an effect on the behavior of the mangabeys. If baboons are present in an area, the mangabeys spend less time there. They also forage in a tighter unit and are not as spread out as normal.

Social Structure and Organization: Cercocebus galeritus

Tana mangabeys live in multi-male/multi-female groups (Wieczkowski & Kinnaird, 2008). There tends to be twice as many adult females as males, and solitary or peripheral males are frequently observed. However, the group that Gust (1994) studied had 2 adult males and 8 adult females (each carrying an infant) and 20 juveniles of various ages. At 3 months an infant will begin to move away from its mother and is often cared for by juvenile females (Gust, 1994). By the time it is 6 months old, the mother will begin to cycle again; the interbirth interval is in the range of 18–24 months. Sexual maturity is attained around 4–5 years of age. Female cycle length is 28–30 days and gestation lasts 6 months. Even though females cycle throughout the year, there is a birth peak during December–January (range November–February) (Kinnaird, 1990).

Tana mangabey males form linear dominance hierarchies, and the alpha male holds a central position within the group. During male-male aggression, group females tend to be wary. Homewood (1998) observed lower-ranking males picking up infants and holding them ventrally, which she interprets to be a submissive gesture.

Like other mangabey species, the Tana mangabeys use a range of vocalizations (Gust, 1994). Tana mangabeys are fairly noisy monkeys, but some vocalizations are used only during specific activities. Alarm calls are staccato-barks, which can be given by any group member, and "wherr" calls are given by an aggressor during agonistic interactions (Gust, 1994). In addition, there are a range of grunts and screams emitted by all group members during various interactions.

The mangabey loud call, the whoop-gobble (a shared feature of *Lophocebus* and *Cercocebus* mangabeys that members of the two groups have converged on), is considered to be a group spacing vocalization. It is a call given by all adult males, who tend to call more often in the early part of the day and more frequently during the dry season (Homewood, 1998). Males appear to call from very distinct locations that tend to be close to core areas. These calls do not usually illicit replies. Upon hearing a loud call from another group, the group under observation will quietly retreat. As a result, groups seldom come into contact with each other. The travel patterns suggest that groups attempt to avoid each other as much as possible.

Kinnaird (1992a) found that behavior toward neighboring groups varies throughout the year. During specific months, the Tana mangabey groups are close and often cross each other's paths without any reactions on either side. Sometimes they even form nonaggressive associations that may last for hours. During these associations, they forage side by side, groom, and sexual-present to each other. However, at other times the two groups confront each other if their paths cross. At such times, all members of each group (except infants) line up facing each other. The adult females will be on the ground and lunge at each other with their eyelids down and tails arched forward over the body while making "cackling" vocalizations. The adult males are more often in the surrounding trees shaking branches and engaging each other in chases. Yet, neither females nor males usually engage in physical contacts during these contests.

The variation in responses to neighboring groups during different times of the year may be due to resource access. This hypothesis is supported by the fact that 63% of the conflicts occur close to fruiting doum palms (*Hyphaene compressa*), and fruit biomass is significantly lower at times when there are no interactions (Kinnaird, 1992a). However, biomass was the same when groups mixed peacefully or when they fought. The only difference was that when fights broke out, the mangabeys were feeding on clumped food resources, while at other times they were feeding on uniformly dispersed food resources.

Even though Tana mangabeys may temporarily form polyspecific associations, for example with red colobus or guenons, they seldom interact. When yellow baboons and Tana mangabeys encounter each other, there is little interaction—rather the mangabeys appear to avoid baboons. Baboons, in turn, displace the mangabeys, especially if there is a contest over sleeping trees (Wahungu, 1998).

Morphology and Ecology: Cercocebus sanjei

The highly endangered Sanje mangabey, found in upland habitats of the Udzungwa Mountains, south-central Tanzania (Ehardt et al., 2008), was first described in the scientific literature by Homewood and Rogers (1981). The species is distributed in two highly restricted populations, separated by more than 100 km of farm and residential areas, along with annually burned grassland (Ehardt & Butynski, 2006b).

C. sanjei have a slender build with a long body and legs. Hair on the body is speckled olive; the tail is long and kept curved over the body with the white tip often touching the head when walking. The hair on the head is long and swept up into two crests on either side; eyelids are light colored but do not stand out as in other *Cercocebus* mangabeys due to the light color around the eyes. Very young infants have pink faces and ears. Sexual dimorphism is evident; the body mass of males is estimated to be around 10 kg while that of females is around 5–6 kg. In addition, males have larger muzzles, bigger canines, more luxuriant hair over the shoulders, and bigger head crests.

The Udzungwa Mountains are part of the southernmost and largest block of the Eastern Arc Mountains in Tanzania and Kenya, an area which is one of the most important biodiversity hotspots in the world. The Sanje mangabeys are among the most important of a wide range of endemic species found in this region (Rovero et al., 2009, 2012). The Udzungwa Mountains cover an area of 10,000 km² and are unusual since the eastern escarpment has continuous forest cover from 300 m up to 2,250 m (Rovero et al., 2006). The natural forest is highly fragmented, and only about 27% of the forest has closed canopy (Ehardt et al., 2005). Due to the high degree of endemism, the Udzungwa Mountains National Park (UMNP) was established in 1992 and covers approximately one-fifth of the mountain area.

Most of the Sanje mangabeys are found at high elevations (1,400–1,650 m), but a few groups have also been recorded lower at 300–700 m (Ehardt et al., 2005). Their habitat is tall primary rainforest located on steep slopes that are dissected by deep ravines with many permanent streams and rivers. The forest ranges from lowland deciduous to montane evergreen but is fragmented and interspersed by degraded forest, woodland, and grassland (Ehardt et al., 2005; Rovero et al., 2006). The mangabeys are mainly found in evergreen wet forest (Rovero et al., 2009). Climatic stability due to persistent Indian Ocean monsoons brings high rainfall along the eastern slopes. Annual rainfall at higher elevations is as high as 2,000 mm, while the western plateau receives around 900 mm (Ehardt et al., 2005). There is a bimodal distribution of rainfall, with most precipitation falling during March–May and a lesser peak during November–January (Jones et al., 2005). Temperature varies depending on elevation; the mean monthly temperature at 300 m is 24°C in June–July and 28°C in November–December (Rovero et al., 2006). At higher elevations the temperature can drop below freezing (Jones et al., 2005).

Sanje mangabeys live in rough and inhospitable terrain that makes observations difficult; in addition, the population is spread over two disjunct forests within the Udzungwa Mountains. Based on repeated censusing, total population size is estimated to be no more than 1,300 individuals. Of these about 60% live within the Mwanihana Forest in UMNP, although there may have been recent local extinctions especially in the lowland populations (Ehardt et al., 2005). The remaining 40% live within poorly protected habitats in the Udzungwa Scarp Forest Reserve. This reserve comprises a series of forest fragments; 18 of the 26 forest fragments are less than 25 km² (Ehardt et al., 2005). Mwanihana Forest is located in the northeast part of UMNP, situated on the slope of the east-facing escarpment and covers 177 km². It has continuous forest cover from 300–2,100 m (Rovero et al., 2006). At lower altitudes there is dry deciduous woodland which gives way to tropical moist montane forest at higher altitudes and eventually to a mosaic of bamboo and alpine flora (Lovett et al., 2006).

There are several habituated groups of Sanje mangabeys within the Mwanihana Forest from which detailed behavioral information has been collected (Ehardt et al., 2005; Mwamende, 2009). During transect walks, 51 groups have been encountered in addition to solitary males. Group size ranges from 27–51 individuals with a mean of 40 based on censusing of three groups (Mwamende, 2009). These estimates indicate there are less than 700 Sanje mangbeys present within the Mwanihana Forest. However, increased protection may be yielding results as one group that has been followed for over 10 years has expanded from 32 individuals in 2000 to 49 in 2005, and further increased to 62 in 2008 (Jones et al., 2006; Mwamende, 2009). The home range of one group is 301 ha (Mwamende, 2009) with seasonal differences—174 ha is used in the wet season and 343 ha in the dry season. The day range (1,760 ± 8 m) also varies seasonally with longer day ranges during the dry season than during the wet season. Continuous monitoring by Mwamende (2009) over 6.5 months shows that his study group has four areas they use more extensively than others, but they spend most of their time close to the river and its gallery forest habitat.

This species has occasionally been sighted in polyspecific associations with Sykes guenon, red colobus, and black-and-white colobus (Ehardt et al., 2005). The main predator threats, aside from humans, are the crowned hawk-eagle, leopards, and pythons (Rovero et al., 2006, 2009). These predators are common in more pristine habitats but less common in disturbed areas (Topp-Jør-gensen et al., 2009). Formation of polyspecific groups may be due to the high density of crowned hawk-eagles (Jones et al., 2006).

Sanje mangabeys feed and forage at all levels of the forest but spend a significant proportion of their time on the forest floor. Detailed diet studies have shown that *C. sanjei* is a true omnivore eating fruits, seeds, nuts, flowers, leaves, shoots, roots, gum, bark, ferns, fungi, vertebrates (lizards), and invertebrates (millipedes, snails, termites, grasshoppers, and spiders). Of the plant resources identified, the Sanje mangabeys consume 99 species, although 68% of the feeding time is on less than 34 species (Mwamende, 2009). Fruits make up the largest single food source (43%), which together with seeds, flowers, and roots make up close to half (47%) of food intake. Vertebrates, invertebrates, and fungi make up another 29%, with shoots and leaves contributing the remaining 24% of the diet (Mwamende, 2009). Adult females spend significantly more time feeding on fruits, seeds, and flowers than other group members, while subadults and juveniles spend more time feeding on shoots and stems.

Social Structure and Organization: Cercocebus sanjei

Sanje mangabeys live in multi-male/multi-female groups, although occasional solitary males are observed. McCabe and Thompson's (2013) study groups comprise 62–65 members with approximately twice as many adult females as males. Both males and females form stable linear dominance hierarchies and, in general, adult males are dominant over females and younger males. The dominance hierarchy is steep with a high degree of power differential between individuals, and aggression tends to be directed down the rank scale. Adult males engage in chasing and biting much more than any other group members but are less aggressive and more tolerant toward estrous females with maximum swellings (Mwamende, 2009).

Female Sanje mangabeys associate mostly with their own offspring and with adult males rather than with other group females. Close to 8% of each day is spent grooming (Mwamende, 2009). Females are the most active groomers, and they spend significantly more time grooming adult males than other females, which results in the females receiving very little grooming attention themselves. However, this grooming attention bestowed on adult males may pay off since females who groom specific males frequently are allowed to feed undisturbed in close proximity to the males. Males do not often reciprocate the grooming attention adult females give them, rather they associate slightly more with subadult females than with the adult females. Subadult females also groom adult males frequently.

Male-male interactions are not common and, when they occur, they are mostly antagonistic. Nonetheless, Mwamende (2009) found that occasionally males will groom each other, although in his study group the alpha male receives more grooming from lower-ranking subadult males while he is never groomed by the second- or third-ranked males.

Infants and juveniles receive social attention from all members of the group. All females and adult males engage in allomothering; adult females also spend much time grooming infants and juveniles. There is little aggression shown toward infants by any group member, although males show more aggression against immatures than females do. Adult males also pay protective attention to immatures, especially infants. During agonistic interactions, adult males will pick up and carry infants away. Males retrieve infants when mothers leave them or when the mother is involved in an agonistic event. Males also take infants away from playing juveniles and return them to their mothers.

The female reproductive cycle is 30 ± 3.0 days in Sanje mangabeys (Fernández et al., 2014). Female anogenital swellings include larger areas than seen in other mangabeys but not as extensive as seen in baboons (Mwamende, 2009). As with other mangabey species, female *C. sanjei* experience swelling about two months after conception. These swellings are dark red like deflating sex skin; the swellings are not turgid and may decrease and increase in size throughout the pregnancy. Gestation lasts 170.5 ± 2.4 days, and the interbirth interval is 570.8 ± 126.8 days (Mwamende, 2009; Fernández, 2011). Even though females tend to conceive throughout the year, a distinct birth peak occurs during the dry season accounting for close to 80% of all births (McCabe & Thompson, 2013).

Matings occur at all stages of a female's estrus cycle, but frequency peak when a female has maximum swelling. Males initiate most copulations, but females initiate a little over one-third of the copulations. A soliciting estrous female will approach a male making soft "precopulation" calls that end with eye contact. She will then present her hindquarters. The male may respond by inspecting her genital area (sniff, touch) or he may mate with her (Mwamende, 2009). Females make copulation calls either during or after mating. There is seldom any interaction between the pair once the copulation is over. The female may solicit the same male again or move off to another male. Re-solicitation can be explained by the fact that more than one mount is required before ejaculation. On average, females will mate with 2.6 ± 0.1 males during the time when they are most likely to conceive. Estrous females seem to have no compunction about mating with solitary males, as 29% of such encounters end with copulations.

Morphology and Ecology: Mandrillus

Mandrills (*Mandrillus sphinx*) and drills (*M. leucophaeus*) exhibit striking sexual dimorphism. Mandrills, in fact, are the most sexually dimorphic of all extant primate species, with adult males nearly 3.5 times the size of females. Male drills are slightly more than twice the size of females (similar to gorillas and orangutans). For the genus, adult male weight ranges between 19–35 kg, while adult female body mass is between 10–15 kg (Marty et al., 2009; Swedell, 2011). Males that have reached alpha position with a fully developed body have the highest weights (Setchell et al., 2001). Females reach adult size around 7 years of age, while males do not attain full size until they are 9–10 years old (Setchell et al., 2001).

The physical appearance of mandrills is quite spectacular. It is the adult males that are especially decorative, and the alpha male has the most developed secondary sexual characteristics. The most striking feature is the elongated muzzle (Figure 16.10), which is covered with bare skin in vivid colors; although present in both sexes, it is less prominent in females. Inflated bony ridges on either side

Figure 16.10 The elongated muzzle of the adult male mandrill (*Mandrillus sphinx*) has bare skin in vivid colors of blue and red with a yellow goatee. Alpha males have the brightest coloration.

(Photo credit: Pete Oxford/naturepl.com/arkive.org)

of the nose, called paranasal swellings, are covered by bright blue skin, while the top of the muzzle is scarlet red. The blue coloration is due to presence of dermal melanin granules and superficial collagen fibers that reflect light (Hill, 1970). The blue coloration, therefore, is not influenced by hormone levels (Setchell & Dixon, 2001). The red coloration of the muzzle is due to hemaglobin present in superficial capillary blood vessels (Hill, 1970). When a male acquires the alpha position, his red facial and genital coloration becomes even brighter, which is correlated with rapid increase in circulating testosterone (Setchell & Dixon, 2001). The head is proportionally large in males, supported by massive neck muscles. Brow ridges are prominent, making the eyes appear deep set. The muzzle is long and ends with expanded fleshy nostrils. Lips are prominent, but display much individual variation in color ranging from dark brown to scarlet red. A brilliant yellow goatee beard, which may extend up around the cheeks, surrounds the face of adult males. Females also have a yellow beard but not colored cheek tufts. Both sexes have a hairy crest on top of the head, which in the male is further enhanced by white hair around pale pink ears. The crest is used during social affiliative communication possibly as a prelude to grooming (Laidre & Yorzinski, 2005). Rather rare among Old World monkeys, both male and female mandrills have scent glands on their chest (Hill, 1970). These glands produce secretions that they rub against tree trunks or branches (Feistner, 1991). For both males and females, scent marking is correlated with dominance rank. Setchell et al. (2010, 2011) suggest that the marks give information on an individual's age, sex, and dominance, which has been supported by a study on captive mandrills (Vaglio et al., 2016).

The size of canine teeth in mandrills is highly sexually dimorphic. In fully adult males the maxillary canines measure 44 mm, while in adult females they are only 9 mm long (Leigh et al., 2005). In males the perineum is brightly colored blue and red. The perianal region, testes, and penis are bright pink, whereas the skin of the rump has blue to violet hues and narrow callosities that are pink. Wispy yellow hair may enhance this color scheme. Females are less colorful; even their sexual swellings are moderate in size and never get brilliant pink as seen, for example, in baboons. A study by Setchell and Dixon (2001) suggests that the coloration in males closely follows maturation and dominance rank and likely signals social status. It also appears that females prefer the most brightly colored males as mating partners (Setchell, 2005). Adult males, and especially top-ranking males, have extra fat deposits in the rump and flanks, often referred to as a "fattened" state (Wickings & Dixon, 1992). In contrast to the colorful face and sexual skin, mandrill body hair is dark olive-green with an almost white belly. The tail is short and held upright over the hindquarters.

Drills (*M. leucophaeus*) lack some of the flamboyant colors seen in mandrills, but high-ranking males can present quite a color palate (Marty et al., 2009). Their almost hairless faces are black, and the paranasal ridges are smooth and very prominent in adult males. Both males and females have a white beard and cheek hair, but males also have bright red lower lips. Like mandrills, drills have prominent brow ridges, but because the hair surrounding the face is shorter, the eyes do not have a deep-set appearance. The ears are dark and not very prominent. Aside from the light ventral region, the hair is a gray-brown and lighter than mandrills. Drills have short tails held upright over the hindquarters. The male perineum is almost iridescent blue and red, with bright pink scrotal skin, penis, and narrow callosities. The bright coloration seen in male drills correlates with dominance rank and attractiveness to females, just as seen in mandrill males (Marty et al., 2009). Females have moderate sexual swellings that are dark pink-brown in color (Hill, 1970; Ankel-Simons, 2000).

Postcranially, both drills and mandrills exhibit adaptations for terrestrial quadrupedal locomotion with an intermembral index of 95 (Fleagle, 2013), along with short stubby fingers used in digitigrade walking. *Mandrillus* and *Cercocebus* (but not *Lophocebus* and *Papio*) share a number of features of their limbs (and teeth) which enhance foraging and processing hard seeds and invertebrates from the forest floor (Fleagle & McGraw, 1999, 2002).

Newborn mandrills and drills do not have any distinctive coloration but look very much like small copies of the adults. The juvenile period in females is short, since females begin to cycle as early as 3.5 years and soon after give birth for the first time. Males have a much more extended development period: Testicular descent occurs around 3–4 years of age, and males have a major growth spurt between 5 years and when they reach physical maturity at 9–10 years (weight increasing from 10 kg to 30–35 kg) (Setchell & Dixson, 2002).

Both mandrills and drills are forest dwellers, found mainly in dense mature rainforest. The drill is found in lowland, submontane rainforests, mostly in mature secondary forest and infrequently in young secondary forests (Oates & Butynski, 2008a). Drills have been studied in Korup National Park, located in southwestern Cameroon. Korup and the surrounding protected area is part of the Cameroon-Gabon lowland rainforest zone (Astaras et al., 2008). Annual rainfall is approximately 5,000 mm, most of which falls during a single wet season (May–October) followed by a distinct dry season (December–February). In this biome, temperature variation is minimal throughout the year; most variation occurs during a daily 24-hour cycle—daily maximum is close to 35°C and daily minimum around 25°C. Most of the area retains fairly complete closed-canopy lowland moist forest dominated by Ceasalpinioideae tree species (part of the legume family), with flowering and fruiting showing a strong seasonal pattern (Astaras et al., 2008).

Mandrills are found in evergreen rainforests, but some populations are known to incorporate more arid semideciduous forest and savanna habitats into their range (Tutin et al., 1997b). Mandrills have been studied at Campo Reserve in southern Cameroon (Hoshino, 1985; Kudo, 1987) and in Lékédi Park in southern Gabon (Brockmeyer et al., 2015; Akoue et al., 2017). The Campo Reserve is located close to the coast at an elevation of 50–230 m. At both the Cameroon and Gabon locations, four distinct seasons occur—a major rainy season, a major dry season, a minor rainy season, and a minor dry season. Temperature varies more on a daily than yearly basis (range 20–30°C). Rainfall is over 2,000 mm annually. Most of the study area in Campo Reserve comprises a dense, continuously humid lowland forest with thick undergrowth resulting in poor visibility. The forest contains many tall and emergent tree species (>25 m), e.g., *Sacoglottis gabonensis*, which provides an important fruit resource for mandrills. The Lékédi Park comprises a mosaic of savanna grassland and evergreen forest (Akoue et al., 2017). Mandrills have also been studied in the Lopé Reserve in central Gabon (Abernethy et al., 2002), a continuous semideciduous Guineo-Congolian lowland rainforest and also mosaic, ancient savanna grassland areas interspersed by forest fragments (Tutin et al., 1997b). The annual rainfall is around 1,500 mm distributed bimodally. There are two dry seasons—a short two- to three-week-long dry spell during December–February, and a longer dry season lasting from mid-June to mid-September (Abernethy et al., 2002).

Little is known about the home range size of either drills or mandrills, although at Lékédi Park a mandrill group ranges over 867 ha (Brockmeyer et al., 2015). Mandrill groups tend to leave their sleeping trees around 7:00 am and travel continuously while foraging until 5:00 pm when they ascend new sleeping trees (Hoshino, 1985). The distance traveled varies between seasons. Since preferred fruit trees are more widely distributed, more movement occurs during the major fruiting period (4.5 km/day) compared with the minor fruiting period (2.5 km/day) when the mandrills rely more on other food resources. The extent of home range overlap of adjacent groups and whether groups are territorial is not known, although dominant males are noted to defend their group against other males (Rogers et al., 1996).

To date, habituation of either drills or mandrills has met with limited success. Both drills and mandrills are very shy, possibly because they are hunted throughout their ranges (Fa et al., 2005). What is known about the diet of drills and mandrills comes mainly from examination of fecal samples and, to a lesser extent, from direct observations. More recently, camera trapping has been used at specific sites to determine potential seasonal changes in habitat use and diet (Hongo et al., 2018). Astaras et al. (2008) investigated the diet of drills in the Korup National Park, Cameroon, during a five-month period which covered part of the dry season and the beginning of the rainy season; seeds, fruits, insects, green leaves, and mushrooms were found in fecal samples. During certain months, the diet includes some specific species. For example, during May, close to all fecal samples collected contained evidence of wild mango fruits (*Irvingia gabonensis*) which were not observed during the previous three months. Fruits and seeds from the parasol tree (*Musanga cecropioides*), however, were eaten throughout the study period. The drills do not include much leaf matter in their diet; only grass blades were recorded. Insects consumed are mainly nonflying species such as ants, termites, and crickets.

Hoshino (1985) studied the diet of mandrills over an intermittent period of 25 months at the Campo Reserve. He found that fruit from a wide range of species make up the major part of the diet, but leaves (mainly young), shoots, pith, and animal matter (mostly ants, termites, and crickets, but also snails, frogs, mice, birds, and eggs) are also part of the diet. At least 27 different fruit species are eaten, and in some cases both fruits and seeds are consumed. During the major fruiting period, fruit from *Sacoglottis gabonensis* (Humiriaceae), which is also a keystone food species for forest elephants (Morgan & Lee, 2007), is an important source between August and October. During March and April, fruits from *Grewia coriacea* (Tiliaceae) make up an important part of the diet. Between April and August, fruits are scarce and more leaves and seeds are consumed, and mushrooms are eaten extensively during the rainy season. Hongo et al. (2018) also found that mandrills in Moukalaba-Doudou National Park, Gabon, have distinctive seasonal preferences, with seeds making up a large portion of their diet, although fruits are preferred and eaten whenever available. Hoshino (1985) found that 41 species (47%) of the plants eaten come from the lowest forest strata (<5m) but, in general, the mandrills forage and eat mostly fallen fruits or seeds from the ground.

Lahm (1986) studied the diet of mandrills in central Gabon, including the Lopé Reserve, and found the same results as those from the Campo Reserve. A wide variety of fruits and seeds make up the majority of the mandrill diet, but they diversify during periods of fruit scarcity to include bark, leaves, young shoots, and stems. Ripe fruits that are consumed come mostly from large trees; however, sometimes unripe wild mangoes (*Irvingia gabonensis*) are consumed. The mandrills also eat the nut-like fruit *Coula edulis*, which is also part of the chimpanzee diet (Boesch & Boesch-Achermann, 2000). Like baboons and *Cercocebus* mangabeys, drills and mandrills occasionally include animal matter in their diets, especially invertebrates (Lahm, 1986; Hongo et al., 2018). Astaras et al. (2008) noted four plant species included in the drill diet that also have been recorded as eaten by mandrills—*Oncoba glauca, Pentadesma butyracae, Irvingia gabonensis,* and *Morinda morindoides;* such overlap in diet may preclude drills and mandrills from being sympatric.

Molecular phylogeny pointing to a link between *Mandrillus* and *Cercocebus* is supported by skeleto-dental features identified as crucial to a shared foraging scheme of the clade (Fleagle & McGraw, 1999, 2002). McGraw and Fleagle (2006: p.201) review the comparative anatomy of these genera which indicates a foraging regimen "characterized by a reliance on hard object foods and habitual aggressive use of the forelimbs during foraging." Food acquisition behaviors, distinguished by unique adaptations for gleaning insects, hard nuts, and seeds from the forest floor, are facilitated by limb anatomy and dentition that represent unambiguous links defining the *Cercocebus-Mandrillus* clade (Fleagle & McGraw, 1999, 2002).

Both drills and mandrills have been observed in mixed foraging parties with mangabeys and guenons and, occasionally, chimpanzees. Drills show the highest rates of association with red-capped mangabeys (*Cercocebus torquatus*), followed by white-nosed guenons (*Cercopithecus nictitans*) (Astaras et al., 2011). Gartlan and Struhsaker (1972) found that drills join up sporadically with a range of guenon species on a temporary basis. At Campo Reserve, mandrills form temporary associations with red-capped mangabeys and crowned guenons (*Cercopithecus pogonias*) (Mitani, 1991). There are certain foods shared by many sympatric primate species (e.g., *Grewia coriacea*), but, in general, feeding competition appears limited because mandrills tend to feed off the ground or from low trees and herbaceous plants, while guenons feed in taller trees. During the minor fruiting season, divergence in diet becomes more obvious between mandrills and guenons since mandrills feed less on fresh fruit and guenons feed more on broad-leaved species not included in the diet of mandrills (Hoshino, 1985).

Social Structure and Organization: Mandrillus

The social grouping patterns of drills and mandrills have received much attention. Yet, accurate census and group structure information is difficult to achieve in dense rainforest habitats. Both drills and mandrills live in multi-male/multi-female groups. More recent observations of mandrill groups support a fluid male membership; the sex ratio is usually heavily biased toward females with only a few adult males present at any one time (Abernethy et al., 2002; Hongo, 2014). Solitary males and small bands of adult males have also been encountered away from main groups.

A study in the Lopé Reserve in Gabon yielded the highest census to date for mandrills. Abernethy et al. (2002) recorded a group size of 620.2 ± 166.3 (range 340–845), which was based on 20 filmed observations collected over four years. The female-to-male ratio is 9:1, with solitary males frequently observed. Abernethy and colleagues also found that these large groups are stable over time, except for adult males that appear to come and go depending on female reproductive conditions. The number of adult males is not correlated to group size as would be expected if the core unit was one-male units. In addition, the fluidity in adult male presence belies the possibility that group structure is based on harems. Rather, it appears that mandrills have a matrilineal-based society, where females remain in their natal group throughout their life while males disperse (Abernethy et al., 2002). The conclusions from this study have been supported by research carried out at Moukalaba-Doudou National Park, Gabon, where Hongo (2014) found slightly smaller group sizes (169–442), although he suspected that on two occasions it was only a subgroup being observed. The two larger groups had 25–29 adult females per adult male.

Astaras et al. (2008) collected census data on four groups of drills in Korup National Park. The group size ranges from 25–77 individuals (with a mean of 52.3). Most of the groups they encountered had more than one adult male, but there is a tendency for subgroups to form when foraging. A second census (Wild et al., 2005) at a study site to the east of Korup found larger group sizes (93.1 ± 8.4, median 70), with solitary males encountered away from the main study groups.

Female *Mandrillus* have a dominance hierarchy with rank passed from mother to daughter, and social interactions occur preferentially within the matriline (Charpentier et al., 2007; Bret et al., 2013). Under specific circumstances, female mandrills have even been observed forming coalitions to attack an alpha male (Setchell et al., 2006). Mandrill males form a strict dominance hierarchy, and usually it is the top three males that are "fatted," but only the top-ranking male displays the most vivid colors

(Setchell & Wickings, 2005). This color display informs all group members of a male's rank. Competition for rank position can be highly aggressive and sometimes lethal. Adult males spend very little time in close proximity to each other, and only the alpha male spends all his time associated with the group. The other males tend to come and go or have a peripheral position. Males use the "grin" display (pulling back the upper lip over the canines) when encountering each other. Lower-ranking males grin when they approach a higher-ranking male who then returns the grin display; however, grins do not correlate significantly with male rank (Setchell & Wickings, 2005). Of the 22 semi-free-ranging adult males observed at the International Medical Research Center in Franceville, Gabon, only 9 (41%) attained alpha position when they reached 9–14 years of age (Setchell et al., 2005). The length of time a male retains the alpha position ranges from a month to 6 years (mean 34 ± 9 months). The potential cost of competition is outweighed by the benefits, since rank plays an important role in reproductive success. Not only does the top-ranking male mate and ejaculate at a higher rate than other males, the alpha male tends to sire the majority of offspring (70–100%; Charpentier et al., 2007). Low-ranking and peripheral males have few mating opportunities and, if detected, the alpha male will interrupt these matings (Setchell et al., 2005).

Even though mandrills and drills do not have a mating season in the strict sense, female mandrills show sexual swellings and proceptive behaviors during a limited time of the year (Setchell et al., 2002; Setchell & Wickings, 2004). Females display estrus cycles during June through November both in the wild and in semi-free conditions (Abernethy et al., 2002; Setchell & Wickings, 2004). The alpha male follows and guards females when their swellings are at peak tumescence, which usually coincides with ovulation and highest probability of conception. While mate guarding, a male invests heavily in visual, olfactory, and auditory displays and scent-marks trees and vegetation with his chest glands. The male almost continuously emits grunt vocalizations when with an estrous female. If more than one peak-estrous female is present, other males—usually the second-ranking male—may have mating access. Males appear to prefer high-ranking, multiparous females and mate/guard them more intensely (Setchell & Wickings, 2006). Females can, however, exert choice during the mating season by either running away from an undesirable male, or approaching other males to solicit mating. It appears that females prefer to mate with males sporting the most brilliant colors. More often than not, this is the alpha male, but deposed alpha males who retain their colors have higher than expected mating success (Setchell, 2005).

Vocal communication is almost continuous within a mandrill group while foraging in the dense understory, especially when they form subgroups and may be long distances apart. Adult males emit "two-phase grunt" long-distance calls when the group is calmly foraging (Kudo, 1987). This type of call is also emitted by drills (Gartlan, 1970). Females and immature mandrills use a different long-distance vocalization ("crowing"), which, again, has also been recorded for drills (Struhsaker, 1969). There are many close-range vocalizations taking place in social groups, but adult males have a different vocabulary compared with females and immatures (Kudo, 1987).

CONSERVATION

Many of the terrestrial species, in particular baboons, exist in areas inhabited by humans. Agricultural crops are highly attractive to nonhuman primates due to the concentration of nutritional food. Crop-raiding primates are considered pests, and conflicts between humans and wild primates are common (Strum, 2010; Hill & Wallace, 2012). Baboons, who are serious crop raiders, are considered "gluttonous, arrogant, and stubborn" and intent on spoiling crops for their own entertainment (Hill & Webber, 2010: p.920). Because they are considered in such a negative way, trapping and killing them is carried out indiscriminately (Abernethy et al., 2013; Knapp et al., 2017). Conflicts between humans and crop-raiding primates can have severe consequences when endangered species are involved. The first kipunji specimen that came to scientific attention was a juvenile that had been killed while raiding crops (Davenport et al., 2006).

In East Africa most primates are not considered edible; however, in West and Central Africa, primates in forested areas are hunted for their meat. Since primates in general have a low reproductive rate, many species are being harvested far beyond the rate they can be replenished. As a result, many are being driven to the brink of extinction, and many local populations have been recorded as extinct (Rovero et al., 2012; Cronin et al., 2017). *Cercocebus lunulatus* could well be the most endangered African papionine. According to the International Union for the Conservation of Nature (IUCN), this species is classified as Endangered and decreasing (Dempsey et al., 2020).

In many African countries, hunting of wildlife is restricted, or even banned, and certainly not allowed within protected areas. Still, hunting takes place because so-called bushmeat is not only a source of protein (e.g., Baya & Storch, 2010; Cronin et al., 2015) but also has become increasingly common in urban markets as a luxury food for those who can afford it. Professional hunters appear to specifically target monkeys (Refisch & Koné, 2005).

Larger species, such as drills and mandrills, in particular, are under increased hunting pressure throughout their range (Linder & Oates, 2011; Cronin et al., 2015). At Korup National Park, Cameroon, there has been a rapid decline in the drill population, partly due to hunters using dogs to chase primates up trees where they are easy targets (Waltert et al., 2002).

The hunting, sale, and consumption of bushmeat also has created a public health problem. Simian immunodeficiency virus (SIV) is a retrovirus responsible for persistent infections in 45 species of primates (Peeters et al., 2001; Peeters & Courgnaud, 2002). SIVsmm, found in wild sooty mangabeys (*C. atys*), is recognized as the progenitor of human immunodeficiency virus type 2 (HIV-2) and has crossed the species barrier to humans on multiple occasions due to the bushmeat trade in West Africa (Apetrei et al., 2005; Santiago et al., 2005). Extensive research on instances of zoonotic disease transfer through exploitation of wild primates is ongoing.

Habitat destruction due to extensive logging is also a major problem affecting primates and their food resources (Matthews & Matthews, 2002). Some primate species appear to react negatively to habitat disturbances in general. For example, the red-capped mangabeys (*C. torquatus*) have been shown to be sensitive to logging, and there has been a rapid decline in population density in logged areas (Waltert et al., 2002). Logging also opens up access to areas previously hidden from bushmeat hunters (Struhsaker, 1997). In addition, transport of hunted meat by lorry drivers into urban centers is made easier with logging roads in place, and primates, especially larger species, fetch a handsome price in urban markets (Cronin et al., 2015). The red-capped mangabeys suffer from hunting pressure, in part because they are highly vocal and easy to locate and, by being semiterrestrial, they are often caught in snares (Maisels et al., 2007).

Species with low population density or restricted distribution are under more severe threat due to habitat loss, hunting, or trapping. The small island of Bioko, Equatorial Guinea, is home to 11 different species of primates, including the endemic Bioko drill (*Mandrillus leucophaeus poensis*). Most of these species are under threat due to habitat loss, yet hunting pressure remains strong despite much effort by organizations such as Bioko Biodiversity Protection Program (Cronin et al., 2015).

In East Africa there are two highly endangered mangabey species (*C. galeritus* and *C. sanjei*), in addition to the endangered kipunji (*Rungwecebus kipunji*). They exist in small isolated populations and are threatened by human agricultural encroachment as well as hunting pressure. Even though reserves and national parks have been established to provide protection, they are not always effective.

The Tana River Primate National Reserve (TRPNR) was established to protect the endemic Tana mangabey (*C. galeritus*) and the Tana River red colobus (*Piliocolobus rufomitratus*) (Karere et al., 2004). Although problems remain with destruction of primate habitat, some areas are showing a slow but steady increase in forest cover (Wieczkowski & Kinnaird, 2008). It is estimated that about half of the mangabey population is located within the TRPNR (Karere et al., 2004); despite this protection, the Tana mangabeys are highly endangered, and it is estimated there are only 1,000–1,200 individuals left (Butynski et al., 2000). In the TRPNR many of the plant resources used by humans are also important primate foods. For example, *Phoenix reclinate*, a palm that is an important food resource for the Tana mangabeys, is used extensively by humans. People cut down the palm trees to use the trunk as building material and the root to make palm wine. However, a more serious threat to the Tana River primate habitat is the interference with water flow in the river by the construction of dams. Dams can have devastating effects on wildlife (Wieczkowski & Kinnaird, 2008) and are especially dangerous since most of the gallery forest vegetation (including the fruits eaten by primates) exists due to a constant water flow and periodic flooding.

Small population sizes that are widely distributed are factors that make both the Sanje mangabeys (*C. sanjei*) and the kipunji (*R. kipunji*) extremely vulnerable to human exploitation and influence. At present the total kipunji population is estimated at approximately 1,000 individuals, making it among the 25 most endangered primate species in the world (Mittermeier et al., 2009). The two kipunji populations are found within forest reserves which are legally protected. Since the 1990s, the Tanzanian government has stopped all logging in government reserves, but illegal logging still takes place and subsistence hunting is widespread (Nielsen, 2006). Some kipunji groups may be fairly well protected since they are far away from human settlements and there appears to be little evidence of human activity in their habitats (Topp-Jørgensen et al., 2009). Nonetheless, there are potential threats from bushfires started at reserve border areas to clear land for farming (Jones, 2006). The fact that the kipunji are shy and wary of humans may suggest that they were hunted in the recent past, which also would explain the low population density. In the Southern Highlands, where the largest population of kipunji can be found, there are severe threats to their habitat. The forest is highly degraded already due to logging and human encroachment (Jones et al., 2005). Even though it is not legal to hunt bushmeat in national parks or reserves (Davenport & Jones, 2008), there is evidence of poaching—snares are frequently used because they are more likely to go undetected and are cheap (Nielsen, 2006; Rovero et al., 2012).

Not all members of the subtribe Papionina are threatened with extinction. The highly successful baboons are, throughout most of their distribution, considered vermin and are heavily persecuted. In some areas, baboons are also trapped and exported for biomedical research (Oates et al., 2008a). Currently, populations of *Papio* are stable except for decreasing numbers of the chacma and Guinea species (IUCN, 2020). Geladas, despite inhabiting a restricted range, have a healthy population and are listed in the IUCN Red List as Least Concern except for the possible northern gelada subspecies, which is listed as Vulnerable (Gippoliti, 2010; IUCN, 2020). Within the Simien National Park, the geladas should in theory be totally protected; however, outside the park boundaries—with ever-increasing expansion of agriculture—there are numerous conflicts between humans and geladas (Yihune et al., 2008). Farmers use dogs to keep the monkeys away from their crops and when all else fails, they shoot them. This is a common scenario throughout Africa where most nonhuman primates have an uneasy time with humans and more often than not come into conflict over agricultural crops.

SUMMARY

Mangabeys, drills, mandrills, baboons, geladas, and kipunjis are relatively large Old World monkeys exhibiting complex, but varied, social systems and distinctive dietary strategies. These arboreal, terrestrial, and semiterrestrial quadrupeds inhabit a wide range of environments within the political boundaries of many African nations.

The *Lophocebus* and *Cercocebus* mangabeys have some shared features, such as adaptations to forests (most species inhabit tropical rainforests), living in mixed age and sex groups, and forming polyspecific associations with other primate species that may last a few hours or the whole day. Nevertheless, these shared features are due to evolutionary convergence; homoplasy is rampant in the papionins, a fact that has obscured their true phylogeny in the past (Lockwood & Fleagle, 1999; Collard & Wood, 2001). *Cercocebus* mangabeys are sister taxa to drills and mandrills, while *Lophocebus* mangabeys are sister taxa to *Papio*, the baboons. *Rungwecebus*, the kipunji, is a *Papio* + *Lophocebus* hybrid that—based on mitochondrial evidence—has led to a distinct and long-surviving taxon (Burrell et al., 2009).

Mandrillus spp. are striking examples of sexual dimorphism. Male mandrills are particularly colorful with massive maxillary and mandibular canines. Both mandrills and drills inhabit dense mature rainforest which has inhibited collection of scientific data on diet and group structure. Baboons (*Papio anubis, P. cynocephalus, P. hamadryas, P. kindae, P. papio,* and *P. ursinus*) may well be the most extensively studied genus of all nonhuman primates, save for possibly the chimpanzee. Baboons have been a logical choice as study animals, not just because they are large, terrestrial, and easy to habituate, but from an evolutionary point of view they have been considered good models based on analogy for early hominin adaptations, especially from an ecological point of view (Jolly, 2001). They appear to have experienced an adaptive radiation at the same time, and very likely in the same type of habitat, as did the early hominins. In the cool highlands of Ethiopia, the gelada (*Theropithecus gelada*) replaces both the anubis and hamadryas baboons, although at lower altitudes either species may be sympatric with the gelada (Dunbar & Dunbar, 1974). The dietary strategy of the grass-eating geladas is specialized and unique among papionins.

Primates throughout the world, including Africa, face an uncertain future (Estrada et al., 2017). The threats they are exposed to come almost exclusively from our own species. An expanding human population demands more space and resources, with the result that wildlife is pushed into increasingly smaller areas and usually less suitable or productive environments. Habitat is lost or degraded by logging, clearing for agriculture, collection of firewood, and charcoal production. In addition, many papionin species are hunted for meat or killed because they are perceived as pests. A solution is not always easy to find, but surely it is easier for us humans to modify our behavior than for nonhuman primates to adapt. If not, then what future does any primate species have?

REFERENCES CITED—CHAPTER 16

Abelló, M. T, ter Meulen, T., & Prins, E. F. (Eds.). (2018). *European Association of Zoos and Aquaria: Mangabey best practice guidelines*—Cercocebus lunulatus (p. 22). Parc Zoològic de Barcelona.

Abernethy, K. A., Coad, L., Taylor, G., Lee, M. E., & Maisel, F. (2013). Extent and ecological consequences of hunting in Central African rainforests in the twenty-first century. *Philosophical Transactions of the Royal Society* B, 368, 20120303.

Abernethy, K. A., White, L. J. T., & Wickings, E. J. (2002). Hordes of mandrills (*Mandrillus sphinx*): Extreme group size and seasonal male presence. *Journal of Zoology*, 258(1), 131-137.

Akoue, N. G., Mbading-Mbading, W., Willaume, E., Souza, A., Mbatchi, B., & Charpentier, M. J. E. (2017). Seasonal and individual predictors of diet in a free-ranging population of mandrills. *Ethology*, 123, 600-613.

Alba, D., Madurell-Malapiera, J., Delson, E., Vinuesa, V., Susanna, I., Patrocinio Espigares, M., Ros-Montoya, S., & Martínez-Navarro, B. (2016). First record of macaques from the Early Pleistocene of Incarcal (NE Iberian Peninsula). *Journal of Human Evolution*, 96, 139-144.

Alberts, S. C., & Altmann, J. (1995). Balancing costs and opportunities: Dispersal in male baboons. *American Naturalist*, 145, 279-306.

Alberts, S. C., Buchan, J. C., & Altmann, J. (2006). Sexual selection in wild baboons: From mating opportunities to paternity success. *Animal Behaviour*, 72, 1177-1196.

Albiach-Serrano, A., Guillén-Salazar, F., & Call, J. (2007). Mangabeys (*Cercocebus torquatus lunulatus*) solve the reverse contingency task without a modified procedure. *Animal Cognition*, 10, 387-396.

Aldrich-Blake, F. P. G., Bunn, T. K., Dunbar, R. I. M, & Headley, P. M. (1971). Observations on baboons, *Papio anubis*, in an arid region of Ethiopia. *Folia Primatologica*, 15, 1-35.

Alea, V., Baldellou, M., Vea, J., & Perez, A. P. (1999). Cost-benefit analysis of allogrooming behaviour in *Cercocebus torquatus lunulatus*. *Behaviour*, 136(2), 243-256.

Altmann, J. (1980). *Baboon mothers and infants*. Harvard University Press.

Altmann, S., & Altmann, J. (1970). *Baboon ecology*. Chicago University Press.

Ankel-Simons, F. (2000). *Primate anatomy: An introduction* (2nd edition). Academic Press.

Apetrei, C., Metzger, M. J., Richardson, D., Ling, B., Telfer, P. T., Reed, P., Robertson, D. L., & Marx, P. A. (2005). Detection and partial characterization of simian immunodeficiency virus SIVsm strains from bush meat samples from rural Sierra Leone. *Journal of Virology*, 79, 2631-2636.

Arlet, M. E., & Isbell, L. A. (2009). Variation in behavioral and hormonal responses of adult male gray-cheeked mangabeys (*Lophocebus albigena*) to crowned eagles (*Stephanoaetus coronatus*) in Kibale National Park. *Behavioural Ecology and Sociobiology*, 63, 491-499.

Arlet, M. E., Isbell, L. A., Kaasik, A., Molleman, F., Chancellor, R. L., Chapman, C. A., Mand, R., & Carey, J. R. (2015). Determinants of reproductive performance among female gray-cheeked mangabeys (*Lophocebus albigena*) in Kibale National Park, Uganda. *International Journal of Primatology*, 36, 55-73.

Arlet, M. E., Isbell, L. A., Molleman, F., Kaasik, A., Chancellor, R. L., Chapman, C. A., Mänd, R., & Carey, J. R. (2014). Maternal investment and infant survival in gray-cheeked mangabeys (*Lophocebus albigena*). *International Journal of Primatology*, 35, 476-490.

Arlet, M. E., Molleman, F., & Chapman, C. A. (2008). Mating tactics in male grey-cheeked mangabeys (*Lophocebus albigena*). *Ethology*, 114, 851-862.

Astaras, C., Krause, S., Mattner, L., Rehse, C., & Waltert, M. (2011). Associations between the drill (*Mandrillus leucophaeus*) and sympatric monkeys in Korup National

Park, Cameroon. *American Journal of Primatology, 71,* 1-8.

Astaras, C., Mühlenberg, M., & Waltert, M. (2008). Note on drill (*Mandrillus leucophaeus*) ecology and conservation status in Korup National Park, southwestern Cameroon. *American Journal of Primatology, 70,* 306-310.

Barrett, L., Henzi, S. P., & Lycett, J. E. (2006). Whose life is it anyway? Maternal investment, developmental trajectories, and life history strategies in baboons. In L. Swedell, & S. R. Leigh (Eds.), *Reproduction and fitness in baboons: Behavioural, ecological, and life history perspectives* (pp. 199-224). Springer.

Baya, L., & Storch, I. (2010). Status of diurnal primate populations at the former settlement of a displaced village in Cameroon. *American Journal of Primatology, 72,* 645-652.

Beehner, J. C., & Bergman, T. J. (2008). Infant mortality following male takeovers in wild geladas. *American Journal of Primatology, 70,* 1152-1159.

Belay, G., & Shotake, T. (1998). Blood protein variation of a new population of gelada baboons (*Theropithecus gelada*), in the southern Rift Valley, Arsi Region, Ethiopia. *Primates, 39*(2), 183-198.

Bentley-Condit, V. K. (2009). Food choices and habitat use by the Tana River yellow baboons (*Papio cynocephalus*): A preliminary report on five years of data. *American Journal of Primatology, 71,* 432-436.

Bergman, T., Ho, L., & Beehner, J. (2009). Chest color and social status in male geladas. *International Journal of Primatology, 30,* 791-806.

Biquand, S., Biquand-Guyot, V., Boug, A., & Gautier, J.-P. (1992). Group composition in wild and commensal hamadryas baboons: A comparative study in Saudi Arabia. *International Journal of Primatology, 13*(5), 533-543.

Boesch, C., & Boesch-Achermann, H. (2000). *The chimpanzees of the Taï Forest: Behavioural ecology and evolution.* Oxford University Press.

Bouchet, H., Pellier, A.-S., Blois-Heulin, C., & Lemanson, A. (2010). Sex differences in the vocal repertoire of adult red-capped mangabeys (*Cercocebus torquatus*): A multi-level acoustic analysis. *American Journal of Primatology, 72,* 360-375.

Bracebridge, C. E., Davenport, T. R. B., & Marsden, S. J., (2011). Can we extend the area of occupancy of the kipunji, a critically endangered African primate? *Animal Conservation, 14,* 687-696.

Bracebridge, C. E., Davenport, T. R. B., Mbofu, V. F., & Marsden, S. J., (2013). Is there a role for human-dominated landscapes in the long-term conservation management of the critically endangered kipunji (*Rungwecebus kipunji*). *International Journal of Primatology, 34,* 1122-1136.

Brain, C. (1990). Aspects of drinking by baboons (*Papio ursinus*) in a desert environment. In M. K. Sealy (Ed.), *Namib ecology: 25 years of Namib research* (pp. 169-172). Transvaal Museum.

Brain, C., & Mitchell, D. (1999). Body temperature changes in free-ranging baboons (*Papio hamadryas ursinus*) in the Namib Desert, Namibia. *International Journal of Primatology, 20*(4), 585-598.

Bret, C., Sueur, C., Ngoubangoye, B., Verrier, D., Deneubourgh, J.-L., & Petit, O. (2013). Social structure of a semi-free ranging group of mandrills (*Mandrillus sphinx*): A social network analysis. *PLoS ONE, 8*(12), e83015.

Brockmeyer, T., Kappeler, P. M., Willaume, E., Benoit, L., Mboumba, S., & Charpentier, M. J. E. (2015). Social organization and space use of a wild mandrill (*Mandrillus sphinx*) group. *American Journal of Primatology, 77,* 1036-1048.

Bronikowski, A. M., & Altmann, J. (1996). Foraging in a variable environment: Weather patterns and the behavioral ecology of baboons. *Behavioral Ecology and Sociobiology, 39,* 11-25.

Brown, M. (2014). Patch occupation time predicts responses of grey-cheeked mangabeys (*Lophocebus albigena*) to real and simulated neighboring groups. *International Journal of Primatology, 35,* 491-508.

Brown, M., & Waser, P. M. (2018). Group movements in response to competitors' calls indicate conflicts of interest between male and female grey-cheeked mangabeys. *American Journal of Primatology, 80*(11), e22918.

Brugiere, D., Gautier, J.-P., Moungazi, A., & Gautier-Hion, A. (2002). Primate diet and biomass in relation to vegetation composition and fruiting phenology in a rain forest in Gabon. *International Journal of Primatology, 3*(5), 999-1024.

Brugiere, D., Sakom, D., & Gautier-Hion, A. (2005). The conservation significance of the proposed Mbaéré-Bodingué National Park, Central African Republic, with special emphasis on its primate community. *Biodiversity and Conservation, 14,* 505-522.

Buchan, J. C., Alberts, S. C., Silk, J. B., & Altmann, J. (2003). True paternal care in a multi-male primate society. *Nature, 425,* 179-181.

Burrell, A. S., Jolly, C. J., Tosi, A. J., & Disotell, T. R. (2009). Mitochondrial evidence for the hybrid origins of the kipunji, *Rungwecebus kipunji* (Primates: Papionini). *Molecular Phylogenetics and Evolution, 51,* 340-348.

Busse, C. D., & Gordon, T. P. (1983). Attacks on neonate by a male mangabey (*Cercocebus atys*). *American Journal of Primatology, 5,* 345-356.

Butynski, T. M. (1990). Comparative ecology of blue monkeys (*Cercopithecus mitis*) in high- and low-density subpopulations. *Ecological Monographs, 60,* 1-26.

Butynski, T. M., Mbora, D. M., Kirathe, J. N., & Wieszcowski, J. (2000). *Group sizes and composition of the Tana River red colobus and Tana River crested mangabey.* Report for KWS, NMK and GEF.

Chancellor, R. L., & Isbell, L. A. (2009). Food site residence time and female competitive relationships in wild gray-cheeked mangabeys (*Lophocebus albigena*). *Behavioral Ecology and Sociobiology, 63,* 1447-1458.

Chapman, C. A., & Chapman, L. J. (2000). Interdemic variation in mixed-species association patterns: Common diurnal primates of Kibale National Park, Uganda. *Behavioral Ecology and Sociobiology*, 47, 129-139.

Chapman, C. A., Struhsaker, T. T., Skorupa, J. P., Snaith, T. V., & Rothman, J. M. (2010). Understanding long-term primate community dynamics: Implications of forest change. *Ecological Applications*, 20(1), 179-191.

Charpentier, M., Peignot, P., Hossaert-McKey, M., & Wickings, E. J. (2007). Kin discrimination in juvenile mandrills, *Mandrillus sphinx. Animal Behaviour*, 73, 37-45.

Charpentier, M. J. E., van Horn, R. C., Altmann, J., & Alberts, S. C. (2008). Paternal effects on offspring fitness in a multimale primate society. *Proceedings of the National Academy of Science*, 105(6), 1988-1992.

Cheney, D. L., & Seyfarth, R. M. (1997). Reconciliatory grunts by dominant female baboons influence victims' behavior. *Animal Behaviour*, 54, 409-418.

Cheney, D. L., & Seyfarth, R. M. (2007). *Baboon metaphysics: The evolution of a social mind*. University of Chicago Press.

Cheney, D. L., Seyfarth, R. M., Fischer, J., Beehner, J., Bergman, T., Johnson, S. E., Kitchen, D. M., Palombit, R. A., Rendall, D., & Silk. J. B. (2004). Factors affecting reproduction and mortality among baboons in the Okavango Delta, Botswana. *International Journal of Primatology*, 25, 401-428.

Collard, M., & Wood, B. (2001). Homoplasy and the early hominid masticatory system: Inferences from analyses of extant hominoids and papionins. *Journal of Human Evolution*, 41, 167-194.

Colmenares, F., Hofer, H., & East, M. L. (2000). Greeting ceremonies in baboons and hyenas. In F. Aureli, & F. B. M. de Waal (Eds.), *Natural conflict resolution* (pp. 94-96). University of California Press.

Cowlishaw, G. (1999). Ecological and social determinants of spacing behaviour in desert baboon groups. *Behavioral Ecology and Sociobiology*, 45, 67-77.

Cronin, D. T., Sesink Clee, P. R., Mitchell, M. W., Mene, D. B., Fernández, D., Riaco, C., Mene, M. F., Echube, J. M. E., Hearn, G. W., & Gonder, M. K. (2017). Conservation strategies for understanding and combating the primate bushmeat trade on Bioko Island, Equatorial Guinea. *American Journal of Primatology*, 79, e22663.

Cronin, D. T., Woloszynek, S., Morra, W. A., Honarvar, S., Linder, J. M., Gonder, M. K., O'Connor, M. P., & Hearn, G. W. (2015). Long-term urban market dynamics reveal increased bushmeat carcass volume despite economic growth and protective environmental legislation on Bioko Island, Equatorial Guinea. *PLoS ONE*, 10(8), e0137470.

Dal Pesco, F., & Fischer, J. (2018). Greetings in male Guinea baboons and the function of rituals in complex social groups. *Journal of Human Evolution*, 125, 87-98.

Davenport, T. R. B., de Luca, D. W., Bracebridge, C. E., Machaga, S. J., Mpunga, N. E., Kibure, O., & Abeid, Y. S. (2010). Diet and feeding patterns in the kipunji (*Rungwecebus kipunji*) in Tanzania's southern highlands: A first analysis. *Primates*, 51, 213-220.

Davenport, T. R. B., & Jones, T. (2008). *Rungwecebus kipunji. The International Union for the Conservation of Nature Red List of Threatened Species*. Version 2014.3. www.iucnredlist.org.

Davenport, T. R. B., Stanley, W. T., Sargis, E. J., de Luca, D. W., Mpunga, N. E., Machaga, S. J., & Olson, L. E. (2006). A new genus of African monkey, *Rungwecebus*: Morphology, ecology, and molecular phylogenetics. *Science*, 312, 1378-1381.

Davidge, C. (1978). Ecology of baboons (*Papio ursinus*) at Cape Point. *Zoologica Africana*, 13(2), 329-350.

De Luca, D. W., Picton Phillipps, G., Machaga, S. J., & Davenport, T. R. B. (2009). Home range, core areas and movement in the "critically endangered" kipunji (*Rungwecebus kipunji*) in southwest Tanzania. *African Journal of Ecology*, 48, 895-904.

Dempsey, A., Gonedelé Bi, S., Matsuda Goodwin, R., & Koffi, A. (2020). *Cercocebus lunulatus. The International Union for the Conservation of Nature Red List of Threatened Species* 2020. https://dx.doi.org/10.2305/IUCN.UK.2020-2.RLTS.T4206A92247733.en.

Devreese, L., Huynen, M.-C., Stevens, J. M. G., & Todd, A. (2013). Group size of a permanent large group of agile mangabeys (*Cercocebus agilis*) at Baï Hokou, Central African Republic. *Folia Primatologica*, 84, 67-73.

Dolado, R., Cooke, C., & Beltran, F. S. (2016). How many for lunch today? Seasonal fission-fusion dynamics as a feeding strategy in wild red-capped mangabeys (*Cercocebus torquatus*). *Folia Primatologica*, 87, 197-212.

Dunbar, R. I. M. (1984). *Reproductive decisions: An economic analysis of gelada baboon social strategies*. Princeton University Press.

Dunbar, R. I. M. (1992). Time: A hidden constraint on the behavioural ecology of baboons. *Behavioural Ecology and Sociobiology*, 31, 35-49.

Dunbar, R. I. M. (1993). Social organization of the gelada. In N. G. Jablonski (Ed.), *Theropithecus: The rise and fall of a genus*. Cambridge University Press.

Dunbar, R. I. M., & Bose, U. (1991). Adaptation to grass-eating in gelada baboons. *Primates*, 32(1), 1-7.

Dunbar, R. I. M., & Dunbar, P. (1974). Ecological relations and niche separation between sympatric terrestrial primates in Ethiopia. *Folia Primatologica*, 21, 36-60.

Dunbar, R. I. M., & Dunbar, P. (1975). *Social dynamics of gelada baboons*. Karger.

Dunbar, R. I. M., Hannah-Stewart, L., & Dunbar, P. (2002). Forage quality and the costs of lactation for female gelada baboons. *Animal Behaviour*, 64, 801-805.

Ehardt, C. L. (1988a). Absence of strongly kin-preferential behaviors by adult female sooty mangabeys (*Cercocebus atys*). *American Journal of Physical Anthropology*, 76, 233-243.

Ehardt, C. L. (1988b). Affiliative behavior of adult female sooty mangabeys (*Cercocebus atys*). *American Journal of Primatology*, 15, 115-127.

Ehardt, C. L., & Butynski, T. M. (2006a). The recently described highland mangabey, *Lophocebus kipunji* (Cercopitheciodea,

Cercopithecinae): Current knowledge and conservation assessment. *Primate Conservation*, 21, 81-87.

Ehardt, C. L., & Butynski, T. M. (2006b). Sanje River mangabey. In R. A. Mittermeier, A. Valladares-Padua, A. B. Rylands, A. A. Eudey, T. M. Bitynski, J. U. Ganzhorn, R. Kormos, J. M. Aguiar, & S. Walker (Eds.), *Primates in peril: The world's 25 most endangered primates 2004-2006. Primate Conservation*, 20, 1-28.

Ehardt, C. L., Butynski, T. M., & Struhsaker, T. (2008). *Cercocebus sanjei. The International Union for the Conservation of Nature Red List of Threatened Species.* Version 2014.3. www.iucnredlist.org.

Ehardt, C. L., Jones, T. P., & Butynski, T. M. (2005). Protective status, ecology and strategies for improving conservation of *Cercocebus sanjei* in the Udzungwa Mountains, Tanzania. *International Journal of Primatology*, 26(3), 557-583.

Eley, R. M., Strum, S. C., Muchemi, G., & Reid, G. D. F. (1989). Nutrition, body condition, activity patterns and parasitism of free-ranging baboons (*Papio anubis*) in Kenya. *American Journal of Primatology*, 18, 209-219.

Eppley, T. M., Hickey, J. R., & Nibbelink, N. P. (2010). Observation of albinistic and leucistic black mangabeys (*Lophocebus aterrimus*) within the Lomako-Yokokala Faunal Reserve, Democratic Republic of Congo. *African Primates*, 7(1), 50-54.

Estrada, A., Garber, P. A., Rylands, A. B., Roos, C., Fernandez-Duque, E., Di Fiore, A., Nekaris, K. A. I., Nijman, V., Heymann, E. W., Lambert, J. E., Rovero, F., Barelli, C., Setchell, J. M., Gillespie, T. R., Mittermeier, R. A., Arregoitia, L. V., de Guinea, M., Gouveia, S., Dobrovolski, R.,... & Li, B. (2017). Impending extinction crisis of the world's primates: Why primates matter. *Science Advances*, 3(1), e1600946.

Fa, J. E., Ryan, S., & Bell, D. J. (2005). Hunting vulnerability, ecological characteristics and harvest rates of bushmeat species in Afrotropical forests. *Biological Conservation*, 121, 167-176.

Fàbregas, M., & Guillén-Salazar, F. (2007). Social compatibility in a newly formed all-male group of white crowned mangabeys (*Cercocebus atys lunulatus*). *Zoo Biology*, 26(1), 63-69.

Fashing, P. J., & Nguyen, N. (2009). Gelada feeding ecology in a tall grass ecosystem: Influence of body size on diet. *American Journal of Primatology*, 71, 60.

Fashing, P. J., Nguyen, N., Fashing, N. J. (2010). Behavior of geladas and other endemic wildlife during a desert locust outbreak at Guassa, Ethiopia: Ecological and conservation implications. *Primates*, 51(3), 193-197.

Feistner, A. T. C. (1991). Scent marking in mandrills, *Mandrillus sphinx. Folia Primatologica*, 57, 42-47.

Fernandez, D. (2011). Life history traits of wild Sanje mangabeys (*Cercocebus sanjei*) in the Udzungwa Mountains National Park, Tanzania. *American Journal of Primatology*, 73(S1), 105.

Fernandez, D., Doran-Sheehy, D., Borries, C., & Brown, J. L. (2014). Reproductive characteristics of wild Sanje mangabeys (*Cercocebus sanjei*). *American Journal of Primatology*, 76, 1163-1174.

Fischer, J., Kopp, G. H., Dal Pesco, F., Goffe, A., Hammerschmidt, K., Kalbitzer, U., Klapproth, M., Maciej, P., Ndao, I., Patzelt, A., & Zinner, D. (2017). Charting the neglected west: The social system of Guinea baboons. *American Journal of Physical Anthropology*, 162, 15-31.

Fleagle, J. G. (2013). *Primate adaptation and evolution* (3rd Ed.). Academic Press.

Fleagle, J. G., & McGraw, W. S. (1999). Skeletal and dental morphology supports diphyletic origin of baboons and mandrills. *Proceedings of the National Academy of Science*, 96, 1157-1161.

Fleagle, J. G., & McGraw, W. S. (2002). Skeletal and dental morphology of African papionins: Unmasking a cryptic clade. *Journal of Human Evolution*, 42, 267-292.

Forthman Quick, D. L. (1986). Activity budgets and the consumption of human food in two troops of baboons, *Papio anubis*, at Gilgil, Kenya. In J. G. Else, & P. C. Lee (Eds.), *Primate ecology and conservation* (pp. 221-228). Cambridge University Press.

Galat, G., & Galat-Luong, A. (2006). Hope for the survival of the critically endangered white-naped mangabey *Cercocebus atys lunulatus*: A new primate species for Burkina Faso. *Oryx*, 40(3), 355-357.

Galat-Luong, A., Galat, G., & Hagell, S. (2006). The social and ecological flexibility of Guinea baboons: Implications for Guinea baboon social organization and male strategies. In L. Swedell, & S. R. Leigh (Eds.), *Reproduction and fitness in baboons: Behavioral, ecological, and life history perspectives* (pp.105-121). Springer.

Gartlan, J. S. (1970). Preliminary notes on the ecology and behaviour of the drill *Mandrillus leucophaeus* Ritgen 1824. In J. R. Napier, & P. H. Napier (Eds.), *Old World monkeys* (pp. 445-480). Academic Press.

Gartlan, J. S., & Struhsaker, T. T. (1972). Polyspecific associations and niche separation of rain-forest anthropoids in Cameroon, West Africa. *Journal of Zoology*, 168, 221-266.

Gautier-Hion, A., Colyn, M., & Gautier, J.-P. (1999). *Histoire naturelle des primates d'Afrique Central.* ECOFAC.

Gilbert, C. C. (2007). Craniomandibular morphology supporting the diphyletic origins of mangabeys and a new genus of the *Cercocebus/Mandrillus* clade, *Procercocebus. Journal of Human Evolution*, 53, 69-102.

Gilbert, C. C. (2013). Cladistic analysis of extant and fossil African papionins using craniodental data. *Journal of Human Evolution*, 64, 399-433.

Gippoliti, S. (2008). *Theropithecus gelada* ssp. *The International Union for the Conservation of Nature Red List of Threatened Species.* www.iucnredlist.org.

Gippoliti, S. (2010). *Theropithecus gelada* distribution and variations related to taxonomy: History, challenges and implications for conservation. *Primates*, 51, 291-297.

Goffe, A., Zinner, D., & Fischer, J. (2016). Sex and friendship in a multilevel society: Behavioural patterns and associations between female and male Guinea baboons. *Behavioral Ecology and Sociobiology, 70*, 323-336.

Gordon, T. P., Gust, D. A., Busse, C. D., & Wilson, M. E. (1991). Hormones and sexual behavior associated with postconception perineal swelling in the sooty mangabey (*Cercocebus torquatus atys*). *International Journal of Primatology, 12*(6), 585-597.

Groves, C. (2001). *Primate taxonomy.* Smithsonian Institution Press.

Grubb, P., Butynski, T. B., Oates, J. F., Bearder, S. K., Disotell, T. R., Groves, C. P., & Struhsaker, T. T. (2003). Assessment of the diversity of African primates. *International Journal of Primatology, 24*(6), 1301-1357.

Gust, D. A. (1994). A brief report on the social behavior of the crested mangabey (*Cercocebus galeritus galeritus*) with a comparison to the sooty mangabey (*C. torquatus atys*). *Primates, 35*(3), 375-383.

Gust, D. A., Busse, C. D., & Gordon, T. P. (1990). Reproductive parameters in the sooty mangabey (*Cercocebus torquatus atys*). *American Journal of Primatology, 22*, 241-250.

Gust, D. A., & Gordon, T. P. (1994). The absence of matrilineally based dominance system in sooty mangabeys, *Cercocebus torquatus atys. Animal Behaviour, 47*, 589-594.

Hamilton, W. J., III. (1985). Demographic consequences of a food and water shortage to desert chacma baboons, *Papio ursinus. International Journal of Primatology, 6*, 451-462.

Hamilton, W. J., III, Buskirk, R. E., & Buskirk, W. L. (1976). Defense of space and resources by chacma (*Papio ursinus*) baboon troops in African desert and swamp. *Ecology, 52*, 1264-1271.

Hapke, A., Zinner, D., & Zischler, H. (2001). Mitochondrial DNA variation in Eritrean hamadryas baboons (*Papio hamadryas hamadryas*): Life history influences population genetic structure. *Behavioral Ecology and Sociobiology, 50*, 483-492.

Harding, R. S. O. (1976). Ranging patterns of a troop of baboons (*Papio anubis*) in Kenya. *American Journal of Primatology, 25*, 143-185.

Harding, R. S. O. (1984). Primates of the Killini area, northwest Sierra Leone. *Folia Primatologica, 42*, 96-114.

Harrison, T., & Harris, E. E. (1996). Plio-Pleistocene cercopithecids from Kanam, western Kenya. *Journal of Human Evolution, 30*, 539-561.

Hart, D., & Sussman, R. W. (2005). *Man the hunted: Primates, predators and human evolution.* Westview Press.

Hart, J., Butynski, T. M., & Kingdon, J. (2008a). *Cercocebus agilis. The International Union for the Conservation of Nature Red List of Threatened Species.* www.iucnredlist.org.

Hart, J., Groves, C. P., & Erhardt, C. (2008b). *Lophocebus aterrimus. The International Union for the Conservation of Nature Red List of Threatened Species.* www.iucnredlist.org.

Henzi, S. P., Byrne, R. W., & Whiten, A. (1992). Patterns of movement by baboons in the Drakensberg Mountains: Primary responses to the environment. *International Journal of Pimatology, 13*(6), 601-628.

Henzi, S. P., & Lycett, J. E. (1995). Population structure, dynamics and demography of mountain baboons. *American Journal of Primatology, 35*, 155-163.

Higham, J. P., Warren, Y., Adanu, J., Umaru, B. N., Maclaron, A. M., Sommer, V., & Ross, C. (2009). Living on the edge: Life history of olive baboons at Gashaka-Gumti National Park, Nigeria. *American Journal of Primatology, 71*, 293-304.

Hill, C. M., & Wallace, G. E. (2012). Crop protection and conflict mitigation: Reducing the costs of living alongside non-human primates. *Biodiversity and Conservation, 21*, 2569-2587.

Hill, C. M., & Webber, A. D. (2010). Perceptions of nonhuman primates in human-wildlife conflict scenarios. *American Journal of Primatology, 72*, 919-924.

Hill, R. A. (2006). Thermal constraint on activity scheduling and habitat choice in baboons. *American Journal of Physical Anthropology, 129*, 242-249.

Hill, R. A., Barrett, L., Gaynor, D., Weingrill, T., Dixon, P., Payne. H., & Henzi, S. P. (2003). Day length, latitude and behavioural (in)flexibility in baboons (*Papio cynocephalus ursinus*). *Behavioural Ecology and Sociobiology, 53*, 278-286.

Hill, W. C. O. (1970). *Primates. Comparative anatomy and taxonomy. VIII. Cynopithecinae.* Edinburgh University Press.

Hoffman, M., & Hilton-Taylor, C. (2008). *Papio ursinus. The International Union for the Conservation of Nature Red List of Threatened Species.* www.iucnredlist.org.

Homewood, K. M. (1998). Monkey on a riverbank. In R. L. Ciochon, & R. A. Nisbett (Eds.), *The primate anthology. Essays on primate behavior, ecology, and conservation from natural history* (pp. 93-98). Prentice Hall.

Homewood, K. M., & Rodgers, W. A. (1981). A previously undescribed mangabey from southern Tanzania. *International Journal of Primatology, 2*(1), 47-55.

Hongo, S. (2014). New evidence from observations of progressions of mandrills (*Mandrillus sphinx*): A multilevel or non-nested society? *Primates, 55*, 473-481.

Hongo, S., Nakashima, Y., Akomo-Okoue, E. F., & Mindonga-Nguelet, F. L. (2018). Seasonal changes in diet and habitat use in wild mandrills (*Mandrillus sphinx*). *International Journal of Primatology, 39*, 27-48.

Horn, A. D. (1987). The socioecology of the black mangabey (*Cercocebus aterrimus*) near Lake Tumba, Zaire. *American Journal of Primatology, 12*, 165-180.

Hoshino, J. (1985). Feeding ecology of mandrills (*Mandrillus sphinx*) in Campo Animal reserve, Cameroon. *Primates, 26*(3), 248-273.

Inogwabini, B.-I., & Thompson, J. A. M. (2013). The golden-bellied mangabey *Cercocebus chrysogaster* (Primates: Cercopithecidae): Distribution and conservation status. *Journal of Threatened Taxa, 5*(7), 4069-4075.

Isbell, L. A, Bidner, L. R., van Cleave, E. K., Matsumoto-Oda, A., & Crofoot, M. C. (2018). GPS-identified vulnerabilities of savannah-woodland primates to leopard predation and their implications for early hominins. *Journal of Human Evolution, 118*, 1-13.

IUCN. (2020). *International Union for the Conservation of Nature Red List of Threatened Species.* https://www.iucnredlist.org/.

Iwamoto, T. (1979). Feeding ecology. In M. Kawai (Ed.), *Ecological and sociological studies of gelada baboons* (pp. 280-330). Karger.

Iwamoto, T. (1993a). The ecology of *Theropithecus gelada.* In N. G. Jablonski (Ed.), *Theropithecus. The rise and fall of a primate genus* (pp. 441-452), Cambridge University Press.

Iwamoto, T. (1993b). Food digestion and energetic conditions in *Theropithecus* gelada. In N. G. Jablonski (Ed.), *Theropithecus: The rise and fall of a primate genus* (pp. 453-463). Cambridge University Press.

Iwamoto, T., & Dunbar, R. I. M. (1983). Thermoregulation, habitat quality and the behavioural ecology of gelada baboons. *Journal of Animal Ecology,* 52, 357-366.

Iwamoto, T., Mori, A., Kawai, M., & Bekele, A. (1996). Anti-predator behavior in gelada baboons. *Primates,* 37(4), 389-397.

Jablonski, N. G. (1994). Convergent evolution in the dentitions of grazing macropodine marsupials and the grass-eating primate *Theropithecus gelada. Journal of the Royal Society of Western Australia,* 77, 37-43.

Jablonski, N. G. (2002). Fossil Old World monkeys: The late Neogene radiation. In W. C. Hartwig (Ed.), *The primate fossil record* (pp. 255-299). Cambridge University Press.

Janmaat, K. R. L., Byrne, R. W., & Zuberbühler, K. (2006). Evidence of spatial memory of fruiting states of rainforest trees in wild mangabeys. *Animal Behaviour,* 72, 797-807.

Janmaat, K. R. L., Olupot, W., Chancellor, R. L., Atlet, M. E., & Waser, P. M. (2009). Long-term site fidelity and individual home range shifts in *Lophocebus albigena. International Journal of Primatology,* 30, 443-466.

Jolly, C. J. (1993). Species, subspecies, and baboon systematics. In W. H. Kimble, & L. B. Martin (Eds.), *Species, species concepts, and primate evolution* (pp. 67-107). Plenum Press.

Jolly, C. J. (2001). A proper study for mankind: Analogies from the papionin monkeys and their implications for human evolution. *Yearbook of Physical Anthropology,* 44, 177-204.

Jolly, C. J. (2003). Cranial anatomy and baboon diversity. *Anatomical Record Part A,* 275A, 1043-1047.

Jolly, C. J. (2007). Baboons, mandrills, and mangabeys. Afro-papionin socioecology in a phylogenetic perspective. In C. J. Campbell, A. Fuentes, K. C. Mackinnon, M. Panger, & S. K. Bearder (Eds.), *Primates in perspective* (pp. 240-251). Oxford University Press.

Jolly, C. J. (2020). Philopatry at the frontier: A demographically driven scenario for the evolution of multilevel societies in baboons (*Papio*). *Journal of Human Evolution,* 146, 102819. https://doi.org/10.1016/j.jhevol.2020.102819.

Jolly, C. J., Burrell, A. S., Phillips-Conroy, J. E., Bergey, C., & Rogers, J. (2011). Kinda baboons (*Papio kindae*) and grayfoot chacma baboons (*P. ursinus griseipes*) hybridize in the Kafue River Valley, Zambia. *American Journal of Primatology,* 73, 291-303.

Jolly, C. J., & Phillips-Conroy, J. E. (2003). Testicular size, mating systems, and maturation schedules in wild anubis and hamadryas baboons. *International Journal of Primatology,* 24, 125-142.

Jones, T. (2006). *Kipunji in Ndundulu Forest, Tanzania. Distribution, abundance and conservation status.* Unpublished report for the Critical Ecosystem Partnership Fund, Fauna and Flora International and Wildlife Conservation Society.

Jones, T., Ehardt, C. L., Butynski, T. M., Davenport, T. R. B., Mpunga, N. E., Machaga, S. J., & de Luca, D. W. (2005). The highland mangabey *Lophocebus kipunji*: A new species of African monkey. *Science,* 308, 1161-1164.

Jones, T., Laurent, S., Mselewa, F., & Mtui, A. (2006). Sanje mangabey *Cercocebus sanjei* kills an African crowned eagle *Stephanoaetus coronatus. Folia Primatologica,* 77(5), 359-363.

Kamilar, J. M. (2006). Geographic variation in savannah baboon (*Papio*) ecology and its taxonomic and evolutionary implications. In S. M. Lehman, & J. G. Fleagle (Eds.), *Primate biogeography* (pp. 169-200). Springer.

Karere, G. M., Oguge, N. O., Kirathe, J., Muoria, P. K., Moinde, N. N., & Suleman, M. A. (2004). Population size and distribution of primates in the lower Tana River forests, Kenya. *International Journal of Primatology,* 25(2), 351-365.

Kawai, M., Dunbar, R., Ohsawa, H., & Mori, U. (1983). Social organization of gelada baboons: Social units and definitions. *Primates,* 24(1), 13-24.

Kawai, M., & Iwamoto, T. (1979). Nomadism and activities. In M. Kawai (Ed.), *Ecological and sociological studies of gelada baboons* (pp. 251-278). Karger.

Kingdon, J. (1997). *The Kingdon field guide to African mammals.* Academic Press.

Kinnaird, M. F. (1990). Pregnancy, gestation and parturition in free-ranging Tana River crested mangabeys (*Cercocebus galeritus galeritus*). *American Journal of Primatology,* 22, 285-289.

Kinnaird, M. F. (1992a). Variable resource defense by the Tana River crested mangabey. *Behavioral Ecology and Sociobiology,* 31, 115-122.

Kinnaird, M. F. (1992b). Phenology of flowering and fruiting of an East African riverine forest ecosystem. *Biotropica,* 24(2a), 187-194.

Kinnaird, M. F. (1992c). Competition for forest palm: Use of *Phoenix reclinata* by humans and nonhuman primates. *Conservation Biology,* 6(1), 101-107.

Kitchen, D. M., Cheney, D. L., & Seyfarth, R. M. (2005). Male chacma baboons (*Papio hamadryas ursinus*) discriminate loud call contests between rivals of different relative ranks. *Animal Cognition,* 8, 1-6.

Kleindorfer, S., & Wasser, S. K. (2004). Infant handling and mortality in yellow baboons (*Papio cynocephalus*): Evidence for female reproductive competition? *Behavioural Ecology and Sociobiology,* 56, 328-337.

Knapp, E. J., Peace, N., & Bechtel, L. (2017). Poachers and poverty: Assessing objective and subjective measures of poverty among illegal hunters outside Ruaha National Park, Tanzania. *Conservation and Society, 15*(1), 24-32.

Knights, K. A., Cipolletta, C., Buren, L., Santochirico, M., Wearn, O., Todd, A. F., & Djimbele, O. (2008). The predatory behaviour of agile mangabeys (*Cercocebus agilis*) on blue duikers (*Cephalophus monticola*) and other vertebrates at Dzanga-Ndoki National Park, Central African Republic. *Abstracts: XXII Congress of the International Primatological Society*, 407.

Kopp, G., Fischer, J., Patzelt, A., Roos, C., & Zinner, D. (2015). Population genetic insights into social organization of Guinea baboons (*Papio papio*): Evidence for female-biased dispersal. *American Journal of Primatology, 77*, 878-889.

Kudo, H. (1987). The study of vocal communication of wild mandrills in Cameroon in relation to their social structure. *Primates, 28*(3), 289-308.

Kummer, H. (1968). *Social organization of hamadryas baboons: A field study.* University of Chicago Press.

Kummer, H. (1995). *In quest of the sacred baboon. A scientist's journey.* Princeton University Press.

Kunz, B. K., & Linsenmair, K. E. (2008). The disregarded West: Diet and behavioural ecology of olive baboons in the Ivory Coast. *Folia Primatologica, 79*, 31-51.

Lahm, S. A. (1986). Diet and habitat preference of *Mandrillus sphinx* in Gabon: Implications of foraging strategy. *American Journal of Primatology, 11*, 9-26.

Laidre, M. E., & Yorzinski, J. L. (2005). The silent bared-teeth face and the crest-raise of the mandrill (*Mandrillus sphinx*): A contextual analysis of signal function. *Ethology, 111*, 143-157.

Lambert, J. E. (2005). Competition, predation, and the evolutionary significance of the cercopithecine cheek pouch: The case of *Cercocebus* and *Lophocebus*. *American Journal of Physical Anthropology, 126*, 183-192.

Leigh, S., Setchell, J. M., & Buchanan, L. S. (2005). Ontogenetic basis of canine dimorphism in anthropoid primates. *American Journal of Physical Anthropology, 129*, 296-311.

le Roux, A., Beehner, J. C., & Bergman, T. J. (2011). Female philopatry and dominance patterns in wild geladas. *American Journal of Primatology, 73*, 422-430.

Lewis, M. C., & O'Riain, M. J. (2017). Foraging profile, activity budget and spatial ecology of exclusively natural-foraging chacma baboons (*Papio ursinus*) on the Cape Peninsula, South Africa. *International Journal of Primatology, 38*, 751–779.

Lilly, A. A., Mehlman, P. T., & Doran, D. (2002). Intestinal parasites in gorillas, chimpanzees, and humans in Mondika research site, Dzanga-Ndoki National Park, Central African Republic. *International Journal of Primatology, 23*(3), 555-573.

Linder, J. M., & Oates, J. F. (2011). Differential impact of bushmeat hunting on monkey species and implications for primate conservation in Korup National Park, Cameroon. *Biological Conservation, 144*, 738-745.

Lockwood, C. A., & Fleagle, J. G. (1999). The recognition and evaluation of homoplasy in primate and human evolution. *Yearbook of Physical Anthropology, 42*, 189-232.

Lovett, J. C., Marshall, A. R., & Jeff, C. (2006). Changes in tropical forest vegetation along an altitudinal gradient in the Udzungwa Mountains National Park, Tanzania. *African Journal of Ecology, 44*(4), 478-490.

Maisels, F., Makaya, Q. P., & Onononga, J.-R. (2007). Confirmation on the presence of the red-capped mangabey (*Cercocebus torquatus*) in Mayumba National Park, southern Gabon and Conkouati-Douli National Park, southern Republic of Congo. *Primate Conservation, 22*, 111-115.

Martinez, F. I., Capelli, C., Ferreira de Silva, M., Aldeias, V., Alemseged, Z., Archer, W., Bamford, M., Biro, D., Bobe, R., Braun, D. R., Habermann, J. M., Ludecke, T., Madiquida, H., Mathe, J., Negash, E., Paulo, L. M., Pinto, M., Stalmans, M., Tata, F., & Carvalho, S. (2019). A missing piece of the *Papio* puzzle: Gorongosa baboon phenostructure and intrageneric relationships. *Journal of Human Evolution, 130*, 1-20.

Marty, J. S, Higham, J. P., Gadsby, E. L., & Ross, C. (2009). Dominance, coloration, and social and sexual behavior in male drills *Mandrillus leucophaeus*. *International Journal of Primatology, 30*, 807-823.

Matthews, A., & Matthews, A. (2002). Distribution, population density, and status of sympatric cercopithecids in the Campo-Ma'an area, southwestern Cameroon. *Primates, 43*(3), 155-168.

McCabe, G. M., & Thompson, M. E. (2013). Reproductive seasonality in wild Sanje mangabeys (*Cercocebus sanjei*) in Tanzania: Relationship between the capital breeding strategy and infant survival. *Behaviour, 150*, 1399-1429.

McGraw, W. S. (1994). Census, habitat preferences and polyspecific association of six monkeys in the Lomako Forest, Zaire. *American Journal of Primatology, 34*, 295-307.

McGraw, W. S. (1998). Comparative locomotion and habitat use of six monkeys in the Taï Forest, Ivory Coast. *American Journal of Physical Anthropology, 105*, 493-510.

McGraw, W. S., & Bshary, R. (2002). Association of terrestrial mangabeys (*Cercocebus atys*) with arboreal monkeys: Experimental evidence for the effects of reduced ground predator pressure on habitat use. *International Journal of Primatology, 23*(2), 311-325.

McGraw, W. S., Cooke, C., & Shultz, S. (2006). Primate remains from African crowned eagle (*Stephanoaetus coronatus*) nests in Ivory Coast's Tai Forest: Implications for primate predation and early hominid taphonomy in South Africa. *American Journal of Physical Anthropology, 131*, 151-165.

McGraw, W. S., & Fleagle, J. G. (2006). Biogeography and evolution of the *Cercocebus-Mandrillus* clade: Evidence from the face. In S. M. Lehman, & J. G. Fleagle (Eds.), *Primate biogeography. Progress and prospects* (pp. 201-224). Springer.

McGraw, W. S., Pampush, J. D., & Daegling, D. J. (2012). Enamel thickness and durophagy in mangabeys revisited. *American Journal of Physical Anthropology, 147*, 326–333.

McGraw, W. S., Vick, A. E., & Daegling, D. J. (2011). Sex and age differences in the diet and ingestive behaviors of sooty mangabeys (*Cercocebus atys*) in the Tai Forest, Ivory Coast. *American Journal of Physical Anthropology, 144*, 140-153.

Mitani, J. C., Sanders, W. J., Lwanga, J. S., & Windfelder, T. L. (2001). Predatory behavior of crowned hawk-eagles (*Stephanoaetus coronatus*) in Kibale National Park, Uganda. *Behavioral Ecology and Sociobiology, 49*, 187-195.

Mitani, M. (1989). *Cercocebus torquatus*: Adaptive feeding and ranging behaviors related to seasonal fluctuations of food resources in the tropical rain forest of south-western Cameroon. *Primates, 30*(3), 307-323.

Mitani, M. (1991). Niche overlap and polyspecific associations among sympatric cercopithcids in the Campo Animal Reserve, southwestern Cameroon. *Primates, 32*(2), 137-151.

Mittermeier, R. A., Rylands, A. B., & Wilson, D. E. (Eds.). (2013). *Handbook of the mammals of the world, Vol. 3, Primates: Cercocebus lunulatus* (p. 951). Lynx Edicions.

Mittermeier, R. A., Wallis, J., Rylands, A. B., Ganzhorn, J. U., Oates, J. F., Williamson, E. A., Palacios, E., Heymann, E. W., Kierulff, C. M., Yongcheng, L., Suprinata, J., Roos, C., Walker, S., Cortés-Ortiz, L., & Schwitzer, C. (2009). Primates in peril: The world's 25 most endangered primates, 2008-2010. *Primate Conservation, 24*(1), 1-57.

Moinde-Fockler, N. N., Oguge, N. O., Karere, G. M., Otina, D., & Suleman, M. A. (2007). Human and natural impacts on forests along lower Tana River, Kenya: Implications towards conservation and management of endemic primate species and their habitat. *Biodiversity and Conservation, 16*, 1161-1173.

Morgan, B. J., & Lee, P. C. (2007). Forest elephant group composition, frugivory and coastal use in the Réserve de Faune du Petit Loango, Gabon. *African Journal of Ecology, 45*, 519-526.

Mori, A., & Belay, G. (1990). The distribution of baboon species and a new population of gelada baboons along the Wabi-Shebeli River, Ethiopia. *Primates, 31*, 495-508.

Mori, U. (1979). Unit formation and the emergence of a new leader. In M. Kawai (Ed.), *Ecological and sociological studies of gelada baboons* (pp. 155-181). Karger.

Muoria, P. K., Karere, G. M., Moinde, N. N., & Suleman, M. A. (2003). Primate census and habitat evaluation in the Tana Delta region, Kenya. *African Journal of Ecology, 41*, 157-163.

Mwamende, K. A. (2009). *Social organisation, ecology and reproduction in the Sanje mangabey* (Cercocebus sanjei) *in the Udzungwa Mountains National Park, Tanzania.* [Master's thesis, Victoria University of Wellington].

Nagel, U. (1973). A comparison of anubis baboons, hamadryas baboons, and their hybrids at a species border in Ethiopia. *Folia Primatologica, 19*, 104-165.

Napier, P. H. (1981). *Catalogue of primates in the British Museum (Natural History) and elsewhere in the British Isles. Part II: Family Cercopithecidae, subfamily Cercopithecinae.* Intercept Scientific.

Newman, T. K., Jolly, C. J., & Rogers, J. (2004). Mitochondrial phylogeny and systematics of baboons (*Papio*). *American Journal of Physical Anthropology, 124*, 17-27.

Nguyen, N., van Horn, R. C., Alberts, S. C., & Altmann, J. (2009). "Friendships" between new mothers and adult males: Adaptive benefits and determinants in wild baboons (*Papio cynocephalus*). *Behavioural Ecology and Sociobiology, 63*, 1331-1344.

Nielson, M. R. (2006). Importance, cause and effect of bushmeat hunting in the Udzungwa Mountains, Tanzania: Implications for community based wildlife management. *Biological Conservation, 128*, 509-516.

Noë, R., & Sluijter, A. A. (1990). Reproductive tactics in male savannah baboons. *Behaviour, 113*, 117-170.

Nolan, R., Welsh, A., Geary, M., Hartley, M., Dempsey, A., Cudjoe Mono, J., Osei, D., & Stanley, C. (2019). Camera traps confirm the presence of the white-naped mangabey *Cercocebus lunulatus* in Cape Three Points Forest Reserve, western Ghana. *Primate Conservation, 33*, 37-41.

Norton, G. W., Rhine, R. J., Wynn, G. M., & Wynn, R. D. (1987). Baboon diet: A five-year study of stability and variability in the plant feeding and habitat of the yellow baboons (*Papio cynocephalus*) of Mikumi National Park, Tanzania. *Folia Primatologica 48*(1-2), 78-120.

Nystrom, P. (1992). *Mating success of hamadryas, anubis and hybrid male baboons in a "mixed" social group in the Awash National Park.* [Doctoral dissertation, Washington University].

Oates, J. F. (1986). *Action plan for African primate conservation.* World Wildlife Fund.

Oates, J. F. (2006). *Primate conservation in the forests of western Ghana: Field survey results, 2005-2006. A report to the Wildlife Division, Forestry Commission, Ghana.* Unpublished report from Department of Anthropology, Hunter College. http://static1.1.sqspcdn.com/static/f/1200343/27604591/1498324782877/JFOsurveyReport_FINAL_16Sep2006.docx.pdf?token=v5S2iT2C4uV3Lq1ym9g0tQvjJj8%3D.

Oates, J. F., & Butynski, T. M. (2008a). *Mandrillus leucophaeus. The International Union for the Conservation of Nature Red List of Threatened Species.* www.iucnredlist.org.

Oates, J. F., & Butynski, T. M. (2008b). *Mandrillus sphinx. The International Union for the Conservation of Nature Red List of Threatened Species.* www.iucnredlist.org.

Oates, J. F., Gippoliti, S., & Groves, C. P. (2008a). *Papio papio. The International Union for the Conservation of Nature Red List of Threatened Species.* www.iucnredlist.org.

Oates, J. F., Gippoliti, S., & Groves, C. P. (2016). *Cercocebus lunulatus. The International Union for the Conservation of Nature Red List of Threatened Species.* http://dx.doi.org/10.2305/IUCN.UK.2016-1.RLTS.T4206A92247225.en.

Oates, J. F., Groves, C. P., & Ehardt, C. (2008b). *Lophocebus albigena. The International Union for the Conservation of Nature Red List of Threatened Species.* www.iucnredlist.org.

O'Driscoll, C. W., & Chapman, C. A. (2006). Density of two frugivorous primates with respect to forest and fragment tree species composition and fruit availability. *International Journal of Primatology, 27*(1), 203-225.

Olupot, W. (2000). Mass differences among male mangabey monkeys inhabiting logged and unlogged forest compartments. *Conservation Biology, 14*(3), 833-843.

Olupot, W., Chapman, C. A., Brown, C. H., & Waser, P. M. (1994). Mangabey (*Cercocebus albigena*) population density size and ranging: A twenty-year comparison. *American Journal of Primatology, 32*, 197-205.

Olupot, W., & Waser, P. M. (2001a). Activity patterns, habitat use and mortality risks of mangabey males living outside social groups. *Animal Behaviour, 61*, 1227-1235.

Olupot, W., & Waser, P. M. (2001b). Correlates of intergroup transfer in male grey-cheeked mangabeys. *International Journal of Primatology, 22*(2), 169-187.

Olupot, W., & Waser, P. M. (2005). Patterns of male residency and intergroup transfer in gray-cheeked mangabeys (*Lophocebus albigena*). *American Journal of Primatology, 66*, 331-349.

Olupot, W., Waser, P. M., & Chapman, C. A. (1998). Fruit finding by mangabeys (*Lophocebus albigena*): Are monitoring of fig trees and use of sympatric frugivore calls possible strategies? *International Journal of Primatology, 19*(2), 339-353.

Onderdonk, D. A., & Chapman, C. A. (2000). Coping with forest fragmentation: The primates of Kibale National Park, Uganda. *International Journal of Primatology, 21*(4), 589-611.

Page, S. L., & Goodman, M. (2001). Catarrhine phylogeny: Noncoding DNA evidence for a diphyletic origin of the mangabeys and for a human-chimpanzee clade. *Molecular Phylogenetics and Evolution, 18*(1), 14-25.

Patzelt, A., Kopp, G. H., Ndao, I., Kalbitzer, U., Zinner, D., & Fischer, J. (2014). Male tolerance and male-male bonds in a multilevel primate society. *Proceedings of the National Academy of Sciences, 111*(41), 14740-14745.

Patzelt, A., Zinner, D., Fickenscher, G., Diedhiou, S., Camara, B., Stahl, D., & Fischer, J. (2011). Group composition of Guinea baboons (*Papio papio*) at a water place suggests a fluid social organization. *International Journal of Primatology, 32*, 652-668.

Peeters, M., & Courgnaud, V. (2002). Overview of primate lentiviruses and their evolution in non-human primates in Africa. In C. Kuiken, B. Foley, E. Freed, B. Hahn, B. Korber, P. A. Marx, F. E. McCutchan, J. W. Mellors, & S. Wolinski (Eds.), *HIV Sequence Compendium* (pp. 2-23). Los Alamos National Laboratory.

Peeters, M., Courgnaud, V., & Abela, B. (2001). Genetic diversity of lentiviruses in non-human primates. *AIDS Reviews, 3*, 3-10.

Perelman, P., Johnson, W. E., Roos, C., Seuanez, H. N., Horvath, J. E., Moreira, M. A. M., Kessing, B., Pontius, J., Roelke, M., Rumpler, Y., Schneider, M. P. C., Silva, A., O'Brien, S. J., & Pecon-Slattery, J. (2011). A molecular phylogeny of living primates. *PLoS Genetics, 7*, e1001342.

Pérez Pérez, A., & Veà Baró, J. (1999). Does allogrooming serve a hygienic function in *Cercocebus torquatus lunulatus*? *American Journal of Primatology, 49*, 223-242.

Petersdorf, M., Weyher, A. H., Kamilar, J. M., Dubuc, C., & Higham, J. P. (2019). Sexual selection in the Kinda baboon. *Journal of Human Evolution, 135*, 102635. https://doi.org/10.1016/j.jhevol.2019.06.006.

Pines, M., Saunders, J., & Swedell, L. (2011). Alternative routes to the leader male role in a multi-level society: Follower vs solitary male strategies and outcomes in hamadryas baboons. *American Journal of Primatology, 73*, 679-691.

Post, D. G. (1981). Activity patterns of yellow baboons (*Papio cynocephalus*) in the Amboseli National Park, Kenya. *Animal Behaviour, 29*, 357-374.

Post, D. G. (1982). Feeding behavior of yellow baboons (*Papio cynocephalus*) in the Amboseli National Park, Kenya. *International Journal of Primatology, 3*, 403-430.

Poulsen, J. R., Clark, C. J., & Smith, T. B. (2001a). Seasonal variation in the feeding ecology of the grey-cheeked mangabey (*Lophocebus albigena*) in Cameroon. *American Journal of Primatology, 54*, 91-105.

Poulsen, J. R., Clark, C. J., & Smith, T. B. (2001b). Seed dispersal by a diurnal primate community in the Dja Reserve, Cameroon. *Journal of Tropical Ecology, 17*, 787-808.

Pugh, K. D., & Gilbert, C. C. (2018). Phylogenetic relationships of living and fossil African papionins: Combined evidence from morphology and molecules. *Journal of Human Evolution, 123*, 35-51.

Pusey, A. E., & Packer, C. (1987). Dispersal and philopatry. In B. B. Smuts, D. L. Cheney, R. M. Seyfarth, R. W. Wrangham, & T. T. Struhsaker (Eds.), *Primate societies* (pp. 250-266). University of Chicago Press.

Quris, R. (1975). Ecologie et organisation sociale de *Cercocebus galeritus agilis*. *Terre et Vie, 29*, 337-398.

Range, F. (2005). Female sooty mangabeys (*Cercocebus torquatus atys*) respond differently to males depending on the male's residence status: Preliminary data. *American Journal of Primatology, 65*, 327-333.

Range, F. (2006). Social behavior of free-ranging juvenile sooty mangabeys (*Cercocebus torquatus atys*). *Behavioral Ecology and Sociobiology, 59*, 511-520.

Range, F., & Noë, R. (2002). Familiarity and dominance relations among female sooty mangabeys in the Taï National Park. *American Journal of Primatology, 56*, 137-153.

Ransom, T. W. (1981). *Beach troop of the Gombe*. Bucknell University Press.

Refisch, J., & Koné, I. (2005). Impact of commercial hunting on monkey populations in the Taï region, Côte d'Ivoire. *Biotropica, 37*(1), 136-144.

Rendall, D., Seyfarth, R. M., & Cheney, D. L. (2000). Proximate factors mediating "contact" calls in adult female baboons (*Papio cynocephalus ursinus*) and their infants. *Journal of Comparative Psychology, 114*, 36-46.

Rendall, D., Seyfarth, R. M., Cheney, D. L., & Owren, M. J. (1998). The meaning and function of grunt variants in baboons. *Animal Behaviour*, 57, 583-592.

Roberts, E. K., Lu, A., Bergman, T. J., & Beehner, J. C. (2017). Female reproductive parameters in wild geladas (*Theropithecus gelada*). *International Journal of Primatology*, 38, 1-20.

Rogers, J., Raveendran, M., Harris, R. A., Mailund, T., Leppälä, K., Athanasiadis, G., Schieerup, M. H., Cheng, J., Munch, K., Walker, J. A., Konkel, M. K., Jordan, V., Steely, C. J., Beckstrom, T. O., Bergey, C., Burrell, A., Schrempf, D., Noll, A., Kothe, M.,... & Worley, K. C. (2019). The comparative genomic and complex population history of *Papio* baboons. *Science Advances*, 5, eaau6947.

Rogers, M. E., Abernethy, K., Bermejo, M., Cipolletta, C., Doran, D., McFarland, K., Nishihara, T., Remis, M., & Tutin, C. E. G. (1994). Western gorilla diet: A synthesis from six sites. *American Journal of Primatology*, 64, 173-192.

Rogers, M. E., Abernethy, K. A., Fontaine, B., Wickings, E. J., White, L., & Tutin, C. E. G. (1996). Ten days in the life of a mandrill horde. *American Journal of Primatology*, 40, 297-313.

Rook, L., Martínez-Navarro, B., & Clark Howell, F. (2004). Occurrence of *Theropithecus* sp. in the late Villafranchian of southern Italy and implication for early Pleistocene "out of Africa" dispersals. *Journal of Human Evolution*, 47, 267-277.

Rovero, F., & de Luca, D. W. (2007). Checklist of mammals of the Udzungwa Mountains of Tanzania. *Mammalia*, 71(1-2), 47-55.

Rovero, F., Marshall, A. R., Jones, T., & Perkin, A. (2009). The primates of the Udzungwa Mountains: Diversity, ecology and conservation. *Journal of Anthropological Science*, 87, 93-126.

Rovero, F., Mtui, A. S., Kitegile, A. S., & Nielsen, M. R. (2012). Hunting or habitat degradation? Decline of primate populations in Udzungwa Mountains, Tanzania: An analysis of threats. *Biological Conservation*, 146, 89-96.

Rovero, F., Struhsaker, T. T., Marshall, A. R., Rinne, T. A., Pedersen, U. B., Butynski, T. M., Egardt, C. L., & Mtui, A. (2006). Abundance of diurnal primates in Mwanihana Forest, Udzungwa Mountains, Tanzania; A multi-observer comparison of line-transect data. *International Journal of Primatology*, 27(3), 675-697.

Rowe, N. (1996). *The pictorial guide to the living primates*. Pogonias Press.

Rowell, T. E. (1966). Forest living baboons in Uganda. *Journal of Zoology*, 149, 344-364.

Santiago, M., Range, F., Keele, B., Li, Y., Bailes, E., Bibollet-Ruche, F., Fruteau, C., Noë, R., Peeters, M., Brookfield, J., Shaw, G., Sharp, P., & Hahn, B. (2005). Simian immunodeficiency virus infection in free-ranging sooty mangabeys (*Cercocebus atys atys*) from the Taï Forest, Côte d'Ivoire: Implications for the origin of epidemic human immunodeficiency virus type 2. *Journal of Virology*, 79(19), 12515-12527.

Schreier, A. (2009). *The influence of resource distribution on the social structure and travel patterns of wild hamadryas baboons* (Papio hamadryas) *in Filoha, Awash National Park, Ethiopia*. [Doctoral dissertation, New York City University].

Schreier, A., & Swedell, L. (2008). Use of palm trees as a sleeping site for hamadryas baboons (*Papio hamadryas hamadryas*) in Ethiopia. *American Journal of Primatology*, 7, 107-113.

Setchell, J. M. (2005). Do female mandrills prefer brightly colored males? *International Journal of Primatology*, 26(4), 715-735.

Setchell, J. M., Charpentier, M., & Wickings, E. J. (2005). Sexual selection and reproductive careers in mandrills (*Mandrillus sphinx*). *Behavioral Ecology and Sociobiology*, 58, 474-485.

Setchell, J. M., & Dixson, A. F. (2001). Changes in the secondary sexual adornments of male mandrills (*Mandrillus sphinx*) are associated with gain and loss of alpha status. *Hormones and Behavior*, 39, 177-184.

Setchell, J. M., & Dixson, A. F. (2002). Developmental variables and dominance rank in male mandrills (*Mandrillus sphinx*). *American Journal of Primatology*, 56, 9-25.

Setchell, J. M., Knapp, L. A., & Wickings, E. J. (2006). Violent coalitionary attack by female mandrills against an injured alpha male. *American Journal of Primatology*, 68, 411-418.

Setchell, J. M., Lee, P. C., Wickings, E. J., & Dixson, A. F. (2001). Growth and ontogeny of sexual size dimorphism in the mandrill (*Mandrillus sphinx*). *American Journal of Physical Anthropology*, 115, 349-360.

Setchell, J. M., Lee, P. C., Wickings, E. J., & Dixson, A. F. (2002). Reproductive parameters and maternal investment in mandrills (*Mandrillus sphinx*). *International Journal of Primatology*, 23, 51-68.

Setchell, J. M., Vaglio, S., Abbott, K. M., Moggi-Cecchi, J., Boscaro, F., Pieraccini, G., & Knapp, L. A. (2011). Odour signals MHC genotype in an Old World monkey. *Proceedings of Biological Science*, 278, 274-280.

Setchell, J. M., Vaglio, S., Moggi-Cecchi, J., Boscaro, F., Calamai, L., & Knapp, L. A. (2010). Chemical composition of scent-gland secretions in an Old World monkey (*Mandrillus sphinx*): Influence of sex, male status, and individual identity. *Chemical Senses*, 35, 205-220.

Setchell, J. M., & Wickings, E. J. (2004). Sexual swellings in mandrills (*Mandrillus sphinx*): A test of the reliable indicator hypothesis. *Behavioral Ecology*, 15(3), 438-445.

Setchell, J. M., & Wickings, E. J. (2005). Dominance, status signals and coloration in male mandrills (*Mandrillus sphinx*). *Ethology*, 111, 25-50.

Setchell, J. M., & Wickings, E. J. (2006). Mate choice in male mandrills (*Mandrillus sphinx*). *Ethology*, 112, 91-99.

Shah, N. F. (2001). Ranging and habitat use of sympatric agile and grey-cheeked mangabeys (*Cercocebus agilis* and *Lophocebus albigena*) in the Dzanga-Ndoki National Park, Central African Republic. *American Journal of Physical Anthropology*, 114(S32), 135.

Shah, N. F. (2003). *Foraging strategies in two sympatric mangabey species* (Cercocebus agilis *and* Lophocebus albigena). [Doctoral dissertation, State University of New York].

Sharman, M. (1981). *Feeding, ranging, and social organization of the Guinea baboon.* [Doctoral dissertation, University of St. Andrews].

Shotake, T., & Nozawa, K. (1984). Blood protein variation in baboon: II. Genetic variability within and among the herds of gelada baboon in central Ethiopia Plateau. *Journal of Human Evolution, 13,* 265-274.

Sigg, H., & Stolba, A. (1981). Home range and daily march in a hamadryas baboon troop. *Folia Primatologica, 36,* 40-75.

Silk, J. B., Beehner, J. C., Bergman, T. J., Crockford, C., Engh, A. L., Moscovice, L. R., Wittig, R. M., Seyfarth, R. M., & Cheney, D. L. (2009). The benefits of social capital: Close social bonds among female baboons enhance offspring survival. *Proceedings of the Royal Society, B, 276*(1670), 3099-3104.

Silk, J. B., Rendall, D., Cheney, D. L., & Seyfarth, R. M. (2003). Natal attraction in adult female baboons (*Papio cynocephalus ursinus*) in the Moremi Reserve, Botswana. *Ethology, 109,* 627-644.

Smuts, B. B. (1985). *Sex and friendship in baboons.* Aldine de Gruyter.

Snyder-Mackler, N., Beehner, J. C., & Bergman, T. J. (2012). Defining higher levels in the multilevel societies of geladas (*Theropithecus gelada*). *International Journal of Primatology, 33,* 1054-1068.

Souquière, S., Bibollet-Ruche, F., Robertson, D. L., Makuwa, M., Apetrei, C., Onanga, R., Kornfeld, C., Plantier, J. C., Gao, F., Abernethy, K., White, L. J., Karesh, W., Telfer, P., Wickings, E. J., Mauclere, P., Marx, P. A., Barre-Sinoussi, F., Hahn, B. H., Muller-Trutwin, M. C., & Simon, F. (2001). Wild *Mandrillus sphinx* are carriers of two types of lentaviruses. *Journal of Virology, 75,* 7086-7096.

Stacey, P. B. (1986). Group size and foraging efficiency in yellow baboons. *Behavioral Ecology and Sociobiology, 18,* 175-187.

Städele, V., Van Doren, V., Pines, M., Swedell, L., & Vigilant, L. (2015). Fine-scale genetic assessment of sex-specific dispersal patterns in a multilevel primate society. *Journal of Human Evolution, 78,* 103-113.

Struhsaker, T. T. (1969). Correlates of ecology and social organization among African cercopithecines. *Folia Primatologica, 11,* 80-118.

Struhsaker, T. T. (1997). *Ecology of an African rain forest: Logging in Kibale and the conflict between conservation and exploitation.* University Press of Florida.

Struhsaker, T. T., & Leakey, M. (1990). Prey selectivity by crowned hawk-eagles on monkeys in the Kibale Forest, Uganda. *Behavioral Ecology and Sociobiology, 26,* 435-443.

Strum, S. C. (1981). Processes and products of change: Baboon predatory behavior at Gilgil, Kenya. In G. Teleki, & R. Harding (Eds.), *Omnivorous primates* (pp. 255-302). Columbia University Press.

Strum, S. C. (1984). Why males use infants. In D. M. Taub (Ed.), *Primate paternalism* (pp. 146-185). Van Nostrand Reinhold.

Strum, S. C. (2010). The development of primate raiding: Implications for management and conservation. *International Journal of Primatology, 31*(1), 133-156.

Surbeck, M., Fowler, A., Deimel, C., & Hohmann, G. (2009). Evidence for the consumption of arboreal, diurnal primates by bonobos (*Pan paniscus*). *American Journal of Primatology, 71,* 171-174.

Swedell, L. (2006). *Strategies for sex and survival in hamadryas baboons. Through a female lens.* Pearson.

Swedell, L. (2011). African papionins. Diversity of social organization and ecological flexibility. In C. J. Campbell, A. Fuentes, K. C. Mackinnon, M. Panger, & S. K. Bearder (Eds.), *Primates in perspective* (pp. 241-277). Oxford University Press.

Swedell, L., Hailemeskel, G., & Schreier, A. (2008). Composition and seasonality of diet in wild hamadryas baboons: Preliminary findings from Filoha. *Folia Primatologica, 79,* 476-490.

Swedell, L., Saunders, J., Schreier, A., Davis, B., Tesfaye, T., & Pines, M. (2011). Female "dispersal" in hamadryas baboons: Transfer among social units in a multilevel society. *American Journal of Physical Anthropology, 145,* 360-370.

Szalay, F., & Delson, E. (1979). *Evolutionary history of the primates.* Academic Press.

Tefler, P. T., Souquière, S., Clifford, S. L., Abernethy, K. A., Bruford, M. W., Disotell, T. R., Sterner, K. N., Roques, P., Marx, P. A., & Wickings, E. J. (2003). Molecular evidence for deep phylogenetic divergence in *Mandrillus sphinx*. *Molecular Ecology, 12,* 2019-2024.

Tinsley Johnson, E., Snyder-Mackler, N., Beehner, J. C., & Bergman, T. J. (2014). Kinship and dominance rank influence the strength of social bonds in female geladas (*Theropithecus gelada*). *International Journal of Primatology, 35,* 288-304.

Topp-Jørgensen, E., Nielsen, M. R., Marshall, A. R., & Pedersen, U. (2009). Relative densities of mammals in response to different levels of bushmeat hunting in the Udzungwa Mountains, Tanzania. *Tropical Conservation Science, 2*(1), 70-87.

Tung, J., Charpentier, J. E., Garfield, D. A., Altman, J., & Alberts, S. C. (2008). Genetic evidence reveals temporal changes in hybridization patterns in a wild baboon population. *Molecular Ecology, 17,* 1998-2011.

Tutin, C. E. G., Ham, R. M., White, L. J. T., & Harrison, M. J. S. (1997a). The primate community of the Lopé Reserve, Gabon: Diets, responses to fruit scarcity, and effects on biomass. *American Journal of Primatology, 42,* 1-24.

Tutin, C. E. G., White, L. J. T., & Mackanga-Missandzou, A. (1997b). The use by rain forest mammals of natural forest fragments in an Equatorial African savanna. *Conservation Biology, 11*(5), 1190-1203.

Vaglio, S., Minicozzi, P., Romoli, R., Boscaro, F., Pieraccini, G., Moneti, G., & Moggi-Cecchi, J. (2016). Sternal gland scent-marking signals sex, age, rank, and group identity in captive mandrills. *Chemical Senses,* 41, 177-186.

van Doorn, A. C., O'Riain, M. J., & Swedell, L. (2010). The effects of extreme seasonality of climate and day length on the activity budget and diet of semi-commensal chacma baboons (*Papio ursinus*) in the Cape Peninsula of South Africa. *American Journal of Primatology,* 71, 1-9.

van Pinxteren, B. O. C. M., Siriani, G., Gratton, P., Després-Einspenner, M.-L., Egas, M., Kühl, H., Lapuente, J., Meier, A. C., & Janmaat, K. R. L. (2018). Sooty mangabeys scavenge on nuts cracked by chimpanzees and red river hogs: An investigation of inter-specific interactions around tropical nut trees. *American Journal of Primatology,* 80, e22895. https://doi.org/10.1002/ajp.22895.

Wahungu, G. M. (1998). Diet and habitat overlap in two sympatric primate species, the Tana crested mangabey (*Cercocebus galeritus*) and yellow baboon (*Papio cynocephalus*). *African Journal of Ecology,* 36, 159-173.

Wahungu, G. M. (2001). Common use of sleeping sites by two primate species in Tana River, Kenya. *African Journal of Ecology,* 39, 18-23.

Walker, S. E., Strasser, M. E., & Field, L. P. (2004). Reproductive parameters and life-history variables in captive golden-bellied mangabeys (*Cercocebus agilis chrysogaster*). *American Journal of Primatology,* 64, 123-131.

Wallis, S. J. (1983). Sexual behavior and reproduction of *Cercocebus albigena johnstonii* in Kibale Forest, western Uganda. *International Journal of Primatology,* 4(2), 153-166.

Waltert, M., Faber Lien, K., & Mühlenberg, M. (2002). Further declines of threatened primates in the Korup Project Area, south-west Cameroon. *Oryx,* 36, 257-265.

Waser, P. M. (1982). The evolution of male loud calls among mangabeys and baboons. In C. T. Snowdon, C. H. Brown, & M. Petersen (Eds.), *Primate communications* (pp. 117-143). Cambridge University Press.

Weingrill, T., Lycett, J. E., & Henzi, S. P. (2000). Consortship and mating success in chacma baboons (*Papio cynocephalus ursinus*). *Ethology,* 106, 1033-1044.

Whitham, J. C., & Maestripieri, D. (2003). Primate rituals: The function of greetings between male Guinea baboons. *Ethology,* 109, 847-859.

Wickings, E. J., & Dixon, A. F. (1992). Testicular function, secondary sexual development and social status in male mandrills (*Mandrillus sphinx*). *Physiology and Behavior,* 52, 909-916.

Wieczkowski, J. (2004). Ecological correlates of abundance in the Tana mangabey (*Cercocebus galeritus*). *American Journal of Primatology,* 63, 125-138.

Wieczkowski, J., & Kinnaird, M. (2008). Shifting forest composition and primate diets: A 13-year comparison of the Tana River mangabey and its habitat. *American Journal of Primatology,* 70, 339-348.

Wild, C., Morgan, B. J., & Dixson, A. (2005). Conservation of drill populations in Bakossiland, Cameroon: Historical trends and current status. *International Journal of Primatology,* 26(4), 759-773.

Yihune, M., Bekele, A., & Tefera, Z. (2008). Human-gelada baboon conflict in and around the Simien Mountains National Park, Ethiopia. *African Journal of Ecology,* 47(3), 276-282.

Zinner, D., Arnold, M. L., & Roos, C. (2009). Is the new primate genus *Rungwecebus* a baboon? *PLoS ONE,* 4(3), e4859.

Zinner, D., Peláez, F., & Torkler, F. (2001). Distribution and habitat associations of baboons (*Papio hamadryas*) in central Eritrea. *International Journal of Primatology,* 22, 397-413.

Zinner, D., Wertheimer, J., Liedigk, R., Groeneveld, L. F., & Roos, C. (2013). Baboon phylogeny as inferred from complete mitochondrial genomes. *American Journal of Physical Anthropology,* 150, 133-140.

CHAPTER 17

Macaques
Ecological Plasticity in Primates

PAMELA C. ASHMORE, DONNA HART, AND CHRISTOPHER A. SHAFFER

MACAQUES ARE QUINTESSENTIAL, GENERIC OLD WORLD MONKEYS THAT exhibit a diverse range of behavior and ecology (Table 17.1). The genus *Macaca* represents one of the most successful primate adaptive radiations, and the geographic distribution of *Macaca* is second only to that of our own genus *Homo*. Extant macaque species have a limited distribution in northwestern Africa and in the Middle East, but inhabit the entire Indian subcontinent and are widely distributed throughout eastern Asia, southern Asia, and numerous islands off the coast of Asia. Within this extensive geographic distribution, macaques occur in a wide variety of habitat types located in tropical, subtropical, and temperate climatic zones. Macaque species live in both pristine rainforest habitats and human-populated urban centers, and depending on the habitat, they may be preferentially arboreal or terrestrial.

As a whole, member species of this genus exhibit neither great morphological nor behavioral specialization.

The adaptive success of the genus is most likely related to the fact that several member species (especially *M. fascicularis*, *M. mulatta*, *M. radiata*, and *M. sinica*) are ecological generalists, also described as "weed macaques," "edge species," or as "adaptable opportunists" (MacKinnon & MacKinnon, 1978; Richard et al., 1989). These species demonstrate a propensity to adapt to secondary and disturbed areas, actually seeking out and colonizing these types of habitats, thereby exhibiting a maximum degree of ecological plasticity. The ability of some macaque species to live in such variable and novel environments is the result of adaptive phenotypic plasticity (Snell-Rood, 2013). This concept refers to the ability of a genotype to vary its phenotype in a range of different environments. It may be expressed as variability in behavior since behavior is directly responsive to environment, and this is especially relevant to large-brained primates (Snell-Rood, 2013; van Schaik, 2013). Changes in ranging behavior,

TABLE 17.1 Species of Genus *Macaca*

Species	Common Name	Taxonomic Description References
M. arctoides	Stump-tailed (bear) macaque	Geoffroy (1831)
M. assamensis	Assam (Assamese) macaque	McClelland (1839)
M. brunnescens [a]	Muna-Butung macaque	Matschie (1901)
M. cyclopis	Taiwanese (Formosan rock) macaque	Swinhoe (1862)
M. fascicularis	Long-tailed (crab-eating, kra, or cynomolgus) macaque	Raffles (1821)
M. fuscata	Japanese macaque (snow monkey)	Blythe (1875)
M. hecki [a]	Heck's macaque	Matschie (1901)
M. leonina	Northern pig-tailed macaque	Blyth (1863)
M. leucogenys	White-cheeked macaque	Li et al. (2015)
M. maura [a]	Moor macaque	Cuvier (1823)
M. mulatta	Rhesus macaque	Zimmerman (1780)
M. munzala	Arunachal macaque	Sinha et al. (2005a)
M. nemestrina	Pig-tailed (Sunda or Beruk) macaque	Linnaeus (1766)
M. nigra [a]	Sulawesi crested macaque (black "ape")	Desmarest (1822)
M. nigrescens [a]	Gorontalo macaque	Temminck (1849)
M. ochreata [a]	Booted macaque	Ogilby (1840)
M. pagensis [b]	Mentawai macaque	Miller (1903)
M. radiata	Bonnet macaque	Geoffroy (1812)
M. siberu [b]	Siberut macaque	Kitchener and Groves (2002)
M. silenus	Lion-tailed macaque	Linnaeus (1758)
M. sinica	Toque macaque	Meyer (1899)
M. sylvanus	Barbary macaque	Linnaeus (1758)
M. thibetana	Tibetan (Milne-Edwards') macaque	Milne-Edwards (1870)
M. tonkeana [a]	Tonkean macaque	Meyer (1899)

[a] Sulawesi species.
[b] Mentawai species.

activity patterns, and dietary flexibility all relate to behavioral plasticity. Consequently, a large degree of ecological plasticity is directly associated with the potential for behavioral plasticity.

Phenotypic plasticity or flexibility in response to external stimuli (Kappeler & Kraus, 2010) is now recognized as an important factor in the adaptability of species. In fact, it may be a main strategy enabling individuals and species to cope with environmental change (Murren et al., 2015). It is also an important factor contributing to intra- and interspecific variation (Strier, 2017). The ability for fluid or plastic behavior may, in fact, be initiated by unique and novel behaviors promoted by selective pressures that influence the potential behavior of an individual (Fuentes, 1999). Within a social group setting, individual behaviors may then promote a shift in the behavior of the group, allowing for the successful adaptation within a new or different set of environmental parameters. It is important to note, however, that not every macaque species exhibits an equal potential for ecological and behavioral plasticity (Richard et al., 1989; Ashmore-DeClue, 1992).

Macaques are cheek-pouched monkeys that have ischial callosities and molar teeth with four cusps (bilophodont). Taxonomically, macaques belong to the superfamily Cercopithecoidea and the family Cercopithecidae. They are members of the Cercopithecinae subfamily, and they are associated with the Asian branch of the tribe Papionini (Groves, 2001). Macaques have robust, heavyset bodies and compact limbs of almost equal length (see Figure 17.1). Delson (1980) described the genus as "conservative" because member species share the following characteristics: 1) a moderate muzzle, 2) a smooth or slightly concave facial profile, 3) lack of a steep anteorbital drop, 4) high-crowned bilophodont teeth, 5) relatively large incisors, and 6) a general lack of maxillary or mandibular *corpus fossae*—all traits reminiscent of ancestral cecopithecines. Males of the smallest species (*M. sinica*) have a mean weight of 5 kg, while the largest (*M. thibetana*) average 18.3 kg (Fooden, 1979; Smith & Jungers, 1997). Macaques are sexually dimorphic, and the degree of sexual differences in body weight (mean adult male body weight/mean adult female body weight) varies from 1.17–1.81 kg with an overall average of 1.54 (Ashmore-DeClue, 1992). The intermembral index is relatively high (92), but within the genus there is considerable range (84–100) (Napier, 1981; Fleagle, 1988). The majority of macaque species have short, stocky tails; nonetheless, some species exhibit no visible tail (e.g., *M. sylvanus*), while others have very long tails (e.g., *M. sinica*) (Hill, 1974; Fooden, 1979). Pelage color varies from light yellowish-brown to black with large variation in pigment coloration and facial hair ornamentation.

Macaques consist of a monophyletic group (Morales & Melnick, 1998), i.e., extant species can be traced

Figure 17.1 *Macaca fascicularis* **showing the relatively equal limb proportions characteristic of macaques.**
(Photo credit: Jeffrey V. Peterson)

back to the most recent common ancestor of the genus. Twenty-four member species are currently recognized, with different strategies having been previously employed to sort species into groups representing four to seven different phyletic lineages. In this chapter, we discuss the genus as consisting of five different species groups, as shown in Table 17.2, that generally reflect an amalgamation of the different phyletic lineages proposed by various authors. Overall, we are preserving the species groupings identified by Fooden (1976), simply adding newly recognized species and acknowledging the primitive status (morphologically and genetically) of *M. sylvanus*.

SYLVANUS SPECIES GROUP

The *sylvanus* species group now consists of a single species. *M. sylvanus* was originally included in the *sylvanus-silenus* species group; however, studies of nucleotide sequences of mtDNA (Hayasaka et al., 1996; Morales & Melnick, 1998) and nuclear DNA (Deinard & Smith, 2001; Ziegler et al., 2007) support putting this species in its own monotypic group. Fossil evidence indicates that extant *M. sylvanus* most closely represents the original evolutionary stock of macaques which originated on the African continent some six to seven mya (Stewart & Disotell, 1998; Elton & O'Regan, 2014).

TABLE 17.2 *Macaca* **Species Groups**

Species Group	Member Species
Sylvanus	*M. sylvanus*, Barbary macaque
Silenus	*M. silenus*, Lion-tailed macaque
	M. brunnescens, Muna-Butung macaque
	M. hecki, Heck's macaque
	M. leonina, Northern pig-tailed macaque
	M. maura, Moor macaque
	M. nemestrina, Pig-tailed macaque
	M. nigra, Sulawesi crested macaque
	M. nigrescens, Gorontalo macaque
	M. ochreata, Booted macaque
	M. pagensis, Mentawai macaque
	M. siberu, Siberut macaque
	M. tonkeana, Tonkean macaque
Sinica	*M. sinica*, Toque macaque
	M. assamensis, Assamese macaque
	M. leucogenys, White-cheeked macaque
	M. munzala, Arunachal macaque
	M. radiata, Bonnet macaque
	M. thibetana, Tibetan macaque
Fascicularis	*M. fascicularis*, Long-tailed macaque
	M. cyclopis, Formosan rock macaque
	M. fuscata, Japanese macaque
	M. mulatta, Rhesus macaque
Arctoides	*M. arctoides*, Stump-tailed macaque

Compared to others in the genus, the Barbary macaque is unique for two reasons: First, it is the only macaque that currently inhabits Africa, and second, its tail is either a boneless vestige or is entirely absent (Fooden, 2007). *M. sylvanus* is a heavyset macaque with thick and coarse variegated pelage that is yellowish brown to grayish black with lighter ventral hair. Individuals have lightly pigmented faces covered with scattered, short black-tipped yellowish hairs, bushy cheek whiskers, short beards, and short, erect, and brightly colored crown hair (see Figure 17.2) (Napier & Napier, 1967; Hill, 1974; Fooden, 2007). As reported in Fooden (2007), mean body weight for selected wild adult females is 9.9 kg; female head and body length is 556.8 mm. Males have a mean weight of 14.5 kg and body length of 634.3 mm.

The geographic distribution of *M. sylvanus* consists of discontinuous and highly fragmented regions of Morocco and Algeria (see Figure 17.3), but in the recent past their range likely extended into Tunisia (McDermott, 1938). In 1740, the British introduced these macaques to Gibraltar (Fa, 1981). The existence of an entire range of *sylvanus*-like fossils throughout the circum-Mediterranean region including the Black Sea and as far north as West Runton in Norfolk, England, suggests that the past distribution of this species was much more extensive than it is today (Elton & O'Regan, 2014). A gross reduction in geographic range can be attributed to large-scale climatic changes in Europe and extensive anthropogenic habitat alteration in North Africa (Thirgood, 1984; Elton & O'Regan, 2014).

Within its modern restricted geographic range, *M. sylvanus* occupies many different habitat types indicative of the fact that this species demonstrates a degree of ecological plasticity (Ménard, 2002; Maibeche et al., 2015). In the wild, Barbary macaques are found in deciduous oak forests, mixed cedar-oak forests, cedar forests, high-altitude (>1,500 m) cedar-fir evergreen-oak forests, and on rocky ridges and thermophillous scrub and grasslands (Deag, 1974; Taub, 1977; Fa, 1982, 1984, 1986a; Mehlman, 1985, 1988; Ménard et al., 1985; Ménard & Vallet, 1996, 1997; Ménard, 2002; Maibeche et al., 2015). The high-altitude forests are subject to severe winters, with 2 m of snow cover and temperatures as low as -18°C (Taub, 1977). The death of 30 macaques living in the Middle Atlas Mountains of Morocco in 2008–2009 was the result of an exceptionally harsh winter consisting of 20–90 cm of snow covering much of the home range of two study groups for a little over three months. Deaths were attributed to starvation; McFarland and Majolo (2013) found that more time spent feeding and having larger numbers of social relationships, i.e., individuals to huddle with, were significant positive indicators of survival.

Figure 17.2 Adult female *Macaca sylvanus*, member of the Rock of Gibraltar colony.
(Photo credit: Agustín Fuentes)

Diversity of habitat includes forests with various degrees of tree and shrub densities, isolated pockets of forests with steep ravines and gorges, and denuded rocky ridges at altitudes of 1,600–2,100 m. Many of these habitats undergo striking seasonal variation in resource availability, thus a high degree of dietary flexibility obviously incorporates much behavioral plasticity (i.e., the ability to shift activity cycles) and physiological plasticity (i.e., generalized dentition and gut tract morphology). This flexibility has enabled the Barbary macaques to occupy and exploit diverse habitat types (El Alami et al., 2012; Maibeche et al., 2015).

Population densities range from 1–70 individuals per km² with the lowest densities (~1.0) reported in scrub habitats and the highest (28.2) found in cedar and cedar/oak forests (Deag, 1974, 1984; Fa, 1982; Drucker, 1984; Ménard & Vallet, 1996). In a survey of macaques living in the Middle Atlas Mountains, home to the largest wild population of *M. sylvanus*, mean population density was 9 per km² with human pressure having a negative effect on density and forest patch size having a positive effect (Ménard et al., 2014). During daylight hours, Barbary macaques spend the majority of their time on the ground.

Fooden (2007) reported mean frequencies of >50% to approximately 100% for daytime terrestriality. In cedar forests, *M. sylvanus* use trees for sleeping, resting, feeding, and predator avoidance, but most traveling is done on the ground (Deag, 1985). When fleeing from humans, they typically do so on the ground but will use arboreal pathways to escape dogs (Taub, 1977; Mehlman, 1984; Deag, 1985). Adults spend more time being terrestrial than do infants or juveniles.

Potential predators, besides humans and domesticated dogs, are large eagles (family Accipitridae), jackals (*Canis aureus*), red fox (*Vulpes vulpes*), and the rare Barbary leopard (*Panthera pardus barbarus*) (Mehlman, 1984; Fa, 1986b). Ménard and Vallet (1996) noted that the macaques respond to dogs and eagles with alarm calls but did not respond in a similar fashion to jackals or foxes.

Barbary macaques, like all macaques, are opportunistic feeders; however, *M. sylvanus* exhibits extreme tendencies toward "dietary eclecticism" (Ménard & Vallet, 1996: p.106). This ability to shift dependence widely between food species likely allows them to compensate for living in habitats that undergo extreme seasonal changes in food availability and promotes their ability to colonize

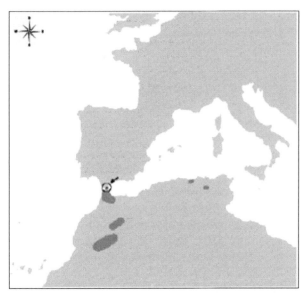

Figure 17.3 Geographic distribution of *Macaca sylvanus* (dark gray areas), based on spatial data from IUCN (2020). The small circled area represents the introduced *M. sylvanus* population on Gibraltar.
(Map credit: MasukudoMahn; image licensed under Creative Commons 3.0 Unported)

a large range of habitat types (Deag, 1974; Drucker, 1984; Fa, 1984; Ménard & Vallet, 1996; Ménard, 2002). They exhibit dietary selectivity on a seasonal basis (Drucker, 1984; Ménard & Vallet, 1996). In a comparison of macaques living in four locations in three different habitat types (deciduous oak forest, cedar-oak forest, and rocky mountains), Mènard (2002) found no clear predominance of any single food category in mean annual diets. The macaques were seen to shift between folivorous, graminivorous, and insectivorous phases in which a particular food category could comprise 83–90% of their monthly feeding time. In two forests of Algeria—a highly seasonal cedar-oak forest (Djurdjura) and a deciduous oak forest (Akfadou)—Barbary macaques spend up to 86% of monthly feeding time consuming acorns and herbaceous seeds during September–March, but in the spring, caterpillars constituted 83% of their monthly feeding time (Ménard, 2002). During times of extreme food scarcity, an ability to harvest rare foods, or those typically ignored, is a critical adaptive strategy of these macaques. Barbary macaques will then concentrate on low-quality food sources such as lichens or the needles, cones, and cambium of cedar trees (Drucker, 1984; Ménard, 2002). In the Middle Atlas Mountains, when food was scarce, an adult male was observed eating a rabbit and other adult males harvesting bird eggs and chicks (Young et al., 2012).

Certainly, related to the diverse and flexible dietary strategies of these macaques, is an ability to shift time budgets or activity cycles in response to resource availability. Ménard (2002) reports that when resources are abundant, Barbary macaques maximize time spent feeding (e.g., 5–6 hours per day). When resources are not as abundant, they reduce feeding time to 1.7 hours per day in deciduous oak forest and 2.6 hours in cedar-oak forest and spend additional time engaging in social behaviors. In the dry season, when resources in the cedar-oak forests and deciduous oak forests are most limited, they reduce feeding and social time but spend more time foraging. During times of moderate to heavy snow cover, macaques spend more time feeding than when there is no snow cover; this is likely related to an increase in thermoregulatory demands, and the poor quality of foods being consumed (Majolo & McFarland, 2013).

In a relictual Mediterranean thermophilous scrub habitat located in the Gouraya National Park, Algeria, a population of *M. sylvanus* lives near sea level and in close proximity to Bejaia, a growing city. Compared to Barbary macaques living elsewhere in Morocco and Algeria, the macaques found here undergo less seasonal constraints in food availability and depend less on the herbaceous layer (Maibeche et al., 2015). The diets of two groups in Gouraya, a peri-urban group and a nonurban group, vary on a monthly basis with leaves, seeds, and fruits consumed year-round. However, the peri-urban macaques eat more leaves, insects, and foods provided by people than the nonurban group. Adult males in the peri-urban group were especially receptive to handouts provided by tourists. Habitat restoration and conservation efforts have benefitted these macaques, but the expansion of the city has resulted in one group living in much closer proximity to it. The high dietary flexibility exhibited by this species may actually allow it to benefit from closer albeit restricted access to humans.

Ménard (2002) cites a modal group size of 40 individuals and notes that 80 may be a maximum group size. In forest habitats, group size remains somewhat stable. However, groups living in marginal mountain-ridge habitats exhibited instability (Ménard et al., 1990). The smaller groups living in marginal habitats (mode of 15–20) display a fission-fusion social structure.

Barbary macaques are seasonal breeders; interbirth intervals of 20 months have been reported for groups living in the fir forests of Morocco (Mehlman, 1989) and 12–24 months in cedar-oak and deciduous oak forests (Ménard & Vallet, 1996) of Algeria. *M. sylvanus* have a highly developed system of alloparenting with frequent interactions observed between older males and infants (Fa, 1984; Taub, 1985; Maestripieri, 1998; Thierry, 2000; Ménard, 2002). Deag and Crook (1971) refer to this as "agonistic-buffering" in which males employ infant-holding as a means to ward off aggression by other males. In favorable habitats, groups exhibit fast population growth (14.6% mean annual intrinsic growth rate in cedar-oak forest), which results in group division along matrilineal lines and promotes male emigration. In poorer habitats, population growth is suppressed (4.8% mean annual intrinsic growth rate in deciduous oak forest) (Ménard & Vallet, 1993). Although no one demographic parameter was identified to explain the difference in growth rates, in the poorer-quality habitats the proportion of female births was lower and infant mortality, particularly for females, was higher. Ménard and Vallet's (1993) seven-year study of the demographic trends of groups living in two different habitats confirmed what was first speculated by Taub (1977) that cedar forests are the preferred habitat of the Barbary macaque. Despite the fact that Barbary macaques can colonize marginal habitats, it is suggested that the demographic characteristics of this species may limit their adaptive success in such habitats (Ménard, 2002; Maibeche et al., 2015).

M. sylvanus exhibit considerable variation in the type of habitats that they occupy, however, it does not approximate the variability in habitat type exhibited by *M. mulatta* or *M. fascicularis*. The limited ecological plasticity of the Barbary macaque appears to be related to the fact that, unlike the more ecologically generalized species, *M. sylvanus* does not invade or colonize human-altered habitats. These macaques also do not respond well to living in more open habitats. Ménard (2002) reports that Barbary macaques move only with great care into the open and are rarely seen more than 100 m from cover. They are not found in environments where trees or cliffs are lacking, and their tendency to maintain stable home ranges and a high level of group sociality limits their dispersal ability (Van Lavieren, 2012; Ménard et al., 2014). Increased competition with sheep and goats for herb and shrub plants, the pruning of cedar trees for livestock feed, and the taking of young macaques for the pet trade are serious and ongoing threats to the remnant populations of Barbary macaques and especially to those living in the Middle Atlas Mountains (Mènard et al., 2014). As human encroachment pushes these macaques into more marginal and isolated habitats, populations will become more and more fragmented and the limits to their plasticity will likely be tested. The International Union for Conservation of Nature and Natural Resources (IUCN) Red List catalogs Barbary macaques as Endangered with a decreasing population trend (Butynski et al., 2008; IUCN, 2020).

SILENUS SPECIES GROUP

The *silenus* group of macaques includes 12 member species making it the largest of the macaque species groups. The timing of the divergence of the *silenus* group from a proto-*fascicularis* group has been placed at 5.1 (+/- 0.4) mya (Ziegler et al., 2007); consequently, the *silenus* group contains species that are the direct ancestors of the original colonizers of the Asian continent. The phylogenetic relationship between member species is complicated and to date not fully resolved. The Plio-Pleistocene dispersal of these species across the Asian continent and the associated environmental and geographical changes linked to this period make any reconstruction of the evolution of this group particularly complex. Fluctuating sea levels produced land connections or caused landmasses to become isolated and separated. During periods of maximum glaciations, the distribution of rainforest habitats that covered much of Sundaland (Maya Peninsula, Borneo, Sumatra, Java, Bali, Palawan, the Mentawai Islands, and numerous smaller islands) was significantly reduced (Eudey, 1980; Abegg & Thierry, 2002; Ziegler et al., 2007). A majority of member species exhibit a more limited range in ecological strategies that reflects their fragmented geographical distribution, encompassing insular as well as mainland habitats (see Figure 17.4). Today, *M. silenus* has a restricted geographic range in the Western Ghats Mountain range of India, *M. nemestrina* and *M. leonina* have much larger distributions across much of Southeast Asia, two species inhabit the Mentawai Islands, and seven species are endemic to the island of Sulawesi, Indonesia.

Member species also exhibit a large range of morphological characteristics. Perhaps the most beautiful of all of the macaque species is the lion-tailed macaque, *M. silenus*. Their long, tufted tails and lion-like silver-colored manes are in striking contrast to silky black pelage and black facial skin (see Figure 17.5). The pelage of other species belonging to this group ranges from yellowish brown (*M. nemestrina*) to glossy black (*M. nigra*). Facial skin is pale brown to black. Crown hairs may form an erectile crest (*M. nigra* and *M. nigrescens*) or radiate to form a whorl and a thick cap of hair (*M. nemestrina*). With the exception of *M. silenus*, these macaques have short tails. *M. nemestrina*, for example, are called pig-tailed macaques because of their short, thin tails. *M. silenus* have the smallest body weight with males averaging 6.7 kg (Napier & Napier, 1967), and *M. nemestrina* are the heaviest with males averaging 11.2 kg (Fooden, 1975). Sulawesi macaques exhibit prognathic (baboon-like) muzzles.

Member species vary in degree of ecological and behavioral plasticity. Despite the fact that *M. silenus* have a very limited geographic distribution and *M. nemestrina* a much broader one, both species demonstrate minimal and limited ecological plasticity (Richard et al., 1989; Ashmore-DeClue, 1992). The tendency for the majority of species in this group to be ecologically restricted to broadleaf evergreen habitats may be linked to the historical biogeography associated with this group (Eudey, 1980; Harrison et al., 2006).

*Lion-tailed Macaque (*Macaca silenus*)*

It is hypothesized that the ancestor of *M. silenus* arrived in India approximately 1.1–1.5 mya (Ziegler et al., 2007). As available habitats shrank and became more fragmented, *M. silenus* became restricted to the broadleaf evergreen tropical rainforests of southern India. Once forests began to re-expand, competition with other more ecologically opportunistic macaques, such as *M. radiata*, contributed to the sustained isolation of this species in forest habitats (Fooden, 1975, 1980; Abegg & Thierry,

2002). Today *M. silenus* occupy a narrow and discontinuous distribution throughout the Western Ghats Mountains of southwest India. They are primarily restricted to *shola* forests that are associated with altitudes of 1,500 m. In fact, this area has been deemed an ecological "hotspot" due to its rich species diversity and large number of endemic plants and animals (Myers, 1988). Extensive tea and cardamom plantations are found in close proximity to the *shola* forests, but the lion-tails do not frequent these plantations and enter them only if there are available patches of undistributed forest (Karr, 1973; Johnson, 1985; Krishnamurthy & Kiester, 1998). Green and Minkowski (1977: p.302) noted a preference for undisturbed rainforest habitats: "In over 1,700 hours observing groups whose range borders on regions of tea, coffee, and eucalyptus, we never saw them cross these areas." A similar aversion to venturing into treeless tea estates was observed by Umapathy and Kumar (2000). Their tendency to avoid deforested and disturbed areas (Singh et al., 2000), and the fact that they predominantly occupy areas of mature forest, defines *M. silenus* as an obligate

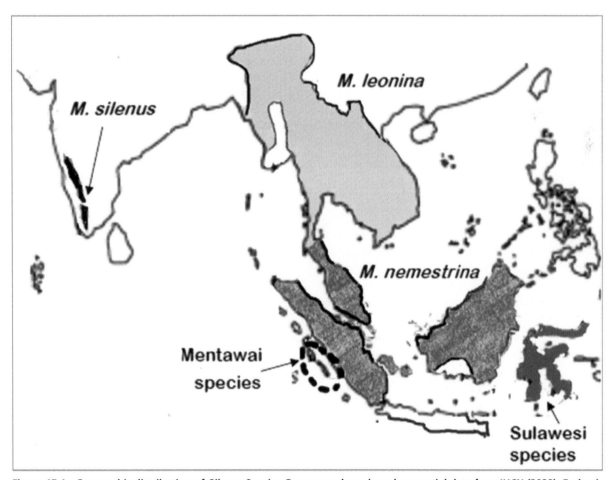

Figure 17.4 Geographic distribution of *Silenus* Species Group members, based on spatial data from IUCN (2020). Endemic Mentawai macaques include *Macaca pagensis* and *M. siberu*; endemic Sulawesi macaques include *Macaca brunnescens*, *M. hecki*, and *M. maura*, *M. nigra*, *M. nigrescens*, *M. ochreata*, and *M. tonkeana*. (Map credit: Ian Colquhoun)

Figure 17.5   Lion-tailed macaque with its silver mane and lion-tufted tail.
(Photo credit: Rajiv Bajaj, https://unsplash.com/photos/TlYiC_srvz8)

evergreen forest species (Krishnamurthy & Kiester, 1998; Kumara et al., 2014).

Home ranges tend to be relatively large (>4.0 km²) (Johnson, 1985), but the range size depends on the quality of available habitat. Daily ranging behavior and sleeping site use is also strongly influenced by the seasonal availability of tree fruits (Erinjery et al., 2015). More than 50% of the groups living in the Western Ghats live in fragmented forest habitats smaller than 20 km² in size (Kumar, 1985).

M. silenus are likely the most arboreal of all macaque species spending less than 1% of their time at ground level (Green & Minkowski, 1977). However, in selectively logged forests of Sirsi-Honnavara in Karnatak State, reduced contiguous canopy cover, and greater habitat heterogeneity contributed to the lion-tails using the lower plant strata and ground level approximately 24% of the time (Santhosh et al., 2015). This finding did not vary between dry and wet seasons. Lion-tails are opportunistic omnivores that are primarily frugivorous (Green & Minkowski, 1977; Johnson, 1985; Krishnadas et al., 2011; Erinjery et al., 2015; Santhosh et al., 2015). In the Anamalai Hills of Kerala State, the annual diet of a group of lion-tails consisted of 79% fruits, 14% insects, and 7% flowers (Erinjery et al., 2015). They may also spend up to 30–45% of their time foraging for insects (Green & Minkowski, 1977; Johnson, 1985) as insects are an important source of protein (Sushma & Singh, 2006). Habitat integrity affects how lion-tails cope with periods of fruit scarcity during the dry season. In a contiguous tract of rainforest habitat, they eat fewer fruit species than they do in the wet season. However, they tend to be more selective and feed on less abundant, high-quality fleshy fruits (Krishnadas et al., 2011). In selectively logged forests in Karnataka State, lion-tails eat a more diverse diet of plants during the dry season and a less diverse diet in the wet season (Santhosh et al., 2015). In the dry season, they spend less time eating fruits and more time feeding on stem-leaves and resins. However, time spent feeding on fauna did not significantly differ between seasons, ranging from 8–11% of total time spent feeding (Santhosh et al., 2015). Lion-tails have been observed eating flying squirrels and taking juvenile Indian giant squirrels from their nests (Umapathy & Kumar, 2000).

Sushma and Singh (2006) studied the resource partitioning and interspecific interactions of sympatric lion-tails, *M. radiata*, Nilgiri langurs (*Semnopithecus johnii*), and Indian giant squirrels (*Ratufa indica*). Using Levins's standard measure of niche breadth to estimate the diversity of food resources in the diets, it was discovered that *M. silenus* had the narrowest niche breadth. On average they consumed 41 food plant species, ate 5 different plant parts, and used a total of 45 different food resources.

Consequently, lion-tails demonstrate limited dietary flexibility.

Groups living in the Anamalai Wildlife Sanctuary of Tamil Nadu were found to spend over 50% of their time feeding and foraging, approximately 33% of the time resting, and 15% of their time moving (Kurup & Kumar, 1993). In Kurup and Kumar's study, seasonal variation in activity patterns indicated an inverse relationship between time spent feeding or foraging and time spent resting; the same relationship was reported by Menon and Poirier (1996).

Potential or verified predators include tigers (*Panthera tigris*), leopards (*P. pardus*), dholes (Asian wild dogs, *Cuon alpinus*), and Indian pythons (*Python molurus*); the low flight of eagles will also elicit alarm calls (Sugiyama, 1968; Pruett, 1973; Green & Minkowski, 1977; Hohmann & Herzog, 1985; Erinjey et al., 2015). Two incidences of eagles (a mountain hawk-eagle, *Nisaetus nipalensis,* and a crested serpent eagle, *Spilornis cheela*) attacking and killing immature lion-tails are reported in Erinjery et al. (2015). Lion-tails also respond to the alarm calls of the Indian giant squirrel (Sushma & Singh, 2006), which may be beneficial when lion-tails forage for insects and are spread out over a wide area with individuals spaced far from one another. They also react to alarm calls given by barking deer (*Muntiacus muntjak*), which feed on jackfruits dropped by lion-tails (Erinjery et al., 2015). Overall, however, the most threatening predator of *M. silenus* is humans.

These macaques live in multi-male/multi-female groups and also in smaller uni-male groups (Green & Minkowski, 1977; Hohmann & Herzog, 1985; Johnson, 1985; Kumar, 1987). Mean group sizes range from 7–38 (Ashmore-DeClue, 1992; Kurup & Kumar, 1993; Kumar, 1995; Ramachandran & Joseph, 2001; Singh et al., 2002; Kumara & Singh, 2004; Erinjery et al., 2015). An average group size of approximately 18 individuals was reported for lion-tails living in contiguous forest (Kumar, 1995) but less for those living in highly fragmented forests (Singh et al., 2002). Solitary males have also been sighted (Singh et al., 2000; Kumar et al., 2001). An extremely high sex ratio in favor of females (F:M = 9.9:1) has been reported for this species (Singh & Sinha, 2004; Thierry, 2007; Kumara et al., 2014). Lion-tails are nonseasonal breeders that exhibit peaks in the frequency of mating behaviors (Sugiyama, 1968; Kumar & Kurup, 1985), possibly related to ecological factors (Sharma et al., 2006). Data on life history from Kumar (1987) illustrate that this species may be particularly vulnerable to the threat of extinction. Females tend to have their first birth at 6.6 years of age, which is somewhat higher than that for other macaque species; they exhibit a long interbirth interval of 29.7 months and a low birth rate (0.28/female/year).

Loud calls are associated with adult males and are produced during intergroup encounters or when males are at a distance from other group members. Loud calls may be related to their arboreal tendencies and rainforest habitats.

Umapathy and Kumar (2000) examined the effects of habitat fragmentation on *M. silenus* living in the Anamalai Hills in the southern Western Ghats. In the small forest fragments, lion-tails were often seen on the ground. Further, demographic patterns in 11 groups living in 8 rainforest fragments indicated that groups living in the smaller fragments had lower birth rates and a lower number of immatures. In addition, these groups exhibited less dietary diversity. The reduced canopy cover in the fragmented forests also required the lion-tails to leap greater distances between trees, and one death as the result of a fall was reported. The reduced canopy cover also likely exposed them to a greater risk of predation by raptors. In the Anamalai Hills, where tracks of evergreen forest still remain but are interspersed with logged areas and monoculture teak plantations, Kumara et al. (2014) calculated a "human disturbance index" and determined that the higher the index value, the greater a negative influence it had on the occurrence of lion-tails. In contrast, a positive effect was correlated with the greater the height of the tallest trees.

M. silenus are one of the most endangered species of macaques with an estimated <2,500 animals remaining in the wild and no subpopulation numbering more than 250 mature individuals (Kumar et al., 2008a). Lion-tails are habitat specialists and obligate frugivores that exhibit a minimal amount of ecological plasticity (Krishnadas et al., 2011; Kumara et al., 2014). Studies on the parasite loads of lion-tails (Hussain et al., 2013; Kumar et al., 2019) additionally indicate that lion-tails do not fare well when living in close association with humans. Groups living in close proximity to human settlement (i.e., livestock) exhibit higher parasite loads than those living farther away from human settlements. In addition, immatures were found to have the highest endoparasite loads, and novel endoparasites may have been introduced to lion-tails through the relocation of *M. radiata* to the same forests.

The IUCN lists the species as Endangered with a decreasing population trend (Kumar et al., 2008a; IUCN, 2020). Hunting by humans, loss of habitat, and degradation of the forest pose very real threats to the continued existence of this species (Singh et al., 2000; Kumara & Singh, 2008; Anitha et al., 2013; Santhosh et al., 2013; Kumara et al., 2014; Sushma et al., 2014). These threats are particularly serious given the limited propensity of *M. silenus* for ecological and behavioral plasticity perpetuated by its evolutionary history and dispersal. As suggested by

Eudey (1980) and Fooden (1975), members of the ancestral *silenus*-stock likely sought out forest refuges during a major cooling phase of the Pliocene. After the cooling phase was over, these ancestral macaques simply could not compete with other more ecologically diverse ancestral macaque species, thus limiting the ability of lion-tails to disperse into different habitat types.

Northern Pig-tailed Macaque (Macaca leonina)

Species status for the northern pig-tail macaque is relatively recent (Mittermeier et al., 2013; Roos et al., 2014) as it was traditionally recognized as a subspecies of *M. nemestrina* (Fooden, 1975; Gippoliti, 2001). Differences in morphology and mtDNA analysis (Roos et al., 2007; Malaivijitnond et al., 2012) support the species level distinction between *M. leonina* and *M. nemestrina*, the southern pig-tail macaque. *M. leonina* are large macaques with a short tail. Adult females weigh on average 5.1 kg and adult males 8.9 kg (Malaivijitnond et al., 2012). They have a dark brown crown patch of hair and a white patch above the eyes. A red streak of pelage extends from the eyes to the ears. The muzzle is shorter than that of the southern pig-tail and the body pelage is reddish to dark in color with a black streak in the middle of the back. Ischial callosities are oval in shape, sexual swelling does not occur around the ischial callosities, but swollen skin does redden (Gippoliti, 2001; Feeroz, 2003; Malaivijitnond et al., 2012).

Northern pig-tail macaques are found in southwestern China, eastern Bangladesh, Cambodia, Lao PDR, Myanmar, Thailand, southern Vietnam, and India. Although predominantly allopatric with *M. nemestrina*, the two species are partially sympatric along the Khlong Marui fault where the Indochinese and Sundaic subregions meet (Malaivijitnond et al., 2012). This suggests that the separation and divergence of southern and northern pig-tails may have been precipitated when sea levels rose, dividing these subregions during the Pleistocene (Woodruff & Turner, 2009).

Long-term field studies have been conducted in seasonally wet evergreen forests in Thailand including areas of human activity, grasslands, shrublands, and secondary and primary forests (Albert et al., 2013a,b; José-Domínguez et al., 2015). Northern pig-tails have also been studied in tropical wet semi-evergreen forests in Bangladesh (Feeroz, 2003). Albert et al. (2013b) describes them as semiterrestrial with most movement occurring on the ground. Northern pig-tails living around a visitor center in the Khao Yai National Park, Thailand, use both energy-minimizing and energy-maximizing strategies in their ranging patterns and dietary choices to cope with periods of low fruit availability (Albert et al., 2013b). When fruit is less available, pig-tails reduce their mean monthly daily path length and home range size and offset this by spending more time near human settlements consuming human-produced foods. *M. leonina* is highly frugivorous (76% of feeding observations) during periods of high fruit abundance but becomes omnivorous when fruit is less abundant (8% of time eating fruit). Conversely, when fruit is most available, they eat human foods 2% of the time, but 69% of the time when fruit is least available. In 1,029 hours of study time, pig-tails were found to spend 46% of the time in the forest and 54% in human settlements. When in the forest, they were found in primary forest 69% of the time, secondary forest 30%, and less than 1% in grasslands, shrublands, and small tree areas (Albert et al., 2013b). Selection of sleeping trees was associated with the location of the last feeding tree and its proximity to a river.

In a study of a non-provisioned group in a seasonally wet evergreen forest, José-Domínguez et al. (2015) found that size of the home range, core area, and daily path length shifted on a monthly basis. Over a 16-month period, home range size was on average 142 ha, ranging between 44–354 ha; monthly core areas were on average 22 ha, ranging between 7–77 ha; and daily path length was on average 2,212 m, ranging between 1,520–3,112 m. Daily path lengths and daily home range size were largest in months of high fruit abundance. Pig-tails exhibit variability in ranging behaviors and what José-Domínguez et al. (2015) describe as low site fidelity. This may be related to the fact that the northern pig-tails are not highly territorial and readily move according to where the next fruit patch is located.

As a highly frugivorous primate, *M. leonina* may be very effective seed dispersers (Albert et al., 2013a,c). The number of fruit species pig-tails consume is positively correlated with the availability of fruit, and they prefer ripe fruits (92%) to unripe fruits (8%). They disperse seeds by swallowing, spitting, and dropping, and may disperse seeds from primary to secondary forests and into grasslands and small tree areas as well. Moreover, by having large home ranges and daily path lengths, these macaques have the potential for long-distance seed dispersal.

In 16 months of observation, Feeroz (2003) found pig-tails to breed throughout the year, to exhibit asynchronous sexual swellings, and to give birth throughout the year with no obvious birth peak. Northern pig-tails live in multi-male/multi-female groups, and reports of group sizes range from 30–67 (Albert et al., 2013b; José-Domínguez et al., 2015). In these social groups, adult females outnumbered adult males by as much as 4:1.

IUCN categorizes this species as Vulnerable with a decreasing population trend (Booratana et al., 2008; IUCN, 2020). Throughout their geographic distribution, forests are being replaced by monocultures and plantations, resulting in a significant loss of feeding and sleeping trees. The pig-tails are hunted for food and immatures are taken for the pet trade. *M. leonina* is susceptible to human immunodeficiency virus type 1 (HIV-1) making it a highly desirable primate model for the study of human AIDS (Lei et al., 2014; Lian et al., 2016).

M. leonina displays a limited degree of ecological and dietary flexibility. In the Khao Yai National Park, they spend almost an equal amount of time in areas of human settlement and forest environments. The human settlements are located in the park and consist of a visitor center, staff housing, restaurants, bungalows, roads, and other housing. The macaques do not appear to attempt to remain in these areas, returning to the forest to forage and to find sleeping trees. In addition, when fruits are readily available, they significantly reduce their use of areas associated with human activity.

Southern Pig-tailed Macaque (Macaca nemestrina)

Compared to the northern pig-tails, the southern pig-tailed macaque is larger-bodied with a little thicker and slightly longer tail, typically arched rearward (Malaivijitnond et al., 2012). Adult females weigh between 5.4–7.6 kg and adult males 10.0–13.6 kg (Fooden, 1975). The upper body tends to be olive to golden-brown in color with lighter ventral parts and a dark brown crown patch similar to the northern form but darker. The ischial callosities are distinctly butterfly-shaped.

Southern pig-tailed macaques are found in Brunei Darussalam, Indonesia, Malaysia, and southern peninsular Thailand (Richardson et al., 2008). Fooden (1975, 1982) identified the geographic distribution of *M. nemestrina* as restricted to the occurrence of tropical broadleaf evergreen habitats. Sundaic pig-tails, those that occupy the biogeographical region of Southeastern Asia, are associated with dipterocarp-dominated lowland forests that exhibit irregular patterns in mast-fruiting (Caldecott, 1986a; Caldecott et al., 1996). The seeds that these trees produce are large, hard, and rich in lipids. However, due to the irregular fruiting cycles and highly variable rainfall patterns, the forests have limited carrying capacities and are considered food-poor, or marginal, habitats (Caldecott et al., 1996). In the Sundaic dipterocarp forests, pig-tails are the only long-term cercopithecine residents. In non-dipterocarp Sundaic forests, where there is greater diversity in fruit-producing species, *M. nemestrina* are sympatric with *M. fascicularis*. Nevertheless, distinctive differences in habitat preferences and locomotion tendencies ecologically differentiate the distribution of these two macaque species. North of the Kra Isthmus, *M. nemestrina* are marginally sympatric with *M. assamensis* and broadly sympatric with the predominantly terrestrial *M. arctoides* (Fooden, 1982). South of the Kra Isthmus they are partially sympatric with *M. leonina* (Malaivijitnond et al., 2012). Southern pig-tails are typically found well within the forest interior, tending to avoid habitats that are highly disturbed. They have also been described as shy and elusive animals that are slow to habituate for study (McCann, 1933; Bernstein, 1967a; Caldecott, 1986a).

Pig-tails are found in higher densities (33.2–36.7 individuals/km^2) in primary than in secondary forests (5.6–22.5 individuals per km^2) (Crockett & Wilson, 1980; Caldecott, 1986b). The use of secondary forests may be related to the proximity of these forests to agricultural crops since *M. nemestrina* are known crop raiders. Overall pig-tails seem to prefer non-riverine, hilly primary forest habitats.

The distribution of food resources in the dipterocarp forests strongly influences the terrestrial locomotion patterns observed for *M. nemestrina*. In Sumatra and East Kalimantan, they were observed by Rodman (1991) to be terrestrial 75% of the time. Terrestrial travel between widely dispersed arboreal food sources may be a conservative adaptation for long-distance travel (Rodman, 1979; Caldecott, 1986a; Caldecott et al., 1996) by this fairly large-bodied macaque. Although highly frugivorous, pig-tails are typically opportunistic foragers that spend time foraging for insects, fungi, and fallen fruit on the forest floor (Caldecott et al., 1996). Caldecott (1986a, 1986b) observed that these macaques may spend 9 to 10 hours per day traveling, or 61.0% of their daily activity engaged in travel. Studies of the activity cycle of *nemestrina* in captive and resource-rich habitats similarly suggest that this species spends a majority of time traveling (Bernstein, 1970; Crocket & Wilson, 1980). In dipterocarp forests home ranges tend to be characteristically large (8–10 km^2) (Rodman, 1978a,b; Caldecott, 1986a,b). These macaques frequently travel over large distances in habitats with poor visibility, with individuals highly dispersed from one another, so contact-calling may be very important (Caldecott et al., 1996).

In the Segari Melintang Forest Reserve, Peninsular Malaysia, oil palm plantations border the forest and the reserve consists of dipterocarp and freshwater swamp forests. Ruppert et al. (2018) compared the activity budgets and habitat use by pig-tails occupying primary forest and the oil palm plantations. In 2013–2014, 17% of the study group's core home range included the oil palm plantation, increasing to 28% in 2014–2015. In both the forest and plantation, pig-tails were most often seen on the ground level. In the forest, fruits made up 32% of the feeding scans, arthropods (not including ants) 24%, shoots and stems 15%, leaves 11%, and the remainder of the diet consisted of ants, fungi, flowers, and tree bark. In the plantation, 85% of the feeding scans involved the macaques consuming oil palm parts with fresh shoots, ants, palm bark, and vertebrates making up the remainder of the scans. Overall, the macaques were found to spend approximately 20% of the day in the plantation, visiting it on each observation day. Activity budgets were significantly quite different in the two habitats: In the plantation, pig-tails spend more time feeding and foraging and less time resting. Conversely, they spend more time traveling, resting, and in social activities in the forest. Sleeping sites were always located in the forest.

M. nemestrina live in groups ranging in size from 15–65 individuals with the largest group sizes observed in rich food resource habitats (Oi, 1986; Caldecott, 1986b; Ruppert et al., 2018). Pig-tail groups are composed of a relatively small number of adult males and a large number of adult females. Three focal groups studied by Oi (1987, 1990) at Mt. Kerinci in dipterocarp forests of West Sumatra had a mean adult sex ratio of 1:6.3 (M:F). Another group of 53 individuals in the Segari Melintang Forest Reserve had a mean adult sex ratio of 1:2.2 (M:F) (Ruppert et al., 2018). Adult males may act aggressively toward other group members, repelling peripheral males (as well as human observers) and will guard groups when they are in open habitats (Bernstein 1967a,b; Dazey et al., 1977; Erwin, 1978; Caldecott, 1986b; Caldecott et al., 1996). Their social organization has been described as consisting of different hierarchical groupings including "one-male harem units" (Caldecott et al., 1996: p.78) and solitary males; such variability in social structure and a tendency to form subgroups was conjectured to be related to unpredictable food resources. However, stable multi-male/multi-female groups have also been reported (Bernstein, 1967a; Oi, 1990). In a study of pig-tails in West Sumatra (Oi, 1990), three groups were observed for three different periods over a three-year span. The last period involved the provisioning of groups for close observation of group structure, but evidence of a multi-leveled social organization or a harem type structure was not seen. Although bimodal peaks in births have been reported (Bernstein, 1967a; Caldecott, 1986b), pig-tails are considered to be nonseasonal breeders.

IUCN categorizes this species as Vulnerable with a decreasing population trend (Richardson et al., 2008; IUCN, 2020). Despite the widespread geographic distribution of this species, pig-tails are considered to be at risk due to severe reduction in numbers from hunting and habitat loss. Throughout Southeast Asia, pig-tails are captured and trained to pick coconuts, which may further depress the male population (Oi, 1990). Pig-tails prefer undisturbed evergreen forest habitats and seem dependent on

them, suggesting a minimal amount of ecological plasticity; however, those living near oil palm plantations do exhibit behavioral plasticity by incorporating these habitats into their daily ranging, foraging, and feeding behaviors. Ruppert et al. (2018: p.239) refer to them as "ecologically flexible crop feeders." The ability to use human-modified habitats may in the long term provide some hope for the continued existence of this species, but their long-term survival will depend on their having access to intact forest habitats.

*Mentawai Island Macaque Species (*Macaca pagensis *and* M. siberu*)*

M. pagensis and *M. siberu* are endemic to the Mentawai Islands, which are located 85–135 km off the west coast of Sumatra and consist of four main islands: Siberut, Sipora, North Pagai, and South Pagai. Paleontological, molecular, and geographic evidence presented by Harrison et al. (2006) suggest that an ancestral population of *M. nemestrina* colonized these deep-water islands during the Middle Pliocene approximately 2.8 mya, concurrent with the onset of a major cold phase and increased forest fragmentation (Prentice & Denton, 1998; Harrison et al., 2006; Meijaard & Groves, 2006; Ziegler et al., 2007).

Both species are short-tailed macaques, but those located on Siberut have darker pelage and differ in hair ornamentation from those found on the other islands. The taxonomic status of the macaques inhabiting the Mentawai Islands has been problematic (Roos & Zinner, 2015). Originally considered to be a subspecies of *M. nemestrina* (Fooden, 1975), a number of short- term studies (Wilson & Wilson, 1976; Tenaza, 1987; Fuentes & Olson, 1995; Groves, 1996) proposed that these macaques were a distinctive species with two subspecific divisions (*M. p. pagensis* found on the Pagai Islands and Sipora, and *M. p. siberu* restricted to the island of Siberut). In 2002, Kitchener and Groves concluded (based predominantly on the morphology of two specimens) that the Siberut macaque was distinct, and this was later confirmed by divergence in the mitochondrial haplotypes of the two species (Roos et al., 2003).

The Mentawai Islands were likely colonized between 1.5–2.6 mya and may have included two separate colonization events (Roos et al., 2003; Ziegler et al., 2007). It is possible that a proto–*M. nemestrina* stock colonized the islands via land bridges, and an independent colonization of Siberut occurred either via an available land bridge from Sumatra or by rafting. In either scenario it is feasible that the Mentawai Islands served as ecological refuges for the early macaque colonizers. These islands are surrounded by deep waters which may have provided a buffer against climatic changes, making the Mentawai

Islands a warm refuge for primates (Gathorne-Hardy et al., 2002).

Although it may be hypothesized that the ancestral form(s) of the Mentawai macaques sought refuge from the changing paleoenvironments of Southeast Asia, the Mentawai Islands in more recent times have been subjected to large-scale habitat alteration. Conversion of forests to oil palm plantations, gardens or other crops, commercial logging, forest clearing, and the harvesting of forest goods have drastically reduced the distribution of forests. On the island of Sipora, only 10–15% of the original forest remains (Fuentes, 1997), and by 2005, 60% of the forest cover on Siberut had disappeared (Richter et al., 2013). *M. pagensis* have been observed in mangrove forest, primary mixed forest, primary dipterocarp forest, secondary forest, riverine forest, swamp forest, logged forest, nipa palm groves, and in gardens; however, it has been suggested that primary riverine lowland forest is the preference for this species (Whitten & Whitten, 1982; Tenaza & Tilson, 1985). Population densities average 7–12 individuals/km² with higher densities associated with logged forests versus unlogged forests, and the highest densities (12.3 individuals/km²) found in mature secondary forests (Paciulli, 2004). A study by Paciulli (2004) on the effects of selective logging on *M. pagensis* looked at densities of macaques living in forests logged 10–12 years ago, forests logged between 20–23 years ago, and forests that were never logged; analysis determined that densities of macaques did not significantly differ between the three forest types. Interestingly, Paciulli (2004) also discovered that macaque densities were higher in areas where there was a greater number of local pathways, i.e., areas closer to human habitation. Findings such as these suggest that *M. pagensis* exhibits a degree of tolerance to the effects of selective logging, is opportunistic in its use of habitat that is in close proximity to villages, and exploits a variety of habitat types.

M. pagensis has most frequently been observed traveling and fleeing on ground level (Kawamura & Megantara, 1986; Paciulli, 2004). Preliminary observations of foraging behaviors indicate that they are primarily frugivorous but, like most macaques, are opportunistic foragers (Whitten & Whitten, 1982). Group sizes range from 5–25, with only one adult male observed in smaller groups and multiple males in groups containing 20 or more individuals. Groups may also split up into smaller foraging units (Tenaza, 1987; Fuentes & Olson, 1995). Additionally, solitary males have been reported (Kawamura & Megantara, 1986; Fuentes & Olson, 1995).

A year-long study of the ecology of *M. siberu* (Richter et al., 2013) confirmed that this species spend the majority of their time (95.7%) in dense, continuous forest and that they tend to be semiterrestrial, spending 29% of time on the ground and 36% in the lower strata of the forest. These macaques travel on the ground and overall spend a large amount of time traveling. Monthly activity budgets reveal they spend 57% of the time traveling, 12% resting, 10% foraging, and only 6% in social activities. Time spent feeding is variable and related to the availability of fruit in the forest. This macaque is primarily frugivorous (76%), but the proportion of fruit in their diet shifts from 43–96%. Dietary diversity is low when the amount of fruit in the diet exceeds 80% and increases as less fruit is available. The macaques also eat insects (ants, termites, and spiders), comprising 11.9% of the diet, 4.5% of the diet are mushrooms, and only 4.4% are leaves. On one occasion, the macaques were observed catching and consuming crabs and shrimp from a river. The macaques had an average home range size of 80.6 ha, with a core area of 26.1 ha, and daily travel distance ranged between 1,054–3,360 m. Daily travel distances were greatest when fruit made up a large portion of their diet. Trees were extensively used as sleeping sites. The group studied by Richter et al. (2013) consisted of 29 individuals including 3 adult males and 8 adult females.

Richter et al. (2013) conclude that ecologically and behaviorally *M. siberu* is most similar to *M. nemestrina*. Both species spend a large portion of time traveling on the ground for long distances, and forage on small and dispersed fruit patches. The distribution of feeding trees promotes foraging as single individuals or in small groups.

Crested serpent eagles (*Spilornis cheela sipora*) and reticulated pythons (*Python reticulatus*) are potential predators of these macaques, although no incidences of predation have been observed. Hunting of Mentawai primates by aboriginal inhabitants of these islands has a long history, but traditional bows and arrows are no longer the weapons of choice as hunters now use high-caliber air rifles (Tenaza, 1987, 1988; Whittaker, 2006). However, Metawians consider macaque meat to be "unsavory and tough," so it is not preferred game (Paciulli, 2004: p.159). Nevertheless, macaques are the most common Mentawai primate to be caught in traps. *M. pagensis* are captured and sold as pets. They are considered pests because of their tendency to raid crops and gardens; consequently, farmers build traps specifically to catch them. In fact, the intensive garden-raiding behavior of macaques living on the island of Berikopek led to their complete extirpation by 1981 (Tenaza, 1987). It is estimated that only ~2,100–3,700 individuals remain on the Pagai Islands and Sipora (an 80% decline from 1980; Setiawan et al, 2020), causing IUCN (2020) to list *M. pagensis* as a Critically

Endangered species. Population estimates gathered by IUCN for *M. siberu* place it in an Endangered category with a decreasing population trend (Traeholt et al., 2020).

Due to the dwindling number of macaques and the widespread habitat alteration that has occurred on the Mentawai Islands, it is difficult to determine the capacity of these species for behavioral and ecological plasticity. *M. siberu's* preference to occupy dense, continuous forest and *M. pagensis'* use of habitat in close proximity to humans may indicate that compared to *M. pagensis*, *M. siberu* is capable of a lesser degree of ecological plasticity. As noted by Richter et al. (2013), the future of these macaques is inextricably linked to the future extent of habitat degradation and loss that may occur on these islands.

Sulawesi Macaque Species *(Macaca brunnescens,* M. hecki, M. maura, M. nigra, M. nigrescens, M. ochreata, *and* M. tonkeana)

The deep-water island of Sulawesi lies east of Borneo (Kalimantan); it has four irregular arms that make up approximately 227,000 km^2 (Whitten et al., 2002). High levels of endemism are characteristic of the island's flora and fauna (e.g., 62% of the island's mammals are endemic) (Whitten et al., 2002). It is hypothesized that an eastern ancestral population of *M. nemestrina* colonized the island via natural rafting or possibly through a land bridge connection via Borneo (Fooden, 1969; Meijaard, 2003; Ziegler et al., 2007). The extant Sulawesi macaques represent a unique adaptive radiation facilitated by a lack of cercopithecine primate competitors and natural predators. The only other primates found on Sulawesi are the tarsiers that underwent a similar radiation.

The macaques of Sulawesi are a taxonomic conundrum and the phylogenetic relationships between them have not been fully resolved (Roos & Zinner, 2015). The number of species recognized by various authorities varies from one species (*M. nigra*) (Thorington & Groves, 1970) to seven (Zinner et al., 2013) or eight (Froehlich & Supriatana, 1996). Typically, however, the standard practice has been to distinguish the seven species identified by Fooden (1969). Traditionally, species have been identified based on unique traits of pelage, ischial callosities, craniofacial dimensions, tail lengths, and dermatoglyphics, along with molecular data.

Similar to the Mentawai Islands, Sulawesi and the neighboring Togian Islands reflect a complex biogeographical history. During the Pleistocene, these islands may have provided important forest refuge to ancestral macaques. Today, coral reefs are evident on Sulawesi, suggesting that during the Pleistocene, contemporary lowlands were submerged, thus creating an archipelago of separate islands which promoted the allopatric speciation of the seven different species of macaques (Fooden, 1969).

The macaques on Sulawesi are predominantly parapatric in their distribution and a number of hybrid zones have been reported (Groves, 1980; Ciani et al., 1989; Watanabe & Matsumura, 1991; Watanabe et al., 1991; Supriatna et al., 1992; Froehlich & Supriatna, 1996; Bynum et al., 1997a,b; Bynum, 2002; Riley et al., 2007, 2013). Gene flow has likely resulted from decreased geographic isolation (once sea levels receded) and secondary intergradation promoted by human-induced habitat alteration (Bynum, 2002). Documented hybrid zones are narrow and may more accurately reflect hybrid sinks rather than expanding areas of gene introgression (Ciani et al., 1989; Evans et al., 1999).

Adult male Sulawesi macaques range from 5.5–16.6 kg and adult females 4.2–9.6 kg (Watanbe et al., 1987). They have short tails with brown to glossy black pelage (some species have contrasting pale, ochraceous gray to buff-colored patches on the throat or limbs), short, almost projecting muzzles, and moderate to pronounced supraorbital ridges; some species sport erect crests of hair on the top of their heads.

Sulawesi macaques are generally described as forest dwelling, preferring lowland rainforest habitats and primary to secondary forests (Supriatna et al., 1992; Kohlhaas & Southwick, 1996; Lee, 1997; O'Brien & Kinnaird, 1997). They have been reported living in a number of different forests types, including humid tropical lowland and upland rainforests, seasonal forests, beach forests, savanna swamp forests, and forests that border agricultural areas (Matsumura, 1991; Lee, 1997; Riley, 2005; Schillaci & Stallmann, 2005; Riley et al., 2007; Nailufar et al., 2015). Compared to other Southeast Asian habitats occupied by macaques, the island of Sulawesi offers a reduced degree of floral diversity (Lee, 1997), and similar to the island of Mentawai, Sulawesi has experienced large-scale anthropogenic change.

Riley (2007a, 2008) compared the habitat use, ranging behaviors, use of forest strata, diet, and activity patterns between two groups of *M. tonkeana* living in minimally altered (2.1% of home range) and heavily altered (66% of home range) lowland and hill forests in central Sulawesi. In the highly altered habitat, group size ranged from 6–9 individuals but averaged 26–>28 in the less-altered habitat. The macaques had similar day ranges, but in the heavily altered habitat they had a greater home range size per individual (4.83 versus 2.99 ha) and more intensely used a limited area of their home range where favored sleeping trees and a garden site were located. Also, in the more heavily altered habitat, the macaques spent significantly more time on the ground (~22%,), while the group living in the less-altered habitat spent 86% of their time in the upper forest strata levels. In heavily altered habitat, *M. tonkeana* spent more time foraging, less time moving, and more time resting, reflecting the reduced availability of resources. Time engaged in feeding and social behaviors was relatively similar between the two groups. Variability in locomotion patterns has been reported for *M. nigra, M. nigrescens*, and *M. maura* (Watanabe & Brotoisworo, 1982; Kohlhass, 1993), but increased terrestriality is often associated with habitat disturbance.

Sulawesi macaques have been described as being highly frugivorous with a strong preference for figs (*Ficus* spp.) (Kohlhass, 1993; Lee, 1997; O'Brien & Kinnaird 1997; Riley, 2007b). Despite a preference for fruits, *M. nigra* include more than 145 species in their annual diet (O'Brien & Kinnaird, 1997) and *M. tonkeana* feed on ~56 different food species (Riley, 2007b). Consequently, at least in these two species, a considerable amount of dietary diversity is evident. Two *M. tonkeana* groups studied by Riley (2007b) exhibited a preference for ripe fruit; however, the group living in a highly disturbed forest ate less fruit than the group found in a minimally altered habitat (66.8% versus 79.8%) and consumed more alternative food items, especially insects (14.6% versus 5.6%). Overall, dietary diversity was less for the group living in the highly disturbed habitat, and 52% of their diet was composed of a single species of palm fruit (*Arenga pinnata*).

Lee (1997) recorded similar findings for *M. nigra* living in North Sulawesi. Key food species were *Ficus* and *Palaquium* spp. Groups living in the more disturbed areas of the Tangkoko-DuaSudara (TDS) Nature Reserve included more insects in their diet than did those groups living in less-disturbed areas of the forest. Insect consumption was inversely related to the amount of fruit consumed. The macaques also spent more time on the ground in the more highly disturbed areas and had larger home ranges and day range lengths than did those living in less-disturbed areas. The ranging behavior of *M. nigra* living in TDS was most affected by the distribution and availability of key food species (O'Brien & Kinnaird, 1997). In fact, feeding on the top four food species of fruits and insects accounted for 74–92% of the observed variations in the patterns of home range use. No relationship between group size and home range size was detected, but the group with the most primary forest in its range spent more time in the trees, had the shortest day range, and spent less time feeding.

The highest reported densities for *M. nigra* come from primary forest on the island of Bacan (170.3 individuals/km², with a mean group size of 24.9) where the macaques are not hunted, while the lowest figures (8.9 individuals/km², with a mean group size of 32) are from groups living at the periphery of secondary regenerating forest in the Manembonembo Nature Reserve where the macaques are hunted (Lee, 1997; Rosenbaum et al., 1998; Palacios et al., 2011; Kyes et al., 2012). Based on a survey of the ecological capacity of specific habitats to support macaques, Lee (1997) found that population densities did not necessarily reflect a 1:1 relationship between habitat quality and density figures but were more a reflection of the impact of hunting on the macaques.

Mean group size for the majority of the Sulawesi macaques approximates that reported for most other macaques (20–30), but home range size is highly variable across species (Riley & Priston, 2010). However, at the Tangkoko Reserve, Marty et al. (2017) recorded considerably larger group sizes (60–90 individuals) for *M. nigra*. In a six-year period, 16 alpha male replacements in 4 groups of *M. nigra* were observed, and Marty et al. (2017) were able to correlate male body weight with rank and witnessed intense male-male fighting over mating access to females. A distinctive male reproductive skew is apparent; however, this population remains genetically variable and viable (Engelhardt et al., 2017).

In a study by Kohlhaas (1993) *M. nigrescens* consumed 60 plant species with 49% of their diet comprised of fruits from *Ficus* species. Their diet consisted of 85.1% fruits, 8.9% arthropods, 3.5% leaves, and 2.5% other plant parts. They were observed spending 10.1% of their day feeding, 19.7% moving, 22.7% engaged in social behaviors, and 47.6% of their time resting. Diet and activity were affected by seasonal factors, and affiliative behaviors were more frequent during times of cooler temperatures. *M. nigrescens* traveled on the ground, but ground use was less when temperatures were high. They used all forest strata, spending 70.5% of the time in the middle to upper canopy at 15–30 m above the ground. Average group size ranged from 9.4–18.1 individuals, although solitary individuals also were observed. Kohlhaas (1993) noted that group size increased with the amount of rainfall. In the Dumoga-Bone National Park of North Sulawesi, where Kohlhaas studied, she noted that pythons, dogs, and humans were predators. She also noted that these macaques were not observed in grassy areas and that the availability of fruits determined habitat use. According to Kohlhass (1993), the future survival of *M. nigrescens* is inherently linked to the lowland forests encompassed by this park.

Although primary forest may be the optimum habitat for Sulawesi macaques, *M. maura* (Supriatna et al., 1992), *M. tonkeana* (Riley, 2008), and *M. brunnescens* (Priston et al., 2012) demonstrate an ability to persist in heavily disturbed habitats because of their ability to appropriate human-produced foods, i.e., engage in crop- and garden-raiding behaviors.

The crop raiding of *M. maura*, *M. tonkeana*, and *M. brunnescens* is an obstacle to their preservation (Supriatna et al., 1992; Riley et al., 2007, 2013; Priston et al., 2012; Hardwick et al., 2017; Zak & Riley, 2017). In southern and central Sulawesi, many of the forests have been illegally converted into agricultural lands, so macaques are found in discontinuous and disturbed forest tracks that are in close proximity to human settlements. Karst landscapes at higher altitudes (>1,000 m) may provide some protection for *M. tonkeana* in central Sulawesi, but in southern Sulawesi, low population densities for *M. maura* have been reported (Supriatna et al., 1992), with groups containing more mature than immature individuals.

Camera traps and farmer reports are methods now used to more accurately access the crop-feeding behavior of *M. maura* (Zak & Riley, 2017). Although predominantly frugivorous, *M. maura* are known to eat a wide variety of crops including cacao and watermelon. These macaques and wild pigs (*Sus scrofa*) are considered by farmers to cause the most severe crop damage. Camera-trap data confirm this; however, the timing of actual macaque crop-feeding events (CFE) and the frequency of CFEs did not match farmers' reports. A majority of macaque crop-feeding events were recorded in early afternoons when farmers were frequently in their gardens, instead of occurring in the mornings as farmer reports indicated and when farmers were not in their fields. In addition, while farmers reported that multiple CFEs were a weekly occurrence, camera traps recorded fewer weekly events. More accurate assessment of the timing and frequency of macaque CFEs may aid in efforts to reduce conflict between farmers and macaques.

Tonkean macaques are primarily frugivorous with fruits comprising between 40–93% of their diet, averaging 81% of feeding records (Riley, 2007a). However, compared to forest fruits, cacao pulp, which they regularly exploit, is higher in digestible carbohydrates and energy content (Riley et al., 2013). In addition, unlike fruits found in the forest, cultivated cacao is densely distributed, making it highly attractive to the macaques.

A study of the actual damage caused through crop raiding by *M. tonkeana* in Lindu, central Sulawesi revealed that fruit loss caused by macaques ranged from 0–6%, while fruit loss caused by the "forest mouse" ranged between 6–24% (Riley, 2005). Interestingly, in four groups of macaques having home ranges that either included or were in close proximity to cacao gardens, only one group raided the gardens. Crop loss and distance of a garden to

a village were positively correlated, suggesting that macaques were most prone to raiding gardens located farthest from a village (Riley, 2007b).

M. brunnescens on the island of Buton has a long history of crop raiding. An eight-year study of their crop-raiding behavior (Priston et al., 2012) revealed that individual macaques most frequently raided close (0–10 m) to farm perimeters; increased group numbers tended to prolong the length of time that the macaques foraged and fed at the farms. In addition, 32–38% of the macaques' daily activity consisted of crop-raiding behaviors. Hardwick et al. (2017) determined that 40–80% of the macaque's diet consisted of agricultural crops.

Habitat preference of the Sulawesi macaques is difficult to determine due to the fact that undisturbed forest habitats are rapidly disappearing from the islands. All species, however, seem to require some forest in their habitats; consequently, they have been described as "noncommensal macaques" (Kohlhaas & Southwick, 1996: p.133). The crop-raiding behavior and use of multiple forest strata by Sulawesi macaques, especially *M. tonkeana*, *M. maura*, and *M. brunnescens*, suggests that these species do demonstrate a potential for moderate ecological plasticity (Riley et al., 2013).

The greatest threat to the Sulawesi macaques is the loss of forest habitats due to large-scale conversion of forests to plantations and expanding areas of human settlement. In the case of *M. nigra*, hunting is also a major threat. In northern Sulawesi, the potential for increasing human-macaque conflict is most critical at elevations between 400–800 m (Nailufar et al., 2015), i.e., areas intensely used by humans. Ownership of macaques as pets in Sulawesi is common and presents concerns about zoonotic and anthropozoonotic disease transmission (Jones-Engel et al., 2001, 2004). Some Sulawesi macaques are also killed for their crop-feeding behaviors (Hardwick et al., 2017), and especially so when they raid crops in less-wealthy communities or where cultural taboos that discourage such killings are absent (Riley & Priston, 2010). The range-restricted *M. brunnescens*, endemic to the small islands of Buton and Muna, is particularly vulnerable to being killed because of its perceived crop-raiding behaviors. However, villlagers often overestimate the crop damage that these macaques cause (Riley, 2007b; Zak & Riley, 2017).

There is great concern regarding the conservation and future status of Sulawesi macaques. Table 17.3 is a listing of the IUCN status for these seven species.

SINICA SPECIES GROUP

The *sinica* group of macaques contains six Asian species: *M. sinica*, *M. radiata*, *M. assamensis*, *M. thibetana*, *M. munzala*, and the newly discovered *M. leucogenys*.

TABLE 17.3 IUCN Red List Status of Sulawesi Macaques

Species	IUCN Red List	Population Trend	References
Macaca brunnescens [a]	Vulnerable	Decreasing	Lee et al. (2021)
M. hecki	Vulnerable	Decreasing	Lee et al. (2020a)
M. maura	Endangered	Decreasing	Riley et al. (2020a)
M. nigra	Critically Endangered	Decreasing	Lee et al. (2020b)
M. nigrescens	Vulnerable	Decreasing	Lee et al. (2020c)
M. ochreata	Vulnerable	Decreasing	Riley et al. (2021)
M. tonkeana	Vulnerable	Decreasing	Riley et al. (2020b)

[a] *M. brunnescens* was previously considered to be a subspecies of *M. ochreata* (i.e., *M. ochreata brunnescens*).

Geographic range for the group encompasses a fragmented South Asian distribution (Thierry, 2007), extending throughout much of Sri Lanka, southern India, Nepal, the mountainous areas of eastern Tibet, and from Szechwan east to Kwangtun and Fukien in China (Fooden, 1971a; Eudey, 1980) (see Figure 17.6). *M. sinica* inhabits an island, while *M. radiata, M. assamensis,* and *M. thibetana* live in tropical and subtropical continental areas (Thierry, 2007). *M. munzala* has been found at a higher altitude than most other macaques (Holden, 2004) and *M. leucogenys* has been photographed in southeastern Tibet (Li et al., 2015).

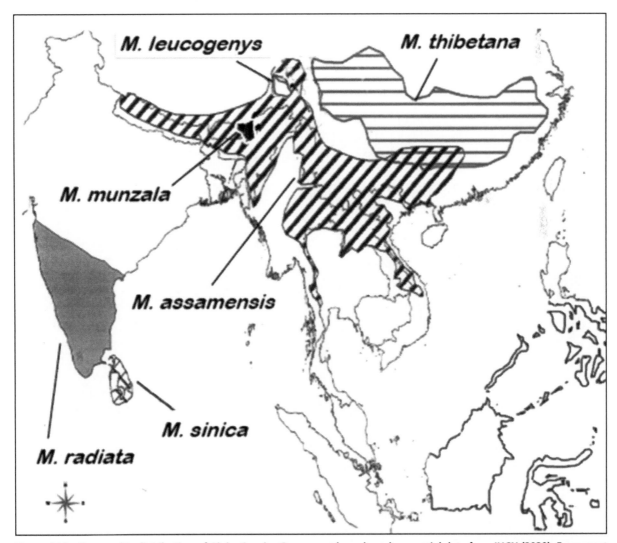

Figure 17.6   Geographic distribution of *Sinica* Species Group members, based on spatial data from IUCN (2020). From west to east: *Macaca radiata, M. sinica* (endemic to the island of Sri Lanka), *M. assamensis, M. munzala* (black polygon within *M. assamensis* distribution), *M. leucogenys* (white polygon within *M. assamensis* distribution), and *M. thibetana*.
(Map credit: Ian Colquhoun)

The *sinica* group is thought to have dispersed later than the *silenus* or *sylvanus* groups, but earlier than the *fascicularis* group. The "moderately disjunct" distribution of the *sinica* group lends support to a theory of dispersal earlier than *fascicularis* (Fooden, 1976: p.228). Fooden (1976) also noted a possible historic or ecological interrelationship between *sinica* and the more southerly *arctoides* species group, a view supported by Roos and Zinner (2015). Eudey (1980) pointed out that *M. sinica* in Sri Lanka and *M. radiata* in southern India are the most representative of the ancestral population of the *sinica* group. She noted that 18,000 years ago (the interlude associated with the last glacial period), northern India consisted of savannas, and this arid period may have directly affected the distribution of the *sinica* group species. Despite theories established in the 1970s and 1980s, Chakraborty et al. (2007) maintain that the specifics of phylogenetic relationships and evolutionary history of the *sinica* species group are basically unknown.

Morphologically, the *sinica* group includes the smallest macaque (*M. sinica*) and the largest (*M. thibetana*). Pelage varies from the relatively bright golden color found in *M. sinica*, to dark brown in *M. assamensis*, *M. thibetana*, and *M. munzala*, and to dull brown in *M. radiata* (Ashmore-Declue, 1992; Sinha et al., 2005a,b). Tail length is also variable among the six species. Both *M. sinica*, the toque macaque, and *M. radiata*, the bonnet macaque—as

their common names imply—have distinctive whorls of hairs that radiate from a central point on their heads (see Figure 17.7). *M. leucogenys* has distinctive pale-to-white long side and chin hairs (Li, et al., 2015).

Species in the *sinica* group are generally allopatric with a couple of exceptions: The Arunachal macaque (*M. munzala*) is believed to be sympatric with the Assamese macaque in the lower elevations of its range (Chakraborty et al., 2007; Kumar et al., 2007). *M. thibetana* may also occupy areas inhabited by *M. leucogenys* (Li et al., 2015). *M. assamensis* and *M. thibetana* are restricted to subtropical broadleaf evergreen forests, while *M. radiata* is often found in disturbed habitat (Fooden, 1982). *M. sinica*, isolated from other macaque species on the island of Sri Lanka, shows evidence of ecological release and populates all the major forest zones, i.e., broadleaf evergreen forest and a variety of other semi-evergreen, transitional, and arid habitats (Fooden, 1982). *M. munzala* inhabits a highly seasonal environment of subtropical to temperate broadleaf forests where cold winters at high altitudes are the norm (Mendiratta et al., 2009). *M. leucogenys* occupies primary and secondary evergreen broadleaf forests (Li et al., 2015).

All members of the *sinica* species group are listed by IUCN as either Endangered or Near-Threatened, except *M. radiata*, which is in the Vulnerable category (IUCN, 2020). The majority of species are known to raid crops and, therefore, face human-induced mortality (Dittus

Figure 17.7 *Macaca radiata* with distinctive "bonnet" of hair.
(Photo credit: Richard Saunders, https://unsplash.com/s/photos/bonnet-macaque)

et al., 2008). The level of ecological plasticity is unknown for *M. munzala* and *M. leucogeny*. However, *M. radiata* and *M. sinica* thrive where, as Richard et al. (1989: p.573) noted, they "often live in and alongside towns and villages and they exploit the fields of farmers and secondary growth nearby. Indeed, they often depend directly or indirectly on human activities for a substantial portion of their diet." Conversely, these authors categorized *M. assamensis* and *M. thibetana* as species that are not adaptable to human disturbance based on variables of diet, habitat, and avoidance of people. Obviously, conservation status and plasticity are independent of each other since *M. sinica* has been categorized by IUCN (2020) as Endangered, yet qualifies under Richard's et al. (1989) criteria as an ecologically adaptable species.

Toque Macaque *(Macaca sinica)*

The toque macaque is the smallest member of the genus; adult males weigh 3.97–6.12 kg, and adult females 2.49–4.30 kg (Fooden, 1979). Unique features include a bright golden coloration, a symmetrical head cap of hair, and an average tail length greater than other macaques (Fooden, 1979).

M. sinica is endemic to the island of Sri Lanka. Ancestors of this species most likely reached Sri Lanka before the middle Pleistocene. Rising sea levels then isolated the founder population from the macaques on the Indian mainland. The geographic distribution of toque macaques is not continuous but limited to areas where there are forests, permanent water sources, or where temperatures do not go below freezing (Dittus, 1974). The species has been found in all major climatic zones, i.e., low-country dry zone, wet zone and lower portions of the hill zone, and the highlands (Wijeyamohan et al., 1996). Elevations at which *M. sinica* are found range from sea level to 2,100 m (Fooden, 1979). At the edge of forests, they live close to human settlements (Mendis & Dangolla, 2016).

Toque macaques are primarily frugivorous, preferring to eat the fruit of fig trees. This species does not usually live in urban areas, but dependence on crop raiding or garbage dumps is highly variable among populations (Richard et al., 1989). Much of the ecological and behavioral data have been gathered at the Polonnaruwa Sacred Reserve, a semi-evergreen forest surrounding an archaeological site located in the north-central dry zone. Toque macaques consume parts of 41 out of 46 tree species found in Polonnaruwa Reserve, fruits from 3 species of figs making up approximately 51% of the total diet (Ashmore-DeClue, 1992). Only a small percentage of the diet consists of insects or small animal prey; however, 20% of the daily foraging time may be spent searching for these items (Dittus, 1974, 1977a). During times of food scarcity, herbs and shrubs become an important component of the diet (Dittus, 1974). At Udawattakelle Sanctuary, a forested area with annual rainfall of 1,750 mm, Vandercone and Santiapillai (2003) found that macaques utilized more than 30 species of plants for fruit, flowers, leaves, bark, or sap. They were observed to spend a large portion of foraging time securing animal matter on the forest floor, leading researchers to classify the Udawattakelle macaques as omnivores. At Udawattakelle, groups emerged from the forest and fed on a steady supply of discarded rice and other carbohydrates throughout the year, suggesting adaptability to different habitats and food items (Vandercone & Santiapillai, 2003).

Toque macaques live in multi-male/multi-female groups, but one all-male group was observed moving in conjunction with the largest of Dittus's study groups at Polonnaruwa. Females are philopatric, but adolescent males migrate an average of once every 3.5 years and adult males migrate once every 5.8 years (Ashmore-DeClue, 1992). At Polonnaruwa the mating season occurs during the months of July–September, with a narrow birth peak during December–February (Dittus, 1974). The timing of reproductive behaviors is highly variable between groups, but the timing of births seems to be generally related to local rainfall patterns (Fooden, 1979). Infants are thus ready to eat solid foods when new vegetation and fruits are available. The average annual birth rate of adult females is 0.69 (Dittus, 1977b). High-ranking females produce more female offspring than mid- or low-ranking females, and a greater proportion of infants born to high-ranking females survive into adulthood (Dittus, 1979). Daughters inherit social rankings from their mothers and obtain a rank just below that of their mothers.

Group size ranged between 20–76 monkeys at Udawattakelle, with a mean of 13% adult males, 30% adult females, 8% subadult males, 31% juveniles, and 18% infants (Vandercone et al., 2006). This compares to smaller group sizes (8–43 animals) observed at Polonnaruwa (Dittus, 1974) and Ruhuna National Park (10–20 animals) in thorn scrub dry zone (Wijeyamohan et al., 1996). There is likely a positive correlation between group size and home range. Typically, *M. sinica* tend to have small home ranges, around 0.4 km² (Hladik & Hladik, 1972; Dittus, 1974), but groups having richer resources tend to have smaller ranges than those occupying resource-poor habitats. Groups that do not fit this pattern are those that have a unique food resource, such as a rice mill or garbage dump, within the confines of their home range (Dittus, 1974).

Toque macaques are well suited for arboreal life, but like most other macaques, they are equally adept at foraging and traveling terrestrially. Dittus (1974) found that individuals at Polonnaruwa spent 76% of the time in trees and 24% on the ground; when startled, they took flight through the trees, and feeding on terminal branches was common. Dittus and Ratnayeke (1989) found that animals spend 46% of the time foraging, 10% traveling, 34% resting, and 10% involved in other activities. Foraging stops at nightfall when a group is close to its sleeping trees. Each group has several preferred sleeping sites, and only 15% of the trees are used on successive nights. The tendency to use different sleeping trees on successive nights may have an adaptive value in lowering predation risks (Dittus, 1974, 1977a).

In the Lower Hanthana woodlands of Peradeniya University, the overall activity budget of the macaques consists of 32% resting, 26% moving, 20% feeding, 10% social activities, and 12% other (Weerasekara & Ranawana, 2018). As monthly temperatures increase, the macaques decrease the amount of time they spend moving. Average monthly home range size of 0.64 km² also negatively correlated with monthly average temperatures (Weerasekara & Ranawana, 2017). This variability in activity budgets and ranging behaviors represent flexible coping responses made by the macaques to offset thermoregulatory costs created by high tropical temperatures.

Potential predators on the island of Sri Lanka include leopards, jackals, fishing cats (*Felis viverrina*), jungle cats (*F. chaus*), crocodiles (*Crocodylus palustris*), Indian pythons, eagles, and hawks (Fooden, 1979). Domestic dogs are also common predators. Predation from carnivores, reptiles, and raptors may inflict significant mortality on some populations of toque macaques since certain age/sex classes are of a size that wild predators seek as prey. Thirteen predations have been observed by researchers (Cheney & Wrangham, 1987). Estimated annual predation rates of 0.5%–1.0% have been calculated based on field studies in Sri Lanka (Hart, 2000). Due to an increased threat of predation in times of drought, toque macaques become extremely vigilant while drinking at water holes where large groups break into smaller groups that alternate between drinking and attentive watching.

Hindu and Buddhist religious beliefs gave M. *sinica* protection in the past, and the toque macaque has never been hunted by humans for food. Nonetheless, much mortality derives from poison, traps, snares, and guns that villagers use in an attempt to prevent damage to cultivated fields. This purposeful killing, and particularly habitat loss due to plantations and destruction of trees

for fuel, has resulted in a 50% decline in the population of toque macaques (Molur et al., 2003), leading to an Endangered designation by IUCN (2020) but no legal protection from the Sri Lankan government (Dittus et al., 2008). Macaques living in the city of Kandy are considered a menace and sterilization programs have been enacted (Mendis & Dangolla, 2016). The toque macaque exhibits a high level of ecological plasticity as evidenced by its ability to exploit many habitat types and to live in the vicinity of human settlements. However, no amount of plasticity can withstand devastating habitat loss and the additive mortality due to extermination by humans. According to Dittus et al. (2008), during the period 1956–1993, Sri Lanka lost 50% of its forest cover, and habitat destruction continues up to the present time. Since there is a 1:1 relationship between loss of critical habitat and population numbers of this primate, *M. sinica* face a bleak future.

Bonnet Macaque (*Macaca radiata*)

Bonnet macaques are drab grayish-brown monkeys with a distinctive head cap of crown hairs radiating from a central whorl; facial skin is usually pinkish but occasionally scarlet in females (Fooden, 1981). Bonnets are larger than toque macaques but smaller than rhesus. Male weight ranges from 5.44–8.85 kg and female weight from 2.93–4.99 kg; adult males average 75% heavier than adult females and 15% greater in length (Fooden, 1981).

The geographic range of *M. radiata* is restricted to southern India; the northernmost limit of *radiata* distribution coincides with the northern limit of the Western Ghat mountain range and the southernmost with the tip of Cape Comorin (Fooden et al., 1981). *M. radiata* living in coastal areas have virtually been eliminated from these habitats, so its current geographic distribution is vastly shrinking (Erinjery et al., 2017). It is conjectured that *M. radiata* range expansion over time is directly correlated with range reduction of *M. silenus*, likely as a result of changes in climate and vegetation (Fooden, 1976). Presently, the geographic distribution of bonnet macaques is also being reduced by the appropriation of their habitats by *M. mulatta*. In fact, in less than 40 years rhesus have invaded nearly 28,000 km² of the range of the bonnet macaque (Erinjery et al., 2017). Fooden (1981) described the monsoonal environment in which bonnet macaques are found as a broad range of wooded and partly wooded habitats from sea level to an elevation of 2,100 m. Forest types include dry and moist deciduous, bamboo, evergreen, and shola. Groups also often are found in disturbed and man-made habitats (cultivated fields, outskirts of villages, temple compounds, urban parks, and market areas). Two common characteristics of the habitats occupied by bonnet macaques are the presence of large trees (such as banyan trees) and a degree of habitat disturbance. Erinjery et al. (2017) describe *M. radiata* as a commensal, habitat generalist.

In roadside habitat, macaques consume over 40 plant species (Makwana, 1980) but since banyan trees (*Ficus* spp.) are commonly found along roadsides throughout India, wild figs may be the most important single item (Simonds, 1965). Fooden's (1981) review of the species found that the diet of *M. radiata* consists of a combination of wild fruits, flowers, leaves, and seeds, combined with foods that are obtained from humans, such as crops, garbage, and temple offerings. Animal food includes insects and spiders (much terrestrial foraging centers on arthropod prey), lizards, and bird eggs. In agricultural areas, as much as 40% of food comes from crop raiding. Groups in temple and market areas may be provisioned due to Hindu religious beliefs; additionally, the macaques may raid houses, shops, and refuse dumps. More recently, the number of macaques living in temple/tourist areas has drastically declined, as these no longer appear to be stable habitats (Erinjery et al., 2017). In addition, roadside habitats have changed dramatically with vegetation eliminated by increasing urbanization. Erinjery et al. (2017) predict that many roadside populations will be nearly extinct within the next 10 years.

Bonnet macaques inhabiting a relatively protected tropical dry evergreen forest without villages were studied by Krishnamani (1994) who found that fruits were the most important diet component, followed by leaves and invertebrates. When a wild population that lived exclusively in the forest was compared to an entirely commensal group inhabiting a village, analysis revealed a significant difference in activity patterns between the two groups (Schlotterhausen, 1998). The wild macaques moved and fed more, suggesting higher survival demands, while the commensal group ate a non-varied diet consisting of human foods. Rahaman and Parthasarathy (1967) estimated that a group of bonnets living in a garden sector of Bangalore, where they did not need to range far for food, would spend 50% of their daily activity budget sleeping and grooming, 40% feeding, and 10% moving. These data may explain the "larger, stronger, and more terrestrial" description assigned by Richard et al. (1989: p.583) to *M. radiata* living close to human settlements in comparison to their forest-living conspecifics. Fooden (1982) categorized bonnet macaques as highly arboreal, i.e., 90% arboreal and 10% terrestrial in forest groups but 70% arboreal in commensal groups; sleeping sites were exclusively arboreal in the forest, but on roofs or building ledges in urban areas.

Bonnet macaques live in large multi-male/multi-female associations with females forming strong linear dominance hierarchies typical for cercopithecine species (Mehu et al., 2006). Extensive social interactions between individuals of both sexes, all ages, all ranks, and kinship groups are the norm in bonnet macaque society. Forest groups consist of only 3–12 indiviuals (Richard et al., 1989), while groups that inhabit roadsides may number as large as 52 animals (Simonds, 1965); normally, however, social groups are in the 30–40 range (Thierry, 2007). In the northern states of Karnataka and Telangana, a study by Erinjery et al. (2017) surveyed 35 groups that ranged in size from 5–21 with a mean group size of 12. Data indicate that the adult male–to–adult female sex ratios of *M. radiata* range from 1:1.3 in forest groups and 1:1.7 in urban areas (Ménard, 2004). Simonds (1974) suggested that males migrate from one group to another without undergoing a long period of isolation. The ability of a bonnet male to shift into a new group with little male-male aggression may be reflective of a highly integrated social organization. Both uni-male groups and female transfer from their natal groups have been observed in bonnet macaques (Sinha et al., 2005b).

Wild *M. radiata* are seasonal breeders (93% of breeding occurs in September–November), with the majority of births in February–March (Rahaman & Parthasarathy, 1969). Most births occur in the dry season, which precedes the spring monsoons, thus newborn bonnets are several months old before the rains revitalize food and water resources (Fooden, 1981). New mothers and their infants form clusters that do not seem dependent on kinship (Rahaman & Parthasarathy, 1969; Sugiyama, 1971; Simonds, 1974). This cross-cutting of kin and dominance lines may provide infants with extended social contacts (Simonds, 1974). Thierry (2000) categorized *M. radiata* as a comparatively socially tolerant member of the genus, reflecting less dominance and kin-bias than many other macaque species.

There have been numerous observations of predation on bonnet macaques by domestic dogs (Makwana, 1980; Simonds, 1965). Typical wild predators are tigers, leopards, dholes, and Indian pythons but may also include large raptors that prey on immature animals. Leopard scat analyzed at Mudumalai Wildlife Sanctuary confirmed that bonnet macaques comprised 0.92% of total prey consumed (Ramakrishnan et al., 1999). Estimated annual predation rate due to nonhuman predators is calculated at 3.0% (Hart, 2000). Alarm calls are given by bonnets in response to terrestrial predators such as tigers, leopards, dholes, jackals, dogs, pythons, and humans (Hohmann, 1989; Ramakrishnan & Coss, 2000; Coss et al., 2007). One aspect of predator avoidance is the selection of sleeping sites in emergent trees on terminal branches that overhang water (Ramakrishnan & Coss, 2001).

Considering their wide distribution and habitat adaptability, combined with a presumed large population, bonnet macaques were not considered in danger until the last decade. However, decreasing numbers now have placed the species in the IUCN Vulnerable category (Singh et al., 2008; IUCN, 2020). Numerous researchers have issued alarms about the dwindling number of *M. radiata* and determined that declining vegetation and increases in urban areas were good predictors of population decline (Singh & Rao, 2004; Erinjery et al., 2017). In fact, from 2003–2015 populations of *M. radiata* have decreased by more than 50% (Erinjery et al., 2017). Although Southwick and Siddiqi (1994a, 1994b) cited cultural and religious protection in the vicinity of Hindu holy places as integral to the survival of primates in India, during a later survey Kumara et al. (2010: p.37) found that bonnet macaques "have been eliminated from about 48% [of the] temples/tourist spots where they occurred in the recent past." Increased intolerance by humans is causing a declining trend in many parts of the bonnet macaque range (Singh et al., 2008; Erinjery et al., 2017).

M. radiata have been classified as a primate species exhibiting a high degree of behavioral flexibility (Anderson, 1998) and as a "weed" species (Richard et al., 1989). Further evidence of the behavioral and social plasticity of bonnet macaques was discussed by Sinha (2005a), in particular the possible connection between food provisioning and the evolution of uni-male groups and female dispersal. Because it is largely commensal and occupies a wide range of habitat types, *M. radiata* do exhibit a maximum range of ecological and behavioral plasticity.

Assamese Macaque (Macaca assamensis)

M. assamensis (the Assamese macaque) are relatively large monkeys; males average 11.3 kg and females 6.9 kg (Thierry, 2007). Pelage is golden brown to dark brown on the back with paler fur on the abdomen (Chalise, 2003a), although geographic differences in morphology have been observed (Chalise, 2003b). Tails are fairly long, ranging in length from 170–250 mm (Fooden, 1988). Assamese macaques have chin tufts, chin whiskers, and variable arrangements of head hair that are either tuft-like or cap-like, but not a complete crown cap. Chalise (2003a) described the species as sluggish, very shy, and less quarrelsome than other macaques.

This shyness, coupled with the rocky terrain and hill slopes, make Assamese macaques difficult to study (Fooden, 1982; Wada, 2005; Zhou et al., 2014b). Assamese macaque range includes Bangladesh, northeast India, Nepal, Bhutan, northeast Myanmar, southern China,

northwestern Thailand, Lao PDR, Myanmar, and northern Vietnam. The northern limit stands at approximately 30° N latitude in the Meridianan Mountains of southwest China (Zhang et al., 1981). According to Wada (2005), Assamese macaques expanded their geographic range in the Early Pleistocene (approximately 1.1 mya); the discontinuous distribution of Assamese macaques in the northern part of their range is probably due to the expansion of rhesus macaques in the late Pleistocene (0.6 mya).

Where they are sympatric along the flanks of the Himalayas, rhesus and Assamese macaques occupy separate ecological niches (Ménard, 2004; Timmins & Duckworth, 2013; Zhou et al., 2014a). *M. assamensis* populations are found from the floodplains to the high mountains, usually associated with rough and steep terrain at an elevation of about 1,000 m., although habitat niches vary depending on the region (Srivastava & Mohnot, 2001; Duckworth, 2013; Timmins & Zhou et al., 2014b; Huang et al., 2015). In a limestone seasonal rainforest in Guangxi Province, China, an average of 71% of *M. assamensis* sightings were located on cliffs, 28% on hillsides and hilltops, and they were never encountered in the valley basin (Zhou et al., 2014b). Fooden (1982) described this species as having a preference for broadleaf evergreen rainforest and surveys of *M. assamensis* in Lao PDR confirm this (Timmins & Duckworth, 2013). They are also found in Lao karst habitat down to 200 m in altitude; however, at such low elevation they were only seen in very steep terrain (Timmins & Duckworth, 2013) and cliffs may serve as sleeping sites (Chalise, 2003a). *M. assamensis* is associated with dense evergreen forest but has been observed entering adjacent secondary forests or agricultural fields; there is no evidence to indicate an attraction to disturbed habitat.

Depending on the habitat, Assamese macaques may be predominantly frugivorous or folivorous. In the limestone seasonal rainforests of Guangxi, China, where Assamese macaques are sympatric with the highly fruigivorous *M. mulatta*, they are predominantly folivorous and consume a less diverse diet (Zhou et al., 2014a). In this habitat, the macaques spend 76% of their annual mean feeding time on young leaves, 2% on mature leaves, and only 20% on fruit, 1% on flowers, and 1% on other food items (Huang et al., 2015). Bamboo (*Bonia saxatilis*) makes up 71% of the diet, and because bamboo contains cyanide, selecting young leaves that are lower in cyanide content is an advantageous strategy. The macaques spend successive days on the same hill thereby reducing time spent foraging; Huang et al. (2015) describe this as an energy-saving strategy. In dry evergreen forests of northeastern Thailand, adult females spend 59% of feeding time on fruits and only 24% on leaves (Heesen et al., 2013). In evergreen and mixed broadleaved forests of central Nepal, they are predominantly frugivorous, spending 66% of total feeding time on fruits (including nuts and seeds) and only 12% on leaves (Koirala et al., 2017). Assamese macaques also consume mushrooms, bark, caterpillars, birds, reptiles, amphibians, mollusks, and spiders (Fooden, 1971b; Roonwal & Mohnot, 1977; Koirala et al., 2017).

In a comparison between a wild group (WG) and a semi-provisioned group (SPG) living in the Shivapuri-Nagarjun National Park in central Nepal, Koirala et al. (2017) were able to document dietary and activity flexibility in these macaques. The WG was predominantly frugivorous but the SPG was heavily dependent on human food with 59% of their total diet being retrieved from an army camp's kitchen garbage. However, the SPG group still foraged on forest food, spending 23% of feeding time on fruits. Mitra (2002) found that a group living close to human settlements and provisioned, still utilized 74% wild foods. *M. assamensis* is known to raid crops and in the highlands of Nepal this has had a significant economic impact on farmers (Chalise, 2003a; Paudel, 2016).

Activity budgets also vary greatly depending on habitat—including the distribution and availability of food (Mitra, 2002; Zhou et al., 2007). Koirala et al. (2017) report that the WG spend on average 55% of its daily activity budget feeding, 19% moving, 14% resting, and 12% in social activities. In contrast, the SPG spend 37% feeding, 14% moving, 31% resting, and 18% in social activities. Activity cycles for both groups also varied on a seasonal basis (also reported by Zhou et al., 2007). In limestone habitats they were observed spending 18.3% of their daily activity budget feeding; 39.6% of their time was devoted to resting, and 33.2% to moving, while the remainder included play, grooming, and other behaviors (Zhou et al., 2007). Two groups (n = 15, n = 17) living in this habitat were found to have average home ranges of 53 ha and 65 ha, with mean daily path lengths of 590 m and 782 m (Zhou et al., 2014b).

Reports of average group size range from 7–50 individuals (Chalise, 2003a) with a sex ratio of 1:1.0 (M:F). On average, smaller group sizes (12–16) have been observed in limestone forests (Zhou et al., 2014a; Huang et al., 2015) and larger average group sizes (34–37) in broad-leaved evergreen forests (Koirala et al., 2017). Assamese macaques, like others in the genus, reside in multi-male/multi-female groups with female philopatry and male dispersal at sexual maturity (Fürtbauer et al., 2010); nevertheless, all-male groups and lone males have been observed. Resident adult males exhibit sentry-like behaviors (Fooden, 1982) and do not possess the despotic social relationships found in females (Cooper & Bernstein, 2008).

Fürtbauer et al. (2010) found that females had their first infant at approximately five years of age and reproduction was strictly seasonal; timing of births was geared to the onset of the rainy season and matings occurred during the dry season. In the same dry evergreen forests in northeastern Thailand, peaks in birth were found to occur a few months prior to a peak in the availability of food (Heesen et al., 2013). Fooden (1982) noted that mothers of wounded infants were known to make desperate attempts to protect their offspring. Males often have affiliative contact with infants, resembling the triadic agonistic buffering of *M. sylvanus* (Cooper & Bernstein, 2008).

In broadleaf evergreen forests, this species is highly arboreal, traveling and foraging in trees (Richard et al., 1989). Koirala et al. (2017) observed a wild group of Assamese macaques in the trees 99% of the time. In contrast, those living in limestone seasonal rainforest are mostly terrestrial (Zhou et al., 2014a). Possible predators include dogs, leopards, and humans (Mitra, 2002; Koirala et al., 2017).

There is a decline in wild populations in certain parts of the species' range but not throughout their entire distribution (Kumar et al., 2001). For example, *M. assamensis* is probably the least common primate species in Nepal (Kumar et al., 2001; Chalise, 2003a), and its fragmented distribution in that country is highly problematic (Paudel, 2016). The Nepalese population is under particular pressure from rapid logging for timber, both as firewood for residents and lodges for tourists (Regmi & Kandel, 2008). In Lao PDR, it is the most common macaque (Duckworth et al., 1999) and is found in many protected areas that are extremely rugged, making hunting difficult (Timmins & Duckworth, 2013). An assessment by IUCN listed this species as Near-Threatened because "it is experiencing declines due to hunting and habitat degradation and fragmentation. These declines are significant...." (Boonratana et al., 2008: p.2; IUCN, 2020). While habitat loss is paramount, other threats include selective logging, human activities, invasive alien animals, hunting and trapping for sport, medicine, food, and the pet trade (Boonratana et al., 2008).

Their preference for undisturbed and often steep, inaccessible terrain suggests that *M. assamensis* have limited propensity for ecological and behavioral plasticity. Recent studies do suggest that some degree of behavioral change may be occurring in populations that are now living in closer proximity to humans (Ménard, 2004; Koirala et al., 2017). Some groups have slowly developed a habit of utilizing agricultural foodstuffs rather than their natural food (Chalise, 1999; Kumar et al., 2001; Mitra, 2002, 2003; Regmi & Kandel, 2008; Paudel, 2016); however, this may denote human encroachment as opposed to suggesting a propensity of these macaques to live commensally with humans.

Tibetan Macaque (Macaca thibetana)

The Tibetan macaque is the largest species in the genus *Macaca*. While body weights of both males and females fluctuate seasonally, mean male weight equals 18.3 kg and mean female weight 12.8 kg (Zhao, 1994). Tibetan macaques have long, dense, brown-colored pelage and stump tails. Adult males have brown faces, facial beards, and buff side whiskers; adult female faces are pink (Fooden, 1983). The species lives in small isolated populations restricted to the mountainous areas of eastern Tibet and east-central China in forest remnants unsuitable for cultivation (Fooden et al., 1985; Berman et al., 2007). IUCN reported it is found in approximately 60 nature reserves (Yongcheng & Richardson, 2008). Preferred habitats of *M. thibetana* are a variety of forests that exist in cold, high mountains ranging from 1,000–2,500 m in elevation (Corbet & Hill, 1992). These forest refuges may differ in floristic composition and include subtropical broadleaf evergreen, temperate deciduous, and subtemperate mixed coniferous and deciduous forest (Fooden, 1983; Fooden et al., 1985). According to Richard et al. (1989), the natural diet of this species includes leaves, shoots, fruits and roots, along with some small invertebrates, snakes, birds, eggs, mushrooms, and maize. At Mt. Emei, China, where a long-term study of a seasonally provisioned population at a Buddhist temple has taken place, a total of 177 plant species were foraged by the macaques (Zhao, 1996). Food handouts from visitors, plus invertebrates, were central dietary items in late spring and summer; bamboo shoots and fruits were main food items during autumn, and mature leaves and bark were resources during the remainder of the year. A dietary emphasis on leaves led Zhao (1996) to conclude that *thibetana* may be more folivorous than other Asian macaques (a foraging strategy shared with *M. sylvanus*). Takahashi and Pan (1994) found that the mandible of Tibetan macaques corresponds to some aspects of folivory in other primates but also varies in some other ways. While Tibetan macaques are predominantly terrestrial (Corbet & Hill, 1992) and feed on the ground, they also move through trees as they forage. When food is plentiful, the macaques eat throughout the day and then retire to sleeping sites on rock ledges or in trees located on cliffs (Zhao & Deng, 1988a).

At Mt. Emei, group size ranged from 28–65 individuals, with a male-to-female ratio of 1:2.0–1:6.5 and a mean population density of 13 individuals/km² (Zhao & Deng, 1988b,c). Mt. Huangshan, a scenic tourist destination in east-central China, is the location of another long-term study of *M. thibetana*. Between 1986–2004, the size of one study group at Mt. Huangshan ranged from 21–52 individuals, the higher levels corresponding to years with regular provisioning by visitors (Berman et al., 2007). There have been aggressive interactions with tourists who readily feed the monkeys (Usui et al., 2014).

The average age of females at first pregnancy is five years (Xiong, 1984; Zhao & Deng, 1988b). There is an increase in mating and aggressive behaviors in late September–December at Huangshan with births occurring between April–June (Xiong, 1984). At Mt. Emei, births also show discrete seasonality, occurring from early January–early May so that females can channel stored nutrients to fetuses and newborns during the winter and spring (Zhao, 1996). As with *M. sylvanus*, intensive male caretaking of infants has been observed in Tibetan macaques (Zhao, 1996). Infants up to the age of 23 weeks were cared for 10% of the time by alloparents, mainly males. Ritualized triads of male-infant-male interactions were specific to the birth season, exhibiting passive agonistic buffering and active spatial cohesion between males. Both the initiator and the recipient of the triads were young adult males, the most sexually active component of the male population, which may reflect a mating strategy. Zhao (1996) observed that the implications of triads may not be the same for *thibetana* as for *sylvanus*, and Thierry (2007) categorized Tibetan macaques as demonstrating less social tolerance overall than Barbary macaques. In a provisioned group of 32 members at Mt. Huangshan National Reserve all 8 adult males were equally able to initiate group movements, indicative of a more tolerant dominance style (Wang et al., 2016).

Potential natural predators at Mt. Emei include leopards, the large Indian civet (*Viverra zibetha*), Asian golden cats (*Felis temmincki*), and dholes (Ashmore-DeClue, 1992). Estimated annual predation rate from wild predators was calculated at 3.0% (Hart, 2000), based on researchers' observations of individuals that disappeared under suspicious circumstances. IUCN (2020) lists the Tibetan macaque as Near Threatened and notes significant habitat loss and population decline during the past 25–30 years. "There has been serious deforestation across [the species'] range, but recent measures seem to have stabilized the situation and the future decline of the species will probably not be as serious as its past decline" (Yongcheng & Richardson, 2008: p.4). Hunting and trapping are considered minor threats. Richard et al. (1989) included *M. thibetana* in their "non-weed" category, indicating limits to plasticity. The fact that they are specialized for living in forest refuges in the mountains of Tibet and China suggests they may be capable of only a minimal level of ecological plasticity.

Arunachal Macaque (*Macaca munzala*)

In 2005 the Arunachal macaque was identified and placed in the genus *Macaca*. Described in the scientific literature by Sinha et al. (2005a), this species was discovered at high altitude in western Arunachal Pradesh, the northeastern portion of India. Recently it has been sighted in Bhutan (Tobgay et al., 2019) and may possibly occur in adjoining areas of Tibet. The Arunachal macaque is unique in its distribution at altitudes of 2,000–3,500 m. Colonization of high elevations may have been promoted after the Last Glacial Maximum at about 20,000 years ago (Chakraborty et al., 2014). The species has been sighted in degraded broadleaf forest, degraded open scrub forest, agricultural areas, undisturbed oak forest, and undisturbed conifer forest. In a survey of 41 groups, Sarania, et al. (2017) found a majority of groups (62%) in human-modified habitats with 38% sighted in different types of forests.

Initially, it was thought there was a resemblance to Assamese and Tibetan macaques, however, distinct morphological features exist among all three species. Sinha et al. (2005a) described the Arunachal monkeys as generally large and heavyset, of dark brown color dorsally shading to paler brown on the upper part of the torso and the limbs. Pelage is long and dense; the tail is distinctively short and dark in color. The head crowns of both males and females contain a predominantly dark patch and pale yellow hair around the neck. A single male specimen weighed 15.0 kg with a head–body length of 575 mm (Mishra & Sinha, 2008).

The Arunachal macaque is largely frugivorous (Kumar et al., 2007) and terrestrial (Kumar et al., 2008c). The rugged terrain and harsh environment inhabited by *M. munzala* undergoes high seasonal fluctuations in food availability. A six-month study carried out by Mendiratta et al. (2009) at an elevation of 2,180 m found that these monkeys spend more time feeding in the winter (41–66%) than in spring (33–51%). Just two species of plants formed 75% of the winter diet, and the macaques spend less time moving in the winter than in the spring when availability of fruits and young leaves increases. Although there is a paucity of data on social organization, the group observed by Mendiratta et al. (2009) was composed of 24 individuals (5 adult females, 4 adult males, 13 subadults, and 2 infants). Sarania et al. (2017) recorded a mean group size of 24 and a range of 12–44 individuals. They also observed 2 solitary males. Initial field studies point to *M. munzala* having a typical macaque matrifocal society, although tolerant relationships and low levels of intragroup aggression were observed (Kumar et al., 2007).

Widespread hunting has reduced the natural predators in western Arunachal Pradesh that might prey on *M. munzala*; the tiger is now extinct in the area, but small populations of snow leopards (*Uncia uncia*) and dhole still exist (Mishra et al., 2006). *M. munzala* is considered to be Endangered by IUCN with a population of 971 individuals (Kumar et al., 2008a,b; Sarania et al., 2017; IUCN, 2020). A distribution model indicates that only 2.4% of the total landmass in the Tawang and West Kameng Districts where this species is found are areas of potential habitat (Sarania et al., 2017). The habitat for this species is greatly threatened by logging and human settlements (Holden, 2004), including severe hill cutting for the expansion of roads (Sarania et al., 2017). In addition, the monkeys are persecuted in response to crop damage and they are sometimes captured to keep as pets (Kumar et al., 2008c). The hunting of monkeys is not practiced by the Monpa people of the Tawang District, but it is a pervasive practice in West Kameng (Sarania et al., 2017). While no specific research has been undertaken concerning the ecological or behavioral plasticity of *M. munzala*, its high-altitude specificity seems to preclude a moderate or large degree of ecological plasticity.

White-Cheeked Macaque (*Macaca leucogenys*)

The white-cheeked macaque, first described by Li et al. (2015), is found in Medog, southeastern Tibet. This macaque has been identified by 738 photos and DNA extracted from four skin samples. It is a heavyset macaque with a short tail, prominent pale to white side and chin hairs with light-colored, long and thick hair around its neck. It has dark facial skin and light to dark brown dorsal pelage. The sequencing of a complete mitochondrion genome (Hou et al., 2016) confirmed species status, and mitochondrial DNA sequences show a close phylogenetic relationship with *M. munzala* and *M. radiata* (Fan et al., 2017). When the monkeys come into contact with humans, they flee by climbing into trees and admit a distinctive high-pitched squeal (Li et al., 2015). *M. leucogenys* has been photographed in tropical primary and secondary evergreen forests at altitudes of 1,395–2,420 m. They have also been seen moving across mixed broadleaf-conifer forests at 2,700 m. A group of 20 individuals was observed by Chetry et al., (2015).

Its conservation status is currently unknown (IUCN, 2020). The macaques are traditionally hunted and also killed for their raiding of cornfields. However, the biggest threat appears to be great loss of habitat through the construction of hydropower stations in Medog (Li et al., 2015).

FASCICULARIS SPECIES GROUP

Four species are included in the *fascicularis* species group: *M. fascicularis* (long-tailed macaque), *M. mulatta* (rhesus macaque), *M. cyclopis* (Taiwan macaque), and *M. fuscata* (Japanese macaque).

Geographically, the *fascicularis* group has the most broadly continuous distribution of all macaque groups, with species that inhabit tropical, subtropical, and temperate regions (Thierry, 2007). Nonetheless, the species populate widely different ranges. *M. cyclopis* is found only on the island of Taiwan, *M. fuscata* only on the Japanese islands, *M. mulatta* from eastern Afghanistan to China, and *M. fascicularis* throughout the Southeast Asia mainland plus the Philippines, Indonesia, and Timor (see Figure 17.8). This group, therefore, includes species with very limited distribution and species with extensive distribution. Species are allopatric with one exception: *mulatta* and *fascicularis* are sympatric with an extensive hybrid zone extending from Vietnam to Lao PDR, and

Thailand and possibly into Myanmar (Fooden, 1964, 1997; Eudey, 1980; Hamada et al., 2005).

The four *fascicularis* species may have initially diverged from one another more than 500,000 years ago (Eudey, 1980). Ross and Zinner (2015) note that *M. cyclopis* and *M. fuscata* were derived from eastern populations of *M. mulatta* approximately 1 mya when rising sea levels resulted in their allopatric distributions on the islands of Japan and Taiwan. The *fascicularis* group likely dispersed in Asia more recently than other macaque species groups as evidenced by its broadly continuous range and may, thereby, have contributed to the reduction/disjunction of geographic ranges of earlier dispersals by successfully competing with them (Fooden, 1976, 1980). Eudey (1980), however, points out that Pleistocene climatic changes in the tropics (i.e., cooler, drier, and more seasonal weather caused by glacial periods in temperate regions) were successfully utilized by the *fascicularis* species group. The arid open-country conditions that drove

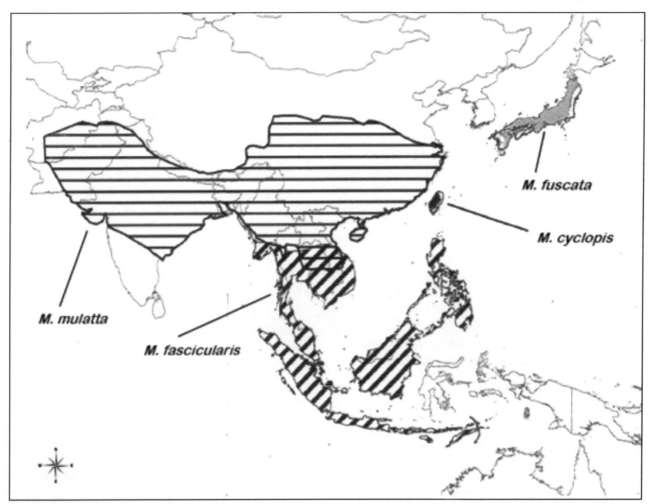

Figure 17.8 Geographic distribution of *Fascicularis* Species Group members, based on spatial data from IUCN (2020). From west to east: *Macaca mulatta*, *M. fascicularis*, *M. cyclopis* (endemic to the island of Taiwan), and *M. fuscata* (endemic to the central and southern Japanese islands of Honshu, Shikoku, and Kyushu).

(Map credit: Ian Colquhoun)

other species groups to forest refugia were tolerated by the more adaptable *fascicularis* group. Richard et al. (1989) suggested that within the past 10,000 years, due to the spread of human agriculture, disturbed habitat has facilitated the "ubiquity" of adaptable macaques more than Pleistocene glacial episodes.

The general morphology of the members of this species group is not exceptionally diverse (Ashmore-Declue, 1992). Face color is pink to pinkish-brown; pelage color varies from gray-brown to yellow-brown to olive-brown. There is some diversity in pelage texture: *M. cyclopis* and *M. fuscata* have denser and thicker hair than the other two species. Facial beards and cheek whiskers are found in adult *M. fascicularis, M. cyclopis,* and *M. fuscata*, but *M. mulatta* has a bare face. *M. fascicularis* has the smallest body and longest tail of the species group (Eudey, 1980); *M. cyclopis* is more robust than *M. fascicularis*, while *M. fuscata* and *M. mulatta* weigh approximately twice as much as *M. fascicularis* (Thierry, 2007).

This species group demonstrates a range of interactions with humans. *M. mulatta* displays "perhaps the most intense relationship between human and nonhuman primates anywhere in the world" (Roonwal & Mohnot, 1977: p.99), while *M. cyclopis* is an elusive primate that avoids humans due to extensive hunting in the twentieth century (Richard et al., 1989). IUCN (2020) currently lists *M. fascicularis* in the Vulnerable category. *M. cyclopis* and *M. fuscata* have been upgraded from Vulnerable and Endangered, respectively, on the basis of population increases over the past decade and a half to species of Least Concern; *M. mulatta* is also listed as Least Concern (IUCN, 2020). The range of plasticity in the species group is diverse. *M. fascicularis* has a high degree of environmental flexibility, demonstrated in particular by its successful colonization of Mauritius, an island with entirely dissimilar flora to Southeast Asia, after a small number of individuals were released there by Portuguese sailors in the 1500s (Sussman et al., 2011). *M. mulatta* has similarly demonstrated a remarkable range of adaptability as an introduced population now lives in the riverine woodlands of Florida (Riley & Wade, 2016). While the two former species show remarkable behavioral and ecological plasticity, *M. cyclopis* and *M. fuscata* do not (Richard et al., 1989).

Long-Tailed Macaque (Macaca fascicularis)

Long-tailed macaques are slender, small monkeys (mean male weight is 5.36 kg; mean female weight is 3.59 kg) with tails that exceed combined head and body length (Fooden, 1995, 1997) (see Figure 17.9). Variation in dorsal pelage color (buffy to yellowish-gray to golden brown to reddish-brown to black), as well as variation in cheek whiskers, mustaches, and facial beards, occur across the populations inhabiting its large Southeast Asian range (Ashmore-DeClue, 1992; Fooden, 1995). Long-tailed macaques are the third most widely distributed primates after humans and rhesus macaques (Fooden, 1995). The species is found in Bangladesh, Brunei, Cambodia, India (Nicobar Islands), Indonesia, Lao PDR, Malaysia, Myanmar, the Philippines, Singapore, Thailand, Timor-Leste, and Vietnam; introduced wild populations occur on the islands of Mauritius, Palau, and in New Guinea (Ong & Richardson, 2008). In addition to being able to live in human-inhabited and -modified environments, *M. fascicularis* is able to exploit a wide variety of natural habitats. They are found in evergreen rainforest, riverine forest, mangrove forest, monsoon forest, bamboo forest, and areas of scrub and grassland (Southwick & Cadigan, 1972; Kurland, 1973; Wilson & Wilson, 1975; Crockett & Wilson, 1980; Eudey, 1980; Fooden, 1991; Hambali et al., 2012a; Brotcorne et al., 2014; Nila et al., 2014). Throughout its range, *M. fascicularis* flourishes in areas that are disturbed and in close proximity to humans (Klegarth et al., 2017). Even in their selection of sleeping trees, long-tailed macaques preferentially select trees located within or in close proximity to human-modified zones (Brotcorne et al., 2014).

Long-tailed macaques are mainly frugivorous with seasonal dependence on leaves, flowers, and insects; during times of low fruit availability, the species reveals an eclectic diet and opportunistic exploitation of resources (Yeager, 1996); fishing even has been observed in groups of long-tailed macaques in northern Sumatra and East Kalimantan, Indonesia (Stewart et al., 2008). In intertidal habitats of southwestern Thailand and Myanmar, long-tailed macaques feed throughout the year on marine prey and have learned to use stone tools to open the hard shells of sessile rock oysters, nonsessile mollusks, crustaceans, and sea almonds (*Terminalia catappa*) (Gumert et al., 2009, 2013; Gumert & Malaivijitnond, 2012). They have also been observed using stone hammers and anvils to open palm oil nuts (Luncz et al., 2017). Long-tailed macaques display a wide range in their use and degree of exploitation of anthropogenic food (Fuentes et al., 2011; Sha & Hanya, 2013; Nila et al., 2014). They are known crop raiders and have been observed eating sugarcane, corn, taro, yams, tapioca, rice, plam oil nuts, and rubber tree fruits (Fooden, 1971b; Poirier & Smith, 1974; Crockett & Wilson, 1980; Sussman et al., 2011; Luncz et al., 2017).

As described above, flexibility in the diet of *M. fascicularis* has been well documented and reflects how these macaques readily adjust to differences in food type,

Figure 17.9 Subadult male *Macaca fascicularis* displaying characteristic lateral cheek whiskers.
(Photo credit: Jeffrey V. Peterson)

availability, and abundance (Nila et al., 2014). Long-tailed macaques living in a high-altitude rainforest of Telaga Warna, West Java, Indonesia, consume both foods from the rainforest and foods provided by visitors to the nature center. Foods from the rainforest were consumed 60% of the time and human-provided foods 40%. In contrast, in secondary and mangrove forests of the Kuala Selangor Nature Park in Malaysia, the macaques were provisioned on a daily basis and demonstrated a clear preference for human-provided food (Hambali et al., 2012b). Here the macaques would also enter the homes of neighboring residents and exploit garbage bins. Compared to the macaques living in Telaga Warna, these macaques spend less time moving and feeding and more time resting, grooming, engaging in agonistic behavior, and mating. Sha and Hanya (2013) note that at least 50% of long-tailed macaques living in Singapore eat anthropogenic food and found within-population variability in behavior associated with the use of anthropogenic food resources.

Strata occupied and locomotion are features that demonstrate the adaptability of long-tailed macaques. The species is nearly exclusively arboreal in Southeast Asia but is able to exploit terrestrial habitats if a continuous canopy is not available (Sussman et al., 2011). The activity

cycle of *M. fascicularis* is affected by food resources, habitat, and group size. Where resources are patchily distributed, MacKinnon and MacKinnon (1980) and Sussman and Tattersall (1981) observed patterns of feeding and moving at a fairly constant rate throughout the day. A selection for riparian habitat is a rare characteristic that has been observed in long-tailed macaques throughout the species' range and has implications for its daily activity cycle. Described as a riverine "refuging" species by Hamilton and Watt (1970), the macaques depart and return to a central location near a river each day (Sussman et al., 2011). Favored sleeping trees of *M. fascicularis* are found overhanging water, and swimming is not unusual.

Long-tailed macaques live in multi-male groups that exhibit fission-fusion into subgroups (Kurland, 1973; van Schaik et al., 1983). The sex ratio in groups averages about one male to three females with males acting as sentinels for predator detection and deterrence (Sussman et al. 2011). Field research has indicated that predator detection is a major determinant of group size and social organization in this species (van Schaik et al., 1983; van Schaik & van Noordwijk, 1985). Group size and subgroup composition differed between the long-tailed macaques living on the predator-free island of Simeulue and those living in the Ketambe Reserve in Sumatra, where a relatively full complement of felid predators exist (van Schaik et al., 1983). On predator-free Simeulue, long-tailed macaques lived in small groups that subdivided into even smaller foraging parties. These small foraging parties often contained infants and juveniles, the age groups most vulnerable to predators. Conversely, groups living in the Ketambe Reserve along with predators were large; foraging parties were likewise large and contained a high proportion of subadult and adult males. Supriatna et al. (1996) encountered long-tailed macaques at widely ranging densities from a low of 3.9 animals/km^2 to a high of 122 individuals/km^2. The highest density was found in a lowland forest with a mean group size of 12 individuals.

Long-tailed macaques are frequent prey items for predators—mammals, raptors, and reptiles (Hart, 2000). For example, on the island of Mindanao, 3.0–6.3% of Philippine eagle (*Pithecophaga jefferyi*) diet consists of long-tailed macaques. In Indonesia, false gharials (*Tomistoma schlegeli*) have been observed preying on *M. fascicularis*, and the monkeys are also prey for the largest of the monitors, Komodo dragons (*Varanus komodoensis*). Long-tailed macaques were one of the identifiable food remains in the alimentary tracts of reticulated pythons caught for the commercial trade in Southeast Asia. In Meru-Betiri Reserve, Indonesia, silver leaf monkeys (*Trachypithecus villosus*) and *M. fascicularis* have become substitutes for a range of other prey normally available to large Asian carnivores. After ungulates were extirpated by human hunting in Meru-Betiri, the two primates became the predominant food of tigers, leopards, and dholes. Based on extrapolation from 46 suspected predations, over 10% of the long-tailed macaque population is estimated to be taken by predators annually at Ketambe, Sumatra (Cheney & Wrangham, 1987).

IUCN (2020) lists the species in the Vulnerable category with a decreasing trend; yet, its wide distribution, and tolerance for a variety of habitats causes population status to vary dramatically from one part of its range to another (Ong & Richardson, 2008). Bangladesh has a population that numbers under 100 individuals and some of the Nicobar Islands populations may have been decimated by the 2004 tsunami (Ong & Richardson, 2008), yet long-tailed macaques were the most frequently encountered primate in Thailand in the early 2000s (Malaivijitnond & Hamada, 2008). Hunting is the major threat to *M. fascicularis* according to Ong and Richardson (2008). Besides killing for meat, sport, and trophies, individuals, particularly females, are removed from the wild into breeding facilities to satisfy a large international trade in primates for medical research. Ninety-four percent (n = 26,512) of the primates imported into the United States in 2008 were long-tailed macaques (Sussman et al., 2011). After China (the departure point for most of the current trade from Southeast Asia), Mauritius is the second-leading exporter of long-tailed macaques to laboratories, a major factor in the drastic decline on the island from 35,000–40,000 animals in 1980–1990 to 8,000 animals in 2010 (Sussman et al., 2011).

In residential areas near the Kuala Selangor Nature Park, Malaysia, long-tailed macaques are considered pests that enter and steal items from homes, cause property damage, and litter (Hambali et al., 2012a). At this location and others where long-tailed macaques live in close proximity to humans, increasing levels of human-macaque conflict are apparent.

Because of their preference for disturbed and/or successional habitats, long-tailed macaques have been described as a "pioneer species" (Angst, 1975), an "edge species" (Medway, 1970), an "adaptable opportunist" (MacKinnon & MacKinnon, 1980), a "quasi-specialized edge species" (Sussman & Tattersall, 1986), and a "weed species" (Richard et al., 1989). Each of these terms depict the ability to survive in habitats that have undergone environmental change and emphasize the adaptability consistently noted for this species. *M. fascicularis* exhibit a large degree of behavioral and dietary flexibility indicative of a maximum degree of ecological plasticity.

Rhesus Macaque (Macaca mulatta)

If vast geographical distribution can be taken as a measure of success for a species, *M. mulatta* "is probably the most successful non-human primate species surviving today" (Melnick, 1981: p.22). The rhesus macaque is widely distributed throughout South, Southeast, and East Asia in eastern Afghanistan, Bangladesh, Bhutan, central and southern China, northern and central India, Lao PDR, Myanmar, Nepal, northern Pakistan, northern Thailand, and Vietnam (Singh et al., 2020). Based on the diversity within its geographic range, the rhesus macaque is credited with pronounced, maximal ecological plasticity (Southwick & Siddiqi, 1988); it resides in temperate coniferous forests, moist and dry deciduous forests, bamboo and mixed forests, mangroves, scrub, and rainforests—from sea level to altitudes exceeding 4,000 m, in locations that receive deep snowfall and in deserts, in the subtropics and tropics, and in semi-arid and sparsely vegetated habitats (Ashmore-DeClue, 1992; (Singh et al., 2020). Populations of rhesus live both in forests where little human intervention occurs and in close association with humans in cities and temples, as well as in the highly disturbed environments of cultivated fields and roadsides (Richard et al., 1989). Conflict due to the crop-raiding behavior of *M. mulatta*, especially in northern India has become a major problem, but Hindu religious beliefs are partly responsible for the ability of this species to survive in India and elsewhere due to the cultural association with Hanuman, the monkey god (Saraswat et al., 2015).

The rhesus monkey is a medium-sized macaque; mean male weight is 11.0 kg, and mean female weight is 8.8 kg (Thierry, 2007). Body weight, as well as pelage colors (yellowish-gray to golden brown to burnt orange on the upper back), vary throughout its range (Fooden, 2000). *M. mulatta* are highly omnivorous, capable of ingesting a wide variety of foods, but individual animals are also able to change their diet to include new food types (Ashmore-DeClue, 1992; Singh et al., 2020). The food preferences of urban rhesus are different from those living in forests; when given a choice between natural or human foods, urban rhesus selected cooked and spicy foods while forest rhesus chose raw fruits and vegetables (Singh, 1969). Local environments influence food habits, resulting in different diets, feeding patterns, activity patterns, and ranging or social behaviors that are linked to foraging strategies. Urbanized rhesus readily replace naturally occurring foods with those they can harvest more quickly and with less energy, even though greater risks may be involved. Those rhesus that depend on humans for a majority of their food have learned to obtain it in a variety of ways. Although rhesus may be fed a great deal by tourists, temple priests, and worshippers, these monkeys also steal food directly from homes, shops, and markets.

Fooden (2000) calculated that *M. mulatta* on average spend about 72% of their daylight hours on the ground and 28% in trees. According to Lindburg (1971) during the cool season (November–February) in the Asarori Forest, India, morning temperatures are as low as 0°C with frosts and heavy dew. Early morning feeding and traveling behaviors are infrequent, and rhesus remain in the trees stretching, sunning, and grooming for as long as two hours. After leaving sleeping trees, concentrated feeding begins and remains fairly constant for the entire day, broken only by a short rest period at midday. Resources tend to be dispersed in winter so rate of feeding activity is intensified, but overall food consumption is reduced (Makwana, 1979). Mating behavior predominates in winter. Time budgets in the hot, dry season (March–June) are influenced by an increase in daylight and warm temperatures, so foraging and social behaviors begin early in the morning. Water is harder to find in the dry season, and a group travels to permanent water sources in the late morning. Many tree species flower at this time of the year, and food resources are highly concentrated. During the warm, wet season (July–October), foraging begins at dawn and by mid-morning social behaviors become frequent. Grasshopper hunts may be conducted in the late morning. At midday, feeding and traveling are reduced and the animals rest, play, and groom. All activities cease during heavy rains, and the rhesus huddle in groups near the base of trees. After the rains subside, drying off and grooming ensues in the trees. Group feeding resumes for the remaining hours of the late afternoon.

In winter and early spring, macaques living in the temperate forest of Mt. Wangwushan, Jiyuan, China, were observed to use sunny slopes at 1,000–1,300 m with slope gradients of 15–40° and less than 60% of canopy cover (Xie et al., 2012). The macaques maximize body temperatures by seeking out increased exposure to the sun.

Jaman and Huffman (2013) compared the activity budgets of a group of rhesus living in the center of urban Dhaka, Bangladesh, to a group living in close proximity to a rural village. Time spent feeding was significantly more (36%) in the rural than in the urban group (22%); time spent grooming and in object manipulation/play was greater (20% versus 15%) in the urban group. Seasonal variations also influenced the activities of both groups. The urban group spent more time feeding on provisioned food (69%), food from gardens and crops (5%), and food from houses or shops (4%) than they spent eating wild plants (22%). In contrast, the rural group spent more time feeding on foods from gardens and crops (40%), foods from houses and shops (17%), wild plants (33%), animal food (3%), and only 6% on provisioned food. Behavioral and dietary patterns were influenced by habitat type, seasonality/availability of food, and level of stability of access to anthropogenic foods.

The behavioral and ecological plasticity of *M. mulatta* may best be portrayed by how this species has adapted to living in a riparian woodland in Florida (Riley & Wade, 2016). In the 1930s, a small number of rhesus were released on an island in the Silver River. A population of 118 individuals living in four social groups (ranging in size from 20–48 individuals) with an estimated population density of 6.2 individuals/km² now inhabit the island. Wild foods are consumed in 88% of feeding records and 12% of food is provisioned by boaters. Samaras (fruit from ash trees) make up 30% of their wild food diet, leaves 29%, flowers and buds 21%, and other plant parts and bark 19%, with 1% insects. This dietary strategy is similar to congeners living in temperate broadleaf forests in Asia. Moreover, the macaques spend up to eight hours a day close to the river's edge—a strategy similar to that of riverine forest–dwelling *M. fascicularis*. The Silver River macaques have adopted temperate-dwelling feeding strategies (Hanya et al., 2011), incorporated unique foods like sedges that make up 10% of their diet, and learned to associate boaters as potential sources of food, thereby successfully exploiting a completely new niche.

For non- or minimally provisioned groups, the mean group size is 32.2 individuals, with extremes of 2–250 animals (Fooden, 2000). In both non-provisioned and provisioned groups, the sex ratio averages one sexually mature male to three sexually mature females (Fooden, 2000). Hasan et al. (2013) found a similar sex ratio in forest groups but a lower sex ratio of one adult male to 1.93 females in urban groups. Mean population density is 37.2 individuals/km^2 in forest habitats, but rises to 201.1 individuals/km^2 in non-forest habitats (Fooden, 2000). A five-year survey of the distribution of rhesus macaque populations in Bangladesh (Hasan et al., 2013) found that group sizes in urban areas ranged from 22–91 individuals with a mean of 41 and those living in forested habitats varied from 10–78 individuals with a mean of 30. Evergreen rainforests supported larger groups than did deciduous forests.

Matrilineal kin groups form the basis of rhesus macaque society. Long-term research revealed that distinct and persistent relationships were formed between adult females and their offspring (identified by grooming bouts) and that incest avoidance was accomplished by male offspring emigration as they neared sexual maturity. Rhesus matrilineage was thus defined as consisting of a stable core of related females through which social behavior is mediated (Sade, 1972). Within the matrilines, strict rules determine acquisition of dominance rank with female offspring achieving a rank below their mothers and in inverse order to age, i.e., younger daughters outranking their elder sisters (Datta, 1992).

Recognizable adult male and female dominance hierarchies are an ordering force to social interactions. However, in contrast to the female social nexus, group males are unstable extensions of the social core. The lines for group fission and social interaction are also based on matrilines (Melnick, 1981; Melnick & Kidd, 1983). Thierry (2000) categorized *M. mulatta* as one of the least socially tolerant of the macaques based on research dealing with asymmetry of conflicts, dominance, and kin bias. Matrilineage and dominance hierarchies—along with age, sex, infant play partners, and adult friendships—all play important roles in rhesus social organization and behavior.

Lindburg (1971) identified a shrill alarm bark with which rhesus macaques warned of predators; Lindburg (1977) also noted that a group of rhesus shifted 12 km after hearing the roar of a tiger. Tigers, leopards, raptors, domestic dogs, and weasels (preying on infants) have been observed attacking rhesus macaques (Fooden, 2000). Captive rhesus react differently to the posture of snake models, indicating that they use this as a cue in threat assessment (Etting & Isbell, 2014). In their natural habitats they coexist with pythons (*Python* spp.), cobras (*Naja* spp.), and kraits (*Bungarus* spp.). Fecal samples gathered in Kanha, India, confirmed that rhesus monkeys were commonly consumed by tigers (Schaller, 1967). Rhesus macaques that have colonized the mangrove swamps in the Sunderbans, West Bengal, India, are preyed upon by Indian pythons, estuarine crocodiles (*Crocodylus porosus*), wolf sharks (*Alopias vulpinus*), and requiem sharks (*Carcharhinus gangeticus*) (Mukherjee & Gupta, 1965). Estimated annual predation rates of 1.2% and 3.0% on rhesus macaque populations have been calculated by field researchers (Hart, 2000).

The 1978 cessation of rhesus exports from India for international lab animal trade resulted in a resurgence of the species in that nation (Richard et al., 1989). Nonetheless, one population of rhesus macaques at the northern edge of the range in China became extinct in the 1980s (Yongzu et al., 1989), and macaques in China have "neither a commensal relationship with people nor high population densities" (Richard et al., 1989: p.576). However, while the population trend for *M. mulatta* is unknown but unlikely to be declining (Singh et al., 2020; IUCN, 2020), the species has been placed in the Least Concern category based on wide distribution and tolerance of a broad range of habitats. As noted by Sengupta et al. (2014), *M. mulatta* are major seed dispersers in tropical forests and as such have the potential to play a critical role in the revitalization and reforestation of fruit trees. As one of the foremost "weed" species (Richard et al., 1989), rhesus macaques are among the hardiest and most adaptable of nonhuman primates (Southwick & Siddiqi, 1988; Ménard, 2004).

*Taiwanese Macaque (*Macaca cyclopis*)*

The Taiwanese macaque (also known as the Formosan rock macaque), endemic to the island of Taiwan, is one of the least known macaques due to the rugged topography of their habitat and extreme wariness of humans. This species is found mainly in forests of the island's central mountain range with some scattered populations in lowland forest remnants (Wu & Richardson, 2008). Primary habitat is broadleaf evergreen forest, but these monkeys are also found in mixed broadleaf-needleleaf, needleleaf, and bamboo forests at elevations ranging from 100–3,400 m (Fooden & Wu, 2001).

Taiwanese macaques have flat faces, prominent foreheads, and long, full facial hair (Peng et al., 1973). Pelage is long, dense, and grayish-brown; the tail has thick hair with a black line on the dorsal surface. The color of the facial skin ranges from white to red with variations relating to age, sex, and regional location (Poirier & Davidson, 1979). Male body weight (based on 1 wild-caught individual) is 8.1 kg and female weight (based on 7 captive animals) varies between 4.0–6.2 kg (Peng et al., 1973; Fooden & Wu, 2001).

Taiwanese macaques consume 78 plant species, along with insects, crustaceans, and mollusks (Lee, 1991). Su and Lee (2001) detailed the diet of non-provisioned *M. cyclopis* by percentage of feeding time as follows: 54% fruits, 29% leaves and buds, 7% roots/flowers/herbs, and 10% invertebrates. They noted that during periods of fruit shortage, *M. cyclopis* may become totally folivorous. Additionally, up to 33 species of cultivated crops are consumed opportunistically (Chang, 2000). While data from field studies confirming strata use by Taiwanese macaques are minimal, one population in central Taiwan averaged 16.8% terrestrial activity and 83.2% arboreal (Lu et al., 1991). The degree of terrestriality, however, may be influenced by local habitat; for example, more ground travel was observed in disturbed areas but more arboreal travel in mature forests (Fooden & Wu, 2001). *M. cyclopis* inhabiting the Yushan National Park remain in the trees feeding, resting, and grooming from sunrise until coming to the ground between 0700–0800 hours (Ashmore-DeClue, 1992). Most social behaviors occur in the late afternoon by which time the group has returned to the trees.

Lu et al. (1990) reported a small mean group size; however, this situation may be analogous to that of the Sulawesi macaques where subgroups or foraging parties are small. An alternative explanation is that hunting pressure (prior to a government ban in 1989) may have been the major cause of small group size (Masui et al., 1986; Wu & Lin, 1992). Post–hunting ban, Hsu and Lin (2001) estimated average group size as 45 individuals. Population density is based on few studies. Hsu and Lin (2001) reported 26 individuals/km² at Mt. Longevity, southern Taiwan, very similar to the mean of 25 individuals/km² for 6 natural populations throughout the range of the species cited in Fooden and Wu (2001). Taiwan macaques live in multi-male/multi-female groups; reproduction is highly seasonal for this species, coinciding with the early rainy season (April–June) when new plant growth is abundant (Wu & Lin, 1992; Hsu et al., 2001). Male–female sex ratios of 1:1.8 were noted during the mating season at Mt. Longevity (Lin et al., 2008); small groups of Taiwanese macaques may contain only one adult male (Wu & Lin, 1992). According to Birky (2001), females initiate grooming with males during the breeding season (interpreted as female mate choice), but outside of the mating season, males are socially peripheral. Within matrilines, female offspring acquire their social rank below and adjacent to their mothers (Su, 2003).

Two reported nonhuman predators of *M. cyclopis* are the clouded leopard (*Neofelis nebulosa*) and the mountain hawk-eagle (*Spitzaetus nipalensis*) (Poirier & Davidson, 1979); however, the presence of clouded leopards in central Taiwan is unconfirmed (Ashmore-DeClue, 1992). Alarm calls are given in the presence of hawk-eagles, and a male macaque once was observed shaking a branch at an eagle perch (Fooden & Wu, 2001). Although *M. cyclopis* was considered a Vulnerable species by IUCN from 1988–2000, justification for an upgrade to Least Concern was based on calculations that population status is stable or even increasing under the 1989 protection conferred by Taiwan's Wildlife Conservation Law (Wu & Richardson, 2008; IUCN, 2020). Nevertheless, no concrete data on population growth are available and the upward trend "could be an artifact due to increased survey work in natural areas" (Hai Yin Wu pers.comm. in Wu & Richardson, 2008: p.3). Continuing habitat loss for agriculture and development, illegal hunting, and non-target trapping constitute current threats to the Taiwan macaque (Wu & Richardson, 2008). Conservation measures may be effectively working to increase numbers of *M. cyclopis*, but any increase may also contribute to escalation of monkey-human conflicts (e.g., crop raiding and occasional attacks on people) (Wu & Richardson, 2008).

The Taiwanese macaque is an elusive animal and avoids contact with humans (Richard et al., 1989). There is only one site in Taiwan, the Shou-Shan Nature Park, where macaques can be easily viewed in their natural habitat. Apprehension of humans is even apparent at Shou-Shan: "The Formosan macaques—unlike the long-tailed macaques in Thailand or rhesus and bonnet macaques in India—neither venture out frequently to explore human settlements nor roam around the bustling roads" (Hsu et al., 2009: p.221). The question of whether or not any tendency toward ecological or behavioral plasticity exists in this species is difficult to decipher.

Japanese Macaque (*Macaca fuscata*)

The Japanese macaque (often called the snow monkey) is a large, stump-tailed monkey endemic to the Japanese archipelago. Adult males weigh an average of 11.3 kg and adult females 8.4 kg (Fooden & Aimi, 2005), but body weight within this species conforms to Bergmann's rule (i.e., weight increases with latitude) (Hamada et al., 1996). The adult pelage is bushy and dense, primarily light brown to gray in color but often including golden shades. The genitalia of both sexes is pink. Facial skin is also pink on both sexes and prominent side whiskers are present (Fooden & Aimi, 2005). The geographic distribution of *M. fuscata* is the northernmost of any nonhuman primate (between 41°30′N and 30°30′N latitude). Macaques are found on the large Japanese islands of Honshu, Shikoku, and Kyushu, as well as six of the small, adjacent islands (Fooden & Aimi, 2005) but have a patchy distribution due to the high density of human occupation (Richard et al., 1989).

M. fuscata are found in a wide range of habitats that encompass three different ecological zones: Zone I) temperate and subalpine mountainous areas of central and northern Honshu with both deciduous broadleaf forests and subalpine conifer forests. The latter is subject to prolonged periods of deep snow cover (110–140 days of snow per year) (Suzuki, 1965). As a consequence of five leafless months, population densities of Japanese macaques are low in the deciduous forest (<10 individuals/km²) (Iwamoto, 1978; Takasaki, 1981). Zone II) warm-temperate habitats of central Honshu which are a mixture of deciduous and coniferous trees. Snow is variable in this area, depending on altitude and the east versus west faces of mountains (Suzuki, 1965). Zone III) warm-temperate habitats of western Honshu, Shikoku, Kyushu, and small islands where the forests are broadleaf evergreen but deciduous trees also occur. The broadleaf evergreen forests are more productive and able to support larger densities of macaques (33 individuals/km²) than the deciduous forest of Zone I (Iwamoto, 1978; Maruhashi, 1980, 1982).

Despite the fact that *M. fuscata* exist in a number of provisioned parks, they are predominantly forest-dwelling animals. This species does not exhibit any selection for disturbed sites (Richard et al., 1989), and groups living in disturbed habitats are found at lower densities than are groups found in relatively undisturbed areas (see Takasaki, 1981). Provisioning does, however, enable unusually large groups of *M. fuscata* to exist and in some cases has eased the strain of habitat disturbance.

Japanese monkeys are highly opportunistic yet selective in their feeding choices, tailoring their diet in response to the wide variety of ecological zones they inhabit (Agetsuma & Nakagawa, 1998). The nutritional intake for macaques living in deciduous forests is lower than those in evergreen forests throughout the entire year, but particularly in winter (Nakagawa et al., 1996). Hill (1997) found that his study animals in the warm, temperate broadleaf forest of Yakushima, in the southern portion of the range, were dietary generalists that could not be assigned either a frugivore or folivore category; foraging time was divided into 35% leaves and shoots, 30% fruit, 13% seeds, 6% flowers, 10% animal matter, 5% fungi, and 1% bark, pitch, and sap. The findings for macaques living in a coniferous forest on the same southern island showed more concentration on animal matter and fungi (Hanya, 2004a). Composition of diet and time spent foraging in a cool, temperate habitat in the northern portion of the Japanese macaque range was markedly different from the south. Agetsuma and Nakagawa (1998) calculated that time spent feeding was 1.7 times greater in the north than in the south due to lower food quality and energy requirements for thermoregulation. Temperature, as well as resources, determine activity budgets in this temperate zone-adapted primate (Hanya 2004a,b).

Specific locality may also affect strata occupied by Japanese macaques. Based on one population at Tsubaki Wild Monkey Park in Wakayama, Chatani (2003) characterized *M. fuscata* as semiterrestrial. However, in the northern habitats, Japanese macaques travel nearly always on the ground in single file through heavy snows (Izawa & Nishida, 1963; Suzuki, 1965). The macaques move slowly, "with one hand stretched forward, and with another hand buried deeply to the elbow in the snow" (Izawa & Nishida, 1963: p.79). While Japanese macaques usually sleep in trees, on two small islands that are devoid of predators the monkeys reportedly sleep on the ground (Mori, 1979; Takahashi, 1997). Use of the ground rather than trees as sleeping sites, in turn, allows the formation of larger "huddling" groups, which is a functional behavior enhancing protection from low temperatures (Takahashi, 1997). Long-distance swimming from island to island has also been observed (Mito, 1980).

In the deciduous broadleaf forests of the Shirakami Mountains (40°32′9′′N) the macaques respond to heavy snow conditions by changing their feeding behavior and exhibit risk-averse foraging and energy-saving strategies (Enari & Sakamaki-Enari, 2013). In this habitat there are 0.15 groups/km² or 5.19 individuals/km². In March 2012, at least 1 m of snow fell in the lowlands with more in upper elevations. During this time the macaques did not dig beneath the snow to feed on shrubs, as they do when there is less snow cover, but instead depended more on tree species. During periods of heavy snow cover, the ranging behavior of the macaques is influenced more by the distribution of refuges, i.e., closed-canopy evergreen conifers at lower elevations as opposed to the distribution of food resources. Thermoregulatory demands also cause the macaques to spend more time resting in the sun.

In the warmer coastal forests of Yakushima where group density is high (4.8 groups/km²), intragroup competition is the major force that affects feeding choices and ranging behavior. Compared to smaller group, larger ones have bigger home ranges and consume more mature leaves (Kurihara & Hanya, 2015).

Habitat quality seems to define natural group size and home range area (Izumiyama et al., 2003). Initial research into the social organization of this species used provisioning on a daily basis to habituate animals, so groups numbering well over 100 individuals have been recorded (Itani et al., 1963; Koyama et al., 1975); however, such large group sizes are a function of artificial access to resources, which results in high birthrates and low infant mortality (Fooden & Aimi, 2005). Average group size of wild macaques in the northern Japan Alps is 48.4 individuals, but even without deliberate provisioning, the groups that frequently raid cultivated land for food are twice as large as those that do not (Izumiyama et al., 2003). Wild groups have a mean composition of 18% adult males, 32% adult females, 35% juveniles, and 15% infants (Fooden & Aimi, 2005). *M. fuscata* is a seasonal breeder, and biennial parturition is the norm.

Fission of groups occurs along matrilines (Koyama, 1970), and matrilineal kin groups provide a focus for the social interactions of females and their offspring (Mori 1977a,b). The inherited rank of young Japanese macaques is similar to *M. mulatta*, i.e., rank is influenced by the relative ranks of their mothers (Koyama, 1970; Furuichi, 1983).

Although grooming interactions most frequently involve related females and their offspring and least frequently involve unrelated females (Furuichi, 1983; Koyama, 1991), huddling at night for warmth, one of the adaptations to cold weather, may increase social bonds in Japanese macaques. "[T]he social function of huddling is that group integration might increase at sleeping sites in winter as close social relationships among adults are extended more widely than those in daytime" (Takahashi, 1997: p.57).

Like many other primate species, Japanese macaques exhibit specificity in alarm calling when they perceive a threat from predators (Fedigan, 1974). Small canids (red fox and raccoon dogs, *Nyctereutes procyonoides*) have been cited as predators on wild Japanese macaques (Iwamoto, 1974), and the extinct Japanese subspecies of wolf (*Canis lupus hodophilax*) undoubtedly was a predator in the past (Fooden & Aimi, 2005). A mountain hawk-eagle (*Nisaetus nipalensis*) was observed attacking an adult female *M. fuscata* (Hamada & Iida, 2000). Another potential avian predator is the golden eagle (*Aquila chrysaetus*) (Izawa & Nishida, 1963). Hart (2000) cited an estimated 5% annual removal rate of Japanese macaques by non-human predators. Human predation for food and medicinal uses was extensive until *M. fuscata* came under the protection of the government in 1947, but presently over 10,000 monkeys perceived as agricultural pests are killed annually (Abe et al., 2005). IUCN considered the Japanese macaque to be an endangered species in 1996, but more recently changed the listing to Least Concern (IUCN, 2020). Rationale for the upgrade stated: "This species is widespread and found in many protected areas. The distribution is expanding and the population is increasing in many areas… although there are some local populations under threat" (Watanabe & Tokita, 2020: p.4).

M. fuscata are likely the most thoroughly studied of all macaque species with the majority of the research focusing on aspects of social structure and social organization; however, most systematic data have been gathered at sites with provisioned feeding stations. Because the wild macaques of northern Japan live in habitats that are inaccessible and subject to severe winters, study of these populations has been hindered. While the species exhibits dietary and activity adaptations to the temperate region in which it lives (Hanya, 2004a,b), speculations concerning ecological plasticity are rare. The species raids crops, but is not commensal with humans, so it was not classified as a weed species by Richard et al. (1989). Although much of the habitat within its range has been disturbed, this species does not exhibit any selection for disturbed sites.

If plasticity can be illustrated by the tendency of a species to exploit new resources which have significant ecological consequences, then the origin and spread of so-called traditions observed in provisioned, free-ranging Japanese macaques should be noted (Chalmers, 1980). As an example, sweet potato washing was initiated by a juvenile female on the island of Koshima. The habit spread to her age-mates and was picked up by their mothers who passed it on to infants. Eventually, all but a few older animals were following the practice. "The monkeys started to wash the potatoes in the sea rather than in the stream that ran across the beach. Some of the younger animals learned to swim and to eat seaweed. This newly acquired behavior had therefore opened up a new diet and new part of the habitat to the monkeys" (Chalmers, 1980: p.224). The ability for innovative, novel behavior may then afford Japanese macaques with a moderate degree of ecological plasticity.

ARCTOIDES SPECIES GROUP

This macaque species group consists of a single species, *Macaca arctoides* (see Figure 17.10). Morales and Melnick (1998) and Abegg and Thierry (2002) proposed that a wave of macaque distribution across the Asian continent saw *M. arctoides* separating from the *M. fascicularis* lineage approximately 1.6 mya. Genomic data suggest that *M. arctoides* is the result of bidirectional hybridization between ancestral members of the *mulatta group* and the *sinica* group (Roos & Zinner, 2015).

The stump-tailed, or bear macaque, is a heavy-set monkey with chestnut pelage, a stubby somewhat hairless tail, and a hairless face mottled with pink, red, or

Figure 17.10 Adult female *Macaca arctoides*.
(Photo credit: Erik Karits, https://pixabay.com/photos/stump-tailed-macaque-macaca-arctoides-4819602/)

black skin colors. Adult males weigh 9.9–15.5 kg and adult females 7.5–9.1 kg (Fooden, 1985). Little is known about the ecology and behavior of this species in the wild. Geographically, it is distributed from north-eastern India to southern China and Malaysia. It is also found in Myanmar, Thailand, Lao PDR, Cambodia, and Vietnam (see Figure 17.11). It was formerly found in eastern Bangladesh but may now be extinct in that country (Htun et al., 2008). *M. arctoides* is found in tropical and subtropical regions and is broadly sympatric with *M. nemestrina*, *M. assamensis*, *M. mulatta*, and *M. fascicularis*, and is allopatric or parapatric with *M. thibetana*.

Fooden (1976) determined that *M. arctoides*, like *M. silenus* and *M. nemestrina*, are predominantly restricted to broadleaf evergreen forests. The species has been encountered in primary evergreen forest, secondary evergreen forest, lowland and montane forests, mixed deciduous forest, dry forest with rocky outcrops in Thailand (Bertrand, 1969; Fooden, 1971b; Fooden et al., 1985; Treesucon, 1988), and in karst forests consisting of dense primary and secondary forests that cover steep limestone hills in Vietnam (Haus et al., 2009).

The preferred habitat was previously described as consisting of secondary forest in close proximity to village crops (Treesucon, 1988); however, a more recent assessment by Choudhury (2001) reported that preference is for dense evergreen forest. In their survey of primates

living in the karst forests of central Vietnam, Haus et al. (2009) found no correlation between vegetation type and primate density but did find a significant negative correlation between primate abundance and the number

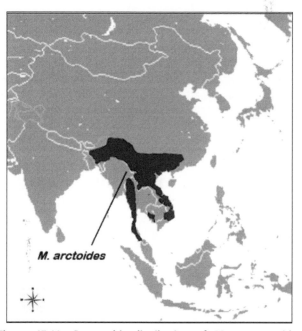

Figure 17.11 Geographic distribution of *Macaca arctoides* (black areas), based on spatial data from IUCN (2020).
(Map credit: Chermundy; image licensed under Creative Commons Attribution-Share Alike 3.0 Unported)

of loggers encountered in survey transects. When compared to *M. assamensis* and two species of langurs, *M. arctoides* had the smallest group sizes and was found at the lowest density in karst forests.

Haus et al. (2009) reported observing 9 groups of *M. arctoides* in Vietnam with a mean group size of 5.89 (ranging from 2–11 individuals) and a density of 1.1 individuals/100 ha. Larger group sizes of >30 individuals have been reported from India (McCann, 1933; Gupta, 2001). Accurate group counts of 22 and 31 individuals were reported by Treesucon (1988) in Thailand, and Fooden et al. (1985) recorded 12 groups in China ranging in size from 12–50 individuals.

M. arctoides use both arboreal and terrestrial modes of locomotion, although they have most frequently been observed foraging and traveling terrestrially (Bertrand, 1969; Fooden, 1971b; Gupta, 2001; Haus et al., 2009). In the Phong Nha-Ke Bang National Park of central Vietnam, stump-tailed macaques are the "most frequently confiscated species in the Rescue Centre... often showing limb injuries caused by snare traps" (Haus et al., 2009: p.310). This suggests that a tendency toward terrestrial foraging and travel may subject them to capture in snares.

Although no comprehensive study has been done on the diet of stump-tailed macaques, they are known to prefer figs (Bertrand, 1969; Gupta, 2001). In southern Thailand they consumed 13 plant foods during two months of the dry season; the fruits of *Ficus* spp. were the most commonly eaten (Bertrand, 1969). In central and southern Yunnan they ate 19 principal plant species over a six-month period (Fooden et al., 1985), eating the leaves from all of the plant species, the fruits from 12 species, and the flowers of one. These macaques also eat insects, small vertebrates, and are known crop raiders, consuming potatoes and rice as well as other agricultural plantings (Treesucon, 1988), which they have been reported raiding in "absolute silence" (Gupta, 2001: p.129). *M. arctoides* may spend up to 50% of their daytime activity foraging and feeding (Bertrand, 1969). Their day ranges seem to be associated with the distribution of fruiting trees (Bertrand, 1969; Estrada & Estrada, 1976; Gupta, 2001).

Reproductive behavior and social organization of *M. arctoides* has been studied in captivity extensively, including 20 years of data collected on stable dominance by a female matriline in Riverside, California (Rhine & Maryanski, 1996). A wide range of submissive signals and tactile behaviors are characteristic of stump-tailed macaques, such as a "mock-bite" that has been described as a ritualized aggressive behavior used to resolve uncertain dominance relationships (Maestripieri, 1996). Overall,

M. arctoides have been described as a socially tolerant species (Thierry, 2010), although third-party harassment of a mating couple has been reported (Fooden, 1990).

In India, Myanmar, Thailand, and China, these macaques are hunted for food and used in traditional medicines (Gupta, 2001; Srivastava & Mohnot, 2001; Chetry et al., 2003). Presently, *M. arctoides* are considered relatively rare throughout much of India and, due to the large-scale loss of habitat, have become extinct in many parts of that nation (Singh, 2001; Chetry et al., 2003; Sharma et al., 2012). Despite the fact that the species lives in the heart of *Macaca* distribution and in broad sympatry with many other species, it is rarely observed and, when a sighting *is* reported, the group sizes and densities are low. IUCN places *M. arctoides* in the Vulnerable category based on past population declines and a projected future decline of 30% over the next 30 years due to hunting and logging (Htun et al., 2008; IUCN, 2020). This species exhibits a limited degree of opportunistic feeding behavior (i.e., crop raiding). While found in a considerable range of different habitat types, *M. arctoides* seem to avoid direct contact with humans and prefer primary evergreen forests. These characteristics associate it with a minimal level of ecological plasticity.

ECOLOGICAL PLASTICITY IN MACAQUES: A SUMMARY

Plasticity is exhibited by both humans and macaques, the two primate genera with the most widespread geographic colonization. Nevertheless, in the case of the majority of macaque species, there appears to be inherent limits to the degree of plasticity they are capable of exhibiting. Almost two decades ago, Muroyama and Eudey (2004) gathered population data that showed declining trends for the majority of macaque species despite their widespread geographic distribution. Population trends for two species that are considered to exhibit maximal plasticity (*M. mulatta* and *M. radiata*) indicate that urban populations are increasing while forest populations are either declining or are unknown. Those macaques living in very close proximity to humans are susceptible to greater parasite loads, nutritional imbalances, and greater stress. Macaques living in provisioned locations where religious beliefs may offer some protection are to a degree safeguarded (e.g., Hindu temple grounds). Similarly, if the monkeys play a major role attracting tourists (e.g. monkey parks), thereby, providing a tangible economic benefit to a region, they are also tolerated. However, in such habitats macaques may be living in very large social groups and experience increased agonistic behavior and competition between individuals. In addition, macaques that live commensally with humans in urban locations experience increased competition with humans for access to food and space.

Those species designated as moderately flexible/plastic in ecology and behavior face increasing conflict with humans due to their crop- and garden-raiding behavior. In these areas, macaques may be trapped and sold as pets or local populations may be totally eradicated. Those species exhibiting limited and minimal degrees of plasticity are being subjected to increased habitat degradation and hunting.

Based on a review of the literature, we propose that the 24 species included in the genus *Macaca* exhibit differing degrees of ecological plasticity. In Table 17.4 we have placed 19 species in categories ranging from maximal to minimal ecological plasticity, with 5 species so little studied in the wild that their ecological flexibility is unknown and cannot be assessed with any reliability. The rigid categories to which we have assigned species, in reality, undoubtedly overlap; however, we offer this classification structure as one way to compare and contrast species in this large nonhuman primate genus.

We concur with the identification originally made by Richard et al. (1989) that four species (*M. fasicularis, M. mulatta, M. radiata,* and *M. sinica*) demonstrate a high degree of ecological plasticity. In this regard, maximal plasticity is indicated by the fact that these species have evolved strategies allowing them to live commensally with humans in highly disturbed habitats or human-created habitats. In addition, evidence exists that they are not restricted to a particular habitat type. It has also been observed that these macaques may actually seek out areas of disturbed habitat to colonize, a capacity most certainly linked to a high degree of behavioral plasticity which allows them to adjust ranging behaviors and activity cycles appropriate for a new or different habitat.

Four species (*M. brunnescens, M. fuscata, M. maura,* and *M. tonkeana*) reflect a moderate degree of ecological plasticity. These species raid agricultural crops on a regular basis and utilize disturbed habitats. They do not, however, live commensally with humans. Nevertheless, in many cases they may seek out secondary forests in close proximity to human settlements and villages, thus providing the macaques with ready access to agricultural fields and garden plots.

Six species (*M. assamensis, M. leonina, M. nigra, M. ochreata, M. pagensis,* and *M. sylvanus*) exhibit limited ecological plasticity. These species, like all macaques, are dietary opportunists. They raid crops on occasion but overall tend to avoid disturbed habitats. They do occupy a range of habitat types, but their limited use of anthropogenic habitats is directly related to the decreased availability of preferred habitat.

Five species (*M. arctoides, M. nemestrina, M. nigrescens, M. silenus,* and *M. thibetana*) give evidence of minimal ecological plasticity. These species are restricted to relatively undisturbed broadleaf evergreen forests and tend to avoid human habitats; when forced to live in more marginal habitats, their demographic responses are not favorable. Within this category, *M. silenus* represents an extreme lack of plasticity, since it is an obligate mature evergreen rainforest species.

The groupings presented in Table 17.4 crosscut phylogenetic boundaries, thus negating phylogeny as a key to the various degrees of plasticity exhibited within the macaque genus. Geographic location is also not a key, nor is island versus continental status. One island species (*M. sinica*) exhibits maximum plasticity. In the case of 11 other island species (*M. cyclopis, M. fuscata,* 7 Sulawesi species, and 2 Mentawai species), the environments in which they live exhibit little floral and faunal diversity compared to mainland East and Southeast Asia. (Of course, it is difficult to reconstruct the original ecological adaptations of these species since many of them live on islands that have undergone much anthropogenic change.)

An evolutionary predisposition or propensity to occupy transitional habitats may have been selected for during the Pleistocene when widespread climatic fluctuation and change promoted ecological edges, as first proposed by Eudey (1980). This may be an inherent adaptation of the four maximally plastic macaque species. The geographic distribution and associated robust

TABLE 17.4 Proposed Degrees of Ecological Plasticity in *Macaca* Species

Maximal Plasticity	Moderate Plasticity	Limited Plasticity	Minimal Plasticity	Unknown Degree of Plasticity
M. fasicularis	*M. brunnescens*	*M. assamensis*	*M. arctoides*	*M. cyclopis*
M. mulatta	*M. fuscata*	*M. leonina*	*M. nemestrina*	*M. hecki*
M. radiata	*M. maura*	*M. nigra*	*M. nigrescens*	*M. leucogenys*
M. sinica	*M. tonkeana*	*M. ochreata*	*M. silenus*	*M. munzala*
		M. pagensis	*M. thibetana*	*M. siberu*
		M. sylvanus		

habitat versatility of these macaques—which Sussman et al. (2011) refer to as "camp-followers" of humans—was then additionally influenced by large-scale anthropogenic change in recent times. In contrast, the less plastic species may have had access to more of their preferred habitat in the past but have now been constrained by human habitat disturbance and an inability to compete against other more successful macaque species that were ecologically more robust.

It is difficult to assess accurately the degree of ecological plasticity exhibited by macaques. They are not highly specialized primates, but they are capable of much behavioral plasticity allowing them to adapt to anthropogenic change. Since the time of Roman and Arab conquerors, when much of the low-altitude forests of North Africa were destroyed, species like *M. sylvanus* have been subjected to human-induced habitat alteration. Consequently, we do not have a deep view of the history of adaptations that this species has made but can only look at surviving populations in isolated pockets of Algeria and Morocco. Human enchroachment and habitat alteration are now regular components and influences on the ecology of macaques and primates in general (McLennan et al., 2017). Due to this fact, McKinney (2015) proposed a new way to classify and identify types and levels of anthropogenic influences. As such, humans have increasingly become a major component of the macaque ecological and even social landscape.

The developing field of ethnoprimatology now seeks to understand human-nonhuman primate social and ecological interconnections (Fuentes & Hockings, 2010; Fuentes, 2012; Riley, 2020). The degree of behavioral and ecological plasticity that macaques exhibit must now be examined within an evolutionary framework that scrutinizes how various species of macaque (as per Riley, 2020) respond to anthropogenic change. This is a fluid situation and humans are now part of the niche construction in which many macaques are found. Undoubtedly there will be inter- and intraspecific differences in how macaques respond to human-induced environmental change, and while some degree of plasticity may be adaptive, in other cases it may not be. Increased proximity to humans results in changes in the quality and type of foods eaten, greater exposure to pathogens, changes in activity patterns, and increased tolerance to humans—all of these result in an increased potential of conflict with humans. The future of this very diverse genus will be explicitly linked, in the case of the limited and minimally ecologically flexible species, to the preservation of their preferred habitats. In the case of the moderate and maximally plastic species, their future will depend on humans adopting strategies that will allow humans and macaques to cohabitate.

REFERENCES CITED—CHAPTER 17

Abe, H., Ishii, N., Ito, T., Kaneko, Y., Maeda, K., Miura, S., & Yoneda, M. (2005). *A Guide to the Mammals of Japan*. Tokai University Press.

Abegg, C., & Thierry, B. (2002). The phylogenetic status of Siberut macaques: Hints from the bared-teeth display. *Primate Report*, 63, 73-78.

Agetsuma, N., & Nakagawa, N. (1998). Effects of habitat differences on feeding behaviors of Japanese monkeys: Comparison between Yakushima and Kinkazan. *Primates*, 39(3), 275-289.

Albert, A., Hambuckers, A., Culot, L., Savini, T., & Huynen, M. (2013a). Frugivory and seed dispersal by northern pigtailed macaques (*Macaca leonina*), in Thailand. *International Journal of Primatology*, 34, 170-193.

Albert, A., Huynen, M., Savini, T., & Hambuckers, A. (2013b). Influence of food resources on the ranging pattern of northern pig-tailed macaques (*Macaca leonina*). *International Journal of Primatology*, 34, 696-713.

Albert, A., Savini, T., & Huynen, M. (2013c). The role of *Macaca* spp. (Primates: Cercopithecidae) in seed dispersal networks. *The Raffles Bulletin of Zoology*, 61(1), 423-434.

Anderson, J. (1998). Home range (core area) squatting in bonnet macaques (*Macaca radiata*). *Mammalia*, 62(4), 595-599.

Angst, W. (1975). Basic data and concepts on the social organization of *Macaca fascicularis*. In L. A. Rosenblum (Ed.), *Primate behavior: Developments in field and laboratory research* (Vol. 4, pp. 325-388). Academic Press.

Anitha, K., Aneesh, A., Raghavan, R., Kanagavel, A., Augustine, T., & Shijo, J. (2013). Identifying habitat connectivity for isolated populations of lion-tailed macaque (*Macaca silenus*) in Valparai Plateau, Western Ghats, India. *Primate Conservation*, 27, 91-97.

Ashmore-DeClue, P. (1992). *Macaques: An adaptive array (A summary and synthesis of the literature of the genus* Macaca *from an ecological perspective)* [Doctoral dissertation, Washington University].

Berman, C. M., Li, J., Ogawa, H., Ionica, C., & Yin, H. (2007). Primate tourism, range restriction, and infant risk among *Macaca thibetana* at Mt. Huangshan, China. *International Journal of Primatology*, 28, 1123-1141.

Bernstein, I. (1967a). A field study of the pigtail monkey (*Macaca nemestrina*). *Primates*, 8, 217-228.

Bernstein, I. (1967b). Intertaxa interactions in a Malayan primate community. *Folia Primatologica*, 7, 198-207.

Bernstein, I. S. (1970). Activity patterns in pigtail monkey groups. *Folia Primatologica*, 12, 187-198.

Bertrand, M. (1969). *The behavioral repertoire of the stumptail macaque: A descriptive and comparative study*. S. Karger.

Birky, W. (2001). Mating season effects on male-female relationships in wild Formosan macaques (*Macaca cyclopis*). *American Journal of Physical Anthropology*, 114(S32), 40.

Boonratana, R., Chalise, M., Das, J., Htun, S. & Timmins, R. J. 2008. *Macaca assamensis. The International Union for the Conservation of Nature Red List of Threatened Species*

2008: e.T12549A3354977. http://dx.doi.org/10.2305/IUCN .UK.2008.RLTS.T12549A3354977.en.

Brotcorne, F., Maslarov, C., Wandia, I. N., Fuentes, A., Beudels-Jamar, R. C., & Huynen, M. (2014). The role of anthropic, ecological, and social factors in sleeping site choice by long-tailed macaques (*Macaca fascicularis*). *American Journal of Primatology, 76*, 1140-1150.

Butynski, T. M., Cortes, J., Waters, S., Fa, J., Hobbelink, M. E., van Lavieren, E., Camperio-Ciani, A. (2008). *Macaca sylvanus. The International Union for the Conservation of Nature Red List of Threatened Species* 2008: e. T12561A3359140. http://dx.doi.org/10.2305/IUCN.UK .2008.RLTS.T12561A3359140.en.

Bynum, E. L., Bynum, D. Z., & Supriatna, J. (1997a). Confirmation and location of the hybrid zone between wild populations of *Macaca tonkeana* and *Macaca hecki* in Central Sulawesi, Indonesia. *American Journal of Primatology, 43*, 181-209.

Bynum, E. L., Bynum, D. Z., Froehlich, J. W., & Supriatna, J. (1997b). Revised geographic ranges and hybridization in *Macaca tonkeana* and *Macaca hecki. Tropical Biodiversity, 4*, 275-283.

Bynum, N. (2002). Morphological variation within a macaque hybrid zone. *American Journal of Physical Anthropology, 118*, 45-49.

Caldecott, J. (1986a). An ecological and behavioural study of the pig-tailed macaque. In F. S. Szalay (Ed.), *Contributions to primatology* (Vol. 21, pp. 1-259). Karger.

Caldecott, J. (1986b). A summary of the ranging and activity patterns of the pig-tailed macaque (*Macaca nemestrina*) in relation to those of sympatric primates in Peninsular Malaysia. In D. M. Taub, & F. A. King (Eds.), *Current perspectives in primate social dynamics* (pp. 152-158). Van Nostrand Reinhold Co.

Caldecott, J. O., Feistner, A. T. C., & Gadsby, E. L. (1996). A comparison of ecological strategies of pig-tailed macaques, mandrills and drills. In J. E. Fa, & D. G. Lindburg (Eds.), *Evolution and ecology of macaque societies* (pp. 73-94). Cambridge University Press.

Chakraborty, D., Ramakrishnan, U., Panor, J., Mishra, C., & Sinha, A. (2007). Phylogenetic relationships and morphometric affinities of the Arunachal macaque *Macaca munzala*, a newly described primate from Arunachal Pradesh, northeastern India. *Molecular Phylogenetics and Evolution, 44*, 838-849.

Chakraborty, D., Sinha, A., & Ramakrishnan, U. (2014). Mixed fortunes: Ancient expansion and recent decline in population size of a subtropical montane primate, the Arunachal macaque *Macaca munzala. PLoS ONE, 9*(7), e97061.

Chalise, M. (1999). Report on the Assamese monkeys (*Macaca assamensis*) of Nepal. *Asian Primates, 7*(1-2), 7-11.

Chalise, M. (2003a). Assamese macaques (*Macaca assamensis*) in Nepal. *Primate Conservation, 19*, 99-107.

Chalise, M. (2003b). Characteristics of the Assamese monkey (*Macaca assamensis*) of Nepal. *American Journal of Primatology, 66*(S1), 195.

Chalmers, N. (1980). *Social behaviour in primates.* University Park Press.

Chang, S. (2000). A survey of crop raiding by the Formosan macaque (*Macaca cyclopis*) in central Taiwan. *Endemic Species Research, 2*, 1-12.

Chatani, K. (2003). Positional behavior of free-ranging Japanese macaques (*Macaca fuscata*). *Primates, 44*, 13-23.

Cheney, D., & Wrangham, R. (1987). Predation. In B. Smuts, D. Cheney, R. Seyfarth, R. Wrangham, & T. Struhsaker (Eds.), *Primate societies* (pp. 227-239). University of Chicago Press.

Chetry, D., Borthakur, U., & Das, R. K. (2015). A short note on a first distribution record of white-cheeked macaque *Macaca leucogenys* from India. *Asian Primates Journal, 5*(1), 45-47.

Chetry, D., Medhi, R., Biswas, J., Das, D., & Bhattacharjee, P. C. (2003). Nonhuman primates in the Namdapha National Park, Arunachal Pradesh, India. *International Journal of Primatology, 24*(2), 383-388.

Choudhury, A. (2001). Primates in northeast India: An overview of their distribution and conservation. *ENVIS Bulletin: Wildlife and Protected Areas, 1*(1), 92-101.

Ciani, A., Stanyon, R., Scheffran, W., & Sampurno, B. (1989). Evidence of gene flow between Sulawesi macaques. *American Journal of Primatology, 9*, 347-364.

Cooper, M., & Bernstein, I. (2008). Evaluating dominance styles in Assamese and rhesus macaques. *International Journal of Primatology, 29*, 225-243.

Corbet, G., & J. Hill. 1992. *Mammals of the Indo-Malayan Region: A systematic review.* Oxford University Press.

Coss, R., McCowan, B., & Ramakrishnan, U. (2007). Threat-related acoustical differences in alarm calls by wild bonnet macaques (*Macaca radiata*) elicited by python and leopard models. *Ethology, 112*, 352-367.

Crockett, C., & Wilson, W. (1980). The ecological separation of *Macaca nemestrina* and *M. fascicularis* in Sumatra. In D. G. Lindburg (Ed.), *The macaques: Studies in ecology, behavior and evolution* (pp. 148-181). Van Nostrand Reinhold Co.

Datta, S. (1992). Effects of availability of allies on female dominance structure. In A. Harcourt & F. de Waal (Eds.), *Coalitions and alliances in humans and other animals* (pp. 61-82). Oxford University Press.

Dazey, J., Kuyk, K., Oswald, M., Martenson, J., & Erwin, J. (1977). Effects of group composition on agonistic behaviour of captive pigtail macaques, *Macaca nemestrina. American Journal of Physical Anthropology, 46*, 73-76.

Deag, J. (1974). *A study of the social behavior and ecology of the wild Barbary macaque* (Macaca sylvanus). *L.)* [Doctoral dissertation, University of Bristol].

Deag, J. (1984). Demography of the Barbary macaque at Ain Kahla in the Moroccan Moyen Atlas. In J. E. Fa (Ed.), *The Barbary macaque: A case study in conservation* (pp. 113-133). Plenum Press.

Deag, J. (1985). The diurnal patterns of behavior of the wild Barbary macaque (*Macaca sylvanus*). *Journal of Zoology, 206*, 403-413.

Deag, J., & Crook, J. (1971). Social behavior and "agonistic buffering" in the wild Barbary macaque, *Macaca sylvana* L. *Folia Primatologica*. 15, 183-200.

Deinard, A., & Smith, D. G. (2001). Phylogenetic relationships among the macaques: Evidence from the nuclear locus *NRAMP*. *Journal of Human Evolution*, 41, 45-59.

Delson, E. (1980). Fossil macaques, phyletic relationships and a scenario of development. In D. G. Lindburg (Ed.), *The macaques: Studies in ecology, behavior and evolution* (pp. 10-30). Van Nostrand Reinhold Co.

Dittus, W. P. J. (1974). *The ecology and behavior of the toque monkey*, Macaca sinica. [Doctoral dissertation, University of Maryland].

Dittus, W. P. J. (1977a). The socioecological basis for the conservation of the toque monkey (*Macaca sinica*) of Sri Lanka (Ceylon). In H. R. H. Rainier III, & G. H. Bourne (Eds.), *Primate conservation*. Academic Press.

Dittus, W. P. J. (1977b). The social regulation of population density and age-sex distribution in the toque monkey. *Behaviour*, 63, 281-322.

Dittus, W. P. J. (1979). The evolution of behavior regulating density and age-specific sex ratios in a primate population. *Behaviour*, 69, 265-302.

Dittus, W. P. J., & S. M. Ratnayeke. (1989). Individual and social behavioral responses to injury in wild toque macaques (*Macaca sinica*). *International Journal of Primatology*, 10, 215- 234.

Dittus, W., Watson, A., & Molur, S. (2008). *Macaca sinica. The International Union for the Conservation of Nature Red List of Threatened Species* 2008: e.T12560A3358720. http://dx.doi .org/10.2305/IUCN.UK.2008.RLTS.T12560A3358720.en.

Drucker, G. (1984). The feeding ecology of the Barbary macaque and cedar forest conservation in the Moroccan Moyen Atlas. In J. E. Fa (Ed.), *The Barbary macaque: A case study in conservation* (pp. 135-64). Plenum Press.

Duckworth, J., Salter, R., & Khounbline, K. (1999). Wildlife in Lao PDR. *1999 status report, IUCN*. Vientiane, Laos.

el Alami, A., van Lavieren, E., Rachida, A., & Chait, A. (2012). Difference in activity budgets and diet between semiprovisioned and wild feeding groups of the endangered Barbary macaque (*Macaca sylvanus*) in the Central High Atlas Mountains, Morocco. *American Journal of Primatology*, 74, 210-216.

Elton, S., & O'Regan, H. J. (2014). Macaques at the margins: The biogeography and extinction of *Macaca sylvanus* in Europe. *Quaternary Science Reviews*, 96, 117-130.

Enari, H., & Sakamaki-Enari, H. (2013). Influence of heavy snow on the feeding behavior of Japanese macaques (*Macaca fuscata*) in northern Japan. *American Journal of Primatology*, 75, 534-544.

Engelhardt, A., Muniz, L., Perwitasari-Farajallah, D., & Widdig, A. (2017). Highly polymorphic microsatellite markers for the assessment of male reproductive skew and genetic variation in critically endangered crested macaques (*Macaca nigra*). *International Journal of Primatology*, 38, 672-691.

Erinjery, J. J., Kavana, T. S., & Singh, M. (2015). Food resources, distribution and seasonal variations in ranging in lion-tailed macaques, *Macaca silenus* in the Western Ghats, India. *Primates*, 56 (1), 45-54.

Erinjery, J. J., Kumara, H., Mohan, K., & Singh, M. (2017). Interactions of lion-tailed macaque (*Macaca silenus*) with non-primates in the Western Ghats, India. *Current Science* 112(10), 2129.

Erwin, J. (1978). Factors contributing to intragroup aggression in captive pig-tailed macaque groups. In D. J. Chivers & J. Herbert (Eds.), *Recent advances in primatology*, (Vol. 1, pp. 581-584). Academic Press.

Estrada, A., & Estrada, R. (1976). Establishment of a free-ranging colony of stumptail macaques (*Macaca arctoides*): Relations to the ecology I. *Primates*, 17, 337-356.

Etting, S. F., & Isbell, L. A. (2014). Rhesus macaques (*Macaca mulatta*) use posture to assess level of threat from snakes. *Ethology*, 120, 1177-1184.

Eudey, A. (1980). Pleistocene glacial phenomena and the evolution of Asian macaques. In D. G. Lindburg (Ed.), *The macaques: Studies in ecology, behavior and evolution*. Van Nostrand Reinhold Co.

Evans, B. J., Morales, J. C., Supriatna, J., & Melnick, D. J. (1999). Origin of the Sulawesi macaques (Cercopithecidae: *Macaca*) as suggested by mitochondrial DNA phylogeny. *Biological Journal of the Linnean Society*, 66(4), 539-560.

Fa, J. E. (1981). The apes on the rock. *Oryx*, 16, 73-76.

Fa, J. E. (1982). A survey of population and habitat of the Barbary macaque *Macaca sylvanus*, L. in north Morocco. *Biological Conservation*, 24(1), 45-66.

Fa, J. E. (1984). Habitat distribution and habitat preference in Barbary macaques (*Macaca sylvanus*). *International Journal of Primatology*, 5(3), 273-286.

Fa, J. E. (1986a). On the ecological status of the Barbary macaque *Macaca sylvanus* L. in north Morocco: Habitat influences versus human impact. *Biological Conservation*, 35(3), 215-258.

Fa, J. E. (1986b). An important new locality for the Barbary macaque (*Macaca sylvanus*) in Morocco. *Primate Conservation*, 7, 31-34.

Fa, J. E. (1989). The genus *Macaca*: A review of taxonomy and evolution. *Mammal Review*, 19, 45-81.

Fan, P., Liu, Y., Zhang, Z., Zhao, C., Li, C., Liu, W., Li, M. (2017). Phylogenetic position of the white-cheeked macaque (*Macaca leucogenys*), a newly described primate from southeastern Tibet. *Molecular Phylogenetics and Evolution*, 107, 80-89.

Fedigan, L. (1974). The classification of predators by Japanese macaques (*Macaca fuscata*) in the mesquite chaparral habitat of south Texas. *American Journal of Physical Anthropology*, 40(1), 135.

Feeroz, M. M. (2003). Wildlife diversity in Satchari Forest of northeastern region of Bangladesh. *Bangladesh Journal of Life Science*, 15, 61-76.

Fleagle, J. G. (1988). *Primate adaptation and evolution*. Academic Press.

Fooden, J. (1964). Rhesus and crab-eating macaques: Intergradation in Thailand. *Science, 143*, 363-365.

Fooden, J. (1969). Taxonomy and evolution of the monkeys of Celebes (Primates: Cercopithecidae). *Bibliotheca Primatologica, 10*, 1-148.

Fooden, J. (1971a). Female genitalia and taxonomic relationships of *Macaca assamensis*. *Primates, 12*, 63-73.

Fooden, J. (1971b). Reports on primates collected in western Thailand Jan–April 1967. *Fieldiana Zoology, 59*, 1-62.

Fooden, J. (1975). Taxonomy and evolution of lion-tailed and pigtail macaques (Primates: Cercopithecidae). *Fieldiana Zoology, 67*, 1-169.

Fooden, J. (1976). Provisional classification and key to living species of macaques (Primates: *Macaca*). *Folia Primatologica, 25*, 225-236.

Fooden, J. (1979). Taxonomy and evolution of the *sinica* group of macaques: I. Species and subspecies accounts of *Macaca sinica*. *Primates, 20*(1), 109-140.

Fooden, J. (1980). Classification and distribution of living macaques (*Macaca* 1799 Lacepède) In D. Lindberg (Ed.), *The macaques: Studies in ecology, behavior and evolution* (pp. 1-9). Van Nostrand Reinhold Co.

Fooden, J. (1981). Taxonomy and evolution of the *sinica* group of macaques: 2. Species and subspecies accounts of the Indian bonnet macaque, *Macaca radiata*. *Fieldiana Zoology, 9*, 1-52.

Fooden, J. (1982). Ecogeographic segregation of macaque species. *Primates, 23*, 574-579.

Fooden, J. (1983). Taxonomy and evolution of the *sinica* group of macaques: 4. Species account of *Macaca thibetana*. *Fieldiana Zoology, 17*, 1-20.

Fooden, J. (1988). Taxonomy and evolution of the *sinica* group of macaques: 6. Interspecific comparisons and synthesis. *Fieldiana Zoology, 45*, 1-44.

Fooden, J. (1990). The bear macaque, *Macaca arctoides*: A systematic review. *Journal of Human Evolution, 19*, 607-686.

Fooden, J. (1991). Systematic review of Philippine macaques (Primates, Cercopithecidae: *Macaca fasicularis* ssp.). *Fieldiana Zoology, 64*, 1-44.

Fooden, J. (1995). Systematic review of southeast Asian longtail macaques. *Fieldiana Zoology, 81*, 1-206.

Fooden, J. (1997). Tail length variation in *Macaca fascicularis* and *Macaca mulatta*. *Primates, 38*, 221-231.

Fooden, J. (2000). Systematic review of the rhesus macaque, *Macaca mulatta* (Zimmermann 1780). *Fieldiana Zoology, 96*, 1-180.

Fooden, J. (2007). Systematic review of the Barbary macaque, *Macaca sylvanus* (Linnaeus, 1758). *Fieldiana Zoology, 113*, 1-58.

Fooden, J., & Aimi, M. (2005). Systematic review of Japanese macaques, *Macaca fuscata* (Gray, 1870). *Fieldiana Zoology, 104*, 1-200.

Fooden, J., Mahabal, A., & Saha, S. S. (1981). Redefinition of rhesus macaque—bonnet macaque boundary in Peninsular India (Primates: *Macaca mulatta, M. radiata*). *Journal of the Bombay Natural History Society, 78*, 463-474.

Fooden, J., Quan, G., Wang, Z., & Wang, X. (1985). The stumptail macaques of China. *American Journal of Primatology, 8*, 11-30.

Fooden, J., & Wu, H. (2001). Systematic review of the Taiwanese macaque, *Macaca cyclopis* Swinhoe, 1863. *Fieldana Zoology, 98*, 1-70.

Froehlich, J., & Supriatna, J. (1996). Secondary intergradation between *Macaca maurus* and *M. tonkeana* in south Sulawesi, and the species status of *M. togeanus*. In J. E. Fa, & D. L. Lindburg (Eds.), *Evolution and ecology of macaque societies* (pp. 43-70). Cambridge University Press.

Fuentes, A. (1997). Current status and future viability for the Mentawai primates. *Primate Conservation, 17*, 111-116.

Fuentes, A. (1999). Variable social organization: What can looking at primate groups tell us about the evolution of plasticity in primate societies? In P. Dolhinow & A. Fuentes (Eds.), *The nonhuman primates* (pp. 183-188). Mayfield Publishing Company.

Fuentes, A. (2012). *Race, monogamy, and other lies they told you: Busting myths about human nature*. University of California Press.

Fuentes, A., & Hockings, K. J. (2010). The ethnoprimatological approach in primatology. *American Journal of Primatology, 72*, 841-847.

Fuentes, A., & Olson, M. (1995). Preliminary observations and status of the Pagai macaque. *Asian Primates, 4*(4), 1-4.

Fuentes, A., Rompis, A. L. T., Arta Putra, I. G. A., Watiniasih, N. L., Suartha, I. N., Wandia, I. N., & Selamet, W. (2011). Macaque behavior at the human-monkey interface: The activity and demography of semi-free ranging, *Macaca fascicularis* at Padangtegal, Bali, Indonesia. In M. Gumert, A. Fuentes, & L. Jones-Engel (Eds.), *Monkeys on the edge: Ecology and management of long-tailed macaques and their interface with humans* (pp. 159-179). Cambridge University Press.

Fürtbauer, I., Schülke, O., Heistermann, M., & Ostner. J. (2010). Reproductive and life history parameters of wild female *Macaca assamensis*. *International Journal of Primatology, 31*, 501-517.

Furuichi, T. (1983). Interindividual distance and influence of dominance on feeding in a natural Japanese macaque troop. *Primates, 24*, 445-455.

Gathorne-Hardy, F. J., Syaukani, Davies, R. G., Eggleton, P., & Jones, D. T. (2002). Quaternary rainforest refugia in southeast Asia: Using termites (Isoptera) as indicators. *Biological Journal of the Linnean Society, 75*, 453-466.

Gippoliti, S. (2001). Notes on the taxonomy of *Macaca nemestrina leonina* Blyth, 1863 (Primates: Cercopithecidae). *Hystrix the Italian Journal of Mammalogy, 12*, 51-54.

Green, S., & Minkowski, K. (1977). The lion-tailed monkey and its south Indian rain forest habitat. In H. R. H. Rainer III, & H. Bourne (Eds.), *Primate conservation* (pp. 289). Academic Press.

Groves, C. (1980). Speciation in *Macaca*: The view from Sulawesi. In D. G. Lindburg (Ed.), *The macaques: Studies in ecology, behavior and evolution* (pp. 84-124). Van Nostrand Reinhold Co.

Groves, C. (1996). The nomenclature of the Tanzanian mangabey and the Siberut macaque. *Australian Primatology,* 10, 2-5.

Groves, C. (2001). *Primate taxonomy.* Smithsonian Institution Press.

Gumert, M., Hamada, Y., & Malaivijitnond, S. (2013). Human activity negatively affect wild stone tool-using Burmese long-tailed macaques *Macaca fascicularis aurea* in Laemson National Park, Thailand. *Oryx,* 14, 1-9.

Gumert, M., Kluck, M., & Malaivijitnond, S. (2009). The physical characteristics and usage patterns of stone axe and pounding hammers used by long-tailed macaques in the Andaman Sea region of Thailand. *American Journal of Primatology,* 71, 594-608.

Gumert, M. D., & Malaivijitnond, S. (2012). Marine prey processed with stone tools by Burmese long-tailed macaques (*Macaca fasciculari aurea*) in intertidal habitats. *American Journal of Physical Anthropology,* 149, 447-457.

Gupta, A. K. (2001). Non-human primates of India: An introduction. *Envis Bulletin: Wildlife and Protected Areas,* 1, 1-29.

Hamada, Y., & Iida, T. (2000). A hawk eagle attacked a Japanese macaque? *Monkey,* 289/290, 16-19.

Hamada, Y., Urasopon, N., Hadi, I., & Malaivijitnond, S. (2005). Body size and proportions and pelage color of free-ranging *Macaca mulatta* from a zone of hybridization in northeastern Thailand. *International Journal of Primatology,* 27, 497-513.

Hamada, Y., Watanabe, T., & Iwamoto, M. (1996). Morphological variations among local populations of Japanese macaque (*Macaca fuscata*). In T. Shotake, & K. Wada (Eds.), *Variations in the Asian macaques* (pp. 97-115). Tokai University Press.

Hambali, K., Ismail, A., & Md-Zain, B. M. (2012a). Daily activity budget of long-tailed macaques (*Macaca fascicularis*) in Kuala Selangor Nature Park. *International Journal of Basic & Applied Sciences,* 12 (04), 47-52.

Hambali, K., Ismail, A., Zulkifli, S. Z., Md-Zain, B. M., & Amir, A. (2012b). Human-macaque conflict and pest behaviors of long-tailed macaques (*Macaca fascicularis*) in Kuala Selangor Nature Park. *Tropical Natural History,* 1(2), 189-205.

Hamilton, W., & Watt, K. (1970). Refuging. *Annual Review of Ecology, Evolution and Systematics,* 1, 263-286.

Hanya, G. (2004a). Diet of a Japanese macaque troop in the coniferous forest of Yakushima. *International Journal of Primatology,* 25(1), 55-69.

Hanya, G. (2004b). Seasonal variations in the activity budget of Japanese macaques in the coniferous forest of Yakushima: Effects of food and temperature. *American Journal of Primatology,* 63, 165-177.

Hanya, G., Mènard, N., Qarro, M., Ibn Tattou, M., Fuse, M., Vallet, D., & Wada, K. (2011). Dietary adaptations of temperate primates: Comparisons of Japanese and Barbary macaques. *Primates,* 52, 187-198.

Hardwick, J. L., Priston, N. E. C., Martin, T. E., Tosh, D. G., Mustari, A. H., & Abernethy, K. E. (2017). Community perceptions of the crop-feeding Buton macaque (*Macaca ochreata brunnescens*): An ethnoprimatological study on Buton Island, Indonesia. *International Journal of Primatology,* 38, 1102-1119.

Harrison, T., Krigbaum, J., & Manser, J. (2006). Primate biogeography and ecology on the Sunda Shelf Islands: A paleontological and zooarchaeological perspective. In S. M. Lehman, & J. G. Fleagle (Eds.), *Primate biogeography: Progress and prospects* (pp. 331-374). Springer.

Hart, D. (2000). *Primates as prey: Ecological, morphological, and behavioral relationships of primate species and their predators.* [Doctoral dissertation, Washington University].

Hasan, M. K., Aziz, M. A., Rabiul, S. M., Kawamoto, Y., Jones-Engel, L., Kyes, R. C., Feeroz, M. M. (2013). Distribution of rhesus macaques (*Macaca mulatta*) in Bangladesh: Interpopulation variation in group size and composition. *Primate Conservation,* 26(1), 125-132.

Haus, T., Vogt, M., Forster, B., Vu, N. T., & Ziegler, T. (2009). Distribution and population densities of diurnal primates in the karst forests of Phong Nha–Ke Bang National Park, Quang Binh Province, Central Vietnam. *International Journal of Primatology,* 30, 301-312.

Hayasaka, K., Fujii, K., & Horai, S. (1996). Molecular phylogeny of macaques: Implications of nucleotide sequences from an 896-base pair region of mitochondrial DNA. *Molecular Biology and Evolution,* 13, 1044-1053.

Heesen, M., Rogahn, S., Ostner, J., & Schülke, O. (2013). Food abundance affects energy intake and reproduction in frugivorous female Assamese macaques. *Behavioral Ecology and Sociobiology,* 67, 1053-1066.

Hill, D. (1997). Seasonal variation in the feeding behavior and diet of Japanese macaques (*Macaca fuscata yakui*) in lowland forest of Yakushima. *American Journal of Primatology,* 43, 305-322.

Hill, W. C. O. (1974). *Primates: Comparative anatomy and taxonomy. VI. Cercopithecoidae, Cercopithecinae.* Edinburgh University Press.

Hladik, C. M., & Hladik, A. (1972). Disponibilités alimentaires et domaines vitaux des primates à Ceylan. *Terre et Vie,* 26, 149-215.

Hohmann, G. M. (1989). Vocal communication of wild bonnet macaques (*Macaca radiata*). *Primates,* 30, 325-345.

Hohmann, G. M., Herzog, M. (1985). Vocal communication in lion-tailed macaques (*Macaca silenus*). *Folia Primatologica,* 45, 148-178.

Holden, C. (2004). New primate discovered in India. *Science,* 306, 2184.

Hou, W., Liu, S., Jiang, J., Fan, Z., Fan, P., & Li, J. (2016). The complete mitochondrial genome of white-cheeked macaque (*Macaca leucogenys*). *Mitochondrial DNA, Part B: Resources,* 1(1), 374-375.

Hsu, M., & Lin, J. (2001). Troop size and structure in free-ranging Formosan macaques (*Macaca cyclopis*) at Mt. Longevity, Taiwan. *Zoological Studies,* 40(1), 49-60.

Hsu, M., Kao, C., & Agoramoorthy, G. (2009). Interactions between visitors and Formosan macaques (*Macaca cyclopis*) at Shou-Shan Nature Park, Taiwan. *American Journal of Primatology, 71*, 214-222.

Hsu, M., Lin, J., & Agoramoorthy, G. (2001). Birth seasonality and interbirth intervals in free-ranging Formosan macaques, *Macaca cyclopis*, at Mt. Longevity, Taiwan. *Primates, 42*(1), 15-25.

Htun, S., Timmins, R., Boonratana, R., & Das, J. (2008). *Macaca arctoides. The International Union for the Conservation of Nature Red List of Threatened Species* 2008: e.T12548A3354519. http://dx.doi.org/10.2305/IUCN.UK .2008.RLTS.T12548A3354519.en.

Huang, Z., Huang, C., Tang, C., Huang L., Tang, H., Ma, G., & Zhou, Q. (2015). Dietary adaptations of Assamese macaques (*Macaca assamensis*) in limestone forests in southwest China. *American Journal of Primatology, 77*, 171-185.

Hussain, S., Ram, M. S., Kumar, A., Shivaji, S., & Umapathy, G. (2013). Human presence increases parasitic load in endangered lion-tailed macaques (*Macaca silenus*) in its fragmented rainforest habitats in southern India. *PLoS ONE, 8*(5), e63685.

IUCN. (2020). *International Union for the Conservation of Nature Red List of Threatened Species*. Version 2020-3. www .iucnredlist.org.

Itani, J., Tokuda, K., Furuya, Y., Kano, K., & Shin, Y. (1963). The social construction of natural troops of Japanese monkeys in Takasakiyama. *Primates, 4*, 1-42.

Iwamoto, T. (1974). A bioeconomic study on a provisioned troop of Japanese monkeys (*Macaca fuscata*) at Koshima Islet Miyazaki. *Primates, 15*, 241-262.

Iwamoto, T. (1978). Food availability as a limiting factor on population density of the Japanese monkey and gelada baboon. In D. J. Chivers, & J. Herbert (Eds.), *Recent advances in primatology* (Vol. 1, 287-303). Academic Press.

Izawa, K., & Nishida, T. (1963). Monkeys living in the northern limits of their distribution. *Primates, 4*, 67-88.

Izumiyama, S., Mochizuki, T., & Shiraishi, T. (2003). Troop size, home range area and seasonal range use of the Japanese macaque in the northern Japan Alps. *Ecological Research, 18*, 465-474.

Jaman, M. F., & Huffman, M. A. (2013). The effect of urban and rural habitats and resource type on activity budgets of commensal rhesus macaques (*Macaca mulatta*) in Bangladesh. *Primates, 54*, 49-59.

Johnson, T. (1985). Lion-tailed macaque behavior in the wild. In P. G. Heltne (Ed.), *The lion-tailed macaque: Status and conservation*. Alan R. Liss.

Jones-Engel, L., Engel, G. A., Schillaci, M. A., Babo, R., & Froehlich, J. (2001). Detection of antibodies to selected human pathogens among wild and pet macaques (*Macaca tonkeana*) in Sulawesi, Indonesia. *American Journal of Primatology, 54*, 171-178.

Jones-Engel, L., Engel, G. A., Schillaci, M. A., Froehlich, J., Paputungan, U., & Kyes, R. C. (2004). Prevalence of enteric parasites in pet macaques in Sulawesi, Indonesia. *American Journal of Primatology, 62*, 71-82.

José-Dominguez, J. M., Savini, T., & Asensio, N. (2015). Ranging and site fidelity in northern pigtailed macaques (*Macaca leonine*) over different temporal scales. *American Journal of Primatology, 77*, 841-853.

Kappeler, P. M., & Kraus, C. (2010). Levels and mechanisms of behavioural variability. In P. M. Kappeler (Ed.), *Animal behavior: Evolution and mechanisms* (pp. 655-684). Springer.

Karr, J. (1973). Ecological and behavioural notes on the lion-tailed macaque (*Macaca silenus*) in South India. *Journal of the Bombay Natural History Society, 70*, 191-193.

Kawamura, S., & Megantara, E. N. (1986) Observation of primates in logged forest on Sipora Island, Mentawai. *Kyoto University Overseas Research Report of Studies on Asian Non-Human Primates, 5*, 1-12.

Kitchener, A. C., & Groves, C. (2002). New insights into the taxonomy of *Macaca pagensis* of the Mentawai Islands, Sumatra. *Mammalia, 66*(4), 533-542.

Klegarth, A. R., Hollocher, H., Jones-Engel, L., Shaw, E., Lee, B. P. Y., Feeney, T., & Fuentes, A. (2017). Urban primate ranging patterns: GPS-collar deployments for *Macaca fascicularis* and *M. sylvanus*. *American Journal of Primatology, 79*(5), e22633.

Kohlhaas, A. K. (1993). *Behavior and ecology of* Macaca nigrescens *in Taman Nasional Dumoga-Bone, Sulawesi Utara, Indonesia*. [Doctoral dissertation, University of Colorado, Boulder].

Kohlhaas, A., & Southwick, C. H. (1996). *Macaca nigrescens*: Grouping patterns and group composition in a Sulawesi macaque. In J. E. Fa, & D. L. Lindburg (Eds.), *Evolution and ecology of macaque societies* (pp. 132-145). Cambridge University Press.

Koirala, S., Chalise, M. K., Katuwal, H. B., Gaire, R., Pandey, B., & Ogawa, H. (2017). Diet and activity of *Macaca assamensis* in wild and semi-provisioned groups in Shivapuri Nagarjun National Park, Nepal. *Folia Primatologica, 88*, 57-74.

Koyama, N. (1970). Changes in dominance rank and division of a wild Japanese monkey troop in Arashiyama. *Primates, 11*, 335-390.

Koyama, N. (1991). Grooming relationships in the Arashiyama group of Japanese monkeys. In L. M. Fedigan, & P. J. Asquith (Eds.), *The monkeys of Arashiyama. Thirty-five years of research in Japan and the West* (pp. 211-226). University of New York Press.

Koyama, N., Norikoshi, K., & Mano, T. (1975). Population dynamics of Japanese monkeys at Arashiyama. In M. Kawai, S. Kondo, & A. Ehara (Eds.), *Contemporary primatology: Proceedings of the Fifth Congress of the International Primatological Society*. Karger.

Krishnadas, M., Chandrasekhara, K., & Kumar, A. (2011). The response of the frugivorous lion-tailed macaque (*Macaca silenus*) to a period of fruit scarcity. *American Journal of Primatology, 73*, 1250-1260.

Krishnamani, R. (1994). Diet composition of the bonnet macaque (*Macaca radiata*) in a tropical dry evergreen forest of southern India. *Tropical Biodiversity, 2*(2), 285-297.

Krishnamurthy, R. S., & Kiester, A. R. (1998). Analysis of lion-tailed macaque habitat fragmentation using satellite imagery. *Current Science, 75,* 283-291.

Kumar, A. (1985). Patterns of extinction in India, Sri Lanka, and elsewhere in southeast Asia: Implications for lion-tailed macaque wildlife management and the Indian conservation system. In P. G. Heltne (Ed.), *The lion-tailed macaque: Status and conservation* (pp. 65-89). Alan R. Liss.

Kumar, A. (1987). *The ecology and population dynamics of lion-tailed monkeys* (Macaca silenus*) in south India.* [Doctoral dissertation, Cambridge University Press].

Kumar, A. (1995). The life history, ecology, distribution and conservation problems in the wild. In A. Kumar, S. Molur, & S. Walker (Eds.), *The lion-tailed macaque: Population and habitat viability assessment workshop Zoo Outreach Organization* (pp. 1-11). Coimbatore.

Kumar, A., & Kurup, G. U. (1985). Sexual behavior of the lion-tailed macaque, *Macaca silenus.* In P. G. Heltne (Ed.), *Lion-tailed macaque: Status and conservation* (pp. 109-130). Alan R. Liss.

Kumar, A., Singh, M., & Molur. S. (2008a). *Macaca silenus. The International Union for the Conservation of Nature Red List of Threatened Species* 2008: e.T12559A3358033. http://dx.doi.org/10.2305/IUCN.UK.2008.RLTS.T12559A3358033.en.

Kumar, A., Sinha, A., & Kumar, S. (2008b). *Macaca munzala. The International Union for the Conservation of Nature Red List of Threatened Species* 2008: e.T136569A4311929. http://dx.doi.org/10.2305/IUCN.UK.2008.RLTS.T136569A4311929.en.

Kumar, M., Karki, J., & Ghimire, M. (2001). Survey of Assamese monkey in Langtang National Park, Nepal. *American Society of Primatology Bulletin, 25,* 4-5.

Kumar, R., Gama, N., Raghunath, R., & Mishra, C. (2008). In search of the munzala: Distribution and conservation status of the newly-discovered Arunachal macaque (*Macaca munzala). Oryx, 42,* 360-366.

Kumar, R., Mishra, C., & Sinha, A. (2007). Foraging ecology and time-activity budget of the Arunachal macaque *Macaca munzala*: A preliminary study. *Current Science, 93*(4), 532-539.

Kumar, S., Kumara, H. N., Santhosh, K., & Sundararaj, P. (2019). Prevalence of gastrointestinal parasites in lion-tailed macaque *Macaca silenus* in central Western Ghats, India. *Primates,* 1-10.

Kumara, H. N., Kumar, S., & Singh, M. (2010). Of how much concern are the "least concern" species? Distribution and conservation status of bonnet macaques, rhesus macaques and Hanuman langurs in Karnataka, India. *Primates, 51,* 37-42.

Kumara, H. N., Sasi, N., Suganthasakthivel, R., Singh, M., Sushma, H. S., Ramachandran, K. K., & Kaumanns, W. (2014). Distribution, demography, and conservation of lion-tailed macaques (*Macaca silenus*) in the Anamalai Hills Landscape, Western Ghats, India. *International Journal of Primatology, 35*(5), 976-989.

Kumara, H. N., & Singh, M. (2004). Distribution and abundance of primates in rain forests of the Western Ghats, Karnataka, India and the conservation of *Macaca silenus. International Journal of Primatology, 25*(5), 1001-1018.

Kumara, H. N., & Singh, V. R. (2008). Status of *Macaca silenus* in the Kudremukh Forest Complex, Karnataka, India. *International Journal of Primatology, 29*(3), 773-781.

Kurihara, Y., & Hanya, G. (2015). Comparison of feeding behavior between two different-sized groups of Japanese macaques (*Macaca fuscata yakui*). *American Journal of Primatology, 77,* 986-1000.

Kurland, J. A. (1973). A natural history of Kra macaques (*Macaca fasciularis* Raffles, 1821) at the Kutai Reserve, Kalimantan, Timur, Indonesia. *Primates, 14,* 245-262.

Kurup, G. U., & Kumar, A. (1993). Time budget and activity patterns of the lion-tailed macaque (*Macaca silenus*). *International Journal of Primatology, 14,* 27-39.

Kyes, R. C., Iskandar, E., Onibala, J., Paputungan, U., Laatung, S., & Huettmann, F. (2013). Long-term population survey of the Sulawesi black macaques (*Macaca nigra*) at Tangkoko Nature Reserve, north Sulawesi, Indonesia. *American Journal of Primatology 75,* 88-94.

Lee, L. (1991). A review of the recent research on *Macaca cyclopis.* In Y. Lin, & K. Chang (Eds.), *Proceedings of the First International Symposium on Wildlife Conservation, Republic of China* (pp. 290-304). Government of Taiwan.

Lee, R. (1997). *The impact of hunting and habitat disturbance on the population dynamics and behavioral ecology of the crested black macaque (*Macaca nigra). [Doctoral dissertation, University of Oregon, Eugene].

Lee, R., Riley, E., Sangermano, F., Cannon, C., and Shekelle, M. (2020a). *Macaca hecki. The IUCN Red List of Threatened Species* 2020: e.T12570A17948969. https://dx.doi.org/10.2305/IUCN.UK.2020-3.RLTS.T12570A17948969.en.

Lee, R., Riley, E., Sangermano, F., Cannon, C., and Shekelle, M. (2020b). *Macaca nigra. The IUCN Red List of Threatened Species* 2020: e.T12556A17950422. https://dx.doi.org/10.2305/IUCN.UK.2020-3.RLTS.T12556A17950422.en.

Lee, R., Riley, E., Sangermano, F., Cannon, C., and Shekelle, M. (2020c). *Macaca nigrescens. The IUCN Red List of Threatened Species* 2020: e.T12568A17948400. https://dx.doi.org/10.2305/IUCN.UK.2020-3.RLTS.T12568A17948400.en.

Lee, R., Riley, E., Sangermano, F., Cannon, C. & Shekelle, M. (2021). *Macaca brunnescens. The IUCN Red List of Threatened Species* 2021: e.T12569A17985924. https://dx.doi.org/10.2305/IUCN.UK.2021-1.RLTS.T12569A17985924.en.

Lei, A., Zhang, G., Tian, R., Zhu, J., Zheng, H., Pang, W., & Zheng, Y. (2014). Replication potentials of HIV-1/HSIV in PBMCs from northern pigtailed macaque (*Macaca leonina*). *Dongwuxue Yanjiu, 35*(3), 186-195.

Li, C., Zhao, C., & Fan, P. (2015). White-cheeked macaque (*Macaca leucogenys*): A new macaque species from Medog, southeastern Tibet. *American Journal of Primatology, 77,* 753-766.

Lian, X., Zhang, X., Dai, Z., & Zheng, Y. (2016). Cloning, sequencing, and polymorphism analysis of novel classical MHC class I alleles in northern pig-tailed macaques (*Macaca leonina*). *Immunogenetics, 68,* 261-274.

Lin, T. J., Agoramoorthy, G., Huang, C. C., & Hsu, M. J. (2008). Effects of troop size on social relations among male Formosan macaques, *Macaca cyclopis. Zoological Studies, 47*(3), 237-246.

Lindburg, D. G. (1971). The rhesus monkey in north India: An ecological and behavioral study. In L. A. Rosenblum (Ed.), *Primate behavior: Developments in field and laboratory research, Vol. 2.* Academic Press.

Lindburg, D. G. (1977). Feeding behavior and diet of rhesus (*Macaca mulatta*) in a Siwalik forest in north India. In T. Clutton-Brock (Ed.), *Primate ecology* (pp. 223-249). Academic Press.

Lu, J., Lin, Y., & Lee, L. (1991). Troop composition, activity pattern and habitat utilization of Formosan macaque (*Macaca cyclopis*) at Nanshi logging road in Yushan National Park. In A. Ehara, T. Kimura, O. Takenaka, & M. Iwamoto (Eds.), *Primatology Today* (pp. 93-96). Elsevier Science Publishers.

Luncz, L. V., Svensson, M. S., Haslam, M., Malaivijitnond, S., Proffitt, T., & Gummert, M. (2017). Technological response of wild macaques (*Macaca fascicularis*) to anthropogenic change. *International Journal of Primatology, 38,* 872-880.

MacKinnon, J., & MacKinnon, K. S. (1978). Comparative feeding ecology of six sympatric primates in west Malaysia. In D. Chivers, & J. Herberts (Eds.), *Recent advances in primatology* (Vol. 1, pp. 305-322). Academic Press.

MacKinnon, J. R., & MacKinnon, K. S. (1980). Niche differentiation in a primate community. In D. J. Chivers (Ed.), *Malayan forest primates.* Plenum Press.

Maestripieri, D. (1996). Gestural communication and its cognitive implications in pigtail macaques (*Macaca nemestrina*). *Behaviour, 133,* 997-1022.

Maestripieri, D. (1998). The evolution of male-infant interactions in the tribe Papionini (Primates: Cercopithecidae). *Folia Primatologica, 69,* 247-251.

Maibeche, Y., Moali, A., Yahi, N., & Ménard, N. (2015). Is diet flexibility an adaptive life trait for relictual and peri-urban populations of the endangered primate *Macaca sylvanus*? *PLoS ONE, 10*(2), e0118596.

Majolo, B., & McFarland, R. (2013). The effect of climate factors on the activity budgets of Barbary macaques (*Macaca sylvanus*). *International Journal of Primatology, 34,* 500-514.

Makwana, S. C. (1979). Field ecology and behaviour of the rhesus macaque, *Macaca mulatta.* II. Food, feeding and drinking in Dehra Dun forests. *Indian Journal of Forestry, 2*(3), 242-253.

Makwana, S. C. (1980). Observations of population and behaviour of the bonnet monkey, *Macaca radiata. Comparative Physiology and Ecology, 5*(1), 9-12.

Malaivijitnond, S., Arsaithamkul, V., Tanaka, H., Pomchote, P., Jaroenporn, S., Suryobroto, B., & Hamada, Y. (2012). Boundary zone between northern and southern pig-tailed macaques and their morphological differences. *Primates, 53,* 377-389.

Malaivijitnond, S., & Hamada, Y. (2008). Current situation and status of long-tailed macaques (*Macaca fascicularis*) in Thailand. *The Natural History Journal of Chulalongkorn University, 8*(2), 185-204.

Marty, P. R., Hodges, K., Agil, M., & Engelhardt, A. (2017). Alpha male replacements and delayed dispersal in crested macaques (*Macaca nigra*). *American Journal of Primatology, 79:* e22448.

Maruhashi, T. (1980). Feeding behavior and diet of the Japanese monkey (*Macaca fuscata yakui*) on Yakushima Island, Japan. *Primates, 21*(2), 141-160.

Maruhashi, T. (1982). An ecological study of troop fissions of Japanese monkeys (*Macaca fuscata yakui*) on Yakushima Island, Japan. *Primates, 23*(3), 317-337.

Masui, K., Narita, Y., & Tanaka, S. (1986). Information on the distribution of Formosan monkeys (*Macaca cyclopis*). *Primates, 27,* 123-191.

Matsumura, S. (1991). A preliminary report of the ecology and social behavior of moor macaques (*Macaca maurus*) in Sulawesi, Indonesia. *Kyoto University Overseas Report of Studies on Asian Nonhuman Primates, 8,* 27-41.

McCann, C. (1933). Notes on some Indian macaques. *Journal of the Bombay Natural History Society, 6,* 796-810.

McDermott, W. C. (1938). *The ape in antiquity.* Johns Hopkins University Press.

McFarland, R., & Majolo, B. (2013). Coping with the cold: Predictors of survival in wild Barbary macaques, *Macaca sylvanus. Biology Letters, 9*(4), 20130428.

McKinney, T. (2015). A classification system for describing anthropogenic influence on nonhuman primate populations. *American Journal of Primatology, 77*(7), 715-726.

McLennan, M. R., Spagnoletti, N., & Hockings, K. J. (2017). The implications of primate behavioral flexibility for sustainable human-primate coexistence in anthropogenic habitats. *International Journal of Primatology, 38,* 105-121.

Medway, L. (1970). The monkeys of Sundaland. In J. R. Napier, & P. H. Napier (Eds.), *Old World monkeys: Evolution, systematics and behavior* (pp. 513-553). Academic Press.

Mehlman, P. (1984). Aspects of the ecology and conservation of the Barbary macaque in the fir forest habitat of the Moroccan Rif Mountains. In J. E. Fa (Ed.), *The Barbary macaque: A case study in conservation* (pp. 67-81). Plenum Press.

Mehlman, P. (1985). Intergroup dynamics of the Barbary macaque (*Macaca sylvanus* L.), Ghomaran Rif Mountains, Morocco. *American Journal of Physical Anthropology, 66,* 204.

Mehlman, P. (1988). Food resources of the wild Barbary macaque (*Macaca sylvanus*) in high altitude fir forest, Ghomaran Rif, Morocco. *Journal of Zoology, 214,* 469-490.

Mehlman, P. (1989). Comparative density, demography, and ranging behavior of Barbary macaques (*Macaca sylvanus*) in marginal and prime conifer habitats. *International Journal of Primatology*, 4, 269-292.

Mehu, M., Huynen, M., & Agoramoorthy, G. (2006). Social relationships in a free-ranging group of bonnet macaques in Tamil Nadu, India. *Primate Report*, 73, 49-55.

Meijaard, E. (2003). Mammals of south-east Asian islands and their late Pleistocene environments. *Journal of Biogeography*, 30, 1245-1257.

Meijaard, E., & Groves, C. P. (2006). The geography of mammals and rivers in mainland Southeast Asia. In S. M. Lehman, & J. G. Fleagle (Eds.), *Primate biogeography: Progress and prospects* (pp. 305-329). Springer.

Melnick, D. J. (1981). *Microevolution in a population of Himalayan rhesus monkeys* (Macaca mulatta). [Doctoral dissertation, Yale University].

Melnick, D. J., & Kidd, K. K. (1983). Genetic consequences of demographic processes in a population of rhesus monkeys in northern Pakistan. *American Journal of Physical Anthropology*, 54, 252-253.

Ménard, N. (2002). Ecological plasticity of Barbary macaques (*Macaca sylvanus*). *Evolutionary Anthropology*, 11(S1), 95-100.

Ménard, N. (2004). Do ecological factors explain variation in social organization? In B. Thierry, M. Singh, & W. Kaumanns (Eds.), *Macaque societies: A model for the study of social organization* (pp. 237-262). Cambridge University Press.

Ménard, N., Hecham, R., Vallet, D., Chikhi, H., & Gautier-Hion, A. (1990). Grouping patterns of a mountain population of *Macaca sylvanus* in Algeria: A fission-fusion system? *Folia Primatologica*, 55, 160-175.

Ménard, N., Rantier, Y., Foulquier, A., Qarro, M., Chillasse, L., Vallet, D.,... & Butet, A. (2014). Impact on human pressure and forest fragmentation on the endangered Barbary macaque *Macaca sylvanus* in the Middle Atlas of Morocco. *Oryx*, 48(2), 276-284.

Ménard, N., & Vallet, D. (1993). Population dynamics of *Macaca sylvanus* in Algeria: An 8 year study. *American Journal of Primatology*, 30, 101-118.

Ménard, N., & Vallet, D. (1996). Demography and ecology of Barbary macaques (*Macaca sylvanus*) in two different habitats. In J. E. Fa, & D. G. Lindburg (Eds.), *Evolution and ecology of macaque societies* (pp. 106-131). Cambridge University Press.

Ménard, N., & Vallet, D. (1997). Behavioral responses of Barbary macaques (*Macaca sylvanus*) to variations in environmental conditions in Algeria. *American Journal of Primatology*, 43, 285-304.

Ménard, N., Vallet, D., & Gautier-Hion, A. (1985). Dèmographie et reproduction de *Macaca sylvanus* dans différents habitats en Algérie. *Folia Primatologica*, 44, 65-81.

Mendiratta, U., Kumar, A., Mishra, C., & Sinha, A. (2009). Winter ecology of the Arunachal macaque *Macaca munzala* in Pangchen Valley, Western Arunachal Pradesh, Northeastern India. *American Journal of Primatology*, 71, 939-947.

Mendis, B. C. G., & Dangolla, A. (2016). Human-monkey (*Macaca sinica*) conflict in Sri Lanka, *Sri Lanka Veterinary Journal*, 63(2), 35-37.

Menon, S., & Poirier, F. E. (1996). Lion-tailed macaque (*Macaca silenus*) in a disturbed forest fragment: Activity patterns and time budgets. *International Journal of Primatology*, 17, 969-985.

Mishra, C., Madhusudan, M. D., & Datta, A. (2006). Mammals of the high altitudes of western Arunachal Pradesh, eastern Himalaya: An assessment of threats and conservation needs. *Oryx*, 40, 29-35.

Mishra, C., & Sinha, A. (2008). A voucher specimen for *Macaca munzala*: Interspecific affinities, evolution, and conservation of a newly discovered primate. *International Journal of Primatology*, 29, 743-756.

Mito, S. (1980). *Bosuzaru heno Michi* [The way of monkey to become a boss]. Popura-sha, Tokyo.

Mitra, S. (2002). Diet and feeding behavior of Assamese macaque (*Macaca assamensis*). *Asian Primates*, 8(1-2), 12-14.

Mitra, S. (2002–2003). Foods consumed by Assamese macaques in West Bengal, India. *Asian Primates*, 8(3&4), 17-20.

Mittermeier, R. A., Rylands, A. B., & Wilson, D. E. (2013). *Handbook of the mammals of the world III: Primates.* Lynx Press, Barcelona.

Molur, S., Brandon-Jones, D., Dittus, W., Eudey, A., Kumar, A., Singh, M., & Walker, S. (2003). Status of South Asian primates: Conservation assessment and management plan report. *Workshop report 2003.* Zoo Outreach Organization/CBSG-South Asia, Coimbatore, India.

Morales, J. C., & Melnick, D. J. (1998). Phylogenetic relationships of the macaques (Cercopithecidae: *Macaca*), as revealed by high resolution restriction site mapping of mitochondrial ribosomal genes. *Journal of Human Evolution*, 34, 1-23.

Mori, A. (1977a). Intra-troop spacing mechanism of the wild Japanese monkeys of the Koshima troop. *Primates*, 18(2), 331-357.

Mori, A. (1977b). The social organization of the provisioned Japanese monkey troop which have extraordinary large population sizes. *Journal of the Anthropological Society of Nippon*, 85, 325-345.

Mori, A. (1979). The role of sex as a centripetal factor in a Japanese monkey troop. In M. Kawai, & S. Azuma (Eds.), *Biosociological studies on the role of sex in the Japanese monkeys society* (pp. 41-49). Inuyama: Primate Research Institute, Kyoto University.

Mukherjee, A., & Gupta, S. (1965). Habits of the rhesus macaque *Macaca mulatta* (Zimmermann) in the Sunderbans, 24-Paganas, West Bengal. *Journal of the Bombay Natural History Society*, 62, 145-146.

Muroyama, Y., & Eudey, A. A. (2004). Do macaque species have a future? In B. Thierry, M. Singh, & W. Kaumanns (Eds.), *Macaque societies: A model for the study of social organization* (pp. 328-332). Cambridge University Press.

Murren, C. J., Auld, J. R., Callahan, H., Ghalambor, C. K., Handelsman, C. A., Heskel, M. A., & Schlichting, C. D. (2015). Constraints on the evolution of phenotypic plasticity: Limits and costs of phenotype and plasticity. *Heredity* 115, 293-301.

Myers, N. (1988). Threatened biotas: "Hot spots" in tropical forests. *Environmentalist,* 8(3), 187-208.

Nailufar, B., Syartinilia, & Perwitasari, D. (2015). Landscape modeling for human–Sulawesi crested black macaques conflict in North Sulawesi. *Procedia Environmental Sciences,* 24, 104-110.

Nakagawa, N., Iwamoto, T., Yokota, N., & Soumah, A. (1996). Inter-regional and inter-seasonal variations of food quality in Japanese macaques: Constraints of digestive volume and feeding time. In J. Fa, & D. Lindburg (Eds.), *Evolution and ecology of macaque societies* (pp. 207-234). Cambridge University Press.

Napier, J., & Napier, P. (1967). *A handbook of living primates.* New York: Academic Press.

Napier, P. H. (1981). *Catalogue of primates in the British Museum (Natural History) and elsewhere in the British Isles. Part II: Family Cercopithecidae, subfamily Cercopithecinae.* London, British Museum (Natural History).

Nila, S., Suryobroto, B., & Widayati, K. A. (2014). Dietary variation in long-tailed macaques (*Macaca fascicularis*) in Telaga Warn, Bogor, West Java. *HAYATI Journal of Biosciences,* 21(1), 8-14.

O'Brien, T. G., & Kinnaird, M. F. (1997). Behavior, diet, and movements of the Sulawesi crested black macaque (*Macaca nigra*). *International Journal of Primatology,* 18(3), 321-351.

Oi, T. (1986). Socio-ecological study on pig-tailed macaques (*Macaca nemestrina*) in Sumatra. Kyoto Univ. *Overseas Research Report of Studies of Asian Non-Human Primates,* 5, 71-78.

Oi, T. (1987). Sexual behavior of the wild pig-tailed macaques in West Sumatra. Kyoto Univ. *Overseas Research Report of Studies of Asian Non-Human Primates,* 6, 67-80.

Oi, T. (1990). Mating systems of macaques. [Abstract]. *13th Congress of the International Primatological Society,* Nagoya and Kyoto, Japan.

Ong, P., & Richardson, M. (2008). *Macaca fascicularis. The International Union for the Conservation of Nature Red List of Threatened Species* 2008: e.T12551A3355536. http://dx.doi.org/10.2305/IUCN.UK.2008.RLTS.T12551A3355536.en.

Paciulli, L. M. (2004). *The effects of logging, hunting, and vegetation on the densities of the Pagai, Mentawai Island primates.* [PhD thesis, Stony Brook University].

Palacios, J. F. G., Engelhardt, A., Agil, M., Hodges, K., Bogia, R., & Waltert, M. (2011). Status of, and conservation recommendations for, the critically endangered crested black macaque in Tangkoko, Indonesia. *Oryx,* 46(2), 290-297.

Paudel, P. K. (2016). Conflict due to Assamese Macaques (*Macaca assamensis* McClelland 1840) and crop protection

114. C.S. strategies in Kaligandaki River Basin, Western Nepal. *Our Nature,* 14(1), 107-114.

Peng, M., Lai, Y., Yang, C., Chiang, H., New, A. E., & Chang, C. (1973). Reproductive parameters of the Taiwan monkey (*Macaca cyclopis*). *Primates,* 14, 201-213.

Poirier, F. E., & Davidson, D. M. (1979). A preliminary study of the Taiwan macaque (*Macaca cyclopis*). *Quarterly Journal of the Taiwan Museum,* 32, 123-192.

Poirier, F. E., & Smith, E. O. (1974). The crab-eating macaques of Angaur Island, Palau, Micronesia. *Folia Primatologica.* 22, 258-306.

Prentice, M. L., & Denton, G. H. (1988). The deep-sea oxygen isotope record, the global ice sheet system and hominid evolution. In F. E. Grine (Ed.), *Evolutionary history of the "robust" australopithecines* (pp. 383-403). New York: Aldine de Gruyter.

Priston, N. E. C., Wyper, R. M., & Lee, P. C. (2012). Buton macaques *Macaca ochreata brunnescens*: Crops, conflict, and behavior on farms. *American Journal of Primatology,* 74, 29-36.

Pruett, C. (1973). A trip to Silent Valley. *Journal of the Bombay Natural History Society,* 70, 544-548.

Rahaman, H., & Parthasarathy, M. D. (1967). A population survey of the bonnet monkey (*Macaca radiata* Geoffroy) in Bangalore, South India. *Journal of the Bombay Natural History Society,* 64, 251-255.

Rahaman, H., & Parthasarathy, M. D. (1969). Studies on the social behaviour of bonnet monkeys. *Primates,* 10, 149-162.

Ramachandran K. K., & Joseph, G. K. (2001). Distribution and demography of diurnal primates in Silent Valley National Park and adjacent areas, Kerala, India. *Journal of the Bombay Natural History Society,* 98, 191-196.

Ramakrishnan, U., & Coss, R. (2000). Recognition of heterospecific alarm vocalizations by bonnet macaques (*Macaca radiata*). *Journal of Comparative Psychology,* 114(1), 3-12.

Ramakrishnan, U., & Coss, R. (2001). Strategies used by bonnet macaques (*Macaca radiata*) to reduce predation risk while sleeping. *Primates,* 42(3), 193-206.

Ramakrishnan, U., Coss, R., & Pelkey, N. (1999). Tiger decline caused by the reduction of large ungulate prey: Evidence from a study of leopard diets in southern India. *Biological Conservation,* 89, 113-120.

Regmi, G., & Kandel, K. (2008). Population status, threats and conservation measures of Assamese macaque (*Macaca assamensis*) in Langtang National Park, Nepal. *Primate Eye,* 96, 19-20.

Rhine, R. J., & Maryanski, A. (1996). A twenty-one-year history of a dominant stump-tail matriline. In J. E. Fa, & D. L. Lindburg (Eds.), *Evolution and ecology of macaque societies* (pp. 473-499). Cambridge University Press.

Richard, A. F., Goldstein, S. J., & Dwar, R. E. (1989). Weed macaques: The evolutionary implications of macaque feeding ecology. *International Journal of Primatology,* 10, 569-594.

Richardson, M., Mittermeier, R. A., Rylands, A. B., & Konstant, B. (2008). *Macaca nemestrina. The International Union for the Conservation of Nature Red List of Threatened Species* 2008: e.T12555A3356892. http://dx.doi.org/10.2305/IUCN.UK.2008.RLTS.T12555A3356892.en.

Richter, C., Taufig, A., Hodges, K., Ostner, J., & Schulke, O. (2013). Ecology of an endemic primate species (*Macaca siberu*) on Siberut Island, Indonesia. *Springer Plus*, 2, 137.

Riley, E. P. (2005). *Ethnoprimatology of* Macaca tonkeana: *The interface of primate ecology, human ecology, and conservation in Lore Lindu National Park, Sulawesi, Indonesia.* [PhD Thesis, University of Georgia].

Riley, E. P. (2007a). Flexibility in diet and activity patterns of *Macaca tonkeana* in response to anthropogenic habitat alteration. *International Journal of Primatology*, 28(1), 107-133.

Riley, E. P. (2007b). The human-macaque interface: Conservation implications of current and future overlap and conflict in Lore Lindu National Park, Sulawesi, Indonesia. *American Anthropologist*, 109(3), 473-484.

Riley, E. P. (2008). Ranging patterns and habitat use of Sulawesi tonkean macaques (*Macaca tonkeana*) in a human-modified habitat. *American Journal of Primatology*, 70, 670-679.

Riley, E. P. (2020). *The promise of contemporary primatology.* New York: Routledge.

Riley, E., Lee, R., Sangermano, F., Cannon, C. & Shekelle, M. (2020). *Macaca maura* (errata version published in 2021). *The IUCN Red List of Threatened Species* 2020: e.T12553A197831931. https://dx.doi.org/10.2305/IUCN.UK.2020-3.RLTS.T12553A197831931.en..

Riley, E., Lee, R., Sangermano, F., Cannon, C., & Shekelle, M. (2020). *Macaca tonkeana. The IUCN Red List of Threatened Species* 2020: e.T12563A17947990. https://dx.doi.org/10.2305/IUCN.UK.2020-3.RLTS.T12563A17947990.en..

Riley, E. P., & Priston, N. C. (2010). Macaques in farms and folklore: Exploring the human nonhuman primate interface in Sulawesi, Indonesia. *American Journal of Primatology* 72, 848-854.

Riley, E. P., Suryobroto B., & Maestripieri, D. (2007). Distribution of *Macaca ochreata* and identification of mixed *ochreata-tonkeana* groups in South Sulawesi, Indonesia. *Primate Conservation*, 22, 75-79.

Riley, E. P., Tolbert, B., & Farida, W. R. (2013). Nutritional content explains the attractiveness of cacao to crop raiding tonkean macaques. *Current Zoology*, 59(2), 160-169.

Riley, E. P., & Wade, T. W. (2016). Adapting to Florida's riverine woodlands: The population status and feeding ecology of the Silver River rhesus macaques and their interface with humans. *Primates*, 57, 195-210.

Rodman, P. (1978a). Diets, densities, and distributions of Bornean primates. In G. G. Montgomery (Ed.), *The ecology of arboreal folivores.* Washington, D.C.: Smithsonian Institute Press.

Rodman, P. (1978b). Food distribution and terrestrial locomotion of crab-eating and pig-tailed macaques in the wild. [Abstract]. *American Journal of Physical Anthropology*, 47, 157.

Rodman, P. (1979). Skeletal differentiation of *Macaca fascicularis* and *Macaca nemestrina* in relation to arboreal and terrestrial quadrupedalism. *American Journal of Physical Anthropology*, 12(4), 357-375.

Rodman, P. (1991). Structural differentiation of microhabitats of sympatric *Macaca fascicularis* and *M. nemestrina* in east Kalimantan, Indonesia. *International Journal of Primatology*, 12(4), 357-375.

Roonwal, M., & Mohnot, S. (1977). *Primates of South Asia: Ecology, sociobiology and behavior.* Harvard University Press.

Roos, C., Boonratana, R., Supriatna, J., Felowes, J. R., Groves, C. P., Nash, S. D.,... & Mittermeier, R. A. (2014). An updated taxonomy and conservation status review of Asian primates. *Asian Primates Journal*, 4(1), 2-38.

Roos, C., Thanh, V. N., Walter, L., & Nadler, T. (2007). Molecular systematics of Indochinese primates. *Vietnam Journal of Primatology*, 1, 41-53.

Roos, C., Ziegler, T., Hodges, J. K., Zischler, H., & Abegg, C. (2003). Molecular phylogeny of Mentawai macaques: Taxonomic and biogeographical implications. *Molecular Phylogenetics and Evolution*, 29(1), 139-150.

Roos, C., & Zinner, D. (2015). Diversity and evolutionary history of macaques with special focus on *Macaca mulatta* and *Macaca fascicularis.* In J. Bluemel, S. Korte, E. Schenck, & G. F. Weinbauer (Eds.), *The Nonhuman Primate in Nonclinical Drug Development and Safety Assessment* (pp. 3-16). Academic Press. https://doi.org/10.1016/B978-0-12-417144-2.00001-9.

Rosenbaum, B. O., Brien, T. G., Kinnaird, M., & Supriatna, J. (1998). Population densities of Sulawesi crested black macaques (*Macaca nigra*) on Bacan and Sulawesi, Indonesia: Effects of habitat disturbance and hunting. *American Journal of Primatology*, 44(2), 89-106.

Ruppert, N., Holzner, A., See, K. W., Gisbrecht, A., & Beck, A. (2018). Activity budgets and habitat use of wild southern pig-tailed macaques (*Macaca nemestrina*) in oil palm plantation and forest. *International Journal of Primatology*, 39, 237-251.

Sade, D. S. (1972). A longitudinal study of social behavior of rhesus monkeys. In R. H. Tuttle (Ed.), *The functional and evolutionary biology of primates.* Chicago: Aldine Atherton Publishing, Inc.

Santhosh, K., Kumara, H. N., Velankar, A. D., & Sinha, A. (2015). Ranging behavior and resource use by lion-tailed macaques (*Macaca silenus*) in selectively logged forests. *International Journal of Primatology*, 36, 288.

Santhosh, K., Raj, V. M., & Kumara, H. N. (2013). Conservation prospects for the lion-tailed macaque (*Macaca silenus*) in the forests of Sirsi-Honnavara, Western Ghats, India. 2013. *Conservation International*, 27, 125-131.

Sarania, B., Devi, A., Kumar, A., Sarma, K., & Gupta, A. K. (2017). Predictive distribution modeling and population status of the endangered *Macaca munzala* in Arunachal

Pradesh, India. *American Journal of Primatology*, 79(2), 1-10.

Saraswat, R., Sinha, A., & Radhakrishna, S. (2015). A god becomes a pest? Human-rhesus macaque interactions in Himachal Pradesh, northern India. *European Journal of Wildlife Research*, 61, 435-443.

Schaller, G. (1967). *The deer and the tiger*. University of Chicago Press.

Schillaci, M. A., & Stallmann, R. R. (2005). Ontogeny and sexual dimorphism in booted macaques (*Macaca ochreata*). *Journal of Zoology*, 267, 19-29.

Schlotterhausen, L. (1998). A comparison of the social and feeding behaviors between a wild and commensal group of bonnet macaques (*Macaca radiata*) in the Indira Gandhi Wildlife Sanctuary, South India. [Abstract]. *American Journal of Primatology*, 45(2), 206.

Sengupta, A., McConkey, K. R., & Radhakrishna, S. (2014). Seed dispersal by rhesus macaques *Macaca mulatta* in Northern India. *American Journal of Primatology*, 76, 1175-1184.

Setiawan, A., Mittermeier, R. A., & Whittaker, D. (2020). *Macaca pagensis. The IUCN Red List of Threatened Species* 2020: e.T39794A17949995. https://dx.doi.org/10.2305/IUCN.UK.2020-2.RLTS.T39794A17949995.en.

Sha, J. C. M., & Hanya, G. (2013). Diet, activity, habitat use, and ranging of two neighboring groups of food-enhanced long-tailed macaques (*Macaca fascicularis*). *American Journal of Primatology*, 75, 581-592.

Sharma, A. K., Singh, M., Kaumanns, W., Krebs, E., Singh, M., Kumar, M. A., & Kumara, H. N. (2006). Birth patterns in wild and captive lion-tailed macaques (*Macaca silenus*). *International Journal of Primatology*, 27, 1429-1439.

Sharma, N., Madhusudan, M. D., Sarkar, P., Bawri, M., & Sinha, A. (2012). Trends in extinction and persistence of diurnal primates in the fragmented lowland rainforests of the upper Brahmaputra Valley, north-eastern India. *Oryx*, 46(2), 308-311.

Simonds, P. E. (1965). The bonnet macaque of South India. In I. E. DeVore (Ed.), *Primate behavior: Field studies of monkeys and apes*. Holt, Rinehart and Winston.

Simonds, P. E. (1974). Sex differences in bonnet macaque networks and social structure. *Archives of Sexual Behavior*, 3, 151-166.

Singh, D. N. (2001). Status and distribution of primates in Arunachal Pradesh. *Envis Bulletin: Wildlife and Protected Areas*, 1, 113-119.

Singh, M., Kumar, A. & Kumara, H. N. (2020). *Macaca mulatta. The IUCN Red List of Threatened Species* 2020: e.T12554A17950825. https://dx.doi.org/10.2305/IUCN.UK.2020-2.RLTS.T12554A17950825.en.

Singh, M., Kumar, A., & Molur, S. (2008). *Macaca radiata. The International Union for the Conservation of Nature Red List of Threatened Species* 2008: e.T12558A3357748. http://dx.doi.org/10.2305/IUCN.UK.2008.RLTS.T12558A3357748.en.

Singh, M., Kumara, H. N., Kumar, M. A., & Sharma, A. (2000). Status and conservation of lion-tailed macaques and other arboreal mammals in tropical rainforests of Sringeri Forest Range, Western Ghats, Karnataka, India. *Primate Report*, 58, 5-16.

Singh, M. & Rao, N. (2004). Population dynamics and conservation of commensal bonnet macaques. *International Journal of Primatology*, 25(4), 847-859.

Singh, M., Singh, M., Kumar, M. A., Kumara, H. N., Sharma, A. K., & Kaumanns, W. (2002). Distribution, population structure and conservation of lion-tailed macaque (*Macaca silenus*) in Anaimalai Hills, Western Ghats, India. *American Journal of Primatology*, 57, 91-102.

Singh, M., & Sinha, A. (2004). Life-history traits: Ecological adaptations or phylogenetic relics? In B. Thierry, M. Singh, & W. Kaumanns (Eds.), *Macaque Societies: A model for the study of social organization* (pp. 80-83). Cambridge University Press.

Singh, S. D. (1969). Urban monkeys. *Scientific American*, 221, 108-115.

Sinha, A. (2005). Not in their genes: Phenotypic flexibility, behavioural traditions and cultural evolution in wild bonnet macaques. *Journal of Bioscience*, 30(1), 51-64.

Sinha, A., Datta, A., Madhusudan, M., & Mishra, C. (2005a). *Macaca munzala*: A new species from Western Arunachal Pradesh, Northeastern India. *International Journal of Primatology*, 26(4), 977-989.

Sinha, A., Mukhopadhyay, K., Datta-Roy, A., & Ram, S. (2005b). Ecology proposes, behavior disposes: Ecological variability in social organization and male behavioural strategies among wild bonnet macaques. *Current Science*, 89(7), 1166-1179.

Smith, R. J., & Jungers, W. L. (1997). Body mass in comparative primatology. *Journal of Human Evolution*, 32, 523-559.

Snell-Rood, E. C. (2013). An overview of the evolutionary causes and consequences of behavioural plasticity. *Animal Behaviour*, 85, 1004-1011.

Southwick, C., & Cadigan, F. C. (1972). Population studies of Malaysian primates. *Primates*, 13(1), 1-18.

Southwick, C. H., & Siddiqi, M. F. (1988). Partial recovery and a new population estimate of rhesus monkey populations in India. *American Journal of Primatology*, 16, 187-197.

Southwick, C. H., & Siddiqi, M. F. (1994a). Population status of nonhuman primates in Asia, with emphasis on rhesus macaque in India. *American Journal of Primatology*, 34, 51-59.

Southwick, C. H., & Siddiqi, M. F. (1994b). Primate commensalism: The rhesus monkey in India. *Revue d'écologie* (*Terre et Vie*), 49, 223-231.

Srivastava, A., & Mohnot, S. (2001). Distribution, conservation status and priorities for primates in northeast India. *ENVIS Bulletin: Wildlife and Protected Areas*, 1(1), 102-108.

Stewart, A., Gordon, C., Wich, S., Schroor, P., & Meijaard, E. (2008). Fishing in *Macaca fascicularis*: A rarely observed innovative behavior. *International Journal of Primatology*, 29, 543-548.

Stewart, C. B., & Disotell, T. R. (1998). Primate evolution—in and out of Africa. *Current Biology,* 8: R582-R588.

Strier, K. B. (2017). What does variation in primate behavior mean? *American Journal of Physical Anthropology,* 162, 4-14.

Su, H. (2003). Acquirement of social ranks in females in one group of Taiwanese macaques (*Macaca cyclopis*) at Fushan Experimental Forest, Taiwan. *American Journal of Physical Anthropology,* (suppl 36), 203.

Su, H., & Lee, L. (2001). Food habits of Formosan rock macaques (*Macaca cyclopis*) in Jentse, northeastern Taiwan, assessed by fecal analysis and behavioral observation. *International Journal of Primatology,* 22(3), 359-377.

Sugiyama, Y. (1968). The ecology of the lion-tailed macaque (*Macaca silenus* Linnaeus): A pilot study. *Journal Bombay Natural History Society,* 65, 283-292.

Sugiyama, Y. (1971). Characteristics of the social life of bonnet macaques (*Macaca radiata*). *Primates,* 12(3-4), 247-266.

Supriatna, J., Froehlich, J. W., Erwin, J. M., & Southwick, C. H. (1992). Population, habitat and conservation status of *Macaca maurus, Macaca tonkeana* and their putative hybrids. *Tropical Biodiversity,* 1, 31-48.

Supriatna, J., Yanuar, A., Martarinza, Wibisono, H., Sinaga, R., Sidik, I., & Iskandar, S. (1996). A preliminary survey of long-tailed and pig-tailed macaques (*Macaca fascicularis* and *Macaca nemestrina*) in Lampung, Bengkulu, and Jambi Provinces, Southern Sumatra, Indonesia. *Tropical Biodiversity,* 3(2), 131-140.

Sushma, H. S., Mann, R., Kumara, H. N., & Udhayan, A. (2014). Population status of the endangered lion-tailed macaque *Macaca silenus* in Kalakad-Mundanthurai Tiger Reserve, Western Ghats, India. *Conservation International,* 28, 171-178.

Sushma, H. S., & Singh, M. (2006). Resource partitioning and interspecific interactions among sympatric arboreal mammals of the Western Ghats, India. *Behavioral Ecology,* 17, 479-490.

Sussman, R., Shaffer, C., & Guidi, L. (2011). *Macaca fascicularis* in Mauritius: Implications for macaque-human interactions and for future research on long-tailed macaques. In M. Gumert, A. Fuentes, & L. Jones-Engel (Eds.), *Monkeys on the edge: Ecology and management of long-tailed macaques and their interface with humans* (pp. 207-235). Cambridge University Press.

Sussman, R. W., & Tattersall, I. (1981). Behavior and ecology of *Macaca fascicularis* in Mauritius: A preliminary study. *Primates,* 22, 192-205.

Sussman, R. W., & Tattersall, I. (1986). Distribution, abundance, and putative ecological strategy of *Macaca fascicularis* on the island of Mauritius, Southwestern Indian Ocean. *Folia Primatologica,* 64, 28-43.

Suzuki, A. (1965). An ecological study of wild Japanese monkeys in snowy areas-focused on their food habits. *Primates,* 6(1), 31-72.

Takahashi, L. (1997). Huddling relationships in night sleeping groups among wild Japanese macaques in Kinkazan Island during winter. *Primates,* 38, 57-68.

Takahashi, L., & Pan, R. (1994). Mandibular morphometrics among macaques: The case of *Macaca thibetana*. *International Journal of Primatology,* 15(4), 597-621.

Takasaki, H. (1981). On the deciduous-evergreen zonal gap in the per capita range area of the Japanese macaque troop from north to south: A preliminary note. *Physiological Ecology Japan,* 18, 1-5.

Taub, D. (1977). Geographic distribution and habitat diversity of the Barbary macaque, *M. sylvanus* L. *Folia Primatologica,* 27, 108-133.

Taub, D. M. (1985). Male-infant interactions in baboons and macaques: A critique and reevaluation. *American Zoology,* 25, 861-871.

Tenaza, R. (1987). The status of primates and their habitats in the Pagai, Islands, Indonesia. *Primate Conservation,* 8, 104-113.

Tenaza, R. (1988). The status of primates and their habitats in the Pagai Islands, Indonesia: A progress report. *Primate Conservation,* 9, 146-149.

Tenaza, R., & Tilson, R. (1985). Human predation and Kloss's gibbon (*Hylobates klossi*) sleeping trees in Siberut Island, Indonesia. *American Journal of Primatology,* 8, 229-308.

Thierry, B. (2000). Covariation of conflict management patterns across macaque species. In F. Aureli, & F. de Waal (Eds.), *Natural conflict resolution* (pp. 106-128). University of California Press.

Thierry, B. (2007). The macaques: A double-layered social organization. In C. J. Campbell, A. Fuentes, K. C. MacKinnon, M. Panger, & S. K. Bearder (Eds.), *Primates in perspective* (pp. 224-239). Oxford University Press.

Thierry, B. (2011). The macaques: A double-layered social organization. In C. J. Campbell, A. Fuentes, K. C. MacKinnon, S. K. Bearder, & R. M. Stumpf (Eds.), *Primates in perspective* (pp. 229-41). Oxford University Press.

Thirgood, J. (1984). The demise of Barbary macaque habitat—past and present forest cover of the Maghreb. In J. E. Fa (Ed.), *The Barbary macaque: A case study in conservation* (pp. 19-69). Plenum Press.

Thorington, R., & Groves, C. (1970). An annotated classification of the Cercopithecoidea. In J. R. Napier, & P. H. Napier (Eds.), *Old World monkeys: Evolution, systematics, and behavior.* Academic Press.

Timmins, R. J., & Duckworth, J. W. (2013). Distribution and habitat of Assamese macaque *Macaca assamensis* in Lao PDR, including its use of low-altitude karsts. *Primate Conservation* 26, 103-114.

Tobgay, S., Dorji, K., & Yangdon, N. (2019). Sighting of arunachal macaque *Macaca munzala* Sinha et al., 2005 (Mammalia: Primates: Ceropithecidae) in Sakteng Wildlife Santuary, Bhutan. *Journal of Threatened Taxa, 11*(6), 13805-13807.

Traeholt, C., Setiawan, A., Quinten, M, Cheyne, S. M., Whittaker, D. & Mittermeier, R. A. (2020). *Macaca siberu. The IUCN Red List of Threatened Species* 2020: e.T39795A17949710. https://dx.doi.org/10.2305/IUCN.UK.2020-2.RLTS.T39795A17949710.en.

Treesucon, U. (1988). A survey of stump-tailed macaques (*Macaca arctoides*) in Thailand. *Natural History Bulletin of the Siam Society, 36,* 61-70.

Umapathy G., & Kumar, A. (2000). The occurrence of arboreal mammals in the rain forest fragments in the Anamalai Hills, South India. *Biological Conservation, 92,* 311-319.

Usui, R., Sheeran, L. K., Li, J., Sun, L., Wang, X., Pritchard, A. J.,... & Wagner, R. S. (2014). Park rangers' behaviors and their effects on tourists and Tibetan macaques (*Macaca thibetana*) at Mt. Huangshan, China. *Animals, 4,* 546-561.

van Lavieren, E. (2012). The Barbary macaque (*Macaca sylvanus*): A unique endangered primate species struggling to survive. *Eubacteria, 30.*

van Schaik, C. P. (2013). The cost and benefits of flexibility as an expression of behavioural plasticity: A primate perspective. *Philosophical Transactions of the Royal Society B, 368,* 20120339. http://dx.doi.org/10.1098/rstb.2012.0339.

van Schaik, C., & van Noordwijk, M. (1985). Evolutionary effect of the absence of felids on the social organization of macaques on the island of Simeulue (*Macaca fascicularis fusca*, Miller, 1903). *Folia Primatologica, 44,* 138-147.

van Schaik, C. P., van Noordwijk, M., Warsono, B., & Sutriono, E. (1983). Party size and early detection of predators in Sumatran forest primates. *Primates, 24,* 211-221.

Vandercone, R., & Santiapillai, C. (2003). Feeding ecology and factors influencing the range of the dusky toque monkey (*Macaca sinica aurifrons*) at Udawattakelle Sanctuary, Sri Lanka. *Tigerpaper, 30*(3), 20-27.

Vandercone, R., Santiapillai, C., & Rasmussen, D. (2006). Group composition and reproduction of toque macaques (*Macaca sinica*) inhabiting the Udawattakelle Sanctuary, Sri Lanka. [Abstract]. *American Journal of Physical Anthropology, 42,* 181.

Wada, K. (2005). The distribution pattern of rhesus and Assamese monkeys in Nepal. *Primates* 46, 115-119.

Wang, X., Sun, L., Sheeran, L. K., Sun, B. H., Zhang Q. X., Zhang, D.,... & Li, J. H. (2016). Social rank versus affiliation: Which is more closely related to leadership of group movements in Tibetan macaques (*Macaca thibetana*)? *American Journal of Primatology, 78*(8), 816-824.

Watanabe, K., Lapasere, H., & Tanto, R. (1991). Associated developmental changes in two species of Sulawesi macaques, *Macaca tonkeana* and *M. hecki,* with special reference to hybrids and the borderland between the species. *Primates, 32,* 61-76.

Watanabe, K., & Matsumura, S. (1991). The borderlands and possible hybrids between three species of macaques, *M. nigra, M. nigrescens,* and *M. hecki,* in the northern peninsula of Sulawesi. *Primates, 32*(3), 365-370.

Watanabe, T., & Brotoisworo, E. (1982). Field observation of Sulawesi Macaques. Kyoto University. *Overseas Research Report of Studies of Asian Non-human Primates, 2,* 3-9.

Watanabe, T., Hamada, Y., Suryobroto, B., & Iwamoto, M. (1987). Somatometrical data of Sulawesi macaques and Sumatran pig-tails collected in 1984 and 1986. *Kyoto University Overseas Research Report of Studies on Asian Non-Human Primates, 6,* 49-56.

Weerasekara, W. M. L. S., & Ranawana, K. B. (2017). Ranging pattern of dusky toque macaques (*Macaca sinica auurifrons*) inhabiting Peradeniya University Premises, Sri Lanka. *Wildlanka, 5*(4), 179-192.

Weerasekara, W. M. L. S., & Ranawana, K. B. (2018). Effect of temperature on activity budgets of free ranging dusky toque macaques (*Macaca sinica aurifrons*): A case study from Peradeniya University premises, Sri Lanka. *Ceylon Journal of Science, 47*(1), 69-75.

Whittaker, D. (2006). A conservation action plan for the Mentawai primates. *Primate Conservation, 20,* 95-105.

Whitten, A., Mustafa, M., & Henderson, G. (2002). *The ecology of Sulawesi.* Periplus Editions.

Whitten, A., & Whitten, J. (1982). Preliminary observations of the Mentawai macaque on Siberut Island, Indonesia. *International Journal of Primatology, 3*(4), 445-459.

Wijeyamohan, S., Alagoda, T., & Santiapillai, C. (1996). Population structure and dynamics of the dusky toque monkey (*Macaca sinica aurifrons*) in the Udawattekelle Sanctuary, Sri Lanka. *TigerPaper, 23*(2), 14-19.

Wilson, C., & Wilson, W. (1975). The influence of selective logging on primates and some other mammals in East Kalimantan. *Folia Primatologica, 23,* 245-274.

Wilson, C., & Wilson, W. (1976). Behavioral and morphological variation among primate populations in Sumatra. *Yearbook of Physical Anthropology, 20,* 207-233.

Woodruff, D. S., & Turner, L. M. (2009). The Indochinese-Sundaic zoogeographic transition: A description and analysis of terrestrial mammal species distributions. *Journal of Biogeography, 36,* 1-19.

Wu, H., & Lin, Y. (1992). Life history variables of wild troops of Formosan macaques (*Macaca cyclopis*) in Kenting, Taiwan. *Primates, 33*(1), 85-97.

Wu, H. Y., & Richardson, M. (2008). *Macaca cyclopis. The International Union for the Conservation of Nature Red List of Threatened Species* 2008: e.T12550A3355290. http://dx.doi.org/10.2305/IUCN.UK.2008.RLTS.T12550A3355290.en.

Xie, D. M., Lu, J. Q., Sichilima, A. M., & Wang, B. S. (2012). Patterns of habitat selection and use by *Macaca mulatta tcheliensis* in winter and early spring in temperate forest, Jiyuan, China. *Biologia, 67*(1), 234-239.

Xiong, C. (1984). Ecological studies of the stump-tailed macaque. [English Summary]. *Acta Theriologica Sinica, 4,* 1-9.

Yeager, C. (1996). Feeding ecology of the long-tailed macaque (*Macaca fascicularis*) in Kalimantan Tengah, Indonesia. *International Journal of Primatology, 17*(1), 51-62.

Yongcheng, L., & Richardson, M. (2008). *Macaca thibetana. The International Union for the Conservation of Nature Red List of Threatened Species* 2008: e.T12562A3359510. http://dx.doi.org/10.2305/IUCN.UK.2008.RLTS.T12562A3359510.en.

Yongzu, Z., Guogiang, Q., Yonglei, L., & Southwick, C. (1989). Extinction of rhesus monkeys (*Macaca mulatta*) in Xinglung, North China. *International Journal of Primatology, 10*(4), 375-381.

Young, C., Schulke, O., Ostner, J., & Majolo, B. (2012). Consumption of unusual prey items in the Barbary macaque (*Macaca sylvanus*). *African Primates, 7*(2), 224-229.

Zak, A. A., & Riley, E. P. (2017). Comparing the use of camera traps and farmer reports to study crop feeding behavior of moor macaques (*Macaca maura*). *International Journal of Primatology, 38,* 224-242.

Zhang, Y., Wang, S., & Quan, G. (1981). On the geographical distribution of primates in China. *Journal of Human Evolution, 10*(3), 215-226.

Zhao, Q. (1994). Seasonal changes in body weight of *Macaca thibetana* at Mt. Emei, China. *American Journal of Primatology, 32,* 223-226.

Zhao, Q. (1996). Etho-ecology of Tibetan macaques at Mount Emei, China. In J. E. Fa, & D. G. Lindburg (Eds.), *In evolution and ecology of macaque societies* (pp. 263-289). Cambridge University Press.

Zhao, Q., & Deng, Z. (1988a). Ranging behavior of *Macaca thibetana* at Mt. Emei, China. *International Journal of Primatology, 9*(1), 37-47.

Zhao, Q., & Deng, Z. (1988b). *Macaca thibetana* at Mt. Emei, China: III. Group composition. *American Journal of Primatology, 16,* 269-273.

Zhao, Q., & Deng, Z. (1988c). *Macaca thibetana* at Mt. Emei, China: II. Birth seasonality. *American Journal of Primatology, 16,* 251-268.

Zhou, Q, Wei, H., Huang, Z., Li, Y., Lu, M., & Huang, C. (2007). Activity patterns and time budgets of the Assamese macaque *Macaca assamensis* in the Longgang Nature Reserve, China. *Acta Zoological Sinica, 53*(5), 791-799.

Zhou, Q., Wei, H., Tang, H., Huang, Z., Krzton, A., & Huang, C. (2014a). Niche separation of sympatric macaques, *Macaca assamensis* and *M. mulatta*, in limestone habitats of Nonggang, China. *Primates, 55,* 125-137.

Zhou, Q., Wei, H., Tang, H., Huang, Z., Krzton, A., & Huang, C. (2014b). Ranging behavior and habitat use of the Assamese macaque (*Macaca assamensis*) in limestone habitats of Nonggang, China. *Mammalia, 78*(2), 171-176.

Ziegler, T., Abegg, C., Meijaard, E., Perwitasari-Farajallah, D., Walter, L., Hodges, J. K., & Roos, C. (2007). Molecular phylogeny and evolutionary history of Southeast Asian macaques forming the *M. Silenus* group. *Molecular Phylogenetics and Evolution, 42,* 807-816.

Zinner, D., Fickenscher, G., & Roos C. (2013). Family Cercopithecidae (Old World monkeys). In R. A. Mittermeier, A. B. Rylands, & D. E. Wildon (Eds.), *The handbook of the mammals of the world—primates* (Vol. 3, pp. 550-627). Lynx Ediciones.

CHAPTER 18

The Ecology and Social Structure of the Asian Colobines

Rajnish Vandercone, Camille Coudrat, and Patricia T. Ormond

The Colobinae derive their name from the word *Kolobos*, the Greek term for "mutilated" in reference to the reduced or absent thumbs of the African species; however, Asian colobines differ in that all species have small thumbs (Oates & Davies, 1994). The most diagnostic feature of the colobines is their large multichambered stomach, which contains rich anaerobic microbial fauna comprised of bacteria, protozoa, and fungi (Bauchop & Martucci, 1968; Kay & Davies, 1994). Colobine premolars and molars are high crowned and possess pointed cusps linked by ridges and separated by deep lateral notches (Oates & Davies, 1994). The sharper crests and higher cusps fold and slice leafy food (Oates & Davies, 1994). These dietary adaptations enable colobines to efficiently ingest and digest foliage, and hence the species in this primate subfamily are popularly referred to as "leaf-eating monkeys."

TAXONOMY

The true phylogeny of the extant colobines is uncertain and, unlike the cercopithecines, relatively little information is available on the evolutionary history of Asian colobines (Osterholz et al., 2008). Based on distribution and morphology, the colobines are divided into an African and an Asian clade (Oates et al., 1994). Based on fossil evidence, it is estimated that the two clades diverged from each other between 10 mya (Stewart & Disotell, 1998) and 13 mya (Delson, 1994). The Asian colobines, which are more diverse than their African cousins, are further split into odd-nosed monkeys (*Rhinopithecus*, *Pygathrix*, *Nasalis*, and *Simias*) and langurs (*Semnopithecus*, *Trachypithecus*, and *Presbytis*) (Osterholz et al., 2008). Both the odd-nosed monkey and the langur groups are considered to be monophyletic. While the monophyly of the odd-nosed monkeys is supported by molecular evidence (Sterner et al., 2006; Whittaker et al., 2006; Wang et al., 2012), the evidence for monophyly of the langur group is still lacking. The mitochondrial and nuclear gene tree discordance in the langurs is likely a result of ancient hybridization events (Ting et al., 2008; Roos et al., 2011; Wang et al., 2012).

Semnopithecus traditionally has included a single species, the Hanuman langur, *S. entellus* (Groves, 2001). Subsequently, Brandon-Jones et al. (2004) reclassified two other langur species (Nilgiri langur, *Trachypithecus johnii*, and purple-faced langur, *T. vetulus*) into *Semnopithecus*. This classification scheme, which was originally supported only by mitochondrial genetic material (Zhang & Ryder, 1998), is now supported by nuclear DNA relationships as well (Karanth et al., 2008; Osterholz et al., 2008).

The taxonomic position of some species of *Trachypithecus* (the capped langur, *T. pileatus*, and golden langur, *T. geei*) is uncertain. Mitochondrial and nuclear DNA relationships show that both these species are closely related to each other and fall into an intermediate position between *Semnopithecus* and the *Trachypithecus* clade (Karanth et al., 2008; Osterholz et al., 2008; Karanth, 2010). It is thought that this *capped-golden* group might have evolved through past hybridization events between *Trachypithecus* and *Semnopithecus* (Karanth et al., 2008; Osterholz et al., 2008; Karanth, 2010). This assertion is further strengthened by the geographic distribution of the *capped-golden* group, which occurs in an area that is sandwiched between the distributions of *Semnopithecus* and *Trachypithecus* (Karanth, 2010).

Since the classification of *Rhinopithecus*, *Pygathrix*, *Simias*, and *Nasalis* in one clade (Groves, 1970), the taxonomic positions between the odd-nosed group and the other Asian colobines have been reviewed in different studies and remain controversial (Brandon-Jones et al., 2004; Sterner et al., 2006); support for its monophyly has not reached a consensus (Jablonski, 1998). In this review, *Rhinopithecus*, *Pygathrix*, *Simias*, and *Nasalis* genera (referred to as the odd-nosed monkeys) are treated separately from *Trachipithecus*, *Semnopithecus*, and *Presbytis*, and characteristics of their socioecology are compared and discussed in terms of this taxonomic grouping.

Due to the uncertain nature of the taxonomy of Asian colobines, for the sake of clarity, the classification proposed by Brandon-Jones et al. (2004) has been adopted in this chapter (Table 18.1).

Table 18.1 Asian Colobine Genera and Geographic Distribution of Species and Subspecies

Species and Subspecies (by Genus) [a]	Common Name	Current Distribution
Presbytis		
P. comata	Javan grizzled surili	Indonesia (W Java)
P. femoralis femoralis	Raffles' banded surili	Malaysia (Johor, Pahang); Singapore
P. femoralis batuana	North Sumatran banded surili	Indonesia (Batu, NC Sumatra)
P. femoralis chrysomelas	Bornean banded surili	Brunei; Indonesia (W Kalimantan); Malaysia (Sarawak)
P. femoralis cruciger	Tricolored surili	Indonesia (NC Kalimantan); Malaysia (W Sabah, C Sarawak)
P. femoralis robinsoni	Robinson's banded surili	Peninsular Thailand; adjacent areas in Malaysia
P. fredericae	Javan fuscous surili	Indonesia (C Java)
P. frontata	White-fronted surili	Indonesia (E and C Kalimantan); Malaysia (E Sarawak)
P. hosei hosei	Hose's grizzled surili	Malaysia (coastal N Sarawak); [Brunei]
P. hosei canicrus	Miller's grizzled surili	Indonesia (EC Kalimantan)
P. hosei everetti	Everett's grizzled surili	E Brunei; Indonesia (N Kalimantan); Malaysia (W Sabah and NE Sarawak)
P. hosei sabana	Crested or Saban grizzled surili	Malaysia (E Sabah)
P. melalophos bicolor	Bicolored mitered surili	Indonesia (SE part of C Sumatra)
P. melalophos mitrata	Depigmented mitered surili	Indonesia (SE Sumatra)
P. melalophos nobilis	Ferruginous mitered surili	Indonesia (inland SW Sumatra)
P. potenziani potenziani	Golden-bellied Mentawai surili	Indonesia (N Pagai, S Pagai, and Sipura)
P. potenziani siberu	Sombre-bellied Mentawai surili	Indonesia (Siberut)
P. rubicunda carimatae	Red-naped red surili	Indonesia (S Kalimantan, Karimata)
P. rubicunda chrysea	Orange-backed red surili	Malaysia (E Sabah)
P. rubicunla ignata	Orange-naped red surila	Malaysia (N Sarawak); [W Brunei]
P. siamensis siamensis	Malayan pale-thighed surili	C and NE peninsular Malaysia; adjacent areas of Thailand
P. siamensis cana	Riau pale-thighed surili	Indonesia (Batam, Galang, Kundur, and adjacent areas of Sumatra)
P. siamensis natunae	Natuna pale-thighed surili	Indonesia (N Natuna)
P. siamensis paenulata	Mantled pale-thighed surili	Indonesia (C Sumatra)
P. thomasi	Sumatran grizzled surili	Indonesia (N Sumatra)
Semnopithecus		
S. entellus entellus	Bengal langur	W Bangladesh; India (SW Bengal, S Bihar, S Chhatisgarh, NE Maharashtra, Orissa)
S. entellus achates	Satpura langur	India (N Chhatisgarh, Gujarat, W Karnataka, Madhya Padesh, W Maharashtra, SE Rajasthan)
S. entellus ajax	Dark-armed Himalayan langur	India (N Himachal Pradesh, S Jammu, and Kashmir); NE Nepal
S. entellus anchises	Deccan langur	India (N Andhra Pradesh, NE Karnataka, SE Maharashtra)
S. entellus dussumieri	Dark-armed Malabar langur	India (SW Karnataka, W Kerala)
S. entellus hector	Lesser hill langur	India (S Uttaranchal, NE Uttar Pradesh, [NW Bengal]), S Nepal
S. entellus hypoleucos	Dark-legged Malabar langur	India (S Karnataka, [NE Karala])
S. entellus priam	Coromandel gray langur	India (S Andhra Pradesh, Tamil Nadu, [SE Karnataka])
S. schistaceus	Nepal sacred langur	NW Pakistan; N India (Jammu and Kashmir, Himachal Pradesh, Uttarakhand, and NW Bengal states, and Sikkim); S China (Tibetan regions of Bo Qu, Ji Long Zang Bu and Chumbi Valleys); Nepal; W Bhutan
S. johnii	Nilgiri black langur	India (S Kanataka, E Kerala, SW Tamil Nadu)
S. vetulus vetulus	Southern purpled-faced langur	SW Sri Lanka
S. vetulus monticola	Highland purple-faced langur	C Sri Lanka
S. vetulus nestor	Western purple-faced langur	W Sri Lanka
S. vetulus philbricki	Northern purple-faced langur	NC and N Sri Lanka
Trachypithecus		
T. auratus auratus	Spangled ebony leaf monkey	Indonesia (Bali, Bangka, Belitung, N Java, SE Kalimantan, Lombok, Riau-lingga, Serasan, S Sumatra)
T. auratus mauritius	West Javan ebony leaf monkey	Indonesia (SW Java)
T. auratus pyrrhus	Ebony leaf monkey/Javan langur	Indonesia (SE Java)
T. barbei	Tenasserim lutung	SW Thailand; adjacent Myanmar
T. delacouri	Delacour's langur	NC Vietnam (Hoa Binh, Ha Nam, Ninh Binh, and Thanh Hoa)
T. francoisi	Francois' langur	SC China (Chongqing, Guizhou, and Guangxi); N Vietnam (Ha Giang, Cao Bang, Tuyen Quang, Bac Kan, and Thai Nguyen)
T. ebenus	Black langur	EC Laos (Khammouane, N Savannakhet); C Vietnam (Quang Binh)

(continued)

Table 18.1 (Continued)

Species and Subspecies (by Genus) [a]	Common Name	Current Distribution
T. hatinhensis	Hantinh langur	NC Vietnam (Quang Binh and Quang Tri Provinces); EC Laos (Khammouane)
T. geei	Golden leaf monkey	Bhutan; India (NW Assam)
T. laotum	Lao langur	WC Laos (Bolikhamsai, Khammouane)
T. obscurus obscurus	Reid's dusky leaf monkey	Peninsular Malaysia; adjacent Thailand
T. obscurus carbo	Taruto dusky leaf monkey	Thailand (Tarutao)
T. obscurus corax	Dark-belllied dusky leaf monkey	Myanmar; N Thailand
T. obscurus flavicauda	Blond-tailed dusky leaf monkey	S Thailand; adjacent Malaysia (?)
T. obscurus halonifer	Cantor's dusky leaf monkey	Malaysia (Dayang Bunting, Langkawi, Penang, adjacent mainland); marginal Thailand (?)
T. obscurus sanctorum	Zadetkyi dusky leaf monkey	Myanmar (Zadetkyi Kyun)
T. obscurus seimundi	Phangan dusky leaf monkey	Thailand (Phangan, adjacent areas in mainland [?])
T. obscurus styx	Perhentian dusky leaf monkey	Malaysia (E Perhentian, adjacent coast [?])
T. phayrei phayrei	Phayre's langur	E Bangladesh; NE India (Assam, Mizoram, and Tripura states); W Myanmar (SE through Arakan to Pegu)
T. phayrei shanicus	Shan States Langur	SW China (Yingjiang-Namting River and Tunchong-Homushu Pass districts, W Yunnan); N and E Myanmar (Shan States and neighboring dry zone of N Myanmar)
T. pileatus pileatus	Blond-bellied capped leaf monkey	W Myanmar; India (Manipur, Meghalaya, Nagaland)
T. pileatus durga	Orange-bellied capped leaf monkey	Bangladesh; W Myanmar; India (C Assam)
T. pileatus tenebricus	Tenebrous capped leaf monkey	India (NC Assam)
T. shortridgei	Shortridge's langur	NE Myanmar (E of the Chindwin River, Kachin State N to Myitkyina District); SW China (Dulong River Valley in Gongshan County, NW Yunnan)
T. poliocephalus	Cat Ba langur	N Vietnam (Cat Ba Island)
T. leucocephalus	White-headed langur	S China (Fusui, Chongzuo Ningming, and Longzhou Counties, SW Guangxi Autonomous Region)
T. villosus	Griffith's silver leaf monkey	Brunei; Indonesia (N Kalimantan, N Sumatra); Malaysia (W peninsular, Sabah, N Sarawak)
T. germaini	Germain's langur	S Myanmar; S Thailand; S Laos; Cambodia (W of Mekong River); S tip of Vietnam
T. margarita	Annamese langur	S Laos; SC Vietnam; E Cambodia (Ratanakiri and Mondulkiri)
T. crepusculus	Indochinese gray langur	SW China (E of Salween River and S of Xishuangbanna, S Yunnan); S Myanmar; N Thailand (S to Raheng, and W to the coast of the Bay of Bengal); N and C Laos; N Vietnam

Nasalis

N. larvatus larvatus	Stripe-naped proboscis monkey	Brunei, Indonesia (Kalimantan); Malaysia (Sabah, Sarawak)
N. larvatus orientalis	Plain-naped proboscis monkey	Indonesia (NE Kalimantan)

Simias

S. concolor concolor	Pagai pig-tailed snub-nosed monkey	Indonesia (N & S Pagai, Sipura)
S. concolor siberu	Siberut pig-tailed snub-nosed monkey	Indonesia (Siberut)

Pygathrix

P. nemaeus	Red-shanked douc	SC Laos; C Vietnam; NE Cambodia
P. nigripes	Black-shanked douc	E Cambodia; SW Vietnam
P. cinerea	Gray-shanked douc	SC Vietnam

Rhinopithecus

R. avunculus	Tonkin snub-nosed monkey	NW Vietnam (Ha Giang, Tuyen Quang, Bac Kan, and Thai Nguyen)
R. bieti	Yunnan snub-nosed monkey	SW China (SE Xizang Autonomous Region [Tibet]; Yun Ling Mts, NW Yunnan)
R. brelichi	Guizhou snub-nosed monkey	SC China (Guizhou, Fanjingshan, Wuling Mts)
R. roxellana roxellana	Moupin golden snub-nosed monkey	WC China (S Gansu, S Shaanxi, W Sichuan)
R. roxellana hubeiensis	Hubei golden snub-nosed monkey	WC China (Shennongjia, W Hubei, NW Sichuan)
R. roxellana qinlingensis	Qinling golden snub-nosed monkey	WC China (Qinling Mts, S Shaanxi)
R. strykeri	Burmese, or black, snub-nosed monkey	NE Myanmar (Salween–N'mai Hka divide); S China (Gaoligongshan, Yunnan)

[a] Taxonomy after Brandon-Jones et al. (2004), with updates from Roos et al. (2014) and IUCN (2014); numerous Asian colobine taxonomies exist in the literature—for some species, the taxonomy presented here is not necessarily accepted as consensus classification.

ASIAN PRIMATE COMMUNITIES

Interaction between sympatric Asian colobine species and interaction between colobines and simple-stomached primates, such as macaques, are poorly understood as relatively few studies have addressed this issue. Sympatric colobines are able to coexist through a number of mechanisms. Studies on sympatric *Semnopithecus entellus* and *S. vetulus* living in disturbed habitats show that the two species are able to coexist by having contrasting dietary preferences (Hladik, 1977), i.e., *S. vetulus* is more folivorous than *S. entellus* (Hladik, 1977). In undisturbed habitats, the food preferences of *S. entellus* and *S. vetulus* are similar, and they coexist by having differential selectivity for shared plant species (Vandercone et al., 2012) and possibly by having contrasting foraging efficiencies (Vandercone, 2011). Similarly, in sympatric populations of *Presbytis potenziani* and *Simias concolor* on Siberut Island, Indonesia, *P. potenziani* has a greater preference for fruit while *Simias concolor* has a greater preference for leaves (Hadi et al., 2012). Colobines are also able to coexist with other simple-stomached primates, such as macaques, by having different microhabitat or dietary preferences. In the southern Westen Ghats, India, *Semnopithecus johnii* has very little dietary overlap with *Macaca silenus* and *M. radiata* and prefers to feed at lower strata in comparison to the macaques (Sushma & Singh, 2006). In Baimaxueshan National Nature Reserve, China, where *Rhinopithecus beiti* and *M. mulatta* are sympatric, *R. beiti* occupies high-altitude habitats and is largely confined to mixed forest, while *M. mulatta* occurs at lower altitudes and occupies a wide range of habitats (Grueter et al., 2010). Direct interaction between both sympatric Asian colobines and colobines sympatric with simple-stomached primates appears to be both tolerant (Sushma & Singh, 2006; Singh et al., 2011; Hadi et al., 2012) *and* intolerant (Sushma & Singh, 2006; Grueter et al., 2010; Vandercone, 2011). However, these interactions seem to be infrequent and avoidance of confrontation appears to be the norm (Sushma & Singh, 2006; Singh et al., 2011; Vandercone, 2011; Hadi et al., 2012). Although interaction between Asian colobines and other primates appears to be sparse, some Asian colobines (e.g., *S. entellus*) are known to form "associations" with other non-primate vertebrates like spotted deer, *Axis axis* (Newton, 1989). During these associations, the spotted deer benefit from the vegetation dropped by langurs and also respond to the anti-predator alarm calls given by langurs. *S. schistaceus* in India seems to have an interspecific feeding association with the Himalayan black bear (*Ursus thibetanus*) based on observations of bears "gleaning" nuts under oak trees where langurs have been feeding on acorns (Nautiyal & Huffman, 2018).

The ability of colobine monkeys to ingest and digest foliage when other food items are scarce permits them to attain high biomass. Generally, in primate communities the colobine biomass tends to be higher than all other primates in the community combined (Chapman & Chapman, 1999). Interestingly, the biomass of Asian colobine species varies drastically from site to site. The combined density of *S. entellus* and *S. vetulus* at Polonnaruwa in Sri Lanka is the highest of all colobine population densities reported from different study localities (Rudran, 1973a). The density of *S. entellus* ranges between 100–200 animals/km², while the density of *S. vetulus* ranges between 150–200 animals/km², and the density of *Presbytis melalophos* in Kuala Lompat, Malaysia, reaches 108 animals/km² (Davies, 1994). Other Asian sites have much lower population densities of colobines. *S. johnii* in the evergreen Shola forests of Kakachi, India, has a density of 71 animals/km² (Oates et al., 1980), and *Trachypithecus pileatus* in Madhupur, Bangladesh, exists at a density of 52 animals/km² (Stanford, 1991). The densities of odd-nosed monkeys tend to be much lower than the densities of other Asian colobines. Both *Rhinopithecus beiti* at Wuyapiya in the Baimaxueshan Nature Reserve, China (Kirkpatrick et al., 1998), and *R. roxellana* at Zhouzhi National Nature Reserve, China, occur at densities of ~7 animals/km². Similar variation in density exists among *populations* of the same species. For example, *S. entellus* at Polonnaruwa in Sri Lanka occur at densities ranging from 100–200 animals/km², while the same species in the forests of Kanha (Davies, 1994) and Dhawar (Sugiyama, 1964) in India have 46 and 85 animals/km², respectively.

Identifying the ecological factors that give rise to the variation in the abundance of primates at different localities is a fundamental question in primatology. Studies addressing this issue have demonstrated a positive relationship between colobine biomass and the index of leaf quality at localities in Asia and Africa (Waterman et al., 1988; Oates et al., 1990; Chapman et al., 2002). In addition, human disturbances, such as hunting (Freese et al., 1982) and logging (Johns & Skorupa, 1987), have been identified as factors that negatively impact primate biomass.

LANGURS: *SEMNOPITHECUS, TRACHYPITHECUS, PRESBYTIS*
Morphology

The genus *Semnopithecus, sensu stricto* (i.e., excluding *S. vetulus* and *S. johnii* and considering only *S. entellus*) has heavy horizontal brow ridges with a distinct depression posteriorly (Brandon-Jones et al., 2004) and infants with blackish-brown pelage, as opposed to the orange pelage of infant *Trachypithecus*. Generally, *S. entellus* does not display a sexually dichromatic pubic integument (Brandon-Jones et al., 2004). (It should be noted again that the decision to include *S. vetulus* and *S. johnii* in the genus *Semnopithecus* is based entirely on molecular evidence.) Cranial morphology, neonatal pelage color (gray for *S. vetulus* and reddish-brown for *S. johnii*), and sexually dichromatic pubic integument integrate them with *Trachypithecus* (Brandon-Jones et al., 2004).

Members of the genus *Trachypithecus* are characterized by having heteromorphic incisors with an edge-to-edge bite and sexually dichromatic pubic integument; neonates of the genus are generally orange, brown, or gray in color (Oates et al., 1994). *Trachypithecus* also have relatively shorter hindlimbs in comparison with *Prebytis* and engage in more quadrupedal walking and running, along with relatively larger stomachs in relation to their body size (Oates et al., 1994).

Some features that characterize the genus *Presbytis* include weakly developed brow ridges, a comparatively short face with a convex nasal profile, underjetted lower incisors, and relatively broad homomorphic incisors (Oates et al., 1994). They also have comparatively thick dental enamel and relatively small stomachs. Neonates are whitish in coloration and, as they develop, pass through a phase where the pelage on the back and upper head displays a cruciform pattern (Pocock, 1928). Furthermore, these monkeys have relatively longer hindlimbs and, consequently, leap frequently and are less quadrupedal than *Trachypithecus*. Adult *Prebytis* are also less sexually dimorphic than *Trachypithecus* (Oates et al., 1994).

Distribution and Habitat

Unlike African colobines, the Asian colobine genera have marked geographic distributions (Figure 18.1). The

genus *Semnopithecus* is restricted to the Indian subcontinent and is distributed from Pakistan through India into Nepal, Bhutan, Bangladesh, and Sri Lanka. Members of the genus *Semnopithecus* are found in a variety of habitats, which include dry tropical scrub jungle, semi-evergreen dry forest, tropical rainforest, and high-altitude coniferous and cloud forest.

The genus *Trachypithecus* is the most diverse of all the langur genera and has a broad distribution ranging from northeast India to mainland Southeast Asia to the Sunda Islands. *Trachypithecus* species occur predominantly in lowland tropical rainforests, dry deciduous forests, and coastal mangrove swamps (Oates et al., 1994). However, some *Trachypithecus* species occupy habitats characterized by karst topography in Southeast Asia (Zhou et al., 2013). The genus *Prebytis* is restricted to rainforests of southern Thailand, the Malay Peninsula, Sumatra, Java, Borneo, Bali, and the Lomboks (Oates et al., 1994).

Some Asian colobines, such as the Hanuman langur (*S. entellus*; see Figure 18.2) and Phayre's leaf monkey (*T. obscurus phayrei*), have wide distribution. *S. entellus* ranges from Sri Lanka north to the Himalayas, from sea

Figure 18.2 Male *Semnopithecus entellus* in Maharashtra, India.
(Photo credit: Shantanu Kuveskar; licensed under Creative Commons Attribution-Share Alike 4.0 International license)

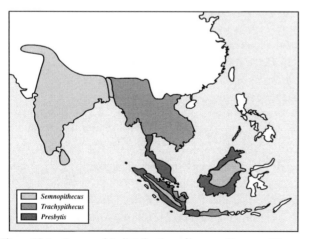

Figure 18.1 Geographic distribution of langurs (*Semnopithecus*, *Trachypithecus*, and *Presbytis*).
(Redrawn from Ting et al., 2008)

level to an altitude of 4,000 m and is found in northeastern India, southern China, northern Vietnam, and much of Laos and Thailand. In comparison, several Asian colobines, such as the white-sideburned black leaf monkey (*T. delacouri*) and the Tonkin snub-nosed monkey (*Rhinopithecus avunculus*), have very limited geographic distributions. *T. delacouri* is restricted to north-central Vietnam, while *R. avunculus* is restricted to central-northern Vietnam. Such restricted distributions may be artifacts of habitat refugia resulting from climatic fluctuations during the Pleistocene (Brandon-Jones, 1996).

Diet and Feeding Behavior

Though the annual dietary profiles of many Asian colobines depict them as obligate folivores, studies on temporal variation in diet, in relation to resource availability and nutritional ecology, lend support to the view that the Asian colobines are not obligate folivores but rather primates with a complex dietary ecology that prefer to feed on fruits and flowers and also selectively utilize leaves as an alternative food source when preferred items are in short supply. Additionally, when comparing the dietary ecology of Asian colobines, it is apparent that intraspecific variation among the colobines can be as great or greater than interspecific variation. This suggests that the differences in feeding ecology observed between populations of the same species are most likely driven by local habitat conditions.

Semnopithecus

The dietary ecology of the genus *Semnopithecus* exemplifies the dietary variation among Asian colobines. A number of long-term studies have examined the feeding ecology of *S. entellus* at different localities across the Indian subcontinent (Table 18.2). Though these studies represent a wide range of habitats across the geographic range of the species, the overall contribution of primary items to their diet appears to be similar across study sites. In the Kanha Tiger Reserve, India, a single group of *S. entellus*, consisting of 18–23 members, fed on 53 species of plants and spent 24.4% of their annual feeding time on fruit, 9.5% on flowers, 34.9% on mature leaves, and 16.7% on immature leaves (Newton, 1992). Studies of *S. entellus* at Ramnagar in southern Nepal (Koenig & Borries, 2001) and Rajaji National Park in India (Kar-Gupta & Kumar, 1994) have also produced similar results. Mature leaves constituted the bulk of the leaf intake by *S. entellus* at all three study sites (Newton, 1992; Koenig & Borries, 2001). The langurs at Kanha and Ramnagar also supplemented their diet with invertebrate prey in the form of insects. However, langurs at other locations, such as Langtang

National Park in Nepal (Sayers & Norconk, 2008) and Polonnaruwa (Hladik, 1977) and Kaludiyapokuna Forest Reserve in Sri Lanka (Vandercone et al., 2012), consumed more immature leaves than mature leaves and rarely consumed insects. Hanuman langurs at Polonnaruwa in Sri Lanka were more frugivorous, while the population at the Kaludiyapokuna Forest Reserve was more florivorous in comparison to other study populations of *S. entellus*. The *S. entellus* at Langtang National Park in Nepal also consumed underground storage organs (Sayers & Norconk, 2008).

Semnopithecus also exhibit considerable temporal variation in plant part consumption. At Kanha in India, the langurs fed on fruit, flowers, and immature leaves according to their availability but did not consume mature leaves according to their availability (Newton, 1992). The langurs at the Kaludiyapokuna Forest Reserve in Sri Lanka also showed a tendency to consume fruit and flowers when they were available but never consumed immature leaves according to their availability (Vandercone et al., 2012). In this population, immature leaves were consumed only when fruits were in short supply, and mature leaves were consumed only when immature leaf availability was low (Vandercone et al., 2012). Hanuman langurs in highland sites also fed on bark and soft underground storage organs when other dietary items were scarce (Sayers & Norconk, 2008).

When fruit availability was high at Kanha, fruits accounted for over 60% of *S. entellus* feeding time during several months of the year (Newton, 1992); fruits consumed by the langurs were fleshy and mature. The flower consumption also accounted for ~60% of their feeding time during one month. Similar patterns have also been observed at other study localities (Hladik, 1977; Vandercone et al., 2012). In addition, the langurs at Kanha also specialized on insects during the monsoon season, which accounted for ~25% of their feeding time during one month (Newton, 1992). The langurs specialized on slow-moving insects, such as caterpillars, that occurred at high densities (Newton, 1992). This pattern is a stark contrast to insectivory by cercopithecines that have a tendency to feed on cryptic and dispersed insects (Newton, 1992). Additionally, studies examining the food plants utilized by langurs have found them to be highly selective, often selectively feeding extensively on plant species that are relatively rare in the environment (Newton, 1992; Vandercone et al., 2012). In addition, studies on plant selection showed most plant species utilized by Hanuman langurs have clumped spatial distribution patterns (Koenig et al., 1998; Vandercone et al., 2012) and vegetative parts high in carbohydrates and proteins and low in fiber (Hladik, 1977; Kar-Gupta & Kumar, 1994; Koenig et al., 1998).

Table 18.2   Dietary Comparison of Some Langur Taxa

Species	Fruit	Seeds	Fruit + Seeds	Flowers	Leaves	Other	No. spp.	Study Site	References
Semnopithecus:									
S. entellus (P 91/92)[a]	-	-	15.1	6.3	59.8	3.1	-	Ramnagar, India	Koenig and Borries (2001)
S. entellus (P 92/93)[a]	-	-	21.9	12.8	52.4	3.8	-	Ramnagar, India	Koenig and Borries (2001)
S. entellus (O)[a]	-	-	23.2	4.4	60.9	0.4	-	Ramnagar, India	Koenig and Borries (2001)
S. entellus	-	-	24.4	9.5	51.6	14.5	53.0	Kanha, India	Newton (1992)
S. entellus	-	-	21.0	6.8	6.8	15.4	43.0	Langtang, Nepal	Sayers and Norconk (2008)
S. entellus	-	-	45.0	7.0	48.0	0.0	-	Polonnaruwa, Sri Lanka	Hladik (1977)
S. entellus (A)[a]	14.6	8.4	23.0	23.9	52.6	0.4	58.0+	Kaludiyapokuna, Sri Lanka	Vandercone et al. (2012)
S. entellus (B)[a]	17.7	11.4	29.1	13.4	57.4	-	32.0	Kaludiyapokuna, Sri Lanka	Vandercone et al. (2012)
S. johnii	-	-	25.1	9.3	62.2	3.4	107.0+	Kakachi, India	Oates et al. (1980)
S. vetulus	-	-	28.0	12.0	60.0	0.0	-	Polonnaruwa, Sri Lanka	Hladik (1977)
S. vetulus (1)[a]	52.3	1.4	53.7	7.6	31.7	-	-	Panadura, Sri Lanka	Dela (2007)
S. vetulus (2)[a]	53.9	6.2	60.1	4.0	29.4	-	-	Piliyandala, Sri Lanka	Dela (2007)
S. vetulus	19.5	6.3	25.8	11.3	62.8	0.0	31.0+	Kaludiyapokuna, Sri Lanka	Vandercone et al. (2012)
Trachypithecus:									
T. pileatus	24.4	9.3	33.7	7.0	57.8	1.5	35.0	Madhupur, Bangladesh	Stanford (1991)
T. pileatus	-	-	16	16	68	-	52.0	Pakhui, India	Solanki et al. (2008)
T. obscurus phayrei	-	-	14	16	70	0.0	29.0	Lawachara, Bangladesh	Aziz and Feeroz (2009)
T. delacouri	-	-	9	5	80	6.0	42.0	Van Long, Vietnam	Workman (2010)
T. leucocephalus	5.7	0.4	6.1	2.7	91.2	-	50.0	Fusui, China	Li, Z., and Rogers (2006)
T. francoisi	-	-	3.1	0.5	94.5	2.0	37.0	Fusui, China	Huang et al (2008)
T. francoisi	17.2	14.2	31.4	7.5	52.8	8.4	90.0	Nonggang, China	Zhou et al. (2006)
T. poliocephalus	6.0	-	6.0	8.0	83.0	3.0	-	Cat Ba Island, Vietnam	Hendershott et al. (2016)
T. germaini	22.7	-	22.7	4.7	67.5	4.5	58	Kien Luong Karst Area, Vietnam	Le et al. (2019)
Presbytis:									
P. rubucunda	-	-	49.6	11.1	37.6	2.0	103.0+	Sepilok, Malaysia	Davies (1991)
P. rubucunda	33.8	26	59.8	6.4	31.1	-	-	Gunung Palung National Park, West Kalimantan, Indonesia	Clink et al. (2017)
P. potenziani	-	-	32	-	55.0	13.0	42.0	Betumonga, North Pagai, Indonesia	Fuentes (1996)
P. siamensis	-	-	56	6.0	35.0	2.0	137.0	Kuala Lompat, Malaysia	Curtin (1980)

[a] Parenthetic alpha/numeric labels are individual group identifiers corresponding to data sets on different study groups occurring at the same research sites (Koenig and Borrie, 2001; Dela, 2007; Vandercone et al., 2012).

Unlike the Hanuman langurs, relatively few long-term studies have examined the dietary ecology of the purple-faced langur (*S. vetulus*) in Sri Lanka (Hladik, 1977; Dela, 2007; Vandercone et al., 2012) and the Nilgiri langur (*S. johnii*) in the Western Ghats, India (Oates et al., 1980). Unlike Hanuman langurs, the dietary ecology of *S. vetulus* varies drastically between study sites. At Polonnaruwa in Sri Lanka, *S. vetulus* spent 28% of their annual feeding time on fruit, 12% on flowers, and 60% on leaves; mature leaves accounted for two-thirds of the leaves consumed (40% of overall diet). A more recent study at the Kaludiyapokuna Forest Reserve showed immature leaf consumption to be high and mature leaf consumption to be extremely low. A study of *S. vetulus* in a human-dominated landscape of the Western Province, Sri Lanka, found the langurs to be extremely frugivorous and include a high proportion of fleshy fruit in their diet year-round (Dela, 2007). In addition, mature and immature leaves were found to be a less important component in the diet of these langurs (Dela, 2007). Dietary item selection by purple-faced langurs appears to be similar to the Hanuman langur. At Kaludiyapokuna, *S. vetulus* consumed fruit and flowers according to their availability but did not consume immature leaves according to their availability, suggesting that, like *S. entellus*, *S. vetulus* also preferred fruits and flowers to immature leaves (Vandercone et al., 2012). As with Hanuman langurs, *S. vetulus* also shows considerable seasonal differences in plant part consumption. At Polonnaruwa, Sri Lanka, during periods of fruit availability, fruit consumption was comparatively high during several months of the year and accounted for ~50% of their feeding time during one month. In the study carried out in the Western Province of Sri Lanka, fruit was the most frequently consumed dietary item year-round and accounted for ~65% of the monthly feeding time of one group and 74% of the monthly feeding time of another group (Dela, 2007). Flowers and seeds were also found to be important components of the *S. vetulus* diet in Sri Lanka. At Kaludiyapokuna, flower and seed consumption were relatively high during some months and accounted for ~27% and 33% of monthly feeding time, respectively (Vandercone et al., 2012). *S. vetulus* was never observed to feed on insects (Hladik, 1977; Dela, 2007; Vandercone et al., 2012). A study on food plant selection showed *S. vetulus* to prefer relatively rare plant species with clumped spatial distributions (Vandercone et al., 2012). At Polonnaruwa, biochemical analyses of food plants showed that the plants consumed by *S. vetulus* were lower in protein in comparison to dietary items consumed by *S. entellus* (Hladik 1977).

A study on the dietary ecology of the Nilgiri langur (*S. johnii*) in an evergreen forest at Kakachi, India, showed the contribution of primary dietary items to be similar to *S. vetulus* (Oates et al., 1980). The Nilgiri langur showed a preference for immature leaves, flowers, and fruits as opposed to mature leaves (Oates et al., 1980). A biochemical analysis of food items consumed by these langurs showed that the most heavily used items were relatively low in fiber and condensed tannins (Oates et al., 1980).

Trachypithecus

Many studies have investigated the dietary ecology of *Trachypithecus* species throughout their ranges (see Table 18.2). A study of the dietary ecology of the capped langur (*Trachypithecus pileatus*) in the moist deciduous forest of Bangladesh, found that they spend the majority of their feeding time on leaves (24.4% on fruits; 9.3% on seeds, 7.0% on flowers, and 42% on mature leaves, which accounted for the bulk of the leaves consumed) (Stanford, 1991). However, brief observations made on *T. pileatus* in wet semi-evergreen forests in India, show the species to be highly frugivorous with fruit making up ~54% of their diet (Stanford, 1992). In comparison, *Trachypithecus pileatus* at Pakhui Wildlife Sanctuary in India was less frugivorous, and the bulk of their diet was made up of immature leaves (Solanki et al., 2008). Further, a study on Phayre's leaf monkey (*T. obscurus phayrei*) at Gumti Wildlife Sanctuary, India, found immature leaves and seeds to account for ~48% and 23% of their feeding time, respectively (Gupta & Kumar, 1994). These langurs rarely utilized fruit and mature leaves. A similar study on silver leaf monkeys (*T. auratus*) also found immature leaves made up the bulk of their diet, although unripe fruit was also an important component (Kool, 1993). Unlike *S. entellus*, *T. auratus* rarely consumed insect prey. A more recent study on silver leaf monkeys found that they feed predominantly on immature leaves (69%) and fruit (21.2%) (Tsuji et al., 2019).

Trachypithecus species associated with limestone hills in China and Vietnam appear to be the most folivorous of all the Asian colobine species. A study on the white-headed langur (*T. leucocephalus*) associated with the limestone hills in Fusui Precious Animal Reserve, China, found the monkeys subsisted predominantly on immature leaves (89%) (Li, Z., & Rogers, 2006). Fruit, flowers, and seeds were rarely consumed (Li, Z., & Rogers, 2006). A study on Delacour's langur (*T. delacouri*) at Van Long Nature Reserve, Vietnam, also found the langurs feed extensively on leaves (Workman, 2010). Studies on Francois' and white-headed langurs in China have also produced similar results (Huang et al., 2008; Hu, 2011). The extremely low fruit consumption of species such as *T. leucocephalus* has been attributed to the lack of suitable fruit in their environment (Li, Z., & Rogers, 2006). A more recent study on Cat Ba langurs (*T. poliocephalus*; Hendershott et al., 2016) and the Indochinese silvered langur (*T. germaini*; Le et al., 2019) find the species to be highly folivorous.

These langur species display significant temporal variation in plant part consumption. *T. pileatus* feed on leaves and fruits according to their availability and utilized mature leaves only when immature leaves were low in availability (Stanford, 1991). Although fruit comprises less that 25% of the annual diet of the species, during several months fruit consumption exceeds 40% of their feeding time. Similar patterns of seasonal plant part use have also been observed in *T. pileatus* at other study localities (Solanki et al., 2008). In the case of *T. leucocephalus*, the monkeys consume immature leaves according to their availability but do not consume mature leaves or fruit according to their availability (Li, Z., & Rogers, 2006). Immature leaf consumption by *T. leucocephalus* remains high, but fruit and flower consumption remain low throughout the year (Li, Z., & Rogers, 2006). Seasonal variation in plant part consumption has also been observed in *T. francoisi* (Hu, 2011). Fruit consumption is comparatively high during the winter (~43%) and immature leaf consumption is highest during the spring. Food plants that are highly preferred by the langurs are relatively rare in the environment (Hu, 2011). Although there is considerable inter- and intraspecific variation in the proportion of plant parts consumed, species of the genus *Trachypithecus* as a whole appear to be the most folivorous of all Asian colobine groups.

Presbytis

In comparison to *Semnopithecus* and *Trachypithecus*, relatively few studies have examined the dietary ecology of *Presbytis* species (see Table 18.2). Red leaf monkeys (*P. rubicunda*) in the Sepilok Virgin Jungle Reserve of East Malaysia were found to spend ~36% of their feeding time on immature leaves, 30% on seeds, 19% on fruit, and 11% on flowers (Davies et al., 1988). A study of red leaf monkeys in a tropical peat-swamp forest found they spend ~76% of their feeding time on seeds, 7.3% on other fruit parts, 7.7% on immature leaves, 2.5% on mature leaves, and 2.8% on flowers (Ehlers Smith et al., 2013a). A recent study of red leaf monkeys in Gunung Palung National Park, West Kalimantan, Indonesia, also found the species consumes a comparatively high proportion of seeds (Clink et al., 2017). Although seeds appear to be an important part of the diet of these langurs, red leaf monkeys show greater selectivity when feeding on leaves than when eating seeds (Hanya & Bernard, 2015). Red leaf monkeys select young leaf species with more digestible

protein and crude ash but less crude lipid (Hanya & Bernard, 2015). In addition to the consumption of plant parts, these langurs are also known to descend to the ground to consume terrestrial fungi (Cheyne et al., 2019). In fact, Cheyne et al. (2018) have suggested that ground use by *P. rubicunda* may be an adaptation to exploit additional food sources in human-disturbed habitat.

Similarly, banded leaf monkeys (*P. melalophos*) in the Krau Game Reserve of western Malaysia also were found to have a diet dominated by fruits and seeds (~50%) (Davies et al., 1988). A study on *P. potenziani* in the Mentawai Islands found they feed predominantly on fruit (Hadi et al., 2012). In contrast, *P. potenziani* at Betumonga, Indonesia, subsist mainly on leaves (55%) (Fuentes, 1996). In the case of the banded and red leaf monkeys, young leaves are a consistent part of the monthly diet in all months except during the fruiting season, during which seeds and fruits account for over 85% of the monthly diet (Davies et al., 1988). Flowers, which contributed relatively little to the overall diet of these langurs, are eaten largely during the peak of the flowering season (Davies et al., 1988). The leaves consumed by the langurs have higher nitrogen, higher digestibility, and lower fiber in comparison to leaves they avoid. The seeds eaten are low in fiber and low in compounds such as tannins and phenolic and hence highly digestible (Davies et al., 1988). Seeds, however, are low in nitrogen in comparison with leaves (Davies et al., 1988).

Ranging Behavior and Activity Patterns

As with dietary ecology, there is considerable intraspecific and interspecific variation in the ranging and activity budgets of Asian colobines. Activity budgets are directly related to metabolism and energetic constraints on these primates. In general, frequent inactivity or rest and infrequent grooming characterize the activity budgets of langurs. However, in most species, there is marked seasonal differences in activity patterns. In some, activity seems to be influenced by diet and the availability of resources, and resource availability also influences the ranging behavior of langurs. The daily range length of several species is influenced by diet and the availability of preferred foods found in their environment. Species with lower than average densities tend to have larger home ranges (Yeager & Kirkpatrick, 1998), which suggests that resource availability may regulate the number of individuals in an area.

Semnopithecus

The annual activity budget of Hanuman langurs at Kanha in India is composed of feeding (25.7%), inactivity (41.8%), moving (13.1%), clinging (7.9%), and allogrooming (6.0%). Activity budgets are similar between adult males and females, except for the fact that adult females engage in significantly more allogrooming and social behavior (Newton, 1992). Subadult females move more and spend less time inactive in comparison to adult females, while juvenile females move more, allogroom less, and spend less time inactive (Newton, 1992). In comparison, S. entellus living in the highlands of Nepal spend more time feeding and traveling and less time resting (Sayers & Norconk, 2008). The proportion of time devoted to different activities also varies between months. For example, more time is devoted to feeding during the winter months (Newton, 1992).

A wide range of home range sizes has been reported for S. entellus (Table 18.3). The home range and average day range of a study group at Kanha was 74.5 ha and 1083 m, respectively (Newton, 1992). Range size tends to be the smallest in winter and mid-monsoon months and comparatively larger in the late monsoon (Newton, 1992); day range length is shorter during the cold weather months than during the rest of the year (Newton, 1992). The home range and average day range of S. entellus in the conifer/broadleaf forests of Junbesi, Nepal, is 1,250 ha and 2,990 m, respectively (Curtin, 1982). The home range of the species in the alpine forests of Machiara, Pakistan, is 235–328 ha in extent, and the average distance covered

per day ranges from 1,230–1,750 m (Minhas et al., 2013). Hanuman langurs in the alpine forests of Pakstan also show altitudinal variation in habitat use (Minhas et al., 2013). In comparison, the home range of S. entellus at Polonnaruwa and Kaludiyapokuna, Sri Lanka, is 10–15 ha and 7.8–9.4 ha in extent, respectively (Hladik, 1977; Vandercone, 2011). In the case of S. entellus, there is a general trend for home range to decrease with increasing population density (Newton, 1984). However, such a trend should be interpreted with caution, as there is little consistency in the methods employed to determine home ranges. At Kaludiyapokuna, the day range of S. entellus was observed to increase during periods of low fruit availability (Vandercone, 2011). S. entellus also displays a tendency to favor certain habitats over others within their home range. For example, at Kanha, S. entellus occupies mixed forest more, and meadows less, than expected; sal forest (a forest type dominated by Shorea robusta) is occupied in proportion to its abundance (Newton, 1992). In some populations, overlap between adjacent groups appears to be substantial (Newton, 1992), while in others, overlap tends to be minimal (Vandercone, 2011).

Trachypithecus

The annual activity budgets of Trachypithecus species are similar to the activity patterns of S. entellus. T. obscurus

Table 18.3 Comparison of Ranging Data Across Selected Langur Species

Genus and Species	Mean Day Range (m)	Day Range (Range of m)	Home Range (ha)	Study Site	References
Semnopithecus					
S. entellus	1083	-	75	Kanha, India	Newton (1992)
S. entellus	2990	-	1250 [a]	Junbesi, Nepal	Curtin (1982)
S. entellus	1230	-	235	Machiara, Pakistan	Minhas et al. (2013)
S. entellus	1750	-	328	Machiara, Pakistan	Minhas et al. (2013)
S. entellus	-	-	10-15	Polonnaruwa, Sri Lanka	Hladik (1977)
S. entellus	441	189-650	9.4	Kaludiyapokuna, Sri Lanka	Vandercone (2011)
S. entellus	348	111-632	7.8	Kaludiyapokuna, Sri Lanka	Vandercone (2011)
S. entellus	-	-	2-3	Polonnaruwa, Sri Lanka	Hladik (1977)
S. vetulus	-	-	11	Kaludiyapokuna, Sri Lanka	Vandercone (2011)
S. vetulus	251	131-409	24	Kakachi, India	Oates et al. (1980)
S. johnii	-	-			
Trachypithecus					
T. pileatus	325	50-700	22	Madhupur, Bangladesh	Stanford (1991)
T. leucocephalus	491	-	23.8	Fusui Nature Reserve, China	Zhou et al. (2011)
T. leucocephalus	512	-	33.8	Fusui Nature Reserve, China	Zhou et al. (2011)
T. francoisi	438	-	19	Fusui Nature Reserve, China	Zhou et al. (2007b)
T. crepusculus	1011	958-1051	357-395	Wuliangshan, Yunnan, China	Pengfei et al. (2014)
Presbytis					
P. siamensis	703	300-1360	30	Kuala Lompat, Malaysia	Bennett (1986)
P. potenziani	540	60-1120	33	Betumonga, Indonesia	Fuentes (1996)
P. potenziani	-	-	20.2	Siberut, Indonesia	Hadi et al. (2012)
P. potenziani	-	-	30.3	Siberut, Indonesia	Hadi et al. (2012)
P. rubricunda	850	225-1670	85	Sepilok, Malaysia	Bennett and Davies (1994)

[a] Home range computed using 0.4 x 0.4 ha grid.

phayrei in the Gumti Wildlife Sanctuary, India, spent 34.9% of their time feeding, 21.1% resting, 14.4% traveling, and 29.5% playing and grooming during the study period (Gupta & Kumar, 1994). There are also significant differences between the seasons as far as the time devoted to various types of activity. Time devoted to feeding in winter is higher than in summer and the monsoon season, and the time devoted to resting is lower in winter in comparison to summer and the monsoon period (Gupta & Kumar, 1994). In comparison, the highly folivorous *Trachypithecus* species associated with limestone hills in China and Vietnam spend a large proportion of their time resting. For example, *T. francoisi* in the Nonggang Nature Reserve, China, spend ~52% of their time resting, while moving and feeding account for 17% and 23% of their time, respectively (Zhou et al., 2007a). In comparison, only about 2% of their time is devoted to grooming. The activity pattern of these langurs was also correlated with their diet and the availability of food resources in the environment (Zhou et al., 2007a). The time spent feeding is negatively correlated with young leaves and flowers and positively correlated with the consumption of seeds (Zhou et al., 2007a). The time spent grooming is affected by the seasonal changes in the availability of food. When fruit and immature leaf availability decline during the dry season, the langurs increase feeding time and reduce grooming time (Zhou et al., 2007a). A study of the activity pattern of *T. leucocephalus* in the Fusui Reserve, China, also showed the langurs spend a greater proportion of their time resting (Li, Z., & Rogers, 2004). In addition, a group living in high-quality habitat spends less time feeding and more time in play than a group living in a low-quality habitat (Li, Z., & Rogers, 2004).

The interspecific variation in home range and daily range in *Trachypithecus* appears to be quite limited in comparison to the variation in the genus *Semnopithecus*. However, as with other Asian colobines, the ranging behavior of *Trachypithecus* varies seasonally. The home range and the average day range of *T. pileatus* in Madhupur, Bangladesh, is 22 ha and 325 m, respectively (Stanford, 1991). Seasonal differences in ranging behavior were also observed in these langurs. Groups of *T. pileatus* were observed to travel farthest in months when the diet is based on fruit and shortest when feeding on mature leaves (Stanford, 1991). Similarly, the home range use of *T. leucocephalus* in the Fusui Nature Reserve is influenced by the distribution of resources within their home range (Zhou et al., 2011). For two study groups in Fusui, the home range was 23.8–33.8 ha and the day range was 215–970 m (Zhou et al., 2011). The most heavily used parts of the home range are located near the most frequently used sleeping sites, and the core area's home range includes more than one permanent water source which the langurs utilize during the dry season (Zhou et al., 2011). A study of a group of *T. francoisi* in Fusui Nature Reserve also produced similar results (Zhou et al., 2007b). The home range of the group is 19 ha and the monthly average day range varies from 341–577 m. The monthly range of the group tends to be larger during the dry season than during the rainy season (Zhou et al., 2007b).

Presbytis

In comparison to *Semnopithecus* and *Trachypithecus*, relatively few studies have focused on the activity budgets of *Presbytis*. As with other Asian colobines, the activity patterns of *Presbytis* are also characterized by prolonged periods of rest (Johns & Skorupa, 1987; Hadi et al., 2012). For example, *P. potenziani* in Siberut, Indonesia, spend ~46% of their daily time resting, 32% feeding, 5% traveling, and 0.6% on social behavior (Hadi et al., 2012). The home range size reported for *Presbytis* fluctuates between 20–108 ha and the daily range length is between 700–1,645 m.

As in other Asian colobines, ranging behavior of *Presbytis* species is also influenced by the availability of resources. For example, the day range length of *P. melalophos* shows correlation with availability of preferred food resources (Bennett, 1986). In addition, the rarity of food plants is considered to be the main reason for large home range size and low population density of *P. rubicunda* in Sepilok, Malaysia (Davies et al., 1988). Further, for *P. thomasi*, larger groups were shown to have longer day ranges and larger home ranges than small groups, which suggests home range size and day range length are indicative of the energetic demands of a group (Steenbeek & van Schaik, 2001).

Social Organization and Reproduction

The most common form of bisexual group observed in *Semnopithecus* (Poirier, 1970; Rudran, 1973a, 1973b; Newton, 1988; Koenig & Borries, 2001), *Trachypithecus* (Stanford, 1991; Solanki et al., 2007; Jin et al., 2009), and *Presbytis* (Davies, 1984; Supriatna et al., 1986) is the one-male unit (OMU). In such groups, a single male associates with multiple females and their offspring. However, there is much variation around this modal pattern. In a number of Asian langur species, bisexual groups with more than one adult male have been observed (Poirier, 1969; Gupta & Kumar, 1994; Jin et al., 2009). In addition, for *S. entellus* in northern Indian and Sri Lanka, the multi-male bisexual social unit predominates (Ripley, 1965; Boggess, 1980; Borries, 1997). Bisexual groups vary widely in their size and the number of females within them (Table 18.4). In the case of *S. entellus*, an analysis by Newton (1988) of 24 wild populations of Hanuman langurs found that the number of adult males per troop is positively correlated with troop size and the number of adult females present. However, in the majority of other langur species, multi-male bisexual groups do not have a greater number of females than OMUs, which suggests that the multi-male bisexual group is not a strategy purely to gain access to females (Kirkpatrick, 2007; Sterck & Van Hooff, 2000).

In bisexual groups, adult males tend to remain aloof from troop activities and are reported to be mainly responsible for intertroop spacing (Ripley, 1965; Stanford, 1991). Langurs devote relatively little time to grooming, and the majority of this activity takes place between females. For example, *S. entellus* at Kanha, India, devoted only 6% of their time toward grooming during Newton's (1992) study. In comparison, *T. francoisi* spend 2% of their time engaged in grooming (Zhou et al., 2007a). In some species, such as *S. johnii*, grooming is almost absent (Poirier, 1970). Although interaction between males and females is sparse in most langurs, during 50% of intergroup encounters observed by Stanford (1991), resident male *T. pileatus* pushed and bit their group females if the females strayed from the immediate proximity of the male.

Nonmaternal caring for infants, or alloparenting, has also been documented in a number of langur species (Davies, 1984; Stanford, 1991; Newton & Dunbar, 1994). *T. pileatus* females of all ages show interest in neonates, but the frequency of female interaction varies greatly (Stanford, 1991). Some neonates have intensive contact with an allomother during the first week of life, while others have little contact with any animals other than the mother (Stanford, 1991). On average, neonates spend one-third of their daylight time during the first month of life in contact with allomothers. After the first month, the time spent with an allomother declines drastically (Stanford, 1991). Interactions between adult males and infants are infrequent and are typically "passively tolerant" (Kirkpatrick, 2007). In the case of *S. entellus*, males that were genetic fathers or resident when infants were conceived, were observed to defend infants from attacks by other males (Borries et al., 1999a). However, *S. entellus* males that immigrated after a female conceived were never involved in protecting the infant (Borries et al., 1999a).

In most langur species, there is considerable interspecific and intraspecific variation in home range overlap between conspecific groups. In Kanha, adjacent groups were frequently observed within the home range of Newton's (1992) study group and occupied half (50.7%) of the quadrats in the study group's annual range. In comparison, overlap between two adjacent *S. entellus* groups in the Kaludiyapokuna Forest Reserve in Sri Lanka was only 1.8 ha (Vandercone, 2011). Overlap between adjacent *T. pileatus* groups in Bangladesh was shown to be extensive (Stanford, 1991), but the home range overlap between *P. rubicunda* groups was shown to be minimal (Davies, 1984). Although most langur species are territorial, intergroup encounters are relatively rare (Davies, 1984; Stanford, 1991). In *S. entellus* at Kanha, intergroup

Table 18.4 Social Organization Parameters of Selected Langur Species

Genus and Species	Study Site	Density	Group Size	Mean No. Adult Males/Group	Mean No. Adult Females/Group	One-Male Units (% Groups)	References
Semnopithecus							
S. entellus	Orchha, India	4.4	19.0	3.7	6.0	0.0	Newton (1988)
S. entellus	Khana, India	46.2	21.7	1.1	9.4	93.0	Newton (1988)
S. entellus	Jodhpur, India	18.0	35.0	1.0	20.0	95.0	Newton (1988)
S. entellus	Ramnagar, India	26.0	16.9	-	-	23.5	Borries (1997)
S. vetulus	Polonnaruwa, Sri Lanka	215.0	8.6	1.1	4.4	93.1	Rudran (1973b)
S. vetulus	Horton Plains, Sri Lanka	92.6	9.0	1.0	3.7	100.0	Rudran (1973b)
S. johnii	Nilgiri Hills, India	-	8.9	1.6	-	71.4	Poirier (1970)
Trachypithecus							
T. pileatus	Madhupur, India	52.0	8.5	1.0	3.6	100.0	Stanford (1991)
T. auratus	Java, Indonesia	345.0	6.0-21.0	1.2	6.6	77.8	Kool (1989)
Presbytis							
P. siamensis	Kuala Lompat, Malaysia	108.0	15.0	1.0	7.7	100.0	Bennett (1983)
P. rubicunda	Sabah (Borneo), Malaysia	19.0	7.0	1.0	2.0	100.0	Davies (1984)

encounters occur with a frequency of 0.68/day (Newton 1992). These encounters are aggressive and involve both sexes (Newton 1992).

Resident adult male replacement and subsequent infantcide has also been reported in langurs (Rudran, 1973b; Sommer et al., 1992; Borries, 1997). The question of whether infanticide in primates is an adaptive strategy to enhance male reproductive success is highly debated (Hrdy et al., 1995; Sussman et al., 1995). Infanticide has been rarely observed and is primarily reported from some populations of *S. entellus* (Kirkpatrick, 2007). In one-male groups that have been studied, residency of an adult male varied between 3 days and 74 months with a mean of 26 months (Sommer & Rajpurohit, 1989). However, the tenure of adult males in multi-male groups changes less drastically (Laws & Laws, 1984). In Hanuman langurs, adult replacement can either be rapid or gradual. Rapid male replacement is an aggressive and acute change often occurring in days, while gradual male replacement involves a staggered pattern of male introductions and exclusions between bisexual groups and nonreproductive units (Newton, 1988). The disappearance and death of infants and juvenile langurs has also been reported during the course of adult replacement and has been attributed to infanticide by the new male (Rudran, 1973b; Sommer et al., 1992; Borries et al., 1999b). Infanticide is estimated to account for approximately 20–30% of infant mortality in Hanuman langur populations in Ramnagar, Nepal, and at Kanha and Jodhpur in India (Borries, 1997).

In general, immature male langurs emigrate from their natal groups and join together to form nonreproductive units, while females remain within their natal groups. *S. entellus* males living outside their natal troops suffer high levels of mortality in comparison to philopatric females (Rajpurohit et al., 1995). However, there is considerable variation around this pattern since *S. vetulus* and *T. leucocephalus* juvenile females have also been observed in nonreproductive units (Rudran, 1973a; Jin et al., 2009). In the case of *S. vetulus*, juvenile females were observed in nonreproductive units after the adult male in their natal group was replaced by a new male (Rudran, 1973a). The transfer of females between groups has also been recorded (Newton, 1987).

Most langur species are seasonal breeders, and copulation is typically initiated by females. Various modes of female solicitation have been observed among different species. In Hanuman langur females, lowering and presenting of the anogenital region, and simultaneous head shaking, were observed during solicitation (Sommer et al., 1992). Almost all matings are preceded by solicitation, but solicitation does not always lead to mating (Borries et al., 2001). Head shaking and presenting have also been observed in *S. vetulus* (Rudran, 1973a), *S. johnii* (Poirier, 1970), and *T. pileatus* (Solanki et al., 2007). In most langur species, mating and births occur during certain times of the year, but for Hanuman langurs at Ramnagar, Nepal, copulations with nonpregnant females occurred throughout the year with the number of copulations rising during July–November (Borries et al., 2001). In the case of one-male Hanuman langur groups, it is only the resident adult male that breeds (Sommer & Rajpurohit, 1989). In multi-male groups, breeding is not entirely monopolized by the dominant male (Launhardt et al., 2001). In one study of multi-male groups, the dominant male sired 57% of the infants in the group (Launhardt et al., 2001). At Ramnagar, infants are born during January–June with a peak in March (Borries et al., 2001). Similar patterns of birth seasonality have also been observed in *S. vetulus* (Rudran, 1973b), *S. johnii* (Poirier, 1970,), and *T. pileatus* (Stanford, 1991; Solanki et al., 2007). At Jodhpur, India, where Hanuman langurs were provisioned, births and mating occurred throughout the year and less seasonality was observed (Sommer et al., 1992). Life history parameters, such as interbirth intervals, lactation periods, and female age at first birth vary between populations (Rudran, 1973b; Borries et al., 2001). For instance, the birth interval of *S. vetulus* in Sri Lanka is 2 years at Polonnaruwa and 1.3 years at Horton Plains (Rudran, 1973b). At least for Hanuman langurs, nutrition has been identified as a key factor influencing reproduction seasonality and variation in life history parameters between populations (Borries et al., 2001). For example, the mean lactation period and interbirth interval in a wild Hanuman langur population in Ramnagar, Nepal, was 24.9 and 28.8 months, respectively, while in a provisioned population of langurs in Jodhpur, India, the mean lactation period was 12.8 months and the interbirth interval was 16.7 months. More recently, phytochemicals in plants have also been shown to influence reproductive function in primates by elevating progestin levels and consequently increasing cycle lengths and follicular phases (Lu et al., 2011).

Predation

Information about predation and its impact on langur populations is poorly understood. This is largely due to the fact that predation events are rarely observed. Documented cases of predation have been largely opportunistic observations. For example, Thapar (1986) photographed the entire sequence of a tiger (*Panthera tigris*) locating her prey, then stalking, capturing, and killing a Hanuman langur in India. Also, in India, several unsuccessful predation attempts on adult *S. entellus* by golden jackals (*Canis aureus*), along with a successful attack on an infant langur, were seen at Kanha Tiger Reserve (Newton, 1985). In Bangladesh, two instances of attacks on *T. pileatus* by jackals were noted; both attacks took place during rare instances of ground travel by the langurs (Stanford, 1989). In addition, a marbled cat (*Pardofelis marmorata*) was observed attacking a juvenile Phayre's leaf monkey (*T. phayrei*) (Borries et al., 2014).

Studies of carnivore dietary ecology indicate that langurs are an important part of the diet of a number of carnivore species in Asia. An analysis of fecal material from leopards (*Panthera pardus*) in the Parambikulam Wildlife Sanctuary in India, found *S. johnii* and *S. entellus* to be an important part of their diet (Joseph et al., 2007). In this study, based on the frequency of primate remains in feces as an estimate of proportion in diet, the langurs collectively accounted for 19% of the leopard diet in the national park. Similarly, *S. entellus* was also found to be an important part of leopard and tiger diets in Kanha, India (Dolhinow, 1972). In Sri Lanka, *S. vetulus* was the second most frequently utilized prey item by leopards in the high-altitude montane forests (Ranawana et al., 1998). A study of the dietary ecology of carnivores in Pakke Tiger Reserve in India showed that the capped langur (*T. pileatus*) was part of the leopard diet (Selvan et al., 2013). Leopards also elicit alarm calls from *S. entellus* (Starin, 1978), which suggests that predation is a significant ecological factor influencing the behavior and population ecology of langurs. This assertion is further strengthened by the presence of anti-predator strategies in some langur species, such as *P. frontata*, indicated by their dull coloration, small groups, freezing upon detection, and vocalizations mostly at night (Nijman & Nekaris, 2012).

ODD-NOSED MONKEYS: *RHINOPITHECUS, PYGATHRIX, NASALIS, SIMIAS*

Research on wild populations of the odd-nosed monkeys started in the late 1970s (*Pygathrix, Simias*) and early 1980s (*Rhinopithecus* in China, *Nasalis*) and for some species, the first data were not collected until the late 1990s (*Rhinopithecus* in Vietnam). These studies collected preliminary information and anecdotal observations of ecology. Armed conflicts in the region prevented any biological surveys for a long period of time; therefore, it was not until the last two decades that research on some species of odd-nosed monkeys increased and permitted a better understanding of their ecology and behavior. Certain species still remain little studied, such as *Pygathrix* and *Simias*, so their ecology and social behavior are still unclear. Previous reviews of Asian colobines note that the odd-nosed monkeys are "some of the most unstudied primates in the Old World" (Bennett & Davies, 1994: p.164); as evidence of this, two new species

of odd-nosed monkeys (*P. cinerea* and *R. strykeri*) were discovered since that review. Kirkpatrick (1998: p.157), in a review of snub-nosed monkeys and doucs, states that "at this stage in research on the snub-nosed and douc langurs, we should not view the results of any particular field or captive study as indicative of species norms. Instead, we should use all available reports to generate testable hypotheses." Most recent advances in the study of Asian primates (notably the odd-nosed monkeys) enabled the establishment of ecological theories to explain the behavior observed in these species (Yeager & Kool, 2000; Kirkpatrick, 2007), and both ongoing and future research on wild populations allow comparison of ecological theories to observed behavior to fill the gaps in our knowledge about this unique group of monkeys.

Morphology

The odd-nosed monkeys get their name from unusual nose morphologies that range from a nasal flap in *Pygathrix*, to small upturned noses in *Simias*, to "absent" upturned noses in *Rhinopithecus*, to tremendously long and large noses in *Nasalis* as shown in Figure 18.3. In addition to the colobine-shared characteristics of the gastrointestinal track and tooth morphology related to their specific diet, Chaplin and Jablonski (1998) identified their hair characteristics and bright or patchy coat coloration (with the most striking one found in *Pygathrix*), as a derived, shared characteristic of the odd-nosed monkeys, which can be explained as camouflage, species recognition strategies, or thermal insulation (e.g., *Rhinopithecus*) adaptations. In addition, Su and Jablonski (2009: p.207) observed that odd-nosed monkeys differ from other Old World monkeys in "the shape of their scapula, their relative olecranon length, intermembral index, clavicular and humeral lengths." These morphological characteristics may explain the specific locomotion of the odd-nosed monkeys compared to the other Asian colobines, which includes vertical climbing, arm-swinging, brachiation, and suspensory behaviors.

Distribution and Habitat

The odd-nosed species are distributed from China and northeastern Myanmar, to the Indochinese peninsula of Laos, Vietnam, and Cambodia, to the Indo-Malaysian island of Borneo and the Mentawei Islands of Indonesia (Figure 18.4). They all have restricted ranges; although their historical range may have been much larger, anthropomorphic activities (e.g., habitat loss, hunting) have led to dramatic reduction in the species' ranges, often followed by the isolation of small populations. The latitudinal variation in distribution implies different habitat characteristics

Figure 18.3   Male *Nasalis larvatus*.
(Photo credit: Charles James Sharp; licensed under Creative Commons Attribution-Share Alike 4.0 International license)

between the different species that have direct impacts on ecology and social behavior. *Rhinopithecus* species in China and Myanmar inhabit extremely cold temperate forests with temperatures below zero and snow during winter. Within the genus, altitudinal and latitudinal range correlates with different habitat types. In contrast, *Simias* and *Nasalis* inhabit the opposite habitat type in lowland tropical forests associated with mangroves and large rivers, whereas *Pygathrix* and Vietnamese *Rhinopithecus* are found in tropical mountainous evergreen forests. While the habitat of the Chinese and Burmese *Rhinopithecus* species is marked by four distinct seasons, the habitats of the other species exhibit only a dry and a wet season. These seasonal changes are responsible for phenological variations in vegetation, with availability of particular foods fluctuating throughout the year which, in turn, influences the ecology and social behavior of the odd-nosed monkeys.

Diet and Feeding Behavior

Feeding ecology studies of the odd-nosed monkeys, although they remain few, allow discussion of some

trends. Odd-nosed monkeys are selective feeders and, as such, prefer to feed on certain foods that maximize their energy requirements and minimize risk of toxicity. Their diet is directly related to their environment; the availability of different foods (species and plant parts) varies with the latitude and altitude of the different habitats and seasonal changes in phenology; these environmental characteristics, in turn, determine the seasonal diet changes. All odd-nosed monkey species seem to have a particular preference for young leaves and fruits whenever available. The tropical species (*Rhinopithecus* of Vietnam, *Pygathrix*, *Nasalis*, and *Simias*) inhabit more productive and diverse habitats than the temperate species (*Rhinopithecus* of China) and, therefore, have a more diverse diet; young leaves and fruits are more often abundant, and consequently they have a larger proportion of these items in their diet than temperate species.

Rhinopithecus

First studies on the feeding ecology of *Rhinopithecus* in China found diverging results. Li, Z., et al. (1982)

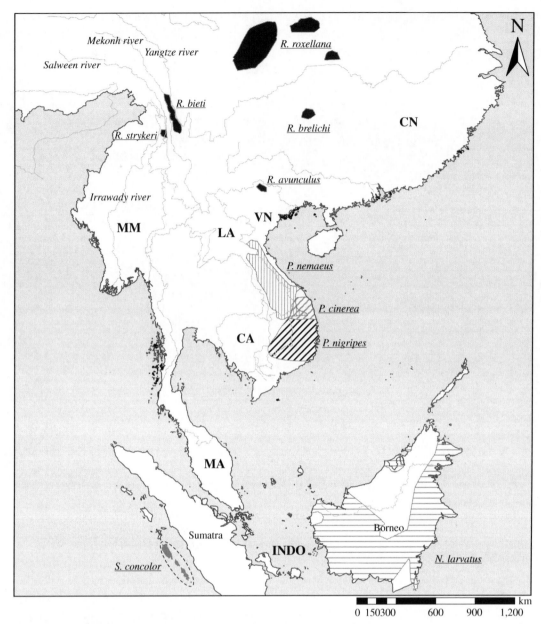

Figure 18.4 Geographic distribution of odd-nosed monkeys. CN: China; MM: Myanmar; LA: Laos; VN: Vietnam; CA: Cambodia; MA: Malysia; INDO: Indonesia.
(Redrawn from Liedigk et al., 2012)

reported that *R. bieti* fed mainly on young leaves, flowers, and buds of conifers and on lichen throughout the year, while Wu and He (1989) from their analysis of food remains in feces, found that the primary food of *R. bieti* was lichen followed by grass and only a small portion of conifer leaves. The latter finding was also confirmed by a long-term study of the same species which depicted the diet of *R. bieti* as primarily composed of lichen (Kirkpatrick, 1996). Bennett and Davies (1994), in their review of Asian colobines, were not able to draw ecological patterns for odd-nosed species given the scarcity of information available at that time. In the last two decades, however, studies on *Rhinopithecus* species—despite

disparity in data collection methods—have helped gain a better understanding of feeding ecology, especially to explain the intra- and interspecies variations of feeding habits in relation to environments. Most recently, the whole-genome sequencing of snub-nosed monkeys provided insights into evolutionary history and adaptation to folivory (Zhou et al., 2014). Specific genes are directly involved in the digestion of toxic secondary compounds found in leaves, the process of fatty acids and lipids to enhance metabolism and energy production, and the digestion of high concntrations of symbiotic foregut bacterial RNA (Zhou et al., 2014).

Within the genus, a wide variety of food habits can be explained by the geographic distribution of each species. From the tropical moist evergreen forest of northern Vietnam, where *R. avunculus* has been recorded at ~100–1,300 masl, to the temperate conifer and mixed deciduous forests of China, where *R. bieti* habitat is at ~2,500–4,600 masl, the diets of the different species seem largely influenced by latitude, elevation, vegetation characteristics, and seasonal variations of their respective habitats.

The harsh environments of the Chinese and Burmese species of *Rhinopithecus* have pushed the animals to adaptations in diet and behavior. *Rhinopithecus* in China are particularly known for feeding on large amounts of lichen, making them one of a few primates relying on nonplant species (Grueter et al., 2009b). *R. bieti* habitat varies with the geographic location of the different groups along the narrow Yuling mountain range delimited by the Yangzte River to the east and the Mekong River to the west (Figure 18.5). Several studies have focused on *R. bieti* feeding ecology; data were collected on its diet either from fecal analysis (Yang & Zhao, 2001; Liu et al., 2004) or from direct observation (Kirkpatrick, 1996; Ding & Zhao, 2004; Xiang et al., 2007; Grueter et al., 2009a), with results that usually did not coincide (e.g., Wu & He, 1989,

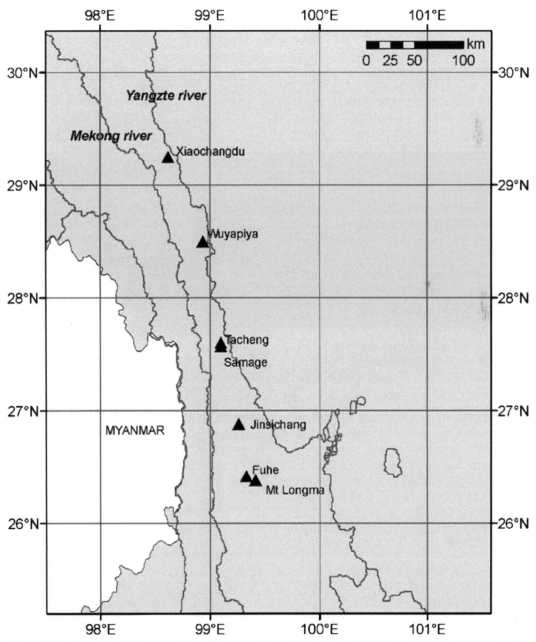

Figure 18.5 Location of sites mentioned in text where *Rhinopithecus bieti* have been studied. Elevation ranges in masl for each site: Xiaochangdu = 3,500–4,250; Wuyupiya = 3,300–4,300; Tacheng = 2,700–3,100; Samage = 2,500–4,000; Jinsichang = 3,200–3,700; Mt. Fuhe = 2,900–3,300; Mt. Longma = 2,260–3,600.

versus Kirkpatrick, 1996). The conflicting results were likely due to an underestimation of ground foods (e.g., bamboo and grass) from direct observation studies or dissimilar digestibility of the different food items (lichen as opposed to plant foods) in fecal analyses (Kirkpatrick, 1996; Yang & Zhao, 2001; Liu et al., 2004). Kirkpatrick et al. (2001) show that the relative proportion of lichen and leaves ingested versus what was defecated were accurately represented through fecal analysis, which is not the case for other food types such as fruits and leaves. In studies of wild monkeys using both methodologies (direct observation and fecal analysis), different results were found for the relative amount of lichen in the diet (Bleisch et al., 1993; Liu et al., 2004). Further studies, therefore, are needed to contrast the two methods.

A gradient has been observed in the diet of different *R. bieti* populations from north to south. The most extreme habitat of the species' distribution range is found in the northern sites of Xiaochangdu and Wuyapiya, characterized by less diversified habitat, lowest temperatures, and highest average elevations. Habitat becomes more clement and more diverse in the southern sites, such as Samage and Mt. Fuhe (Table 18.5). As an adaptation to severe conditions, *R. bieti* rely primarily on year-round available lichens (Kirkpatrick, 1996; Xiang et al., 2007; Grueter et al., 2009a, 2009b). This resource is found abundantly in the coniferous forests and is nutritionally more interesting than mature evergreen broadleaves or conifer needles, both of which contain more fiber and toxins than lichens (Grueter et al., 2009a); conifer needles, in addition, are not expected to be consumed by colobines since they may inhibit bacterial activities in the forestomach of the monkeys (Freeland & Janzen, 1974; Brattsten, 1979; Xiang et al., 2007). Historical reports of *R. bieti* feeding primarily on conifer needles are most likely erroneous (Zhixiang et al., 1982). The percentage of lichen in feeding records decreases from north to south, and other food items (leaves, fruits, and flowers) either increase in proportion or are added to the diet (e.g., bamboo) in the southern sites. For example, while bamboo does not seem to be part of the species' diet repertoire in the northern sites (Kirkpatrick, 1996; Xiang et al., 2007), it is the main component of their diet (59 %) at Jinsichang compared to only 5% lichen at that site (Yang & Zhao, 2001). Bamboo (20%) was also found to be consumed farther south at Mt. Fuhe, after leaves at 50.1%, with lichens constituting only 5.5% of the diet (Liu et al., 2004). Therefore, populations in the north are living in poor habitat with fewer dietary choices and are more reliant on lichen than the southern populations living in richer habitats where a larger proportion of leaves and fruits can be incorporated into their diet. Lichens (*Bryoria* and *Usnea* being the main types consumed) have no fibrous components and are highly digestible, rich in carbohydrates and vitamin D, and have a low content of toxins (Grueter at al., 2009a); these characteristics make lichens an important food in low productivity/diversity environments and can be complemented by other items when available (Grueter et al., 2009a). Reports of feeding on invertebrates and small mammals (e.g., Yang & Zhao, 2001; Xiang et al., 2007; Grueter et al., 2009b; Ren et al., 2010) may be adaptations to supplement a low-protein diet. Research from studies that lasted 10–21 months shows that plant species diversity, extending the diet repertoire further, also increases southward as the latitude and altitude decrease (Ding & Zhao, 2004; Huo, 2005; Xiang et al., 2007; Grueter et al., 2009a, 2009b).

R. roxellana inhabits comparable temperate habitats to *R. bieti*, characterized by mixed evergreen conifer and deciduous broadleaf forests and cold winters; however, *R. roxellana* is found at lower altitudes (Kirkpatrick, 1995; Figure 18.6, and see Table 18.5). Studies that provide details about the diet of *R. roxellana* were conducted at different sites ranging in latitude and altitude. In contrast with *R. bieti*, the northern group of *R. roxellana* was found to have a smaller amount of lichen in its annual diet compared to the group at the southern site (29.0% versus 43.8%, respectively). The animals at Quinmuchuan site (midway between the northern and southern sites) were not observed feeding on lichen at all. These observations can be explained by the contrasting habitat of the sites. Conifers, the main lichen hosts, are less common at the northern site, lichens are less available and replaced by readily available bark and leaf/flower buds (Guo et al., 2007; Hou et al., 2018), and the central site lies at lower elevations where evergreen vegetation provides alternative foods to the monkeys (Li, Y., et al., 2010). In the Qinling Mountains of China during the summer and autumn, *R roxellana* supplement their diet with cicadas (*Karenia caelatata*), which are high in protein and fat (Yang et al., 2016).

R. brelichi and *R. avunculus* differ from the two previously discussed species in that they do not feed on lichen. A preliminary study of *R. brelichi* showed that their diet is composed mainly of young leaves (including petiole and buds), flower buds, fruits, seeds, and bark (Bleisch et al., 1993; Bleisch & Xie, 1998). A recent quantitative study on the species' diet reports that the animals mostly feed on young leaves, followed by mature leaves, fruits/seeds, buds, and flowers (Xiang et al., 2012). *R. brelichi* feed primarily on young leaves in spring, on fruits/seeds in summer and autumn, and on buds, mature leaves, and bark in winter (Xiang et al., 2012).

Table 18.5 Diets of Odd-Nosed Monkey Species

Genus and Species	Study Site [a]	Food Items [b] (%)															No. Plant Species	References
		TL	YL	ML	Li	Bu	Se	T Se+Fr	T Fr	R-Fr	Un-Fr	Flo	Bam	Ba/He	Inv	U/O		
Rhinopithecus																		
R. bieti	Xiaochangdu, Tibet	15.9	—	—	74.8	—	—	—	2.5	—	—	—	—	4.9	2	—	25	Xiang et al. (2007)
R. bieti	Tacheng, Yunnan, China	—	—	—	60	—	—	—	—	—	—	—	—	—	—	—	59	Ding and Zhao (2004)
R. bieti	Jinsichang, Linjiang, China	28	—	—	5	4	—	—	—	—	—	—	59	—	—	0.2	—	Yang and Zhao, (2001) [e]
R. bieti	Samage, Yunnan, China	16.5	12.4	4.1	63.9	3.5	—	—	11.4	—	—	0.2	—	1.2	—	0.5	94	Grueter et al. (2009a)
R. bieti	Mt. Fuhe, Yunnan, China	50.1	—	—	5.5 (63) [c]	—	—	—	38.1 [d]	—	—	—	20	—	—	—	—	Liu et al. (2004) [e]
R. roxellana	Qianjiapin, Hubei, China	32.2	28.7	3.5	43.3	5.4	—	14.6	—	—	—	1.1	—	3.5	—	—	23	Li, Y. (2006)
R. roxellana	Zhouzhi East, Qinling, China	24	—	—	29	—	—	29.4	—	—	—	—	—	11	—	2.3	>84	Guo et al. (2007)
R. brelichi	Fanjingshan, Guizhou, China	47.5	25.5	21.8	—	15.3	—	21.6	—	—	—	9.4	—	—	—	6.3	107	Xiang et al. (2012)
R. brelichi	Fanjingshan, Guizhou, China	33.3	—	—	—	29.1	—	—	24.3	—	—	6.9	—	—	—	6.4	—	Nie et al. (2009)
R. avunculus	Tat Ke/Nam Trang-Ban Bung, Vietnam	38	38	—	—	—	15	—	—	—	47	—	—	—	—	—		Boonratana and Le (1998)
R. avunculus	Tat Ke/Khau Ca, Vietnam	64.5	52.5	11.6	—	5.8	—	—	—	3.4	12.6	13.6	—	—	—	—	—	Thanh (2007)
R. avunculus	Khau Ca, Vietnam	33.3	33.3	—	—	—	5.6	—	—	22.2	25	8.3	—	—	—	2.8	31	Quyet et al. (2007)
Pygathrix																		
P. nemaeus	Many forest areas, Vietnam	82	[75	7]	—	—	—	14	—	—	—	4	—	—	—	—	—	Lippold (1998)
P. nemaeus	Bach Ma NP, Vietnam	63	—	—	—	—	—	—	37	—	—	—	—	—	—	—	50	Pham (1993)
P. nemaeus	Hin Namno NPA, Lao PDR	54.7	—	—	—	—	—	—	33.8	—	—	3.9	—	—	—	7.6	112	Phiapalath et al. (2011)
P. nemaeus	Son Tra NR, Vietnam	87.8	66.6	31.4	—	—	—	10.2	1.6	—	—	—	—	0.4	—	—	79	Ulibarri (2013)
P. nigripes	Seima, Mondulkiri, Cambodia	29.9	24	5.9	—	—	39.7	—	11.4	—	—	8.8	—	—	—	10.2	190	Rawson (2009)
P. nigripes	Nui Chua/Phuoc Binh NPs, Vietnam	54.6	—	—	—	—	—	—	29.3	—	—	14.6	—	—	—	1.5	152	Hoang et al. (2009)
P. cinerea	Kon Ka Kinh NP, Vietnam	58.9	49.6	9.3	—	—	—	—	—	21.9	19.1	—	—	—	—	0.1	166	Ha (2009)

Location															Reference[b]
Nasalis															
N. larvatus															
Natai Lengkuas, Indonesia	51.9	41.2	2.7	—	—	—	40	—	—	—	3	—	<1	4.7	47	Yeager (1989)
Samunsam, Indonesia	41	38	3	—	—	—	—	—	50	—	3	—	—	—	—	Bennett and Sebastian (1988)
Sukau, Sabah, Malaysia	73	72.7	0.3	—	2.4	0.5	—	1.7	—	6.6	7.8	—	8	—	—	Boonratana (1993)
Abai, Sabah, Malaysia	49.7	49.7	—	—	11.6	—	—	—	—	20.6	15.5	—	2.6	—	—	Boonratana (1993)
Menangul, Sabah, Malaysia	65.9	65.9	—	—	—	—	25.9	—	—	—	7.7	—	—	—	127	Matsuda et al. (2009)
Simias															
S. concolor															
Peleonan, Siberut, Indonesia	15.8	—	—	—	—	—	6.3	—	—	—	4.9	—	0.6	99	Hadi et al. (2012)[f]

[a] Length of studies range from 6-36 months.
[b] Li=lichen, TL=total leaves, YL=young leaves, ML=mature leaves, Bu=buds, Se=seeds, T Se+Fr=total seeds+fruits, R-Fr=ripe fruits, Un-Fr=unripe fruits, T Fr=total fruits, Flo=flowers, Bam=bamboo, Ba/He=bark and herbs, Inv=invertebrates, U/O=unknown and other.
[c] Entry from direct observation feeding record.
[d] Summer and autumn only.
[e] Percentages derived from fecal analysis.
[f] Percentages given as events/hr of observation.
– Absent from diet or not indicated in data.

Although research on *R. avunculus* is preliminary and ongoing, the findings so far indicate that the species falls into the known ecological patterns of most colobines with a diet mainly composed of young leaves, mature leaves, and fruits/seeds. This is expected given its tropical habitat where most other colobine species are found (Kirkpatrick, 2007). Nonetheless, different studies on *R. avunculus* give divergent amounts of food items in the diet, either dominated by unripe fruits (Boonratana & Le, 1998) or young leaves (Quyet et al., 2007; Thanh, 2007), although differences may be due to different methodology and study length. The preliminary research on *R. avunculus* has not yet presented dietary selection across the tropical seasons (wet and dry), but it can be expected to follow the colobine feeding behavior pattern and be comparable to other tropical odd-nosed monkeys.

The diet of *R. strykeri* has yet to be studied (initial discovery of the species was in 2010), and there is no complete information on its feeding habits due to the difficulty of following the monkeys across steep mountainous terrain (Yang et al., 2019). Given its distribution, it can be expected that the diet repertoire of *R. strykeri* is similar to *R. bieti* and/or *R. roxellana*. Preliminary assessments identified 14 plant species and 4 lichen species consumed by the monkeys (Yang et al., 2019).

The diet of all *Rhinopithecus* species, like other colobines, is characterized by clear shifts throughout the year following seasonal variations and availability of foods and demonstrates adaptations to habitat and food selectivity. Overall, *Rhinopithecus* species across their range select food items from a variety of species (plant and lichen) that maximize their energy requirements, preferably choosing foods with low fiber and tannin content and high protein content (Kay & Davies, 1994; Huang et al., 2010) whenever these are available in the environment. Negative correlations are apparent between feeding times on less rich foods (e.g., lichen, mature leaves, bark) and feeding time on preferred foods (e.g., young leaves, bamboo leaves and shoots, fruits/seeds), and positive correlations are seen between feeding time on preferred foods and availability of the latter in the habitat (e.g. Li, Y., 2006; Xiang et al., 2007; Grueter et al., 2009a). Such observations demonstrate the plasticity of these species and place them in the category of generalist feeders, which allows them to cope with periods (seasonal or sporadic) of food restrictions. For example, *R. roxellana* was observed to shift its diet to lichen in autumn instead of fruits after a heavy snow fall which had limited fruit availability (Li, B., et al., 2003).

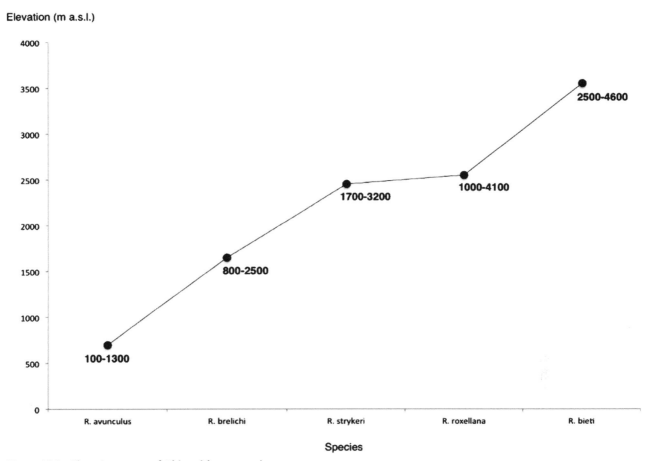

Elevation (m a.s.l.)

100-1300

800-2500

1700-3200

1000-4100

2500-4600

R. avunculus   R. brelichi   R. strykeri   R. roxellana   R. bieti

Species

Figure 18.6   Elevation range of *Rhinopithecus species*.

Pygathrix

Information on the diet of douc langurs remains scarce at present, but at least one intensive ecological study exists for each of the three species—*P. nemaeus*: (Phiapalath, 2009) in Laos and (Lippold, 1998; Ulibarri, 2013) in Vietnam; *P. cinerea*: (Ha, 2009) in Vietnam; and *P. nigripes*: (Hoang et al., 2009; O'Brien, 2014) in Vietnam and (Rawson, 2009) in Cambodia. Different data collection and analysis methodologies were used in these studies making comparisons difficult. The earliest research on the diet of *P. nemaeus* reported that they mainly feed on leaves followed by fruits/seeds (Lippold, 1998; Pham, 1993, 1994; cf. Table 18.5). *P. nemaeus* in Bach Ma National Park, Vietnam, were observed eating leaves, fruits, buds, flowers, and bamboo shoots; the proportion of these items in their diet was not recorded, but the species diversity for fruits and leaves was high (they feed on fruits from 26 tree species and on leaves from 24 tree species) (Le, 2009). In the single study of *P. nemaeus* in Laos, in a limestone forest of Hin Namno National Protected Area, it was found these monkeys feed mainly on leaves, fruits, and flowers from 112 plant species in 25 taxonomic families (Phiapalath, 2009; Phiapalath et al., 2011). *P. cinerea* has so far only been studied in Khon Kha Kin National

Park, Vietnam, and was found to eat a majority of young leaves, followed by fruits and mature leaves but never was observed eating flowers (cf. Table 18.5; Ha, 2009).

P. nigripes in Nui Chua and Phuoc Binh National Parks in Vietnam feed primarily on leaves also (cf. Table 18.5; Hoang et al., 2009); in contrast, the diet of a group in Cambodia was dominated by a large amount of seeds (Rawson, 2009). Thus, the species is highly frugivorous (~50% of time is spent on consuming fruits, when seeds and fruits are combined). *P. nigripes* in Cat Tien National Park, Vietnam, select foods with the highest mean protein content (O'Brien, 2014), which is consistent with what is observed in other colobine monkeys (Kay & Davies, 1994). There is a comparable and distinct seasonal variation in *P. nigripes*' diet with an increasing amount of fruits and a reduced amount of leaves in the diet during the wet season (May–October), while leaves are eaten more during the dry season (November–April) (Ha, 2009; Hoang et al., 2009; Phiapalath et al., 2011; Ulibarri, 2013). Seasonal variation in the diet of a population of *P. nigripes* in Cambodia was not, however, significant. The monkeys mainly feed on fruits/seeds year-round (Rawson, 2009). This difference, of course, may be due to the variation in food availability in different environments.

There is overlap between the tree families eaten by *Pygathrix* spp. across their range: *Moraceae*, *Fagaceae*, *Myrtaceae*, *Sapindaceae*, *Euphorbiaceae*, and *Meliaceae*. *Ficus* spp. (*Moraceae*) is often observed as particularly important in the diet of the species. Results of initial studies demonstrate that *Pygathrix* are selective feeders, preferably choosing to feed on fruits (and young leaves) whenever available. This confirms the conclusion of recent reviews of Asian colobine feeding ecology (Yeager & Kool, 2000; Kirkpatrick, 2007).

Water consumption is rarely observed in the field, but a study in captivity showed that *Pygathrix* obtain water from food (leaves) along with direct water consumption (Kullik, 2010). *P. nigripes* was observed drinking from water holes in the wild in Vietnam (Nadler, 2008) and likely descends occasionally to the ground to drink from water sources. *P. nigripes* was recently recorded coming regularly to the ground to consume soil from salt licks in Cambodia (Rawson & Bach, 2011).

Nasalis

Early studies of *Nasalis* ecology described the species as mainly folivorous (Kern, 1964; Kawabe & Mano, 1972; Macdonald, 1982), but this was contradicted by subsequent long-term studies which classified the species as folivorous/frugivorous and highly selective (Bennett & Sebastian, 1988; Yeager, 1989; Boonratana, 1993, 2003; Matsuda et al., 2009). In Indonesia, the species feed on fruits whenever available, although they continue to feed on young leaves year-round, and during times of preferred foods shortage, the monkeys increase the diversity of their diet by including more food items that are widely available (cf. Table 18.5; Yeager, 1989). In Malaysia, two groups of *N. larvatus* from two different sites were studied in the lower Kinabatangan (the Sukau riverine site and the Abai mangrove site). Comparison of the two groups showed a difference in diet, although both were dominated by young leaves (cf. Table 18.5; Boonratana, 1993). The group inhabiting a more productive habitat (Abai) was found to eat larger proportions of fruits and seeds than the Sukau group (Boonratana, 1993). Matsuda's et al. (2009) study confirms previous findings; groups studied along the river in Sabah, Malaysia, feed primarily on young leaves. Anecdotal observations of feeding on invertebrates and insect larvae also have been reported (Salter et al., 1985; Yeager, 1989). There is an overall negative correlation between amount of young leaves in the diet and amount of fruits/seeds, indicating the preference for these latter food items whenever available.

Simias

Among the odd-nosed monkeys, *Simias* is the least-studied genus. Research providing quantitative data on the species' diet focused on the niche overlap with *Presbytis potenziani* and found that *S. concolor* feed mainly on leaves (15.8 events/hour), followed by fruits (6.3 events/hour), and flowers (4.9 events/hour) (Hadi et al., 2012).

Ranging Behavior and Activity Patterns
Rhinopithecus

Chinese *Rhinopithecus* species show consistent seasonal variation in their activity budget. During the winter months (November–February), the monkeys spend more time feeding than during the other seasons. In general, in summer (June–August), time spent moving and resting is higher than during other seasons. Seasonal shifts are closely connected to environmental variables (e.g., day length and temperature) and food availability. Xiang et al. (2010) suggest that time spent foraging (feeding + moving) across seasons will determine time allocated to all other activities. Xiang et al. (2010) found that resting time was positively correlated with seasonal variation in day length, ambient temperature, and food availability. Thus, intra- and interspecific variation in habitat characteristics of *Rhinopithecus* species across their range (especially in China) may induce variations in adaptive strategies, leading to the differences often observed between sites.

Another interesting aspect of seasonal adaptation by *Rhinopithecus* in China is their spatio-temporal vertical ranging. Movement of the monkeys at each site is largely influenced by their distinctive habitat, primarily by food availability rather than temperature or human disturbance (Kirkpatrick & Long, 1994; Tan et al., 2007; Li, D. et al., 2008). Quan et al. (2011) suggest that seasonal solar radiation may drive *R. bieti* to range at high altitudes during winter. Kirkpatrick et al. (1998) and Li, D., et al. (2008) argue that the use of mid-high altitudes throughout the year by *R. bieti* is due to higher concentration of lichen at these altitudes. That the same species used low altitudinal range at a different site (Jinsichang) was explained by the availability of bamboo leaves at lower elevations (Yang, 2003).

Rhinopithecus have been observed to descend occasionally to lower elevations during heavy snowfalls (Kirkpatrick & Long, 1994; Li, D., et al., 2008). At one site (Samage), *R. bieti* ranged at the highest altitudes in summer to escape from warm temperatures and to feed on the seasonal bamboo shoots available; they descended to lower altitudes in spring during the lush period of leaves and fruits (Li, D., et al., 2008). *R. roxellana* in Zhouzhi range at lower elevations in autumn where their main diet (fruits/seeds from *Quercus aliena*) is mostly available (Tan et al., 2007). In contrast, *R. brelichi* in Yangaoping do not show significant vertical movement across seasons, though they do tend occasionally to descend to lower altitudes during periods of snow or freezing rainfall (Niu et al., 2010). Normally, however, they remain at elevations where their annual staple foods (i.e., *Prunus vaniotii*, *Dendrobenthamia angustata*, *Magnolia sprengeri*, *Acer flabellatum*, and *Sorbus xanthoneura*) are concentrated within a zone of mixed evergreen and deciduous broadleaf forest. The monkeys at this site show a clear daily altitudinal movement pattern from lower to higher and back to lower elevations every day. This may be due to a combination of food availability at higher elevations during the day and predation avoidance at lower elevations at night, although this hypothesis has yet to be confirmed (Niu et al., 2010).

In a comparative review, Kirkpatrick et al. (1998) showed that odd-nosed monkeys (*Rhinopithecus*, *Nasalis*, and *Simias*) have longer daily path lengths (DPL) and larger home ranges than other Asian colobines (such as *Presbytis*) or African colobines. In particular, *Rhinopithecus* species have the longest reported DPLs and largest home ranges among all colobines—home ranges as large as ~30 km² (Table 18.6). It has been consistently confirmed, across studies, that home range size and DPL are strongly influenced by diet and food resource availability/distribution, habitat heterogenetity being the primary factor (Kirkpatrick et al., 1998; Li, B., et al., 2000; Liu et al., 2004; Li, D., et al., 2008). Annual diet is directly influenced by seasonal phenology of plant foods, which, in turn, determines the specific areas used within the total home range of the monkeys. Therefore, the seasonal variation in home range and DPL varies across species and sites according to plant species and phenology. For example, in the Qinling Mountains, home range size of *R. roxellana* does not significantly change from summer (8 km²) to winter (7 km²), but the areas used within the total home range differ (showing an overlap of 0.13 km²) and follow the distribution of seasonal foods (Li, D., et al., 2008). In contrast, *R. bieti* at Mt. Fuhe has a larger home range and DPL in summer–autumn than in winter–spring since its preferred food (fruit of *Acanthopanax evodiaefolius*) becomes available in summer but is patchily distributed, forcing the monkeys to travel longer distances over a larger area (Liu et al., 2004). *R. roxellana*, similarly, has their largest home range in spring, compared to other seasons when their preferred foods (young leaves and buds) become available but are patchily distributed (Tan et al., 2007). In Samage Forest, Yunnan, the home range of *R. bieti* is similar from winter through summer but is

much smaller in autumn during the time preferred foods (fruits from *Acanthopanax evodiaefolius*, *Sorbus thibetica*, *Sorbus* spp., and *Cornus macrophylla*) are concentrated over a small area. Regular visits to the different clumped areas increase their autumn DPL (Li, D., et al., 2008). Ren et al. (2009a, 2009b) found that, although DPL of *R. bieti* does not vary significantly across seasons, it is positively correlated with ambient temperature and day length.

Pygathrix

Available data on *Pygathrix* activity budgets come from different countries and habitat types with various degrees of anthropogenic pressure and studies have been conducted only on unhabituated groups; any comparison or patterns within the genus must, therefore, be seen as preliminary. *Pygathrix* species spend a large amount of time resting, which is consistent with their diet and physiology. The large amount of leaves ingested needs time to be processed through bacterial fermentation in their forestomachs (Kay & Davies, 1994). On average, *Pygathrix* species were observed spending more time feeding and traveling during the dry season and more time resting during the wet season. This may be explained by the fact that high-quality foods (fruits and seeds) become available during the wet season. The monkeys receive a net energy gain more quickly from these items and, therefore, can decrease their amount of feeding time while they increase their time resting. In contrast, they need to increase their food intake during the dry season to compensate for the lower-quality foods available (leaves) and, accordingly, increase their feeding and traveling time. There is, however, no significant difference in any activities between wet and dry seasons for *P. nemaeus* in Son Tra National Reserve, Vietnam (Ulibarri, 2013). Human disturbance has been mentioned by researchers as a potential factor that may influence activitiy patterns of the monkeys between seasons; notably travel time can increase when monkeys are aware of observers (Hoang, 2007; Ha, 2009; Ha et al., 2010). Future studies on activities of *Pygathrix* in relation to diet and food distribution/availability are needed.

Data on home range and DPL for the genus *Pygathrix* are inconsistent due to the difficulty of following unhabituated monkeys in arduous terrain, which results in small samples and/or incomplete data sets (Hoang, 2007; Rawson, 2009). Daily path length varies on average between 500–1,070 m for the genus, and home range was found to vary from 20–940 ha (see Table 18.6). The lowest home range estimates come from either incomplete data (Rawson, 2009) or from an area surrounded by agricultural land, which may affect ranging of the animals (Hoang, 2007); the largest estimate is for several groups

Table 18.6 Odd-nosed Monkey Group Size, Daily Path Length (DPL), Home Range (HR), and Elevation Range (ER)

Genus and Species	Study Site	Group Size	DPL (m)	HR (km²)	ER (m)	References
Rhinopithecus:						
R. bieti	Wuyapiya, Yunnan, China	~150	-	-	(3750-4500)	Kirkpatrick and Long (1994)
R. bieti	Wuyapiya, Yunnan, China	≥175	1310 (300-2400)	25.25 [a]	-	Kirkpatrick et al. (1998)
R. bieti	Samage, Yunnan, China	-	-	-	3200 (2600-4000)	Li, D., et al. (2008)
R. bieti	Samage, Yunnan, China	>400	162 [b]	24.75	-	Li, D., et al. (2008)
R. bieti	Nanren Baima, Yunnan, China	~300	-	-	3907 (3700-4240)	Cui et al. (2006)
R. bieti	Jinsichang, Lijiang, China	-	-	-	3630-3500	Yang (2003)
R. bieti	Laojunshan, Yunnan, China	-	909 (180-3626)	18-33	-	Ren et al. (2009a, 2009b)
R. bieti	Mt Fuhe, Yunnan, China	~80	800	11	3105	Liu et al. (2004)
R. roxellana	Shennongjia, China	100+	71	40	1800-2700	Su et al. (1998)
R. roxellana	Zhouzhi West, Shaanxi, China	~90	-	22.5	-	Li, B., et al. (2000)
R. roxellana	Zhouzhi East, Shaanxi, China	112	2100 (750-5000)	18.3	2137 (1500-2600)	Tan et al. (2007)
R. roxellana	Qingmuchuan, Shaanxi, China	100-120	-	20.35	-	Li, Y., et al. (2010)
R. brelichi	Yangaoping, Guizhou, China	-	935 (523-1672)	-	1660 (1350-1870)	Niu et al. (2010)
Pygathrix:						
P. nemaeus	Him Namno, Lao PDR	~30	-	2.9	-	Phiapalath (2009)
P. nemaeus	Son Tra, Vietnam	21(three units)	509 (137-987)	0.36	-	Ulibarri (2013)
P. nigripes	Nui Chua/Phuoc Binh, Vietnam	10	976	0.4-0.5	200-800	Hoang (2007)
P. nigripes	Seima, Mondulkiri, Cambodia	-	514	0.2	-	Rawson (2009)
P. cinerea	Kon Ka Kinh, Vietnam	88(band)	1068 (50-4080) [c]	9.4(several groups)	-	Ha (2009)
Nasalis:						
N. larvatus	Sukau, Sabah, Malaysia	~20 (OMU)	910 (370-1810)	2.2	-	Boonratana (2000a)
N. larvatus	Abai, Sabah, Malaysia	~15 (OMU)	-	3.2 [d]	-	Boonratana (2000a)
N. larvatus	Samunsam, Sarawak, Malaysia	16 (OMU)	483 [e] (300-590)	9	-	Bennett and Sebastian (1988)
N. larvatus	Samunsam, Sarawak, Malaysia	-	1312 (?-1550)	-	-	Salter et al. (1985)
N. larvatus	Sekonyer, Kalimantan, Indonesia	~50(4 OMUs)	-	1.3 [f]	-	Yeager (1989)
Simias:						
S. concolor	Siberut, Indonesia	-	-	0.06	-	Hadi et al. (2012)
S. concolor	Peleonan, Siberut, Indonesia	-	-	0.05 (0.04-0.1)	-	Hadi et al. (2012)
S. concolor	Pagai Islands, Indonesia	-	-	0.07-0.2	-	Teneza and Fuentes (1995)
S. concolor	Sirimuri, Siberut, Indonesia	-	-	~0.25-0.3	-	Tilson (1977)

[a] HR not corrected for slope.
[b] DPL only calculated over September–November.
[c] DPL calculated regardless of group size followed; based on 26 full-day follows.
[d] HR estimated from river follows and inland movement from river.
[e] DPL likely underestimate—includes only days with complete follows; some incomplete follows indicated DPLs of at least 2000 m.
[f] Values are means for 4 groups studied (group sizes=6, 7, 15, 14).

traveling together (Ha, 2009), which makes comparisons difficult. Group size does not affect home range in *P. nemaeus* and *P. nigripes* (Hoang, 2007; Phiapalath, 2009). While home range is larger during the dry season in two different studies for *P. nemaeus* and *P. nigripes* (Hoang, 2007; Phiapalath, 2009), the opposite was observed for a group of *P. nemaeus* in Vietnam (Ulibarri, 2013). Home range overlap (1–3 ha) between two adjacent groups was observed in *P. nemaeus*, and DPL was reported as longer during the dry season (Ulibarri, 2013).

Nasalis

Along the Menanggul River (Sabah, Malaysia), Matsuda et al. (2009) found that proboscis monkeys devote a striking amount of time to resting (77%), much higher than the other odd-monkeys. In Bako National Park (Sarawak, Malaysia), during a four-month study, the species spent 26% of the time resting, and in Samunsam Wildlife Sanctuary (also in Sarawak), they were found to spend the majority of their time (58%) resting (Salter et al., 1985). While the sample of studies is small and their length and methodologies differ, *N. larvatus* seems to spend a large portion of their time being inactive. However, the behavioral category of "resting" probably includes time when the monkeys are inactive but vigilant (e.g., sitting while scanning surroundings). Boonratana (2000b) found that *N. larvatus* in Sabah spend 30% of their time in vigilance, which the author proposes may function as predator, food, and/or male competitor detection, besides being a rest, during which the animals can digest their food.

Similar to other odd-nosed monkeys, home range of *N. larvatus* varies across sites, mainly due to habitat productivity and seasonal change in food availability (Boonratana, 2000a). In Samunsam Wildlife Sanctuary, monkeys have to travel farther to feed on their preferred food, resulting in larger home ranges than in the Sukau region (Sabah, Malaysia) (Bennett & Sebastian, 1988; Boonratana, 2000a). In Sukau, the intensity with which the monkeys use some areas of their home range is positively correlated with the presence of trees producing fruits and flowers (Boonratana, 2000a). Monthly shifts in home range have been observed at some sites (Bennett & Sebastian, 1988) but not at others (Boonratana, 2000a) and is related to habitat productivity. Home ranges between groups completely overlap (Bennett & Sebastian, 1988; Boonratana, 2000a), which is consistent with their multilevel social organization. There seems to be a clear daily range pattern that is consistently observed across sites. This is strongly affected by the habit of returning to sleeping sites along main watercourses (if present) at the end of each day (Salter et al., 1985; Bennett & Sebastian, 1988; Yeager, 1989; Boonratana, 1993, 2000a; Onuma, 2002),

which, in turn, can limit their maximum home range size (Bennett & Sebastian, 1988). The species typically moves from a sleeping site along a river in the morning and travels inland (~400–700 m from river), mostly within mangrove swamp or riverine forest, and returns to the river in late afternoon (Salter et al., 1985; Bennett & Sebastian, 1988; Boonratana, 1993, 2000a; Onuma, 2002). Feilen and Marshall (2017) found that *Nasalis* in Indonesian Borneo sleep farther inland from rivers when biting sandflies are abundant and/or less food is available along the river. In Bako National Park (Sarawak, Malaysia), where no major watercourse occurs, distinctly different daily ranges dependent on habitat types were observed, probably influenced by regular variations in tidal levels (Salter et al., 1985; Onuma, 2002). During low tides, monkeys sleep in mixed riverine and dipterocarp/high *kerangas* (heath) forests and progressively move to mangrove forest through the morning and early afternoon, returning to the mixed forest in mid-afternoon (Salter et al., 1985; Onuma, 2002). During periods of high tide, the monkeys stay in the higher elevations (80 masl) of the mixed riverine and dipterocarp/high *kerangas* forests (Onuma, 2002). Since small and large rivers are a major characteristic of their habitat, *Nasalis* often swim across rivers during their daily travel, sometimes more than once a day and at points over 100 m wide (Salter et al., 1985; Bennett & Sebastian, 1988; Boonratana, 2000a).

Simias

Home ranges of *S. concolor* are small compared to other odd-nosed monkeys, from 2–10 ha (see Table 18.6) and barely overlap with other groups (Tenaza & Fuentes, 1995; Hadi et al., 2012). *Simias* behavior is similar to other odd-nosed monkeys with large amounts of time spent resting, followed by traveling/moving and feeding (Paciulli & Holmes, 2008; Hadi et al., 2012).

Social Organization and Reproduction

At present, data on odd-nosed monkeys remain scarce and comparisons within and between species are preliminary; nonetheless, recent advances on some ecological aspects of the species have led to a better understanding of their behavior. The multilevel society observed in the clade has been the subject of numerous debates seeking to explain the evolutionary factors behind this form of social organization. In nonhuman primates, multilevel societies are found in only a few species (e.g., Hamadryas baboons, *Papio hamadryas*, and geladas, *Theropithecus gelada*); nonetheless, *Rhinopithecus*, *Pygathrix*, and *Nasalis* all exhibit multilevel societies, but it remains to be confirmed whether *Simias* shares a multilevel social organization. Evolutionarily speaking, it is assumed that multilevel societies have evolved independently—either two (*Papio-Theropithecus*/odd-nosed) or three (*Papio*/*Theropithecus*/odd-nosed) times due to the phylogenetic distances of these different groups (Grueter & Zinner, 2004). Authors have tentatively analyzed both the proximate and ultimate factors that may have led to the appearance of multilevel societies observed in odd-nosed monkeys. To date, the determinants that best explain the evolution of multilevel societies in odd-nosed monkeys may be a combination of factors discussed below.

Sexual conflict between males ("conspecific threat," Kirkpatrick & Grueter, 2010; "bachelor threat hypothesis," Grueter & van Schaik, 2009b) involves the presence of nonreproductive males forming all-male units (AMU) at the periphery of one-male units (OMU) in a "band." Coalition or parallel chasing behavior of the resident males in an OMU band can decrease risks (in the form of infanticide, takeover, and sexual coercion) from bachelor males. The evolution of sexual dimorphism in body weight (Grueter & van Schaik, 2009a) and canine size (Plavcan & van Schaik, 1992; Jablonski & Pan, 1995) observed in odd-nosed monkeys is consistent with other primate species exhibiting multilevel societies.

The multilevel form of social organization in the odd-nosed monkeys may also have evolved due to weak food competition based on an abundance of staple foods in the habitat, such as lichen and foliage (Grueter & van Schaik, 2009a). Other ecological factors have been examined and cannot be ruled out due to the difficulty of testing in wild conditions and the overall lack of data. For example, predation risk may explain formation of bands (at least for some species) (Yeager, 1992; Grueter & van Schaik, 2009b; Kirkpatrick & Grueter, 2010). Further, since the odd-nosed monkeys are considered to be phylogenetically related and form one clade (Sterner et al., 2006; Whittaker at al., 2006), phylogenetic inertia of modular society in this clade also may have played a role in multilevel societies (Kirkpatrick & Grueter, 2010).

Rhinopithecus

Despite early reports on *Rhinopithecus* species that gave inconsistent observations of their social organization (reviewed in Kirkpatrick, 1998), it is now widely accepted that they are organized in multilevel societies (Grueter & van Schaik, 2009a; Kirkpatrick & Grueter, 2010). Even the recently discovered *R. strykeri* is thought to live in a multilevel society (Meyer et al., 2017). The basic social organization of all *Rhinopithecus* species is the OMU, which consists of one male, several females, and their offspring (infants, juveniles, and subadults). A number of these units gather to travel and forage together and form

a band (Kirkpatrick & Grueter, 2010). A few (one or two) all-male units (AMU) are also commonly reported to be part of the band, traveling at its periphery, e.g., *R. bieti* (Kirkpatrick et al., 1998); *R. roxellana* (Zhang et al., 2006; Tan et al., 2007; Qi et al., 2008, 2009); *R. brelichi* (Bleisch et al., 1993; Nie et al., 2009); and *R. avunculus* (Boonratana & Le, 1998; Thanh, 2007). Reports of multi-male/multi-female units and male-female pairs are rare (cf. Ren et al., 1998; Kirkpatrick et al., 1999; Liu et al., 2007; Cui et al., 2008; Table 18.7) and mainly considered to be transitory units derived from OMUs or in the process of forming new OMUs (Liu et al., 2007; Cui et al., 2008). Solitary individuals (usually males) are also observed but typically are in the process of joining a unit by replacing the resident male of an OMU or forming a new one by attracting females (cf. Zhang et al., 2006; Cui et al., 2008; Zhao et al., 2008b; Qi et al., 2008, 2009).

Although the multilevel society is common to all *Rhinopithecus* species, band size, unit size, adult male-to-female ratio, and cohesiveness varies within and between species (see Table 18.7). Band size can vary from ≥400 in *R. brelichi* (Bleisch et al., 1993; Xiang et al., 2009) and *R. bieti* (Li, D. et al., 2008) to ~20 in *R. avunculus* (Thanh, 2007) but is often believed to be underestimated due to poor visibility in the field. Studies at one site may result in different counts due to the variation in population dynamics over time; often some units appearing or disappearing in/from the band over the course of the research (e.g., Zhao et al., 2008a). Within the genus, the number of OMUs forming a band varies from 3–30; the average size of an OMU is 9.6 individuals (calculated from Table 18.7), with variation in the number of females, subadults, juveniles, and infants. Group size of units and bands has been suggested to vary seasonally. Overall, the Chinese species of *Rhinopithecus* form much larger bands than the Vietnamese species (*R. avunculus*), which may reflect differences in habitat productivity and food availability (Kirkpatrick & Grueter, 2010). The species inhabiting temperate/subtropical habitats may have larger food patches, therefore, allowing formation of larger groups compared to the tropical habitat of *R. avunculus*. Cohesiveness of the bands, as observed through fission-fusion events, varies among *Rhinopithecus* species. *R. avunculus* bands are the least cohesive, with units regularly observed splitting or joining at sleeping sites, feeding trees, and for travel (Boonratana & Le, 1998; Thanh, 2007). Fission-fusion in *R. roxellana* and *R. brelichi* is also common (cf. Table 18.7), although bands may stay cohesive or split for several months at a time. *R. bieti* form the most cohesive bands, with units that rarely split, and if they do, the split is for less than a day (Kirkpatrick et al., 1998). These differences in band cohesiveness among species are

Table 18.7 Social Organization Parameters of Odd-Nosed Monkeys

Genus and Species	Study Site[a]	Social Units Observed					Fission-Fusion?[b]	No. of Units	Group/OMU Size (Mean or Range)	Band Size[c]	AM:AF (Band or OMU)	References
		MMU	OMU	AMU	MF pair	Solitary						
Rhinopithecus												
R. bieti	Nanren/Bamei, Yunnan, China	X	X	X		X		15-31	7.5	70 ->175	1:4.3	Cui et al. (2008)
R. bieti	Wuyupiya, Yunnan, China		X	X			yes	15-18		175-200	1:3.1	Kirkpatrick et al. (1998)
R. bieti	20 sites, Yunnan, China									50-200		Long et al. (1994)
R. bieti	Tacheng, Yunnan, China	X	X					26	11.3	366	1:4.7	Liu et al. (2007)
R. roxellana	Zhouzhi West, Qinling, China		X					7	18.6	<300	1:7	Guo et al. (2010)
R. roxellana	Zhouzhi West, Qinling, China		X					6	6.5		1:1.5	Li, B, and Zhao (2007)
R. roxellana	Shennongjia, Hubei, China		X							270	1:0.79	Li, Y., et al. (2009)
R. roxellana	Zhouzhi West, Qinling, China		X	X		X		6-12	9.42	132	1:4.25	Qi et al. (2008, 2009)
R. roxellana	Zhouzhi East, Qinling, China		X	X						≥112	1:3.7	Tan et al. (2007)
R. roxellana	Shennongjia, Hubei, China		X	X				3-4	10.1	34-53		Yao et al. (2011)
R. roxellana	Zhouzhi West, Qinling, China		X	X		X	yes	6-8	9	45-82	1:3.3	Zhang et al. (2006)
R. roxellana	Zhouzhi West, Qinling, China		X	X		X	yes	11	8.73	61-116	1:0.29	Zhao et al. (2008a)
R. roxellana	Baihe, China	X	X	X			yes			>200	1:2.5	Kirkpatrick et al. (1999)
R. roxellana	Shennongjia, Hubei, China	X	X	X				11-21	12	95-340	1:4.3	Ren et al. (1998)
R. brelichi	Fanjingshan, Guizhou, China		X	X					6.13	>400	1:0.93-1.27	Bleisch et al. (1993)
R. brelichi	Fanjingshan, Guizhou, China		X	X			yes					Nie et al. (2009)
R. avunculus	Tat Ke/Nam Trang-Ban Bung, Vietnam		X	X			yes	≥3-5	15.2	50-400	1:2.5	Boonratana and Le (1998)
R. avunculus	Khau Ca, Vietnam		X				yes		4.50	29		Covert et al. (2008)
R. avunculus	Khau Ca, Vietnam		X	X			yes	5	12.1	17		Thanh (2007)
R. avunculus	Tat Ke, Vietnam		X	X			yes	2	5.6	22-81		Thanh (2007)
R. strykeri	Myanmar (Burma)	------	------	No information available	------	------						Geissmann et al. (2011)
Pygathrix												
P. nemaeus	Son Tra, Vietnam	X							6-17		1:2	Lippold (1998)
P. nemaeus	Pu Mat, Vietnam	X							25-35		1:2.6	Lippold (1998)
P. nemaeus	Bach Ma, Vietnam	X							5-20		1:2	Lippold (1998)
P. nemaeus	Kon Cha Rang, Vietnam	X							51		1:2.3	Lippold (1998)
P. nemaeus	Son Tra, Vietnam								6-24		1:2	Lippold and Thanh (2008)
P. nemaeus	Son Tra, Vietnam								6-40			Dinh et al. (2010)
P. nemaeus	Phong Na Ke Bang, Vietnam								6.25			Haus et al. (2009)
P. nemaeus	Hin Namno NPA, Lao PDR	X				X	yes		24.5 (17-45)		1:1.89	Phiapalath et al. (2011)
P. nemaeus	Son Tra, Vietnam	X	X			X	yes	2.7	6.8(3-12)	18	1:1.63	Ulibarri (2013)
P. nigripes	Seima, Mondulkiri, Cambodia		X	X		X	yes		6.9[d] (1-26)		1:2.09	Rawson (2009)
P. nigripes	Nui Chua/Phuoc Binh, Vietnam		X	X		X	yes		11.6(1-55)	13-45	1:2.33	Hoang (2007)[e]
P. cinerea	Kon Ka Kinh, Vietnam		X	X		X	yes	2-6	5-6	16-88	1:2.02	Ha (2009)
Nasalis												
N. larvatus	Tajung Puting, Kalimantan, Indonesia		X			X	yes	2-6	12.6 (3-23)		1:5	Yeager (1991a, 1992)

Species	Location					Band/Unit	Group size	M:F ratio	Reference
N. larvatus	Samunsam, Indonesia	X		yes	2-6		9 (5-16)	1:3.6	Bennett and Sebastian (1988)
N. larvatus	Sukau, Sabah, Malaysia	X	X	yes		X	17 (14-20)	1:7.3	Boonratana (1993)
N. larvatus	Abai, Sabah, Malaysia	X	X	yes		X	14.1 (8-22)	1:7.1	Boonratana (1993)
N. larvatus	Menangul, Sabah, Malaysia								Matsuda et al. (2009)
N. larvatus	Padas Bay, Sabah, Malaysia	X			11-32			1:1.8	Kawabe and Mano (1972)
Simias									
S. concolor	Sirimuri, Siberut, Indonesia	X		X			3.6 (2-5)	1:1	Tilson (1977)
S. concolor	Pagai Islands, Indonesia	X	X	X			4.1 (1-9)	1:1.8	Teneza and Fuentes (1995)
S. concolor	Sarabua, Siberut, Indonesia	X		X			3 (2-5)	1:1	Watanabe (1981)
S. concolor	Grukna, Siberut, Indonesia	X		X			7.1 (2-20)	1:2	Watanabe (1981)
S. concolor	Peleonan, Siberut, Indonesia	X					8.7 (8-10)	1:3	Hadi et al. (2012)
S. concolor	Peleonan, Siberut, Indonesia						2.6		Waltert et al. (2008)
S. concolor	Peleonan, Siberut, Indonesia	X	X	X				1:3	Erb et al. (2011)

a Lengths of studies cited ranged from 3 months to >60 months.
b When blank, either not observed or not indicated in study.
c Density (Ind/km2) estimates in studies ranged from 0.18 (Thanh, 2007) to 220 (Watanabe, 1981).
d Includes solitary individuals; mean group size with no solitary individual=7.5.
e Presence of MMUs (multi-male/multi-female units) recorded but considered by author to be grouping of several OMUs.

believed to be primarily influenced by the seasonal availability of foods (Bleisch et al., 1993; Kirkpatrick et al., 1998; Kirkpatrick & Grueter, 2010). *R. bieti* relies year-round on a widely available food source—lichen—which allows large bands to remain cohesive without experiencing food competition (Grueter et al., 2009a; Kirkpatrick & Grueter, 2010).

Social interactions, and inter- versus intra-unit dynamics within the band and between bands, can be distinguished. Within the unit, social affiliation between members is highly tolerant and is relatively strong between the females (although not hierarchial), and maintained through allogrooming, reconciliation, food sharing, and huddling behaviors, e.g., *R. avunculus* (Boonratana & Le, 1998; Thanh, 2007); *R. bieti* (Kirkpatrick et al., 1998; Grueter, 2004; Li, D., et al., 2010); *R. roxellana* (Zhang, P., et al., 2006, 2008a, 2008b; Zhang, J., et al., 2010). Alloparental care has been reported regularly in the genus (Xi et al., 2008; Kirkpatrick & Grueter, 2010), another evidence of tolerant relationships among unit members. In addition, direct paternal care, although infrequent, is recorded in *R. bieti* (Xiang et al., 2009) suggesting male tolerance to infants, which is commonly observed in Asian colobines (Kirkpatrick, 2007). Female association with the resident male (e.g., participating in inter-unit aggressions) is found to be a crucial component in the case of inter-unit contests in *R. roxellana* (Zhao et al., 2008b; Zhao & Tan, 2010). When a new male takes over an OMU, *R. roxellana* females show less affiliation to him than to the previous male for a period of time (Zhao et al., 2008a).

Members of different units generally maintain a relatively neutral relationship. Although infrequent, even grooming has been observed between members of different *R. roxellana* units (Zhang et al., 2006). The boundary between units usually is maintained through aggression between the resident males of OMUs (*R. bieti*: Kirkpatrick et al., 1998); cooperative aggression by OMU members occurs if units are less than 2 m apart (*R. roxellana*: Zhang et al., 2006). Inter-unit contests were studied by Zhao and Tan (2010) in a provisioned band of *R. roxellana*; when aggression was initiated by individuals of two different units, the outcome was determined by the number of participants of each unit joining forces—the higher the number of participants, the better the chance of winning a contest. Furthermore, the number of females participating was a determining factor in the outcome of the contest (Zhao & Tan, 2010).

Clear dominance relationships were observed in *R. roxellana* between the different OMUs within the band, the oldest unit established in the band dominating over newer units (Zhao et al., 2008b). Resident males from the band can join forces to chase away bachelor males from

AMUs (Grueter & van Schaik, 2009a; Zhao & Li, 2009; Kirkpatrick & Grueter, 2010) in order to reduce risks of an OMU takeover. This may not be infrequent since male takeovers are often observed in wild provisioned *R. roxellana* (e.g., Wang et al., 2004; Zhao et al., 2005, 2008b; Zhang et al., 2008b; Zhao & Li, 2009).

Though information is not yet available for all species of the genus and more studies are needed before drawing generalizations, both male and female *R. roxellana* appear to disperse from their unit within as well as outside the band. This dispersal seems to be seasonal (Zhang et al., 2006, 2008a; Zhao et al., 2008a; Qi et al., 2009; Yao et al., 2011).

Although females maintain affiliative relationships with other females in their unit, they are not philopatric (Kirkpatrick, 2007) and disperse voluntarily. Over the course of an intensive six-year study of semi-habituated, food-provisioned *R. roxellana*, adult and subadult female movement between OMUs was common, and the majority (72%) of the females that transferred stayed in the new unit for over three years (Qi et al., 2009). Dispersal has been observed to occur between the birth season and the mating season (Zhao et al., 2008a). It has been suggested that female dispersal in *Rhinopithecus* is promoted by a relatively low feeding competition in the species, familiarity of neighbors, and a low predation risk (Zhang et al., 2008a; Qi et al., 2009). In addition, female dispersal has been explained by two major ultimate factors: (1) an increase of reproductive success by reducing sexual competition, and (2) inbreeding avoidance (Zhao et al., 2008a; Qi et al., 2009). Female *R. roxellana* dispersing from their OMU preferably transfer to a newly formed OMU with a smaller group size of less females, and a high proportion of transferred females have post-transfer successful reproduction (Zhao et al., 2008a; Qi et al., 2009). Moreover, females tend to transfer from their unit if their father is still the resident male when they reach sexual maturity (Qi et al., 2009).

Dispersal has also been observed in male *R. roxellana*, the males often joining an AMU before emigrating again (secondary dispersal) to form a new OMU or to take over an established OMU. Yao et al. (2011) propose that male dispersal in *R. roxellana* is determined by two main ultimate factors: (1) mating competition, and (2) inbreeding avoidance. Indeed, they observed that males disperse more during the mating season, resulting in male takeovers and evictions from OMUs. It was also observed that no male remained in the same OMU for more than five years and never returned to his previous one. Males usually are replaced before their daughter(s) reaches sexual maturity (Qi et al., 2009; Yao et al., 2011). Yao et al. (2011) observed that familiar males (juveniles/subadults) from

the same OMU tend to transfer together while staying in close proximity to other units; factors that may influence the timing and the collective nature of the transfer include predation avoidance and low food competition since dispersal occurs mostly when preferred foods are available.

Bands seem to never meet; none of the studies on *Rhinopithecus* species has recorded contact between two different bands. Band home ranges normally do not overlap (Tan et al., 2007; Kirkpatrick & Grueter, 2010); movement of individuals between bands does, however, occur (Zhang et al., 2006; Zhao et al., 2008a; Qi et al., 2009).

Mating season (autumn) and birth season (end of winter/start of spring in China and the wet season in Vietnam) are relatively consistent between species and correspond to periods of preferred-food availability (Qi et al., 2008). *R. bieti*, however, seems to have an earlier birth season, which may be a strategy to increase infant survival; this gives the mother a longer period of time (all spring and summer) with access to preferred foods, since the species relies on poorer-quality lichen the rest of the year (Xiang et al., 2007). Copulations within the unit are usually initiated by females. Although most of the copulations occur within the unit, extra-unit copulations, mainly initiated by females, have been recorded in *R. roxellana*. Zhao et al. (2008a) recorded one instance of an extra-unit copulation initiated by a new OMU resident male, which they suggest is a strategy to stimulate fertilization of resident females. Guo et al. (2010) reported that 6.5% of female-initiated copulations are extra-unit copulations, while over 50% of analyzed DNA samples from immatures reveal they were sired by extra-unit males. During mating season, female interference with other copulation events within the unit have been observed, but those instances may be hormonally induced, rather than evidence of sexual competition, since interference does not result in the decrease of reproductive success of other females (Qi et al., 2011). A significant number of *R. roxellana* females solicit the resident male within OMUs that contain more females, and success of solicitation decreases as the number of females in the OMU increases (Li, B., & Zhao, 2007).

Pygathrix

The social organization of *Pygathrix* remains unclear since too few studies have been conducted on habituated groups and long-term ecological studies have been rare, e.g., *P. nigripes* (Hoang, 2007; Rawson, 2009); *P. cinerea* (Ha, 2009); and *P. nemaeus* (Phiapalath, 2009; Ulibarri, 2013). Early data on the species were collected mainly from short-term surveys in Vietnam (Lippold, 1977,

1998; also reviewed in Kirkpatrick, 1998; Nadler et al., 2003). These surveys mostly reported a social organization of multi-male/multi-female groups (referred to as MMUs in Table 18.7 and hereafter), with group size ranging from 5 to~50 individuals (Lippold, 1998), although Lippold (1977) gave accounts of OMUs, MMUs, and solitary individuals. It cannot be ruled out that the MMUs may have been a combination of several OMUs, as is phylogenetically expected for the genus (Kirkpatrick, 2007). Surveys in Son Tra National Park, Vietnam, report group sizes of *P. nemaeus* ranging from 6–40 individuals, but information on social organization is not mentioned (Lippold & Thanh, 2008; Dinh et al., 2010). Long-term observations (≥1 yr) of *P. cinerea* in Vietnam at Kon Kah Kin National Park (Ha, 2009), *P. nigripes* in Vietnam at Nui Chua and Phuoc Binh National Parks (Hoang, 2007), and Cambodia at Seima Conservation Forest (Rawson, 2009) suggest *Pygathrix* social organization consists of a basic level of OMUs regularly gathering to form a secondary level of large bands. Solitary individuals and AMUs were also observed during these studies (see Table 18.7). Phiapalath et al. (2011) studied *P. nemaeus* in Hin Namno National Protected Area, Laos, and reported a multi-male/multi-female group organization based on the fact that at least two adult males were observed at any one time. They observed solitary males on three occasions but never AMUs. The authors, nevertheless, do not rule out a multilevel society for *P. nemaeus* and urge further studies. The fact that they observed fission-fusion in the species (through group size change over time) still favors an interpretation of multilevel society in the genus *Pygathrix*. A long-term study of three semi-habituated groups of *P. nemaeus* in Son Tra National Reserve, Vietnam, provides a better understanding of social organization. Ulibbari (2013) describes three levels of organization: the "unit" (one-male or multi-male family units with an average of 6.5 individuals), the "group" (several adjacent units, 2.7 on average, fused together with an average of 18 individuals), and the "super-troop" (occasional fusion/encounter of 2–3 groups with a total of 40–55 individuals). Fission-fusion events increase during the wet season, i.e., during periods of leaf-flush and increased precipitation, and groups are less stable with the result of a higher degree of fission-fusion events. In addition, units are always observed fused into groups at sleeping sites, suggesting units may split during daytime and fuse at night (Ulibarri, 2013). In all studies, variation in group size was observed across seasons. While *P. nigripes* in Vietnam and Cambodia (Hoang 2007; Rawson, 2009) and *P. nemaeus* in Laos (Phiapalath, 2009) have larger groups during the wet season in April–October, *P. cinerea* form larger groups during the dry season in November–March (Ha, 2009). At each site, the wet season was characterized by an increase of fruits, the preferred food, and wider availability of fruits may allow formation of larger groups (Rawson, 2009; Phiapalath et al., 2011).

Social affiliation between members occurs through grooming, although it is infrequent relative to the total activity budget (Hoang, 2007; Rawson, 2009). *P. cinerea* females are the most affiliative members of the unit, and individuals are observed most of the time less than 2 m apart (Ha, 2009). Over the course of his study, Ha (2009) observed 13 instances of fission-fusion events; when fusion occurs, the band travels and forages together for one to three days. Fissions, of course, may have been induced by the presence of the observers.

Nasalis

From brief early studies, the social organization of *Nasalis* was misinterpreted as multi-male/multi-female (Kawabe & Mano, 1972; Macdonald, 1982; Salter et al., 1985). Subsequent long-term studies at different sites consistently found that the species forms multilevel societies with the OMU as the basic social unit. Several of these OMUs can come together to form a band, which travels together or gathers at sleeping sites along a river (Bennett & Sebastian, 1988; Yeager, 1990a, 1991b, 1992; Boonratana, 1993; Yeager, 1995). In addition to the OMUs, all-male units and solitary individuals (mainly males) are often encountered around the band. Boonratana (1993: p.14) also reports the existence of nonreproductive groups defined as "loosely bonded... predominantly males with at least one female," but this form of grouping was not observed in other studies. Yeager (1990a) observed one "monogamous" (adult female and adult male) group, but all these unusual groups may be transitory.

Members of a *Nasalis* unit stay closer to each other, generally no more than 15 m apart, when compared to separation from members of other units (Yeager, 1990a). OMUs are female-centered; females direct the majority of their affiliative and agonistic behaviors toward other females (and offspring), and the females maintain the social integrity of the unit (Yeager, 1990a; Matsuda et al., 2012). The majority of social grooming is performed by females and is directed toward other females or youngsters of the group but rarely, if ever, toward the male (Matsuda et al., 2012). Although a rigid female dominance hierarchy does not occur in this species, some individuals are thought to be dominant over others due to the variation in relationships within the unit (Yeager, 1990b; Matsuda et al., 2012). Resident males are regarded as the mediators when agonistic behaviors occur (Yeager, 1990b).

OMUs meet other OMUs at least once every day and occasionally travel together for no less than two days (Bennett & Sebastian, 1988). The OMUs and AMUs in a band usually stay 50–100 m apart when gathered at a site (Bennett & Sebastian, 1988; Boonratana, 1993). Despite range overlap of several units, the association of units that form bands was not random in Tanjung Puting National Park, Kalimantan Tengah, Indonesia (Yeager, 1991b). Therefore, in opposition to what has been found in *Rhinopithecus*, *Nasalis* bands have overlapping ranges. Yet, there seems to be temporal avoidance between the bands via morning and evening vocalizations, branch shaking displays, and coordination of unit movements in each band between sleeping sites (Yeager, 1991b). Fission-fusion of the band is commonly observed with OMUs spending some time alone (but OMUs and AMUs are more often observed in association than alone (Yeager, 1991b). Although the number of units associated together varies, there are typically two to six (cf. Table 18.7). Inter-unit interactions are typically neutral, with some exceptions of high-intensity agonistic behaviors that are rapidly mediated by the resident males or between males of OMUs and AMUs (Yeager, 1992). One of the factors that may be determinant in the formation of bands in *Nasalis* is predation risk. *Nasalis* units seem to coordinate their associations along rivers and when crossing rivers, during which time they are the most vulnerable to crocodilian predators (Yeager, 1991a).

Both males and females (most commonly juveniles and subadults) are known to transfer voluntarily from their unit (Bennett & Sebastian, 1988; Yeager, 1990a; Boonratana, 1999; Murai et al., 2007); males generally dispersing before adolescence (Yeager, 1990b). Murai et al. (2007) estimates the average tenure of a resident male is six years. Female maturity is reached around five years old; female dispersal, therefore, plays a role in inbreeding avoidance. Furthermore, females were observed to disperse in large groups (Murai et al., 2007). Female transfer occurs when groups are associated in bands, which makes female transfer easier due to familiarity with the habitat and neighbors (Boonratana, 1999; Murai et al., 2007). Resident males are indifferent to subadult females leaving the unit, however, they may call out nonaggressively to adult females who are transferring (Murai et al., 2007).

Hollihn (1973), Yeager (1990b), Gorzitze (1996), and Boonratana (2011) report that females typically initiate copulations, in contrast to Murai et al. (2007) who report that solicitation by males is more frequent than by females. The reasons for this difference in field observations remains unclear. The large nose in male *Nasalis* plays a distinct role in attracting females; Koda et al. (2018) found significant correlation between nose size and number of females in an OMU, which supports theories of both female choice and male-male competition in the species. Harassment of a copulating pair by juveniles and infants was consistently observed in different studies, although the harassments were never successful in stopping the copulation (Yeager, 1990b; Murai, 2006; Boonratana, 2011). Similarly, nonsexual mounts (female-female, male-male, juvenile/infant-adult, juvenile/infant-juvenile/infant) are commonly observed (Yeager, 1990b; Murai, 2006; Boonratana, 2011). The function of this behavior is unknown but could have social consequences, such as dominance or bonding (Boonratana, 2011). Boonratana (2011) investigated birth seasonality over a 23-month period in the lower Kanabatangan, Sabah, Malaysia, using the monthly ratio of young infants per adult female, and concluded that *N. larvatus* have no particular birth season.

Simias

Simias is the least-studied genus of the odd-nosed monkeys. Social organization of *S. concolor* varies across the few long-term studies on the species. In most of the studies, however, the basic social unit of *Simias* is composed of 1 male, ≥1 female(s), and their offspring (Tenaza & Fuentes, 1995), differing, therefore, from the OMUs observed in other odd-nosed monkeys. Solitary males and all-male groups (of no more than 2 males) are also reported traveling and feeding in the vicinity of other social units (Tilson, 1977; Watanabe, 1981; Tenaza & Fuentes, 1995). Several groups with only one bisexual pair of adults (plus offspring) have been reported, which has previously led to categorizing the species as monogamous (Tilson, 1977; Watanabe, 1981; Tenaza & Fuentes, 1995). One group with two males was also observed at a site on Pagai Island, Indonesia (Tenaza & Fuentes, 1995). It is a consensus that these discrepancies and the social units observed are the results of human hunting pressure (Tilson, 1977; Hadi et al., 2009; Erb et al., 2011). In a recent survey of an undisturbed site, Erb et al. (2011) found that the species is organized into OMUs with an average of three adult females; all-male units were also observed, but bisexual adult pairs were never seen. By comparing different populations of *S. concolor* with differing hunting levels, Erb et al. (2011) demonstrate that both the number of adult females and the number of immatures is negatively correlated with hunting pressure. Similarly, in an undisturbed site on northeast Siberut, Indonesia, Hadi et al. (2009) report a higher average group size than at other sites and a social organization made up of OMUs.

Information on intragroup interactions comes mostly from anecdotal observations in the course of Tenaza and Fuentes's (1995) study, during which they observed no particular dominance hierarchy within a group and that either sex can lead a group when traveling. Group members are sometimes observed splitting into subgroups of two to four individuals. Similarly, intergroup interactions are poorly documented. Ad libitum instances of intergroup contact show that physical contact does not occur. Loud calls, moving through the canopy, and coming to the ground all are observed prior to the departure of one of the groups (Tenaza & Fuentes, 1995). Two foraging groups were also observed ~75 m apart with no particular agonistic behavior, except for vocalizations from the adult males (Tilson, 1977).

Both male and female dispersal occurs in the species, particularly with juveniles (Erb et al., 2011). Birth season is unclear. It was initially thought that births occurred principally between June–July (Tilson, 1977), but a later study shows no particular birth season and includes the observation of births from November–June (Hadi et al., 2009), although these differences may be due to different habitats and study periods. One characteristic of *S. concolor* is the female exhibition of conspicuous sexual swelling (Tenaza, 1989a), which does not occur in other odd-nosed monkeys. The exact function of this remains unknown, although it certainly plays a role in sexual behavior. Loud calls produced by the males are thought to play a role in sexual selection, displaying the quality state of the caller (Erb et al., 2012).

Predation

Potential nonhuman predators of *Rhinopithecus* species include many terrestrial mammals and raptors. Probable mammalian predators are leopards (*Panthera pardus*), leopard cats (*Prionailurus bengalensis*), clouded leopards (*Neofelis nebulosa*), snow leopards (*Uncia uncia*), China tigers (*Panthera tigris amoyensis*), Asian golden cats (*Catopuma temmincki*), black bears (*Selenarctors thibetanus*), lynx (*Lynx lynx*), wolves (*Canis lupus*), dholes (*Cuon alpinus*), and jackals (*Canis aureus*) (Johnson et al., 1993; Kirkpatrick et al., 1998; Liu & Zhao, 2004; Cui et al., 2006; Li, D., et al., 2006; Li, Y., 2007; Niu et al., 2010; Quan et al., 2011; Yao et al., 2011). Birds of prey that may be predators include hawks (*Accipiter* spp.), buzzards (*Buteo* spp.), and eagles (*Aquila* spp.) (Zhang et al., 1999; Cui, 2003; Cui et al., 2006; Li, D., et al., 2006). Nowadays, the density of terrestrial predators in the habitat of *Rhinopithecus* has been reduced from human-induced pressure (e.g., hunting and habitat loss), and although they may have been important predators in the past, they do not seem to present high predation risk

today. Raptors, however, probably prey on *Rhinopithecus* at a high level, especially on juveniles and infants (Zhang et al., 1999; Cui, 2003; Cui et al., 2006; Li, D., et al., 2006), although this has never been quantified and has not often been observed (Zhang et al., 1999; Cui, 2003; Li, Y., 2007). Predation risk (at least in the past) is suggested to be one of the factors that influence sleeping sites (Liu & Zhao, 2004; Cui et al., 2006; Li, D., et al., 2006, 2010) and tree stratum use (Li, Y., et al., 2002; Li, Y., 2007) as anti-predation strategies.

Potential nonhuman predators of *Pygathrix* include leopards (*Panthera pardus*), leopard cats (*Prionailurus bengalensis*), clouded leopards (*Neofelis nebulosa*), Asian golden cats (*Catopuma temmincki*), marble cats (*Pardofelis marmorata*), as well as pythons and raptors (e.g., *Spilornis*, *Haliaeetus*, *Accipiter*, *Milvus*, *Ictinaetus*, and *Aquila*). Large cats may occur at too low a density presently to put strong predation pressure on the genus (e.g., Coudrat et al., 2014), and python and raptor predation may be minimal (Hoang, 2007; Rawson, 2009). No long-term studies on *Pygathrix* have recorded predation events.

The main nonhuman predators of *Nasalis* are clouded leaopards (*Neofelis nebulosa*), false gharials (*Tomistoma*), and other crocodilians (Galdikas, 1985; Yeager, 1991a; Boonratana, 1993; Matsuda et al., 2008), but may also potentially include monitor lizards (*Varanus* spp.), pythons, and raptors (Yeager, 1991a; Boonratana, 1993). Crossing at narrow points of a river and the crossing coordination by several OMUs may be anti-predator strategies (Yeager, 1991a).

There are no nonhuman predators of *Simias* in its present-day habitat (Tenaza & Tilson, 1985; Mitchell & Tilson, 1986; Tenaza, 1989b).

CONSERVATION

As with the vast majority of primates throughout the world, the langurs are an example of wildlife falling victim to human encroachment of their habitat, hunting, and deliberate destruction due to crop raiding. There are only a few langur species that are not threatened, mainly those living comensually with humans. The International Union for the Conservation of Nature (IUCN) lists four Endangered and one Critically Endangered species of *Semnopithecus*; in addition, there are three Vulnerable and two Near-Threatened species in the genus, along with four species assessed to be Least Concern (IUCN, 2020). Habitat conflicts with humans pose a threat to endangered *S. ajax* in India; however, sightings of this primate increased along with a rise in elevation in the northwestern Himalayas (Sharma & Ahmed, 2017).

Trachypithecus and *Presbytis* are both, unfortunately, of tremendous conservation concern. All species in both genera are classified as Near Threatened, Vulnerable, Endangered, or Critically Endangered (IUCN, 2020). *T. leucocephalus* is one of the critically endangered primates existing in the human-altered and fragmented environment of southern China. Wang et al. (2017) identified anthropogenic land modification as having the most negative impact on *T. leucocephalus* and recommended that habitat continuity should be restored and enforced to save the species. Nature preserves have played an important role in the conservation of primates in China, but the sheer size of a reserve has important implications on the recovery of langur populations. *T. francoisi* numbers, for example, increased very slowly over time in small reserves compared to larger ones (Deng et al., 2019). A survey of *T. germaini* in Vietnam estimated a total population of only 362–406 individuals due to hunting and habitat loss (Bang et al., 2017). A similar effort assessed the total population of *T. shortridgei* in China at 250–370 individuals; threats to the species are habitat loss and poaching (Cui et al., 2016). Thinley et al. (2019) discovered that the population of *T. geei* in Bhutan is much lower than current IUCN estimates. They also identified a number of threats that may not have been obvious in prior studies, such as road kills, electrocution from electrical transmission lines, road construction, and hydropower projects.

Predation by other wild animals is not a factor in the declines of the odd-nosed monkeys species; pressure comes mostly from human hunting. For some species, hunting pressure has been so high that it has led to local extinctions and placed the global population in crisis. All species of odd-nosed monkeys are threatened with extinction due to habitat degradation and hunting pressure. As a result, they are all either classified in the IUCN Red List as Endangered or Critically Endangered and in need of immediate intensive protection (IUCN, 2020).

Forest fragmentation has had dire effects on species of *Rhinopithecus* (Nüchel et al., 2018), For example, Huang et al. (2017) found that endangered *R. bieti* populations in southwest China are stressed for food in their isolated and degraded habitat patch. Species of the genus *Rhinopithecus* also are hunted by humans for food and traditional medicine, which poses a great threat to the populations (Nguyen, 2000; Thanh, 2007; Covert et al., 2008; Geissmann et al., 2011). The most recently discovered species, *R. strykeri,* is on the verge of extinction (estimated population is <400 individuals), and strict enforcement of habitat protection regulations in China and Myanmar will be essential for the survival of the species (Yang et al., 2019). According to Meyer et al. (2017: p.30), *R. strykeri* are among the primates that have been hunted and trapped in Myanmar "for their heads (skull and brain) and bones to be sold to nearby Chinese logging and road construction camps, as well as to wildlife traders in Gangfang."

For *Pygathrix*, the main threat is from hunting; their behavioral response of hiding in trees above hunters makes them particularly vulnerable (Coudrat et al., 2013). The *Pygathrix* species are highly sought for food, traditional medicine, or pets and are either used locally and/or traded nationally or internationally (Phiapalath, 2009; Tran, 2010). Hunting has already resulted in local extinctions across the range of the genus (Coudrat et al., 2012). *Nasalis* is threatened throughout its range. Hunting is high at some sites where the monkeys are mainly killed for food by non-Muslims (Boonratana, 1993; Meijaard & Nijman, 2000, 2009; Stark et al., 2012). *Simias* also face a very high level of hunting by humans, which is believed to have modified the social organization of the species (Watanabe, 1981; Hadi et al., 2009; Erb et al., 2011). Pressure on *Simias* is particularly intense due to the low mammalian faunal diversity on the Indonesian islands they inhabit, leading hunters to focus on primates (Tilson, 1977). During a ceremonial hunt of *S. concolor*, three groups (23 individuals) were hunted and killed in the Loh Bajou Bay region of northeast Siberut (Hadi et al., 2009). In a little more than a decade, local areas in Siberut saw the extirpation of *Simias* populations (Tenaza & Fuentes, 1995).

SUMMARY

Asia contains a wide variety of forest habitats, ranging from tropical evergreen and montane forests to seasonal deciduous and temperate forests. These habitats support 7 genera and 31 species of Asian colobines. Asian colobines vary in size from species of the genus *Presbytis,* weighing just over 5 kg, to proboscis monkey males, weighing over 20 kg. Asian colobines also exhibit a broad range of diet and social organization. While leaves comprise a significant proportion of the diet for some species (Hladik, 1977), nevertheless, fruit (Dela, 2007), flowers (Vandercone et al., 2012), and seeds (Davies, 1991) are important components of the diets of Asian colobines. Similar patterns of dietary variation are also present between *populations* of Asian colobine species. The diets of Asian colobine species are often seasonal, with fruits, flowers, and seeds being important components during certain times of the year.

Typically, Asian colobines live in relatively small groups comprised of a few females, their offspring, and one male (Kirkpatrick, 2007). In addition, these groups also have relatively small home ranges and short day range lengths. However, some species (and some populations) of Asian colobines live in multi-male matrilineal groups and large bands similar to cercopithecines. These groupings maintain large home ranges and have long day range lengths as well.

The langurs are divided into three genera: *Semnopithecus*, *Trachypithecus*, and *Presbytis*. Their geographic distribution is extensive, from the temperate Himalayas to tropical Southeast Asia. They are indigenous to Pakistan through India into Nepal, Bhutan, Sri Lanka, Indonesia, Malaysia, Myanmar, and Bangladesh, with areas of sympatry between genera (Kirkpatrick, 2007). While the social organization, behavior, and life history characteristics of some langur species, such as *S. entellus*, have been well documented, similar data on *Trachypithecus* and *Presbytis* species are sparse. As with other aspects of the behavior and ecology of langurs, there is considerable variation in social organization, sexual behavior, and life history between populations of the same species, which seems to be driven by habitat quality.

The *Rhinopithecus-Pygathrix-Nasalis-Simias* clade of the odd-nosed monkeys seems to have evolved with generally similar ecological and social behaviors. However, the determinants of specific variations among the different species appear to be due to their respective habitat characteristics. Consistent patterns among odd-nosed monkeys in relation to ecological theories cannot be established systematically; their behaviors seem to be habitat-specific rather than species-specific, and they will adopt the strategies to best exploit their respective habitats. On average, odd nosed monkeys spend most of their time resting, followed by feeding, traveling, and socializing. Characteristics of their habitat also create a gradient in the feeding ecology of the different species, but, overall, they are selective feeders, preferably choosing fruits and young leaves whenever these are seasonally available. Otherwise, they rely on widely available alternative foods, such as lichen (especially *Rhinopithecus*) and mature foliage. Their selection for specific foods is influenced by the nutrient content, opting for foods with the highest protein and lowest fiber and tannin contents.

Seasonality in mating and birth varies across species of odd-nosed monkeys and depends on habitat characteristics; while some species are seasonal (*Rhinopithecus*), others do not show any particular seasonality (*Nasalis* and *Simias*). They have a basic social unit of one-male and/or multi-male, with several females and offspring, but all-male units and solitary individuals are also observed in all species. With the exception of *Simias* (although this may be the result of intense historical human hunting pressure), they form a second level of social organization called bands/groups, which consists of several one-male, multi-male, or all-male units that regularly (or semipermanently) come together. Social

relationships seem female-centered; females maintain close relationships within the unit, but no strict linear hierarchy exists, although loose dominance can occur between females. Further research on the odd-nosed monkey species (in particular *Pygathrix* and *Simias*) is urgently needed to fill the gaps in our knowledge of their ecology and behavior.

Although basic information on ecology and social organization is available for at least one population of most Asian colobine species, only a few species have been studied across their geographic range over long periods. Thus, for many Asian colobines, the variation in their social organization and ecology at a given site, in relation to temporal variation in ecological conditions, is poorly understood. The concentration of field studies on a few species, together with the relatively poor representation of Asian colobines in the primate literature, have tended to obscure the fact that Asian colobines are an anatomically, ecologically, and socially varied group of primates (Oates & Davies, 1994).

REFERENCES CITED—CHAPTER 18

Aziz, M. A., & Feeroz, M. M. (2009). Utilization of forest flora by Phayre's leaf-monkey *Trachypithecus phayrei* (Primates: Cercopithecidae) in semi-evergreen forests of Bangladesh. *Journal of Threatened Taxa, 1*, 257-262.

Bang, T. V., Nguyen, M. A., Nguyen, D. Q., Truong, Q. B. T., Ang, A., Covert, H. H., & Hoang, D. M. (2017). Current conservation status of Germain's langur (*Trachypithecus germaini*) in Vietnam. *Primates, 58*(3), 435-440.

Bauchop, T., & Martucci, R. W. (1968). Ruminant-like digestion of the langur monkey. *Science, 161*(3842), 698-700.

Bennett, E. L. (1983). *The banded langur: Ecology of a colobine in west Malaysian rain-forest*. [Doctoral dissertation, Cambridge University].

Bennett, E. L. (1986). Environmental correlates of ranging behavior in the banded langur *Presbytis melalophos*. *Folia Primatologica, 47*(1), 26-38.

Bennett, E. L., & Davies, A. G. (1994). The ecology of Asian colobines. In A. G. Davies, & J. F. Oates (Eds.), *Colobine monkeys: Their ecology, behavior and evolution* (pp. 129-172). Cambridge University Press.

Bennett, E. L., & Sebastian, A. C. (1988). Social organization and ecology of proboscis monkeys (*Nasalis larvatus*) in mixed coastal forest in Sarawak. *International Journal of Primatology, 9*(3), 233-255.

Bleisch, W., Cheng, A.-S., Ren, X.-D., & Xie, J.-H. (1993). Preliminary results from a field study of wild Guizhou snub-nosed monkeys (*Rhinopithecus brelichi*). *Folia Primatologica, 60*, 72-82.

Bleisch, W., & Xie, J. (1998). Ecology and behavior of the Guizhou snub-nosed langur (*Rhinopithecus* [*rhinopithecus*] *brelichi*), with a discussion of socioecology in the genus. In N. G. Jablonski (Ed.), *The natural history of the doucs* and snub-nosed monkeys (pp. 217-240). World Scientific Publishing Co.

Boggess, J. (1980). Intermale relations and troop male membership changes in langurs (*Presbytis entellus*) in Nepal. *International Journal of Primatology, 1*(3), 233-274.

Boonratana, R. (1993). *The ecology and behaviour of the proboscis monkey (*Nasalis larvatus*) in the lower Kinabatangan, Sabah.* [Doctoral dissertation, Mahidol University].

Boonratana, R. (1999). Dispersal in proboscis monkeys (*Nasalis larvatus*) in the lower Kinabatangan, northern Borneo. *Tropical Biodiversity, 6*(3), 179-187.

Boonratana, R. (2000a). Ranging behavior of proboscis monkeys (*Nasalis larvatus*) in the lower Kinabatangan, northern Borneo. *International Journal of Primatology, 21*(3), 497-518.

Boonratana, R. (2000b). A short note on vigilance exhibited by proboscis monkey (*Nasalis larvatus*) in the lower Kinabatangan, Sabah, Malaysia. *Tiger Paper, 27*(4), 21-22.

Boonratana, R. (2003). Feeding ecology of proboscis monkeys *Nasalis larvatus*, in the lower Kinabatangan, Sabah, Malaysia. *Sabah Parks Nature Journal, 6*, 1-26.

Boonratana, R. (2011). Observations on the sexual behavior and birth seasonality of proboscis monkey (*Nasalis larvatus*) along the lower Kinabatangan River, northern Borneo. *Asian Primates Journal, 2*(1), 2-9.

Boonratana, R., & Le, X. C. (1998). Preliminary observations of the ecology and behavior of the Tonkin snub-nosed monkey (*Rhinopithecus [presbytiscus] avunculus*) in northern Vietnam. In N. G. Jablonski (Ed.), *The natural history of the doucs and snub-nosed monkeys* (pp. 207-216). World Scientific Publishing Co.

Borries, C. (1997). Infanticide in seasonally breeding multimale groups of Hanuman langurs (*Presbytis entellus*) in Ramnagar (south Nepal). *Behavioral Ecology and Sociobiology, 41*, 139-150.

Borries, C., Koenig, A., & Winkler, P. (2001). Variation of life history traits and mating patterns in female langur monkeys (*Semnopithecus entellus*). *Behavioral Ecology and Sociobiology, 50*, 391-402.

Borries, C., Launhardt, K., Epplen, C., Epplen, J. T., & Winkler, P. (1999a). Males as infant protectors in Hanuman langurs (*Presbytis entellus*) living in multimale groups: Defence pattern, paternity, and sexual behaviour. *Behavioral Ecology and Sociobiology, 46*, 350-356.

Borries, C., Launhardt, K., Epplen, C., Epplen, J. T., & Winkler, P. (1999b). DNA analyses support the hypothesis that infanticide is adaptive in langur monkeys. *Biological Sciences, 266*(1422), 901-904.

Borries, C., Primeau, Z. M., Ossi, K., Dtubpraserit, S., & Koenig, A. (2014). Possible predation attempt by marbled cat on a juvenile Phayre's leaf monkey. *The Raffles Bulletin of Zoology, 62*, 561-565.

Brandon-Jones, D. (1996). The Asian colobinae (Mammalia: Cercopithecidae) as indicators of quaternary climatic change. *Biological Journal of the Linnean Society, 59*(3), 327-350.

Brandon-Jones, D., Eudey, A. A., Geissmann, T., Groves, C. P., Melnick, D. J., Morales, J. C., Shekelle, M., & Stewart, C. B. (2004). Asian primate classification. *International Journal of Primatology, 25*(1), 97-164.

Brattsten, L. B. (1979). Biochemical defense mechanisms in herbivores against plan allelochemics. In G. A. Rosenthal, & D. H. Janzen (Eds.), *Herbivores: Their interaction with secondary plant metabolites* (pp. 199-270). Academic Press.

Chaplin, G., & Jablonski, N. G. (1998). The integument of the odd-nosed colobines. In N. G. Jablonski (Ed.), *The natural history of the doucs and snub-nosed monkeys* (pp. 79-104). World Scientific Publishing Co.

Chapman, C. A., & Chapman, L. J. (1999). Implications of small scale variation in ecological conditions for the diet and density of red colobus monkeys. *Primates, 40*(1), 215-232.

Chapman, C. A., Chapman, L. J., Bjorndal, K. A., & Onderdonk, D. A. (2002). Application of protein-to-fiber ratios to predict colobine abundance on different special scales. *International Journal of Primatology, 23*(2), 283-310.

Cheyne, S. M., Supiansyah, A., Neale, C. J., Thompson, C., Wilcox, C. H., Ehlers Smith, Y. C., & Ehlers Smith, D. A. (2018). Down from the treetops: Red langur (*Presbytis rubicunda*) terrestrial behavior. *Primates, 59*(5), 437-448.

Cheyne, S. M., Supiansyah, S., Adul, A., Wilcox, C. H., Cahyaningrum, E., Ehlers Smith, Y.C., Ehlers Smith, D. A., & Kulu, I. P. (2019). A short cut to mushrooms—red langur (*Presbytis rubicunda*) consumption of terrestrial fungus. *Folia Primatologica, 90*(3), 190-198.

Clink, D. J., Dillis, C., Feilen, K. L., Beaudrot, L., & Marshall, A. J. (2017). Dietary diversity, feeding selectivity, and responses to fruit scarcity of two sympatric Bornean primates (*Hylobates albibarbis* and *Presbytis rubicunda rubida*). *PLoS ONE, 12*(3), e0173369.

Coudrat, C. N. Z., Duckworth, J. W., & Timmins, R. J. (2012). Distribution and conservation status of the red-shanked douc (*Pygathrix nemaeus*) in Lao PDR: An update. *American Journal of Primatology, 74*(10), 874-889.

Coudrat, C. N. Z., Nanthavong, C., & Nekaris, K. A. I. (2013). Conservation of the red-shanked douc *Pygathrix nemaeus* in Lao People's Democratic Republic: Density estimates based on distance sampling and habitat suitability modelling. *Oryx, 48*(4), 540-547.

Coudrat, C. N. Z., Nanthavong, C., Sayavong, S., Johnson, A., Johnston, J. B., & Robichaud, W. G. (2014). Non-panthera cats in Nakai-Nam Theun National Protected Area, Lao PDR. *Cat News, 8*, 45-52.

Covert, H. H., Quyet, L. K., & Wright, B. W. (2008). On the brink of extinction: Research for the conservation of the Tonkin snub-nosed monkey (*Rhinopithecus avunculus*). In J. G. Fleagle, & C. C. Gilbert (Eds.), *Elwin Simons: A search for origins* (pp. 409-427). Springer Science.

Cui, L.-W. (2003). A note on an interaction between *Rhinopithecus bieti* and a buzzard at Baima Snow Mountain. *Folia Primatologica, 74*(1), 51-53.

Cui, L.-W., Huo, S., Zhong, T., Xiang, Z.-F., Xiao, W., & Quan, R.-C. (2008). Social organization of black-and-white snub-nosed monkeys (*Rhinopithecus bieti*) at Deqin, China. *American Journal of Primatology, 70*(2), 169-174.

Cui, L.-W., Li, Y.-C., Ma, C., Scott, M. B., Li, J.-F., He, X.-Y., Li, D.-H., Sun, J., Sun, W.-M., & Xiao, W. (2016). Distribution and conservation status of Shortridge's capped langurs *Trachypithecus shortridgei* in China. *Oryx*, 50(4), 732-741.

Cui, L.-W., Quan, R.-C., & Xiao, W. (2006). Sleeping sites of black-and-white snub-nosed monkeys (*Rhinopithecus bieti*) at Baima Snow Mountain, China. *Journal of Zoology*, 270(1), 192-198.

Curtin, R A. (1982). Range use of gray langurs in highland Nepal. *Folia Primatologica*, 38(1-2), 1-18.

Curtin, S. H. (1980). Dusky and banded leaf monkeys. In D. J. Chivers (Ed.), *Malayan forest primates: Ten years' study in tropical rain forest* (pp. 105-145). Plenum Press.

Davies, A. G. (1984). *An ecological study of the red leaf monkey* (Presbytis rubicunda) *in the dipterocarp forests of Sabah, northern Borneo.* [Doctoral dissertation, University of Cambridge].

Davies, A. G. (1991). Seed-eating by red leaf monkeys (*Presbytis rubicunda*) in dipterocarp forest of northern Borneo. *International Journal of Primatology*, 12(2), 119-144.

Davies, A. G. (1994). Colobine populations. The natural history of African colobines. In A. G. Davies, & J. F. Oates (Eds.), *Colobine monkeys: Their ecology, behaviour and evolution* (pp. 285-310). Cambridge University Press.

Davies, A. G., Bennett, E. L., & Waterman, P. G. (1988). Food selection by two south-east Asian colobine monkeys (*Presbytis rubicunda* and *Presbytis melalophos*) in relation to plant chemistry. *Biological Journal of the Linnean Society*, 34(1), 33-56.

Dela, J. (2007). Seasonal food use strategies of *Semnopithecus vetulus nestor*, at Panadura and Piliyandala, Sri Lanka. *International Journal of Primatology*, 28(3), 607-626.

Delson, E. (1994). Evolutionary history of the colobine monkeys in paleoenvironmental perspective. In A. G. Davies, & J. F. Oates (Eds.), *Colobine monkeys: Their ecology, behaviour and evolution* (pp. 11-44). Cambridge University Press.

Deng, H., Cui, H., Zhao, Q., Pan, R., Zhou, J., & Lan, A. (2019). Constrained François' langur (*Trachypithecus francoisi*) in Yezhong Nature Reserve, Guizhou, China. *Global Ecology and Conservation*, 19, 1-9.

Ding, W., & Zhao, Q.-K. (2004). *Rhinopithecus bieti* at Tacheng, Yunnan: Diet and daytime activities. *International Journal of Primatology*, 25(3), 583-598.

Dinh, T. P. A., Nguyen, D. H. C., & Huynh, T. N. H. (2010). Status and distribution of red- shanked douc langurs (*Pygathrix nemaeus*) and threats to their population at Son Tra Nature Reserve, Danang City. In T. Nadler, B. M. Rawson, & V. N. Thinh (Eds.), *Conservation of primates in Indochina* (pp. 71-78). Frankfurt Zoological Society.

Dolhinow, P. (1972). The north Indian langur. In P. Dolhinow (Ed.), *Primate patterns* (pp. 181-239). Holt, Rhinehart and Winston.

Ehlers Smith, D. A., Ehlers Smith, Y. C., & Cheyne, S. M. (2013b). Home-range use and activity patterns of the red langur (*Presbytis rubicunda*) in Sabangau tropical peat-swamp forest, central Kalimantan, Indonesian Borneo. *International Journal of Primatology*, 34(5), 957-972.

Ehlers Smith, D. A., Husson, S. J., Ehlers Smith, Y. C., & Harrison, M. E. (2013a). Feeding ecology of red langurs in Sabangau tropical peat-swamp forest, Indonesian Borneo: Extreme granivory in a non-masting forest. *American Journal of Primatology*, 75(8), 848-859.

Erb, W. M., Borries, C., Lestari, N. S., & Ziegler, T. E. (2011). Demography and dispersal of Simakobu (*Simias concolor*) and the impact of human disturbance. *American Journal of Primatology*, 73(S1), 16.

Erb, W. M., Borries, C., Lestari, N. S., & Ziegler, T. E. (2012). Demography of Simakobu (*Simias concolor*) and the impact of human disturbance. *American Journal of Primatology*, 74(6), 580-590.

Feilen, K. L., & Marshall, A. J. (2017). Multiple ecological factors influence the location of proboscis monkey (*Nasalis larvatus*) sleeping sites in west Kalimantan, Indonesia. *International Journal of Primatology*, 38(3), 448-465.

Freeland, W. J., & Janzen, D. H. (1974). Strategies in herbivory by mammals: The role of plant secondary compounds. *American Naturalist*, 108(961), 269-289.

Freese, C. H., Heltne, P. G., Napoleon, C. R., & Whitesides, G. (1982). Patterns and determinants of monkey densities in Peru and Bolivia, with notes on distribution. *International Journal of Primatology*, 3(1), 53-90.

Fuentes, A. (1996). Feeding and ranging in the Mentawai Island langur (*Presbytis potenziani*). *International Journal of Primatology*, 17(4), 525-548.

Galdikas, B. M. F. (1985). Crocodile predation on a proboscis monkey in Borneo. *Primates*, 26(4), 495-496.

Geissmann, T., Lwin, N., Aung, S. S., Aung, T. N., Aung, Z. M., Hla, T. H., Grindley, M., & Momberg, F. (2011). A new species of snub-nosed monkey, genus *Rhinopithecus* Milne-Edwards, 1872 (Primates, Colobinae), from northern Kachin State, northeastern Myanmar. *American Journal of Primatology*, 73(1), 96-107.

Gorzitze, A. B. (1996). Birth-related behavior in wild proboscis monkeys (*Nasalis larvatus*). *Primates*, 37(1), 75-78.

Groves, C. P. (1970). The forgotten leaf-eaters, and the phylogeny of the Colobinae. In J. R. Napier, & P. H. Napier (Eds.), *Old World monkeys: Evolution, systematics, and behaviour* (pp. 555-587). Academic Press.

Groves, C. P. (2001). *Primate taxonomy*. Smithsonian Institute Press.

Grueter, C. C. (2004). Conflict and postconflict behaviour in captive black-and-white snub-nosed monkeys (*Rhinopithecus bieti*). *Primates*, 45(3), 197-200.

Grueter, C. C., Li, D., Feng, S.-K., & Ren, B. (2010). Niche partitioning between sympatric rhesus macaques and Yunnan snub-nosed monkeys at Baimaxueshan Nature Reserve, China. *Zoological Research*, 31(5), 516-522.

Grueter, C. C., Li, D., Ren, B., Wei, F., & van Schaik, C. P. (2009a). Dietary profile of *Rhinopithecus bieti* and its socioecological implications. *International Journal of Primatology*, 30(4), 601-624.

Grueter, C. C., Li, D., Ren, B., Wei, F., Xiang, Z., & van Schaik, C. P. (2009b). Fallback foods of temperate-living primates: A case study on snub-nosed monkeys. *American Journal of Physical Anthropology, 140*(4), 700-715.

Grueter, C. C., & van Schaik, C. P. (2009a). Evolutionary determinants of modular societies in colobines. *Behavioral Ecology, 21*(1), 63-71.

Grueter, C. C., & van Schaik, C. P. (2009b). Sexual size dimorphism in Asian colobines revisited. *American Journal of Primatology, 71*(7), 609-616.

Grueter, C. C., & Zinner, D. (2004). Nested societies: Convergent adaptations of baboons and snub-nosed monkeys? *Primate Report, 70*, 1-98.

Guo, S., Ji, W., Ming, L., Chang, H., & Li, B. (2010). The mating system of the Sichuan snub-nosed monkey (*Rhinopithecus roxellana*). *American Journal of Primatology, 72*(1), 25-32.

Guo, S., Li, B., & Watanabe, K. (2007). Diet and activity budget of *Rhinopithecus roxellana* in the Qinling Mountains, China. *Primates, 48*(4), 268-276.

Gupta, A. K., & Kumar, A. (1994). Feeding ecology and conservation of Phayre's leaf monkey *Presbytis phayrei* in northeast India. *Biological Conservation, 69*(3), 301-306.

Ha, T. L. (2009). *Behavioural ecology of grey-shanked douc monkeys* Pygathrix cinerea *in Vietnam.* [Doctoral dissertation, University of Cambridge].

Ha, T. L., Tinh, N. T., Vy, T. H., & Minh, H. T. (2010). Activity budget of grey-shanked douc langurs (*Pygathrix cinerea*) in Kon Ka Kinh National Park, Vietnam. *Vietnamese Journal of Primatology, 4*, 27-39.

Hadi, S., Ziegler, T., & Hodges, J. K. (2009). Group structure and physical characteristics of Simakobu monkeys (*Simias concolor*) on the Mentawai Island of Siberut, Indonesia. *Folia Primatologica, 80*(2), 74-82.

Hadi, S., Ziegler, T., Waltert, M., Syamsuri, F., Mühlenbergb, M., & Hodges, J. K. (2012). Habitat use and trophic niche overlap of two sympatric colobines, *Presbytis potenziani* and *Simias concolor*, on Siberut Island, Indonesia. *International Journal of Primatology, 33*(1), 218-232.

Hanya, G., & Bernard, H. (2015). Different roles of seeds and young leaves in the diet of red leaf monkeys (*Presbytis rubicunda*): Comparisons of availability, nutritional properties, and associated feeding behavior. *International Journal of Primatology, 36*(1), 177-193.

Haus, T., Vogt, M., Forster, B., Vu, N. T., & Ziegler, T. (2009). Distribution and population densities of diurnal primates in the karst forests of Phong Nha–Ke Bang National Park, Quang Binh Province, central Vietnam. *International Journal of Primatology, 30*(2), 301-312.

Hendershott, R., Behie, A., & Rawson, B. (2016). Seasonal variation in the activity and dietary budgets of Cat Ba langurs (*Trachypithecus poliocephalus*). *International Journal of Primatology, 37*(4-5), 586-604.

Hladik, C. M. (1977). A comparative study of the feeding strategies of two sympatric species of leaf monkeys: *Prebytis senex* and *Presbytis entellus*. In T. H. Clutton-Brock (Ed.), *Primate ecology: Studies of feeding and ranging behaviour in lemurs, monkeys and apes* (pp. 323-353). Academic Press.

Hoang, M. D. (2007*). Ecology and conservation status of the black-shanked douc (*Pygathrix nigripes*) in Nui Chua and Phuoc Binh National Parks, Ninh Thuan Province, Vietnam.* [Doctoral dissertation, University of Queensland].

Hoang, M. D., Baxter, G. S., & Page, M. J. (2009). Diet of *Pygathrix nigripes* in southern Vietnam. *International Journal of Primatology, 30*(1), 15-28.

Hollihn, U. (1973). Remarks on the breeding and maintenance of colobus monkeys (*Colobus guereza*), proboscis monkeys (*Nasalis larvatus*), and douc langurs (*Pygathrix nemaeus*) in zoos. *International Zoo Yearbook, 13*(1), 185-188.

Hou, R., He, S., Wu, F., Chapman, C. A., Pan, R., Garber, P. A., Guo, S., & Li, B. (2018). Seasonal variation in diet and nutrition of the northernmost population of *Rhinopithecus roxellana*. *American Journal of Primatology, 80*(4), 1-9.

Hrdy, S. B., Janson, C. H., & van Schaik, C. P. (1995). Infanticide: Let's not throw out the baby with the bath water. *Evolutionary Anthropology, 3*(5), 151-154.

Hu, G. (2011). Dietary breadth and resource use of François' langur in a seasonal and disturbed habitat. *American Journal of Primatology, 73*(11), 1176-1187.

Huang, C., Wu, H., Zhou, Q., Li, Y., & Cai, X. (2008). Feeding strategy of François' langur and white-headed langur at Fusui, China. *American Journal of Primatology, 70*(4), 320-326.

Huang, Z., Huo, S., Yang, S., Cui, L., & Xiao, W. (2010). Leaf choice in black-and-white snub-nosed monkeys *Rhinopithecus bieti* is related to the physical and chemical properties of leaves. *Current Zoology, 56*(6), 643-649.

Huang, Z., Scott, M. B., Li, Y.-P., Ren, G.-P., Xiang, Z.-F., Cui, L.-W., & Xiao, W. (2017). Black-and-white snub-nosed monkey (*Rhinopithecus bieti*) feeding behavior in a degraded forest fragment: Clues to a stressed population. *Primates, 58*(4), 517-524.

Huo, S. (2005). *Diet and habitat use of* Rhinopithecus bieti *at Mt. Longma.* [Doctoral dissertation, Kunming Institute of Zoology].

IUCN. (2014). *The International Union for the Conservation of Nature Red List of Threatened Species.* Version 2014.3. http://www.iucnredlist.org.

IUCN. (2020). *The International Union for the Conservation of Nature Red List of Threatened Species.* Version 2020.3. https://www.iucnredlist.org.

Jablonski, N. G. (1998). The evolution of the doucs and snub-nosed monkeys and the question of the phyletic unity of the odd-nosed colobines. In N. G. Jablonski (Ed.), *The natural history of doucs and snub-nosed monkeys* (pp. 13-52). World Scientific Publishing Co.

Jablonski, N. G., & Pan, R. (1995). Sexual dimorphism in the snub-nosed langurs (Colobinae: *Rhinopithecus*). *American Journal of Physical Anthropology, 96*(3), 251-272.

Ji, W., Zou, R., Shang, E., Zhou, H., Yang, S., & Tian, B. (1998). Maintenance and breeding of Yunnan snub-nosed monkeys

(*Rhinopithecus [rhinopithecus] bieti*) in captivity. In N. G. Jablonski (Ed.), *The natural history of the doucs and snub-nosed monkeys* (pp. 323-336). World Scientific Publishing Co.

Jin, T., Wang, D. Z., Zhao, Q., Yin, L.-J., Qin, D.-G., Ran, W.-Z., & Pan, W.-S. (2009). Social organization of white-headed langurs (*Trachypithecus leucocephalus*) in the Nongguan Karst Hills, Guangxi, China. *American Journal of Primatology*, 71(3), 206-213.

Johns, A. D., & Skorupa, J. P. (1987). Response of rain-forest primates to habitat disturbance: A review. *International Journal of Primatology*, 8(2), 157-191.

Johnson, K. G., Wang, W., Reid, D. G., & Hu, J. C. (1993). Food habits of Asiatic leopards (*Panthera pardus fusea*) in Wolong Reserve Sichuan, China. *Journal of Mammalogy*, 74(3), 646-650.

Joseph, S., Thomas, A., Satheesh, R., & Sugathan, R. (2007). Foraging ecology and relative abundance of large carnivores in Parambikulam Wildlife Sanctuary, southern India. *Zoos Print Journal*, 22(5), 2667-2670.

Karanth, K. P. (2010). Molecular systematics and conservation of the langurs and leaf monkeys of South Asia. *Journal of Genetics*, 89(4), 393-399.

Karanth, K. P., Singh, L., Collura, R. V., & Stewart, C. B. (2008). Molecular phylogeny and biogeography of langurs and leaf monkeys of South Asia (Primates: Colobinae). *Molecular Phylogenetics and Evolution*, 46(2), 683-694.

Kar-Gupta, K., & Kumar, A. (1994). Leaf chemistry and food selection by common langurs (*Presbytis entellus*) in Rajaji National Park, Uttar Pradesh, India. *International Journal of Primatology*, 15(1), 75-93.

Kawabe, M., & Mano, T. (1972). Ecology and behavior of the wild proboscis monkey, *Nasalis larvatus* (von Wurmb), in Sabah, Malaysia. *Primates*, 13(2), 213-228.

Kay, R. N. B., & Davies, A. G. (1994). Digestive physiology. In A. G. Davies, & J. F. Oates (Eds.), *Colobine monkeys: Their ecology, behaviour and evolution* (pp. 229-250). Cambridge University Press.

Kern, J. A. (1964). Observations on the habits of the proboscis monkey, *Nasalis larvatus* (von Wurmb), made in the Brunei Bay area, Borneo. *Zoologica*, 49(11), 183-192.

Kirkpatrick, R. C. (1995). The natural history and conservation of the snub-nosed monkeys (genus *Rhinopithecus*). *Biological Conservation*, 72(3), 363-369.

Kirkpatrick, R. C. (1996). *Ecology and behavior of the Yunnan snub-nosed langur* Rhinopithecus bieti *(Colobinae).* [Doctoral dissertation, University of California, Davis].

Kirkpatrick, R. C. (1998). Ecology and behavior in snub-nosed and douc langurs. In N. G. Jablonski (Ed.), *The natural history of the doucs and snub-nosed monkeys* (pp. 155-190). World Scientific Publishing Co.

Kirkpatrick, R. C. (2007). The Asian colobines: Diversity among leaf-eating monkeys. In C. J. Campbell, A. Fuentes, K. C. Mackinnon, M. Panger, & S. K. Bearder (Eds.), *Primates in perspective* (pp. 186-200). Oxford University Press.

Kirkpatrick, R. C., & Grueter, C. C. (2010). Snub-nosed monkeys: Multilevel societies across varied environments. *Evolutionary Anthropology*, 19(3), 98-113.

Kirkpatrick, R. C., Gu, H. J., & Zhou, X. P. (1999). A preliminary report on Sichuan snub-nosed monkeys (*Rhinopithecus roxellana*) at Baihe Nature Reserve. *Folia Primatologica*, 70(2), 117-120.

Kirkpatrick, R. C., & Long, Y. C. (1994). Altitudinal ranging and terrestriality in the Yunnan snub-nosed monkey (*Rhinopithecus bieti*). *Folia Primatologica*, 63(2), 102-106.

Kirkpatrick, R. C., Long, Y. C., Zhong, T., & Xiao, L. (1998). Social organization and range use in the Yunnan snub-nosed monkey *Rhinopithecus bieti*. *International Journal of Primatology*, 19(1), 13-51.

Kirkpatrick, R. C., Zou, R. J., Dierenfeld, E. S., & Zhou, H. W. (2001). Digestion of selected foods by Yunnan snub-nosed monkey *Rhinopithecus bieti* (Colobinae). *American Journal of Physical Anthropology*, 114(2), 156-162.

Koda, H., Murai, T., Tuuga, A., Goossens, B., Nathan, S., Stark, D., Ramirez, D., Sha, J., Osman, I., Sipangkul, R., Seino, S., & Matsuda, I. (2018). Nasalization by *Nasalis larvatus*: Larger noses audiovisually advertise conspecifics in proboscis monkeys. *Science Advances*, 4(2), 1-6.

Koenig, A., Beise, J., Chalise, M. K., & Ganzhorn, J. U. (1998). When females should contest for food: Testing hypotheses resource density, distribution, size, and quality with Hanuman langurs (*Presbytis entellus*). *Behavioral Ecology and Sociobiology*, 42 (4), 225-237.

Koenig, A., & Borries, C. (2001). Socioecology of Hanuman langurs: The story of their success. *Evolutionary Anthropology*, 10(4), 122-137.

Kool, K. M. (1989). *Behavioural ecology of the silver leaf monkey,* Trachypithecus auratus sondaicus, *in the Pangandaran Nature Reserve, west Java, Indonesia.* [Doctoral dissertation, University of New South Wales].

Kool, K. M. (1993). The diet and feeding behavior of the silver leaf monkey (*Trachypithecus auratus sondaicus*) in Indonesia. *International Journal of Primatology*, 14(5), 667-700.

Kullik, H. (2010). Water consumption of Delacour's langurs (*Trachypithecus delacouri*) and grey-shanked douc langurs (*Pygathrix cinerea*) in captivity. *Vietnamese Journal of Primatology*, 4, 41-47.

Launhardt, K., Borries, C., Hardt, C., Epplen, J., & Winkler, P. (2001). Paternity analysis of alternative male reproductive routes among the langurs (*Semnopithecus entellus*) of Ramnagar. *Animal Behaviour*, 61(1), 53-64.

Laws, J. W., & Laws, J. V. H. (1984). Social interactions among adult male langurs (*Presbytis entellus*) at Rajaji Wildlife Sanctuary. *International Journal of Primatology*, 5(1), 31-50.

Le, H. T., Hoang, D. M., & Covert, H. H. (2019). Diet of the Indochinese silvered langur (*Trachypithecus germaini*) in Kien Luong Karst Area, Kien Giang Province. *American Journal of Primatology*, 81(9), 1-17.

Le, Thi Dien. 2009. *Conservation of the red-shanked douc langur (*Pygathrix nemaeus nemaeus*) in Bach Ma National Park, Vietnam, Bach.* People's Trust for Endangered Species.

Li, B., Chen, C., Ji, W., & Ren, B. (2000). Seasonal home range changes of the Sichuan snub-nosed monkey (*Rhinopithecus roxellana*) in the Qinling Mountains of China. *Folia Primatologica*, 71(6), 375-386.

Li, B., Zhang, P., Watanabe, K., Tan, C. L., Fukuda, F., & Wada, K. (2003). A dietary shift in Sichuan snub-nosed monkeys. *Acta Theriologica Sinica*, 23(4), 358-360.

Li, B., & Zhao, D. (2007). Copulation behavior within one-male groups of wild *Rhinopithecus roxellana* in the Qinling Mountains of China. *Primates*, 48(3), 190-196.

Li, D., Grueter, C. C., Ren, B., Long, Y., Li, M., Peng, Z., & Wei, F. (2008). Ranging of *Rhinopithecus bieti* in the Samage Forest, China. II. Use of land cover types and altitudes. *International Journal of Primatology*, 29(5), 1147-1173.

Li, D., Grueter, C. C., Ren, B., Zhou, Q., Li, M., Peng, Z., & Wei, F. (2006). Characteristics of night-time sleeping places selected by golden monkeys (*Rhinopithecus bieti*) in the Samage Forest, Baima Snow Mountain Nature Reserve, China. *Integrative Zoology*, 1(4), 141-152.

Li, D., Ren, B., Grueter, C. C., Li, B., & Li, M. (2010). Nocturnal sleeping habits of the Yunnan snub-nosed monkey in Xiangguqing, China. *American Journal of Primatology*, 72(12), 1092-1099.

Li, Y. (2006). Seasonal variation of diet and food availability in a group of Sichuan snub-nosed monkeys in Shennongjia Nature Reserve, China. *American Journal of Primatology*, 68(3), 217-233.

Li, Y. (2007). Terrestriality and tree stratum use in a group of Sichuan snub-nosed monkeys. *Primates*, 48(3), 197-207.

Li, Y. (2009). Activity budgets in a group of Sichuan snub-nosed monkeys in Shennongjia Nature Reserve, China. *Current Biology*, 55(3), 173-179.

Li, Y., Jiang, Z., Li, C., & Grueter, C. C. (2010). Effects of seasonal folivory and frugivory on ranging patterns in *Rhinopithecus roxellana*. *International Journal of Primatology*, 31(4), 609-626.

Li, Y., Liu, X., Liao, M., Yang, J., & Stanford, C. B. (2009). Characteristics of a group of Hubei golden snub-nosed monkeys (*Rhinopithecus roxellana hubeiensis*) before and after major snow storms. *American Journal of Primatology*, 71(6), 523-526.

Li, Y., Stanford, C. B., & Yuhui, Y. (2002). Winter feeding tree choice in Sichuan snub-nosed monkeys (*Rhinopithecus roxellanae*) in Shennongjia Nature Reserve, China. *International Journal of Primatology*, 23(3), 657-675.

Li, Z.-X., Ma, S.-L., Hua, C.-H., & Wang, Y.-X. (1982). The distribution and habit of the Yunnan golden monkey, *Rhinopithecus bieti*. *Journal of Human Evolution*, 11(7), 633-638.

Li, Z., & Rogers, M. E. (2004). Habitat quality and activity budgets of white-headed langurs in Fusui, China. *International Journal of Primatology*, 25(1), 41-54.

Li, Z., & Rogers, M. E. (2006). Food items consumed by white-headed langurs in Fusui, China. *International Journal of Primatology*, 27(6), 1551-1567.

Liedigk, R., Yang, M., Jablonski, N. G., Momberg, F., Geissmann, T., Lwin, N.,... & Christian Roos. (2012). Evolutionary history of the odd-nosed monkeys and the phylogenetic position of the newly described Myanmar snub-nosed monkey *Rhinopithecus strykeri*. *PLoS ONE*, 7(5), e37418. https://doi.org/10.1371/journal.pone.0037418.

Lippold, L. K. (1977). The douc langur: A time for conservation. In H. S. H. Prince Rainier, & G. H. Bourne (Eds.), *Primate conservation* (pp. 513-538). Academic Press.

Lippold, L. K. (1998). Natural history of douc langurs. In N. G. Jablonski (Ed.), *The natural history of the doucs and snub-nosed monkeys* (pp. 191-206). World Scientific Publishing Co.

Lippold, L. K., & Thanh, V. N. (2008). The time is now: Survival of the douc langurs of Son Tra, Vietnam. *Primate Conservation*, 23(1), 75-79.

Liu, Z., Ding, W., & Grueter, C. C. (2004). Seasonal variation in ranging patterns of Yunnan snub-nosed monkeys *Rhinopithecus bieti* at Mt. Fuhe, China. *Acta Zoologica Sinica*, 50(5), 691-696.

Liu, Z., Ding, W., & Grueter, C. C. (2007). Preliminary data on the social organization of black-and-white snub-nosed monkeys (*Rhinopithecus bieti*) at Tacheng, China. *Acta Theriologica Sinica*, 27(2), 120-122.

Liu, Z., & Zhao, Q. (2004). Sleeping sites of *Rhinopithecus bieti* at Mt. Fuhe, Yunnan. *Primates*, 45(4), 241-248.

Long, Y., Kirkpatrick, R. C., Xiao, L., & Zhong, T. (1998). Time budgets of the Yunnan snub-nosed monkey (*Rhinopithecus [rhinopithecus] bieti*). In N. G. Jablonski (Ed.), *The natural history of the doucs and snub-nosed monkeys* (pp. 279-289). World Scientific Publishing Co.

Long, Y., Kirkpatrick, C. R., Zhong, T., & Xiao, L. (1994). Report on the distribution, population, and ecology of the Yunnan snub-nosed monkey (*Rhinopithecus bieti*). *Primates*, 35(2), 241-250.

Lu, A., Beehner, J. C., Czekala, N. M., Koenig, A., Larney, E., & Borries, C. (2011). Phytochemicals and reproductive function in wild female Phayre's leaf monkeys (*Trachypithecus phayrei crepusculus*). *Hormones and Behavior*, 59(1), 28-36.

Lu, A., Borries, J. C., Czekala, N. M., & Beehner, J. C. (2010). Reproductive characteristics of wild female Phayre's leaf monkeys. *American Journal of Primatology*, 72(12), 1073-1081.

Lu, J., & Li, B. (2006). Diurnal activity budgets of the Sichuan snub-nosed monkey *Rhinopithecus roxellana* in the Qinling Mountains of China. *Acta Theriologica Sinica*, 26(1), 26-32.

Macdonald, D. W. (1982). Notes on the size and composition of groups of proboscis monkey, *Nasalis larvatus*. *Folia Primatologica*, 37(1-2), 95-98.

Matsuda, I., Tuuga, A., Bernard, H., & Furuichi, T. (2012). Inter-individual relationships in proboscis monkeys: A preliminary comparison with other non-human primates. *Primates*, 53(1), 13-23.

Matsuda, I., Tuuga, A., & Higashi, S. (2008). Clouded leopard (*Neofelis diardi*) predation on proboscis monkeys (*Nasalis larvatus*) in Sabah, Malaysia. *Primates*, 49(3), 227-231.

Matsuda, I., Tuuga, A., & Higashi, S. (2009). The feeding ecology and activity budget of proboscis monkeys. *American Journal of Primatology*, 71(6), 478-492.

Meijaard, E., & Nijman, V. (2000). Distribution and conservation of the proboscis monkey (*Nasalis larvatus*) in Kalimantan, Indonesia. *Biological Conservation*, 92(1), 15-24.

Meijaard, E., & Nijman, V. (2009). The local extinction of the proboscis monkey *Nasalis larvatus* in Pulau Kaget Nature Reserve, Indonesia. *Oryx*, 34(1), 66-70.

Meyer, D., Momberg, F., Matauschek, C., Oswald, P., Lwin, N., Aung, S. S., Yang, T., Xiao, W., Long, Y.-C., Grueter, C., & Roos, C. (2017). *Conservation status of the Myanmar or black snub-nosed monkey* Rhinopithecus strykeri. Fauna & Flora International, Myanmar Primate Conservation Program, Dali University, Institute of Eastern-Himalaya Biodiversity Research, and German Primate Center.

Minhas, R. A., Ali, U., Awan, M. S., Ahmed, K. B., Khan, M. N., Dar, N. I., Qamar, Q. Z., Ali, H., Grueter, C. C., & Tsuji, Y. (2013). Ranging and foraging of Himalayan grey langurs (*Semnopithecus ajax*) in Machiara National Park, Pakistan. *Primates*, 54(2), 147-152.

Mitchell, A. H., & Tilson, R. L. (1986). Restoring the balance: Traditional hunting and primate conservation in the Mentawai Islands, Indonesia. In J. G. Else, & P. C. Lee (Eds.), *Primate ecology and conservation* (pp. 249-260). Cambridge University Press.

Murai, T. (2006). Mating behaviors of the proboscis monkey (*Nasalis larvatus*). *American Journal of Primatology*, 68(8), 832-837.

Murai, T., Maryati, M., Bernard, H., Mahedi, P. A., Saburi, R., & Higashi, S. (2007). Female transfer between one-male groups of proboscis monkey (*Nasalis larvatus*). *Primates*, 48(2), 117-121.

Nadler, T. (2008). Color variation in black-shanked douc langurs (*Pygathrix nigripes*), and some behavioural observations. *Vietnamese Journal of Primatology*, 1(2), 71-76.

Nadler, T., Momberg, F., Nguyen, X. D., & Lormee, N. (2003). *Vietnam primate conservation status review. Part II: Leaf monkeys.* Frankfurt Zoological Society and Fauna & Flora International—Vietnam Programme.

Nautiyal, H., & Huffman, M. (2018). Interspecific feeding association between central Himalayan langurs (*Semnopithecus schistaceus*) and Himalayan black bears (*Ursus thibetanus*) in a temperate forest of the western Indian Himalayas. *Mammal Study*, 43(1), 55-60.

Newton, P. N. (1984). *The ecology and social organisation of Hanuman langurs* (Presbytis entellus, Dufresne 1797) in *Kanha Tiger Reserve, central Indian highlands*, [Doctoral dissertation, University of Oxford].

Newton, P. N. (1985). A note on golden jackals (*Canis aureus*) and their relationship with langurs (*Presbytis entellus*) in Kanha Tiger Reserve. *Journal of the Bombay Natural History Society* 82, 633-636.

Newton, P. N. (1987). The social organization of forest Hanuman langurs (*Presbytis entellus*). *International Journal of Primatology*, 8(3), 199-232.

Newton, P. N. (1988). The variable social organization of Hanuman langurs (*Presbytis entellus*), infanticide, and the monopolization of females. *International Journal of Primatology*, 9(1), 59-77.

Newton, P. N. (1989). Association between langur monkeys (*Presbytis entellus*) and chital deer (*Axis axis*): Chance encounters or a mutualism? *Ethology*, 83(2), 89-120.

Newton, P. N. (1992). Feeding and ranging patterns of forest Hanuman langurs (*Presbytis entellus*). *International Journal of Primatology*, 13(3), 245-285.

Newton, P. N., & Dunbar, R. I. M. (1994). Colobine monkey society. In A. G. Davies, & J. F. Oates (Eds.), *Colobine monkeys: Their ecology, behaviour and evolution* (pp. 311-346). Cambridge University Press.

Nguyen, N. (2000). A survey of Tonkin snub-nosed monkeys (*Rhinopithecus avunculus*) in northern Vietnam. *Folia Primatologica*, 71(3), 157-160.

Nie, S., Xiang, Z., & Li, M. (2009). Preliminary report on the diet and social structure of gray snub-nosed monkeys (*Rhinopithecus brelichi*) at Yangaoping, Guizhou, China. *Acta Theriologica Sinica*, 29(3), 326-331.

Nijman, V., & Nekaris, K. A. I. (2012). Loud calls, startle behaviour, social organisation and predator avoidance in arboreal langurs (Cercopithecidae: *Presbytis*). *Folia Primatologica*, 83(3-6), 274-287.

Niu, K., Tan, C. L., & Yang, Y. (2010). Altitudinal movements of Guizhou snub-nosed monkeys (*Rhinopithecus brelichi*) in Fanjingshan National Nature Reserve, China: Implications for conservation management of a flagship species. *Folia Primatologica*, 81(4), 233-244.

Nüchel, J., Bøcher, P. K., Xiao, W., Zhu, A.-X., & Svenning, J.-C. (2018). Snub-nosed monkeys (*Rhinopithecus*): Potential distribution and its implication for conservation. *Biodiversity and Conservation*, 27(6), 1517-1538.

Oates, J. F., & Davies, A. G. (1994). What are the colobines? In A. G. Davies, & J. F. Oates (Eds.), *Colobine monkeys: Their ecology behaviour and evolution* (pp. 1-10). Cambridge University Press.

Oates, J. F., Davies, A. G., & Delson, E. (1994). The diversity of living colobines. In A.G. Davies, & J. F. Oates (Eds.), *Colobine monkeys: Their ecology behaviour and evolution* (pp. 45-74). Cambridge University Press.

Oates, J. F., Waterman, P. G., & Choo, G. M. (1980). Food selection by south Indian leaf-monkey, *Presbytis johnii*, in relation to leaf chemistry. *Oecologia*, 45(1), 45-56.

Oates, J. F., Whitesides, G. H., Davies, A. G., Waterman, P. G., Green, S. M., Dasilva, G. L., & Mole, S. (1990). Determinants of variation in tropical forest primate biomass: New evidence from West Africa. *Ecology*, 71(1), 328-343.

O'Brien, J. A. (2014). *The ecology and conservation of black-shanked doucs* (Pygathrix nigripes*) in Cat Tien National Park, Vietnam*. [Doctoral dissertation, University of Colorado].

Onuma, M. (2002). Daily ranging patterns of the proboscis monkey, *Nasalis larvatus*, in coastal areas of Sarawak, Malaysia. *Mammal Study, 27*(2), 141-144.

Osterholz, M., Walter, L., & Roos, C. (2008). Phylogenetic position of the langur genera *Semnopithecus* and *Trachypithecus* among Asian colobines, and genus affiliations of their species groups. *BMC Evolutionary Biology, 8,* 58.

Paciulli, L. M., & Holmes, S. (2008). Activity budget of simakobu monkeys (*Simias concolor*) inhabiting the Mentawai Islands, Indonesia. *XXII Congress of the International Primatological Society, Edinburgh, UK. Primate Eye, 96,* Special Issue: 304. [Abstract].

Pengfei, F., Garber, P., Chi, M., Guopeng, R., Changming, L., Xiaoyong, C., & Junxing, Y. (2014). High dietary diversity supports large group size in Indo-Chinese gray langurs in Wuliangshan, Yunnan, China. *American Journal of Primatology, 77*(5), 479-491.

Pham, N. (1993). First results of the diet of the red-shanked douc langur, *Pygathrix nemaeus*. *Australian Primatology, 8,* 5-6.

Pham, N. (1994). Preliminary results on the diet of the red-shanked douc langur (*Pygathrix nemaeus*). *Asian Primates, 4,* 9-11.

Phiapalath, P. (2009). *Distribution, behavior and threat of red-shanked douc langur* Pygathrix nemaeus *in Hin Namno National Protected Area, Khammouane Province, Lao PDR.* [Doctoral dissertation, Suranaree University of Technology].

Phiapalath, P., Borries, C., & Suwanwaree, P. (2011). Seasonality of group size, feeding, and breeding in wild red-shanked douc langurs (Lao PDR). *American Journal of Primatology, 73*(11), 1134-1144.

Pocock, R. I. (1928) The langurs, or leaf monkeys, of British India. *Journal of the Bombay Natural History Society, 32*(3-4), 472-504.

Poirier, F. E. (1969). The Nilgiri langur (*Presbytis johnii*) troop: Its composition, structure, function, and change. *Folia Primatologica, 10*(1), 20-47.

Poirier, F. E. (1970). Dominance structure of the Nilgiri langur (*Prebytis johnii*) of south India. *Folia Primatologica, 12*(3), 161-186.

Qi, X.-G., Li, B., Garber, P. A, Ji, W.-H., & Watanabe, K. (2009). Social dynamics of the golden snub-nosed monkey (*Rhinopithecus roxellana*): Female transfer and one-male unit succession. *American Journal of Primatology, 71*(8), 670-679.

Qi, X.-G., Li, B., & Ji, W.-H. (2008). Reproductive parameters of wild female *Rhinopithecus roxellana*. *American Journal of Primatology, 70*(4), 311-319.

Qi, X.-G., Yang, B., Garber, P. A., Ji, W.-H., Watanabe, K., & Li, B. (2011). Sexual interference in the golden snub-nosed monkey (*Rhinopithecus roxellana*): A test of the sexual competition hypothesis in a polygynous species. *American Journal of Primatology, 73*(4), 366-377.

Quan, R.-C., Ren, G., Behm, J. E., Wang, L., Huang, Y., Long, Y., & Zhu, J. (2011). Why does *Rhinopithecus bieti* prefer the highest elevation range in winter? A test of the sunshine hypothesis. *PLoS ONE, 6*(9), e24449.

Quyet, L. K., Duc, N. A., Tai, V. A., Wright, B. W., & Covert, H. H. (2007). Diet of the Tonkin snub-nosed monkey (*Rhinopithecus avunculus*) in the Khau Ca area, Ha Giang Province, northeastern Vietnam. *Vietnamese Journal of Primatology, 1*(1), 75-83.

Rajpurohit, L. S., Sommer, V., & Mohnot, S. M. (1995). Wanderers between harems and bachelor bands: Male Hanuman langurs (*Presbytis entellus*) at Jodhpur in Rajasthan. *Behaviour, 132*(3-4), 255-299.

Ranawana, K. B., Bambaradeniya, C. N. B., Bogahawatte, T. D., & Amarasinghe, F. P. (1998). A preliminary survey of the food habits of the Sri Lanka leopard (*Panthera pardus fusca*) in three montane wet zone forests of Sri Lanka. *Ceylon Journal of Science, 25,* 65-69.

Rawson, B. M. (2009). *The socio-ecology of the black-shanked douc in Mondulkiri Province, Cambodia.* [Doctoral dissertation, Australian National University].

Rawson, B. M., & Bach, L. T. (2011). Preliminary observations of geophagy amongst Cambodia's Colobinae. *Vietnamese Journal of Primatology, 5,* 41-46.

Ren, B., Li, D., Liu, Z., Li, B., Wei, F., & Li, M. (2010). First evidence of prey capture and meat eating by wild Yunnan snub-nosed monkeys *Rhinopithecus bieti* in Yunnan, China. *Current Zoology, 56*(2), 227-231.

Ren, B., Li, M., Long, Y., & Wei, F. (2009b). Influence of day length, ambient temperature, and seasonality on daily travel distance in the Yunnan snub-nosed monkey at Jinsichang, Yunnan, China. *American Journal of Primatology, 71*(3), 233-241.

Ren, B., Li, M., Long, Y., Wei, F., & Wu, R. (2009a). Home range and seasonality of Yunnan snub-nosed monkeys. *Integrative Zoology, 4*(2), 162-171.

Ren, B., Zhang, S.-Y., Xia, S.-Z., Li, Q.-F., Liang, B., & Lu, M.-Q. (2003). Annual reproductive behavior of *Rhinopithecus roxellana*. *International Journal of Primatology, 24*(3), 575-589.

Ren, R., Su, Y., Yan, K., Li, J., Zhou, Y., Zhu, Z., Hu, Z., & Hu, Y. (1998). Preliminary survey of the social organization of *Rhinopithecus [rhinopithecus] roxellana* in Shennongjia National Natural Reserve, Hubei, China. In N. G. Jablonski (Ed.), *The natural history of the doucs and snub-nosed monkeys* (pp. 269-278). World Scientific Publishing Co.

Ripley, S. (1965). *The ecology and social behavior of the Ceylon gray langur,* Presbytis entellus thersites. [Doctoral dissertation, University of California, Berkeley].

Roos, C., Boonratana, R., Supriatna, J., Fellowes, J. R., Groves, C. P., Nash, S. D., Rylands, A. B., & Mittermeier, R. A. (2014). An updated taxonomy and conservation status review of Asian primates. *Asian Primates, 4*(1), 2-38.

Roos, C., Zinner, D., Kubatko, L. S., Schwarz, C., Yang, M., Meyer, D., Nash, S. D., Xing, J., Batzer, M. A., Brameier, M., Leendertz, F. H., Ziegler, T., Perwitasari-Farajallah, D., Nadler, T., Walter, L., & Osterholz, M. (2011). Nuclear versus mitochondrial DNA: Evidence for hybridization in colobine monkeys. *BMC Evolutionary Biology, 11,* 77.

Rudran, R. (1973a). Adult male replacement in one-male troops of purple-faced langurs (*Presbytis senex senex*) and its effect on population structure. *Folia Primatologica*, 19(2), 166-192.

Rudran, R. (1973b). The reproductive cycles of two subspecies of purple-faced langurs (*Presbytis senex*) with relation to environmental factors. *Folia Primatologica*, 19(1), 41-60.

Salter, R. E., MacKenzie, N. A., Nightingale, N., Aken, K. M., & Chai, P. K. P. (1985). Habitat use, ranging behaviour, and food habits of the proboscis monkey, *Nasalis larvatus* (von Wurmb), in Sarawak. *Primates*, 26(4), 436-451.

Sayers, K., & Norconk, M. A. (2008). Himalayan *Semnopithecus entellus* at Langtang National Park, Nepal: Diet, activity patterns, and resources. *International Journal of Primatology*, 29(2), 509-530.

Selvan, K. M., Gopi, G. V., Lyngdoh, S., Habib, B., & Hussain, S. A. (2013). Prey selection and food habits of three sympatric large carnivores in a tropical lowland forest of the eastern Himalayan biodiversity hotspot. *Mammalian Biology*, 78(4), 296-303.

Sharma, N., & Ahmed, M. (2017). Distribution of endangered Kashmir gray langur (*Semnopithecus ajax*) in Bhaderwah, Jammu and Kashmir, India. *Journal of Wildlife Research*, 5(1), 1-5.

Singh, M., Roy, K., & Singh, M. (2011). Resource partitioning in sympatric langurs and macaques in tropical rainforests of the central western Ghats, south India. *American Journal of Primatology*, 73(4), 335-346.

Solanki, G. S., Kumar, A., & Sharma, B. K. (2007). Reproductive strategies of *Trachypithecus pileatus* in Arunachal Pradesh, India. *International Journal of Primatology*, 28(5), 1075-1083.

Solanki, G. S., Kumar, A., & Sharma, B. K. (2008). Feeding ecology of *Trachypithecus pileatus* in India. *International Journal of Primatology*, 29(1), 173-182.

Sommer, V., & Rajpurohit, L. S. (1989). Male reproductive success in harem troops of Hanuman langurs (*Presbytis entellus*). *International Journal of Primatology*, 10(4), 293-317.

Sommer, V., Srivastava, A., & Borries, C. (1992). Cycles, sexuality, and conception in free-ranging langurs (*Presbytis entellus*). *American Journal of Primatology*, 28(1), 1-27.

Stanford, C. B. (1989). Predation on capped langurs (*Presbytis pileata*) by cooperatively hunting jackals (*Canis aureus*). *American Journal of Primatology*, 19(1), 53-56.

Stanford, C. B. (1991). The capped langur in Bangladesh: Behavioral ecology and reproductive tactics. *Contributions to Primatology*, 26, 1-179.

Stanford, C.B. (1992). Comparative ecology of the capped langur *Presbytis pileata* Blyth in two forest types in Bangladesh. *Journal of Bombay Natural History Society*, 89, 187-193.

Starin, E. D. (1978). A preliminary investigation of home range use in the Gir Forest langur. *Primates*, 19(3), 551-567.

Stark, D. J., Nijman, V., Lhota, S., Robins, J. G., & Goossens, B. (2012). Modeling population viability of local proboscis monkey *Nasalis larvatus* populations: Conservation implications. *Endangered Species Research*, 16(1), 31-43.

Steenbeek, R., & van Shaik, C. P. (2001). Competition and group size in Thomas's langurs (*Presbytis thomasi*): The folivore paradox revisited. *Behavioral Ecology and Sociobiology*, 49(2), 100-110.

Sterck, E. H. M., & van Hooff, J. A. R. A. M. (2000). The number of males in langur groups: Monopolizability of females or demographic processes? In P. M. Kappeler (Ed.), *Primate males: Causes and consequences of variation in group composition* (pp. 120-129). Cambridge University Press.

Sterner, K. N., Raaum, R. L., Zhang, Y. P., Stewart, C. B., & Disotell, T. R. (2006). Mitochondrial data support an odd-nosed colobine clade. *Molecular Phylogenetics and Evolution*, 40(1), 1-7.

Stewart, C. B., & Disotell, T. R. (1998). Primate evolution: In and out of Africa. *Current Biology*, 8(16), 582-588.

Su, D. F., & Jablonski, N. G. (2009). Locomotor behavior and skeletal morphology of the odd-nosed monkeys. *Folia Primatologica*, 80(3), 189-219.

Su, Y., Ren, R., Yan, K., Li, J., Zhou, Y., Zhu, Z., Hu, Z., & Hu, Y. (1998). Preliminary survey of the home range and ranging behavior of golden monkeys (*Rhinopithecus [rhinpithecus] roxellana*) in Hennongjia National Natural Reserve, Hubei, China. In N. G. Jablonski (Ed.), *The natural history of the doucs and snub-nosed monkeys* (pp. 255-268). World Scientific Publishing Co.

Sugiyama, Y. (1964). Group composition, population density and some sociological observations of Hanuman langurs (*Presbytis entellus*). *Primates*, 5(3-4), 7-37.

Supriatna, J., Manullang, B. O., & Soekara, E. (1986). Group composition, home range, an diet of the maroon leaf monkey (*Presbytis rubicunda*) at Tanjung Putting Reserve, central Kalimantan, Indonesia. *Primates*, 27(2), 185-190.

Sushma, H. S., & Singh, M. (2006). Resource partitioning and interspecific interactions among sympatric rain forest arboreal mammals of the western Ghats, India. *Behavioral Ecology*, 17(3), 479-490.

Sussman, R. W., Cheverud, J. M., & Bartlett, T. Q. (1995). Infant killing as an evolutionary strategy: Reality or myth? *Evolutionary Anthropology*, 3(5), 149-151.

Tan, C. L., Guo, S., & Li, B. (2007). Population structure and ranging patterns of *Rhinopithecus roxellana* in Zhouzhi National Nature Reserve, Shaanxi, China. *International Journal of Primatology*, 28(3), 577-591.

Tenaza, R. R. (1989a). Female sexual swellings in the Asian colobine *Simias concolor*. *American Journal of Primatology*, 17(1), 81-86.

Tenaza, R. R. (1989b). Male intergroup vocalizations of the pig-tailed langur (*Simias concolor*). *Primates*, 30(2), 199-206.

Tenaza, R. R., & Fuentes, A. (1995). Monandrous social organization of pigtailed langurs (*Simias concolor*) in the Pagai Islands, Indonesia. *International Journal of Primatology*, 16(2), 295-310.

Tenaza, R. R., & Tilson, R. L. (1985). Human predation and Kloss's gibbon (*Hylobates kiossii*) sleeping trees. *American Journal of Primatology*, 8(4), 299-308.

Thanh, H. D. (2007). *Behavioural ecology and conservation of Rhinopithecus avunculus in Vietnam*. Rufford Small Grants Foundation.

Thapar, V. (1986). *Tiger: Portrait of a predator*. Facts on File.

Thinley, P., Norbu, T., Rajaratnam, R., Vernes, K., Wangchuk, K., Choki, K., Tenzin, J., Tenzin, S., Kinley, Dorji, S., Wangchuk, T., Cheda, K., & Gempa, G. (2019). Population abundance and distribution of the endangered golden langur (*Trachypithecus geei*, Khajuria 1956) in Bhutan. *Primates*, 60(5), 437-448.

Tilson, R. L. (1977). Social organization of Simakobu monkeys (*Nasalis concolor*) in Siberut Island, Indonesia. *Journal of Mammalogy*, 58(2), 202-212.

Ting, N., Tosi, A. J., Li, Y., Zhang, Y.-P., & Disotell, T. R. (2008). Phylogenetic incongruence between nuclear and mitochondrial markers in the Asian colobines and the evolution of the langurs and leaf monkeys. *Molecular Phylogenetics and Evolution*, 46(2), 466-474.

Tran, T. H. (2010). Stopping the trade of Vietnam's primates: Experiences and cases from ENV's Wildlife Crime Unit. In T. Nadler, B. M. Rawson, & V. N. Thinh (Eds.), *Conservation of primates in Indochina* (pp. 233-236). Frankfurt Zoological Society.

Tsujia, Y., Mitani, M., Widayatic, K. A., Suryobroto, B., & Watanabe, K. (2019). Dietary habits of wild Javan lutungs (*Trachypithecus auratus*) in a secondary plantation mixed forest: Effects of vegetation composition and phenology. *Mammalian Biology*, 98, 80-90.

Ulibarri, L. (2013). *The socioecology of red-shanked doucs (Pygathrix nemaeus) in Son Tra Nature Reserve, Vietnam*. [Doctoral dissertation, University of Colorado].

Vandercone, R. (2011). *Dietary shifts, niche relationships and interspecific competition in the sympatric grey langur (Semnopithecus entellus) and the purple-faced langur (Trachypithecus vetulus) in Sri Lanka*. [Doctoral dissertation, Washington University].

Vandercone, R., Dinad, C., Wijethunga, G., Ranawana, K., & Rasmussen, D. T. (2012). Dietary diversity and food selection in Hanuman langurs (*Semnopithecus entellus*) and purple-faced langurs (*Trachypithecus vetulus*) in the Kaludiyapokuna Forest Reserve in the dry zone of Sri Lanka. *International Journal of Primatology*, 33(6), 1382-1405.

Waltert, M., Abegg, C., Ziegler, T., Hadi, S., Priata, D., & Hodges, K. (2008). Abundance and community structure of Mentawai primates in the Peleonan forest, north Siberut, Indonesia. *Oryx*, 42(3), 375-379.

Wang, H., Tan, C. L., Gao, Y., & Li, B. (2004). A takeover of resident male in the Sichuan snub-nosed monkey *Rhinopithecus roxellanae* in Qinling Mountains. *Acta Zoologica Sinica*, 50(5), 859-862.

Wang, W., Qiao, Y., Li, S., Pan, W., & Yao, M. (2017). Low genetic diversity and strong population structure shaped by anthropogenic habitat fragmentation in a critically endangered primate, *Trachypithecus leucocephalus*. *Heredity*, 118(6), 542-553.

Wang, X. P., Yu, L., Roos, C., Ting, N., Chen, C. P., Wang, J., & Zhang, Y. P. (2012). Phylogenetic relationships among the colobine monkeys revisited: New insights from analyses of complete mt genomes and 44 nuclear non-coding markers. *PLoS ONE*, 7(4), e36274.

Watanabe, K. (1981). Variations in group composition and population density of the two sympatric Mentawaian leaf-monkeys. *Primates*, 22(2), 145-160.

Waterman, P. G., Ross, J. A. M., Bennet, E. L., & Davies, A. G. (1988). A comparison of the floristics and leaf chemistry of the tree flora in two Malaysian rain forests and the influence of leaf chemistry on the populations of colobine monkeys in the Old World. *Biological Journal of the Linnean Society*, 34(1), 1-32.

Whittaker, D. J., Ting, N., & Melnick, D. J. (2006) Molecular phylogenetic affinities of the simakobu monkey (*Simias concolor*). *Molecular Phylogenetics and Evolution*, 39(3), 887-892.

Wich, S. A., Steenbeek, R., Sterck, E. H. M., Korstjens, A. H., Willems, E. P., & van Schaik, C. P. (2007). Demography and life history of Thomas langurs (*Presbytis thomasi*). *American Journal of Primatology*, 69(6), 641-651.

Workman, C. (2010). Diet of the Delacour's langur (*Trachypithecus delacouri*) in Van Long Nature Reserve, Vietnam. *American Journal of Primatology*, 72(4), 317-324.

Wu, B. Q., & He, S. J. (1989). A micro-quantitative analysis of types of residuary diets among excrements of a group of *Rhinopithecus bieti* in snowing season. *Zoological Research*, 10, 101-109.

Xi, W., Li, B., Zhao, D., Ji, W., & Zhang, P. (2008). Benefits to female helpers in wild *Rhinopithecus roxellana*. *International Journal of Primatology*, 29(3), 593-600.

Xiang, Z.-F., Huo, S., & Xiao, W. (2010). Activity budget of *Rhinopithecus bieti* at Tibet: Effects of day length, temperature and food availability. *Current Zoology*, 56(6), 650-659.

Xiang, Z.-F., Huo, S., Xiao, W., Quan, R.-C., & Grueter, C. C. (2007). Diet and feeding behavior of *Rhinopithecus bieti* at Xiaochangdu, Tibet: Adaptations to a marginal environment. *American Journal of Primatology*, 69(10), 1141-1158.

Xiang, Z.-F., Liang, W.-B., Nie, S.-G., & Li, M. (2012). Diet and feeding behavior of *Rhinopithecus brelichi* at Yangaoping, Guizhou. *American Journal of Primatology*, 74(6), 551-560.

Xiang, Z.-F., Nie, S.-G., Lei, X.-P., Chang, Z.-F., Wei, F., & Li, M. (2009). Current status and conservation of the gray snub-nosed monkey *Rhinopithecus brelichi* (Colobinae) in Guizhou, China. *Biological Conservation*, 142(3), 469-476.

Xiang, Z.-F., & Sayers, K. (2009). Seasonality of mating and birth in wild black-and-white snub-nosed monkeys (*Rhinopithecus bieti*) at Xiaochangdu, Tibet. *Primates*, 50(1), 50-55.

Yang, B., Zhang, P., Garber, P., Hedley, R., & Li, B. (2016). Sichuan snub-nosed monkeys (*Rhinopithecus roxellana*) consume cicadas in the Qinling Mountains, China. *Folia Primatologica*, 87(1), 11-16.

Yang, M., Sun, D. Y., Zinner, D., & Roos, C. (2009). Reproductive parameters in Guizhou snub-nosed monkeys (*Rhinopithecus brelichi*). *American Journal of Primatology*, 71(3), 266-270.

Yang, S. (2003). Altitudinal ranging of *Rhinopithecus bieti* at Jinsichang, Lijiang, China. *Folia Primatologica*, 74(2), 88-91.

Yang, S., & Zhao, Q. K. (2001). Bamboo leaf–based diet of *Rhinopithecus bieti* at Lijiang, China. *Folia Primatologica*, 72(2), 92-95.

Yang, Y., Groves, C., Garber, P., Wang, X., Li, H., Long, Y., Li, G., Tian, Y., Dong, S., Yang, S., Behie, A., & Xiao, W. (2019). First insights into the feeding habits of the critically endangered black snub-nosed monkey, *Rhinopithecus strykeri* (Colobinae, Primates). *Primates*, 60(2), 143-153.

Yao, H., Liu, X., Stanford, C., Yang, J., Huang, T., Wu, F., & Li, Y. (2011). Male dispersal in a provisioned multilevel group of *Rhinopithecus roxellana* in Shennongjia Nature Reserve, China. *American Journal of Primatology*, 73(12), 1280-1288.

Yeager, C. P. (1989). Feeding ecology of the proboscis monkey (*Nasalis larvatus*). *International Journal of Primatology*, 10(6), 497-530.

Yeager, C. P. (1990a). Proboscis monkey (*Nasalis larvatus*) social organization: Group structure. *American Journal of Primatology*, 20(2), 95-106.

Yeager, C. P. (1990b). Notes on the sexual behavior of the proboscis monkey (*Nasalis larvatus*). *American Journal of Primatology*, 21(3), 223-227.

Yeager, C. P. (1991a). Possible antipredator behavior associated with river crossings by proboscis monkeys (*Nasalis larvatus*). *American Journal of Primatology*, 24(1), 61-66.

Yeager, C. P. (1991b). Proboscis monkey (*Nasalis larvatus*) social organization: Intergroup patterns of association. *American Journal of Primatology*, 23(2), 73-86.

Yeager, C. P. (1992). Proboscis monkey (*Nasalis larvatus*) social organization: Nature and possible functions of intergroup patterns of association. *American Journal of Primatology*, 26(2), 133-137.

Yeager, C. P. (1995). Does intraspecific variation in social systems explain reported differences in the social structure of the proboscis monkey (*Nasalis larvatus*)? *Primates*, 36(4), 575-582.

Yeager, C. P., & Kirkpatrick, R. C. (1998). Asian colobine social structure: Ecological and evolutionary constraints. *Primates*, 39(2), 147-155.

Yeager, C. P., & Kool, K. M. (2000). The behavioral ecology of Asian colobines. In P. Whitehead, & C. Jolly (Eds.), *Old World monkeys* (pp. 496-521). Cambridge University Press.

Zhang, J., Zhao, D., & Li, B. (2010). Postconflict behavior among female Sichuan snub-nosed monkeys *Rhinopithecus roxellana* within one-male units in the Qinling Mountains, China. *Current Zoology*, 56(2), 222-226.

Zhang, P., Watanabe, K., & Li, B. (2008a). Female social dynamics in a provisioned free-ranging band of the Sichuan snub-nosed monkey (*Rhinopithecus roxellana*) in the Qinling Mountains, China. *American Journal of Primatology*, 70(11), 1013-1022.

Zhang, P., Watanabe, K., Li, B., & Qi, X. (2008b). Dominance relationships among one-male units in a provisioned free-ranging band of the Sichuan snub-nosed monkeys (*Rhinopithecus roxellana*) in the Qinling Mountains, China. *American Journal of Primatology*, 70(7), 634-641.

Zhang, P., Watanabe, K., Li, B., & Tan, C. L. (2006). Social organization of Sichuan snub-nosed monkeys (*Rhinopithecus roxellana*) in the Qinling Mountains, central China. *Primates*, 47(4), 374-382.

Zhang, S., Ren, B. P., & Li, B. G. (1999). A juvenile Sichuan golden monkey (*Rhinopithecus roxellana*) predated by a goshawk (*Accipiter gentilis*) in the Qinling Mountains. *Folia Primatologica*, 70(3), 175-176.

Zhang, Y., & Ryder, O. A. (1998). Mitochondrial cytochrome *b* gene sequences of Old World monkeys: With a special reference on evolution of Asian colobines. *Primates*, 39(1), 39-49.

Zhao, D., Ji, W., Li, B., & Watanabe, K. (2008a). Mate competition and reproductive correlates of female dispersal in a polygynous primate species (*Rhinopithecus roxellana*). *Behavioural Processes*, 79(3), 165-170.

Zhao, D., & Li, B. (2009). Do deposed adult male Sichuan snub-nosed monkeys *Rhinopithecus roxellana* roam as solitary bachelors or continue to interact with former band members? *Current Zoology*, 55(3), 235-237.

Zhao, D., Li, B., Groves, C. P., & Watanabe, K. (2008b). Impact of male takeover on intra-unit sexual interactions and subsequent interbirth interval in wild *Rhinopithecus roxellana*. *Folia Primatologica*, 79(2), 93-102.

Zhao, D., Li, B., Li, Y., & Wada, K. (2005). Extra-unit sexual behaviour among wild Sichuan snub-nosed monkeys (*Rhinopithecus roxellana*) in the Qinling Mountains of China. *Folia Primatologica*, 76(3), 172-176.

Zhao, Q., Borries, C., & Pan, W. (2011). Male takeover, infanticide and female countertactics in white-headed leaf monkeys (*Trachypithecus leucocephalus*). *Behavioral Ecology and Sociobiology*, 65(8), 1535-1547.

Zhao, Q., & Tan, C. L. (2010). Inter-unit contests within a provisioned troop of Sichuan snub-nosed monkeys (*Rhinopithecus roxellana*) in the Qinling Mountains, China. *American Journal of Primatology*, 73(3), 262-269.

Zhixiang, L., Shilai, M., Chenhui, H., & Yingxiang, W. (1982). The distribution and habitat of the Yunnan golden monkey, *Rhinopithecus bieti*. *Journal of Human Evolution*, 11(7), 633-638.

Zhou, Q., Huang, C., Li, Y., & Cai, X. (2007b). Ranging behavior of the François' langur (*Trachypithecus francoisi*) in the Fusui Nature Reserve, China. *Primates*, 48(4), 320-323.

Zhou, Q., Luo, B., Wei, F., & Huang, C. (2013). Habitat use and locomotion of the François' langur (*Trachypithecus francoisi*) in limestone habitats of Nonggang, China. *Integrative Zoology*, 8(4), 346-355.

Zhou, Q., Tang, X., Huang, H., & Huang, C. (2011). Factors affecting the rangin behavior of white-headed langurs (*Trachypithecus leucocephalus*). *International Journal of Primatology, 32*(2), 511-523.

Zhou, Q., Wei, F., Huang, C., Li, M., Ren, B., & Luo, B. (2007a). Seasonal variation in the activity patterns and time budgets of *Trachypithecus francoisi* in the Nonggang Nature Reserve, China. *International Journal of Primatology, 28*(3), 657-671.

Zhou, Q. H., Wei, F. W., Li, M., Huang, C. M., & Luo, B. (2006). Diet and food choice of *Trachypithecus francoisi* in the Nonggang Nature Reserve, China. *International Journal of Primatology, 27*(5), 1441-1460.

Zhou, X., Wang, B., Pan, Q., Zhang, J., Kumar, S., Sun, X., Liu, Z., Pan, H., Lin, Y., Liu, G., Zhan, W., Li, M., Ren, B., Ma, X., Ruan, H., Cheng, C., Wang, D., Shi, F., Hui, Y.,... & Li, M. (2014). Whole-genome sequencing of the snub-nosed monkey provides insights into folivory and evolutionary history. *Nature Genetics, 46*(12), 1303-1310.

CHAPTER 19

Regarding Old World Monkeys

Larissa Swedell and W. Scott McGraw

Since the dawn of primatology, Old World monkeys have served as the archetypical primate. Early field studies of apes notwithstanding, it was macaques that emerged as standard bearers of the nonhuman primate flag nearly a century ago. Crucial to this movement were Clarence Ray Carpenter, who founded the Cayo Santiago rhesus macaque colony in 1938, and Kinji Imanishi, who launched field research on Japanese macaques in 1948. That same year, the first animal was sent into space—a macaque of course—and shortly thereafter Japanese macaques starting salting their potatoes on Koshima Island (Kawamura, 1954). Baboons received a fair amount of early attention as well, though few knew about the first baboon field study because it was not published until 60 years after its completion (Marais, 1969). It was instead Zuckerman's (1932) study of captive hamadryas baboons at the London Zoo that established the dominant paradigm regarding the drivers of baboon behavior; unfortunately, we now know that many of his conclusions were misguided, due in no small measure to a flurry of field studies of baboons in their natural habitats. Bolwig's insightful early descriptions of baboon behavior in South Africa (Bolwig, 1959) were closely followed by DeVore and Washburn's observations of olive and yellow baboons in Kenya, Hall's of chacma baboons in southern Africa, Kummer and Kurt's of hamadryas baboons in Ethiopia (Kummer & Kurt, 1963; DeVore & Hall, 1965), and Rowell's (1966) of olive baboons in Uganda. These influential descriptive accounts of baboon society were paralleled by other early fieldwork on Old World monkeys: Buxton's (1952), Haddow's (1952), and Booth's (1956) studies of arboreal African monkeys; Sugiyama's (1965) and Jay's (1963, 1965) of langurs in India; Nolte's (1955) and Simonds's (1965) of bonnet macaques in India; and the first studies of the Cayo Santiago rhesus macaques by Carpenter (1942a,b) and Stuart Altmann (1962). Early field studies such as these paved the collective primatological consciousness, so much so that cercopithecoids, especially baboons and macaques, became the "typical primate" that subsequently reduced our ability to see variation across the rest of the order (Strier, 1994). Our views of cercopithecoids are changing, however; research over the past two decades has enhanced our understanding of their biology via methodological advances, discovery of both new and fossil species, and taxonomic revisions. These developments include both confirmations and contradictions to received wisdom, and even a few paradigm shifts.

The large African papionins in particular occupy a special place in primatology, not only for their use as anthropological windows on our evolutionary history, but also as superb exemplars of adaptation and homoplasy (Jolly, 1970, 2001). For many years it was argued that the largest African papionins—baboons (Papio), geladas (Theropithecus), and mandrills and drills (Mandrillus)—were each other's closest relatives, while the "mangabeys," united by their long limbs, sunken cheekbones, and shared "whoop-gobble" vocalizations, formed a sister clade. Despite early molecular evidence to the contrary (Cronin & Sarich, 1976), many had difficulty accepting the notion of mangabey monophyly or that the larger-bodied African papionins did not form a cohesive radiation (Sarich, 1970). Nevertheless, research conducted in the lab and the field over the last 25 years has confirmed that baboons are in fact the sister taxon of arboreal mangabeys (Lophocebus), while terrestrial mangabeys (Cercocebus) are most closely related to mandrills and drills (Mandrillus) (Disotell, 1994; Fleagle & McGraw, 1999). Referring to "mangabeys" as an entity is therefore incorrect unless one's purpose is to highlight paraphyly; however, the more interesting challenge is to identify the causes of such rampant convergence in both clades.

The close affinities of Lophocebus and Papio are well illustrated by the East African monkey known as the kipunji. Discovered by science in 2004, this enigmatic cercopithecoid looks like an arboreal mangabey (it was originally considered a member of Lophocebus) but is genetically more similar to baboons. The nature of the kipunji continues to be debated, but most authorities contend it is the result of hybridization between Lophocebus and yellow baboons (Papio cyncocephalus) as recently as 0.65 mya (Burrell et al., 2009). With fewer than 1,000 individuals, this monkey at once becomes one of the newest known Old World monkeys and the most endangered.

The dynamic nature of baboon phylogeny is evidenced by inconsistent taxonomy, even in recent literature, which reflects the tension between the clear variation in baboon phenotypes and behavior and the consistent interbreeding at all known hybrid zones. The lesson here is that baboons as we know them are still evolving, perhaps rapidly so (Rogers et al., 2019). The same has been said for the guenons, a clade routinely described as a recent radiation. We often point to *Cercopithecus* as an example of "evolution in action," with the many variations in facial patterns and coat color schemes shaped by recent (< 3 mya) contractions and expansions of the sub-Saharan forest belt responding to climatic change. Guenon populations stranded in forest refugia and cut off from populations on other islands evolved the spots, stripes, chevrons, counter-shadings, and so forth that form the basis for guenon taxonomy. Prevailing wisdom holds that the present distributions of the 20 or so *Cercopithecus* species closely mirror the refuge sites of the parent populations and that the ability of sympatric species to produce fertile hybrids is evidence that interruptions in gene flow were relatively recent. Several new fossil finds, however, call into question not only the youth of the guenon clade, but also the notion that the present distribution of *Cercopithecus* spp. mirrors the locations of the ancestral islands from which modern decedents emerged (Gilbert et al., 2014; Plavcan et al., 2019). It seems that guenon biogeographic history is far more complex than previously envisioned.

A dominant theme for much of primatology has been that Old World monkeys retain more or less the same *baüplan* and display less social and ecological diversity than other radiations. Few authorities argued for behavioral homogeneity; however, there was a general feeling that cercopithecoids occupied fewer niches than other primates and were comparatively invariable generalists. Discovering that there is variation in nature is no cause for fanfare, but what the chapters in this section demonstrate is that cercopithecoids are far more variable in their behavior and biology than they get credit for. Take postcranial anatomy and positional behavior. In his influential 1970 paper, Schultz famously catalogued the skeletal "uniformity of the Cercopithecoidea" (Schultz, 1970), and the notion that monkeys of the Old World were all variants on a generalized quadrupedal theme was more or less taken as a given. Subsequent field and laboratory study has highlighted surprising locomotor diversity including, among other things, brachiating Asian colobines (Byron & Covert, 2004; Wright et al., 2008; Su & Jablonski, 2009; Zhu et al., 2015), bounding African colobines (Mittermeier & Fleagle, 1976; Morbeck, 1977; Rose, 1979;

McGraw, 1998; Dunham, 2015), and a growing number of guenon species found to be moving on or near the ground (Mekonnen et al., 2018; Hart et al., 2019; Arenson et al., 2020). Dietary monikers continue to be revised. Gone is the notion that colobines are dedicated folivores. Studies of cusp morphology and fracture mechanics (e.g., Lucas & Teaford, 1994) suggest colobine teeth are designed as much for seed processing as they are for shearing leaves, a conclusion supported by the multiple field studies highlighting the importance of seeds in colobine diets (Dasilva, 1992; Maisels et al., 1994; Korstjens, 2001; McGraw et al., 2016).

Each field study reveals previously undiscovered variation among populations, which is decreasing our ability to make sweeping generalizations about particular taxa. Indeed, much of our received wisdom about categorical patterns of variation in cercopithecoid biology requires reassessment. The challenge facing the next generation of primatologists will be not merely to document this variation, but to explain its determinants. This challenge has become even greater with, for example, evidence hinting at cultural differences in guenon feeding behavior (e.g., Tournier et al., 2014), mismatches between diets and the cranio-dental morphologies that evolved to process them (McGraw & Daegling, 2019), and home ranges that appear unconstrained by group size or resource distribution (Henriquez et al., in press).

Our view of cercopithecoid social behavior has also undergone paradigm revisions over the past half century. We have shifted from an early focus on aggression (Zuckerman, 1932) and male dominance (Hall & DeVore, 1965) to a view of primate sociality centered on social relationships, kinship, and females as core elements structuring cercopithecoid societies, and these changes have not uncoincidentally paralleled the increase in female primatologists leading field studies and research programs (Rowell, 1966; Altmann, 1980; Strum, 1982, 1987; Hrdy & Williams, 1983). We have also seen a shift from group-level processes to individual-level selection, centered on the rise of sociobiology over four decades ago, with much of this discussion within primatology focusing on whether sexual selection shapes infanticidal behavior by males (Hrdy, 1974; Hausfater & Hrdy, 1984; Hrdy et al., 1995; Sussman et al., 1995). The past two decades in particular have brought a renewed focus on the importance of primate sociality, spearheaded by a series of analyses demonstrating the selective advantages of social bonds and social integration for both males and females (Silk et al., 2003, 2009, 2010; Schülke et al., 2010; Thompson, 2019).

As with many other gregarious animals, dyadic social relationships comprise the core of cercopithecoid social groups (Hinde, 1976), and this is true for even the most complex of social systems, multilevel societies (Colmenares, 2004; Grueter et al., 2020). A few decades ago, the only multilevel societies that had been identified in primates were those of geladas (Dunbar & Dunbar, 1975) and hamadryas baboons (Kummer & Kurt, 1963), but since then we've identified them in several other cercopithecoid taxa, including Guinea baboons (Galat-Luong et al., 2006; Fischer et al., 2017), snub-nosed monkeys (Zhang et al., 2006; Ren et al., 2012), and, most recently, Rwenzori colobus monkeys (Stead & Teichroeb, 2019). The past decade has also shone the spotlight on two previously little-known baboon taxa, the Kinda baboon (Petersdorf et al., 2019) and the Guinea baboon (Fischer et al., 2017), and ongoing genetic work has revealed a previously underappreciated degree of shared evolutionary history and behavioral similarity between Guinea and hamadryas baboons (Kopp et al., 2014; Jolly, 2020). In both Guineas and hamadryas, males depart from the standard cercopithecoid pattern and remain philopatric, whereas females are the primary agents of gene flow (Kopp et al., 2015; Staedele et al., 2015). Males in both taxa are not only tolerant of one another but in fact appear to maintain strong and differentiated social relationships that likely shape patterns of reproduction and fitness over time (Abegglen, 1984; Patzelt et al., 2014; Chowdhury et al., 2015).

No discussion of Old World monkeys would be complete without considering their conservation. Unfortunately, little of the news is good and an increasing number of cercopithecoid taxa are becoming threatened, if they are not already. The monkeys in the most desperate shape are the red colobus, and each of the 20 or so species is categorized by the International Union for the Conservation of Nature (IUCN) as either Vulnerable, Endangered, or Critically Endangered. One of the great tragedies of primate conservation is that as early as the mid-1950s, biologists were warning that red colobus populations were plummeting and that, unless trends were reversed, several species—including one commonly known as Miss Waldron's red colobus (*Piliocolobus waldroni*)—were in danger of extinction (Booth, 1956). Despite this early alarm, we failed to take measures that would have provided adequate protection or halt the forces that threatened its existence, and Miss Waldron's red colobus had the inglorious distinction of becoming the first primate to go extinct in approximately 500 years (Oates et al., 2000). Equally disturbing is the fact that several other red colobus species are similarly imperiled by habitat destruction, increased human hunting, and other pressures from an expanding human population. These factors, combined with an inability to adapt well to humans,

have made red colobus the most threatened group of primates in Africa. A five-year action plan outlining specific conservation priorities for each red colobus species has been developed and was recently published (Linder et al., 2021). We hope the priorities set out in this document can be operationalized and that aggressive efforts are made to help safeguard these and other monkey populations. The need to do so is great: Recent surveys reveal that the conservation status of several guenon species—once common throughout African forests—is dire. Indeed, few African cercopithecoids can be accurately or responsibly described as "common." (A similar action plan is being prepared for members of the *Cercocebus-Mandrillus* clade; Fernandez et al., 2019.)

A growing number of Asian colobines also require immediate conservation attention. Indeed, since creation of the IUCN's list of 25 most endangered primates, nine Asian colobines have appeared on the roster: *Trachypithecus delacouri, T. poliocephalus, T. geei, Semnopithecus vetulus, S. ajax, Presbytis hosei, Pygathrix cinerea, Rhinopithecus avunculus,* and *Simias concolor*. For macaques and baboons, who are more flexible in their adaptation to human encroachment, the challenges are different; conflict between these taxa and humans at points of contact can have adverse effects on the monkeys and humans involved (Beamish, 2009; Chowdhury et al., 2020), which can lead to local extirpation, especially in regions where the monkeys are not legally protected. The good news is that there is a growing awareness among researchers, students, and conservation workers to focus increased conservation efforts on all Old World monkeys, primates that—we argue—have for too long not received the conservation attention or resources they deserve.

It is an exciting time to be working on Old World monkeys. Cercopithecoids have now been cloned (Liu et al., 2018), have been shown to possess syntactic abilities (Arnold & Zuberbühler, 2006), and are performing comparably to apes in several tests of intelligence (e.g., Schmitt et al., 2012). New species continue to be discovered (Jones et al., 2005; Geissmann et al., 2011; Hart et al., 2012), and new methodological and analytical approaches continue to give us a more detailed understanding of links between ecology, physiology, and behavior (Jaeggi et al., 2018; Habig et al., 2019; Fürtbauer et al., 2020; Henriquez et al., in press). One of our challenges moving forward will be to examine some of the unanswered questions about the patterns we see in cercopithecoid biology. For example, to what extent is variation in behavior explained by traditional socioecological models? Do we abandon them given the lack of fit to much of the variation in nature, or do we refine them further? Is local ecological variation and phenotypic plasticity a better explanation in most cases than ecological pressures during evolutionary history? There has been much discussion in the literature of these questions, and there are no easy answers (Janson, 2000; Thierry, 2008; Clutton-Brock & Janson, 2012). One thing we can say for certain is that our understanding and appreciation of the variation in behavior and ecology in cercopithecoids, and primates as a whole, continues to grow.

REFERENCES CITED—CHAPTER 19

Abegglen, J.-J. (1984). *On socialization in Hamadryas baboons.* Associated University Presses.

Altmann, J. (1980). *Baboon mothers and infants.* Harvard University Press.

Altmann, S. A. (1962). A field study of the sociobiology of rhesus monkeys, *Macaca mulatta. Annals of the New York Academy of Science,* 102(2), 338-435.

Arenson, J. L., Sargis, E. J., Hart, J. A., Hart, T. B., Detwiler, K. M., & Gilbert, C. C. (2020). Skeletal morphology of the lesula (*Cercopithecus lomamiensis*) and the evolution of guenon locomotor behavior. *American Journal of Physical Anthropology,* 172, 3-24.

Arnold, K., & Zuberbühler, K. (2006). Semantic combinations in primate calls. *Nature,* 441, 303.

Beamish, E. (2009). *Causes and consequences of mortality and mutilation in the Cape Peninsula baboon population, South Africa.* [Masters thesis, University of Cape Town].

Bolwig, N. (1959). A study of the behaviour of the chacma baboon, *Papio ursinus. Behaviour,* 14 (1-2), 136-163.

Booth, A. H. (1956). The distribution of primates in the Gold Coast. *Journal of the West African Scientific Association,* 2, 122-133.

Burrell, A. S., Jolly, C. J., Tosi, A. J., & Disotell, T. R. (2009). Mitochondrial evidence for the hybrid origin of the kipunji, *Rungwecebus kipunji* (Primates: Papionini). *Molecular Phylogenetics and Evolution,* 51(2), 340-348.

Buxton, A. P. (1952). Observations on the diurnal behaviour of the redtail monkey (*Cercopithecus ascanius schmidti* Matschie) in a small forest in Uganda. *Journal of Animal Ecology,* 21(1), 25-58.

Byron, C. D., & Covert, H. H. (2004). Unexpected locomotor behavior: Brachiation by an Old World monkey (*Pygathrix nemaeus*) in Vietnam. *Journal of Zoology,* 263, 101-106.

Carpenter, C. R. (1942a). Sexual behavior of free-ranging rhesus monkeys (*Macaca mulatta*) I. Specimens, procedures and behavioral characteristics of estrus. *Journal of Comparative Psychology,* 33, 113-142.

Carpenter, C. R. (1942b). Sexual behavior of free-ranging rhesus monkeys (*Macaca mulatta*) II. Periodicity of estrus, homosexual, autoerotic and non-conformist behavior. *Journal of Comparative Psychology,* 33, 143-162.

Chowdhury, S., Brown, J., & Swedell, L. (2020). Anthropogenic effects on the physiology and behaviour of chacma baboons in the Cape Peninsula of South Africa. *Conservation Physiology,* 8(1), coaa066.

Chowdhury, S., Pines, M., Saunders, J., & Swedell, L. (2015). The adaptive value of secondary males in the polygynous multi-level society of hamadryas baboons. *American Journal of Physical Anthropology,* 158, 501-513.

Clutton-Brock, T., & Janson, C. (2012). Primate socioecology at the crossroads: Past, present, and future. *Evolutionary Anthropology,* 21, 136-150.

Colmenares, F. (2004). Kinship structure and its impact on behavior in multi-level societies. In B. Chapais, & C. M. Berman (Eds.), *Kinship and behavior in primates* (pp. 242-270). Oxford University Press.

Cronin, J. E., & Sarich, V. M. (1976). Molecular evidence for dual origin of mangabeys among Old World monkeys. *Nature,* 260, 700-702.

Dasilva, G. L. (1992). The western black-and-white colobus as a low-energy strategist: Activity budgets, energy expenditure and energy intake. *Journal of Animal Ecology,* 61, 79-91.

DeVore, I., & Hall, K. R. L. (1965). Baboon ecology. In I. DeVore (Ed.), *Primate behavior: Field studies of monkeys and apes* (pp. 20-52). Holt, Rinehart and Winston.

Disotell, T. R. (1994). Generic level relationships of the papionini (Cercopithecoidea). *American Journal of Physical Anthropology,* 94, 47-57.

Dunbar, R., & Dunbar, P. (1975). *Social dynamics of gelada baboons: Contributions to primatology.* Karger.

Dunham, N. T. (2015). Ontogeny of positional behavior and support use among *Colobus angolensis palliatus* of the Diani Forest, Kenya. *Primates,* 56, 183-192.

Fernandez, D., Dempsey, A., McCabe, G., & McGraw, W. S. (2019). Mangadrill conservation action plan: A strategy for the survival of *Cercocebus* and *Mandrillus. African Primates,* 13, 66-67.

Fischer, J., Kopp, G. H., Dal Pesco, F., Goffe, A., Hammerschmidt, K., Kalbitzer, U., Klapproth, M., Maciej, P., Ndao, I., Patzelt, A., & Zinner, D. (2017). Charting the neglected west: The social system of Guinea baboons. *American Journal of Physical Anthropology,* 162, 15-31.

Fleagle, J. G., & McGraw, W. S. (1999). Skeletal and dental morphology supports diphyletic origin of baboons and mandrills. *Proceedings of the National Academy of Sciences,* 96, 1157-1161.

Fürtbauer, I., Christensen, C., Bracken, A., O'Riain, M. J., Heistermann, M., & King, A. J. (2020). Energetics at the urban edge: Environmental and individual predictors of urinary C-peptide levels in wild chacma baboons (*Papio ursinus*). *Hormones and Behavior,* 126, 104846.

Galat-Luong, A., Galat, G., & Hagell, S. (2006). The social and ecological flexibility of Guinea baboons: Implications for Guinea baboon social organization and male strategies. In L. Swedell, & S. R. Leigh (Ed.), *Reproduction and fitness in baboons: Behavioral, ecological, and life history perspectives* (pp. 105-121). Springer.

Geissmann, T., Lwin, N., Aung, S. S., Aung, T. N., Aung, Z. M., Hla, T. H., Grindley, M., & Momberg, F. (2011). A new species of snub-nosed monkey, genus *Rhinopithecus* Milne-Edwards, 1872 (Primates, Colobinae), from northern

Kachin State, northeastern Myanmar. *American Journal of Primatology, 73,* 96-107.

Gilbert, C. C., Bibi, F., Hill, A., & Beech, M. J. (2014). Early guenon from the late Miocene Baynunah Formation, Abu Dhabi, with implications for cercopithecoid biogeography and evolution. *Proceedings of the National Academy of Sciences, 111*(28), 10119-10124.

Grueter, C. C., Qi, X., Zinner, D., Bergman, T., Li, M., Xiang, Z., Zhu, P., Migliano, A. B., Miller, A., Krützen, M., Fischer, J., Rubenstein, D. I., Vidya, T. N. C., Li, B.-G., Cantor, M., & Swedell, L. (2020). Multilevel organisation of animal sociality. *Trends in Ecology & Evolution, 35*(9), 835-847.

Habig, B., Jansen, D. A. W. A. M., Akinyi, M. Y., Gesquiere, L. R., Alberts, S. C., & Archie, E. A. (2019). Multi-scale predictors of parasite risk in wild male savanna baboons (*Papio cynocephalus*). *Behavioral Ecology and Sociobiology, 73,* 134.

Haddow, A. J. (1952). Field and laboratory studies on an African monkey, *Cercopithicus ascanius schmidti* Matschie. *Proceedings of the Zoological Society of London, 122*(2), 297-394.

Hall, K. R. L., & DeVore, I. (1965). Baboon social behavior. In I. DeVore (Ed.), *Primate behavior: Field studies of monkeys and apes* (pp. 53-110). Holt, Rinehart and Winston.

Hart, J. A., Detwiler, K. M., Alempijevic, D., Lokasola, A., & Rylands, A. B. (2019). *Cercopithecus dryas.* In *The IUCN Red List of Threatened Species.* Cambridge University Press.

Hart, J. A., Detwiler, K. M., Gilbert, C. C., Burrell, A. S., Fuller, J. L., Emetshu, M., Hart, T. B., Vosper, A., Sargis, E. J., & Tosi, A. J. (2012). Lesula: A new species of *Cercopithecus* monkey endemic to the Democratic Republic of Congo and implications for conservation of Congo's central basin. *PLoS ONE, 7*(9), e44271.

Hausfater, G., & Hrdy, S. B. (Eds.). (1984). *Infanticide: Comparative and evolutionary perspectives.* Aldine.

Henriquez, M. C., Amann, A., Zimmerman, D., Sanchez, C., Murray, S., McCann, C., Tesfaye, T., & Swedell, L. (2021). Home range, sleeping site use, and band fissioning in hamadryas baboons: Improved estimates using GPS collars. *American Journal of Primatology, 83,* e23248. https://doi .org/10.1002/ajp.23248.

Hinde, R. A. (1976). Interactions, relationships and social structure. *Man, New Series, 11*(1), 1-17.

Hrdy, S. B. (1974). Male-male competition and infanticide among the langurs (*Presbytis entellus*) of Abu, Rajasthan. *Folia Primatologica, 22,* 19-58.

Hrdy, S. B., Janson, C., & van Schaik, C. (1995). Infanticide: Let's not throw out the baby with the bathwater. *Evolutionary Anthropology, 3*(5), 151-154.

Hrdy, S. B., & Williams, G. C. (1983). Behavioral biology and the double standard. In S. K. Wasser (Ed.), *Social behavior of female vertebrates* (pp. 3-18). Academic Press.

Jaeggi, A. V., Trumble, B. C., & Brown, M. (2018). Group-level competition influences urinary steroid hormones among wild red-tailed monkeys, indicating energetic costs. *American Journal of Primatology, 80,* e22757.

Janson, C. H. (2000). Primate socio-ecology: The end of a golden age. *Evolutionary Anthropology, 9*(2), 73-86.

Jay, P. (1963). The Indian langur monkey (*Presbytis entellus*). In C. H. Southwick (Ed.), *Primate social behavior* (pp. 114-123). D. Van Nostrand.

Jay, P. (1965). The common langur of north India. In I. DeVore (Ed.), *Primate behavior: Field studies of monkeys and apes* (pp. 197-249). Holt, Rinehart and Winston.

Jolly, C. J. (1970). The large African monkeys as an adaptive array. In J. R. Napier, & P. H. Napier (Eds.), *Old World monkeys: Evolution, systematics, and behavior* (pp. 139-174). Academic Press.

Jolly, C. J. (2001). A proper study for mankind: Analogies from the papionin monkeys and their implications for human evolution. *Yearbook of Physical Anthropology, 44,* 177-204.

Jolly, C. J. (2020). Philopatry at the frontier: A demographically-driven scenario for the evolution of multi-level societies in baboons (*Papio*). *Journal of Human Evolution, 146,* 102819.

Jones, T., Ehardt, C. L., Butynski, T. M., Davenport, T. R. B., Mpunga, N. E., Machaga, S. J., & De Luca, D. W. (2005). The highland mangabey *Lophocebus kipunji*: A new species of African monkey. *Science, 308,* 1161-1164.

Kawamura, S. (1954). A new type of action expressed in feeding behavior of Japanese monkeys in the wild. *Seibutsu Shinka, 2*(1), 11-13.

Kopp, G. H., Ferreira da Silva, M. J., Fischer, J., Brito, J. C., Regnaut, S., Roos, C., & Zinner, D. (2014). The influence of social systems on patterns of mitochondrial DNA variation in baboons. *International Journal of Primatology, 35,* 210-225.

Kopp, G. H., Fischer, J., Patzelt, A., Roos, C., & Zinner, D. (2015). Population genetic insights into the social organization of Guinea baboons (*Papio papio*): Evidence for female-biased dispersal. *American Journal of Primatology, 77*(8), 878-889.

Korstjens, A. H. (2001). *The mob, the secret sorority, and the phantoms.* [Doctoral dissertation, University of Utrecht].

Kummer, H., & Kurt, F. (1963). Social units of a free-living population of hamadryas baboons. *Folia Primatologica, 1,* 4-19.

Linder, J., Cronin, D., Ting, N., Abwe, E., Davenport, T., Detwiler, K., Galat, G., Galat-Luong, A., Hart, J., Ikemeh, R., Kivail, S., Koné, I., Kujirakwinja, D., Maisels, F., McGraw, S., Oates, J., & Struhsaker, T. (2021). *Red colobus (Piliocolobus) conservation action plan 2021-2026.* Gland, Switzerland: IUCN. https://doi.org/10.2305/IUCN.CH.2021.08.en.

Liu, Z., Cai, Y., Wang, Y., Nie, Y., Zhang, C., Xu, Y., Zhang, X., Lu, Y., Wang, Z., Poo, M., & Sun, Q. (2018). Cloning of macaque monkeys by somatic cell nuclear transfer. *Cell, 172*(4), 881-887.

Lucas, P. W., & Teaford, M. F. (1994). Functional morphology of colobine teeth. In A. G. Davies, & J. F. Oates (Eds.), *Colobine monkeys: Their ecology, behaviour and evolution* (pp. 173-203). Cambridge University Press.

Maisels, F., Gautier-Hion, A., & Gautier, J.-P. (1994). Diets of two sympatric colobines in Zaire: More evidence on seed-eating in forests on poor soils. *International Journal of Primatology*, 15, 681-701.

Marais, E. N. (1969). *The soul of the ape*. Anthony Blond Ltd.

McGraw, W. S. (1998). Comparative locomotion and habitat use of six monkeys in the Taï Forest, Ivory Coast. *American Journal of Physical Anthropology*, 105, 493-510.

McGraw, W. S., & Daegling, D. J. (2019). Diet, feeding behavior, and jaw architecture of Taï monkeys: Congruence and chaos in the realm of functional morphology. *Evolutionary Anthropology*, 29(1), 14-28.

McGraw, W. S., van Casteren, A., Kane, E. E., Geissler, E., Burrows, B., & Daegling, D. J. (2016). Feeding and oral processing behaviors of two colobine monkeys in Taï Forest, Ivory Coast. *Journal of Human Evolution*, 98, 90-102.

Mekonnen, A., Fashing, P. J., Sargis, E. J., Venkataraman, V. V., Bekele, A., Hernandez-Aguilar, R. A., Rueness, E. K., & Stenseth, N. C. (2018). Flexibility in positional behavior, strata use, and substrate utilization among bale monkeys (*Chlorocebus djamdjamensis*) in response to habitat fragmentation and degradation. *American Journal of Primatology*, 80(5), e22760. https://doi.org/10.1002/ajp.22760.

Mittermeier, R. A., & Fleagle, J. G. (1976). The locomotor and postural repertoires of *Ateles geoffroyi* and *Colobus guereza*, and reevaluation of the locomotor category semibrachiation. *American Journal of Physical Anthropology*, 45, 235-256.

Morbeck, M. E. (1977). Positional behavior, selective use of habitat substrate and associated non-positional behavior in free-ranging *Colobus guereza* (Rüppel, 1835). *Primates*, 18, 35-58.

Nolte, A. (1955). Field observations on the daily routine and social behaviour of common Indian monkeys, with special reference to the bonnet monkey (*Macaca radiata* Geoffroy). *Journal of the Bombay Natural History Society*, 53, 177-184.

Oates, J. F., Abedi-Lartey, M., McGraw, W. S., Struhsaker, T., & Whitesides, G. H. (2000). Extinction of a West African red colobus monkey. *Conservation Biology*, 14, 1526-1532.

Patzelt, A., Kopp, G. H., Ndao, I., Kalbitzer, U., Zinner, D., & Fischer, J. (2014). Male tolerance and male-male bonds in a multilevel primate society. *Proceedings of the National Academy of Sciences*, 111(41), 14740-14745.

Petersdorf, M., Weyher, A. H., Kamilar, J. M., Dubuc, C., & Higham, J. P. (2019). Sexual selection in the Kinda baboon. *Journal of Human Evolution*, 135, 102635.

Plavcan, J. M., Ward, C. V., Kay, R. F., & Manthi, F. K. (2019). A diminuitive Pliocene guenon from Kanapoi, west Turkana, Kenya. *Journal of Human Evolution*, 135, 102623.

Ren, R., Garber, P. A., & Li, M. (2012). Fission-fusion behavior in Yunnan snub-nosed monkeys (*Rhinopithecus bieti*) in Yunnan, China. *International Journal of Primatology*, 33, 1096-1109.

Rogers, J., Raveendran, M., Harris, R. A., Mailund, T., Leppälä, K., Athanasiadis, G., Schierup, M. H., Cheng, J., Munch, K., Walker, J. A., Konkel, M. K., Jordan, V., Steely, C. J., Beckstrom, T. O., Bergey, C., Burrell, A., Schrempf, D., Noll, A., Kothe, M.,... & Worley, K. Baboon Genome

Analysis Consortium. (2019). The comparative genomics and complex population history of *Papio* baboons. *Science Advances*, 5(1), eaau6947.

Rose, M. D. (1979). Positional behavior of natural populations: Some quantitative results of a field study of *Colobus guereza* and *Cercopithecus aethiops*. In M. E. Morbeck, H. Preuschoft, & G. Gomberg (Eds.), *Environment, behavior and morphology: Dynamic interactions in primates* (pp. 75–93). Gustav Fischer.

Rowell, T. E. (1966). Forest living baboons in Uganda. *Journal of Zoology*, 149, 344-364.

Sarich, V. M. (1970). Primate systematics with special reference to Old World monkeys: A protein perspective. In J. R. Napier, & P. H. Napier (Eds.), *Old World monkeys: Evolution, systematics, and behavior* (pp. 175-226). Academic Press.

Schmitt, V., Pankau, B., & Fischer, J. (2012). Old World monkeys compared to apes in the primate cognition test battery. *PLoS ONE*, 7(4), e32024.

Schülke, O., Bhagavatula, J., Vigilant, L., & Ostner, J. (2010). Social bonds enhance reproductive success in male macaques. *Current Biology*, 20, 2207-2210.

Schultz, A. H. (1970). The comparative uniformity of the cercopithecoidea. In J. R. Napier, & P. H. Napier (Eds.), *Old World monkeys: Evolution, systematics, and behavior* (pp. 39-51). Academic Press.

Silk, J. B., Alberts, S. C., & Altmann, J. (2003). Social bonds of female baboons enhance infant survival. *Science*, 302, 1231-1234.

Silk, J. B., Beehner, J. C., Bergman, T. J., Crockford, C., Engh, A. L., Moscovice, L. R., Wittig, R. M., Seyfarth, R. M., & Cheney, D. L. (2009). The benefits of social capital: Close social bonds among female baboons enhance offspring survival. *Proceedings of the Royal Society of London B*, 276, 3099-3104.

Silk, J. B., Beehner, J. C., Bergman, T. J., Crockford, C., Engh, A. L., Moscovice, L. R., Wittig, R. M., Seyfarth, R. M., & Cheney, D. L. (2010). Strong and consistent social bonds enhance the longevity of female baboons. *Current Biology*, 20(15), 1359-1361.

Simonds, P.E. (1965). The bonnet macaque in south India. In I. DeVore (Ed.), *Primate behavior: Field studies of monkeys and apes* (pp. 175-196). Holt, Rinehart and Winston.

Staedele, V., Van Doren, V., Pines, M., Swedell, L., & Vigilant, L. (2015). Fine-scale genetic assessment of sex-specific dispersal patterns in a multilevel primate society. *Journal of Human Evolution*, 78, 103-113.

Stead, S. M., & Teichroeb, J. A. (2019). A multi-level society comprised of one-male and multi-male core units in an African colobine (*Colobus angolensis ruwenzorii*). *PLoS ONE*, 14 (10), e0217666.

Strier, K. B. (1994). Myth of the typical primate. *Yearbook of Physical Anthropology*, 37, 233-271.

Strum, S. C. (1982). Agonistic dominance in male baboons: An alternative view. *International Journal of Primatology*, 3(2), 175-202.

Strum, S. C. (1987). *Almost human*. W. W. Norton.

Su, D. F., & Jablonski, N. G. (2009). Locomotor behavior and skeletal morphology of the odd-nosed monkeys. *Folia Primatologica*, 80, 189-219.

Sugiyama, Y. (1965). On the social change of Hanuman langurs (*Presbytis entellus*) in their natural condition. *Primates*, 6(3-4), 381-418.

Sussman, R. W., Cheverud, J. M., & Bartlett, T. Q. (1995). Infant killing as an evolutionary strategy: Reality or myth? *Evolutionary Anthropology*, 3(5), 149-151.

Thierry, B. (2008). Primate socioecology, the lost dream of ecological determinism. *Evolutionary Anthropology*, 17, 93-96.

Thompson, N. A. (2019). Understanding the links between social ties and fitness over the life cycle in primates. *Behaviour*, 156(9), 859-908.

Tournier, E., Tournier, V., van de Waal, E., Barrett, A., Brown, L., & Bshary, R. (2014). Differences in diet between six neighboring groups of vervet monkeys. *Ethology*, 120, 471-482.

Wright, K. A., Stevens, N. J., Covert, H. H., & Nadler, T. (2008). Comparisons of suspensory behavior among *Pygathrix cinerea*, *P. nemaeus*, and *Nomascus leucogenys* in Cuc Phuong National Park, Vietnam. *International Journal of Primatology*, 29, 1467-1480.

Zhang, P., Watanabe, K., Li, B., & Tan, C. L. (2006). Social organization of Sichuan snub-nosed monkeys (*Rhinopithecus roxellana*) in the Qinling Mountains, central China. *Primates*, 47, 374-382.

Zhu, W. W., Garber, P. A., Bezanson, M., Qi, X. G., & Li, B.-G. (2015). Age and sex-based patterns of positional behavior and substrate utilization in the golden snub-nosed monkey (*Rhinopithecus roxellana*). *American Journal of Primatology*, 77, 98-108.

Zuckerman, S. (1932). *The social life of monkeys and apes*. Routledge & Kegan Paul.

PART IV

APES

CHAPTER 20

Gibbons

Arboreal Acrobats of Southeast Asia

Thad Q. Bartlett and Robert W. Sussman

Considered the "gentlemen" among the primates by the ancient Chinese (van Gulik, 1967), gibbons are the most acrobatic of all arboreal primate species. The family name Hylobatidae, derived from the Latin *Hylobates*, means "tree walker," though the name *Brachitanytes*, "arm swinger," proposed by Schultz (1932) and now defunct, might be more apt. Indeed, these small apes, which are native to Southeast Asia, are characterized by adaptations for specialized, brachiating locomotion. They have long arms and highly mobile shoulder joints, long hook-like hands with a deep cleft between the fingers and the thumb, and lack a tail. Unlike other apes, male and female gibbons are roughly the same size and both sexes have long saber-like canine teeth. In addition, all gibbons share a number of common features of social organization, typically living in socially, if not sexually, monogamous social groups (Figure 20.1) that exclude conspecifics from their home ranges by calling and chasing. Gibbon loud calls take the form of solos, characteristically sung by adult and subadult males, and coordinated duets, which are sung by mated pairs in all but two species (Kloss's and Javan gibbons; see Raemaekers et al., 1984; Geissmann, 1993). In addition, gibbon calls are species specific, making them an important means of addressing phylogenetic relationships among species and subspecies (Marshall, 1976; Thinh et al., 2011).

TAXONOMY AND DISTRIBUTION

As with many other primates, gibbon taxonomy has undergone extensive revision over the last several decades, marked by a steady growth in the number of recognized species (see Table 20.1). For example, since the comprehensive review of gibbon taxonomy by Groves in 1972, in which he identified 6 species of gibbon in the single genus *Hylobates*, the number of distinct species grew to 8 in 1984 (Groves, 1984), 10 in 1990 (Groves & Wang, 1990), and 14 in 2001 (Groves, 2001). Furthermore, in 2005, Groves (Mootnick & Groves, 2005) endorsed the now common practice of elevating all 4 recognized subgenera, distinguishable by the distinct number of chromosomes in each taxon, to the genus level. As a result, the International Union for the Conservation of Nature Red List of Threatened Species currently identifies 20 gibbon species in 4 genera (IUCN, 2020). Most of this taxonomic growth is a result of an increased appreciation for the taxonomic diversity of the once poorly studied gibbons of the genus *Nomascus*, which until 1990 most investigators placed in a single species, but which has swelled to 7 species today. The most recent inclusions are *Nomascus annamensis,* the northern buff-cheeked gibbon (Thinh et al., 2010) and *Hoolock tianxing*, the skywalker hoolock gibbon (Fan et al., 2017).

The distribution of the four genera can be briefly summarized as follows (see Figure 20.2): *Hoolock* (3 species)—white browed gibbons of Assam India, Bangladesh, Burma (Myanmar), and southern China; *Hylobates* (9 species)—gibbons of the so-called lar group, extending from southern China to the Malay Peninsula, Borneo, Sumatra, Java, and the Mentawai Islands; *Nomascus* (7 species)—crested gibbons of southern China, Laos, Vietnam, and Cambodia; and *Symphalangus* (1 species)—the large black siamang (*S. syndactylus*) of Sumatra and the Malay Peninsula. *Symphalangus* is sympatric with *Hylobates* throughout most of its range, but sympatry between other gibbon species is largely limited to small zones of contact; for example, between white-handed (*Hylobates lar*) and pileated (*Hy. pileatus*) gibbons in central Thailand and between white-handed and agile (*Hy. agilis*) gibbons on the Malay Peninsula. A third area of contact, between white-bearded (*Hy. albibarbis*) and Mueller's gibbons (*Hy. muelleri*) in central Borneo is much more extensive and the occurrence of hybrid individuals is much more common, so much so, in fact, that Mather (1992; see Chivers, 2001) has argued that these taxa should be considered the same species. Nevertheless, pelage and genetic differences between populations outside the hybrid zone are reason enough for an increasing number of investigators to recognize them as distinct species (e.g., Groves, 2001; Hirai et al., 2003; Cheyne, 2010; but see Heller et al., 2010).

Siamangs are by far the largest of all gibbon species, with a mean female body weight of 10.7 kg, much larger than sympatric white-handed gibbons, 5.3 kg, and agile gibbons, 5.8 kg. With the elevation of all subgenera to

Figure 20.1 Family group of yellow-cheeked gibbons (*Nomascus gabriellae*); male (left) and female holding infant (right). The range of *N. gabriellae* includes northeastern Cambodia, southeastern Lao PDR, and southern Vietnam.
(Photo credit: German Primate Center–Leibniz Institute for Primate Research)

the genus level, a shorthand has reemerged to distinguish the larger siamang from all other "small-bodied" gibbons (e.g., Elder, 2009; Bartlett, 2009b); however, Geissmann (1993; see Figure 20.3) argues that any such dichotomy (large vs. small) disappears when body sizes of *Hoolock* (~7 kg) and *Nomascus* (~8 kg) are taken into account because they are intermediate between the other genera (see also Jungers, 1984). Geissmann also notes that there is considerable body size variation within the genus *Hylobates*; nevertheless, he has since come to refer to all members of this genus as "dwarf gibbons" (Geissmann, 2015). Despite small body size, gibbon life-history characteristics—age of first reproduction, interbirth interval, and juvenile period—approach those seen in great apes (Kelley, 1997; Reichard & Barelli, 2008).

CONSERVATION STATUS

According to the most recent IUCN Red List, all gibbon species are threatened with extinction (Table 20.1). Among the most critically endangered are five species of crested gibbons (*Nomascus concolor*, *N. hainanus*, *N. leucogenys*, *N. nasutus*, and *N. siki*). The Hainan gibbon (*N. hainanus*) is likely the single most endangered primate worldwide with fewer than 30 individuals remaining

(Chan et al., 2005; Cressey, 2014), while the cao-vit gibbon (*N. nasutus*) fares little better with just over 100 individuals living in the cross-border region of China and Vietnam (Le et al., 2008; Fan et al., 2013). Given that many of the *Nomascus* species are not held in captivity and therefore entirely reliant on in situ conservation, Melfi (2012: p.239) asserts that gibbons are "probably the most endangered primates in world." In the long term, the greatest threat to all gibbon species is habitat loss and degradation. Gibbon habitat throughout Southeast Asia is on the decline, with an estimated loss of potential habitat of over 10% in the period from 2000–2014 (Fan & Bartlett, 2017). Nevertheless, hunting and the pet trade represent acute threats to some populations (Geissmann, 2007; Hamard et al., 2010) as evidenced by the fact that many intact nature areas support few large animals (Harrison et al., 2016). In such circumstances, community conservation activities offer some hope. For example, Kolasartsanee and Srikosamatara (2014) found that gibbon population density in the Khao Soi Dao wildlife sanctuary increased after they worked with local influencers to establish a network of poachers who agreed to stop poaching gibbons (but not other animals) during foraging trips.

Table 20.1 Species of the Family Hylobatidae [a]

Genus	Species	Common Name	Distribution	Status and Trend
Hoolock (diploid number 38)	*Ho. hoolock*	Western Hoolock gibbon	Bangladesh; India (Assam); Myanmar; China (uncertain)	Endangered, Decreasing
	Ho. leuconedys	Eastern Hoolock gibbon	Myanmar; India (uncertain)	Vulnerable, Decreasing
	Ho. tianxing	Skywalker Hoolock gibbon	China; Myanmar	Endangered, Decreasing
Hylobates (diploid number 44)	*Hy. abbotti*	Abbott's gray gibbon	Indonesia (Kalimantan); Brunei Darussalam	Endangered, Decreasing
	Hy. agilis	Agile, or dark-handed, gibbon	Indonesia (Sumatra); Peninsular Malaysia; Thailand	Endangered, Decreasing
	Hy. albibarbis	Bornean white-bearded gibbon	Indonesia (Kalimantan)	Endangered, Decreasing
	Hy. funereus	Northern gray gibbon	Indonesia (Kalimantan); Malaysia (Sabah)	Endangered, Decreasing
	Hy. klossii	Kloss's gibbon	Indonesia (Mentawai Islands)	Endangered, Decreasing
	Hy. lar	White-handed, or lar, gibbon	Indonesia (Sumatra); Peninsular Malaysia; Myanmar; Thailand; Lao PDR	Endangered, Decreasing
	Hy. moloch	Javan, or silvery, gibbon	Indonesia (Java)	Endangered, Decreasing
	Hy. muelleri	Müller's, gray, or Bornean gibbon	Indonesia (Kalimantan)	Endangered, Decreasing
	Hy. pileatus	Pileated, or capped, gibbon	Cambodia; Lao PDR; Thailand	Endangered, Decreasing
Nomascus (diploid number 52)	*N. annamensis* [b]	Northern buff-cheeked gibbon	Vietnam; Lao PDR; Cambodia	Endangered, Decreasing
	N. concolor	Black crested gibbon	China (Yunnan); Lao PDR; Vietnam	Critically Endangered, Decreasing
	N. gabriellae	Yellow-cheeked crested gibbon	Cambodia; Lao PDR; Vietnam	Endangered, Decreasing
	N. hainanus	Hainan gibbon	China (Hainan)	Critically Endangered, Stable
	N. leucogenys	Northern white-cheeked gibbon	Lao PDR; Vietnam	Critically Endangered, Decreasing
	N. nasutus	Cao-Vit crested gibbon	China; Vietnam	Critically Endangered, Decreasing
	N. siki	Southern white-cheeked gibbon	Lao PDR; Vietnam	Critically Endangered, Decreasing
Symphalangus (diploid number 50)	*S. syndactylus*	Siamang	Indonesia (Sumatra); Peninsular Malaysia; Thailand	Endangered, Decreasing

[a] For details see Rainer et al. (2014), Cheyne et al. (2016), IUCN (2020).
[b] See Thinh et al. (2010).

In the past (as recently as the tenth century), gibbons may have been distributed much farther north in China than they are today, perhaps as far north as the Yellow River, and ancient literature describes gibbons as living in snow-covered mountains (van Gulik, 1967). Gibbons still persist in occasionally snow-covered habitats in China (Fan et al., 2008), but their distribution is restricted to extreme southern China with relic populations in Yunnan Province, Guangxi Province, and Hainan Island (Ma et al., 1988; Zhou et al., 2005; Fan, 2017). While white-handed gibbons once extended into southwestern Yunnan, they are now thought to be extinct in China (Grueter et al., 2009). Eastern hoolock gibbons (*Hoolock leuconedys*) are known to persist within and around the Gaoligongshan Nature Reserve, Yunnan, but their remaining habitat in China is highly fragmented with some subpopulations comprised of only a single group; thus, the long-term prospects for this population are dire (Walker et al., 2009; Fan et al., 2011). Highly fragmented habitat poses a threat to gibbons throughout their range, but perhaps most acutely in Java (Gates, 1998) and northeast India (Kakati et al., 2009; Vasudev & Fletcher, 2015). Given the impact of habitat loss and degradation on gibbon populations, conservation scientists have refocused their efforts on documenting gibbon distribution and population density in order to inform conservation

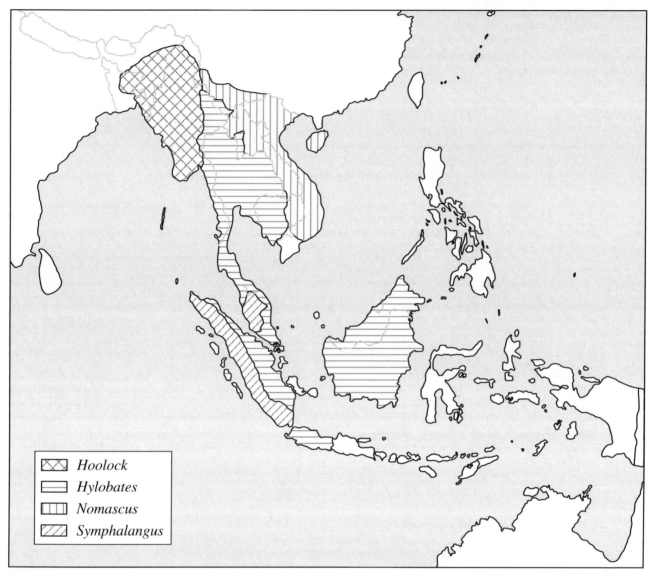

Figure 20.2   Approximate historic distributions of the four gibbon genera (see text for details).

practice throughout their range (e.g., Cheyne et al., 2016; Lappan et al., 2018; Vu et al., 2018).

HISTORY OF GIBBON FIELD RESEARCH

Gibbons were among the first primates to be studied systematically in the wild. In 1937, Carpenter (1940) spent approximately four months in Thailand (then Siam) studying the natural history of white-handed gibbons. Approximately 20 years later, Kawamura (1958) returned to Thailand for a preliminary survey, also of white-handed gibbons, and then Ellefson (1967, 1968) studied the same species in peninsular Malaysia in 1964. Subsequently, Chivers (1974) studied siamangs for two years, from 1968–1970, at two sites where white-handed gibbons and siamangs occurred together in the same forest. Brief studies of Kloss's (*Hy. klossii*; Tenaza & Hamilton, 1971; Tenaza, 1975) and hoolock (*Ho. hoolock*; Tilson, 1979) gibbons were

initiated in 1970 and 1971. During this same period Joe Marshall and Warren Brockelman began recording gibbon songs in Khao Yai National Park, Thailand, using them to document a hybrid zone between pileated and white-handed gibbons (Marshall et al., 1972; Brockelman, 2013).

By the middle of the next decade field research on gibbons had proliferated to include most members of the genus *Hylobates* (see Preuschoft et al., 1984; Leighton, 1987). Only more recently have researchers begun amassing detailed data on the ecology and social structure of the more poorly studied *Hoolock* (e.g., Ahsan, 2001; Sarma & Kumar, 2016) and *Nomascus* (e.g., Fan et al., 2008; Fan et al., 2012); nevertheless, *Hylobates* (e.g., Savini et al., 2008; Cheyne, 2010; Phoonjampa et al., 2010; Kim et al., 2011; Reichard et al., 2012; Clink et al., 2017) and *Symphalangus* (e.g., Lappan 2007, 2009; Yanuar & Chivers, 2010; Morino, 2015) continue to be the best studied taxa.

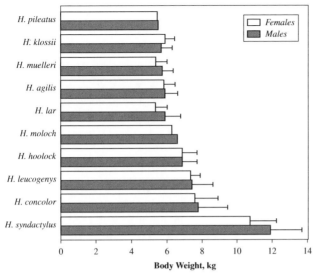

Figure 20.3 Male and female body size in 10 gibbon species.
(Redrawn after Geissmann, 1993)

HABITAT AND LOCOMOTION

Gibbons are completely arboreal, spending nearly all of their time in the forest canopy or in emergent tree crowns that extend above the main canopy (Cannon & Leighton, 1994). Consequently, their distribution is limited predominantly to evergreen forests with a more or less continuous canopy. Even when their habitat is comprised in part of deciduous forest, the available evidence suggests that access to evergreen forest patches is important to gibbon sustainability (Kenyon, 2008; Rawson et al., 2009; Gray et al., 2010). On the other hand, gibbons do appear to tolerate selectively logged and degraded forest, perhaps because the tree most economically valuable to humans (i.e., trees with excellent timber qualities such as dipterocarps) are not those in which gibbons feed (Lee et al., 2015). Furthermore, the degree of climatic variation characteristic of areas occupied by gibbons should not be underestimated. For example, mean annual rainfall for Ketambe, Sumatra, Indonesia (~3° N Latitude) is 3,229 mm (van Schaik, 1986), which is over twice the 1,607 mm measured at Mt. Wuliang, Yunnan, China (~24° N Latitude), some of which falls in the form of snow (Fan et al., 2008). Perhaps just as important in terms of understanding forest dynamics is the distribution of rainfall. It is estimated that once mean monthly precipitation falls below 60 mm in a given month, evaporation exceeds precipitation, leading to water stress and reduced forest productivity (Whitmore, 1984; van Schaik & Pfannes, 2005). At equatorial sites, like Ketambe (see also Kinnaird & O'Brien, 2005; Kim et al., 2011), mean monthly precipitation rarely falls below 60 mm, whereas Fan et al. (2008) report that during their study, 96% of the precipitation occurred from May to October, which means little or no rain fell during the remaining six

months of the year. Khao Yai National Park, Thailand (~14° N Latitude), represents an intermediate condition both geographically and climatically, with mean annual rainfall of 2,326 mm (Kitamura et al., 2004) and a dry season (with < 60 mm precipitation per month) of three to four months in duration (Bartlett, 2009b).

The typical pattern of locomotion by gibbons is brachiation—using both hands while swinging beneath tree branches. This means of travel gives the gibbons a most agile, spectacular, and graceful appearance as they glide through the branches of the upper canopy. During fast locomotion, gibbons use a rapid, ricochet-style arm-swinging in which the animal almost throws itself from branch to branch (Fleagle, 1976). They are probably the fastest flightless animals in the trees (Gittins, 1983).

Fleagle (1976) and Gittins (1983) give the most detailed quantitative description of locomotor behavior of siamangs and agile gibbons, respectively. Both authors recognize four main types of locomotion: brachiation, climbing, bipedal walking, and leaping (Figure 20.4). During travel, agile gibbons were observed to brachiate 74% of the time, climb 14%, walk 7%, and leap 6%. Siamangs brachiate less (51% of the samples) and climb more (37%

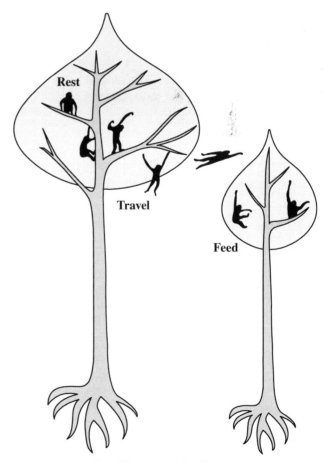

Figure 20.4 Modes of locomotion in gibbons.
(Redrawn after Gittins, 1983)

of the time samples), while walking (6%) and leaping (6%) occurred at the same rates as in agile gibbons. Animals normally brachiate on larger branches and mostly along these branches rather than between them. In climbing, the apes use a variety of supports and employ hands as well as feet. Large branches and boughs are used, but most climbing is on small twigs. Bipedal walking only occurs on large, nearly horizontal branches, and leaps are usually from higher to lower branches, from various size supports often to a mesh of terminal twigs. Fleagle's (1976: p.252) description of leaping in siamangs is applicable to gibbons in general: "Often a siamang simply allows itself to drop from the higher branch. In the air, a leaping siamang looks like a giant furry spider, with a horizontal trunk and four abducted, slightly flexed limbs" (Figure 20.5).

Brachiation is an amazing adaptation for fast locomotion; however, as Prost (1965) has stressed, locomotor and postural adaptations are complementary and the anatomical characteristics of gibbons that are used in brachiation must also be seen as postural adaptations. The suspensory capabilities of gibbons allow them a great deal of mobility during foraging and feeding, especially on small branches (Preuschoft & Demes, 1984; Nowak & Reichard, 2010). Alternatively, in large crowned trees, gibbons sit when possible, especially when feeding on larger fruits that need to be manipulated with both hands (Gittins, 1983).

Gibbons rest, sleep, and are social mainly while sitting on large, stable supports high in the upper canopy, where there are many spreading trees with open branches (Chivers, 1974; Gittins & Raemaekers, 1980; Gittins, 1983). Gibbons rarely travel in the discontinuous lower canopy and compared to sympatric macaques, travel higher in the canopy and are able to cross wider gaps (up to 9 m; Cannon & Leighton, 1994). Thus, the positional

Figure 20.5 According to Fleagle (1976: p.252), in the air a leaping gibbon looks like a "giant furry spider." (Photo credit: Thad Q. Bartlett)

behavior of the hylobatids is a highly efficient foraging *and* feeding adaptation, especially useful in exploiting food resources located on peripheral, terminal branches. It also enables these apes to move rapidly and may be a very important factor in the ability of gibbon groups to defend the boundaries of their home ranges and to visit dispersed feeding trees (Ellefson, 1974).

DIET

All gibbon species display a marked preference for ripe fruits, which account for roughly 60% of the gibbon diet on average (Bartlett, 2011). While a diverse diet is necessary to meet all an animal's nutritional requirements, it is clear that gibbons as a rule target ripe, succulent fruits with high sugar content. Fruits are harvested during distinct feeding bouts of 10–15 minutes in length from a series of known feeding patches distributed throughout their range (Gittins, 1982; Bartlett, 2009b). Aside from *Ficus* spp. (see Janzen, 1979; Leighton & Leighton, 1983), most species exploited by gibbons produce fruit synchronously over a period weeks or months, and the gibbon diet reflects this pattern with a different species dominating feeding time each month (Gittins, 1982; McConkey et al., 2003; Fan et al., 2009; Bartlett, 2009b; but see Dillis et al., 2015). Individual trees within favored species may have a significant impact on feeding patterns, attracting gibbons over multiple days or even multiple times per day (Gittins, 1982; Bartlett, 2009b; Fan et al., 2015). Bartlett (2009b), for example, found that the single most visited fruit patch each month was visited on 79% of observation days; during March, a single large *Prunus javanica* tree was visited on seven consecutive days by a single gibbon group and accounted for a full third of that group's diet during the sample.

To investigate gibbon food preference in a controlled setting, Jildmalm et al. (2008) presented captive white-handed gibbons with a series of food items and calculated their rank order preference. Captive gibbons consistently preferred grapes, bananas, and figs (*Ficus carica*) to all other food types presented (i.e., apple, pear, melon, carrot, tomato, cucumber, and avocado), with grapes as the single most preferred food item. Rank order preference correlated strongly with the total carbohydrate content (e.g., fructose and glucose) of the foods presented, but not with the total energy content of the foods or with protein or lipid levels. For example, avocado, which is high in lipid content, was selected over another food item (a carrot) just once in 290 trials. Extrapolation of these findings to wild gibbons is complicated, in part, by the fact that other foods that round out the gibbon diet, such as young leaves, insects, and flowers were not included in the study design. Nevertheless, a preference for ripe fruit pulp is consistent with field observations. Even in relatively folivorous gibbon species, a dietary preference for ripe fruit is clear. For example, Fan et al. (2009) found that the annual diet of black crested gibbons was comprised of slightly more leaves than fruits (46.5% versus 44.1%) and that fruit feeding in some months fell to just 5% of feeding time; nevertheless, fruit feeding climbed to nearly 83% when preferred species were available.

Despite a demonstrable preference for succulent fruit, molar enamel in at least some gibbon species is intermediate between that observed in chimpanzees and bonobos (*Pan* spp.), and orangutans (*Pongo* spp.). Accordingly, Vogel et al. (2009) found that both orangutans and gibbons (*Hy. albibarbis*) consumed harder and tougher mesocarp tissue (the fleshy middle of the fruit) compared to chimpanzees—though the mesocarp tissue consumed by orangutans is still tougher than that eaten by gibbons. This research highlights the importance of the mechanical properties of food even among highly frugivorous species. Moreover, based on these findings, Vogel et al. (2009) breathed new life into Frisch's (1963: p.723) alternative explanation for monomorphic canine size in gibbons, which is typically attributed to their monogamous social organization: "The adaptive advantage of tall canines for puncturing and tearing protective fruit tissues is readily apparent, and we suggest that natural selection favored large canine size in both sexes to facilitate access to tough-husked fruits." However, as Frisch observes, male gibbons are twice as likely to have damaged or broken canines than females (42% vs. 20%), which in the absence of clear dietary differences points to a different function for canine teeth. Differential fracture rates are most easily explained by the greater role taken by males in intergroup aggression, as discussed below, though that does not rule out a relationship with food processing as well.

Within the broad category of fruit feeding, figs (*Ficus* spp.) make up a full quarter of the gibbon diet on average (Bartlett, 2011). In fact, fig trees have been found to be completely absent at only one study site where gibbons are found, the Ailao Mountains in China. Not surprisingly, the portion of fruit in the diet of gibbons at that site is lower than in any other forest where they have been studied, just 24% (Chen, 1995). Elsewhere, the abundance of fig trees in the environment has been found to correlate positively with social group size (Mather, 1992) and gibbon density (Marshall & Leighton, 2006). O'Brien et al. (2003) found that siamang groups living in fire-damaged areas, where fig density was lower, had higher infant and juvenile mortality compared to groups living in intact forest. In sum, there can be no doubt of the significance

of figs to gibbon foraging behavior and ecology, a fact that has led many authors to argue that hylobatids are specifically adapted to exploit fig trees (e.g., MacKinnon & MacKinnon, 1980). For example, based on their consecutive studies of siamangs and white-handed gibbons on the Malay Peninsula, Chivers and Raemaekers (1986: p.46) conclude, "The plant part most eaten is not necessarily that most sought after. When opportunities taken are expressed as percentages of those provided, it is clear that both gibbon species selected figs more strongly than other fruit and flowers overall, and these more strongly than young leaves overall." This view was subsequently endorsed by Palombit (1992: p.204), who studied the same two species in Sumatra, Indonesia: "The Ketambe hylobatids both show a clear dietary preference for figs." However, a dietary *preference* for figs has been called into question by others who argue that figs are fallback foods; that is, food items that are exploited primarily when more preferred succulent fruits are scarce (Leighton & Leighton, 1983; Bricknell, 1999; Marshall & Leighton, 2006; Vogel et al., 2009; Bartlett, 2009b). For example, during a 12-month study of white-handed gibbons in Khao Yai, Bartlett (2009b: p.122) showed that the contribution of figs to the diet was greatest in the cool season when overall fruit abundance was lowest: "During November when non-fig fruit feeding was at its lowest, figs accounted for 53% of the gibbon diet on average. In fact, fig feeding during the cool season never fell below 30% of feeding time. In contrast figs accounted for 0%–15% of diet during the remaining months, when fruit was more abundant" (see Figure 20.6).

Marshall and Leighton (2006) draw the same conclusion based on a multiyear study of white-bearded gibbons in Gunung Palung, West Kalimantan, Indonesia. Comparing gibbon feeding records collected opportunistically during line transect surveys in seven distinct habitats to an index of food availability over the same 63-month period, the authors showed that "the percentage of gibbon feeding observations on figs was significantly negatively correlated to overall food availability" (Marshall & Leighton, 2006: p.326). These findings are broadly consistent with the finding that fig fruits (technically a fruit-like mass, or syconium, encasing numerous tiny flowers and seeds) are often of lower quality than more succulent fruit species and may contain higher tannin content (which inhibits digestion) as well as more fiber (Leighton, 1993; Wrangham et al., 1993). At the same time, fig trees exploited by gibbons are often among the largest trees in the forest and it is impossible to discount the possibility that such trees are preferred due to the energetic efficiency associated with exploiting large—but

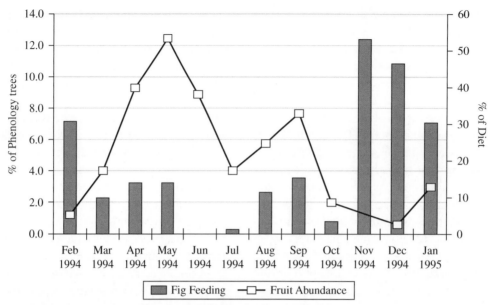

Figure 20.6 Percentage of fig fruit in the diet of white-handed gibbons compared with overall fruit abundance in the forest. (Redrawn after Bartlett, 2009b)

perhaps lower-quality—fruit crops (Gittins & Raemaekers, 1980; Vellayan, 1981; Elder, 2009).

Most gibbons supplement their frugivorous diet primarily with young leaves and shoots. While high in nondigestible fiber, leaves are an important source of dietary protein and young leaves are especially valued for their higher protein/fiber ratio and lower levels of secondary compounds (e.g., tannins). Taken together, leaves and shoots make up about a third of the gibbon diet, but variation between species and between study sites is extreme (Bartlett, 2011). Most notably, the diet of Kloss's gibbons on Siberut, Indonesia, consists of just 2% leaves; in fact, this species fed on the vegetative parts of only two plants (leaf petioles of orchids and shoots of a leguminose vine). Whitten (1982a) speculates that Kloss's gibbons may avoid eating leaves because they live in habitats with soils poor in nutrients and in which the tree leaves contain unusually high levels of secondary compounds. Nevertheless, Kloss's gibbons are not unique in their low rates of folivory. Leaf eating is also relatively low in both white-handed gibbons (4%) and siamangs (17%) at Ketambe (Palombit, 1997). Palombit suggests that the low levels of folivory at Ketambe are the result of the abundance of fig trees, which according to data tabulated by Mather (1992), occur at a density up to 10 times higher there than in Siberut (see Chivers, 2001). In any case, it is significant that the contribution of insects to the diet of these three gibbon populations are the highest known of any study (Kloss's gibbon, 25%; white-handed gibbon, 24%; siamang, 21%), a finding that supports the conclusion that leaves and insects represent alternative sources of protein for gibbons.

To date, the highest rates of folivory in gibbons have been documented among *Nomascus* species (e.g., *N. concolor*, 47% [Fan et al., 2009]; *N. leucogenys*, 53% [Hu et al., 1989, 1990]); however, no long-term studies detailing the feeding behavior of crested gibbons have been completed outside of the extremely seasonal areas of southern China, so it is not possible to know to what extent such high rates of folivory characterize crested gibbons in other habitats (but see Frechette et al., 2017).

On average, flowers are the plant part that contributes least to the gibbon diet (approximately 4% overall; Bartlett, 2011), though again there are exceptions (Ahsan, 2001; McConkey et al., 2003). For example, Lappan (2009) found that siamangs in southern Sumatra devoted 12% of their annual feeding time to flowers and that flower-feeding time reached 43% of feeding during one month. This spike in flower consumption coincided with reduced siamang activity levels, suggesting that flowers, like figs, might in some cases represent foods eaten primarily when other plant parts are not available.

High rates of insectivory are not generally expected among large-bodied primates (Kay, 1975; Lambert, 2011) and, with the exceptions detailed above, insectivory in gibbons is limited (~7% of the diet on average; Bartlett, 2011). Flying insects are taken sporadically when accidentally displaced, but more often gibbons capture social insects, such as ants and termites, by laying the back of their hand over marching columns and then licking the unsuspecting creatures off their hair (Gittins, 1982). Caterpillars, spiders, and galls are slowly searched for on branches and among dead leaves. Occasionally, trees infested with larvae in rolled leaves can result in extended feeding bouts of over an hour or more (Bartlett, pers.obs.).

Evidence of gibbon predation on vertebrate species is quite rare; nevertheless, gibbons will occasionally harvest eggs from birds' nests (Carpenter, 1940; Delacour, 1961; Newkirk, 1973; Traeholt et al., 2007; Light, 2016) and capture small lizards (Traeholt et al., 2007; Fan & Jiang, 2009). In addition, captive gibbons have been known to kill and partially consume birds in or around their enclosures (Newkirk, 1973), and Carpenter (1940) reported finding the skull of a bird among the stomach contents of a white-handed gibbon shot in northern Thailand. To date, the best evidence for predation by gibbons is from Mt. Wuliang, China, where Fan and Jiang (2009) recorded 11 attacks on giant flying squirrels (*Petaurista philippensis*), 4 of which were successful. The authors describe one successful attack as follows:

When the group had finished a feeding session, they began to forage in the forest and encountered a giant flying squirrel nest made of leaves located between branches about 12 m above the ground, at 1233 h. FB-female smashed the nest with her right hand five or six times and pulled from the nest a giant flying squirrel infant about 10 cm in head-body length. She bit the infant, killing it, and began eating. (p.47)

On two other occasions the male of the group ate eggs or chicks from birds' nests and on a third occasion a juvenile gibbon consumed a small lizard. The only other report of a gibbon feeding on a mammal comes from a brief report of a single predation event by *N. annamensis* on an immature variable squirrel in the Veun Sai-Siem Pang Conservation Area (Rawson et al., 2011). It is intriguing that both documented cases of gibbon predation on squirrels have involved crested gibbons, but the preponderance of evidence suggests that predation on small vertebrates is a potential feeding strategy available to all gibbons.

Finally, water is apparently obtained primarily from fruits and young leaves, though gibbons do occasionally drink free-standing water, for example, by putting their arms and hands into natural water bowls in trees and then letting the water drip from their hands into open mouths. They will also collect water on the back of their arms and hands and then lick their hair much like they do to collect ants (Chivers, 1977).

In sum, with the possible exception of some species of the genus *Nomascus*, gibbons supplement a highly frugivorous diet with varying amounts of leaves, flowers, fruits, and insects. Further distinction between gibbon species is complicated by high intraspecific variability suggesting that local ecology plays a significant role in diet choice

(Elder, 2009; Malone & Fuentes, 2009; Bartlett, 2009b). A further illustration of this point can be found in the dietary differences in sympatric gibbons and siamangs. Early studies characterized siamangs as folivorous, devoting more feeding time to leaves than to fruit; however, more recent comparisons suggest that the dietary distinction between co-occurring gibbons and siamangs is more subtle (Elder, 2009). For example, at Kuala Lompat, West Malaysia, siamangs devoted 36% of their diet to fruit and 43% to leaves, while white-handed gibbons devoted 50% of their diet to fruit and 29% to leaves (Raemaekers, 1979). In contrast, at Ketambe both species were found to be highly frugivorous, and the rate of fruit eating by siamangs exceeded that by white-handed gibbons at Kuala Lompat (Palombit, 1997). As described above, the high rates of frugivory at Ketambe are likely a product of the high abundance of figs, which contribute heavily to the diet of both species. Nevertheless, despite broad dietary overlap, it does appear that Ketambe siamangs rely more heavily on leaves then sympatric white-handed gibbons, which eat more non-fig fruit. Thus, their relatively greater reliance on leaves (as well as lower-quality fig fruit) is likely a significant factor in the ability of these two species to occupy the same habitat, a fact that is consistent with observed differences in body size and the inferred differences in metabolic rate (Chivers, 1974).

In light of their highly frugivorous diet, an additional ecological consideration is the ability of gibbons to coexist with other primates that likewise eat large amounts of fruit. While different observers have suggested that gibbons have distinct preferences for fruit size (Ungar, 1995; McConkey et al., 2002) or patch size (Gittins & Raemaekers, 1980; Chivers, 2001), the available data are equivocal (see Bartlett, 2009b, 2011). On the other hand, it does appear that gibbons on the whole rely more heavily on figs than do co-occurring primates (MacKinnon & MacKinnon, 1980; Feeroz et al., 1994; Ungar, 1995). In addition, Vogel et al. (2009: p.716) found that gibbons and sympatric orangutans exhibited different fallback strategies: among gibbons "there was a progressively greater reliance on figs, liana products, and unripe fruit. Orangutans relied heavily on unripe fruit and fracture-resistant bark and pith tissues." Gibbons may also avoid competition with sympatric primates by feeding higher in the canopy (MacKinnon & MacKinnon, 1980; Ungar, 1996) and by making greater use of the small terminal branches of the crown margin (Cant, 1992). Hornbills (Bucerotidae), which are frequently found in sympatry with gibbons, feed heavily on fig fruit and thus represent a potential competitor for gibbons. Nevertheless, based on their analysis of dietary overlap between white-handed gibbons and four species of hornbill in Khao Yai,

Kanwatanakid-Savini et al. (2009: p.189) conclude that the large size of fig trees combined with the low nutritional quality of their fruit makes dietary overlap "energetically inconsequential for either group."

Seed Dispersal

In general, primates are recognized as important seed dispersers, and not surprisingly, frugivorous primates, such as gibbons, are thought to be especially important in this regard as they swallow large quantities of seeds that they then deposit some distance away from the parent tree (Gittins, 1982; Whitington & Treesucon, 1991; Corlett, 1998; McConkey & Chivers, 2007; Russo & Chapman, 2011). In one study, McConkey (2000) calculated that hybrid gibbons in Borneo disperse, through defecation, an average of 5.6 seeds/individual/day, or 19.6 seeds per group, which based on an average home range size of 44.5 ha would be the equivalent of 7,300 seeds/group/per year. Even factoring in seeds lost due to subsequent predation, McConkey estimates that a single gibbon group on an annual basis is responsible for establishing 13 seedlings/ha. Given the overall loss of primates in many Asian forests, the role of gibbons in maintaining biodiversity is an area of growing concern. An important area of future research will be understanding the degree to which seed dispersal by gibbons duplicates that by other dispersers, especially other primates. Work by McConkey and Brockelman (2011) in Khao Yai has shown that macaques are more effective dispersers than gibbons (as measured by the number of seeds dispersed and the likelihood of seedling survival) for the two fruit species they studied (*Prunus javanica* and *Salacia chinensis*). Nevertheless, these two primates were more effective dispersers than other frugivores such as birds and squirrels.

ACTIVITY PATTERNS

Gibbons become active early in the morning, often before first light. One of the first signs of activity are the solo vocalizations of adult or subadult males which are sung in or near their sleeping trees; however, in some species solo calls are rare among mated males and duet songs predominate at or around dawn (e.g., Rawson, 2004). After the calls have subsided, all group members move in a coordinated fashion to a single feeding patch, typically a fruiting tree. The remainder of the day consists of a series of relatively distinct feeding visits interspersed with brief bouts of travel, rest, duetting, and social activity (including both intra- and intergroup interactions). Unlike most primates, gibbons lack a prolonged midday rest period and retire well before sunset (Gittins & Raemaekers, 1980; Cheyne, 2010). In seasonal environments, activity

is further reduced during periods of food scarcity (Chivers, 1974; Islam & Feeroz, 1992; Bartlett, 2009b).

In general, data on affiliative social behavior among gibbons (e.g., grooming and play) have been slow to materialize. It has been argued that the small size of gibbon social groups, and the consequently small number of social partners, results in extremely low rates of pro-social behavior (Leighton, 1987). Gittins and Raemaekers (1980), for example, concluded that social behavior was almost completely absent in agile gibbons. More recent studies, however, have shown that low rates of sociality are not universal. At Khao Yai, for example, rates of affiliative social behavior are comparable to levels reported for primates in general (Sussman et al., 2005). Bartlett (2003) found that social activity accounted for 11% of the activity budget in white-handed gibbons, and that during periods of fruit abundance, play and grooming accounted for as much as 20% of the activity time. Bouts of mutual grooming between adults tend to parallel juvenile play bouts, though adults do engage in various forms of play with immature group members (Ellefson, 1974; Bartlett, 2003). Furthermore, juvenile gibbons will sometimes play and even groom with juveniles from neighboring groups (Treesucon, 1984; Reichard & Sommer, 1997; Ahsan, 2001; Bartlett, 2003; Chan et al., 2005). According to Bartlett (2003), 17% of intergroup encounters occurring over a 12-month period included chase play between juveniles, and on two occasions, a neighbor male groomed the adolescent and juvenile of the neighboring group. Affiliative encounters between neighboring adults were not observed; nevertheless, this finding demonstrates that neighboring adults exhibit unappreciated levels of tolerance, which is likely a product of the complex social and genetic relationships that characterize gibbons living in the same neighborhood (Brockelman et al., 1998; Bartlett, 2009b).

PREDATION ON GIBBONS

As with all primates, predator pressure on gibbons is difficult to measure and direct observations of predation on gibbons are almost entirely absent (Uhde & Sommer, 2002). Schneider (1905) discovered the body of a full-grown siamang in the belly of a python. However, pythons will scavenge carrion so it is conceivable the siamang was consumed after it died from some other cause. To date, the only unequivocal (though also not witnessed) predation event involved the case of a young juvenile siamang and a clouded leopard (*Neofelis nebulosa*) at Way Canguk. Drawn to the area by siamang alarm calls, Morino (2010) startled a clouded leopard that was standing over the dead body of the siamang. The leopard fled the area and Morino recovered the body, which was still warm. The profound wounds to the head and neck were consistent with a leopard attack. Morino notes that the attack occurred near a known sleeping tree just at dawn, the time at which gibbons are thought to be most vulnerable.

Despite the paucity of direct evidence of predation, the alarm calls triggered by the presence of predators including birds of prey, snakes, and tigers (Reichard, 1998; Uhde & Sommer, 2002) make it unlikely that gibbons are "virtually immune from cursorial and arboreal predators" as has been suggested by some (van Schaik & Dunbar, 1990: pp.40–41). Gibbons are cautious about coming to the ground, and several aspects of their sleep behavior are indicative of an adaptive response to predation pressure (Tenaza & Tilson, 1985; Reichard, 1998; Fan & Jiang, 2008a; Phoonjampa et al., 2010). First, gibbons typically select tall sleeping trees with few low branches and few lianas (Whitten, 1982b; Ahsan, 2001). Second, gibbons rarely reuse sleeping trees on consecutive nights, which may limit the ability of predators to locate them based on olfactory cues. And finally, it has also been suggested that one of the reasons gibbons enter night trees so much earlier than sympatric species is to limit the ability of crepuscular predators to track gibbons to their night trees (Caine, 1987; Reichard, 1998). Indeed, despite entering sleep trees hours before sunset, individual gibbons remain remarkably inactive and, with the exception of mother-infant pairs, avoid social interactions.

Dooley and Judge (2014) point out that anti-predator strategies that are effective at reducing the risk of predation by snakes, cats, and raptors may not be effective against human hunters. Humans have been the primary predator on gibbons in the Mentawai Islands for an estimated 2,000 years, which the authors argue is sufficient time for Kloss's gibbons to have evolved specific strategies to avoid human hunters, such as altered singing and activity patterns as well as "decoy" behavior wherein one adult approaches the hunters and appears to distract them while the other animals flee. Dooley and Judge's work highlights the fact that the role of human hunters on the evolution of primate behavior is too often overlooked.

HOME RANGE, RANGE USE, AND TERRITORIALITY

The average home range size across all gibbon species is approximately 40 ha but varies from 15 to well over 100 ha (Bartlett, 2011; but see Sarma & Kumar, 2016). It is likely that home range size for a given group or population is the result of a complex interplay between resource density and the density of competitors. At one extreme of the range-size continuum are western black crested gibbons inhabiting the mountainous regions of southern China (Bleisch & Chen, 1991; Sheeran, 1993; see also Fan et al., 2010). For example, Sheeran et al. (1998) estimated

home range size based on four study groups to be from 100–200 ha. Though imprecise, this estimate is supported by more recent work by Fan and Jiang (2008b) who found that the home range size for the only habituated group of black crested gibbons studied to date is 151 ha. Such extreme range sizes are almost certainly a consequence of marked seasonality in resource abundance.

Though limited, intraspecific comparisons also point to the influence of resource density on home range size. For example, Kim et al. (2011) argue that altitudinal differences in resource abundance explain differences in range size among Javan (*Hy. moloch*) gibbons. Despite broad similarities in terms of activity budget, diet, and group size, the authors found that gibbons living in montane forest (950–1,100 masl) had substantially larger home ranges (37 ha) compared to counterparts living in lowland forest (15–17 ha). Resource density has also been shown to account for differences in home range size within the gibbon population at Khao Yai. Savini et al. (2008) found that home range size among white-handed gibbons (range 14.8–49.7 ha) correlated negatively with home range quality (i.e., fruit productivity/ha). Interestingly they also found a significant correlation between home range size and infant mortality, which they attribute to increased travel costs associated with traversing a larger home range.

Patterns of movement within a group's range appears to be influenced most by the location of specific food sources (Chivers, 1977; Whitten, 1982a; Fan & Jiang, 2008b; Asensio et al., 2011, 2014), especially large fruiting trees that are visited repeatedly over several consecutive days. For example, Khao Yai gibbons will sometimes visit a single tree two or even three times in a single day, which is reflected in day ranges that double back or cross themselves (Bartlett, 2009b). To better document the impact of feeding trees on gibbon travel routes, Asensio et al. (2011) mapped gibbon movements using GIS and then applied a "change point test" to determine the specific location where the direction of travel changed significantly; they determined that most directional changes occurred at feeding trees. The authors conclude, "Because consecutive travel change points were far from the gibbons' sight, *planned movement* between preferred food sources was the most parsimonious explanation for the observed travel patterns" (Asensio et al., 2011: p.395, emphasis added). Group leadership appears to fall to females in most cases (*S. syndactylus* [Chivers, 1974]; *Hy. klossii* [Tenaza, 1975]; *Hy. lar* [Reichard & Sommer, 1997]; *Ho. hoolock* [Ahsan, 2001]; *Hy. albibarbis* [Cheyne, 2010, but see Gittins, 1980]), which is likely due to the increased nutritional demands associated with pregnancy, lactation, and carrying infants (Barelli et al., 2008).

Gibbons do not markedly alter the areas of the range that they use between seasons (Whitten, 1982b), though as with activity levels in general, gibbons tend to travel least when resources are scarce (Chivers, 1974; Raemaekers, 1980; Bartlett, 2009b).

Territoriality and Intergroup Encounters

The most conspicuous aspect of gibbon range use is the fact that they actively defend a fixed range or territory, which they do via a host of behaviors that include duets, male and female solos, and protracted chases between adults, especially males (Ellefson, 1974; Gittins, 1984; Bartlett, 2009a). Disputes are initiated when two groups come into visual contact or when the male of one group approaches a territorial boundary in response to the calls of a rival male. Adult males take the lead role in virtually all disputes; however, subadult males and adult females also participate in chases and calls (Brockelman & Srikosamatara, 1984; Bartlett, 2009b; Cheyne, 2010). During a dispute, rival males sit or hang within view of one another for periods at a time interspersed by rapid chases back and forth across the area of overlap between the two ranges. Encounters tend to end rather suddenly with the males quickly moving to rejoin their respective groups. In rare cases, encounters may be lethal to one of the participants (Palombit, 1993; Brockelman et al., 1998; Reichard, 2003; Cheyne, 2010).

It is generally accepted that gibbon loud calls also function in range defense. This is supported by the fact that calls are loud enough to be heard from a considerable distance and also that duets are often sung in the context of intergroup encounters. Comparing data from eight species, Gittins (1984) concludes that gibbon groups sing on average once a day and engage in territorial disputes on average once every five days. Yet reported encounter rates between study populations may vary due to a myriad of factors, including degree of habituation, the presence of neighboring groups, extent of forest fragmentation, and population density. Cheyne (2010), for example, found that the rate of encounters among Bornean white-bearded gibbons living in peat swamp forest was just once every 22 days, a finding that she attributes to low population densities and relatively large territories. Rates of encounters among black crested gibbons, which also have large home ranges and low population densities, are similarly infrequent (Fan, pers.comm.).

Assessing the evolutionary function of gibbon encounters is complicated and a consensus view is lacking. Anecdotally, it is clear that some cases of agonism between neighbor males represent a form of mate defense; nevertheless, in other instances (even instances involving the same groups), encounters may involve intersexual aggression, mutual indifference, or even cooperation (e.g., during juvenile play). Consequently, attempts to link gibbon territoriality to a single causal factor are bound to fail. What is clear is that any attempt to model the adaptive benefits of territoriality in gibbons must account for the fact that gibbons maintain fixed, temporally stable territories, a phenomenon not required for effective mate defense (Bartlett & Light, 2017). Observations of the gibbon population at Khao Yai have continued on and off since at least 1981. Since that time, the locations of most gibbon home ranges have remained remarkably stable (Savini et al., 2008; Bartlett et al., 2016). This supports the view that one of the principal benefits of territoriality in gibbons is that it promotes foraging efficiency for individuals living in small, well-known ranges (Raemaekers & Chivers, 1980; Brockelman, 2009; Bartlett, 2009b). Furthermore, for these benefits to be fully realized, social group size must be small. Consequently, most discussions of the evolution of gibbon social structure begin with the premise that intragroup feeding competition, tied to a diet dominated by ripe fruit, limits gibbon group size to a single adult pair (Wrangham, 1979; Raemaekers & Chivers, 1980; Leighton, 1987; Bartlett, 2009b; but see Fan et al., 2015).

SOCIAL STRUCTURE AND ORGANIZATION

Gibbons are distinct among apes, and among primates generally, in that they live in small social groups that are made up of a mated adult pair and associated offspring. Mean group size is approximately four individuals (*Hy. lar*, 4.2 [Reichard & Barelli, 2008]; *Hy. albibarbis*, 4.0 [Mitani, 1990]; *S. syndactylus*, 3.9 [O'Brien et al., 2004]; *N. gabriellae*, 4.5 [Kenyon, 2008]; *Ho. hoolock*, 2.9 [Ahsan, 2001]) and rarely exceeds six even at sites where supernumerary males are common (e.g., Khao Yai [Reichard, 2009]; Way Canguk [Lappan, 2007]). The only well-documented exceptions to this pattern occur in crested gibbons, where bi-female groups are common. For example, in black crested gibbons, group size exceeds four group members on average (4.8 [Jiang et al., 1999]; 5.5 [Sheeran et al., 1998]; 6.2 [Fan et al., 2006]; see also Zhou et al., 2005; Fan et al., 2010), and one group of cao-vit gibbons maintained a group size of eight to nine members over a several-year period (Fan et al., 2015). Small group size in gibbons is generally maintained via mutual intolerance between same-sex adults (Ellefson, 1974; Mitani, 1984, 1985). In crested gibbons, however, co-resident females are tolerant to the point of engaging in mutual grooming (Fan et al., 2006; Guan et al., 2013).

Heterosexual pair-bonds between resident adults are signaled by a host of affiliative interactions that include joint defense of territory, duetting, and affiliative social interactions such as grooming and physical proximity (Palombit, 1994a; Geissmann & Orgeldinger, 2000). Available data suggest that the effort contributed to grooming relationships differs between male and female pair mates. For example, in white-handed gibbons (Palombit, 1996), hoolock gibbons (Ahsan, 2001), and white-bearded gibbons (Cheyne, 2010), males groom females more often than the reverse, while in siamangs it appears that grooming relationships are more even (Gittins & Raemaekers, 1980; Palombit, 1996). Palombit (1996) argues that differences in grooming effort indicate that white-handed gibbon males contribute more to the maintenance of the pair-bond than females because the costs of bonding to females are higher in smaller-bodied species relative to larger-bodied siamangs, which have lower relative energy requirements and a lower-quality diet. A potential test of this hypothesis would be to examine grooming parity in crested gibbons, which are intermediate with respect to body size. Fan and Jiang (2010) report only that the spatial relationship in the polygynous (1 male, 2 females) group they studied was equal among all adults.

REPRODUCTION

Females give birth to a single young after a gestation length of approximately seven months (Napier & Napier, 1985; Geissmann, 1991). In multiparous females, the interbirth interval (IBI) is typically between two and four years (*Hy. albibarbis*, 2.4 years [Cheyne, 2010]; *Hy. agilis*, 3.2 years [Mitani, 1990]; *Hy. klossii*, 3.3 years [Tilson, 1981]; *Hy. lar*, 3.4 years [Reichard & Barelli, 2008]; *S. syndactylus*, 2.8 years [Lappan, 2008]). Infants are carried ventrally by females until they are weaned during their second year of life (Treesucon, 1984; Reichard & Barelli, 2008). In siamangs alone, adult males and older juveniles are known to carry infants (Chivers, 1974; Dielentheis et al., 1991) and data from both captive (Alberts, 1987) and wild populations (Lappan, 2009) suggest that male siamangs become increasingly responsible for infant carrying after the infant's first year. Energetically, lactation is the most costly period for gibbon mothers, so it is interesting that male care should increase at the same time that infants are beginning to wean rather than when female investment is highest. In light of this finding, Lappan (2009: p.99) concludes that "any energetic savings the female experienced as a result of male 'help' were likely directed to somatic investment, rather than the current reproductive attempt." This conclusion is consistent with speculation, based on data from Khao Yai, that "gibbon females must attain a certain threshold of physical condition necessary to start cycling and to be able to conceive" (Savini et al., 2008: p.9).

Data from captivity indicate that male gibbons can breed as early as four years of age and females as early as five (Geissmann, 1991; Hodgkiss et al., 2010); however, because mating behavior in wild gibbons is generally delayed until after dispersal, maturity is more readily defined as having reached adult body size, which occurs at approximately eight years of age (Brockelman et al., 1998). Even so, age at first reproduction (i.e., age at first surviving offspring) is greatly delayed relative to maturity in both males (12.9 years; Brockelman et al., 1998) and females (11.1 years; Reichard & Barelli, 2008). In contrast to many other studies (e.g., Tilson, 1981; Ahsan, 2001; Burns et al., 2011), Brockelman et al. (1998) did not find evidence that subadults are forced out of their natal group via intrasexual aggression upon reaching sexual maturity. They speculate that extra subadults may provide a beneficial service to their parents by acting as social partners for younger infants and contributing to territorial defense, and their presence is tolerated as a result.

Monogamy and Pair-bonds

Gibbon social groups are regularly described as monogamous family units; however, the use of such terms as applied to gibbons must be done so with caution. Given the myriad ways in which gibbon groups are formed (see below), so-called *family* members may not always be close genetic relatives. Similar caution must be taken when applying the term *monogamy*, which can imply an exclusive social bond with a member of the opposite sex, sexual fidelity to a single mate, or both. Consequently, most authors are now careful to distinguish social (or grouping) monogamy from sexual (or mating) monogamy; although, others argue that such imprecise labels should be jettisoned altogether in favor of terminology less encumbered by connotative baggage (Fuentes, 2000; Sommer & Reichard, 2000). While warnings about the imprecision of the term *monogamy* are not new (see Carpenter, 1940; Wickler, 1976; Wickler & Seibt, 1983), there can be no doubt that the last two decades have witnessed a sea change in how we view gibbon social systems. The most important developments can be summarized as follows:

First, it is now well documented that bonded adults engage in extra-pair copulations (EPC), or "sexual contacts between individuals who [do] not maintain a close spatiosocial pair bond" (Reichard, 2009: p.349), and in some cases extra-pair paternity is the result (Kenyon et al., 2011; Barelli et al., 2013). EPCs were first documented in siamangs at Ketambe (Palombit, 1994b) and then soon after among white-handed gibbons in Khao Yai (Reichard, 1995). Based on data collected on white-handed gibbons over 14 years in Khao Yai, Reichard (2009) reports that of the adults who were seen to copulate, 42%

of males (8 of 19) and 69% of females (9 of 13) engaged in EPCs. In a subsequent study, Barelli et al. (2013) were able to use short tandem repeats (STRs) to successfully assign paternity to 41 gibbon offspring in the same population. Interestingly, despite high reported rates of EPC, the overwhelming majority of offspring were sired by resident males (38/41 or 92.7%). Nevertheless, in two cases the authors were able to confirm extra-pair paternity by known neighbor males. In a separate study of crested gibbons (*N. gabriellae*), Kenyon (2008; Kenyon et al., 2011) successfully genotyped 10 putative offspring from six social groups in Cat Tien National Park, Vietnam. In all cases, the offspring were confirmed to belong to the resident female and in all but one case to the resident male. In the single case of extra-pair paternity, paternity was successfully assigned to a lone male who was not a territory holder. In this case, EPC was inferred based on the long-term stability of the resident pair, which had been under observation for two years and which were genetically confirmed as parents of an older offspring in the group.

Second, long-term studies show that adult membership changes regularly as a result of death (Palombit, 1994a), mate desertion (Palombit, 1994a), and displacement (Palombit, 1994a; Brockelman et al., 1998; Ahsan, 2001; Fan & Jiang, 2010), or by replacement of one pair mate by a maturing offspring (Chivers & Raemaekers, 1980; Tilson, 1981; Raemaekers & Raemaekers, 1984; Palombit, 1994a; Ahsan, 2001). In fact, it now appears that bonds formed by two maturing subadults, once thought to be the dominant pathway to group formation, are relatively rare. Long-term research on the population dynamics at Khao Yai determined that just one of seven documented pair formations was the result of two dispersing subadults jointly entering an unoccupied territory (Brockelman et al., 1998; see also Reichard, 2009). At Ketambe, Indonesia, just one of six new pair-bonds was established through simultaneous dispersal events (Palombit, 1994a).

Third, when adults or subadults enter new groups, they are occasionally accompanied by one or more immatures, including putative siblings (Palombit, 1994a; Brockelman et al., 1998; Fan et al., 2010). Consequently, the structure of genetic relationships within groups will frequently diverge from that implied by a nuclear family model. Furthermore, secondary dispersal by maturing animals may result in close genetic relationships—or non-genetic social ties—between individuals living in neighboring groups. Thus, it is likely that the complex pattern of familial and affiliative relationships transcends the level of the individual group, forming a social superstructure recognizable in the pattern of interactions (e.g.,

play and grooming) among neighbors (Bartlett, 2003; Bartlett et al., 2016; see also Rowell, 2000).

Fourth, while demographic data indicate that two-adult groups are the most common pattern in gibbons (Mitani, 1990; Fuentes, 2000; Reichard, 2009; Cheyne, 2010), at some study sites social groups with supernumerary adults, especially males, are common (Lappan, 2007; Reichard, 2009; see also Malone & Fuentes, 2009). At Khao Yai, for example, 18% of groups censused over 14 years contained more than two adults, the vast majority of which were multi-male–single female groups (Reichard, 2009). In some cases, it is clear that multi-male social structure is quite stable, with females engaging in sexual contact with both males (i.e., socio-sexual polyandry; but see Morino, 2015). By comparison, multi-female groups, though present, were rare (< 3%) and temporary: "The longest known multi-male group existed for 12 years; the longest known pair-living groups existed for 14 years. In contrast, the longest known multi-female group existed for about two years" (Reichard, 2009: p.357).

Given the stability of multi-male groups at Khao Yai, Reichard concludes that multi-male groups are an alternative grouping strategy with fitness benefits both to females (e.g., increased paternal investment, increased fertilization probability, decreased infanticide risk) and to males. While standard socioecological theory assumes rigid competition between males, Reichard (2009)—building on the same logic Brockelman et al. (1998) use to explain delayed dispersal by subadult males—suggests that under some circumstances cooperative male defense of female territories may be more successful than defense by a single male operating alone (see also Lappan, 2007). Elsewhere, Bartlett (2009b) has summarized the benefit of joint male defense as follows:

> Although the addition of a second male should increase intragroup feeding competition, it is reasonable to expect that the loss of resources would be offset by his contribution to territorial defense. Groups with more males should be able to maintain larger or more-exclusive ranges. The energetic benefits might be realized in terms of reduced interbirth intervals, higher survival rates, or healthier offspring. (p.142)

Alternatively, when the available habitat is saturated, as may be the case at Khao Yai, polyandry may not be so much an evolved *strategy* (c.f. Lappan, 2007; Reichard, 2009) as it is a case of males making the best of a bad situation by gaining a position as secondary male rather than having no position at all. At the same time, primary males will tolerate secondary males simply because the cost of excluding them is too high (Lappan, 2007).

Finally, in contrast to the situation described for Khao Yai and Way Canguk, where stable multi-male groups are common but multi-female groups rare and short lived, crested gibbons in China and northern Vietnam form *stable* multi-female (or polygynous) groups. This has been best documented in black crested gibbons at Mt. Wuliang (Fan & Jiang, 2010); however, social groups with two adult females, each with dependent offspring, have also been documented in Hainan gibbons (*N. hainanus*; Chan et al., 2005), in cao-vit gibbons (*N. nasutus*; Fan et al., 2010), and in yellow-cheeked crested gibbons (*N. gabriellae*; Barca et al., 2016). In such cases, the consequences for territorial economics are quite different than in the case of extra-males because the addition of a female in combination with one or more offspring significantly increases group size and net nutritional demand without significantly increasing resource holding potential. Indeed, it is likely that gibbon territories at Mt. Wuliang, for example, are at the limits of defendability (Mitani & Rodman, 1979). That is, home range size (at least for some groups) is so large that residents are just barely able to monitor territorial borders effectively. In this population, at least, the synergistic relationship between frugivory, territoriality, and social monogamy that characterizes gibbons in more stable habitat (Bartlett, 2009b, 2011) is less apparent. Whether this is a species-specific adaptive response to a markedly more variable resource base or simply an expression of the ecological flexibility common to all hylobatids must await further study.

SUMMARY

Gibbons are small, highly acrobatic, arboreal apes found primarily in the forests of Southeast Asia, with remnant populations reaching north into China and west into India and Bangladesh. Their behavior and ecology are characterized by an interrelated constellation of traits that include frugivory, social monogamy, territoriality, and loud complex calls, or songs. In forests with reliable access to ripe fruit, small gibbon groups are able to maintain access to fixed ranges that males, in particular, actively defend against encroachment by neighbors. Notably, these same traits, along with slow life history, make gibbons vulnerable to hunting and habitat loss. For example, loud vocal displays make it easier for human hunters to find groups, and fragmented habitat robs gibbons of the arboreal highways they rely on to traverse their range. In light of these threats, fieldworkers have intensified efforts to document current gibbon distribution in hopes of targeting conservation action.

REFERENCES CITED—CHAPTER 20

Ahsan, M. F. (2001). Socio-ecology of the hoolock gibbon (*Hylobates hoolock*) in two forests of Bangladesh. *The apes: Challenges for the 21st century, Conference Proceedings* (pp. 286-299). Chicago Zoological Society.

Alberts, S. (1987). Parental care in captive siamangs (*Hylobates syndactylus*). *Zoo Biology, 6,* 401-406.

Asensio, N., Brockelman, W. Y., Malaivijitnond, S., & Reichard, U. (2011). Gibbon travel paths are goal oriented. *Animal Cognition, 14*(3), 395-405.

Asensio, N., Brockelman, W. Y., Malaivijitnond, S., & Reichard, U. (2014). White-handed gibbon (*Hylobates lar*) core area use over a short-time scale. *Biotropica, 46*(4), 461-469.

Barca, B., Vincent, C., Soeung, K., Nuttall, M., & Hobson, K. (2016). Multi-female group in the southernmost species of *Nomascus*: Field observations in eastern Cambodia reveal multiple breeding females in a single group of southern yellow-cheeked crested gibbon *Nomascus gabriellae*. *Asian Primates Journal, 6*(1), 15-19.

Barelli, C., Boesch, C., Heistermann, M., & Reichard, U. H. (2008). Female white-handed gibbons (*Hylobates lar*) lead group movements and have priority of access to food resources. *Behaviour, 145,* 965-981.

Barelli, C., Matsudaira, K., Wolf, T., Roos, C., Heistermann, M., Hodges, K., Ishida, T., Malaivijitnond, S., & Reichard, U. H. (2013). Extra-pair paternity confirmed in wild white-handed gibbons. *American Journal of Primatology, 75*(12), 1185-1195.

Bartlett, T. Q. (2003). Intragroup and intergroup social interactions in white-handed gibbons. *International Journal of Primatology, 24,* 239-259.

Bartlett, T. Q. (2009a). Seasonal home range use and defendability in white-handed gibbons (*Hylobates lar*) in Khao Yai National Park, Thailand. In S. M. Lappan, & D. Whittaker (Eds.), *The gibbons: New perspectives on small ape socioecology and population biology* (pp. 265-275). Springer.

Bartlett, T. Q. (2009b). *The gibbons of Khao Yai: Seasonal variation in behavior and ecology.* Pearson Prentice Hall.

Bartlett, T. Q. (2011). The Hylobatidae: Small apes of Asia. In C. J. Campbell, A. Fuentes, K. C. MacKinnon, R. M. Stumpf, & S. K. Bearder (Eds.), *Primates in perspective* (2nd Ed., pp. 300-312). Oxford University Press.

Bartlett, T. Q., & Light, L. E. O. (2017). Territory. In A. Fuentes (Ed.), *The international encyclopedia of primatology.* John Wiley & Sons, Inc.

Bartlett, T. Q., Light, L. E. O., & Brockelman, W. Y. (2016). Long-term home range use in white-handed gibbons (*Hylobates lar*) in Khao Yai National Park. *American Journal of Primatology, 78*(2), 192-203.

Bleisch, W. V., & Chen, N. (1991). Ecology and behavior of wild black-crested gibbons (*Hylobates concolor*) in China with a reconsideration of evidence for polygyny. *Primates, 32,* 539-548.

Bricknell, S. J. (1999). *Hybridisation and behavioral variation: A socio-ecological study of hybrid gibbons (Hylobates agilis* albibarbis x H. muelleri) *in central Kalimantan, Indonesia.* [Doctoral dissertation, Australian National University].

Brockelman, W. Y. (2009). Ecology and the social system of gibbons. In D. Whittaker, & S. Lappan (Eds.), *The gibbons. Developments in primatology: Progress and prospects* (pp. 211-239). Springer.

Brockelman, W. Y. (2013). Gibbon studies in Khao Yai National Park: Some personal reminiscences. *Natural History Bulletin of the Siam Society, 59*(2), 109-135.

Brockelman, W. Y., Reichard, U., Treesucon, U., & Raemaekers, J. J. (1998). Dispersal, pair formation and social structure in gibbons (*Hylobates lar*). *Behavioral Ecology and Sociobiology, 42,* 329-339.

Brockelman, W. Y., & Srikosamatara, S. (1984). Maintenance and evolution of social structure in gibbons. In H. Preuschoft, D. J. Chivers, W. Y. Brockelman, & N. Creel (Eds.), *The lesser apes: Evolutionary and behavioural biology* (pp. 298-323). Edinburgh University Press.

Burns, B. L., Dooley, H. M., & Judge, D. S. (2011). Social dynamics modify behavioural development in captive white-cheeked (*Nomascus leucogenys*) and silvery (*Hylobates moloch*) gibbons. *Primates, 52*(3), 271.

Caine, N. G. (1987). Vigilance, vocalizations, and cryptic behavior at retirement in captive groups of red-bellied tamarins. *American Journal of Primatology,* 12, 241-250.

Cannon, C. H., & Leighton, M. (1994). Comparative locomotor ecology of gibbons and macaques: Selection of canopy elements for crossing gaps. *American Journal of Physical Anthropology, 93,* 505-524.

Cant, J. G. H. (1992). Positional behavior and body size of arboreal primates: A theoretical framework for field studies and an illustration of its application. *American Journal of Physical Anthropology, 88,* 273-284.

Carpenter, C. R. (1940). *A field study in Siam of the behavior and social relations of the gibbon (Hylobates lar).* Johns Hopkins University Press.

Chan, B. P. L., Fellowes, J. R., Geissmann, T., & Zhang, J. (Eds.). (2005). *Status survey and conservation action plan for the Hainan gibbon.* Kadoorie Farm and Botanic Garden.

Chen, N. (1995). *Ecology of the black-crested gibbon (*Hylobates concolor) *in the Ailao Mt. Reserve, Yunnan, China.* [Master thesis, Mahidol University].

Cheyne, S. M. (2010). Behavioural ecology and socio-biology of gibbons (*Hylobates albibarbis*) in a degraded peat-swamp forest. In J. Supriatna, & S. L. Gursky (Eds.), *Indonesian primates* (pp. 121-156). Springer.

Cheyne, S. M., Gilhooly, L. J., Hamard, M. C., Höing, A., Houlihan, P. R., Kursani, Loken, B., Phillips, A., Rayadin, T., Capilla, B. R., Rowland, D., Sastramidjaja, W. J., Spehar, S., Thompson, C. J. H., & Zrust, M. (2016). Population mapping of gibbons in Kalimantan, Indonesia: Correlates of gibbon density and vegetation across the species' range. *Endangered Species Research, 30,* 133-143.

Chivers, D. J. (1974). The siamang in Malaya: A field study of a primate in tropical rain forest. *Contributions to Primatology,* 4, 1-331.

Chivers, D. J. (1977). The feeding behaviour of siamang (*Symphalangus syndactylus*). In T. H. Clutton-Brock (Ed.), *Primate ecology* (pp. 355-382). Academic Press.

Chivers, D. J. (2001). The swinging singing apes: Fighting for food and family in the far-east forests. In *The apes: Challenges for the 21st century, Conference Proceedings* (pp. 1-28). Chicago Zoological Society.

Chivers, D. J., & Raemaekers, J. J. (1980). Long-term changes in behavior. In D. J. Chivers, & K. A. Joysey (Eds.), *Malayan forest primates* (pp. 209-260). Academic Press.

Chivers, D. J., & Raemaekers, J. J. (1986). Natural and synthetic diets of Malayan gibbons. In J. G. Else, & P. C. Lee (Eds.), *Primate ecology and conservation* (pp. 39-56). Cambridge University Press.

Clink, D. J., Bernard, H., Crofoot, M. C., & Marshall, A. J. (2017). Investigating individual vocal signatures and small-scale patterns of geographic variation in female Bornean gibbon (*Hylobates muelleri*) great calls. *International Journal of Primatology, 38*(4), 656-671.

Corlett, R. T. (1998). Frugivory and seed dispersal by vertebrates in the oriental (Indomalayan) region. *Biological Reviews, 73*(4), 413-448.

Cressey, D. (2014). Time running out for rarest primate. *Nature News, 508*(7495), 163.

Delacour, J. (1961). Gibbons at liberty. *Zoologische Garten,* 26, 96-99.

Dielentheis, T. F., Zaiss, E., & Geissmann, T. (1991). Infant care in a family of siamangs (*Hylobates syndactylus*) with twin offspring at Berlin Zoo. *Zoo Biology,* 10, 309-317.

Dillis, C., Beaudrot, L., Feilen, K. L., Clink, D. J., Wittmer, H. U., & Marshall, A. J. (2015). Modeling the ecological and phonological predictors of fruit consumption by gibbons (*Hylobates albibarbis*). *Biotropica, 47*(1), 85-93.

Dooley, H. M., & Judge, D. S. (2014). Kloss gibbons (*Hylobates klossii*) behavior facilitates the avoidance of human predation in the Peleonan forest, Siberut Island, Indonesia. *American Journal of Primatology, 77*(3), 196-308.

Elder, A. A. (2009). Hylobatid diets revisited: The importance of body mass, fruit availability, and interspecific competition. In S. Lappan, & D. J. Whittaker (Eds.), *The gibbons: New perspectives on small ape socioecology and population biology* (pp. 133-159). Springer.

Ellefson, J. O. (1967). *A natural history of gibbons in the Malay Peninsula.* [Doctoral dissertation, University of California].

Ellefson, J. O. (1968). Territorial behavior in the common white-handed gibbon, *Hylobates lar* Linn. In P. C. Jay (Ed.), *Primates: Studies in adaptation and variability* (pp. 180-199). Holt, Rinehart and Winston.

Ellefson, J. O. (1974). A natural history of white-handed gibbons in the Malayan Peninsula. In *Gibbon and Siamang,* 3, 1-136.

Fan, P. (2017). The past, present, and future of gibbons in China. *Biological Conservation,* 210, 29-39.

Fan, P. F., & Bartlett, T. Q. (2017). Overlooked small apes need more attention! *American Journal of Primatology, 79*(6), e22658.

Fan, P. F., Bartlett, T. Q., Fei, H. L., Ma, C. Y., & Zhang, W. (2015). Understanding stable bi-female grouping in gibbons: Feeding competition and reproductive success. *Frontiers in Zoology, 12,* 5.

Fan, P. F., Fei, H. L., & Ma, C. Y. (2012). Behavioral responses of Cao Vit gibbon (*Nomascus nasutus*) to variations in food abundance and temperature in Bangliang, Jingxi, China. *American Journal of Primatology, 74,* 632-641.

Fan, P. F., Fei, H. L., Xiang, Z. F., Zhang, W., Ma, C. Y., & Huang, T. (2010). Social structure and group dynamics of the Cao Vit gibbon (*Nomascus nasutus*) in Bangliang, Jingxi, China. *Folia Primatologica, 81,* 245-253.

Fan, P. F., He, K., Chen, X., Ortiz, A., Zhang, B., Zhao, C., Li, Y.-Q., Zhang, H.-B., Kimock, C., Wang, W.-Z., Groves, C., Turvey, S. T., Roos, C., Helgen, K. M., & Jiang, X.-L. (2017). Description of a new species of hoolock gibbon (Primates: Hylobatidae) based on integrative taxonomy. *American Journal of Primatology, 79*(5), e22631.

Fan, P. F., & Jiang, X. L. (2008a). Sleeping sites, sleeping trees and sleep-related behavior of black crested gibbons (*Nomascus concolor jingdongensis*) at Mt. Wuliang, Yunnan, China. *American Journal of Primatology, 70*(2), 153-160.

Fan, P. F., & Jiang, X. L. (2008b). Effects of food and topography on ranging behavior of black crested gibbon (*Nomascus concolor jingdongensis*) in Wuliang Mountain, Yunnan, China. *American Journal of Primatology, 70*(9), 871-878.

Fan, P. F., & Jiang, X. L. (2009). Predation on giant flying squirrels (*Petaurista philippensis*) by black crested gibbons (*Nomascus concolor jingdongensis*) at Mt. Wuliang, Yunnan, China. *Primates, 50*(1), 45-49.

Fan, P. F., & Jiang, X. L. (2010). Maintenance of multifemale social organization in a group of *Nomascus concolor* at Wuliang Mountain, Yunnan, China. *International Journal of Primatology, 31*(1), 1-13.

Fan, P. F., Jiang, X. L., Liu, C. M., & Luo, W. S. (2006). Polygynous mating system and behavioural reason of black crested gibbon (*Nomascus concolor jingdongensis*) at Dazhaizi, Mt. Wuliang, Yunnan, China. *Zoological Research, 27*(2), 216-220.

Fan, P. F., Ni, Q. Y., Sun, G. Z., Huang, B., & Jiang, X. L. (2008). Seasonal variations in the activity budget of black-crested gibbons (*Nomascus concolor jingdongensis*) at Mt. Wuliang, central Yunnan, China: Effects of diet and temperature. *International Journal of Primatology, 29*(4), 1047-1057.

Fan, P. F., Ni, Q. Y., Sun, G. Z., Huang, B., & Jiang, X. L. (2009). Gibbons under seasonal stress: The diet of the black crested gibbons (*Nomascus concolor*) on Mt. Wuliang, central Yunnan. *Primates, 50*(1), 37-44.

Fan, P. F., Ren, G. P., Wang, W., Scott, M. B., Ma, C. Y., Fei, H. L., Wang, L., Xhao, W., & Zhu, J. G. (2013). Habitat evaluation and population viability analysis of the last population of Cao Vit gibbon (*Nomascus nasutus*): Implications for conservation. *Biological Conservation, 161,* 39-47.

Fan, P. F., Xiao, W., Huo, S., Ai, H. S., Wang, T. C., & Lin, R. T. (2011). Distribution and conservation status of *Hoolock leuconedys* in China. *Oryx, 45*(1), 129-134.

Feeroz, M. M., Islam, M. A., & Kabir, M. M. (1994). Food and feeding behaviour of hoolock gibbon (*Hylobates hoolock*), capped langur (*Presbytis pileata*) and pigtailed macaque (*Macaca nemestrina*) of Lawachara. *Bangladesh Journal of Zoology, 22,* 123-132.

Fleagle, J. G. (1976). Locomotion and posture of the Malayan siamang and implications for hominoid evolution. *Folia Primatologica, 26,* 245-269.

Frechette, J. L., Hon, N., Behie, A. M., & Rawson, B. M. (2017). Seasonal variation in the diet and activity budget of the northern yellow-cheeked crested gibbon *Nomascus annamensis*. *Cambodian Journal of Natural History,* 168-178.

Frisch, J. E. (1963). Sex-differences in the canines of the gibbon (*Hylobates lar*). *Primates, 4,* 1-10.

Fuentes, A. (2000). Hylobatid communities: Changing views on pair bonding and social organization in hominoids. *Yearbook of Physical Anthropology, 43,* 33-60.

Gates, R. (1998). *In situ* and *ex situ* conservation of the silvery or Moloch gibbon *Hylobates moloch*. *International Zoo Yearbook, 36,* 81-84.

Geissmann, T. (1991). Reassessment of age of sexual maturity in gibbons. *American Journal of Primatology, 23,* 11-22.

Geissmann, T. (1993). *Evolution of communication in gibbons (Hylobatidae).* [Doctoral dissertation, Universitaet Zuerich].

Geissmann, T. (2007). First field data on the Laotian black crested gibbon (*Nomascus concolor lu*) of the Nam Kan area of Laos. *Gibbon Journal, 3,* 56-65.

Geissmann, T. (2015). *The gibbons (Hylobatidae): An introduction.* Gibbon Research Lab. http://www.gibbons.de /main/index.html.

Geissmann, T., & Orgeldinger, M. (2000). The relationship between duet songs and pair bonds in siamangs, *Hylobates syndactylus*. *Animal Behaviour, 60,* 805-809.

Gittins, S. P. (1980). Territorial behavior in the agile gibbon. *International Journal of Primatology, 1,* 381-399.

Gittins, S. P. (1982). Feeding and ranging in the agile gibbon. *Folia Primatologica, 38,* 39-71.

Gittins, S. P. (1983). Use of the forest canopy by the agile gibbon. *Folia Primatologica, 40,* 134-144.

Gittins, S. P. (1984). Territorial advertisement and defense in gibbons. In H. Preuschoft, D. J. Chivers, W. Y. Brockelman, & N. Creel (Eds.), *The lesser apes: Evolutionary and behavioural biology* (pp. 420-424). Edinburgh University Press.

Gittins, S. P., & Raemaekers, J. J. (1980). Siamang, lar and agile gibbons. In D. J. Chivers (Ed.), *Malayan forest primates: Ten years' study in tropical rain forest.* (pp. 63-105). Plenum Press.

Gray, T. N. E., Phan, C., & Long, B. (2010). Modelling species distribution at multiple spatial scales: Gibbon habitat preferences in a fragmented landscape. *Animal Conservation*, 1, 1-9.

Groves, C. P. (1984). A new look at the taxonomy and phylogeny of the gibbons. In H. Preuschoft, D. J. Chivers, W. Y. Brockelman, & N. Creel (Eds.), *The lesser apes: Evolutionary and behavioural biology* (pp. 452-561). Edinburgh University Press.

Groves, C. P. (2001). *Primate taxonomy*. Smithsonian Institution Press.

Groves, C. P., & Wang, Y. (1990). The gibbons of the subgenus *Nomascus* (Primates, Mammalia). *Zoological Research*, 11, 147-154.

Grueter, C. C., Jiang, X., Konrad, R., Fan, P. F., Guan, Z., & Geissmann, T. (2009). Are *Hylobates lar* extirpated from China? *International Journal of Primatology, 30*, 553-567.

Guan, Z. H., Huang, B., Ning, W. H., Ni, Q. Y., Sun, G. Z., & Jiang, X. L. (2013). Significance of grooming behavior in two polygynous groups of western black crested gibbons: Implication for understanding social relationships among immigrant and resident group members. *American Journal of Primatology*, 75, 1165-1173.

Hamard, M., Cheyne, S. M., & Nijman, V. (2010). Vegetation correlates of gibbon density in the peat swamp forest of the Sabangau catchment, central Kalimantan, Indonesia. *American Journal of Primatology*, 72(7), 607-616.

Harrison, R. D., Sreekar, R., Brodie, J. F., Brook, S., Luskin, M., O'Kelly, H., Rao, M., Scheffers, B., & Velho, N. (2016). Impacts of hunting on tropical forests in southeast Asia. *Conservation Biology*, 30(5), 972-981.

Heller, R., Sander, A. F., Wang, C. W., Usma, F., & Dabelsteen, T. (2010). Macrogeographical variability in the great call of *Hylobates agilis*: Assessing the applicability of vocal analysis in studies of fine-scale taxonomy of gibbons. *American Journal of Primatology*, 72(2), 142-151.

Hirai, H., Mootnick, A. R., Takenaka, O., Suryobroto, B., Mouri, T., Kamanaka, Y., Katoh, A., Kimura, N., Katoh, A., & Maeda, N. (2003). Genetic mechanism and property of a whole-arm translocation (WAT) between chromosomes 8 and 9 of agile gibbons (*Hylobates agilis*). *Chromosome Research*, 11(1), 37-50.

Hodgkiss, S., Thetford, E., Waitt, C. D., & Nijman, V. (2010). Female reproductive parameters in the Javan gibbon (*Hylobates moloch*). *Zoo Biology*, 29(4), 449-456.

Hu, Y., Xu, H., & Yang, D. (1989). The studies on ecology of *Hylobates leucogenys*. *Zoological Research*, 10(zk), 61-66.

Hu, Y., Xu, H., & Yang, D. (1990). Feeding ecology of the white-cheek gibbon (*Hylobates concolor leucogenys*). *Acta Ecologica Sinica*, 10(2), 155-159.

Islam, M. A., & Feeroz, M. M. (1992). Ecology of hoolock gibbon of Bangladesh. *Primates*, 33(4), 451-464.

IUCN. (2020). *International Union for the Conservation of Nature Red List of Threatened Species*. Version 2020-3. http://www.iucnredlist.org.

Janzen, D. H. (1979). How to be a fig. *Annual Review of Ecology and Systematics*, 10, 13-51.

Jiang, X., Wang, Y., & Wang, Q. (1999). Coexistence of monogamy and polygyny in black-crested gibbon (*Hylobates concolor*). *Primates*, 40(4), 607-611.

Jildmalm, R., Amundin, M., & Laska, M. (2008). Food preferences and nutrient composition in captive white-handed gibbons, *Hylobates lar*. *International Journal of Primatology*, 29(6), 1535-1547.

Jungers, W. L. (1984). Aspects of size and scaling in primate biology with special reference to the locomotor skeleton. *Yearbook of Physical Anthropology*, 27, 73-97.

Kakati, K., Raghavan, R., Chellam, R., Qureshi, Q., & Chivers, D. J. (2009). Status of western hoolock gibbon (*Hoolock hoolock*) populations in fragmented forests of eastern Assam. *Primate Conservation*, 24(1), 127-137.

Kanwatanakid-Savini, C. O., Poonswad, P., & Savini, T. (2009). An assessment of food overlap between gibbons and hornbills. *Raffles Bulletin of Zoology*, 57, 189-198.

Kawamura, S. (1958). The preliminary survey on the white-handed gibbon in Thailand. *Primates*, 1, 157-158.

Kay, R. F. (1975). The functional adaptations of primate molar teeth. *American Journal of Physical Anthropology*, 43(2), 195-215.

Kelley, J. (1997). Paleobiological and phylogenetic significance of life history in Miocene hominoids. In D. R. Begun, C. V. Ward, & M. D. Rose (Eds.), *Function, phylogeny and fossils: Miocene hominoid evolution and adaptations* (pp. 173-208). Plenum Press.

Kenyon, M. A. (2008). *Ecology of the golden-cheeked gibbon (*Nomascus gabriellae*) in Cat Tien National Park, Vietnam* [Doctoral dissertation, University of Cambridge].

Kenyon, M., Roos, C., Binh, V. T., & Chivers, D. (2011). Extra-pair paternity in golden-cheeked gibbons (*Nomascus gabriellae*) in the secondary lowland forest of Cat Tien National Park, Vietnam. *Folia Primatologica*, 82, 154-164.

Kim, S., Lappan, S., & Choe, J. C. (2011). Diet and ranging behavior of the endangered Javan gibbon (*Hylobates moloch*) in a submontane tropical rainforest. *American Journal of Primatology*, 73(3), 270-280.

Kinnaird, M. F., & O'Brien, T. G. (2005). Fast foods of the forest: The influence of figs on primates and hornbills across Wallace's line. In J. L. Dew, & J. P. Bouble (Eds.), *Tropical fruits and frugivores: The search for strong predictors* (pp. 155-184). Springer.

Kitamura, S., Suzuki, S., Yumoto, T., Poonswad, P., Chuailua, P., & Plongmai, K. (2004). Dispersal of *Aglaia spectabilis*, a large-seeded tree species in a moist evergreen forest in Thailand. *Journal of Tropical Ecology*, 20(04), 421-427.

Kolasartsanee, I., & Srikosamatara, S. (2014). Applying "diffusion of innovation" theory and social marketing for the recovery of pileated gibbon *Hylobates pileatus* in north Ta-riu watershed, Khao Soi Dao Wildlife Sanctuary, Thailand. *Conservation Evidence*, 11, 61-65.

Lambert, J. E. (2011). Primate nutritional ecology: Feeding biology and diet at ecological and evolutionary scales. In C. J. Campbell, A. Fuentes, K. C. MacKinnon, R. M. Stumpf, &

S. K. Bearder (Eds.), *Primates in perspective* (pp. 512-522). Oxford University Press.

Lappan, S. (2007). Patterns of dispersal in Sumatran siamangs (*Symphalangus syndactylus*): Preliminary mtDNA evidence suggests more frequent male than female dispersal to adjacent groups. *American Journal of Primatology, 69*, 692-698.

Lappan, S. (2008). Male care of infants in a siamang (*Symphalangus syndactylus*) population including socially monogamous and polyandrous groups. *Behavioral Ecology and Sociobiology, 62*, 1307-1317.

Lappan, S. (2009). The effects of lactation and infant care on adult energy budgets in wild siamangs (*Symphalangus syndactylus*). *American Journal of Physical Anthropology, 140*, 290-301.

Lappan, S., Ruppert, N., Mohd Rameli, N. I. A., Kamaruzaman, A. S., Heng, P. Y., Sharul, A. M. S., Quilter, D., & Bartlett, T. Q. (2018). Join the chorus! *Malaysian Naturalist, 31-33.*

Le, T. D., Fan, P. F., Yan, L., Le, H. O., & Josh, K. (2008). *The global Cao Vit gibbon (*Nomascus nasutus*) population.* Fauna and Flora International.

Lee, D. C., Powell, V. J., & Lindsell, J. A. (2015). The conservation value of degraded forests for agile gibbons *Hylobates agilis. American Journal of Primatology, 77*, 76-85.

Leighton, M. (1987). Gibbons: Territoriality and monogamy. In B. B. Smuts, D. L. Cheney, R. M. Seyfarth, R. W. Wrangham, & T. T. Struhsaker (Eds.), *Primate societies* (pp. 135-145). University of Chicago Press.

Leighton, M. (1993). Modeling dietary selectivity by Bornean orangutans: Evidence for integration of multiple criteria in fruit selection. *International Journal of Primatology, 14*, 257-313.

Leighton, M., & Leighton, D. R. (1983). Vertebrate responses to fruiting seasonality within a Bornean rain forest. In S. L. Sutton, T. C. Whitmore, & A. C. Chadwick. (Eds.), *Tropical rain forests: Ecology and management* (pp. 181-196). Blackwell Scientific Publications.

Light, L. E. O. (2016). *Life at the extreme: The behavioral ecology of white-handed gibbons (*Hylobates lar*) living in a dry forest in Huai Kha Khaeng Wildlife Sanctuary, western Thailand.* [Doctoral dissertation, University of Texas, San Antonio].

Ma, S., Wang, Y., & Poirier, F. E. (1988). Taxonomy, distribution, and status of gibbons (*Hylobates*) in southern China and adjacent areas. *Primates, 29*(2), 277-286.

MacKinnon, J. R., & MacKinnon, K. S. (1980). Niche differentiation in a primate community. In D. J. Chivers (Ed.), *Malayan forest primates: Ten years' study in tropical rain forest* (pp. 167-191). Plenum Press.

Malone, N., & Fuentes, A. (2009). The ecology and evolution of hylobatid communities: Causal and contextual factors underlying inter- and intraspecific variation. In S. Lappan, & D. J. Whittaker (Eds.), *The gibbons: New perspectives on small ape socioecology and population biology* (pp. 241-264). Springer.

Marshall, A. J., & Leighton, M. (2006). How does food availability limit the population density of agile gibbons?

In G. Hohmann, M. Robbins, & C. Boesch (Eds.), *Feeding ecology of the apes* (pp. 313-335). Cambridge University Press.

Marshall, J. T., Jr., & Marshall, E. R. (1976). Gibbons and their territorial songs. *Science, 193*, (4249), 235-237.

Marshall, J. T., Jr., Ross, B. A., & Chantharojvong, S. (1972). The species of gibbons in Thailand. *Journal of Mammalogy, 53*(3), 479-486.

Mather, R. J. (1992). *A field study of hybrid gibbons in central Kalimantan, Indonesia.* [Doctoral dissertation, University of Cambridge].

McConkey, K. R. (2000). Primary seed shadow generated by gibbons in the rain forests of Barito Ulu, central Borneo. *American Journal of Primatology, 52*, 13-29.

McConkey, K. R., Aldy, F., Ario, A., & Chivers, D. J. (2002). Selection of fruit by gibbons (*Hylobates muelleri x agilis*) in the rain forests of central Borneo. *International Journal of Primatology, 23*, 123-145.

McConkey, K. R., Ario, A., Aldy, F., & Chivers, D. J. (2003). Influence of forest seasonality on gibbon food choice in the rain forests of Barito Ulu, central Kalimantan. *International Journal of Primatology, 24*, 19-32.

McConkey, K. R., & Brockelman, W. Y. (2011). Nonredundancy in the dispersal network of a generalist tropical forest tree. *Ecology, 92*(7), 1492-1502.

McConkey, K. R., & Chivers, D. J. (2007). Influence of gibbon ranging patterns on seed dispersal distance and deposition in a Bornean forest. *Journal of Tropical Ecology, 23*, 269-275.

Melfi, V. A. (2012). Gibbons: Probably the most endangered primates in the world. *International Zoo Yearbook, 46*(1), 239-240.

Mitani, J. C. (1984). The behavioral regulation of monogamy in gibbons (*Hylobates muelleri*). *Behavioral Ecology and Sociobiology, 15*(3), 225-229.

Mitani, J. C. (1985). Location-specific responses of gibbons (*Hylobates muelleri*) to male songs. *Zeitschrift für Tierpsychology, 70*, 219-224.

Mitani, J. C. (1990). Demography of agile gibbons (*Hylobates agilis*). *International Journal of Primatology, 11*(5), 411-424.

Mitani, J. C., & Rodman, P. S. (1979). Territoriality: The relation of ranging pattern and home range size to defendability, with an analysis of territoriality among primate species. *Behavioral Ecology and Sociobiology, 5*, 241-251.

Mootnick, A. R., & Groves, C. P. (2005). A new generic name for the hoolock gibbon (Hylobatidae). *International Journal of Primatology, 26*, 971-976.

Morino, L. (2010). Clouded leopard predation on a wild juvenile siamang. *Folia Primatologica, 81*(6), 362-368.

Morino, L. (2015). Dominance relationships among siamang males living in multimale groups. *American Journal of Primatology, 78*(3), 288-297.

Napier, J. R., & Napier, P. H. (1985). *The natural history of the primates.* Cambridge University Press.

Newkirk, J. B. (1973). A possible case of predation in the gibbon. *Primates, 14*, 301-304.

Nowak, M. G., & Reichard, U. H. (2010). *Hylobatid positional behavior and postcranial anatomy: Terminal branch feeding and the evolution of hominoid orthogrady*. XXIII Congress of the International Primatological Society.

O'Brien, T. G., Kinnaird, M. F., Nurcahyo, A., Iqbal, M., & Rusmanto, M. (2004). Abundance and distribution of sympatric gibbons in a threatened Sumatran rain forest. *International Journal of Primatology, 25*(2), 267-284.

O'Brien, T. G., Kinnaird, M. F., Nurcahyo, A., Prasetyaningrum, M., & Iqbal, M. (2003). Fire, demography and the persistence of siamang (*Symphalangus syndactylus*: Hylobatidae) in a Sumatran rainforest. *Animal Conservation, 6*, 115-121.

Palombit, R. A. (1992). *Pair bonds and monogamy in wild siamang (*Hylobates syndactylus*) and white-handed gibbon (*Hylobates lar*) in northern Sumatra*. [Doctoral dissertation, University of California, Davis].

Palombit, R. A. (1993). Lethal territorial aggression in a white-handed gibbon. *American Journal of Primatology, 31*, 311-318.

Palombit, R. A. (1994a). Dynamic pair bonds in hylobatids: Implications regarding monogamous social systems. *Behaviour, 128*(1-2), 65-101.

Palombit, R. A. (1994b). Extra-pair copulations in a monogamous ape. *Animal Behavior, 47*, 721-723.

Palombit, R. A. (1996). Pair bonds in monogamous apes: A comparison of the siamang *Hylobates syndactylus* and the white-handed gibbon *Hylobates lar*. *Behaviour, 133*, 321-356.

Palombit, R. A. (1997). Inter- and intraspecific variation in the diets of sympatric siamang (*Hylobates syndactylus*) and lar gibbons (*Hylobates lar*). *Folia Primatologica, 68*, 321-337.

Phoonjampa, R., Koenig, A., Borries, G., Gale, G. A., & Savini, T. (2010). Selection of sleeping trees in pileated gibbons (*Hylobates pileatus*). *American Journal of Primatology, 72*, 617-625.

Preuschoft, H., Chivers, D. J., Brockelman, W. Y., & Creel, N. (Eds.). (1984). The lesser apes: Evolutionary and behavioural biology. Edinburgh University Press.

Preuschoft, H., & Demes, B. (1984). Biomechanics of brachiation. In H. Preuschoft, D. J. Chivers, W. Y. Brockelman, & N. Creel (Eds.), *The lesser apes: Evolutionary and behavioural biology* (pp. 96-118). Edinburgh University Press.

Prost, J. H. (1965). A definitional system for the classification of primate locomotion. *American Anthropology, 67*, 1198-1214.

Raemaekers, J. J. (1979). Ecology of sympatric gibbons. *Folia Primatologica, 31*, 227-245.

Raemaekers, J. J. (1980). Causes of variation between months in the distance traveled daily by gibbons. *Folia Primatologica, 34*, 46-60.

Raemaekers, J. J., & Chivers, D. J. (1980). Socio-ecology of Malayan forest primates. In D. J. Chivers (Ed.), *Malayan forest primates: Ten years' study in tropical rain forest* (pp. 279-316). Plenum Press.

Raemaekers, J. J., & Raemaekers, P. M. (1984). Vocal interactions between two male gibbons, *Hylobates lar*. *Natural History Bulletin of the Siam Society, 32*(2), 95-106.

Raemaekers, J. J., Raemaekers, P. M., & Haimoff, E. H. (1984). Loud calls of the gibbon (*Hylobates lar*): Repertoire, organisation and context. *Behaviour, 91*, 146-189.

Rainer, H., White, A. R., & Lanjouw, A. (Eds.). (2014). *Extractive industries and ape conservation*. Cambridge University Press.

Rawson, B. M. (2004). Vocalisation patterns in the yellow-cheeked crested gibbon (*Nomascus gabriellae*). In T. Nadler, U. Streicher, & H. T. Long (Eds.), *Conservation of primates in Vietnam* (pp. 130-136). Haki Publishing.

Rawson, B. M., Clements, T., & Meng, H. (2009). Status and conservation of yellow-cheeked crested gibbons (*Nomascus gabriellae*) in the Seima Biodiversity Conservation Area, Mondulkiri Province, Cambodia. In S. Lappan, & D. Whittaker (Eds.), *The gibbons: New perspectives on small ape socioecology and population biology* (pp. 387-408). Springer.

Rawson, B. M., Insua-Cao, P., Nguyen, M. H., Thinh, V. N., Minh, D. H., Mahood, S., Geissmann, T., & Roos, C. (2011). *The conservation status of gibbons in Vietnam*. Fauna & Flora International.

Reichard, U. H. (1995). Extra-pair copulation in a monogamous gibbon (*Hylobates lar*). *Ethology, 100*, 99-112.

Reichard, U. H. (1998). Sleeping sites, sleeping places, and presleep behavior of gibbons (*Hylobates lar*). *American Journal of Primatology, 46*, 35-62.

Reichard, U. H. (2003). Social monogamy in gibbons: The male perspective. In U. H. Reichard, & C. Boesch (Eds.), *Monogamy: Mating strategies and partnerships in birds, humans, and other mammals* (pp. 190-213). Cambridge University Press.

Reichard, U. H. (2009). The social organization and mating system of Khao Yai white-handed gibbons: 1992-2006. In S. Lappan, & D. J. Whittaker (Eds.), *The gibbons: New perspectives on small ape socioecology and population biology* (pp. 347-384). Springer.

Reichard, U. H., & Barelli, C. (2008). Life history and reproductive strategies of Khao Yai white-handed gibbon females (*Hylobates lar*). *International Journal of Primatology, 29*, 823-844.

Reichard, U. H., & Boesch, C. (Eds.). (2003). *Monogamy mating strategies and partnerships in birds, humans and other mammals*. Cambridge University Press.

Reichard, U. H., Ganpanakngan, M., & Barelli, C. (2012). White-handed gibbons of Khao Yai: Social flexibility, reproductive strategies, and a slow life history. In P. M. Kappeler, & D. Watts (Eds.), *Long-term field studies of primates* (pp. 237-258). Springer.

Reichard, U. H., & Sommer, V. (1997). Group encounters in wild gibbons (*Hylobates lar*): Agonism, affiliation, and the concept of infanticide. *Behaviour, 134*, 1135-1174.

Rowell, T. (2000). A few peculiar primates. In S. C. Strum, & L. M. Fedigan (Eds.), *Primate encounters: Models of science, gender, and society* (pp. 57-70). University of Chicago Press.

Russo, S. S., & Chapman, C. A. (2011). Primate seed dispersal: Linking behavioural ecology and forest community structure. In C. J. Campbell, A. F. Fuentes, K.C. MacKinnon, M. Panger, & S. Bearder (Eds.), *Primates in perspective* (pp. 523-534). Oxford University Press.

Sarma, K., & Kumar, A. (2016). The day range and home range of the eastern hoolock gibbon *Hoolock leuconedys* (Mammalia: Primates: Hylobatidae) in lower Dibang Valley District in Arunachal Pradesh, India. *Journal of Threatened Taxa*, 8(4), 8641-8651.

Savini, T., Boesch, C., & Reichard, U. H. (2008). Home-range characteristics and the influence of seasonality on female reproduction in white-handed gibbons (*Hylobates lar*) at Khao Yai National Park, Thailand. *American Journal of Physical Anthropology*, 135, 1-12.

Schneider, G. (1905). Ergebnisse zoologischer Forschungsreisen auf Sumatra. Erster Teil: Säugetiere (Mammalia). *Zoologisches Jahrbuch*, 23, 1-172.

Schultz, A. H. (1932). The generic position of *Symphalangus klossi. Journal of Mammalogy*, 13, 368-369.

Sheeran, L. K. (1993). *A preliminary study of the behavior and socio-ecology of black gibbons* (Hylobates concolor). [Doctoral dissertation, Ohio State University].

Sheeran, L. K., Yongzu, Z., Poirier, F. E., & Dehua, Y. (1998). Preliminary report on the behavior of the Jingdong black gibbon (*Hylobates concolor jingdongensis*). *Tropical Biodiversity*, 5, 113-125.

Sommer, V., & Reichard, U. H. (2000). Rethinking monogamy: The gibbon case. In P. M. Kappeler (Ed.), *Primate males: Causes and consequences of variation in group composition* (pp. 159-168). Cambridge University Press.

Sussman, R. W., Garber, P. A., & Cheverud, J. M. (2005). Importance of cooperation and affiliation in the evolution of primate sociality. *American Journal of Physical Anthropology*, 128, 84-97.

Tenaza, R. R. (1975). Territory and monogamy among Kloss' gibbons (*Hylobates klossii*) in Siberut Island, Indonesia. *Folia Primatologica*, 24, 60-80.

Tenaza, R. R., & Hamilton, W. J., III. (1971). Preliminary observations of the Mentawai Islands gibbon, *Hylobates klossii. Folia Primatologica*, 15, 201-211.

Tenaza, R. R., & Tilson, R. L. (1985). Human predation and Kloss's gibbon (*Hylobates klossii*) sleeping trees in Siberut Island, Indonesia. *American Journal of Primatology*, 8, 299-308.

Thinh, V. N., Hallam, C., Roos, C., & Hammerschmidt, K. (2011). Concordance between vocal and genetic diversity in crested gibbons. *BMC Evolutionary Biology*, 11, 36-44.

Thinh, V. N., Mootnick, A. R., Thanh, V. N., Nadler, T., & Roos, C. (2010). A new species of crested gibbon, from the central Annamite mountain range. *Vietnamese Journal of Primatology*, (4), 1-12.

Tilson, R. L. (1979). Behavior of hoolock gibbons (*Hylobates hoolock*) during different seasons in Assam, India. *Journal of Bombay Natural History Society*, 76, 1-16.

Tilson, R. L. (1981). Family formation strategies of Kloss's gibbons. *Folia Primatologica*, 35, 259-287.

Traeholt, C., Bunthoen, R., Vuthin, S., Samuth, M., & Virak, C. (2007). *Habitat utilization and food preference of the yellow-cheeked crested gibbon,* Nomascus gabriellae, *in Cambodia*. Fauna and Flora International.

Treesucon, U. (1984). *Social development of young gibbons* (Hylobates lar) *in Khao Yai National Park Thailand*. [Master's thesis, Mahidol University].

Uhde, N. L., & Sommer, V. (2002). Antipredatory behavior in gibbons (*Hylobates lar*, Khao Yai, Thailand). In L. E. Miller (Ed.), *Eat or be eaten: Predator sensitive foraging among primates* (pp. 268-291). Cambridge University Press.

Ungar, P. S. (1995). Fruit preferences of four sympatric primate species at Ketambe, northern Sumatra, Indonesia. *International Journal of Primatology*, 16, 221-245.

Ungar, P. S. (1996). Feeding height and niche separation in sympatric Sumatran monkeys and apes. *Folia Primatologica*, 67, 163-168.

van Gulik, R. H. (1967). *The gibbons in China: An essay in Chinese animal lore*. E. J. Brill.

van Schaik, C. P., & Dunbar, R. I. M. (1990). The evolution of monogamy in large primates: A new hypothesis and some crucial tests. *Behaviour*, 115, 30-62.

van Schaik, C. P., & Pfannes, K. R. (2005). Tropical climates and phenology: A primate perspective. In D. K. Brockman, & C. P. van Schaik (Eds.), *Seasonality in primates: Studies of living and extinct human and non-human primates* (pp. 24-54). Cambridge University Press.

Vasudev, D., & Fletcher, R. J., Jr. (2015). Incorporating movement behavior into conservation prioritization in fragmented landscapes: An example of western hoolock gibbons in Garo Hills, India. *Biological Conservation*, 181, 124-132.

Vellayan, S. (1981). The nutritive value of *Ficus* in the diet of lar gibbon (*Hylobates lar*). *Malaysian Applied Biology*, 10, 177-182.

Vogel, E. R., Haag, L., Mitra-Setia, T., Schaik, C. P., & Dominy, N. J. (2009). Foraging and ranging behavior during a fallback episode: *Hylobates albibarbis* and *Pongo pygmaeus wurmbii* compared. *American Journal of Physical Anthropology*, 140, 716-726.

Vu, T. T., Tran, L. M., Nguyen, M. D., Van Tran, D., Doherty, P. F., Jr., Giang, T. T., & Dong, H. T. (2018). A distance sampling approach to estimate density and abundance of gibbon groups. *American Journal of Primatology*, 80(9), e22903.

Walker, S., Molur, S., Brockelman, W. Y., Das, J., Islam, A., Geissmann, T., & Fan, P.-F. (2009). Western hoolock gibbon *Hoolock hoolock* (Harlan, 1831). In R. A. Mittermeier, J. Wallis, A. B. Rylands, J. U. Ganzhorn, J. F. Oates, E. A. Williamson, E. Palacios, E. W. Heymann, M. C. M. Kierulff, Y. Long, J. Supriatna, C. Roos, S. Walker, L. Cortés-Ortiz, & C. Schwitzer (Eds.), *Primates in peril: The world's 25 most endangered primates 2008-2010* (pp. 62-64). IUCN Primate Specialist Group.

Whitington, C., & Treesucon, U. (1991). Selection and treatment of food plants by white-handed gibbons (*Hylobates lar*) in Khao Yai National Park, Thailand. *Natural History Bulletin of the Siam Society, 39*, 111-122.

Whitmore, T. C. (1984). *Tropical rain forests of the Far East.* (2nd Ed.). Oxford University Press.

Whitten, A. J. (1982a). Diet and feeding behaviour of Kloss gibbons on Siberut Island, Indonesia. *Folia Primatologica, 37*, 177-208.

Whitten, A. J. (1982b). Home range use by Kloss gibbons (*Hylobates klossii*) on Siberut Island, Indonesia. *Animal Behavior, 30*, 182-198.

Wickler, W. (1976). The ethological analysis of attachment: Sociometric, motivational and sociophysiological aspects. *Zeitschrift für Tierpsychologie, 42*, 12-28.

Wickler, W., & Seibt, U. (1983). Monogamy: An ambiguous concept. In P. Bateson (Ed.), *Mate choice* (pp. 33-50). Cambridge University Press.

Wrangham, R. W. (1979). On the evolution of ape social systems. *Social Science Information, 18*, 335-368.

Wrangham, R. W., Conklin, N. L., Etot, G., Obua, J., Hunt, K. D., Hauser, M. D., & Clark, A. P. (1993). The value of figs to chimpanzees. *International Journal of Primatology, 14*(2), 243-256.

Yanuar, A., & Chivers, D. J. (2010). Impact of forest fragmentation on ranging and home range of siamang (*Symphalangus syndactylus*) and agile gibbons (*Hylobates agilis*). In S. Gursky-Doyen, & J. Supriatna (Eds.), *Indonesia primates* (pp. 97-120). Springer.

Zhou, J., Wei, F. W., Li, M., Zhang, J. F., Wang, D. L., & Pan, R. L. (2005). Hainan black crested gibbon is headed for extinction. *International Journal of Primatology, 26*, 453-465.

CHAPTER 21

The Orangutans

Asia's Endangered Great Apes

ROBERTO ANTONIO DELGADO JR.

ASIA'S ONLY EXTANT GREAT APE, THE ORANGUTAN (FIGURE 21.1), was first described in the early seventeenth century but not assigned its modern taxonomic name, *Pongo pygmaeus*, until the International Commission on Zoological Nomenclature did so in the early twentieth century. The evolutionary history of orangutans (*Pongo* spp.) is marked by drastic changes in biogeography over the last few million years (Rijksen & Meijaard, 1999; Delgado & van Schaik, 2000), and the subfossil record indicates that ancestral populations inhabited a significantly larger area, spanning southern China to northeast India, mainland Southeast Asia, and most of Sundaland (Delgado & van Schaik, 2000; Goosens et al., 2009). As of 2017, there are three extant species of orangutans distributed across a limited geographic range on two islands (Figure 21.2).

Orangutans (*Pongo* spp.) have been the subject of long-term field studies across their range revealing substantial variation in their behavioral ecology. Notably, ecological differences spanning their range, in particular the relative abundance and temporal variance in food availability and the presence or absence of predators, have been important factors driving behavioral variation. For example, Sumatran forests tend to have higher fruiting rates than Bornean forests, in comparable habitat types and altitude, and less variation in fruit availability (Marshall et al., 2009). Sumatra lies on the edge of active volcanic plates, leading to ongoing mineral deposition and erosion, and higher plant nutrients (MacKinnon et al., 1996, 2009). In addition, there is greater climatic variation in Borneo than Sumatra, resulting in higher seasonal variation in fruit availability (Wich & van Schaik, 2000; Marshall et al., 2009). In general, there is a decrease in forest productivity across a west-to-east cline (Sumatra > West Borneo/Central Borneo > East Borneo), which is made worse by droughts and fires in East Borneo (Wich et al., 2009; Russon et al., 2015).

DISTRIBUTION AND TAXONOMY

An account by Nicolaes Tulp, a Dutch surgeon and anatomist in the seventeenth century, is thought to be the first detailed description of the orangutan and includes an engraving; but there is some dispute regarding the subject's provenance and its nomenclature (Cribb et al., 2014). Although Tulp used "ourang-outang" and accurately translated this term as "forest person," the use of "ourang-outang" is also regarded as having been a catch-all term for all great apes until the mid-nineteenth century and may have emerged from earlier writings by a Dutch physician, Jacobus Bontius, based in Batavia during the 1630s (Conniff, 2011; Cribb et al., 2014). Though widely used today in reference to the lone extant Asian great ape, some scholars now argue that the name "orangutan" was not an indigenous term; in many communities in Borneo and Sumatra, *mawas*, or sometimes *mias*, is the most common local name.

Until 2017, most taxonomists recognized two species of orangutans; *Pongo abelii* in Sumatra, and *P. pygmaeus* in Borneo, with three subspecies (*P. p. pygmaeus*, *P. p. wurmbi*, and *P. p. morio*) across Borneo (Brandon-Jones et al., 2004; Singleton et al., 2004) (Table 21.1). These island species are characterized principally by anatomical and genetic differences, though behavioral features can also be distinctive, with the general pattern being that Sumatran orangutans are more gregarious and social than their Bornean counterparts (Rodman, 1988; Delgado & van Schaik, 2000; Wich et al., 2009). However, Bornean orangutans tend to have a more robust craniodental morphology (Taylor, 2006), possibly resulting from greater load resistance when consuming harder and tougher foods; accordingly, the hypothesis that mandibular robusticity in Bornean populations is functionally linked to feeding on tough foods has been supported (Vogel et al., 2014). Orangutans are the largest living habitually arboreal vertebrates. Adult males appear to occur in two morphs, and flanged adult males can reach twice the body mass of adult females (Markham & Groves, 1990; Smith & Jungers, 1997). Flanged males also have long cape-like hair, laterally projecting fatty cheek pads (i.e., flanges), a pendulous throat sack, and they periodically emit loud and conspicuous vocalizations known as "long calls." Unflanged males may also produce distinct long calls, though at lower rates, and they lack the throat sac, flanges, and large body size, but are sexually mature

Figure 21.1   Flanged adult male Sumatran orangutan (*Pongo abelii*).
(Photo credit: James Askew)

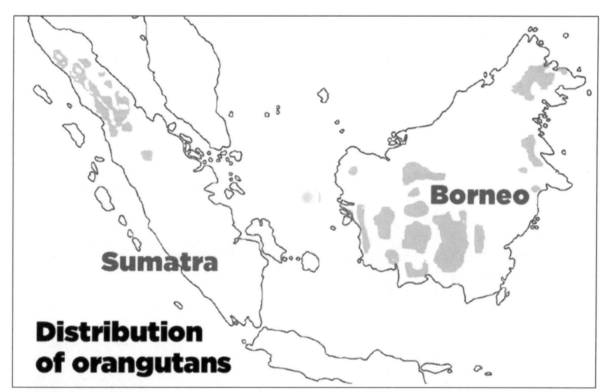

Figure 21.2   Geographic distributions of orangutan species on the islands of Sumatra (*Pongo abelii* in northern Sumatra; *P. tapanuliensis* in isolated range in central Sumatra) and Borneo (*P. pygmaeus*).
(Map based on orangutan distribution image from orangutans.org)

Table 21.1   Orangutan Taxonomy and Distribution[a]

Island	Genus	Species	Subspecies	Range
Borneo	*Pongo*	*pygmaeus*	*morio*	Northeast Borneo
			pygmaeus	Western Borneo
			wurmbii	Southern Borneo
Sumatra	*Pongo*	*abelii*	-	Northern Sumatra
	Pongo	*tapanuliensis*	-	Batang Toru

[a] Data from Brandon-Jones et al. (2004) and Nater et al. (2016).

and have been known to sire more than 50% of offspring in at least some wild populations (Utami et al., 2002).

In 2017, Nater and colleagues described a second orangutan species in Sumatra, the Tapanalui orangutan (*P. tapanuliensis*), separated by the Lake Toba caldera. The Tapanuli orangutan is found 200 km south of Lake Toba in the highlands of the Batang Toru ecosystem (Nater et al., 2017). The Bornean orangutan (*P. pygmaeus*) still comprises three subspecies separated geographically by a central mountain range and the Kapuas, Makam, and Barito Rivers: (1) the northwest Bornean orangutan (*P. p. pygmaeus*), (2) the southwest Bornean orangutan (*P. p. wurmbii*), and (3) the northeast Bornean orangutan (*P. p. morio*). All three orangutan species are classified as Critically Endangered by the International Union for Conservation of Nature (IUCN) Red List due to habitat loss and hunting (see Ancrenaz et al., 2016; IUCN, 2020). Continued deforestation has forced orangutans into fragmented populations, including isolated pockets unsuitable for sustaining viable populations (Goosens et al., 2009).

In Sumatra, the Gunung Leuser Ecosystem to the north is the main stronghold for orangutans, though smaller, genetically distinct populations are also found south of Lake Toba in West Batang Toru and East Sarulla (Singleton et al., 2004). In Borneo, the orangutan is found throughout the Indonesian province of Kalimantan and the Malaysian states of Sabah and Sarawak, across diverse habitats including, but not limited to, mixed hill and lowland dipterocarp forests, peat swamps, and karst outcrops (Singleton et al., 2004; Marshall et al., 2007; Wich et al., 2009). In recent years, orangutans have also been observed feeding and nesting within *Acacia* plantations established by the paper and pulp industry, but the extent to which these communities are viable is not yet known (Meijaard et al., 2010a). Accurate and reliable estimates of orangutan population size are difficult to obtain given the arboreal, semi-solitary, and wide-ranging nature of the species, but the consistent trend is toward dramatic declines across recent decades in most, if not all, habitats (van Schaik et al., 2001; Wich et al., 2003, 2009; Goosens et al., 2006; Meijaard et al., 2010b).

HABITAT SELECTION AND LOCOMOTION

Orangutans are generally characterized as preferring lowland tropical rainforest. Throughout Southeast Asia, relatively little primary forest remains undisturbed, and orangutans are found in such ecologically diverse swamp and lowland forests from sea level up to 1,800 m above sea level (Singleton et al., 2004; Wich et al., 2009). Local orangutan population densities appear to be correlated positively with local habitat productivity or sources of fruit (Buij et al., 2003; Morrogh-Bernard et al., 2003; Knop et al., 2004), often coinciding with a negative relationship along an altitudinal gradient (Johnson et al., 2005). For several sites in Sumatra, however, soil pH, and strangler fig density are also strong predictors of local orangutan density (Wich et al., 2004a).

Orangutans are mainly arboreal, though large "past-prime" flanged adult males may spend more time on the ground than females; alternatively, studies from some sites have reported, based on focal animal observations (i.e., at Tuanan; Ashbury et al., 2015) and camera-trap data (at Wehea Forest; Loken et al., 2013), that terrestriality is relatively common for orangutans and represents a regular strategy employed by individuals of all age-sex classes. Orangutans have a long trunk, longer arms than legs, and their post-cranial anatomy seems to be highly adapted for arboreal suspensory locomotion, sometimes described as "quadrumanous scrambling." Arboreal locomotion among orangutans tends to be dominated by "orthograde clambering" (see Cant, 1987) and, though often described as slow and cautious, can also be fast and acrobatic (Thorpe & Crompton, 2005, 2006). Orangutans prefer to travel in the canopy level with the most continuity, typically less than 20 m above the ground (Sugardjito & van Hooff, 1986; Thorpe & Crompton, 2005). Given their large body size, the orangutan's postural repertoire is quite diverse and relies on multiple supports whose diameter and type influence specific locomotor modes (Thorpe & Crompton, 2005). In short, for sites both in Borneo and Sumatra, forest structure and support availability influence orangutan locomotion (Manduell et al., 2012). Furthermore, increased balance and stability for orangutans are achieved through long contact times between multiple limbs and supports as well as with a combination of horizontal and vertical postures (Thorpe et al., 2009).

DIET

Orangutans are predominantly ripe-fruit specialists but regularly include leaves, flowers, bark, and insects in their diet to varying degrees with evident geographic variation (Fox et al., 2004; Wich et al., 2009). A comprehensive study by Leighton (1993) concluded that orangutan foraging decisions are strongly influenced by patch size and that the energy content of foods is the primary determinant of dietary selectivity, though protein content may also be pertinent for some populations (Hamilton & Galdikas, 1994). However, given the wide geographic range of orangutans, there is also considerable variation in diets (Russon et al., 2009), suggesting that populations may respond differentially to fluctuations in fruit availability (Knott, 1998; Wich et al., 2006; Harrison et al., 2010; Kanamori et al., 2010) or nutritional differences (and thereby impacting local population density; e.g., Vogel et al., 2015). In some cases, diet and food selection traditions may emerge that are best explained by a preferential social learning model (Bastian et al., 2010). Overall, orangutans feed on a large number of plants, more than 200 species at some sites, but they can be very selective on a seasonal basis.

Several orangutan populations live in masting habitats, where fruit availability can be very low for periods of between 2 and up to 10 years, but then increase suddenly and dramatically. During such mast-fruiting periods for a population in West Kalimantan, Indonesia, up to 100% of the orangutan diet may comprise fruit, whereas this proportion may fall to as low as 21% during non-masting periods (Knott, 1998). In contrast, other populations, namely, in North Sumatra, that also face fluctuations in fruit availability may not differ significantly in the proportion of fruit in their diets (Wich et al., 2006). However, at some Bornean sites, orangutans have been observed to rely heavily on unripe fruit and fracture-resistant bark and pith tissues, leading to negative protein balance (Vogel et al., 2010, 2011). The relative proportion of non-fruit items in the diet also varies; for example, at Suaq Balimbing, a peat swamp forest in North Sumatra, an abundance of social insects forms a staple in the local orangutan diet (see Figure 21.3), sometimes obtained by using extraction tools (Fox et al., 1999, 2004). Despite relatively similar habitat types between Bornean and Sumatran populations, tool use has thus far only been observed in Sumatran sites (van Schaik & Knott, 2001). Tools are used to extract the lipid-rich, nutritious seeds of *Neesia* fruit at several Sumatran swamp sites, and the technique probably emerges when there are suitable conditions (i.e., high population density and social

Figure 21.3   Flanged adult male Sumatran orangutan feeding on termites.
(Photo credit: James Askew)

tolerance) for social learning and transmission to occur (van Schaik & Knott, 2001). There appears to be greater reliance on bark as a seasonal food for Bornean orangutans (Rodman, 1988; Delgado & van Schaik, 2000; Wich et al., 2009) and, to a lesser extent, insects (Galdikas, 1988; Knott, 1998), though work at low-resource sites in Sumatra (e.g., Batang Toru & Sikundur) may demonstrate some increased similarities between Bornean and Sumatran orangutan feeding ecology (Askew, 2019).

ACTIVITY BUDGETS

Orangutan activity profiles tend to vary within and between individuals on a monthly basis (Mitani, 1989), as well as between sites (Fox et al., 2004; Wich et al., 2009). Patterns of feeding, resting, and traveling may differ across months as a function of age and sex (but not within the same age-sex class), with adult females and unflanged males typically traveling and feeding for longer than flanged males (Mitani, 1989; Fox et al., 2004). At Danum Valley in Sabah, Malyasia, resting generally increases as feeding decreases in the late stages of fruiting seasons, suggesting that the orangutans adopt an energy-minimizing strategy to withstand periods of fruit shortage (Kanamori et al., 2010).

From a series of long-term studies (for a review, see Wich et al., 2009), it is now clear that the proportion of time spent by orangutans in different activities may vary considerably. Depending on the site, feeding can account for anywhere between 36%–60% of observation time, resting 18%–56%, moving 10%–19%, while any remaining time is spent socializing, nesting, or engaging in other behaviors (MacKinnon, 1974; Rijksen, 1978; Galdikas, 1988; Rodman, 1988; Mitani, 1989; Delgado & van Schaik, 2000; Fox et al., 2004; Wich et al., 2009; Kanamori et al., 2010). Orangutans will be active between 10–12 hours each day, and all independent individuals build resting and sleeping platforms called nests (see Figure 21.4).

Although nests are sometimes re-used, orangutans typically construct a new sleeping platform every night. Nests are often located near major travel routes or feeding sites (MacKinnon, 1974), but rarely *within* fruit trees (Rijksen, 1978). Nest positions vary but are generally at stable locations on large, strong branches with varying degrees of exposure. Nest size and location can be used to determine the age-structure of a population (Rayadin & Saitoh, 2009). Orangutans also construct resting platforms during the day, and adult females will sometimes build nests to avoid sexual harassment from adult males. One unique aspect of orangutan nest building is the use of covers or "umbrellas" during rainy periods (Davenport, 1967; MacKinnon, 1974; Rijksen, 1978).

Figure 21.4 Nest positions in tree: (1) on horizontal branch near trunk (often where two horizontal branches emerge), (2) on horizontal branch(es) away from trunk, (3) in acute angle of large branches/trunk split, and (4) branches from two trees woven together.
(Adapted from Merrill, 2004)

Presently, researchers and conservationists utilize nests as a way to estimate local orangutan population densities (e.g., Felton et al., 2003; Morrogh-Bernard et al., 2003; Ancrenaz et al., 2004, 2010; Johnson et al., 2005; van Schaik et al., 2005; Marshall et al., 2006; Spehar et al., 2010). However, underlying assumptions are not always supported and population size can be overestimated (Mathewson et al., 2008; Marshall & Meijaard, 2009). Nonetheless, with long-term biological monitoring, population trends can still be documented to provide important data for conservation and management strategies.

PREDATION

Today, the Sumatran tiger (*Panthera tigris sumatrae*) occurs only on the island of Sumatra and is thought to cause increased group size and reduced terrestriality in orangutan populations due to predation pressure (Mitra Setia et al., 2009; Thorpe & Crompton, 2009), but rigorous tests of this hypothesis have not been conducted to date. Tigers are regionally extinct on Borneo, and orangutans at Sabangau and Wehea frequently use the ground for travel (Loken et al., 2013; Ancrenaz et al., 2014). Adult orangutans appear to face a relatively low predation risk from tigers, leopards, and other nonhuman predators (see Rikjsen, 1978), though, historically, humans may have been a significant cause of mortality. The caves at Niah in Sarawak and at Madai in Sabah retain extensive evidence of prehistoric hunting in the form of orangutan remains that had been left behind by tribal hunters (Hooijer, 1948, 1961; Medway, 1977). In historical times, seven major gatherer-hunter societies have been identified and the early ethnographers noted that the favorite prey of all these tribes comprised "monkeys and apes," which were hunted with poisoned arrows, by means of a blowgun, or with dogs and spears (Rijksen & Meijaard, 1999). The indications of human habitation and the high frequency of orangutan remains relative to that of other taxa suggest selective targeting of orangutans.

SOCIAL SYSTEM AND COMMUNITY STRUCTURE

The orangutan is the only known diurnal primate species that does not live in permanent male-female associations; however, this pattern does not preclude the presence of long-lasting relationships or broader social networks. Though long considered solitary, and similar to many nocturnal prosimians in this respect, orangutan gregariousness and sociality vary among populations (Wich et al., 2009). The most common social unit tends to be either solitary flanged adult males or adult females with dependent offspring. However, depending on the population, mixed-sex parties, "nursery" groups (multiple adult females and their offspring), as well as same-sex associations of unflanged males are not uncommon (Galdikas, 1985a,b), suggesting an individual-based fission-fusion social system at least at some sites (van Schaik, 1999; Delgado, & van Schaik, 2000; Singleton & van Schaik, 2001, 2002).

Orangutan party size tends to be constrained by the distribution and relative abundance of food resources, but regular feeding aggregations and travel bands occur during periods of relative fruit abundance (Rijksen, 1978; Sugardjito et al., 1987; Galdikas, 1988; te Boekhorst et al., 1990; Mitani et al., 1991). As with other aspects of their behavioral ecology, there is variation across sites in the proportion of time that orangutans spend in feeding aggregations and travel bands (Delgado & van Schaik, 2000; Wich et al., 2009).

Adult female orangutans are now known to occupy ranges of up to 900 ha or more with extensive overlap (Horr, 1975; Rijksen, 1978; Galdikas, 1984, 1995; Singleton & van Schaik, 2001; Knott et al., 2008; Wich et al., 2009; Wartmann et al., 2010). In turn, adult male home ranges may exceed 2,000–3,000 ha, also with high overlap, encompassing several females, and males do not appear to defend territorial boundaries (Singleton & van Schaik, 2001). The spatio-temporal availability of resources, such as cycling females and fruit trees, probably drives variation in male ranging patterns (MacKinnon, 1974; Rodman & Mitani, 1987; Sugardjito et al., 1987; te Boekhorst et al., 1990; Mitani et al., 1991; van Schaik, 1999; Delgado & van Schaik, 2000; Singleton & van Schaik, 2002).

Mean party size among orangutans tends to remain small because of the high costs of feeding competition (Mitani et al., 1991; van Schaik & van Hooff, 1996; van Schaik, 1999; Delgado & van Schaik, 2000). However, preferred associations among certain subjects indicate established relationships between different individuals (van Schaik & van Hooff, 1996; Singleton & van Schaik, 2002). At several Sumatran sites, clusters of females associate preferentially (Singleton & van Schaik, 2002), suggesting a scenario in which orangutans form a dispersed society with individualized relationships (MacKinnon, 1974; Sugardjito et al., 1987; van Schaik, 1999; Delgado & van Schaik, 2000). Consequently, in orangutans, the absence of cohesive social units does not prevent the presence of a community structure. If social units are defined as a group of animals that associate more with each other than with individuals of other such groups (see Struhsaker, 1969), then constant spatial proximity is not required. Long-distance calling, as well as regular contacts, may facilitate coordinated travel as well as mate choice and, hence, lead to a dispersed social network.

At least three lines of evidence from field observations by Delgado (2006) in northern Sumatra suggest that relative loud call production remained the best indicator of male status and attractiveness (Delgado, 2006). First, typically, only the locally dominant male was observed to respond upon hearing the loud calls of other males; specifically, the locally dominant male vocalized and aggressively chased subordinate individuals whenever the distance between them was less than 200 m at the time of the subordinate's long call. Second, the resident dominant was the only male to whom the cycling females approached and with whom they consorted. Lastly, females initiated matings, and all copulations by the locally dominant male were unforced. Throughout this mating period, the male's calling rate was more than twice as high as the two other males who were present in the study area. Hence, because of the orangutan's dispersed social system, there is a strong reason to believe that long calls play a crucial role in social organization and both male and female reproductive strategies, perhaps especially so during contested mating periods. Such a system would necessarily require a high degree of home range overlap and enduring relationships, as previously reported among orangutans.

Given the extent of geographic variation thus far observed across orangutan sites (Rodman, 1988; Delgado & van Schaik, 2000; Wich et al., 2009), at least two possible models for social structure emerge: one is a system characterized by roving male promiscuity (Galdikas, 1985a,b; Rodman & Mitani, 1987; van Schaik & van Hooff, 1996; Delgado, 2010) and the other is a network of loose associations within a socially distinct and open community organized around resident flanged males or related female clusters (van Schaik, 1999; Delgado & van Schaik, 2000; Singleton & van Schaik, 2001; Delgado, 2010). Female clusters have been reported at several sites (e.g., Galdikas, 1984; Singleton & van Schaik, 2002; Knott et al., 2008), but more variable is the degree to which there is coordinated travel with the locally dominant male. Thus, one reasonable interpretation is that behavioral flexibility in social organization could well be favored by selection in response to variation in local resource distribution and abundance. For example, in habitats with relatively low productivity, social structure is better characterized by roving male promiscuity, whereas a dispersed community is more likely to develop at sites with high habitat productivity.

REPRODUCTIVE STRATEGIES AND DEVELOPMENT

In the wild, adult males are thought to compete for access to fertile females (Galdikas, 1981; Rodman & Mitani, 1987; van Schaik & van Hooff, 1996). Matings between the locally dominant flanged male and adult females occur typically within the context of a consortship whose duration varies (Rijksen, 1978; Schürmann, 1981, 1982; Schürmann & van Hooff, 1996; Fox, 1998). During consortships, the male and female travel together and coordinate activities from a few minutes up to several days. Copulations tend to be cooperative and occur a number of times over the duration of the consortship. Cooperative matings are normally initiated by the female, and almost invariably involve flanged males (Rijksen, 1978; Schürmann, 1982; Galdikas, 1984; Mitani, 1985a). Sexual encounters between adult females and unflanged males are more often coercive and forcibly resisted, though can sometimes lead to successful fertilizations in some populations (Utami et al., 2002). Females tend to initiate consortships, suggesting that, under some circumstances, they express mate choice (Schürmann, 1982; Nadler, 1988; Fox, 1998).

Gestation length in orangutans is approximately 260 days, and interbirth intervals can range anywhere between 5–9 years, depending on the particular population (Wich et al., 2004b, 2009). Infants may reach full locomotor independence by 3 years of age, continue to share their mother's nest until weaning (approximately 6–7 years of age), and do not become completely ecologically competent until the birth of the next infant (van Noordwijk & van Schaik, 2005; van Andrichem et al., 2006). In the wild, males and females vary in the age they reach sexual maturity; males at approximately 14 years (Wich et al., 2004b) and females between 10–11 years (Knott, 2001), with age at first reproduction anywhere between 12–16 years of age, and estimated life span in the wild for females approaching 60 years (Knott, 2001; Wich et al., 2004b). For males, reaching full physical maturity and achieving "flanged" status is a long and variable, yet poorly understood, process. Several studies indicate that chronic stress does not fully explain bimaturism, but the male developmental "arrest" could be part of an alternative reproductive strategy (Maggioncalda et al., 1999, 2002; Harrison & Chivers, 2007).

POPULATION DIFFERENCES

Similar to chimpanzees, orangutans have well-documented population differences involving learned skill behaviors that vary geographically and are likely maintained within populations by social learning and geographic barriers to dispersal (van Schaik et al., 1999, 2003). Alternatively, the presence of population-specific skilled behaviors may indicate population differences in genetic expression or local ecological factors, rather than patterns of innovation and the appropriate conditions for social learning and diffusion. However, when comparing the tool-using skills at one site each on Borneo and Sumatra, where tool use was absent at the Bornean site (Cabang Panti) and present at the Sumatran site (Suaq Balimbing), van Schaik and Knott (2001) demonstrated that neither genetic nor ecological differences were sufficient to explain the documented patterns of tool use. The high local population density and extent of social tolerance observed at Suaq Balimbing implies that favorable demographic conditions are responsible for the emergence and spread of learned tool-use behavior among orangutans (van Schaik et al., 1999, 2003; van Schaik & Knott, 2001; van Schaik, 2004). High densities of orangutan populations residing in other Sumatran field sites with observed tool use further corroborate this interpretation of learned population differences (van Schaik & Knott, 2001). Finally, a wider, multisite comparison found correlations between geographic distance and population differences, as well as between the opportunities for social learning and the size of the local tool-using repertoire (van Schaik et al., 1999).

LONG CALLS AND ORANGUTAN SOCIETIES REVISITED

Already well accepted as serving a spacing function between adult males (MacKinnon, 1974; Galdikas, 1983; Mitani, 1985b; van't Land, 1990), long calls are also the most likely mechanism by which orangutans maintain a socially distinct network of associations within their loosely knit community. Long calls tend to be given by the dominant resident male three to four times a day, although there is variation between individuals and between sites that is likely dependent on local population density (Mitani, 1985b; Delgado & van Schaik, 2000; Delgado, 2006, 2007). Interestingly, there is also a positive relationship between mean long call duration and local population density, found from available data at four Bornean sites (Cabang Panti, Sabangau, Tanjung Puting, and Tuanan) and two Sumatran sites (Ketambe and Suaq Balimbing) (Figure 21.5). These calls carry long distances

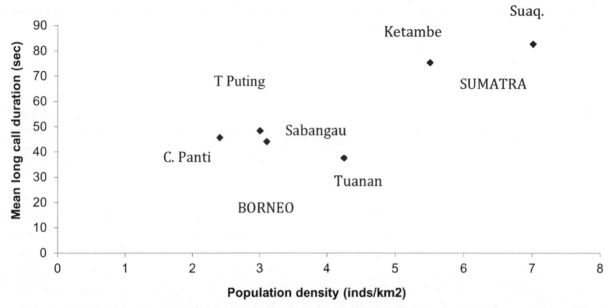

Figure 21.5 Mean long call duration is positively correlated with local orangutan population density across five sites in Borneo and Sumatra (R=0.85, R2=0.72, df=5, p<0.05). (Long call duration data calculated by the author except for Tanjung Putting: Galdikas, 1983; Galdikas & Insley, 1988. Population density data from the author's direct observations and various published sources, i.e., Tanjung Putting: Galdikas, 1983; Galdikas & Insley, 1988; Suaq Balimbing: van Schaik, 1999; Sabangau: Morrogh-Bernard et al., 2003).

in the forest and are audible to human observers on the ground up to 800 m away (Mitani, 1985b), but this distance likely depends on local habitat topography since other investigators claim to hear long calls up to several kilometers away (see MacKinnon, 1974; Galdikas, 1983).

Long calls potentially enable dispersed parties to remain in spatial contact with the dominant male of the area (MacKinnon, 1974) or any other vocalizing males. Further, if long calls communicate a male's travel direction (see van Schaik et al., 2013), these vocalizations may, therefore, act as a coordinating signal for orangutan populations—observations of coordinated seasonal movements by whole "communities" over several kilometers have been noted (Sugardjito et al., 1987; te Boekhorst et al., 1990; Buij et al., 2003; but cf. Singleton & van Schaik, 2001) and could be guided by male long calls.

Among males, previous experimental playbacks and behavioral studies show receiver-specific responses to long calls based on dominance relationships (Mitani, 1985b; van't Land, 1990) in which adult males ignore, avoid, or threaten the caller, depending on their putative relationship with the vocalizer. These behavioral responses suggest individual recognition based on the vocalizations alone or at least some assessment criteria within the long call properties that may include, but are not limited to, the rate, speed, and duration of the long calls. In fact, Galdikas and Insley (1988) report that long calls delivered at a faster tempo (i.e., more pulses per unit time) reflect high levels of arousal because "fast" calls occurred most

often following inter-male encounters and were often associated with a signaling male pushing over a dead tree.

In addition to responses to disturbances, flanged males emit long calls seemingly spontaneously as they move through their home ranges and when resting (MacKinnon, 1974; Horr, 1975; van Schaik & van Hooff, 1996; Delgado & van Schaik, 2000; Mitra Setia & van Schaik, 2007). Adult females may also avoid, approach, or ignore these calls depending on their reproductive condition and social context. Both male and female responses are likely to be relationship-dependent but long-term observations during simultaneous follows and additional field playback experiments are needed to fully understand the role of long-distance vocalizations in orangutan social organization and reproductive strategies (Askew, 2019).

CONSERVATION STATUS

Although Indonesia and Malaysia together have over 50 national parks, orangutan populations tend to occur mainly outside of these protected areas and legal enforcement of violations is virtually nonexistent. In short, the geographic range of orangutans is becoming increasingly fragmented, and all populations face severe conservation threats due principally to habitat loss and related anthropogenic activities (Figure 21.6). Unsustainable human population growth further exacerbates habitat loss. Regional threats also may include deforestation for conversion to oil palm plantations and mining concessions,

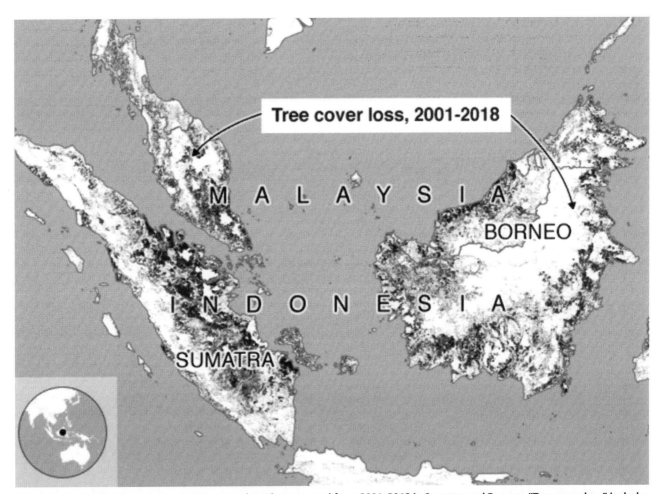

Figure 21.6 Shaded areas represent tree cover loss that occurred from 2001–2018 in Sumatra and Borneo. "Tree cover loss" includes deforestation, forest degradation due to selective mechanical harvesting, fire damage, disease, and damage from tropical cyclones. (Image derived from Global Forest Watch, University of Maryland)

as well as hunting and live capture for the entertainment industry or the trade in exotic pets (Singleton et al., 2004; Nellermann et al., 2007; Nijman, 2009). In 2007, in the face of these clear threats, the Indonesian government launched a 10-year action plan for the conservation of orangutans (Soehartono et al., 2007). In sum, the plan compels the Indonesian government to secure wild orangutan populations within formal protected areas or among compatible land uses such as sustainably managed timber concessions (e.g., Ancrenaz et al., 2010), and to enforce existing conservation laws. However, as of 2014, little progress had been made. Despite a moratorium on new forest concession licenses that went into effect in 2011, immense areas of orangutan habitat have been lost, mostly due to conversion to acacia and oil palm plantations, and rates of primary forest deforestation have increased, with significant negative implications for biodiversity conservation efforts (Margono et al., 2014). The success and effectiveness of the proposed conservation and management strategies were scheduled to be evaluated in 2017 but, as of 2020, the Indonesian

government has not yet conducted its official evaluation of the 2007–2017 action plan and a new 10-year (2019–2029) orangutan conservation and action plan was recently published. Despite the value of studying great apes as potential models for early human evolution and unraveling the selective pressures leading to diverse adaptations, our fascination with, and interest in, the orangutan will be for naught if the species and their habitats are not protected and managed responsibly. Despite collaborative, multisite research and long-term biological monitoring throughout their geographic range, without commitments from the Indonesian and Malaysian governments to uphold law enforcement in protected areas and policies in support of environmental protection, scientists and conservationists alike will struggle to preserve orangutan populations and their habitats.

SUMMARY

Orangutans are among the largest of the primates and show a high degree of sexual dimorphism. Both of these factors are extremely important in understanding the behavioral ecology of these apes and their versatile social structure. They are found only in the fragmented forests of Borneo and Sumatra, habitats that have lost over 6.02 million hectares between 2000–2012 (Margono et al., 2014). Consequently, orangutan populations are being reduced accordingly, with an estimated loss of 100,000 Bornean individuals between 1999–2015 (Voigt et al., 2018). This remarkable great ape is critically endangered throughout its range in Borneo and Sumatra (IUCN, 2020).

Although they are large and slow moving, orangutans can also be fast and agile. They spend most of their time in the trees and require vast tracts of forest to sustain viable populations. The diet of the orangutans is highly frugivorous but supplemented with a variety of other food items such as leaves, bark, and insects, which they access sometimes using tools. For some populations, leaves, bark, and insects are considered seasonal foods, only consumed during periods when preferred fruits are not available. Thus far, tool use is observed only in densely populated sites with high social tolerance, namely, in Sumatran swamp forests. Feeding, traveling, and resting make up over 95% of the activity of orangutans. Gregariousness and sociality vary among orangutan populations and appear to depend on local habitat productivity. Orangutans construct platforms for both resting during the day and for sleeping at night, sometimes adding covers or "umbrellas" for protection against the rain.

The social structure of orangutans is paradoxical—semi-solitary due to feeding constraints, yet still social with dispersed networks. Flanged males or adult females with dependent offspring normally range alone over wide areas approaching tens of square kilometers. Spacing between males is often accomplished by site-independent loud and conspicuous vocalizations known as long calls. These calls, which maintain distance between adult males, may also attract receptive females to the caller, and sexually active females will initiate consort relationships with flanged adult males. Unflanged males attempt to follow and mate with adult females, sometimes forcibly and with resistance, but do manage to inseminate females opportunistically (Galdikas, 1981, 1985a,b), particularly during periods when flanged males are competing for local dominance (Utami, 2002). Females have interbirth intervals of up to nine years, the longest of all living primates, and part of their extended life history. Males, in turn, have a long and variable developmental period that leads to two adult morphs, a phenomenon known as bimaturism. Although the exact mechanism for this developmental arrest is not yet known, the selective pressures leading to the occurrence of flanged and unflanged males are probably related to alternative male reproductive strategies.

Tool use in orangutans for extractive foraging is but one example of population differences that likely emerges when productive habitats are densely populated and high fruit availability allows for increased social tolerance. Vocal signals and comfort skills (e.g., dealing with wet or thorny substrates, creating sun or rain hats) are possibly other local-learned variants that distinguish populations but cannot be explained readily by genetic or ecological differences. Understanding behavioral flexibility and the determinants of geographic variation, perhaps in relation to the function of adult male long calls, coordinated travel, and mate assessment, may provide further clues into the adaptive versatility of the orangutan's social organization.

REFERENCES CITED—CHAPTER 21

Ancrenaz, M., Ambu, L., Sunjoto, I., Ahmad, E., Manokaran, K., Meijaard, E., & Lackman, I. (2010). Recent surveys in the forests of Ulu Segama Malua, Sabah, Malaysia, show that orang-utans (*P. p. morio*) can be maintained in slightly logged forests. *PLoS ONE,* 5(7), e11510. DOI: 10.1371/journal.pone.0011510.

Ancrenaz, M., Calaque, R., & Lackman-Ancrenaz, I. (2004). Orangutan nesting behavior in disturbed forest of Sabah, Malaysia: Implications for nest census. *International Journal of Primatology,* 25(5), 983-1000.

Ancrenaz, M., Gumal, M., Marshall, A. J., Meijaard, E., Wich, S. A., & Husson, S. (2016). *Pongo pygmaeus* (errata version published in 2018). *The IUCN Red List of Threatened Species* 2016: e.T17975A123809220. https://www.iucnredlist.org/species/17975/123809220.

Ancrenaz, M., Sollmann, R., Meijaard, E., Hearn, A. J., Ross, J., Samejima, H., Loken, B., Cheyne, S. M., Stark, D. J., Gardner, P. C., & Goossens, B. (2014). Coming down from the trees: Is terrestrial activity in Bornean orangutans natural or disturbance driven? *Scientific Reports,* (4), 4024.

Ashbury, A. M., Posa, M. R. C., Dunkel, L. P., Spillman, B., Utami Atmoko, S. S., van Schaik, C. P., & van Noordwijk, M. A. (2015). Why do orangutans leave the trees? Terrestrial behavior among wild Bornean orangutans (*Pongo pygmaeus wurmbii*) at Tuanan, central Kalimantan. *American Journal of Primatology,* 77, 1216-1229.

Askew, J. A. (2019). *Geographic variation in the feeding ecology and long-distance vocalizations of orangutans.* [Doctoral dissertation, University of Southern California].

Bastian, M. L., Zweifel, N., Vogel, E. R., Wich, S. A., van Schaick, C. P. (2010). Diet traditions in wild orangutans. *American Journal of Physical Anthropology,* 143, 175-187.

Brandon-Jones, D., Eudey, A. A., Geissmann, T., Groves, C. P., Melnick, D. J., Morales, J. C., Shekelle, M., & Stewart, C. B. (2004). Asian primate classification. *International Journal of Primatology,* 25(1), 97-164.

Buij, R., Singleton, I., Krakauer, E., & van Schaik, C. P. (2003). Rapid assessment of orangutan density. *Biological Conservation,* 114, 103-113.

Cant, J. G. H. (1987). Positional behavior of female Bornean orangutans (*Pongo pygmaeus*). *American Journal of Primatology, 12*, 71-90.

Conniff, R. (2011). *The species seekers: Heroes, fools, and the mad pursuit of life on earth*. W. W. Norton & Company.

Cribb, R., Gilbert, H., & Tiffin, H. (2014). *The wild man from Borneo*. University of Hawaii Press.

Davenport, R. K. (1967). The orang-utan in Sabah. *Folia Primatologica, 5*, 247-263.

Delgado, R. A. (2006). Sexual selection in the loud calls of male primates: Signal content and function. *International Journal of Primatology, 27*(1), 5-25.

Delgado, R. A. (2007). Geographic variation in the long calls of male orangutans (*Pongo* spp.). *Ethology, 113*, 487-498.

Delgado, R. A. (2010). Communication, culture, and conservation in orangutans. In S. Gursky, & J. Priatna (Eds.), *Indonesian primates* (pp. 23-40). Springer.

Delgado, R. A., & van Schaik, C. P. (2000). The behavioral ecology and conservation of the orangutan: A tale of two islands. *Evolutionary Anthropology, 9*(5), 201-218.

Felton, A. M., Engstrom, L. M., Felton, A., & Knott, C. D. (2003). Orangutan population density, forest structure and fruit availability in hand-logged and unlogged peat swamp forests in west Kalimantan, Indonesia. *Biological Conservation, 114*, 91-101.

Fox, E. A. (1998). *The function of female mate choice in the Sumatran orangutan, (Pongo pygmaeus abelii)*. [Doctoral dissertation, Duke University].

Fox, E. A., Sitompul, A. F., & van Schaik, C. P. (1999). Intelligent tool use in wild Sumatran orangutans. In S. T. Parker, R. W. Mitchell, & H. L. Miles (Eds.), *The mentalities of gorillas and orangutans: Comparative perspectives* (pp. 99-116). Cambridge University Press.

Fox, E. A., van Schaik, C. P., Sitompul, A., & Wright, D. N. (2004). Intra- and inter-populational differences in orangutan (*Pongo pygmaeus*) activity and diet: Implications for the invention of tool use. *American Journal of Physical Anthropology, 125*, 162-174.

Galdikas, B. M. F. (1981). Orangutan reproduction in the wild. In C. E. Graham (Ed.), *Reproductive biology of the great apes: Comparative and biomedical perspectives* (pp. 281-300). Academic Press.

Galdikas, B. M. F. (1983). The orangutan long call and snag crashing at Tanjung Puting reserve. *Primates, 24*, 371-384.

Galdikas, B. M. F. (1984). Adult female sociality among wild orangutans at Tanjung Puting Reserve. In M. F. Small (Ed.), *Female primates: Studies by women primatologists* (pp. 217-235). Alan R. Liss.

Galdikas, B. M. F. (1985a). Adult male sociality and reproductive tactics among orangutans at Tanjung Puting. *Folia Primatologica, 45*, 9-24.

Galdikas, B. M. F. (1985b). Subadult male orangutan sociality and reproductive behavior at Tanjung Puting. *American Journal of Primatology, 8*, 87-99.

Galdikas, B. M. F. (1988). Orangutan diet, range, and activity at Tanjung Puting, central Borneo. *International Journal of Primatology, 9*(1), 1-35.

Galdikas, B. M. F. (1995). Social and reproductive behavior of wild adolescent female orangutans. In R. D. Nadler, B. F. M. Galdikas, L. K. Sheeran, & N. Rosen (Eds.), *The neglected ape* (pp. 163-182). Plenum Press.

Galdikas, B. M. F., & Insley, S. J. (1988). The fast call of the adult male orangutan. *Journal of Mammalogy, 69*(2), 371-382.

Global Forest Watch. (2014). *World Resources Institute*. www.globalforestwatch.org.

Goossens, B., Chikhi, L., Ancrenaz, M., Lackman-Ancrenaz, I., Andau, P., & Bruford, M. W. (2006). Genetic signature of anthropogenic population collapse in orang-utans. *PLoS Biology, 4*(2), e25.

Goossens, B., Chikhi, L., Jalil, M. F., James, S., Ancrenaz, M., Lackman-Ancrenaz, I., & Bruford, M. W. (2009). Taxonomy, geographic variation and population genetics of Bornean and Sumatran orangutans. In S. A. Wich, S. S. Utami Atmoko, T. M. Setia, & C. P. van Schaik (Eds.), *Orangutans: Geographic variation in behavioural ecology and conservation* (pp. 1-13). Oxford University Press.

Hamilton, R. A., & Galdikas, B. M. F. (1994). A preliminary study of food selection by the orangutan in relation to plant quality. *Primates, 35*(3), 255-263.

Harrison, M. E., & Chivers, D. J. (2007). The orang-utan mating system and the unflanged male: A product of declining food availability during the late Miocene and Pliocene? *Journal of Human Evolution, 52*, 275-293.

Harrison, M. E., Morrogh-Bernard, H. C., & Chivers, D. (2010). Orangutan energetics and the influence of fruit availability in the nonmasting peat-swamp forest of Sabangau, Indonesian Borneo. *International Journal of Primatology, 31*, 585-607.

Hooijer, D. A. (1948). Prehistoric teeth of man and of the orang-utan from central Sumatra, with notes on the fossil orang-utan from Java and southern China. *Zoologische Mededeelingen, 29*, 175-301.

Hooijer, D. A. (1961). The orang-utan in Niah cave pre-history. *Sarawak Museum Journal, 9*, 408-421.

Horr, D. A. (1975). The Borneo orang-utan: Population structure and dynamics in relationship to ecology and reproductive strategy. In L.A. Rosenblum (Ed.), *Primate behavior: Developments in field and laboratory research* (Vol. 4, pp. 307-323). Academic Press.

IUCN. (2016). *International Union for the Conservation of Nature Red List of Threatened Species*. www.iucnredlist.org.

IUCN. (2020). *International Union for the Conservation of Nature Red List of Threatened Species*. www.iucnredlist.org.

Johnson, A. E., Knott, C. D., Pamungkas, B., Pasaribu, M., & Marshall, A. J. (2005). A survey of the orangutan (*Pongo pygmaeus wurmbii*) population in and around Gunung Palung National Park, west Kalimantan, Indonesia based on nest counts. *Biological Conservation, 121*, 495-507.

Kanamori, T., Kuze, T., Bernard, H., Malim, T. P., & Kohshima, S. (2010). Feeding ecology of Bornean orangutans (*Pongo pygmaeus morio*) in Danum Valley, Sabah, Malaysia: A 3-year record including two mast fruitings. *American Journal of Primatology, 72,* 820-840.

Knop, E., Ward, P. I., & Wich, S. A. (2004). A comparison of orangutan density in a logged and unlogged forest on Sumatra. *Biological Conservation, 120,* 183-188.

Knott, C. D. (1998). Changes in orangutan caloric intake, energy balance, and ketones in response to fluctuating fruit availability. *International Journal of Primatology, 19*(6), 1061-1079.

Knott, C. D. (2001). Female reproductive ecology of the apes: Implications for human evolution. In P. T. Ellison (Ed.), *Reproductive ecology and human evolution* (pp. 429-463). Aldine de Gruyter.

Knott, C., Beaudrot, L., Snaith, T., White, S., Tschauner, H., & Planansky, G. (2008). Female-female competition in Bornean orangutans. *International Journal of Primatology, 29,* 975-997.

Leighton, M. (1993). Modeling dietary selectivity by Bornean orangutans: Evidence for integration of multiple criteria in fruit selection. *International Journal of Primatology, 14*(2), 257-313.

Loken, B., Spehar, S., & Rayadin, Y. (2013). Terrestriality in the Bornean orangutan and implications for their ecology and conservation. *American Journal of Primatology, 75*(11), 1129-1138.

MacKinnon, J. (1974). The behaviour and ecology of wild orang-utans (*Pongo pygmaeus*). *Animal Behaviour, 22,* 3-74.

MacKinnon, K., Hatta, G., Mangalik, A., & Halim, H. (1996). *The ecology of Kalimantan* (Vol. 3). Oxford University Press.

Maggioncalda, A., Czekala, N., & Sapolsky, R. (2002). Male orangutan subadulthood: A new twist on the relationship between chronic stress and developmental arrest. *American Journal of Physical Anthropology, 118*(1), 25-32.

Maggioncalda, A., Sapolsky, R., & Czekala, N. (1999). Reproductive hormone profiles in captive male orangutans: Implications for understanding developmental arrest. *American Journal of Physical Anthropology, 109,* 19-32.

Manduell, K. L., Harrison, M., & Thorpe, S. K. S. (2012). Forest structure and support availability influence orangutan locomotion in Sumatra and Borneo. *American Journal of Primatology, 74,* 1128-1142.

Margono, B. A., Potapov, P. V., Turubanova, S., Stolle, F., & Hansen, M. C. (2014). Primary forest cover loss in Indonesia over 2000-2012. *Nature Climate Change, 4,* 730-735.

Markham, R., & Groves, C. P. (1990). Brief communication: Weights of wild orang-utans. *American Journal of Physical Anthropology, 81,* 1-3.

Marshall, A. J., Ancrenaz, M., Brearley, F. Q., Fredriksson, G. M., Ghaffar, N., Heydon, M., Husson, S. J., Leighton, M., McConkey, K. R., Morrogh-Bernard, H. C., & Proctor, J. (2009). The effects of forest phenology and floristics on populations of Bornean and Sumatran orangutans. In S. A. Wich, S. S. Utami Atmoko, T. M. Setia, & C. P. van Schaik (Eds.), *Orangutans: Geographic variation in behavioural ecology and conservation* (pp. 97-117). Oxford University Press.

Marshall, A. J., & Meijaard, E. (2009). Orangutan nest surveys: The devil is in the details. *Oryx, 43,* 416-418.

Marshall, A. J., Nardiyono, Engstrom, L. M., Pamungkas, B., Palapa, J., Meijaard, E., & Stanley, S. A. (2006). The blowgun is mightier than the chainsaw in determining population density of Bornean orang-utans (*Pongo pygmaeus morio*) in the forests of east Kalimantan. *Biological Conservation, 129,* 566-578.

Marshall, A. J., Salas, L. A., Stephens, S., Nardiyono, Engstrom, L., Meijaard, E., & Stanley, S. A. (2007). Use of limestone karst forests by Bornean orangutans (*Pongo pygmaeus morio*) in the Sangkulirang Peninsula, east Kalimantan, Indonesia. *American Journal of Primatology, 69*(2), 212-219.

Mathewson, P. D., Spehar, S. N., Meijaard, E., Nardiyono, Purnomo, Sasmirul, A., Sudiyanto, Oman, Sulhnudin, Jasary, Jumali, & Marshall, A. J. (2008). Evaluating orangutan census techniques using nest decay rates: Implications for population estimates. *Ecological Applications, 18,* 208-221.

Medway, L. (1977). The Niah excavations and an assessment of the impact of early man on mammals in Borneo. *Asian Perspectives, 20,* 51-69.

Meijaard, E., Albar, G., Nardiyono, Rayadin, Y., Ancrenaz, M., & Spehar, S. (2010a). Unexpected ecological resilience in Bornean orangutans and implications for pulp and paper plantation management. *PLoS ONE, 5*(9), e12813. DOI: 10.1371/journal.pone.0012813.

Meijaard, E., Welsh, A., Ancrenaz, M., Wich, S., Nijman, V., & Marshall, A. J. (2010b). Declining orangutan encounter rates from Wallace to the present suggest the species was once more abundant. *PLoS ONE, 5*(8), e12042. DOI: 10.1371/journal.pone.0012042.

Merrill, M. Y. (2004). *Orangutan cultures? Tool use, social transmission, and population differences.* [Doctoral dissertation, Duke University].

Mitani, J. C. (1985a). Mating behavior of male orangutans in the Kutai Game Reserve, Indonesia. *Animal Behaviour, 33,* 392-402.

Mitani, J. C. (1985b). Sexual selection and adult male orangutan long calls. *Animal Behaviour, 33,* 272-283.

Mitani, J. C. (1989). Orangutan activity budgets: Monthly variations and the effects of body size, parturition, and sociality. *American Journal of Primatology, 18,* 87-100.

Mitani, J. C., Grether, G. F., Rodman, P. S., & Priatna, D. (1991). Associations among wild orang-utans: Sociality, passive aggregations or chance? *Animal Behaviour, 42*(1), 33-46.

Mitra Setia T., Delgado, R. A., Utami Atmoko, S. S., Singleton, I., & van Schaik, C. P. (2009). Social organization and male-female relationships. In S. A. Wich, S. S. Utami Atmoko, T. Mitra Setia, & C. P. van Schaik (Eds.), *Orangutans: Geographic variation in behavioral ecology and conservation* (pp. 245-253). Oxford University Press.

Mitra Setia, T., & van Schaik, C. P. (2007). The response of adult orang-utans to flanged male long calls: Inferences about their function. *Folia Primatologica, 78*, 215-226.

Morrogh-Bernard, H., Husson, S., Page, S. E., & Rieley, J. O. (2003). Population status of the Bornean orang-utan (*Pongo pygmaeus*) in the Sebangau peat swamp forest, central Kalimantan, Indonesia. *Biological Conservation, 110*(1), 141-152.

Nadler, R. D. (1988). Sexual and reproductive behavior. In J. H. Schwartz (Ed.), *Orang-utan biology* (pp. 105-116). Oxford University Press.

Nater, A., Mattle-Greminger, M. P., Nurcahyo, A., Nowak, M., de Manuel, M., Desai, T., Groves, C., Pybus, M., Bilgin Sonay, T., Roos, C., Lameira, A. R., Wich, S. A., Askew, J. A., Davila-Ross, M., Fredriksson, G., de Valles, G., Casals, F., Prado-Martinez, J., Goosens, B.,... & Krützen M. (2017). Morphometric, behavioral, and genomic evidence for a new orangutan species (*P. tapanuliensis*). *Current Biology, 27*(22), 3487-3498.

Nellemann, C., Miles, L., Kaltenborn, B. P., Virtue, M., & Ahlenius, H. (2007). *The last stand of the orangutan, state of emergency: Illegal logging, fire and palm oil in Indonesia's national parks.* United Nations Environment Programme.

Nijman, V. (2009*). An assessment of trade in gibbons and orang-utans in Sumatra, Indonesia.* TRAFFIC Southeast Asia.

Rayadin, Y., & Saitoh, T. (2009). Individual variation in nest size and nest site features of the Bornean orangutans (*Pongo pygmaeus*). *American Journal of Primatology, 71*, 1-7.

Rijksen, H. D. (1978). *A field study on Sumatran orang utans (*Pongo pygmaeus abelii *Lesson 1827): Ecology, behaviour and conservation.* Wageningen.

Rijksen, H. D., & Meijaard, E. (1999). *Our vanishing relative: The status of wild orang-utans at the close of the twentieth century.* Kluwer Academic Publishers.

Rodman, P. S. (1988). Diversity and consistency in ecology and behavior. In J. H. Schwartz (Ed.), *Orang-utan biology* (pp. 31-51). Oxford University Press.

Rodman, P. S., & Mitani, J. C. (1987). Orangutans: Sexual dimorphism in a solitary species. In B. B. Smuts, D. Cheney, R. Seyfarth, R. Wrangham, & T. Struhsaker (Eds.), *Primate societies* (pp. 145-154). University of Chicago Press.

Russon, A. E., Kuncoro, P., & Ferisa, A. (2015). Orangutan behavior in Kutai National Park after drought and fire damage: Adjustments to short- and long-term natural forest regeneration. *American Journal of Primatology, 77*(12), 1276-1289.

Russon, A. E., Wich, S., Ancrenaz, M., Tomoko, K., Knott, C. D., Kuze, N., Morrogh-Bernard, H. C., Pratje, P., Ramlee, H., Rodman, P., Sawang, A., Sidiyasa, K., Singleton, I., & van Schaik, C. P. (2009). Geographic variation in orangutan diets. In S. A. Wich, S. S. Utami Atmoko, T. Mitra Setia, & C. P. van Schaik (Eds.), *Orangutans: Geographic variation in behavioral ecology and conservation* (pp. 135-156). Oxford University Press.

Schürmann, C. L. (1981). Courtship and mating behavior of wild orangutans in Sumatra. In A. B. Chiarelli, & R. S. Corruccini (Eds.), *Primate behavior and sociobiology* (pp. 130-135). Springer.

Schürmann, C. L. (1982). Mating behaviour of wild orang utans. In L. E. M. de Boer (Ed.), *The orang utan: Its biology and conservation* (pp. 269-284). W. Junk Publishers.

Schürmann, C. L., & van Hooff, J. A. R. A. M. (1986). Reproductive strategies of the orang-utan: New data and a reconsideration of existing sociosexual models. *International Journal of Primatology, 7*, 265-287.

Singleton, I., & van Schaik, C. P. (2001). Orangutan home range size and its determinants in a Sumatran swamp forest. *International Journal of Primatology, 22*(6), 877-911.

Singleton, I., & van Schaik, C. P. (2002). The social organisation of a population of Sumatran orang-utans. *Folia Primatologica, 73*, 1-20.

Singleton, I., Wich, S., Husson, S., Stephens, S., Utami Atmoko, S., Leighton, M., Rosen, N., Traylor-Holzer, T., Lacy, R., & Byers, O. (2004). *Orangutan population and habitat viability assessment: Final report.* IUCN/SSC Conservation Breeding Specialist Group.

Smith, R. J., & Jungers, W. L. (1997). Body mass in comparative primatology. *Journal of Human Evolution, 32*(6), 523-559.

Soehartono, T., Susilo, H. D., Anayani, N., Utami-Atmoko, S. S., Sihite, J., Saleh, C., & Sutrisno, A. (2007). *Strategi dan rencana aksi konservasi orangutan Indonesia 2007-2017.* Direktorat Jenderal Perlindungan Hutan dan Konservasi Alam, Departemen Kehutanan.

Spehar, S. N., Mathewson, P. D., Nuzuar, Wich, S. A., Marshall, A. J., Kuhl, H., Nardiyono, & Meijaard, E. (2010). Estimating orangutan densities using the standing crop and marked nest count method: Lessons learned for conservation. *Biotropica, 42*(6), 748-757.

Struhsaker, T. T. (1969). Correlates of ecology and social organisation among African cercopithecines. *Folia Primatologica, 11*, 80-118.

Sugardjito, J., te Boekhorst, I. J. A., & van Hooff, J. A. R. A. M. (1987). Ecological constraints on the grouping of wild orang-utans (*Pongo pygmaeus*) in the Gunung Leuser National Park, Sumatra, Indonesia. *International Journal of Primatology, 8*(1), 17-41.

Sugardjito, J., & van Hooff, J. A. R. A. M. (1986). Age-sex class differences in the positional behavior of the Sumatran orang utan (*Pongo pygmaeus abelii*) in the Gunung Leuser National Park, Indonesia. *Folia Primatologica, 47*, 14-25.

Taylor, A. B. (2006). Feeding behavior, diet, and the functional consequences of jaw form in orangutans, with implications for the evolution of *Pongo*. *Journal of Human Evolution, 50*, 377-393.

te Boekhorst, I. J. A., Schürmann, C. L., & Sugardjito, J. (1990). Residential status and seasonal movements of wild orang-utans in the Gunung Leuser Reserve (Sumatra, Indonesia). *Animal Behaviour, 39*(6), 1098-1109.

Thorpe, S. K. S., & Crompton, R. H. (2005). Locomotor ecology of wild orangutans (*Pongo pygmaeus abelii*) in the Gunung Leuser ecosystem, Sumatra, Indonesia: A multivariate analysis using log-linear modelling. *American Journal of Physical Anthropology, 127,* 58-78.

Thorpe, S. K. S., & Crompton, R. H. (2006). Orangutan positional behavior and the nature of arboreal locomotion in Hominoidea. *American Journal of Physical Anthropology, 131,* 384-401.

Thorpe, S. K. S., Holder, R., & Crompton, R. H. (2009). Orangutans employ unique strategies to control branch flexibility. *Proceedings of the National Academy of Sciences, 106*(31), 12646-12651.

Utami, S. S., Goossens, B., Bruford, M. W., de Ruiter, J. R., & van Hooff, J. A. R. A. M. (2002). Male bimaturism and reproductive success in Sumatran orang-utans. *Behavioral Ecology, 13*(5), 643-652.

van Adrichem, G. G. J., Utami, S. S., Wich, S. A., van Hooff, J. A. R. A. M., & Sterck, E. H. M. (2006). The development of wild immature Sumatran orangutans (*Pongo abelii*) at Ketambe. *Primates, 47,* 300-309.

van Noordwijk, M. A., & van Schaik, C. P. (2005). Development of ecological competence in Sumatran orangutans. *American Journal of Physical Anthropology, 127,* 79-94.

van Schaik, C. P. (1999). The socioecology of fission-fusion sociality in orangutans. *Primates, 40*(1), 69-86.

van Schaik, C. P. (2004). *Among orangutans: Red apes and the rise of human culture.* Harvard University Press.

van Schaik, C. P., Ancrenaz, M., Borgen, G., Galdikas, B., Knott, C. D., Singleton, I., Suzuki, A., Utami, S. S., & Merrill, M. (2003). Orangutan cultures and the evolution of material culture. *Science, 299,* 102-105.

van Schaik, C. P., Damerius, L., & Isler, K. (2013). Wild orangutan males plan and communicate their travel direction one day in advance. *PLoS ONE, 8*(9), e74896.

van Schaik, C. P., Deaner, R. O., & Merrill, M. Y. (1999). The conditions for tool use in primates: Implications for the evolution of material culture. *Journal of Human Evolution, 36*(6), 719-741.

van Schaik, C. P., & Knott, C. D. (2001). Geographic variation in tool use on neesia fruit in orangutans. *American Journal of Physical Anthropology, 114,* 331-342.

van Schaik, C. P., Monk, K. A., & Robertson, J. M. Y. (2001). Dramatic decline in orang-utan numbers in the Leuser Ecosystem, north Sumatra. *Oryx, 35*(1), 14-25.

van Schaik, C. P., & van Hooff, J. A. R. A. M. (1996). Toward an understanding of the orangutan's social system. In W. C. McGrew, L. F. Marchant, & T. Nishida (Eds.), *Great ape societies* (pp. 3-15). Cambridge University Press.

van Schaik, C. P., Wich, S. A., Utami, S. S., & Odom, K. (2005). A simple alternative to line transects of nests for estimating orangutan densities. *Primates, 46,* 249-254.

van't Land, J. (1990). *Who is calling there? The social context of adult male orangutan long calls.* [Doctoral dissertation, University of Utrecht].

Vogel, E. R., Haag, L., Mitra-Setia, M., van Schaik, C. P., & Dominy, N. J. (2010). Foraging and ranging behavior during a fallback episode: *Hylobates albibarbis* and *Pongo pygmaeus wumbii* compared. *American Journal of Physical Anthropology, 140,* 716-726.

Vogel, E. R., Harrison, M. E., Zulfa, A., Bransford, T. D., Alavi, S. E., Husson, S., Morrogh-Bernard, H., Firtsman, T., Utami-Atmoko, S. S., van Noordwijk, M. A., & Farida, W. R. (2015). Nutritional differences between two orangutan habitats: Implications for population density. *PLoS ONE, 10*(10), e0138612.

Vogel, E. R., Knott, C. D., Crowley, B. E., Blakely, M. D., Larsen, M. D., & Dominy, N. J. (2011). Bornean orangutans on the brink of protein bankruptcy. *Biology Letters, 8*(3), 333-336.

Vogel, E. R., Zulfa, A., Hardus, M., Wich, S. A., Dominy, N. J., & Taylor, A. B. (2014). Food mechanical properties, feeding ecology, and the mandibular morphology of wild orangutans. *Journal of Human Evolution, 75,* 110-124.

Voigt, M., Wich, S. A., Ancrenaz, M., Meijaard, E., Abram, N., Banes, G. L., Campbell-Smith, G., d'Arcy, L. J., Delgado, R. A., Erman, A., Gaveau, D., Goossens, B., Heinicke, S., Houghton, M., Husson, S. J., Leiman, A., Sanchez, K. L., Makinuddin, N., Marshall, A. J.,... & Kuhl, H. S. (2018). Global demand for natural resources eliminated more than 100,000 Bornean orangutans. *Current Biology, 28,* 761-769.

Wartmann, F. M., Purves, R. S., & van Schaik, C. P. (2010). Modeling ranging behavior of female orangutans: A case study in Tuanan, central Kalimantan, Indonesia. *Primates, 51,* 119-130.

Wich, S. A., Buij, R., & van Schaik, C. P. (2004a). Determinants of orangutan density in the dryland forests of the Leuser Ecosystem. *Primates, 45,* 177-182.

Wich, S. A., Geurts, M. L., Mitra Setia, T., & Utami Atmoko, S. S. (2006). Influence of fruit availability on Sumatran orangutan sociality and reproduction. In G. Hohmann, M. M. Robbins, & C. Boesch (Eds.), *Feeding ecology in apes and other primates: Ecological, physical and behavioral aspects* (pp. 335-356). Cambridge University Press.

Wich, S. A., Singleton, I., Utami-Atmoko, S. S., Geurts, M. L., Rijksen, H. D., & van Schaik, C. P. (2003). The status of the Sumatran orang-utan *Pongo abelii*: An update. *Oryx, 37*(1), 49-54.

Wich, S. A., Utami-Atmoko, S. S., Mitra Setia, T., Schurmann, C., van Hooff, J. A. R. A. M., & van Schaik, C. P. (2004b). Life history of wild Sumatran orangutans (*Pongo abelii*). *Journal of Human Evolution, 47*(6), 385-398.

Wich, S. A., Utami Atmoko, S. S., Mitra Setia, T., & van Schaik, C. P. (2009). *Orangutans: Geographic variation in behavioral ecology and conservation.* Oxford University Press.

Wich, S. A., & van Schaik, C. V. (2000). The impact of El Nino on mast fruiting in Sumatra and elsewhere in Malesia. *Journal of Tropical Ecology, 16*(4), 563-577.

CHAPTER 22

Gorillas

David B. Morgan, Robert W. Sussman, Kristena E. Cooksey, Kathryn Judson, and Crickette M. Sanz

Over the last 15 years, gorilla systematics have been debated considerably and gone through several revisions. Previous examinations split the genus *Gorilla* into two separate subspecies: eastern and western lowland populations. More informed interpretations and studies based on mtDNA sequences divide the genus *Gorilla* into two species: the eastern gorilla (*G. beringei*: Matschie, 1903) and the western gorilla (*G. gorilla*: Savage, 1847) (Sarmiento & Butynski, 1996; Seaman et al., 1999; Sarmiento & Oates, 2000; Groves, 2001). Each species contains two subspecies (Groves, 2003). The eastern species consists of the mountain gorilla (*G. b. beringei*: Matschie, 1903) and eastern lowland or Grauer's gorilla (*G. b. graueri*: Matschie, 1914). The western species (see Figure 22.1) consists of the western lowland gorilla (*G. g. gorilla*: Savage, 1847) and the Cross River gorilla (*G. g. diehli*: Matschie, 1904). The findings of a gorilla population by Morgan and colleagues (2003) at Ebo, in northern Cameroon, have yet to be assigned to the subspecies level (Maisels et al., 2018).

Based on genetic evidence, *G. gorilla* and *G. beringei* diverged between 1.20–3.01 mya with gene flow occurring as late as 200,000 or as recently as 80,000 years ago (Thalmann et al., 2006; Langergraber et al., 2012). A largely forested gap of about 1,000 km now serves as a barrier to gene flow between the two species. Within the western subspecies, populations of *G. g. diehli* diverged from ancestral *G. g. gorilla* populations roughly 17,800 years ago and gene flow ceased as recently as 420 years ago (Rouquet et al., 2005; Thalmann et al., 2011). Though the western population is highly fragmented, gene flow may still occur between localities (Bergl & Vigilant, 2007). Within the eastern subspecies, gene flow occurred approximately 400,000 years ago. Eastern lowland and mountain gorilla populations reside as close as 100 km apart; however, there is not strong evidence of recent gene flow between them (Das et al., 2014).

Gorillas are the largest and one of the most sexually dimorphic species of living primates. On average, males are almost double the size of females (Taylor, 1997). *G. beringei* are somewhat larger than *G. gorilla;* adult male *G. beringei* weigh upward of 220 kg. Females reach 80 kg in *G. b. graueri* and 97.7 kg in *G. b. beringei. G. beringei* adult males may reach a height of 1.8 m and females 1.4 m. Western males may reach 1.7 m in height, and the average height of females ranges from 1.2–1.4 m. Eastern and mountain female gorillas grow to an average of 1.5 and 1.4 m, respectively (Wood, 1979; Leigh & Shea, 1995; Miller-Schroeder, 1997).

GEOGRAPHIC DISTRIBUTION

The genus *Gorilla* has a wide and varied distribution across a diverse range of habitat and sub-habitat types, from the high-altitude alpine forests in the Albertine Rift Valley to the moist lowland coastal forests of Western Africa (Figure 22.2). The western extent of *G. gorilla* occurs on the Nigeria-Cameroon boundary with small fragmented populations occurring in the Cross River region where the subspecies (*G. g. diehli*) derived its common name. To the east, more extensive and continuous populations of *G. g. gorilla* occur in Cameroon, mainland Equatorial Guinea, Gabon, Republic of Congo, and the southern tip of the Central African Republic. In this region, the mosaic of suitable forest covers an estimated area of over 700,000 km^2 (Maisels et al., 2018) with gorilla populations divided into mostly distinct genetic subpopulations by major rivers (Anthony et al., 2007; Fünfstück, 2014). The overall population of *G. g. gorilla* is unevenly distributed. Approximately 60% of the population reside in the Republic of Congo, with the majority persisting in northern Republic of Congo (Strindberg et al., 2018). Reports also suggest gorillas exist in the Cabinda Province of Angola to the south, but the size and status of these ape populations remain unconfirmed (Ferriss, 2005). The eastern boundary of *G. gorilla* distribution is demarcated by the Congo and Oubangui Rivers.

At present, mountain gorillas (*G. b. beringei*—see Figure 22.3) inhabit three countries in two small, isolated populations. One population inhabits Rwanda, the Democratic Republic of Congo (DRC), and Uganda within the Virunga Volcano region. The second population inhabits the Bwindi Impenetrable Forest located in the Kigezi Highlands of southwest Uganda, on the edge of the Albertine Rift Valley bordering the DRC to the west. This population

Figure 22.1 Many of the first insights into western lowland gorilla behavior were from observations of a silverback named "Kingo" and his group at the Mondika study site in the Djéké Triangle, Republic of Congo.
(Photo Credit: Crickette Sanz, WCS-Congo)

is narrowly separated by 35 km of varied terrain from the Virunga population. The tri-national protection system, consisting of Virunga National Park, Volcanoes National Park, Mgahinga Gorilla National Park, and Bwindi Impenetrable National Park, form the core of the mountain gorilla population distribution (see Figure 22.2).

The Grauer's gorilla (*G. b. graueri*) occurs farther to the west in eastern DRC and within the Lualaba River and the Burundi-Rwanda-Uganda border. As a result, the subspecies has a wider geographical distribution and larger overall population size than those of the *G. b. beringei* (Figure 22.2). While observations of gorillas have occurred in the more peripheral limits of *G. b. graueri* range, the core populations occur in Maiko National Park, Kahuzi-Biega National Park, Itombwe Forest, and North Kivu (Mehlman, 2008).

GORILLA FIELD STUDIES

Until the late 1950s most attempts to encounter gorillas in their natural habitat were motivated by the demand for display specimens in museums or zoological gardens.

The first intensive fieldwork on gorillas began in 1959 by Schaller and Emlen on *G. b. beringei* (Emlen & Schaller, 1960; Schaller, 1963). Following Schaller's pioneering work, Dian Fossey established the Karisoke Research Center in the Virunga Mountains, which began one of the longest-running field studies of any mammal to date. In addition to this site, other research projects focusing on the behavior and ecology of Grauer's gorilla populations in DRC in the Mt. Kahuzi region were initiated by Casimir (1975, 1979), Goodall (1977), and subsequently Yamagiwa and colleagues (1993, 1996). Most recently, studies of the mountain gorilla population in Uganda's Bwindi Impenetrable Forest National Park have been undertaken by Robbins (1995), Robbins and McNeilage (2003), and Robbins and Sawyer (2007). Together, occurring at varying altitudes and habitat types, these research sites provide an important cross-section from which to base intra- and interspecific comparisons on gorilla behavior and ecology.

In contrast, detailed and long-term studies on western lowland gorillas were not successfully implemented until the late 1980s. Initial attempts were made with investigations by Jones and Sabater (1971) in the Central African state of Rio Muni (Equatorial Guinea). It was not until Tutin and Fernandez (1984) conducted a census of gorillas in Gabon, and established the Lopé Great Ape Research in central Gabon, that information about the ecology of *G. g. gorilla* began to emerge. Results from these preliminary studies indicated that gorilla distribution was not dictated by secondary or regenerating forest, but that western gorillas occurred across a range of diverse forest types (Jones & Sabater, 1971). The initial studies of dietary overlap between sympatric western lowland gorillas and chimpanzees at Lopé contributed tremendously to our understanding of ape ecology in a Central African context. These studies were monumental in inspiring further investigations into western lowland gorillas, while also highlighting the difficulty of habituating western lowland gorillas.

Compared to mountain and Grauer's gorillas, western lowland gorillas have been drastically more challenging to habituate for multiple reasons, e.g., thicker vegetation impacting both visibility and tracking ability (Tutin & Fernandez, 1991). While habituation success has increased since these preliminary attempts, the effect of delayed habituation persists in recent investigations (Cipolleta, 2003; Greer & Cipolleta, 2006; Doran-Sheehy et al., 2007). Given the challenges of habituation, a majority of published studies on western lowland gorillas only account for their behavior at swampy forest clearings (locally known as *bais*), as discussed later in the chapter (Magliocca et al., 1999; Parnell, 2002b). Even less

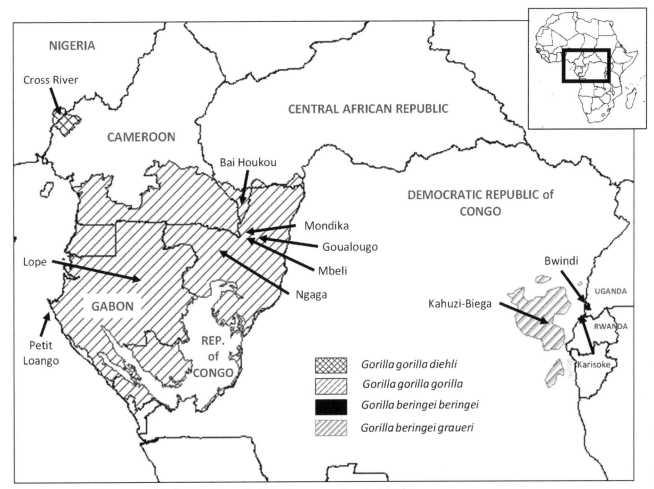

Figure 22.2   Gorilla geographic distribution and study sites across Equatorial Africa.

information is available on Cross River gorillas, given the fragmented forests in which they live and scientists' reliance on indirect behavioral inferences (Sunderland-Groves et al., 2009).

Currently, there are more than 10 sites across Africa dedicated to studying the behavior and ecology of wild gorillas (see Figure 22.2). The long-term studies of East African apes have produced a large body of influential works by multiple generations of scientists (Harcourt & Stewart, 2007; Robbins et al., 2011a), which has no parallel in western Equatorial Africa. This trend is especially true for gorillas, where despite major scientific advancements in studying gorilla ecology over the last 30 years, there is still a sizeable gap in knowledge of the eastern *G. gorilla*. Specifically, in regard to behavior and reproduction, there is far more published information on the eastern gorilla subspecies than the western gorilla subspecies.

Several of the ongoing studies of gorillas have shown convincingly how detailed monitoring programs based on known individuals and groups provide important insights into the ecology and behavior of wild apes. Such a deeper understanding of these species also aids in

informing the public, policymakers, and governments, bringing awareness to their plight and addressing the conservation threats they face. While some general aspects of gorilla behavior are consistent among taxa, the high degree of ecological variation, phenology, and other landscape features between adjacent populations implies that there is great potential for subtle but meaningful differences in social and foraging strategies within and between gorilla subspecies. Most notably, these behavioral repertoires differ along a gradient in terms of diet, nesting, and arboreality.

RANGING AND HABITAT USE

Among the primates, gorillas are particularly noteworthy for their wide distribution across Equatorial Africa and the broad spectrum of climatically and floristically diverse regions they inhabit. Gorilla populations can be found from the high elevations of the east, down to and within the lowland forests of the Congo Basin to coastal habitats lining the Atlantic Ocean, highlighting the ecological flexibility of these great apes. Population numbers vary markedly across the geographical distribution and

Figure 22.3 Mountain gorilla research has been ongoing at several sites, including Karisoke Research Center in Volcanoes National Park (Rwanda) and Bwindi Impenetrable National Park (Uganda). (Photo Credit: Martha Robbins, MPI)

habitat types of the genus (see Table 22.1). A common environmental characteristic has emerged from these patterns: the positive relationship between availability and abundance of staple herbaceous vegetation and gorilla population densities.

Among the most striking aspects of eastern gorilla ecology are the extreme elevations and correspondingly lower temperatures at which these populations are found (especially mountain gorillas). Some groups range up to elevations of >3,000 m and habitat types unique to such altitude. Overall, mountain gorillas are typically described as inhabiting an environment relatively uniform in food availability, but groups will use bamboo, moorland, and swamp forest habitats at particular times (Schaller, 1963; Casimir, 1975; Harcourt & Stewart, 2007). Habitat use is, in part, linked to dietary needs; as the suitability and attractiveness of particular habitats vary, so do the foraging efforts of gorillas (Waterman et al., 1983). Alpine forest predominates at the highest altitudes of the mountain gorilla range, and it seems that groups tend to limit the

amount of time foraging in this extreme climate. The impacts of habitat change and seasonality are most conspicuous via the growth of new shoots (Fossey & Harcourt, 1977; Vedder, 1989).

In contrast to the higher elevations where mountain gorillas occur, the vast majority of western gorillas inhabit lower-lying forests relying on a mosaic of different habitats. Most western gorillas are found in mixed-species forests characterized by gradual habitat changes, although other populations exist with abrupt and notable habitat shifts. Perhaps most striking is the Likouala region in the northern Republic of Congo, where there is a high prevalence of both seasonally and permanently inundated forests that gorillas utilize (Fay & Agnagna, 1992; Rainey et al., 2010). While there is little detail on how western lowland gorillas exist in these largely inundated conditions, there could be markedly different aspects to their behavior and ecology in comparison to conspecifics inhabiting drier terra firma forests. Most obvious is travel, which is likely compromised. At Mondika in the Central African Republic, terra firma forests are more abundant, and gorillas spend 26% of their time in swamps where, energetically, the travel costs are steep. Western lowland gorilla daily path lengths are 50% longer on days spent traversing swamp compared to days spent in terra firma forest (Doran-Sheehy et al., 2004). Not surprisingly, at certain times of the year, when flooding occurs, these forests may be abandoned for drier areas (Poulsen & Clark, 2004) where travel, nesting, and locomotion are less constrained. While some of the highest densities of western lowland gorillas are found in regions with comparable herb densities to that of the eastern species habitat, western lowland gorillas are prevalent in areas with much lower herb densities than their eastern counterparts, yet maintain similar group sizes.

Gorilla diets are largely comprised of high-quality herbaceous vegetation in higher altitudes, such as the Virunga Volcanoes of DRC and Rwanda and Kahuzi-Biega National Park in DRC (Watts, 1984; Yamagiwa et al., 1993; Goldsmith, 2003; Kalpers et al., 2003; Nkurunungi et al., 2004). Studies across the western gorilla range indicate a preference for mixed-species forests (both terra firma and inundated) with closed understory habitat (Carroll, 1986; Fay et al., 1989; Tutin et al., 1995; Fay, 1997; Brugiere & Sakom, 2001; Mehlman & Doran, 2002; Morgan et al., 2006; Stokes et al., 2010). Across a few geographical locations in Central Africa, environmental conditions are comparable in structure to mountainous eastern Africa—namely, low tree densities and expansive stands of terrestrial vegetation. Nest surveys across these regions indicate gorilla abundance reaches its highest levels in areas where the understory stratum is dominated by

Table 22.1 Density Estimates of Eastern and Western Gorilla Populations from Nest-Count Surveys and Direct Observations

Species Country/Study Site	Subspecies	Density (indiv/km²)	Survey (km)	Habitat	References
Eastern Gorillas (*Gorilla beringei*)					
Rwanda					
Karisoke	*G. b. beringei*	604 [a]	-	High montane forest	Hickey et al., 2019
Uganda					
Bwindi	*G. b. beringei*	430 [a]	-	High montane forest	Roy et al., 2014
Democratic Republic Congo					
North Maiko	*G. b. graueri*	160- 1440 [a]	-	High montane forest	Mehlman, 2008
South Maiko	*G. b. graueri*	200-1880	-	High montane forest	Mehlman, 2008
Kahuzi-Biega (Highlands)	*G. b. graueri*	60- 536 [a]	-	High montane forest	Plumptre et al., 2019
Kahuzi-Biega (Lowland)	*G. b. graueri*	292- 2945 [a]	-	Lowland montane forest	Plumptre et al., 2019
Western Gorillas (*Gorilla gorilla*)					
Gabon					
Nationwide	*G. g. gorilla*	0.18 (0.01- 0.44)	783	Semi-evergreen forest	Tutin and Fernandez, 1984
Lopé Reserve	*G. g. gorilla*	0.3- 1.0	700 [b]	Semi-evergreen forest	White, 1992
Petit Loango	*G. g. gorilla*	0.59	98	Coastal, Interior	Furuichi et al., 1997
Cameroon					
Dja	*G. g. gorilla*	1.71 (1.02- 2.86)	95	Semi-evergreen forest	Williamson and Usongo, 1995
Campo	*G. g. gorilla*	0.2	404	Semi-evergreen forest	Matthews and Matthews, 2004
Ntonga	*G. g. gorilla*	3.87	58	Coastal, Interior	Dupain et al., 2004
Republic of Congo					
Conkouati	*G. g. gorilla*	3.90 (2.41- 5.39)	19	Semi-deciduous forest	Maisels and Cruikshank, 1996
Goualougo	*G. g. gorilla*	2.34 (1.83- 2.99)	222	All habitats	Morgan et al., 2006
Lac Tele	*G. g. gorilla*	2.91 (1.6- 5.6)	234	All habitats	Poulsen and Clark, 2004
Northern Congo	*G. g. gorilla*	1.65 (1.24- 2.21)	330	All habitats	Stokes et al., 2010
Nigeria					
Cross River	*G. g. diehli*	1.65 (1.24- 2.21)	-	All habitats	Oates et al., 2003

[a] Encounter rate.
[b] Smaller subset of transects were surveyed several times.

herbaceous vegetation (White, 1992, 1994; Devos et al., 2008). Likewise, the *Raphia* swamps of the Likouala region in northern Republic of Congo also show a similar phenomenon where significant gorilla numbers occur in areas dominated by herbaceous vegetation (Fay & Agnagna, 1992; Blake et al., 1995; Poulsen & Clark, 2004; Rainey et al., 2010).

LOCOMOTIVE AND POSITIONAL BEHAVIORS

Given differences in *G. gorilla* and *G. beringei* habitat types and foraging strategies, morphological features related to positional repertoires and behavioral adaptions to habitat utilization have been observed (Groves & Stott, 1979). During ground travel, gorillas are quadrupedal and use a knuckle-walking posture. Their hind feet are flat on the ground, and the upper torso is supported on the knuckles. Physical characteristics associated with terrestrial travel, such as thickened skin with callouses on the middle joints of the second to fifth fingers, as well as modifications to the hand and wrist, have been observed

(Tuttle & Basmajian, 1974). There is a great deal of interest in elucidating functional morphological conditions to respective local ecologies, specifically for *G. b. beringei* and *G. g. gorilla*. A reexamination of foot morphology contrasting these subspecies indicates dissimilarity in their feet (Tocheri et al., 2011). Mountain gorillas have shorter metatarsals compared to lowland gorillas, supporting claims of adaptations to greater terrestriality in the mountain gorillas (Inouye, 1994, 2003) and lending support to the separation of *G. gorilla* and *G. beringei* based on distinct adaptations to different ecological conditions (Tocheri et al., 2011; Jabbour & Pearman, 2016). More detailed studies of the talar morphology also have found recognizable differences distinguishing *G. gorilla* from *G. beringei* (Dunn et al., 2014). Fast-speed locomotion or running have also been observed in the knuckle-walking posture, though individuals will often crouch close to the ground when moving through the brush stratum. About one-third of the gorilla's time is spent in an upright position with the hands freed, mostly while sitting and feeding and less so during social interactions (such as rare grooming sessions; Schaller, 1963). Bipedal locomotion is performed by juveniles during play or the first few steps when an animal is surprised or frightened and during aggressive, chest-beating displays. Prolonged bipedal locomotion is rare and occurs mostly when traveling in swamps or crossing deeper streams with no option for using overhanging vegetation. It is in these situations that gorillas have been observed to seek assistance from conspecifics (e.g., dependents and their mothers) or occasionally to use tools (Breuer et al., 2005).

In the early stages of gorilla research, scientists indicated gorillas were predominantly terrestrial, although they live in complex three-dimensional forested environments. These assumptions were based on Schaller's (1963) initial research that showed mountain gorillas in the Virunga region spent 80%–90% of their waking hours on the ground. Tree use by the gorillas was considered to be rare. When they did venture into the canopy, gorillas spent time feeding, nesting, resting, and using trees as vantage points to scan the surrounding environment. While in the canopy, they moved very cautiously and deliberately and were not observed to brachiate (Schaller, 1963).

Shortly after these initial findings, the impression that gorillas avoided tree use began to change. It has now been shown that some eastern and all of the studied western gorilla populations spend considerable amounts of time climbing and locomoting in the canopy (Goodall, 1977; Tutin et al., 1991; Masi, 2008). Even the largest silverbacks, which can weigh at least 200 kg, were observed in the canopy on occasion. There are, however, age-sex class differences in climbing abilities and propensities. There is some indication that male arboreality is affected by the surrounding substrate, such as smaller trees, to allow easier access to the core of the main tree (Remis et al., 1999). Remis et al. (1999) also found that males tended to be more seasonal in their arboreality, climbing more when there was more fruit availability, while female arboreality did not vary throughout the year. Infants and juveniles are stronger climbers by far than subadults and adults, displaying somewhat acrobatic postures. Female gorillas, with or without infants, climb much farther per day than adult males and immatures (Masi, 2008).

Gorillas have a distinct shape, possessing long muscular arms that tend to be longer than their legs, and a broad chest. *G. g. gorilla*, in particular, spends considerably more time arboreally than conspecifics, and this aspect of their lifestyle is thought to be reflected anatomically. Aspects such as vertebral formula differ between eastern and western species and suggest *G. beringei* has elevated levels of terrestrial locomotion while *G. gorilla*

has a more arboreal existence (Williams, 2012). As well, western species have stronger forelimb structures than eastern gorillas. Grauer's gorillas demonstrate intermediary strength comparable to either mountain or western gorillas, based on altitude, which correlates with arboreality in this subspecies (Ruff et al., 2018). Studies found certain talar characteristics were best explained by taxon, while other characteristics were best explained by ecological habitat, specifically convergent traits for improved arboreality (Knigge, 2015). These locomotor and habitat specialization differences can be attributed to the varied forest structure and floral diversities. The lowland Itebero region of Kahuzi-Biega National Park in eastern DRC and the Buhoma and Ruhija regions of Bwindi Impenetrable National Park, Uganda, offer a richer and more three-dimensional habitat (of which gorilla populations in these regions take advantage) than the Virungas (Goldsmith & Moles, 2003; Yamagiwa et al., 2003; Ganas et al., 2004).

The trend of increasing forest structure, diversity, and complexity extends from the lower-lying eastern regions to the lowland forests across Central Africa and has links to increased levels of not only suspensory locomotion and terrestrial travel (Yamagiwa et al., 2003; Doran-Sheehy et al., 2004; Ganas & Robbins, 2005; Masi et al., 2009) but also arboreality (Doran & McNeilage, 1998a; Masi, 2008). The mixed-species forests of Equatorial Africa contain emergent trees reaching heights of just over 50 m, while most trees forming the upper canopy reach heights of 40 m. Western lowland gorillas make use of all levels of the terrestrial and canopy stratum to nest and forage (Tutin et al., 1995; Doran, 2002; Morgan & Sanz, 2006). Differences in gorilla resource preferences and acquisition among these varied forest types have important implications on foraging strategies, grouping patterns, and social dynamics.

DIETARY DIVERSITY

The location of resources affects where gorillas search for food and spend their time. Among the parallels between eastern and western gorillas is their incorporation of large amounts of staple food resources, mainly high-quality herbaceous vegetation obtained terrestrially (see Table 22.2) (Fossey & Harcourt, 1977; Watts, 1984; Tutin & Fernandez, 1985; Vedder, 1989; Plumptre, 1991, 1995; McNeilage, 1995; Doran et al., 2002; Ganas et al., 2004; Rogers et al., 2004). The gorilla's physiological adaptations for dealing with plant defenses and fiber include large body size, longer digestive retention times, thick dental enamel, and molars with shearing crests (Groves, 2003). These specializations afford advantages for gorillas to procure and consume plant material low in toxins and digestibility-reducers but high in protein content and

Table 22.2 **Diets of Eastern and Western Gorilla Populations Based on Fecal Surveys and Direct Observations**

Species Country/Study Site	Subspecies	Total No. of Food Items	Habitat Description	References
Eastern Gorilla (*Gorilla beringei*)				
Rwanda				
Karisoke	*G. b. beringei*	36	High montane forest	Vedder 1984; Watts 1984, 1991a
Uganda				
Bwindi (Mubare)	*G. b. beringei*	205	Low altitude	Ganas et al., 2004
Bwindi (Habinyanja)	*G. b. beringei*	187	Low altitude	Ganas et al., 2004
Bwindi (Kyaguriro)	*G. b. beringei*	106	High altitude	Ganas et al., 2004
Bwindi	*G. b. beringei*	96	High montane forest	Stanford and Nkurunungi, 2003
Democratic Republic Congo				
Kahuzi-Biega	*G. b. graueri*	79	High montane forest	Mehlman, 2008
Western Gorilla (*Gorilla gorilla*)				
Gabon				
Petit Loango	*G. g. gorilla*	203	Moist forest	Head et al., 2011
Lopé Reserve	*G. g. gorilla*	213	Mixed-deciduous	Tutin and Fernandez, 1993
Belinga	*G. g. gorilla*	89	Mixed-deciduous	Tutin and Fernandez, 1985
Republic of Congo				
Goualougo	*G. g. gorilla*	80	Semi-deciduous	Morgan and Sanz, 2006
Mondika	*G. g. gorilla*	127	Semi-deciduous	Doran et al., 2002
Ndoki	*G. g. gorilla*	79	Semi-deciduous	Kuroda et al., 1992
Ndoki	*G. g. gorilla*	182	Semi-deciduous	Kuroda et al., 1996; Nishihara, 1992, 1995
Central African Republic				
Bai Hokou	*G. g. gorilla*	230	Semi-deciduous	Remis, 1997; Remis et al., 2001
Nigeria				
Cross River	*G. g. diehli*			Oates et al., 2003

tannins (Waterman et al., 1983; Watts, 1984; Rogers et al., 1990; Plumptre, 1995). When comparing body size and dietary flexibility, gorillas fall at the far end of the primate body mass spectrum, illustrating that larger species are more capable of coping with less digestible items, such as terrestrial herbs and leaves (Kay, 1984; Janson & Chapman, 1999).

The breadth and selectivity of any gorilla population and their diet seems, in a large sense, to be a functional measure of altitude, local ecology, climate, and floral diversity, rather than a result of dichotomous food preferences or gorilla phylogeny (Goldsmith, 2003; Robbins & McNeilage, 2003; Ganas et al., 2004). The genus *Gorilla* is characterized by a variable diet, with their taxa representing extremes along the primate diet continuum. Species-typical and specialized ecological conditions and associated dietary strategies are argued to have led to important differences in the species' brain structures (Barks et al., 2015). Along the primate diet continuum, *Gorilla* is notable, with western species considered a ripe-fruit specialist while the mountain gorilla is among the most folivorous of all primates (Clutton-Brock, 1977; Harcourt & Stewart, 2007; Doran-Sheehy et al., 2009).

At the highest elevations, between 2,700–3,800 m in the alpine forests of the Virunga ecosystem, mountain gorillas encounter an environment largely colonized by herbaceous vegetation, and it is hard to underestimate the centrality of their subsistence on monocotyledonous species (Fossey & Harcourt, 1977; Vedder, 1989). However, with over 30 years of expanded research on gorillas across the continent, the Virunga ecosystem and the mountain gorilla population can be considered an outlier in terms of diet, given that most other gorilla populations live at lower elevations and in floristically richer habitats that provide more succulent fruit options. The consumption of terrestrial herbaceous vegetation is common throughout their distribution at lower altitudes as well, but gorillas complement this staple food item with a variety of other leaves from medium- to emergent-sized trees, bark, and roots. Simultaneously there is an increasing trend for greater frugivory and overall dietary diversity with decreasing elevation (Robbins, 2011). In comparison to initial studies of *G. b. beringei*, the consumption and preference for high-quality fruit by most populations of gorillas is striking.

The foraging options for eastern gorilla populations change rather dramatically between neighboring areas within the region. In the southeast DRC, increased numbers of plant species, including greater numbers of fruit items, are common in Grauer's gorilla diets at Kahuzi-Biega National Park (Casimir, 1975; Yamagiwa et al., 1994). Yamagiwa and colleagues (1996) found subtle but greater fruit consumption by gorillas within the park at the Itebero lowland site (600–1,300 m) than those of groups inhabiting increased altitudinal sites in the Kahuzi highland at 1,800–3,300 m. Research at highland and lowland locations indicate that in Bwindi National Park, mountain gorillas have more in common in terms of diet to mountain gorillas found in Kahuzi-Biega National Park than conspecific *G. b. beringei* of the Virungas (Yamagiwa et al., 1996a; Robbins & McNeilage, 2003; Ganas et al., 2004; Yamagiwa, 2004). Floristically and phenologically, Bwindi has been shown to be a richer habitat with greater fruit availability when compared to higher elevation habitats (Sarmiento et al., 1996; Robbins & McNeilage, 2003; Nkurunungi et al., 2004).

Across much of Central Africa, a diverse and heterogeneous mosaic of habitats exists owing to a combination of environmental and climatic drivers (Ernst et al., 2012). Central African forests, characterized as having biomass high above ground and very low tree stem densities, typically contain greater mean tree sizes than other tropical forests (Lewis et al., 2013). The occurrence of medium- to large-sized tree classes belonging to families such as Sapotaceae and Irvingiaceae provides apes with important fruit-bearing species. Western lowland gorilla populations' dietary regimes reflect their diverse environments, with some populations consuming greater than 100 species of fruit across the lowland landscape (Tutin, 1993; Doran et al., 2002; Rogers et al., 2004; Head et al., 2011). It is here at these sites that fruit production can span months, with some tree species fruiting for long durations and across a brief dry season (Doran et al., 2002). This preference has led some to suggest western lowland gorillas are "pursuers" of particular fruit items (Doran & McNeilage, 1998a; Goldsmith, 1999; Doran et al., 2002; Rogers et al., 2004). Resulting inferences suggest that this subspecies has a foraging strategy which follows a more typical "ape-pattern," featuring prevalence of succulent ripe fruits, rather than demonstrating an Old World primate foraging strategy typified by a wider selection of diverse fruits at all phenological stages (Doran-Sheehy et al., 2009). When in pursuit of highly valuable food (such as preferred fruits, or within swamps), gorillas have been found to take longer steps, faster, and travel more directly versus when pursuing terrestrial herbaceous vegetation (Salmi et al., 2020). As preferred fruit availability increases, gorillas will spend up to 70% of their time feeding on selected species at one site (Doran-Sheehy et al., 2009). One of the benefits gorillas gain from fruit consumption is increased accessibility to free simple sugars that are easily digestible and provide energy (Rogers et al., 1990). This preference and selectivity also mean their foraging is subject to seasonal fluctuations as tropical tree fruit production varies in time and space (Chapman et al., 1999; van Schaik & Pfannes, 2005). There may also be dramatic multiyear fruiting cycles or interannual variation in fruit production, which can have meaningful impacts on aspects of consumption by gorillas (Cipolletta et al., 2007; Doran-Sheehy et al., 2009; Masi et al., 2009).

At some point, the costs of searching for high-quality food resources outweigh the benefits and cause gorilla foraging strategy and diet shifts to focus on more nutritionally marginal foods or items more abundant but not typically selected for. This is particularly apparent in the highly seasonal forest located in the Cross River region. It is here that gorillas are believed to rapidly transition between foraging strategies, incorporating both high- and low-quality food resources. Initial dietary studies of *G. g. diehli* indicated one of the most diverse diets documented in wild gorillas, including a high number of fruit species (Oates et al., 2003).

It is commonly assumed that gorillas capitalize on their physiological adaptations to effectively metabolize vegetative items low in sugar, protein content, and fiber (Tutin & Fernandez, 1985; Williamson, 1990; Tutin, 1993; Remis, 1994; Nishihara, 1995; Kuroda et al., 1996; Fay, 1997; Doran et al., 2002; Oates et al., 2003; Ganas et al., 2004; Rogers et al., 2004). Though considerable variation in nutritional quality and chemical composition within food types has been documented, several researchers suggest caution should be used when making generalizations about selection of many foods by gorillas (Barton et al., 1993). While the timing and dietary changes in the ratio of frugivory and folivory are consistent with subspecies' ecology, our understanding of gorilla fallback foods is still tentative. During preferred food-limiting periods, gorillas spend 50%–70% of their feeding time on leaves followed by upward of 20% of their foraging time consuming herbs (Doran-Sheehy et al., 2009). Nutritional analysis indicates the two most commonly preferred leaf species are considerably high in protein content, which could explain the benefits of such focused foraging (Doran-Sheehy et al., 2009). But for most terrestrial herbaceous vegetation items used as fallbacks, aspects of nutritional attributes provide little indication as to how and why gorillas differentiate between those preferentially fed upon as opposed to those consumed during low resource periods (Doran-Sheehy et al., 2009).

Bais and Swamps

Throughout the year, there are herb species such as *Hydrocharis* (a species found in aquatic environments) and *Haumania danckelmaniana* (inhabiting terra firma forests) that are consumed preferentially by gorillas (Kuroda et al., 1996). To feed at patches of aquatic herbs, individuals must navigate heavily inundated forests or swamps to arrive in open canopy clearings, which can range up to tens of hectares in size. In the most detailed study yet, Doran and colleagues (2004) found gorillas at Mondika spent roughly 27% of their time in swampy settings accessing aquatic herbs and other food resources. *G. g. gorilla*, across much of Central Africa, have access to another type of forest clearing referred to as a *bai*. The longer-term and more detailed observations of gorilla groups in bai settings indicate groups spend roughly 1% of their daily light hours in such clearings (Magliocca et al., 1999; Parnell, 2002b). In the southwest Central African Republic at Bai Hokou, swamps are less prevalent but gorillas do have access to bais that are relatively dry (Cipolletta, 2004). The amount of time spent in the bai setting in comparison to other habitat types remains unknown at Bai Hokou, but individuals were most often observed foraging in the bai setting during times of low fruit availability (Shelly Masi, pers.comm.). Gorilla use of bais at Mbeli was most frequent during the months associated with increased availability of fruit in the forest (Parnell, 2002b).

Gorilla group visitation can also be influenced dramatically depending on the phenology of particular tree species located at the periphery of such clearings and in inundated forest (Walsh et al., 2007). High-return foods (e.g., fruits) may play a large role in attracting gorillas to forage in and around swamps at Mondika, where travel costs are 50% higher than costs associated with traveling in neighboring terra firma forest (Doran-Sheehy et al., 2004). In a study across multiple bais, gorillas consumed diverse food species based on macronutrient availability, with mineral content varying across sites; however, all consumed bai species demonstrated higher concentrations of Na, K, and Ca compared to other consumed species, suggesting a possible metabolic benefit to bais (Sienne et al., 2014). The results here indicate swamp and bai use is likely driven by many complex and interlinked factors such as local flora within and beyond those gorilla subspecies' ranges. Only through continued monitoring of gorilla foraging dynamics in association with phenology and nutritional analysis will a clearer understanding emerge of fallback versus preferred foraging in western lowland gorillas or of the underlying role floristic composition of neighboring habitats plays in social dynamics, such as female transfer (Guschanski et al., 2008).

Diet and Ranging Behavior

The factors that shape gorilla home range size, spatial configuration, and overlap is as much an ecological question as a social one. Gorilla ranging is intricately related to habitat type, resource base, the density of apes in a given area, male-female mating strategies, and social relationships. Initial accounts predicted that gorilla dependence on terrestrial herbs as stable food items served to buffer groups against high levels of feeding competition given the lack of easily monopolized feeding patches (Fossey & Harcourt, 1977; Watts, 1998a; Robbins & McNeilage, 2003). In such ecological and dietary conditions, home ranges of neighboring groups can overlap extensively and food patches often can be shared between groups, although likely not at the same time (Schaller, 1963; Jones & Sabater, 1971; Fossey & Harcourt, 1977; Yamagiwa et al., 1996).

However, in more recent accounts, this predicted pattern has been shown to be complicated. Even in more stable gorilla environments (e.g., *G. b. beringei* in the Virunga ecosystem), where resources are abundant, groups have been observed to modify ranging behavior in response to changes in abundance of preferred shoots, piths, and leaves. With intense foraging on favored herbs, overexploitation may exhaust these food sources and require *G. b. beringei* groups to forage elsewhere (Watts, 1998a). For *G. beringei*, home range sizes vary between 3 km² up to 40 km² with *G. b. beringei* generally navigating smaller home ranges than those at lower elevations (Ganas & Robbins, 2005; Caillaud et al., 2014) (see Table 22.3). For some groups, the dynamics of navigating extensively overlapping home ranges can intensify intergroup interactions. Over a 12-year period, Caillaud and colleagues (2014) found that an increase in the overlap between groups that once monopolized particular areas for their use, resulted in a 50% reduction in the exclusive use of such zones (Caillaud et al., 2014). Mountain gorilla groups in Bwindi Impenetrable National Park were found to constrict their range with increased local gorilla density, suggesting avoidance between groups, possibly due to either intergroup feeding competition or male mating strategies (Seiler et al., 2018). Areas of high-quality food resources are generally visited more frequently and constitute a core area of use by a group (Watts, 1998a; Robbins & McNeilage, 2003; Caillaud et al., 2014; Seiler et al., 2017). The location of core areas is most often associated as being concentrated and centrally located within a group's home range. Caillaud and colleagues (2014) found *G. b. beringei* groups have high home range overlap and core areas of approximately 2.5 km² in size (50% Kernel Analysis: a spatial analysis technique accounting for the relative location of variables to one another). While home range overlap can be quite high between neighboring groups, groups shared a much smaller proportion of their core area. Although gorillas are not considered territorial, some aspects of territorial behavior have been observed (Seiler et al., 2017).

Table 22.3 Ranging and Group Size of Eastern and Western Gorilla Populations Based on Direct Observations

Species Country/Study Site	Subspecies	Day Range (m)	Home Range (km²)	Group Size	References
Eastern Gorilla (*Gorilla beringei*)					
Rwanda					
Karisoke (Pablo)	*G. b. beringei*	-	12.6	34.3	Caillaud et al., 2014
Karisoke (Shinda)	*G. b. beringei*	-	4.99	16.8	Caillaud et al., 2014
Karisoke (Beetsme)	*G. b. beringei*	-	8.62	17.7	Caillaud et al., 2014
Uganda					
Bwindi (Kyaguriro)	*G. b. beringei*	1034	31.3	14	Ganas and Robbins, 2005
Bwindi (Rushegura)	*G. b. beringei*	633	13.7	8	Ganas and Robbins, 2005
Bwindi (Habinyanja)	*G. b. beringei*	978	37.6	30	Ganas and Robbins, 2005
Bwindi (Mubare)	*G. b. beringei*	547	22.9	12	Ganas and Robbins, 2005
Democratic Republic Congo					
Kahuzi-Biega (Kahuzi)	*G. b. graueri*	1800-3300	23-31	13.5	Yamagiwa et al., 1992, 1994, 1996
Kahuzi-Biega (Itebero)	*G. b. graueri*	2155	23-31	-	Yamagiwa et al., 1992, 1994, 1996
Western Gorilla (*Gorilla gorilla*)					
Gabon					
Lopé Reserve	*G. g. gorilla*	213	-	-	Tutin and Fernandez, 1993
Republic of Congo					
Maya	*G. g. gorilla*	-	-	5.4, 25.4	Bermejo, 1999
Mbeli	*G. g. gorilla*	-	-	2-13	Parnell, 2002
Mondika	*G. g. gorilla*	2014	15.4	10, 13	Doran et al., 2004
Central African Republic					
Bai Hokou (Munye)	*G. g. gorilla*	1527	18.3	8	Cipolletta, 2004

In the lowland forests where widely dispersed fruit-bearing tree species are more prevalent, *G. g. gorilla* spatial movements, and the resulting home range size and use, are different from their eastern counterparts. Current studies indicate that *G. gorilla* has larger home ranges compared to eastern species, but most studies are based on singular group studies and more intra- and intersite comparisons are needed (Bermejo, 2004; Cipolleta, 2004; Doran et al., 2004). In one of the first studies of intergroup encounters in western gorillas, Bermejo (2004) found low home range overlap between groups. Most interactions occurred in the periphery of group home ranges and at particular fruit-bearing tree species (Bermejo, 2004). Findings from Mondika in the Central African Republic indicate *G. g. gorilla* groups can have multiple disconnected core areas located even at far ends of their range and corresponding with concentrated patches of preferred aquatic herbs (Doran-Sheehy et al., 2004). At Bai Hokou, one study group was observed to have a smaller range during the dry season than the wet season, which coincides with a higher herbaceous diet versus a more frugivorous diet (Remis, 1997). Such an effect may be widespread in the Ndoki forests with its high prevalence of small- to medium-sized rivers and associated swamps. These results indicate the availability and importance of bais and swamps to gorillas and lead to interesting questions as to how groups incorporate predictable terrestrial herbaceous or fruiting food patches within their home range. Daily tracking of preferred resources may also contribute to home range use. For the

folivorous mountain gorilla of the Virunga ecosystem, resources are abundantly and evenly distributed, and the median overall travel distance for groups ranges between 570–756 m (Vedder, 1989; Watts, 1991b; McNeilage, 1995). Daily path lengths (DPL) for mountain gorillas have subsequently been found to be significantly shorter than most other gorilla populations; this is likely due to lowland gorilla increased frugivory (Barton et al., 1992; Hemingway & Bynum, 2005). In line with observations from other primate studies, gorillas inhabiting environments with fruit-producing trees have been observed to navigate longer DPL that are subject to seasonal variation (Robbins & McNeilage, 2003; Cipolletta, 2004; Doran-Sheehy et al., 2004; Ganas & Robbins, 2005; Masi et al., 2009). At Bwindi, mountain gorilla DPL ranged between 547 m to >1,034 m (Robbins & McNeilage, 2003; Ganas et al., 2004; Ganas & Robbins, 2005). These values are rather similar to the DPL of Grauer's gorillas ranging from 813–1,057 m (Goodall, 1977; Yamagiwa et al., 1996). A similar study at Mondika found that western lowland gorillas navigate a mean DPL of 2,014 m (Doran-Sheehy et al., 2004). With increased daily travel distances, *G. g. gorilla* were estimated to locomote two to three times farther than eastern conspecifics with half of the variation observed in DPL, which is likely explained by fruit availability and consumption (Doran-Sheehy et al., 2004). At Mondika, low herb abundance along with patchy, dispersed, high-quality fruit-bearing species may force gorillas on long daily searches. As a result, western lowland gorillas can spend nearly double the amount of time traveling compared to mountain gorillas at Karisoke (Masi et al., 2009).

While there are commonalities in ranging behavior shared between groups of gorillas living in similar habitat types, there is additional variation in daily ranging not related to preferred resource availability but to social circumstances (Robbins & McNeilage, 2003). Social stressors, such as the external threat of male-male competition or within-group tension can drive range use up or down (Watts, 1991b, 1994b, 1998b; Robbins & McNeilage, 2003). A clear example was documented at Bwindi when a *G. g. beringei* group was rejoined by a former silverback male. In this situation, the large observed increase in home range was believed to result from the resident dominant male's avoidance strategy during the reintegration period (Robbins & McNeilage, 2003). Instead of confrontation or acceptance of the arrival, the dominant male sought to avoid the rival and maintain as much distance from him as possible. This tactic led to long travel distances, often into areas outside the group's normal home range (Robbins & McNeilage, 2003). A similar phenomenon involving western lowland gorillas was recorded by Cipolleta (2004) who, over a three-year period, saw a silverback male expand his home range after a significant number of females emigrated from his group and then to contract his range a year later back to the previous size before the females emigrated. During this time, the male moved to areas he had never previously visited during the study and obtained a new female for a period of time when in this novel area. Once that female left, the male reduced his range back to the previous size, suggesting that the range expansion was more likely due to mating versus feeding strategies (Cipolleta, 2004). Considering such observations and the complex and interrelated nature of gorilla range use, a long-standing question within the field of gorilla research deals with whether annual home range size is a reflection of the number of individuals within a group, since some research suggests positive trends in group size and mean home range size (McNeilage, 1995; Watts, 1998b; Caillaud et al., 2014). More information is needed to understand how ecology and social dynamics affect home range size and travel patterns in all gorilla species, particularly the western species.

SOCIAL ORGANIZATION

Social organization and group structure vary within primate species and populations and the genus *Gorilla* is no exception (Strier, 1994; Thierry, 2007). What is clear is that females are always associated with at least one mature male and likely other females, while males show slightly greater variation. Such variation is often described via combinations of group type and adult sex ratios, e.g., solitary (one male), breeding group (multi-/uni-male, multi-female), or nonbreeding (typically, all male). When groups form, they may be composed of as few as 2 to as many as 34 independent adults along with dependent offspring (Yamagiwa et al., 1993; Parnell, 2002b; Kalpers et al., 2003). In summarizing available data, Harcourt and Stewart (2007) found a median number of 1 silverback male, 3.25 juveniles/subadults, and 2 adult females per group in western lowland gorillas; whereas in mountain gorillas a median of 1.95 silverback males, 2.5 juveniles/subadults, and 5.2 adult females were observed (N = 43 groups). Originally, mountain gorillas were believed to reach larger group sizes than conspecific western lowland gorillas due to an increased abundance of low-quality food sources and decreased competition (Wrangham, 1979). Yet, in certain localities where herbaceous food resources reach high density and abundance, western gorilla groups can attain similarly sized social units as observed at Rwandan, Ugandan, and DRC research sites (Bermejo, 1999; Magliocca et al., 1999; Kalpers et al., 2003). Based on long-term monitoring of eastern populations and several sites in the western gorilla range, Harcourt and Stewart (2007) came to the consensus that there are no significant differences in group sizes between species.

Male Sociality

The various social groups observed in gorillas are often composed of the same type of base unit: a mature (i.e., silverback) male, at least one parous adult female, and dependent offspring. Groups may also include male offspring at various stages of maturity that tend to be less socially integrated within their natal groups as they age. At lower elevations and in the DRC, multi-male social units are also observed in *G. b. graueri* but are far less prevalent than in mountain gorilla groups (Yamagiwa et al., 1993; Yamagiwa & Kahekwa, 2001). Based on the common occurrence of more than one mature adult male in *G. b. beringei* groups, it was speculated that *G. g. gorilla* behavioral strategies could follow a similar multi-male social unit pattern. Field observations have not supported this assumption, as single male groups are the norm and multi-male social units have rarely been observed in *G. g. gorilla* populations (Magliocca et al., 1999; Parnell, 2002b; Robbins et al., 2004).

Silverback or blackback males will sometimes attempt to evict maturing and potential male rivals but such aggression is not always present (Watts, 1996; Robbins & Robbins, 2005; Stoinski et al., 2009a). Sex ratio, dominance rank, and opportunities to copulate likely play an important role in male natal dispersal decisions (Robbins, 1995; Watts, 2000; cf. Stoinski et al., 2009b). As well, some evidence indicates that maternal support may influence whether a male disperses or remains philopatric (Robbins et al., 2016). Little is known about exactly when and why newly adult males decide to transfer, where they decide to emigrate to, or whether they remain in their natal group. Robbins (1996) and subsequent research (Stoinski et al., 2009c) on mountain gorillas, found that nearly half of all silverback males did not emigrate from their natal groups. In many cases, the young males are presumably the offspring of the silverback and the resulting dispersal events were often considered a form of inbreeding avoidance. Nevertheless, exactly why males decide to either remain or disperse is intriguing and complex; research in this area has important implications for predictive reproductive success. Maturing males who remain philopatric are effectively opting to compete for mating opportunities in their natal group with, likely, their fathers (Harcourt & Stewart, 1981). In philopatric cases, natal males forgo an opportunity to develop a new group and instead will remain and stand to inherit the group's resident females, most of which are not their kin. A male that remains in his natal group also retains a home range for which he has detailed ecological knowledge regarding topography and location of food resources, presumably a benefit to foraging efficiency.

The reproductive opportunities for philopatric *G. b. beringei* males may be higher still, since groups with multiple males not only attract more females but are more successful in maintaining stable long-term group membership (Robbins, 1995). Extending the tenure length of a dominant male has the potential to increase his reproductive output, as multi-male groups experience far less infant mortality than uni-male groups (Robbins et al., 2007). While infant mortality is likely when lead males in uni-male groups die, such infant mortality events are less common in multi-male groups that suffer such a loss (Watts, 1989; Robbins & Robbins, 2005; Harcourt & Stewart, 2007; Robbins et al., 2013). Coalitionary support in intergroup interactions and protection of offspring may outweigh the potential reproductive costs that dominant males experience when subordinate counterparts remain within the group (Sicotte, 1993; Robbins, 1995; Robbins, 2001). The existence of the multi-male system may be an indication that in many cases, a single male is not able to adequately protect his reproductive resources. However, even though there are many potential benefits to male philopatry, almost all males emigrate, with only males that obtain dominance in their natal group remaining philopatric since males that remain in their natal group have a jump-start on breeding opportunities

compared to those males who disperse to create their own group (Robbins, 1995; Robbins et al., 2019). It is also not guaranteed that every mature male leaving the natal group will become a harem holder (Watts, 2000; Gatti et al., 2004; Robbins & Robbins, 2005). In these cases, males may spend their adult lives in a semi-solitary lifestyle with few breeding opportunities (Sicotte, 2001).

While bisexual groups are the norm, nonreproductive groups also occur and are typically composed of a combination of immature individuals as well as adult males (Yamagiwa, 1987; Robbins, 1995, 1996; Gatti et al., 2004). The formation of a nonreproductive group can be predicted from several social and ecological factors; however, most typically, resident females decide to emigrate from a group leaving only juveniles and the dominant male behind (Robbins et al., 2004). Otherwise, "bachelor groups" originate through the social grouping of individuals from various natal groups that are likely in a transitional phase (i.e., blackback males; Robbins, 1996; Yamagiwa, 1987). Within these bachelor groups, maturing or past-prime males presumably gain the benefits of protection from predation or risky intergroup interactions. A small percentage of mature males live a solitary existence. In *G. b. beringei*, approximately 1.8% of the population is comprised of lone silverback males; this can be compared to *G. b. graueri* with 3.5% and *G. g. gorilla* where the population is composed of approximately 5.4% solitary males (Gatti et al., 2004). At Mbeli Bai in the northern Republic of Congo, a total of 26 solitary males between 1995–2014 were identified (Greenway, 2015). The increased number of *G. g. gorilla* solitary males may reflect a slightly different male mating strategy and/or grouping dynamic (Parnell, 2002b). The occurrence of multi-male groups has long been established in *G. b. beringei* (Schaller, 1963; Weber & Vedder, 1983). It is this harem-type grouping pattern along with associated male reproductive behaviors that contribute to what amounts to a multi-male group (Watts, 1990a; Sicotte, 1994; Robbins, 1999). However, remaining in nonreproductive groups is not a viable long-term option for young males; therefore, most disperse from nonreproductive groups upon reaching maturity.

The prevalence of uni-male groups in *G. g. gorilla* can likely be explained by increased rates of maturing male emigration and female preferences for immigrating to smaller groups (Stokes et al., 2003; Robbins et al., 2004). Most often *G. g. gorilla* males choose a solitary status rather than remaining in their natal group or joining a formed group. Young males become dominant males in uni-male groups by competing for access to females of other established groups or by attracting young females who are dispersing from their natal group. This transition from a nonreproductive to a reproductive group can be a long and potentially risky process. To attract females, solitary males must confront and/or shadow reproductive groups with defensive silverback males. Even if males successfully acquire females, a proportion of these adult males will not sire offspring (Breuer et al., 2008). While the prevalence of multi-male groups is vastly different between *G. gorilla* and *G. beringei*, the overall generalized process seems to be that the majority of male gorillas will leave their natal group (Yamagiwa et al., 1993; Yamagiwa & Kahekwa, 2001; Stokes et al., 2003; Robbins et al., 2004, 2019).

More information on social and ecological factors that predict male emigration are needed to understand the formation and prevalence of multi-male groups and why they seem to be subspecies specific. The existence of differing male mating strategies suggests intense male-male competition within the genus *Gorilla*. In addition, the low proportion of males that acquire breeding females suggests that the potential for considerable variation in reproductive success is high. In accordance with sexual selection theory, the extreme dimorphism of adult male gorillas being twice the size of adult females and exhibiting secondary sexual traits supports this assertion (Smith & Jungers, 1997). Achieving reproductive success can be accomplished in a variety of ways by male gorillas. Studies of male reproductive success at Mbeli found that males leading larger social units had lower offspring mortality than the smaller groups (Breuer et al., 2010). Reproductive skew or bias toward mating success does occur in mountain gorillas and by attaining dominance or high rank, males in multi-male groups improve their reproductive potential, as rank is a strong predictor of copulations (Stoinski et al., 2009b).

Few investigations of mating conflict and female choice in gorillas have successfully linked observations to differences in male competitive abilities. The studies of western lowland gorillas at Mbeli and Lokoue indicate the variation observed in group size, and reproductive success of males is dictated by male quality as assessed by phenotypic traits such as body length, sagittal crest size, and gluteal muscles (Caillaud et al., 2008; Breuer et al., 2012). Meaningful physical differences between adult male gorillas may allow them to physically out-compete male rivals in contests over females. These findings also support the assertion that *G. g. gorilla* males face elevated levels of intense male-male competition. Such aggression may play a role in influencing the turnover of groups within a population. When differences in observation hours are accounted for, there are higher levels of group disintegration in *G. g. gorilla* than in mountain gorillas (Robbins, 1995; Kalpers et al., 2003; Robbins et al., 2004).

Female Sociality

Gorilla females spend their entire lives in heterogeneous and cohesive groups. Previously, female social relationships and their role in shaping primate society had been overlooked, and female gorilla social dynamics were no exception. However, recently more studies have begun to systematically explore female social dynamics in different environmental settings. Most investigations within the gorilla genus base the cohesiveness of groups on the strong relationships formed between females and the male(s). In mountain gorilla groups, females are often within close proximity to other females, but this is in part an artifact of their tendency to stay spatially close to the dominant male or forage in the same resource patch, rather than seeking one another's company (Harcourt 1979a,b; Watts 1992, 1994c; Harcourt & Stewart, 2007).

Robbins et al. (2007) suggest *G. b. beringei* female dominance relations to be best described as "egalitarian." Subtle dominance hierarchies between resident females have been detected in groups at Karisoke and Bwindi (Harcourt, 1979a; Watts, 1994c; Robbins et al., 2005, 2009). Mountain gorilla female dominance dyads are described as relatively fluid with bi-directionality in the exchange of supplants and displacements between female individuals (Watts, 1994a). Based on 30 years of observational data, it is now evident that female mountain gorillas do form long-term, stable dominance hierarchies but nepotism does not play an influential role in dominance dyads (Robbins et al., 2005). These observations lead to interesting questions as to what extent within-group female-female dynamics in western lowland gorilla differ from those of mountain gorillas.

Considering the vastly dissimilar distribution, diversity, and abundances of food items in the context of lowland forests compared to higher altitude gorilla habitats, there may be important variations in competitive strategies and female social relations. As has already been noted, group spread in western lowland gorillas and range use differs from eastern lowland gorillas (Doran et al., 2002; Masi et al., 2009). There is just one published investigation on female-female competition and relationships in western lowland gorillas, which demonstrated stronger linear dominance hierarchies in western lowland gorillas compared to mountain gorillas (Lodwick, 2015). There has not been a documented case of females remaining in their natal group in *G. g. gorilla*. However, many studies have demonstrated that females may reside in groups with closely related kin, which at some point may have belonged to their natal group, at a frequency greater than that predicted by chance (Bradley et al., 2007; Arandjelovic et al., 2014; Hagemann et al., 2018). Along with the variability of the relationships and associated costs and benefits of social grouping, group size dynamics also play an important role. Increases in social group size may also invite elevated levels of within-group feeding competition (Janson & Goldsmith, 1995).

The ecological costs of group living for female gorillas are likely minimal and those residing in larger groups do not experience decreased levels of reproductive success (Watts, 1984; Doran et al., 2002; Rogers et al., 2004; Robbins et al., 2006, 2007). Social status may play a moderate role as female gorillas in multi-male groups with higher rank had elevated reproductive success when considering measures such as interbirth interval, success rearing offspring, and surviving birth rate per mother, but the relationships are not well supported (Robbins et al., 2007). Presumably, females in these larger groups spend a greater amount of time in closer proximity to other females while foraging on the same resource patches. However, transfers of female mountain gorillas at the Karisoke site indicate intergroup movements are not directional and female transfer decisions are independent of group sizes (Watts, 1990a).

Dispersal decisions by female mountain gorillas are influenced by the number of adult males within the group. These females also show a preference for multi-male groups as opposed to single-male groups or solitary individuals (Watts, 2000; Robbins et al., 2009). Analyzing over 40 years of data on female relations and reproduction in the Virunga population, however, did not find any significant fitness benefits for females in selecting multi-male over single-male units (Robbins et al., 2013). In contrast, G. g. gorilla females do tend to transfer to smaller-sized groups with single males based on observations at one forest clearing (Stokes et al., 2003). This is hypothesized to be related to female mate choice, selecting males younger in their silverback tenure, as they may be physically more equipped to protect periphery individuals from risks such as predation, infanticide, or other groups encroaching on resources. However, additional intersite studies are needed (Breuer et al., 2010).

Young, maturing females leave their natal group to join another group or a solitary male (Harcourt et al., 1976; Harcourt, 1978). Female transfer events occur while groups directly interact or during spatial overlap periods (Sicotte, 2001; Robbins & Sawyer, 2007). During these interactions, forced transfer may take place in which a neighboring male enters the group and seizes a female (Michael Stucker, pers.comm.). However, female mate choice is believed to be an important factor in most dispersal events (Stokes et al., 2003; Harcourt & Stewart, 2007; Robbins et al., 2007). In mountain gorillas, initial transfers occur around eight years of age (Harcourt et al., 1976; Harcourt & Stewart, 1981). A female may transfer

several times throughout her lifetime, and in general, female gorillas from all subspecies are assumed to emigrate at least twice (Harcourt, 1978; Stokes et al., 2003; Harcourt & Stewart, 2007). These secondary dispersal events suggest not all transfers are an inbreeding avoidance strategy in gorillas. Determining if and how G. g. gorilla behave in comparison to G. b. beringei conspecifics in patterns of residency and dispersal will require further observations from the field. Similarly, more investigations are needed on female mate choice in gorillas and how this differs among the species, as differences in female mate choice could potentially explain differences in group dynamics and gorilla society (Stokes et al., 2003; Breuer et al., 2012).

Male-Female Relationships

While gorilla groups are cohesive, social interactions among adult group members are infrequent and can be subtle. Although the bonds between the dominant male and reproducing females are stable and can be long-lasting, there is little overt social interaction among individuals. Grooming is only rarely observed between G. b. beringei adults (Schaller, 1963; Harcourt, 1979b). Apart from mothers grooming their infants, this social behavior seems to be absent in G. g. gorilla (Masi et al., 2009). Although mountain gorillas spend less time traveling and more time resting than western lowland gorillas, such differences in activity budgets do not, alone, explain these subspecific differences in affiliative tendencies (Masi et al., 2009).

Younger females without relatives and less stable relationships tend to engage in affinitive behaviors with the silverback male regularly (Harcourt & Stewart, 1987; Watts, 1994c), although aggressive interactions from the dominant male to resident females are common (Harcourt, 1979b). Males will intervene to end quarrels between female group members. These interactions may be intense but, generally, do not entail the kind of severe wounds observed in extra-group aggressive contacts (Harcourt, 1979b). Aggression also occurs when males herd females during group movements or in the context of a male asserting his dominance over an immigrant female (Parnell, 2002b). The integration of a female into a new group can be intense with repeated confrontations with the dominant male, as well as resident females, during the socialization process. Silverback males repeatedly assert their authority over group members, but there is wide variation among individual males in their dominance styles.

The "male protection model" posits that infanticide is a primary force underlying sexually mature females seeking the protective services of males through permanent association (Dunbar, 1988). The model places the threat of infanticide as a primary cause of observed associations and heterogeneous group formation. The killing of infants by non-group males is reasoned to accrue a near-term reproductive benefit through increased mating opportunities by way of the female's resumption of cycling (Hrdy, 1974). The lethal male harassment observed in *G. b. beringei* has been argued to support this view of sexual competition. While infant gorillas face many life-threatening risks, the main cause of infant mortality before the age of weaning is the targeted killing of infants by rival males (Watts, 1990b; Robbins et al., 2007). Long-term monitoring of infant mortality of the Virunga mountain gorillas indicates infanticide accounted for over 20% of infant deaths (Robbins et al., 2013). As an alternative or counterstrategy to this loss of reproductive effort, *G. b. beringei* females are thought to choose groups with multiple protector males rather than single-male units to decrease the risk of infanticide. The first studies of infant mortality and survival in *G. b. beringei* indicated infants are less likely to be victims of infanticide with two male protectors as opposed to one. Understanding of infanticide in this population is evolving, and while ongoing monitoring supports previous studies indicating infants in single-male groups do suffer overall higher rates of infanticide (Robbins, 1995), more detailed analysis also finds they are not significantly different from multi-male group rates (Robbins et al., 2013).

Social Strategies

Contrary to previous interpretations, there are no detectable fitness benefits to females in maintaining membership in multi-male social units (Robbins et al., 2013). In light of these findings and empirical tests, the "male strength model" suggests there may be real disparities in individual male mountain gorilla capabilities as protectors (Pradhan & van Schaik, 2008; Robbins et al., 2013). In some cases, single-male social units may have higher offspring survival as a result of a lower threat of infanticide than local multi-male groups because of a formidable dominant male protector (Robbins et al., 2013), and other evidence supports the assertion that variability in male strength is important. According to the sexual selection hypothesis, an infanticidal event alters the behavior of the dependent victims' mother, such that the infanticidal male will likely sire her next offspring (van Schaik & Janson 2000). Female mountain gorillas that have just lost a dependent to an outsider male were found to immigrate into the perpetrator male's group, which Watts (1989) concluded is evidence of the female's need for a capable protector male. Temporary forays by western lowland females with solitary males at Mbeli Bai

are also argued to support the "protection/infanticide" model (Parnell, 2002a). While infanticide has been deduced as present at Mbeli Bai, some observed cases of females transferring with dependent offspring did not lead to the disappearance or death of offspring, suggesting that infanticide is not universal (Stokes et al., 2003). More recent evidence from the nearby Ngaga study site as well implies resident silverbacks, in some instances, are tolerant of unrelated immature offspring (Forcina et al., 2019). These observations suggest more research into the prevalence of infanticide in western lowland gorillas is necessary before drawing firm conclusions on infanticidal behavior in gorillas.

A second mainstream hypothesis as to why some primate species, including gorillas, evolved male-female or polyandrous associations, argues that the potential risk of predation is an underlying cause of group association (van Schaik & van Hooff, 1983). Female gorillas seek permanent associations with mature adult male(s) capable of fending off potential attacks as a counterstrategy to the risk of predation. The fact that dispersing female gorillas without infants make direct transfers from either their natal group or from one harem to the next implies that the risk of predation could be substantial. They spend close to no time on their own or just with other females during transfers. Residing within a group with multiple members affords the added benefit of a cooperative warning system that vigilant group members provide (Arnold & Zuberbuhler, 2006). Considering why group formation occurs in taxa such as *Gorilla* requires disentangling the influence of infanticide from predation, which is not easily accomplished. However, van Schaik and Kappeler (1997), in a phylogenetic test of primates, concluded that the infanticide prevention hypothesis drew more support than predation. Part of the issue is that eyewitness observations of predation attempts on primates are rare since much predation occurs at night. While direct evidence of predation exists, it is often available through study of the predators (e.g., fecal analyses, den remains) rather than the prey, which leaves an understanding of this common risk incomplete (Hart & Sussman, 2009). The large size of gorillas (particularly adult males) has historically been suggested to protect these apes from predation by carnivores, such as leopards (*Panthera pardus*) and in rare cases, lions (*P. leo*) (Fay et al., 1995).

The potential for life-threatening leopard attacks exists since gorillas spend a substantial amount of time traveling and foraging on the ground, and the effectiveness of vigilance in detecting a predator depends on many factors. Large group spread in western lowland gorillas results in individuals spending considerable amounts of time spatially distant from others and potentially out of visual or auditory contact (Masi et al., 2009). In addition, western lowland gorillas are likely to spatially overlap extensively with leopards as a consequence of being attracted to many of the same food items that the felids' main prey base (ungulates and diurnal primates) feed upon (Henschel et al., 2005).

While predation attempts on gorillas are overall likely low, the intensity of potential predation varies among populations. As local extinctions of top carnivores like lions and leopards become more commonplace throughout sub-Saharan Africa, the potential threat of predation changes. Gorillas inhabiting the Virunga Volcano region are currently not at risk of leopard predation as this species is now absent in the ecosystem (Robbins et al., 2013). The Congolian forest-savanna mosaic in the Odzala region of northern Republic of Congo was one of the few areas where lions and gorillas historically shared overlap in Central Africa, but the populations of lions that once roamed the mosaic of transition forests are also considered locally extinct (Henschel et al., 2014). Leopard density and the abundance of their main prey base also influence the degree of predation risk to gorillas in different areas. Populations of apes inhabiting forests where human hunting of the ungulate populations is intense may end up facing higher levels of threat as leopards are forced to shift their foraging to feed on other less-preferred resources such as apes. A detailed study on leopard predation in Lopé, Gabon, found gorillas are indeed a prey item (Henschel et al., 2005), which supports anecdotal observations in the region (Fay et al., 1995).

A further advantage of social grouping as a means of protecting against predation is increased reproductive output. The "predation-protection" model predicts that large groups will have improved infant survivorship because of enhanced anti-predation efforts (Dunbar, 1988). As a result, the selection of larger social units by females for protection purposes could ensue (Hamilton, 1971). However, studies of mountain gorillas thus far indicate predation may not be a primary driver in gorilla group formation or variability in group size (Watts, 1990b, 1996). Similarly, observations of western lowland gorillas also do not support the model, as females select for smaller groups when dispersing rather than larger units (Stokes et al., 2003). Though evidence from G. g. gorilla at Mbeli Bai in the Nouabalé-Ndoki National Park, Republic of Congo, provides some support of this prediction, as harem-holding males with larger social units have lower offspring mortality (Breuer et al., 2010).

In addition to benefiting from a protector, male night nest building may be another tactic employed by some primates, such as gorillas, to avoid predation. In support of this model, all independent individuals construct nests,

which implies that both sexes are subject to predation risks while sleeping. However, the importance of nesting behavior as a gorilla survival strategy is not well understood. Rarely do gorillas nest high in the tree canopy where the risk of predation attempts is likely lower. More often, gorillas nest on or near the ground. At Mondika, 79% of gorilla nests were found located on the ground (Mehlman & Doran, 2002), presumably vulnerable to leopard attacks during the night despite the presence of a protector male. Observations of a suspected predation on a juvenile gorilla at a nest site (Patrice Mongo, pers .comm.) and the fact that gorillas forage before dawn and after sunset imply that the conditions under which events take place and the risks incurred as a result of predation may be greater than previously supposed. Studies on the spatial distribution and group nest cohesion dynamics of western lowland gorillas in relation to leopard activity will allow researchers to more fully understand the overall impact of the risk of predation on gorillas.

Gorilla Mating System

Gorillas are long-lived and over the course of a female's life she will invest significantly in her offsprings' development (Nowell & Fletcher, 2007). Gorillas are not seasonal breeders and the likelihood of conception depends in part on the female's overall health condition. Because there is great variation in the quality and abundance of food resources, and thus energy available between populations and species, the reproductive ecology (cycling, conception, and interbirth intervals) can be expected to vary between groups. Sexual activity in *G. b. beringei* females starts before an individual is fully mature, with labial swelling occurring at 7–7.5 years of age. In contrast, female *G. g. gorilla* reach this stage later in life, between 9.6 years old and 10.3 years old (Breuer et al., 2009). Over the course of a two- to three-day estrous period (Harcourt et al., 1980), receptive females show labial swelling.

Typically, receptive adult females approach the dominant male and solicit copulations with him (Watts, 1990b). Silverbacks are vigilant during this period and, on occasion, actively block any potential rival's attempts at mating with the parous female (Harcourt & Stewart, 2007). Within the multi-male harem system documented among mountain gorillas, such mate-guarding tactics may require a considerable amount of time and effort. Maintaining close proximity or vocal communication with cycling females are other tactics often employed by dominant males (Sicotte, 1994). Given the complexity of the forested environment and large group sizes with multiple males, dominant males may not be able to monopolize all mating opportunities within their group. Dominant males do, however, allow younger subordinate males to mate with adolescent and young nulliparous females within the group (Harcourt et al., 1980; Watts, 1992; Robbins, 1995). Results thus far indicate extragroup copulations are uncommon (Sicotte, 2001).

Gorilla gestation ranges from 237–270 days, with a median of 252 days (Harcourt et al., 1981). The resumption of sexual cycling in adult females depends on the duration of the infant-weaning process. Length of interbirth intervals in the genus are variable (Robbins et al., 2004), and most likely correlated with available resources. Immature mountain gorillas stop suckling sooner than western lowland gorillas and females have shorter interbirth intervals (approximately 3.5 years versus 4.9, respectively) (Breuer et al., 2009). Mountain gorilla interbirth intervals are relatively short compared to all other nonhuman great apes (Wich et al., 2004). The shorter interbirth intervals documented in mountain gorillas are likely the result of groups inhabiting environments with a greater abundance of food resources than conspecifics at lower elevations. Importantly, these findings indicate that western lowland gorilla females have a reduced reproductive output compared to mountain gorillas. Breeding opportunities of dominant western lowland males during their tenure are also likely lower than those documented in mountain gorillas. These new insights imply there may be biologically meaningful differences between the two species of gorilla.

The observations that western lowland gorillas differ from mountain gorillas in stages of development and maturation across all age-sex classes support findings on patterns of primate development. Growth patterns of folivorous primates are accelerated compared to non-folivores, such as ripe-fruit specialists (Leigh, 1994). Male mountain gorillas are physically mature between 12–16 years of age (Watts, 1990b; Robbins, 2007). Breuer and colleagues (2009) found that maturity in male western gorillas does not occur until 18 years of age. Based on these observations, mountain gorillas have a faster life-history trajectory than that observed in western lowland gorillas (Breuer et al., 2009). When considering the slow maturation of male western lowland gorillas, Breuer and colleagues (2009) reasoned that the interplay between social structure and demographics of this subspecies effectively reduces the chance of forming multiple male groups that are so typical among the mountain gorillas; the relatively delayed maturation and short tenure time result in a deficit of mature males to form multi-male groups. Gorilla life-history parameters have, until recently, largely been based on long-term data sets restricted to mountain gorilla groups at Karisoke. As findings from the lowland forests of Equatorial Africa continue to emerge, our understanding of how different ecologies impact important milestones in gorilla life-history trajectories and population fluctuations will certainly continue to expand.

Intergroup Dynamics

Gorilla groups and solitary individuals probably have a long history of interacting. Familiarity is one aspect that may factor into the quality of interactions between these apes. Recent findings at the Ngaga study site indicate neighboring western gorilla groups in a forested environment interact nonaggressively at times (Forcina et al., 2019). Male relatedness may also play an important role in intergroup relations. Genetic profiles of neighboring gorilla groups have shown that kin relationships between non-group males may influence the quality of such group interactions (Bradley et al., 2005), though relatedness among adult males residing in a region is not always a prerequisite for elevated levels of social tolerance (Forcina et al., 2019). There are large variations in data among research sites concerning whether males preferentially stay close to kin (Bradley et al., 2004; Douadi et al., 2007; Inoue et al., 2013), and whether there are sex-biased dispersal distances (Douadi et al., 2007; Fünfstück et al., 2014), but a lack of comparable methodology prevents conclusive results. Encounters between established groups and nonbreeding groups or solitary individuals can vary from tolerant to highly agonistic (Parnell, 2002a; Sicotte, 1993; Yamagiwa, 1986, 1987). Threat displays involve chest-beating, running, and branch breaking, even dynamic splash displays (Parnell & Buchanan-Smith, 2001), and physical contact may occur with fatal consequences (Watts, 1989). Males bear scars of previous encounters and wounds often take weeks to heal (Harcourt & Stewart, 2007). The intensity of aggressive interactions at Karisoke depends in part on the number of potential migrant females within interacting groups (Sicotte, 1993). Groups with greater numbers of potential dispersing females had more intense interactions than groups with fewer such females. Interactions between groups at Bwindi occurred at an average rate of 0.78 per month (Robbins & Sawyer, 2007), while Sicotte (2001) found Virunga mountain gorilla groups averaged slightly more contacts with one encounter per month.

While the aggressive intergroup interactions observed in mountain gorillas are less frequent in western lowland gorillas, this discrepancy could be in part due to differences in sample sizes. There are fewer habituated groups of western lowland gorillas, and we may not yet have an accurate portrayal of intergroup encounters among these apes, particularly in the context of the forested environment (e.g., Southern and colleagues [2021] recently reported on two observations of protracted intergroup aggression between large parties of chimpanzees [n = 27] and two much smaller, and outnumbered, gorilla groups [n = 5, and n = 7]; each encounter resulted in chimpanzees killing a gorilla infant). Based on long-term monitoring at Mbeli Bai clearing, male life-history patterns and tenure length imply considerable levels of male competition in this species (Breuer, 2008). Severe intergroup aggression can result in a dominant male's demise, which leads to the disintegration of his group if a male offspring does not immediately inherit the group. Rates of group disintegration in western lowland gorillas are higher than those of mountain gorillas (Robbins et al., 2004). When the breakup of a group occurs, long-term fitness consequences, such as survival prospects of offspring, decrease (Crockett, 2000; Steenbeek & van Schaik, 2001), and ultimately the dominant male's reproductive success suffers (Robbins & Robbins, 2005; Greenway, 2015).

However, the type of intergroup encounters in western lowland gorillas can vary dramatically, from agonistic conflict to affiliation. Over half of the group encounters observed at Mbeli Bai resulted in one group seemingly ignoring the other's advances (Parnell, 2002a,b). Similar results were reported from another bai in central Republic of Congo, where researchers indicated that 62% of contacts between groups resulted in indifference (Magliocca & Gautier-Hion, 2004). Some of these differences may be in part due to the identity of the inter-actors. Long-term monitoring at Mbeli Bai and Lokoue Bai has demonstrated that some between-group interactions are based on the identity of the groups, and a variation in tolerance and proximity models that of multilevel societies (Morrison et al., 2019). Affiliative interactions between immature nongroup members frequently occurred in the bai setting (Parnell, 2002a), indicating that in some environmental settings, opportunities for less tense social engagement may occur.

GORILLA CONSERVATION

One factor limiting conservation efforts is the fact that population estimates of gorillas throughout most of the genus range were not available until recently, and those that were available were out of date. Recent surveys have provided drastic changes in our perception of gorilla population numbers. For the first time in more than two decades, *G. b. beringei* are no longer considered Critically Endangered, and have been reclassified as Endangered on the International Union for the Conservation of Nature Red List (IUCN) (Hickey et al., 2019). Based on the most recent long-term systematic survey throughout the entire Virunga range, there are at minimum 604 individuals (Hickey et al., 2019), more than double the estimates from the early 1980s (Gray et al., 2013). Increases, however, were not equally shared between social groups

in the Virunga Mountains due to disparities in human intervention among groups. During the 2010 survey of the entire Virunga gorilla population, annual increases in growth rate occurred in habituated gorilla groups, while annual declines were documented from non-habituated social units over the same period (Robbins et al., 2011b). In the most recent survey, it appears that the growth rate for non-habituated gorillas is still decreasing (Hickey et al., 2019). Occupying an area slightly smaller in geographic extent, the *G. b. beringei* population at Bwindi is roughly 430 individuals (Roy et al., 2014), up from earlier estimates of 320 individuals (McNeilage et al., 2006).

G. b. graueri has an overall much larger geographic range than the Virunga and Bwindi mountain gorilla populations, but it also occupies a region rife with sporadic instability and human occupation of remote forests, plus intense exploitation of natural resources, and, as a consequence, the population appears to be facing rapid decline. Most recent estimates suggest there are 3,800 individuals remaining, a 77% decline in one generation, with some populations decreasing 81.7%–100% (Plumptre et al., 2019). *G. b. graueri* has been upgraded from Endangered to Critically Endangered by IUCN due to this drastic decrease in population size (Maisels et al., 2018).

Both subspecies of *G. gorilla* are listed as Critically Endangered by IUCN (Maisels et al., 2018). Studies also suggested the population of *G. g. gorilla* to be considerably higher than other taxa with estimates of 302,973–460,093 individuals (Strindberg et al., 2018). Yet, while the *G. g. gorilla* population is in decline (Strindberg et al., 2018), their genetic diversity is still well intact, having higher genetic diversity compared to all other subspecies of gorillas (Fünfstück & Vigilant, 2015). The geographically isolated and smaller range of the Cross River gorilla (*G. g. diehli*) supports around 200–250 individuals according to Oates and colleagues (2003). Recent efforts to verify and analyze new and old ape survey data from the region created a predicted density map for *Gorilla gorilla* (IUCN, 2014). Results produced 18 priority conservation landscapes for the western lowland gorilla along with associated action points believed required for the species' long-term protection (IUCN, 2014).

Major Threats to Gorilla Populations

Among the threats to gorilla survival are health risks such as *Zaïre ebolavirus* (EBOV), a hemorrhagic fever virus, as well as a variety of other respiratory diseases. Disease outbreaks in Central Africa over the last 10 years are believed to have significantly reduced western lowland gorilla populations (Huijbregts et al., 2003; Walsh et al., 2003; Rouquet et al., 2005; Bermejo et al., 2006).

While the extent of gorilla population declines remains largely unknown, gorilla-density estimates at Ebola outbreak sites in Gabon and Republic of Congo indicate that this virus has the potential to cause local ape extinction in some areas (Strindberg et al., 2018). Ebola virus has become a major concern for the future viability of African ape populations and was a driving force behind the reclassification of the *G. g. gorilla* by IUCN. The repeated emergence of Ebola outbreaks in gorillas and the potential rapid spread of the virus to healthy gorilla populations are thus of great conservation concern. Transmission dynamics of the Ebola virus in relation to gorilla ecology and sociality remain largely unknown, though several bat species are considered putative reservoirs, including the hammer-headed fruit bat (*Hypsignathus monstrosus*), Franquet's epauletted fruit bat (*Epomops franqueti*), and little collared fruit bat (*Myonycteris torquata*) (Leroy et al., 2005). Recent studies demonstrate that an Ebola outbreak affected the social dynamics of gorilla groups and reduced group cohesion; however, the rapid recovery of certain social dynamics after the event demonstrates some resilience to immediate environmental change (Genton et al., 2015). Identifying ecological or infrastructural drivers (i.e., access via roads) underpinning Ebola virus emergence and transmission dynamics in apes is critical to creating better predictive models to guide wildlife management, develop potential protective measures for wildlife, and reduce transmission to humans. Modeling disease spread among gorillas across the Odzala region in the central Republic of Congo indicated roads were associated with attenuation of modeled virus spread (Cameron et al., 2016).

There are also disease risks inherent in observing or working in areas with wild gorillas. Several documented cases of human viruses and bacteria being transmitted by ape researchers and tourists have occurred (Homsy, 1999; Goldberg et al., 2007). Findings of human respiratory syncytial virus (HRSV) infection both in local workers and western lowland gorillas at a research and tourism site in Dzanga Sangha Protected Areas in the Central African Republic underscore the risk of interspecies disease exchanges (Grutzmacher et al., 2016). The HRSV pathogen causes acute lower respiratory disease (ALRI) in young human children and is a leading cause of child mortality worldwide (Nair et al., 2010). Such transmission events, even of common human pathogens, can prove fatal to apes who have yet to develop antibodies to many common human diseases. Thus, monitoring of ape health at gorilla research and tourism sites has expanded, and many informative collaborations among field researchers, veterinarians, and health-care professionals have developed to address such concerns and to develop evidence-based measures to decrease risks. For example, the Mountain Gorilla Veterinarian Project (MGVP) in partnership with the Dian Fossey Gorilla Fund, has developed a detailed and standardized health-monitoring system for gorilla groups that are observed by researchers and tourists (Cranfield, 2004).

The most significant threats to gorillas are linked to the destruction of habitat and illegal poaching (Fa et al., 2005; Mehlman, 2008; Nellemann et al., 2010). Regarding habitat destruction, alteration of gorilla habitats occurs in a variety of ways with differing degrees of severity, which has resulting implications for gorilla survival prospects. Permanent deforestation (such as agriculture parcels, clearance for livestock grazing, and human settlements) eliminates these areas for gorilla use. Such anthropogenic disturbance has had a severe impact on both eastern and mountain gorillas, leaving populations fragmented and isolated in many areas as humans continue to intensively compete for land occupied by gorillas (Mehlman, 2008; Nellemann et al., 2010). In the 1970s, nearly one-third of 10,000 ha in the Rwandan Volcanoes National Park was converted to crop cultivation (Nellemann et al., 2010). Habitat loss in and around the park continues, particularly as the demand for a variety of natural resources has significantly increased. Grauer's gorilla habitat loss has been drastic, with the Kahuzi-Biega National Park–Reserve des Gorilles de Punia region, DRC, decreasing from 15,870 km^2 in 1959, when initial surveys were conducted by Elmen and Schaller (1960), to 9,005 km^2 currently (Hall et al., 1998; Plumptre et al., 2016). Further, increased levels of mineral extraction are predicted in the lowlands of the eastern DRC and Albertine Rift Mountains and in regions outside of protected areas (Edwards et al., 2014). These regions are noted for corruption, political instability, and weak governance, which further promote activities such as mining and illegal timber and charcoal production, and compromise gorilla habitat and survival prospects (Nellemann et al., 2010).

The isolation, fragmentation, and loss of habitat documented in the eastern DRC and Rwanda are rare in Central Africa, with the exception of the Cross River gorilla population. The geographical range covering suitable western lowland gorilla habitats is estimated at over 700,000 km^2, of which 22% falls within the boundaries of formally protected areas (Maisels et al., 2018). Thus, there is a considerable need to devise strategies to protect gorillas existing outside of parks and in multiuse forest zones. Most western lowland gorilla populations are impacted by selective timber exploitation or other forms of natural resource extraction. Large-scale industrial logging is widespread in the region, and the forestry sectors in Cameroon, Gabon, and the Republic of Congo generate significant revenue (Perez et al., 2005). Forestry operations also increase development in remote forested regions as well as foster the emergence of once nonexistent markets. Although rates of deforestation range from 0.02–1.3% of forest loss per year in the range states of gorillas, it is predicted (but perhaps underestimated) that total forest cover in Central Africa will decline by more than 30% in the next 50 years (Nellemann et al., 2010). Given the expansion of resource exploitation in the region, it is becoming exceedingly apparent that few intact forest blocks will remain in the Congo Basin outside the protected area networks. Between 2000–2013, selective logging led to the loss of 77% of identified Intact Forest Landscape (IFL) in the Congo Basin (Potapov et al., 2017).

Gorillas are typically found in mixed-species forest, which may result in direct competition and conflict with timber-extraction operations, as well as with local inhabitants (Tutin & Fernandez, 1984). Although mechanized logging has received considerable attention and is often blamed as the primary cause of faunal decline in tropical forests, knowledge of the impact of logging on complex species and forest dynamics is still poorly understood (Skorupa, 1988; Putz et al., 2001). What *is* apparent is that species richness of invertebrates, amphibians, and mammals decreases as timber outtake rates increase (Burivalova et al., 2014). Several studies imply that gorillas show initial declines within logged habitats but may recover to more normative levels as time passes (Tutin & Fernandez, 1984; Clark et al., 2009; Stokes et al., 2010). However, caution needs to be exercised when making any firm conclusions on the consequences of this form of habitat alteration on the long-term survival prospects of gorillas. The increasing demand for additional varieties of tree species for the timber trade will continue, and it is inevitable that more species of trees, important in the gorilla diet, will be exploited as the world markets incorporate lesser-known varieties of trees. Unless these food resources are replaced with similarly preferred resources, a negative impact on gorilla survival is likely to occur with further timber-harvesting cycles. Increased understanding of the nutritional needs of forest-dwelling primates in relation to future timber-extraction cycles will be necessary to ensure gorilla existence in production forests (Felton et al., 2010).

Other land conversion threats to gorillas include the palm oil industry. Although oil palm is native to Africa, the palm oil industry has not been as large or had as much of an effect in African countries as in Southeast Asia. However, 10.7% (*G. beringei*) and 73.8% (*G. gorilla*) of gorilla distribution overlaps with suitable palm oil areas (Wich et al., 2014). This poses a bigger threat to western gorillas as only about 13% of their distribution is in protected areas (Strindberg et al., 2018). Oil palm has led to the loss of Intact Forest Landscape (IFL) throughout the tropics (Potapov et al., 2017). Demand for palm oil may at least double by 2050 (Corely, 2009), with most of this conversion expected to occur in Africa (Wich et al., 2014). Estimates of exactly how much growth will take place in Africa range from 22 Mha to 53 Mha, a more than 100% increase from the current amount of land (Strona et al., 2018). When Strona et al. (2018) tried to find areas for growing palm oil that would have minimal impact on primate conservation, only 3.3 Mha were found, highlighting how massive an issue this will be going forward. More research needs to be done on sustainable palm oil to mitigate the effects of the industry on gorilla populations.

Associated with the expansion in timber exploitation and palm oil agriculture is a significant increase in the demand and availability of non-timber forest products such as bushmeat (Wilkie & Carpente, 1999; Poulsen et al., 2009). The expansion of commercial hunting is believed to have increased beyond sustainable levels (Fa & Brown, 2009). Gorillas are primarily killed as a food source but occasionally for traditional medicines or cultural fetishes (Nellemann et al., 2010). Assessments on the availability of ape meat in markets, or along control points of travel routes, have been conducted and indicate large forest-dwelling primates, such as apes, may play only a minor role in the trade (Wilkie & Carpenter, 1999; Fa & Brown, 2009; Poulsen et al., 2009). However, traditional survey methods in markets or interviews may not adequately detect the availability and trade of the meat. There is also great variation among close population centers in the demand and subsequent availability of bushmeat. Recent studies from Central Africa highlight the important and complex influence local and nonlocal hunters have on the bushmeat trade (Kuehl et al., 2009; Poulsen et al., 2009). Regions where indigenous people have historically considered hunting apes taboo are now undergoing an influx of foreigners who do not hold similar beliefs. The changes in cultural dynamics and significant expansion of logging roads in Central Africa (Laporte et al., 2007; Kleinscroth et al., 2019) have increased the threat of hunting to remaining ape populations (Tutin & Fernandez, 1984; Kuehl et al., 2009; Morgan et al., 2019).

SUMMARY

This review highlights the behavioral and ecological flexibility that is the result of gorillas adapting to life in very dynamic environments, which have repeatedly provided many opportunities and posed many challenges. The resilience of these great apes is unquestionable, but the current human impact on this genus has no historical analogue. At the same time, the long-term monitoring and expanding efforts in the Virunga Mountains aimed at the science and conservation of protecting gorillas also showcases how interventions can have beneficial impacts not only on safeguarding apes but facilitating population increases (Robbins et al., 2011b). While each population of gorillas faces their own unique set of environmental and social circumstances, threats such as human poaching, settlements, cultivation, mining, and removal of timber have combined with dramatic effect on most remaining gorilla populations. Further, many ape populations are at increased risks of emerging diseases (Huijbregts et al., 2003; Walsh et al., 2003; Caillaud et al., 2006). To combat the current trends, a synergistic approach that includes more informed and measurable conservation strategies, devised by government officials in partnership with scientists, conservationists, nongovernment organizations, and industry are urgently needed (Tutin, 2001).

REFERENCES CITED—CHAPTER 22

Anthony, N. M., Johnson-Bawe, M., Jeffery, K., Clifford, S. L., Abernethy, K. A., Tutin, C. E., Lahm, S. A., White, L. J., Utley, J. F., Wickings, E. J., & Bruford, M. W. (2007). The role of Pleistocene refugia and rivers in shaping gorilla genetic diversity in central Africa. *Proceedings of the National Academy of Sciences*, 104(51), 20432-20436.

Arandjelovic, M., Head, J. S., Boesch, C., Robbins, M. M., & Vigilant, L. (2014). Genetic inference of group dynamics and female kin structure in a western lowland gorilla population (*Gorilla gorilla gorilla*). *Primate Biology*, 1(1), 29-38.

Arnold, K., & Zuberbühler, K. (2006). The alarm-calling system of adult male putty-nosed monkeys, *Cercopithecus nictitans martini*. *Animal Behaviour*, 72(3), 643-653.

Barks, S. K., Calhoun, M. E., Hopkins, W. D., Cranfield, M. R., Mudakikwa, A., Stoinski, T. S., Patterson, F. G., Erwin, J. M., Hecht, E. E., Hof, P. R., & Sherwood, C. C. (2015). Brain organization of gorillas reflects species differences in ecology. *American Journal of Physical Anthropology*, 156(2), 252-262.

Barton, R. A., Whiten, A., Byrne, R. W., & English, M. (1993). Chemical composition of baboon plant foods: Implications for the interpretation of intra- and interspecific differences in diet. *Folia Primatologica*, 61(1), 1-20.

Barton, R. A., Whiten, A., Strum, S. C., Byrne, R. W., & Simpson, A. J. (1992). Habitat use and resource availability in baboons. *Animal Behaviour*, 43(5), 831-844.

Bergl, R. A., & Vigilant, L. (2007). Genetic analysis reveals population structure and recent migration within the highly fragmented range of the Cross River gorilla (*Gorilla gorilla diehli*). *Molecular Ecology*, 16(3), 501-516.

Bermejo, M. (1999). Status and conservation of primates in Odzala National Park, Republic of the Congo. *Oryx*, 33(4), 323-331.

Bermejo, M. (2004). Home-range use and intergroup encounters in western gorillas (*Gorilla g. gorilla*) at Lossi Forest, north Congo. *American Journal of Primatology*, 64(2), 223-232.

Bermejo, M., Rodríguez-Teijeiro, J. D., Illera, G., Barroso, A., Vilà, C., & Walsh, P. D. (2006). Ebola outbreak killed 5000 gorillas. *Science*, 314(5805), 1564.

Blake, S., Rogers, E., Fay, J. M., Ngangoué, M., & Ebéké, G. (1995). Swamp gorillas in northern Congo. *African Journal of Ecology*, 33(3), 285-290.

Bradley, B. J., Doran-Sheehy, D., Boesch, C., & Vigilant, L. (2005). Related dyads of females are common in western gorilla groups despite routine female dispersal. *American Journal of Physical Anthropology*, 126(S40), 77.

Bradley, B. J., Doran-Sheehy, D. M., Lukas, D., Boesch, C., & Vigilant, L. (2004). Dispersed male networks in western gorillas. *Current Biology*, 14(6), 510-513.

Bradley, B. J., Doran-Sheehy, D. M., & Vigilant, L. (2007). Potential for female kin associations in wild western gorillas despite female dispersal. *Proceedings of the Royal Society B: Biological Sciences*, 274(1622), 2179-2185.

Breuer, T., Hockemba, M. B. N., Olejniczak, C., Parnell, R. J., & Stokes, E. J. (2009). Physical maturation, life-history classes and age estimates of free-ranging western gorillas—insights from Mbeli Bai, Republic of Congo. *American Journal of Primatology*, 71(2), 106-119.

Breuer, T., Ndoundou-Hockemba, M., & Fishlock, V. (2005). First observation of tool use in wild gorillas. *PLoS Biology*, 3(11), 380.

Breuer, T., Robbins, A. M., Boesch, C., & Robbins, M. M. (2012). Phenotypic correlates of male reproductive success in western gorillas. *Journal of Human Evolution*, 62(4), 466-472.

Breuer, T., Robbins, A. M., Olejniczak, C., Parnell, R. J., Stokes, E. J., & Robbins, M. M. (2010). Variance in the male reproductive success of western gorillas: Acquiring females is just the beginning. *Behavioral Ecology and Sociobiology*, 64(4), 515-528.

Breuer, T., Stokes, E. J., Parnell, R. J., Robbins, A. M., & Robbins, M. M. (2008). Male life history patterns and reproductive success in western gorillas: Insights from Mbeli Bai, Republic of Congo. *Folia Primatologica*, 79(5), 314-315.

Brugiere, D., & Sakom, D. (2001). Population density and nesting behaviour of lowland gorillas (*Gorilla gorilla gorilla*) in the Ngotto forest, Central African Republic. *Journal of Zoology*, 255(2), 251-259.

Burivalova, Z., Şekercioğlu, Ç. H., & Koh, L. P. (2014). Thresholds of logging intensity to maintain tropical forest biodiversity. *Current Biology*, 24(16), 1893-1898.

Caillaud, D., Levréro, F., Cristescu, R., Gatti, S., Dewas, M., Douadi, M., Gautier-Hion, A., Raymond, M., & Ménard, N. (2006). Gorilla susceptibility to Ebola virus: The cost of sociality. *Current Biology*, 16(13), 489-491.

Caillaud, D., Levréro, F., Gatti, S., Menard, N., & Raymond, M. (2008). Influence of male morphology on male mating status and behavior during interunit encounters in western lowland gorillas. *American Journal of Physical Anthropology*, 135(4), 379-388.

Caillaud, D., Ndagijimana, F., Giarrusso, A. J., Vecellio, V., & Stoinski, T. S. (2014). Mountain gorilla ranging patterns: Influence of group size and group dynamics. *American Journal of Primatology*, 76(8), 730-746.

Cameron, K. N., Reed, P., Morgan, D. B., Ondzié, A. I., Sanz, C. M., Kühl, H. S., Olson, S. H., Leroy, E., Karesh, W. B., & Mundry, R. (2016). Spatial and temporal dynamics of a mortality event among central African great apes. *PLoS ONE*, 11(5), e0154505.

Carroll, R. W. (1986). Status of the lowland gorilla and other wildlife in the Dzangha-Sangha region of southwestern Central African Republic. *Primate Conservation*, 7, 38-41.

Casimir, M. J. (1975). Feeding ecology and nutrition of an eastern gorilla group in the Mt. Kahuzi region (Republique du Zaire). *Folia Primatologica*, 24(2), 81-136.

Casimir, M. J. (1979). An analysis of gorilla nesting sites of the Mt. Kahuzi region (Zaire). *Folia Primatologica*, 32(4), 290-308.

Chapman, C. A., Wrangham, R. W., Chapman, L. J., Kennard, D. K., & Zanne, A. E. (1999). Fruit and flower phenology at two sites in Kibale National Park, Uganda. *Journal of Tropical Ecology*, 15(2), 189-211.

Cipolletta, C. (2003). Ranging patterns of a western gorilla group during habituation to humans in the Dzanga-Ndoki National Park, Central African Republic. *International Journal of Primatology*, 24(6), 1207-1226.

Cipolletta, C. (2004). Effects of group dynamics and diet on the ranging patterns of a western gorilla group (*Gorilla gorilla gorilla*) at Bai Hokou, Central African Republic. *American Journal of Primatology*, 64(2), 193-205.

Cipolletta, C., Spagnoletti, N., Todd, A., Robbins, M. M., Cohen, H., & Pacyna, S. (2007). Termite feeding by *Gorilla gorilla gorilla* at Bai Hokou, Central African Republic. *International Journal of Primatology*, 28(2), 457.

Clark, C. J., Poulsen, J. R., Malonga, R., & Elkan, P. W. (2009). Logging concessions can extend the conservation estate for central African tropical forests. *Conservation Biology*, 23(5), 1281-1293.

Clutton-Block, T. H. (1977). Species differences in feeding and ranging behaviour in primates. In T. H. Clutton-Block (Ed.), *Primate ecology: Studies of feeding and ranging behaviour in lemurs, monkeys, and apes* (pp. 557-584). Academic Press.

Corley, R. H. V. (2009). How much palm oil do we need? *Environmental Science & Policy*, 12(2), 134-139. http://dx.doi.org/10.1016/j.envsci.2008.10.011.

Cranfield, M. R. (2004). Standardised health monitoring system for the mountain gorilla (*Gorilla beringei beringei*). *Folia Primatologica*, 75, 104.

Crockett, C. (2000). Infanticide in red howlers: Female group size, male membership, and a possible link to folivory. In C. Van Schaik, & C. Janson (Eds.), *Infanticide by males and its implications* (pp. 75-98). Cambridge University Press. DOI: 10.1017/CBO9780511542312.006.

Das, R., Hergenrother, S. D., Soto-Calderón, I. D., Dew, J. L., Anthony, N. M., & Jensen-Seaman, M. I. (2014). Complete mitochondrial genome sequence of the eastern gorilla (*Gorilla beringei*) and implications for African ape biogeography. *Journal of Heredity*, 105(6), 846-855.

Devos, C., Sanz, C., Morgan, D., Onononga, J. R., Laporte, N., & Huynen, M. C. (2008). Comparing ape densities and habitats in northern Congo: Surveys of sympatric gorillas and chimpanzees in the Odzala and Ndoki regions. *American Journal of Primatology*, 70(5), 439-451.

Doran, D. M., & McNeilage, A. (1998a). Gorilla ecology and behavior. *Evolutionary Anthropology*, 6(4), 120-131.

Doran, D. M., McNeilage, A., Greer, D., Bocian, C., Mehlman, P., & Shah, N. (2002). Western lowland gorilla diet and resource availability: New evidence, cross-site comparisons, and reflections on indirect sampling methods. *American Journal of Primatology*, 58(3), 91-116.

Doran-Sheehy, D. M., Derby, A. M., Greer, D., & Mongo, P. (2007). Habituation of western gorillas: The process and factors that influence it. *American Journal of Primatology*, 69(12), 1354-1369.

Doran-Sheehy, D. M., Greer, D., Mongo, P., & Schwindt, D. (2004). Impact of ecological and social factors on ranging in western gorillas. *American Journal of Primatology*, 64(2), 207-222.

Doran-Sheehy, D., Mongo, P., Lodwick, J., & Conklin-Brittain, N. L. (2009). Male and female western gorilla diet: Preferred foods, use of fallback resources, and implications for ape versus Old World monkey foraging strategies. *American Journal of Physical Anthropology*, 140(4), 727-738.

Douadi, M. I., Gatti, S., Levrero, F., Duhamel, G., Bermejo, M., Vallet, D., Menard, N., & Petit, E. J. (2007). Sex-biased dispersal in western lowland gorillas (*Gorilla gorilla gorilla*). *Molecular Ecology*, 16(11), 2247-2259.

Dunbar, R. I. M. (1988). *Primate social systems*. Cornell University Press.

Dunn, R. H., Tocheri, M. W., Orr, C. M., & Jungers, W. L. (2014). Ecological divergence and talar morphology in gorillas. *American Journal of Physical Anthropology*, 153(4), 526-541.

Dupain, J., Guislain, P., Nguenang, G. M., De Vleeschouwer, K., & Van Elsacker, L. (2004). High chimpanzee and gorilla densities in a non-protected area on the northern periphery of the Dja Faunal Reserve, Cameroon. *Oryx*, 38, 209-216.

Edwards, D. P., Sloan, S., Weng, L., Dirks, P., Sayer, J., & Laurance, W. F. (2014). Mining and the African environment. *Conservation Letters*, 7(3), 302-311.

Emlen, J. T., & Schaller, G. B. (1960). Distribution and status of the mountain gorilla (*Gorilla gorilla beringei*). *Zoologica*, 45(1), 41-52.

Ernst, C., Verhegghen, A., Mayaux, P., Hansen, M., Defourny, P., Kondjo, K., Makak, J. S., Biang, J. D. M., Musampa, C., Motogo, R. N., & Neba, G. (2012). Central African forest cover and cover change mapping. In C. de Wasseige, P. de Marcken, N. Bayol, F. Hiol Hiol, Mayaux Ph., B. Desclee, R. Nasi, A. Billand, P. Defourny, & A. R. Eba'a (Eds.), *The forests of the Congo Basin: State of the forest 2010* (pp. 23-41). Observatoire des Forêts d'Afrique Centrale.

Fa, J. E., & Brown, D. (2009). Impacts of hunting on mammals in African tropical moist forests: A review and synthesis. *Mammal Review*, 39(4), 231-264.

Fa, J. E., Ryan, S. F., & Bell, D. J. (2005). Hunting vulnerability, ecological characteristics and harvest rates of bushmeat species in afrotropical forests. *Biological Conservation*, 121(2), 167-176.

Fay, J. M. (1997). *The ecology, social organization, populations, habitat and history of the western lowland gorilla (*Gorilla gorilla gorilla*)*. [Doctoral dissertation, Washington University].

Fay, J. M., & Agnagna, M. (1992). Census of gorillas in northern Republic of Congo. *American Journal of Primatology*, 27(4), 275-284.

Fay, J. M., Agnagna, M., Moore, J., & Oko, R. (1989). Gorillas (*Gorilla gorilla gorilla*) in the Likouala swamp forests of north central Congo: Preliminary data on populations and ecology. *International Journal of Primatology*, 10(5), 477-486.

Fay, J. M., Carroll, R., Peterhans, J. K., & Harris, D. (1995). Leopard attack on and consumption of gorillas in the Central African Republic. *Journal of Human Evolution*, 29, 93-99.

Felton, A. M., Felton, A., Foley, W. J., & Lindenmayer, D. B. (2010). The role of timber tree species in the nutritional ecology of spider monkeys in a certified logging concession, Bolivia. *Forest Ecology and Management*, 259(8), 1642-1649.

Ferriss, S. (2005). Western gorilla (*Gorilla gorilla*). In J. Caldecott, & L. Miles (Eds.), *World atlas of great apes and their conservation* (pp. 105-127). University of California Press.

Forcina, G., Vallet, D., Le Gouar, P. J., Bernardo-Madrid, R., Illera, G., Molina-Vacas, G., Dréano, S., Revilla, E., Rodríguez-Teijeiro, J. D., Ménard, N., & Bermejo, M., (2019). From groups to communities in western lowland gorillas. *Proceedings of the Royal Society B*, 286(1896), 1-9.

Fossey, D., & Harcourt, A. H. (1977). Feeding ecology of free-ranging mountain gorilla (*Gorilla gorilla beringei*). In T. H. Clutton-Brock (Ed.), *Primate ecology* (pp. 415–447). Academic Press.

Fünfstück, T., Arandjelovic, M., Morgan, D., Sanz, C., Breuer, T., Stokes, E., Reed, P., Olson, S., Cameron, K., Ondzie, A., Peeters, M., Kühl, H., Cipolletta, C., Todd, A., Masi, S., Doran-Sheehy, D., Bradley, B., & Vigilant, L. (2014). The genetic population structure of wild western lowland gorillas (*Gorilla gorilla gorilla*) living in continuous rain forest. *American Journal of Primatology*, 76(9), 868–878. DOI: 10.1002/ajp.22274.

Fünfstück, T., & Vigilant, L. (2015). The geographic distribution of genetic diversity within gorillas. *American Journal of Primatology*, 77(9), 974-985.

Furuichi, T., Inagaki, H., & Angoue-Ovono, S. (1997). Population density of chimpanzees and gorillas in the Petit Loango Reserve, Gabon: Employing a new method to distinguish between nests of the two species. *International Journal of Primatology*, 18(6), 1029-1046.

Ganas, J., & Robbins, M. M. (2005). Ranging behavior of the mountain gorillas (*Gorilla beringei beringei*) in Bwindi Impenetrable National Park, Uganda: A test of the ecological constraints model. *Behavioral Ecology and Sociobiology*, 58(3), 277-288.

Ganas, J., Robbins, M. M., Nkurunungi, J. B., Kaplin, B. A., & McNeilage, A. (2004). Dietary variability of mountain gorillas in Bwindi Impenetrable National Park, Uganda. *International Journal of Primatology*, 25(5), 1043-1072.

Gatti, S., Levréro, F., Ménard, N., & Gautier-Hion, A. (2004). Population and group structure of western lowland gorillas (*Gorilla gorilla gorilla*) at Lokoue, Republic of Congo. *American Journal of Primatology*, 63(3), 111-123.

Genton, C., Pierre, A., Cristescu, R., Lévréro, F., Gatti, S., Pierre, J., Ménard, N., & Gouar, P. (2015). How Ebola impacts social dynamics in gorillas: A multistate modelling approach. *Journal of Animal Ecology*, 84(1), 166-176.

Goldberg, T. L., Gillespie, T. R., Rwego, I. B., Wheeler, E., Estoff, E. L., & Chapman, C. A. (2007). Patterns of gastrointestinal bacterial exchange between chimpanzees and humans involved in research and tourism in western Uganda. *Biological Conservation*, 135(4), 511-517.

Goldsmith, M. L. (1999). Ecological constraints on the foraging effort of western gorillas (*Gorilla gorilla gorilla*) at Bai Hoköu, Central African Republic. *International Journal of Primatology*, 20(1), 1-23.

Goldsmith, M. L. (2003). Comparative behavioral ecology of a lowland and highland gorilla population: Where do Bwindi gorillas fit. In A. B. Taylor, & M. L. Goldsmith (Eds.), *Gorilla biology: A multidisciplinary perspective* (pp. 358-384). Cambridge University Press.

Goldsmith, M. L., & Moles, H. (2003). Does topography affect the foraging effort of mountain gorillas in Bwindi Impenetrable National Park, Uganda? *American Journal of Physical Anthropology*, 77, 102.

Goodall, A. G. (1977). Feeding and ranging behavior of a mountain gorilla group (*Gorilla gorilla beringei*) in the Tshibinda-Kahuzi region (Zaire). In T. H. Clutton-Brock (Ed.), *Primate ecology* (pp. 449-479). Academic Press.

Gray, M., Roy, J., Vigilant, L., Fawcett, K., Basabose, A., Cranfield, M., Uwingeli, P., Mburanumwe, I., Kagoda, E., & Robbins, M. M. (2013). Genetic census reveals increased but uneven growth of a critically endangered mountain gorilla population. *Biological Conservation*, 158, 230-238.

Greenway, K. (2015). *Threat and display: Reproductive competition in wild male western gorillas (*Gorilla gorilla*)*. [Doctoral dissertation, University of Kent].

Greer, D., & Cipolletta, C. (2006). Western gorilla tourism: Lessons learned from Dzanga-Sangha. *Gorilla Journal*, 33, 16–19.

Groves, C. (2001). *Primate taxonomy*. Smithsonian Institution Press.

Groves, C. (2003). A history of gorilla taxonomy. In A. B. Taylor, & M. L. Goldsmith (Eds.), *Gorilla biology:*

A multidisciplinary perspective (pp. 15-24). Cambridge University Press.

Groves, C. P., & Stott, K. W., Jr. (1979). Systematic relationships of gorillas from Kahuzi, Tshiaberimu and Kayonza. *Folia Primatologica*, 32(3), 161-179.

Grützmacher, K. S., Köndgen, S., Keil, V., Todd, A., Feistner, A., Herbinger, I., Petrzelkova, K., Fuh, T., Leendertz, S. A., Calvignac-Spencer, S., & Leendertz, F. H. (2016). Codetection of respiratory syncytial virus in habituated wild western lowland gorillas and humans during a respiratory disease outbreak. *EcoHealth*, 13(3), 499-510.

Guschanski, K., Caillaud, D., Robbins, M. M., & Vigilant, L. (2008). Females shape the genetic structure of a gorilla population. *Current Biology*, 18(22), 1809-1814.

Hagemann, L., Boesch, C., Robbins, M. M., Arandjelovic, M., Deschner, T., Lewis, M., Froese, G., & Vigilant, L. (2018). Long-term group membership and dynamics in a wild western lowland gorilla population (*Gorilla gorilla gorilla*) inferred using non-invasive genetics. *American Journal of Primatology*, 80(8), e22898.

Hall, J. S., White, L. J., Inogwabini, B. I., Omari, I., Morland, H. S., Williamson, E. A., Saltonstall, K., Walsh, P., Sikubwabo, C., Bonny, D., & Kiswele, K. P. (1998). Survey of Grauer's gorillas (*Gorilla gorilla graueri*) and eastern chimpanzees (*Pan troglodytes schweinfurthi*) in the Kahuzi-Biega National Park lowland sector and adjacent forest in eastern Democratic Republic of Congo. *International Journal of Primatology*, 19(2), 207-235.

Hamilton, W. D. (1971). Geometry for the selfish herd. *Journal of Theoretical Biology*, 31(2), 295-311.

Harcourt, A. H. (1978). Strategies of emigration and transfer by primates, with particular reference to gorillas. *Zeitschrift für Tierpsychologie*, 48(4), 401-420.

Harcourt, A. H. (1979a). Social relationships among adult female mountain gorillas. *Animal Behaviour*, 27, 251-264.

Harcourt, A. H. (1979b). Social relationships between adult male and female mountain gorillas in the wild. *Animal Behaviour*, 27, 325-342.

Harcourt, A. H., Fossey, D., & Sabater-Pi, J. (1981). Demography of *Gorilla gorilla*. *Journal of Zoology*, 195(2), 215-233.

Harcourt, A. H., Fossey, D., Stewart, K. J., & Watts, D. P. (1980). Reproduction in wild gorillas and some comparisons with chimpanzees. *Journal of Reproduction and Fertility*, 28, 59-70.

Harcourt, A. H., & Stewart, K. J. (1981). Gorilla male relationships: Can differences during immaturity lead to contrasting reproductive tactics in adulthood? *Animal Behaviour*, 29(1), 206-210.

Harcourt, A. H., & Stewart, K. J. (1987). The influence of help in contests on dominance rank in primates: Hints from gorillas. *Animal Behaviour*, 35(1), 182-190.

Harcourt, A. H., & Stewart, K. J. (2007). *Gorilla society: Conflict, compromise, and cooperation between the sexes*. University of Chicago Press.

Harcourt, A. H., Stewart, K. S., & Fossey, D. (1976). Male emigration and female transfer in wild mountain gorilla. *Nature*, 263(5574), 226.

Hart, D., & Sussman, R. W. (2009). *Man the hunted: Primates, predators, and human evolution. (Expanded Ed.)*. Westview Press.

Head, J. S., Boesch, C., Makaga, L., & Robbins, M. M. (2011). Sympatric chimpanzees (*Pan troglodytes troglodytes*) and gorillas (*Gorilla gorilla gorilla*) in Loango National Park, Gabon: Dietary composition, seasonality, and intersite comparisons. *International Journal of Primatology*, 32(3), 755-775.

Hemingway, C. A., & Bynum, N. (2005). The influence of seasonality on primate diet and ranging. In D. K. Brockman, & C. P. van Schaik (Eds.), *Seasonality in primates: Studies of living and extinct human and non-human primates* (pp. 57-104). Cambridge University Press.

Henschel, P., Abernethy, K. A., & White, L. J. T. (2005). Leopard food habits in the Lope National Park, Gabon, central Africa. *African Journal of Ecology*, 43(1), 21-28.

Henschel, P., Malanda, G. A., & Hunter, L. (2014). The status of savanna carnivores in the Odzala-Kokoua National Park, northern Republic of Congo. *Journal of Mammalogy*, 95(4), 882-892.

Hickey, J. R., Granjon, A. C., Vigilant, L., Eckardt, W., Gilardi, K. V., Cranfield, M., Musana, A., Masozera, A. B., Babaasa, D., Ruzigandekwe, F., & Robbins, M. M. (2019). *Virunga 2015–2016 surveys: Monitoring mountain gorillas, other select mammals and illegal activities*. International Gorilla Conservation Programme.

Homsy, J. (1999). *Ape tourism and human diseases: How close should we get?: A critical review of rules and regulations governing park management and tourism for the wild mountain gorilla*. International Gorilla Conservation Programme.

Hrdy, S. B. (1974). Male-male competition and infanticide among the langurs (*Presbytis entellus*) of Abu, Rajasthan. *Folia Primatologica*, 22(1), 19-58.

Huijbregts, B., De Wachter, P., Obiang, L. S. N., & Akou, M. E. (2003). Ebola and the decline of gorilla *Gorilla gorilla* and chimpanzee *Pan troglodytes* populations in Minkebe Forest, north-eastern Gabon. *Oryx*, 37(4), 437-443.

Inoue, E., Akomo-Okoue, E., Ando, C., Iwata, Y., Judai, M., Fujita, S., Hongo, S., Nze-Nkogue, C., Inoue-Murayama, M., & Yamagiwa, J. (2013). Male genetic structure and paternity in western lowland gorillas (*Gorilla gorilla gorilla*). *American Journal of Physical Anthropology*, 151(4), 583-588.

Inouye, S. E. (1994). Ontogeny of knuckle-walking hand postures in African apes. *Journal of Human Evolution*, 26(5-6), 459-485.

Inouye, S. E. (2003). Intraspecific and ontogenetic variation in the forelimb morphology of gorilla. In A. Taylor, & M. Goldsmith (Eds.), *Gorilla biology: A multidisciplinary perspective* (pp. 194-235). Cambridge University Press.

IUCN. (2014). *Regional action plan for the conservation of western lowland gorillas and central chimpanzees 2015-2025*.

IUCN SSC Primate specialist group. www.primate-sg.org/WEA2014.pdf.

Jabbour, R. S., & Pearman, T. L. (2016). Geographic variation in gorilla limb bones. *Journal of Human Evolution*, 95, 68-79.

Janson C. H., & Chapman, C. A. (1999). Resouces and the determination of primate community structure. In J. G. Fleagle, C. H. Janson, & K. Reed (Eds.), *Primate communities* (pp. 237-267). Cambridge University Press.

Janson, C. H., & Goldsmith, M. L. (1995). Predicting group size in primates: Foraging costs and predation risks. *Behavioral Ecology*, 6(3), 326-336.

Jones, C., & Sabater J. (1971). Comparative ecology of *Gorilla gorilla* (Savage and Wyman) and *Pan troglodytes* (Blumenbach) in Rio Muni, West Africa. *Bibliotheca Primatologica*, 13, 1-96.

Kalpers, J., Williamson, E. A., Robbins, M. M., McNeilage, A., Nzamurambaho, A., Lola, N., & Mugiri, G. (2003). Gorillas in the crossfire: Population dynamics of the Virunga mountain gorillas over the past three decades. *Oryx*, 37(3), 326-337.

Kay, R. F. (1984). On the use of anatomical features to infer foraging behavior in extinct primates. In P. Rodman, & J. G. H. Cant (Eds.), *Adaptations for foraging in nonhuman primates: Contributions to an organismal biology of prosimians, monkeys, and apes* (pp. 21-53). Columbia University Press.

Kleinschroth, F., Garcia, C., & Ghazoul, J. (2019). Reconciling certification and intact forest landscape conservation. *Ambio*, 48, 153-159.

Knigge, R. P., Tocheri, M. W., Orr, C. M., & Mcnulty, K. P. (2015). Three-dimensional geometric morphometric analysis of talar morphology in extant gorilla taxa from highland and lowland habitats. *Anatomical Record*, 298(1), 277-290.

Kuehl, H. S., Nzeingui, C., Yeno, S. L. D., Huijbregts, B., Boesch, C., & Walsh, P. D. (2009). Discriminating between village and commercial hunting of apes. *Biological Conservation*, 142(7), 1500-1506.

Kuroda, S. (1992). Ecological interspecies relationships between gorillas and chimpanzees in the Ndoki-Nouabale reserve, northern Congo. In N. Itoigawa, Y. Sugiyama, G. P. Sackett, & R. K. R. Thompson (Eds.), *Topics in primatology: Behavior, ecology, and conservation* (Vol. 2, pp. 385-394). University of Tokyo Press.

Kuroda, S., Nishihara, T., Suzuki, S., & Oko, R. A. (1996). Sympatric chimpanzees and gorillas in the Ndoki Forest, Congo. In W. C. McGrew, L. F. Marchant, & T. Nishida (Eds.), *Great ape societies* (pp. 71-81). Cambridge University Press.

Langergraber, K. E., Pruefer, K., Rowney, C., Boesch, C., Crockford, C., Fawcett, K., Inoue, E., Inoue-Muruyama, M., Mitani, J. C., Muller, M. N., Robbins, M. M., Schubert, G., Stoinski, T. S., Viola, B., Watts, D., Wittig, R. M., Wrangham, R. W., Zuberbuhler, K., Paabo, S., & Vigilant, L. (2012). Generation times in wild chimpanzees and gorillas suggest earlier divergence times in great ape and human evolution. *Proceedings of the National Academy of Sciences*, 109(39), 15716-15721.

Laporte, N. T., Stabach, J. A., Grosch, R., Lin, T. S., & Goetz, S. J. (2007). Expansion of industrial logging in central Africa. *Science*, 316(5830), 1451.

Leigh, S. R. (1994). Ontogenetic correlates of diet in anthropoid primates. *American Journal of Physical Anthropology*, 94(4), 499-522.

Leigh, S. R., & Shea, B. T. (1995). Ontogeny and the evolution of adult body size dimorphism in apes. *American Journal of Primatology*, 36(1), 37-60.

Leroy, E. M., Kumulungui, B., Pourrut, X., Rouquet, P., Hassanin, A., Yaba, P., Délicat, A., Paweska, J. T., Gonzalez, J. P., & Swanepoel, R. (2005). Fruit bats as reservoirs of Ebola virus. *Nature*, 438(7068), 575-577.

Lewis, S. L., Sonké, B., Sunderland, T., Begne, S. K., Lopez-Gonzalez, G., van der Heijden, G. M. F., Phillips, O. L., Affum-Baffoe, K., Baker, T. R., Banin, L., & Bastin, J. F. (2013). Above-ground biomass and structure of 260 African tropical forests. *Philosophical Transactions of the Royal Society B: Biological Sciences*, 368(1625). https://doi.org/10.1098/rstb.2012.0295.

Lodwick, J. L. (2015). *Links between foraging strategies, feeding competition, and female agonistic relationships in wild western gorillas (Gorilla gorilla).* [Doctoral dissertation, Stony Brook University].

Magliocca, F., & Gautier-Hion, A. (2004). Intergroup encounters in western lowland gorillas at a forest clearing. *Folia Primatologica*, 75(6), 379-382.

Magliocca, F., Querouil, S., & Gautier-Hion, A. (1999). Population structure and group composition of western lowland gorillas in north-western Republic of Congo. *American Journal of Primatology*, 48(1), 1-14.

Maisels, F., Bergl, R. A., & Williamson, E. A. (2018). *Gorilla gorilla (amended version of 2016 assessment). The IUCN Red List of Threatened Species* 2018: e.T9404A136250858. https://dx.doi.org/10.2305/IUCN.UK.2018-2.RLTS.T9404A136250858.en.

Maisels F., & Cruickshank, A. (1996). *Inventaire et recensement des grands mammifères dans le Reserve de Faune de Conkouati.* UICN/ Projet Conkouati.

Masi, S. (2008). *Seasonal influence on foraging [strategies], activity and energy budgets of western lowland gorillas (Gorilla gorilla gorilla) in Bai-Hokou, Central African Republic.* [Doctoral dissertation, Università Sapienza Roma].

Masi S., Cipolletta C., & Robbins, M. M. (2009). Western lowland gorillas (Gorilla gorilla gorilla) change their activity patterns in response to frugivory. *American Journal of Primatology*, 71(2), 91-100.

Matschie, P. (1903). Über einen gorilla aus Deutsch-Ostafrika. *Sitzungsberichte der Gesellschaft Naturforschender Freunde, Berlin*, 1903, 253-259.

Matschie, P. (1904). Bermerkungen über die Gattung gorilla. *Sitzungsberichte der Gesellschaft Naturforschender Freunde, Berlin*, 1904, 45-53.

Matschie, P. (1914). Neue Affen aus Mittelafrika. *Sitzungsber der Naturforschender Freunde Berlin, 1914,* 323-342.

Matthews, A., & Matthews, A. (2004). Survey of gorillas (*Gorilla gorilla gorilla*) and chimpanzees (*Pan troglodytes troglodytes*) in southwestern Cameroon. *Primates, 45,* 15-24.

McNeilage, A. J. (1995). *Mountain gorillas in the Virunga Volcanoes: Ecology and carrying capacity.* [Doctoral dissertation, University of Bristol].

McNeilage, A. J., Robbins, M., Gray, M., & Olupot, W. (2006). Census of the mountain gorilla *Gorilla beringei beringei* population in Bwindi Impenetrable National Park, Uganda. *Oryx, 40*(4), 419-427.

Mehlman, P. T. (2008). Current status of wild gorilla populations and strategies for their conservation. In T. S. Stoinski, H. D. Steklis, & P. T. Mehlman (Eds.), *Conservation in the 21st century: Gorillas as a case study* (pp. 3-54). Springer.

Mehlman, P. T., & Doran, D. M. (2002). Influencing western gorilla nest construction at Mondika Research Center. *International Journal of Primatology, 23*(6), 1257-1285.

Miller-Schroeder, P. (1997). *Gorillas.* Diane Publishing Inc.

Morgan, B. J., Wild, C., & Ekobo, A. (2003). Newly discovered gorilla population in the Ebo forest, Littoral Province, Cameroon. *International Journal of Primatology, 24*(5), 1129-1137.

Morgan, D., & Sanz, C. (2006). Chimpanzee feeding ecology and comparisons with sympatric gorillas in the Goualougo Triangle, Republic of Congo. *Cambridge Studies in Biological and Evolutionary Anthropology, 48,* 97-122.

Morgan, D., Sanz, C., Onononga, J. R., & Strindberg, S. (2006). Ape abundance and habitat use in the Goualougo Triangle, Republic of Congo. *International Journal of Primatology, 27*(1), 147-179.

Morgan, D., Strindberg, S., Winston, W., Stephens, C. R., Traub, C., Ayina, C. E., Ebika, N., Thony, S., Mayoukou, W., Koni, D., & Iyenguet, F. (2019). Impacts of selective logging and associated anthropogenic disturbance on intact forest landscapes and apes of northern Congo. *Frontiers in Forests and Global Change, 2*(28), 1-13.

Morrison, R. E., Groenenberg, M., Breuer, T., Manguette, M. L., & Walsh, P. D. (2019). Hierarchical social modularity in gorillas. *Proceedings of the Royal Society B, 286*(1906), 1-7.

Nair, H., Nokes, D. J., Gessner, B. D., Dherani, M., Madhi, S. A., Singleton, R. J., O'Brien, K.L., Roca, A., Wright, P. F., Bruce, N., & Chandran, A. (2010). Global burden of acute lower respiratory infections due to respiratory syncytial virus in young children: A systematic review and meta-analysis. *The Lancet, 375*(9725), 1545-1555.

Nellemann, C., Redmond, I., & Refisch, J. (Eds.). (2010). *The last stand of the gorilla: Environmental crime and conflict in the Congo Basin.* United Nations Environment Programme.

Nishihara, T. (1992). A preliminary report on the feeding habits of western lowland gorillas (*Gorilla gorilla gorilla*) in the Ndoki Forest, northern Congo. In N. Itoigawa, Y. Sugiyama, G. P. Sackett, & R. K. R. Thompson (Eds.), *Topics in primatology: Behavior, ecology, and conservation* (pp. 225-240). University of Tokyo Press.

Nishihara, T. (1995). Feeding ecology of western lowland gorillas in the Nouabale-Ndoki National Park, Congo. *Primates, 36*(2), 151-168.

Nkurunungi, J. B., Ganas, J., Robbins, M. M., & Stanford, C. B. (2004). A comparison of two mountain gorilla habitats in Bwindi Impenetrable National Park, Uganda. *African Journal of Ecology, 42*(4), 289-297.

Nowell, A. A., & Fletcher, A. W. (2007). Development of independence from the mother in *Gorilla gorilla gorilla. International Journal of Primatology, 28*(2), 441-455.

Oates, J. F., McFarland, K. L., Groves, J. L., Bergl, R. A., Linder, J. M., & Disotell, T. R. (2003). The Cross River gorilla: Natural history and status of a neglected and critically endangered subspecies. In A. B. Taylor, & M. L. Goldsmith (Eds.), *Gorilla biology: A multidisciplinary perspective* (pp. 472-497). Cambridge University Press.

Parnell, R. J. (2002a). *The social structure and behaviour of western lowland gorillas* (Gorilla gorilla gorilla) *at Mbeli Bai, Republic of Congo.* [Doctoral dissertation, University of Stirling].

Parnell, R. J. (2002b). Group size and structure in western lowland gorillas (*Gorilla gorilla gorilla*) at Mbeli Bai, Republic of Congo. *American Journal of Primatology, 56*(4), 193-206.

Parnell, R. J. & Buchanan-Smith, H. M. (2001). An unusual social display by gorillas. *Nature, 412,* 294.

Pérez, M. R., de Blas, D. E., Nasi, R., Sayer, J. A., Sassen, M., Angoué, C., Gami, N., Ndoye, O., Ngono, G., Nguinguiri, J. C., & Nzala, D. (2005). Logging in the Congo Basin: A multi-country characterization of timber companies. *Forest Ecology and Management, 214*(1-3), 221-236.

Plumptre, A. J. (1991). *Plant-herbivore dynamics in the Virungas.* [Doctoral dissertation, University of Bristol].

Plumptre, A. J. (1995). The chemical composition of montane plants and its influence on the diet of the large mammalian herbivores in the Parc National des Volcans, Rwanda. *Journal of Zoology, 235*(2), 323-337.

Plumptre, A., Robbins, M. M., & Williamson, E. A. (2019). *Gorilla beringei. The IUCN Red List of Threatened Species* 2019: e.T39994A115576640. https://dx.doi.org/10.2305/IUCN .UK.2019-1.RLTS.T39994A115576640.en.

Plumptre, A. J., Nixon, S., Kujirakwinja, D. K., Vieilledent, G., Critchlow, R., Williamson, E. A., Kirkby, A. E., & Hall, J. S. (2016). Catastrophic decline of world's largest primate: 80% loss of Grauer's gorilla (*Gorilla beringei graueri*) population justifies critically endangered status. *PLoS ONE, 11*(10), e0162697.

Potapov, P., Hansen, M. C., Laestadius, L., Turubanova, S., Yaroshenko, A., Thies, C., Smith, W., Zhuravleva, I., Komarova, A., Minnemeyer, S., & Esipova, E. (2017). The last frontiers of wilderness: Tracking loss of intact forest landscapes from 2000 to 2013. *Science Advances, 3*(1), e1600821.

Poulsen, J. R., & Clark, C. J. (2004). Densities, distributions, and seasonal movements of gorillas and chimpanzees in swamp forest in northern Congo. *International Journal of Primatology*, 25(2), 285-306.

Poulsen, J. R., Clark, C. J., Mavah, G., & Elkan, P. W. (2009). Bushmeat supply and consumption in a tropical logging concession in northern Congo. *Conservation Biology*, 23(6), 1597-1608.

Pradhan, G. R., & van Schaik, C. (2008). Infanticide-driven intersexual conflict over matings in primates and its effects on social organization. *Behaviour*, 145(2), 251-275.

Putz, F. E., Blate, G. M., Redford, K. H., Fimbel, R., & Robinson, J. (2001). Tropical forest management and conservation of biodiversity: An overview. *Conservation Biology*, 15(1), 7-20.

Rainey, H. J., Iyenguet, F. C., Malanda, G. A. F., Madzoké, B., Dos Santos, D., Stokes, E. J., Maisels, F., & Strindberg, S. (2010). Survey of raphia swamp forest, Republic of Congo, indicates high densities of critically endangered western lowland gorillas *Gorilla gorilla gorilla*. *Oryx*, 44(1), 124-132.

Remis, M. J. (1994). *Feeding ecology and positional behavior of lowland gorillas in the Central African Republic.* [Doctoral dissertation, Yale University].

Remis, M. J. (1997). Ranging and grouping patterns of a western lowland gorilla group at Bai Hokou, Central African Republic. *American Journal of Primatology*, 43(2), 111-133.

Remis, M. J. (1999). Tree structure and sex differences in arboreality among western lowland gorillas (*Gorilla gorilla gorilla*) at Bai Hokou, Central African Republic. *Primates*, 40(2), 383-396.

Remis, M. J., Dierenfeld, E. S., Mowry, C. B., & Carroll, R. W. (2001). Nutritional aspects of western lowland gorilla (*Gorilla gorilla gorilla*) diet during seasons of fruit scarcity at Bai Houkou, Central African Republic. *International Journal of Primatology*, 22, 807-836.

Robbins, A. M., Gray, M., Basabose, A., Uwingeli, P., Mburanumwe, I., Kagoda, E., & Robbins, M. M. (2013). Impact of male infanticide on the social structure of mountain gorillas. *PLoS ONE*, 8(11), 1-10.

Robbins, A. M., Gray, M., Breuer, T., Manguette, M., Stokes, E. J., Uwingeli, P., Mburanumwe, I., Kagoda, E., & Robbins, M. M. (2016). Mothers may shape the variations in social organization among gorillas. *Royal Society Open Science*, 3(10), 1-10.

Robbins, A. M., & Robbins, M. M. (2005). Fitness consequences of dispersal decisions for male mountain gorillas (*Gorilla beringei beringei*). *Behavioral Ecology and Sociobiology*, 58(3), 295-309.

Robbins, A. M., Robbins, M. M., Gerald-Steklis, N., & Steklis, H. D. (2006). Age-related patterns of reproductive success among female mountain gorillas. *American Journal of Physical Anthropology*, 131(4), 511-521.

Robbins, A. M., Stoinski, T., Fawcett, K., & Robbins, M. M. (2009). Leave or conceive: Natal dispersal and philopatry of female mountain gorillas in the Virunga Volcano region. *Animal Behaviour*, 77(4), 831-838.

Robbins, A. M., Stoinski, T., Fawcett, K., & Robbins, M. M. (2011a). Lifetime reproductive success of female mountain gorillas. *American Journal of Physical Anthropology*, 146(4), 582-593.

Robbins, M. M. (1995). A demographic analysis of male life history and social structure of mountain gorillas. *Behaviour*, 132(1), 21-47.

Robbins, M. M. (1996). Male-male interactions in heterosexual and all-male wild mountain gorilla groups. *Ethology*, 102(7), 942-965.

Robbins, M. M. (1999). Male mating patterns in wild multimale mountain gorilla groups. *Animal Behaviour*, 57(5), 1013-1020.

Robbins, M. M. (2001). Variation in the social system of mountain gorillas: The male perspective. In M. M. Robbins, P. Sicotte, & K. G. Stewart (Eds.), *Mountain gorillas: Three decades of research at Karisoke* (pp. 29-58). Cambridge University Press.

Robbins, M. M. (2007). Gorillas: Diversity in ecology and behavior. In C. J. Campbell, A. Fuentes, K. C. MacKinnon, M. Panger, & S. K. Bearder (Eds.), *Primates in perspective* (pp. 305-321). Oxford University Press.

Robbins, M. M. (2011). Gorillas: Diversity in ecology and behavior. In C. J. Campbell, A. Fuentes, K. C. MacKinnon, S. K. Bearder, & R. M. Stumpf (Eds.), *Primates in perspective* (2nd Ed., pp. 326-339). Oxford University Press.

Robbins, M. M., Akantorana, M., Arinaitwe, J., Kabano, P., Kayijamahe, C., Gray, M., Guschanski, K., Richardson, J., Roy, J., Tindimwebwa, V., & Vigilant, L. (2019). Dispersal and reproductive careers of male mountain gorillas in Bwindi Impenetrable National Park, Uganda. *Primates*, 60(2), 133-142.

Robbins, M. M., Bermejo, M., Cipolletta, C., Magliocca, F., Parnell, R. J., & Stokes, E. (2004). Social structure and life-history patterns in western gorillas (*Gorilla gorilla gorilla*). *American Journal of Primatology*, 64(2), 145-159.

Robbins, M. M., Gray, M., Fawcett, K. A., Nutter, F. B., Uwingeli, P., Mburanumwe, I., Kagoda, E., Basabose, A., Stoinski, T. S., Cranfield, M. R., Byamukama, J., Spelman, L. H., & Robbins, A. M. (2011b). Extreme conservation leads to recovery of the Virunga mountain gorillas. *PLoS ONE*, 6(6), e19788.

Robbins, M. M., & McNeilage, A. (2003). Home range and frugivory patterns of mountain gorillas in Bwindi Impenetrable National Park, Uganda. *International Journal of Primatology*, 24(3), 467-491.

Robbins, M. M., Robbins, A. M., Gerald-Steklis, N., & Steklis, H. D. (2005). Long-term dominance relationships in female mountain gorillas: Strength, stability and determinants of rank. *Behaviour*, 142(6), 779-810.

Robbins, M. M., Robbins, A. M., Gerald-Steklis, N., & Steklis, H. D. (2007). Socioecological influences on the reproductive success of female mountain gorillas (*Gorilla beringei beringei*). *Behavioral Ecology and Sociobiology*, 61(6), 919-931.

Robbins, M. M., & Sawyer, S. (2007). Intergroup encounters in mountain gorillas of Bwindi Impenetrable National Park, Uganda. *Behaviour*, 144(12), 1497-1519.

Rogers, M. E., Abernethy, K., Bermejo, M., Cipolletta, C., Doran, D., McFarland, K., Nishihara, T., Remis, M., & Tutin, C. E. (2004). Western gorilla diet: A synthesis from six sites. *American Journal of Primatology*, 64(2), 173-192.

Rogers, M. E., Maisels, F., Williamson, E. A., Fernandez, M., & Tutin, C. E. (1990). Gorilla diet in the Lope Reserve, Gabon. *Oecologia*, 84(3), 326-339.

Rouquet, P., Froment, J. M., Bermejo, M., Kilbourn, A., Karesh, W., Reed, P., Kumulungui, B., Yaba, P., Délicat, A., Rollin, P. E., & Leroy, E. M. (2005). Wild animal mortality monitoring and human Ebola outbreaks, Gabon and Republic of Congo, 2001–2003. *Emerging Infectious Diseases*, 11(2), 283.

Roy, J., Vigilant, L., Gray, M., Wright, E., Kato, R., Kabano, P., Basabose, A., Tibenda, E., Kühl, H. S., & Robbins, M. M. (2014). Challenges in the use of genetic mark–recapture to estimate the population size of Bwindi mountain gorillas (*Gorilla beringei beringei*). *Biological Conservation*, 180, 249-261.

Ruff, C. B., Burgess, M. L., Junno, J. A., Mudakikwa, A., Zollikofer, C. P., Ponce de León, M. S., & McFarlin, S. C. (2018). Phylogenetic and environmental effects on limb bone structure in gorillas. *American Journal of Physical Anthropology*, 166(2), 353-372.

Salmi, R., Presotto, A., Scarry, C. J., Hawman, P., & Doran–Sheehy, D. M. (2020). Spatial cognition in western gorillas (*Gorilla gorilla*): An analysis of distance, linearity, and speed of travel routes. *Animal Cognition*, 23, 545-557.

Sarmiento, E. E., Butynski, T. M., & Kalina, J. (1996). Gorillas of Bwindi-Impenetrable Forest and the Virunga volcanoes: Taxonomic implications of morphological and ecological differences. *American Journal of Primatology*, 40(1), 1-21.

Sarmiento, E. E., & Oates, J. F. (2000). The Cross River gorillas: A distinct subspecies, *Gorilla gorilla diehli* Matschie 1904. *American Museum Novitates*, 3304, 1-55.

Savage, T. S., & Wyman, J. (1847). *Notice of the external characters and habits of Troglodytes gorilla, a new species of orang from the Gaboon River*. Little Brown and Company.

Schaller, G. E. (1963). *The mountain gorilla: Ecology and behavior*. University of Chicago Press.

Seaman, M. I., Deinard, A. S., & Kidd, K. K. (1999). Incongruence between mitochondrial and nuclear DNA estimates of divergence between *Gorilla* subspecies. *American Journal of Physical Anthropology*, 108(S28), 247.

Seiler, N., Boesch, C., Mundry, R., Stephens, C., & Robbins, M. M. (2017). Space partitioning in wild, non-territorial mountain gorillas: The impact of food and neighbours. *Royal Society Open Science*, 4(11), 1-13.

Seiler, N., Boesch, C., Stephens, C., Ortmann, S., Mundry, R., & Robbins, M. M. (2018). Social and ecological correlates of space use patterns in Bwindi mountain gorillas. *American Journal of Primatology*, 80(4), 1-13.

Sicotte, P. (1993). Inter-group encounters and female transfer in mountain gorillas: Influence of group composition on male behavior. *American Journal of Primatology*, 30(1), 21-36.

Sicotte, P. (1994). Effect of male competition on male-female relationships in bi-male groups of mountain gorillas. *Ethology*, 97(1-2), 47-64.

Sicotte, P. (2001). Female mate choice in mountain gorillas. In M. M. Robbins, P. Sicotte, & K. G. Stewart (Eds.), *Mountain gorillas: Three decades of research at Karisoke* (pp. 59-87). Cambridge University Press.

Sienne, J., Buchwald, R., & Wittemyer, G. (2014). Plant mineral concentrations related to foraging preferences of western lowland gorilla in Central African forest clearings. *American Journal of Primatology*, 76(12), 1115-1126.

Skorupa, J. P. (1988). *The effect of selective timber harvesting on rain-forest primates in Kibale Forest, Uganda*. [Doctoral dissertation, University of California].

Smith, R. J., & Jungers, W. L. (1997). Body mass in comparative primatology. *Journal of Human Evolution*, 32(6), 523-559.

Southern, L. M., Deschner, T., & Pika, S. (2021). Lethal coalitionary attacks of chimpanzees (*Pan troglodytes troglodytes*) on gorillas (*Gorilla gorilla gorilla*) in the wild. *Scientific Reports* 11, 14673. https://doi.org/10.1038/s41598-021-93829-x.

Stanford, C. B., & Nkurunungi, J. B. (2003). Behavioral ecology of sympatric chimpanzees and gorillas in Bwindi Impenetrable National Park, Uganda: Diet. *International Journal of Primatology*, 24(4), 901-918.

Steenbeek, R., & van Schaik, C. P. (2001). Competition and group size in Thomas's langurs (*Presbytis thomasi*): The folivore paradox revisited. *Behavioral Ecology and Sociobiology*, 49(2), 100-110.

Stoinski, T. S., Perdue, B. M., & Legg, A. M. (2009a). Sexual behavior in female western lowland gorillas (*Gorilla gorilla gorilla*): Evidence for sexual competition. *American Journal of Primatology*, 71(7), 587-593.

Stoinski, T. S., Rosenbaum, S., Ngaboyamahina, T., Vecellio, V., Ndagijimana, F., & Fawcett, K. (2009b). Patterns of male reproductive behaviour in multi-male groups of mountain gorillas: Examining theories of reproductive skew. *Behaviour*, 146(9), 1193-1215.

Stoinski, T. S., Vecellio, V., Ngaboyamahina, T., Ndagijimana, F., Rosenbaum, S., & Fawcett, K. A. (2009c). Proximate factors influencing dispersal decisions in male mountain gorillas, *Gorilla beringei beringei*. *Animal Behaviour*, 77(5), 1155-1164.

Stokes, E. J., Parnell, R. J., & Olejniczak, C. (2003). Female dispersal and reproductive success in wild western lowland gorillas (*Gorilla gorilla gorilla*). *Behavioral Ecology and Sociobiology*, 54(4), 329-339.

Stokes, E. J., Strindberg, S., Bakabana, P. C., Elkan, P. W., Iyenguet, F. C., Madzoké, B., Malanda, G. A. F., Mowawa, B. S., Moukoumbou, C., Ouakabadio, F. K., & Rainey, H. J. (2010). Monitoring great ape and elephant abundance at large spatial scales: Measuring effectiveness of a conservation landscape. *PLoS ONE*, 5(4), e10294.

Strier, K. B. (1994). Myth of the typical primate. *American Journal of Physical Anthropology*, 37(S19), 233-271.

Strindberg, S., Maisels, F., Williamson, E. A., Blake, S., Stokes, E. J., Aba'a, R., Abitsi, G., Agbor, A., Ambahe, R. D., Bakabana, P. C., & Bechem, M. (2018). Guns, germs, and trees determine density and distribution of gorillas and chimpanzees in western Equatorial Africa. *Science Advances*, 4(4), eaar2964.

Strona, G., Stringer, S. D., Vieilledent, G., Szantoi, Z., Garcia-Ulloa, J., & Wich, S. A. (2018). Small room for compromise between oil palm cultivation and primate conservation in Africa. *Proceedings of the National Academy of Sciences*, 115(35), 8811-8816.

Sunderland-Groves, J., Ekinde, A., & Mboh, H. (2009). Nesting behavior of *Gorilla gorilla diehli* at Kagwene Mountain, Cameroon: Implications for assessing group size and density. *International Journal of Primatology*, 30(2), 253-266.

Taylor, A. B. (1997). Relative growth, ontogeny, and sexual dimorphism in gorilla (*Gorilla gorilla gorilla* and *G. g. beringei*): Evolutionary and ecological considerations. *American Journal of Primatology*, 43(1), 1-31.

Thalmann, O., Fischer, A., Lankester, F., Pääbo, S., & Vigilant, L. (2006). The complex evolutionary history of gorillas: Insights from genomic data. *Molecular Biology and Evolution*, 24(1), 146-158.

Thalmann, O., Wegmann, D., Spitzner, M., Arandjelovic, M., Guschanski, K., Leuenberger, C., Bergl, R. A., & Vigilant, L. (2011). Historical sampling reveals dramatic demographic changes in western gorilla populations. *BMC Evolutionary Biology*, 11(1), 85.

Thierry, B. (2007). Unity in diversity: Lessons from macaque societies. *Evolutionary Anthropology*, 16(6), 224-238.

Tocheri, M. W., Solhan, C. R., Orr, C. M., Femiani, J., Frohlich, B., Groves, C. P., Harcourt-Smith, W. E., Richmond, B. G., Shoelson, B., & Jungers, W. L. (2011). Ecological divergence and medial cuneiform morphology in gorillas. *Journal of Human Evolution*, 60(2), 171-184.

Tutin, C. E. G. (2001). Saving the gorillas (*Gorilla g. gorilla*) and chimpanzees (*Pan t. troglodytes*) of the Congo Basin. *Reproduction, Fertility and Development*, 13(8), 469-476.

Tutin, C. E. G., & Fernandez, M. (1984). Nationwide census of gorilla and chimpanzee populations in Gabon. *American Journal of Primatology*, 6, 313-336.

Tutin, C. E. G., & Fernandez, M. (1985). Foods consumed by sympatric populations of *Gorilla g. gorilla* and *Pan t. troglodytes* in Gabon: Some preliminary data. *International Journal of Primatology*, 6(1), 27-43.

Tutin, C. E. G., & Fernandez, M. (1991). Responses of wild chimpanzees and gorillas to the arrival of primatologists: Behaviour observed during habituation. In H. O. Box (Ed.), *Primates responses to environmental change* (pp. 187-197). Chapman & Hall.

Tutin, C. E. G., & Fernandez, M. (1993). Composition of the diet of chimpanzees and comparisons with that of sympatric lowland gorillas in the Lopé Reserve, Gabon. *American Journal of Primatology*, 30(3), 195-211.

Tutin, C. E. G., Fernandez, M., Rogers, M. E., Williamson, E. A., & McGrew, W. C. (1991). Foraging profiles of sympatric lowland gorillas and chimpanzees in the Lope Reserve, Gabon. *Philosophical Transactions of the Royal Society. Series B: Biological Sciences*, 334(1270), 179-186.

Tutin, C. E. G., Parnell, R. J., White, L. J., & Fernandez, M. (1995). Nest building by lowland gorillas in the Lopé Reserve, Gabon: Environmental influences and implications for censusing. *International Journal of Primatology*, 16(1), 53-76.

Tuttle, R. H., & Basmajian, J. V. (1974). Electromyography of forearm musculature in *Gorilla* and problems related to knuckle walking. In F. A. Jenkins (Ed.), *Primate Locomotion* (pp. 293-347). Academic Press.

van Schaik, C. P., & Janson, C. H. (Eds.). (2000). *Infanticide by males and its implications*. Cambridge University Press.

van Schaik, C. P., & Kappeler, P. M. (1997). Infanticide risk and the evolution of male–female association in primates. *Proceedings of the Royal Society of London. Series B: Biological Sciences*, 264(1388), 1687-1694.

van Schaik, C. P., & Pfannes, K. R. (2005). Tropical climates and phenology: A primate perspective. *Cambridge Studies in Biological and Evolutionary Anthropology*, 44, 23-54.

van Schaik, C. P., & Van Hooff, J. (1983). On the ultimate causes of primate social systems *Behaviour*, 1, 91-117.

Vedder, A. L. (1984). Movement patterns of a group of free-ranging mountain gorillas (*Gorilla gorilla beringei*) and their relation to food availability. *American Journal of Primatology*, 7(2), 73-88.

Vedder, A. L. (1989). *Feeding ecology and conservation of the mountain gorilla (*Gorilla gorilla beringei). [Doctoral dissertation, University of Wisconsin].

Walsh, P. D., Abernethy, K. A., Bermejo, M., Beyers, R., De Wachter, P., Akou, M. E., Huijbregts, B., Mambounga, D. I., Toham, A. K., Kilbourn, A. M., & Lahm, S. A., (2003). Catastrophic ape decline in western equatorial Africa. *Nature*, 422(6932), 611-614.

Walsh, P. D., Breuer, T., Sanz, C., Morgan, D., & Doran-Sheehy, D. (2007). Potential for Ebola transmission between gorilla and chimpanzee social groups. *American Naturalist*, 169(5), 684-689.

Waterman, P. G., Choo, G. M., Vedder, A. L., & Watts, D. (1983). Digestibility, digestion-inhibitors and nutrients of herbaceous foliage and green stems from an African montane flora and comparison with other tropical flora. *Oecologia*, 60(2), 244-249.

Watts, D. P. (1984). Composition and variability of mountain gorilla diets in the central Virungas. *American Journal of Primatology*, 7(4), 323-356.

Watts, D. P. (1989). Infanticide in mountain gorillas: New cases and a reconsideration of the evidence. *Ethology*, 81(1), 1-18.

Watts, D. P. (1990a). Ecology of gorillas and its relation to female transfer in mountain gorillas. *International Journal of Primatology*, 11(1), 21-45.

Watts, D. P. (1990b). Mountain gorilla life histories, reproductive competition, and sociosexual behavior and some implications for captive husbandry. *Zoo Biology*, 9(3), 185-200.

Watts, D. P. (1991a). Harassment of immigrant female mountain gorillas by resident females. *Ethology*, 89(2), 135-153.

Watts, D. P. (1991b). Strategies of habitat use by mountain gorillas. *Folia Primatologica*, 56(1), 1-16.

Watts, D. P. (1992). Social relationships of immigrant and resident female mountain gorillas. I. Male-female relationships. *American Journal of Primatology*, 28(3), 159-181.

Watts, D. P. (1994a). Agonistic relationships between female mountain gorillas (*Gorilla gorilla beringei*). *Behavioral Ecology and Sociobiology*, 34(5), 347-358.

Watts, D. P. (1994b). The influence of male mating tactics on habitat use in mountain gorillas (*Gorilla gorilla beringei*). *Primates*, 35(1), 35-47.

Watts, D. P. (1994c). Social relationships of immigrant and resident female mountain gorillas, II. Relatedness, residence, and relationships between females. *American Journal of Primatology*, 32(1), 13-30.

Watts, D. P. (1996). Comparative socio-ecology of gorillas. In W. C. McGrew, L. F Marchant, & T. Nishida (Eds.), *Great ape societies* (pp. 16-28). Cambridge University Press.

Watts, D. P. (1998a). Long-term habitat use by mountain gorillas (*Gorilla gorilla beringei*). Reuse of foraging areas in relation to resource abundance, quality, and depletion. *International Journal of Primatology*, 19(4), 681-702.

Watts, D. P. (1998b). Long-term habitat use by mountain gorillas (*Gorilla gorilla beringei*). Consistency, variation, and home range size and stability. *International Journal of Primatology*, 19(4), 651-680.

Watts, D. P. (2000). Causes and consequences of variation in male mountain gorilla life histories and group membership. In P. M. Kappeler (Ed.), *Primate males* (pp. 169-179). Cambridge University Press.

Weber, A. W., & Vedder, A. (1983). Population dynamics of the Virunga gorillas: 1959–1978. *Biological Conservation*, 26(4), 341-366.

White, L. J. T. (1992). *The effects of mechanised selective logging on the flora and mammalian fauna of the Lope Reserve, Gabon.* [Doctoral dissertation, University of Edinburgh].

White, L. J. T. (1994). The effects of commercial mechanised logging on forest structure and composition on a transect in the Lopé Reserve, Gabon. *Journal of Tropical Ecology*, 10, 309-318.

Wich, S. A., Garcia-Ulloa, J., Kühl, H. S., Humle, T., Lee, J. S., & Koh, L. P. (2014). Will oil palm's homecoming spell doom for Africa's great apes? *Current Biology*, 24(14), 1659-1663.

Wich, S. A., Utami-Atmoko, S. S., Setia, T. M., Rijksen, H. D., Schürmann, C., van Hooff, J., & van Schaik, C. P. (2004). Life history of wild Sumatran orangutans (*Pongo abelii*). *Journal of Human Evolution*, 47(6), 385-398.

Wilkie, D. S., & Carpenter, J. F. (1999). Bushmeat hunting in the Congo Basin: An assessment of impacts and options for mitigation. *Biodiversity and Conservation*, 8(7), 927-955.

Williams, S. A. (2012). Variation in anthropoid vertebral formulae: Implications for homology and homoplasy in hominoid evolution. *Journal of Experimental Zoology Part B: Molecular and Developmental Evolution*, 318(2), 134-147.

Williamson, E. A., Tutin, C. E., Rogers, M. E., & Fernandez, M. (1990). Composition of the diet of lowland gorillas at Lopé in Gabon. *American Journal of Primatology*, 21(4), 265-277.

Williamson, L., & Usongo, L. (1995). *Survey of elephants, gorillas, and chimpanzees. Reserve de Fauna du Dja, Cameroun Project.* Ecosystèmes Forestiers d'Afrique Centrale.

Wood, B. A. (1979). Relationship between body size and long bone lengths in *Pan* and *Gorilla*. *American Journal of Physical Anthropology*, 50(1), 23-25.

Wrangham, R. (1979). On the evolution of ape social systems. *Social Science Information*, 18(3), 336-368.

Yamagiwa, J. (1986). Activity rhythm and the ranging of a solitary male mountain gorilla (*Gorilla gorilla beringei*). *Primates*, 27(3), 273-282.

Yamagiwa, J. (1987). Intra- and inter-group interactions of an all-male group of Virunga Mountains gorillas (*Gorilla gorilla beringei*). *Primates*, 28, 1-30.

Yamagiwa, J. (2004). Foraging strategies of eastern lowland gorillas in Kahuzi-Biega National Park. *Folia Primatologica*, 75, 180-181.

Yamagiwa, J., Basabose, A. K., Kaleme, K., & Yumoto, T. (2003). Within group feeding competition and socioecological factors influencing social organization of gorillas in the Kahuzi-Biega National Park, Democratic Republic of Congo. In A. B. Taylor, & M. L. Goldsmith (Eds.), *Gorilla biology: A multidisciplinary perspective* (pp. 328-357). Cambridge University Press.

Yamagiwa, J., & Kahekwa, J. (2001). Dispersal patterns, group structure, and reproductive parameters of eastern lowland gorillas at Kahuzi in the absence of infanticide. In A. M. Robbins, P. Sicotte, & K. J. Stewart (Eds.), *Mountain gorillas: Three decades of research at Karisoke* (pp. 89-122). Cambridge University Press.

Yamagiwa, J., Maruhashi, T., Yumoto, T., & Mwanza, N. (1996). Dietary and ranging overlap in sympatric gorillas and chimpanzees in Kahuzi-Biega National Park, Zaire. In W. C. McGrew, L. F. Marchant, & T. Nishida (Eds.), *Great ape societies* (pp. 82-98). Cambridge University Press.

Yamagiwa, J., Mwanza, N., Spangenberg, A., Maruhashi, T., Yumoto, T., Fischer, A., & Steinhauer-Burkart, B. (1993). A census of the eastern lowland gorillas *Gorilla gorilla graueri* in Kahuzi-Biega National Park with reference to mountain gorillas *G. g. beringei* in the Virunga region, Zaire. *Biological Conservation*, 64(1), 83-89.

Yamagiwa, J., Mwanza, N., Spangenberg, A., Maruhashi, T., Yumoto, T., Fisher, A., Steinhauer-Burkart, B., & Refish, J. (1992). Population density and ranging pattern in Kahuzi-Biega National Park, Zaire: A comparison with sympatric population of gorillas. *African Study Monographs*, 13(4), 217-230.

Yamagiwa, J., Mwanza, N., Yumoto, T., & Maruhashi, T. (1994). Seasonal change in the composition of the diet of eastern lowland gorillas. *Primates*, 35(1), 1-14.

CHAPTER 23

Chimpanzees and Bonobos

CRICKETTE M. SANZ, ROBERT W. SUSSMAN, STEPHANIE MUSGRAVE, JUAN SALVADOR ORTEGA PERALEJO, AND DAVID B. MORGAN

THE GENUS *PAN* IS COMPOSED OF TWO SPECIES, the chimpanzee *Pan troglodytes* and the bonobo *Pan paniscus*. These allopatric species are characterized by distinct morphology, ecology, and behavior (see Table 23.1; Zihlman & Cramer, 1978; Jungers & Susman, 1984; Coolidge & Shea, 1992; White, 1996). Chimpanzees have been allocated into four genetically and geographically distinct subspecies: *P. t. verus* in West Africa, *P. t. troglodytes* in Central Africa, *P. t. schweinfurthii* in East Africa, and *P. t. ellioti* (Oates et al., 2009) (previously *P. t. vellerosus*) in Nigeria and northwestern Cameroon (Gonder et al., 1997, 2006; Gagneux et al., 1999; Groves, 2001).

Studies of genetic diversity have provided insight into the phylogeny of the chimpanzee subspecies, historic population sites, and patterns of gene flow (Fischer et al., 2004). Using rates of mutation per generation and direct observations of generation times in living apes, scientists have recently recalibrated population split times of African apes (Langergraber et al., 2012). The estimated time for the split of the *Pan* species is between 1.82–2.55 and 1.45–2.37 mya (Becquet et al., 2007; Pruefer et al., 2012). Subspecies divergence times based on complete mitochondrial DNA sequences have yielded estimates of 1.1–0.76 mya for western and eastern/central chimpanzees and 0.25–0.18 mya for eastern and central chimpanzees (Stone et al., 2010). Chimpanzees have maintained relatively long-distance gene flow, and it has been reported that chimpanzees separated by up to 900 km shared nearly identical sequences from two different mitochondrial DNA haplotype groups (Morin et al., 1993, 1994). The extent of genetic differentiation among groups of chimpanzees may not actually support the idea that subspecies are genetically distinct entities (Fischer et al., 2006). This conclusion is supported by a recent landscape-scale analysis of genetic differentiation of the central (*P. t. troglodytes*) and the eastern (*P. t. schweinfurthii*) chimpanzees, which shows clinal genetic variation rather than a strong subspecies break (Fuenfstueck et al., 2015).

Historically, the geographic range of *P. troglodytes* was continuous across the African equatorial rainforest zone and neighboring areas of forest–savanna mosaic and woodland (see Figure 23.1). Currently, populations of chimpanzees are located in patchy, and sometimes isolated forest blocks across this region. The western limit of the chimpanzee is the Atlantic Ocean at Senegal, and the eastern limit is in Uganda. The ecological flexibility of the chimpanzee is reflected in this expansive range, which is much larger and more structurally diverse than the range of other great apes (Kano, 1984). In contrast, the distribution of *P. paniscus* is restricted to the region south of the Congo River.

FIELD STUDIES PAST AND PRESENT

The chimpanzee was first described by Western science in 1641 by the Dutch physician Nicolaes Tulp. In 1699 Edward Tyson, often considered to be the father of primatology because of his excellent anatomical studies, completed the first anatomical treatise on the chimpanzee. Because the size of these two animals differed between paintings, it has been claimed that the animal described by Tulp was a bonobo and that Tyson's description was of a chimpanzee (Reynolds, 1967).

The first attempt to study chimpanzees in their natural habitat came approximately 200 years later. In the late 1800s, Robert Garner traveled to Central Africa and built a cage in the forest with the hopes of observing wild chimpanzees and gorillas (Garner, 1896). A few apes approached the cage, but little was learned about the behavior or ecology of chimpanzees. In 1930, Robert Yerkes sent an expedition to West Africa to capture wild chimpanzees for a research colony of chimpanzees in Orange Park, Florida. Also, he had hoped the mission would gather information on the natural history of wild apes. Henry Nissen led the expedition and succeeded in observing the apes for 49 days, dispelling a number of fictitious ideas about wild chimpanzees (Nissen, 1931).

Modern research on free–ranging chimpanzees did not begin until the early 1960s with the initiation of two longitudinal studies of East African chimpanzee populations on the shores of Lake Tanganyika in Tanzania. Under the guidance of Louis Leakey, Jane Goodall founded the Gombe site in 1960 (Goodall, 1965, 1986). Leakey was a paleontologist interested in ape behavior, particularly in habitats similar to those of ancestral hominins.

Table 23.1 Overview of Chimpanzee and Bonobo Socioecology

Socioecological Variable	Chimpanzee (*Pan troglodytes*)	Bonobo (*Pan paniscus*)
Geographic location	Across Equatorial Africa Example sites: Gombe, Kibale, Budongo, Goualougo Triangle, Taï, Bossou, Fongoli	Democratic Republic of Congo *only* Example sites: Wamba, Lomako, Lukuru, LuiKotale
Conservation status	Endangered (*P.t. verus*: Critically Endangered) IUCN Red List Population Trend: Decreasing [a]	Endangered IUCN Red List Population Trend: Decreasing [a]
Sympatry with other apes	Overlap with gorillas in Central Africa	Do *not* overlap with gorillas or chimpanzees
Morphology	Dark lips, shorter hair; light-faced infants; more prominent brow ridge	Pink lips, longer hair parted in the middle; dark-faced infants; lighter build; longer legs; paedomorphism
Sexual dimorphism	Male/female bodyweight ratio 1.29; males have sagittal crest and larger canines	Male/female body weight ratio 1.36; less cranio-dental sexual dimorphism than in chimpanzees; males lack sagittal crest
Feeding ecology	Frugivorous; highly diverse diet, including meat; hunting and tool use observed across populations; diverse foraging tool kits	Frugivorous diet similar to chimpanzees, but consume more terrestrial herbaceous vegetation; some vertebrate hunting observed; minimal foraging tool use
Inter-community interactions	Territorial; boundary patrols; lethal aggression can occur	Displays common, but lethal aggression rarely if ever occurs; extra-group mating occurs regularly; food sharing observed
Social structure	Individual-based fission fusion; mixed parties are most common; higher estrus sex ratio (fewer estrus females in proportion to males); smaller parties	Individual-based fission-fusion; mixed parties are most common; lower estrus sex ratio; larger, more stable parties, with higher female attendance
Dominance	Male dominance	Female-male co-dominance; highest ranks in mixed-sex hierarchies often held by females and the sons of high-ranking females
Female-Female sociality	Females maintain dominance hierarchies. Sociability varies between East and West Africa, but females generally have weaker social bonds compared to males.	Females are highly gregarious and form strong relationships, which are maintained through sociosexual contact (genital-genital rubbing); female coalitions protect against male aggression; food sharing common.
Male-Female sociality	More male to female contact aggression. Affiliation occurs less and is more dependent on estrus cycle. Mother-son bond wanes at adolescence	More socializing between males and females, less male to female aggression; mother-son bond is primary and continues throughout life
Male-Male sociality	Dominance hierarchies— Clearly defined linear hierarchies; males have strong social bonds, form alliances, and cooperate in multiple contexts such as inter-community interactions	Dominance hierarchies— Ranks less clearly defined; weaker social bonds; more agonism occurs and males are less likely to support each other in conflicts
Reproductive strategies	Promiscuous mating, sperm competition; male rank affects reproductive success— Depending on population, a spectrum ranging from male sexual coercion of females to female choice; possessive mating, consortships, mate guarding, infanticide observed	Promiscuous mating, sperm competition; male rank affects reproductive success— More estrus cycles, pseudo-estrus, reduced predictability of ovulation; cycle characteristics give females leverage over males and minimize potential effectiveness of male-male competition
Dispersal	Males remain in natal group and females disperse	Males remain in natal group and females disperse

[a] IUCN (2022).

In 1961, the Japanese primatologist Kinji Imanishi organized the Kyoto University Africa Primatological Expedition to study wild chimpanzees. This eventually led to the establishment of the Mahale chimpanzee study site (Nishida, 1968, 1990, 2012). These are two of the longest-running primate field studies, with study groups representing third-generation descendants of the original study animals.

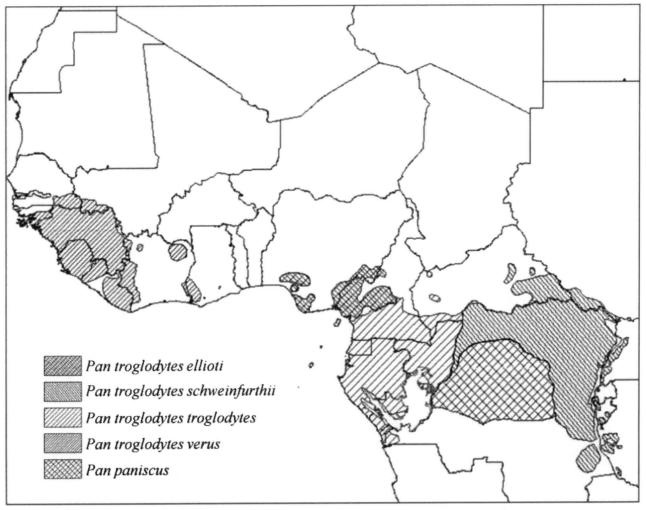

Figure 23.1 Geographic distributions of chimpanzees and bonobos.

Soon after, long-term studies of wild chimpanzees were also established in the Budongo and Kibale Forests of Uganda. In 1962, Vernon Reynolds spent 10 months observing chimpanzees in the Budongo Forest and established a permanent study site in 1994 (Reynolds, 1965, 1967, 2005; Reynolds & Reynolds 1965). Ghiglieri (1984) spent three field seasons observing chimpanzees at Kibale between 1976 and 1981, but a chimpanzee study site was not established at this site until 1987 by Richard Wrangham. Currently, there are two active research sites within the Kibale National Park: Kanyawara and Ngogo (Potts et al., 2011).

Bossou (Guinea) was the first research site initiated in West Africa dedicated to the study of *P. t. verus*. Albrecht and Dunnett conducted a study of chimpanzee behavior at this site between 1968 and 1969 (Albrecht & Dunnett, 1971). Beginning in 1976, Yukimaru Sugiyama began conducting research at Bossou for a few months each year, establishing a continuous research presence in 1987 (Sugiyama, 1989). Tetsuro Matsuzawa expanded the scope of research at Bossou by installing an outdoor

laboratory to study tool use and cognition (Matsuzawa et al., 2011). In 1976, Boesch and Boesch-Achermann began studying chimpanzees in the Taï Forest of Côte d'Ivoire and established a permanent research presence in 1979 (Boesch & Boesch-Achermann, 2000). The Fongoli chimpanzee community in Senegal has been studied by Jill Pruetz since 2001.

Until recently, relatively little was known about the central and Nigerian subspecies of chimpanzee despite valiant attempts to habituate chimpanzee communities at Lopé Reserve in Gabon and Guga in the Republic of Congo (Tutin & Fernandez, 1991; Tutin et al., 1991; Kuroda et al., 1996). In 1999, David Morgan began studying sympatric chimpanzees and gorillas in the Goualougo Triangle of northern Republic of Congo. It was possible to identify and observe the apes during initial contacts at this site because of their limited experience with and lack of disturbance by humans (Morgan & Sanz, 2003). A new study of the central subspecies of chimpanzee was initiated on the coast of Gabon at Loango in 2005 by Christophe Boesch. The site is currently directed by

Tobias Deschner and Simone Pika. In 2000, Volker Sommer established a site in Gashaka to study the Nigerian subspecies. Researchers have documented the feeding ecology and tool-using behavior of these chimpanzees through indirect traces, but detailed behavioral reports are still forthcoming (Sommer et al., 2004). In reflecting upon the insights generated from comparative studies, it is clear that continued research on different subspecies and populations is imperative to understanding the behavioral diversity of *Pan troglodytes*.

Whereas there are a number of chimpanzee communities habituated to human observations, scientists have struggled to establish and maintain long-term studies of wild bonobos due to political turmoil and lack of infrastructure in the Democratic Republic of Congo (DRC). Most of our current knowledge of wild bonobo behavioral ecology has been informed by reports from Takeshi Furuichi's research team at Wamba (Furuichi et al., 1998, 2012) and Gottfried Hohmann's research at LuiKotale (Hohmann & Fruth, 2008). Other bonobo study sites include Lomako, Lukuru, and Kokolopori (Surbeck et al., 2017a). Lomako is now part of the DRC's first community-managed nature reserve, the Lomako-Yokokala

Faunal Reserve (Dupain et al., 2000). In 1998, Jo Thompson bought 34 km² of pristine forest in Lukuru to create the Bososandja Faunal Reserve (Myers Thompson, 2002). Kokolopori Bonobo Reserve was officially established in 2009 and is a community-managed project in collaboration with Vie Sauvage and the Bonobo Conservation Initiative.

LIFE HISTORY

Gestation length of chimpanzees averages 228–229 days (range: 202–248 days, n = 56) (Nissen & Yerkes, 1943; Martin et al., 1978). The infant is born altricial and helpless (Plooij, 1984; Vanderijtplooij & Plooij, 1987). The first two months of life are spent on the mother's ventral surface (Figure 23.2), and infants may be awake for only about 30% of the time. While in the earliest months, the infant may sleep for more than half of the day, by four months of age napping may be limited to less than an hour per day. During the fourth month, infants begin to manipulate objects and chew solid foods. The infant may climb onto its mother's back and make initial excursions beyond her reach. Siblings are close to the mother and infant for over 90% of the time, whereas other individuals

Figure 23.2 Adult female chimpanzee with infant in the Moto Community of the Goualougo Triangle, Republic of Congo.
(Photo credit: Crickette Sanz, WCS-Congo)

of the community are seldom near the infant during the first three to six months. By eight months onward, the infant begins to approach and interact with other community members (Plooij, 1984).

With increasing age, the infant spends more and more time at greater distances from its mother. At 20 months, over 50% of the time is spent off the mother, but in close proximity. Toward the end of the second year, the infant begins to move farther away (Plooij, 1984; Vanderijtplooij & Plooij, 1987). At Gombe, male infants, compared to female infants, begin to travel independently at younger ages and show increased distance from their mothers by three years of age (Lonsdorf et al., 2014). A three-year-old is allowed to suckle, ride on its mother's back and share her food, but by four to five years of age the infant is increasingly forced to move and feed on its own. A comparison of immature female bonobos and chimpanzees from LuiKotale and Gombe, respectively, indicates that these milestones occur at similar ages between species (Figure 23.3), suggesting conserved patterns of maturation within *Pan* (Lee et al., 2019).

Weaning is a gradual process in which the mother takes an active part, but there is very little aggression involved (Clark, 1977). Cessation of suckling occurs between four and seven years of age (Lonsdorf et al., 2019). In a review of the life history of 65 offspring of 29 mothers at Gombe, Lonsdorf and colleagues (2019) found that male offspring were less likely than female offspring to

wean by a given age, and the weaned age of males varied more than the weaned age of females. There is variation in time that females require to resume cycling, as ovarian activity is closely correlated with energetic condition. Research at Kibale has shown that mothers inhabiting lower-quality areas of the Kanyawara community home range experienced higher metabolic loads of lactation than mothers in the higher-quality areas of the community range (Thompson et al., 2012). At Ngogo, alloparenting is associated with accelerated weaning. This could be because infant handling led to longer intervals between nursing bouts and reduced rates of nursing, which may be associated with earlier resumption of ovulation (Boesch-Achermann & Boesch, 1994; Badescu et al., 2016).

Males associate with their mother until approximately 8 years of age (Pusey, 1983). They then progressively spend less time with their mother and increasing amounts of time with nonfamily members, primarily adult males and estrus females. A male's weight increases sharply between 9–12 years of age, with maximum weight reached at 16–17 years. Adult males weigh between 35–45 kg. Sexual maturity is estimated to be achieved at 12–14 years of age (Goodall, 1986). By late adolescence, and for the rest of his life, a male may spend considerable time alone. However, adult males are generally more sociable with nonfamily members than are anestrus females.

Females remain with their mother until about 10 years of age, when they begin to show their first sexual

Figure 23.3 Adult female bonobo with infant in LuiKotale, Democratic Republic of Congo.
(Photo credit: LKBP, Kathryn Judson)

swellings (Pusey, 1983; Goodall, 1986). During this period of late adolescence, the female often leaves her mother during estrus to travel with males. At this time, she may visit neighboring communities and even transfer to one of these. Females normally have a period of adolescent sterility, with first birth occurring between 13–15 years of age. After her first infant is born, the female generally settles into a community. Adult weight of a female is between 32–37 kg and is reached at about 19 years of age (Goodall, 1986). Mean interbirth interval averages between 66–72 months, with a much shorter interval of 12–16 months until the next conception if an infant dies. Fecundity varies throughout a female's lifetime but tends to be highest between 20–35 years of age. Females observed from a young age produced an average of 3.9 offspring (with 1.4 surviving past age of weaning). This is in contrast to females observed from middle age (18–33 years old) until death (31–48 years old) who gave birth to an average of 2.7 offspring (with 2.0 surviving past weaning) (Nishida et al., 2003). These results suggest that although older mothers produce fewer offspring, they have a higher rate of success in raising infants to the age of weaning.

Long-term research is revealing some significant differences in life-history characteristics between chimpanzee populations. A multisite examination of chimpanzee life-history reported a mean mortality rate of 7% across four sites (Hill et al., 2001), but subsequent reports indicate that this may be overestimated. For example, Kanyawara reported a mean mortality rate of 4% per year (Muller & Wrangham, 2014). Further research is needed to determine the ecological and potential anthropogenic factors shaping these differences. There is consistency in the fact that life expectancy for females tends to be longer than for males among chimpanzees. In the Kanyawara community of Kibale, life expectancy at birth of females is 4.5 years higher than for males (21.6 versus 17.1) (Muller & Wrangham, 2014). Fifty percent of females in this community are expected to survive to 15 years of age compared to only 34% of males (Muller & Wrangham, 2014).

HABITAT

Chimpanzees were not restricted to forest refugia during climatic shifts in the Pleistocene and have been referred to as "historical ecological generalists" (Goldberg, 1998). They are capable of maintaining gene flow across many different habitat types, including montane forests, mid-altitude rainforests, lowland rainforests, deciduous dry forests, evergreen riverine forests, thickets, grasslands, savannas, and swamp forest (Itani, 1979; McGrew et al.,

1981; Collins & McGrew, 1988; Moore, 1996; Goldberg, 1998; Pruetz et al., 2002).

Chimpanzees are known for their behavioral plasticity and have adopted unique behaviors to cope with diverse environments. Within dry habitats, chimpanzees tend to be found in gallery forests, thickets, and grasslands (McGrew et al., 1981; Hunt & McGrew, 2002; Lanjouw, 2002). They have also adopted particular behavior patterns that enable them to cope with extreme environments. At Fongoli, for example, cortisol levels are positively correlated with environmental temperature, and urinary creatinine levels peak during periods of lowest water availability and highest temperatures, suggesting susceptibility to heat stress and dehydration (Wessling et al., 2018). These chimpanzees may cope through the use of pools and caves as well as through adoption of nocturnal activity (Pruetz, 2007, 2018).

Bonobos also show a high degree of flexibility with regard to habitat utilization. They have been observed in lowland rainforest, primary semi–deciduous Leguminosae forest, swamp forest, and dry forest/savanna mosaic (Hashimoto et al., 1998; Terada et al., 2015). Bonobos thrive in primary forest, but also successfully exploit secondary forest and abandoned fields (Badrian & Malenky, 1984; Kano & Mulavwa, 1984). Continued research will illuminate the extent to which bonobos rely on other habitats and cope with changing environments.

LOCOMOTION

Studies of free-ranging chimpanzee and bonobo populations show that they divide their time between the ground and the trees (Reynolds & Reynolds, 1965; Susman, 1984; Doran, 1993a; Doran & Hunt, 1994; Pontzer & Wrangham, 2004). Chimpanzees have been reported to spend over 60% of the day on the ground (Wrangham, 1977), and the vast majority of their locomotion time is spent knuckle-walking (Doran & Hunt, 1994). Further, chimpanzees spend approximately 10 times more energy per day on terrestrial travel than on vertical climbing. This suggests that their postcranial adaptations for arboreality are not a strategy to reduce daily energy expenditures but may reflect pressures to securely navigate arboreal environments to exploit food resources (Pontzer & Wrangham, 2004). Just as chimpanzees are "flexible" in the range of habitats they exploit, so are they flexible in their locomotion:

> Bonobos, chimpanzees, and gorillas in the wild are skilled climbers as well as adept terrestrial quadrupeds. Arboreal adaptations are evidenced by grasping halluces and long, curved toes, powerful long fingers, and

relatively long upper limbs. A major terrestrial specialization is seen in knuckle–walking adaptations.... Most would agree that knuckle–walking quadrupedalism and terrestriality in African apes is a secondary adaptation, one that is "superimposed" on the arboreal common ancestor of chimpanzees and gorillas... the diverse locomotor repertoire of chimpanzees is probably unequalled among other primates. (Susman, 1984: p.389)

Chimpanzees and bonobos differ in their locomotor anatomy in a number of subtle but important ways. Bonobos have more curved finger bones, longer arms in relation to legs, a long and narrow shoulder blade, and generally smaller body size. Correlated with these differences, currently available data suggest that bonobos may be more arboreal and use suspensory locomotion more frequently than chimpanzees (Kortlandt, 1962; Goodall, 1968; Doran & Hunt, 1994).

Doran and Hunt (1994) compared locomotor behavior between *Pan* species, subspecies, and populations. Hunt (1992) observed *P. t. schweinfurthii* at Gombe Stream National Park and the Mahale Mountains in Tanzania. Meanwhile, Doran (1993a,b) observed *P. t. verus* at the Taï National Park in Côte d'Ivoire and *P. paniscus* at Lomako Forest in the DRC. At all three of these sites, the chimpanzees or bonobos spend close to 90% of their overall locomotor time in quadrupedalism and around 97% in either quadrupedal or quadrumanous climb/scramble locomotion. Thus, less than 3% of travel time is spent in suspensory, bipedal, or leaping locomotor activities. Even while in the trees, chimpanzees spend most of their time (over 90%) in quadrupedal locomotion and little time using other types of locomotion. Researchers have also noted differences in locomotor behavior between populations of bonobos. Such differences could be the result of habitat differences, effects of provisioning, or degree of habituation. Locomotor behavior of bonobos residing in the Lomako Forest changed as they became more habituated to human observers (Susman, 1984; Doran, 1993a). As habituation increased, the frequency of bipedalism, leaping, and diving decreased while quadrumanous climbing/scrambling increased.

Locomotion patterns are also related to sex and body size. Chimpanzee and bonobo males at all three sites studied by Doran and Hunt (1994) spend more than 50% of their time on the ground, and females spend around 50% or less of their time on the ground. Larger males feed in the lower canopy and on the ground more often than smaller males at the same research site (Hunt, 1994). Interestingly, no difference has been observed between male and female chimpanzees in time spent on the ground or in the trees during travel or feeding activities

in the more closed canopy habitat of Taï (Doran, 1993b). Females, however, spend more time resting in the trees, and males are more terrestrial while resting. The greater tendency of females to rest above the ground in more protected environments could be related to their smaller body size, as well as the fact that females are often accompanied by infants and juveniles, which makes them more susceptible to predation. Females also sit more and lie down less than males when they rest. This might be a potential predator avoidance behavior, as sitting is a more alert way to rest.

The locomotor anatomy and behavior of great apes has long been used to inform reconstructions of early hominin locomotion. Early models of the evolution of bipedalism suggested different origins, including ancestral roots in knuckle-walking or climbing hominoid ancestors (Fleagle et al., 1981). Further, it has been posited that postural bipedalism may have preceded and been important in the evolution of bipedalism in locomotion, as stated by Hunt (1998: p.415): "The synthesis of chimpanzee ecology and australopithecine functional morphology yields a postural feeding hypothesis that suggests that australopithecines were semi-arboreal postural bipeds that specialized on gathering small, hard-husked fruit in short statured trees." More recently, interest has shifted from assessing morphological and behavioral homologies to developing biomechanical models that link locomotor performance to anatomy (Pontzer et al., 2014). While no living primate is a perfect model for the locomotor behavior of ancestral hominins, consistencies in the efficiency of various types of locomotion across taxa can provide insights on patterns within primate evolution. Analysis of energetics and biomechanics of chimpanzees and humans has shown that human bipedalism is much less costly than both quadrupedal and bipedal walking in chimpanzees (Sockol et al., 2007).

DIET

Specialization in Ripe Fruit

Chimpanzees and bonobos are ripe fruit specialists who regularly supplement their diets with vegetative plant parts and animal protein (*P. paniscus*: Badrian & Malenky 1984; Kano & Mulavwa, 1984; *P. t. verus*: Goodall, 1986; Sugiyama & Koman 1987, 1992; *P. t. schweinfurthii*: Isabirye-Basuta, 1989; Newton-Fisher, 1999a; Watts et al., 2012; Wrangham et al., 1998; *P. t. troglodytes*: Tutin & Fernandez, 1993; Morgan & Sanz 2006; Oelze et al., 2011). The chimpanzee diet is comprised of 60–80% fruit (Boesch & Boesch-Achermann, 2000; Watts et al., 2012), with lower proportions of fruit consumed by populations living in dryer savanna habitats (Hunt & McGrew, 2002). According to early reports, bonobos may spend a slightly higher proportion of their time foraging on fruit, between 70–85% (Badrian & Malenky, 1984; Kano & Mulavwa, 1984).

As with the majority of frugivorous primates, chimpanzees and bonobos are important seed dispersers. In the Kanyawara community at Kibale, 98.5% of chimpanzee fecal specimens contain seeds and these seeds have a higher germination rate than undigested seeds (Wrangham et al., 1994). Fig seeds are the most common seed type in feces, but large seeds also frequently pass through the digestive system intact. It has been estimated that chimpanzees at Kibale disperse 369 large seeds per kilometer per day (Wrangham et al., 1994). Due to their large home ranges, chimpanzees could be more important primary dispersers than monkeys. At LuiKotale (DRC), bonobos disperse 40% of the tree species at that site an average distance of 1.2 km from the parent tree (Beaune et al., 2013a). As an example, bonobos specialize in *Dialium* spp. which they can disperse to distances averaging 1.25 km within 24 hours of ingestion (Beaune et al., 2013b).

Chimpanzees pursue ripe fruit of rarely occurring species and opportunistically exploit figs (several species of *Ficus*) when they are available. Mahale chimpanzees spend most of their feeding time concentrating on fruits with high caloric content, and fibrous foods are consumed when fruit is rare (Matsumoto-Oda & Hayashi, 1999). Kanyawara chimpanzees concentrate on non-fig fruit when available but consume figs throughout the year. Terrestrial fibrous foods provide calories in the form of fermentable fiber during dry periods and times of fruit shortage. Immature leaves are eaten when available, but intake is not influenced by fruit abundance (Wrangham et al., 1991).

Densities of a particular species of *Ficus* have been shown to influence ape abundance in Uganda (Potts et al., 2009; Watts et al., 2012). Throughout 29 months of study, chimpanzees at Kibale relied heavily on figs and consumed more than 40 calories per minute when harvesting large figs (Wrangham et al., 1993). In contrast, fig species are rare or absent at some field sites in West and Central Africa. At Lopé in Gabon, it has been reported that figs are too rare to be a dependable food source for chimpanzees (Tutin et al., 1997).

The basic foraging pattern of the chimpanzee involves intensive search and use of isolated patches of fruit, as described by Ghiglieri (1984: pp.86–87): "By specializing on relatively rare patches of fruit, chimpanzees have taken a narrow ecological path which requires superior abilities to locate and exploit such resources." In his use of the term "specializing," Ghiglieri is not referring to the number of items or species in the diet of the chimpanzee but to their propensity to search for widely scattered, rare patches of food, i.e., a "patch" specialist. Although this adaptation is rare in large–sized animals, the advanced spatial abilities of chimpanzees may enable them to be successful in this niche (e.g., Janmaat et al., 2013). Chimpanzees may select foraging areas by considering the presence of multiple resource types. For example, at Ngogo, chimpanzee choice of a highly preferred fig species (*F. mucoso*) was determined by presence of other available resources in the area (Potts et al., 2016). Chimpanzees have also been documented to exploit agricultural habitats, as crops provide clumped food sources (Bryson-Morrison et al., 2017). Including crops may be an efficient foraging strategy, as crops provide high levels of sugars, are low in antifeedants and insoluble fiber, and provide a predictable and concentrated resource (McLennan & Ganzhorn, 2017).

Omnivory and Dietary Generalism

Although Ghiglieri considers chimpanzees "patch" specialists, they must be considered dietary generalists or omnivores in that they feed on many types of foods from many species of plants and animals. Both chimpanzees and bonobos supplement their frugivorous diet with leaves, shoots, buds, blossoms, grains, seeds, bark, wood, cambium, gum, and with invertebrate and vertebrate prey (Hladik, 1973, 1977; Wrangham, 1977; Nishida & Uehara, 1983; Isabirye-Basuta, 1989; Newton-Fisher, 1999a; Morgan & Sanz, 2006; Watts et al., 2012). In fact, chimpanzees have one of the most diverse diets of any primate species. Over a 15-year study period, the chimpanzees at Mahale fed upon 328 plant food types from 198 species and at least 42 animal foods (Nishida & Uehara, 1983; Nishida et al., 1983). Also, along the shores of Lake Tanganyika, chimpanzees at Gombe fed upon 203 plant foods from 147 species and 25 animal foods during a similar time frame (Wrangham, 1977). Chimpanzees at Ngogo ate 167 identified plant foods (with at least 24 additional unidentified plant foods) during a 15-year period (Watts et al., 2012). In an 8-year study at Lopé in Gabon, chimpanzees ate 161 plant foods from at least 132 species and 9 invertebrate and 3 mammal species (Tutin & Fernandez, 1993). In a less diverse savanna habitat, only 68 chimpanzee foods were documented after 5 years of study (McGrew et al., 1988). However, it was estimated that a maximum of only 122 food items were available to primates in this marginal environment. Similarly, in the open savanna habitat of the Issa Valley (Tanzania), 69 food items were documented in the diet of a group of semi-habituated chimpanzees during a 4-year period (Piel et al., 2017). Interestingly, comparative studies have also revealed local differences in plant-feeding habits. Only 59% of the 286 plant foods available at both

Gombe and Mahale were eaten by chimpanzees at both sites (Nishida et al., 1983).

Although less diverse than chimpanzees, the diet of bonobos also includes a wide variety of plant species. Researchers at Wamba and Lomako reported 113 and 133 food items from 81 and 114 species, respectively (Badrian et al., 1981; Kano & Mulavwa, 1984). In an attempt to determine factors driving differences in the dietary patterns of *Pan*, Hohmann and colleagues (2010) examined variation of nutritional ecology across chimpanzee and bonobo study sites. Although plant samples collected at each site differ in terms of macronutrient content, the nutritious quality and gross energy content of food samples are similar. Therefore, dietary quality between populations and species of *Pan* is more likely a reflection of dietary selectivity rather than habitat ecology (Hohmann et al., 2010).

Inter-community variation in diet could be maintained in part through cultural transmission, as has been suggested for feeding techniques. For example, Beaune and colleagues (2017) examined consumption of tannin-rich fruit across bonobo study sites. Bonobos in Wamba, Lomako, and Manzano crunch and ingest only the pulp of a particular fruit species, *Canarium schweinfurthii*, while bonobos at LuiKotale ingest the whole fruit, subsequently extracting the seeds from the feces and re-ingesting the pulp (Beaune et al., 2017).

Even with such a diversity of plants in their diet, chimpanzees and bonobos spend most of their feeding time on a few foods during any specific period. The Gombe chimpanzees eat an average of 15 plant foods per day and 60 per month (Wrangham, 1977). Similarly, chimpanzees eat 14 plant foods per day in the Kasakati Basin, a savanna-woodland environment in western Tanzania (Suzuki, 1969). For bonobos at Wamba, the average number of food types eaten per day and per month are 10 and 40, respectively (Kano & Mulavwa, 1984).

In recent decades, analysis of stable isotope ratios and use of mass spectrometry have significantly enhanced our understanding of the composition of primate diets and enabled more systematic comparisons of diets within and between great ape taxa. Stable carbon isotopic data indicate that despite local availability of grasses and scarcity of fruits, chimpanzees inhabiting savanna habitats in West Africa (Fongoli) and East Africa (Ishasha and Ugalla) consume little C_4 vegetation or animals that consume such resources (Sponheimer et al., 2006). Isotopic data have also corroborated behavioral observations of niche partitioning and seasonal dietary variation among sympatric chimpanzees and gorillas in Central Africa (Oelze et al., 2014). At Ngogo (Uganda), pronounced isotopic differences between ground and canopy resources (4.2%

versus 2.2%) have been found (Carlson & Crowley, 2016). Stable isotope analyses have also informed estimation of age-related feeding transitions at Budongo (Boesch-Achermann & Boesch, 1994; Badescu et al., 2017). Stable carbon and nitrogen isotope ratios in hair samples of bonobos at LuiKotale over a period of five months showed a high degree of homogeneity in diet over time among members of the focal group and between sexes (Oelze et al., 2011). Using plasma mass spectrometry to assess plant food mineral content at the same site, Hohmann et al. (2019) documented high iodine concentrations in aquatic herbs consumed by bonobos, suggesting this vital nutrient may be more available than previously thought. Similar approaches have revealed that termite mound soil consumed by chimpanzees at Budongo, Gombe, and Mahale contains high concentrations of minerals, including iron and aluminum as well as manganese and copper (Reynolds et al., 2019).

The diet of the chimpanzee, both in major food items and in supplements, may shift radically between seasons and years (Hladik, 1977; Wrangham, 1977; Nishida & Uehara, 1983; Wrangham et al., 1991; Watts et al., 2012). In Gabon, for example, fruits may form up to 90% of the daily intake and never less than 40% (usually between 55–80%) (Hladik, 1977). During the major dry season (July–September) and the minor rainy season (April–June), a few species of fruit are available, and leaves and shoots are eaten in large proportions (up to 40–50% of the diet). During the major rainy season (October–December) and especially during the minor dry season (January–March), fruits are plentiful and the proportion of leaves in the diet falls to as low as 15%. Over a two-year period at Gombe, fruit was the item most eaten every month, but leaves were also eaten frequently in all months (Goodall, 1986). Seeds made up a high proportion of the diet in three months during one year and insects were consumed in large amounts in November of both years. A small amount of meat was eaten in approximately half of the months. Specific foods eaten also change from year to year. Researchers at Mahale discovered that some foods neglected in one year might be eaten in large amounts in another year (Nishida & Uehara, 1983).

Chimpanzees also incorporate certain plants into their diet for medicinal purposes (Huffman, 2001; Huffman & Caton, 2001; Krief et al., 2006), as is suspected of the Awash, Ethiopia, baboons and of Mauritian macaques, as well as by some New World monkeys (Garber & Kitron, 1997). All have been observed eating the leaves and pith of a number of plants that are used medicinally by humans as effective antibiotics and anthelminthics (Huffman, 2001; Krief et al., 2006). Evidence for self-medication has also been reported among bonobos at Lomako (Dupain et al., 2002) as well as at LuiKotale, where bonobos swallow unchewed pieces of stems as well as whole leaves (Fruth et al., 2014). A recent report from Fongoli has shown that chimpanzees recurrently ingest ethanol from raffia palm using tools, although the relative importance of this resource in their diet is currently unknown (Hockings et al., 2015).

Faunivory

Fruits in the chimpanzee diet are low in protein, and chimpanzees typically compensate by ingesting young leaves throughout the year. If a large primate (over 3 kg) does not obtain most of its protein or essential amino acids from leaves, however, it must include animal foods in its diet. These foods typically comprise small vertebrates, or social insects, as hunting for individual insects does not yield adequate protein for large primates. Chimpanzees routinely incorporate animal prey into their diet, although it comprises only 4% of the diet throughout the year (2.5–6% of daily intake). Bonobos have also been reported to occasionally incorporate vertebrates and invertebrates in their diet (Badrian & Malenky, 1984; Kano & Mulavwa, 1984; Ihobe, 1992; Hohmann & Fruth, 1993, 2008; Surbeck & Hohmann, 2008; Surbeck et al., 2009; Oelze et al., 2011).

Insects, specifically social ants and termites account for the greatest proportion of the animal prey of chimpanzees. On a per-gram basis, insect prey eaten by Gombe chimpanzees are comparable to vertebrate meat in macronutrients and overall energy (O'Malley & Power, 2012). Examinations of pre-agriculture human diets also suggest that insects could serve as equivalents not only of wild meat, but of a range of other foods including fruits, vegetables, nuts, and even some shellfish (Raubenheimer et al., 2014). Some termite species (such as *Macrotermes*) likely provide significant amounts of protein as well as high levels of lipids, vitamins, and minerals such as manganese, iron, and zinc (Deblauwe & Janssens, 2008). At the Fongoli site in Senegal, male chimpanzees spend 7.4% of their daily activity budget termite-fishing throughout the year (Bogart & Pruetz, 2008, 2011). During the late dry and early wet seasons, consumption increases with these males spending up to 19.5% of their time gathering termites, which is second only to fruit in their overall dietary intake.

Hunting

The role of vertebrate prey in the chimpanzee diet, as well as the frequency of hunting behavior, varies across populations. Hunting and meat eating have been extensively investigated at Gombe (Goodall, 1963, 1965, 1968,

1986; Teleki 1981; Stanford et al., 1994; Stanford, 1995, 1999), where the average consumption of meat is approximately 22 g per day per chimpanzee (Wrangham, 1975). Boesch and Boesch-Achermann reported a higher frequency of both hunting behaviors and meat consumption in the chimpanzees of Taï Forest (Boesch & Boesch-Achermann, 2000). They estimate that Taï chimpanzees capture 125 colobus monkeys per year, with males consuming an average of 186 grams and females 25 grams of meat per day. Using this same approach at both Taï and Gombe, they calculated that Gombe males eat 55 grams and females 7 grams of meat per day. Chimpanzees at Taï also hunt colobus monkeys twice as frequently as chimpanzees in Gombe (Taï = 125 colobus per year, Gombe = 66 colobus per year). Hunting is observed at least once a month at Taï and Gombe, while it was observed less than five times at both Bossou and Budongo over a period of several years. A total of 413 hunts were observed in a 12-year period at Taï, with 267 of these resulting in prey capture.

Historically, there were few observations of meat consumption by bonobos at Wamba and Lomako (Badrian & Malensky, 1984; Kano & Mulavwa, 1984; Ihobe, 1992; Fruth & Hohmann, 2002). However, there is increasing evidence for behavioral variation in meat eating and hunting among bonobos as research efforts expand across different study populations. Over a four-year study period at LuiKotale, 18 cases of vertebrate consumption were documented. Duikers comprised the majority of prey, but bonobos also consumed rodents, birds, and other mammals (Hohmann & Fruth, 2008). Bonobos at this site also hunt and consume other primate species, including a galago (Hohmann & Fruth, 2008; Surbeck & Hohmann, 2008; Surbeck et al., 2009). Similarly, at Iyondji, a site adjacent to Wamba, bonobos have been observed eating duikers as well as red-tailed monkeys (Sakamaki et al., 2016). Hunting by bonobos, particularly for monkeys, has been hypothesized to occur less often than for chimpanzees as a result of selection against aggressive behavior in bonobos (Wrangham, 1999). However, it is also possible that observations of bonobos hunting have historically been precluded by other factors, such as reduced observation time relative to that for chimpanzees (Stanford, 1998). The intriguing variation could indicate that interactions between bonobos and sympatric primate species reflect different environmental conditions and/or cultural differences in hunting behaviors across populations (Surbeck & Hohmann, 2008; Sakamaki et al., 2016).

Consistent seasonal differences in hunting have been documented among chimpanzees at some study sites. During the short rainy season months of

September–October, Taï chimpanzees hunt significantly more often (hunts are conducted almost every day) than during other times of the year (when hunting averages once per week) (Boesch & Boesch-Achermann, 2000). This seasonal pattern is attributed to three factors: (1) the rainy season makes it more difficult for colobus to escape on slippery substrates, (2) reduced availability of other food sources, and (3) birth season of the colobus monkeys. In contrast, hunting peaks strongly during the dry season and is rare during the wettest months at Gombe. The end of the hunting season at Taï coincides with the beginning of the nut-cracking season, which provides an alternate source of protein.

Mammals comprise most of the vertebrate prey for chimpanzees, but specific preferences for prey or prey characteristics may vary between populations. Gombe chimpanzees are reported to kill 50–75 mammalian prey per year, comprised of 44% red colobus, 31% bush pig, 18% bushbuck, 3% redtail monkey, 3% blue monkey, and 1% baboon (Wrangham, 1975). At both Gombe and Mahale, chimpanzees more frequently capture immature colobus monkeys than chimpanzee hunters at Taï, who focus on adult monkeys (Goodall, 1986; Stanford et al., 1994; Boesch, 1994a,b; Boesch & Boesch-Achermann, 2000). Population differences in preferred prey have also been reported. At Budongo forest, the Sonso community shows a strong preference for hunting colobus monkeys (*Colobus guereza occidentalis*: 74.9% hunts), while the neighboring community at Waibira tend to hunt other primate species, as well as duikers (Hobaiter et al., 2017). The size of prey targeted may also vary across habitat types. Savanna chimpanzees tend to hunt smaller vertebrates, while forest chimpanzees tend to target larger prey (Moore et al., 2017).

Approximately 8–13% of the colobus population at Gombe is killed each year by chimpanzees. Current levels of chimpanzee predation on red colobus at Gombe are not sustainable and will lead to the extinction of these monkeys at the site. Regarding this situation, one possibility is that habitat fragmentation has limited migration of red colobus into the park, contributing to negative growth rates of the colobus population (Fourrier et al., 2008). Data from Gombe have shown that particular males can have a large impact on hunting rates and success; for example, researchers observed a particularly keen hunter named "Frodo" who killed up to 10% of the red colobus population living in his home range each year (Stanford, 1999).

McGrew (1983) explains differences across chimpanzee populations in hunting behavior in terms of the following environmental influences: (1) presence or absence of prey species, (2) competing predators, (3) characteristics of habitat, and (4) human interference. For example, the difference in frequency of predation on red colobus at Gombe and Mahale can be explained by the structure of the habitat. At Gombe, the broken canopy makes the red colobus vulnerable to the mode of capture employed by chimpanzees. At Mahale, the vegetation is primary forest or closed woodland with a more continuous canopy which gives the monkeys more opportunity to escape. Differences in hunting success between chimpanzees at Gombe and Taï can also be attributed to forest conditions (Boesch, 1994a,b). At Taï, the forest canopy reaches over 40–50 m high and colobus have many arboreal escape routes. The forest canopy in the woodland-savanna habitat at Gombe is much lower at 15 m high, which provides limited escape routes for monkeys.

Understanding the factors driving hunting behavior of nonhuman primates helps inform models of hominin evolution. These include the possibilities that males hunt for social reasons, so as to share meat with sexually receptive females in exchange for mating (the Meat for Sex model) or to partition meat among males in exchange for social support (Male Bonding) (Mitani & Watts, 2001; Gilby et al., 2006). Chimpanzees may also hunt more frequently during food shortages (Nutrient Shortfall), or when there is a surplus energy reserve (Nutrient Surplus), or when they encounter monkeys in habitat types where prey are most vulnerable (Mitani & Watts, 2001; Gilby et al., 2006; Gilby & Wrangham, 2007). Gilby and colleagues examined 25 years of hunting behavior by the Gombe chimpanzees and found that ecological factors account for more of the hunting behavior than social factors, such as exchanging meat for sex or social support (Gilby et al., 2006). However, there is strong evidence that male chimpanzees at Taï exchange meat for sex (Gomes & Boesch, 2009). Over a 22-month period, female chimpanzees at Taï copulated more frequently with those males who shared meat with them. Sharing of resources, including meat among Taï chimpanzees, is best predicted by enduring, mutual grooming relationships, suggesting that long-term social bonds underlie sharing behavior (Samuni et al., 2018a). Male chimpanzees are generally considered to hunt more frequently than do females, and there are also sex differences in prey type. For example, a recent comparison of hunting behavior between chimpanzees at Gombe and Kanyawara showed that females follow a lower-risk strategy of targeting smaller vertebrates, while males choose larger prey (Gilby et al., 2017).

The question of what drives hunting in groups, and whether chimpanzees actually hunt cooperatively, has been debated (Busse, 1978; Teleki, 1981; Boesch, 1994b, 2002). It has generally been assumed that cooperative hunting should maximize per capita caloric intake. More recently, it has been posited that group hunting may be advantageous because it maximizes an individual's likelihood of obtaining important micronutrients that are found in even small quantities of meat, as described by the Meat Scrap Hypothesis (Tennie et al., 2009). In some cases, male chimpanzees at Gombe in groups ranging from two to nine adults were observed to chase prey (Goodall, 1968; Teleki, 1973; Goodall, 1986) and on occasion to position themselves "to effectively anticipate and cut off all potential escape routes of the prey" (Teleki, 1981: p.332). Cooperative hunts at Gombe are infrequent, however, comprising only 36% of hunts, and they are less successful than individual hunting. Thus, hunting in groups does not seem to benefit individuals at Gombe on a prey–capture basis. In contrast, 72% of hunts at Mahale and 85% at Taï are group hunts, and the hunting success of chimpanzees at Taï increases as they hunt in groups (Boesch & Boesch, 1989; Boesch, 1994a,b). Further, at Taï, participants in group hunts can be attributed specific roles, such as driver, blocker, and ambusher (Boesch & Boesch-Achermann, 2000; Boesch, 2002). Group hunts may be more efficient than solitary hunts in closed canopy forests where multiple individuals are needed to block arboreal escape routes. A cooperative basis for hunting at Taï is supported by the relationship between hunt participation and meat sharing (Samuni et al., 2018b), as well as by increased oxytocin secretion in association with hunting and sharing (Samuni et al., 2018a).

There is considerable variation in hunting behaviors across species and populations of *Pan* species. A number of ecological factors (such as seasonality, presence or absence of prey, prey size, predators, and habitat type) in combination with social dynamics may explain the different patterns of hunting behavior in chimpanzees and bonobos.

Food Sharing

There are several descriptions of chimpanzee mothers sharing food with their young (Goodall 1968, 1986; Sugiyama, 1972; Silk, 1978; Boesch & Boesch-Achermann, 2000). In this context, the exchange of food is normally preceded by a particular gesture in which the young animal holds its hand, palm upward, with an extended arm. Adult chimpanzees also use this gesture to request food, including meat, from other individuals. Meat is not evenly distributed between group members after a hunt, and more meat is consumed by males than females. At Gombe, males have access to meat in 36% of all captures and females in 12% of all captures (Goodall, 1986). Male chimpanzees at Taï have access to meat in 48% of all captures and females in 15% of captures (Boesch & Boesch-Achermann, 2000). When begging effort is considered, meat is preferentially shared with hunt participants and

with older individuals (Samuni et al., 2018b). Foods acquired by tools are rarely shared between adults, but sharing is sometimes observed between mothers and infants (Goodall, 1986). For example, infant chimpanzees scrounge termites at Gombe (Lonsdorf, 2006), and mothers share freshly cracked nuts at Taï (Boesch & Boesch-Achermann, 2000) and Bossou (Biro et al., 2006). A mother may share up to 20% of her nut quarry with her infant, the amount decreasing with age as the infant acquires the nut-cracking skills (Boesch & Boesch-Achermann, 2000). Food sharing events among adult chimpanzees are associated with higher urinary oxytocin levels than other types of social feeding. These events may activate neurobiological pathways similar to those involved in mother-infant bonding, thus facilitating cooperation between unrelated adults (Wittig et al., 2014).

Bonobos often share meat and some plant foods, especially large fruits (Hohmann & Fruth, 1993; Fruth & Hohmann, 2002). Food sharing among bonobos differs from that in chimpanzees in the following ways: (1) adults often share plant food, (2) females share food with individuals other than their infants, (3) males seldom share food among themselves whereas females often do so, and (4) sexual behavior often occurs in connection with food sharing (Kuroda, 1984). Female bonobos are most often the possessors of meat and large fruits at Lomako, and food sharing is not explained by kin relations and very rarely by reciprocity (Fruth & Hohmann, 2002). Rather, mutualistic explanations most often account for food exchanges. Another possibility is that subordinate female bonobos beg for food to gauge the tolerance of more dominant individuals, and to strengthen social bonds (Yamamoto, 2015; Goldstone et al., 2016). Food sharing has also been documented between bonobo communities. During an intergroup encounter between two communities at LuiKotale, an adult male from one community caught a duiker and allowed females from both communities to take parts of the prey. The communities' home ranges showed increased spatial overlap in the weeks following this sharing event (Fruth & Hohmann, 2018).

It is clear that kinship plays an important role in the sharing of food among chimpanzees. This is in striking contrast with food sharing by female bonobos, who share food with non-kin in a sexually affiliative context that is remarkably different to that of chimpanzees. The mechanisms underlying these differences remain largely unknown.

ACTIVITY CYCLES

Similar to other primates, chimpanzees adjust their activity patterns to different ecological and demographic conditions. The fission-fusion social structure of chimpanzees allows a particularly high degree of flexibility in budgeting time for feeding, traveling, and social behaviors throughout the typical day. This has enabled chimpanzees to adeptly cope with fluctuating resources in a wide variety of habitats across Africa.

Chimpanzees are diurnal primates, with consistent active periods that coincide with daylight hours. Compared to males, sexually nonreceptive females have shorter active periods and sexually receptive females have longer active periods at Gombe (Lodwick et al., 2004). Chimpanzees begin to move at dawn, usually about 10–15 minutes after awakening. Most of their day is spent feeding (25–66%) and inactive (10–50%), with an average of around 20% of the time spent traveling and 16% grooming (see Table 23.2). Goodall (1968) reports that Gombe chimpanzees normally spend 1–2 hours at one food source and then move 2–16 km to another. At Gombe, chimpanzees move from feeding site to site throughout the day, stopping at various intervals to rest until late afternoon. Morning and afternoon peaks of feeding have been reported from most sites (Izawa & Itani, 1966; Wrangham, 1977; Doran, 1997; Anderson et al., 2002; Sanz, 2004). Social interactions are frequent during day rest periods (Goodall, 1986; Plumptre & Reynolds, 1997).

Chimpanzees are also active at night. Recent cameratrap data obtained from the Pan-Africa Program indicate that terrestrial nocturnal activity is infrequent but occurs at 17 out of 22 sites studied, comprising 1.8% of the total chimpanzee activity budget. Nocturnal activity is associated with higher daytime temperatures (Tagg et al., 2018). At Fongoli, direct observations of chimpanzee nocturnal activity provide evidence for foraging, traveling, and social behaviors during the hottest period of the year, suggesting a thermoregulatory benefit for nocturnal activity (Pruetz, 2018).

All great apes build nests each night for sleeping. Young animals up to the age of five or six years sleep in their mother's nest, but infants as young as nine months of age begin to build rudimentary nests (Goodall, 1968). Nests normally take from 5 to 10 minutes to construct. Goodall (1965) describes nest construction as follows:

> The chimpanzee takes up a central position on a suitable "foundation" (such as a horizontal fork or two adjacent parallel branches), takes hold of a fairly thick branch and bends it down across the feet, and a second branch is bent across it. From four to six main cross-pieces are bent in, and then between six and ten smaller branches are bent across to form secondary crosspieces. The branches are bent so that their leafy ends form part of the main structure.... Finally, all the small twigs projecting round the next are bent in and normally a number of additional twigs are picked and laid loose on top of the nest. (p.447)

Nest construction on the ground has been reported, but night nests are typically built at heights between 5–20 m (Goodall, 1968; Sanz et al., 2007). Nest height is directly related to habitat, available vegetation, and presence of nocturnal predators (Pruetz et al., 2008). Chimpanzees routinely nest in particular tree species, branches of which enable the construction of stable, firm, and resilient sleeping platforms. This could enhance sleep quality and minimize fall risk, complementing other potential anti-predator, thermoregulatory, and pathogen avoidance functions of nests (Samson & Hunt, 2014). Chimpanzees sometimes make day nests for their midday rest period, though day nests are usually not as elaborate as night nests (Plumptre & Reynolds, 1997). In general terms, bonobo nest building is similar to that of the chimpanzee (Kano, 1992; Fruth & Hohmann, 1993, 1996).

Chimpanzees typically build sleeping nests earlier in the rainy season (about 1.5 hours before sunset) than in the dry season (about 45 minutes before sunset), although the time of settling might vary due to number of individuals nesting close together and to individual idiosyncrasies. Generally, two to six chimpanzees build nests in a single tree or in adjacent trees (Goodall, 1968). Females with their dependent offspring often form small nesting groups, whereas single adolescent and adult males frequently sleep alone. In the Goualougo Triangle, average nesting group sizes are relatively small, with only 2.75 ± 1.88 individuals (Morgan et al., 2006). The largest group observed nesting together by Goodall (1968) at Gombe was 17 animals and the largest in one tree was 10. On some occasions, a number of small groups that are separate during the day come together to sleep in a nesting group. However, at other times a large feeding group might break up into smaller nesting groups.

Unlike chimpanzees, bonobo foraging parties have been reported to gather at night nest sites (Fruth & Hohmann, 1996). When feeding on preferred fruits, bonobos at Wamba tend to form larger parties and aggregate into even larger groups for night nesting (Mulavwa et al., 2010). Fruit availability at nesting sites may also positively influence cohesion among bonobos living in forest-savanna mosaic habitat in western DRC (Serckx et al., 2014).

RANGING BEHAVIOR

The home range of an individual or a group is the area where essential resources, such as food, mates, and habitat, are obtained and a majority of activities are conducted. Clutton-Brock and Harvey (1979) defined home

Table 23.2 Activity Budgets Reported from Chimpanzee Study Sites

Site	Subjects	% Feed	% Inactive	% Travel	% Social
Budongo	Males and females [a]	53	25	8	14 (groom)
Gombe	Males [b]	56	30	14	-
	Males and females [c]	42.8	18.9	13.4	24.9
	Dry season [c]	44.4	14.1	15.7	25.8
	Short wet [c]	44.0	15.6	21.3	19.1
	Long wet season [c]	40.2	9.3	18.8	31.7
	Male (Figan) [d]	56.5±14.6	18.8±8.9	14.5±4.2	9.9±7.2 (groom)
Kibale	Males [e]	62	26	12	-
	Females [e]	52	38	10	-
Mahale	Males (1985) [f]	27	35	16	23 (groom)
	Males (1987) [f]	39	31	24	6 (groom)
	Females (1985) [f]	28±4	41±4	20±4	11±5 (groom)
	Females (1985) [f]	31±9	39±10	20±2	10±3 (groom)
	Estrous females [g]	31±4	34±3	26±2	9±2 (groom)
	Anestrous females [g]	26±6	32±10	32±4	10±2 (groom)
	Early dry season [h]	36	32	32	-
	Late dry season [h]	26	41	33	-
	Early wet season [h]	24	51	25	-
	Late wet season [h]	25	51	24	-
Taï	Wet season, 1987 [i]	49	25	23	-
	Dry season, 1987 [i]	66	16	17	-
	Wet season, 1988 [i]	57	20	23	-
	Dry season, 1988 [i]	50	31	22	-
	Males and females [j]	43	39	12	-
	Summary Data Mean (±SD):	42.0 (±12.8)	29.7 (±11.2)	19.9 (±6.9)	16.2 (±8.4)

[a] Fawcett, 2000; [b] Wrangham, 1977; [c] Teleki, 1981; [d] Riss and Busse, 1977; [e] Ghiglieri, 1984; [f] Huffman, 1990; [g] Matsumoto-Oda and Oda, 1998; [h] Matsumoto-Oda, 2002; [i] Boesch and Boesch-Achermann, 2000; [j] Doran, 1997.
- Indicates no data.

range as the total area shared by the social group. Consideration discussion continues over how to define such an area, as range use can vary with time and intensity. If defended against conspecifics over time, such an area can be defined as a territory. Levels of spatial structure in chimpanzee home ranges have been documented at the community, subgroup, and individual levels.

Home ranges obviously have strong ecological functions, and in chimpanzees, long-term site fidelity has been linked to local resource distribution and availability. Long-term observations in East and West Africa have shown that chimpanzee communities maintain their ranges, even after dramatic changes in group demography (Halperin, 1979; Kawanaka, 1984; Williams et al., 2002a; Lehmann & Boesch, 2004). Scientists have suggested that chimpanzees should be reluctant to share established resources, such as food and/or mates, which require large amounts of energy and risk to protect from conspecifics in neighboring social units. Significant social gains may be associated with territorial behavior that outweigh the energetic costs of patrolling territorial boundaries and the potentially high costs of antagonistic inter-community interactions (Amsler, 2010). Territorial expansion to increase access to valuable resources has been posited as an explanation for such risky behavior (Mitani et al., 2010).

Sex Differences in Ranging Behavior

For adult female and male chimpanzees, critical resources differ, which affects ranging and sociality. Food is the limiting resource for females, and optimization of foraging efficiency is a female priority. Daily travel distances and home ranges of adult females are typically smaller than those of adult males, especially when in female-only parties (Wrangham & Smuts, 1980; Fawcett, 2000; Williams et al., 2002a; Newton-Fisher, 2003; Lehmann & Boesch, 2005). Consistent use of core areas increases familiarity with localized resources and enhances foraging efficiency, although females shift their ranges to maintain access to food and ensure their safety. Reproductive state adds an additional dimension to ranging; cycling females are noticeably more gregarious and wide-ranging than females with dependent offspring (Nishida, 1979; Tutin, 1979; Goodall, 1986).

In contrast to females, the ranging behavior of males seems to be driven by social factors more than by ecological parameters. They range over the entire community to maintain regular contact with all community females to monitor and assess potential mating opportunities (Wrangham, 2000). Males in several populations conduct patrols to ensure the integrity of their community range (Goodall et al., 1979; Boesch & Boesch-Achermann, 2000; Watts & Mitani, 2001; Amsler, 2010; Mitani et al.,

2010). Patrols tend to originate within exclusive community ranges and generally cover portions of a shared boundary, with occasional incursions into a neighboring group's range. Behavior of males on such patrols is consistent across sites. Adult males, accompanied by subadults and juveniles, travel slowly and silently, occasionally in single file with patrollers inspecting the environment for signs of apes from neighboring groups. Socializing and foraging are not high priorities on patrols. Frequency of boundary patrols differs among chimpanzee communities, which seems to be a function of the number of resident males in the communities (Watts & Mitani, 2001). Patrols occurred at Ngogo every 9.7 days from 1998–1999 (Watts & Mitani, 2001) and every 9.3 days from 2004–2006 (Amsler, 2010). At Taï, patrols were observed every 14 days from 1984–1991 (Boesch & Boesch-Achermann, 2000). From 1977–1982, patrols occurred at Gombe every 22 days (Goodall, 1986).

Structured ranges serve important functions within chimpanzee communities. Males of the Sonso community of Budongo have core areas, with multi-nuclear patterning, in which particular areas are more heavily visited than others (Newton-Fisher 1997, 2003). Males at Budongo form long-term alliances with other males and structure their ranges to associate with important allies. In this sense, preferred areas of use by males support the "social cores" hypothesis, which suggests that spatial structuring facilitates the maintenance of relationships between individuals, as they may be more "locatable" to resident allies and resident females (Newton-Fisher, 1997).

Ranging patterns in chimpanzees are thus sex-specific, with localized ranging of females and more extensive ranging of males. As female chimpanzees are highly dispersed, it may be difficult for them to develop relationships with same-sex conspecifics, and also challenging for any one male to monopolize a female for extended periods. These observations, along with a lack of infant care by males, suggest that social organization in such species will be determined by whether the range of females is defendable (Clutton-Brock, 1989). In contrast to females, male chimpanzees develop strong relationships with other adult males, which may increase their mating opportunities and their ability to maintain the integrity of community boundaries—effectively defending females and food resources.

In bonobos, an intriguing strategy of female leadership in collective action has emerged within a male-philopatric society (Tokuyama & Furuichi, 2017). Researchers at Wamba recently investigated the initiation of group departure to determine the distribution of leadership among group members. The frequency of initiation

of group departure was associated with age, affiliative relationships, and male dominance rank. Intriguingly, the three oldest females in the community initiated departures more frequently than expected and were described as "key individuals" who helped to maintain community cohesiveness.

Community Spatial Structure

Three models have been proposed to describe chimpanzee dispersion and sociality. The Male-Only Community Model was originally formulated by Itani and Suzuki and asserts that adult male and female chimpanzees should be evenly dispersed across a landscape with overlapping, but independent ranges (Itani & Suzuki, 1967; Wrangham, 1979). The Male-Bonded Community Model predicts that adult female ranges will be smaller than male home ranges and encompassed within the settlement of a social network of adult males (Wrangham, 1979, 1980). It is further predicted that core areas heavily used by females may or may not overlap extensively but should cluster more toward the center of the community range. Finally, the Bisexually-Bonded Community Model suggests that males and females will be equally social and sex-specific ranging patterns will not be apparent (Lehmann & Boesch, 2005). Males and females of a single community are predicted to range relatively equally throughout a defended range. As a result, high levels of overlap in home range and core area use are expected. Lehmann and Boesch (2005) outline testable predictions for each of these models with regard to home range size and degree of overlap. In the Male-Only Community Model, males will have larger home ranges than females and degree of overlap will be low. Observations that adult female chimpanzees in boundary areas are subject to aggression by adult males contradicts the predictions of this model (Watts & Mitani, 2001). In addition, females will also occasionally participate in boundary patrols, which suggests that resource defense by females may be more important than previously thought (Lehmann & Boesch, 2005). The Male-Bonded Model predicts that overlap between female ranges will be low and that males will show a higher degree of overlap than females. Long-term observations at Gombe have supported the Male-Bonded Model. Females are dispersed within a community home range that is defended mostly by males (Williams et al., 2004). They use identifiable core areas within the home range that overlap with other females (Wrangham & Smuts, 1980; Williams et al., 2002a; Murray et al., 2006). The structuring of the female core areas suggests that feeding competition may have been influential in female settlement. High-ranking females at Gombe tend to reside toward the center of the community range (Williams et al., 2002a). The reproductive success of these females is higher than females residing in more peripheral areas of the range, which suggests that spatial dispersion of females may be in response to feeding competition. At Budongo, in contrast, no relationship has been detected between female dominance and core area usage within the Sonso community range (Fawcett, 2000). The spatial arrangement of the Sonso community aligns more closely with the predictions of the Male-Bonded Model, at least during periods of tension among males over alpha status.

The Bisexually-Bonded Community Model predicts a high degree of overlap between males and females throughout a community range, which results in little or no sex differences in ranging. Support for the Bisexually-Bonded Model of chimpanzee society comes primarily from Taï (Lehmann & Boesch, 2005). Long-term monitoring of the northern community has found that males and females are similar in their use of the community range (Boesch & Boesch-Achermann, 2000; Lehmann & Boesch, 2005). Males and females exhibit high degrees of range overlap, which differentiates this social system from other studied chimpanzee populations and shows similarity to bonobos. It was generally assumed that female chimpanzees avoid boundary areas due to the danger posed by conspecifics of neighboring communities, but higher-ranking adult females at Taï do not show avoidance of boundary areas, which is another indication that this chimpanzee society differs from other populations. However, the increased participation in home range defense efforts by the females could also be in response to changes in group demography, as the number of males has dramatically decreased within the community. It is also possible that the risk posed by the neighboring community has been relaxed due to population decline. Decreased detection of neighboring chimpanzee groups in the local environment could prompt changes in ranging behavior (Lehmann & Boesch, 2005).

Current models of chimpanzee sociality are not well equipped to cope with intra-sexual variation in ranging observed among female chimpanzees across populations. In particular, the models described above are challenged by observations of "peripheral" females, who are spatially and socially removed from the rest of the community. "Peripheral" or "stranger" females have been documented in three chimpanzee communities. Gombe, Kibale, and Budongo have adult females who occupy boundary areas or other regions of the community ranges in what appear to be suboptimal ecological and social situations (Goodall, 1986; Wrangham et al., 1996; Fawcett, 2000; Williams et al., 2002a; Emery Thompson et al., 2007; Murray et al., 2007). Although information about these individuals remains relatively scarce, females and offspring inhabiting peripheral areas may not benefit from an equal amount of protection from resident males as females residing closer to the community's core area. For example, a peripheral female in Kibale is reported to have lost an infant directly through infanticide (Arcadi & Wrangham, 1999). Lack of male protection is also suspected to have caused the decrease in reproductive rates of peripheral females compared to more central females at Gombe (Williams et al., 2002a). Structured use of a community range may be a strategy that chimpanzees have adopted to mitigate this threat. Core areas of a home range are almost exclusively occupied by community members, whereas peripheral or boundary areas may overlap between neighboring communities.

MATING SYSTEM

Chimpanzees breed throughout the year. The estrus cycle lasts for a mean of about 32–37 days, and perineal sex skin swellings serve as a graded signal for ovulation (Deschner et al., 2003, 2004). Mating between reproductively active adults occurs in three contexts: (1) opportunistic matings in mixed parties, (2) possessive matings in mixed parties, and (3) during consortships. Most matings occur promiscuously between sexually receptive females and community males in an opportunistic fashion. At Gombe, in a study of sexual behavior, 73% of 1,137 observed copulations were of this type (Tutin & McGinnis, 1981), as were 92% of 660 copulations at Mahale (Hasegawa & Hirai-Hasegawa, 1983). In some instances, one male might attempt to dominate the attentions of a female and to keep other males from copulating with her. Twenty-five percent of the matings at Gombe and 7% at Mahale were of this possessive type. Most often it is the alpha male of a group who has the greatest success in controlling access to a particular female (McGinnis, 1979; Tutin, 1979; Nishida, 1983; Deschner et al., 2004).

A male may also attempt to form a consort relationship with a female to draw her away from other males in the community. "Once alone, the pair, plus any dependent offspring of the female, cease all loud vocalizations and avoid encounters with other chimpanzees. This avoidance often results in the pair moving to the edge or even outside the normal community range" (Tutin & McGinnis, 1981: p.257). The male initiates a consortship, but he is only successful if the female cooperates by following him. Over a 16-month period, 15 consortships were formed at Gombe lasting a mean of 7 days (ranges: 3–28 days). Only 2% of the copulations observed occurred during consortships (Tutin, 1979).

Although most mating is opportunistic, partner preferences do exist. As in most primates, chimpanzees have an aversion to close kin as sex partners. Mother–adult son matings are virtually unknown, and mature brother and sister matings are rare (Tutin, 1979; Goodall, 1986). Adolescent males at Gombe and Mahale mate more often with nulliparous than with parous females, and mostly outside of periovulatory periods (Watts, 2015). Differences in access to mating opportunities are intricately related to the demographic profile of the social group. Young females around 10–11 years of age migrate out of their natal community (Pusey, 1979). After establishing a stable position within a new community, the female may then spend the rest of her life in this community. However, some instances of temporary transfers have been observed. Almost all adolescent females at Mahale and Taï transferred from their natal communities (Nishida & Hiraiwa-Hasegawa, 1987; Boesch & Boesch-Achermann, 2000). However, half of adolescent females at Gombe have remained in their natal community and maintained long-term relationships with related females (Goodall, 1986; Pusey et al., 1997). It has been suggested that male dispersal occurs at Bossou, due to the lack of female transfers and frequent departure of adolescent males (Sugiyama, 1999). However, these differences could be due to habitat destruction that has limited the choices of emigrating females.

Although bonobos also exhibit promiscuous mating throughout the year, there are important differences between the two species. Chimpanzee and bonobo females spend a similar proportion of days per estrus cycle with tumescent swellings, 12.5 out of 31.5 days for chimpanzees, and 14.6 out of 42 days for bonobos. However, female bonobos also exhibit swellings during non-conceptive time periods. They sometimes show continuous swellings throughout the cycle and may resume estrus within a year after giving birth, which is in contrast to female chimpanzees who resume cycling 55.5 months after birth (Furuichi, 2011). Female bonobos also continue cycling until a month before parturition (Kano, 1992). Therefore, at any given time, there are a greater number of bonobo females exhibiting tumescent swellings than chimpanzees. Also, the ratio of adult males to cycling females is lower in bonobos than chimpanzees. Thus, it is less feasible for male bonobos to control access to receptive females than it is for male chimpanzees (Furuichi, 2011). In addition, maximal swelling periods do not necessarily correspond to probability of ovulation among female bonobos. At LuiKotale, ovulation occurred during the maximal swelling period in only 52.9% of cycles analyzed (Douglas et al., 2016). This pattern likely acts as a constraint against mate-guarding efforts by male bonobos. Prolonged sexual swellings in female bonobos may also function to attract other females and to facilitate female-female bonding. Female bonobos with maximally tumescent swellings more frequently affiliate with other females via grooming, remaining in close proximity, and genito-genital rubbing (Ryu et al., 2015).

SOCIAL ORGANIZATION

Before the pioneering field studies at Gombe and Mahale in the 1960s, the social organization of chimpanzees was unknown. Imanishi had predicted that the social unit of chimpanzees would be the one-male group, identical to gorillas (Imanishi, 1961). However, Nishida reported that groups containing a single male and adult females accounted for only 4.2% of all groups observed during his field observations of wild chimpanzees conducted between 1965–1967 (Nishida, 1968). Although several early researchers documented the temporary grouping patterns of chimpanzees, Itani first recognized the existence of a larger inclusive group (Azuma & Toyoshima, 1961; Goodall, 1963; Reynolds, 1965; Nishida, 1968). Nishida (1968: p.189) stated that "it first became clear by means of artificial feeding... at Kasoge that... the chimpanzee members with whom they make a group are definite; in other words, joining and parting is, an intra-entire-group phenomenon."

In relation to stable groups of other primate societies, Nishida (1979) describes the social structure of chimpanzees in the Mahale Mountains:

> The corresponding group in chimpanzee society proved to be the multi-male bisexual group ranging from 20–100 in size, which I have called the unit-group. This unit-group is involved in a continual process of splitting into several temporary subgroups, which then rejoin, and re-separate into different subgroups. Thus, it is very rare that all members of the unit-group meet together in one group. (p.74)

Although labels have changed, this early description of a fission-fusion society has subsequently been confirmed at all chimpanzee study sites where individuals are identified and community membership is known (Kawanaka, 1984; Goodall, 1986; Wrangham et al., 1992; Boesch & Boesch-Achermann, 2000; Fawcett, 2000; Morgan, 2007).

Nishida identified two distinct levels of grouping in chimpanzee society, the unit-group and the temporary subgroup. A chimpanzee unit-group (Nishida, 1968) or community (Goodall, 1973) consists of all individuals seen together in various subgroups over time. Subgroups or parties are the temporary groups that contain only a portion of the community membership. The neighborhood is a level of chimpanzee society that resides between unit-groups and temporary subgroups. Female neighborhoods have been identified at Gombe based on association patterns and spatial fidelity of individual home ranges (Williams et al., 2002a). In the Ngogo chimpanzee community, researchers have detected stable subgrouping among males along the lines of age and rank (Mitani & Amsler, 2003).

Chimpanzee community sizes range from less than 20 to more than 200 individuals. Community structure is often biased toward females, with sex-ratios ranging from 1:1 to 1:3 (Goodall, 1986; Nishida, 1990; Boesch & Boesch-Achermann, 2000; Nishida et al., 2003; Reynolds, 2005; Mitani, 2006; Nishida, 2012). It has generally been accepted that in chimpanzee and bonobo societies, males are philopatric and young adult females transfer from their natal group (Kawanaka, 1984; Nishida & Hiraiwa-Hasegawa, 1987; Boesch & Boesch-Achermann, 2000; Langergraber et al., 2007a; Stumpf et al., 2009).

Party Size and Composition

Although variation in patterns of party size and composition has been reported among field sites, there are several general characteristics that define chimpanzee parties. Average party size at most sites typically consists of fewer than 10 mature individuals (Goodall, 1968; Nishida, 1968; Riss & Busse, 1977; Wrangham et al., 1992; Chapman et al., 1994; Sakura, 1994; Boesch, 1996; Mitani et al., 2002; Sanz, 2004; Furuichi, 2009). Several changes in party composition occur each day, with average duration of parties ranging from 14 minutes to more than 2 hours (Halperin, 1979; Boesch & Boesch-Achermann, 2000). Recent comparisons have shown that relative party size (assessed as percentage of total community size) is significantly larger for bonobos than chimpanzees (Furuichi, 2009). The prolonged estrus of bonobo females, close associations between mother and adult sons, and strong relationships between females are all likely contributors to the increased cohesiveness of bonobos, but are complemented by ecological conditions that support larger aggregations. Although early studies focused on identifying a single factor that determined grouping patterns, subsequent researchers have found that ecological, demographic, and social factors interact to determine subgroup size and composition (Chapman et al., 1995; Boesch & Boesch-Achermann, 2000; Anderson et al., 2002; Mitani et al., 2002; Furuichi, 2009).

Changes in subgrouping patterns reflect a group's responses to and interactions with environmental and social factors within their immediate surroundings. Ecological factors may include abundance, distribution, and quality of food resources (Symington, 1990; Chapman et al., 1995). Demographic factors include the number of sexually receptive females and population parameters

(Goodall, 1986; Sakura, 1994; Boesch, 1996; Mitani, 2006). Social factors include within- or between-group aggression and transmission of information to group members (Goodall et al., 1979; Goodall, 1986; Hamai et al., 1992; Williams et al., 2002b). Also, it is important to identify other factors (such as degree of habituation and sampling method) that may affect assessment of subgrouping patterns (Chapman et al., 1993).

Influence of Food Availability on Subgrouping

Although methodological differences have made comparisons between studies problematic, a positive relationship between food availability and subgroup size has generally been reported for *Pan* (Nishida, 1974; Goodall, 1986; Chapman, 1990; Wrangham et al., 1992; Chapman et al., 1995; Anderson et al., 2002). Different measures of food abundance include size (patch, crown), density (patches per unit area), distribution (location of food resources), type (fruits, leaves, etc.), and quality (nutritional value, processing effort required). In a comparison of the grouping patterns of chimpanzees and spider monkeys, Chapman and colleagues (1995) measured the size, density, and distribution of food resources and concluded that ecological variables are critical influences on party size. Individuals often join larger parties if preferred items are available (Newton-Fisher, 1999b). For example, subgroup sizes at Gombe are larger when chimpanzees feed on preferred food items that are highly localized and infrequently available, specifically *Dalbergia* spp. and meat (Riss & Busse, 1977). Party sizes at Taï are larger when chimpanzees are cracking nuts or eating meat (Boesch & Boesch-Achermann, 2000; Anderson, 2001). At Taï and Ngogo, party sizes also increase with the formation of hunting parties (Boesch, 1996; Mitani & Watts, 1999).

Rather than focusing on specific indicators of food abundance, some researchers have examined food availability in relation to primate activity patterns (time spent foraging, travel costs between food patches, relative amount of time spent resting or socially interacting). Chimpanzees in the Taï Forest respond to food scarcity in the minor dry season by spending more time feeding, feeding more frequently on lower-quality food items, reducing day range, reducing party size, and spending more time alone and less time in mixed groups than during the rainy season (Doran, 1997). Similar flexibility in diet, grouping patterns, and time budgets has been documented between seasons at Mahale (Matsumoto-Oda, 2002). For example, in the wet season when fruit availability is highest, chimpanzees form large parties and spend more time feeding on animal foods. In the dry seasons, when food availability is intermediate, they remain in large parties but spend more time feeding and traveling than in the wet season. In the late wet season, when fruit is scarce, chimpanzees are found in small groups and travel shorter distances.

Living in social groups provides opportunities for individuals to gather information from other group members, such as the location of food resources and novel foraging techniques. Further, putative cultural variants have been identified among several chimpanzee communities, and it is thought that these specific behavior patterns are socially transmitted between group members (Whiten et al., 1999, 2001). Aggregations also provide important opportunities for the socialization of youngsters (Williams et al., 2002b).

Influence of Sexually Receptive Females on Subgrouping

Social and reproductive interests, which differ between adult males and females, also influence group composition and stability. The presence of sexually receptive females has long been thought to be an important influence on chimpanzee subgrouping patterns. Male chimpanzees show differential attention to the phases of female reproductive cycles, as exhibited by perineal swellings in conjunction with ovulation, and swellings influence probabilities of association between the sexes (Deschner et al., 2003, 2004). Presence of receptive females is a driving force in chimpanzee gregariousness in all intensively studied populations (Goodall, 1986; Chapman et al., 1995; Doran, 1997; Newton-Fisher, 1999; Boesch & Boesch-Achermann, 2000; Wrangham, 2000; Anderson et al., 2002; Mitani et al., 2002). While food availability influences subgroups when chimpanzees are in the core area of their range at Taï and estrous females are not present, grouping patterns are mainly driven by the presence or absence of estrous females (Anderson et al., 2002). Food availability and number of estrous females are positively correlated with monthly party sizes at Ngogo. Together, these independent variables explain 87% of the variation in monthly party sizes (Mitani et al., 2002).

Ecological and reproductive factors influencing party size are not independent, as female reproductive cycles are also affected by food resources. Chimpanzees do not exhibit true birth seasonality, but researchers have shown that conception peaks occur in relation to the availability of specific high-quality foods (Wallis, 2002). There is a peak in cycling at the end of the dry season at Budongo, which coincides with a dependence on *Cynometra alexandrii* seeds, rich in lipids. New leaf flushes coincide with increased sociality of cycling females at Taï (Anderson, 2001). New leaves are presumed to have high nutritional value, and these findings support the notion of food quality influencing the reproductive cycles of females with subsequent effects on sociality (Fawcett, 2000; Anderson, 2001). These findings prompt further investigation into the relationship between food characteristics (i.e., abundance, quality) and reproductive status of females. It has further been suggested that some form of dietary endocrine inhibitor could be at work with plant secondary compounds ingested during the wet season, which block ovulation. Changes in diet that release females from this natural "contraceptive" would allow cycling to resume (Wallis, 2002).

Population Parameters and Subgrouping

The relationship between community structure and group cohesiveness is not well understood in chimpanzees. With regard to demographic influences on subgrouping patterns, Halperin (1979) concluded:

> The findings from the present analysis suggest that, although seasonality of food availability affects chimpanzee grouping patterns (Reynolds & Reynolds, 1965; Wrangham, 1975), the most important element in understanding the overall grouping pattern (size, frequency, and variety) is the specific age and sex composition of that chimpanzee community. (p.499)

Indeed, community composition can be a limiting factor on the associations that occur between individuals (Mitani, 2006). It has been suggested that when chimpanzee and bonobo community sizes decrease, parties become more stable and the fission-fusion structure falters (Boesch & Boesch-Achermann, 2000). In the early 1970s, a community fission was documented at Gombe. Recent analysis on this episode highlights that group composition prior to the split was unusually male-biased and that subsequent communities were formed depending on the affiliative associations of male individuals (Feldblum et al., 2018).

With several neighboring communities, the decision to transfer to another group may be based on the community size or composition. Researchers at Mahale and Taï have reported that female immigration tends be directed toward communities with higher numbers of males (Nishida, 1990; Boesch & Boesch-Achermann, 2000). The comparatively lower rate of female transfer at Gombe (Goodall, 1986; Pusey et al., 1997) and Bossou (Sugiyama, 1984) could be due to extensive habitat alteration around both of these study communities, which may reduce the viable options for inter-community transfer. It may be more advantageous for young females at Gombe and Bossou to remain in their natal communities than transfer to a distant community that resides along the forest edge and closer to human habitation. Such shifts in dispersal alter the demographic composition of the community and subgrouping patterns.

Overall, the extremely dynamic pattern of fission-fusion present in *Pan* can be seen as a response at the population level to many different factors. However, comparisons between species and populations continue to be problematic due to differences in sampling methodology and possible artifacts of incomplete habituation.

SOCIAL STRUCTURE

Male bonding has been reported from all studied chimpanzee communities. Newton-Fisher (1999b) has suggested that males at Budongo frequently associate with other males to gain social benefits and that aligning with particular partners reflects tactical strategies. Although it was previously thought that male bonds formed most often between related individuals, studies have shown that genetic relatedness is not correlated with grooming, levels of cooperation in coalitions, meat sharing, or participation in patrols (Goldberg & Wrangham, 1997; Mitani et al., 2000). At Ngogo, familiar males of similar age and rank are mostly likely to affiliate and cooperate (Mitani et al., 2000). Langergraber and colleagues (2007b) used a combination of molecular genetics and long-term field observations to identify kin relations and degree of cooperation between male chimpanzees at Ngogo. Although maternal brothers clearly prefer to affiliate and cooperate in several behavioral contexts, the impact of kinship is limited. The majority of highly affiliative and cooperative dyads are either distantly or completely unrelated.

Male chimpanzees have stable linear dominance hierarchies, in which subordinate individuals challenge those more dominant. These challenges usually arise from coalitions and alliances of lower-ranking males (Nishida, 1983; Goodall, 1986; Nishida & Hosaka, 1996). However, females at Taï are also frequently involved in coalitions and alliances (Boesch & Boesch-Achermann, 2000). Although male coalitions and alliances are frequent at Mahale and Gombe, females are very rarely involved in aggressive attacks against males. These population-specific differences in coalitions have prompted doubt about broad generalizations of social relationships across chimpanzee populations (Boesch & Boesch-Achermann, 2000).

Based on early reports from East African populations, chimpanzees were not considered female bonded. Males were clearly bonded within their community, whereas females transferred from their natal groups and were consistently less gregarious than males. This is evident in the social relationships observed among adult chimpanzees at Mahale, with males being consistently more sociable than females (Nishida, 1968; Takahata, 1990a,b). Grooming between males accounts for 46% of social bouts, with only 10% occurring between females. However, recent research in East African chimpanzee populations such as Ngogo shows that males and females do not differ in the long-term stability of their party associations (Langergraber et al., 2009).

The quality and degree of female bonds among chimpanzees may differ widely across sites. At Gombe, Mahale, and Kibale, interactions between females occur less often than male-male and male-female affiliation (Wrangham & Smuts, 1980; Goodall, 1986; Nishida, 1989). Although females in the Kanyawara community in Kibale spend as much time grooming as males, females focus much of their grooming attention on offspring rather than on other adults (Wrangham et al., 1992). In the neighboring Ngogo community, females actively maintain social relationships and bond with other females, which results in social "cliques" (Wakefield, 2008, 2013). Reports from West African sites suggest that females bond and form long-lasting relationships (Sugiyama & Koman, 1979; Boesch, 1996; Boesch & Boesch-Achermann, 2000). Although overall female-female association indices at Taï are lower than those between males, 17 of 24 females had at least one stable female associate with whom they shared food and formed coalitions (Boesch & Boesch-Achermann, 2000). Further, three female dyads had the highest association indices within the community, spending 66%, 71%, and 79% of their time with other females over a period of several years. Members of these long-lasting female dyads were not related and held similar positions of rank, often the highest ranks within the female hierarchy (Boesch & Boesch-Achermann, 2000).

Dominance hierarchies among females have been more challenging to characterize than among males. A linear dominance hierarchy among females has been identified in Taï and positively correlated with reproductive success (Wittig & Boesch, 2003). At Gombe, females have been assigned to broader dominance categories (Murray, 2007). It has been suggested that there are different socioecological influences for females of different ranks and reproductive states (Williams et al., 2002a). At Gombe, low-ranking females are most affected by contest competition, whereas females with infants are susceptible to scramble competition. While maternal influence on male rank among bonobos has been well documented, recent evidence suggests that maternal rank influences the outcome of aggressive interactions between immature chimpanzees (Markham et al., 2015).

Dominance rank in both males and females has been related to increased reproductive success at Gombe (Pusey et al., 1997, Constable et al., 2001). Dominant males' direct aggression toward females during times when they are not sexually receptive (sex skins are not swollen), and the long-term patterning of this aggression, is correlated with siring offspring (Feldblum et al., 2014). In concert with their paternity success, the intimidation behavior of these high-ranking males has been described as support for the sexual coercion hypothesis (Muller et al., 2007, 2011). At Budongo, alpha males sire a disproportionate number of offspring, but middle- and low-ranking males also father offspring (Newton-Fisher et al., 2010).

Ranging has a large influence on sociality. The sociospatial relationships between males and females have been shown to play a large role in chimpanzee male reproductive strategies at Ngogo (Langergraber et al., 2013). Males in this community tend to associate with particular females who concentrate their ranging within certain areas of the community territory that coincide with a particular male's range. In the Kanyawara community, male-female dyads associate as a result of overlap in ranging and changes in female reproductive state. Based on differential ranging patterns and asocial habits of female chimpanzees at Gombe, it has been widely concluded that bonds between males and females are not strong (Wrangham & Smuts, 1980). Also, large parties at Kibale typically include all community males and many adolescent females, but these large aggregations have never included all the community's adult females (Wrangham, 2000). Together, these results confirm that associations between male and female chimpanzees are not the result of strong affiliative bonds (Machanda et al., 2013).

The most sociable female chimpanzees tend to be recent nulliparous immigrants who spend much of their time with adult males even when they are not sexually receptive (Goodall, 1986). These females typically become less gregarious after giving birth to their first offspring. Based on these results, female (and possibly group) sociability is related to parity (Anderson et al., 2002). The observation of female "cliques" at Ngogo, however, challenges the notion of asocial females, as these association clusters are not simply an artifact of spatial overlap (Wakefield, 2008, 2013).

Adult females at most sites spend a high proportion of their time feeding and traveling with only dependent offspring, but are also frequently observed in large, mixed-sex parties. At most sites, parties including both males and females (mixed-sex parties and adult parties) are the most common subgroupings (30% at Gombe: Wrangham & Smuts, 1980; Goodall, 1986; 52% at Mahale: Nishida,

1990; 52% at Kibale: Wrangham et al., 1992; Goldberg & Wrangham, 1997; Wrangham, 2000; 41% at Budongo: Fawcett, 2000; 61% at Taï: Boesch & Boesch-Achermann, 2000; 42% at Bossou: Sugiyama, 1988; Sakura, 1994). Potential benefits for non-estrous females to join these aggregations may include protection against harassment by males, increased access to resources, protection against predators, benefit from the food sharing after cooperative hunts, and evaluation of potential mates. In Kibale, females show significantly lower C-peptide levels in insulin (indicative of energetic and reproductive costs) when they associate with more males, which is not the case when they associate with females (Thompson et al., 2014).

Grooming is a critical means by which chimpanzees and bonobos develop and maintain relationships, and both social and economic factors are hypothesized to influence grooming behavior and choice of grooming partners. The grooming trade model (Seyfarth, 1977) hypothesizes that high-ranking individuals will receive more grooming than lower-ranking individuals. Biological Markets Theory (BMT) further posits that partner choice will depend on the "supply and demand" conditions of the specific environment (Noë & Hammerstein, 1994). A study of male chimpanzees at Mahale found that grooming-initiator individuals were less likely to reciprocate in the presence of a bystander, and also more prone to abandon a grooming dyad (Kaburu & Newton-Fisher, 2016). Another study at the same site provided evidence that females trade grooming for sex (Kaburu & Newton-Fisher, 2015b). Among males of the Sonso community at Budongo, the rank of a bystander relative to the rank of a current partner exerts a greater influence on grooming investment than does the number of bystanders (Newton-Fisher & Kaburu, 2017). Consistent with predictions of BMT, high-ranking individuals at Budongo will trade coalitionary support for grooming with lower-ranking individuals; however, this was not observed at Mahale, where the hierarchy was more egalitarian, and thus agonistic support was less valuable (Newton-Fisher & Lee, 2011, 2015a). Grooming reciprocity is also stronger for males of similar dominance rank at Budongo, while at Mahale, grooming is reciprocated independently of dominance rank (Newton-Fisher & Lee, 2011; Kaburu & Newton-Fisher, 2015a). Grooming reciprocity was not found to vary relative to hierarchy steepness, though grooming was more reciprocal at Mahale during a period of social stability (Kaburu & Newton-Fisher, 2015a). Some authors have highlighted methodological challenges in applying BMT, such as defining the precise time frame during which commodities are exchanged, measuring some commodities (e.g., lack of aggression), and determining whether primates can cognitively track instances of reciprocation (Sánchez-Amaro & Amici, 2015). Less is known about what underpins grooming decisions among bonobos, but continued investigation of patterns of exchange over time will help to clarify to what extent bonobos reciprocate grooming or exchange it for other resources, such as feeding tolerance (Surbeck & Hohmann, 2014).

While the social structure of bonobos is similar to chimpanzees, recent studies highlight strong species-specific differences in chimpanzee and bonobo social affiliation. A comparative analysis of seven communities of chimpanzees and two of bonobos concludes that chimpanzee males (but not bonobos) prefer to associate with other males, while in bonobos, both females and males tend to associate more with female individuals (Surbeck et al., 2017b). These social patterns may arise as a result of differential between-group competition and associated differences between species in the possible fitness benefits of affiliation and cooperation among males (Surbeck et al., 2017c). In contrast to chimpanzees, bonobo females are generally considered to be co-dominant or dominant to males (Furuichi, 2011). Furuichi (2017) has suggested that the high social status of female bonobos and their initiative in social, sexual, and ranging behaviors may contribute to the relatively peaceful nature of their society. Females at LuiKotale may even trade feeding opportunities in exchange for intra-sexual cohesion and protection against male aggression (Nurmi et al., 2018). At Wamba, older females typically support younger females in conflicts with males (Tokuyama & Furuichi, 2016).

There are intra-sexual dominance hierarchies in bonobos, and a mother's presence and rank within the group influences the rank of her son (Furuichi, 2011). A mother's presence is also associated with enhanced mating and paternity success for male bonobos, but this is not the case for chimpanzees (Surbeck et al., 2010, 2019). More generally, male bonobos have a greater support network of related females within their social group than do male chimpanzees. Adult male bonobos are twice as likely to live in the same group as their mother and immature male bonobos are three times as likely to reside with their paternal grandmother than are chimpanzees (Schubert et al., 2013).

Increasingly, there is recognition of the role of neuro-endocrinological factors in regulating different aspects of cooperative and coalitionary behavior for both chimpanzees and bonobos. Recent studies on oxytocin patterns suggest that these neuropeptides act as a neural reward mechanism enabling chimpanzees to keep track of interaction histories with different members of the community (Crockford et al., 2013, 2018), mediate post-conflict affiliative interactions (Preis et al., 2018), and share resources (Samuni et al., 2018a). In addition, the oxytocinergic system modulates female bonobo genito-genital interactions, the frequency of which predicts coalitionary support among unrelated females (Moscovice et al., 2019).

While great progress has been made in the past decade to compare wild populations of chimpanzees and bonobos, further research is needed to determine the factors underlying specific differences in these great ape societies.

INTER-COMMUNITY INTERACTIONS

Chimpanzee and bonobo communities are both typically surrounded by several other communities, which are settled in spatially distinct home ranges that overlap in peripheral areas. Interactions between neighboring groups of chimpanzees have been characterized as ranging from avoidance to aggression, while interactions among bonobo groups may be affiliative and have involved sharing of meat between members of different communities (Idani, 1990; Fruth & Hohmann, 2018). Intergroup encounters in chimpanzees have been associated with lethal aggression, the prevalence and evolutionary significance of which have been extensively debated. A recent review of 18 chimpanzee communities and 4 bonobo communities studied over five decades reported 58 observed conspecific killings, 41 inferred killings, and 53 suspected killings by chimpanzees, as well as 1 suspected killing by bonobos (Wilson et al., 2014). Sixty-six percent of killings were associated with intercommunity aggression; no killings were associated with human impact. Conspecific killing may confer adaptive benefits, potentially in accordance with enhancing or maintaining access to food and/or mates. Three hypotheses have been proposed as to why chimpanzees exhibit hostile behavior during intergroup encounters and range maintenance. The first is that males exclude neighboring conspecifics from their territory in defense of local food resources. The second hypothesis also includes that males actively defend food resources but extends this idea by implying that males maximize their reproductive success by increasing the access to resources on the part of their community females. The third hypothesis involves inter-community competition as a result of males defending access to mates. Not only do the types of inter-community encounters differ between bonobos and chimpanzees, but also the forces underlying these interactions between groups. Rather than decrease during times of resource abundance, inter-community encounters among bonobos were highest during the yearly peak in fruit abundance at Wamba (Sakamaki et al., 2018). During periods of low fruit abundance, the probability of an encounter increased when adult females with maximum sexual swellings were present.

While lethal intergroup aggression has been documented in several chimpanzee populations, it is nonetheless a relatively rare phenomenon, and most interactions between communities consist of long-distance communication or bluff displays (Goodall et al., 1979; Boesch & Boesch-Achermann, 2000). For example, of 129 territorial activities that were observed at Taï, none involved physical attacks (Boesch & Boesch-Achermann, 2000). This territorial monitoring consisted of 38 patrols, 32 drumming exchanges, 14 instances of avoidance, and 45 occasions of visual contact.

Aggressive events have been used as evidence for both group cohesiveness and dispersion within discussions of chimpanzee sociality. There are two contradictory hypotheses that are widely cited to explain group living in relation to within- and between-group aggression. First, primates may join groups to protect themselves and their infants from conspecifics. The notion that lone females are vulnerable to the risk of attack has arisen from events at Gombe and Mahale (Goodall et al., 1979; Hamai et al., 1992; Pusey et al., 1997; Watts & Mitani, 2000). An alternate view is that subordinate individuals who are often the subject of harassment may avoid subgroups to reduce risk of aggression. Further research is needed to determine the influence of these factors on not only chimpanzees, but on other primate societies as well.

TOOL USE

Aside from humans, the tool-using skills of chimpanzees surpass those of any other mammal or nonhuman primate. Tool use is defined as, "The manipulation of an object (the tool), not part of the actor's anatomical equipment and not attached to a substrate, to change the position, action, or condition of another object, either directly through the action of the tool on the object or of the object on the tool, or through action at a distance as in aimed throwing" (Parker & Gibson 1977: pp.624–625). Materials used for these functions are often modified and/or transported before use. Jane Goodall (1963, 1964, 1968) was the first to describe the manufacture and use of tools by wild chimpanzees to gather termites:

> Stalks and small twigs were used when the chimpanzees fed on termites.... Some animals inspected several clumps of grass etc. before selecting their tools, sometimes they picked several to carry back to the termite heap and then used them one at a time.... Sometimes tools were carefully prepared: leaves were stripped from stems or twigs, and long strips were sometimes pulled from a piece of grass that was too wide... individuals

> were seen to pick a tool for subsequent use on a heap that was out of sight and as far as 100 yd away. One male twice carried a tool over half a mile whilst inspecting a series of termite heaps, none of which were [sic] ready for working. (Goodall, 1968: pp.204–205)

The tool repertoire of wild chimpanzees is diverse and expressed in a variety of behavioral contexts (McGrew, 1992; Sanz & Morgan, 2007). All studied chimpanzee populations exhibit at least one type of tool used for extraction or social activities, and tool repertoires of a population may include more than 20 different types of tools. A comparison across study sites revealed that 53% of tool behaviors shown by all wild chimpanzee populations are directed toward food gathering. Tools are most often used in gathering otherwise inaccessible items, such as nuts, insects, bone marrow, honey, and water (Sanz & Morgan, 2007). Chimpanzees at the West African site of Fongoli use tools as spears in hunting small primates in tree hollows (Pruetz & Bertolani, 2007; Pruetz et al., 2015). Tools are also used by several populations in self-care (e.g., grooming and comfort: 32%); for example, branches or leaves are used to manufacture "umbrellas" in the case of rain. Chimpanzees also use tools in a variety of social contexts (e.g., display and play: 14%) (Sanz & Morgan, 2007). New tool variants continue to be recorded among wild chimpanzee populations. For instance, in Bakoun Classified Forest (Guinea), chimpanzees were recently documented using woody twigs and sticks to fish for algae (*Spirogyra* spp.; Boesch et al., 2017). At Comoé National Park (Côte d'Ivoire), chimpanzees manufacture water-dipping tools by chewing the tips of sticks to produce a fibrous, absorbent sponge (Lapuente et al., 2017). A recent summary of tool-using behaviors of bonobos indicates a striking paucity of tool-assisted foraging behaviors, with the exception of a single observation of leaf sponging; rather, bonobo tool use seems driven by promoting self-comfort and communicating information within a social context. For example, during play, bonobos may place vegetation in their hand or mouth and chase conspecifics (Furuichi et al., 2015).

Comparisons of different populations of chimpanzees across Africa have revealed intriguing intraspecific differences in tool-using behaviors. Researchers reported variations in tool use to gather termites across the range of chimpanzees (McGrew et al., 1979; McGrew & Rogers, 1983), but also noted that groups within the same deme exhibit variations in tool use and termite capture (McGrew & Collins, 1985). Within the range of the central subspecies of chimpanzee, three different types of tools are used in termite predation (Sanz et al., 2004). For chimpanzees at both Gombe and Goualougo, termite fishing involves inserting a flexible herb stem into a termite nest to extract termites biting the invading object. However, a variation of the termite-fishing technique in Central Africa involves use of a perforating tool to open the termite exit holes on the surface of the nest, rather than (as in East Africa) just picking open the hole with one's fingers. Extracting termites from subterranean (as opposed to elevated) nests involves another type of tool, a puncturing stick. This involves inserting the length of a stout stick into the ground to create a long, narrow tunnel for insertion of the fishing probe. Prior to termite fishing, chimpanzees in Central Africa also apply a set of deliberate, distinguishable actions to modify herb stems to fashion a brush-tipped probe, which is different from the form of fishing tools used by chimpanzees in East and West Africa. Brush-tipped probes are more effective in gathering insects than unmodified fishing probes (Sanz et al., 2009). The development of an improved fishing probe design among these chimpanzees provides some of the best evidence to date of cumulative culture among nonhumans. A recent report from chimpanzees at Budongo Forest documenting a transition from leaf-sponge to moss-sponge in order to drink water provides further evidence for innovation of tool technology and cumulative culture (Lamon et al., 2018).

The application of archaeological methods has provided novel insights into tool raw material procurement and cultural variation in chimpanzees. At Gombe, for instance, availability of plant raw material is comparable between two neighboring chimpanzee communities, yet members of one community produce wider, longer tools and use a broader array of material types (Pascual-Garrido, 2019). In the Issa Valley, Tanzania, chimpanzees manufacture termite-fishing tools only from bark, despite availability of other source materials such as grass or twigs (Almeida-Warren et al., 2017). These findings suggest possible cultural preferences for tool characteristics within specific communities.

Similar sophistication and variation in chimpanzee technology are seen in the use of anvils and hammers to break open hard-shelled foods. Although there were some early reports of nut-cracking by wild apes (Savage & Wyman, 1844; Beatty, 1951), Struhsaker and Hunkeler (1971) provided the first convincing evidence of the use of stones and detached branches to crack the shells of *Coula edulis* and *Panda oleosa* nuts. Subsequently, Sugiyama and Koman (1979) provided detailed descriptions of this behavior. They observed chimpanzees at Bossou placing palm nuts (*Elaeis guinensis*) on relatively large anvil (previously referred to as "platform") stones and lifting smaller "hammer" stones to crack nuts. Each nut-cracking site contained from one to four platforms, and

the tools were used in succession by different individuals over a 20- to 60-minute period.

In the Taï Forest, nut-cracking can provide a major portion of caloric and protein intake. During the four-month season of *Coula edulis*, chimpanzees have been observed spending an average of 2.25 hours per day cracking nuts (Boesch & Boesch 1984a,b; Gunther & Boesch, 1993; Boesch & Boesch-Achermann, 2000). Stone hammers are frequently transported between anvils of different trees, and particular types of tools are used for different species of nuts (Boesch & Boesch, 1984a). For example, *Panda oleosa* nuts are harder than *Coula edulis,* and the chimpanzees transport harder hammers (almost exclusively stones), use hammers of greater weight, and carry the stones over a greater distance for *Panda* nuts than for those of *Coula*. The location of stones used for *Panda* nuts may also be chosen to keep transport distance at a minimum. Even when nut-cracking of only *Coula* nuts is considered, chimpanzees' selection of tools reflects the integrated assessment of multiple variables in order to optimize effort (Sirianni et al., 2015). Sex differences in chimpanzee nut-cracking behavior have been reported in this context (Boesch & Boesch, 1984b). Males collect 12–15 nuts in the canopy and carry them down to a stone anvil and a wooden club to crack the shells, whereas females often carry the hammer into the tree and crack nuts on a horizontal branch, usually eating more nuts per minute than do the males. Females more often crack the harder *Panda* nuts. Sex differences in tool behavior have also been observed in other populations. At Fongoli, females more often use tools to hunt (Pruetz & Bertolani, 2007). Females at Gombe termite-fish more often and for longer durations than do males (McGrew, 1979; Pandolfi et al., 2003). Infant chimpanzees at Gombe also show sex differences in termite-fishing behaviors. Young females spend more time observing their mothers termite-fish and acquire tool-using skills earlier than young males (Lonsdorf et al., 2004).

It was previously held that nut-cracking was limited to chimpanzee populations residing on the west side of the Sassandra River, Côte d'Ivoire (Boesch et al., 1994). Although nut species associated with this behavior are present on both sides of the river, it was hypothesized that this watercourse acted as a barrier to transmission of nut-cracking behavior between populations. This trait may not have spread to Central and East Africa because the classical biogeographical barriers may also have acted as an effective barrier to cultural transmission. However, chimpanzees in the Ebo Forest of Cameroon east of this barrier were discovered to crack nuts, which may indicate a broader historical distribution of this behavior or possible independent invention of this foraging technology (Morgan & Abwe, 2006). Percussive technology has also been documented in Loango National Park in Gabon, where chimpanzees smash tortoises (*Kinixis eroxa*) against anvils; the meat is then typically shared with bystanders (Pika et al., 2019). Between 2014–2016, surveys in the northern DRC revealed a new chimpanzee tool kit, termed the Bili-Uéré Behavioral Realm, including multiple probing and percussive tool variants. These behaviors show consistency across ecologically diverse regions as well as subtle variations in technique and target food items (Hicks et al., 2019).

Chimpanzees show intriguing variation in tool repertoires across multiple habitats, with some differences reflecting cultural specificity. This variability may be linked to a combination of factors, such as differential energy requirements, habitat type, or geographical barriers. However, within-population differences in tool material suggest that cultural preferences may operate in tool-use decision making. Finally, an open question remains: Why do bonobos show such a paucity of tool-use behavior in the wild?

PREDATION

It was previously held that chimpanzees were too large to be susceptible to predation pressure, but this has since been dramatically disproven. In 1985, Boesch noticed two or three sharply cut parallel wounds on an adult male chimpanzee at Taï that could only have been from a leopard (Boesch, 1991). He was soon aware that leopard predation was a constant factor for the chimpanzees: "The tremendous power of the leopard's bite makes him a rapid killer and, if taken by surprise, even an adult individual seems unable to prevent it from the fatal biting. Thus, all age-sex classes may suffer from predation by leopards" (Boesch, 1991: p.228). Boesch found that the presence of humans was not much of a deterrent to the leopards, as a chimpanzee was attacked within 10 m from where researchers were working. Boesch calculated that leopards killed about 5.5% of his study group of chimpanzees each year. In Mahale, lions annually killed an estimated 6.3% of the chimpanzee population (Tsukahara, 1993). Leopard attacks were the main cause of mortality in the Taï chimpanzees between 1988–1991 (Boesch, 1991). Further, lions were suspected of killing at least four apes at Mahale in 1989 (Tsukahara, 1993).

Potential mammalian predators include leopard (*Panthera pardus*), lion (*P. leo*), spotted hyena (*Crocuta crocuta*), and wild dog (*Lycaon pictus*). Other potential predators are the crowned hawk–eagle (*Stephanoaetus coronatus*) and poisonous snakes (Ghiglieri, 1984; Goodall, 1986). Goodall placed an almost dead python in the observation area and the chimpanzees reacted to it with fear and avoidance (Goodall, 1968). She also observed a female chimpanzee react with a protective gesture to her infant when a small hawk flew overhead. Tutin and colleagues examined the reactions of chimpanzees to the calls of potential predators at the Niokolo Koba National Park near Mt. Assirik in Senegal (Tutin et al., 1981). Of the 36 reactions observed, 30 occurred during the night while the chimpanzees were in their nests. Hyena calls never provoked a reaction from the chimpanzees, but leopards were responded to either with loud calls, silent flight, or on one occasion, by a female and her two dependent offspring remaining silent in their nest as the leopard passed below. Lions provoked a variety of responses, ranging from ignoring to charging displays. Chimpanzees have been observed chasing away leopards in the Taï Forest (Boesch & Boesch-Achermann, 2000). Bonobos are also susceptible to predation by leopards (D'Amour et al., 2006). Direct observations of predation attempts are atypical and may underestimate the threat to these primates (Hart & Sussman, 2009).

CONSERVATION OUTLOOK

Wild ape populations have faced an increasing number of threats over the last 50 years. Threats include loss of habitat, poaching pressure, political instability, and disease (Leendertz et al., 2006; Meijaard et al., 2010, 2011; Robbins et al., 2011; Hickey et al., 2013). The combined impact of these threats has resulted in the classification of chimpanzees and bonobos as endangered species. Among the subspecies of chimpanzees, however, there is considerable variation in population status. The western chimpanzee (*P. troglodytes verus*) has experienced significant population decline across its range. In Côte d'Ivoire, a country that once contained a large continuous population of chimpanzees, nearly 90% has disappeared during the last two decades (Campbell et al., 2008). Recent analysis of survey data across multiple sites across the subspecies range collected from 1990–2014 indicate an annual decline of 6% and a total population decline of 80.2% (Kühl et al., 2017). These findings prompted the uplisting of the western chimpanzee to Critically Endangered status by the International Union for Conservation of Nature (IUCN, 2022). Population declines of the eastern subspecies (*P. t. schweinfurthii*) have also been documented in eastern DRC with between 22–45% of the population disappearing over the last two decades (Plumptre et al., 2015).

Habitat loss through encroachment and conversion to crops is one of the most significant threats to all remaining wild ape populations. The number of suitable habitats for great apes declined notably between 1995–2010 (Junker et al., 2012). Chimpanzee habitat reduction has been steepest in West Africa with 20% of the *P. t. verus* subspecies range lost (Kühl et al., 2017). Chimpanzees are typically found in mixed-species forest characterized by taller trees (Strindberg et al., 2018). Such forest stands are also targeted by logging companies because of the relatively high density of valuable timber trees. Although logging is often considered a primary cause of faunal decline in tropical forests, our knowledge of the impact of logging on complex species and tropical forest dynamics is still poorly understood.

The commercial bushmeat trade is another threat to the conservation of great apes, particularly in Central Africa, which contains the largest remaining populations of great apes on the planet (Morgan et al., 2006; Rainey et al., 2010; Stokes et al., 2010; Strindberg et al., 2018). Vast and remote areas of intact forest combined with relatively low human population densities and poor access networks and infrastructure has, until recently, largely safeguarded these populations from the threat of poaching in Central Africa. However, the past decade has been defined by unprecedented expansion of logging roads (Laporte et al., 2007; Kleinschroth et al., 2015), which provides a transport network for the commercial bushmeat trade.

The expansion of roads in Central Africa over the last 16 years has varied over the region depending on land-use. In Central Africa, areas leased by logging firms experienced nearly a doubling in total length of roads since 2003 while those outside such timber production zones saw a 40% increase in route length (Kleinschroth et al., 2019). The degradation and loss of habitat has also coincided in altering other aspects of chimpanzee well-being, including their behavioural diversity. Recent findings throughout Equatorial Africa show the number of behavioral variants documented across chimpanzee communities decline with increasing human disturbance, suggesting unique cultures of social groups are disappearing (Kuehl et al., 2019). Simultaneously, the poaching pressure in and around the forests associated with road development has dramatically increased in this region and poses health risks to local human populations (Hahn et al., 2000; Rouquet et al., 2005).

Disease epidemics have also impacted many chimpanzee populations. Emerging infectious diseases (such as Ebola and anthrax) have had a significant impact on chimpanzees and gorillas across Africa (Walsh et al., 2003; Bermejo et al., 2006; Köndgen et al., 2018). Pathogen transmission from humans is also a threat to wild apes, as they have been shown to contract parasites, respiratory diseases, polio, and scabies from local human populations, research and protection staff, or tourists (Wallis & Lee, 1999; Leendertz et al., 2006; Lonsdorf et al., 2006; Goldberg et al., 2007). Human respiratory syncytial viruses (HRSV) and human metapneumoviruses (HMPV) have played a role in respiratory outbreaks in wild chimpanzees in Côte d'Ivoire (Köndgen et al., 2017). Transmission of human respiratory viruses to wild populations can play an influential role, causing high morbidity and, in some instances, mortality (Köndgen et al., 2017). Chimpanzee communities may be affected differently depending on the circumstances of disease exposure and possible health interventions (Pusey et al., 2008; Lonsdorf et al., 2018). While long-term initiatives to monitor chimpanzee health have been implemented at some sites in Côte d'Ivoire, Cameroon, or Tanzania (Köndgen et al., 2010; Lonsdorf et al., 2018), there is a critical need to expand and conduct systematic monitoring of wild ape populations and health protocols (Grützmacher et al., 2016). Not only will such efforts alert local agencies about potential epidemics, but they could also lead to the discovery of novel pathogens of importance to wild apes and humans alike (Calvignac-Spencer et al., 2012; Köndgen et al., 2017).

SUMMARY

Chimpanzees and bonobos are the closest relatives of humans, sharing 99% of our genetic makeup (Mikkelsen et al., 2005; Prüfer et al., 2012). As such, studies of their behavior and ecology are fundamental in reconstructing our own evolutionary history. Further, these species are integral to their ecosystems. Maintaining a long-term research presence has been shown to have a positive impact on local chimpanzee populations and the local human communities involved in the research and conservation efforts. The following benefits have been attributed to sustained presence at sites across the species range (Pusey et al., 2007; Campbell et al., 2011; Piel et al., 2015; Morgan et al., 2020): (1) discoveries are made of new behaviors, resulting in increased protected status, (2) publication and subsequent media coverage of research findings has increased worldwide awareness of wild chimpanzees and bonobos, which has resulted in increased financial support for specific study sites and local surrounding communities, (3) conservation initiatives and policies have been informed by scientific data on chimpanzee sociality and ecology, and (4) monitoring of population trends has allowed researchers to identify threats to important habitats, population viability, and their cultures. Despite these meaningful benefits, expanding research and analysis directed at the emerging environmental and socioeconomic challenges, facing not only chimpanzees and bonobos but the people and nations where they exist, is urgently needed. More holistic approaches that build awareness on these issues with timely evidence will facilitate more robust engagement with decision-makers at local, regional, and governmental scales as well as those working in industry. The outcomes of such research and partnerships have the potential to facilitate a more viable future for humans as well as chimpanzees and bonobos in Africa.

REFERENCES CITED—CHAPTER 23

Albrecht, H., & Dunnett, S. C. (1971). *Chimpanzees in western Africa*. R. Piper & Co.

Almeida-Warren, K., Sommer, V., Piel, A. K., & Pascual-Garrido, A. (2017). Raw material procurement for termite fishing tools by wild chimpanzees in the Issa Valley, western Tanzania. *American Journal of Physical Anthropology*, 164(2), 292-304.

Amsler, S. J. (2010). Energetic costs of territorial boundary patrols by wild chimpanzees. *American Journal of Primatology*, 72(2), 93-103.

Anderson, D. P. (2001). *Tree phenology and distribution, and their relation to chimpanzee social ecology in the Taï National Park, Côte d'Ivoire.* [Doctoral dissertation, University of Wisconsin].

Anderson, D. P., Nordheim, E. V., Boesch, C., & Moermond, T. C. (2002). Factors influencing fission-fusion grouping in chimpanzees in the Taï National Park, Côte d'Ivoire. In C. Boesch, G. Hohmann, & L. F. Marchant (Eds.), *Behavioral diversity in chimpanzees and bonobos* (pp. 90-101). Cambridge University.

Arcadi, A. C., & Wrangham, R. W. (1999). Infanticide in chimpanzees: Review of cases and a new within-group observation from the Kanyawara study group in Kibale National Park. *Primates*, 50, 337-351.

Azuma, S., & Toyoshima, A. (1961). Progress report of the survey of chimpanzees in their natural habitat, Kabogo Point area, Tanganyika. *Primates*, 3, 61-70.

Badescu, I., Katzenberg, M. A., Watts, D. P., & Sellen, D. W. (2017). A novel fecal stable isotope approach to determine the timing of age-related feeding transitions in wild infant chimpanzees. *American Journal of Physical Anthropology*, 162(2), 285-299.

Badrian, N., Badrian, A., & Randall, L. (1981). Preliminary observations on the feeding behavior of *Pan paniscus* in the Lomako Forest of central Zaire. *Primates*, 22(2), 173-181.

Badrian, N., & Malenky, R. K. (1984). Feeding ecology of Pan paniscus in the Lomako Forest, Zaire. In R. W. Susman (Ed.), *The pygmy chimpanzee* (pp. 275-299). Plenum Press.

Beatty, H. (1951). A note on the behavior of the chimpanzee. *Journal of Mammalogy*, 32, 118.

Beaune, D., Bretagnolle, F., Bollache, L., Bourson, C., Hohmann, G., & Fruth, B. (2013a). Ecological services performed by the bonobo (*Pan paniscus*): Seed dispersal effectiveness in tropical forest. *Journal of Tropical Ecology*, 29(5), 367-380.

Beaune, D., Bretagnolle, F., Bollache, L., Hohmann, G., Surbeck, M., Bourson, C., & Fruth, B. (2013b). The bonobo-dialium positive interactions: Seed dispersal mutualism. *American Journal of Primatology*, 75(4), 394-403.

Beaune, D., Hohmann, G., Serckx, A., Sakamaki, T., Narat, V., & Fruth, B. (2017). How bonobo communities deal with tannin rich fruits: Re-ingestion and other feeding processes. *Behavioural Processes*, 142, 131-137.

Becquet, C., Patterson, N., Stone, A. C., Przeworski, M., & Reich, D. (2007). Genetic structure of chimpanzee populations. *PLoS Genetics*, 3(4), e66.

Bermejo, M., Rodriguez-Teijeiro, J. D., Illera, G., Barroso, A., Vila, C., & Walsh, P. D. (2006). Ebola outbreak killed 5000 gorillas. *Science*, 314(5805), 1564.

Biro, D., Sousa, C., & Matsuzawa, T. (2006). Ontogeny and cultural propagation of tool use by wild chimpanzees in Bossou, Guinea: Case studies in nut cracking and leaf folding. In T. Matsuzawa, M. Tomonaga, & M. Tanaka (Eds.), *Cognitive development in chimpanzees* (pp. 476-508). Springer.

Boesch, C. (1991). The effect of leopard predation on grouping patterns in forest chimpanzees. *Behaviour*, 117(3-4), 220-242.

Boesch, C. (1994a). Hunting strategies of Gombe and Taï chimpanzees. In R. W. Wrangham, W. C. McGrew, F. B. M. de Waal, & P. G. Heltne (Eds.), *Chimpanzee cultures* (pp. 77-91). Harvard University Press.

Boesch, C. (1994b). Cooperative hunting in wild chimpanzees. *Animal Behaviour*, 48, 653-667.

Boesch, C. (1996). Social grouping in Taï chimpanzees. In W. C. McGrew, L. F. Marchant, & T. Nishida (Eds.), *Great ape societies* (pp. 101-113). Cambridge University Press.

Boesch, C. (2002). Cooperative hunting roles among Taï chimpanzees. *Human Nature*, 13(1), 27-46.

Boesch, C., & Boesch, H. (1984a). Mental map in wild chimpanzees: An analysis of hammer transports for nut cracking. *Primates*, 25, 160-170.

Boesch, C., & Boesch, H. (1984b). Possible causes of sex differences in the use of natural hammers by wild chimpanzees. *Journal of Human Evolution*, 13, 415-440.

Boesch, C., & Boesch, H. (1989). Hunting behavior of wild chimpanzees in the Taï National Park. *American Journal of Physical Anthropology*, 78, 547-573.

Boesch, C., & Boesch-Achermann, H. (2000). *The chimpanzees of the Taï forest: Behavioural ecology and evolution*. Oxford University Press.

Boesch, C., Kalan, A. K., Agbor, A., Arandjelovic, M., Dieguez, P., Lapeyre, V., & Kuhl, H. S. (2017). Chimpanzees routinely fish for algae with tools during the dry season in Bakoun, Guinea. *American Journal of Primatology*, 79(3), 1-7.

Boesch, C., Marchesi, N., Fruth, B., & Joulian, F. (1994). Is nut cracking in wild chimpanzees a cultural behaviour? *Journal of Human Evolution*, 26(4), 325-338.

Boesch-Achermann, H., & Boesch, C. (1994). Hominization in the rainforest: The chimpanzee's piece of the puzzle. *Evolutionary Anthropology*, 3(1), 9-16.

Bogart, S. L., & Pruetz, J. D. (2008). Ecological context of savanna chimpanzee (*Pan troglodytes verus*) termite fishing at Fongoli, Senegal. *American Journal of Primatology*, 70, 1-8.

Bogart, S. L., & Pruetz, J. D. (2011). Insectivory of savanna chimpanzees (*Pan troglodytes verus*) at Fongoli, Senegal. *American Journal of Physical Anthropology, 145*, 11-20.

Bryson-Morrison, N., Tzanopoulos, J., Matsuzawa, T., & Humle, T. (2017). Activity and habitat use of chimpanzees (*Pan troglodytes verus*) in the anthropogenic landscape of Bossou, Guinea, West Africa. *International Journal of Primatology, 38*(2), 282-302.

Busse, C. D. (1978). Do chimpanzees hunt cooperatively? *American Naturalist, 112*, 767-770.

Calvignac-Spencer, S., Leendertz, S. A. J., Gillespie, T. R., & Leendertz, F. H. (2012). Wild great apes as sentinels and sources of infectious disease. *Clinical Microbiology and Infection, 18*(6), 521-527.

Campbell, G., Kuehl, H., Diarrassouba, A., N'Goran, P. K., & Boesch, C. (2011). Long-term research sites as refugia for threatened and over-harvested species. *Biology Letters, 7*(5), 723-726.

Campbell, G., Kuehl, H., N'Goran Kouame, P., & Boesch, C. (2008). Alarming decline of West African chimpanzees in Côte d'Ivoire. *Current Biology, 18*(19), 903-904.

Carlson, B. A., & Crowley, B. E. (2016). Variation in carbon isotope values among chimpanzee foods at Ngogo, Kibale National Park and Bwindi Impenetrable National Park, Uganda. *American Journal of Primatology, 78*(10), 1031-1040.

Chapman, C. A. (1990). Association patterns of spider monkeys: The influence of ecology and sex on social organization. *Behavioral Ecology and Sociobiology, 26*, 409-414.

Chapman, C. A., Chapman, L. J., & Wrangham, R. W. (1995). Ecological constraints on group size: An analysis of spider monkey and chimpanzee subgroups. *Behavioral Ecology and Sociobiology, 36*, 59-70.

Chapman, C. A., White, F. J., & Wrangham, R. W. (1993). Defining subgroup size in fission-fusion societies. *Folia Primatologica, 61*, 31-34.

Chapman, C. A., White, F. J., & Wrangham, R. W. (1994). Party size in chimpanzees and bonobos: A reevaluation of theory based on two similarly forested sites. In R. W. Wrangham, W. C. McGrew, F. B. M. de Waal, & P. Heltne (Eds.), *Chimpanzee cultures* (pp. 41-57). Harvard University Press.

Clark, C. B. (1977). Preliminary report on weaning among chimpanzees of Gombe National Park, Tanzania. *American Journal of Physical Anthropology, 47*(1), 123-124.

Clutton-Brock, T. H. (1989). Mammalian mating systems. *Proceedings of the Royal Society B Biological Sciences, 236*(1285), 339-372.

Clutton-Brock, T. H., & Harvey, P. H. (1979). Home range size, population density and phylogeny in primates. In I. Bernstein, & E. O. Smith (Eds.), *Primate ecology and human origins* (pp. 201-214). Garland.

Collins, A., & McGrew, W. C. (1988). Habitats of three groups of chimpanzees (*Pan troglodytes*) in western Tanzania compared. *Journal of Human Evolution, 17*, 553-574.

Constable, J., Ashley, M., Goodall, J., & Pusey, A. (2001). Noninvasive paternity assignment in Gombe chimpanzees. *Molecular Ecology, 10*, 1279-1300.

Coolidge, H. J., & Shea, B. T. (1992). External body dimensions of *Pan paniscus* and *Pan troglodytes* chimpanzees. *Primates, 23*, 245-251.

Crockford, C., Deschner, T., & Wittig, R. M. (2018). The role of oxytocin in social buffering: What do primate studies add? *Behavioral Pharmacology of Neuropeptides: Oxytocin, 35*, 155-173.

Crockford, C., Wittig, R. M., Langergraber, K., Ziegler, T. E., Zuberbuhler, K., & Deschner, T. (2013). Urinary oxytocin and social bonding in related and unrelated wild chimpanzees. *Proceedings of the Royal Society B Biological Sciences, 280*(1755), 20122765.

D'Amour, D. E., Hohmann, G., & Fruth, B. (2006). Evidence of leopard predation on bonobos (*Pan paniscus*). *Folia Primatologica, 77*(3), 212-217.

Deblauwe, I., & Janssens, G. P. J. (2008). New insights in insect prey choice by chimpanzees and gorillas in southeast Cameroon: The role of nutritional value. *American Journal of Physical Anthropology, 135*(1), 42-55.

Deschner, T., Heistermann, M., Hodges, K., & Boesch, C. (2003). Timing and probability of ovulation in relation to sex skin swelling in wild West African chimpanzees, *Pan troglodytes troglodytes*. *Animal Behaviour, 66*, 551-560.

Deschner, T., Heistermann, M., Hodges, K., & Boesch, C. (2004). Female sexual swelling size, timing of ovulation, and male behavior in wild West African chimpanzees. *Hormones and Behavior, 46*, 204-215.

Doran, D. M. (1993a). Comparative locomotor behavior of chimpanzees and bonobos: The influence of morphology on locomotion. *American Journal of Physical Anthropology, 91*, 83-98.

Doran, D. M. (1993b). Sex-differences in adult chimpanzee positional behavior: The influence of body size on locomotion and posture. *American Journal of Physical Anthropology, 91*(1), 99-115.

Doran, D. M. (1997). Influence of seasonality on activity patterns, feeding behaviour, ranging and grouping patterns in Taï chimpanzees. *International Journal of Primatology, 19*(2), 183-206.

Doran, D. M., & Hunt, K. D. (1994). Comparative locomotor behavior of chimpanzees and bonobos: Species and habitat differences. In. R. W. Wrangham, W. C. McGrew, F. B. M. de Waal, & P. G. Heltne (Eds.), *Chimpanzee cultures* (pp. 93-108). Harvard University Press.

Dupain, J., Van Elsacker, L., Nell, C., Garcia, P., Ponce, F., & Huffman, M. A. (2002). New evidence for leaf swallowing and *Oesophagostomum* infection in bonobos (*Pan paniscus*). *International Journal of Primatology, 23*(5), 1053-1062.

Dupain, J., Van Krunkelsven, E., Van Elsacker, L., & Verheyen, R. F. (2000). Current status of the bonobo (*Pan paniscus*) in the proposed Lomako Reserve (Democratic Republic of Congo). *Biological Conservation, 94*(3), 265-272.

Emery Thompson, M., Kahlenberg, S. M., Gilby, I. C., & Wrangham, R. W. (2007). Core area quality is associated with variance in reproductive success among female chimpanzees at Kibale National Park. *Animal Behaviour*, 73, 501-512.

Estienne, V., Cohen, H., Wittig, R. M., & Boesch, C. (2019). Maternal influence on the development of nut-cracking skills in the chimpanzees of the Taï forest, Côte d'Ivoire (*Pan troglodytes verus*). *American Journal of Primatology*, 81(7), ee23022.

Fawcett, K. (2000). *Female relationships and food availability in a forest community of chimpanzees*. [Doctoral dissertation, University of Edinburgh].

Feldblum, J. T., Manfredi, S., Gilby, I. C., & Pusey, A. E. (2018). The timing and causes of a unique chimpanzee community fission preceding Gombe's "four-year war." *American Journal of Physical Anthropology*, 166(3), 730-744.

Feldblum, J. T., Wroblewski, E. E., Rudicell, R. S., Hahn, B. H., Paiva, T., Cetinkaya-Rundel, M., Pusey, A. E., & Gilby, I. C. (2014). Sexually coercive male chimpanzees sire more offspring. *Current Biology*, 24(23), 2855-2860.

Fischer, A., Pollack, J., Thalmann, O., Nickel, B., & Paeaebo, S. (2006). Demographic history and genetic differentiation in apes. *Current Biology*, 16(11), 1133-1138.

Fischer, A., Wiebe, V., Paabo, S., & Przeworski, M. (2004). Evidence for a complex demographic history of chimpanzees. *Molecular Biology and Evolution*, 21(5), 799-808.

Fleagle, J. G., Stern, J. T., Jungers, W., Susman, R. L., Vangor, A. K., & Wells, J. P. (1981). Climbing: A biomechanical link with brachiation and with bipedalism. In M. H. Day (Ed.), *Vertebrate locomotion* (pp. 359-375). Symposium of the Zoological Society of London.

Fourrier, M., Sussman, R. W., Kippen, R., & Childs, G. (2008). Demographic modeling of a predator-prey system and its implication for the Gombe population of *Procolobus rufomitratus tephrosceles*. *International Journal of Primatology*, 29(2), 497-508.

Fruth, B., & Hohmann, G. (1993). Ecological and behavioral aspects of nest-building in wild bonobos (*Pan paniscus*). *Ethology*, 94(2), 113-126.

Fruth, B., & Hohmann, G. (1996). Nest building behavior in the great apes: The great leap forward? In W. C. McGrew, L. F. Marchant, & T. Nishida (Eds.), *Great ape societies* (pp. 225-240). Cambridge University Press.

Fruth, B., & Hohmann, G. (2002). How bonobos handle hunts and harvests: Why share food? In C. Boesch, G. Hohmann, & L. F. Marchant (Eds.), *Behavioral diversity in chimpanzees and bonobos* (pp. 231-243). Cambridge University Press.

Fruth, B., & Hohmann, G. (2018). Food sharing across borders: First observation of intercommunity meat sharing by bonobos at LuiKotale, DRC. *Human Nature*, 29, 91–103.

Fruth, B., Ikombe, N. B., Matshimba, G. K., Metzger, S., Muganza, D. M., Mundry, R., & Fowler, A. (2014). New evidence for self-medication in bonobos: *Manniophyton fulvum* leaf and stem-strip swallowing from LuiKotale,

Salonga National Park, DR Congo. *American Journal of Primatology*, 76(2), 146-158.

Fuenfstueck, T., Arandjelovic, M., Morgan, D. B., Sanz, C., Reed, P., Olson, S. H., Cameron, K., Ondzie, A., Peeters, M., & Vigilant, L. (2015). The sampling scheme matters: *Pan troglodytes troglodytes* and *P.t. schweinfurthii* are characterized by clinal genetic variation rather than a strong subspecies break. *American Journal of Physical Anthropology*, 156(2), 181-191.

Furuichi, T. (2009). Factors underlying party size differences between chimpanzees and bonobos: A review and hypotheses for future study. *Primates*, 50(3), 197-209.

Furuichi, T. (2011). Female contributions to the peaceful nature of bonobo society. *Evolutionary Anthropology*, 20(4), 131-142.

Furuichi, T. (2017). Female contributions to the peaceful nature of bonobo society. In B. Hare, & S. Yamamoto (Eds.), *Bonobos: Unique in mind, brain, and behavior* (pp. 17-34). Oxford University Press.

Furuichi, T., Idani, G., Ihobe, H., Hashimoto, C., Tashiro, Y., Sakamaki, T., Mulavwa, M., Yangozene, K., & Kuroda, S. (2012). Long-term studies on wild bonobos at Wamba, Luo Scientific Reserve, D. R. Congo: Towards the understanding of female life history in a male-philopatric species. In P. M. Kappeler, & D. P. Watts (Eds.), *Long-term Field Studies of Primates* (pp. 413-433). Springer.

Furuichi, T., Idani, G., Ihobe, H., Kuroda, S., Kitamura, K., Mori, A., Enomoto, T., Okayasu, N., Hashimoto, C., & Kano, T. (1998). Population dynamics of wild bonobos (*Pan paniscus*) at Wamba. *International Journal of Primatology*, 19(6), 1029-1043.

Furuichi, T., Sanz, C., Koops, K., Sakamaki, T., Ryu, H., Tokuyama, N., & Morgan (2015). Why do wild bonobos not use tools like chimpanzees do? *Behaviour*, 152(3-4), 425-460.

Gagneux, P., Wills, C., Gerloff, U., Tautz, D., Morin, P. A., Boesch, C., Fruth, B., Hohmann, G., Ryder, O. A., & Woodruff, D. S. (1999). Mitochondrial sequences show diverse evolutionary histories of African hominoids. *Proceedings of the National Academy of Sciences*, 96, 5077-7082.

Garber, P. A., & Kitron, U. (1997). Seed swallowing in tamarins: Evidence of a curative function or enhanced foraging efficiency? *International Journal of Primatology*, 18(4), 523-538.

Garner, R. L. (1896). *Gorillas and chimpanzees*. Osgood McIlvaine.

Ghiglieri, M. P. (1984). *The chimpanzees of Kibale Forest*. Columbia University Press.

Gilby, I. C., Eberly, L. E., Pintea, L., & Pusey, A. E. (2006). Ecological and social influences on the hunting behaviour of wild chimpanzees, *Pan troglodytes schweinfurthii*. *Animal Behaviour*, 72, 169-180.

Gilby, I. C., Machanda, Z. P., O'Malley, R. C., Murray, C. M., Lonsdorf, E. V., Walker, K., Mjungu, D. C., Otali, E., Muller, M. N., Emery Thompson, M., Pusey, A. E., & Wrangham,

R. W. (2017). Predation by female chimpanzees: Toward an understanding of sex differences in meat acquisition in the last common ancestor of *Pan* and *Homo*. *Journal of Human Evolution*, 110, 82-94.

Gilby, I. C., & Wawrzyniak, D. (2018). Meat eating by wild chimpanzees (*Pan troglodytes schweinfurthii*): Effects of prey age on carcass consumption sequence. *International Journal of Primatology*, 39(1), 127-140.

Gilby, I. C., & Wrangham, R. W. (2007). Risk-prone hunting by chimpanzees (*Pan troglodytes schweinfurthii*) increases during periods of high diet quality. *Behavioral Ecology and Sociobiology*, 61(11), 1771-1779.

Goldberg, T. (1998). Biogeographic predictors of genetic diversity in populations of eastern African chimapnzees (*Pan troglodytes schweinfurthii*). *International Journal of Primatology*, 19(2), 237-254.

Goldberg, T. L., Gillespie, T. R., Rwego, I. B., Wheeler, E., Estoff, E. L., & Chapman, C. A. (2007). Patterns of gastrointestinal bacterial exchange between chimpanzees and humans involved in research and tourism in western Uganda. *Biological Conservation*, 135, 511-517.

Goldberg, T. L., & Wrangham, R. W. (1997). Genetic correlates of social behaviour in wild chimpanzees: Evidence from mitochondrial DNA. *Animal Behaviour*, 54, 559-570.

Goldstone, L. G., Sommer, V., Nurmi, N., Stephens, C., & Fruth, B. (2016). Food begging and sharing in wild bonobos (*Pan paniscus*): Assessing relationship quality? *Primates*, 57, 367-376.

Gomes, C. M., & Boesch, C. (2009). Wild chimpanzees exchange meat for sex on a long-term basis. *PLoS ONE*, 4(4), e5116.

Gonder, M. K., Disotell, T. R., & Oates, J. F. (2006). New genetic evidence on the evolution of chimpanzee populations and implications for taxonomy. *International Journal of Primatology*, 27(4), 1103-1127.

Gonder, M. K., Oates, J. F., Disotell, T. R., Forstner, M. R. J., Morales, J. C., & Melnick, D. J. (1997). A new West African chimpanzee subspecies? *Nature*, 388(6640), 337.

Goodall, J. (1963). Feeding behaviour of wild chimpanzees: A preliminary report. *Symposium of the Zoological Society of London*, 10, 39-48.

Goodall, J. (1964). Tool-using and aimed throwing in a community of free-living chimpanzees. *Nature*, 201, 1264-1266.

Goodall, J. (1965). Chimpanzees of the Gombe Stream Reserve. In I. DeVore (Ed.), *Primate behavior: Field studies of monkeys and apes* (pp. 425-447). Holt, Rinehart, and Winston.

Goodall, J. (1968). The behaviour of free-living chimpanzees in the Gombe Stream Reserve. *Animal Behaviour Monographs*, 1, 161-311.

Goodall, J. (1973). Cultural elements in the chimpanzee community. In W. Montagna (Ed.), *Precultural primate behaviour* (pp. 144-184). E. W. Menzel.

Goodall, J. (1986). *The chimpanzees of Gombe: Patterns of behavior*. Belknap Press.

Goodall, J., Bandora, A., Bergmann, E., Busse, C., Matama, H., Mpongo, E., Pierce, A., & Riss, A. (1979). Intercommunity interactions in the chimpanzee population of the Gombe National Park. In D. A. Hamburg, & E. R. McCown (Eds.), *The great apes* (pp. 13-53). Benjamin/Cummings Publishing Company.

Groves, C. (2001). *Primate taxonomy*. Smithsonian Institution Press.

Grützmacher, K., Keil, V., Leinert, V., Leguillon, F., Henlin, A., Couacy-Hymann, E., Kondgen, S., Lang, A., Deschner, T., Wittig, R. M., & Leendertz, F. H. (2018). Human quarantine: Toward reducing infectious pressure on chimpanzees at the Taï Chimpanzee Project, Côte d'Ivoire. *American Journal of Primatology*, 80(1), e22619.

Gunther, M., & Boesch, C. (1993). Energetic costs of nut-cracking behavior in wild chimpanzees. In D. Chivers, & H. Preuschoft (Eds.), *Evolution of hands* (pp. 109-129). Gustav Fisher Verlag.

Hahn, B. H., Shaw, G. M., de Cock, K. M., & Sharp, P. M. (2000). AIDS as a zoonosis: Scientific and public health implications. *Science*, 287(607), 607-614.

Halperin, S. (1979). Temporary association patterns in free ranging chimpanzees: An assessment of individual grouping preferences. In D. A. Hamburg, & E. R. McCown (Eds.), *The great apes* (pp. 491-499). Benjamin/Cummings Publishing Company.

Hamai, M., Nishida, T., Takasaki, H., & Turner, L. (1992). New records of within-group infanticide and cannibalism in wild chimpanzees. *Primates*, 33(2), 151-162.

Hart, D., & Sussman, R. W. (2009). *Man the hunted: Primates, predators, and human evolution*. Westview Press.

Hasegawa, T., & Hirai-Hasegawa, M. (1983). Opportunistic and restrictive matings among wild chimpanzees in the Mahale Mountains, Tanzania. *Journal of Ethology*, 1, 75-85.

Hashimoto, C., Tashiro, Y., Kimura, D., Enomoto, T., Ingmanson, E. J., Idani, G., & Furuichi, T. (1998). Habitat use and ranging of wild bonobos (*Pan paniscus*) at Wamba. *International Journal of Primatology*, 19(6), 1045-1060.

Hickey, J. R., Nackoney, J., Nibbelink, N. P., Blake, S., Bonyenge, A., Coxe, S., Dupain, J., Emetshu, M., Furuichi, T., Grossmann, F., Guislain, P., Hart, J., Hashimoto, C., Ikembelo, B., Ilambu, O., Inogwabini, B.-I., Liengola, I., Lokasola, A. L., Lushimba, A.,... & Kuehl, H. S. (2013). Human proximity and habitat fragmentation are key drivers of the rangewide bonobo distribution. *Biodiversity and Conservation*, 22(13-14), 3085-3104.

Hicks, T. C., Kuhl, H. S., Boesch, C., Dieguez, P., Ayimisin, A. E., Fernandez, R. M., Zungawa, D. B., Kambere, M., Swinkels, J., Menken, S. B. J., Hart, J., Mundry, R., & Roessingh, P. (2019). Bili-Uere: A chimpanzee behavioural realm in northern Democratic Republic of Congo. *Folia Primatologica*, 90(1), 3-64.

Hill, K., Boesch, C., Goodall, J., Pusey, A., Williams, J., & Wrangham, R. (2001). Mortality rates among wild chimpanzees. *Journal of Human Evolution*, 40, 437-450.

Hladik, C. M. (1973). Alimentation et activité d'un groupe de chimpanzés réintroduits en forêt Gabonaise. *La Terre et la Vie*, 27, 343-413.

Hladik, C. M. (1977). Chimpanzees of Gabon and chimpanzees of Gombe: Some comparative data on diet. In T. H. Clutton-Brock (Ed.), *Primate ecology: Studies of feeding and ranging behaviour in lemurs, monkeys, and apes* (pp. 481-501). Academic Press.

Hobaiter, C., Samuni, L., Mullins, C., Akankwasa, W. J., & Zuberbuhler, K. (2017). Variation in hunting behaviour in neighbouring chimpanzee communities in the Budongo Forest, Uganda. *PLoS ONE,* 12(6), e0178065.

Hockings, K. J., Bryson-Morrison, N., Carvalho, S., Fujisawa, M., Humle, T., McGrew, W. C., Nakamura, M., Ohashi, G., Yamanashi, Y., Yamakoshi, G., & Matsuzawa, T. (2015). Tools to tipple: Ethanol ingestion by wild chimpanzees using leaf-sponges. *Royal Society Open Science,* 2(6), 150150.

Hohmann, G., & Fruth, B. (1993). Field observations on meat sharing among bonobos (*Pan paniscus*). *Folia Primatologica,* 60, 225-229.

Hohmann, G., & Fruth, B. (2008). New records on prey capture and meat eating by bonobos at LuiKotale, Salonga National Park, Democratic Republic of Congo. *Folia Primatologica,* 79(2), 103-110.

Hohmann, G., Ortmann, S., Remer, T., & Fruth, B. (2019). Fishing for iodine: What aquatic foraging by bonobos tells us about human evolution. *BMC Zoology,* 4(1), 5.

Hohmann, G., Potts, K., N'Guessan, A., Fowler, A., Mundry, R., Ganzhorn, J. U., & Ortmann, S. (2010). Plant foods consumed by *Pan*: Exploring the variation of nutritional ecology across Africa. *American Journal of Physical Anthropology,* 141(3), 476-485.

Huffman, M. A. (2001). Self-medicative behavior in the African great apes: An evolutionary perspective into the origins of human traditional medicine. *Bioscience,* 51(8), 651-661.

Huffman, M. A., & Caton, J. M. (2001). Self-induced increase of gut motility and the control of parasitic infections in wild chimpanzees. *International Journal of Primatology,* 22(3), 329-346.

Hunt, K. D. (1992). Positional behavior of *Pan troglodytes* in the Mahale Mountains and Gombe Stream National Parks, Tanzania. *American Journal of Physical Anthropology,* 87, 83-105.

Hunt, K. D. (1994). Body size effects on vertical climbing among chimpanzees. *International Journal of Primatology,* 15(6), 855-865.

Hunt, K. D., & McGrew, W. C. (2002). Chimpanzees in the dry habitats of Assirik, Senegal and Semliki Wildlife Reserve, Uganda. In C. Boesch, G. Hohmann, & L. F. Marchant (Eds.), *Behavioural diversity of chimpanzees and bonobos* (pp. 35-51). Cambridge University Press.

Idani, G. (1990). Relations between unit-groups of bonboos at Wamba, Zaire: Encounters and temporary fusions. *African Study Monographs,* 11, 153-186.

Ihobe, H. (1992). Observations on the meat-eating behavior of wild bonobos (*Pan paniscus*) at Wamba, Republic of Zaire. *Primates,* 33(2), 247-250.

Imanishi, K. (1961). The origin of the human family: A primatological approach. *Japanese Journal of Ethnology,* 25, 119-130.

IUCN. (2022). *The International Union for the Conservation of Nature Red List of Threatened Species.* Version 2021-3. https://www.iucnredlist.org.

Isabirye-Basuta, G. (1989). Feeding ecology of chimpanzees in the Kibale Forest, Uganda. In P. Heltne, & L. Marquardt (Eds.), *Understanding chimpanzees* (pp. 116-127). Harvard University Press.

Itani, J. (1979). Distribution and adaptation of chimpanzees in arid area. In D. A. Hamburg, & E. R. McCown (Eds.), *The great apes* (pp. 55-71). Benjamin/Cummings Publishing.

Itani, J., & Suzuki, A. (1967). The social unit of wild chimpanzees. *Primates,* 8, 355-381.

Izawa, I., & Itani, J. (1966). Chimpanzees in the Kasakati Basin, Tanganyika. 1. Ecological study of the rainy season. *Kyoto University African Studies,* 1, 73-156.

Janmaat, K. R., Ban, S. D., & Boesch, C. (2013). Chimpanzees use long-term spatial memory to monitor large fruit trees and remember feeding experiences across seasons. *Animal Behaviour,* 86(6), 1183-1205.

Jungers, W. L., & Susman, R. L. (1984). Body size and skeletal allometry in African apes. In R. L. Susman (Ed.), *The pygmy chimpanzee: Evolutionary biology and behavior* (pp. 131-177). Plenum.

Junker, J., Blake, S., Boesch, C., Campbell, G., du Toit, L., Duvall, C., Ekobo, A., Etoga, G., Galat-Luong, A., Gamys, J., Ganas-Swaray, J., Gatti, S., Ghiurghi, A., Granier, N., Hart, J., Head, J., Herbinger, I., Hicks, T. C., Huijbregts, B.,... & Kuehl, H. S. (2012). Recent decline in suitable environmental conditions for African great apes. *Diversity and Distributions,* 18(11), 1077-1091.

Kaburu, S. S., & Newton-Fisher, N. E. (2013). Social instability raises the stakes during social grooming among wild male chimpanzees. *Animal Behaviour,* 86(3), 519-527.

Kaburu, S. S., & Newton-Fisher, N. E. (2015a). Egalitarian despots: Hierarchy steepness, reciprocity and the grooming-trade model in wild chimpanzees, *Pan troglodytes. Animal Behaviour,* 99, 1-154.

Kaburu, S. S., & Newton-Fisher, N. E. (2015b). Trading or coercion? Variation in male mating strategies between two communities of East African chimpanzees. *Behavioral Ecology and Sociobiology,* 69(6), 1039-1052.

Kaburu, S. S., & Newton-Fisher, N. E. (2016). Bystanders, parcelling, and an absence of trust in the grooming interactions of wild male chimpanzees. *Scientific Reports,* 6, 20634.

Kano, T. (1984). Distribution of pygmy chimpanzees (*Pan paniscus*) in the central Zaire Basin. *Folia Primatologica,* 43, 36-52.

Kano, T. (1992). *The last ape: Pygmy chimpanzee behavior and ecology.* Stanford University Press.

Kano, T., & Mulavwa, M. (1984). Feeding ecology of the pygmy chimpanzees (*Pan paniscus*) of Wamba. In R. L. Susman (Ed.), *The pygmy chimpanzee* (pp. 233-274). Plenum.

Kawanaka, K. (1984). Association, ranging, and the social unit in chimpanzees of the Mahale Mountains, Tanzania. *International Journal of Primatology*, 5, 411-434.

Kleinschroth, F., Gourlet-Fleury, S., Sist, P., Mortier, F., & Healey, J. R. (2015). Legacy of logging roads in the Congo Basin: How persistent are the scars in forest cover? *Ecosphere*, 6(4), art64.

Kleinschroth, F., Laporte, N., Laurance, W. F., Goetz, S. J., & Ghazoul, J. (2019). Road expansion and persistence in forests of the Congo Basin. *Nature Sustainability*, 2(7), 628-634.

Köndgen, S., Calvignac-Spencer, S., Grutzmacherl, K., Keil, V., Matz-Rensing, K., Nowak, K., Metzger, S., Kiyangs, J., Becker, A. L., Deschner, T., Wittig, R. M., Lankester, F., & Leendertz, F. H. (2017). Evidence for human *Streptococcus pneumoniae* in wild and captive chimpanzees: A potential threat to wild populations. *Scientific Reports*, 7, 14581.

Köndgen, S., Kuehl, H., N'Goran, P. K., Walsh, P. D., Schenk, S., Ernst, N., Biek, R., Formenty, P., Maetz-Rensing, K., Schweiger, B., Junglen, S., Ellerbrok, H., Nitsche, A., Briese, T., Lipkin, W. I., Pauli, G., Boesch, C., & Leendertz, F. H. (2008). Pandemic human viruses cause decline of endangered great apes. *Current Biology*, 18(4), 260-264.

Köndgen, S., Schenk, S., Pauli, G., Boesch, C., & Leendertz, F. H. (2010). Noninvasive monitoring of respiratory viruses in wild chimpanzees. *Ecohealth*, 7(3), 332-341.

Kortlandt, A. (1962). Chimpanzees in the wild. *Scientific American*, 206(5), 128-138.

Krief, S., Huffman, M. A., Sevenet, T., Hladik, C. M., Grellier, P., Loiseau, P. M., & Wrangham, R. W. (2006). Bioactive properties of plant species ingested by chimpanzees (*Pan troglodytes schweinfurthii*) in the Kibale National Park, Uganda. *American Journal of Primatology*, 68(1), 51-71.

Kuehl, H. S., Boesch, C., Kulik, L., Haas, F., Arandjelovic, M., Dieguez, P., Bocksberger, G., McElreath, M. B., Agbor, A., Angedakin, S., Ayimisin, E. A., Bailey, E., Barubiyo, D., Bessone, M., Brazzola, G., Chancellor, R., Cohen, H., Coupland, C., Danquah, E.,... & Kalan, A. K. (2019). Human impact erodes chimpanzee behavioral diversity. *Science*, 363(6434), 1453-1455.

Kühl, H. S., Sop, T., Williamson, E. A., Mundry, R., Brugiere, D., Campbell, G., Cohen, H., Danquah, E., Ginn, L., Herbinger, I., Jones, S., Junker, J., Kormos, R., Kouakou, C. Y., N'Goran, P. K., Normand, E., Shutt-Phillips, K., Tickle, A., Vendras, E. G.,... & Boesch, C. (2017). The critically endangered western chimpanzee declines by 80%. *American Journal of Primatology*, 79(9), e22681.

Kuroda, S. (1984). Interactions over food among pygmy chimpanzees. In R. L. Susman (Ed.), *The pygmy chimpanzee* (pp. 301-324). Plenum.

Kuroda, S., Nishihara, T., Suzuki, S., & Oko, R. A. (1996). Sympatric chimpanzees and gorillas in the Ndoki Forest, Congo. In W. C. McGrew, L. F. Marchant, & T. Nishida (Eds.), *Great ape societies* (pp. 71-81). Cambridge University Press.

Lamon, N., Neumann, C., Gier, J., Zuberbuhler, K., & Gruber, T. (2018). Wild chimpanzees select tool material based on efficiency and knowledge. *Proceedings of the Royal Society B Biological Sciences*, 285(1888), 20181715.

Langergraber, K. E., Mitani, J. C., Vigilant, L. (2007b). The limited impact of kinship on cooperation in wild chimpanzees. *Proceedings of the National Academy of Sciences*, 104(19), 7786-7790.

Langergraber, K., Mitani, J., & Vigilant, L. (2009). Kinship and social bonds in female chimpanzees (*Pan troglodytes*). *American Journal of Primatology*, 71, 1-12.

Langergraber, K. E., Mitani, J. C., Watts, D. P., & Vigilant, L. (2013). Male-female socio-spatial relationships and reproduction in wild chimpanzees. *Behavioral Ecology and Sociobiology*, 67(6), 861-873.

Langergraber, K. E., Pruefer, K., Rowney, C., Boesch, C., Crockford, C., Fawcett, K., Inoue, E., Inoue-Muruyama, M., Mitani, J. C., Muller, M. N., Robbins, M. M., Schubert, G., Stoinski, T. S., Viola, B., Watts, D., Wittig, R. M., Wrangham, R. W., Zuberbuehler, K., Paeaebo, S., & Vigilant, L. (2012). Generation times in wild chimpanzees and gorillas suggest earlier divergence times in great ape and human evolution. *Proceedings of the National Academy of Sciences*, 109(39), 15716-15721.

Langergraber, K. E., Siedel, H., Mitani, J. C., Wrangham, R. W., Reynolds, V., Hunt, K., & Vigilant, L. (2007a). The genetic signature of sex-biased migration in patrilocal chimpanzees and humans. *PLoS ONE*, 2(10), e973.

Lanjouw, A. (2002). Behavioural adaptations to water scarcity in Tongo chimpanzees. In C. Boesch, G. Hohmann, & L. F. Marchant (Eds.), *Behavioural diversity in chimpanzees and bonobos* (pp. 52-60). Cambridge University Press.

Laporte, N., Stabach, J. A., Grosch, R., Lin, T. S., & Goetz, S. J. (2007). Expansion of industrial logging in central Africa. *Science*, 316, 1451.

Lapuente, J., Hicks, T. C., & Linsenmair, K. E. (2017). Fluid dipping technology of chimpanzees in Comoe National Park, Ivory Coast. *American Journal of Primatology*, 79(5), ee22628.

Lee, S. M., Murray, C. M., Lonsdorf, E. V., Fruth, B., Stanton, M. A., Nichols, J., & Hohmann, G. (2019). Wild bonobo and chimpanzee females exhibit broadly similar patterns of behavioral maturation but some evidence for divergence. *American Journal of Physical Anthropology*, 171(1), 100-109.

Leendertz, F. H., Pauli, G., Maetz-Rensing, K., Boardman, W., Nunn, C., Ellerbrok, H., Jensen, S. A., Junglen, S., & Boesch, C. (2006). Pathogens as drivers of population declines: The importance of systematic monitoring in great apes and other threatened mammals. *Biological Conservation*, 131, 325-337.

Lehmann, J., & Boesch, C. (2004). To fission or to fusion: Effects of community size on wild chimpanzee (*Pan troglodytes verus*) social organization. *Behavioral Ecology and Sociobiology*, 56, 207-216.

Lehmann, J., & Boesch, C. (2005). Bisexually bonded ranging in chimpanzees (*Pan troglodytes verus*). *Behavioral Ecology and Sociobiology*, 57, 525-535.

Lodwick, J. L., Borries, C., Pusey, A. E., Goodall, J., & McGrew, W. C. (2004). From nest to nest: Influence of ecology and reproduction on the active period of adult Gombe chimpanzees. *American Journal of Primatology*, 64(3), 249-260.

Lonsdorf, E. V. (2006). What is the role of mothers in the acquisition of termite-fishing behaviors in wild chimpanzees (*Pan troglodytes schweinfurthii*)? *Animal Cognition*, 9(1), 36-46.

Lonsdorf, E. V., Eberly, L. E., & Pusey, A. E. (2004). Sex differences in learning in chimpanzees: Young females quickly learn how to fish for termites—but young males prefer to play. *Nature*, 428, 715.

Lonsdorf, E. V., Gillespie, T. R., Wolf, T. M., Lipende, I., Raphael, J., Bakuza, J., Murray, C. M., Wilson, M. L., Kamenya, S., Mjungu, D., Collins, D. A., Gilby, I. C., Stanton, M. A., Terio, K. A., Barbian, H. J., Li, Y. Y., Ramirez, M., Krupnick, A., Seidl, E.,... & Travis, D. A. (2018). Socioecological correlates of clinical signs in two communities of wild chimpanzees (*Pan troglodytes*) at Gombe National Park, Tanzania. *American Journal of Primatology*, 80(1), 22562.

Lonsdorf, E. V., Markham, A. C., Heintz, M. R., Anderson, K. E., Ciuk, D. J., Goodall, J., & Murray, C. M. (2014). Sex differences in wild chimpanzee behavior emerge during infancy. *PLoS ONE*, 9(6), e99099.

Lonsdorf, E. V., Stanton, M. A., Pusey, A. E., & Murray, C. M. (2019). Sources of variation in weaned age among wild chimpanzees in Gombe National Park, Tanzania. *American Journal of Physical Anthropolology*, 171(3), 419-427.

Lonsdorf, E. V., Travis, D., Pusey, A. E., & Goodall, J. (2006). Using retrospective health data from the Gombe Chimpanzee Study to inform future monitoring efforts. *American Journal of Primatology*, 68, 897-908.

Machanda, Z. P., Gilby, I. C., & Wrangham, R. W. (2013). Male-female association patterns among free-ranging chimpanzees (*Pan troglodytes schweinfurthii*). *International Journal of Primatology*, 34(5), 917-938.

Markham, A. C., Lonsdorf, E. V., Pusey, A. E., & Murray, C. M. (2015). Maternal rank influences the outcome of aggressive interactions between immature chimpanzees. *Animal Behaviour*, 100, 192-198.

Martin, D. E., Graham, C. E., & Gould, K. G. (1978). Successful artificial insemination in the chimpanzee. *Symposium Zoological Society of London*, 43, 249-260.

Matsumoto-Oda, A. (2002). Behavioral seasonality in Mahale chimpanzees. *Primates*, 43(2), 103-117.

Matsumoto-Oda, A., & Hayashi, Y. (1999). Nutritional aspects of fruit choice by chimpanzees. *Folia Primatologica*, 70, 154-162.

Matsuzawa, T., Humle, T., & Sugiyama, Y. (Eds.). (2011). *The chimpanzees of Bossou and Nimba*. Springer.

Maynard Smith, J. M., & Parker, G. A. (1976). The logic of asymmetric contests. *Animal Behaviour*, 24, 159-175.

McGinnis, P. R. (1979). Sexual behavior in free-living chimpanzees: Consort relationships. In D. A. Hamburg, & E. R. McCown (Eds.), *The great apes* (pp. 429-440). Benjamin/Cummings Publishing.

McGrew, W. (1979). Evolutionary implications of sex differences in chimpanzee predation and tool use. In D. A. Hamburg, & E. R. McCown (Eds.) *The great apes* (pp. 440-463). Benjamin/Cummings Publishing.

McGrew, W. C. (1983). Animal foods in the diets of wild chimpanzees (*Pan troglodytes*): Why cross-cultural variation? *Journal of Ethology*, 1, 46-61.

McGrew, W. C. (1992). *Chimpanzee material culture: Implications for human evolution*. Cambridge University Press.

McGrew, W. C., Baldwin, P. J., & Tutin, C. E. G. (1981). Chimpanzees in a hot, dry and open habitat: Mt. Assirik, Senegal, West Africa. *Journal of Human Evolution*, 10, 227-244.

McGrew, W. C., Baldwin, P. J., & Tutin, C. E. G. (1988). Diet of wild chimpanzees (*Pan troglodytes verus*) at Mt. Assirik, Senegal: 1. Composition. *American Journal of Primatology*, 16, 213-226.

McGrew, W. C., & Collins, D. A. (1985). Tool use by wild chimpanzees (*Pan troglodytes*) to obtain termites (*Macrotermes herus*) in the Mahale Mountains, Tanzania. *American Journal of Primatology*, 9, 47-62.

McGrew, W. C., & Rogers, M. E. (1983). Chimpanzees, tools, and termites: New record from Gabon. *American Journal of Primatology*, 5, 171-174.

McGrew, W. C., Tutin, C. E. G., & Baldwin, P. J. (1979). Chimpanzees, tools and termites: Cross-cultural comparisons of Senegal, Tanzania, and Rio Muni. *Man*, 14, 185-214.

McLennan, M. R., & Ganzhorn, J. U. (2017). Nutritional characteristics of wild and cultivated foods for chimpanzees (*Pan troglodytes*) in agricultural landscapes. *International Journal of Primatology*, 38(2), 122-150.

Meijaard, E., Buchori, D., Hadiprakarsa, Y., Utami-Atmoko, S. S., Nurcahyo, A., Tjiu, A., Prasetyo, D., Nardiyono, Christie, L., Ancrenaz, M., Abadi, F., Antoni, I. N. G., Armayadi, D., Dinato, A., Ella, Gumelar, P., Indrawan, T. P., Kussaritano, Munajat, C.,... & Mengersen, K. (2011). Quantifying killing of orangutans and human-orangutan conflict in Kalimantan, Indonesia. *PLoS ONE*, 6(11), e27491.

Meijaard, E., Welsh, A., Ancrenaz, M., Wich, S., Nijman, V., & Marshall, A. J. (2010). Declining orangutan encounter rates from Wallace to the present suggest the species was once more abundant. *PLoS ONE*, 5(8), e12042.

Mikkelsen, T., Hillier, L., Eichler, E., Zody, M., Jaffe, D., Yang, S. P., Enard, W., Hellmann, I., Lindblad-Toh, K., Altheide, T., & Archidiacono, N. (2005). Initial sequence of the chimpanzee genome and comparison with the human genome. *Nature*, 437(7055), 69-87.

Mitani, J. C. (2006). Demographic influences on the behavior of chimpanzees. *Primates, 47*(1), 6-13.

Mitani, J. C., & Amsler, S. J. (2003). Social and spatial aspects of male subgrouping in a community of wild chimpanzees. *Behaviour, 140*, 869-884.

Mitani, J. C., Merriwether, D. A., & Zhang, C. B. (2000). Male affiliation, cooperation and kinship in wild chimpanzees. *Animal Behaviour, 59*, 885-893.

Mitani, J. C., & Watts, D. P. (1999). Demographic influences on the hunting behavior of chimpanzees. *American Journal of Physical Anthropology, 109*(4), 439-454.

Mitani, J. C., & Watts, D. P. (2001). Why do chimpanzees hunt and share meat? *Animal Behaviour, 61*, 915-924.

Mitani, J. C., Watts, D. P., & Amsler, S. J. (2010). Lethal intergroup aggression leads to territorial expansion in wild chimpanzees. *Current Biology, 20*(12), R507-R508.

Mitani, J. C., Watts, D. P., & Lwanga, J. S. (2002). Ecological and social correlates of chimpanzee party size and composition. In C. Boesch, G. Hohmann, & L. F. Marchant (Eds.), *Behavioral diversity of chimpanzees and bonobos* (pp. 102-111). Cambridge University Press.

Moore, J. (1996). Savanna chimpanzees, referential models and the last common ancestor. In W. C. McGrew, L. F. Marchant, & T. Nishida (Eds.), *Great ape societies* (pp. 275-292). Cambridge University Press.

Moore, J., Black, J., Hernandez-Aguilar, R. A., Idani, G., Piel, A., & Stewart, F. (2017). Chimpanzee vertebrate consumption: Savanna and forest chimpanzees compared. *Journal of Human Evolution, 112*, 30-40.

Morgan, B. J., & Abwe, E. E. (2006). Chimpanzees use stone hammers in Cameroon. *Current Biology, 16*(16), R632-R633.

Morgan, D., & Sanz, C. (2003). Naive encounters with chimpanzees in the Goualougo Triangle, Republic of Congo. *International Journal of Primatology, 24*(2), 369-381.

Morgan, D., & Sanz, C. (2006). Chimpanzee feeding ecology and comparisons with sympatric gorillas in the Goualougo Triangle, Republic of Congo. In G. Hohmann, M. Robbins, & C. Boesch (Eds.), *Primate feeding ecology in apes and other primates: Ecological, physiological, and behavioural aspects* (pp. 97-122). Cambridge University Press.

Morgan, D., Sanz, C., Onononga, J. R., & Strindberg, S. (2006). Ape abundance and habitat use in the Goualougo Triangle, Republic of Congo. *International Journal of Primatology, 27*(1), 147-179.

Morgan, D., Winston, W., Eyana Ayina, C., Mayoukou, W., Lonsdorf, E., & Sanz, C. (2020). Forest certification and the high conservation value concept: Protecting great apes in the Sangha Trinational landscape in an era of industrial logging. In L. M. Hopper, & S. R. Ross (Eds.), *Chimpanzee in context* (pp. 644-670). University of Chicago Press.

Morgan, D. B. (2007). *Socio-ecology of chimpanzees (*Pan troglodytes troglodytes*) in the Goualougo Triangle, Republic of Congo.* [Doctoral dissertation, Cambridge University].

Morin, P. A., Moore, J. J., Chakraborty, R., Jin, L., Goodall, J., & Woodruff, D. S. (1994). Kin selection, social structure, gene flow, and the evolution of chimpanzees. *Science, 265,* 1193-1201.

Morin, P. A., Wallis, J., Moore, J. J., Chakraborty, R., & Woodruff, D. S. (1993). Non-invasive sampling and DNA amplification for paternity exclusion, community structure, and phylogeography in wild chimpanzees. *Primates, 34*(3), 347-356.

Moscovice, L. R., Surbeck, M., Fruth, B., Hohmann, G., Jaeggi, A. V., & Deschner, T. (2019). The cooperative sex: Sexual interactions among female bonobos are linked to increases in oxytocin, proximity and coalitions. *Hormones and Behavior, 116*, 104581.

Mulavwa, M. N., Yangozene, K., Yamba-Yamba, M., Motema-Salo, B., Mwanza, N. N., & Furuichi, T. (2010). Nest groups of wild bonobos at Wamba: Selection of vegetation and tree species and relationships between nest group size and party size. *American Journal of Primatology, 72*(7), 575-586.

Muller, M. N., Kahlenberg, S. M., Thompson, M. E., & Wrangham, R. W. (2007). Male coercion and the costs of promiscuous mating for female chimpanzees. *Proceedings of the Royal Society B Biological Sciences, 274*(1612), 1009-1014.

Muller, M. N., Thompson, M. E., Kahlenberg, S. M., & Wrangham, R. W. (2011). Sexual coercion by male chimpanzees shows that female choice may be more apparent than real. *Behavioral Ecology and Sociobiology, 65*(5), 921-933.

Muller, M. N., & Wrangham, R. W. (2014). Mortality rates among Kanyawara chimpanzees. *Journal of Human Evolution, 66*, 107-114.

Murray, C. M. (2007). Method for assigning categorical rank in female *Pan troglodytes schweinfurthii* via the frequency of approaches. (2007). *International Journal of Primatology, 28*, 853-864.

Murray, C. M., Eberly, L. E., & Pusey, A. E. (2006). Foraging strategies as a function of season and rank among wild female chimpanzees (*Pan troglodytes*). *Behavioral Ecology, 17*, 1020-1028.

Murray, C. M., Mane, S. V., & Pusey, A. E. (2007). Dominance rank influences female space use in wild chimpanzees, *Pan troglodytes*: Towards an ideal despotic distribution. *Animal Behaviour, 74*, 1795-1804.

Myers Thompson, J. A. (2002). Bonobos of the Lukuru Wildlife Research Project. In C. Boesch, G. Hohmann, & L. F. Marchant (Eds.), *Behavioural diversity in chimpanzees and bonobos* (pp. 61-70). Cambridge University Press.

Nakamura, M., Hosaka, K., Itoh, N., & Zamma, K. (Eds.). (2015). *Mahale chimpanzees: 50 years of research.* Cambridge University Press.

Newton-Fisher, N. (1997). *Tactical behaviour and decision making in wild chimpanzees.* [Doctoral dissertation, University of Cambridge].

Newton-Fisher, N. E. (1999a). The diet of chimpanzees in the Budongo Forest Reserve, Uganda. *African Journal of Ecology, 37*, 344-354.

Newton-Fisher, N. E. (1999b). Association by male chimpanzees: A social tactic? *Behaviour, 136,* 705-730.

Newton-Fisher, N. E. (2002). Relationships of male chimpanzees in the Budongo Forest, Uganda. In C. Boesch, G. Hohmann, & L. F. Marchant (Eds.), *Behavioral diversity in chimpanzees and bonobos* (pp. 125-137). Cambridge University Press.

Newton-Fisher, N. E. (2003). The home range of the Sonso community of chimpanzees from the Budongo Forest, Uganda. *African Journal of Ecology, 41,* 150-156.

Newton-Fisher, N. E., & Kaburu, S. S. K. (2017). Grooming decisions under structural despotism: The impact of social rank and bystanders among wild male chimpanzees. *Animal Behaviour, 128,* 153-164.

Newton-Fisher, N. E., & Lee, P. C. (2011). Grooming reciprocity in wild male chimpanzees. *Animal Behaviour, 81*(2), 439-446.

Newton-Fisher, N. E., Thompson, M. E., Reynolds, V., Boesch, C., & Vigilant, L. (2010). Paternity and social rank in wild chimpanzees (*Pan troglodytes*) from the Budongo Forest, Uganda. *American Journal of Physical Anthropology, 142*(3), 417-428.

Nishida, T. (1968). The social group of wild chimpanzees in the Mahale Mountains. *Primates, 9,* 167-224.

Nishida, T. (1974). The ecology of wild chimpanzees. In. R. Ohtsuka, J. Tanaka, & T. Nishida (Eds.), *Human ecology* (pp. 15-160). Kyoritsu-Shuppan.

Nishida, T. (1979). The social structure of chimpanzees of the Mahale Mountains. In D. A. Hamburg, & E. R. McCown (Eds.), *The great apes* (pp. 73-121). Benjamin/Cummings Publishing.

Nishida, T. (1983). Alpha status and agonistic alliance in wild chimpanzees (*Pan troglodytes schweinfurthii*). *Primates, 24,* 318-336.

Nishida, T. (1989). Social interactions between resident and immigrant female chimpanzees. In P. Heltne, & L. Marquardt (Eds.), *Understanding chimpanzees* (pp. 68-89). Harvard University Press.

Nishida, T. (1990). *The chimpanzees of the Mahale Mountains: Sexual and life history strategies.* University of Tokyo Press.

Nishida, T. (2012). *Chimpanzees of the lakeshore: Natural history and culture at Mahale.* Cambridge University Press.

Nishida, T., Corp, N., Hamai, M., Hasegawa, T., Hiraiwa-Hasegawa, M., Hosaka, K., Hunt, K. D., Itoh, N., Kawanaka, K., Matsumoto-Oda, A., Mitani, J. C., Nakamura, M., Norikoshi, K., Sakamaki, T., Turner, L., Uehara, S., & Zamma, K. (2003). Demography, female life history, and reproductive profiles among the chimpanzees of Mahale. *American Journal of Primatology, 59*(3), 99-121.

Nishida, T., & Hiraiwa-Hasegawa, M. (1987). Chimpanzees and bonobos: Cooperative relationships among males. In B. B. Smuts, D. L. Cheney, R. M. Seyfarth, R. W. Wrangham, & T. T. Struhsaker (Eds.), *Primate societies* (pp. 165-177). University of Chicago Press.

Nishida, T., & Hosaka, K. (1996). Coalition strategies among adult male chimapnzees of the Mahale Mountains, Tanzania.

In W. C. McGrew, L. F. Marchant, & T. Nishida (Eds.), *Great ape societies* (pp. 114-134). Cambridge University Press.

Nishida, T., & Uehara, S. (1983). Natural diet of chimpanzees (*Pan troglodytes schweinfurthii*): Long-term record from the Mahale Mountains, Tanzania. *African Study Monographs, 3,* 109-130.

Nishida, T., Wrangham, R. W., Goodall, J., & Uehara, S. (1983). Local differences in plant-feeding habits of chimpanzees between the Mahale Mountains and Gombe National Park, Tanzania. *Journal of Human Evolution, 12,* 467-480.

Nissen, H. W. (1931). A field study of the chimpanzee. *Comparative Psychology Monographs, 8,* 1-121.

Nissen, H. W., & Yerkes, R. M. (1943). Reproduction in the chimpanzee: Report on forty-nine births. *Anatomical Record, 86*(4), 567-578.

Noë, R., & Hammerstein, P. (1994). Biological markets: Supply and demand determine the effect of partner choice in cooperation, mutualism and mating. *Behavioral Ecology and Sociobiology, 35*(1), 1-11.

Nurmi, N. O., Hohmann, G., Goldstone, L. G., Deschner, T., & Schülke, O. (2018). The "tolerant chimpanzee": Towards the costs and benefits of sociality in female bonobos. *Behavioral Ecology, 29*(6), 1325-1339.

Oates, J. F., Groves, C. P., & Jenkins, P. D. (2009). The type locality of *Pan troglodytes vellerosus* (Gray, 1862), and implications for the nomenclature of West African chimpanzees. *Primates, 50*(1), 78-80.

Oelze, V. M., Fuller, B. T., Richards, M. P., Fruth, B., Surbeck, M., Hublin, J.-J., & Hohmann, G. (2011). Exploring the contribution and significance of animal protein in the diet of bonobos by stable isotope ratio analysis of hair. *Proceedings of the National Academy of Sciences, 108*(24), 9792-9797.

Oelze, V. M., Head, J. S., Robbins, M. M., Richards, M., & Boesch, C. (2014). Niche differentiation and dietary seasonality among sympatric gorillas and chimpanzees in Loango National Park (Gabon) revealed by stable isotope analysis. *Journal of Human Evolution, 66,* 95-106.

O'Malley, R. C., & Power, M. L. (2012). Nutritional composition of actual and potential insect prey for the Kasekela chimpanzees of Gombe National Park, Tanzania. *American Journal of Physical Anthropology, 149*(4), 493-503.

O'Malley, R. C., & Power, M. L. (2014). The energetic and nutritional yields from insectivory for Kasekela chimpanzees. *Journal of Human Evolution, 71,* 46-58.

Pandolfi, S., van Schaik, C., & Pusey, A. (2003). Sex differences in termite fishing among Gombe chimpanzees. In F. B. M. de Waal, & P. L. Tyack (Eds.), *Animal social complexity: Intelligence, culture and individualized societies* (pp. 414-418). Harvard University Press.

Parker, G. A. (1974). Assessment strategy and the evolution of fighting behaviour. *Journal of Theoretical Biology, 47,* 223-243.

Parker, S. T., & Gibson, K. R. (1977). Object manipulation, tool use, and sensorimotor intelligence as feeding adaptations in

Cebus monkeys and great apes. *Journal of Human Evolution,* 6, 623-641.

Pascual-Garrido, A. (2019). Cultural variation between neighbouring communities of chimpanzees at Gombe, Tanzania. *Scientific Reports,* 9(1), 8260.

Pascual-Garrido, A., Umaru, B., Allon, O., & Sommer, V. (2013). Apes finding ants: Predator–prey dynamics in chimpanzee habitat in Nigeria. *American Journal of Primatology* 75(12), 1231-1244.

Piel, A. K., Lenoel, A., Johnson, C., & Stewart, F. A. (2015). Deterring poaching in western Tanzania: The presence of wildlife researchers. *Global Ecology and Conservation,* 3, 188-199.

Piel, A. K., Strampelli, P., Greathead, E., Hernandez-Aguilar, R. A., Moore, J., & Stewart, F. A. (2017). The diet of open-habitat chimpanzees (*Pan troglodytes schweinfurthii*) in the Issa Valley, western Tanzania. *Journal of Human Evolution,* 112, 57-69.

Pika, S., Klein, H., Bunel, S., Baas, P., Theleste, E., & Deschner, T. (2019). Wild chimpanzees (*Pan troglodytes troglodytes*) exploit tortoises (*Kinixys erosa*) via percussive technology. *Scientific Reports,* 9(1), 7661.

Plooij, F. X. (1984). *The behavioral development of free-living chimpanzee babies and infants.* Praeger.

Plumptre, A., Nixon, S., Critchlow, R., Vielledent, G., Nishuli, R., Kirkby, A., Williamson, E. A., Hall, J. S., & Kujirakwinja, D. (2015). *Status of Grauer's gorilla and chimpanzees in eastern Democratic Republic of Congo: Historical and current distribution and abundance.* Wildlife Conservation Society.

Plumptre, A. J., & Reynolds, V. (1997). Nesting behavior of chimpanzees: Implications for censuses. *International Journal of Primatology,* 18(4), 475-485.

Pontzer, H., Raichlen, D. A., & Rodman, P. S. (2014). Bipedal and quadrupedal locomotion in chimpanzees. *Journal of Human Evolution,* 66, 64-82.

Pontzer, H., & Wrangham, R. W. (2004). Climbing and the daily energy cost of locomotion in wild chimpanzees: Implications for hominoid locomotor evolution. *Journal of Human Evolution,* 46(3), 317-335.

Potts, K. B., Baken, E., Levang, A., & Watts, D. P. (2016). Ecological factors influencing habitat use by chimpanzees at Ngogo, Kibale National Park, Uganda. *American Journal of Primatology,* 78(4), 432-440.

Potts, K. B., Chapman, C. A., & Lwanga, J. S. (2009). Floristic heterogeneity between forested sites in Kibale National Park, Uganda: Insights into the fine-scale determinants of density in a large-bodied frugivorous primate. *Journal of Animal Ecology,* 78(6), 1269-1277.

Potts, K. B., Watts, D. P., & Wrangham, R. W. (2011). Comparative feeding ecology of two communities of chimpanzees (*Pan troglodytes*) in Kibale National Park, Uganda. *International Journal of Primatology,* 32(3), 669-690.

Preis, A., Samuni, L., Mielke, A., Deschner, T., Crockford, C., & Wittig, R. M. (2018). Urinary oxytocin levels in relation to post-conflict affiliations in wild male chimpanzees (*Pan troglodytes verus*). *Hormones and Behavior,* 105, 28-40.

Pruefer, K., Munch, K., Hellmann, I., Akagi, K., Miller, J. R., Walenz, B., Koren, S., Sutton, G., Kodira, C., Winer, R., Knight, J. R., Mullikin, J. C., Meader, S. J., Ponting, C. P., Lunter, G., Higashino, S., Hobolth, A., Dutheil, J., Karakoc, E.,... & Paeaebo, S. (2012). The bonobo genome compared with the chimpanzee and human genomes. *Nature,* 486(7404), 527-531.

Pruetz, J. D. (2007). Evidence of cave use by savanna chimpanzees (*Pan troglodytes verus*) at Fongoli, Senegal: Implications for thermoregulatory behavior. *Primates,* 48, 316-319.

Pruetz, J. D. (2018). Nocturnal behavior by a diurnal ape, the West African chimpanzee (*Pan troglodytes verus*), in a savanna environment at Fongoli, Senegal. *American Journal of Physical Anthropology,* 166(3), 541-548.

Pruetz, J. D., & Bertolani, P. (2007). Savanna chimpanzees, *Pan troglodytes verus,* hunt with tools. *Current Biology,* 17, 1-6.

Pruetz, J. D., Bertolani, P., Ontl, K. B., Lindshield, S., Shelley, M., & Wessling, E. G. (2015). New evidence on the tool-assisted hunting exhibited by chimpanzees (*Pan troglodytes verus*) in a savannah habitat at Fongoli, Sénégal. *Royal Society Open Science,* 2(4).

Pruetz, J. D., Fulton, S. J., Marchant, L. F., McGrew, W. C., Schiel, M., & Waller, M. (2008). Arboreal nesting as anti-predator adaptation by savanna chimpanzees (*Pan troglodytes verus*) in southeastern Senegal. *American Journal of Primatology,* 70(4), 393-401.

Pruetz, J. D., Marchant, L. F., Arno, J., & McGrew, W. C. (2002). Survey of savanna chimpanzees (*Pan troglodytes verus*) in southeaster Senegal. *American Journal of Primatology,* 58, 35-43.

Pusey, A. (1979). Intercommunity transfer of chimpanzees in the Gombe National Park. In D. A. Hamburg, & E. R. McCown (Eds.), *The great apes* (pp. 465-479). Benjamin/Cummings Publishing.

Pusey, A. (1983). Mother-offspring relationships in chimpanzees after weaning. *Animal Behaviour,* 28, 543-552.

Pusey, A., Williams, J., & Goodall, J. (1997). The influence of dominance rank on the reproductive success of female chimpanzees. *Science,* 277, 828-831.

Pusey, A. E., Pintea, L., Wilson, M. L., Kamenya, S., & Goodall, J. (2007). The contribution of long-term research at Gombe National Park to chimpanzee conservation. *Conservation Biology,* 21(3), 623-634.

Pusey, A. E., Wilson, M. L., & Collins, D. A. (2008). Human impacts, disease risk, and population dynamics in the chimpanzees of Gombe National Park, Tanzania. *American Journal of Primatology,* 70(8), 738-744.

Rainey, H. J., Iyenguet, F. C., Malanda, G. A. F., Madzoke, B., Dos Santos, D., Stokes, E. J., Maisels, F., & Strindberg, S. (2010). Survey of *Raphia* swamp forest, Republic of Congo, indicates high densities of critically endangered western lowland gorillas *Gorilla gorilla gorilla. Oryx,* 44(1), 124-132.

Raubenheimer, D., Rothman, J. M., Pontzer, H., & Simpson, S. J. (2014). Macronutrient contributions of insects to the diets of hunter-gatherers: A geometric analysis. *Journal of Human Evolution,* 71, 70-76.

Reynolds, V. (1965). *Budongo: An African forest and its chimpanzees*. Natural History Press.

Reynolds, V. (1967). *The apes: The gorilla, chimpanzee, orangutan, and gibbon—their history and their world*. E. P. Dutton.

Reynolds, V. (2005). *The chimpanzees of the Budongo Forest: Ecology, behaviour, and conservation*. Oxford University Press.

Reynolds, V., Pascual-Garrido, A., Lloyd, A. W., Lyons, P., & Hobaiter, C. (2019). Possible mineral contributions to the diet and health of wild chimpanzees in three East African forests. *American Journal of Primatology, 81*(6), e22978.

Reynolds, V., & Reynolds, F. (1965). Chimpanzees of the Budongo Forest. In I. DeVore (Ed.), *Primate behavior* (pp. 368-424). Holt, Rinehart, and Winston.

Riss, D. C., & Busse, C. D. (1977). Fifty-day observation of a free-ranging adult male chimpanzee. *Folia Primatologica, 28*, 283-297.

Robbins, M. M., Gray, M., Fawcett, K. A., Nutter, F. B., Uwingeli, P., Mburanumwe, I., Kagoda, E., Basabose, A., Stoinski, T. S., Cranfield, M. R., Byamukama, J., Spelman, L. H., & Robbins, A. M. (2011). Extreme conservation leads to recovery of the Virunga mountain gorillas. *PLoS ONE, 6*(6), e19788.

Ross, S. R., Lukas, K. E., Lonsdorf, E. V., Stoinski, T. S., Hare, B., Shumaker, R., & Goodall, J. (2008). Science priorities: Inappropriate use and portrayal of chimpanzees. *Science, 319*(5869), 1487.

Rouquet, P., Froment, J. M., Bermejo, M., Kilbourn, A., Karesh, W., Reed, P., Kumulungui, B., Yaba, P., Delicat, A., Rollin, P. E., & Leroy, E. M. (2005). Wild animal mortality monitoring and human Ebola outbreaks, Gabon and Republic of Congo, 2001-2003. *Emerging Infectious Diseases, 11*(2), 283-290.

Ryu, H., Hill, D. A., & Furuichi, T. (2015). Prolonged maximal sexual swelling in wild bonobos facilitates affiliative interactions between females. *Behaviour, 152*(3-4), 285-311.

Sakamaki, T., Maloueki, U., Bakaa, B., Bongoli, L., Kasalevo, P., Terada, S., & Furuichi, T. (2016). Mammals consumed by bonobos (*Pan paniscus*): New data from the Iyondji Forest, Tshuapa, Democratic Republic of the Congo. *Primates, 57*(3), 295-301.

Sakamaki, T., Ryu, H., Toda, K., Tokuyama, N., & Furuichi, T. (2018). Increased frequency of intergroup encounters in wild bonobos (*Pan paniscus*) around the yearly peak in fruit abundance at Wamba. *International Journal of Primatology, 39*(4), 685-704.

Sakura, O. (1994). Factors affecting party size and composition of chimpanzees (*Pan troglodytes verus*) at Bossou, Guinea. *International Journal of Primatology, 15*, 167-183.

Samson, D. R., & Hunt, K. D. (2014). Chimpanzees preferentially select sleeping platform construction tree species with biomechanical properties that yield stable, firm, but compliant nests. *PLoS ONE, 9*(4), e95361.

Samuni, L., Preis, A., Deschner, T., Crockford, C., & Wittig, R. M. (2018b). Reward of labor coordination and hunting success in wild chimpanzees. *Communications Biology, 1*(1), 138, 1-9.

Samuni, L., Preis, A., Mielke, A., Deschner, T., Wittig, R. M., & Crockford, C. (2018a). Social bonds facilitate cooperative resource sharing in wild chimpanzees. *Proceedings of the Royal Society B Biological Sciences, 285*(1888), 20181643.

Samuni, L., Preis, A., Mundry, R., Deschner, T., Crockford, C., & Wittig, R. M. (2017). Oxytocin reactivity during intergroup conflict in wild chimpanzees. *Proceedings of the National Academy of Sciences, 114*(2), 268-273.

Sánchez-Amaro, A., & Amici, F. (2015). Are primates out of the market? *Animal Behaviour, 110*, 51-60.

Sanz, C. M. (2004). *Behavioral ecology of chimpanzees in a Central African forest:* Pan troglodytes troglodytes *in the Goualougo Triangle, Republic of Congo*. [Doctoral dissertation, Washington University].

Sanz, C., Call, J., & Morgan, D. (2009). Design complexity in termite-fishing tools of chimpanzees (*Pan troglodytes*). *Biology Letters, 5*(3), 293-296.

Sanz, C. M., & Morgan, D. B. (2007). Chimpanzee tool technology in the Goualougo Triangle, Republic of Congo. *Journal of Human Evolution, 52*(4), 420-433.

Sanz, C., Morgan, D., & Gulick, S. (2004). New insights into chimpanzees, tools, and termites from the Congo Basin. *American Naturalist, 164*(5), 567-581.

Sanz, C., Morgan, D., Strindberg, S., & Onononga, J. R. (2007). Distinguishing between the nests of sympatric chimpanzees and gorillas. *Journal of Applied Ecology, 44*(2), 263-272.

Savage, T. S., & Wyman, J. (1844). Observation on the external characters and habits of the *Troglodytes niger*, Geoff., and on its organization. *Boston Journal of Natural History, 4*, 362-386.

Sayers, K., & Lovejoy, C. O. (2008). The chimpanzee has no clothes: A critical examination of *Pan troglodytes* in models of human evolution. *Current Anthropology, 49*(1), 87-114.

Sayers, K., Raghanti, M. A., & Lovejoy, C. O. (2012). Human evolution and the chimpanzee referential doctrine. *Annual Review of Anthropology, 41*, 119-138.

Schubert, G., Vigilant, L., Boesch, C., Klenke, R., Langergraber, K., Mundry, R., Surbeck, M., & Hohmann, G. (2013). Co-residence between males and their mothers and grandmothers is more frequent in bonobos than chimpanzees. *PLoS ONE, 8*(12), e83870.

Serckx, A., Huynen, M. C., Bastin, J. F., Hambuckers, A., Beudels-Jamar, R. C., Vimond, M., Raynaud, E., & Kühl, H. S. (2014). Nest grouping patterns of bonobos (*Pan paniscus*) in relation to fruit availability in a forest-savannah mosaic. *PLoS ONE, 9*(4), e93742.

Seyfarth, R. (1977). Model of social grooming among adult female monkeys. *Journal of Theoretical Biology, 65*, 671-698.

Silk, J. B. (1978). Patterns of food sharing among mother and infant chimpanzees at Gombe National Park, Tanzania. *Folia Primatologica, 29*, 129-141.

Sirianni, G., Mundry, R., & Boesch, C. (2015). When to choose which tool: Multidimensional and conditional selection of nut-cracking hammers in wild chimpanzees. *Animal Behaviour, 100*, 152-165.

Sockol, M. D., Raichlen, D. A., & Pontzer, H. (2007). Chimpanzee locomotor energetics and the origin of human bipedalism. *Proceedings of the National Academy of Sciences,* 104(30), 12265-12269.

Sommer, V., Adanu, J., Faucher, I., & Fowler, A. (2004). Nigerian chimpanzees (*Pan troglodytes vellerosus*) at Gashaka: Two years of habituation efforts. *Folia Primatologica,* 75, 295-316.

Sponheimer, M., Loudon, J. E., Codron, D., Howells, M. E., Pruetz, J. D., Codron, J., de Ruiter, D. J., & Lee-Thorp, J. A. (2006). Do "savanna" chimpanzees consume C-4 resources? *Journal of Human Evolution,* 51(2), 128-133.

Stanford, C. B. (1995). The influence of chimpanzee predation on group-size and antipredator behavior in red colobus monkeys. *Animal Behaviour,* 49(3), 577-587.

Stanford, C. B. (1998). The social behavior of chimpanzees and bonobos: Empirical evidence and shifting assumptions. *Current Anthropology,* 39(4), 399-420.

Stanford, C. B. (1999). *The hunting apes: Meat eating and the origins of human behavior.* Princeton University Press.

Stanford, C. B., Wallis, J., Matama, H., & Goodall, J. (1994). Patterns of predation by chimpanzees on red colobus monkeys in Gombe-National-Park, 1982-1991. *American Journal of Physical Anthropology,* 94(2), 213-228.

Stokes, E. J., Strindberg, S., Bakabana, P. C., Elkan, P. W., Iyenguet, F. C., Madzoke, B., Malanda, G. A. F., Mowawa, B. S., Moukoumbou, C., Ouakabadio, F. K., & Rainey, H. J. (2010). Monitoring great ape and elephant abundance at large spatial scales: Measuring effectiveness of a conservation landscape. *PLoS ONE,* 5(4), e10294.

Stone, A. C., Battistuzzi, F. U., Kubatko, L. S., Perry, G. H., Trudeau, E., Lin, H. M., & Kumar, S. (2010). More reliable estimates of divergence times in *Pan* using complete mtDNA sequences and accounting for population structure. *Philosophical Transactions of the Royal Society B Biological Sciences,* 365(1556), 3277-3288.

Strindberg, S., Maisels, F., Williamson, E. A., Blake, S., Stokes, E. J., Aba'a, R., Abitsi, G., Agbor, A., Ambahe, R. D., Bakabana, P. C., Bechem, M., Berlemont, A., de Semboli, B. B., Boundja, P. R., Bout, N., Breuer, T., Campbell, G., De Wachter, P., Akou, M. E.,... & Wilkie, D. S. (2018). Guns, germs, and trees determine density and distribution of gorillas and chimpanzees in western Equatorial Africa. *Science Advances,* 4(4), eaar2964.

Struhsaker, T. T., & Hunkeler, P. (1971). Evidence of tool-using by chimpanzees in the Ivory Coast. *Folia Primatologica,* 15, 212-219.

Stumpf, R. M., Thompson, M. E., Muller, M. N., & Wrangham, R. W. (2009). The context of female dispersal in Kanyawara chimpanzees. *Behaviour,* 146, 629-656.

Sugiyama, Y. (1972). Social characteristics and socialization of wild chimpanzees. In F. E. Poirer (Ed.), *Primate socialization* (pp. 143-165). Random House.

Sugiyama, Y. (1984). Population dynamics of wild chimpanzees at Bossou, Guinea, between 1976 and 1983. *Primates,* 23, 391-400.

Sugiyama, Y. (1988). Grooming interactions among adult chimpanzees at Bossou, Guinea, with special reference to social structure. *International Journal of Primatology,* 9(5), 393-407.

Sugiyama, Y. (1989). Population dynamics of chimpanzees at Bossou, Guinea. In P. G. Heltne, & L. A. Marquardt (Eds.), *Understanding chimpanzees* (pp. 134-145). Harvard University Press.

Sugiyama, Y. (1999). Socioecological factors of male chimpanzee migration at Bossou, Guinea. *Primates,* 40, 61-68.

Sugiyama, Y., & Koman, J. (1979). Social structure and dynamics of wild chimpanzees at Bossou, Guinea. *Primates,* 20, 323-329.

Sugiyama, Y., & Koman, J. (1987). A preliminary list of chimpanzees' alimentation at Bossou, Guinea. *Primates,* 28(1), 133-147.

Sugiyama, Y., & Koman, J. (1992). The flora of Bossou: Its utilization by chimpanzees and humans. *African Studies Monographs,* 13, 127-169.

Surbeck, M., Boesch, C., Crockford, C., Thompson, M. E., Furuichi, T., Fruth, B., Hohmann, G., Ishizuka, S., Machanda, Z., Muller, M. N., Pusey, A., Sakqamaki, T., Tokuyama, N., Walker, K., Wrangham, R., Wroblewski, E., Zuberbuhler, K., Vigilant, L., & Langergraber, K. (2019). Males with a mother living in their group have higher paternity success in bonobos but not chimpanzees. *Current Biology,* 29(10), R354-R355.

Surbeck, M., Boesch, C., Girard-Buttoz, C., Crockford, C., Hohmann, G., & Wittig, R. M. (2017c). Comparison of male conflict behavior in chimpanzees (*Pan troglodytes*) and bonobos (*Pan paniscus*), with specific regard to coalition and post-conflict behavior. *American Journal of Primatology,* 79(6), 22641.

Surbeck, M., Coxe, S., & Lokasola, A. L. (2017a). Lonoa: The establishment of a permanent field site for behavioural research on bonobos in the Kokolopori Bonobo Reserve. *Pan Africa News,* 24(2), 13-15.

Surbeck, M., Fowler, A., Deimel, C., & Hohmann, G. (2009). Evidence for the consumption of arboreal, diurnal primates by bonobos (*Pan paniscus*). *American Journal of Primatology,* 71(2), 171-174.

Surbeck, M., Girard-Buttoz, C., Boesch, C., Crockford, C., Fruth, B., Hohmann, G., Langergraber, K. E., Zuberbuhler, K., Wittig, R. M., & Mundry, R. (2017b). Sex-specific association patterns in bonobos and chimpanzees reflect species differences in cooperation. *Royal Society Open Science,* 4(5), 161081.

Surbeck, M., & Hohmann, G. (2008). Primate hunting by bonobos at LuiKotale, Salonga National Park. *Current Biology,* 18(19), R906-R907.

Surbeck, M., & Hohmann, G. (2014). Social preferences influence the short-term exchange of social grooming among male bonobos. *Animal Cognition,* 18(2), 573-579.

Surbeck, M., Mundry, R., & Hohmann, G. (2010). Mothers matter! Maternal support, dominance status and mating success in male bonobos (*Pan paniscus*). *Proceedings of the Royal Society B Biological Sciences*, 278(1705), 590-598.

Susman, R. L. (1984). *The pygmy chimpanzee*. Plenum.

Suzuki, A. (1969). An ecological study of chimpanzees in savanna woodland. *Primates, 10*, 103-148.

Symington, M. M. (1990). Fission-fusion social organization in *Ateles* and *Pan*. *International Journal of Primatology*, 11(1), 47-61.

Tagg, N., McCarthy, M., Dieguez, P., Bocksberger, G., Willie, J., Mundry, R., Stewart, F., Arandjelovic, M., Widness, J., Landsmann, A., Agbor, A., Angedakin, S., Ayimisin, A. E., Bessone, M., Brazzola, G., Corogenes, K., Deschner, T., Dilambaka, E., Eno-Nku, M.,... & Boesch, C. (2018). Nocturnal activity in wild chimpanzees (*Pan troglodytes*): Evidence for flexible sleeping patterns and insights into human evolution. *American Journal of Physical Anthropology*, 166(3), 510-529.

Takahata, Y. (1990a). Adult males' social relations with adult females. In T. Nishida (Ed.), *The chimpanzees of the Mahale Mountains: Sexual and life history strategies* (pp. 133-148). University of Tokyo Press.

Takahata, Y. (1990b). Social relationships among adult males. In T. Nishida (Ed.), *The chimpanzees of the Mahale Mountains: Sexual and life history strategies* (pp. 149-170). University of Tokyo Press.

Teleki, G. (1973). The wild chimpanzee: A hunter of mammals. *Scientific American*, 228, 32-42.

Teleki, G. (1981). The omnivorous diet and eclectic feeding habits of chimpanzees in Gombe National Park, Tanzania. In R. S. O. Harding, & G. Teleki (Eds.), *Omnivorous primates* (pp. 303-343). Columbia University Press.

Tennie, C., Gilby, I., & Mundry, R. (2009). The meat-scrap hypothesis: Small quantities of meat may promote cooperative hunting in wild chimpanzees (*Pan troglodytes*). *Behavioral Ecology and Sociobiology*, 63(3), 421-431.

Terada, S., Nackoney, J., Sakamaki, T., Mulavwa, M. N., Yumoto, T., & Furuichi, T. (2015). Habitat use of bonobos (*Pan paniscus*) at Wamba: Selection of vegetation types for ranging, feeding, and night-sleeping. *American Journal of Primatology*, 77(6), 701-713.

Thompson, M. E., Muller, M. N., & Wrangham, R. W. (2012). The energetics of lactation and the return to fecundity in wild chimpanzees. *Behavioral Ecology*, 23(6), 1234-1241.

Thompson, M. E., Muller, M. N., & Wrangham, R. W. (2014). Male chimpanzees compromise the foraging success of their mates in Kibale National Park, Uganda. *Behavioral Ecology and Sociobiology, 68*(12), 1973-1983.

Tokuyama, N., & Furuichi, T. (2016). Do friends help each other? Patterns of female coalition formation in wild bonobos at Wamba. *Animal Behaviour*, 119, 27-35.

Tokuyama, N., & Furuichi, T. (2017). Leadership of old females in collective departures in wild bonobos (*Pan paniscus*) at Wamba. *Behavioral Ecology and Sociobiology*, 71(3), 55.

Tsukahara, T. (1993). Lions eat chimpanzees: The first evidence of predation by lions on wild chimpanzees. *American Journal of Primatology*, 29(1), 1-11.

Tutin, C. E. G. (1979). Mating patterns and reproductive strategies in a community of wild chimpanzees (*Pan troglodytes schweinfurthii*). *Behavioral Ecology and Sociobiology*, 44, 225-282.

Tutin, C. E. G., & Fernandez, M. (1991). Responses of wild chimpanzees and gorillas to the arrival of primatologists: Behaviour observed during habituation. In H. O. Box (Ed.), *Primate responses to environmental change* (pp. 187-197). Chapman and Hall.

Tutin, C. E. G., & Fernandez, M. (1993). Composition of the diet of chimpanzees and comparisons with that of sympatric lowland gorillas in the Lopé Reserve, Gabon. *American Journal of Primatology*, 30, 195-211.

Tutin, C. E. G., Fernandez, M., Rogers, M. E., Williamson, E. A., & McGrew, W. C. (1991). Foraging profiles of sympatric lowland gorillas and chimpanzees in the Lopé Reserve, Gabon. *Philosophical Transactions of the Royal Society of London*, 334, 178-186.

Tutin, C. E. G., Ham, R. M., White, L. J. T., & Harrison, M. J. S. (1997). The primate community of the Lopé Reserve, Gabon: Diets, responses to fruit scarcity and effects on biomass. *American Journal of Primatology*, 42, 1-24.

Tutin, C. E. G., & McGinnis, P. R. (1981). Chimpanzee reproduction in the wild. In C. E. Graham (Ed.), Reproductive biology of the great apes (pp. 239-264). Academic Press.

Tutin, C. E. G., McGrew, W. C., & Baldwin, P. A. (1981). Responses of wild chimpanzees to potential predators. In A. B. Chiarelli, & R. S. Corruccini (Eds.), *Primate behavior and sociobiology* (pp. 136-141). Springer.

Vanderijtplooij, H. H. C., & Plooij, F. X. (1987). Growing independence, conflict and learning in mother-infant relations in free-ranging chimpanzees. *Behaviour*, 101, 1-86.

Wakefield, M. (2008). Grouping patterns and competition among female *Pan troglodytes schweinfurthii* at Ngogo, Kibale National Park, Uganda. *International Journal of Primatology*, 29, 907-929.

Wakefield, M. L. (2013). Social dynamics among females and their influence on social structure in an East African chimpanzee community. *Animal Behaviour*, 85(6), 1303-1313.

Wallis, J. (2002). Seasonal aspects of reproduction and sexual behavior in two chimpanzee populations: A comparison of Gombe (Tanzania) and Budongo (Uganda). In C. Boesch, G. Hohmann, & L. F. Marchant (Eds.), *Behavioural diversity in chimpanzees and bonobos* (pp. 181-191). Cambridge University Press.

Wallis, J., & Lee, D. R. (1999). Primate conservation: The prevention of disease transmission. *International Journal of Primatology*, 20(6), 803-826.

Walsh, P. D., Abernathy, K. A., Bermejo, M., Beyers, R., de Wachter, P., Akou, M. E. Huijbregts, B., Mambounga, D. K., Toham, A. K., Kilbourn, A. M., Lahm, S. A., Latour, S., Maisels, F., Mbina, C., Mihindou, Y., Obiang, S. N., Effa, E. N., Starkey, M. P., Telfer, P.,... & Wilkie, D. S. (2003). Catastrophic ape decline in western Equatorial Africa. *Nature*, 422, 611-613.

Watts, D. P. (2015). Mating behavior of adolescent male chimpanzees (*Pan troglodytes*) at Ngogo, Kibale National Park, Uganda. *Primates, 56*(2), 163-172.

Watts, D. P., & Mitani, J. C. (2000). Infanticide and cannibalism by male chimpanzees at Ngogo, Kibale National Park, Uganda. *Primates, 41*(4), 357-365.

Watts, D. P., & Mitani, J. C. (2001). Boundary patrols and intergroup encounters in wild chimpanzees. *Behaviour, 138*, 299-327.

Watts, D. P., Potts, K. B., Lwanga, J. S., & Mitani, J. C. (2012). Diet of chimpanzees (*Pan troglodytes schweinfurthii*) at Ngogo, Kibale National Park, Uganda, 2. Temporal variation and fallback foods. *American Journal of Primatology, 74*(2), 130-144.

Wessling, E. G., Kuhl, H. S., Mundry, R., Deschner, T., & Pruetz, J. D. (2018). The costs of living at the edge: Seasonal stress in wild savanna-dwelling chimpanzees. *Journal of Human Evolution, 121*, 1-11.

Wessling, E. G., Oelze, V. M., Eshuis, H., Pruetz, J. D., & Kuhl, H. S. (2019). Stable isotope variation in savanna chimpanzees (*Pan troglodytes verus*) indicate avoidance of energetic challenges through dietary compensation at the limits of the range. *American Journal of Physical Anthropology, 168*(4), 665-675.

White, F. J. (1996). *Pan paniscus* 1973 to 1996: Twenty-three years of field research. *Evolutionary Anthropology, 5*(1), 86-92.

Whiten, A., Goodall, J., McGrew, W. C., Nishida, T., Reynolds, V., Sugiyama, Y., Tutin, C. E. G., Wrangham, R. W., & Boesch, C. (1999). Cultures in chimpanzees. *Nature, 399*, 682-686.

Whiten, A., Goodall, J., McGrew, W. C., Nishida, T., Reynolds, V., Sugiyama, Y., Tutin, C. E. G., Wrangham, R., & Boesch, C. (2001). Charting cultural variation in chimpanzees. *Behaviour, 138*, 1481-1516.

Whiten, A., McGrew, W. C., Aiello, L. C., Boesch, C., Boyd, R., Byrne, R. W., Dunbar, R. I. M., Matsuzawa, T., Silk, J. B., Tomasello, M., Van Schaik, C. P., & Wrangham, R. (2010). Studying extant species to model our past. *Science, 327*(5964), 410-410.

Williams, J. M., Lui, H., & Pusey, A. E. (2002b). Costs and benefits of grouping for female chimpanzees at Gombe. In C. Boesch, G. Hohmann, & L. F. Marchant (Eds.), *Behavioural diversity in chimpanzees and bonobos* (pp. 192-203). Cambridge University Press.

Williams, J. M., Oehlert, G. W., Carlis, J. V., & Pusey, A. E. (2004). Why do male chimpanzees defend a group range? *Animal Behaviour, 68*, 523-532.

Williams, J. M., Pusey, A. E., Carlis, J. V., Farms, B. P., & Goodall, J. (2002a). Female competition and male territorial behaviour influence female chimpanzees' ranging patterns. *Animal Behaviour, 63*, 347-360.

Wilson, M. L., Boesch, C., Fruth, B., Furuichi, T., Gilby, I. C., Hashimoto, C., Hobaiter, C. L., Hohmann, G., Itoh, N., Koops, K., Lloyd, J. N., Matsuzawa, T., Mitani, J. C., Mjungu, D. C., Morgan, D., Muller, M. N., Mundry, R.,

Nakamura, M., Pruetz, J.,... & Wrangham, R. W. (2014). Lethal aggression in *Pan* is better explained by adaptive strategies than human impacts. *Nature, 513*(7518), 414-417.

Wittig, R. M., & Boesch, C. (2003). Food competition and linear dominance hierarchy among female chimpanzees of the Taï National Park. *International Journal of Primatology, 24*(4), 847-867.

Wittig, R. M., Crockford, C., Deschner, T., Langergraber, K. E., Ziegler, T. E., & Zuberbuehler, K. (2014). Food sharing is linked to urinary oxytocin levels and bonding in related and unrelated wild chimpanzees. *Proceedings of the Royal Society B Biological Sciences, 281*(1778), 20133096.

Wrangham, R. W. (1975). *The behavioral ecology of chimpanzees in the Gombe National Park, Tanzania.* [Doctoral dissertation, Cambridge University].

Wrangham, R. W. (1977). Feeding behaviour of chimpanzees in Gombe National Park, Tanzania. In T. H. Clutton-Brock (Ed.), *Primate ecology* (pp. 503-538). Academic Press.

Wrangham, R. W. (1979). On the evolution of ape social systems. *Social Science Information, 18*(3), 335-368.

Wrangham, R. W. (1980). An ecological model of female-bonded primate groups. *Behaviour, 75*, 262-299.

Wrangham, R. W. (1999). Evolution of coalitionary killing. *American Journal of Physical Anthropology, 110*(S29), 1-30.

Wrangham, R. W. (2000). Why are male chimpanzees more gregarious than mothers? A scramble competition hypothesis. In P. M. Kappeler (Ed.), *Primate males: Causes and consequences of variation in group composition* (pp. 248-258). Cambridge University Press.

Wrangham, R. W., Chapman, C. A., & Chapman, L. J. (1994). Seed dispersal by forest chimpanzees in Uganda. *Journal of Tropical Ecology, 10*, 355-368.

Wrangham, R. W., Chapman, C. A., Clark-Arcadi, A. P., & Isabirye-Basuta, G. (1996). Social ecology of Kanywara chimpanzees: Implications for understanding the costs of great ape groups. In W. C. McGrew, L. F. Marchant, & T. Nishida (Eds.), *Great ape societies* (pp. 45-57). Cambridge University Press.

Wrangham, R. W., Clark, A. P., & Isabirye-Basuta, G. (1992). Female social relationships and social organization of Kibale Forest chimpanzees. In T. Nishida, W. C. McGrew, P. Marler, M. Pickford, & F. B. M. de Waal (Eds.), *Human origins* (pp. 81-98). University of Tokyo Press.

Wrangham, R. W., Conklin-Brittain, N. L., & Hunt, K. D. (1998). Dietary response of chimpanzees and cercopithecines to seasonal variation in fruit abundance. I. Antifeedants. *International Journal of Primatology, 19*(6), 949-970.

Wrangham, R. W., Conklin, N. L., Chapman, C. A., & Hunt, K. D. (1991). The significance of fibrous foods for Kibale chimpanzees. *Philosophical Transactions of the Royal Society of London, 334*, 171-178.

Wrangham, R. W., Conklin, N. L., Etot, G., Obua, J., Hunt, K. D., Hauser, M. D., & Clark, A. P. (1993). The value of figs to chimpanzees. *International Journal of Primatology, 14*(2), 243-256.

Wrangham, R. W., & Smuts, B. (1980). Sex differences in the behavioural ecology of chimpanzees in the Gombe National Park, Tanzania. *Journal of Reproduction and Fertility*, 28, 13-31.

Yamamoto, S. (2015). Non-reciprocal but peaceful fruit sharing in wild bonobos in Wamba. *Behaviour*, 152(3-4), 335-357.

Zihlman, A. (1996). Reconstructions reconsidered: Chimpanzee models and human evolution. In W. C. McGrew, L. F. Marchant, & T. Nishida (Eds.), *Great ape societies* (pp. 293-304). Cambridge University Press.

Zihlman, A. L., & Cramer, D. L. (1978). Skeletal differences between pygmy (*Pan paniscus*) and common (*Pan troglodytes*) chimpanzees. *Folia Primatologica*, 29, 86-94.

Zihlman, A. L., Cronin, J. E., Cramer, D. L., & Sarich, V. M. (1978). Pygmy chimpanzee as a possible prototype for common ancestor of humans, chimpanzees, and gorillas. *Nature*, 275(5682), 744-746.

CHAPTER 24

The Wonderful World of Apes... and the Adventures We Have Studying Them

MICHELE GOLDSMITH

WE'VE COME A LONG WAY FROM MY PHD qualifying exam circa 1993 when one of the questions was to resolve the trichotomy of *Homo*, *Gorilla*, and *Pan*. Historically, *Homo* has consistently been classified in a group of their own to the exclusion of other apes. For example, the family Hominidae included *Homo* but excluded all other apes, which were in the family Pongidae (e.g., Simpson, 1945). Changes to ape and human systematics continued for decades. First, the recognition of "lesser-apes" (*Hylobates*) as an out-group—therefore, excluded from Pongidae (Goodman, 1964; Sarich & Wilson, 1967). Next, *Pongo* was determined to be the out-group to both great apes and humans and placed within their own subfamily, Ponginae (Goodman, 1964). This resulted in the trichotomy Stony Brook graduate students were tasked with resolving to achieve ABD (all but dissertation). No small feat and one not resolved until only recently. All through these changes, however, *Homo* remained within a distinct tribe (Hominini). One can argue this was due, in part, to the inherent challenge of humans to see themselves as great apes, let alone animals. Finally, humans could avoid it no longer. Advances in molecular genetics finally revealed we are not a "special" species after all. Gorillas were actually found to be the out-group, solidifying that our most recent common ancestor was shared with *Pan*. Figure 24.1 shows the ancestral relationships among all apes within the superfamily Hominoidea.

Within the wonderful world of apes, there are as many social systems as genera. Even more interesting is that we are still discovering variances within these systems. For example, a quick history of gibbon and siamang field research reveals how they were first thought to be monogamous (pair-bonded for life). As a result, all captive exhibits kept them as mated pairs. It was not until field studies extended beyond decades that researchers discovered a type of serial monogamy. Most recently, studies have verified cases of extra-pair copulations (Reichard, 2009). The longer we study them, the more we unravel these variances. The chapters in this section do just that by providing a complete natural history of nonhuman apes along with recent research findings that often challenge the status quo.

In this section, we pay homage to the brilliant primatologist, Bob Sussman, by reviewing the natural history of apes, including the Southeast Asian gibbons and orangutans, the Equatorial African gorillas, chimpanzees, and bonobos. A graduate student of Bob's, Thad Bartlett, is his co-author on the gibbon chapter. Thad has been studying and publishing research on the behavioral ecology of gibbons since 2003. This chapter is wonderfully rich, well cited (with over 200 sources), and includes some of Thad's personal observations from the field. The chapter on orangutans presents an interesting look and investigation into the lives of these amazing great apes with a cautionary note from the author, Roberto Antonio Delgado, on their imminent risk of extinction. Bob has a number of co-authors on the gorilla and chimpanzee/bonobo chapters. Lead author on the gorilla chapter, David Morgan, is co-director with Crickette Sanz of the Goualougo Triangle Ape Research Project where long-term research is conducted on the behavioral ecology of sympatric gorillas and chimpanzees in the Congo Basin (including Cameroon, Central African Republic, and the Republic of Congo). Kristena Cooksey and Kathyrn Judson, both bioanthropology graduate students of Crickette's at Washington University in St. Louis, just published their western lowland gorilla research on intergroup encounters, and home range and site fidelity, respectively (Cooksey et al., 2020; Judson et al., 2020). Crickette is lead author on the chimpanzee and bonobo chapter, again with David Morgan, and two more Washington University graduate students: Juan Salvador Ortega Peralejo and Stephanie Musgrave. Stephanie studies how chimpanzee mothers teach offspring to use tools (Musgrave et al., 2016, 2020). It is clear: the enlightening contributions in this section were truly a Sussman/Wash U family affair!

The natural history of apes goes back further and is more complete than most other primates due to early efforts of one man who never even studied extant species: Louis Leakey. He was responsible for finding and hiring the "Trimates." Jane Goodall was the first and was sent to study chimpanzees in Tanzania in 1960 at what is now known as Gombe Stream Research Center. She discovered

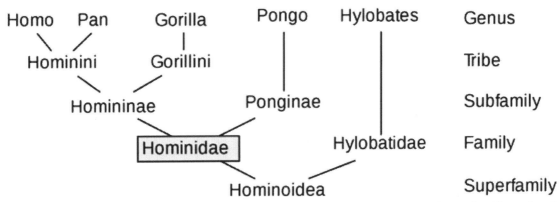

Figure 24.1 A generalized representation of phylogenetic relationships among apes (e.g., cf. with Bartlett & Sussman, this volume).
(from "Apes" Wikipedia, 2020)

some novel behaviors among chimpanzees, such as the fact that they hunted and ate monkeys and other mammals, but perhaps the most pivotal finding was their use of tools. The latter forced us to redefine what it meant to be human. A few years after Jane, Leakey met Dian Fossey who left for the Virunga volcanoes to study mountain gorillas in 1967. George Schaller (1963) had been there earlier and published what we first learned about this population. Interestingly, no mountain gorillas exist in captivity with almost all individuals being western lowland gorillas, yet it was not until the 1980s that we would come to learn about their behavior in the wild. The third trimate was Birute Galdikas, who set off for Borneo in 1971 to study the orangutan. For an interesting review of women in primatology, especially the trimates, see Morell (1993).

One of these three women would change my life!

It was the second time that week I ran from the lab where the frogs lay screaming. Like most kids who grow up loving animals, I wanted to become a veterinarian. It is only after you get to college you realize this means invasive experimentation on the very animals you love (as well as a year of organic chemistry and physics). Fortunately, when I was a junior in college, Dian Fossey presented a guest lecture at my university, and I learned there were other ways to work with animals that did not involve experimentation and euthanasia. Dian paced in the front of the lecture hall sharing her tales about Digit and other Virunga gorillas. I sat mesmerized. Daydreaming of my life as a gorilla researcher left me completely unprepared for her last image; that of Digit with his head, hands, and feet cut off. I guess this is what people refer to as an "aha" moment. My future path was clear; I was going to study gorillas in Africa. Sadly, the year I graduated, Dian was beheaded, much like her beloved Digit who she lays next to in eternity.

So, you might be surprised to learn the last entry in my field journal during my maiden primate research voyage read, "at least I know field work is not for me!" I had

joined Warren Kinzey (another exemplar primatologist who left us too soon) in Venezuela to survey islands, once mountaintops, in a lake outside Caracas for a new field site. We found a lot of howler monkeys but few *Pithecia* and *Chiropotes*. As such, the post-doc and I started tagging turtles. Warren told me not to be discouraged—I am so glad I took his advice.

For my PhD, I decided to compare the behavioral ecology of mountain gorillas and western lowland gorillas as a natural experiment. My work with Charles Janson on foraging effort and group size (Janson & Goldsmith, 1995) led me to hypothesize that herbivory in mountain gorillas versus a more frugivorous diet in lowland gorillas would increase lowland foraging effort, in turn, affecting their grouping patterns. After a short pilot study at Bai Hokou, Central African Republic (CAR) with Melissa Remis, I returned home and started submitting grants. As I prepared to leave for CAR, the genocide in Rwanda started with a vengeance. As a result, funding partners, such as Fulbright and the Wildlife Conservation Society, did not want to support my comparative study on Rwandan mountain gorillas. So, instead, I spent two years deep in the Dzanga-Sangha Forest in CAR studying non-habituated gorilla groups.

Although Savage and Wyman described western lowland gorillas in 1847, first referred to as *Troglodytes gorilla*, behavioral studies did not start until more than a century later. Jones and Sabater Pi (1971) and Sabater Pi (1977) were pioneers studying lowland gorilla ecology in Rio Muni (now known as Equatorial Guinea). Prior to their work, most information came from often-exaggerated stories of "great white hunters" (e.g., Sharpe, 1921; Allen, 1931 as cited in Blomfield, 2002) and explorers (e.g., Blancou, 1951). The first field researcher was a Virginia native, Richard Lynch Garner (1896), who studied gorillas and chimpanzees in Gabon and was a pioneer in chimpanzee language and the use of playback experiments. In one study, he sat in a cage for 112 days waiting, unsuccessfully, for gorillas to approach.

One problem standing in the way of research on western lowland gorillas is that they are very difficult to habituate. The lowland forest, throughout their range, is so dense you can walk within 5 feet of an elephant and not even know it's there! And, you better watch out if you do, as they are one of the most aggressive animals in the forest. Being a group of three (two B'Aka trackers and me), if an elephant emerged, we would take off in different directions running blindly through the forest looking for a tree to climb. But, I digress.

To habituate, that is, remove the fear of humans, you must be able to approach from a distance and then slowly get closer over time. The dense forest structure makes this quite impossible. In addition, western lowland gorillas demonstrate little group cohesion while foraging within the forest. When you come upon one individual, others are rarely nearby. This explains why the first observable behavior comes from large, swampy clearings within the forest, such as the Mbeli Bai in the Republic of Congo run by the Wildlife Conservation Society for over 20 years. In contrast, mountain gorillas live in more open forest, rich in their primary food source with increased group cohesion, which eases habituation attempts—some groups acclimating to human presence in less than a year.

Needless to say, my advisors were skeptical. A couple of years earlier a student went to study bonobos and came back empty-handed. One professor even said to me, something along the lines of, "Don't come back from the field with just a photo of a gorilla's butt as it runs away from you." Not very encouraging. Nevertheless, I persisted. I returned to Bai Hokou in 1992 and never looked back. What is this "field work is not for me" stuff? I had the most amazing experience living in the middle of the rainforest tracking gorillas every day with extremely talented B'Aka pygmies who can see a green snake in a green tree from 20 m away. They saved my life more than once—but those are stories for another time.

After 24 years of gorilla fieldwork, I was never able to re-create that particular time in the field. Showering in a cave under a waterfall with black-and-white colobus monkeys playing in the trees overhead. Bats coming out and flying around me. The smell of fresh gorilla dung in the morning. Sitting alone in the forest while the B'Aka headed out in different directions to patrol the area for gorilla signs. I never did see much of the gorillas, except for accidental interactions where, more often than not, the silverback charged, beating his chest and baring his teeth. Why were friends so compelled to send me Michael Creighton's *Congo* when it came out? I did not go into the forest for a week after reading it. Alas, I came

home with no photos (not even one of a gorilla's butt). Yet, it is truly amazing the amount of detailed feeding, ranging, and grouping data that can be collected using only nest sites, hair, dung, and gorilla tracks! In fact, it made me wonder, if we could learn so much about a species by indirect means, maybe we should consider the ethics of *reducing* their fear toward humans—an often-deadly predator (Goldsmith, 2004).

Findings from my dissertation (1996) added one more piece to the western lowland gorilla puzzle following pioneer work by Tutin and Fernandez (earliest publication, 1984) and Williamson (dissertation, 1988) at Lopé in Gabon; Carroll (unpub. findings, 1986) and Remis (dissertation, 1994) at Bai Hokou; Mitani (earliest publication, 1992) at Ndoki in CAR; and Olejniczak (earliest publication, 1994) at Mbeli Bai in the northern Republic of Congo.

After studying western lowland gorillas, I was finally able conduct the comparative study on mountain gorillas as a post-doctoral fellow. To the dismay of my family, whom I had promised I would not go back after a near-fatal battle with cerebral malaria, I packed my bags and left for Uganda. In 1996, I was one of the first primatologists to study mountain gorillas in this region and co-established the Bwindi Impenetrable Great Ape Project (BIGAPE) with Craig Stanford.

Soon after arriving in Bwindi, I discovered the mountain gorillas spent most of their time foraging and nesting outside of the park; making comparisons with lowland behavior ecology challenging, but achievable (Goldsmith, 2003). The habituation of mountain gorilla groups for ecotourism left them unafraid of coming out the park to raid farmers' fields (Goldsmith et al., 2006); they became pests and nuisances and threatened the livelihood of local human populations. My interest in this phenomenon grew, and in 2001, as a *National Geographic* researcher, I focused the rest of my time in Uganda examining the impact of ecotourism on the behavioral ecology and well-being of this small (less than 450 individuals; Hickey et al., 2018), isolated gorilla population in Uganda (Goldsmith, 2014).

To be sure, many adventures await a primatologist in the field. Each project brings us closer to understanding our kin and protecting their future. Take advantage of unexpected opportunities, be flexible, and persevere. Certainly, it's a wonderful world studying apes!

REFERENCES CITED—CHAPTER 24

Blancou, L. (1951). La protection de la faune sauvage en Afrique Equatoriale Francaise. *Mammalia*, 15(3), 157.

Blomfield, A. (2002, February 14). *Death of last white hunter who chased wildlife and women*. The Telegraph. https://www

.telegraph.co.uk/news/worldnews/africaandindianocean/kenya/1384865/Death-of-last-white-hunter-who-chased-wildlife-and-women.html.

Carroll, R. W. (1986). *The status, distribution, and density of the lowland gorilla (*Gorilla gorilla gorilla *[Savage and Wyman]), forest elephant (*Loxodonta africana cyclotis*) and associated dense forest fauna in southwestern Central African Republic: Research towards establishment of a reserve for their protection.* [Unpublished report]. Yale University.

Cooksey, K., Sanz, C., Ebombi, T. F., Massamba, J. M., Teberd, P., Magema, E., Abea, G., Ortega Peralejo, J. S., Kienast, I., Stephens, C., & Morgan, D. (2020). Socioecological factors influencing intergroup encounters in western lowland gorillas (*Gorilla gorilla gorilla*). *International Journal of Primatology, 41,* 181-202.

Garner, R. L. (1896). *Gorilla and chimpanzee.* Osgood, McIlvaine.

Goldsmith, M. L. (1996). *Ecological influences on the ranging and grouping behavior of western lowland gorillas in Bai Hokou, Central African Republic.* [Doctoral dissertation, State University of New York].

Goldsmith, M. L. (2003). Comparative behavioral ecology of lowland and highland gorilla populations: Where do the Bwindi gorillas fit? In A. B. Taylor, & M. L. Goldsmith (Eds.), *Gorilla biology: A multidisciplinary perspective* (pp. 358-384). Cambridge University Press.

Goldsmith, M. L. (2004). Habituating primates for field study: Ethical considerations for African great apes. In T. L. Turner (Ed.), *Ethics in biological anthropology* (pp. 49-64). State University of New York Press.

Goldsmith, M. L. (2014). Mountain gorilla tourism as a conservation tool: Have we tipped the balance? In A. Russon, & J. Wallis (Eds.), *Primate tourism: A conservation tool* (pp. 177-198). Cambridge University Press.

Goldsmith, M. L., Glick, J., & Ngabirano, E. (2006). Gorillas living on the edge: Literally and figuratively. In N. E. Newton-Fisher, H. Notman, J. D. Paterson, & V. Reynolds (Eds.), *Primates of western Uganda* (pp. 405-422). Springer.

Goodman, M. (1964). Man's place in the phylogeny of the primates as reflected in serum proteins. In S. L. Washburn (Ed.), *Classification and human evolution* (pp 204-234). Aldine.

Hickey, J. R., Basabose, A., Gilardi, K. V., Greer, D., Nampindo, S., Robbins, M. M., & Stoinski, T. S. (2018). *Gorilla beringei* ssp. *beringei. The International Union for the Conservation of Nature Red List of Threatened Species.* e.T39999A17989719. https://dx.doi.org/10.2305/IUCN.UK.2018-2.RLTS.T39999A17989719.en.

Janson, C. H., & Goldsmith, M. L. (1995). Predicting group size in primates: Foraging costs and predation risks. *Behavioral Ecology, 6(3),* 326-336.

Jones, C., & Sabater Pi, J. (1971). Comparative ecology of *Gorilla gorilla* (Savage and Wyman) and *Pan troglodytes* (Blumenbach) in Rio Muni, West Africa. *Bibliotheca Primatologica, 13,* 1-96.

Judson, K., Morgan, D., Massamba, J. M., Ebombi, F., Teberd, P., Abea, G., Magema, E., Mbeboute, G., Ortega, J., Kienast, I., & Sanz, C. (2020). Home range estimations and site fidelity of western lowland gorillas (*Gorilla gorilla gorilla*) in Ndoki Forest. *American Journal of Physical Anthropology, 171,* 134-135.

Mitani, M. (1992). Preliminary results of the studies on wild western lowland gorillas and other sympatric diurnal primates in the Ndoki Forest, northern Congo. In N. Itoigawa, Y. Sugiyama, G. P. Sackett, & R. K. R. Thompson (Eds.), *Topics in primatology behavior, ecology and conservation* (pp. 215-225). University of Tokyo Press.

Morell, V. (1993). Called "Trimates," three bold women shaped their field. *Science, 260(5106),* 420-425.

Musgrave, S., Lonsdorf, E., Morgan, D., Prestipino, M., Bernstein-Kurtycz, L., Mundry, R., & Sanz, C. (2020). Teaching varies with task complexity in wild chimpanzees. *Proceedings of the National Academy of Sciences, 117(2),* 969-976.

Musgrave, S., Morgan, D., Lonsdorf, E., Mudry, R., & Sanz C. (2016). Tool transfers are a form of teaching among chimpanzees. *Scientific Reports, 6,* 34783.

Olejniczak, C. L. (1994). Report on pilot study of western lowland gorillas at Mbeli Bai. *Gorilla Conservation News, 8,* 9-11.

Reichard, U. H. (2009). The social organization and mating system of Khao Yai white-handed gibbons: 1992-2006. In S. Lappan, & D. J. Whittaker (Eds.), *The gibbons: New perspectives on small ape socioecology and population biology* (pp. 347-384). Springer.

Remis, M. J. (1994). *Feeding ecology and positional behavior of western lowland gorillas (*Gorilla gorilla gorilla*) in the Central African Republic.* [Doctoral dissertation, Yale University].

Sabater Pi, J. (1977). Contribution to the study of alimentation of lowland gorillas in the natural state, in Rio Muni, Republic of Equatorial Guinea (West Africa). *Primates, 18,* 183-204.

Sarich, V. M., & Wilson, A. C. (1967). Immunological time scale for hominid evolution. *Science, 158(3805),* 1200-1203.

Savage, T. S., & Wyman, J. (1847). *Notice of the external characters and habits of troglodytes gorilla, a new species of orang from the Gaboon River.* Little Brown Press.

Schaller, G. E. (1963). *The mountain gorilla: Ecology and behavior.* University of Chicago Press.

Sharpe, A. (1921*), The backbone of Africa: A record of travel during the Great War with some suggestions for administrative reform.* Witherby.

Simpson, G. G. (1945). The principles of classification and a classification of mammals. *Bulletin American Museum of Natural History, 85,* 1-350.

Tutin, C. E. G., & Fernandez, M. (1984). Nationwide census of gorilla (*Gorilla gorilla gorilla*) and chimpanzee (*Pan troglodytes troglodytes*) populations in Gabon: Some preliminary data. *International Journal of Primatology, 6,* 27-43.

Williamson, E. A. (1988). *Behavioral ecology of western lowland gorillas in Gabon.* [Doctoral dissertation, University of Stirling].

Index

Page numbers in *italics* and **bold** refer to photographs and tables, respectively.

golden-bellied mangabeys, 406–7
golden-brown mouse lemurs, 39, 49
golden cats, 341
golden-headed lion tamarins, 144, 145
golden jackals, 493–94
golden lion tamarins, 144, 145, 161
Goldsmith, Michele, 642–44
Gombe, Tanzania, 322
Gombe chimpanzees, 601, 608
Gombe Stream National Park, 607
Gombe Stream Research Center, 642–43
Goodall, Jane, 601, 642
gorilla groups
    anti-predation efforts, 585–86
    in bai settings, 579
    bisexual groups, 582
    body mass, 571
    conservation status, 588–590
    conservation strategies, 589–590
    Cross River. *See* Cross River gorillas
    dietary diversity, 576–581
    diets/dietary adaptations, 574
    eastern. *See* eastern gorillas
    evolution, 571
    female social relationships, 583–84
    field studies, 572–73
    geographic distribution, 571–72, *573*
    Gorilla beringei. See eastern gorillas
    Gorilla beringei beringei. See mountain gorillas
    Gorilla beringei graueri. See Grauer's gorilla
    Gorilla gorilla. See western gorillas
    Gorilla gorilla diehli. See Cross River gorillas
    Gorilla gorilla gorilla. See western lowland gorillas
    health-monitoring system for, 589
    health risks, 588–89
    infanticidal behavior, 584–85
    infant mortality, 582
    intergroup dynamics, 587–88
    land conversion threats, 590
    locomotor behavior, 575–76
    male emigration, 583
    male-female relationships, 584–85
    male protection model, 584–85
    male social behavior, 581–83
    mating system, 586–87
    morphological features, 575
    mountain. *See* mountain gorillas
    nonreproductive groups, 582
    physiological adaptations, 576–77
    population density, 573–74
    predators, 585–86
    ranging behavior, 573–75, 579–581
    as sexually dimorphic, 571
    social organization, 581–88
    social status, 584
    social strategies, 585–86
    in swamps, 579
    territorial behaviors, 579

    threat displays, 587
    western. *See* western gorillas
    western lowland. *See* western lowland gorillas
Gorontalo macaques, 442–44
Goualougo Triangle, Republic of Congo, 603
gouging behavior, 11, 20, 42, 133, 141–42, 151–53
Gouraya National Park, Algeria, 433
Gran Chaco forests, 180, 182
Grauer's gorilla, 572, 575, 577–78, 580
gray-cheeked mangabeys
    activity budgets, 387–88
    arboreal feeding, 386
    breeding system, 388
    diets/dietary adaptations, 386
    dominance hierarchies, 387, 388
    habitat, 386
    morphological characteristics, 385–86
    polyspecific associations, 386
    predators, 386–87
    ranging behavior, 386
    social structure/organization, 387–88
gray gentle lemurs, 78, 82–83
gray mouse lemurs, 39–41, 48–49
gray slender loris, 6–7, 18
greater bamboo lemurs, 73, 79, 83, 132
greater dwarf lemur. *See* dwarf lemurs
greater galagos, locomotion, 7
green monkeys, 355
grivets, 355
group mobbing, 114
Guangxi Province, China, 450
Guassa Plateau, 400
guenons
    alarm calls, 341
    blue monkeys, 13, 341, 342, 346
    clades, 335
    classification, 335
    commercial trade, 348
    conservation threats, 347–48
    diets/dietary adaptations, 337–39, 349
    dominance hierarchies, 342–43
    evolution, 335
    feeding flexibility, 339
    in a fruiting *Ficus natalensis* tree, *340*
    geographic distribution, **336–37**
    habitat preferences, 335–37, 348–49
    home ranges, 346–47
    hovering males, 345
    infant development, 342
    interbirth intervals, 342
    mating systems, 344–45
    mixed-species groups, 347
    nonresident males, 345
    polyspecific associations, 345–47, **346**
    population decline, 348
    population density, 343
    as potential umbrella species, 348–49
    predator avoidance, 341–42

umbrella species, defined, 348–49

União Biological Reserve, Brazil, 166

uni-male groups, 272, 316, 320–21, 324, 342, 437, 449–450, 581–82

United Nations World Health Organization (WHO), 288

Upper Guinean rainforest of West Africa, 405

urine-washing, 12, 186–87

Varecia. See black-and-white ruffed lemurs

Varecia species, 71–72

venomous primates, 14

Verreaux's eagle-owl, 368

Verreaux's sifaka, 68, *76*

vervet monkeys

    activity budgets, 364

    adult, *358*

    alloparenting, 360

    anatomy, 357

    cohabitation with humans, 357

    crop-raiding behavior, 357

    diets/dietary adaptations, 365–67, **366**

    dominance hierarchies, 370–71, 373

    female-female alliances, 371

    female social relationships, **372**

    field studies, 360–61

    geographic distribution, 355–56, *356*

    group size, **362**

    growth and development, 359

    guenons and, 370

    habitat destruction, 361

    habitat preferences, 361–62

    infant mortality, 368

    intergroup interactions, 363–64

    interspecies relationship, 363

    juvenile, *357*

    mating system, 359–360

    morphological characteristics, **373**

    mortality, 368

    population decline, 361

    predators, 368–69

    ranging behavior, 362, **362**, 367–68

    sexual maturity, 359

    sleeping sites, 364

    social behavior, **373**

    social structure/organization, 370–73, **373**

    taxonomic classification, 355

    as territorial, 363

    warning vocalizations, 368–69

    water sources, 367–68

    See also Bale mountain monkeys

Veun Sai-Siem Pang Conservation Area, Cambodia, 543

Vie Sauvage, 604

Vietnam

    Cat Tien National Park, 501, 548

    Kon Kah Kin National Park, 510

    Nui Chua National Park, 501, 510

    Phong Nha-Ke Bang National Park, 464

    Phuoc Binh National Park, 501, 510

Son Tra National Reserve, 503, 510

Virunga gorilla population, 588

Virunga National Park, DRC, 572

Virunga Volcano region, 586

visual predation theory, 134

vocal communication. *See* acoustic communication

Volcanoes National Park, Rwanda, 572, 589

V. rubra. See red-ruffed lemurs

Wallace, Alfred Russel, 289

wandering males, 345

Way Canguk, Thailand, 544, 549

weasel lemurs, 34, *37*, 43, 51

Weddell's saddleback tamarins, 159, 160

West African patas monkeys. *See* patas monkeys

Western Amazonian forests, 211

Western Amazonian saki, 212

western dwarf galagos, 7

western gentle lemurs, 73

Western Ghats Mountains, India, 435, 488

western gorillas

    arboreal locomotion, 576

    behavioral strategies, 581–83

    conservation status, 588

    foot morphology, 575

    geographic distribution, 571, 572

    group size, **580**

    habitat use, 574

    home ranges, 580

    population density, **575**

    ranging behavior, **580**

    See also eastern gorillas; gorilla groups; western lowland gorillas

western lowland gorillas

    breeding opportunities, 587

    conservation status, 588

    development stages, 587

    diets/dietary adaptations, 578

    female dominance, 583

    field studies, 572, 643–44

    foot morphology, 575

    group size, 581

    home range, 581

    intergroup interactions, 587–88

    locomotion, 576

    male mating strategy, 582

    photograph, *572*

    population decline, 588

    ranging behavior, 580–81

    sexual activity, 586

    uni-male groups, 582–83

western tarsiers

    competitors, 115–16

    diets/dietary adaptations, 112

    geographic distribution, 105–7

    gestation period, 120

    morphological characteristics, 107–8

    olfactory communication, 119